Applied Engineering Mathematics

Dr. Madan Mohan Singh
Chairman
Board of Studies (Applied Mathematics)
Chhattisgarh Swami Vivekanand Technical University,
Bhilai (Chhattisgarh).

BSP BS Publications
A unit of **BSP Books Pvt. Ltd.**
4-4-309/316, Giriraj Lane, Sultan Bazar,
Hyderabad - 500 095.
Phone : 040 - 23445605, 23445688

Applied Engineering Mathematics by *Dr. Madan Mohan Singh*

Published by:

BSP BS Publications
A unit of **BSP Books Pvt. Ltd.**
4-4-309/316, Giriraj Lane, Sultan Bazar,
Hyderabad - 500 095.
Phone : 040 - 23445605, 23445688
e-mail : info@bspbooks.net
www.bspbooks.net

ISBN: 978-98-83635-79-5

Price : Rs. 995.00

Dedicated to

My Father

D.B. Singh

who served the cause of education for over six decades

Preface

There are several books on Engineering Mathematics available today. Then one may ask, "why did you venture to add one more to the list?" The answer is quite simple and obvious.

As I critically analyze the books on Engineering Mathematics, two things strike my mind. While some books claim to cover principles, numerical problems are shabbily treated. On the other hand books containing sufficient numerical problems do not cover theoretical background adequately. Between these extremes this textbook is an earnest endeavour to blend both the principles and numerical problems together. It is my strong conviction that solving numerical problems requires both sound knowledge of principles and their applications in an intelligent manner. My main concern in this book has been to impart concepts of fundamentals and approach to the problems i.e., how to think and proceed for solution.

Once the student starts proceeding in the right direction with sound knowledge of fundamentals and with confidence, there is no question that he will fail to solve any problem. The number of questions is not important, what matters most is the approach and conception.

I am thankful to Shri I.P Mishra, President, Shri Gangajali Education Society, Bhilai who has always inspired and encouraged me in my endeavour. I am indebted to various authors whose masterly texts have been freely consulted during the preparation of this book. I am also thankful to my son Mr. Kumar Anand, Asst. Professor (CSE Dept), SSTC, who has always motivated me during the course of writing.

Finally I would like to thank my publisher and printers for their kind cooperation.

Suggestions for improvement will be gratefully accepted.

Janmashtmi
18.8.2014

Dr. M. M. Singh
Shri Shankaracharya Technical Campus,
Bhilai (Chhattisgarh).

Contents

Part II

CHAPTER 4

CHAPTER 5

Part IV

CHAPTER 16

CHAPTER 18

CHAPTER 19

Part I

CHAPTER 1

Matrices

1.1 Introduction

Matrix algebra is a powerful mathematical formulation in the context of solving linear algebraic equations, linear transformations and the solutions of various types of differential equations. It has also become a powerful tool to tackle the problems of rigid body motions, oscillations, transformation of different coordinate systems, development and mathematical formulations of Quantum mechanics and the theory of representation of groups.

Let us consider the following system of equations

$$x + 2y + 3z = 11$$

$$2x - y - z = -3$$

$$3x + 4y + 2z = 17$$

If we arrange the coefficients of x, y and z in the order in which they occur in the given equations and enclose them in brackets, we get the following rectangular array of numbers

$$\begin{pmatrix} 1 & 2 & 3 \\ 2 & -1 & -1 \\ 3 & 4 & 2 \end{pmatrix} \text{ or } \begin{bmatrix} 1 & 2 & 3 \\ 2 & -1 & -1 \\ 3 & 4 & 2 \end{bmatrix} \text{ or } \begin{Vmatrix} 1 & 2 & 3 \\ 2 & -1 & -1 \\ 3 & 4 & 2 \end{Vmatrix}$$

This type of rectangular array of numbers has been given the name matrix. The horizontal lines are called rows and the vertical lines are called columns.

1.2 Definition of Matrix

Matrix is a rectangular array of real or complex numbers in rows and columns. A matrix is denoted by the capital letters A, B, C etc. If there are m rows and n columns in the matrix, then the matrix is called a $m \times n$ matrix.

Example: $A = \begin{pmatrix} 1 & 2 & 3 \\ 3 & 2 & -4 \end{pmatrix}$ or $\begin{bmatrix} 1 & 2 & 3 \\ 3 & 2 & -4 \end{bmatrix}$

Here A is a 2×3 matrix because A has two rows and three columns.

Note:

(i) Although matrix is just a rectangular array of numbers but these numbers are enclosed within big or small brackets to make it good looking.

(ii) Matrix is just an array of numbers and has no numerical value as in case of a determinant.

(iii) Matrices are not necessarily square as determinants must be. A matrix is simply a display or table of values.

Elements of a matrix: The numbers occurring in the rectangular array (matrix) are called the elements of the matrix. The elements of the matrix denoted by the capital letters are usually denoted by the corresponding small letters with lower suffixes. Thus the element of the ith row and j^{th} column of the matrix denoted by the capital letters A is usually denoted by the corresponding small letter a_{ij}. The matrix A is sometimes also written as (a_{ij}) or $[a_{ij}]$.

Types of Matrices

1. ***Square matrix:*** A matrix having equal number of rows and columns is called a square matrix. If the matrix A has n rows and n columns it is said to be a square matrix of order n.

   Example: $\begin{bmatrix} 1 & 2 & 3 \\ 2 & 3 & 4 \\ 2 & 0 & 5 \end{bmatrix}$ is a square matrix of order 3.

2. ***Horizontal matrix:*** A $m \times n$ matrix is called a horizontal matrix if $m < n$.

   Example: $A = \begin{pmatrix} 1 & 0 & 2 \\ 3 & 4 & 5 \end{pmatrix}$. Here A is a horizontal matrix.

3. ***Vertical matrix:*** A $m \times n$ matrix is called a vertical matrix if $m > n$.

   Example: $B = \begin{bmatrix} 1 & 0 \\ 2 & 3 \\ 4 & 5 \end{bmatrix}$. Here B is a vertical matrix.

4. ***Row matrix and Row vector:*** A matrix having only one row is called a row matrix or a row vector.

   Example: $A = \begin{pmatrix} 1 & 2 & 3 \end{pmatrix}$. Here A is a row matrix.

5. ***Column matrix or Column vector:*** A matrix having only one column is called a column matrix or column vector.

Example: $B = \begin{bmatrix} 1 \\ 2 \\ 0 \\ 4 \end{bmatrix}$. Here B is a column matrix

6. ***Zero matrix or Null matrix:*** A matrix whose each element is a zero is called a zero matrix or null matrix.

Example: (i) $A = \begin{pmatrix} 0 & 0 & 0 \\ 0 & 0 & 0 \end{pmatrix}$, (ii) $B = \begin{bmatrix} 0 & 0 & 0 \\ 0 & 0 & 0 \\ 0 & 0 & 0 \end{bmatrix}$

Here A and B are zero matrices.

Diagonal of matrix

Let A be a square matrix of order n.

Let $$A = \begin{bmatrix} a_{11} & a_{12} & a_{13} & \cdots & a_{1n} \\ a_{21} & a_{22} & a_{23} & \cdots & a_{2n} \\ a_{31} & a_{32} & a_{33} & \cdots & a_{3n} \\ \cdots & \cdots & \cdots & \cdots & \cdots \\ a_{n1} & a_{n2} & \cdots & \cdots & a_{nn} \end{bmatrix}$$

Here the elements $a_{11}, a_{22}, a_{33}, ..., a_{nn}$ are called the diagonal elements and the line along which these elements lie is called the principal diagonal of the matrix.

Example: Let $A = \begin{bmatrix} 1 & 2 & 3 \\ 2 & 0 & 5 \\ 3 & 2 & 4 \end{bmatrix}$

Here 1, 0, 4 are the diagonal elements of the matrix A.

7. ***Diagonal matrix:*** A square matrix having all elements not occurring along the principal diagonal equal to zero is called a diagonal matrix. Thus the square matrix (a_{ij}) is called the diagonal matrix if $a_{ij} = 0$ for $i \neq j$

Examples: (i) $A = \begin{bmatrix} 2 & 0 & 0 \\ 0 & 1 & 0 \\ 0 & 0 & 0 \end{bmatrix}$, $B = \begin{bmatrix} 0 & 0 & 0 \\ 0 & 0 & 0 \\ 0 & 0 & 0 \end{bmatrix}$

Here A and B are diagonal matrices

8. ***Scalar matrix:*** A square matrix is said to be a scalar matrix if all elements along the principal diagonal are equal and all elements not occurring along the principal diagonal are zero.

Examples: (i) $A = \begin{bmatrix} 2 & 0 & 0 \\ 0 & 2 & 0 \\ 0 & 0 & 2 \end{bmatrix}, B = \begin{bmatrix} 0 & 0 & 0 \\ 0 & 0 & 0 \\ 0 & 0 & 0 \end{bmatrix}$

Here A and B are scalar matrices.

9. ***Unit matrix:*** A square matrix is said to be a unit matrix if all the elements along the principal diagonal are unity (1) and all elements not occurring along the principal diagonal are zero.

Examples: (i) $A = \begin{pmatrix} 1 & 0 \\ 0 & 0 \end{pmatrix}$ (ii) $B = \begin{bmatrix} 1 & 0 & 0 \\ 0 & 1 & 0 \\ 0 & 0 & 1 \end{bmatrix}$

A unit matrix of order n is denoted by I_n.

Thus in the examples given above $A = I_2, B = I_3$.

10. ***Triangular matrix***

(a) *Lower triangular matrix*: A square matrix (a_{ij}) is called a lower triangular matrix if $a_{ij} = 0$ when $i < j$.

(b) *Upper triangular matrix*: A square matrix (a_{ij}) is called an upper triangular matrix if $a_{ij} = 0$ when $i > j$.

Examples: (i) $A = \begin{bmatrix} 1 & 0 & 0 \\ 2 & 2 & 0 \\ 0 & 5 & 3 \end{bmatrix}, B = \begin{bmatrix} 1 & 0 & 0 \\ 0 & 0 & 0 \\ 0 & 2 & 0 \end{bmatrix}$

(ii) $C = \begin{bmatrix} 1 & -1 & 4 \\ 0 & 2 & 5 \\ 0 & 0 & 3 \end{bmatrix}, D = \begin{bmatrix} 1 & 3 & 5 \\ 0 & 0 & 0 \\ 0 & 0 & 2 \end{bmatrix}$

Here A and B are lower triangular matrices.

Here C and D are upper triangular matrices

Submatrix of a matrix: Any matrix obtained by omitting some rows and some columns from the matrix A is called a submatrix of the matrix A.

Example: Let $A = \begin{bmatrix} 1 & 3 & 4 & 5 \\ 2 & 0 & 3 & -1 \\ 4 & 3 & 2 & 0 \end{bmatrix}, B = \begin{pmatrix} 1 & 3 & 4 \\ 4 & 3 & 2 \end{pmatrix}$

Here B has been obtained by omitting 4th column and 2nd row from A. Hence B is a submatrix of A.

Note: A itself is a submatrix of A.

Equality of two matrices: Two m × n matrices A and B are said to be equal if the corresponding elements of the two matrices are equal.

i.e., if $a_{ij} = b_{ij}$ for i = 1, 2, …, m; j = 1, 2, 3, …, n

i.e., if $a_{11} = b_{11}$, $a_{12} = b_{12}$ etc.

Example: Let $A = \begin{pmatrix} 1 & 2 & 0 \\ 3 & 2 & 4 \end{pmatrix}$, $B = \begin{pmatrix} 1 & 2 & 0 \\ 3 & 2 & 4 \end{pmatrix}$

Here A and B are equal matrices and we write A = B.

Addition of matrices: Let A and B be two m × n matrices. The m × n matrix obtained by adding the corresponding elements of the matrices A and B is called the sum of the matrices A and B and is denoted by A + B.

Thus if $A = (a_{ij}),\ B = (b_{ij}),\ \begin{matrix} i = 1,2,\ldots,m \\ j = 1,2,\ldots,n \end{matrix}$

then $A + B = (a_{ij}) + b_{ij}),\ \begin{matrix} i = 1,2,\ldots,m \\ j = 1,2,\ldots,n \end{matrix}$

Example: Let $A = \begin{pmatrix} 2 & 3 & 4 \\ 0 & 2 & -1 \end{pmatrix}$, $B = \begin{pmatrix} -3 & 4 & 5 \\ 6 & 2 & 9 \end{pmatrix}$

Then $A + B = \begin{pmatrix} 2-3 & 3+4 & 4+5 \\ 0+6 & 2+2 & -1+9 \end{pmatrix} = \begin{pmatrix} -1 & 7 & 9 \\ 6 & 4 & 8 \end{pmatrix}$

Note: A + B is defined only when they are matrices if the same type i.e., when the number of rows in A is equal to the number of rows in B and the number of columns in A is equal to the number of columns in B.

Properties of Matrix Addition

Property I: Matrix addition is commutative i.e., if A and B be any two m × n matrices, then $A + B = B + A$.

Proof: Let $A = (a_{ij})$ and $B = (b_{ij});\ \begin{matrix} i = 1,2,\ldots,m \\ j = 1,2,\ldots,n \end{matrix}$

Then $A + B = (a_{ij} + b_{ij});\ \begin{matrix} i = 1,2,\ldots,m \\ j = 1,2,\ldots,n \end{matrix}$

$$= (b_{ij} + a_{ij}); \quad \begin{matrix} i = 1, 2, ..., m \\ j = 1, 2, ..., n \end{matrix}$$

$$= (b_{ij}) + (a_{ij}); \quad \begin{matrix} i = 1, 2, ..., m \\ j = 1, 2, ..., n \end{matrix}$$

$$= B + A$$

Property II: Matrix addition is associative i.e., if A, B and C be three $m \times n$ matrices, then

$$A + (B + C) = (A + B) + C$$

Proof: Let $A = (a_{ij})$, $B = (b_{ij})$ and $C = (c_{ij})$ be three $m \times n$ matrices.

$$\text{Now} \quad A + (B + C) = (a_{ij}) + \{(b_{ij}) + (c_{ij})\}; \quad \begin{matrix} i = 1, 2, ..., m \\ j = 1, 2, ..., n \end{matrix}$$

$$= (a_{ij}) + (b_{ij} + c_{ij})$$

$$= (a_{ij} + \{b_{ij} + c_{ij}\})$$

$$= (\{a_{ij} + b_{ij}\} + c_{ij})$$

$$= (a_{ij} + b_{ij}) + (c_{ij})$$

$$= \{(a_{ij}) + (b_{ij})\} + (c_{ij})$$

$$= (A + B) + C$$

Property III: Cancellation laws hold good for addition of matrices i.e., if A, B, C, be any three $m \times n$ matrices, then

(i) $A + B = A + C \Rightarrow B = C$ (left cancellation law)

(ii) $B + A = C + A \Rightarrow B = C$ (right cancellation law)

Negative of a matrix: Let $A = (a_{ij})$ be $m \times n$ matrix. Then the negative of the matrix A is denoted by $-A$ and is defined as $(-a_{ij})$.

In order to find $-A$ sign of each element of A must be changed.

Example: Let $A = \begin{pmatrix} 1 & -2 & 0 \\ 3 & 2 & -5 \end{pmatrix}$. Then $-A = \begin{pmatrix} -1 & 2 & 0 \\ -3 & -2 & 5 \end{pmatrix}$

Note: If A is a $m \times n$ matrix, then $-A$ is also a $m \times n$ matrix

Subtraction of two matrices: Let A and B be two $m \times n$ matrices.

Then the difference of A and B is denoted by $A - B$ and is defined by

$$A - B = A + (-B)$$

$A - B$ will be also a $m \times n$ matrix. In order to find $A - B$, the elements of B must be subtracted from the corresponding elements of A.

Examples: Let $A = \begin{pmatrix} 1 & 2 & 3 \\ 3 & -1 & 0 \end{pmatrix}$, $B = \begin{pmatrix} 2 & -1 & 4 \\ 5 & 0 & 6 \end{pmatrix}$

then $$A - B = \begin{pmatrix} 1-2 & 2+1 & 3-4 \\ 3-5 & -1-0 & 0-6 \end{pmatrix} = \begin{pmatrix} -1 & 3 & -1 \\ -2 & -1 & -6 \end{pmatrix}$$

Multiplication of a matrix by a scalar: Let A be any m × n matrix and k be any scalar (real or complex number), then the scalar multiple of matrix A by k is denoted by kA or Ak and is defined as the m × n matrix obtained by multiplying each element of A by k.

Thus if $A = (a_{ij});\quad \begin{matrix} i = 1,2,\ldots,m \\ j = 1,2,\ldots,n \end{matrix}$

then $kA = (ka_{ij});\quad \begin{matrix} i = 1,2,\ldots,m \\ j = 1,2,\ldots,n \end{matrix}$

Example: Let $A = \begin{bmatrix} 1 & 2 \\ 3 & 0 \\ 5 & 4 \end{bmatrix}$, then $3A = \begin{bmatrix} 3 & 6 \\ 9 & 0 \\ 15 & 12 \end{bmatrix}$

Properties of scalar multiplication

(i) If A and B are any two m × n matrices and k is any scalar, then

$$k(A + B) = kA + kB$$

(ii) If A is any m × n matrix and a and b are any two scalars, then

$$(a + b)A = aA + bA$$

(iii) If A be any m × n matrix and k be any scalar, then

$$(-k)A = -(kA) = k(-A)$$

Multiplication of two matrices

Let $A = [a_{ij}]$ be a m × n matrix and $B = [b_{jk}]$ be a n × p matrix such that the number of columns in A is equal to the number of rows in B, then the m × p matrix $C = [c_{ik}]$ such that

$$C_{ik} = \sum_{j=1}^{n} a_{ij} b_{jk}$$

$$= a_{i1} b_{1k} + a_{i2}b_{2k} + a_{i3}b_{3k} + \ldots + a_{in}b_{nk}$$

is said to the product of the matrices A and B in that order and is denoted by AB.

Note:

(i) Product AB is defined only when the number of columns in A is equal to the number of rows in B.

(ii) If A is a m × n matrix and B is a n × p matrix, then AB will be a m × p matrix.

$$\begin{aligned} A &: m \times \boxed{n} \\ B &: \boxed{n} \times p \\ \hline AB &: m \times p \end{aligned}$$

(iii) Element of ith row and k^{th} column of the product matrix is

$$a_{i1}b_{1k} + a_{i2}b_{2k} + a_{i3}b_{3k} + \ldots + a_{in}b_{nk}$$

i^{th} row of A is $a_{i1}\, a_{i2}\, a_{i3} \ldots b_{in}$

k^{th} column of B is $b_{1k} b_{2k} b_{3k} \ldots b_{nk}$

Element of ith row and k^{th} column of AB

$$= a_{i1}\, b_{1k} + a_{i2}\, b_{2k} + \ldots + a_{in}b_{nk}$$

Thus in order to write down the element of the ith row and k^{th} column of AB, we take the elements of ith row of the first matrix A and multiply them with the corresponding elements of the kth column of the second matrix B and then add them.

(iv) In the product AB, A is called the pre-factor and B is called the post-factor.

(v) If the product AB is possible then it is not necessary that the product BA is also possible.

(vi) If A be a $m \times n$ matrix and both AB and BA are defined then B will be a $n \times m$ matrix.

Examples:

(i) $A = 3 \times 2 \text{ matrix} \begin{bmatrix} a_{11} & a_{12} \\ a_{21} & a_{22} \\ a_{31} & a_{32} \end{bmatrix}$, $B = 2 \times 2 \text{ matrix} \begin{bmatrix} b_{11} & b_{12} \\ b_{21} & b_{22} \end{bmatrix}$

Then $AB = 3 \times 2 \text{ matrix} \begin{bmatrix} a_{11}b_{11} + a_{12}b_{21} & a_{11}b_{12} + a_{12}b_{22} \\ a_{21}b_{11} + a_{22}b_{12} & a_{21}b_{21} + a_{22}b_{22} \\ a_{31}b_{11} + a_{32}b_{21} & a_{31}b_{12} + a_{32}b_{22} \end{bmatrix}$

(ii) $A = 3 \times 3 \text{ matrix} \begin{bmatrix} 1 & 0 & 5 \\ -1 & 2 & 4 \\ 3 & -2 & 6 \end{bmatrix}$, $B = 3 \times 2 \text{ matrix} \begin{bmatrix} 4 & -1 \\ 2 & -2 \\ 5 & 3 \end{bmatrix}$

$$\begin{bmatrix} 1.4 + 0.2 + 5.5 & 1(-1) + 0(-2) + 5.3 \\ (-1)4 + 2.2 + 4.5 & (-1)(-1) + 2(-2) + 4.3 \\ 3.4 + (-2)2 + 6.5 & 3(-1) + (-2)(-2) + 6.3 \end{bmatrix} = \begin{bmatrix} 29 & 14 \\ 20 & 9 \\ 38 & 19 \end{bmatrix}$$

Here BA is not defined because number of column in B is not equal to the number of rows in A.

Properties of matrix multiplication

(i) ***Matrix multiplication is associative:***

i.e., If A, B and C be m × m, n × p and p × q matrices, then

$$A(BC) = (AB)C$$

Proof: Let $A = [a_{ij}]_{m \times n}$, $B = [b_{jk}]_{n \times p}$ and $C = [c_{kl}]_{p \times q}$

B is a n × p matrix and C is a p × q matrix, therefore, BC will be a n × q matrix and since A is a m × n matrix, therefore, A (BC) will be a m × q matrix.

Again A is a m × n matrix and B is n × p matrix, therefore, AB will be m × p matrix and since C is a p × q matrix, therefore (AB) C is a m × q matrix.

Now element of ith row and k^{th} column of AB

$$u_{ik} = \sum_{j=1}^{n} a_{ij} b_{jk}$$

$$= a_{i1}b_{1k} + a_{i2}b_{2k} + ... + a_{in}b_{nk} \qquad .....(1.1)$$

Element of ith row and l^{th} column of (AB) C

$$= \sum_{k=1}^{p} u_{ik} c_{kl} = \sum_{k=1}^{p}\left\{\left(\sum_{j=1}^{n} a_{ij} b_{jk}\right) c_{kl}\right\} \qquad .....(1.2)$$

Again element of jth row and l^{th} column of BC

$$v_{jl} = \sum_{k=1}^{p} b_{jk} c_{kl} \qquad .....(1.3)$$

Element of ith row and l^{th} column of A (BC)

$$= \sum_{j=1}^{n} a_{ij} v_{jl} = \sum_{j=1}^{n}\left\{ a_{ij} \sum_{k=1}^{p} b_{jk} c_{kl}\right\}$$

$$= \sum_{k=1}^{p}\left\{\left(\sum_{j=1}^{n} a_{ij}.b_{jk}\right) c_{kl}\right\}$$

From eq. 1.2 and 1.4 it follows that the element of ith row and l^{th} column of A(BC) = element of ith row and i^{th} column of (AB) C for all admissible values of i and l.

Hence $\quad A(BC) = (AB)C$

(ii) Multiplication of matrices is distributive with respect to addition of matrices i.e., if A is a m × n matrix and B and C are both n × p matrices.

then $\quad A (B + C) = AB + AC$

Proof: Let $A = [a_{ij}]_{m \times n}$, $B = [b_{jk}]_{n \times p}$ and $C = [c_{jk}]_{n \times p}$

Elements of ith row of A are

$$a_{i1}, a_{i2}, \ldots, a_{in} \quad \text{.....(1.4)}$$

Elements of kth column of B are

$$b_{1k}, b_{2k}, \ldots, b_{nk} \quad \text{.....(1.5)}$$

Elements of kth column of C are

$$c_{1k}, c_{2k}, \ldots, c_{nk} \quad \text{.....(1.6)}$$

$\therefore$ Elements of kth column of (B + C) are

$$b_{1k} + c_{1k}, b_{2k} + c_{2k}, \ldots, b_{nk} + c_{nk} \quad \text{.....(1.7)}$$

Elements of ith row of A are

$$a_{i1}, a_{i2}, \ldots, a_{in} \quad \text{.....(1.8)}$$

Element of ith row and k^{th} column of AB

$$= a_{i1}b_{1k} + a_{i2}b_{2k} + \ldots + a_{in}b_{nk} \qquad \text{[From (1.4) and (1.5)]} \quad \text{.....(1.9)}$$

Element of ith row and k^{th} column of AC

$$= a_{i1}c_{1k} + a_{i2}c_{2k} + \ldots + a_{in}c_{nk} \qquad \text{[From (1.4) and (1.6)]} \quad \text{.....(1.10)}$$

$\therefore$ Element of ith row and k^{th} column of AB + AC

$$= (a_{il}b_{1k} + a_{il}c_{1k}) + (a_{i2}b_{2k} + a_{i2}c_{2k}) + \ldots + (a_{in}b_{nk} + a_{in}c_{nk}) \quad \text{.....(1.11)}$$

Element of ith row and k^{th} column of A (B + C)

$$= a_{il}(b_{1k} + a_{il}c_{1k}) + a_{i2}(b_{2k} + c_{2k}) + \ldots + a_{in}(b_{nk} + a_{in}c_{nk})$$

$$\text{[From (1.4) to (1.7)]} \quad \text{.....(1.12)}$$

$$= (a_{il}b_{1k} + a_{il}c_{1k}) + (a_{i2}b_{2k} + a_{i2}c_{2k}) + \ldots + (a_{in}b_{nk} + a_{in}c_{nk}) \quad \text{.....(1.13)}$$

From (1.12) and (11.3), we have

$$A(B + C) = AB + AC$$

Note: It can be also proved that

(i) if A and B are $m \times n$ matrices and C is a $n \times p$ matrix, then

$$(A + B)C = AC + BC$$

(ii) if A be a $m \times n$ matrix and I_n be the unit matrix of order n, then

$$AI_n = I_nA = A$$

Transpose of a matrix: Let A be any matrix then the matrix obtained by interchanging its rows and columns is called the transpose of A and is denoted by A' or A^T. If A is a $m \times n$ matrix then A' will be a $n \times m$ matrix.

Example: $A = \begin{pmatrix} 1 & 2 & 3 \\ 2 & 0 & 5 \end{pmatrix}$, then $A' = \begin{bmatrix} 1 & 2 \\ 2 & 0 \\ 3 & 5 \end{bmatrix}$

Note: If $A = [a_{ij}]; \quad i = 1, 2, ..., m, j = 1, 2, ..., n$

then $A' = [a_{ji}]; \quad j = 1, 2, ..., n, \ i = 1, 2, .., m$

Properties of Transpose of Matrices

Property I: $(A + B)' = A' + B'$

Proof: Let $A = [a_{ij}]_{m \times n}$ and $B = [b_{ij}]_{m \times n}$

$\therefore$ $A + B = [a_{ij} + b_{ij}]$, then $(A + B)' = [a_{ji} + b_{ji}]_{n \times m}$ .....(1.14)

$A' = [a_{ij}]_{n \times m}$ and $B' = [b_{ji}]_{n \times m}$

$\therefore$ $A' + B' = [a_{ij} + b_{ji}]_{n \times m}$ .....(1.15)

From (1.14) and (1.15) it follows that $(A + B) = A' + B'$

Property II: If A is any matrix, then $(A')' = A$

Proof: Let $A = [a_{ij}]_{m \times n}$, then

$A' = [a_{ij}]_{n \times m} \quad \therefore (A')' = [a_{ij}]_{m \times n} = A$

Property III: If k is any number real or complex and A be any matrix, then

$(kA)' = kA'$

Proof: Let $A = [a_{ij}]_{m \times n}$, then $A' = [a_{ji}]_{n \times m}$

Now $kA = [ka_{ij}]_{m \times n} \therefore (kA)' = [ka_{ji}]_{n \times m}$ .....(1.16)

Also $kA' = [ka_{ji}]_{n \times m}$ .....(1.17)

From (1.16) and (1.17) it follows that $(kA)' = kA'$

Property IV: If A be a $m \times n$ matrix and B be a $n \times p$ matrix, then

$$(AB)' = B'A'$$

Proof: Let $A = [a_{ij}]_{m\times n},\ B = [b_{jk}]_{n\times p}$

Then $A' = [a_{ji}]_{n\times m},\ B' = [b_{kj}]_{n\times p}$

AB will be a $m \times p$ matrix, $B'A'$ will be a $p \times m$ matrix.

Element of the kth row and i^{th} column of $(AB)'$

$$= \text{element of } i^{th} \text{ row and } k^{th} \text{ column of AB}$$

$$= \sum_{j=1}^{n} a_{ij} b_{jk} \qquad \text{.....(1.18)}$$

Element of kth row and i^{th} column of $B'A'$

$$= \sum_{j=1}^{n} b_{jk} a_{ij} = \sum_{j=1}^{n} a_{ij} b_{jk} \qquad \text{.....(1.19)}$$

$\therefore$ [elements of kth row of B' = elements of kth column of B]

They are : $b_{1k}, b_{2k}, ..., b_{nk}$

and elements of ith column of A' = element of ith row of A

They are: $a_{i1}, a_{i2}, ..., a_{in}$

From (1.18) and (1.19) it follows that $(AB) = B'A'$

Symmetric, Skew Symmetric and Orthogonal Matrices

(i) ***Symmetric matrix:*** A square matrix $A = [a_{ij}]$ is said to be a symmetric matrix if

$$a_{ij} = a_{ji} \text{ for all i and j}$$

Example: $A = \begin{bmatrix} 1 & 3 & 5 \\ 3 & 2 & 0 \\ 5 & 0 & -1 \end{bmatrix},\ B = \begin{bmatrix} 1 & 3 \\ 3 & 0 \end{bmatrix} \because$

Note: A square matrix A is symmetric if and only if $A' = A$.

(ii) ***Skew symmetric matrix:*** A square matrix $A = [a_{ij}]$ is said to be a skew symmetric matrix if $a_{ij} = -a_{ji}$ for all i and j.

$\because \quad a_{ij} = -a_{ji}$ for all i and j

$\therefore \quad a_{ii} = -a_{ii}$ [putting j = i]

or $\quad 2a_{ii} = 0$ or $a_{ii} = 0$

Thus in a skew symmetric matrix all elements along the principal diagonal are zero.

Examples: $A = \begin{pmatrix} 0 & 2 \\ -2 & 0 \end{pmatrix}, B = \begin{bmatrix} 0 & 2 & 1 \\ -2 & 0 & -3 \\ -1 & 3 & 0 \end{bmatrix}$

***Note*:** A square matrix A is skew symmetric if and only if $A' = -A$.

(iii) ***Orthogonal matrix:*** A matrix A is said to be orthogonal If $A'A = I$ where A' is the transpose of A.

Some results related to symmetric and skew symmetric matrices:

(i) If A is any square matrix, then $A + A'$ is a symmetric matrix and $A - A'$ is a skew symmetric matrix.

Proof: $(A + A')' = A' + (A')' = A' + A = A + A' \quad [\because (A+B)' = A' + B']$

Hence $A + A'$ is a symmetric matrix.

Again $(A - A')' = A' - (A')' = A' - A = -(A - A')$

Hence $A - A'$ is a skew symmetric matrix.

(ii) Every square matrix can be uniquely expressed as the sum of a symmetric matrix and a skew symmetric matrix.

Proof: Let A be any square matrix. Then as in (i) $\frac{1}{2}(A + A')$ will be a symmetric matrix and $\frac{1}{2}(A - A')$ will be a skew symmetric matrix.

Let $B = \frac{1}{2}(A + A')$ and $C = \frac{1}{2}(A - A')$

Then $A = \frac{1}{2}(A + A') + \frac{1}{2}(A - A') = B + C$

where B is a symmetric matrix and C is a skew symmetric matrix.

To prove that the representation is unique: If possible let A = D + E where D is a symmetric and E is a skew symmetric matrix. Then

$$D' = D \text{ and } E' = -E \qquad \ldots..(1.20)$$

Now $\quad A = D + E \Rightarrow A' = D' + E' = D - E \qquad$ [From (1.20)]

Thus $\quad A = D + E$
and $\quad A' = D - E$
$\therefore \quad D = \frac{1}{2}(A + A') = B$ and $E = \frac{1}{2}(A - A') = C$

Hence representation of A is unique.

Determinant of a square matrix: Let A be a square of order n then the determinant of the matrix A is the value of the determinant whose elements are the corresponding elements of the matrix A and is denoted by |A| or Det. A

Thus if $A = [a_{ij}]$ be a square matrix of order n, then the number

$$\begin{vmatrix} a_{11} & a_{12} & \cdots & a_{1n} \\ a_{21} & a_{22} & \cdots & a_{2n} \\ \cdots & \cdots & \cdots & \cdots \\ \cdots & \cdots & \cdots & \cdots \\ a_{n1} & an_2 & \cdots & a_{nn} \end{vmatrix}$$

is the determinant of the matrix A.

Example: $A = \begin{bmatrix} 1 & 2 & 3 \\ 0 & -1 & 2 \\ 3 & 4 & 5 \end{bmatrix}$, then $|A| = \begin{vmatrix} 1 & 2 & 3 \\ 0 & -1 & 2 \\ 3 & 4 & 5 \end{vmatrix} = 1(-5-8) - 0(10-12) + 3(4+3) = 8$

Minor and cofactor of an element of a determinant

Let $A = [a_{ij}]$ be a square matrix, then

(i) The minor of the element a_{ij} of |A| is the value of the determinant obtained by deleting its ith row and j^{th} column and it is denoted by M_{ij}.

(ii) The cofactor of the element a_{ij} of |A| is denoted by the corresponding capital letter A_{ij} and $A_{ij} = (-1)^{i+j} M_{ij}$

Example: Let $A = \begin{bmatrix} 1 & 2 & 3 \\ -2 & 4 & 5 \\ 0 & -3 & 6 \end{bmatrix}$

In det. A, minor of 1 $= \begin{vmatrix} 4 & 5 \\ -3 & 6 \end{vmatrix} = 24 + 15 = 39$

Cofactor of 1 $= (-1)^{1+1}\, 39 = 39 \qquad$ [$\because$ 1 lies in the 1st row and 1st column]

$$\text{Minor of } 2 = \begin{vmatrix} -2 & 5 \\ 0 & 6 \end{vmatrix} = -12 - 0 = -12$$

Cofactor of 2 = $(-1)^{1+2}(-12) = 12$ [$\because$ 2 occurs in the 1st row and 2nd column]

Adjoint of a square matrix

Let $A = [a_{ij}]$ be a square matrix.

Let $B = [A_{ij}]$ where A_{ij} is the cofactor of the element a_{ij} in the det. A. The transpose B' of the matrix B is called the adjoint of the matrix A and is denoted by adj. A.

Example: Let $A = \begin{bmatrix} 1 & 2 & 3 \\ 2 & 3 & 4 \\ 2 & 0 & 5 \end{bmatrix}$, then $B = \begin{bmatrix} 15 & -2 & -6 \\ -10 & -1 & 4 \\ -1 & 2 & -1 \end{bmatrix}$

$$\text{adj. } A = B' = \begin{bmatrix} 15 & -10 & -1 \\ -2 & -1 & 2 \\ -6 & 4 & -1 \end{bmatrix}$$

Theorem: If A is any square matrix of order n, then

$$A.\,(\text{adj } A) = (\text{adj } A).\,A = |A|\,I_n$$

Proof: Let $A = \begin{bmatrix} a_{11} & a_{12} & \cdots & a_{1n} \\ a_{21} & a_{22} & \cdots & a_{2n} \\ \dots & \dots & \dots & \dots \\ \dots & \dots & \dots & \dots \\ a_{n1} & a_{n2} & & a_{nn} \end{bmatrix}$

Then $\text{adj. } A = \begin{bmatrix} A_{11} & A_{21} & \cdots & A_{n1} \\ A_{21} & A_{22} & \cdots & a_{n2} \\ \dots & \dots & \dots & \dots \\ \dots & \dots & \dots & \dots \\ A_{1n} & A_{2n} & \dots & A_{nn} \end{bmatrix}$

Now $A.\,(\text{adj } A) = \begin{bmatrix} |A| & 0 & 0 & \cdots & 0 \\ 0 & |A| & 0 & \cdots & 0 \\ 0 & 0 & |A| & \cdots & 0 \\ \dots & \dots & \dots & \dots & \dots \\ 0 & 0 & 0 & \cdots & |A| \end{bmatrix}$ [a square matrix of order n]

$$= |A|\begin{bmatrix} 1 & 0 & 0 & \cdots & 0 \\ 0 & 1 & 0 & \cdots & 0 \\ \cdots & \cdots & \cdots & \cdots & \cdots \\ 0 & 0 & 0 & \cdots & 1 \end{bmatrix} = |A|I_n$$

$$\because \sum_{j=1}^{n} a_{ij}A_{ij} = |A| \qquad \text{for } i = 1, 2, \ldots, n$$

And $$\sum_{j=1}^{n} a_{ij}A_{kj} = 0, \qquad i \neq k$$

Similarly we can show that (adj A). A = $|A|I_n$

Thus A(adj. A) = (adj. A) A = $|A|\ I_n$

Note: (i) $|A\ (\text{adj. } A)| = |A|^n$ $[\because\ |I_n| = 1]$

(ii) If $|A| \neq 0$, then $|\text{adj. } A| = |A|^{n-1}$

Non-singular and singular matrices

(i) A square matrix A is said to be a non-singular matrix if $|A| \neq 0$

(ii) A square matrix A is said to be a singular matrix if $|A| = 0$

Inverse or reciprocal of a square matrix: Let A be a square matrix of order n. Then a matrix B (if such a matrix exists) is called the inverse of A if

$$AB = BA = I_n$$

Inverse of the square matrix A is denoted by A^{-1}.

Existence of the inverse: The inverse of a square matrix A exists if and only if A is a non-singular matrix.

If part: Let A be non-singular square matrix of order n. Then $|A| \neq 0$.

Let $$B = \frac{\text{adj.}A}{|A|}$$

Then $$AB = \frac{A(\text{adj } A)}{|A|} = \frac{|A|I_n}{|A|} = I_n \qquad [\because A.\ (\text{adj } A) = |A|I_n] \qquad \ldots..(1.21)$$

Hence B i.e., $\frac{\text{adj } A}{|A|}$ is the inverse of matrix A (by definition of inverse)

Only if part: Let A be a square matrix of order n. Let inverse of A exist. Let B be the inverse of A.

Then by definition of inverse

$$AB = I_n \Rightarrow |AB| = |I_n| = 1$$

or $\quad |A|\,|B| = 1 \qquad [\because |AB| = |A|\,|B|]$

$\therefore \quad |A| \neq 0, \qquad$ because product $|A|\,|B|$ is non-zero

Hence A is non singular

Note: (i) $A^{-1} = \dfrac{\text{adj } A}{|A|}$ (ii) $AA^{-1} = I_n$ [From (1.21)]

Theorem:

(i) If A and B be any two non-singular matrices, then AB is also a non-singular matrix and $(AB)^{-1} = B^{-1}A^{-1}$

$\because \quad$ A, B are non-singular $\therefore |A| \neq 0, |B| \neq 0$

$|AB| = |A|\,|B| \neq 0$. Hence AB is non-singular

Non $AB(B^{-1}A^{-1})$

$$= A\{B(B^{-1}A^{-1})\} = A\{(BB^{-1})A^{-1}\} \quad \text{[by associative law]}$$

$$= A\{I_n A^{-1}\} \qquad [\because BB^{-1} = I_n]$$

$$= AA^{-1} \qquad [\because I_n A^{-1} = A^{-1}]$$

$$= I_n$$

Hence $B^{-1}A^{-1}$ is the inverse of AB

$$(AB)^{-1} = B^{-1}A^{-1}$$

(ii) If A is a non singular matrix, then

$(A^{-1})^{-1} = A$

Let A be a square matrix of order n,

Then $A^{-1}A = I_n \therefore$ inverse of $A^{-1} = A \qquad \because (A^{-1})^{-1} = A$

$$I_n^{-1} = I_n \text{ as } I_n^{-1} = I_n$$

Elementary Operations or Elementary Transformations of a Matrix

Any of the following operations is called an elementary transformation.

(i) The interchange of any two rows (or columns)

(ii) The multiplication of the elements of any row (or column) by any non-zero number.

(iii) The addition to the elements of any row (or column) the corresponding elements of any other row (or column) multiplied by any number.

Note: An elementary transformation is called a row transformation or column transformation according as it applies to rows or columns.

1.3 Echelon Form of a Matrix

A matrix A is said to be in echelon form if

(i) every row of A which has all its elements 0, occurs below row which has a non zero element.

(ii) the first non-zero element in each non-zero row is 1 .

(iii) the number of zeros before the first non-zero element in a row is less than the number of such zeros in the next row.

Examples: (i) $\begin{bmatrix} 1 & 1 & 3 \\ 0 & 1 & 2 \\ 0 & 0 & 1 \end{bmatrix}$ (ii) $\begin{bmatrix} 1 & 3 & 4 \\ 0 & 1 & 2 \\ 0 & 0 & 1 \end{bmatrix}$

***Note*:** A row of a matrix is said to be a zero row if all its elements are zero.

1.4 Rank of a Matrix

Definition: Let A be a matrix of order m × n. If atleast one of its minors of order r is different from zero and all minors of order (r +1) are zero, then the number r is called the rank of the matrix A and is denoted by ρ(A).

***Note*:**

(i) The rank of a zero matrix is zero and the rank of an identity matrix of order n is n.

(ii) The rank of a non-singular matrix of order n is n.

(iii) The rank of a matrix in echelon form is equal to the number of non-zero rows of the matrix.

***Example* 1.1:** Reduce the following matrix to upper triangular form (Echelon form)

$$\begin{bmatrix} 1 & 2 & 3 \\ 2 & 5 & 7 \\ 3 & 1 & 2 \end{bmatrix}$$

Solution: Upper triangular matrix. If in a square matrix, all the elements below the principal diagonal are zero, the matrix is called an upper triangular matrix.

$$\begin{bmatrix} 1 & 2 & 3 \\ 2 & 5 & 7 \\ 3 & 1 & 2 \end{bmatrix} \sim \begin{bmatrix} 1 & 2 & 3 \\ 0 & 1 & 1 \\ 0 & -5 & -7 \end{bmatrix} \begin{matrix} R_2 - 2R_1 \\ R_3 - 3R_1 \end{matrix} \quad \textbf{Ans.} \overset{R_3 + 5R_2}{\sim \begin{bmatrix} 1 & 2 & 3 \\ 0 & 1 & 1 \\ 0 & 0 & -2 \end{bmatrix}}$$

***Example* 1.2:** Transform $\begin{bmatrix} 1 & 3 & 3 \\ 2 & 4 & 10 \\ 3 & 8 & 4 \end{bmatrix}$ into a unit matrix

Solution: $\begin{bmatrix} 1 & 3 & 3 \\ 2 & 4 & 10 \\ 3 & 8 & 4 \end{bmatrix} \sim \begin{bmatrix} 1 & 3 & 3 \\ 0 & -2 & 4 \\ 0 & -1 & -5 \end{bmatrix} \begin{matrix} R_2 - 2R_1 \\ R_3 - 3R_1 \end{matrix}$

$$\sim \begin{bmatrix} 1 & 3 & 3 \\ 0 & 1 & -2 \\ 0 & -1 & -5 \end{bmatrix} -\frac{1}{2}R_2 \quad \sim \begin{bmatrix} 1 & 0 & 9 \\ 0 & 1 & -2 \\ 0 & 0 & -7 \end{bmatrix} \begin{matrix} R_1 - 3R_2 \\ R_3 + R_2 \end{matrix}$$

$$\sim \begin{bmatrix} 1 & 0 & 9 \\ 0 & 1 & -2 \\ 0 & 0 & 1 \end{bmatrix} -\frac{1}{7}R_3 \quad \sim \begin{bmatrix} 1 & 0 & 0 \\ 0 & 1 & 0 \\ 0 & 0 & 1 \end{bmatrix} \begin{matrix} R_1 - 9R_3 \\ R_2 + 2R_3 \end{matrix}$$ **Ans.**

Elementary Matrices

A matrix obtained from a unit matrix by a single elementary transformation is called elementary matrix.

$$I = \begin{bmatrix} 1 & 0 & 0 \\ 0 & 1 & 0 \\ 0 & 0 & 1 \end{bmatrix}$$

Consider the matrix obtained by $R_2 + 3\,R_1$

$\begin{bmatrix} 1 & 0 & 0 \\ 3 & 1 & 0 \\ 0 & 0 & 1 \end{bmatrix}$ is called the elementary matrix

***Example* 1.3:** Use Gauss-Jordan reduction method (Elementary matrices method) to compute the inverse of the matrix.

$\begin{bmatrix} 3 & -3 & 4 \\ 2 & -3 & 4 \\ 0 & -1 & 1 \end{bmatrix}$ by applying elementary row transformation

Solution: $A = \begin{bmatrix} 3 & -3 & 4 \\ 2 & -3 & 4 \\ 0 & -1 & 1 \end{bmatrix}$

Elementary row transformation, which will reduce A = IA to I = PA, then matrix P will be in inverse of matrix A.

$$\begin{bmatrix} 3 & -3 & 4 \\ 2 & -3 & 4 \\ 0 & -1 & 1 \end{bmatrix} = \begin{bmatrix} 1 & 0 & 0 \\ 0 & 1 & 0 \\ 0 & 0 & 1 \end{bmatrix} A, \frac{1}{3}R_1 \begin{bmatrix} 1 & -1 & \frac{4}{3} \\ 2 & -3 & 4 \\ 0 & -1 & 1 \end{bmatrix} = \begin{bmatrix} \frac{1}{3} & 0 & 0 \\ 0 & 1 & 0 \\ 0 & 0 & 1 \end{bmatrix} A$$

$$R_2 - 2R_1 \begin{bmatrix} 1 & -1 & \frac{4}{3} \\ 0 & -1 & \frac{4}{3} \\ 0 & -1 & 1 \end{bmatrix} = \begin{bmatrix} \frac{1}{3} & 0 & 0 \\ -\frac{2}{3} & 1 & 0 \\ 0 & 0 & 1 \end{bmatrix} A - R_2 \begin{bmatrix} 1 & -1 & \frac{4}{3} \\ 0 & 1 & -\frac{4}{3} \\ 0 & -1 & 1 \end{bmatrix} = \begin{bmatrix} \frac{1}{3} & 0 & 0 \\ \frac{2}{3} & -1 & 0 \\ 0 & 0 & 1 \end{bmatrix} A$$

$$\begin{matrix} R_1 + R_2 \\ R_3 + R_2 \end{matrix} \begin{bmatrix} 1 & 0 & 0 \\ 0 & 1 & -\frac{4}{3} \\ 0 & 0 & -\frac{1}{3} \end{bmatrix} = \begin{bmatrix} 1 & -1 & 0 \\ \frac{2}{3} & -1 & 0 \\ \frac{2}{3} & -1 & 1 \end{bmatrix} A - 3R_3 \begin{bmatrix} 1 & 0 & 0 \\ 0 & 1 & -\frac{4}{3} \\ 0 & 0 & 1 \end{bmatrix} = \begin{bmatrix} 1 & -1 & 0 \\ \frac{2}{3} & -1 & 0 \\ -2 & 3 & -3 \end{bmatrix} A$$

$$R_2 + \frac{4}{3}R_3 \begin{bmatrix} 1 & 0 & 0 \\ 0 & 1 & 0 \\ 0 & 0 & 1 \end{bmatrix} = \begin{bmatrix} 1 & -1 & 0 \\ -2 & 3 & -4 \\ -2 & 3 & -3 \end{bmatrix} A$$

$$\therefore I = PA \qquad \therefore P = A^{-1} \text{ or } A^{-1} = \begin{bmatrix} 1 & -1 & 0 \\ -2 & 3 & -4 \\ -2 & 3 & -3 \end{bmatrix}$$ **Ans.**

***Example* 1.4:** Compute the inverse of the following matrix by using elementary transformations.

$$\begin{bmatrix} 2 & -6 & -2 & -3 \\ 5 & -13 & -4 & -7 \\ -1 & 4 & 1 & 2 \\ 0 & 1 & 0 & 1 \end{bmatrix}$$

Solution:

$$\begin{bmatrix} 2 & -6 & -2 & -3 \\ 5 & -13 & -4 & -7 \\ -1 & 4 & 1 & 2 \\ 0 & 1 & 0 & 1 \end{bmatrix} = \begin{bmatrix} 1 & 0 & 0 & 0 \\ 0 & 1 & 0 & 0 \\ 0 & 0 & 1 & 0 \\ 0 & 0 & 0 & 1 \end{bmatrix} A$$

$$\begin{bmatrix} -1 & 4 & 1 & 2 \\ 2 & -6 & -2 & -3 \\ 5 & -13 & -4 & -7 \\ 0 & 1 & 0 & 1 \end{bmatrix} = \begin{bmatrix} 0 & 0 & 1 & 0 \\ 1 & 0 & 0 & 0 \\ 0 & 1 & 0 & 0 \\ 0 & 0 & 0 & 1 \end{bmatrix} A \qquad \begin{matrix} R_1 \to R_3 \\ R_2 \to R_1 \\ R_3 \to R_2 \end{matrix}$$

$R_2 + 2R_1$, $R_3 + 5R_1$

$$\begin{bmatrix} -1 & 4 & 1 & 2 \\ 0 & 2 & 0 & 1 \\ 0 & 7 & 1 & 3 \\ 0 & 1 & 0 & 1 \end{bmatrix} = \begin{bmatrix} 0 & 0 & 1 & 0 \\ 1 & 0 & 2 & 0 \\ 0 & 1 & 5 & 0 \\ 0 & 0 & 0 & 1 \end{bmatrix} A$$

$-R_1$, $\frac{1}{2}R_2$

$$\begin{bmatrix} 1 & -4 & -1 & -2 \\ 0 & 1 & 0 & \frac{1}{2} \\ 0 & 7 & 1 & 3 \\ 0 & 1 & 0 & 1 \end{bmatrix} = \begin{bmatrix} 0 & 0 & -1 & 0 \\ \frac{1}{2} & 0 & 1 & 0 \\ 0 & 1 & 5 & 0 \\ 0 & 0 & 0 & 1 \end{bmatrix} A$$

$R_3 - 7R_2$, $R_4 - R_2$

$$\begin{bmatrix} 1 & -4 & -1 & -2 \\ 0 & 1 & 0 & \frac{1}{2} \\ 0 & 0 & 1 & -\frac{1}{2} \\ 0 & 0 & 0 & \frac{1}{2} \end{bmatrix} = \begin{bmatrix} 0 & 0 & -1 & 0 \\ \frac{1}{2} & 0 & 1 & 0 \\ -\frac{7}{2} & 1 & -2 & 0 \\ -\frac{1}{2} & 0 & -1 & 1 \end{bmatrix} A$$

$R_3 + R_4$, $R_2 - R_4$, $R_1 + 4R_4$

$$\begin{bmatrix} 1 & -4 & -1 & 0 \\ 0 & 1 & 0 & 0 \\ 0 & 0 & 1 & 0 \\ 0 & 0 & 0 & \frac{1}{2} \end{bmatrix} = \begin{bmatrix} -2 & 0 & -5 & 4 \\ 1 & 0 & 2 & -1 \\ -4 & 1 & -3 & 1 \\ -\frac{1}{2} & 0 & -1 & 1 \end{bmatrix} A$$

$2R_4$

$$\begin{bmatrix} 1 & -4 & -1 & 0 \\ 0 & 1 & 0 & 0 \\ 0 & 0 & 1 & 0 \\ 0 & 0 & 0 & 1 \end{bmatrix} = \begin{bmatrix} -2 & 0 & -5 & 4 \\ 1 & 0 & 2 & -1 \\ -4 & 1 & -3 & 1 \\ -1 & 0 & -2 & 2 \end{bmatrix} A$$

$R_1 + R_3$

$$\begin{bmatrix} 1 & -4 & 0 & 0 \\ 0 & 1 & 0 & 0 \\ 0 & 0 & 1 & 0 \\ 0 & 0 & 0 & 1 \end{bmatrix} = \begin{bmatrix} -6 & 1 & -8 & 5 \\ 1 & 0 & 2 & -1 \\ -4 & 1 & -3 & 1 \\ -1 & 0 & -2 & 2 \end{bmatrix} A$$

$R_1 + 4R_2$

$$\begin{bmatrix} 1 & 0 & 0 & 0 \\ 0 & 1 & 0 & 0 \\ 0 & 0 & 1 & 0 \\ 0 & 0 & 0 & 1 \end{bmatrix} = \begin{bmatrix} -2 & 1 & 0 & 1 \\ 1 & 0 & 2 & -1 \\ -4 & 1 & -3 & 1 \\ -1 & 0 & -2 & 2 \end{bmatrix} A$$

Inverse matrix = $\begin{bmatrix} -2 & 1 & 0 & 1 \\ 1 & 0 & 2 & -1 \\ -4 & 1 & -3 & 1 \\ -1 & 0 & -2 & 2 \end{bmatrix}$ **Ans.**

***Example* 1.5:** Find the inverse of the matrix A = $\begin{bmatrix} 2 & 4 & 3 & 2 \\ 3 & 6 & 5 & 2 \\ 2 & 5 & 2 & -3 \\ 4 & 5 & 14 & 14 \end{bmatrix}$

Solution:

$$\begin{bmatrix} 2 & 4 & 3 & 2 \\ 3 & 6 & 5 & 2 \\ 2 & 5 & 2 & -3 \\ 4 & 5 & 14 & 14 \end{bmatrix} = \begin{bmatrix} 1 & 0 & 0 & 0 \\ 0 & 1 & 0 & 0 \\ 0 & 0 & 1 & 0 \\ 0 & 0 & 0 & 1 \end{bmatrix} A$$

$\frac{1}{2}R_1$

$$\begin{bmatrix} 1 & 2 & 3/2 & 1 \\ 3 & 6 & 5 & 2 \\ 2 & 5 & 2 & -3 \\ 4 & 5 & 14 & 14 \end{bmatrix} = \begin{bmatrix} 1/2 & 0 & 0 & 0 \\ 0 & 1 & 0 & 0 \\ 0 & 0 & 1 & 0 \\ 0 & 0 & 0 & 1 \end{bmatrix} A,$$

$R_2 - 3R_1, R_3, -2R_1, R_4 - 4R_1$

$$\begin{bmatrix} 1 & 2 & 3/2 & 1 \\ 0 & 0 & \frac{1}{2} & -1 \\ 0 & 1 & -1 & -5 \\ 0 & -3 & 8 & 10 \end{bmatrix} = \begin{bmatrix} 1/2 & 0 & 0 & 0 \\ -\frac{3}{2} & 1 & 0 & 0 \\ -1 & 0 & 1 & 0 \\ -2 & 0 & 0 & 1 \end{bmatrix} A$$

$R_2 \leftrightarrow R_3$

$$\begin{bmatrix} 1 & 2 & 3/2 & 1 \\ 0 & 1 & -1 & -5 \\ 0 & 0 & \frac{1}{2} & -1 \\ 0 & -3 & 8 & 10 \end{bmatrix} = \begin{bmatrix} 1/2 & 0 & 0 & 0 \\ -1 & 0 & 1 & 0 \\ -\frac{3}{2} & 1 & 0 & 0 \\ -2 & 0 & 0 & 1 \end{bmatrix} A,$$

$R_1 - 2R_2, R_4 + 3R_2, R_3 \to 2R_3$

$$\begin{bmatrix} 1 & 0 & \frac{7}{2} & 11 \\ 0 & 1 & -1 & -5 \\ 0 & 0 & 1 & -2 \\ 0 & 0 & 5 & -5 \end{bmatrix} = \begin{bmatrix} \frac{5}{2} & 0 & -2 & 0 \\ -1 & 0 & 1 & 0 \\ -3 & 2 & 0 & 0 \\ -5 & 0 & 3 & 1 \end{bmatrix} A$$

$R_1 - \frac{7}{2} R_3, R_2 + R_3, R_4 - 5R_3$

$$\begin{bmatrix} 1 & 0 & 0 & 18 \\ 0 & 1 & 0 & -7 \\ 0 & 0 & 1 & -2 \\ 0 & 0 & 0 & 5 \end{bmatrix} = \begin{bmatrix} 13 & -7 & -2 & 0 \\ -4 & 2 & 1 & 0 \\ -3 & 2 & 0 & 0 \\ 10 & -10 & 3 & 1 \end{bmatrix} A,$$

$\frac{1}{5} R_4$

$$\begin{bmatrix} 1 & 0 & 0 & 18 \\ 0 & 1 & 0 & -7 \\ 0 & 0 & 1 & -2 \\ 0 & 0 & 0 & 1 \end{bmatrix} = \begin{bmatrix} 13 & -7 & -2 & 0 \\ -4 & 2 & 1 & 0 \\ -3 & 2 & 0 & 0 \\ 2 & -2 & \frac{3}{5} & \frac{1}{5} \end{bmatrix} A$$

$R_1 - 18R_4, R_2 + 7R_4, R_3 + 2R_4$

$$\begin{bmatrix} 1 & 0 & 0 & 0 \\ 0 & 1 & 0 & 0 \\ 0 & 0 & 1 & 0 \\ 0 & 0 & 0 & 1 \end{bmatrix} = \begin{bmatrix} -23 & 29 & -\frac{64}{5} & -\frac{18}{5} \\ 10 & -12 & \frac{26}{5} & \frac{7}{5} \\ 1 & -2 & \frac{6}{5} & \frac{2}{5} \\ 2 & -2 & \frac{3}{5} & \frac{1}{5} \end{bmatrix} A,$$

$$A^{-1} = \begin{bmatrix} -23 & 29 & -\frac{64}{5} & -\frac{18}{5} \\ 10 & -12 & \frac{26}{5} & \frac{7}{5} \\ 1 & -2 & \frac{6}{5} & \frac{2}{5} \\ 2 & -2 & \frac{3}{5} & \frac{1}{5} \end{bmatrix}$$ **Ans.**

Exercise 1.1

Find the inverse of the following matrices:

1. $\begin{bmatrix} 1 & 3 & 3 \\ 1 & 4 & 3 \\ 1 & 3 & 4 \end{bmatrix}$ **Ans.** $\begin{bmatrix} 7 & -3 & -3 \\ -1 & 1 & 0 \\ -1 & 0 & 1 \end{bmatrix}$

2. $\begin{bmatrix} 1 & -1 & -1 \\ 4 & 1 & 0 \\ 8 & 1 & 1 \end{bmatrix}$ **Ans.** $\begin{bmatrix} 1 & 2 & -1 \\ -4 & -7 & 4 \\ -4 & -9 & 5 \end{bmatrix}$

3. $\begin{bmatrix} 1 & 2 & 5 \\ 2 & 3 & 1 \\ -1 & 1 & 1 \end{bmatrix}$ **Ans.** $\frac{1}{21}\begin{bmatrix} 2 & 3 & -13 \\ -3 & 6 & 9 \\ 5 & -3 & -1 \end{bmatrix}$

4. $\begin{bmatrix} 7 & 6 & 2 \\ -1 & 2 & 4 \\ 3 & 6 & 8 \end{bmatrix}$ **Ans.** $\frac{1}{20}\begin{bmatrix} -4 & -18 & 10 \\ 10 & 25 & -15 \\ -6 & -12 & 10 \end{bmatrix}$

5. $\begin{bmatrix} 2 & 3 & 1 \\ 1 & 2 & 3 \\ 3 & 1 & 2 \end{bmatrix}$ **Ans.** $\frac{1}{18}\begin{bmatrix} 1 & -5 & 7 \\ 7 & 1 & -5 \\ -5 & 7 & 1 \end{bmatrix}$

6. $\begin{bmatrix} 2 & 1 & -1 & 2 \\ 1 & 3 & 2 & -3 \\ -1 & 2 & 1 & -1 \\ 2 & -3 & -1 & 4 \end{bmatrix}$ **Ans.** $\frac{1}{18}\begin{bmatrix} 2 & 5 & -7 & 1 \\ 5 & -1 & 5 & -2 \\ -7 & 5 & 11 & 10 \\ 1 & -2 & 10 & 5 \end{bmatrix}$

***Example* 1.6:** Determine the rank of a matrix

$$\begin{bmatrix} 1 & 4 & 5 \\ 2 & 6 & 8 \\ 3 & 7 & 22 \end{bmatrix}$$

Solution :

$$\begin{bmatrix} 1 & 4 & 5 \\ 2 & 6 & 8 \\ 3 & 7 & 22 \end{bmatrix} \sim \begin{bmatrix} 1 & 4 & 5 \\ 0 & -2 & -2 \\ 0 & -5 & 7 \end{bmatrix} \begin{matrix} R_2 - 2R_1 \\ R_3 - 3R_1 \end{matrix}$$

$$\sim \begin{bmatrix} 1 & 4 & 5 \\ 0 & 1 & 1 \\ 0 & -5 & 7 \end{bmatrix} -\frac{1}{2}R_2 \sim \begin{bmatrix} 1 & 4 & 5 \\ 0 & 1 & 1 \\ 0 & 0 & 12 \end{bmatrix} R_3 \times 5R_2$$

Rank = Number of non-zero rows = 3 **Ans.**

***Example* 1.7:** Find the rank of the matrix

$$\begin{bmatrix} -1 & 2 & 3 & -2 \\ 2 & -5 & 1 & 2 \\ 3 & -8 & 5 & 2 \\ 5 & -12 & -1 & 6 \end{bmatrix}$$

Solution:

$$\begin{bmatrix} -1 & 2 & 3 & -2 \\ 2 & -5 & 1 & 2 \\ 3 & -8 & 5 & 2 \\ 5 & -12 & -1 & 6 \end{bmatrix} \sim \begin{bmatrix} -1 & 2 & 3 & -2 \\ 0 & -1 & 7 & -2 \\ 0 & -2 & 14 & -4 \\ 0 & -2 & 14 & -4 \end{bmatrix} \begin{matrix} R_2 + 2R_1 \\ R_3 + 3R_1 \\ R_4 + 5R_1 \end{matrix}$$

$$\sim \begin{bmatrix} -1 & 2 & 3 & -2 \\ 0 & -1 & 7 & -2 \\ 0 & 0 & 0 & 0 \\ 0 & 0 & 0 & 0 \end{bmatrix} \begin{matrix} \\ R_3 - 2R_2 \\ R_4 - 2R_2 \\ \\ \end{matrix}$$

Here the 4th order and 3rd order minors are zero. But a minor of second order

$$\begin{vmatrix} 3 & -2 \\ 7 & -2 \end{vmatrix} = -6 + 14 = 8 \neq 0$$

Rank = Number of non-zero rows = 2 **Ans.**

***Example* 1.8:** Find the rank of the matrix

$$\begin{bmatrix} 3 & -4 & -1 & 2 \\ 1 & 7 & 3 & 1 \\ 5 & -2 & 5 & 4 \\ 9 & -3 & 7 & 7 \end{bmatrix}$$

Solution:

$$\begin{bmatrix} 1 & 7 & 3 & 1 \\ 3 & -4 & -1 & 2 \\ 5 & -2 & 5 & 4 \\ 9 & -3 & 7 & 7 \end{bmatrix} R_1 \leftrightarrow R_2$$

$$R_2 - 3R_1,\ R_3 - 5R_1,\ R_4 - 9R_1,\ R_3 - \frac{37}{25}R_2,\ R_4 - \frac{66}{25}R_2\ R_4 - \frac{4}{3}R_3$$

$$\begin{bmatrix} 1 & 7 & 3 & 1 \\ 0 & -25 & -10 & -1 \\ 0 & -37 & -10 & -1 \\ 0 & -66 & -20 & -2 \end{bmatrix} = \begin{bmatrix} 1 & 7 & 3 & 1 \\ 0 & -25 & -10 & -1 \\ 0 & 0 & \frac{24}{5} & \frac{12}{25} \\ 0 & 0 & \frac{32}{5} & \frac{16}{25} \end{bmatrix} \sim \begin{bmatrix} 1 & 7 & 3 & 1 \\ 0 & -25 & -10 & -1 \\ 0 & 0 & \frac{24}{5} & \frac{12}{25} \\ 0 & 0 & 0 & 0 \end{bmatrix}$$

$\therefore$ Rank = 3 **Ans.**

***Example* 1.9:** Find the rank of

$$\begin{bmatrix} 2 & 3 & -1 & -1 \\ 1 & -1 & -2 & -4 \\ 3 & 1 & 3 & -2 \\ 6 & 3 & 0 & -7 \end{bmatrix}$$

Solution:

$$\overset{R_1 \leftrightarrow R_2}{\sim}\begin{bmatrix} 1 & -1 & -2 & -4 \\ 2 & 3 & -1 & -1 \\ 3 & 1 & 3 & -2 \\ 6 & 3 & 0 & -7 \end{bmatrix} \overset{R_2-2R_1, R_3-3R_1, R_4-6R_1}{\sim} \begin{bmatrix} 1 & -1 & -2 & -4 \\ 0 & 5 & 3 & -7 \\ 0 & 4 & 9 & 10 \\ 0 & 9 & 12 & 17 \end{bmatrix}$$

$$\overset{R_3-\frac{4}{5}R_2, R_4-\frac{9}{5}R_2}{\sim}\begin{bmatrix} 1 & -1 & -2 & -4 \\ 0 & 5 & 3 & 7 \\ 0 & 0 & \frac{33}{5} & \frac{22}{5} \\ 0 & 0 & \frac{33}{5} & \frac{22}{5} \end{bmatrix} \overset{R_4-R_3}{-} \begin{bmatrix} 1 & -1 & -2 & -4 \\ 0 & 5 & 3 & 7 \\ 0 & 0 & \frac{33}{5} & \frac{22}{5} \\ 0 & 0 & 0 & 0 \end{bmatrix}$$

Rank = Number of non-zero rows = 3 **Ans.**

Exercise 1.2

Find the rank of the following matrices:

1. $\begin{bmatrix} 1 & 2 & 3 \\ 2 & 4 & 7 \\ 3 & 6 & 10 \end{bmatrix}$ **Ans.** 2

2. $\begin{bmatrix} 3 & -1 & 2 \\ -6 & 2 & 4 \\ -3 & 1 & 2 \end{bmatrix}$ **Ans.** 2

3. $\begin{bmatrix} 1 & 2 & 1 \\ -1 & 0 & 2 \\ 2 & 1 & -3 \end{bmatrix}$ **Ans.** 3

4. $\begin{bmatrix} 1 & 2 & 1 & 0 \\ -2 & 4 & 3 & 0 \\ 1 & 0 & 2 & -8 \end{bmatrix}$ **Ans.** 3

5. $\begin{bmatrix} 0 & 1 & 2 & -2 \\ 4 & 0 & 2 & 6 \\ 2 & 1 & 3 & 1 \end{bmatrix}$ **Ans.** 2

6. $\begin{bmatrix} 2 & 4 & 3 & -2 \\ -3 & -2 & -1 & 4 \\ 6 & -1 & 7 & 2 \end{bmatrix}$ **Ans.** 3

7. $\begin{bmatrix} 3 & 4 & 1 & 1 \\ 2 & 4 & 3 & 6 \\ -1 & -2 & 6 & 4 \\ 1 & -1 & 2 & -3 \end{bmatrix}$ **Ans.** 4

8. $\begin{bmatrix} 9 & 3 & 1 & 0 \\ 3 & 0 & 1 & -6 \\ 1 & 1 & 1 & 1 \\ 0 & -6 & 1 & 9 \end{bmatrix}$ **Ans.** 4

1.5 Normal Form or Canonical Form

(i) A matrix of the form $\begin{bmatrix} I_r & 0 \\ 0 & 0 \end{bmatrix}$, where I_r is a unit matrix of order 'r' and 0 is the null matrix is called the normal form or first canonical form.

(ii) Matrices of the form $[I_r \quad 0]$, $[I_r]$, $\begin{bmatrix} I_r \\ 0 \end{bmatrix}$, are also known as normal forms.

(iii) Every m × n matrix can be reduced to the normal form by a series of elementary transformations.

(iv) The rank of the above canonical form is 'r'.

(v) *Algorithm to reduce a matrix to the normal form :*

***Step* 1 :** Make $a_{11} = 1$, by applying suitable row or column operations or both.

***Step* 2 :** Make all the elements below a_{11} in 1st column and on the right of a_{11}, in the 1st row equal to zero with the help of suitable row and column operations.

***Step* 3 :** If the resulting matrix is in the normal form, the process ends. Otherwise repeat the above steps regarding a_{22} such that no row or column operation should involve 1st row or 1st column. If the matrix is still not in the normal form proceed to a_{33} and apply the same process. The process is continued till the normal form is obtained.

***Example* 1.10:** Reduce the matrix

$$A = \begin{bmatrix} 3 & 1 & 4 & 6 \\ 2 & 1 & 2 & 4 \\ 4 & 2 & 5 & 8 \\ 1 & 1 & 2 & 2 \end{bmatrix}$$ to an Echelon form and hence find its rank.

Solution:

Step 1 : To get '1' in the first position of 1st row, (i.e., a_{11} position).

Applying R14 ($R_1 \leftrightarrow R_4$), we get

$$A \sim \begin{bmatrix} 1 & 1 & 2 & 2 \\ 2 & 1 & 2 & 4 \\ 4 & 2 & 5 & 8 \\ 3 & 1 & 4 & 6 \end{bmatrix}$$

Step 2 : To get zeros below a_{11}.

Applying $R_2 \to R_2 - 2R_1$; $R_3 \to R_3 - 4R_1$, $R_4 \to R_4 - 3R_1$,

$$\sim \begin{bmatrix} 1 & 1 & 2 & 2 \\ 0 & -1 & -2 & 0 \\ 0 & -2 & -3 & 0 \\ 0 & -2 & -2 & 0 \end{bmatrix}$$

Step 3 : To get 1 in a_{22} position.

Applying $R_2 \to -R_2$,

$$\sim \begin{bmatrix} 1 & 1 & 2 & 2 \\ 0 & 1 & 2 & 0 \\ 0 & -2 & -3 & 0 \\ 0 & -2 & -2 & 0 \end{bmatrix}$$

Step 4 : To get zeros below a_{22} :

Applying $R_3 \to R_3 + 2R_2$, $R_4 \to R_4 + 2R_2$

$$\sim \begin{bmatrix} 1 & 1 & 2 & 2 \\ 0 & 1 & 2 & 0 \\ 0 & 0 & 1 & 0 \\ 0 & 0 & 2 & 0 \end{bmatrix}$$

Step 5 : $a_{33} = 1$; so we have to get zeros below it. ∴ Applying $R_4 \to R_4 - 2R_3$;

$$\sim \begin{bmatrix} 1 & 1 & 2 & 2 \\ 0 & 1 & 2 & 0 \\ 0 & 0 & 1 & 0 \\ 0 & 0 & 0 & 0 \end{bmatrix},$$

which is in the Echelon form which has 3 non-zero rows in it.

$$\therefore \qquad \rho(A) = 3$$

***Example* 1.11:** Apply elementary transformations to find the rank of

$$A = \begin{bmatrix} 1 & -7 & 3 & -3 \\ 7 & 20 & -2 & 25 \\ 5 & -2 & 4 & 7 \end{bmatrix}$$

Solution: Applying $R_2 \to R_2 - 7R_1$; $R_3 \to R_3 - 5R_1$, we get,

$$A \sim \begin{bmatrix} 1 & -7 & 3 & -3 \\ 0 & 69 & -23 & 46 \\ 0 & 33 & -11 & 22 \end{bmatrix}$$

$$\sim \begin{bmatrix} 1 & -7 & 3 & -3 \\ 0 & 3 & -1 & 2 \\ 0 & 3 & -1 & 2 \end{bmatrix} \quad \left(R_2 \to \frac{1}{23}R_2; R_3 \to \frac{1}{11}R_3\right)$$

$$\sim \begin{bmatrix} 1 & -7 & 3 & -3 \\ 0 & 3 & -1 & 2 \\ 0 & 0 & 0 & 0 \end{bmatrix} = B \text{ (say) } [R_3 \to R_3 - R_2],$$

which is in Echelon form which has two non-zero rows,

$$\therefore \quad \rho(A) = \rho(B) = 2$$

(or) every 3-rowed minor of B = 0

and $\quad \begin{vmatrix} 1 & -7 \\ 0 & 3 \end{vmatrix} \neq 0.$

$$\therefore \quad \rho(A) = 2$$

Example **1.12:** Find the rank of the matrix

$$A = \begin{bmatrix} 2 & 3 & -1 & -1 \\ 1 & -1 & -2 & -4 \\ 3 & 1 & 3 & -2 \\ 6 & 3 & 0 & -7 \end{bmatrix}$$

Applying $R_1 \leftrightarrow R_2$,

$$A \sim \begin{bmatrix} 1 & -1 & -2 & -4 \\ 2 & 3 & -1 & -1 \\ 3 & 1 & 3 & -2 \\ 6 & 3 & 0 & -7 \end{bmatrix}$$

$$\sim \begin{bmatrix} 1 & -1 & -2 & -4 \\ 0 & 5 & 3 & 7 \\ 0 & 4 & 9 & 10 \\ 0 & 9 & 12 & 17 \end{bmatrix} \quad \begin{Bmatrix} R_2 \to R_2 - 2R_1 \\ R_3 \to R_3 - 3R_1 \\ R_4 \to R_4 - 6R_1 \end{Bmatrix}$$

$$\sim \begin{bmatrix} 1 & -1 & -2 & -4 \\ 0 & 5 & 3 & 7 \\ 0 & 4 & 9 & 10 \\ 0 & 0 & 0 & 0 \end{bmatrix} \quad (R_4 \to R_4 - (R_2 + R_3))$$

= B ; In B, the 4 rowed minor = 0

$\therefore \quad \rho(A) \neq 4$

The leading 3 rowed minor,

$$= \begin{vmatrix} 1 & -1 & -2 \\ 0 & 5 & 3 \\ 0 & 4 & 9 \end{vmatrix} = 45 - 12 \neq 0$$

$\therefore \quad \rho(A) = 3.$

(**Note :** The problem can also be solved by reducing to Echelon form).

***Example* 1.13:** Find the constants 'l' and 'm' such that the rank of the matrix $\begin{bmatrix} 1 & -2 & 3 & 1 \\ 2 & 1 & -1 & 2 \\ 6 & -2 & l & m \end{bmatrix}$ is (i) 3 (ii) 2.

Solution:

$$A = \begin{bmatrix} 1 & -2 & 3 & 1 \\ 2 & 1 & -1 & 2 \\ 6 & -2 & l & m \end{bmatrix}$$

$$\sim \begin{bmatrix} 1 & -2 & 3 & 1 \\ 0 & 5 & -7 & 0 \\ 0 & 10 & (l-18) & (m-6) \end{bmatrix} \begin{Bmatrix} R_2 \to R_2 - 2R_1 \\ R_3 \to R_3 - 6R_1 \end{Bmatrix}$$

$$\sim \begin{bmatrix} 1 & -2 & 3 & 1 \\ 0 & 5 & -7 & 0 \\ 0 & 0 & (l-4) & (m-6) \end{bmatrix} \quad (R_3 \to R_3 - 2R_2) = B$$

Matrix B is in Echelon form.

(i) If $\rho(A) = 3$, then $\rho(B) = 3$, $\therefore$ B should have 3 non zero rows.

$\therefore \quad l \neq 4$ (or) $m \neq 6$.

(ii) If $\rho(A) = 2$, then $\rho(B) = 2$, $\therefore$ B should have 2 non-zero rows.

$\therefore \quad l = 4$ and $m = 6$.

***Example* 1.14:** Find the rank of the matrix $\begin{bmatrix} 0 & 1 & 2 & 2 \\ 1 & 1 & 2 & 3 \\ 3 & 4 & 8 & 11 \\ 1 & 3 & 6 & 7 \end{bmatrix}$.

Solution: The given matrix be A. Then,

$$A \sim \begin{bmatrix} 1 & 1 & 2 & 3 \\ 0 & 1 & 2 & 2 \\ 3 & 4 & 8 & 11 \\ 1 & 3 & 6 & 7 \end{bmatrix} (R_1 \leftrightarrow R_2)$$

$$\sim \begin{bmatrix} 1 & 1 & 2 & 3 \\ 0 & 1 & 2 & 2 \\ 0 & 1 & 2 & 2 \\ 0 & 2 & 4 & 4 \end{bmatrix}, \begin{Bmatrix} R_3 \to R_3 - 3R_1 \\ R_4 \to R_4 - R_1 \end{Bmatrix}$$

$$\sim \begin{bmatrix} 1 & 1 & 2 & 3 \\ 0 & 1 & 2 & 2 \\ 0 & 0 & 0 & 0 \\ 0 & 0 & 0 & 0 \end{bmatrix} \begin{matrix} R_3 \to R_3 - R_2 \\ R_4 \to R_4 - 2R_2 \end{matrix} = B \text{ (say)}$$

B is in Echelon form with two non zero rows.

$\therefore \quad \rho(A) = \rho(B) = 2.$

***Example* 1.15:** Reduce the matrix $\begin{bmatrix} 1 & 2 & 3 \\ 4 & 6 & 8 \\ 3 & 4 & 5 \end{bmatrix}$ to normal form and hence find its rank.

Solution: Given matrix be A. Then,

$$A \sim \begin{bmatrix} 1 & 2 & 3 \\ 0 & -2 & -4 \\ 0 & -2 & -4 \end{bmatrix}, \begin{Bmatrix} R_2 \to R_2 - 4R_1 \\ R_3 \to R_3 - 3R_1 \end{Bmatrix}$$

$$\sim \begin{bmatrix} 1 & 0 & 0 \\ 0 & -2 & -4 \\ 0 & -2 & -4 \end{bmatrix}, \begin{Bmatrix} C_2 \to C_2 - 2C_1 \\ C_3 \to C_3 - 3C_1 \end{Bmatrix}$$

$$\sim \begin{bmatrix} 1 & 0 & 0 \\ 0 & 1 & 2 \\ 0 & 0 & 0 \end{bmatrix}, \begin{Bmatrix} R_2 \to R_2 \times -\dfrac{1}{2} \\ R_3 \to R_3 - R_2 \end{Bmatrix}$$

$$\sim \begin{bmatrix} 1 & 0 & 0 \\ 0 & 1 & 0 \\ 0 & 0 & 0 \end{bmatrix}, (C_3 \to C_3 - 2C_2),$$

which is of the normal form $\begin{bmatrix} I_2 & 0 \\ 0 & 0 \end{bmatrix}$.

Hence $\rho(A) = 2.$

***Example* 1.16:** Reduce the matrix $\begin{bmatrix} 1 & -2 & 1 & 2 \\ 2 & -2 & 0 & 6 \\ 4 & 2 & 0 & 2 \\ 1 & -1 & 0 & 3 \end{bmatrix}$ into the normal form and hence find its rank.

Solution: Given matrix be A. Then

$$A \sim \begin{bmatrix} 1 & -2 & 1 & 2 \\ 0 & 2 & -2 & 2 \\ 0 & 10 & -4 & -6 \\ 0 & 1 & -1 & 1 \end{bmatrix}, \begin{matrix} R_2 \to R_2 - 2R_1 \\ R_3 \to R_3 - 4R_1 \\ R_4 \to R_4 - R_1 \end{matrix}$$

$$\sim \begin{bmatrix} 1 & 0 & 0 & 0 \\ 0 & 2 & -2 & 2 \\ 0 & 10 & -4 & -6 \\ 0 & 1 & -1 & 1 \end{bmatrix}, \begin{matrix} C_2 \to C_2 + 2C_1 \\ C_3 \to C_3 - C_1 \\ C_4 \to C_4 - 2C_1 \end{matrix}$$

$$\sim \begin{bmatrix} 1 & 0 & 0 & 0 \\ 0 & 1 & -1 & 1 \\ 0 & 10 & -4 & -6 \\ 0 & 1 & -1 & 1 \end{bmatrix} R_2 \to \frac{1}{2}R_2$$

$$\sim \begin{bmatrix} 1 & 0 & 0 & 0 \\ 0 & 1 & 0 & 1 \\ 0 & 0 & 6 & -6 \\ 0 & 1 & 0 & 1 \end{bmatrix} \begin{matrix} C_2 \to C_2 + C_3 + C_4 \\ C_3 \to C_3 + C_2 \end{matrix}$$

$$\sim \begin{bmatrix} 1 & 0 & 0 & 0 \\ 0 & 1 & 0 & 1 \\ 0 & 0 & 1 & -1 \\ 0 & 1 & 0 & 1 \end{bmatrix} R_3 \to \frac{1}{6}R_3$$

$$\sim \begin{bmatrix} 1 & 0 & 0 & 0 \\ 0 & 1 & 0 & 0 \\ 0 & 0 & 1 & -1 \\ 0 & 1 & 0 & 0 \end{bmatrix} C_4 \to C_4 - C_2$$

$$\sim \begin{bmatrix} 1 & 0 & 0 & 0 \\ 0 & 1 & 0 & 0 \\ 0 & 0 & 1 & -1 \\ 0 & 0 & 0 & 0 \end{bmatrix} R_4 \to R_4 - R_2$$

$$\sim \begin{bmatrix} 1 & 0 & 0 & 0 \\ 0 & 1 & 0 & 0 \\ 0 & 0 & 1 & 0 \\ 0 & 0 & 0 & 0 \end{bmatrix} C_4 \to C_4 + C_3$$

which is of the normal form $\begin{bmatrix} I_3 & 0 \\ 0 & 0 \end{bmatrix}$. Hence $\rho(A) = 3$.

***Example* 1.17:** Reduce the matrix $\begin{bmatrix} 1 & 0 & -3 & 2 \\ 0 & 1 & 4 & 5 \\ 1 & 3 & 2 & 0 \\ 1 & 1 & -2 & 0 \end{bmatrix}$, to normal form and find its rank.

Solution: Given matrix be A. Then

$$A \sim \begin{bmatrix} 1 & 0 & -3 & 2 \\ 0 & 1 & 4 & 5 \\ 0 & 3 & 5 & -2 \\ 0 & 1 & 1 & -2 \end{bmatrix}, \begin{matrix} R_3 \to R_3 - R_1 \\ R_4 \to R_4 - R_1 \end{matrix}$$

$$\sim \begin{bmatrix} 1 & 0 & 0 & 0 \\ 0 & 1 & 4 & 5 \\ 0 & 3 & 5 & -2 \\ 0 & 1 & 1 & -2 \end{bmatrix}, \begin{matrix} C_3 \to C_3 + 3C_1 \\ C_4 \to C_4 - 2C_1 \end{matrix}$$

$$\sim \begin{bmatrix} 1 & 0 & 0 & 0 \\ 0 & 1 & 4 & 5 \\ 0 & 0 & -7 & -17 \\ 0 & 0 & -3 & -7 \end{bmatrix}, \begin{matrix} R_3 \to R_3 - 3R_2 \\ R_4 \to R_4 - R_2 \end{matrix}$$

$$\sim \begin{bmatrix} 1 & 0 & 0 & 0 \\ 0 & 1 & 0 & 0 \\ 0 & 0 & -7 & -17 \\ 0 & 0 & -3 & -7 \end{bmatrix}, \begin{matrix} C_3 \to C_3 - 4C_2 \\ C_4 \to C_4 - 5C_2 \end{matrix}$$

$$\sim \begin{bmatrix} 1 & 0 & 0 & 0 \\ 0 & 1 & 0 & 0 \\ 0 & 0 & 1 & \frac{17}{7} \\ 0 & 0 & -3 & -7 \end{bmatrix}, R_3 \to \frac{-1}{7} R_3$$

$$\sim \begin{bmatrix} 1 & 0 & 0 & 0 \\ 0 & 1 & 0 & 0 \\ 0 & 0 & 1 & \frac{17}{7} \\ 0 & 0 & 0 & \frac{2}{7} \end{bmatrix}, R_4 \to R_4 + 3R_3$$

$$\sim \begin{bmatrix} 1 & 0 & 0 & 0 \\ 0 & 1 & 0 & 0 \\ 0 & 0 & 1 & 0 \\ 0 & 0 & 0 & \frac{2}{7} \end{bmatrix}, C_4 \to C_4 - \frac{17}{7} C_3$$

$$\sim \begin{bmatrix} 1 & 0 & 0 & 0 \\ 0 & 1 & 0 & 0 \\ 0 & 0 & 1 & 0 \\ 0 & 0 & 0 & 1 \end{bmatrix}, R_4 \to \frac{7}{2} R_4$$

which is in the normal form I_4. $\therefore$ $\rho(A) = 4$

***Example* 1.18:** If $A = \begin{bmatrix} 1 & -1 & -1 & 2 \\ 4 & 2 & 2 & -1 \\ 2 & 2 & 0 & -2 \end{bmatrix}$, find two non-singular matrices P and Q such that PAQ is in the normal form.

Method of finding two non-singular matrices P and Q such that PAQ is in the normal form where A is any given $m \times n$ matrix.

Step 1 :

Write $\quad A = I_m A I_n, \quad$ .....(1.22)

where I_m, I_n are unit matrices of orders m and n respectively.

Step 2 :

Subject the matrix A on the L.H.S of (1.22) to elementary row and column transformations to reduce it to the normal form. Perform each of the same row transformations on I_m of R.H.S and each of the same column transformations on I_n of R.H.S. In all the steps keep the matrix A on R.H.S as it is.

Solution: Given matrix be A. Since it is of order 3×4, we write

$$A = I_3 A I_4$$

i.e.,
$$\begin{bmatrix} 1 & -1 & -1 & 2 \\ 4 & 2 & 2 & -1 \\ 2 & 2 & 0 & -2 \end{bmatrix} = \begin{bmatrix} 1 & 0 & 0 \\ 0 & 1 & 0 \\ 0 & 0 & 1 \end{bmatrix} A \begin{bmatrix} 1 & 0 & 0 & 0 \\ 0 & 1 & 0 & 0 \\ 0 & 0 & 1 & 0 \\ 0 & 0 & 0 & 1 \end{bmatrix}$$

Applying $R_2 \to R_2 - 4R_1$, $R_3 \to R_3 - 2R_1$ on L.H.S as well as the prefactor (I_3) of A on R.H.S, we get,

$$\begin{bmatrix} 1 & -1 & -1 & 2 \\ 0 & 6 & 6 & -9 \\ 0 & 4 & 2 & -6 \end{bmatrix} = \begin{bmatrix} 1 & 0 & 0 \\ -4 & 1 & 0 \\ -2 & 0 & 1 \end{bmatrix} A \begin{bmatrix} 1 & 0 & 0 & 0 \\ 0 & 1 & 0 & 0 \\ 0 & 0 & 1 & 0 \\ 0 & 0 & 0 & 1 \end{bmatrix}$$

Applying $C_2 \to C_2 + C_1$, $C_3 \to C_3 + C_1$, $C_4 \to C_4 - 2C_1$, on L.H.S as well as the postfactor (I_4) of A on R.H.S, we get,

$$\begin{bmatrix} 1 & 0 & 0 & 0 \\ 0 & 6 & 6 & -9 \\ 0 & 4 & 2 & -6 \end{bmatrix} = \begin{bmatrix} 1 & 0 & 0 \\ -4 & 1 & 0 \\ -2 & 0 & 1 \end{bmatrix} A \begin{bmatrix} 1 & 1 & 1 & -2 \\ 0 & 1 & 0 & 0 \\ 0 & 0 & 1 & 0 \\ 0 & 0 & 0 & 1 \end{bmatrix}$$

Applying $R_2 \to \frac{1}{6} R_2$ on L.H.S as well as the prefactor on R.H.S, we get,

$$\begin{bmatrix} 1 & 0 & 0 & 0 \\ 0 & 1 & 1 & \frac{-3}{2} \\ 0 & 4 & 2 & -6 \end{bmatrix} = \begin{bmatrix} 1 & 0 & 0 \\ \frac{-2}{3} & \frac{1}{6} & 0 \\ -2 & 0 & 1 \end{bmatrix} A \begin{bmatrix} 1 & 1 & 1 & -2 \\ 0 & 1 & 0 & 0 \\ 0 & 0 & 1 & 0 \\ 0 & 0 & 0 & 1 \end{bmatrix}$$

Applying $R_3 \to R_3 - 4R_2$ on L.H.S as well as prefactor on R.H.S, we get,

$$\begin{bmatrix} 1 & 0 & 0 & 0 \\ 0 & 1 & 1 & \frac{-3}{2} \\ 0 & 0 & -2 & 0 \end{bmatrix} = \begin{bmatrix} 1 & 0 & 0 \\ \frac{-2}{3} & \frac{1}{6} & 0 \\ \frac{2}{3} & \frac{-2}{3} & 1 \end{bmatrix} A \begin{bmatrix} 1 & 1 & 1 & -2 \\ 0 & 1 & 0 & 0 \\ 0 & 0 & 1 & 0 \\ 0 & 0 & 0 & 1 \end{bmatrix}$$

Applying $C_3 \to C_3 - C_2$ on L.H.S as well as the postfactor on R.H.S, we get,

$$\begin{bmatrix} 1 & 0 & 0 & 0 \\ 0 & 1 & 0 & \frac{-3}{2} \\ 0 & 0 & -2 & 0 \end{bmatrix} = \begin{bmatrix} 1 & 0 & 0 \\ \frac{-2}{3} & \frac{1}{6} & 0 \\ \frac{2}{3} & \frac{-2}{3} & 1 \end{bmatrix} A \begin{bmatrix} 1 & 1 & 0 & -2 \\ 0 & 1 & -1 & 0 \\ 0 & 0 & 1 & 0 \\ 0 & 0 & 0 & 1 \end{bmatrix}$$

Applying $R_3 \to \frac{-1}{2} R_3$ on L.H.S as well as the prefactor on R.H.S, we get,

$$\begin{bmatrix} 1 & 0 & 0 & 0 \\ 0 & 1 & 0 & \frac{-3}{2} \\ 0 & 0 & 1 & 0 \end{bmatrix} = \begin{bmatrix} 1 & 0 & 0 \\ \frac{-3}{2} & \frac{1}{6} & 0 \\ \frac{-1}{3} & \frac{1}{3} & \frac{-1}{2} \end{bmatrix} A \begin{bmatrix} 1 & 1 & 0 & -2 \\ 0 & 1 & -1 & 0 \\ 0 & 0 & 1 & 0 \\ 0 & 0 & 0 & 1 \end{bmatrix}$$

Applying $C_4 \to C_4 + \left(\frac{3}{2}\right) C_2$ on L.H.S as well as postfactor on R.H.S, we get,

$$\begin{bmatrix} 1 & 0 & 0 & 0 \\ 0 & 1 & 0 & 0 \\ 0 & 0 & 1 & 0 \end{bmatrix} = \begin{bmatrix} 1 & 0 & 0 \\ \frac{-2}{3} & \frac{1}{6} & 0 \\ \frac{-1}{3} & \frac{1}{3} & \frac{-1}{2} \end{bmatrix} A \begin{bmatrix} 1 & 1 & 0 & \frac{-1}{2} \\ 0 & 1 & -1 & \frac{3}{2} \\ 0 & 0 & 1 & 0 \\ 0 & 0 & 0 & 1 \end{bmatrix}$$

The L.H.S which is now of the form $[I_3\ 0]$ is in the normal form. Hence the two required matrices are

$$P = \begin{bmatrix} 1 & 0 & 0 \\ \frac{-2}{3} & \frac{1}{6} & 0 \\ \frac{-1}{3} & \frac{1}{3} & \frac{-1}{2} \end{bmatrix} \text{ and } Q = \begin{bmatrix} 1 & 1 & 0 & \frac{-1}{2} \\ 0 & 1 & -1 & \frac{3}{2} \\ 0 & 0 & 1 & 0 \\ 0 & 0 & 0 & 1 \end{bmatrix}$$

***Example* 1.19:** Find two non-singular matrices P and Q such that P AQ will be in the normal form where

$$A = \begin{bmatrix} 1 & 2 & 3 \\ 2 & -1 & 0 \\ 3 & 1 & 2 \end{bmatrix}$$

Solution: Here A is 3 × 3 matrix, hence we write $A = I_3 AI_3$

i.e.,
$$\begin{bmatrix} 1 & 2 & 3 \\ 2 & -1 & 0 \\ 3 & 1 & 2 \end{bmatrix} = \begin{bmatrix} 1 & 0 & 0 \\ 0 & 1 & 0 \\ 0 & 0 & 1 \end{bmatrix} A \begin{bmatrix} 1 & 0 & 0 \\ 0 & 1 & 0 \\ 0 & 0 & 1 \end{bmatrix}$$

Applying $R_2 \to R_2 - 2R_1$, $R_3 \to R_3 - 3R_1$ on L.H.S as well as the prefactor of A on R.H.S, we get,

$$\begin{bmatrix} 1 & 2 & 3 \\ 0 & -5 & -6 \\ 0 & -5 & -7 \end{bmatrix} = \begin{bmatrix} 1 & 0 & 0 \\ -2 & 1 & 0 \\ -3 & 0 & 1 \end{bmatrix} A \begin{bmatrix} 1 & 0 & 0 \\ 0 & 1 & 0 \\ 0 & 0 & 1 \end{bmatrix}$$

Applying $C_2 \to C_2 - 2C_1$, $C_3 \to C_3 - 3C_1$ on L.H.S as well as the postfactor on R.H.S, we get,

$$\begin{bmatrix} 1 & 0 & 0 \\ 0 & -5 & -6 \\ 0 & -5 & -7 \end{bmatrix} = \begin{bmatrix} 1 & 0 & 0 \\ -2 & 1 & 0 \\ -3 & 0 & 1 \end{bmatrix} A \begin{bmatrix} 1 & -2 & -3 \\ 0 & 1 & 0 \\ 0 & 0 & 1 \end{bmatrix}$$

Applying $R_2 \to \frac{-1}{5} R_2$ on L.H.S as well as the prefactor of R.H.S, we get

$$\begin{bmatrix} 1 & 0 & 0 \\ 0 & 1 & \frac{6}{5} \\ 0 & -5 & -7 \end{bmatrix} = \begin{bmatrix} 1 & 0 & 0 \\ \frac{2}{5} & \frac{-1}{5} & 0 \\ -3 & 0 & 1 \end{bmatrix} A \begin{bmatrix} 1 & -2 & -3 \\ 0 & 1 & 0 \\ 0 & 0 & 1 \end{bmatrix}$$

Applying $R_3 \to R_3 + 5R_2$ on L.H.S and the prefactor on R.H.S, we get

$$\begin{bmatrix} 1 & 0 & 0 \\ 0 & 1 & \frac{6}{5} \\ 0 & 0 & -1 \end{bmatrix} = \begin{bmatrix} 1 & 0 & 0 \\ \frac{2}{5} & \frac{-1}{5} & 0 \\ -1 & -1 & 1 \end{bmatrix} A \begin{bmatrix} 1 & -2 & -3 \\ 0 & 1 & 0 \\ 0 & 0 & 1 \end{bmatrix}$$

Applying $C_3 \to C_3 - \frac{6}{5}C_2$, on L.H.S and the postfactor on R.H.S, we get

$$\begin{bmatrix} 1 & 0 & 0 \\ 0 & 1 & 0 \\ 0 & 0 & -1 \end{bmatrix} = \begin{bmatrix} 1 & 0 & 0 \\ \frac{2}{5} & \frac{-1}{5} & 0 \\ -1 & -1 & 1 \end{bmatrix} A \begin{bmatrix} 1 & -2 & \frac{-3}{5} \\ 0 & 1 & \frac{-6}{5} \\ 0 & 0 & 1 \end{bmatrix}$$

Applying $R_3 \to -R_3$ on L.H.S and the prefactor of R.H.S, we get,

$$\begin{bmatrix} 1 & 0 & 0 \\ 0 & 1 & 0 \\ 0 & 0 & 1 \end{bmatrix} = \begin{bmatrix} 1 & 0 & 0 \\ \frac{2}{5} & \frac{-1}{5} & 0 \\ 1 & 1 & -1 \end{bmatrix} A \begin{bmatrix} 1 & -2 & \frac{-3}{5} \\ 0 & 1 & \frac{-6}{5} \\ 0 & 0 & 1 \end{bmatrix}$$

Now the L.H.S is in the normal form (I_3). Hence the required matrices are

$$P = \begin{bmatrix} 1 & 0 & 0 \\ 2/5 & -1/5 & 0 \\ 1 & 1 & -1 \end{bmatrix} \text{ and } Q = \begin{bmatrix} 1 & -2 & \frac{-3}{5} \\ 0 & 1 & \frac{-6}{5} \\ 0 & 0 & 1 \end{bmatrix}$$

Exercise - 1.3

I. Reduce the following matrices to Echelon form and hence find their ranks.

(a) $\begin{bmatrix} 1 & 3 & 5 \\ 2 & -1 & 0 \\ 3 & 1 & 4 \end{bmatrix}$ **[Ans : 3]**

(b) $\begin{bmatrix} 1 & -1 & 0 & 2 \\ 2 & -3 & 1 & 4 \\ 3 & 4 & -2 & 6 \end{bmatrix}$ **[Ans : 3]**

(c) $\begin{bmatrix} 3 & 4 & 5 & 2 \\ -1 & 2 & 0 & -3 \\ 1 & -1 & 1 & 4 \end{bmatrix}$. **[Ans : 3]**

(d) $\begin{bmatrix} 1 & 2 & 3 & 4 \\ 2 & 3 & 4 & 5 \\ 3 & 4 & 5 & 6 \\ 4 & 5 & 6 & 7 \end{bmatrix}$ **[Ans : 2]**

II Reduce the following matrices to the normal form and hence find their ranks.

(a) $\begin{bmatrix} 0 & 1 & -3 & -1 \\ 1 & 0 & 1 & 1 \\ 3 & 1 & 0 & 2 \\ 1 & 1 & -2 & 0 \end{bmatrix}$ **[Ans : 2]**

(b) $\begin{bmatrix} 1 & 2 & -1 & 3 \\ 2 & 4 & -4 & 7 \\ -1 & -2 & -1 & -2 \end{bmatrix}$ **[Ans : 2]**

(c) $\begin{bmatrix} 1 & 2 & 3 & 2 \\ 2 & 3 & 5 & 1 \\ 1 & 3 & 4 & 5 \end{bmatrix}$ **[Ans : 2]**

(d) $\begin{bmatrix} 8 & 1 & 3 & 6 \\ 0 & 3 & 2 & 2 \\ -8 & -1 & -3 & -4 \end{bmatrix}$ **[Ans : 3]**

III Find two non-singular matrices P and Q such that P AQ is in the normal form where A is given by the matrix.

(a) $\begin{bmatrix} 1 & 2 & 3 \\ 3 & 5 & 7 \\ 4 & 6 & 8 \end{bmatrix}$ **Ans :** $P = \begin{bmatrix} 1 & 0 & 0 \\ 3 & -1 & 0 \\ 2 & -2 & 1 \end{bmatrix}$ $Q = \begin{bmatrix} 1 & -2 & 1 \\ 0 & 1 & -2 \\ 0 & 0 & 1 \end{bmatrix}$

(b) $\begin{bmatrix} 1 & -1 & 0 & 2 \\ 2 & 1 & 3 & -2 \\ 0 & 1 & 2 & 4 \end{bmatrix}$ **Ans :** $P = \begin{bmatrix} 1 & 0 & 0 \\ \frac{-2}{3} & \frac{1}{3} & 0 \\ \frac{2}{3} & \frac{-1}{3} & 1 \end{bmatrix}$ $Q = \begin{bmatrix} 1 & 1 & -1 & 3 \\ 0 & 1 & -1 & 4 \\ 0 & 0 & 1 & -3 \\ 0 & 0 & 0 & \frac{1}{2} \end{bmatrix}$

IV **Find the constants 'a' and 'b' such that the matrix** $\begin{bmatrix} 1 & 2 & 3 & 4 \\ -1 & 0 & 2 & -3 \\ 1 & 1 & a & b \end{bmatrix}$ **is of rank**

(i) 3 and (ii) 2.

Ans : (i) $a \neq \frac{1}{2}$ **(or)** $b \neq \frac{7}{2}$

(ii) $a = \frac{1}{2}$ and $b = \frac{7}{2}$

V **If the rank of the matrix** $\begin{bmatrix} 1 & -1 & 3 \\ 2 & 1 & -1 \\ 3 & 0 & a \end{bmatrix}$ **is '2', find the value of 'a'.**

[Ans : 2]

1.6 Simultaneous Linear Equations

In this section, the theory of matrices is applied to determine the existence of solutions of simultaneous linear equations and obtain the solutions if they exist.

Matrix form of a system of non-homogeneous Linear Equations

A system of 'm' simultaneous non-homogeneous linear equations in 'n' unknowns $x_1, x_2, x_3, \ldots x_n$ is of the form

$$\left.\begin{array}{l} a_{11}x_1 + a_{12}x_2 + a_{13}x_3 + \cdots + a_{1n}x_n = b_1 \\ a_{21}x_1 + a_{22}x_2 + a_{23}x_3 + \cdots + a_{2n}x_n = b_2 \\ \ldots\ldots\ldots\ldots\ldots\ldots\ldots\ldots\ldots\ldots \\ \ldots\ldots\ldots\ldots\ldots\ldots\ldots\ldots\ldots\ldots \\ a_{m1}x_1 + a_{m2}x_2 + a_{m3}x_3 + \cdots a_{mn}x_n = b_m \end{array}\right\} \quad \ldots\ldots(1.23)$$

The above system can be written in a matrix form (using the concepts of multiplication of matrices and equality of matrices) as

$$\begin{bmatrix} a_{11} & a_{12} & \ldots\ldots & a_{1n} \\ a_{21} & a_{22} & \ldots\ldots & a_{2n} \\ \ldots & \ldots & \ldots\ldots & \ldots \\ a_{m1} & a_{m2} & \ldots\ldots & a_{mn} \end{bmatrix} \begin{bmatrix} x_1 \\ x_2 \\ \vdots \\ x_n \end{bmatrix} = \begin{bmatrix} b_1 \\ b_2 \\ \vdots \\ b_m \end{bmatrix} \quad \ldots\ldots(1.24)$$

or simply, $AX = B$ .....(1.25)

where $B \neq 0$,

$$A = \begin{bmatrix} a_{11} & a_{12} & \cdots & a_{1n} \\ a_{21} & a_{22} & \cdots & a_{2n} \\ \cdots & \cdots & \cdots & \cdots \\ \cdots & \cdots & \cdots & \cdots \\ a_{m1} & a_{m2} & \cdots & a_{mn} \end{bmatrix}$$, which is called the coefficient matrix,

$$X = \begin{bmatrix} x_1 \\ x_2 \\ \vdots \\ \vdots \\ x_n \end{bmatrix}$$ which is the solution matrix

and $$B = \begin{bmatrix} b_1 \\ b_2 \\ \vdots \\ \vdots \\ b_m \end{bmatrix}$$ is the matrix of the constants $b_1, b_2, \ldots b_m$ of the R.H.S of (1.23)

The set of values $x_1, x_2, \ldots x_n$ which satisfy the system (1.23) is called the solution of the system.

Consistency and Inconsistency

The system of equations (1.23) is said to be consistent if it has at least one solution.

(1.23) is said to be inconsistent if it has no solution.

Augmented matrix (A/B)

The $m \times (n + 1)$ matrix given by

$$\begin{bmatrix} a_{11} & a_{12} & \cdots & a_{1n} & b_1 \\ a_{21} & a_{22} & \cdots & a_{2n} & b_2 \\ \cdots & \cdots & \cdots & \cdots & \cdots \\ \cdots & \cdots & \cdots & \cdots & \cdots \\ a_{m1} & a_{m2} & \cdots & a_{mn} & b_m \end{bmatrix}$$

(which is obtained by annexing the elements of B to those of (A) is called the augmented matrix and is denoted by (A/B) or (A, B).

Condition for the consistency of (1)

1. The condition for the consistency of (1) is given by the following theorem (with out proof).
2. ***Theorem*** : The system of simultaneous linear equations AX = B is consistent if and only if the rank of A = rank of (A/B).

Note: 1. Thus the ranks of the coefficient matrix and the augmented matrix play a vital role in determining the consistency of the system (1).

2. We apply the elementary transformations to the augmented matrix (A/B) since they do not alter its rank. This reduced equivalent system will enable us to test for consistency and also to find the solution.

Working Rules

Step 1 : Write the system in the form AX = B.

Step 2 : Find ρ(A) and ρ(A/B) by applying elementary transformations on (A/B).

Conclusions :

Case (i) : If ρ(A) = ρ(A/B) = n, where 'n' is the number of unknowns, the system is consistent and it has a unique solution [in this case A is non-singular]. This unique solution (given by $X = A^{-1}B$) can be obtained by the reduced system of equations.

Case (ii) : If ρ(A) = ρ(A/B) < n, then the system is consistent and has an infinite number of solutions.

Case (iii) : If ρ(A) ≠ ρ(A/B), the system is inconsistent and has no solution.

Note : ρ(A) ≯ ρ(A/B)

Summary

The nature of solutions of the system AX = B with respect to its consistency or inconsistency is shown in the following table.

Consistent	ρ(A, B) = ρ(A) = n	unique solution
	ρ(A, B) = ρ(A) < n	Infinite solutions
Inconsistent	ρ(A, B) ≠ ρ(A)	No solution
	i.e, ρ(A, B) > ρ(A)	

Some other methods of solving non-homogenous linear equations

In this section, certain methods of solving non-homogeneous equations (in which the concept of the rank of a matrix need not be used) are discussed.

Note that these are applicable only when the coefficient matrix of the system is a square matrix.

Method 1 : *Matrix inversion method :*

The solution of the system AX = B is given by $X = A^{-1}B$ provided A^{-1} exists.

∴ This method is applicable only when A is nonsingular.

Method 2 : *CRAMER'S RULE (using determinants) :*

To apply this method A must be non-singular.

i.e., $|A| \neq 0$

Let A be 3 × 3 matrix.

Let $|A| = \Delta$

we find the values of 3 other determinants which are obtained by replacing the 1st, 2nd and 3rd columns of Δ with the column matrix 'B' of the given system respectively.

Let Δ_1, Δ_2, Δ_3 be the values of these determinants respectively.

Then, $x = \dfrac{\Delta_1}{\Delta}, y = \dfrac{\Delta_2}{\Delta}, z = \dfrac{\Delta_3}{\Delta}$

Method 3 : *Gauss-Jordan method:*

Take the matrix (A, B) and subject it to a series of elementary row transformations till it is reduced to the form $(I_n X)$ where I_n is unit matrix of order n and X is the column matrix of the order of B. Then 'X' is the solution matrix.

Homogeneous Linear equations

The system of equations AX = 0 is said to be homogeneous where

$$A = \begin{bmatrix} a_{11} & a_{12} & \cdots & a_{1n} \\ a_{21} & a_{22} & \cdots & a_{2n} \\ \cdots & \cdots & \cdots & \cdots \\ \cdots & \cdots & \cdots & \cdots \\ a_{m1} & a_{m2} & \cdots & a_{mn} \end{bmatrix}$$

$$X = \begin{bmatrix} x_1 \\ x_2 \\ \vdots \\ x_n \end{bmatrix}_{n\times 1} \text{ and } 0 = \begin{bmatrix} 0 \\ 0 \\ \vdots \\ 0 \end{bmatrix}_{m\times 1}$$

Here the augmented matrix is

$$(A/B) = (A/0)$$

Hence $\rho(A/B) = \rho(A)$

Conclusions :

$\therefore$ 1. The system AX = 0 is always consistent. [Obviously, the trivial solution $x_1 = x_2 = ... = x_n = 0$ always exists].

2. If $\rho(A/B) = \rho(A) = n$ (i.e., is $|A| \neq 0$) then the trival solution is the unique solution.

3. If $\rho(A/B) = \rho(A) = r < n$, the system has non-trivial solutions. (In this case $|A| = 0$).

   Let $\rho(A/B) = \rho(A) = r < n$. Then the system will have $(n - r)$ linearly independent solutions. The values of 'r' unknowns can be expressed in these arbitrarily chosen $(n - r)$ unknowns.

Solved Examples

***Example* 1.20:** Test the system of equations

$$2x + y + 5z = 4;\ 3x - 2y + 2z = 2\ ;\ 5x - 8y - 4z = 1$$

for consistency. If consistent solve them.

Solution: The given system of equations can be put in the matrix form AX = B where,

$$A = \begin{bmatrix} 2 & 1 & 5 \\ 3 & -2 & 2 \\ 5 & -8 & -4 \end{bmatrix}, X = \begin{bmatrix} x \\ y \\ z \end{bmatrix}, B = \begin{bmatrix} 4 \\ 2 \\ 1 \end{bmatrix}$$

Augmented matrix

$$(A/B) = \begin{bmatrix} 2 & 1 & 5 & 4 \\ 3 & -2 & 2 & 2 \\ 5 & -8 & -4 & 1 \end{bmatrix}$$

$$\sim \begin{bmatrix} 1 & \frac{1}{2} & \frac{5}{2} & 2 \\ 3 & -2 & 2 & 2 \\ 5 & -8 & -4 & 1 \end{bmatrix}, R_1 \rightarrow \frac{1}{2} R_1$$

$$\sim \begin{bmatrix} 1 & \frac{1}{2} & \frac{5}{2} & 2 \\ 0 & \frac{-7}{2} & \frac{-11}{2} & -4 \\ 0 & \frac{-21}{2} & \frac{-33}{2} & -9 \end{bmatrix}, \begin{matrix} R_2 \rightarrow R_2 - 3R_1 \\ R_3 \rightarrow R_3 - 5R_1 \end{matrix}$$

$$\sim \begin{bmatrix} 1 & \frac{1}{2} & \frac{5}{2} & 2 \\ 0 & 1 & \frac{11}{7} & \frac{8}{7} \\ 0 & \frac{-21}{2} & \frac{-33}{2} & -9 \end{bmatrix} R_2 \to \frac{-2}{7} R_2$$

$$\sim \begin{bmatrix} 1 & 0 & \frac{12}{7} & \frac{10}{7} \\ 0 & 1 & \frac{11}{7} & \frac{8}{7} \\ 0 & 0 & 0 & 3 \end{bmatrix} \begin{matrix} R_1 \to R_1 - \frac{1}{2} R_2 \\ R_3 \to R_3 + \frac{21}{2} R_2 \end{matrix}$$

From the above equivalent matrix, we find that rank of A = 2; rank of (A/B) = 3

$\therefore \quad \rho(A) \neq \rho(A/B)$

Hence the system is not consistent

$\therefore$ It has no solution.

***Example* 1.21:** Solve the system $2x - y + 4z = 12$; $3x + 2y + z = 10$; $x + y + z = 6$; if it is consistent.

Solution: The matrix form of the system is

$$\begin{bmatrix} 2 & -1 & 4 \\ 3 & 2 & 1 \\ 1 & 1 & 1 \end{bmatrix} \begin{bmatrix} x \\ y \\ z \end{bmatrix} = \begin{bmatrix} 12 \\ 10 \\ 6 \end{bmatrix} \qquad \text{.....(1.26)}$$

or $\quad AX = B$

where $\quad A = \begin{bmatrix} 2 & -1 & 4 \\ 3 & 2 & 1 \\ 1 & 1 & 1 \end{bmatrix}, X = \begin{bmatrix} x \\ y \\ z \end{bmatrix}, B = \begin{bmatrix} 12 \\ 10 \\ 6 \end{bmatrix}$

Augmented matrix

$$(A, B) = \begin{bmatrix} 2 & -1 & 4 & 12 \\ 3 & 2 & 1 & 10 \\ 1 & 1 & 1 & 6 \end{bmatrix}$$

$$\sim \begin{bmatrix} 1 & 1 & 1 & 6 \\ 3 & 2 & 1 & 10 \\ 2 & -1 & 4 & 12 \end{bmatrix}, R_1 \leftrightarrow R_3$$

$$\sim\begin{bmatrix}1 & 1 & 1 & 6\\ 0 & -1 & -2 & -8\\ 0 & -3 & 2 & 0\end{bmatrix}\begin{matrix}R_2 = R_2 - 3R_1\\ R_3 = R_3 - 2R_1\end{matrix}$$

$$\sim\begin{bmatrix}1 & 1 & 1 & 6\\ 0 & 1 & 2 & 8\\ 0 & -3 & 2 & 0\end{bmatrix} R_2 \to -R_2$$

$$\sim\begin{bmatrix}1 & 0 & -1 & -2\\ 0 & 1 & 2 & 8\\ 0 & 0 & 8 & 24\end{bmatrix}\begin{matrix}R_1 \to R_1 - R_2\\ R_3 \to R_3 + 3R_2\end{matrix}$$

$$\sim\begin{bmatrix}1 & 0 & -1 & -2\\ 0 & 1 & 2 & 8\\ 0 & 0 & 1 & 3\end{bmatrix} R_3 \to \frac{1}{8}R_3$$

$$\sim\begin{bmatrix}1 & 0 & 0 & 1\\ 0 & 1 & 0 & 2\\ 0 & 0 & 1 & 3\end{bmatrix}, \begin{matrix}R_1 \to R_1 + R_3\\ R_2 \to R_2 - 2R_3\end{matrix} \qquad .....(1.27)$$

which is in Echelon form.

$$\rho(A, B) = \rho(A) = 3$$

So the system is consistent and has unique solution.

Further (1.27) is equivalent to

$$\begin{bmatrix}1 & 0 & 0\\ 0 & 1 & 0\\ 0 & 0 & 1\end{bmatrix}\begin{bmatrix}x\\ y\\ z\end{bmatrix} = \begin{bmatrix}1\\ 2\\ 3\end{bmatrix}$$

$\therefore$ $x = 1 ; y = 2 ; z = 3$

***Example* 1.22:** If consistent, solve the system of equations,

$$x + y + z + t = 4$$
$$x - z + 2t = 2$$
$$y + z - 3t = -1$$
$$x + 2y - z + t = 3$$

Solution: The system can be written as

$$\begin{bmatrix}1 & 1 & 1 & 1\\ 1 & 0 & -1 & 2\\ 0 & 1 & 1 & -3\\ 1 & 2 & -1 & 1\end{bmatrix}\begin{bmatrix}x\\ y\\ z\\ t\end{bmatrix} = \begin{bmatrix}4\\ 2\\ -1\\ 3\end{bmatrix} \qquad .....(1.28)$$

The augmented matrix

$$(A, B) = \begin{bmatrix} 1 & 1 & 1 & 1 & 4 \\ 1 & 0 & -1 & 2 & 2 \\ 0 & 1 & 1 & -3 & -1 \\ 1 & 2 & -1 & 1 & 3 \end{bmatrix}$$

$$\sim \begin{bmatrix} 1 & 1 & 1 & 1 & 4 \\ 0 & -1 & -2 & 1 & -2 \\ 0 & 1 & 1 & -3 & -1 \\ 0 & 1 & -2 & 0 & -1 \end{bmatrix} \begin{matrix} R_2 \to R_2 - R_1 \\ R_4 \to R_4 - R_1 \end{matrix}$$

$$\sim \begin{bmatrix} 1 & 1 & 1 & 1 & 4 \\ 0 & 1 & 2 & -1 & 2 \\ 0 & 1 & 1 & -3 & -1 \\ 0 & 1 & -2 & 0 & -1 \end{bmatrix} R_2 \to -R_1$$

$$\sim \begin{bmatrix} 1 & 0 & -1 & 2 & 2 \\ 0 & 1 & 2 & -1 & 2 \\ 0 & 0 & -1 & -2 & -3 \\ 0 & 0 & -4 & 1 & -3 \end{bmatrix} \begin{matrix} R_1 \to R_1 - R_2 \\ R_3 \to R_3 - R_2 \\ R_4 \to R_4 - R_2 \end{matrix}$$

$$\sim \begin{bmatrix} 1 & 0 & -1 & 2 & 2 \\ 0 & 1 & 2 & -1 & 2 \\ 0 & 0 & 1 & 2 & 3 \\ 0 & 0 & -4 & 1 & -3 \end{bmatrix} R_3 \to -R_3$$

$$\sim \begin{bmatrix} 1 & 0 & 0 & 4 & 5 \\ 0 & 1 & 0 & -5 & -4 \\ 0 & 0 & 1 & 2 & 3 \\ 0 & 0 & 0 & 9 & 9 \end{bmatrix} \begin{matrix} R_1 \to R_1 + R_3 \\ R_2 \to R_2 - 2R_3 \\ R_4 \to R_4 + 4R_3 \end{matrix}$$

$$\sim \begin{bmatrix} 1 & 0 & 0 & 4 & 5 \\ 0 & 1 & 0 & -5 & -4 \\ 0 & 0 & 1 & 2 & 3 \\ 0 & 0 & 0 & 1 & 1 \end{bmatrix} R_4 \to \frac{1}{9} R_1$$

$$\sim \begin{bmatrix} 1 & 0 & 0 & 0 & 1 \\ 0 & 1 & 0 & 0 & 1 \\ 0 & 0 & 1 & 0 & 1 \\ 0 & 0 & 0 & 1 & 1 \end{bmatrix} \begin{matrix} R_1 \to R_1 - 4R_4 \\ R_2 \to R_2 + 5R_4 \\ R_3 \to R_3 - 2R_4 \end{matrix} \qquad .....(1.29)$$

which is in Echelon form

$$\rho(A) = \rho(A/B) = 4 = \text{number of unknowns}$$

∴ The system (1.28) is consistent and has a unique solution.

Again from the reduced form (1.29) of (A/B), the system becomes,

$$\begin{bmatrix} 1 & 0 & 0 & 0 \\ 0 & 1 & 0 & 0 \\ 0 & 0 & 1 & 0 \\ 0 & 0 & 0 & 1 \end{bmatrix} \begin{bmatrix} x \\ y \\ z \\ t \end{bmatrix} = \begin{bmatrix} 1 \\ 1 \\ 1 \\ 1 \end{bmatrix}$$

⇒ $x = 1,\ y = 1,$
$z = 1,$ and $t = 1;$

***Example* 1.23:** Test for consistency and if consistent solve the system,

$$5x + 3y + 7t = 4$$
$$3x + 26y + 2t = 9$$
$$7x + 2y + 10t = 5$$

Solution: The given system can be written as

$$\begin{bmatrix} 5 & 3 & 7 \\ 3 & 26 & 2 \\ 7 & 2 & 10 \end{bmatrix} \begin{bmatrix} x \\ y \\ t \end{bmatrix} = \begin{bmatrix} 4 \\ 9 \\ 5 \end{bmatrix} \qquad .....(1.30)$$

or $AX = B,$

where $A = \begin{bmatrix} 5 & 3 & 7 \\ 3 & 26 & 2 \\ 7 & 2 & 10 \end{bmatrix}, X = \begin{bmatrix} x \\ y \\ t \end{bmatrix}$ and $B = \begin{bmatrix} 4 \\ 9 \\ 5 \end{bmatrix}$

The augmented matrix

$$(A, B) = \begin{bmatrix} 5 & 3 & 7 & 4 \\ 3 & 26 & 2 & 9 \\ 7 & 2 & 10 & 5 \end{bmatrix}$$

$$\sim \begin{bmatrix} 5 & 3 & 7 & 4 \\ 3 & 26 & 2 & 9 \\ 2 & -1 & 3 & 1 \end{bmatrix} R_3 \to R_3 - R_1$$

$$\sim \begin{bmatrix} 5 & 3 & 7 & 4 \\ 1 & 27 & -1 & 8 \\ 2 & -1 & 3 & 1 \end{bmatrix} R_2 \to R_2 - R_3$$

$$\sim \begin{bmatrix} 1 & 27 & -1 & 8 \\ 5 & 3 & 7 & 4 \\ 2 & -1 & 3 & 1 \end{bmatrix} R_1 \leftrightarrow R_2$$

$$\sim \begin{bmatrix} 1 & 27 & -1 & 8 \\ 0 & -132 & 12 & -36 \\ 0 & -55 & 5 & -15 \end{bmatrix} \begin{matrix} R_2 \to R_2 - 5R_1 \\ R_3 \to R_3 - 2R_1 \end{matrix}$$

$$\sim \begin{bmatrix} 1 & 27 & -1 & 8 \\ 0 & 1 & -\frac{1}{11} & \frac{3}{11} \\ 0 & -55 & 5 & -15 \end{bmatrix} R_2 \to \frac{-1}{132} R_2$$

$$\sim \begin{bmatrix} 1 & 0 & 16/11 & 7/11 \\ 0 & 1 & -1/11 & 3/11 \\ 0 & 0 & 0 & 0 \end{bmatrix} \begin{matrix} R_1 \to R_1 - 27R_2 \\ R_3 \to R_3 + 55R_2 \end{matrix} \qquad .....(1.31)$$

which is in Echelon form.

$\therefore$ $\rho(A, B) = \rho(A) = 2 <$ the number of unknowns.

$\therefore$ The system (1.30) is consistent with infinite number of solutions.

$\rho(A) = r = 2$, number of unknowns $n = 3$.

Hence we have to consider $(3 - 2 = 1)$ one variable as an arbitrary constant and the other two variables are expressed in terms of this constant.

The final reduced system (1.31) can be written as

$$\left.\begin{aligned} x + \frac{16t}{11} = \frac{7}{11} \\ y - \frac{t}{11} = \frac{3}{11} \end{aligned}\right\} \qquad .....(1.32)$$

$$\Rightarrow \qquad y = \frac{3+t}{11}; \; x = \frac{7-16t}{11}$$

If t = c, where c is an arbitrary constant, we have, the solution of the system (1.30) as,

$$x = \frac{1}{11}(7-16c), \; y = \frac{1}{11}(c+3),$$

For different values of '*c*' we get different solutions. Thus we get infinite number of solutions.

***Example* 1.24:** Solve the system $x - z - 2 = 0$; $y - z + 2 = 0$; $x + y - 2z = 0$; if it is consistent.

***Solution*:** The system is AX = B, where

$$A = \begin{bmatrix} 1 & 0 & -1 \\ 0 & 1 & -1 \\ 1 & 1 & -2 \end{bmatrix}, X = \begin{bmatrix} x \\ y \\ z \end{bmatrix}, B = \begin{bmatrix} 2 \\ -2 \\ 0 \end{bmatrix}$$

Augmented matrix

$$(A, B) = \begin{bmatrix} 1 & 0 & -1 & 2 \\ 0 & 1 & -1 & -2 \\ 1 & 1 & -2 & 0 \end{bmatrix}$$

$$\sim \begin{bmatrix} 1 & 0 & -1 & -2 \\ 0 & 1 & -1 & -2 \\ 0 & 1 & -1 & -2 \end{bmatrix} R_3 \to R_3 - R_1$$

$$\sim \begin{bmatrix} 1 & 0 & -1 & 2 \\ 0 & 1 & -1 & -2 \\ 0 & 0 & 0 & 0 \end{bmatrix} R_3 \to R_3 - R_2$$

$$\rho(A, B) = \rho(A) = 2 < \text{number of unknowns } 3.$$

So the system being consistent, will have infinite number of solutions which involve 3 – 2 = 1 arbitrary constant.

The final reduced matrix ~ (A,B) makes the system equivalent to,

$$\begin{bmatrix} 1 & 0 & -1 \\ 0 & 1 & -1 \end{bmatrix} \begin{bmatrix} x \\ y \\ z \end{bmatrix} = \begin{bmatrix} 2 \\ -2 \end{bmatrix}$$

$\Rightarrow \quad x - z = 2,\ y - z = -2$

$\Rightarrow \quad x = z + 2,\ y = z - 2$

Taking z = c (an arbitrary constant) we get the solution as,

$$x = c + 2,\ y = c - 2,\ z = c.$$

***Example* 1.25:** Investigate for what values of α, β, the system of equations given by

$$\begin{bmatrix} 1 & 1 & 1 \\ 1 & 2 & 3 \\ 1 & 2 & \alpha \end{bmatrix} \begin{bmatrix} u \\ v \\ w \end{bmatrix} = \begin{bmatrix} 6 \\ 10 \\ \beta \end{bmatrix}, \text{ has}$$

(i) no solution (ii) unique solution (iii) an infinite number of solutions.

***Solution*:** The given system can be put in the form

$$AX = B, \text{ where } A = \begin{bmatrix} 1 & 1 & 1 \\ 1 & 2 & 3 \\ 1 & 2 & \alpha \end{bmatrix}, x = \begin{bmatrix} u \\ v \\ w \end{bmatrix}, B = \begin{bmatrix} 6 \\ 10 \\ \beta \end{bmatrix}$$

Augmented matrix

$$[A, B] = \begin{bmatrix} 1 & 1 & 1 & 6 \\ 1 & 2 & 3 & 10 \\ 1 & 2 & \alpha & \beta \end{bmatrix}$$

$$\sim \begin{bmatrix} 1 & 1 & 1 & 6 \\ 0 & 1 & 2 & 4 \\ 0 & 1 & (\alpha-1) & (\beta-6) \end{bmatrix} \begin{matrix} R_2 \to R_2 - R_1 \\ R_3 \to R_3 - R_1 \end{matrix}$$

$$\sim \begin{bmatrix} 1 & 0 & -1 & 2 \\ 0 & 1 & 2 & 4 \\ 0 & 0 & (\alpha-3) & (\beta-10) \end{bmatrix} \begin{matrix} R_1 \to R_1 - R_2 \\ R_3 \to R_3 - R_2 \end{matrix}$$

$$= C \text{ (say)}$$

Case (i) :

If the system has no solution,

$$\rho(A/B) \neq \rho(A)$$

i.e., $\quad \alpha = 3,\ \beta \neq 10$

Explanation :

If $\alpha = 3 ; \beta \neq 10$;

$$A \sim \begin{bmatrix} 1 & 0 & -1 \\ 0 & 1 & 2 \\ 0 & 0 & 0 \end{bmatrix} \text{ and } \rho(A) = 2 \text{ and } (A, B) \sim \begin{bmatrix} 1 & 0 & -1 & 2 \\ 0 & 1 & 2 & 4 \\ 0 & 0 & 0 & \gamma \end{bmatrix}$$

(where $\gamma = \beta - 10 \neq 0$)

$$\therefore \quad \begin{vmatrix} 0 & -1 & 2 \\ 1 & 2 & 4 \\ 0 & 0 & \gamma \end{vmatrix} \neq 0, \text{ so } \rho(A,B) = 3$$

Hence $\rho(A) \neq \rho(A, B)$

$\therefore$ System has no solution.

Case (ii) :

If the system has a unique solution, $\rho(A) = \rho(A,B) =$ number of unknowns = 3

$\therefore$ $\alpha - 3 \neq 0$, and $(\beta - 10)$ can take any value

i.e., $\alpha \neq 3$, and β can have any value

Explanation : $\alpha \neq 3 \Rightarrow C = \begin{bmatrix} 1 & 0 & -1 & 2 \\ 0 & 1 & 2 & 4 \\ 0 & 0 & \delta & \beta-10 \end{bmatrix}$ (where $\delta = \alpha - 3 \neq 0$)

$$\therefore \quad \begin{vmatrix} 1 & 0 & -1 \\ 0 & 1 & 2 \\ 0 & 0 & \delta \end{vmatrix} = \delta \neq 0$$

$\therefore$ $\rho(A) = \rho(A, B) = 3$

Case (iii) :

The system has infinite number of solutions (say).

Then $\rho(A) = \rho(A, B) < 3$ number of unknowns.

$$\therefore \quad \text{If } \alpha = 3, \text{ and } \beta = 10, C = \begin{bmatrix} 1 & 0 & -1 & 2 \\ 0 & 1 & 2 & 4 \\ 0 & 0 & 0 & 0 \end{bmatrix}, \text{ so that } \rho(A) = \rho(A, B) = 2 < 3$$

$\therefore$ $\alpha = 3$ and $\beta = 10$.

Hence the required answers are

(i) $\alpha = 3, \beta \neq 10$

(ii) $\alpha \neq 3$, β can have any value

(iii) $\alpha = 3$ and $\beta = 10$.

***Example* 1.26:** Test the following system for consistency and if consistent solve it.

$$u + 2v + 2w = 1,\ 2u + v + w = 2,\ 3u + 2v + 2w = 3,\ v + w = 0.$$

Solution: The system is in the form AX = B, where

$$A = \begin{bmatrix} 1 & 2 & 2 \\ 2 & 1 & 1 \\ 3 & 2 & 2 \\ 0 & 1 & 1 \end{bmatrix},\ X = \begin{bmatrix} u \\ v \\ w \end{bmatrix},\ B = \begin{bmatrix} 1 \\ 2 \\ 3 \\ 0 \end{bmatrix}$$

Augmented matrix

$$(A,B) = \begin{bmatrix} 1 & 2 & 2 & 1 \\ 2 & 1 & 1 & 2 \\ 3 & 2 & 2 & 3 \\ 0 & 1 & 1 & 0 \end{bmatrix}$$

$$\sim \begin{bmatrix} 1 & 2 & 2 & 1 \\ 0 & -5 & -5 & 0 \\ 0 & -4 & -4 & 0 \\ 0 & 1 & 1 & 0 \end{bmatrix} \begin{matrix} R_2 \to R_2 - 2R_1 \\ R_3 \to R_3 - 3R_1 \end{matrix}$$

$$\sim \begin{bmatrix} 1 & 2 & 2 & 1 \\ 0 & 1 & 1 & 0 \\ 0 & 1 & 1 & 0 \\ 0 & 1 & 1 & 0 \end{bmatrix} \begin{matrix} R_2 \to \frac{-1}{5} R_2 \\ R_3 \to \frac{-1}{4} R_3 \end{matrix}$$

$$\sim \begin{bmatrix} 1 & 2 & 2 & 1 \\ 0 & 1 & 1 & 0 \\ 0 & 0 & 0 & 0 \\ 0 & 0 & 0 & 0 \end{bmatrix} \begin{matrix} R_3 \to R_3 - R_2 \\ R_4 \to R_4 - R_2 \end{matrix}$$

Which is in Echelon form with two nonzero rows.

$\therefore \qquad \rho(A) = \rho(A, B) = 2 < 3$ (number of unknowns)

$\therefore$ The system is consistent and has infinite solutions containing $3 - 2 = 1$ arbitrary constant 'c' say.

The system is reduced to

$$\left.\begin{array}{l} u + 2v + 2w = 1 \\ v + w = 0 \end{array}\right\} \text{[from final equivalent matrix]}$$

Let $\qquad w = c$, then $v = -c$

and $\qquad u = 1 \; (\because 2v + 2w = 0)$

$$\therefore \qquad X = \begin{bmatrix} u \\ v \\ w \end{bmatrix} = \begin{bmatrix} 1 \\ -c \\ c \end{bmatrix},$$

where c is arbitrary, is the solution.

***Example* 1.27:** Solve the following system completely.

$x + y + z = 1$; $x + 2y + 4z = \alpha$

$x + 4y + 10z = \alpha^2$

***Solution*:** The system can be put in the matrix form $AX = B$, as,

$$\begin{bmatrix} 1 & 1 & 1 \\ 1 & 2 & 4 \\ 1 & 4 & 10 \end{bmatrix} \begin{bmatrix} x \\ y \\ z \end{bmatrix} = \begin{bmatrix} 1 \\ \alpha \\ \alpha^2 \end{bmatrix} \qquad .....(1.33)$$

Here $\qquad A = \begin{bmatrix} 1 & 1 & 1 \\ 1 & 2 & 4 \\ 1 & 4 & 10 \end{bmatrix}, X = \begin{bmatrix} x \\ y \\ z \end{bmatrix}, B = \begin{bmatrix} 1 \\ \alpha \\ \alpha^2 \end{bmatrix}$

Augmented matrix

$$(A,B) = \begin{bmatrix} 1 & 1 & 1 & 1 \\ 1 & 2 & 4 & \alpha \\ 1 & 4 & 10 & \alpha^2 \end{bmatrix}$$

$$\sim \begin{bmatrix} 1 & 1 & 1 & 1 \\ 0 & 1 & 3 & \alpha - 1 \\ 0 & 3 & 9 & \alpha^2 - 1 \end{bmatrix} R_2 \to R_2 - R_1, R_3 \to R_3 - R_1$$

$$\sim \begin{bmatrix} 1 & 0 & -2 & 2-\alpha \\ 0 & 1 & 3 & \alpha-1 \\ 0 & 0 & 0 & \alpha^2-3\alpha+2 \end{bmatrix} R_1 \to R_1 - R_2,\ R_3 \to R_3 - 3R_2$$

$= C$ (say)

For the system to be consistent, $\rho(A)$ and $\rho(A, B)$ must be equal.

$\because$ $$A \sim \begin{bmatrix} 1 & 0 & -2 \\ 0 & 1 & 3 \\ 0 & 0 & 0 \end{bmatrix},\ \rho(A) = 2$$

$\therefore$ $\rho(A, B) = \rho(C) = 2 \Rightarrow \alpha^2 - 3\alpha + 2 = 0 \Rightarrow \alpha = 1$ or $\alpha = 2$.

$\therefore$ The system is consistent when $\alpha = 1$ or 2.

Case (i) :

$$\alpha = 1 \Rightarrow C = \begin{bmatrix} 1 & 0 & -2 & 1 \\ 0 & 1 & 3 & 0 \\ 0 & 0 & 0 & 0 \end{bmatrix}.$$

$\Rightarrow$ $x - 2z = 1,\ y + 3z = 0$ .....(1.34)

$\because$ $\rho(A) = \rho(A, B) = 2 <$ number of unknowns 3, the system has infinite solutions involving one arbitrary constant.

$\therefore$ Let $z = c_1$, so that (1.34) gives $y = -3z$

$\Rightarrow$ $y = -3c_1$ and $x = 1 + 2z$

$\Rightarrow$ $x = 1 + 2c_1$

$\therefore$ when $\alpha = 1$, the solution is,

$x = 1 + c_1$,

$y = -3c_1$,

$z = c_1$,

where c_1 is an arbitrary constant.

Case (ii) :

$$\alpha = 2 \Rightarrow C = \begin{bmatrix} 1 & 0 & -2 & 0 \\ 0 & 1 & 3 & 1 \\ 0 & 0 & 0 & 0 \end{bmatrix}$$

$\Rightarrow \quad x - 2z = 0, \; y + 3z = 1,$

Taking $z = c_2$ an arbitrary constant, we get the solution as,

$$x = 2c_2,$$
$$y = 1 - 3c_2,$$
$$z = c_2$$

***Example* 1.28:** Show that the system

$$3x + 4y + 5z = \alpha,$$
$$4x + 5y + 6z = \beta,$$
$$5x + 6y + 7z = \gamma,$$

is consistent only when α, β, γ are in A . P.

***Solution*:** The given system, when put in matrix form is

$$\begin{bmatrix} 3 & 4 & 5 \\ 4 & 5 & 6 \\ 5 & 6 & 7 \end{bmatrix} \begin{bmatrix} x \\ y \\ z \end{bmatrix} = \begin{bmatrix} \alpha \\ \beta \\ \gamma \end{bmatrix}$$

Which when compared with AX = B gives

$$A = \begin{bmatrix} 3 & 4 & 5 \\ 4 & 5 & 6 \\ 5 & 6 & 7 \end{bmatrix}, X = \begin{bmatrix} x \\ y \\ z \end{bmatrix}, B = \begin{bmatrix} \alpha \\ \beta \\ \gamma \end{bmatrix}$$

Augmented matrix

$$(A, B) = \begin{bmatrix} 3 & 4 & 5 & \alpha \\ 4 & 5 & 6 & \beta \\ 5 & 6 & 7 & \gamma \end{bmatrix}$$

$$\sim \begin{bmatrix} 3 & 4 & 5 & \alpha \\ 1 & 1 & 1 & \beta-\alpha \\ 1 & 1 & 1 & \gamma-\beta \end{bmatrix} \begin{matrix} R_2 \to R_2 - R_1 \\ R_3 \to R_3 - R_2 \end{matrix}$$

$$\sim \begin{bmatrix} 1 & 1 & 1 & \beta-\alpha \\ 3 & 4 & 5 & \alpha \\ 1 & 1 & 1 & \gamma-\beta \end{bmatrix}, R_1 \leftrightarrow R_2$$

$$\sim \begin{bmatrix} 1 & 1 & 1 & \beta-\alpha \\ 0 & 1 & 2 & 4\alpha-3\beta \\ 0 & 0 & 0 & \gamma-2\beta+\alpha \end{bmatrix} \begin{matrix} R_2 \to R_2 - 3R_1 \\ R_3 \to R_3 - R_1 \end{matrix}$$

$\Rightarrow$ $\rho(A) = 2$ for any α, β, γ, and ρ (A, B) = 2 when $\alpha - 2\beta + \gamma = 0$.

But the system is consistent if and only if

$$\rho(A) = \rho(A, B)$$

i.e., if $\rho(A, B) = 2$,

i.e., if $\alpha - 2\beta + \gamma = 0$

i.e., if $\alpha + \gamma = 2\beta$,

i.e., when α, β, γ are in A.P.

***Example* 1.29:** Find the totality of solutions of the system given by,

$$5x + 2y - 6z + 2t + 1 = 0$$
$$x - y + z - t + 2 = 0$$

***Solution*:** The given system is,

$$\begin{bmatrix} 5 & 2 & -6 & 2 \\ 1 & -1 & 1 & -1 \end{bmatrix} \begin{bmatrix} x \\ y \\ z \\ t \end{bmatrix} = \begin{bmatrix} -1 \\ -2 \end{bmatrix}$$

Which is in the form AX = B, where

$$A = \begin{bmatrix} 5 & 2 & -6 & 2 \\ 1 & -1 & 1 & -1 \end{bmatrix}, X = \begin{bmatrix} x \\ y \\ z \\ t \end{bmatrix}, B = \begin{bmatrix} -1 \\ -2 \end{bmatrix}$$

Augmented matrix *X*,

$$(A, B) = \begin{bmatrix} 5 & 2 & -6 & 2 & -1 \\ 1 & -1 & 1 & -1 & -2 \end{bmatrix}$$

$$\sim \begin{bmatrix} 5 & 2 & -6 & 2 & -1 \\ 0 & \frac{-7}{5} & \frac{11}{5} & \frac{-7}{5} & \frac{-9}{5} \end{bmatrix} \left(R_2 \to R_2 - \frac{1}{5} R_1 \right)$$

$\because$ $\begin{vmatrix} 5 & 2 \\ 0 & \frac{-7}{5} \end{vmatrix} \neq 0$ is a minor of order 2 of A and

$\begin{vmatrix} 5 & -1 \\ 0 & \frac{-9}{5} \end{vmatrix} \neq 0$ is a minor of order 2 of (A, B).

$\rho(A) = \rho(A, B) = 2 <$ number of unknowns,

$\therefore$ The system is consistent and has infinite solutions which involve 4 – 2 = 2 arbitrary constants.

The final - reduced matrix makes the system as,

$$5x + 2y - 6z + 2t = -1 \qquad .....(1.35)$$

and
$$\frac{-7}{5}y + \frac{11}{5}z \frac{-7}{5}t = \frac{-9}{5}$$

(or)
$$-7y + 11z - 7t = -9 \qquad .....(1.36)$$

Taking $z = k_1$, $t = k_2$, (k_1, k_2 being arbitrary constants)

we get, from (1.36),

$$y = \frac{1}{7}(11k_1 - 7k_2 + 9)$$

$$(1.35) \Rightarrow \quad x = \frac{1}{5}(-2y + 6z - 2t - 1)$$

$$= \frac{1}{5}\left[\frac{-2}{7}(11k_1 - 7k_2 + 9) + 6k_1 - 2k_2 - 1\right]$$

$$= \frac{1}{7}(4k_1 - 5).$$

$\therefore$ The totality of solutions of the system is

$$x = \frac{1}{7}(4k_1 - 5),\ y = \frac{1}{7}(11k_1 - 7k_2 + 9),\ (z = k_1,\ t = k_2$$

where k_1, k_2 are arbitrary constants).

***Example* 1.30:** Solve the system

$$x + y + z = 6$$
$$2x - 3y + 4z = 8$$
$$x - y + 2z = 5, \text{ by the methods}$$

(i) Matrix inversion method

(ii) Using cramer's rule

(iii) Gauss Jordan method

***Solution*:** The system is given by

$$\begin{bmatrix} 1 & 1 & 1 \\ 2 & -3 & 4 \\ 1 & -1 & 2 \end{bmatrix}\begin{bmatrix} x \\ y \\ z \end{bmatrix} = \begin{bmatrix} 6 \\ 8 \\ 5 \end{bmatrix}$$

If it is put in the form AX = B, we have

$$A = \begin{bmatrix} 1 & 1 & 1 \\ 2 & -3 & 4 \\ 1 & -1 & 2 \end{bmatrix}, X = \begin{bmatrix} x \\ y \\ z \end{bmatrix} \quad B = \begin{bmatrix} 6 \\ 8 \\ 5 \end{bmatrix}$$

(i) Matrix inversion method :

We find A^{-1}

Let $$A = \begin{bmatrix} a_{11} & a_{12} & a_{13} \\ a_{21} & a_{22} & a_{23} \\ a_{31} & a_{32} & a_{33} \end{bmatrix} = \begin{bmatrix} 1 & 1 & 1 \\ 2 & -3 & 4 \\ 1 & -1 & 2 \end{bmatrix}$$

Adjoint of $$A = \begin{bmatrix} A_{11} & A_{21} & A_{31} \\ A_{12} & A_{22} & A_{32} \\ A_{13} & A_{23} & A_{33} \end{bmatrix}$$

where $A_{11}, A_{12}, \ldots$ are cofactors of $a_{11}, a_{12}, \ldots$ respectively.

$$A_{11} = (-1)^{1+1}\begin{vmatrix} -3 & 4 \\ -1 & 2 \end{vmatrix} = -2 \; ; \; A_{12} = (-1)^{1+2}\begin{vmatrix} 2 & 4 \\ 1 & 2 \end{vmatrix} = 0;$$

$$A_{13} = (-1)^{1+3}\begin{vmatrix} 2 & -3 \\ 1 & -1 \end{vmatrix} = +1;$$

$$|A| = \det A = a_{11} A_{11} + a_{12} A_{12} + a_{13} A_{13}$$
$$= 1.(-2) + 1(0) + 1\,(1) = -1$$

$$A_{21} = (-1)^{2+1}\begin{vmatrix} 1 & 1 \\ -1 & 2 \end{vmatrix} = -3,$$

$$A_{22} = (-1)^{2+2}\begin{vmatrix} 1 & 1 \\ -1 & 2 \end{vmatrix} = 1 \; ; \; A_{23} = (-1)^{2+3}\begin{vmatrix} 1 & 1 \\ 1 & -1 \end{vmatrix} = 2$$

$$A_{31} = (-1)^{1+3}\begin{vmatrix} 1 & 1 \\ -3 & 4 \end{vmatrix} = 7 \; ; \; A_{32} = (-1)^{3+2}\begin{vmatrix} 1 & 1 \\ 2 & 4 \end{vmatrix} = -2 \; ;$$

$$A_{33} = (-1)^{3+3}\begin{vmatrix} 1 & 1 \\ 2 & -3 \end{vmatrix} = -5$$

$$\therefore \qquad \text{Adj}\,A = \begin{bmatrix} -2 & -3 & 7 \\ 0 & 1 & -2 \\ -1 & 2 & -5 \end{bmatrix}$$

$$A^{-1} = \frac{1}{|A|} \text{ Adj } A = -\begin{bmatrix} -2 & -3 & 7 \\ 0 & 1 & -2 \\ 1 & 2 & -5 \end{bmatrix}$$

Solution matrix

$$X = A^{-1} B$$

$$\therefore \quad X = \frac{+1}{-1}\begin{bmatrix} -2 & -3 & 7 \\ 0 & 1 & -2 \\ 1 & 2 & -5 \end{bmatrix}\begin{bmatrix} 6 \\ 8 \\ 5 \end{bmatrix}$$

$$= -\begin{pmatrix} -2.6 - 3.8 + 7.5 \\ 0.6 + 1.8 - 2.5 \\ 1.6 + 2.8 - 5.5 \end{pmatrix}$$

$$= -\begin{bmatrix} -1 \\ -2 \\ -3 \end{bmatrix} = \begin{bmatrix} 1 \\ 2 \\ 3 \end{bmatrix}$$

$\therefore$ x = 1, y = 2, z = 3 is the solution.

(ii) Cramer's Rule :

$$\Delta = |A| = -1$$

$$\Delta_1 = \begin{vmatrix} 6 & 1 & 1 \\ 8 & -3 & 4 \\ 5 & -1 & 2 \end{vmatrix} = 6(-6+4) - 1(16-20) + 1(-8+15)$$

$$= -12 + 4 + 7 = -1$$ Replacing 1st column of Δ by B

Similarly, $$\Delta_2 = \begin{vmatrix} 1 & 6 & 1 \\ 2 & 8 & 4 \\ 1 & 5 & 2 \end{vmatrix} = 1(16-20) - 6(4-4) + 1(10-8)$$

$$= -4 - 0 + 2 = -2$$

and $$\Delta_3 = \begin{vmatrix} 1 & 1 & 6 \\ 2 & -3 & 8 \\ 1 & -1 & 5 \end{vmatrix} = 1(-15+8) - 1(10-8) + 6(-2+3)$$

$$= -7 - 2 + 6 = -3$$

$\therefore$ solution is

$$x = \frac{\Delta_1}{\Delta} = 1;\ y = \frac{\Delta_2}{\Delta} = 2;\ z = \frac{\Delta_3}{\Delta} = 3$$

(iii) Gauss-Jordan method :

$$[A, B] = \begin{bmatrix} 1 & 1 & 1 & 6 \\ 2 & -3 & 4 & 8 \\ 1 & -1 & 2 & 5 \end{bmatrix}$$

$$\sim \begin{bmatrix} 1 & 1 & 1 & 6 \\ 0 & -5 & 2 & -4 \\ 0 & -2 & 1 & -1 \end{bmatrix} \begin{matrix} R_2 \to R_2 - 2R_1 \\ R_3 \to R_3 - R_1 \end{matrix}$$

$$\sim \begin{bmatrix} 1 & 1 & 1 & 6 \\ 0 & 1 & -2/5 & \frac{4}{5} \\ 0 & -2 & 1 & -1 \end{bmatrix} \quad R_2 \to -\frac{1}{5}R_2$$

$$\sim \begin{bmatrix} 1 & 0 & 7/5 & 26/5 \\ 0 & 1 & -2/5 & 4/5 \\ 0 & 0 & 1/5 & 3/5 \end{bmatrix} \begin{matrix} R_1 \to R_1 - R_2 \\ R_3 \to R_3 + 2R_2 \end{matrix}$$

$$\sim \begin{bmatrix} 1 & 0 & 7/5 & 26/5 \\ 0 & 1 & -2/5 & 4/5 \\ 0 & 0 & 1 & 3 \end{bmatrix} R_3 \to 5R_3$$

$$\sim \begin{bmatrix} 1 & 0 & 0 & 1 \\ 0 & 1 & 0 & 2 \\ 0 & 0 & 1 & 3 \end{bmatrix} \begin{matrix} R_1 \to R_1 - \frac{7}{5}R_3 \\ \\ R_2 \to R_2 + \frac{2}{5}R_3 \end{matrix}$$

$$= [I_3 \cdot X]$$

where $X = \begin{bmatrix} 1 \\ 2 \\ 3 \end{bmatrix}$

$\therefore$ solution is

$$x = 1, y = 2, z = 3$$

***Example* 1.31:** Show that the system

$$2x - 3y + z = 0,$$
$$4x + 9y + z = 0,$$
$$8x - 27y + z = 0,$$

has no non-trivial solution.

***Solution*:** The given system can be written as

$$\begin{bmatrix} 2 & -3 & 1 \\ 4 & 9 & 1 \\ 8 & -27 & 1 \end{bmatrix}\begin{bmatrix} x \\ y \\ z \end{bmatrix} = \begin{bmatrix} 0 \\ 0 \\ 0 \end{bmatrix},$$

which is of the form

$$AX = 0,$$

where $A = \begin{bmatrix} 2 & -3 & 1 \\ 4 & 9 & 1 \\ 8 & -27 & 1 \end{bmatrix}$, $X = \begin{bmatrix} x \\ y \\ z \end{bmatrix}$, and $0 = \begin{bmatrix} 0 \\ 0 \\ 0 \end{bmatrix}$

$$A = \begin{bmatrix} 2 & -3 & 1 \\ 4 & 9 & 1 \\ 8 & -27 & 1 \end{bmatrix}$$

$$\sim \begin{bmatrix} 1 & -3/2 & 1/2 \\ 4 & 9 & 1 \\ 8 & -27 & 1 \end{bmatrix} R_1 \to \frac{1}{2}R_1$$

$$\sim \begin{bmatrix} 1 & -3/2 & 1/2 \\ 0 & 15 & -1 \\ 0 & -15 & -3 \end{bmatrix} \begin{matrix} R_2 \to R_2 - 4R_1 \\ R_3 \to R_3 - 8R_1 \end{matrix}$$

$$\sim \begin{bmatrix} 1 & -3/2 & 1/2 \\ 0 & 1 & -1/15 \\ 0 & -15 & -3 \end{bmatrix} R_2 \to \frac{1}{15}R_2$$

$$\sim \begin{bmatrix} 1 & 0 & 2/5 \\ 0 & 1 & -1/15 \\ 0 & 0 & -4 \end{bmatrix} \begin{matrix} R_1 \to R_1 + \frac{3}{2}R_2 \\ R_3 \to R_3 + 15R_2 \end{matrix}$$

$$\sim \begin{bmatrix} 1 & 0 & 2/5 \\ 0 & 1 & -1/15 \\ 0 & 0 & 1 \end{bmatrix} R_3 \to \frac{-1}{4}R_3$$

$$\sim \begin{bmatrix} 1 & 0 & 0 \\ 0 & 1 & 0 \\ 0 & 0 & 1 \end{bmatrix} \begin{matrix} R_1 \to R_1 - 2/5\, R_3 \\ R_2 \to R_2 + 1/15\, R_3 \end{matrix}$$

which is in the canonical form I_3.

$\therefore$ $\rho(A) = 3$ = number of unknowns.

$\therefore$ The system is consistent and has unique solution which is the trivial solution $x = 0, y = 0, z = 0$.

(ALITER : Since A is a 3×3 matrix, it is easy to find $\rho(A)$ by showing that $|A| \neq 0$.)

***Example* 1.32:** Solve the system of homogeneous equations given by

$$2x + y + 2z = 0,$$
$$x + y + 3z = 0,$$
$$4x + 3y + 8z = 0.$$

Solution: The given system is of the form AX = 0 where

$$A = \begin{bmatrix} 2 & 1 & 2 \\ 1 & 1 & 3 \\ 4 & 3 & 8 \end{bmatrix}, X = \begin{bmatrix} x \\ y \\ z \end{bmatrix}, 0 = \begin{bmatrix} 0 \\ 0 \\ 0 \end{bmatrix}$$

$$A = \begin{bmatrix} 2 & 1 & 2 \\ 1 & 1 & 3 \\ 4 & 3 & 8 \end{bmatrix}$$

$$\sim \begin{bmatrix} 1 & 1 & 3 \\ 2 & 1 & 2 \\ 4 & 3 & 8 \end{bmatrix} R_1 \leftrightarrow R_2$$

$$\sim \begin{bmatrix} 1 & 1 & 3 \\ 0 & -1 & -4 \\ 0 & -1 & -4 \end{bmatrix} \begin{matrix} R_2 \to R_2 - 2R_1 \\ R_3 \to R_3 - 4R_1 \end{matrix}$$

$$\sim \begin{bmatrix} 1 & 1 & 3 \\ 0 & 1 & 4 \\ 0 & 1 & 4 \end{bmatrix} \begin{matrix} R_2 \to -R_2 \\ R_3 \to -R_3 \end{matrix}$$

$$\sim \begin{bmatrix} 1 & 1 & 3 \\ 0 & 1 & 4 \\ 0 & 0 & 0 \end{bmatrix} R_3 \to R_3 - R_2,$$

$\therefore$ $\rho(A) = 2$,

Hence the system is consistent and has infinite solutions with one arbitrary constant.

The final reduced matrix reduces the system to

$$x + y + 3z = 0$$
$$y + 4z = 0$$

Taking z = k, we get y = – 4k, and x = k (k being arbitrary constant)

$\therefore$ The solution of the system is

$$X = \begin{bmatrix} x \\ y \\ z \end{bmatrix} = \begin{bmatrix} k \\ -4k \\ k \end{bmatrix}$$

***Example* 1.33:** Solve the system,

$$x + y = 0,$$
$$x + 2y + 3z = 0,$$
$$u + v + w = 0,$$
$$x + u + w = 0.$$

Solution: In this problem, we have to find 6 variables x, y, z, u, v, w.

The given system is

$$\begin{bmatrix} 1 & 1 & 0 & 0 & 0 & 0 \\ 1 & 2 & 3 & 0 & 0 & 0 \\ 0 & 0 & 0 & 1 & 1 & 1 \\ 1 & 0 & 0 & 1 & 0 & 1 \end{bmatrix} \begin{bmatrix} x \\ y \\ z \\ u \\ v \\ w \end{bmatrix} = \begin{bmatrix} 0 \\ 0 \\ 0 \\ 0 \end{bmatrix}$$

If written in the form AX = 0, we have,

$$\text{we find that, } \begin{vmatrix} 1 & 0 & 0 & 0 \\ 2 & 3 & 0 & 0 \\ 0 & 0 & 1 & 1 \\ 0 & 0 & 1 & 0 \end{vmatrix} = \begin{vmatrix} 3 & 0 & 0 \\ 0 & 1 & 1 \\ 0 & 1 & 0 \end{vmatrix}$$

$$= 3 \begin{vmatrix} 1 & 1 \\ 1 & 0 \end{vmatrix} \neq 0, \text{ is the largest non- vanishing minor of A.}$$

$\therefore$ $\rho(A) = 4$

Hence the system has infinite solutions and involves 6 – 4 = 2 arbitrary constants.

Taking $v = k_1, w = k_2$ (k_1, k_2 being arbitrary constants)

We have, $u = -k_1 - k_2$

$x = k_1, y = -k_1$

$z = \frac{1}{3} k_1$

$\therefore$ The solution is

$x = +k_1,$

$y = -k_1,$

$z = \frac{1}{3} k_1,$

$u = -k_1 - k_2,$

$v = k_1,$

$w = k_2$

Note : $\rho(A)$ can also be found by any alternate method.

***Example* 1.34:** If a, b, c, are distinct non-zero numbers, show that the homogeneous system

$$\begin{bmatrix} a & b & c \\ a^2 & b^2 & c^2 \\ a^3 & b^3 & c^3 \end{bmatrix} \begin{bmatrix} x \\ y \\ z \end{bmatrix} = \begin{bmatrix} 0 \\ 0 \\ 0 \end{bmatrix}$$

has no non-trivial solution.

Solution: Here AX = 0 is the system where

$$A = \begin{bmatrix} a & b & c \\ a^2 & b^2 & c^2 \\ a^3 & b^3 & c^3 \end{bmatrix}$$

$$|A| = \begin{vmatrix} a & b & c \\ a^2 & b^2 & c^2 \\ a^3 & b^3 & c^3 \end{vmatrix} w$$

$$= abc \begin{vmatrix} 1 & 1 & 1 \\ a & b & c \\ a^2 & b^2 & c^2 \end{vmatrix}$$

$$= abc\begin{vmatrix} 1 & 0 & 0 \\ a & b-a & c-a \\ a^2 & b^2-a^2 & c^2-a^2 \end{vmatrix} \quad \begin{matrix} c_2 \to c_2 - c_1 \\ c_3 \to c_3 - c_1 \end{matrix}$$

$$= abc\,.\,(b-a)\,(c-a)\begin{vmatrix} 1 & 0 & 0 \\ a & 1 & 1 \\ a^2 & b+a & c+a \end{vmatrix} \quad \begin{matrix} c_2 \to \dfrac{1}{b-a}c_2 \\ c_3 \to \dfrac{1}{c-a}c_3 \end{matrix}$$

$$= abc\,(b-a)\,(c-a)\begin{vmatrix} 1 & 1 \\ b+a & c+a \end{vmatrix}$$

$= abc\,(b-a)\,(c-a)\,(c-b) \neq 0$, since a, b, c are distinct non-zero numbers.

$\therefore$ $\rho(A) = 3$ = number of unknowns. Hence the system has a unique trivial solution $x = 0, y = 0, z = 0$.

Hence the problem ($\rho(A)$ can also be found by any other method).

Exercise - 1.4

I **Test for consistency the following systems. If consistent, solve them.**

(i) $x - 2y + t = 4$; $3x + 5y + t = 6$, $6x - y + 4t = 2$

(Ans : not-consistent)

(ii) $4x - y + 3z = 11$, $2x + y - 3z = -5$, $x + y + z = 6$

(Ans : $x = 1, y = 2, z = 3$)

(iii) $x - y + z = 3$, $2x - 3y + 5z = 10$, $x + y + 4z = 4$

(Ans : $x = 1, y = -1, z = 1$)

(iv) $x_1 + x_2 - x_3 + x_4 = 2$; $3x_1 - x_2 + 2x_3 + 5x_4 = 9$

$4x_1 + x_2 - x_3 - x_4 = 3$; $2x_1 - x_2 + 3x_3 + x_4 = 5$

(Ans : $x_1 = x_2 = x_3 = x_4 = 1$)

1.7 Eigen Values and Eigen Vectors

Matrix Polynomial

Let P_0, P_1, P_2 ..... P_n ($P_n \neq 0$) be square matrices of same order, then an expression of the form

$$P_0 + P_1x + P_2x^2 + \ldots\ldots + P_nx^n,$$

is called a matrix polynomial of order n.

Example 1

Let $P_0 = \begin{bmatrix} 1 & 2 \\ 3 & 0 \end{bmatrix}, P_1 = \begin{bmatrix} 1 & 2 \\ 3 & 1 \end{bmatrix}, P_2 = \begin{bmatrix} 2 & 3 \\ 1 & 4 \end{bmatrix},$

Then, $P_0 + P_1x + P_2x^2 = \begin{bmatrix} 1 & 2 \\ 3 & 0 \end{bmatrix} + \begin{bmatrix} 1 & 2 \\ 3 & 1 \end{bmatrix}x + \begin{bmatrix} 2 & 3 \\ 1 & 4 \end{bmatrix}x^2$

$$= \begin{bmatrix} (1+x+2x^2) & (2+2x+3x^2) \\ (3+3x+x^2) & (x+4x^2) \end{bmatrix}$$

is a matrix polynomial of order 2.

Characteristic matrix

Let (1) A be a square matrix of order '*n*' (2) I be a unit matrix of order *n*. Then the matrix polynomial given by $(A - \lambda I)$ is called the characteristic matrix of *A*.

Example 2

If $A = \begin{bmatrix} 1 & 2 \\ 2 & 1 \end{bmatrix}$, then the polynomial

$$A - \lambda I = \begin{bmatrix} 1 & 2 \\ 2 & 1 \end{bmatrix} - \lambda \begin{bmatrix} 1 & 0 \\ 0 & 1 \end{bmatrix} = \begin{bmatrix} 1-\lambda & 2 \\ 2 & 1-\lambda \end{bmatrix}$$

is the characteristic matrix of A.

Characteristic equation of A

For a square matrix A the equation $|A - \lambda I| = 0$ is called the characteristic equation.

Example 3

In the example 2, the characteristic equation is given by

$$\begin{vmatrix} 1-\lambda & 2 \\ 2 & 1-\lambda \end{vmatrix} = 0 \Rightarrow \lambda^2 - 2\lambda - 3 = 0.$$

Eigen Values

The roots of the characteristic equation are called the characteristic values or roots or Eigen values or Latent roots of the square matrix.

In example 3, the eigen values of $\begin{bmatrix} 1 & 2 \\ 2 & 1 \end{bmatrix}$ are the roots of the equation $\lambda^2 - 2\lambda - 3 = 0$

i.e., $\lambda = 3$ and -1 are the eigen values.

Eigen vectors

Let λ be an eigen value of the square matrix A.

Then, a non-zero column vector X = $\begin{bmatrix} x_1 \\ x_2 \\ \vdots \\ x_n \end{bmatrix}$ such that

$$(A - \lambda I)X = 0_{n\times 1}$$

(or) $AX = \lambda IX$

(or) $AX = \lambda X$, is called an eigen vector of A corresponding to λ.

Properties of Eigen Values and Eigen Vectors

1. The sum of the eigen values of a square matrix A is its trace and their product is $|A|$.

[JNTU 2003s, 2004s]

Proof :

Let $A = (a_{ij})_{n\times n}$ be a square matrix of order n and I be the unit matrix of same order.

The eigen values λ_i, $i = 1, 2, \ldots n$, are the roots of the equation

$$|A - \lambda I| = \begin{vmatrix} (a_{11}-\lambda) & a_{12} & \cdots & a_{1n} \\ a_{21} & (a_{22}-\lambda) & \cdots & a_{2n} \\ \cdots & \cdots & \cdots & \cdots \\ \cdots & \cdots & \cdots & \cdots \\ a_{n1} & a_{n2} & \cdots & (a_{nn}-\lambda) \end{vmatrix} = 0 \qquad \text{.....(1.37)}$$

Expanding the determinant let

$$|A - \lambda I| = p_0\lambda^n + p_1\lambda^{n-1} + \ldots + p_n \quad .....(1.38)$$

where $p_0, p_1, p_2, \ldots p_n$ are constants.

Taking $\lambda = 0$ in (1.38), we get, $|A| = p_n$ .....(1.39)

Expanding $|A - \lambda I|$ by first row, we get from (1.37),

$$|A - \lambda I| = (-1)^n \lambda^n + (-1)^{n-1} \lambda^{n-1}(a_{11} + a_{22} + \ldots + a_{nn}) + \\ + \ldots\ldots\ldots \quad .....(1.40)$$

Comparing (1.38) and (1.40), we have,

$$p_0 = (-1)^n, \; p_1 = (-1)^{n-1} (a_{11} + a_{22} + \ldots + a_{nn}), \ldots\ldots\ldots$$

Sum of eigen values $= \lambda_1 + \lambda_2 + \ldots + \lambda_n$

$$= \frac{-p_1}{p_0}$$

$$= \frac{(-1)^n (a_{11} + a_{22} + \ldots + a_{nn})}{(-1)^n}$$

$$= a_{11} + a_{22} + \ldots + a_{nn}$$

= Trace of A.

Product of eigen values $= \lambda_1 \lambda_2 \ldots\ldots \lambda_n$

$$= (-1)^n . \frac{p_n}{p_0}$$

$$= \frac{(-1)^n p_n}{(-1)^n} = p_n = |A| \quad \text{from (1.39).}$$

Note 1 :

$\because \quad |A| = \lambda_1 \lambda_2 \ldots \lambda_n,$

(i) If A is non-singular,

$|A| \neq 0$

$\Rightarrow \quad \lambda_1 \lambda_2 \ldots \lambda_n \neq 0 \quad \Rightarrow$ all eigen values are non-zero numbers.

(ii) If A is singular, $|A| = 0 = \lambda_1 \lambda_2 \ldots \lambda_n \Rightarrow$ at least one eigen value of A must be equal to zero.

2. The eigen values of A and A^T are equal.

Proof :

For any square matrix B, $|B| = |B^T|$.

$\therefore \quad |A - \lambda I| = |(A - \lambda I)^T| = |A^T - (\lambda I)^T| = |A^T - \lambda I|$ which proves the result.

3. If A is a non-singular matrix and λ is an eigen value of A, then $\frac{1}{\lambda}$ is an eigen value of A^{-1}. Hence deduce that $\frac{|A|}{\lambda}$ is an eigen value of adj A **[JNTU Nov 2005]**

Proof :

Let X be the eigen vector corresponding to λ.

Then, $AX = \lambda X$

$\therefore \quad A^{-1}(AX) = A^{-1}(\lambda X)$

$\Rightarrow \quad IX = \lambda A^{-1}X$

$\Rightarrow \quad A^{-1}X = \left(\frac{1}{\lambda}\right)X$

$\Rightarrow \quad \frac{1}{\lambda}$ is an eigen value of A^{-1}.

Therefore we have

$$|A|A^{-1}X = \frac{|A|}{\lambda}X$$

$$\Rightarrow \quad (\text{Adj } A)X = \frac{|A|}{\lambda}X \qquad .....(1.41)$$

$\Rightarrow \quad \frac{|A|}{\lambda}$ is eigen value of (AdjA)

Corollary : If λ_i, $i = 1, 2, ..., n$ are eigen values of A ($|A| \neq 0$), then $\frac{1}{\lambda_i}$, $i = 1, 2, ..., n$ are eigen values of A^{-1}.

4. If λ is an eigen value of A, then $\mu\lambda$ is an eigen value of μA where μ is a non-zero scalar

   Proof :

   X be an eigen vector of A corresponding to λ.

   Then $$AX = \lambda X$$
   $$(\mu A)X = \mu(AX) = \mu(\lambda X) = (\mu\lambda)X$$

   which prove the result

5. If λ is an eigen value of A, then λ^m is an eigen value of A^m, m being any positive integer

   [JNTU 2000, 2003]

   Proof :

   Let X be an eigen vector of A corresponding to λ.

   Then $$AX = \lambda X \qquad .....(1.42)$$
   $$A^2X = A(AX) = A(\lambda X) = \lambda(AX)$$
   $$= \lambda(\lambda X) \qquad \text{from (1.42)}$$
   $$= \lambda^2 X$$

   $\therefore$ λ^2 is an eigen value of A^2. Proceeding the same way, the result can be proved.

   Corollary : If λ_i, $i = 1, 2, \ldots\ldots n$ are eigen values of A, λ_i^m, $i = 1, 2 \ldots\ldots n$, are eigen values of A^m where m is any positive integer.

6. The eigen values of a diagonal matrix are its diagonal elements

   Proof :

   Let A = Diagonal $(a_{11}, a_{22} \ldots\ldots a_{nn})$.

   Then $$|A - \lambda I| = 0$$

   $$\Rightarrow \begin{vmatrix} a_{11}-\lambda & 0 & 0 & \cdots & 0 \\ 0 & a_{22}-\lambda & 0 & \cdots & 0 \\ 0 & 0 & a_{33}-\lambda & \cdots & 0 \\ \cdots & \cdots & \cdots & \cdots & \cdots \\ 0 & 0 & 0 & \cdots & a_{nn}-\lambda \end{vmatrix} = 0$$

   $$\Rightarrow (a_{11}-\lambda)(a_{22}-\lambda)\cdots(a_{nn}-\lambda) = 0$$

   $$\Rightarrow \lambda = a_{11}, a_{22}, \cdots a_{nn}$$; which completes the proof.

7. If B be a non-singular matrix, and A, B are matrices of same order, then A and $B^{-1}AB$ have same eigen values

Proof :

A and B are of same order.

Let $\qquad C = B^{-1} AB$

Then
$$C - \lambda I = B^{-1}AB - \lambda I = B^{-1}AB - \lambda B^{-1}B$$
$$= B^{-1}AB - B^{-1}\lambda B$$
$$= B^{-1}(AB - \lambda IB)$$
$$= B^{-1}(A - \lambda I)B$$
$$|C - \lambda I| = |B^{-1}||A - \lambda I||B|$$
$$|C - \lambda I| = |A - \lambda I| \qquad \left(\because |B^{-1}||B| = |B^{-1}B| = |I| = 1\right)$$

$\therefore$ $\quad A$ and $C = B^{-1}AB$ have same eigen values.

8. λ is a characteristic root of a square matrix A if and only if there exists a non-zero vector X such that $AX = \lambda X$.

Proof :

Let λ be a characteristic root of A.

$\therefore \qquad |A - \lambda I| = 0$

$\Rightarrow \qquad (A - \lambda I)$ is a singular matrix.

$\therefore$ $\quad$ The homogeneous system of equations given by $(A - \lambda I)X = 0$ possesses non-zero solutions.

i.e., $\quad$ there exists a non-zero vector X such that

$$(A - \lambda I)X = 0$$

$\Rightarrow \qquad AX = \lambda IX = \lambda X$

conversely, $\qquad AX = \lambda X$

$\Rightarrow \qquad (A - \lambda I)X = 0$

where X is a non-zero vector.

$\therefore$ $\quad$ The system of homogeneous equations $(A - \lambda I)X = 0$ has a non-zero solution.

Hence the coefficient matrix $(A - \lambda I)$ is singular.

i.e., $|A - \lambda I| = 0$

which shows that λ is an eigen value of A.

9. If X be an eigen vector of A corresponding to the eigen value λ, then cX is also an eigen vector of A corresponding to λ, c being a non-zero scalar

Proof :

X is an eigen vector of A corresponding to λ.

$\therefore \quad AX = \lambda X$

$\Rightarrow \quad cAX = c\lambda X$

$\Rightarrow \quad A(cX) = \lambda(cX)$

which proves the result.

10. If X is an eigen vector of a square matrix A, then X cannot correspond to more than one eigen value of A.

Proof :

If possible, let X correspond to two eigen values λ_1 and λ_2 of A.

Then, we have,

$$AX = \lambda_1 X \quad \text{and} \quad AX = \lambda_2 X$$

$\therefore \quad \lambda_1 X = \lambda_2 X \Rightarrow (\lambda_1 - \lambda_2)X = 0$

$\Rightarrow \quad \lambda_1 = \lambda_2 \quad (\because X \neq 0)$

Hence the result.

11. Zero is an eigen value of a matrix if and only if it is singular

Proof :

Let $\lambda = 0$ be an eigen value of a matrix 'A'.

Then it satisfies the equation

$$|A - \lambda I| = 0 \qquad \text{.....(1.43)}$$

$\Rightarrow \quad |A| = 0 \quad \Rightarrow \quad A$ is singular.

Conversely, A is singular

$\Rightarrow \quad |A| = 0$

$\Rightarrow \quad \lambda = 0$ satisfies the equation (1.43)

$\Rightarrow \quad \lambda = 0$ is an eigen value of A.

12. If λ is an eigen value of a non-singular matrix A, show that $\frac{|A|}{\lambda}$ is an eigen value of *Adj A*.

Proof :

A is non-singular $\Rightarrow \lambda \neq 0$.

$X \neq 0$ be the eigen vector corresponding to λ.

Then $\quad AX = \lambda X$

$\Rightarrow \quad (Adj\ A)\ AX = (Adj\ A)\lambda X$

$\Rightarrow \quad |A|\ X = (Adj\ A)\lambda X \qquad [\because (Adj\ A)A = |A|\ I]$

$\Rightarrow \quad \frac{|A|}{\lambda} X = (Adj\ A)X$

(or) $\quad (Adj\ A)X = \frac{|A|}{\lambda} X$ which proves the result.

Solved Examples

***Example* 1.35:** If $A = \begin{bmatrix} 3 & 0 & 0 \\ 5 & 7 & 0 \\ 2 & 6 & 1 \end{bmatrix}$, find the eigen values of (i) A^{-1} (ii) A^3 and (iii) A^T.

Solution: Characteristic equation of A is

$$|A - \lambda I| = 0$$

$$\Rightarrow \quad \begin{vmatrix} 3-\lambda & 0 & 0 \\ 5 & 7-\lambda & 0 \\ 2 & 6 & 1-\lambda \end{vmatrix} = 0$$

$$\Rightarrow \quad (3-\lambda)\ (7-\lambda)\ (1-\lambda) = 0$$

$\Rightarrow \quad \lambda = 3, 7, 1$ are eigen values of A.

∴ (i) Eigen values of A^{-1} are $\frac{1}{1}, \frac{1}{3}, \frac{1}{7}$,

i.e., $1, \frac{1}{3}, \frac{1}{7}$.

(ii) Eigen values of A^3 are $1^3, 3^3, 7^3$.

(iii) Eigen values of A^T are same as those of A,

i.e., 1, 3, 7.

***Example* 1.36:** Without finding the actual eigen values of $A = \begin{bmatrix} 1 & 2 & 6 \\ 0 & 3 & 4 \\ 0 & 0 & 5 \end{bmatrix}$, find their sum and product.

Solution: $|A| = 15$,

sum of eigen vectors of A = Trace of A

$= 1 + 3 + 5 = 9$

Product = |A| = 15

***Example* 1.37:** Find the characteristic values of the matrices,

(i) $\begin{pmatrix} \cos\alpha & -\sin\alpha \\ -\sin\alpha & \cos\alpha \end{pmatrix}$ and (ii) $\begin{pmatrix} \cos\alpha & -\sin\alpha \\ -\sin\alpha & -\cos\alpha \end{pmatrix}$.

Solution: The given matrix be A. Then

(i) Characteristic equation of A is

$$|A - \lambda I| = 0$$

$$\Rightarrow \quad (\cos\alpha - \lambda)^2 - \sin^2\alpha = 0$$

$$\Rightarrow \quad (\cos\alpha - \lambda + \sin\alpha)\ (\cos\alpha - \lambda - \sin\alpha) = 0$$

∴ $\lambda = \cos\alpha + \sin\alpha, \cos\alpha - \sin\alpha$ are eigen values.

(ii)
$$|A - \lambda I| = 0$$

$$\Rightarrow \quad \begin{vmatrix} \cos\alpha - \lambda & -\sin\alpha \\ -\sin\alpha & -\cos\alpha - \lambda \end{vmatrix} = 0$$

$$-\left(\cos^2\alpha - \lambda^2\right) - \sin^2\alpha = 0$$

$$\Rightarrow \quad \lambda^2 = \cos^2\alpha + \sin^2\alpha = 1$$

$\lambda = \pm 1$ are eigen values.

Example **1.38:** Find the eigen values of the matrix A = $\begin{bmatrix} -9 & 4 & 4 \\ -8 & 3 & 4 \\ -16 & 8 & 4 \end{bmatrix}$.

Solution: $|A-\lambda I| = \begin{vmatrix} -9-\lambda & 4 & 4 \\ -8 & 3-\lambda & 4 \\ -16 & 8 & 4-\lambda \end{vmatrix}$

$$= (-9-\lambda)\{\lambda^2-7\lambda-20\}-4(-32+8\lambda+64)\ +4(-64+48-16\lambda)$$

$$= -9\lambda^2+63\lambda+180-\lambda^3+7\lambda^2+20\lambda\ -32\lambda-128-64\lambda-64$$

$$= -\lambda^3-2\lambda^2-13\lambda-12$$

Characteristic equation of A is

$|A-\lambda I| = 0$

$\Rightarrow \quad \lambda^3+2\lambda^2+13\lambda+12=0$

By inspection, $\lambda=-1$ is a root of this equation.

The above equation can be written as

$(\lambda+1)\ (\lambda^2+\lambda+12)=0$

$\lambda^2+\lambda+12=0$

$$\Rightarrow \lambda = \frac{-1\pm\sqrt{1-48}}{2}$$

$$= \frac{-1\pm\sqrt{47}i}{2}$$

$\therefore$ Eigen values of A are -1, $\dfrac{-1\pm\sqrt{47}i}{2}$

Example **1.39:** Find the eigen values of the matrix A = $\begin{bmatrix} 3 & -9 & -12 \\ 1 & 3 & 4 \\ 0 & 0 & 1 \end{bmatrix}$

Solution: $A-\lambda I = \begin{bmatrix} 3-\lambda & -9 & -12 \\ 1 & 3-\lambda & 4 \\ 0 & 0 & 1-\lambda \end{bmatrix}$

$$|A-\lambda I| = (1-\lambda)\left[(3-\lambda)^2+9\right] \qquad \text{(expanding by last row)}$$

$$= (1-\lambda)\left(\lambda^2-6\lambda+18\right)$$

Characteristic equation of A is

$$|A-\lambda I| = 0$$

$$\Rightarrow \qquad (1-\lambda)(\lambda^2-6\lambda+18) = 0$$

$\therefore$ Eigen values of A are

$$\lambda = 1, \frac{6 \pm \sqrt{36-72}}{2}$$

i.e., $\lambda = 1, 3+3i, 3-3i$

Example **1.40:** Find the eigen values and corresponding eigen vectors of the unit matrix

$$I = \begin{bmatrix} 1 & 0 & 0 \\ 0 & 1 & 0 \\ 0 & 0 & 1 \end{bmatrix}.$$

Solution: If A is a square matrix, its eigen values 'λ' are given by the characteristic equation $|A-\lambda I| = 0$.

$\therefore$ In this problem, the characteristic equation is $\begin{vmatrix} 1-\lambda & 0 & 0 \\ 0 & 1-\lambda & 0 \\ 0 & 0 & 1-\lambda \end{vmatrix} = 0.$

$$\Rightarrow \qquad (1-\lambda)^3 = 0,$$

$\therefore$ Eigen values are $\lambda = 1, 1, 1$

Eigen vectors are given by,

$$AX = \lambda X$$

$$\Rightarrow \qquad IX = X \qquad \text{.....(1.44)}$$

$\because$ (1.44) is satisfied by all non-zero vectors X, all non-zero vectors are the eigen vectors of the given unit matrix.

Example **1.41:** Determine the eigen values and corresponding eigen vectors of A = $\begin{bmatrix} 2 & 4 \\ 1 & 5 \end{bmatrix}$.

Solution: $|A-\lambda I| = \begin{vmatrix} 2-\lambda & 4 \\ 1 & 5-\lambda \end{vmatrix} = \lambda^2 - 7\lambda + 6$

Characteristic equation is $|A - \lambda I| = 0$

i.e., $\lambda^2 - 7\lambda + 6 = 0$

$\Rightarrow$ $\lambda = 1, 6$

Let $X = \begin{bmatrix} x_1 \\ x_2 \end{bmatrix}$ be the eigen vector.

Case (i): $\lambda = 1$

$$AX = \lambda X$$

$$\Rightarrow \quad AX = X$$

$$\Rightarrow \quad \begin{bmatrix} 2 & 4 \\ 1 & 5 \end{bmatrix} \begin{bmatrix} x_1 \\ x_2 \end{bmatrix} = \begin{bmatrix} x_1 \\ x_2 \end{bmatrix}$$

$$\Rightarrow \quad x_1 + 4x_2 = 0 \ ; \ x_1 + 4x_2 = 0$$

$$\therefore \quad x_2 = k$$

$$\Rightarrow \quad x_1 = -4k$$

$\therefore$ $X = \begin{bmatrix} -4k \\ k \end{bmatrix}$ is the eigen vector corresponding to $\lambda = 1$

Case (ii) : $\lambda = 6$

$$\Rightarrow \quad AX = 6X$$

$$\Rightarrow \quad \begin{bmatrix} 2 & 4 \\ 1 & 5 \end{bmatrix} \begin{bmatrix} x_1 \\ x_2 \end{bmatrix} = \begin{bmatrix} 6x_1 \\ 6x_2 \end{bmatrix}$$

$$-4x_1 + 4x_2 = 0$$

$$x_1 - x_2 = 0$$

Both these equations are equivalent to one equation

$$x_1 - x_2 = 0$$

$$\Rightarrow \qquad x_1 = x_2 = c$$

(say) where c is a constant.

$$\therefore \qquad X = \begin{bmatrix} c \\ c \end{bmatrix} \text{ is the eigen vector corresponding to } \lambda = 6.$$

***Example* 1.42:** Find the eigen values and corresponding eigen vectors of the matrix

$$A = \begin{bmatrix} 1 & 1 & 3 \\ 1 & 5 & 1 \\ 3 & 1 & 1 \end{bmatrix}$$

Solution: Characteristic equation is $|A - \lambda I| = 0$

$$\Rightarrow \qquad \begin{vmatrix} 1-\lambda & 1 & 3 \\ 1 & 5-\lambda & 1 \\ 3 & 1 & 1-\lambda \end{vmatrix} = 0$$

which on expansion and simplification gives

$$\lambda^3 - 7\lambda^2 + 36 = 0$$

$$\Rightarrow \qquad (\lambda+2)\,(\lambda-6)\,(\lambda-3) = 0$$

$$\Rightarrow \qquad \lambda = -2, 3, 6 \text{ are eigen values of } A.$$

***Case (i)* :**

To find the eigen vector when $\lambda = -2$:

Let $X = \begin{bmatrix} x_1 \\ x_2 \\ x_3 \end{bmatrix}$ be the eigen vector, so that $AX = -2X$

$$\Rightarrow \qquad \begin{bmatrix} 1 & 1 & 3 \\ 1 & 5 & 1 \\ 3 & 1 & 1 \end{bmatrix} \begin{bmatrix} x_1 \\ x_2 \\ x_3 \end{bmatrix} = \begin{bmatrix} -2x_1 \\ -2x_2 \\ -2x_3 \end{bmatrix}$$

$$\Rightarrow \qquad 3x_1 + x_2 + 3x_3 = 0$$

$$x_1 + 7x_2 + x_3 = 0$$

$$3x_1 + x_2 + 3x_3 = 0,$$

which are equivalent to only two equations. Solving the first two equations by rule of cross-multiplication, we have

$$\frac{x_1}{-20} = \frac{x_2}{0} = \frac{x_3}{20} = c \text{ (say)}$$

$\Rightarrow$ $x_1 = -c,\ x_2 = 0,\ x_3 = c$ (c being arbitrary constant)

$\Rightarrow$ $x = \begin{bmatrix} -c \\ 0 \\ c \end{bmatrix}$ is eigen vector for $\lambda = -2$. If $c = 1$ is a particular value,

$x = \begin{bmatrix} -1 \\ 0 \\ 1 \end{bmatrix}$ is the corresponding eigen vector.

Case (ii) :

When $\lambda = 3$; $AX = 3X$

i.e., $$\begin{bmatrix} 1 & 1 & 3 \\ 1 & 5 & 1 \\ 3 & 1 & 1 \end{bmatrix} \begin{bmatrix} x_1 \\ x_2 \\ x_3 \end{bmatrix} = \begin{bmatrix} 3x_1 \\ 3x_2 \\ 3x_3 \end{bmatrix}$$

$$\Rightarrow \quad \begin{aligned} -2x_1 + x_2 + 3x_3 &= 0 \\ x_1 + 2x_2 + x_3 &= 0 \\ 3x_1 + x_2 - 2x_3 &= 0 \end{aligned}$$

The first two equations give,

$$\frac{x_1}{-5} = \frac{x_2}{5} = \frac{x_3}{-5} = \text{c (say)}$$

$\Rightarrow$ $x_1 = c,\ x_2 = -c,\ x_3 = $ c, (which satisfy the third equation also).

$\therefore$ $X = \begin{bmatrix} c \\ -c \\ c \end{bmatrix}$ is the eigen vector.

Taking $c = 1, X = \begin{bmatrix} 1 \\ -1 \\ 1 \end{bmatrix}$ is eigen vector for $\lambda = 3$.

Case (iii) :

When $\lambda = 6$; $AX = 6X$

$$\Rightarrow \begin{aligned} -5x_1 + x_2 + 3x_3 &= 0 \\ x_1 - x_2 + x_3 &= 0 \\ 3x_1 + x_2 - 5x_3 &= 0 \end{aligned}$$

Solving the first two equations,

$$\frac{x_1}{4} = \frac{x_2}{8} = \frac{x_3}{4} = \text{c (say)}$$

$\Rightarrow$ $x_1 = c, x_2 = 2c, x_3 = c$ (these satisfy third equation)

$\therefore$ $X = \begin{bmatrix} c \\ 2c \\ c \end{bmatrix}$ is the eigen vector and a particular vector is $\begin{bmatrix} 1 \\ 2 \\ 1 \end{bmatrix}$.

***Example* 1.43:** Find the eigen values and corresponding eigen vectors of $A = \begin{bmatrix} 1 & 2 & 3 \\ 0 & -4 & 2 \\ 0 & 0 & 7 \end{bmatrix}$.

Solution: $|A - \lambda I| = \begin{vmatrix} (1-\lambda) & 2 & 3 \\ 0 & (-4-\lambda) & 2 \\ 0 & 0 & 7-\lambda \end{vmatrix} = (1 - \lambda)(-4 - \lambda)(7 - \lambda)$

Characteristic equation of A is

$$(1 - \lambda)(-4 - \lambda)(7 - \lambda) = 0$$

$\therefore$ Eigen values are $\lambda = 1, -4, 7$.

To find eigen vectors.

Case (i) : When $\lambda = 1$

Let $X = \begin{bmatrix} x_1 \\ x_2 \\ x_3 \end{bmatrix}$ be the eigen vector corresponding to $\lambda = 1$.

$$\therefore \quad AX = X \Rightarrow \begin{bmatrix} 1 & 2 & 3 \\ 0 & -4 & 2 \\ 0 & 0 & 7 \end{bmatrix} \begin{bmatrix} x_1 \\ x_2 \\ x_3 \end{bmatrix} = \begin{bmatrix} x_1 \\ x_2 \\ x_3 \end{bmatrix}$$

$$\Rightarrow \quad \left.\begin{aligned} 2x_2 + 3x_3 &= 0 \\ -5x_2 + 2x_3 &= 0 \\ 6x_3 &= 0 \end{aligned}\right\} \qquad .....(1.45)$$

The solution of the system (1.45) is

$$x_3 = 0,\ x_2 = 0,\ x_1 = c$$

$$\therefore \quad \text{Eigen vector} = X = \begin{bmatrix} c \\ 0 \\ 0 \end{bmatrix}, \text{ (c being constant)}$$

In particular $X = \begin{bmatrix} 1 \\ 0 \\ 0 \end{bmatrix}$.

Case (ii) : Let $X = \begin{bmatrix} x_1 \\ x_2 \\ x_3 \end{bmatrix}$ be the eigen vector corresponding to $\lambda = -4$.

$$\therefore \quad AX = -4X$$

$$\Rightarrow \quad 5x_1 + 2x_2 + 3x_3 = 0,\ 2x_3 = 0,\ 11x_3 = 0$$

$$\Rightarrow \quad x_3 = 0,\ x_2 = k,\ x_1 = \frac{-2k}{5}$$

A particular solution is $x_1 = \dfrac{-2}{5}, x_2 = 1, x_3 = 0$

Case (iii) : Let $X = \begin{bmatrix} x_1 \\ x_2 \\ x_3 \end{bmatrix}$ be the eigen vector corresponding to $\lambda = 7$.

$$\therefore \quad AX = 7X$$

$$\Rightarrow \quad \begin{bmatrix} 1 & 2 & 3 \\ 0 & -4 & 2 \\ 0 & 0 & 7 \end{bmatrix} \begin{bmatrix} x_1 \\ x_2 \\ x_3 \end{bmatrix} = \begin{bmatrix} 7x_1 \\ 7x_2 \\ 7x_3 \end{bmatrix}$$

$$\Rightarrow \qquad \left.\begin{aligned} -6x_1 + 2x_2 + 3x_3 &= 0 \\ -11x_2 + 2x_3 &= 0 \end{aligned}\right\} \qquad .....(1.46)$$

Taking $x_3 = c$, c being constant, the solution of (1.46) is

$$x_3 = c,\ x_2 = \frac{2}{11}\ c,\ x_1 = \frac{1}{6}\left[\frac{4c}{11} + 3c\right]$$

$$\therefore \qquad X = \begin{bmatrix} \frac{1}{6}\left(\frac{4c}{11} + 3c\right) \\ \frac{2c}{11} \\ c \end{bmatrix} = \begin{bmatrix} \frac{37c}{66} \\ \frac{2c}{11} \\ c \end{bmatrix} \text{ is the eigen vector for } \lambda = 7.$$

***Example* 1.44:** Determine the characteristic roots and vectors of $A = \begin{bmatrix} 8 & -6 & 2 \\ -6 & 7 & -4 \\ 2 & -4 & 3 \end{bmatrix}$

[CSVTU 2005, 2007, 2008]

Solution:

$$|A - \lambda I| = \begin{vmatrix} 8-\lambda & -6 & 2 \\ -6 & 7-\lambda & -4 \\ 2 & -4 & 3-\lambda \end{vmatrix}$$

$$= (8 - \lambda)[\lambda^2 - 10\lambda + 5] + 6\,[6\lambda - 10] + 2[10 + 2\lambda]$$

$$= -\lambda^3 + 18\lambda^2 - 45\lambda$$

Characteristic equation is $|A - \lambda I| = 0$

$$\Rightarrow \qquad -\lambda^3 + 18\lambda^2 + 45\lambda = 0$$

$$\Rightarrow \qquad \lambda(\lambda - 3)(\lambda - 15) = 0$$

$\therefore$ Characteristic values are $\lambda = 0, 3, 15$

Let us find the corresponding characteristic vectors

Let $X = \begin{bmatrix} x_1 \\ x_2 \\ x_3 \end{bmatrix}$ be the required vector.

(i) when $\lambda = 0$, $\therefore AX = \lambda X = 0$

$$\Rightarrow \qquad \begin{bmatrix} 8 & -6 & 2 \\ -6 & 7 & -4 \\ 2 & -4 & 3 \end{bmatrix} \begin{bmatrix} x_1 \\ x_2 \\ x_3 \end{bmatrix} = 0$$

$$\Rightarrow \left.\begin{array}{rl} 8x_1 - 6x_2 + 2x_3 &= 0 \\ -6x_1 + 7x_2 - 4x_3 &= 0 \\ 2x_1 - 4x_2 + 3x_3 &= 0 \end{array}\right\}$$

Solving the first two equations, we have,

$$\frac{x_1}{10} = \frac{x_2}{20} = \frac{x_3}{20}$$

$$\Rightarrow \quad x_1 = \frac{x_2}{2} = \frac{x_3}{2} = c \text{ (say)}$$

$\Rightarrow \quad x_1 = c\,;\ x_2 = 2c,\ x_3 = 2c$ (these values satisfy the third equation also)

$$\therefore \quad X = \begin{bmatrix} c \\ 2c \\ 2c \end{bmatrix}$$ where c is a constant.

A particular value $= \begin{bmatrix} 1 \\ 2 \\ 2 \end{bmatrix}$.

(ii) when $\lambda = 3\,;\ AX = 3X$

$$\Rightarrow \begin{bmatrix} 8 & -6 & 2 \\ -6 & 7 & -4 \\ 2 & -4 & 3 \end{bmatrix} \begin{bmatrix} x_1 \\ x_2 \\ x_3 \end{bmatrix} = \begin{bmatrix} 3x_1 \\ 3x_2 \\ 3x_3 \end{bmatrix}$$

$$\Rightarrow \begin{array}{rl} 5x_1 - 6x_2 + 2x_3 &= 0 \\ -6x_1 + 4x_2 - 4x_3 &= 0 \\ 2x_1 - 4x_2 &= 0 \end{array}$$

Solving first two equations, we get

$$\frac{x_1}{16} = \frac{x_2}{8} = \frac{x_3}{-16} \Rightarrow \frac{x_1}{2} = \frac{x_2}{1} = \frac{x_3}{-2} = c \text{ (say)}$$

$\therefore \quad x_1 = 2c,\ x_2 = c,\ x_3 = -2c$ and these values satisfy the 3rd equation.

$$\therefore \qquad X = \begin{bmatrix} 2c \\ c \\ -2c \end{bmatrix} \text{ is the required vector where } c \text{ is a constant.}$$

A particular vector can be $\begin{bmatrix} 2 \\ 1 \\ -2 \end{bmatrix}$.

(iii) When $\lambda = 15$; $AX = 15X$

i.e., $$\begin{bmatrix} 8 & -6 & 2 \\ -6 & 7 & -4 \\ 2 & -4 & 3 \end{bmatrix} \begin{bmatrix} x_1 \\ x_2 \\ x_3 \end{bmatrix} = \begin{bmatrix} 15x_1 \\ 15x_2 \\ 15x_3 \end{bmatrix}$$

By multiplication and equality of matrices, we have

$$-7x_1 - 6x_2 + 2x_3 = 0$$
$$-6x_1 - 8x_2 - 4x_3 = 0$$
$$2x_1 - 4x_2 - 12x_3 = 0$$

From the first two equations, we have,

$$\frac{x_1}{40} = \frac{x_2}{-40} = \frac{x_3}{20} \Rightarrow \frac{x_1}{2} = \frac{x_2}{-2} = \frac{x_3}{1} = c \text{ (say)}$$

$\therefore$ $x_1 = 2c$, $x_2 = -2c$, $x_3 = c$, where c is a constant.

These values satisfy the 3rd equation also

$$\therefore \qquad X = \begin{bmatrix} 2c \\ -2c \\ c \end{bmatrix} \text{ is the required vector and a particular value can be taken as } \begin{bmatrix} 2 \\ -2 \\ 1 \end{bmatrix}$$

***Example* 1.45:** Find the latent roots and the corresponding latent vectors of the matrix

$$A = \begin{bmatrix} 2 & -2 & 2 \\ 1 & 1 & 1 \\ 1 & 3 & -1 \end{bmatrix}.$$

Solution: $|A - \lambda I| = \begin{vmatrix} 2-\lambda & -2 & 2 \\ 1 & 1-\lambda & 1 \\ 1 & 3 & -1-\lambda \end{vmatrix}$

$$= (2-\lambda)\left\{-(1+\lambda)(1-\lambda)-3\right\} + 2(-1-\lambda-1) + 2(3-1+\lambda)$$

$$= (2-\lambda)(\lambda^2-4) - 2(2+\lambda) + 2(2+\lambda)$$

$$= -(\lambda-2)(\lambda^2-4)$$

Characteristic equation is $|A - \lambda I| = 0$

i.e., $(\lambda - 2)(\lambda - 2)(\lambda + 2) = 0$

$\Rightarrow$ Latent roots of A are $\lambda = 2, 2, -2$

Case (i) : $\lambda = 2$

Let $X = \begin{bmatrix} x_1 \\ x_2 \\ x_3 \end{bmatrix}$ be the corresponding latent vector.

$\therefore$ $AX = 2X$

$$\Rightarrow \begin{bmatrix} 2 & -2 & 2 \\ 1 & 1 & 1 \\ 1 & 3 & -1 \end{bmatrix}\begin{bmatrix} x_1 \\ x_2 \\ x_3 \end{bmatrix} = \begin{bmatrix} 2x_1 \\ 2x_2 \\ 2x_3 \end{bmatrix}$$

Applying matrix multiplication and equality of matrices, we get on simplification,

$$-x_2 + x_3 = 0$$

$$x_1 - x_2 + x_3 = 0$$

$$x_1 + 3x_2 - 3x_3 = 0$$

From the first two equations

$$\frac{x_1}{0} = \frac{x_2}{1} = \frac{x_3}{1} = c \text{ (say)}$$

$\therefore$ $x_1 = 0, x_2 = c, x_3 = c$ which satisfy the 3rd equation also

$X = \begin{bmatrix} 0 \\ c \\ c \end{bmatrix}$ is the required vector.

case (ii) when $\lambda = -2$, we have AX = – 2X

$$\Rightarrow \quad \begin{aligned} 4x_1 - 2x_2 + 2x_3 &= 0 \\ x_1 + 3x_2 + x_3 &= 0 \\ x_1 + 3x_2 + x_3 &= 0 \end{aligned}$$

$$\Rightarrow \quad \frac{x_1}{-8} = \frac{x_2}{-2} = \frac{x_3}{14}.$$

$$\Rightarrow \quad \frac{x_1}{-4} = \frac{x_2}{-1} = \frac{x_3}{7} = c$$

Required vector is $\begin{bmatrix} -4c \\ -c \\ 7c \end{bmatrix}$ or in particular $\begin{bmatrix} -4 \\ -1 \\ 7 \end{bmatrix}$.

***Example* 1.46:** Determine the constants p, q, r, s, t, u so that [1 1 1]', [1 0 –1]' and [1 –1 0]' are the eigen vectors of the matrix $\begin{bmatrix} 1 & 1 & 1 \\ p & q & r \\ s & t & u \end{bmatrix}$.

Solution: Let $\qquad A = \begin{bmatrix} 1 & 1 & 1 \\ p & q & r \\ s & t & u \end{bmatrix}$, let l_1, l_2 l_3 be the eigen values of A.

[1 1 1]' = $\begin{bmatrix} 1 \\ 1 \\ 1 \end{bmatrix}$ be the eigen vector corresponding to λ_1.

$$\therefore \quad AX_1 = \lambda_1 X$$

$$\Rightarrow \quad \begin{bmatrix} 1 & 1 & 1 \\ p & q & r \\ s & t & u \end{bmatrix}\begin{bmatrix} 1 \\ 1 \\ 1 \end{bmatrix} = \lambda_1 \begin{bmatrix} 1 \\ 1 \\ 1 \end{bmatrix} = \begin{bmatrix} \lambda_1 \\ \lambda_1 \\ \lambda_1 \end{bmatrix}$$

$$1.1 + 1.1 + 1.1 = \lambda_1; \qquad \text{i.e., } \lambda_1 = 3$$

$$\Rightarrow \quad p.1 + q.1 + r.1 = \lambda_1 \Rightarrow \quad p + q + r = 3 \qquad .....(1.47)$$

$$s.1 + t.1 + u.1 = \lambda_1 = \lambda_1 \quad s + t + u = 3 \qquad .....(1.48)$$

Let $X = [1\ 0\ -1]' = \begin{bmatrix} 1 \\ 0 \\ -1 \end{bmatrix}$ be the eigen vector corresponding to λ_2.

Then $AX = \lambda_2 X$

$$\Rightarrow \quad \begin{bmatrix} 1 & 1 & 1 \\ p & q & r \\ s & t & u \end{bmatrix}\begin{bmatrix} 1 \\ 0 \\ -1 \end{bmatrix} = \lambda_2 \begin{bmatrix} 1 \\ 0 \\ -1 \end{bmatrix} = \begin{bmatrix} \lambda_2 \\ 0 \\ -\lambda_2 \end{bmatrix}$$

$\Rightarrow \quad 1.1 + 1.0 + 1.-1 = \lambda_2 \quad$ (i.e.,) $\lambda_2 = 0$

$p.1 + q.0 + r.-1 = 0 \quad \Rightarrow \quad p - r = 0$ .....(1.49)

$s.1 + t.0 + u.-1 = -\lambda_2 = 0 \Rightarrow \quad s - u = 0$ .....(1.50)

Similarly, $\begin{bmatrix} 1 & 1 & 1 \\ p & q & r \\ s & t & u \end{bmatrix}\begin{bmatrix} 1 \\ -1 \\ 0 \end{bmatrix} = \lambda_3 \begin{bmatrix} 1 \\ -1 \\ 0 \end{bmatrix} = \begin{bmatrix} \lambda_3 \\ -\lambda_3 \\ 0 \end{bmatrix}$

gives, $\lambda_3 = 0$,

$p - q = 0$ .....(1.51)

and $s - t = 0$ .....(1.52)

To get the values of p, q, r, s, t, u we have to solve the equations (1.47) to (1.52).

(1.47) + (1.49) + (1.51) $\Rightarrow 3p = 3 \Rightarrow p = 1$

$\therefore$ (1.49) $\Rightarrow r = 1$, and (1.51) $\Rightarrow q = 1$.

Similarly, from (1.48), (1.50) and (1.52) we get $s = t = u = 1$.

$\therefore \quad p = q = r = s = t = u = 1$.

***Example* 1.47:** If A and B are non – singular matrices of same order , show that AB and BA have same eigen values.

Solution: $AB = (B^{-1}B)(AB)$ [since $B^{-1}B = I$] $= B^{-1}(BA)B$ have same eigen values. i.e., BA and AB have same eigen values.

***Example* 1.48:** Show that the eigen values of a triangular matrix are just the diagonal elements of the matrix.

Solution: Let $A=\begin{bmatrix} a_{11} & a_{12}\ldots\ldots a_{1m} \\ 0 & a_{22}\ldots\ldots a_{2m} \\ \ldots\ldots\ldots\ldots & \\ \ldots\ldots\ldots\ldots & \\ 0\ldots\ldots 0\ldots\ldots & a_{mm} \end{bmatrix}$ be triangular matrix.

The characteristic equation of A is $|A - \lambda I| = 0$

$$\Rightarrow \begin{bmatrix} (a_{11}-\lambda) & a_{12}\ldots\ldots\ldots a_{1m} \\ 0 & (a_{22}-\lambda)\ldots\ldots a_{2m} \\ \ldots\ldots\ldots\ldots\ldots\ldots & \\ 0\ldots\ldots\ldots\ldots\ldots & (a_{mm}-\lambda) \end{bmatrix} = 0 \Rightarrow (a_{11}-\lambda)(a_{22}-\lambda)\ldots\ldots(a_{mm}-\lambda) = 0$$

$\Rightarrow \lambda = a_{11}, a_{22}, \ldots\ldots a_{mm}$ (diagonal elements of A) are the eigen values of A.

Exercise - 1.5

1. Write the characteristic matrices and the characteristic equations of the following matrices.

(a) $\begin{bmatrix} 3 & 4 \\ 1 & 5 \end{bmatrix}$ **[Ans :** $\begin{bmatrix} 3-\lambda & 4 \\ 1 & 5-\lambda \end{bmatrix}$; $\lambda^2 - 8\lambda + 11 = 0$**]**

(b) $\begin{bmatrix} 1 & 2 & 0 \\ 0 & 3 & 2 \\ 0 & 4 & 5 \end{bmatrix}$ **[Ans :** $\begin{bmatrix} 1-\lambda & 2 & 0 \\ 0 & 3-\lambda & 2 \\ 0 & 4 & 5-\lambda \end{bmatrix}$; $(1-\lambda)(\lambda^2 - 8\lambda + 7) = 0$**]**

2. If $A = \begin{bmatrix} 1 & 2 & 3 \\ 0 & 4 & 0 \\ 0 & 0 & 6 \end{bmatrix}$, find the eigen values of

(i) A^{-1} (ii) A^2 and (iii) A^3

[Ans : (i) $1, \frac{1}{4}, \frac{1}{6}$; **(ii)** 1, 16, 36; **(iii)** 1, 64, 216**]**

3. Find the product and sum of the eigen values of the following matrices without finding the eigen values.

(i) $\begin{bmatrix} 4 & 2 \\ 6 & 5 \end{bmatrix}$ (ii) $\begin{bmatrix} 4 & 2 & 3 \\ 0 & 3 & 2 \\ 0 & 0 & 5 \end{bmatrix}$ **[Ans :** (i) 8 ; 9; (ii) 60 ; 12**]**

4. Find the eigen values of the following matrices.

(i) $\begin{bmatrix} 2 & 3 & 4 \\ 0 & 2 & 1 \\ 0 & 0 & 2 \end{bmatrix}$ (ii) $\begin{bmatrix} 3 & 4 \\ 2 & 6 \end{bmatrix}$ **[Ans :** (i) 2, 2 , 2; (ii) $\frac{1}{2}\left(9 \pm \sqrt{41}\right)$**]**

5. Find the eigen values and the corresponding eigen vectors of the following matrices.

(i) $\begin{bmatrix} 0 & 1 & 0 \\ 1 & 0 & 1 \\ 0 & 1 & 0 \end{bmatrix}$ **[Ans :** $0, \pm\sqrt{2}$; $\begin{bmatrix} -c \\ 0 \\ c \end{bmatrix}, \begin{bmatrix} c \\ \sqrt{2}c \\ c \end{bmatrix}, \begin{bmatrix} -c \\ \sqrt{2}c \\ -c \end{bmatrix}$, $(c \neq 0)$**]**

(ii) $\begin{bmatrix} -2 & 8 & -12 \\ 1 & 4 & 4 \\ 0 & 0 & 1 \end{bmatrix}$ **[Ans :** 0, 1, 2 ; $\begin{bmatrix} 4c \\ -c \\ 0 \end{bmatrix}; \begin{bmatrix} 4c \\ 0 \\ -c \end{bmatrix}, \begin{bmatrix} 2c \\ -c \\ 0 \end{bmatrix}$, $c \neq 0$**]**

(iii) $\begin{bmatrix} 2 & 1 & 1 \\ 2 & 3 & 4 \\ -1 & -1 & -2 \end{bmatrix}$ **[Ans :** 1, –1, 3 ; $\begin{bmatrix} c \\ -c \\ 0 \end{bmatrix}, \begin{bmatrix} 0 \\ c \\ -c \end{bmatrix}, \begin{bmatrix} -2c \\ -3c \\ c \end{bmatrix}$, $c \neq 0$**]**

(iv) $\begin{bmatrix} 2 & 0 & 2 \\ -1 & 3 & 1 \\ 1 & -1 & 3 \end{bmatrix}$ **[Ans :** 2, 2, 4 ; $\begin{bmatrix} c \\ c \\ 0 \end{bmatrix}, \begin{bmatrix} c \\ 0 \\ c \end{bmatrix}$, $c \neq 0$**]**

(v) $\begin{bmatrix} 6 & -2 & 2 \\ -2 & 3 & -1 \\ 2 & -1 & 3 \end{bmatrix}$ **[Ans :** 2, 2, 8 ; $\begin{bmatrix} c_1 \\ c_2 \\ c_2 - 2c_1 \end{bmatrix}, \begin{bmatrix} 2c \\ -c \\ c \end{bmatrix}$**]**

(vi) $\begin{bmatrix} -2 & 2 & -3 \\ 2 & 1 & -6 \\ -1 & -2 & 0 \end{bmatrix}$ **[Ans :** –3, –3, 5 ; $\begin{bmatrix} -2c_1 + 3c_2 \\ c_1 \\ c_2 \end{bmatrix}; \begin{bmatrix} c \\ 2c \\ -c \end{bmatrix}$**]**

1.8 Cayley-Hamilton Theorem

Cayley-Hamilton Theorem

It states that "Every square matrix satisfies its characteristic equation".

Proof : Let A be a square matrix of order n and I be the unit matrix of the same order.

Then $\quad |A - \lambda I| = 0 \quad$ .....(1.53)

is the characteristic equation and 'λ' the characteristic root of A.

$\because$ (1.54) is an equation of degree 'n' in λ, it can be written as,

$$(-1)^n \left[\lambda^n + p_1\lambda^{n-1} + p_2\lambda^{n-2} + \ldots + p_{n-1}\lambda + p_n\right] = 0 \quad \text{.....(1.54)}$$

where $p_1, p_2, \ldots, p_n$ are constants.

$\therefore$ We have to prove that,

$$A^n + p_1A^{n-1} + p_2 A^{n-2} + \ldots\ldots + p_{n-1} A + p_n I = 0 \quad \text{.....(1.55)}$$

The elements of the matrix $(A - \lambda I)$ are atmost of the first degree in λ.

$\therefore$ $Adj\ (A - \lambda I)$ has its elements as ordinary polynomials in λ of degree $(n - 1)$ or less. So $Adj\ (A - \lambda I)$ can be written as a matrix polynomial as,

$$Adj\ (A - \lambda I) = C_0\lambda^{n-1} + C_1\lambda^{n-2} + \ldots. + C_{n-2}\lambda + C_{n-1},$$

where $C_0, C_2, \ldots, C_{n-1}$ are $n \times n$ matrices and their elements are functions of elements of A.

Again, since

$$(A - \lambda I) \,.\, Adj\ (A - \lambda I) = |A - \lambda I| I,$$

we have,

$$(A - \lambda I)\left[C_0\lambda^{n-1} + C_1\lambda^{n-2} + \ldots + C_{n-2}\lambda + C_{n-1}\right]$$

$$= (-1)^n\left[\lambda^n + p_1\lambda^{n-1} + p_2\lambda^{n-2} + \ldots + p_{n-1}\lambda + p_n\right]I$$

[from(1.54)]

Comparing coefficients of like powers of λ on both sides, we get

$$-IC_0 = (-1)^n I$$

$$AC_0 - IC_1 = (-1)^n . p_1 I$$

$$AC_1 - IC_2 = (-1)^n . p_2 I$$

$$\ldots \quad \ldots \quad \ldots \quad \ldots$$

$$AC_{n-2} - IC_{n-1} = (-1)^n . p_{n-1} I$$

$$AC_{n-1} = (-1)^n . p_n I$$

Premultiplying the above equations respectively by A^n, A^{n-1}, ... A, I and adding, we get

$$0 = (-1)^n \left[A^n + p_1 A^{n-1} + p_2 A^{n-2} + ... + p_{n-1} A + p_n I\right]$$

which is equal to (1.55) and hence the theorem is proved.

Corollary 1 :

The inverse of a nonsingular matirx can be found by using the above therom as follows :

A is nonsingular $\Rightarrow |A| \neq 0$.

Then premultiplying (1.55) of above theroem by A^{-1}, we get ;

$$A^{n-1} + p_1 A^{n-2} + ... + p_{n-1} I + p_n A^{-1} = 0$$

$$\Rightarrow \qquad A^{-1} = \frac{-1}{p_n}\left[A^{n-1} + p_1 A^{n-2} + ... + p_{n-1} I\right]$$

(If $\lambda = 0$ (1.53) and (1.54) $\Rightarrow |A| = (-1)^n p_n \Rightarrow p_n \neq 0$)

Corollary 2 :

If m be a positive integer such that $m \geq n$, then multiplying the result (1.55) by A^{m-n}, we get,

$$A^m + p_1 A^{m-1} + ... + p_n A^{m-n} = 0,$$

which shows that any positive integral power A^m $(m \geq n)$ of A can be expressed in terms of powers of lower order.

Solved Examples

***Example* 1.49:** Verify Cayley Hamilton Theorem for the matrix $A = \begin{bmatrix} 3 & 2 \\ 1 & 5 \end{bmatrix}$.

Solution:

$$A = \begin{bmatrix} 3 & 2 \\ 1 & 5 \end{bmatrix}; \; A - \lambda I = \begin{bmatrix} 3-\lambda & 2 \\ 1 & 5-\lambda \end{bmatrix}$$

$$|A - \lambda I| = \begin{vmatrix} 3-\lambda & 2 \\ 1 & 5-\lambda \end{vmatrix} = (3-\lambda)(5-\lambda) - 2$$

$$= \lambda^2 - 8\lambda + 13$$

characteristic equation of A is

$$|A - \lambda I| = 0$$

$$\Rightarrow \qquad \lambda^2 - 8\lambda + 13 = 0 \qquad .....(1.56)$$

Cayley Hamilton theorem states that A satisfies its characteristic equation.

i.e., We have to verify that $A^2 - 8A + 13I = 0$

$$A^2 = \begin{bmatrix} 3 & 2 \\ 1 & 5 \end{bmatrix}\begin{bmatrix} 3 & 2 \\ 1 & 5 \end{bmatrix}$$

$$= \begin{bmatrix} 3.3+2.1 & 3.2+2.5 \\ 1.3+5.1 & 1.2+5.5 \end{bmatrix}$$

$$= \begin{bmatrix} 11 & 16 \\ 8 & 27 \end{bmatrix}$$

$$A^2 - 8A + 13I = \begin{bmatrix} 11 & 16 \\ 8 & 27 \end{bmatrix} - 8\begin{bmatrix} 3 & 2 \\ 1 & 5 \end{bmatrix} + 13\begin{bmatrix} 1 & 0 \\ 0 & 1 \end{bmatrix}$$

$$= \begin{bmatrix} 11-24+13 & 16-16+0 \\ 8-8+0 & 27-40+13 \end{bmatrix} = \begin{bmatrix} 0 & 0 \\ 0 & 0 \end{bmatrix} = 0$$

Hence the theorem is verified.

***Example* 1.50:** Velrify Cayley-Hamilton theorem for the matrix $A = \begin{bmatrix} 1 & 2 & -1 \\ 3 & 1 & 0 \\ -2 & 1 & 4 \end{bmatrix}$.

Solution:

$$|A - \lambda I| = \begin{vmatrix} 1-\lambda & 2 & -1 \\ 3 & 1-\lambda & 0 \\ -2 & 1 & 4-\lambda \end{vmatrix}$$

$$= (1-\lambda)[(1-\lambda)\ (4-\lambda)-0] - 2[3(4-\lambda)-0]$$
$$-1[3+2(1-\lambda)]$$

$$= 4 - 9\lambda + 6\lambda^2 - \lambda^3 - 24 + 6\lambda - 5 + 2\lambda$$

$$= -\lambda^3 + 6\lambda^2 - \lambda - 25$$

Characteristic equation of A is $|A - \lambda I| = 0$

i.e., $\lambda^3 - 6\lambda^2 + \lambda + 25 = 0$

Cayley-Hamilton theorem states that the matrix satisfies its characteristic equation.

$\therefore$ we have to verify that

$$A^3 - 6A^2 + A + 25I = 0$$

$$A^2 = \begin{bmatrix} 1 & 2 & -1 \\ 3 & 1 & 0 \\ -2 & 1 & 4 \end{bmatrix}\begin{bmatrix} 1 & 2 & -1 \\ 3 & 1 & 0 \\ -2 & 1 & 4 \end{bmatrix}$$

$$= \begin{bmatrix} (1.1+2.3+-1.-2) & (1.2+2.1-1.1) & (1.-1+2.0-1.4) \\ (3.1+1.3+0.-2) & (3.2+1.1+0.1) & (3.-1+1.0+0.4) \\ (-2.1+1.3+4.-2) & (-2.2+1.1+4.1) & (-2-1+1.0+4.4) \end{bmatrix}$$

$$= \begin{bmatrix} 9 & 3 & -5 \\ 6 & 7 & -3 \\ -7 & 1 & 18 \end{bmatrix}$$

$$A^3 = A^2 . A = \begin{bmatrix} 9 & 3 & -5 \\ 6 & 7 & -3 \\ -7 & 1 & 18 \end{bmatrix}\begin{bmatrix} 1 & 2 & -1 \\ 3 & 1 & 0 \\ -2 & 1 & 4 \end{bmatrix} = \begin{bmatrix} 28 & 16 & -29 \\ 33 & 16 & -18 \\ -40 & 5 & 79 \end{bmatrix}$$

$$A^3 - 6A^2 + A + 25I = \begin{bmatrix} 28 & 16 & -29 \\ 33 & 16 & -18 \\ -40 & 5 & 79 \end{bmatrix} - 6\begin{bmatrix} 9 & 3 & -5 \\ 6 & 7 & -3 \\ -7 & 1 & 18 \end{bmatrix}$$

$$+ \begin{bmatrix} 1 & 2 & -1 \\ 3 & 1 & 0 \\ -2 & 1 & 4 \end{bmatrix} + 25\begin{bmatrix} 1 & 0 & 0 \\ 0 & 1 & 0 \\ 0 & 0 & 1 \end{bmatrix}$$

$$= \begin{bmatrix} (28-54+1+25) & (16-18+2+0) & (-29+30-1+0) \\ (33-36+3+0) & (16-42+1+25) & (-18+18+0+0) \\ (-40+42-2+0) & (5-6+1+0) & (79-108+4+25) \end{bmatrix}$$

$$= \begin{bmatrix} 0 & 0 & 0 \\ 0 & 0 & 0 \\ 0 & 0 & 0 \end{bmatrix};$$

$\therefore$ The theorem is verified.

***Example* 1.51:** Using Cayley-Hamilton theorem find A^{-1} if $A = \begin{bmatrix} 1 & 2 & 1 \\ 2 & 3 & 5 \\ 0 & -1 & 4 \end{bmatrix}$.

Solution: Characteristic equation of A is $|A - \lambda I| = 0$

$$\Rightarrow \quad \begin{vmatrix} 1-\lambda & 2 & 1 \\ 2 & 3-\lambda & 5 \\ 0 & -1 & 4-\lambda \end{vmatrix} = 0$$

$$\Rightarrow \quad (1-\lambda)(\lambda^2 - 7\lambda + 17) - 2(+8 - 2\lambda) + 1(-2) = 0$$

$$\Rightarrow \quad -\lambda^3 + 8\lambda^2 - 24\lambda + 17 - 16 + 4\lambda - 2 = 0$$

$$\Rightarrow \quad \lambda^3 - 8\lambda^2 - 20\lambda + 1 = 0$$

According to Cayley-Hamilton theorem,

$$A^3 - 8A^2 + 20A + 1I = O_{3\times 3}$$

Multipying each term with A^{-1}, we get,

$$A^2 - 8A + 20I + 1A^{-1} = O_{3\times 3}$$

$$\Rightarrow \quad A^{-1} = -\left[A^2 - 8A + 20I\right] \qquad \text{.....(1.57)}$$

$$A^2 = \begin{bmatrix} 1 & 2 & 1 \\ 2 & 3 & 5 \\ 0 & -1 & 4 \end{bmatrix}\begin{bmatrix} 1 & 2 & 1 \\ 2 & 3 & 5 \\ 0 & -1 & 4 \end{bmatrix} = \begin{bmatrix} 5 & 7 & 15 \\ 8 & 8 & 37 \\ -2 & -7 & 11 \end{bmatrix}$$

$$\therefore \quad A^{-1} = -\left[\begin{pmatrix} 5 & 7 & 15 \\ 8 & 8 & 37 \\ -2 & -7 & 11 \end{pmatrix} - 8\begin{pmatrix} 1 & 2 & 1 \\ 2 & 3 & 5 \\ 0 & -1 & 4 \end{pmatrix} + 20\begin{pmatrix} 1 & 0 & 0 \\ 0 & 1 & 0 \\ 0 & 0 & 1 \end{pmatrix}\right]$$

$$= -\begin{bmatrix} (5-8+20) & (7-16+0) & (15-8+0) \\ (8-16+0) & (8-24+20) & (37-40+0) \\ (-2-0+0) & (-7+8+0) & (11-32+20) \end{bmatrix}$$

$$= -\begin{bmatrix} 17 & -9 & 7 \\ -8 & 4 & -3 \\ -2 & 1 & -1 \end{bmatrix}$$

$$= \begin{bmatrix} -17 & 9 & -7 \\ 8 & -4 & 3 \\ 2 & -1 & 1 \end{bmatrix}$$

***Example* 1.52:** Verify Cayley-Hamilton theorem for the matrix $A = \begin{bmatrix} 2 & -1 & 0 \\ 3 & 1 & -1 \\ 2 & 0 & 3 \end{bmatrix}$. Hence find A^{-1}.

Solution: Characteristic equation of A is $|A - \lambda I| = 0$

$$\Rightarrow \quad \begin{vmatrix} 2-\lambda & -1 & 0 \\ 3 & 1-\lambda & -1 \\ 2 & 0 & 3-\lambda \end{vmatrix} = 0$$

$$\Rightarrow \quad (2-\lambda)(1-\lambda)(3-\lambda) + 3(3-\lambda) + 2 = 0$$

$$\Rightarrow \quad -\lambda^3 + 6\lambda^2 - 11\lambda + 6 + 9 - 3\lambda + 2 = 0$$

$$\Rightarrow \quad \lambda^3 - 6\lambda^2 + 14\lambda - 17 = 0$$

Cayley-Hamilton theorem states that A satisfies its characteristic equation.

i.e., we have to verify that

$$A^3 - 6A^2 + 14A - 17I = O_{3\times 3} \qquad .....(1.58)$$

$$A^2 = \begin{bmatrix} 2 & -1 & 0 \\ 3 & 1 & -1 \\ 2 & 0 & 3 \end{bmatrix}\begin{bmatrix} 2 & -1 & 0 \\ 3 & 1 & -1 \\ 2 & 0 & 3 \end{bmatrix} = \begin{bmatrix} 1 & -3 & 1 \\ 7 & -2 & -4 \\ 10 & -2 & 9 \end{bmatrix}$$

$$A^3 = \begin{bmatrix} 1 & -3 & 1 \\ 7 & -2 & -4 \\ 10 & -2 & 9 \end{bmatrix}\begin{bmatrix} 2 & -1 & 0 \\ 3 & 1 & -1 \\ 2 & 0 & 3 \end{bmatrix} = \begin{bmatrix} -5 & -4 & 6 \\ 0 & -9 & -10 \\ 32 & -12 & 29 \end{bmatrix}$$

$\therefore \quad A^3 - 6A^2 + 14A - 17I$

$$= \begin{bmatrix} -5 & -4 & 6 \\ 0 & -9 & -10 \\ 32 & -12 & 29 \end{bmatrix} - 6\begin{bmatrix} 1 & -3 & 1 \\ 7 & -2 & -4 \\ 10 & -2 & 9 \end{bmatrix} + 14\begin{bmatrix} 2 & -1 & 0 \\ 3 & 1 & -1 \\ 2 & 0 & 3 \end{bmatrix} - 17\begin{bmatrix} 1 & 0 & 0 \\ 0 & 1 & 0 \\ 0 & 0 & 1 \end{bmatrix}$$

$$= \begin{bmatrix} (-5-6+28-17) & (-4+18-14-0) & (6-6+0-0) \\ (0-42+42-0) & (-9+12+14-17) & (-10+24-14-0) \\ 32-60+28-0 & (-12+12+0-0) & (29-54+42-17) \end{bmatrix}$$

$$= \begin{bmatrix} 0 & 0 & 0 \\ 0 & 0 & 0 \\ 0 & 0 & 0 \end{bmatrix}$$

Hence the theorem is verified.

To find A^{-1} : Multiply (1.58) by A^{-1} to get,

$$A^2 - 6A + 14I - 17A^{-1} = O_{3\times 3}$$

$$\Rightarrow \quad A^{-1} = \frac{1}{17}\left[A^2 - 6A + 14I\right]$$

$$= \frac{1}{17}\left[\begin{pmatrix} 1 & -3 & 1 \\ 7 & -2 & -4 \\ 10 & -2 & 9 \end{pmatrix} - 6\begin{pmatrix} 2 & -1 & 0 \\ 3 & 1 & -1 \\ 2 & 0 & 3 \end{pmatrix} + 14\begin{pmatrix} 1 & 0 & 0 \\ 0 & 1 & 0 \\ 0 & 0 & 1 \end{pmatrix}\right]$$

$$= \frac{1}{17}\begin{pmatrix} (1-12+14) & (-3+6+0) & (1-0+0) \\ (7-18+0) & (-2-6+14) & (-4+6+0) \\ (10-12+0) & (-2-0+0) & (9-18+14) \end{pmatrix}$$

$$= \frac{1}{17}\begin{bmatrix} 3 & 3 & 1 \\ -11 & 6 & 2 \\ -2 & -2 & 5 \end{bmatrix}$$

***Example* 1.53:** If $A = \begin{bmatrix} 1 & 1 \\ 2 & 4 \end{bmatrix}$, (i) find its eigen values (ii) verify Cayley-Hamilton theorem for A. (iii) Find A^{-1} and (iv) Express $(A^5 - 4A^4 - 2A^3 - A^2 - I)$ as a linear polynomial in A.

Solution:

(i) Characteristic equation of A is $|A - \lambda I| = 0$

$$\Rightarrow \quad \begin{vmatrix} 1-\lambda & 1 \\ 2 & 4-\lambda \end{vmatrix} = 0$$

$$\Rightarrow \quad \lambda^2 - 5\lambda + 2 = 0 \qquad .....(1.59)$$

$\therefore$ Eigen values of A are roots of (1.59).

They are $\lambda = \dfrac{5+\sqrt{17}}{2}, \dfrac{5-\sqrt{17}}{2}$

(ii) Cayley Hamilton theorem states that a square matrix satisfies its characteristic equation.

i.e., we have to verify that A satisfies (1.59)

i.e., $$A^2 - 5A + 2I = O_{2\times 2} \qquad .....(1.60)$$

$$A^2 = \begin{bmatrix} 1 & 1 \\ 2 & 4 \end{bmatrix}\begin{bmatrix} 1 & 1 \\ 2 & 4 \end{bmatrix} = \begin{bmatrix} 3 & 5 \\ 10 & 18 \end{bmatrix}$$

$$\therefore (1.60) \Rightarrow A^2 - 5A + 2I = \begin{bmatrix} 3 & 5 \\ 10 & 18 \end{bmatrix} - 5\begin{bmatrix} 1 & 1 \\ 2 & 4 \end{bmatrix} + 2\begin{bmatrix} 1 & 0 \\ 0 & 1 \end{bmatrix}$$

$$= \begin{bmatrix} 3-5+2 & 5-5+0 \\ 10-10+0 & 18-20+2 \end{bmatrix}$$

$$= \begin{bmatrix} 0 & 0 \\ 0 & 0 \end{bmatrix}$$

Hence the theorem is verified.

(iii) From (1.60), we have

$$A^2 - 5A + 2I = O_{2\times 2}$$

Multiplying by A^{-1}, we have,

$$A - 5I + 2A^{-1} = O_{2\times 2}$$

$$\Rightarrow \qquad A^{-1} = \frac{1}{2}[-A + 5I]$$

$$= \frac{1}{2}\left[\begin{pmatrix} -1 & -1 \\ -2 & -4 \end{pmatrix} + 5\begin{pmatrix} 1 & 0 \\ 0 & 1 \end{pmatrix}\right]$$

$$= \frac{1}{2}\begin{bmatrix} 4 & -1 \\ -2 & 1 \end{bmatrix}$$

(iv) (1.60) $\Rightarrow$ $A^2 = 5A - 2I$ .....(1.61)

(1.61) $\times A \Rightarrow A^3 = 5A^2 - 2A$

$= 5\,(5A - 2I) - 2A$ (from(1.61))

$= 23A - 10I$ .....(1.62)

(1.62) $\times A \Rightarrow A^4 = 23A^2 - 10A$

$= 23\,(5A - 2I) - 10A$ (from(1.62))

$= 105A - 46I$ .....(1.63)

(1.63) $\times A \Rightarrow A^5 = 105A^2 - 46A$

$= 105(5A - 2I) - 46A$ (from(1.63))

$= 479\,A - 210\,I$ .....(1.64)

Substituting the values of A^2, A^3, A^4, A^5 obtained above (in equations (1.61), (1.62), (1.63), (1.64) respectively) in the given expression, we get,

$$A^5 - 4A^4 - 2A^3 - A^2 - I = (479A - 210I) - 4(105A - 46I)$$
$$- 2\,(23A - 10I) - (5A - 2I) - I$$
$$= (8A - 5I)$$

***Example* 1.54:** If $A = \begin{bmatrix} 2 & 3 \\ 1 & 2 \end{bmatrix}$, express A^4 in lower powers of A.

Solution: $A = \begin{bmatrix} 2 & 3 \\ 1 & 2 \end{bmatrix}$;

The characteristic equation, $|A - \lambda I| = 0$

$$\Rightarrow \qquad (2 - \lambda)^2 - 3 = 0$$

$$\Rightarrow \qquad \lambda^2 - 4\lambda + 1 = 0$$

According to Cayley Hamilton theorem, we have,

$$A^2 - 4A + I = O_{2 \times 2}$$

$$\Rightarrow \quad A^2 = 4A - I \quad .....(1.65)$$

$$(1.65) \times A \Rightarrow \quad A^3 = 4A^2 - A \quad .....(1.66)$$

$$(1.66) \times A \Rightarrow \quad A^4 = 4A^3 - A^2 \quad .....(1.67)$$

(1.67) can be further simplified, if necessary, as

$$A^4 = 4(4A^2 - A) - A^2$$

$$= 15A^2 - 4A \quad \text{from (1.66)}$$

$$= 15(4A - I) - 4A \quad \text{from (1.65)}$$

$$= 56A - 15I$$

Exercise - 1.6

I. Verify Cayley–Hamilton theorem for the following matrices.

(i) $\begin{bmatrix} 2 & 1 \\ 0 & 4 \end{bmatrix}$ (ii) $\begin{bmatrix} 1 & -1 \\ -1 & 4 \end{bmatrix}$

(iii) $\begin{bmatrix} 2 & 3 & 4 \\ 1 & -1 & 0 \\ 0 & 2 & 1 \end{bmatrix}$ (iv) $\begin{bmatrix} 1 & 3 & 5 \\ 2 & -1 & 2 \\ 0 & 1 & 3 \end{bmatrix}$

II. Using Cayley Hamilton theorem, find the inverses of the followig matrices.

(i) $\begin{bmatrix} 2 & 1 \\ 5 & 4 \end{bmatrix}$ **Ans :** $\frac{1}{3}\begin{bmatrix} 4 & -1 \\ -5 & 2 \end{bmatrix}$

(ii) $\begin{bmatrix} 4 & -1 \\ 1 & 2 \end{bmatrix}$ **Ans :** $\frac{1}{9}\begin{bmatrix} 2 & 1 \\ -1 & 4 \end{bmatrix}$

(iii) $\begin{bmatrix} 4 & 1 & 2 \\ 1 & 2 & 0 \\ 0 & -1 & 3 \end{bmatrix}$ **Ans :** $\frac{1}{19}\begin{bmatrix} 6 & -5 & -4 \\ -3 & 12 & 2 \\ -1 & 4 & 7 \end{bmatrix}$

(iv) $\begin{bmatrix} 1 & 3 & 4 \\ 2 & 0 & -1 \\ 3 & 1 & 6 \end{bmatrix}$ **Ans :** $+\frac{1}{36}\begin{bmatrix} -1 & +14 & +3 \\ +15 & +6 & -9 \\ -2 & -8 & +6 \end{bmatrix}$

III Using Cayley – Hamilton theorem, express $(A^5 - 4A^3 + 3A^2 - 2A)$ as a linear polynomial in A, where $A = \begin{bmatrix} 2 & 1 \\ 1 & 2 \end{bmatrix}$

[Ans : $79A - 81I$**]**

IV If $A = \begin{bmatrix} 3 & -1 \\ 1 & 2 \end{bmatrix}$, express A^5 in powers of A of lower order.

[Ans : $149A - 385I$**]**

V If $A = \begin{bmatrix} 0 & 1 \\ 1 & 0 \end{bmatrix}$, find the constants a, b, c, such that $A^3 = aA$, $A^4 = bA$ and $A^5 = cA$.

[Ans : $a = b = c = 1$**]**

VI If $A = \begin{bmatrix} 0 & 2 \\ 2 & 0 \end{bmatrix}$, prove that

(i) $A^4 = 4A^2$ and (ii) $A^5 = 16A$.

(**Hint :** Use Cayley – Hamilton theorem).

***Example* 1.55:** Find the characteristic equation of the matrix

$$A = \begin{bmatrix} 1 & 2 & -2 \\ 1 & 1 & 1 \\ 1 & 3 & -1 \end{bmatrix}$$

Hence find A^{-1}

***Solution*:** Characteristic equation of matrix A is

$$|A - \lambda I| = 0$$

$$\begin{bmatrix} 1-\lambda & 2 & -2 \\ 1 & 1-\lambda & 1 \\ 1 & 3 & -1-\lambda \end{bmatrix} = 0$$

or $(1-\lambda)[(1-\lambda)(-1-\lambda)-3]-2(-1-\lambda-1)-2(3-1+\lambda)=0$

or $(1-\lambda)(-1+\lambda^2-3)-2(-\lambda-2)-2(2+\lambda)=0$

or $(1-\lambda)(\lambda^2-4)+2\lambda+4-4-2\lambda=0$

or $(-\lambda+1)(\lambda^2-4)=0$ or $\lambda^3-\lambda^2-4\lambda+4=0$

By Cayley – Hamilton Theorem

$$A^3-A^2-4A+4I=0$$

or $A^2-A-4I+4A^{-1}=0$

or $A^2-A-4I+4A^{-1}=0$ (Multiplying by A^{-1})

or $4A^{-1}=[-A^2+A+4I]$ .....(1.68)

$$A^2=\begin{bmatrix}1 & 2 & -2\\ 1 & 1 & 1\\ 1 & 3 & -1\end{bmatrix}\begin{bmatrix}1 & 2 & -2\\ 1 & 1 & 1\\ 1 & 3 & -1\end{bmatrix}=\begin{bmatrix}1 & -2 & 2\\ 3 & 6 & -2\\ 3 & 2 & 2\end{bmatrix}$$

From (1.68), we have

$$4A^{-1}=-\begin{bmatrix}1 & -2 & 2\\ 3 & 6 & -2\\ 3 & 2 & 2\end{bmatrix}+\begin{bmatrix}1 & 2 & -2\\ 1 & 1 & 1\\ 1 & 3 & -1\end{bmatrix}=\begin{bmatrix}4 & 0 & 0\\ 0 & 4 & 0\\ 0 & 0 & 4\end{bmatrix}$$

$$=\begin{bmatrix}-1+1+4 & 2+2+0 & -2-2+0\\ -3+1+0 & -6+1+4 & 2+1+0\\ -3+1+0 & -2+3+0 & -2-1+4\end{bmatrix}$$

$$A^{-1}=\frac{1}{4}\begin{bmatrix}4 & 4 & -4\\ -2 & -1 & 3\\ -2 & 1 & 1\end{bmatrix}$$ **Ans.**

***Example* 1.56:** Find the characteristic equation of the symmetric matrix

$$A=\begin{bmatrix}2 & -1 & 1\\ -1 & 2 & -1\\ 1 & -1 & 2\end{bmatrix}$$

and verify that it is satisfied by A and hence obtain A^{-1}.

Express $A^6-6A^5+9A^4-2A^3-12A^2+23A-9I$ in linear polynomial in A.

Solution: Characteristic equation is $|A-\lambda I|=0$

$$\begin{bmatrix} 2-\lambda & -1 & 1 \\ -1 & 2-\lambda & -1 \\ 1 & -1 & 2-\lambda \end{bmatrix} = 0$$

$$(2-\lambda)[(2-\lambda)^2 - 1] + 1[-2+\lambda+1] + 1[1-2+\lambda] = 0$$

or $\quad (2-\lambda)^3 - (2-\lambda) + \lambda - 1 + \lambda - 1 = 0$

or $\quad (2-\lambda)^3 - 2 + \lambda + \lambda - 1 + \lambda - 1 = 0$ or $(2-\lambda)^3 + 3\lambda - 4 = 0$

or $\quad 8 - \lambda^3 - 12\lambda + 6\lambda^2 + 3\lambda - 4 = 0$

or $\quad -\lambda^3 + 6\lambda^2 - 9\lambda + 4 = 0$ or $\lambda^3 - 6\lambda^2 + 9\lambda - 4 = 0$

By Cayley – Hamilton Theorem $A^3 - 6A^2 + 9A - 4I = 0$ .....(1.69)

Verification: $A^2 = \begin{bmatrix} 2 & -1 & 1 \\ -1 & 2 & -1 \\ 1 & -1 & 2 \end{bmatrix}\begin{bmatrix} 2 & -1 & 1 \\ -1 & 2 & -1 \\ 1 & -1 & 2 \end{bmatrix}$

$$= \begin{bmatrix} 4+1+1 & -2-2-1 & 2+1+2 \\ -2-2-1 & 1+4+1 & -1-2-2 \\ 2+1+2 & -1-2-2 & 1+1+4 \end{bmatrix} = \begin{bmatrix} 6 & -5 & 5 \\ -5 & 6 & -5 \\ 5 & -5 & 6 \end{bmatrix}$$

$$A^3 = \begin{pmatrix} 6 & -5 & 5 \\ -5 & 6 & -5 \\ 5 & -5 & 6 \end{pmatrix}\begin{pmatrix} 2 & -1 & 1 \\ -1 & 2 & -1 \\ 1 & -1 & 2 \end{pmatrix}$$

$$= \begin{pmatrix} 12+5+5 & -6-10-5 & 6+5+10 \\ -10-6-5 & 5+12+5 & -5-6-10 \\ 10+5+6 & -5-10-6 & 5+5+12 \end{pmatrix} = \begin{pmatrix} 22 & -21 & 21 \\ -21 & 22 & -21 \\ 21 & -21 & 22 \end{pmatrix}$$

$$A^3 - 6A^2 + 9A - 4I$$

$$= \begin{pmatrix} 22 & -21 & 21 \\ -21 & 22 & -21 \\ 21 & -21 & 22 \end{pmatrix} - 6\begin{pmatrix} 6 & -5 & 5 \\ -5 & 6 & -5 \\ 5 & -5 & 6 \end{pmatrix} + 9\begin{pmatrix} 2 & -1 & 1 \\ -1 & 2 & -1 \\ 1 & -1 & 2 \end{pmatrix} - 4\begin{pmatrix} 1 & 0 & 0 \\ 0 & 1 & 0 \\ 0 & 0 & 1 \end{pmatrix}$$

$$= \begin{pmatrix} 22-36+18-4 & -21+30-9-0 & 21-30+9-0 \\ -21+30-9-0 & 22-36+18-4 & -21+30-9-0 \\ 21-30+9-0 & -21+30-9-0 & 22-36+18-4 \end{pmatrix} = \begin{pmatrix} 0 & 0 & 0 \\ 0 & 0 & 0 \\ 0 & 0 & 0 \end{pmatrix} = 0$$

So it is verified that the characteristic equation (1.69) is satisfied by A.

Inverse of Matrix A,

$$A^3 - 6A^2 + 9A - 4I = 0$$

On multiplying by A^{-1}, we get

$$A^2 - 6A + 9I - 4A^{-1} = 0 \text{ or } 4A^{-1} = A^2 - 6A + 9I$$

or $$4A^{-1} = \begin{pmatrix} 6 & -5 & 5 \\ -5 & 6 & -5 \\ 5 & -5 & 6 \end{pmatrix} - 6\begin{pmatrix} 2 & -1 & 1 \\ -1 & 2 & -1 \\ 1 & -1 & 2 \end{pmatrix} + 9\begin{pmatrix} 1 & 0 & 0 \\ 0 & 1 & 0 \\ 0 & 0 & 1 \end{pmatrix}$$

$$= \begin{pmatrix} 6-12+9 & -5+6+0 & 5-6+0 \\ -5+6+0 & 6-12+9 & -5+6+0 \\ 5-6+0 & -5+6+0 & 6-12+9 \end{pmatrix}, \quad A^{-1} = \frac{1}{4}\begin{pmatrix} 3 & 1 & -1 \\ 1 & 3 & 1 \\ -1 & 1 & 3 \end{pmatrix}$$ **Ans.**

$$A^6 - 6A^5 + 9A^4 - 2A^3 - 12A^2 + 23A - 9I$$

$$= A^3(A^3 - 6A^2 + 9A - 4I) + 2(A^3 - 6A^2 + 9A - 4I) + 5A - I$$

$$= 5A - I$$ **Ans.**

***Example* 1.57:** Find the characteristic equation of the matrix A

$$A = \begin{bmatrix} 4 & 3 & 1 \\ 2 & 1 & -2 \\ 1 & 2 & 1 \end{bmatrix}$$

Hence find A^{-1}

Solution: Characteristic equation is

$$\begin{bmatrix} 4-\lambda & 3 & 1 \\ 2 & 1-\lambda & -2 \\ 1 & 2 & 1-\lambda \end{bmatrix} = 0$$

$$(4-\lambda)[1+\lambda^2 - 2\lambda + 4] - 3(2 - 2\lambda + 2) + 1(4 - 1 + \lambda) = 0$$

or $$(4-\lambda)(\lambda^2 - 2\lambda + 5) - 3(-2\lambda + 4) + (3 + \lambda) = 0$$

$$4\lambda^2 - 8\lambda + 20 - \lambda^3 + 2\lambda^2 - 5\lambda + 6\lambda - 12 + 3 + \lambda = 0$$

or $$-\lambda^3 + 6\lambda^2 - 6\lambda + 11 = 0 \text{ or } \lambda^3 - 6\lambda^2 + 6\lambda - 11 = 0$$

By Cayley Hamilton Theorem

$$A^3 - 6A^2 + 6A - 11I = 0$$

By multiplying by A^{-1} we get

$$A^2 - 6A + 6I - 11\,A^{-1} = 0 \text{ or } 11A^{-1} = A^2 - 6A + 6I$$

$$11A^{-1} = \begin{bmatrix} 4 & 3 & 1 \\ 2 & 1 & -2 \\ 1 & 2 & 1 \end{bmatrix}\begin{bmatrix} 4 & 3 & 1 \\ 2 & 1 & -2 \\ 1 & 2 & 1 \end{bmatrix} - 6\begin{bmatrix} 4 & 3 & 1 \\ 2 & 1 & -2 \\ 1 & 2 & 1 \end{bmatrix} + 6\begin{bmatrix} 1 & 0 & 0 \\ 0 & 1 & 0 \\ 0 & 0 & 1 \end{bmatrix}$$

$$= \begin{bmatrix} 23 & 17 & -1 \\ 8 & 3 & -2 \\ 9 & 7 & -2 \end{bmatrix} + \begin{bmatrix} -24 & -18 & -6 \\ -12 & -6 & 12 \\ -6 & -12 & -6 \end{bmatrix} + \begin{bmatrix} 6 & 0 & 0 \\ 0 & 6 & 0 \\ 0 & 0 & 6 \end{bmatrix}$$

$$= \begin{bmatrix} 5 & -1 & -7 \\ -4 & 3 & 10 \\ 3 & -5 & -2 \end{bmatrix} \therefore A^{-1} = \frac{1}{11}\begin{bmatrix} 5 & -1 & -7 \\ -4 & 3 & 10 \\ 3 & -5 & -2 \end{bmatrix}$$ **Ans.**

***Example* 1.58:** Find the characteristic equation of the matrix [CSVTU 2007]

$$A = \begin{bmatrix} 2 & 1 & 1 \\ 0 & 1 & 0 \\ 1 & 1 & 2 \end{bmatrix}$$

Verify Cayley-Hamilton theorem and hence evaluate the matrix equation.

$$A^8 - 5A^7 + 7\,A^6 - 3\,A^5 + A^4 = 5A^3 - 8\,A^2 + 2A - I$$

Solution: Characteristic equation of the matrix A is

$$\begin{bmatrix} 2-\lambda & 1 & 1 \\ 0 & 1-\lambda & 0 \\ 1 & 1 & 2-\lambda \end{bmatrix} = 0 \text{ or } (2-\lambda)[(1-\lambda)\,(2-\lambda)] - 1(0) + 1\,(0 - 1 + \lambda) = 0$$

or $\lambda^3 - 5\,\lambda^2 + 7\lambda - 3 = 0$

According to Cayley-Hamilton Theorem

$$A^3 - 5\,A^2 + 7A - 3I = 0 \qquad .....(1.70)$$

We have to verify the eq. (1.70).

$$A^2 = \begin{bmatrix} 2 & 1 & 1 \\ 0 & 1 & 0 \\ 1 & 1 & 2 \end{bmatrix}\begin{bmatrix} 2 & 1 & 1 \\ 0 & 1 & 0 \\ 1 & 1 & 2 \end{bmatrix} = \begin{bmatrix} 5 & 4 & 4 \\ 0 & 1 & 0 \\ 4 & 4 & 5 \end{bmatrix}$$

$$A^3 = \begin{bmatrix} 2 & 1 & 1 \\ 0 & 1 & 0 \\ 1 & 1 & 2 \end{bmatrix}\begin{bmatrix} 5 & 4 & 4 \\ 0 & 1 & 0 \\ 4 & 4 & 5 \end{bmatrix} = \begin{bmatrix} 14 & 13 & 13 \\ 0 & 1 & 0 \\ 13 & 13 & 14 \end{bmatrix}$$

$$A^3 - 5A^2 + 7A - 3I = \begin{bmatrix} 14 & 13 & 13 \\ 0 & 1 & 0 \\ 13 & 13 & 14 \end{bmatrix}\begin{bmatrix} 5 & 4 & 4 \\ 0 & 1 & 0 \\ 4 & 4 & 5 \end{bmatrix} + 7\begin{bmatrix} 2 & 1 & 1 \\ 0 & 1 & 0 \\ 1 & 1 & 2 \end{bmatrix} - 3\begin{bmatrix} 1 & 0 & 0 \\ 0 & 1 & 0 \\ 0 & 0 & 1 \end{bmatrix}$$

$$= \begin{bmatrix} 14-25+14-3 & 13-20+7+0 & 13-20+7+0 \\ 0+0+0+0 & 1-5+7-3 & 0-0+0-0 \\ 13-20+7+0 & 13-20+7-0 & 14-25+14-3 \end{bmatrix} = \begin{bmatrix} 0 & 0 & 0 \\ 0 & 0 & 0 \\ 0 & 0 & 0 \end{bmatrix} = 0$$

Hence, Cayley Hamilton Theorem is verified.

Now $\quad A^8 - 5A^7 + 7A^6 - 3A^5 + A^4 - 5A^3 + 8A^2 - 2A + I$

$$= A^5(A^3 - 5A^2 + 7A - 3I) + A(A^3 - 5A^2 + 7A - 3I) + A^2 + A + I$$

$$= A^5 XO + AXO + A^2 + A + I = +A^2 + A + I$$

$$= \begin{bmatrix} 5 & 4 & 4 \\ 0 & 1 & 0 \\ 4 & 4 & 5 \end{bmatrix} + \begin{bmatrix} 2 & 1 & 1 \\ 0 & 1 & 0 \\ 1 & 1 & 2 \end{bmatrix} + \begin{bmatrix} 1 & 0 & 0 \\ 0 & 1 & 0 \\ 0 & 0 & 1 \end{bmatrix}$$

$$= \begin{bmatrix} 8 & 5 & 5 \\ 0 & 3 & 0 \\ 5 & 5 & 8 \end{bmatrix}$$

Exercise 1.7

1. Find the characteristic polynomial of the matrix

$$A = \begin{bmatrix} 3 & 1 & 1 \\ -1 & 5 & -1 \\ 1 & -1 & 3 \end{bmatrix}$$

Verify Cayley Hamilton Theorem for this matrix. Hence find A^{-1}.

Ans. $A^{-1} = \frac{1}{20}\begin{bmatrix} 7 & -2 & -3 \\ 1 & 4 & 1 \\ -2 & 2 & 8 \end{bmatrix}$

2. Use Cayley Hamilton Theorem to find the inverse of the matrix

$$\begin{bmatrix} \cos\theta & \sin\theta \\ -\sin\theta & \cos\theta \end{bmatrix}$$

Ans. $\begin{bmatrix} \cos\theta & -\sin\theta \\ \sin\theta & \cos\theta \end{bmatrix}$

3. Using the Cayley-Hamilton Theorem, find A^{-1}, given that

$$A = \begin{bmatrix} 2 & -1 & 3 \\ 1 & 0 & 2 \\ 4 & -2 & 1 \end{bmatrix}$$

Ans. $-\frac{1}{5}\begin{bmatrix} 4 & -5 & -2 \\ 7 & -10 & -1 \\ -2 & 0 & 1 \end{bmatrix}$.

4. Using Cayley Hamilton Theorem, find the inverse of the matrix

$$\begin{bmatrix} 5 & -1 & 5 \\ 0 & 2 & 0 \\ -5 & 3 & -15 \end{bmatrix}$$

Ans. $\frac{1}{10}\begin{bmatrix} 3 & 0 & 1 \\ 0 & 5 & 0 \\ -1 & 1 & -1 \end{bmatrix}$

5. Find the characteristic equation of the matrix

$$A = \begin{bmatrix} 1 & 3 & 7 \\ 4 & 2 & 3 \\ 1 & 2 & 1 \end{bmatrix}$$

and show that the equation is also satisfied by A. **Ans.** $\lambda^3 - 4\lambda^2 - 20\,\lambda - 35 = 0$

6. Find the eign values of the matrix

$$\begin{bmatrix} 2 & -3 & 1 \\ 3 & 1 & 3 \\ -5 & 2 & -4 \end{bmatrix}$$

Ans. Eigenvalues are 0, +1, –2.

7. Using Cayley-Hamilton relation obtain the inverse of the matrix

$$\begin{bmatrix} 1 & 1 & 3 \\ 1 & 3 & -3 \\ -2 & -4 & -4 \end{bmatrix}$$

Ans. $\frac{1}{8}\begin{bmatrix} 24 & 8 & 12 \\ -10 & -2 & -6 \\ -2 & -2 & -2 \end{bmatrix}$

8. Show that the matrix $A = \begin{bmatrix} 1 & -2 & 2 \\ 1 & 2 & 3 \\ 0 & -1 & 2 \end{bmatrix}$

satisfies its characteristic equation. Hence find A^{-1}. **Ans.** $\frac{1}{9}\begin{bmatrix} 7 & 2 & -10 \\ -2 & 2 & -1 \\ -1 & 1 & 4 \end{bmatrix}$

9. Use Cayley Hamilton Theorem to find the inverse of

$$A = \begin{bmatrix} 1 & 2 & 4 \\ -1 & 0 & 3 \\ 3 & 1 & -2 \end{bmatrix} \qquad \textbf{Ans. } A^{-1} = \frac{1}{7}\begin{bmatrix} -3 & 8 & 6 \\ 7 & -14 & -7 \\ -1 & 5 & 2 \end{bmatrix}$$

10. Verify Cayley-Hamilton Theorem for the matrix

$$A = \begin{bmatrix} 1 & 1 & 2 \\ 3 & 1 & 1 \\ 2 & 3 & 1 \end{bmatrix} \text{ Hence evaluate } A^{-1} \qquad \textbf{Ans. } \frac{1}{11}\begin{bmatrix} -2 & 5 & -1 \\ -1 & -3 & 5 \\ 7 & -1 & -2 \end{bmatrix}$$

11. If $A = \begin{bmatrix} 1 & 4 \\ 2 & 3 \end{bmatrix}$, then express $A^5 - 4A^4 - 7A^3 + 11A^2 - A - 10I$ in terms of A.

Ans. $A + 5I$

12. If λ_1, λ_2 and λ_3 are the eigen values of the matrix

$$\begin{bmatrix} -2 & -9 & 5 \\ -5 & -10 & 7 \\ -9 & -21 & 14 \end{bmatrix} \text{ then } \lambda_1 + \lambda_2 + \lambda_3 \text{ is equal to}$$

(i) –16 (ii) 2 (iii) – 6 (iv) –14 **Ans.** (ii)

13. The matrix $A = \begin{bmatrix} 1 & 0 \\ 2 & 4 \end{bmatrix}$ is given. The eigen values of $4A^{-1} + 3A + 2I$ are

(A) 6, 15; (B) 9, 12 (C) 9, 15; (D) 7, 15 **Ans.** (C)

14. The matrix A is defined as

$$A = \begin{bmatrix} 1 & 2 & 3 \\ 0 & -2 & 6 \\ 0 & 0 & -3 \end{bmatrix}$$

Find the eigen values of $3A^3 + 5A^2 + 6A + 1$ **Ans.** 15, –15, –53

15. If a matrix $A = \begin{pmatrix} 1 & 0 & 0 \\ 0 & -1 & 0 \\ 1 & 0 & 1 \end{pmatrix}$, find the matrix A^{32}, using Cayley Hamilton Theorem.

CHAPTER 2

Complex Numbers

2.1 Introduction

Imaginary quantities are of frequent occurrence in Higher Mathematics and the results obtained by their aid are as valid and important as those deduced from real quantities.

But evaluation of these numbers requires extraction of square roots of negative quantities, which according to the laws of algebra, is an impossible operation, because there is no number defined as yet, of which the square is negative.

Prior to the introduction of negative numbers we were justified in saying that the equation $2x + 4 = 0$ has no solution. But with the introduction of negative numbers this equation $2x + 4 = 0$ became solvable. Our motive for the introduction of complex numbers was our desire to make every algebraic equation solvable. Let us consider the equation $x^2 + 4 = 0$. This equation has no solution in the set of real numbers, for there cannot be a real number x whose square is -4. In order to remedy this situation a new kind of numbers were introduced and were given the name complex numbers. Euler was the first to introduce the symbol i for $\sqrt{-1}$ with the property $i^2 = -1$. i is a symbol used for $\sqrt{-1}$ so that $i^2 = -1$.

Definitely i is not a real number and it was given the name an imaginary number.

Note:

(i) If $a > 0$, then $\sqrt{a}$ denotes the positive square root of a

(ii) $\sqrt{-2} = \sqrt{2}i, \sqrt{-3} = \sqrt{3}i$ etc.

From definition of i it is clear that

(a) $i = \sqrt{-1}$, (b) $i^2 = -1$, (c) $i^3 = i^2 \times i = -i$, (d) $i^4 = i^2 \times i^2 = -1 \times -1 = 1$.

Thus any positive integral powers of i will be equal to $\pm i$ or ± 1. Example

$$i^{0.101} = i^{0.4\times25+1} = i^{0.4\times25} \times i = 1 \times i = i$$

2.2 Complex Number

Definition: An expression of the form x + iy, whre x and y are real numbers and i is a symbol is called a complex number and is usually denoted by z. x is called the real part and y the imaginary part of z. If y = 0, z is called purely real and if x = 0, z is called purely imaginary.

Operations of addition, substraction, multiplication and division involving complex numbers are performed according to the same rules as in case of real numbers taking $i^2 = -1$.

Example:

(i) $z_1 = -3 - 4i$. Here $R(z_1) = -3$ and $I(z_1) = -4$

(ii) $z_2 = -4i = 0 - 4i$. Here $R(z_2) = 0$ and $I(z_2) = -4$

(iii) $z_3 = -3 = -3 + 0i$. Here $R(z_3) = -3$ and $I(z_3) = 0$

(iv) $z_4 = 0 = 0 + i0$. Here $R(z_4) = 0$ and $I(z_4) = 0$

Note:

1. Every real number is a complex number.
2. 0 is both purely real and purely imaginary number.
3. In case of purely imaginary number real part must be zero and imaginary part may or may not be zero.
4. Every real number is purely real and every purely real number is real but an imaginary number may not be purely imaginary and these is a purely imaginary number 0 which is not an imaginary number. An imaginary number is of the form x + iy, where y is a non-zero real number.

Do not think that the word "imaginary" means that these numbers are mystical or fictitious or unreal in the usual sense of the word or that complex means complicated.

Imaginary numbers do exist and have very real application in many branches of physical science. We call them numbers for the obvious reason that the familiar operations can be defined for them and that they can be shown to satisfy the familiar laws like the commutative, distributive and the associative laws as in the case of real numbers.

2.3 Equality of Complex Numbers

Two complex number $x_1 + iy_1$ and $x_2 + iy_2$ are said to be equal if and only if

$$x_1 = x_2 \text{ and } y_1 = y_2$$

i. e., $$z_1 = z_2 \Leftrightarrow \text{Re}(z_1) = \text{Re}(z_2) \text{ and } \text{Im}(z_1) = \text{Im}(z_2).$$

Justification: $x_1 + iy_1 = x_2 + iy_2 \Rightarrow x_1 - x_2 = i(y_2 - y_1)$

$$\Rightarrow (x_1 - x_2)^2 = i^2 (y_2 - y_1)^2$$

$$\Rightarrow (x_1 - x_2)^2 = -(y_2 - y_1)^2 \Rightarrow (x_1 - x_2)^2 + (y_2 - y_1)^2 = 0$$

$$\Rightarrow (x_1 - x_2)^2 = 0 \text{ and } (y_2 - y_1)^2 = 0$$

$$\Rightarrow x_1 = x_2 \text{ and } y_1 = y_2$$

Note: Inequality relation does not hold good in case of complex numbers having non-zero imaginary parts. Hence it can not be decided that out of the two numbers $4 + 3i$ and $-3 + 2i$ which one is greater.

2.4 Algebra of Complex Numbers

(i) Addition: The sum of two complex numbers $x_1 + iy_1$ and $x_2 + iy_2$ is the complex number $(x_1 + x_2) + i(y_1 + y_2)$.

Thus $(x_1 + iy_1) + (x_2 + iy_2) = (x_1 + x_2) + i(y_1 + y_2)$.

(ii) Subtraction: The difference of two complex numbers $x_1 + iy_1$, and $x_2 + iy_2$ is the complex number

$$(x_1 - x_2) + i(y_1 - y_2).$$

Thus $(x_1 + iy_1) - (x_2 + iy_2) = x_1 - x_2 + i(y_1 - y_2)$.

Note: $(x + iy) = 0 + i\,0 - (x + iy) = 0 - x + i(0 - y) = -x + i(-y)$.

(iii) Multiplication: The product of two complex numbers $x_1 + iy_1$ and $x_2 + iy_2$ is the complex number

$$(x_1x_2 - y_1y_2) + i(x_1y_2 + y_1x_2).$$

Thus $(x_1 + iy_1)(x_2 + iy_2) = (x_1x_2 - y_1y_2) + i(x_1y_2 + y_1x_2)$.

Complex numbers as ordered pairs: An ordered pair (x, y), where x and y are real numbers is called a complex number if the following conditions are satisfied

(i) $(x_1, y_1) = (x_2, y_2)$ if $x_1 = x_2$, $y_1 = y_2$

(ii) $(x_1, y_1) + (x_2, y_2) = (x_1 + x_2, y_1 + y_2)$

(iii) $(x_1, y_1)(x_2, y_2) = (x_1x_2 - y_1y_2, x_1y_2 + x_2y_1)$

(iv) Division: If $x_1 + iy_1$ and $x_2 + iy_2$ be two complex numbers and $x_2 + iy_2 \neq 0$, then

$$\frac{x_1 + iy_1}{x_2 + iy_2} = \frac{(x_1 + iy_1)(x_2 - iy_2)}{(x_2 + iy_2)(x_2 - iy_2)}$$

$$= \frac{x_1x_2 + y_1y_2 + i(x_2y_1 - x_1y_2)}{x_2^2 + y_2^2} = \frac{x_1x_2 + y_1y_2}{x_2^2 + y_2^2} + \frac{i(x_2y_1 - x_1y_2)}{x_2^2 + y_2^2}$$

2.5 Properties of Addition of Complex Numbers

(i) Closure Law: Closure law for addition holds in the set of complex numbers i.e., if z_1 and z_2 are any two complex numbers then $z_1 + z_2$ is also a complex number.

Let $z_1 = x_1 + iy_1$, $z_2 = x_2 + iy_2$, then

$$z_1 + z_2 = (x_1 + iy_1) + (x_2 + iy_2) = (x_1 + x_2) + i(y_1 + y_2)$$
$$= \text{a complex number}$$

(ii) Commutative law: Commutative law for addition holds in the set of complex numbers i.e., $z_1 + z_2 = z_2 + z_1$

Let, $z_1 = x_1 + iy_1$ and $z_2 = x_2 + iy_2$

Now,
$$\begin{aligned} z_1 + z_2 &= (x_1 + iy_1) + (x_2 + iy_2) \\ &= (x_1 + x_2) + i(y_1 + y_2) = (x_2 + x_1) + i(y_2 + y_1) \\ &= z_2 + z_1 \end{aligned}$$

(iii) Associative law: Associative law holds in the set of complex numbers.

i.e., $z_1 + (z_2 + z_3) = (z_1 + z_2) + z_3$

Let $z_1 = x_1 + iy_1$, $z_2 = x_2 + iy_2$, $z_3 = x_3 + iy_3$

Now,
$$\begin{aligned} z_1 + (z_2 + z_3) &= (x_1 + iy_1) + \{(x_2 + iy_2) + (x_3 + iy_3)\} \\ &= (x_1 + iy_1) + \{(x_2 + x_3) + i(y_2 + y_3)\} \\ &= \{(x_1 + \{(x_2 + x_3)\} + i\{y_1 + (y_2 + y_3)\} \\ &= \{(x_1 + x_2) + x_3\} + i\{(y_1 + y_2) + y_3\} \\ &= \{(x_1 + x_2) + i(y_1 + y_2)\} + (x_3 + iy_3) \\ &= \{(x_1 + iy_1) + (x_2 + iy_2)\} + (x_3 + iy_3) \\ &= (z_1 + z_2) + z_3 \end{aligned}$$

(iv) Additive identity: There exists a complex number $0 + i0$ such that for every complex number $x + iy$,

$$(x + iy) + (0 + i0) = (x + 0) + i(y + 0) = x + iy$$

Here $0 + i0$ is called the additive identity.

(v) Additive inverse: For every complex number $x + iy$ there exists a complex number $-x - iy$ such that $(x + iy) + (-x - iy) = (x - x) + i(y - y) = 0 + i\,0$ = additive identity

$-x - iy$ is called the additive inverse of $x + iy$.

2.6 Properties of Multiplication of Complex Numbers

(i) Closure property: Closure law of multiplication holds in the set of complex numbers i.e., if z_1 and z_2 are any two complex numbers then $z_1 z_2$ is also a complex number.

Let $z_1 = x_1 + iy_1$, $z_2 = x_2 + iy_2$

Now,
$$\begin{aligned} z_1 z_2 &= (x_1 + iy_1)(x_2 + iy_2) \\ &= (x_1x_2 - y_1y_2) + i(x_1y_2 + x_2y_1), \text{ which is a complex number.} \end{aligned}$$

(ii) **Commutativity:** Commutative law holds in the set of complex numbers i.e.,

$z_1z_2 = z_2z_1$ for any two complex numbers z_1 and z_2.

Let $z_1 = x_1 + iy_1, z_2 = x_2 + iy_2$

Now, $z_1z_2 = (x_1 + iy_1)(x_2 + iy_2)$

$= (x_1x_2 - y_1y_2) + i(x_1y_2 + x_2y_1)$

$= (x_2x_1 - y_2y_1) + i(x_2y_1 + x_1y_2)$

$= (x_2 + iy_2)(x_1 + iy_1) = z_2z_1$

(iii) **Associativity:** Associative law holds in the set of complex numbers i.e.,

$z_1(z_2z_3) = (z_1z_2)z_3$ for any three complex numbers.

Let $z_1 = x_1 + iy_1, z_2 = x_2 + iy_2, z_3 = x_3 + iy_3$

Now, $z_1(z_2z_3) = (x_1 + iy_1)\{(x_2 + iy_2)(x_3 + iy_3)\}$

$= (x_1 + iy_1)\{(x_2x_3 - y_2y_3) + i(x_2y_3 + x_3y_2)\}$

$= x_1(x_2x_3 - y_2y_3) - y_1(x_2y_3 + x_3y_2) + i\{x_1(x_2y_3 + x_3y_2) + y_1(x_2x_3 - y_2y_3)\}$

$= (x_1x_2x_3 - x_1y_2y_3 - x_2y_1y_3 - x_3y_1y_2) + i(x_1x_2y_3 + x_1x_3y_2 + x_2x_3y_1 - y_1y_2y_3)$

Similarly we can show that

$(z_1z_2)z_3 = (x_1x_2x_3 - x_1y_2y_3 - x_2y_1y_3 - x_3y_1y_2) + i(x_1x_2y_3 + x_1x_3y_2 + x_2x_3y_1 - y_1y_2y_3)$

Hence, $z_1(z_2z_3) = (z_1z_2)z_3$

(iv) **Identity:** There exists complex number. $1 + i.0$ such that for every complex number $x + iy$,

$(x + iy)(1 + i.0) = (x.1 - y.0) + i(x.0 + y.1) = x + iy,$

$1 + i.0$ is called the multiplicative identity

(v) **Inverse:** For every complex number $z = x + iy \neq 0 + i.0$; there exists a complex number $\frac{x - iy}{x^2 + y^2}$ such that

$$(x+iy)\left(\frac{x-iy}{x^2+y^2}\right) = (x+iy)\left(\frac{x}{x^2+y^2} - i\frac{y}{x^2+y^2}\right)$$

$$= \left(\frac{x^2}{x^2+y^2} + \frac{y^2}{x^2+y^2}\right) + i\left(\frac{-xy}{x^2+y^2} + \frac{xy}{x^2+y^2}\right)$$

$= 1 + i.0 =$ multiplicative identity

$$\frac{x-iy}{x^2+y^2} \quad \text{i.e.,} \quad \frac{x}{x^2+y^2}-i\frac{y}{x^2+y^2}$$

is called the multiplicative inverse of x + iy.

(vi) Distributive law: Distributive law holds in the set of complex numbers i.e., for any three complex number z_1, z_2, z_3

$$z_1(z_2+z_3) = z_1z_2+z_1z_3$$

Let $z_1 = x_1 + iy_1,\ z_2 = x_2 + iy_2,\ z_3 = x_3 + iy_3$

Now,
$$\begin{aligned}
z_1(z_2+z_3) &= (x_1+iy_1)\{(x_2+iy_2)+(x_3+iy_3)\} \\
&= (x_1+iy_1)\{(x_2+x_3)+i(y_2+y_3)\} \\
&= \{x_1(x_2+x_3)-y_1(y_2+y_3)\}+i\{x_1(y_2+y_3)+y_1(x_2+x_3)\} \\
&= (x_1x_2-y_1y_2)+i(x_1y_2+x_2y_1)+(x_1x_3-y_1y_3)+ \\
&\quad i(x_1y_3+x_3y_1) \\
&= (x_1+iy_1)(x_2+iy_2)+(x_1+iy_1)(x_3+iy_3) \\
&= z_1z_2+z_1z_3.
\end{aligned}$$

Thus we see that the complex numbers obey the fundamental laws of algebra of real numbers and their algebra is identical in form, but not in meaning with the algebra of real numbers.

2.7 Modulus of a Complex Number

The modulus of a complex number x + iy is denoted by $|x + iy|$ and is defined as

$$|x+iy| = \sqrt{x^2+y^2} = \text{non negative square root of } (x^2+y^2)$$

***Example*:**

(i) Let z = 3 – 4i, then $|z| = \sqrt{3^2+(-4)^2} = 5$

(ii) Let z = 4i, then z = 0 + 4i $\therefore |z| = \sqrt{0^2+4^2} = 4$

(iii) Let z = 3, then z = 3 + 0i $\therefore |z| = \sqrt{3^2+0^2} = 3$

2.8 Conjugate of a Complex Number

The conjugate of a complex number x + iy is denoted by $\overline{x+iy}$ and is defined as

$$\overline{x+iy} = x - iy$$

Example:

(i) Let $z = -3 - 4i$, then $\bar{z} = -3 + 4i$

(ii) Let $z = 3 = 3 + 0i$, then $\bar{z} = 3 - 0i = 3$

(iii) Let $z = 4i$, then $\bar{z} = -4i$

Note:

(i) A complex number z is purely real if $\bar{z} = z$

(ii) A complex number z is purely imaginary if and only if $\bar{z} = -z$

2.9 Argument or Amplitude of a Complex Number

The argument or amplitude of a complex number $z = x + iy$ is the value of θ which satisfies the two equations

$$\cos\theta = \frac{x}{\sqrt{x^2 + y^2}} \text{ and } \sin\theta = \frac{y}{\sqrt{x^2 + y^2}}$$

Argument of z is dentoed by arg z or amp z.

There will be infinite number of values of θ satisfying the above equations and all these values will be argument of z but usually we take only that value of θ for which $0 \le \theta < 2\pi$.

Note: Argument of 0 is not defined

Example:

(i) Let $z = -1 - i$, here $x = -1, y = -1$

Now $\cos\theta = \frac{x}{\sqrt{x^2 + y^2}} = -\frac{1}{\sqrt{2}}$; $\quad \therefore \theta = \frac{3\pi}{4}, \frac{5\pi}{4}$

and $\sin\theta = \frac{y}{\sqrt{x^2 + y^2}} = -\frac{1}{\sqrt{2}}$; $\quad \therefore \theta = \frac{5\pi}{4}, \frac{7\pi}{4}$

since $\theta = \frac{5\pi}{4}$ satisfies both the equations

$$\therefore \boxed{\arg z = \frac{5\pi}{4}}$$

And general value of $\arg z = 2n\pi + \frac{5\pi}{4}$, where $n = 0, \pm 1, \pm 2, \ldots$

(ii) Let $z = -1 + \sqrt{3}i$, here $x = -1, y = \sqrt{3}$

Now $\cos\theta = \frac{x}{\sqrt{x^2+y^2}} = -\frac{1}{2}$; $\quad \therefore \theta = \frac{2\pi}{3}, \frac{4\pi}{3}$

and $\sin\theta = \frac{y}{\sqrt{x^2+y^2}} = \frac{\sqrt{3}}{2}$; $\quad \therefore \theta = \frac{\pi}{3}, \frac{2\pi}{3}$

since the value of θ satisfying both the equations is $\frac{2\pi}{3}$

$$\therefore \boxed{\arg z = \frac{2\pi}{3}}$$

General value of $\arg z = 2n\pi + \frac{2\pi}{3}$, $n = 0, \pm 1, \pm 2, \ldots$

Another way of finding argument of a complex number:

***Working Rule*:**

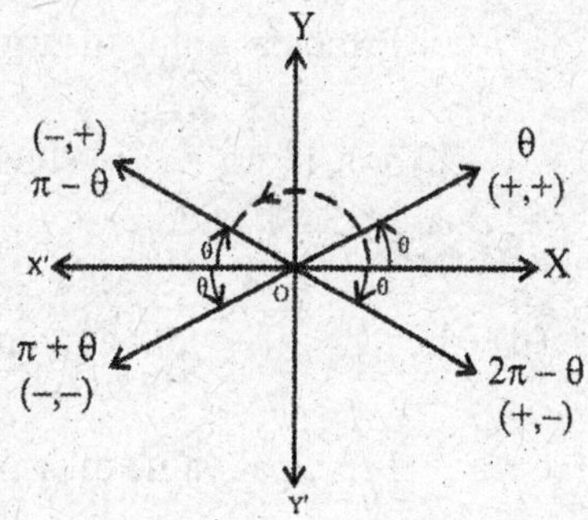

1. Take $\tan\theta = \left|\frac{y}{x}\right|$ and from this find the value of θ lying between 0 and $\frac{\pi}{2}$.
2. Then find in which quadrant the point z lies.
3. Argument of z will be $\theta, \pi - \theta, \pi + \theta$, or $2\pi - \theta$ according as the point z lies in the 1st, 2nd, 3rd, or, 4th quadrants.

Example:

(i) Let $z = -1 - i$, here $x = -1, y = -1$

Now $\tan\theta = \left|\frac{y}{x}\right| = \left|\frac{-1}{-1}\right| = 1$ $\quad \therefore \theta = \frac{\pi}{4}.\left(\text{between } 0 \text{ and } \frac{\pi}{2}\right)$

Since the point $z = -1 - i \equiv (-1, -1)$ lies in the 3rd quadrant

$$\therefore \arg z = \pi + \theta = \pi + \frac{\pi}{4} = \frac{5\pi}{4}$$

(ii) $z = -1 + \sqrt{3}i$, here $x = -1, y = \sqrt{3}$

Now $\tan\theta = \left|\frac{y}{x}\right| = \left|\frac{\sqrt{3}}{-1}\right| = \sqrt{3}$ $\quad \therefore \theta = \frac{\pi}{3}$

Since the point $z=-1+\sqrt{3}i$ i.e. $\left(-1,\sqrt{3}\right)$ lies in the 2nd quadrant

$$\therefore \text{ arg } z=\pi-\theta=\pi-\frac{\pi}{3}=\frac{2\pi}{3}$$

Principal value of the argument

There are infinite number of values of θ satisfying the two equations.

$$\cos\theta=\frac{x}{\sqrt{x^2+y^2}} \text{ and } \sin\theta=\frac{y}{\sqrt{x^2+y^2}}$$

But there will be a unique value of such that $-\pi<\theta\leq\pi$.

The value of argument θ satisfying the inequality $-\pi<\theta\leq\pi$. is called the principal value of the argunment.

Example:

(i) Let $z=-1-i$, here $x=-1$, $y=-1$. As found above $\arg z=\frac{5\pi}{4}$

$$\therefore \text{ Principal value of arg } z=\frac{5\pi}{4}-2\pi=-\frac{3\pi}{4}$$

[***Note:*** If arg $z>\pi$, substract 2π from it to get the principal value of arg z and if arg $z\leq-\pi$, add 2π to it to get the principal value of arg z.]

(ii) Let $z=-1+\sqrt{3i}$, as found above $\arg z=\frac{2\pi}{3}$

$$\therefore \text{ Principal value of arg } z=\frac{2\pi}{3}$$

2.10 Polar Form of a Complex Number

Every complex number $z=x+iy$ can be put in the form $r(\cos\theta+i\sin\theta)$, where $r=|z|$ and $\theta=\arg z$.

This form $r(\cos\theta+i\sin\theta)$ is called the polar or trigonometrical form of the complex number z.

Example:

Let $z=-1-i$, then $|z|=\sqrt{2}$, and $\arg z=\frac{5\pi}{4}$

Polar form of z is $\sqrt{2}\left(\cos\frac{5\pi}{4}+i\sin\frac{5\pi}{4}\right)$

Taking principal value of arg z, $\sqrt{2}\left[\cos\left(-\frac{3\pi}{4}\right)+i\sin\left(-\frac{3\pi}{4}\right)\right]$ is also the polar form of z.

Note: Althoguh $-1-i=\sqrt{2}\left(\cos\frac{3\pi}{4}-i\sin\frac{3\pi}{4}\right)$. But it is not the polar form of z

2.11 Properties of Modulii

(i) $|z_1z_2|=|z_1||z_2|$ (ii) $|z^n|=|z|^n$ (iii) $\frac{|z_1|}{|z_2|}=\frac{|z_1|}{|z_2|}\ (z_2\neq 0)$

(iv) $|z_1+z_2|\leq|z_1|+|z_2|$ (v) $|z_1-z_2|\geq|z_1|\sim|z_2|$ (vi) $z\bar{z}=|z|^2$

(vii) $|z|=|\bar{z}|$

Proof: **First Method:** $|z_1z_2|=|z_1||z_2|$

Let $z_1 = x_1 + iy_1$ and $z_2 = x_2 + iy_2$

Now $|z_1z_2|=|(x_1+iy_1)(x_2+iy_2)|$

$$=|(x_1x_2-y_1y_2)+i(x_1y_2+x_2y_1)|$$

$$=\sqrt{(x_1x_2-y_1y_2)^2+(x_1y_2+x_2y_1)^2}$$

$$=\sqrt{x_1^2x_2^2+y_1^2y_2^2-2x_1x_2y_1y_2+x_1^2y_2^2+x_2^2y_1^2+2x_1x_2y_1y_2}$$

$$=\sqrt{x_1^2x_2^2+x_1^2y_2^2+y_1^2y_2^2+x_2^2y_1^2}$$

$$=\sqrt{(x_1^2+y_1^2)(x_2^2+y_2^2)}=\sqrt{x_1^2+y_1^2}\ \sqrt{x_2^2+y_2^2}=|z_1||z_2|.$$

Second Method: Every complex number z = x + iy can be put in the form r (cos θ + i sin θ), where $r=|z|$ and θ = arg z.

Let $z_1 = r_1(\cos\theta_1 + i\sin\theta_1)$, $z_2 = r_2(\cos\theta_2 + i\sin\theta_2)$,
where $|z_1|=r_1,\ |z_2|=r_2,\ \arg z_1=\theta_1,\ \arg z_2=\theta_2$

Now $z_1z_2 = r_1r_2(\cos\theta_1 + i\sin\theta_1)(\cos\theta_2 + i\sin\theta_2)$

$$= r_1r_2[(\cos\theta_1\cos\theta_2 - \sin\theta_1\sin\theta_2) + i(\sin\theta_1\cos\theta_2 + \cos\theta_1\sin\theta_2)]$$

$$= r_1r_2[\cos(\theta_1+\theta_2) + i\sin(\theta_1+\theta_2)]$$

$$= r(\cos\theta + i\sin\theta),\text{ where } r = r_1r_2,\ \theta = \theta_1+\theta_2$$

$\therefore\quad |z_1z_2| = r = r_1r_2 = |z_1||z_2|$

(ii) $\because \quad |z_1 z_2 \ldots \ldots z_n| = |z_1||z_2|\ldots\ldots|z_n|$

$\therefore \quad |z^n| = |z.z\ldots \text{ to n factors}| = |z||z|\ldots \text{ to n factors} = |z|^n$

(iii) $\left|\dfrac{z_1}{z_2}\right| = \dfrac{|z_1|}{|z_2|}$

Let $z_1 = r_1 (\cos\theta_1 + i \sin\theta_1)$, $z_2 = r_2 (\cos\theta_2 + i \sin\theta_2)$.

Now $\dfrac{z_1}{z_2} = \dfrac{r_1 (\cos\theta_1 + i \sin\theta_1)}{r_2 (\cos\theta_2 + i \sin\theta_2)} = \dfrac{r_1 (\cos\theta_1 + i \sin\theta_1)(\cos\theta_2 - i \sin\theta_2)}{r_2 (\cos\theta_2 + i \sin\theta_2)(\cos\theta_2 - i \sin\theta_2)}$

$$= \frac{r_1}{r_2}\frac{\left[(\cos\theta_1\cos\theta_2 + \sin\theta_1\sin\theta_2) + i(\sin\theta_1\cos\theta_2 - \cos\theta_1\sin\theta_2)\right]}{(\cos^2\theta_2 + \sin^2\theta_2)}$$

$$= \frac{r_1}{r_2}\left[\cos(\theta_1 - \theta_2) + i \sin(\theta_1 - \theta_2)\right]$$

$\therefore \quad \left|\dfrac{z_1}{z_2}\right| = \dfrac{r_1}{r_2} = \dfrac{|z_1|}{|z_2|}$

(iv) $|z_1 + z_2| = |r_1(\cos\theta_1 + i \sin\theta_1) + r_2(\cos\theta_2 + i \sin\theta_2)|$

$$= |(r_1\cos\theta_1 + r_2\cos\theta_2) + i(r_1\sin\theta_1 + r_2\sin\theta_2)|$$

$$= \sqrt{(r_1\cos\theta_1 + r_2\cos\theta_2)^2 + (r_1\sin\theta_1 + r_2\sin\theta_2)^2}$$

$$= \sqrt{r_1^2(\cos^2\theta_1 + \sin^2\theta_1) + r_2^2(\cos^2\theta_2 + \sin^2\theta_2) + 2r_1r_2(\cos\theta_1\cos\theta_2 + \sin\theta_1\sin\theta_2)}$$

$$= \sqrt{r_1^2 + r_2^2 + 2r_1r_2\cos(\theta_1 - \theta_2)}$$

$$\leq \sqrt{r_1^2 + r_2^2 + 2r_1r_2} \qquad [\because \cos(\theta_1 - \theta_2) \leq 1]$$

$\therefore \quad |z_1 + z_2| \leq \sqrt{(r_1 + r_2)^2}$

or $\quad |z_1 + z_2| \leq r_1 + r_2$ Thus $|z_1 + z_2| \leq |z_1| + |z_2|$

(v) $$|z_1 - z_2| = |r_1(\cos\theta_1 + i\sin\theta_1) - r_2(\cos\theta_2 + i\sin\theta_2)|$$

$$= |(r_1\cos\theta_1 - r_2\cos\theta_2) + i(r_1\sin\theta_1 - r_2\sin\theta_2)|$$

$$= \sqrt{(r_1\cos\theta_1 - r_2\cos\theta_2)^2 + (r_1\sin\theta_1 - r_2\sin\theta_2)^2}$$

$$= \sqrt{r_1^2 + r_2^2 - 2r_1r_2(\cos\theta_1\cos\theta_2 + \sin\theta_1\sin\theta_2)}$$

$$= \sqrt{r_1^2 + r_2^2 - 2r_1r_2\cos(\theta_1 - \theta_2)}$$

$$\geq \sqrt{r_1^2 + r_2^2 - 2r_1r_2} = \sqrt{(r_1 - r_2)^2} = |r_1 - r_2| \qquad [\because \cos(\theta_1 - \theta_2) \leq 1]$$

$\therefore$ $$|z_1 - z_2| \geq |r_1 - r_2| = r_1 \sim r_2 = |z_1| \sim |z_2|$$

(vi) Let $z = x + iy$, then $\bar{z} = x - iy$

Now $z\bar{z} = x^2 - i^2y^2 = x^2 + y^2 = |z|^2$

(vii) Let $z = x + iy$, then $\bar{z} = x - iy$

Now $|z| = \sqrt{x^2 + y^2}$ and $|\bar{z}| = \sqrt{x^2 + (-y)^2} = \sqrt{x^2 + y^2}$

2.12 Properties of Conjugates

(i) $\overline{z_1z_2} = \bar{z}_1\,\bar{z}_2$ (ii) $\overline{z^n} = (\bar{z})^n$

(iii) $\overline{\left(\dfrac{z_1}{z_2}\right)} = \dfrac{\bar{z}_1}{\bar{z}_2}$, $z_2 \neq 0$ (iv) $z\bar{z} = |z|^2$ (v) $\overline{z_1 + z_2} = \bar{z}_1 + \bar{z}_2$

(vi) $\overline{z_1 - z_2} = \bar{z}_1 - \bar{z}_2$ (vii) $z + \bar{z} =$ purely real

(viii) $z - \bar{z} =$ purely imaginary (ix) $\bar{\bar{z}} = z$

Proof: (i) $\overline{z_1z_2} = \bar{z}_1\,\bar{z}_2$

Let $z_1 = x_1 + iy_1$, $z_2 = x_2 + iy_2$

$z_1z_2 = (x_1 + iy_1)(x_2 + iy_2) = x_1x_2 - y_1y_2 + i(x_2y_1 + x_1y_2)$

$\therefore$ $\overline{z_1z_2} = x_1x_2 - y_1y_2 - i(x_2y_1 + x_1y_2)$ .....(2.1)

$\bar{z}_1\bar{z}_2 = (x_1 - iy_1)(x_2 - iy_2)$

$= (x_1x_2 - y_1y_2) - i(x_2y_1 + x_1y_2)$ .....(2.2)

From (2.1) and (2.2), $\overline{z_1z_2} = \bar{z}_1\,\bar{z}_2$

Note: $\overline{z_1z_2...z_n} = \bar{z}_1, \bar{z}_3 ..., \bar{z}_n$

(ii) $$\overline{(z)^n} = (\bar{z})^n$$

$$\overline{z^n} = \overline{(z.zz...n \text{ times})} = \bar{z}.\ \bar{z}\ ...\ n \text{ times} = (\bar{z})^n$$

(iii) $$\overline{\left(\frac{z_1}{z_2}\right)} = \frac{\bar{z}_1}{\bar{z}_2}$$

Let $z_1 = x_1 + iy_1$, $z_2 = x_2 + iy_2$

Now $$\frac{z_1}{z_2} = \frac{x_1 + iy_1}{x_2 + iy_2} = \frac{(x_1 + iy_1)(x_2 - iy_2)}{(x_2 + iy_2)(x_2 - iy_2)} = \frac{x_1x_2 + y_1y_2 + i(x_2y_1 - x_1y_2)}{x_2^2 + y_2^2}$$

$$= \frac{x_1x_2 + y_1y_2}{x_2^2 + y_2^2} + \frac{i(x_2y_1 - x_1y_2)}{x_2^2 + y_2^2} \quad .....(2.3)$$

$$\frac{\bar{z}_1}{\bar{z}_2} = \frac{x_1 - iy_1}{x_2 - iy_2} = \frac{(x_1 - iy_1)(x_2 + iy_2)}{(x_2 - iy_2)(x_2 + iy_2)}$$

$$= \frac{x_1x_2 + y_1y_2 - i(x_2y_1 - x_1y_2)}{x_2^2 + y_2^2}$$

$$= \frac{x_1x_2 + y_1y_2}{x_2^2 + y_2^2} - \frac{i(x_2y_1 - x_1y_2)}{x_2^2 + y_2^2} \quad .....(2.4)$$

From (2.3) and (2.4), $\overline{\left(\frac{z_1}{z_2}\right)} = \frac{\bar{z}_1}{\bar{z}_2}$

(iv) $$z\bar{z} = |z|^2$$

Let $z = x + iy$, then $\bar{z} = x - iy$

Now $z\bar{z} = (x + iy)(x - iy) = x^2 - i^2y^2 = x^2 + y^2 = |z|^2$

(v) $$\overline{z_1 + z_2} = \bar{z}_1 + \bar{z}_2$$

Let $z_1 = x_1 + iy_1$ and $z_2 = x_2 + iy_2$

Now $z_1 + z_2 = x_1 + x_2 + i\,(y_1 + y_2)$

$\therefore$ $\overline{z_1 + z_2} = x_1 + x_2 - i(y_1 + y_2)$

$$= (x_1 - iy_1) + (x_2 - iy_2) = \bar{z}_1 + \bar{z}_2$$

(vi) $\overline{z_1 - z_2} = \overline{z_1} - \overline{z_2}$

$$z_1 - z_2 = (x_1 + iy_1) - (x_2 + iy_2)$$

$$= x_1 - x_2 + i(y_1 - y_2)$$

$\therefore$ $\overline{z_1 - z_2} = x_1 - x_2 - i(y_1 - y_2)$

$$= (x_1 - iy_1) - (x_2 - iy_2) = \overline{z_1} - \overline{z_2}$$

(vii) $z + \overline{z}$ is purely real.

Let $z = x + iy$ then $\overline{z} = x - iy$

Now $z + \overline{z} = 2x$ = purely real

(viii) $z - \overline{z} = (x + iy) - (x - iy) = i\ 2y$ = purely imaginary

Note: $R(z) = \dfrac{z + \overline{z}}{2}$, $I(z) = \dfrac{z - \overline{z}}{2i}$

(ix) Let $z = x + iy$, then $\overline{z} = x - iy$

and $\overline{\overline{z}} = \overline{x - iy} = x + iy = z$

2.13 Properties of Argument

(i) $\arg(z_1 z_2) = \arg z_1 + \arg z_2$ (ii) $\arg(z^n) = n(\arg z)$

(iii) $\arg\left(\dfrac{z_1}{z_2}\right) = \arg z_1 - \arg z_2$

Proof:

(i) Let $z_1 = r_1(\cos\theta_1 + i\sin\theta_1), z_2 = r_2(\cos\theta_2 + i\sin\theta_2)$

Now $z_1 z_2 = r_1 r_2\left[(\cos\theta_1\cos\theta_2 - \sin\theta_1\sin\theta_2) + i(\sin\theta_1\cos\theta_2 + \cos\theta_1\sin\theta_2)\right]$

$$= r_1 r_2\left[\cos(\theta_1 + \theta_2) + i\sin(\theta_1 + \theta_1)\right]$$

$= r(\cos\theta + i\sin\theta)$, where $r = r_1 r_2$ and $\theta = \theta_1 + \theta_2$

$\therefore$ $\arg(z_1 z_2) = \theta = \theta_1 + \theta_2 = \arg z_1 + \arg z_2$

(ii) $$\arg(z_1, z_2 \dots z_n) = \arg z_1 + \arg z_2 + \dots + \arg z_n$$

Putting $z_1 = z_2 = \dots = z_n = z_1$ we get

$$\arg\left(z^n\right) = n(\arg z)$$

(iii) $$\frac{z_1}{z_2} = \frac{r_1(\cos\theta_1 + i\sin\theta_1)}{r_2(\cos\theta_2 + i\sin\theta_2)}$$

$$= \frac{r_1(\cos\theta_1 + i\sin\theta_1)(\cos\theta_2 - i\sin\theta_2)}{r_2(\cos\theta_2 + i\sin\theta_2)(\cos\theta_2 - i\sin\theta_2)}$$

$$= \frac{r_1}{r_2}\frac{\left[(\cos\theta_1\cos\theta_2 + \sin\theta_1\sin\theta_2) + i(\sin\theta_1\cos\theta_2 - \cos\theta_1\sin\theta_2)\right]}{\cos^2\theta_2 + \sin^2\theta_2}$$

$$= \frac{r_1}{r_2}\left[\cos\ (\theta_1 - \theta_2) + i\sin(\theta_1 - \theta_2)\right]$$

$$\therefore \quad \arg\left(\frac{z_1}{z_2}\right) = \theta_1 - \theta_2 = \arg\ z_1 - \arg\ z_2$$

2.14 Any Root of an Imaginary Number is an Imaginary Number

Proof: Let x + iy be an imaginary number, where $y \neq 0$

Let $(x + iy)^n = a; \ \therefore\ x + iy = a^n$ .....(2.5)

Now if a is real a^n will also be real and from (eq. 2.5) an imaginary number x + iy is equal to a real number a^n which is not possible. Hence a must be an imaginary number.

***Note*:** Any power of an imaginary number is an imaginary number.

2.15 Square Root of a Complex Number

If $a > 0$ then the symbol $\sqrt{a}$ denotes the positive square root of a.

Thus $\sqrt{16} = 4$

In symbolic form square root of a is written as $\pm\sqrt{a}$, where $a > 0$. The symbol $\sqrt{a}$ denotes the positive square root of a.

To find $\sqrt{a + ib}$: Here the symbol $\sqrt{a + ib}$ denotes the square root of (a + ib)

Let $\sqrt{a + ib} = x + iy$

Then $\sqrt{a+ib} = x+iy \Rightarrow (a+ib) = (x+iy)^2$

$$\Rightarrow (a+ib) = (x^2-y^2)+(2xy)i$$

$$\Rightarrow x^2 - y^2 = a \quad .....(2.6(a))$$

and $2xy = b \quad .....(2.6)$

Now, $x^2+y^2 = \sqrt{(x^2-y^2)^2+4x^2y^2} = \sqrt{a^2+b^2}$. .....(2.7)

Solving the eq. (2.6) and (2.7), we get

$$x = \pm\sqrt{\left(\frac{a+\sqrt{(a^2+b^2)}}{2}\right)} \text{ and } y = \pm\sqrt{\left(\frac{\sqrt{a^2+b^2}-a}{2}\right)}$$

From relation (eq. 2.6(a)) we can determine the sign of xy.

If $xy > 0$, then x and y will have same sign

$$\therefore \quad \sqrt{a+ib} = \pm\left\{\sqrt{\left(\frac{a+\sqrt{a^2+b^2}}{2}\right)} + \sqrt{\frac{\sqrt{a^2+b^2}-a}{2}}\right\}$$

If xy is < 0, then $\sqrt{a+ib} = \pm\left\{\sqrt{\left(\frac{a+\sqrt{a^2+b^2}}{2}\right)} - i\sqrt{\frac{\sqrt{a^2+b^2}-a}{2}}\right\}$

Note: $\sqrt{a}\times\sqrt{b} = \sqrt{ab}$ only when at least one of a and b is non-negative thus if both a and b are negative then. $\sqrt{a}\times\sqrt{b} \neq \sqrt{ab}$.

2.16 Cube Roots of Unity

Let $x = (1)^{\frac{1}{3}}$, then $x^3=1$ or $x^3-1=0$

or $(x-1)(x^2+x+1)=0$

When $x-1=0$, $x=1$ and when $x^2+x+1=0$, $x = \frac{-1\pm\sqrt{-3}}{2} = \frac{-1\pm\sqrt{3}\,i}{2}$

Thus cube roots of unity are 1, $\frac{-1+\sqrt{3}\,i}{2}$ and $\frac{-1-\sqrt{3}\,i}{2}$

Properties of cube roots of unity

1. **Each imaginary cube roots of unity is the square of the other.**

   Proof:

$$\left(\frac{-1+\sqrt{3}\,i}{2}\right)^2=\frac{1}{4}\left(1-2\sqrt{3}\,i+3\,i^2\right)=\frac{1}{4}\left(1-2\sqrt{3}\,i-3\right)=\frac{-1-\sqrt{3}\,i}{2}$$

Again $$\left(\frac{-1-\sqrt{3}\,i}{2}\right)^2=\frac{1}{4}\left(1+2\sqrt{3}\,i+3i^2\right)=\frac{1}{4}\left(1+2\sqrt{3}\,i-3\right)=\frac{-1+\sqrt{3}\,i}{2}$$

Thus if $\frac{-1+\sqrt{3}\,i}{2}=\omega$, then $\frac{-1-\sqrt{3}\,i}{2}=\omega^2$

and if $\frac{-1-\sqrt{3}\,i}{2}=\omega$, then $\frac{-1+\sqrt{3}\,i}{2}=\omega^2$

thus cube roots of unity are 1, ω, ω^2

2. **Sum of all the cube roots of unity is zero**

   i.e., $1+\omega+\omega^2=0$

   Proof: Sum of the cube roots of unity

$$=1+\frac{-1+\sqrt{3}\,i}{2}+\frac{-1-\sqrt{3}\,i}{2}=\frac{2-1+\sqrt{3}\,i-1-\sqrt{3}\,i}{2}=0$$

3. **Product of imaginary cube roots of unity is 1 i.e., $\omega^3=1$**

   Proof: Product of imaginary cube roots of unity $=\left(\frac{-1+\sqrt{3}\,i}{2}\right)\left(\frac{-1-\sqrt{3}\,i}{2}\right)$

$$=\frac{1}{4}\left(1^2-3i^2\right)=\frac{1}{4}(1+3)=1$$

Note: **1.** If n is a positive integer then $\omega^n=\omega^r$, where r is the remainder when n is divided by 3.

Justification: $3\overline{)n}(m$, subtract $3m$, remainder r $\quad\therefore\ n=3m+r$

Now $\omega^n=\omega^{3m+r}=\omega^{3m}\,.\,\omega^r=(\omega^3)^m\,.\,\omega^r=1^m\,.\,\omega^r=\omega^r$

Example: $\omega^{55}=\omega^1$ as when 55 is divided by 3 remainder is 1

Note: **2.** If α be one cube roots of a number then its other cube roots will be $\alpha\omega$ and $\alpha\omega^2$

Example:

(i) $(-8)^{\frac{1}{3}}=-2,\ -2\omega,\ -2\omega^2$ (ii) $(27)^{\frac{1}{3}}=3,\ 3\omega,\ 3\omega^2$

2.17 De Moivre's Theorem

Statement: If n is any integer then

$(\cos\theta + i\sin\theta)^n = \cos n\theta + i\sin n\theta$

If n is any rational number then one of the values of

$(\cos\theta + i\sin\theta)^n = \cos n\theta + i\sin n\theta$

Proof:

***Case* I. When n = 0**

Clearly, $(\cos\theta + i\sin\theta)^0 = 1 = \cos(0.\theta) + i\sin(0.\theta)$

***Case* II. When n is a positive integer**

By actual multiplication, we have

$(\cos\theta_1 + i\sin\theta_1)(\cos\theta_2 + i\sin\theta_2)$

$= (\cos\theta_1\cos\theta_2 - \sin\theta_1\sin\theta_2) + i(\sin\theta_1\cos\theta_2 + \cos\theta_1\sin\theta_2)$

$= \cos(\theta_1 + \theta_2) + i\sin(\theta_1 + \theta_2)$

Similarly, $(\cos\theta_1 + i\sin\theta_1)(\cos\theta_2 + i\sin\theta_2)(\cos\theta_3 + i\sin\theta_3)$

$= [(\cos\theta_1 + i\sin\theta_1)(\cos\theta_2 + i\sin\theta_2)](\cos\theta_3 + i\sin\theta_3)$

$= [\cos(\theta_1 + \theta_2) + i\sin(\theta_1 + \theta_2)](\cos\theta_3 + i\sin\theta_3)$

$= [\cos(\theta_1 + \theta_2 + \theta_3) + i\sin(\theta_1 + \theta_2 + \theta_3)]$

Proceeding in this way, for n factors we have

$(\cos\theta_1 + i\sin\theta_1)(\cos\theta_2 + i\sin\theta_2)\ldots(\cos\theta_n + i\sin\theta_n)$

$= \cos(\theta_1 + \theta_2 + \ldots + \theta_n) + i\sin(\theta_1 + \theta_2 + \ldots + \theta_n)$ ...(2

Putting

$\theta_1 = \theta_2 = \theta_3 = \ldots = \theta_n = \theta$ in (eq. 2.8), we get:

$(\cos\theta + i\sin\theta)^n = (\cos n\theta + i\sin n\theta)$

***Case* III. When n is a negative integer**

Let n = – m, where m is a positive integer

Now, $(\cos\theta + i\sin\theta)^n = (\cos\theta + i\sin\theta)^{-m}$

$$= \frac{1}{(\cos\theta + i\sin\theta)^m} = \frac{1}{(\cos m\theta + i\sin m\theta)} \quad \text{[by case II]}$$

$$=\frac{1}{(\cos m\theta+i\sin m\theta)}\times\frac{(\cos m\theta-i\sin m\theta)}{(\cos m\theta+i\sin m\theta)}$$

$$=\frac{(\cos m\theta-i\sin m\theta)}{(\cos^2 m\theta+i^2\sin^2 m\theta)}=\frac{(\cos m\theta-i\sin m\theta)}{(\cos^2 m\theta+\sin^2 m\theta)}$$

$= \cos m\theta - i \sin m\theta$

$= \cos(-m)\theta + i\sin(-m)\theta$ $\quad [\because \cos(-\theta) = \cos\theta \text{ and } \sin(-\theta) = -\sin\theta]$

$= \cos n\theta + \sin n\theta$ $\quad [\because -m = n]$

***Case* IV. When n is a rational number**

Let $n=\frac{p}{q}$, where q is a positive integer and p is an integer

Now $\left(\cos\frac{p}{q}\theta+i\sin\frac{p}{q}\theta\right)^q=\left(\cos q.\frac{p}{q}\theta+i\sin q.\frac{p}{q}\theta\right)$

$= \cos p\theta + i\sin p\theta$ $\quad [\because q \text{ is a positive integer}]$

$= (\cos\theta + i\sin\theta)^p$ $\quad$ [since p is an integer]

$\therefore \quad \cos\frac{p}{q}\theta+i\sin\frac{p}{q}\theta=(\cos\theta+i\sin\theta)^{p/q}$

$\therefore \quad \cos n\theta+i\sin n\theta=(\cos\theta+i\sin\theta)^n$ $\quad \left[\because n=\frac{p}{q}\right]$

Note: If n is a positive integer then

(i) $(\cos\theta + i\sin\theta)^n = \cos n\theta + i\sin n\theta$

(ii) $(\cos\theta+i\sin\theta)^{\frac{1}{n}}=\cos\frac{2r\pi+\theta}{n}+i\sin\frac{2r\pi+\theta}{n}$,

where $r = 0, 1, 2 \ldots, (n-1)$

2.18 nth Roots of Unity

$1 = \cos 0 + i\sin 0$

$\therefore \quad (1)^{\frac{1}{n}}=(\cos 0+i\sin 0)^{\frac{1}{n}}$

$$= \cos\frac{2r\pi + 0}{n} + i\sin\frac{2r\pi + 0}{n}, \quad \text{where } r = 0, 1, 2, \ldots, (n-1)$$

$$= \cos\frac{2r\pi}{n} + i\sin\frac{2r\pi}{n}, \quad \text{where } r = 0, 1, 2, \ldots (n-1)$$

$$= e^{\frac{i2r\pi}{n}}, \quad \text{where } r = 0, 1, 2, \ldots, (n-1)$$

$$= 1, e^{\frac{i2\pi}{n}}, e^{\frac{i4\pi}{n}}, e^{\frac{i6\pi}{n}}, \ldots e^{\frac{i2(n-1)\pi}{n}}$$

$$= 1, \alpha, \alpha^2, \alpha^3, \ldots \alpha^{n-1}, \quad \text{where } e^{\frac{i2\pi}{n}}$$

Thus nth roots of unity are $1, \alpha, \alpha^2, \ldots, \alpha^{n-1}$, where $\alpha = e^{\frac{i2\pi}{n}}$

2.19 Properties of nth Roots of Unity

(i) $1 + \alpha + \alpha^2 + \ldots + \alpha^{n-1} = 0$

Explanation: $1 + \alpha + \alpha^2 + \ldots + \alpha^{n-1} = \dfrac{1-\alpha^n}{1-\alpha} = \dfrac{1-1}{1-\alpha} = 0$

$$\left[\because \alpha = e^{\frac{i2\pi}{n}} \therefore \alpha^n = e^{i2\pi} = \cos 2\pi + i\sin 2\pi = 1\right]$$

(ii) $1 \ldots \alpha \ldots \alpha^2 \ldots \alpha^{n-1} = (-1)^{n-1}$

Explanation: $1 . \alpha . \alpha^2 \ldots \alpha^{n-1} = \alpha^{1+2+\ldots+(n-1)}$

$$= \alpha^{\frac{(n-1)n}{2}} = \left(\alpha^{\frac{n}{2}}\right)^{n-1} = \left(e^{i\pi}\right)^{n-1} \qquad \left[\because \alpha^n = e^{i2\pi} \therefore \alpha^{\frac{n}{2}} = e^{i\pi}\right]$$

$$= (\cos\pi + i\sin\pi)^{n-1} = (-1)^{n-1}$$

2.20 Argand Plane and Geometrical Representation of Complex Numbers

1. Let O be the origin and OX and OY be the x-axis and y-axis respectively. Corresponding to each complex number x + iy there will be one and only one point P(x, y) in the xy – plane.

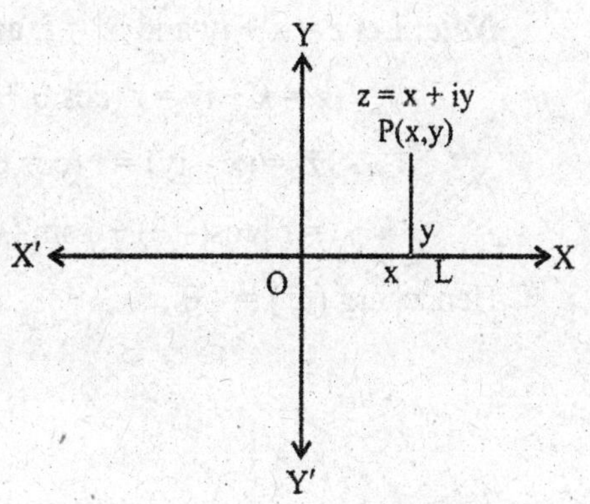

Thus each complex number x + iy can be represented by a point P (x, y) of the xy-plane and conversely corresponding to each point P (x, y) of the xy-plane will be unique complex number x + iy. The xy-plane is called the Argand plane or complex plane, x-axis is called *the real axis* and y-axis is *the imaginary axis.*

2. Each complex number z can be represented by a vector $\overrightarrow{OP}$, where P is the point representing the complex number z.

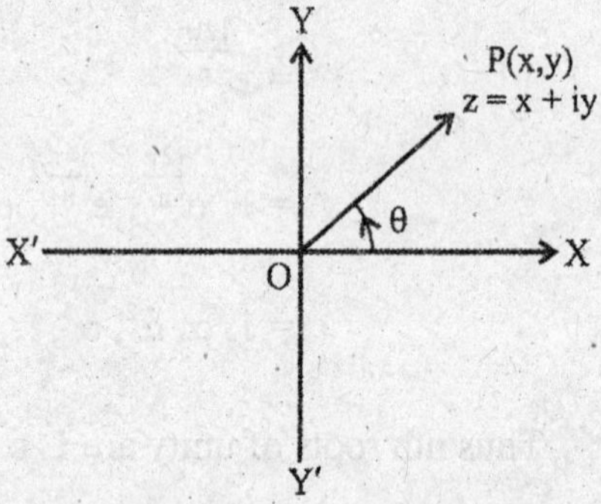

Thus $z = \overrightarrow{OP}$

∴ (i) $|z| = |\overrightarrow{OP}| = OP$

= distance between points representing z_1 and z_2.

(ii) arg z = θ = angle which vector $\overrightarrow{OP}$ makes with the positive direction of x-axis

Note: Any other vector $\overrightarrow{AB}$ which has the same magnitude, direction and sense as that of $\overrightarrow{OP}$ but has a different initial point also represents the complex number z.

Note: Complex numbers can be considered as vectors in case of sum, difference and modulus of complex numbers.

2.21 Geometrical Representation of some Complex Numbers

(i) ***Geometrical representation of conjugate of a complex number***: Let z = x + iy be any complex number. It will be represented by a point P(x, y) in the Argand plane.

Now $\bar{z} = (x - iy)$, $\bar{z}$ will be represented by the point Q(x, – y)

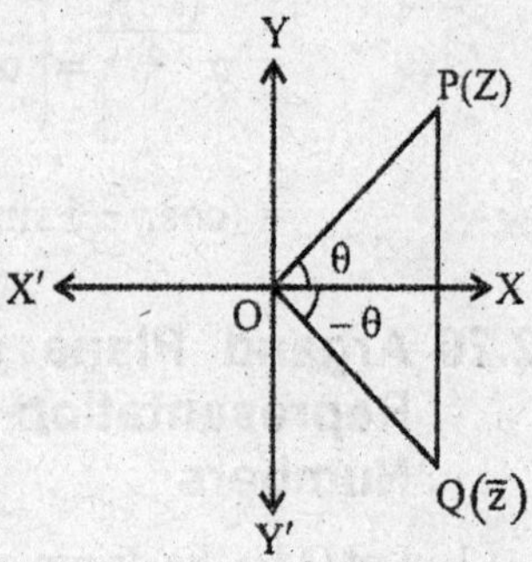

Clearly, Q is the image of the point P in the real axis.

Note: Let z = x + iy and |z| = r, arg z = θ then,

$$z = x + iy = r(\cos\theta + i\sin\theta),$$

$$\bar{z} = (x - iy) = r(\cos\theta - i\sin\theta)$$

$$= r[\cos(-\theta) + i\sin(-\theta)]$$

Hence, arg $(\bar{z}) = -\theta$.

(ii) ***Geometrical representation of the sum of two complex numbers*:** Let the complex numbers z_1 and z_2 be represented by the points P and Q on the Argand plane. We complete the parallelogram OPRQ. The point R will represent the complex number $z_1 + z_2$.

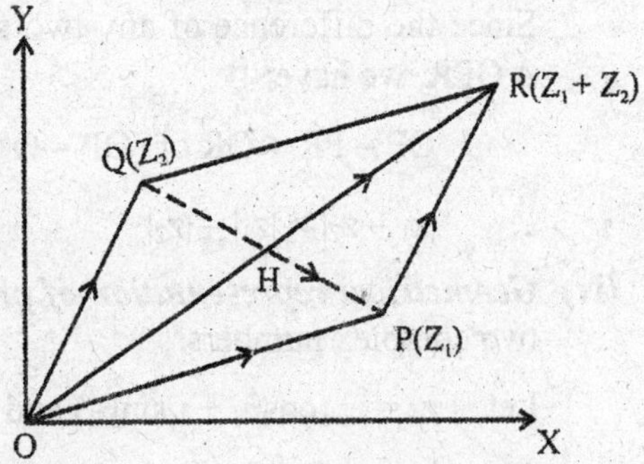

In vector notation

$$z_1 + z_2 = \overline{OP} + \overline{OQ} = \overline{OP} + \overline{PR} = \overline{OR}$$

(iii) ***Geometrical Representation of difference of two complex numbers*:** Let $z_1 = x_1 + iy_1$ and $z_2 = x_2 + iy_2$ be two complex numbers.

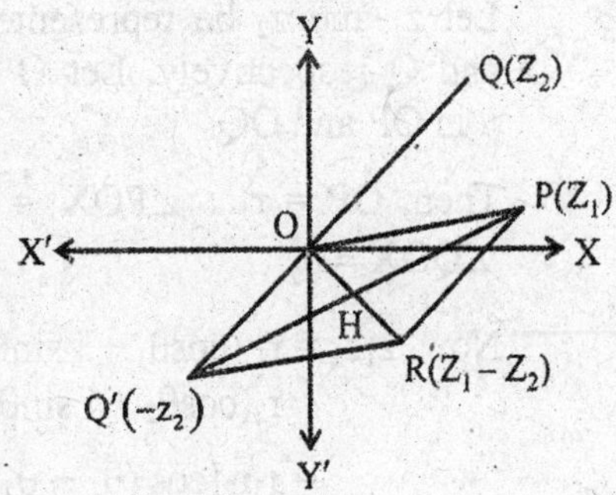

Let z_1 and z_2 be represented by the points P (x_1, y_1) and Q (x_2, y_2) in the Argand plane.

Let O be the origin. Join OP and OQ. Produce QO backwards to Q' such that $OQ' = OQ$.

Then, the co-ordinates of Q' will be $(-x_2, -y_2)$ and therefore Q' will represent the complex number $-z_2 = -x_2 + i(-y_2)$.

Complete the parallelogram $POQ'R$ and let its diagonals intersect at the point H.

Since H is middle point of PQ', its co-ordinates are

$$\left(\frac{x_1 - x_2}{2}, \frac{y_1 - y_2}{2}\right)$$

Now, if (x, y) be the co-ordinates of R, then since H is the middle point of OR, the co-ordinates of H are $\left(\frac{x}{2}, \frac{y}{2}\right)$

$$\therefore \quad \frac{x}{2} = \frac{x_1 - x_2}{2} \text{ and } \frac{y}{2} = \frac{y_1 - y_2}{2}$$

$\Rightarrow \quad x = x_1 - x_2$ and $y = y_1 - y_2$

Thus, the co-ordinates of R are $(x_1 - x_2, y_1 - y_2)$

Hence, the point R represents the complex number

$$(x_1 - x_2) + i(y_1 - y_2) = [(x_1 + iy_1) - (x_2 + iy_2)] = z_1 - z_2$$

Clearly $|z_1| = OP$, $|z_2| = |-z_2| = OQ'$ and $|z_1 - z_2| = OR$

Since the difference of any two sides of a triangle is less than the third side, from Δ OPR, we have:

$$OP - PR < OR \text{ or } (OP - OQ') < OR \text{ or } OR > (OP - OQ')$$

$\therefore \quad |z_1 - z_2| > |z_1| - |z_2|$

(iv) ***Geometrical representation of product of two complex numbers*:** Let z_1 and z_2 be two complex numbers.

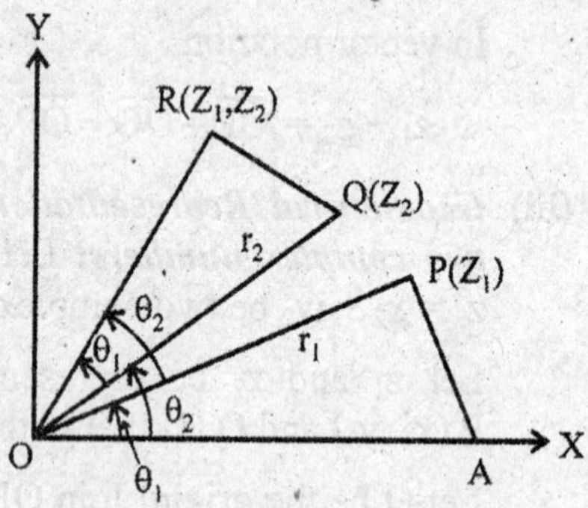

Let $z_1 = r_1 (\cos\theta_1 + i \sin\theta_1)$ and

$z_2 = r_2 (\cos\theta_2 + i \sin \theta_2)$

Let z_1 and z_2 be represented by the points P and Q respectively. Let O be the origin. We join OP and OQ.

Then, $OP = r_1 : \angle POX = \theta_1 : OQ = r_2$ and $\angle QOX = \theta_2$

Now, $z_1z_2 = r_1 (\cos\theta_1 + i \sin\theta_1), r_2(\cos\theta_2 + i \sin\theta_2)$

$= r_1r_2[\cos (\theta_1 + \theta_2) + i \sin (\theta_1 + \theta_2)]$

$= r(\cos \theta + i \sin \theta)$, where $r = r_1r_2$ and $\theta = \theta_1 + \theta_2$

Thus $z_1 z_2$ is a complex number whose modulus is $r_1 r_2$ and argument is $\theta_1 + \theta_2$.

Take a point A on the real axis such that OA = I. Join PA

Now, construct a triangle OQR similar to the triangle OAP

From similar Δ OAP and Δ OQR we have

$$\frac{OR}{OQ} = \frac{OP}{OA} \quad \text{or} \quad \frac{OR}{r_2} = \frac{r_1}{1}$$

Or, $OR = r_1r_2$

And, $\angle XOR = \angle XOP + \angle POR = \theta_1 + \theta_2$

Thus, the point R represents z_1z_2

(v) ***Geometrical representation of*** $\frac{z_1}{z_2}$**:** Let z_1 and z_2 be two given complex numbers.

Let $z_1 = r_1 (\cos \theta_1 + i \sin \theta_1)$

and $z_2 = r_2 (\cos\theta_2 + i \sin \theta_2)$

Let z_1 and z_2 be represented by the points P and Q respectively

Let O be the origin. We join OP and OQ. Then,

$OP = r_1$; $\angle POX = \theta_1$; $OQ = r_2$ and $\angle QOX = \theta_2$

Now, $\dfrac{z_1}{z_2} = \dfrac{r_1(\cos\theta_1 + i\sin\theta_1)}{r_2(\cos\theta_2 + i\sin\theta_2)} \times \dfrac{(\cos\theta_2 - i\sin\theta_2)}{(\cos\theta_2 - i\sin\theta_2)}$;

$= \dfrac{r_1}{r_2}(\cos\theta_1 + i\sin\theta_1)[\cos(-\theta_2) + i\sin(-\theta_2)]$

$= \dfrac{r_1}{r_2}[\cos(\theta_1 - \theta_2) + i\sin(\theta_1 - \theta_2)]$

Thus, $\dfrac{z_1}{z_2}$ is a complex number whose modulus is $\dfrac{r_1}{r_2}$ and argument is $\theta_1 - \theta_2$

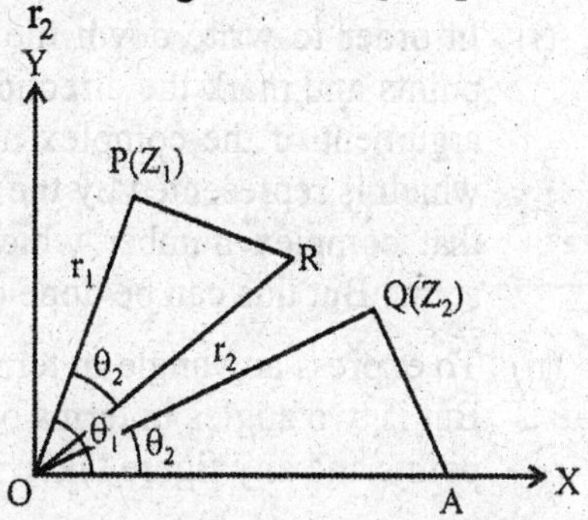

We take a point A on the real axis such that OA = 1. Join QA

Now, construct a Δ OPR, similar to Δ OQA

From similar Δ OPR and Δ OQA, we have

$\dfrac{OP}{OR} = \dfrac{OQ}{OA}$ $\therefore$ $\dfrac{r_1}{OR} = \dfrac{r_2}{1}$ or $OR = \dfrac{r_1}{r_2}$

Also, $\angle XOR = \angle XOP - \angle POR = \theta_1 - \theta_2$

Thus, R represents the complex number (z_1/z_2)

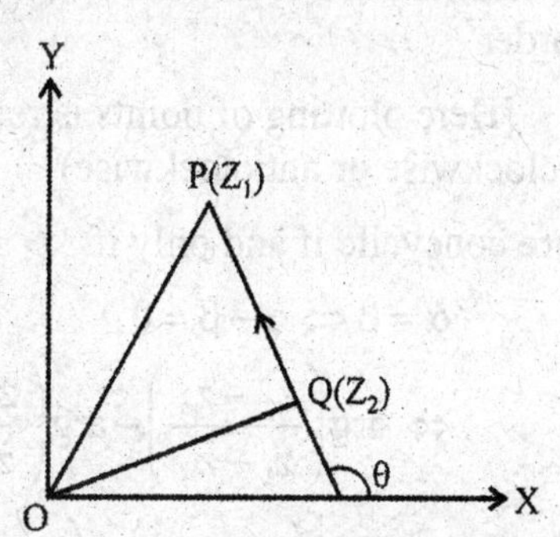

(i) $z_1 - z_2 = \overrightarrow{OP} - \overrightarrow{OQ} = \overrightarrow{QP}$

$\therefore$ $|z_1 - z_2| = |\overrightarrow{QP}| = QP =$ distance between points representing z_1 and z_2

Arg $(z_1 - z_2)$ = angle which $\overrightarrow{QP}$ makes with the positive direction of x-axis = θ,

(ii) $\theta = \arg\left(\dfrac{z_3 - z_1}{z_2 - z_1}\right)$

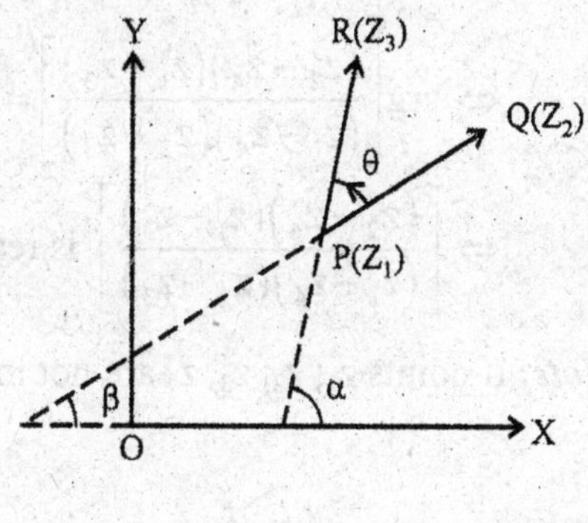

Explanation: Method 1

$\theta = \alpha - \beta$

$= \arg(z_3 - z_1) - \arg(z_2 - z_1)$

$= \arg\left(\dfrac{z_3 - z_1}{z_2 - z_1}\right)$

Method II

(ii) In the given figure z_1, z_2, z_3 are represented by points A, B, C respectively as in the given figure, then

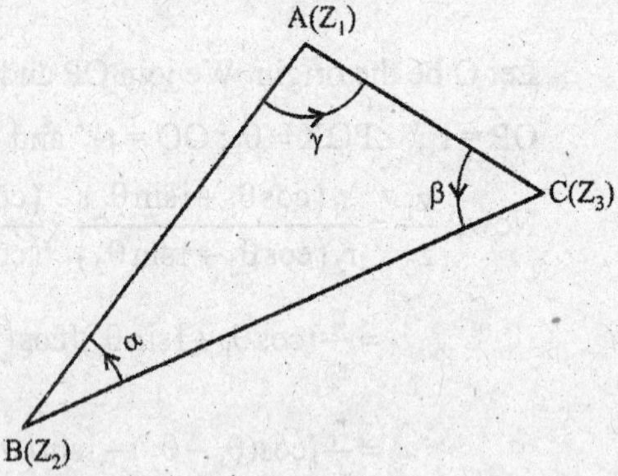

$$\alpha = \arg\left(\frac{z_1 - z_2}{z_3 - z_2}\right),\ \beta = \arg\left(\frac{z_2 - z_3}{z_1 - z_3}\right)$$

$$\text{and } \gamma = \arg\left(\frac{z_3 - z_1}{z_2 - z_1}\right)$$

***Note*:**

(i) In order to write down the angle between two line segments, first of all plot the points and mark the direction of arrow in anticlockwise direction and then take the argument of the complex number keeping that complex number in the numerator which is represented by the vector on which the arrow goes and in the denominator that complex number which is represented by the vector from which the arrow starts. But this can be done only after the points have been actually plotted.

(ii) To express any angle in terms of argument, actual position of points must be taken. But if two angles in terms of arguments are to be equated then we need not plot the points and any figure taken may be considered as the actual figure.

Condition for four points to be concyclic: Points representing z_1, z_2, z_3, z_4 taken in order

[Here plotting of points is required or it should be given that they are in order (clockwise or anticlockwise)]

are concyclic if and only if

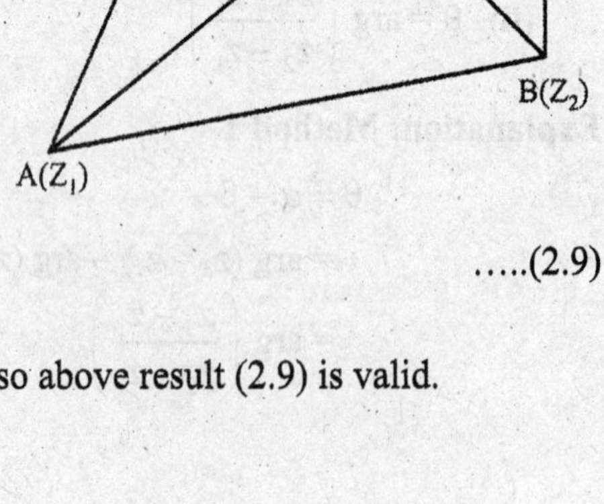

$$\alpha = \beta \Leftrightarrow \alpha - \beta = 0$$

$$\Leftrightarrow \arg\left(\frac{z_2 - z_4}{z_1 - z_4}\right) - \arg\left(\frac{z_2 - z_3}{z_1 - z_3}\right) = 0$$

$$\Leftrightarrow \arg\left[\left(\frac{z_2 - z_4}{z_1 - z_4}\right) \Big/ \left(\frac{z_2 - z_3}{z_1 - z_3}\right)\right] = 0$$

$$\Leftrightarrow \arg\left[\frac{(z_2 - z_4)(z_1 - z_3)}{(z_1 - z_4)(z_2 - z_3)}\right] = 0$$

$$\Leftrightarrow \left[\frac{(z_2 - z_4)(z_1 - z_3)}{(z_1 - z_4)(z_2 - z_3)}\right] \text{ is real} \qquad \text{.....(2.9)}$$

***Note*:** If points z_1, z_2, z_3, z_4 are not in order, then also above result (2.9) is valid.

Problems based on Demoivre's Theorem

***Example* 2.1:** Show that $\left(\dfrac{\cos\theta + i\sin\theta}{\sin\theta + i\cos\theta}\right)^4 = \cos 8\theta + i\sin 8\theta$

***Solution*:** $(\cos\theta + i\sin\theta)^4 = \cos 4\theta + i\sin 4\theta$

$$(\sin\theta + i\cos\theta)^4 = \left[i\left(\cos\theta + \frac{1}{i}\sin\theta\right)\right]^4$$

$$= i^4(\cos\theta - i\sin\theta)^4 = (\cos\theta - i\sin\theta)^4 = \cos 4\theta - i\sin 4\theta$$

Now $\left(\dfrac{\cos\theta + i\sin\theta}{\sin\theta + i\cos\theta}\right)^4 = \dfrac{\cos 4\theta + i\sin 4\theta}{\cos 4\theta - i\sin 4\theta}$

$$= \frac{(\cos 4\theta + i\sin 4\theta)(\cos 4\theta + i\sin 4\theta)}{(\cos 4\theta - i\sin 4\theta)(\cos 4\theta + i\sin 4\theta)}$$

$$= \frac{(\cos 4\theta + i\sin 4\theta)^2}{1} = \cos 8\theta + i\sin 8\theta$$

***Example* 2.2:** If $z^2 - 2z\cos q + 1 = 0$, show that $z^2 + z^{-2} = 2\cos 2\theta$

***Solution*:** Given, $z^2 - 2z\cos\theta + 1 = 0$

$$\therefore \quad z = \frac{2\cos\theta \pm \sqrt{4\cos^2\theta - 4}}{2} = \cos\theta \pm \sqrt{\cos^2\theta - 1}$$

$$= \cos\theta \pm \sqrt{-\sin^2\theta} = \cos\theta \pm \sqrt{i^2\sin^2\theta} = \cos\theta \pm i\sin\theta$$

When $z = \cos\theta + i\sin\theta$,

$$z^2 + z^{-2} = \cos 2\theta + i\sin 2\theta + (\cos 2\theta - i\sin 2\theta) = 2\cos 2\theta$$

when $z = \cos\theta - i\sin\theta$,

$$z^2 + z^{-2} = \cos 2\theta - i\sin 2\theta + \cos 2\theta + i\sin 2\theta = 2\cos 2\theta$$

***Example* 2.3:** If $x_r = \cos\dfrac{\pi}{2^r} + i\sin\dfrac{\pi}{2^r}$, prove that $x_1x_2x_3 \ldots$ to $\infty = -1$

***Solution*:** $x_r = \cos\dfrac{\pi}{2^r} + i\sin\dfrac{\pi}{2^r} = e^{i\frac{\pi}{2^r}}$

Now $x_1x_2x_3 \ldots$ to $\infty = e^{i\frac{\pi}{2}} . e^{i\frac{\pi}{2^2}} . e^{i\frac{\pi}{2^3}} \ldots$

$$= e^{i\frac{\pi}{2}}\left(1+\frac{1}{2}+\frac{1}{2^2}+\text{...to}\infty\right)$$

$$= e^{i\frac{\pi}{2}\left(\frac{1}{1-\frac{1}{2}}\right)} = e^{i\pi} = \cos\pi + i\sin\pi = -1$$

***Example* 2.4:** Prove that $(1 + i)^n + (1 - i)^n = 2^{\frac{n}{2}+1}\cos\frac{n\pi}{4}$.

***Solution*:** $1 + i = \sqrt{2}\left(\cos\frac{\pi}{4}+i\sin\frac{\pi}{4}\right)$ [Polar form of (1 + i)]

$$\therefore \quad (1+i)^n = \left[\sqrt{2}\left(\cos\frac{\pi}{4}+i\sin\frac{\pi}{4}\right)\right]^n = \left(\sqrt{2}\right)^n\left(\cos\frac{n\pi}{4}+i\sin\frac{n\pi}{4}\right)$$

$$= 2^{\frac{n}{2}}\left(\cos\frac{n\pi}{4}+i\sin\frac{n\pi}{4}\right) \quad \text{.....(i)}$$

Taking conjugate, we get

$$\therefore \quad (1-i)^n = 2^{\frac{n}{2}}\left(\cos\frac{n\pi}{4}.-i\sin\frac{n\pi}{4}\right) \quad \text{.....(ii)}$$

$$\text{(i) + (ii)} \Rightarrow (1+i)^n + (1-i)^n = 2^{\frac{n}{2}+1}\cos\frac{n\pi}{4}$$

***Example* 2.5:** Show that the value of $\sum_{k=1}^{6}\left(\sin\frac{2\pi k}{7}-i\cos\frac{2\pi k}{7}\right)$ is i.

***Solution*:** $\sum_{k=1}^{6}\left(\sin\frac{2\pi k}{7}-i\cos\frac{2\pi k}{7}\right)$

$$= -i\sum_{k=1}^{6}\left(\cos\frac{2\pi k}{7}-i\sin\frac{2\pi k}{7}\right) = -i\sum_{k=1}^{6} e^{\frac{i2\pi k}{7}}$$

$$= -i\left[e^{\frac{i2\pi}{7}}+e^{\frac{i4\pi}{7}}+e^{\frac{i6\pi}{7}}\text{...}+e^{\frac{i12\pi}{7}}\right]$$

$$= -i\left[\alpha+\alpha^2+\text{...}\alpha^6\right], \qquad \text{where } \alpha = e^{i\frac{2\pi}{7}}$$

$$= -i\left[\left(1+\alpha+\alpha^2+\text{...}+\alpha^6\right)-1\right]$$

$$= -i\left[\left(\frac{1-\alpha^7}{1-\alpha}\right)-1\right] = -i[0-1] = i$$

$$\because \quad \left[\alpha^7 = \left(e^{\frac{i2\pi}{7}}\right)^7 = e^{i2\pi} = \cos 2\pi + i\sin 2\pi = 1\right]$$

***Example* 2.6:** Show that $e^{2mi} \cot^{-1} p. \left(\frac{pi+1}{pi-1}\right)^m = 1$

***Solution*:** Let $\quad \cot^{-1} p = \theta$, then $\cot\theta = p$

Now L.H.S $= e^{2mi\theta}.\left(\frac{i\cot\theta+1}{i\cot\theta-1}\right)^m = e^{2mi\theta}.\left[\frac{i(\cot\theta-i)}{i(\cot\theta+i)}\right]^m$

$$= e^{2mi\theta}.\left(\frac{\cot\theta-i}{\cot\theta+i}\right)^m = e^{2mi\theta}\left(\frac{\cos\theta-i\sin\theta}{\cos\theta+i\sin\theta}\right)^m$$

$$= e^{2mi\theta}\left(\frac{e^{-i\theta}}{e^{i\theta}}\right)^m = e^{2mi\theta}\left(e^{-2i\theta}\right)^m = e^{2mi\theta} = e^0 = 1$$

***Example* 2.7:** Prove that

$$\left(\frac{1+\sin\phi+i\cos\phi}{1+\sin\phi-i\cos\phi}\right)^n = \cos\left(\frac{n\pi}{2}-n\phi\right)+i\sin\left(\frac{n\pi}{2}-n\phi\right)$$

***Solution*:** Let $\quad 1+\sin\phi+i\cos\phi = r(\cos\theta + i\sin\theta)$

$\therefore \quad 1 + \sin\theta = r\cos\theta \quad$ .....(i)

and $\quad \cos\theta = r\sin\theta \quad$ .....(ii)

now $(1+\sin\phi+i\cos\phi)^n = r^n(\cos\theta+i\sin n\theta)^n = r^n(\cos n\theta+i\sin n\theta) \quad$ .....(1)

Taking conjugates, we get

$$(1+\sin\phi-i\cos\phi)^n = r^n(\cos n\theta-i\sin n\theta) \quad ......(2)$$

From (1) and (2), we get

$$\left(\frac{1+\sin\phi+i\cos\phi}{1+\sin\phi-i\cos\phi}\right)^n = \frac{\cos n\theta+i\sin n\theta}{\cos n\theta-i\sin n\theta} = \frac{e^{in\theta}}{e^{-in\theta}}$$

$$= e^{in\theta}.e^{in\theta} = e^{2in\theta} = \cos 2n\theta + i\sin 2n\theta \quad .....(3)$$

$$\tan\theta = \frac{\cos\phi}{1+\sin\phi} = \frac{\cos^2\frac{\phi}{2} - \sin^2\frac{\phi}{2}}{\left(\cos\frac{\phi}{2} + \sin\frac{\phi}{2}\right)}$$

$$= \frac{\cos\frac{\phi}{2} - \sin\frac{\phi}{2}}{\cos\frac{\phi}{2} + \sin\frac{\phi}{2}} = \frac{1-\tan\frac{\phi}{2}}{1+\tan\frac{\phi}{2}} = \tan\left(\frac{\pi}{4} - \frac{\phi}{2}\right)$$

$$\therefore \qquad \theta = \frac{\pi}{4} - \frac{\phi}{2} \qquad \therefore \; 2n\theta = \left(\frac{n\pi}{2} - n\phi\right)$$

$\therefore$ From (3),

$$\left(\frac{1+\sin\phi + i\cos\phi}{1+\sin\phi - i\cos\phi}\right) = \cos\left(\frac{n\pi}{2} - n\phi\right) + i\sin\left(\frac{n\pi}{2} - n\phi\right)$$

***Example* 2.8:** If $\sin\alpha + \sin\beta + \sin\gamma = \cos\alpha + \cos\beta + \cos\gamma = 0$, show that

$$\cos 3\alpha + \cos 3\beta + \cos 3\gamma = 3\cos(\alpha + \beta + \gamma)$$

and $\quad \sin 3\alpha + \sin 3\beta + \sin 3\gamma = 3\sin(\alpha + \beta + \gamma)$

***Solution*:** Let $a = \cos\alpha + i\sin\alpha$, $b = \cos\beta + i\sin\beta$, $c = \cos\gamma + i\sin\gamma$

Now $\quad a + b + c = (\cos\alpha + \cos\beta + \cos\gamma) + i(\sin\alpha, \sin\beta + \sin\gamma)$

$$= 0 + i.\,0 = 0$$

Also $\quad \frac{1}{a} + \frac{1}{b} + \frac{1}{c} = (\cos\alpha - i\sin\alpha) + (\cos\beta - i\sin\beta) + (\cos\gamma - i\sin\gamma)$

$$= (\cos\alpha + \cos\beta + \cos\gamma) - i(\sin\alpha + \sin\beta + \sin\gamma)$$

$$= 0 - i.0. = 0$$

Now $\quad a^3 + b^3 + c^3 - 3abc = (a + b + c)(a^2 + b^2 - c^2 - ab - bc - ca) = 0$

$$[\because a + b + c = 0]$$

$$\therefore \quad a^3 + b^3 + c^3 = 3abc$$

$$\therefore \quad (\cos\alpha + i\sin\alpha)^3 + (\cos\beta + i\sin\beta)^3 + (\cos\gamma + i\sin\gamma)^3$$

$$= 3(\cos\alpha + i\sin\alpha)(\cos\beta + i\sin\beta)(\cos\gamma + i\sin\gamma)$$

or $\quad \cos 3\alpha + \cos 3\beta + \cos 3\gamma + i(\sin 3\alpha + \sin 3\beta + \sin 3\gamma)$

$$= 3[(\cos(\alpha + \beta + \gamma) + i\sin(\alpha + \beta + \gamma)]$$

Equating real and imaginary parts, we get

$$\cos 3\alpha + \cos 3\beta + \cos 3\gamma = 3\cos(\alpha + \beta + \gamma)$$

and $\quad \sin 3\alpha + \sin 3\beta + \sin 3\gamma = 3\sin(\alpha + \beta + \gamma)$

***Example* 2.9:** If sin x + sin y + sin z = cos x + cos y + cos z = 0, show that

$$\sin 2x + \sin 2y + \sin 2z = \cos 2x + \cos 2y + \cos 2z = 0$$

Solution: Let $a = \cos x + i \sin x,\ b = \cos y + i \sin y$ and $z = \cos z + i \sin z$

Now $a + b + c = (\cos x + \cos y + \cos z) + i(\sin x + \sin y + \sin z)$

$$= 0 + i.0 = 0$$

Also $\frac{1}{a}+\frac{1}{b}+\frac{1}{c} = \cos x - i \sin x + \cos y - i \sin y + \cos z - i \sin z$

$$= (\cos x + \cos y + \cos z) - i(\sin x + \sin y + \sin z) = 0$$

$\therefore$ $a^2 + b^2 + c^2 = (a + b + c)^2 - 2(ab + bc + ca)$

$$= (a + b + c)^2 - 2abc\left(\frac{1}{a}+\frac{1}{b}+\frac{1}{c}\right) = 0^2 - 2abc.\ 0 = 0$$

$\therefore$ $(\cos x + i \sin x)^2 + (\cos y + i \sin y)^2 + (\cos z + i \sin z)^2$

or $(\cos 2x + i \sin 2x) + (\cos 2y + i \sin 2y) + (\cos 2z + i \sin 2z) = 0$

or $(\cos 2x + \cos 2y + \cos 2z) + i(\sin 2x + \sin 2y + \sin 2z) = 0 + i.0$

Equating real and imaginary parts, we get

$$\cos 2x + \cos 2y + \cos 2z = 0 \text{ and } \sin 2x + \sin 2y + \sin 2z = 0.$$

***Example* 2.10:** If α, β are the roots of the equation $t^2 - 2t + 2 = 0$, show that a value of x, satisfying $\frac{(x+\alpha)^n-(x+\beta)^n}{(\alpha-\beta)} = \frac{\sin^n\theta}{\sin^n\theta}$ is $x = (\cot\theta - 1)$

Solution: $t^2 - 2t + 2 = 0 \Leftrightarrow t = \frac{2\pm\sqrt{4-8}}{2} = (1\pm i)$

Let $\alpha = (1 + i)$ and $b = (1 - i)$.

$\therefore$ $x + \alpha = (x + 1) + i,\ x + \beta = (x + 1) - i$ and $\alpha - \beta = 2i$.

Let $x + 1 = r\cos\phi$ and $1 = r\sin\phi$.

Then, $x + \alpha = r(\cos\phi + i\sin\phi)$ and $x + \beta = r(\cos\phi - i\sin\phi)$.

$\therefore$ $(x + a)^n = r^n(\cos n\phi + i \sin n\phi)$

and $(x + \beta)^n = r^n(\cos n\phi - i \sin n\phi)$.

Thus, $\frac{(x+\alpha)^n-(x+\beta)^n}{(\alpha-\beta)} = \frac{\sin n\theta}{\sin^n\theta}$

$$\Leftrightarrow \frac{r^n(\cos n\phi + i\sin n\phi) - r^n(\cos n\phi - i\sin n\phi)}{2i} = \frac{\sin n\theta}{\sin^n\theta}$$

$$\Leftrightarrow r^n \sin n\phi = \frac{\sin n\theta}{\sin^n\theta}$$

$$\Leftrightarrow \frac{\sin n\theta}{\sin^n\theta} = \frac{\sin n\theta}{\sin^n\theta} \qquad \left[\because r\sin\phi = 1 \therefore r^n = \frac{1}{\sin^n\phi}\right]$$

$\Leftrightarrow$ one of the values of ϕ is θ

$\therefore \qquad x + 1 = r\cos\theta$ .....(i)

and $\qquad 1 = r\sin\theta$ .....(ii)

Dividing (i) by (ii) we get, $x + 1 = \dfrac{\cos\theta}{\sin\theta}$ or $x = \cot\theta - 1$.

***Example* 2.11:** If $(1 + x)^n = p_0 + p_1 x + p_2x^2 + \ldots + p_nx^n$, show that

$$p_0 - p_2 + p_4 \ldots = 2^{\frac{n}{2}}\cos\frac{n\pi}{4} \quad \text{and} \quad p_1 - p_3 + p_5 - \ldots = 2^{\frac{n}{2}}\sin\frac{n\pi}{4}$$

***Solution*:** Given, $(1 + x)^n = p_0 + p_1x + p^2x^2 + \ldots + p^nx^n$

Putting $x = i$, we get

$$(1 + i)^n = p_0 + p_1i + p_2i^2 + p_3i^3 + \ldots + p_ni^n$$

$$= (p_0 - p_2 + p_4) + i\,(p_1 - p_3 + \ldots)$$

or
$$\left[\sqrt{2}\left(\cos\frac{\pi}{4} + i\sin\frac{\pi}{4}\right)\right]^n = (p_0 - p_2 + p_{4-}\ldots) + i(p_1 - P_3 + \ldots)$$

or
$$2^{\frac{n}{2}}\left(\cos\frac{n\pi}{4} + i\sin\frac{n\pi}{4}\right) = (p_0 - p_2 + p_{4-}\ldots) + i(p_1 - P_3 + \ldots)$$

Equating real and imaginary parts, we get

$$p_0 - p_2 + p_{4-}\ldots = 2^{\frac{n}{2}}\cos\frac{n\pi}{4}$$

and
$$p_1 - p_3 + p_5^- \ldots = 2^{\frac{n}{2}}\sin\frac{n\pi}{4}$$

***Example* 2.12:** If $(1 - x + x^2)n = a_0 + a_1x + a_2x^2 + \ldots + a_{2n}x^{2n}$ show that

$$a_0 + a_3 + a_6 + \ldots = \frac{1}{3}\left(1 + 2^{n+1}\cos\frac{n\pi}{3}\right)$$

***Solution*:** Given $\qquad (1 - x + x^2) = a_0 + a_1x + a_2x^2 + a_3x^3 + \ldots + a_{2n}x^{2n}$ .....(i)

Putting x = 1, ω and ω^2, we get

$$1 = a_0 + a_1 + a_2 + a_3 + a_4 + \ldots + a_{2n} \qquad \ldots\ldots(ii)$$

$$(-2\omega)^n = a_0 + a_1\omega + a_2\omega^2 + a_3\omega^3 + \ldots + a_{2n}\omega^{2n} \qquad \ldots\ldots(iii)$$

$$(-2\omega^2)^n = a_0 + a_1\omega^2 + a_2\omega^4 + a_3\omega^6 + a_4\omega^8 + \ldots + a_{2n}\omega^{4n} \qquad \ldots\ldots(iv)$$

(i) + (ii) + (iii) $\Rightarrow 3(a_0 + a_3 + a_6 + \ldots) = 1 + (-2)^n(\omega^n + \omega^{2n})$ .....(v)

Now $\omega = \dfrac{1+\sqrt{3}i}{2} = -\dfrac{1}{2} + \dfrac{\sqrt{3}}{2}i = 1.\left(\cos\dfrac{2\pi}{3} + i\sin\dfrac{2\pi}{3}\right)$

∴ $\omega^n = \cos\dfrac{2n\pi}{3} + i\sin\dfrac{2n\pi}{3}$ .....(vi)

$\omega^2 = \dfrac{-1-\sqrt{3}i}{2} = -\dfrac{1}{2} - \dfrac{\sqrt{3}}{2}i = \cos\dfrac{2\pi}{3} - i\sin\dfrac{2\pi}{3}$ .....(vii)

∴ $\omega^{2n} = \omega^{2n} = 2\cos\dfrac{2n\pi}{3} = 2\cos\left(n\pi - \dfrac{n\pi}{3}\right)$

$= 2(-1)^n \cos\dfrac{n\pi}{3}$ $[\because \cos(\pi+\theta) = -\cos\theta, \cos(-\theta) = \cos\theta]$

From (v). $3(a_0 + a_3 + a_6 + \ldots) = 1 + (-2)^n 2.(-1)^n \cos\dfrac{n\pi}{3} = 1 + 2^{n+1}\cos\dfrac{n\pi}{3}$

∴ $a_0 + a_3 + a_6 + \ldots = \dfrac{1}{3}\left(1 + 2^{n+1}\cos\dfrac{n\pi}{3}\right)$

***Example* 2.13:** If n is a positive integer and

$$(1 + x)^n = c_0 + c_1x + c_2x^2 + c_nx^n$$

Show that $c_0 + c_4 + c_8 + \ldots = 2^{n-2} + 2^{\frac{n}{2}-1}\cos\dfrac{n\pi}{4}$

***Solution*:** Given $(1 + x)^n = c_0 + c_1x + c_2x^2 + \ldots + c_nx^n$ .....(i)

Putting x = 1 and –1, we get

$$2^n = c_0 + c_1 + c_2 + c_3 + c_4 + \ldots. + (-1)^n c_n \qquad \ldots\ldots(ii)$$

and $0 = c_0 - c_1 + c_2 - c_3 + c_4 - \ldots. + (-1)^n c_n$ .....(iii)

adding (ii) and (iii), we get

$$2^n = 2(c_0 + c_2 + c_4 + \ldots)$$

or $c_0 + c_2 + c_4 + \ldots = 2^{n-1}$ .....(iv)

putting x = i in (i), we get

$$(1+i)^n = c_0 + c_1 i + c_2 i^2 + c_3 i^3 + c_4 i^4 + \dots + c_n i^n$$

or
$$\left[\sqrt{2}\left(\cos\frac{\pi}{4} + i\sin\frac{\pi}{4}\right)\right]^n = (c_0 - c_2 + c_4 - \dots) + i(c_1 - c_3 + \dots)$$

or
$$2^{\frac{n}{2}}\left(\cos\frac{n\pi}{4} + i\sin\frac{n\pi}{4}\right) = (c_0 - c_2 + c_4 - \dots) + i(c_1 - c_3 + \dots)$$

Equating real parts, we get

$$c_0 - c_2 + c_4 - \dots = 2^{\frac{n}{2}}\cos\frac{n\pi}{4} \qquad \text{.....(v)}$$

Adding (iv) and (v), we get

$$2[c_0 + c_4 + c_8 + \dots] = 2^{n-1} + 2^{\frac{n}{2}}\cos\frac{n\pi}{4}$$

$$\therefore \quad c_0 + c_4 + c_8 + \dots = 2^{n-1} + 2^{\frac{n}{2}-1}\cos\frac{n\pi}{4}$$

***Example* 2.14:** Solve the equation $z^8 + 1 = 0$ and deduce that

$$\cos 4\theta = 8\left(\cos\theta - \cos\frac{\pi}{8}\right)\left(\cos\theta - \cos\frac{3\pi}{8}\right)$$
$$\left(\cos\theta - \cos\frac{5\pi}{8}\right)\left(\cos\theta - \cos\frac{7\pi}{8}\right)$$

Solution:
$$z^8 + 1 = 0 \Rightarrow z^8 = -1 = \cos\pi + i\sin\pi$$

$$\therefore \quad z = (\cos\pi + i\sin\pi)^{\frac{1}{8}}$$

$$= \cos\frac{2r\pi + \pi}{8} + i\sin\frac{2r\pi + \pi}{8}, \qquad r = 0, 1, 2. \dots 7$$

$$\therefore \quad z = \cos\frac{\pi}{8} \pm i\sin\frac{\pi}{8}, \quad \cos\frac{3\pi}{8} \pm i\sin\frac{3\pi}{8}$$
$$\cos\frac{5\pi}{8} \pm i\sin\frac{5\pi}{8}, \quad \cos\frac{7\pi}{8} \pm i\sin\frac{7\pi}{8}$$

These are the roots of eqn. $z^8 + 1 = 0$.

Now quadratic eqn. whose roots are $\cos\frac{\pi}{8} \pm i\sin\frac{\pi}{8}$ is

$$z^2 - 2\cos\frac{\pi}{8}z + 1 = 0$$

Similarly we can find the quadratic equations whose roots are

$$\cos\frac{3\pi}{8} \pm i\sin\frac{3\pi}{8},\ \cos\frac{5\pi}{8} \pm i\sin\frac{5\pi}{8} \text{ and } \cos\frac{7\pi}{8} \pm i\sin\frac{7\pi}{8}$$

Thus
$$z^8 + 1 = \left(z^2 - 2\cos\frac{\pi}{8}z + 1\right)\left(z^2 - 2\cos\frac{3\pi}{8}z + 1\right)$$
$$\left(z^2 - 2\cos\frac{5\pi}{8}z + 1\right)\left(z^2 - 2\cos\frac{7\pi}{8}z + 1\right)$$

Dividing both sides by z^4, we get

$$z^4 + \frac{1}{z^4}\left(z + \frac{1}{z} - 2\cos\frac{\pi}{8}\right)\left(z + \frac{1}{z} - 2\cos\frac{3\pi}{8}\right)$$
$$\left(z + \frac{1}{z} - 2\cos\frac{5\pi}{8}\right)\left(z + \frac{1}{z} - 2\cos\frac{7\pi}{8}\right)$$

Putting $z = \cos\theta + i\sin\theta$, so that $z^n + \frac{1}{z^n} - 2\cos n\theta$, we get

$$2\cos 4\theta = 2\left(\cos\theta - \cos\frac{\pi}{8}\right)2\left(\cos\theta - \cos\frac{3\pi}{8}\right)$$
$$2\left(\cos\theta - \cos\frac{5\pi}{8}\right)2\left(\cos\theta - \cos\frac{7\pi}{8}\right)$$

$\therefore$
$$\cos 4\theta = 8\left(\cos\theta - \cos\frac{\pi}{8}\right)\left(\cos\theta - \cos\frac{3\pi}{8}\right)$$
$$\left(\cos\theta - \cos\frac{5\pi}{8}\right)\left(\cos\theta - \cos\frac{7\pi}{8}\right)$$

***Example* 2.15:** Prove that the roots of the equation $8x^3 - 4x^2 - 4x + 1 = 0$ are $\cos\frac{\pi}{7},\ \cos\frac{3\pi}{7},\ \cos\frac{5\pi}{7}$.

***Solution*:** Let $z = \cos\theta + i\sin\theta$, then $z^7 = \cos 7\theta + i\sin 7\theta$

If $\theta = \frac{\pi}{7}, \frac{3\pi}{7}, \frac{5\pi}{7}, \frac{7\pi}{7}, 9, \frac{9\pi}{7}, \frac{11\pi}{7}, \frac{13\pi}{7}$

Then $z^7 = \cos 7\theta + i\sin 7\theta = -1$ or $z^7 + 1 = 0$ .....(i)

Thus $z = \cos\theta + i\sin\theta$,

where $\theta = \frac{\pi}{7}, \frac{3\pi}{7}, \frac{5\pi}{7}, \pi, \frac{9\pi}{7}, \frac{11\pi}{7}, \frac{13\pi}{7}$ are the roots of eq.(i).

Also when $\theta = \pi$, $z = \cos\pi + i\sin\theta = -1$

Now $z^7 + 1 = 0 \Rightarrow (z+1)(z^6 - z^5 + z^4 - z^3 + z^2 - z + 1) = 0$

Root of equation $z + 1 = 0$ is $\cos\theta + i\sin\theta$, where $\theta = \pi$

$\therefore$ Roots of equation $z^6 - z^5 + z^4 - z^3 + z^2 - z + 1 = 0$ .....(ii)

are $\cos\theta + i\sin\theta$,

where $\theta = \frac{\pi}{7}, \frac{3\pi}{7}, \frac{5\pi}{7}, \frac{7\pi}{7}, \frac{9\pi}{7}, \frac{11\pi}{7}, \frac{13\pi}{7}$

let $x = \cos\theta$, then $z + \frac{1}{z} = \cos\theta + i\sin\theta + \frac{1}{\cos\theta + i\sin\theta} = 2\cos\theta = 2x$

Thus $z + \frac{1}{z} = 2x$, where $x = \cos\theta$ and $\theta = \frac{\pi}{7}, \frac{3\pi}{7}, \frac{5\pi}{7}, \frac{7\pi}{7}, \frac{9\pi}{7}, \frac{11\pi}{7}, \frac{13\pi}{7}$

But $\cos\frac{13\pi}{7} = \cos\left(2\pi - \frac{\pi}{7}\right) = \cos\frac{\pi}{7}$, $\cos\frac{11\pi}{7} = \cos\frac{3\pi}{7}, \cos\frac{9\pi}{7} = \cos\frac{5\pi}{7}$

From (ii), $z^3 - z^2 + z - 1 + \frac{1}{z} - \frac{1}{z^2} + \frac{1}{z^3} = 0$ [Dividing by z^6]

$$\Rightarrow \left(z^3 + \frac{1}{z}\right) - \left(z^2 + \frac{1}{z^2}\right) + \left(z + \frac{1}{z}\right) - 1 = 0$$

$$\Rightarrow \left(z + \frac{1}{z}\right)^3 - 3z.\frac{1}{z}\left(z + \frac{1}{z}\right) - \left[\left(z + \frac{1}{z}\right)^2 - 2z\frac{1}{z}\right] + z + \frac{1}{z} - 1 = 0$$

$$\Rightarrow 8x^3 - 3.2x - (4x^2 - 2) + 2x - 1 = 0$$

$$\Rightarrow 8x^3 - 4x^2 - 4x + 1 = 0 \qquad .....(iii)$$

Roots of equation (iii) are $\cos\frac{\pi}{7}$, $\cos\frac{3\pi}{7}$ and $\cos\frac{5\pi}{7}$.

***Example* 2.16:** Solve the equation $x^{10} - 1 = 0$ and deduce that

$$\sin 5\theta = 5\sin\theta\left(1 - \frac{\sin^2\theta}{\sin^2\frac{\pi}{5}}\right)\left(1 - \frac{\sin^2\theta}{\sin^2\frac{2x}{5}}\right)$$

Solution: Given, $x^{10} - 1 = 0 \Rightarrow x^{10} = 1 = \cos 0 + i \sin 0$

$$\therefore \qquad x = (\cos 0 + i\sin 0)^{\frac{1}{10}} = \cos\frac{2r\pi + 0}{10} + i\sin\frac{2r\pi + 0}{10}$$

$$= \cos\frac{r\pi}{5} + i\sin\frac{r\pi}{5}, \quad r = 0,1,2,....,9$$

$$= 1,\ \cos\frac{\pi}{5} \pm i,\ \sin\frac{\pi}{5},\ \cos\frac{2\pi}{5} \pm i\sin\frac{2\pi}{5},\ \cos\frac{3\pi}{5} \pm i\sin\frac{3\pi}{5},$$

$$\cos\frac{4\pi}{5} \pm i\sin\frac{4\pi}{5}, \cos\frac{5\pi}{5} + i\sin\frac{5\pi}{5}$$

$$= \pm 1, \cos\frac{\pi}{5} \pm i\sin\frac{\pi}{5}, \cos\frac{2\pi}{5} \pm i\sin\frac{2\pi}{5},$$

$$\cos\frac{3\pi}{5} \pm i\sin\frac{3\pi}{5}, \cos\frac{4\pi}{5} \pm i\sin\frac{4\pi}{5}$$

Second part: Quadratic equation whose roots are 1 and -1 is $x^2 - 1 = 0$.

And quadratic equation whose roots are $\cos\frac{\pi}{5} \pm i, \sin\frac{\pi}{5}$ is

$$x^2 - 2\cos\frac{\pi}{5}x + 1 = 0$$

Thus
$$x^{10} - 1 = (x^2 - 1)\left(x^2 - 2\cos\frac{\pi}{5}.x + 1\right)\left(x^2 - 2\cos\frac{2\pi}{5}.x + 1\right)$$

$$\left(x^2 - 2\cos\frac{3\pi}{5}.x + 1\right)\left(x^2 - 2\cos\frac{4\pi}{5}.x + 1\right) \qquad .....(i)$$

Dividing both sides by x^5, we get

$$x^5 - \frac{1}{x^5} = \left(x - \frac{1}{x}\right)\left(x + \frac{1}{x} - 2\cos\frac{\pi}{5}\right)\left(x + \frac{1}{x} - 2\cos\frac{2\pi}{5}\right)$$

$$\left(x + \frac{1}{x} - 2\cos\frac{3\pi}{5}\right)\left(x + \frac{1}{x} - 2\cos\frac{4\pi}{5}\right)$$

Putting $x = \cos\theta + i\sin\theta$ in (i), so that $x^5 - \frac{1}{x^5} = 2i\sin 5\theta$,

We get $2i\sin 5\theta = 2i\sin\theta.2\left(\cos\theta - \cos\frac{\pi}{5}\right)2\left(\cos\theta - \cos\frac{2\pi}{5}\right)$

$$2\left(\cos\theta-\cos\frac{3\pi}{5}\right).2\left(\cos\theta-\cos\frac{4\pi}{5}\right)$$

$$\therefore\quad \sin 5\theta = 16\sin\theta\left(\cos\theta-\cos\frac{\pi}{5}\right).2\left(\cos\theta-\cos\frac{4\pi}{5}\right)$$

$$\left(\cos\theta-\cos\frac{2\pi}{5}\right)\left(\cos\theta-\cos\frac{3\pi}{5}\right)$$

$$= 16\sin\theta\left(\cos\theta-\cos\frac{\pi}{5}\right)\left(\cos\theta+\cos\frac{\pi}{5}\right)$$

$$\left(\cos\theta-\cos\frac{2\pi}{5}\right)\left(\cos\theta+\cos\frac{2\pi}{5}\right)$$

$$= 16\sin\theta\left(\cos^2\theta-\cos^2\frac{\pi}{5}\right)\left(\cos^2\theta+\cos^2\frac{2\pi}{5}\right)$$

$$= 16\sin\theta\left(\sin^2\frac{\pi}{5}-\sin^2\theta\right)\left(\sin^2\frac{2\pi}{5}-\sin^2\theta\right)$$

$$= 16\sin\theta\sin^2\frac{\pi}{5}.\sin^2\frac{2\pi}{5}\left(1-\frac{\sin^2\theta}{\sin^2\frac{\pi}{5}}\right)\left(1-\frac{\sin^2\theta}{\sin^2\frac{2\pi}{5}}\right)$$

$$= 16\sin\theta.\sin^2 36^\circ\sin^2 72^\circ\left(1-\frac{\sin^2\theta}{\sin^2\frac{\pi}{5}}\right)\left(1-\frac{\sin^2\theta}{\sin^2\frac{2\pi}{5}}\right)$$

$$= 16\sin\theta\left(\frac{\sqrt{10-2\sqrt{5}}}{4}\right)^2\left(\frac{10+2\sqrt{5}}{4}\right)^2\left(1-\frac{\sin^2\theta}{\sin^2\frac{\pi}{5}}\right)\left(1-\frac{\sin^2\theta}{\sin^2\frac{2\pi}{5}}\right)$$

$$= 5\sin\theta\left(1-\frac{\sin^2\theta}{\sin^2\frac{\pi}{5}}\right)\left(1-\frac{\sin^2\theta}{\sin^2\frac{2\pi}{5}}\right)$$

***Example* 2.17:** Solve the equation $x^7+1=0$ and deduce that $\cos\frac{\pi}{7}\cos\frac{3\pi}{7}\cos\frac{5\pi}{7}=-\frac{1}{8}$.

***Solution*:** Given, $x^7+1=0$ or $x^7=-1=\cos\pi+i\sin\pi$

$$\therefore x = (\cos\pi + i\sin\pi)^{\frac{1}{7}} = \cos\frac{2r\pi+\pi}{7} + i\sin\frac{2r\pi+\pi}{7}, \quad r = 0,1,2,\dots,6$$

$$= \cos\frac{\pi}{7} \pm i\sin\frac{\pi}{7},\ \cos\frac{2\pi}{7} \pm i\sin\frac{\pi}{7},\ \cos\frac{3\pi}{7} \pm \sin\frac{3\pi}{7},\ \cos\pi + i\sin\pi(=-1)$$

***Second part*:**

$$x^7+1 = (x+1)\left(x^2 - 2\cos\frac{\pi}{7}x+1\right)\left(x^2 - 2\cos\frac{2\pi}{7}x+1\right)\left(x^2 - 2\cos\frac{3\pi}{7}.x+1\right)$$

Putting $x^2 + 1 = 0$ i.e., $x = i$, we get

$$i^7+1 = (1+i)\left(-2i\cos\frac{\pi}{7}\right)\left(-2i\cos\frac{2\pi}{7}\right)\left(-2i\cos\frac{3\pi}{7}\right)$$

or $$1-i = 8(1+i)i\cos\frac{\pi}{7}\cos\frac{2\pi}{7}\cos\frac{3\pi}{7}$$

$$= 8(-1+i)\cos\frac{\pi}{7}\cos\frac{2\pi}{7}\cos\frac{3\pi}{7}$$

$$= -8(1-i)\cos\frac{\pi}{7}\cos\frac{2\pi}{7}\cos\frac{3\pi}{7}$$

or $$\cos\frac{\pi}{7}\cos\frac{2\pi}{7}\cos\frac{3\pi}{7} = -\frac{1}{8}$$

***Example* 2.18:** Prove that

$$(1+\cos\theta+i\sin\theta)^n + (1+\cos\theta-i\sin\theta)^n = 2^{n+1}\cos^n\frac{\theta}{2}\cos\frac{n\theta}{2}$$

Where n is an integer.

***Solution*:** $(1+\cos\theta+i\sin\theta)^n + (1+\cos\theta-i\sin\theta)^n$

$$= \left[1+2\cos^2\frac{\theta}{2}-1+2\,i\sin\frac{\theta}{2}\cos\frac{\theta}{2}\right]^n + \left[1+2\cos^2\frac{\theta}{2}-1-i2\sin\frac{\theta}{2}\cos\frac{\theta}{2}\right]^n$$

$$= \left[2\cos^2\frac{\theta}{2}+2i\sin\frac{\theta}{2}\cos\frac{\theta}{2}\right]^n + \left[2\cos^2\frac{\theta}{2}-2i\sin\frac{\theta}{2}-\cos\frac{\theta}{2}\right]^n$$

$$= \left(2\cos\frac{\theta}{2}\right)^n\left[\cos\frac{\theta}{2}+i\sin\frac{\theta}{2}\right]^n + \left(2\cos\frac{\theta}{2}\right)^n.\left[\cos\frac{\theta}{2}-i\sin\frac{\theta}{2}\right]^n$$

$$= 2^n\cos^n\frac{\theta}{2}\left[\cos\frac{n\theta}{2}+i\sin\frac{n\theta}{2}\right] + 2^n\cos^n\frac{\theta}{2}\left[\cos\frac{n\theta}{2}-i\sin\frac{n\theta}{2}\right]$$

$$= 2^n \cos^n \frac{\theta}{2}\left[\cos\frac{n\theta}{2} + i\sin\frac{n\theta}{2} + \cos\frac{n\theta}{2} - i\sin\frac{n\theta}{2}\right]$$

$$= 2^n \cos^n \frac{\theta}{2}\left(2\cos\frac{n\theta}{2}\right) = 2^{n+1}\cos^n\frac{\theta}{2}\cos\frac{n\theta}{2}$$ **Proved.**

***Example* 2.19:** If n is positive integer, prove that

$$\left(\sqrt{3}+i\right)^n + \left(\sqrt{3}-i\right)^n = 2^{n+1}\cos\frac{n\pi}{6}$$

***Solution*:** Let $\sqrt{3}+i = r(\cos\alpha + i\sin\alpha)$, $r = \sqrt{3+1} = 2$

$$\alpha = \tan^{-1}\frac{1}{\sqrt{3}} = \frac{\pi}{6}$$

$$\left(\sqrt{3}+i\right)^n + \left(\sqrt{3}-i\right)^n = \left[r(\cos\alpha + i\sin\alpha)\right]^n + \left[r(\cos\alpha - i\sin\alpha)\right]^n$$

$$= r^n(\cos n\alpha + i\sin n\alpha) + r^n(\cos n\alpha - i\sin n\alpha)$$

$$= 2r^n\cos n\alpha$$

$$= 2(2)^n\cos\frac{n\pi}{6} = 2^{n+1}\cos\frac{n\pi}{6}$$ **Proved**

***Example* 2.20:** Find all the roots of $x^{12} - 1 = 0$ and identify the roots which are also the roots of $x^4 - x^2 + 1 = 0$.

***Solution*:** $x^{12} - 1 = 0$ or $x^{12} = 1 = \cos 0 + i\sin 0$

$$x^{12} = \cos 2n\pi + i\sin 2n\pi$$

$$x = [\cos 2n\pi + i\sin 2n\pi]^{1/12}$$

$$= \cos\frac{n\pi}{6} + i\sin\frac{n\pi}{6}$$ By De Moivre's Theorem

Putting n = 0, 1, 2, 3, 4, 5, 6, 7, 8, 9, 10, 11, we get tweleve roots

viz.,

$$x_1 = \cos 0 + i\sin 0,\ x_2 = \cos\frac{\pi}{6} + i\sin\frac{\pi}{6}$$

$$x_3 = \cos\frac{2\pi}{6} + i\sin\frac{2\pi}{6},\ x_4 = \cos\frac{3\pi}{6} + i\sin\frac{3\pi}{6}$$

$$x_5 = \cos\frac{4\pi}{6} + i\sin\frac{4\pi}{6},\ x_6 = \cos\frac{5\pi}{6} + i\sin\frac{5\pi}{6}$$

$$x_7 = \cos\frac{6\pi}{6} + i\sin\frac{6\pi}{6},\ x_8 = \cos\frac{7\pi}{6} + i\sin\frac{7\pi}{6}$$

$$x_9 = \cos\frac{8\pi}{6} + i\sin\frac{8\pi}{6},\ x_{10} = \cos\frac{9\pi}{6} + i\sin\frac{9\pi}{6}$$

$$x_{11} = \cos\frac{10\pi}{6} + i\sin\frac{10\pi}{6},\ x_{12} = \cos\frac{11\pi}{6} + i\sin\frac{11\pi}{6} \quad .....(1)$$

By algebra $\quad x^{12} - 1 = (x^2 - 1)(x^6 + 1)(x^4 + x^2 + 1)$

The roots of $x^4 + x^2 + 1 = 0$ are the roots of $x^{12} - 1 = 0$ excluding the roots of $x^2 - 1 = 0$ and $x^6 + 1 = 0$.

$$x^2 - 1 = 0 \text{ or } x^2 = 1 = \cos 0 + i\sin 0 = \cos 2n\pi + i\sin 2n\pi$$

$$x = \cos n\pi + i\sin n\pi$$

where n = 0 and 1. There are two roots.

i.e., The roots are cos 0 + i sin 0 and cos π + i sin π $\quad .....(2)$

$$x^6 + 1 = 0 \text{ or } x^6 = -1 = \cos\pi + i\sin\pi = \cos(2n\pi + \pi) + i\sin(2n\pi + \pi)$$

$$x = [\cos(2n\pi + \pi) + i\sin(2n\pi + \pi)]^{1/16} = \cos(2n+1)\frac{\pi}{6} + i\sin(2n+1)\frac{\pi}{6}$$

where n = 0, 1, 2, 3, 4, 5, we get six roots.

viz.,
1. $\cos\frac{\pi}{6} + i\sin\frac{\pi}{6}$
2. $\cos\frac{3\pi}{6} + i\sin\frac{3\pi}{6}$
3. $\cos\frac{5\pi}{6} + i\sin\frac{5\pi}{6}$
4. $\cos\frac{7\pi}{6} + i\sin\frac{7\pi}{6}$
5. $\cos\frac{9\pi}{6} + i\sin\frac{9\pi}{6}$
6. $\cos\frac{11\pi}{6} + i\sin\frac{11\pi}{6}$ $\quad .....(3)$

On exclusing two roots of (2) and six roots of (3) from twelve roots of (1), we get four roots of

$$x^4 + x^2 + 1 = 0$$

Hence, the required roots are

1. $\cos\frac{\pi}{3} + i\sin\frac{\pi}{3}$
2. $\cos\frac{2\pi}{3} + i\sin\frac{2\pi}{3}$
3. $\cos\frac{4\pi}{3} + i\sin\frac{4\pi}{3}$
4. $\cos\frac{5\pi}{3} + i\sin\frac{5\pi}{3}$

i.e., $\frac{1}{2} + i\frac{\sqrt{3}}{2}$, $\quad -\frac{1}{2} + i\frac{\sqrt{3}}{2}$, $\quad -\frac{1}{2} - i\frac{\sqrt{3}}{2}$, $\quad \frac{1}{2} - i\frac{\sqrt{3}}{2}$ **Ans.**

Exercise 2.1

Find the values of :

1. $(1+i)^{1/5}$

Ans. $2^{1/10}\left[\cos\frac{1}{5}\left(2n\pi+\frac{\pi}{4}\right)+i\sin\frac{1}{5}\left(2n\pi+\frac{\pi}{4}\right)\right]$, where n = 0, 1, 2, 3, 4.

2. $(1+\sqrt{-3})^{3/4}$

Ans. $(2)^{3/4}\left[\cos\frac{3}{4}\left(2n\pi+\frac{\pi}{3}\right)+\sin\frac{3}{4}\left(2n\pi+\frac{\pi}{3}\right)\right]$, where n = 0, 1, 2, 3.

3. $(-i)^{1/6}$

Ans. $\cos(4n+1)\frac{\pi}{12}-i\sin\ (4n+1)\frac{\pi}{12}$, where n = 0, 1, 2, 3, 4, 5.

4. $(1+i)^{2/3}$

Ans. $2^{1/3}\left[\cos\left(\frac{4n\pi}{3}+\frac{\pi}{6}\right)+i\sin\left(\frac{4n\pi}{3}+\frac{\pi}{6}\right)\right]$, where n = 0, 1, 2.

5. Solve the equation with the help of De Moivre's theorem $x^{7-1}=0$.

Ans. $\cos\frac{2n\pi}{7}+i\sin\frac{2n\pi}{7}$, where n = 0, 1, 2, 3, 4, 5, 6.

6. Find the roots of the equation $x^3+8=0$.

Ans. $2\left[\cos\left(\frac{2n\pi+\pi}{3}\right)+i\sin\left(\frac{2n\pi+\pi}{3}\right)\right]$, where n = 0, 1, 2.

7. Use D-Moivre's theorem to solve $x^9-x^5+x^4-1=0$.

Ans. $\left[\cos(2n+1)\frac{\pi}{5}+i\sin(2n+1)\frac{\pi}{5}\right]$, where n = 0, 1, 2, 3, 4.

and $\left[\cos\frac{n\pi}{2}+i\sin\frac{n\pi}{2}\right]$, where n = 0, 1, 2, 3.

8. Show that the roots of $(x+1)^6+(x-1)^6=0$ are given by $i\cot\frac{(2n+1)}{12}$, n = 0, 1, 2, 3, 4, 5.

9. Show that all the roots of $(x+1)^7=(x-1)^7$ are given by $\pm i\cot\left(\frac{n\pi}{7}\right)$ where n =1, 2, 3, why r ≠ 0.

10. Solve the equation $x^5=1+i$ and find the continued product of the roots.

2.22 Hyperbolic Functions

(i) $\sin hx = \frac{e^x - e^{-x}}{2}$ (ii) $\cos hx = \frac{e^x + e^{-x}}{2}$

(iii) $\tan hx = \frac{e^x - e^{-x}}{e^x + e^{-x}}$ (iv) $\cot hx = \frac{e^x + e^{-x}}{e^x - e^{-x}}$

(v) $\text{sec hx} = \frac{2}{e^x + e^{-x}}$ (vi) $\text{cosec hx} = \frac{2}{e^x - e^{-x}}$

(vii) $\cos hx + \sin hx = \frac{e^x + e^{-x}}{2} + \frac{e^x - e^{-x}}{2} = e^x$

(viii) $\cos hx - \sin hx = \frac{e^x + e^{-x}}{2} - \frac{e^x - e^{-x}}{2} = e^{-x}$

(ix) $(\cosh x + \sinh x)^n = \cosh nx + \sinh nx$

Relation between circular and Hyperbolic Functions

$\sin ix = i \sinh x$ $\sinh ix = i \sin x$

$\cos ix = \cosh x$ $\cosh ix = \cos x$

$\tan ix = i \tanh x$ $\tanh ix = i \tan x$

Formulae of Hypoerbolic Functions

(a) 1. $\cosh^2 x - \sinh^2 x = 1$ 2. $\text{sech}^2 x = 1 - \tan h^2 x$

3. $\text{cosech}^2 x = \coth^2 x - 1$

(b) 1. $\sinh(x \pm y) = \sinh x \cosh y \pm \cosh x \sinh y$

2. $\cosh(x \pm y) = \cosh x \cosh y \pm \sinh x \sinh y$

3. $\tanh(x \pm y) = \frac{\tanh x \pm \tanh y}{1 \pm \tanh x \tanh y}$

(c) 1. $\sinh 2x = 2 \sinh x \cosh x$ 2. $\cosh 2x = \cosh^2 x + \sinh^2 x$

3. $\cosh 2x = 2\cosh^2 x - 1$ 4. $\cosh 2x = 1 + 2\sinh^2 x$

5. $\tanh 2x = \frac{2 \tanh x}{1 + \tanh^2 x}$

(d) 1. $\sinh x + \sinh y = 2 \sinh \frac{x+y}{2} \cosh \frac{x-y}{2}$

2. $\sinh x - \sinh y = 2 \cosh \frac{x+y}{2} \sinh \frac{x-y}{2}$

3. $\cosh x + \cosh y = 2 \cosh \frac{x+y}{2} \cosh \frac{x-y}{2}$

4. $\cosh x - \cosh y = 2\sinh\frac{x+y}{2}\sinh\frac{x-y}{2}$

Note: For proof, Put $\sinh x = \frac{e^x - e^{-x}}{2}$ and $\cosh x = \frac{e^x + e^{-x}}{2}$.

Real and Imaginary Parts of Circular and Hyperbolic Functions

***Example* 2.21:** Separate the following into real and imaginary parts:

(i) $\sin(x+iy)$ (ii) $\cos(x+iy)$ (iii) $\tan(x+iy)$ (iv) $\cot(x+iy)$

(v) $\sec(x+iy)$ (vi) $\operatorname{cosec}(x+iy)$

Solution:

(i) $\sin(x+iy) = \sin x \cos iy + \cos x \sin(iy) = \sin x \cosh y + i \cos x \sinh y$

(ii) $\cos(x+iy) = \cos x \cos(iy) - \sin x \sin(iy) = \cos x \cosh y - i \sin x \sinh y$

(iii) $$\tan(x+iy) = \frac{\sin(x+iy)}{\cos(x+iy)} = \frac{2\sin(x+iy)\cos(x-iy)}{2\cos(x+iy)\cos(x+iy)}$$

$\because$ $2\sin A.\cos B = \sin(A+B) + \sin(A-B)$

and $2\cos A.\cos B = \cos(A+B) + \cos(A-B)$

$$= \frac{\sin 2x + \sin(2iy)}{\cos 2x + \cos 2iy} = \frac{\sin 2x + i\sinh 2y}{\cos 2x + \cosh 2y}$$

(iv) $$\cot(x+iy) = \frac{\cos(x+iy)}{\sin(x+iy)} = \frac{2\cos(x+iy)\sin(x-iy)}{2\sin(x+iy)\sin(x-iy)}$$

$$= \frac{\sin 2x - \sin(2\,iy)}{\cos(2iy) - \cos 2x} = \frac{\sin 2x - i\sinh 2y}{\cosh 2y - \cos 2x}$$

(v) $$\sec(x+iy) = \frac{1}{\cos(x+iy)} = \frac{2\cos(x-iy)}{2\cos(x+iy)\cos(x-iy)}$$

$$= \frac{2[\cos x \cos(iy) + \sin x.\sin(iy)]}{\cos 2x + \cos(2iy)}$$

$$= \frac{2[\cos x \cosh y + i \sin x \sinh y]}{\cos 2x + \cosh 2y}$$

(vi) $$\operatorname{cosec}(x+iy) = \frac{1}{\sin(x+iy)} = \frac{2\sin(x-iy)}{2\sin(x+iy)\sin(x-iy)}$$

$$= \frac{2[\sin x \cos(iy) - \cos x \sin(iy)]}{\cos(2iy) - \cos 2x}$$

$$= \frac{2[\sin x.\cosh y - i \cos x.\sinh y]}{\cos 2y - \cos 2x}$$

***Example* 2.22:** If $\sin(\alpha + i\beta) = x + iy$, prove that

(a) $\dfrac{x^2}{\cosh^2\beta} + \dfrac{y^2}{\sinh^2\beta} = 1$ (b) $\dfrac{x^2}{\sinh^2\alpha} + \dfrac{y^2}{\cosh^2\alpha} = 1$

***Solution*:**

(a) $x + iy = \sin(\alpha + i\beta) = \sin\alpha\cosh\beta + i\cos\alpha\sinh\beta$

Equating real and imaginary parts

$$x = \sin\alpha\cosh\beta, \qquad y = \cos\alpha\sinh\beta$$

$$\sin\alpha = \frac{x}{\cosh\beta} \quad \text{and} \quad \cos\alpha = \frac{y}{\sinh\beta}$$

Squaring and adding $\sin^2\alpha + \cos^2\alpha = \dfrac{x^2}{\cosh^2\beta} + \dfrac{y^2}{\sinh^2\beta}$

$$1 = \frac{x^2}{\cosh^2\beta} + \frac{y^2}{\sinh^2\beta}$$

(b) Again $\cosh\beta = \dfrac{x}{\sin\alpha}$ and $\sin\beta = \dfrac{y}{\cos\alpha}$

$$\cosh^2\beta - \sinh^2\beta = \frac{x^2}{\sin^2\alpha} - \frac{y^2}{\cos^2\alpha}$$

$$1 = \frac{x^2}{\sin^2\alpha} - \frac{y^2}{\cos^2\alpha}$$

***Example* 2.23:** If $\tan(\theta + i\phi) = \tan\alpha + i\sec\alpha$, Prove that

(a) $e^{2\phi} = \pm\cot\dfrac{\alpha}{2}$ (b) $2\theta = \left(n + \dfrac{1}{2}\right)\pi + \alpha$

***Solution*:** $\tan(\theta + i\phi) = \tan\alpha + i\sec\alpha$

(a) $$\tan 2i\phi = \frac{\tan(\theta + i\phi) - \tan(\theta - i\phi)}{1 + \tan(\theta + i\phi)\tan(\theta - i\phi)}$$

$$i\tanh 2\phi = \frac{(\tan\alpha + i\sec\alpha) - (\tan\alpha - i\sec\alpha)}{1 + (\tan\alpha + i\sec\alpha)(\tan\alpha - i\sec\alpha)}$$

$$= \frac{2i\sec\alpha}{1+(\tan^2\alpha+\sec^2\alpha)} = \frac{2i\sec\alpha}{1+(\sec^2\alpha-1+\sec^2\alpha)}$$

$$i\frac{e^{2\phi}-e^{-2\phi}}{e^{2\phi}+e^{-2\phi}} = \frac{2i\sec\alpha}{2\sec^2\alpha} = i\cos\alpha$$

$$\frac{e^{2\phi}-e^{-2\phi}}{e^{2\phi}+e^{-2\phi}} = \cos\alpha \text{ or } \frac{e^{4\phi}-1}{e^{4\phi}+1} = \cos\alpha$$

or $$e^{4\phi} = \frac{1+\cos\alpha}{1-\cos\alpha} = \frac{1+2\cos^2\frac{\alpha}{2}-1}{1-\left(1-2\sin^2\frac{\alpha}{2}\right)}$$

or $$e^{4\phi} = \cot^2\frac{\alpha}{2} \quad \text{or} \quad e^{2\phi} = \pm\cot\frac{\alpha}{2}$$

$\therefore$ $$\tan(\theta-i\phi) = \tan\alpha - i\sec\alpha$$

(b) $$\tan 2\theta = \frac{\tan(\theta+i\phi)+\tan(\theta-i\phi)}{1-\tan(\theta+i\phi)\tan(\theta-i\phi)} = \frac{(\tan\alpha+i\sec\alpha)+(\tan\alpha-i\sec\alpha)}{1-(\tan\alpha+i\sec\alpha)(\tan\alpha-i\sec\alpha)}$$

$$= \frac{2\tan\alpha}{1-(\tan^2\alpha+\sec^2\alpha)} = \frac{2\tan\alpha}{1-(\tan^2\alpha+1+\tan^2\alpha)}$$

$$= \frac{2\tan\alpha}{-2\tan^2\alpha} = -\cot\alpha = \tan\left(\frac{\pi}{2}+\alpha\right)$$

$$2\theta = n\pi + \left(\frac{\pi}{2}+\alpha\right) = \left(n+\frac{1}{2}\right)\pi + \alpha$$ **Proved.**

Example 2.24: Separate the following into real and imaginary parts:

(a) $\sinh(x+iy)$ (b) $\cosh(x+iy)$ (c) $\tanh(x+iy)$

Solution:

(a) $\sinh(x+iy) = \sinh x\cosh(iy) + \cosh x\sinh(iy)$

$= \sinh x\cos y + i\sin y\cosh x$ **Ans.**

(b) $\cosh(x+iy)$

$= \cosh x\cosh(iy) - \sinh x\sinh iy = \cosh x\cos y - i\sinh x\sin y$ **Ans.**

(c) $$\tan(x+iy) = \frac{\sinh(x+iy)}{\cosh(x+iy)} = \frac{-i\sin i(x+iy)}{\cos i(x+iy)} = \frac{-i\sin(ix-y)}{\cos(ix-y)}$$

$$= \frac{-i2\sin(ix-y).\cos(ix+y)}{2\cos(ix-y)\cos(ix+y)}$$

$$= -i\frac{\sin 2ix - \sin 2y}{\cos 2ix + \cos 2y} = -i\frac{i\sinh 2x - \sin 2y}{\cosh 2x + \cos 2y}$$

$$= \frac{\sinh 2x + i\sin 2y}{\cosh 2x + \cos 2y}$$

$$= \frac{\sin 2x}{\cosh 2x + \cos 2y} + i\frac{\sin 2y}{\cosh 2x + \cos 2y}$$ **Ans.**

Inverse Hyporbolic Functions

***Example* 2.25:** Prove that $\sinh^{-1} x = \log\left(x + \sqrt{x^2+1}\right)$.

Solution: Let $\sinh^{-1} x = y \Rightarrow x - \sinh y$

$$x = \frac{e^y - e^{-y}}{2} \text{ or } e^y - e^{-y} = 2x \text{ or } e^{2y} - 2x\,e^y - 1 = 0$$

This is quadratic in e^y.

$$e^y = \frac{2x \pm \sqrt{4x^2+4}}{2} = x \pm \sqrt{x^2+1}$$

$$y = \log(x + \sqrt{x^2+1})$$ (Taking positive sign only)

or $$\sinh^{-1} x = \log(x + \sqrt{x^2+1})$$ **Proved.**

***Example* 2.26:** Prove that $\cosh^{-1} x = \log(x + \sqrt{x^2-1})$

***Solution*:** Let $y = \cosh^{-1} x \Rightarrow x = \cosh y$

$$x = \frac{e^y + e^{-y}}{2} \text{ or } 2x = e^y + e^{-y}$$

$$e^{2y} - 2xe^y + 1 = 0$$ (This is quadratic in e^y)

$$e^y = \frac{2x \pm \sqrt{4x^2-4}}{2} = x \pm \sqrt{x^2-1}$$

$$y = \log(x + \sqrt{x^2-1})$$ (Taking positive sign only)

$$\cosh^{-1} x = \log(x + \sqrt{x^2-1})$$ **Proved.**

***Example* 2.27:** Prove that $\tanh^{-1} x = \frac{1}{2}\log\frac{1+x}{1-x}$

***Solution*:** Let $\tanh^{-1} x = y \Rightarrow x = \tanh y$

$$x = \frac{e^y - e^{-y}}{e^y + e^y}$$

Applying componendo and dividenodo, we obtain

$$\frac{1+x}{1-x} = \frac{e^y}{e^{-y}} = e^{2y}$$

$$2y = \log\frac{1+x}{1-x} \quad \text{or} \quad \tanh^{-1} x = \frac{1}{2}\log\frac{1+x}{1-x}$$ **Proved.**

Similarly, $$\coth^{-1} x = \frac{1}{2}\log\frac{x+1}{x-1}$$

***Example* 2.28:** Prove that $\operatorname{sech}^{-1}x = \log\frac{1+\sqrt{1-x^2}}{x}$

***Solution*:** Let $$y = \operatorname{sech}^{-1}x \Rightarrow x = \operatorname{sech} y$$

$$x = \frac{2}{e^y + e^{-y}} \quad \text{or} \quad x = \frac{2e^y}{e^{2y}+1}$$

$$\Rightarrow \quad xe^{2y} - 2e^y + x = 0 \Rightarrow \quad e^y = \frac{2 \pm \sqrt{4-4x^2}}{2x} = \frac{1 \pm \sqrt{1-x^2}}{x}$$

We take only positive sign only

$$e^y = \frac{1+\sqrt{1-x^2}}{x}$$

$$y = \log\frac{1+\sqrt{1-x^2}}{x}$$

$$\operatorname{sech}^{-1}x = \log\frac{1+\sqrt{1-x^2}}{x}$$

Similarly, $$\operatorname{cosech}^{-1}x = \log\frac{1+\sqrt{1+x^2}}{x}$$

***Example* 2.29:** Separate $\tan^{-1}(a+ib)$ into real and imaginary parts.

***Solution*:** Let $$\tan^{-1}(a+ib) = x+iy$$

$$\therefore \quad \tan(x+iy) = a+ib \qquad \text{.....(1)}$$

On both sides for i write – i $\quad \therefore \tan(x - iy) = a - ib$

Now $$\tan 2x = \tan[(x+iy)+(x-iy)] = \frac{\tan(x+iy)+\tan(x-iy)}{1-\tan(x+iy)\tan(x-iy)}$$

$$= \frac{a+ib+a-ib}{1-(a+ib)(a-ib)} = \frac{2a}{1-a^2-b^2}$$

$$2x = \tan^{-1}\left[\frac{2a}{1-a^2-b^2}\right] \quad \therefore \quad x = \frac{1}{2}\tan^{-1}\left[\frac{2a}{1-a^2-b^2}\right] \qquad .....(2)$$

and $$\tan(2yi) = \tan[(x+iy)-(x-iy)]$$

$$= \frac{\tan(x+iy)-\tan(x-iy)}{1+\tan(x+iy)\tan(x-iy)} = \frac{a+bi-a+bi}{1+(a+bi)(a-bi)}$$

$$i\tanh 2y = \frac{2bi}{1+a^2+b^2} \quad \text{so} \quad \tanh^2 y = \frac{2b}{1+a^2+b^2}$$

$$2y = \tanh^{-1}\left[\frac{2b}{1+a^2+b^2}\right]$$

so $$y = \frac{1}{2}\tanh^{-1}\left[\frac{2b}{1+a^2+b^2}\right] \qquad .....(3)$$

From (1), (2) and (3), we have

$$\tan^{-1}(a+ib) = \frac{1}{2}\tan^{-1}\left[\frac{2b}{1-a^2-b^2}\right] + \frac{i}{2}\tanh^{-1}\left[\frac{2b}{1+a^2+b^2}\right] \quad \textbf{Ans}$$

***Example* 2.30:** If $\sinh(\theta + i\phi) = \cos\alpha + i\sin\alpha$, prove that $\sinh^4\theta = \cos^2\alpha = \cos^4\phi$

***Solution*:** $$\sinh(\theta + i\phi) = \cos\alpha + i\sin\alpha$$

$$\sinh\theta\cos\phi + i\sin\phi\cosh\theta = \cos\alpha + i\sin\alpha$$

Equating real and imaginary parts, we have

$$\sinh\theta\cos\phi = \cos\alpha \qquad .....(1)$$

and $$\sin\phi\cosh\theta = \sin\alpha \qquad .....(2)$$

Let us eliminate ϕ from eq. (1) and (2).

$$\cos\phi = \frac{\cos\alpha}{\sinh\theta} \quad \text{and} \quad \sin\phi = \frac{\sin\alpha}{\cosh\theta}$$

Squaring and adding, we get

$$1=\frac{\cos^2\alpha}{\sinh^2\theta}+\frac{\sin^2\alpha}{\cosh^2\theta} \quad \text{or} \quad \frac{\cos^2\alpha}{\sinh^2\alpha}=1-\frac{\sin^2\alpha}{\cosh^2\theta}$$

or $$\frac{\cos^2\alpha}{\sinh^2\theta}=1-\frac{1-\cos^2\alpha}{1+\sinh^2\theta}=\frac{1+\sinh^2\theta-1+\cos^2\alpha}{1+\sinh^2\theta}$$

$$\frac{\cos^2\alpha}{\sinh^2\theta}=\frac{\sinh^2\theta+\cos^2\alpha}{1+\sinh^2\theta}$$

$$\sinh^4\theta+\sinh^2\theta\cos^2\alpha=\cos^2\alpha+\cos^2\alpha\sinh^2\theta$$

or $$\sinh^4\theta=\cos^2\alpha$$ **Proved.**

For second result, let us eliminate θ.

$$\sinh\theta=\frac{\cos\alpha}{\cos\phi} \quad \text{and} \quad \cosh\theta=\frac{\sin\alpha}{\sin\phi}$$

$$\cosh^2\theta-\sinh^2\theta=\frac{\sin^2\alpha}{\sin^2\phi}-\frac{\cos^2\alpha}{\cos^2\phi} \quad \text{or} \quad 1=\frac{1-\cos^2\alpha}{1-\cos^2\phi}-\frac{\cos^2\alpha}{\cos^2\phi}$$

or $$\frac{\cos^2\alpha}{\cos^2\phi}=\frac{1-\cos^2\alpha-1+\cos^2\phi}{1-\cos^2\phi}$$

$$\frac{\cos^2\alpha}{\cos^2\phi}=\frac{\cos^2\phi-\cos^2\alpha}{1-\cos^2\phi}$$

or $$\cos^4\phi-\cos^2\phi\cos^2\alpha=\cos^2\alpha-\cos^2\alpha\cos^2\phi$$

or $$\cos^4\phi=\cos^2\alpha$$ **Proved.**

***Example* 2.31:** If $\sin(\theta+i\phi)=\tan\alpha+i\sec\alpha$ show that $\cos 2\theta\cosh 2\phi=3$

***Solution*:** $$\sin(\theta+i\phi)=\tan\alpha+i\sec\alpha$$

Equating real and imaginary parts, we get

$$\sin\theta\cosh\phi=\tan\alpha \quad \text{.....(1)}$$

$$\cos\theta\sinh\phi=\sec\alpha \quad \text{.....(2)}$$

We know that

$$\sec^2\alpha-\tan^2\alpha=1$$

$$\cos^2\theta\sinh^2\phi-\sin^2\theta\cosh^2\phi=1 \quad \text{[From (1) and (2)]}$$

$$\left(\frac{1+\cos 2\theta}{2}\right)\left(\frac{\cosh 2\phi-1}{2}\right)-\left(\frac{1-\cos 2\theta}{2}\right)\left(\frac{\cos 2\phi+1}{2}\right)=1$$

$$[-1+\cosh 2\phi-\cos 2\theta+\cos 2\theta\cosh 2\phi]-$$

$$[\cosh 2\phi+1-\cos 2\theta\cosh 2\phi-\cos 2\theta]=4$$

$$-2+2\cos 2\theta\cosh 2\phi=4\Rightarrow 2\cos 2\theta\cosh 2\phi=6\Rightarrow \cos 2\theta\cosh 2\phi=3$$

Proved.

***Examle* 2.32:** If $x=2\sin\alpha\cosh\beta$, $y=2\cosh\alpha\sinh\beta$, show that

(i) $\operatorname{cosec}(\alpha-i\beta)+\operatorname{cosec}(\alpha+i\beta)=\dfrac{4x}{x^2+y^2}$

(ii) $\operatorname{cosec}(\alpha-i\beta)-\operatorname{cosec}(\alpha+i\beta)=\dfrac{i4y}{x^2+y^2}$

Solution: $\operatorname{cosec}(\alpha+i\beta)=\dfrac{1}{\sin(\alpha+i\beta)}=\dfrac{1}{\sin\alpha\cos i\beta+\cos\alpha\sin i\beta}$

$$=\frac{1}{\sin\alpha\cosh\beta+i\cos\alpha\sinh\beta}=\frac{1}{\frac{x}{2}+i\frac{y}{2}}=\frac{2}{x+iy}$$

Similarly, $\cos(\alpha-i\beta)=\dfrac{2}{x-iy}$

$$\operatorname{cosec}(\alpha-i\beta)+\operatorname{cosec}(\alpha+i\beta)=\frac{2}{x-iy}+\frac{2}{x+iy}=\frac{4x}{x^2+y^2}$$

$$\operatorname{cosec}(\alpha-i\beta)-\operatorname{cosec}(\alpha+i\beta)=\frac{2}{x-iy}-\frac{2}{x+iy}=\frac{4iy}{x^2+y^2}$$ **Proved.**

***Example* 2.33:** If $\cos(\alpha+i\beta)\cos(\gamma+i\delta)=1$, prove that $\tanh^2\beta\cosh^2\delta=\sin^2\gamma$.

Solution: $\cos(\alpha+i\beta)\cos(\gamma+i\delta)=1$

or $\cos(\gamma+i\delta)=\sec(\alpha+i\beta)$ .....(1)

$$\sin(\gamma+i\delta)=\sqrt{1-\cos^2(\gamma+i\delta)}\sqrt{1-\sec^2(\alpha+i\beta)}$$

$$=\sqrt{-\tan^2(\alpha+i\beta)}$$

$$=i\tan(\alpha+i\beta) \quad .....(2)$$

Dividing eq. (2) by (1), we get

$$\frac{\sin(\gamma+i\delta)}{\cos(\gamma+i\delta)}=\frac{i\tan(\alpha+i\beta)}{\sec(\alpha+i\beta)} \text{ or } \tan(\gamma+i\delta)=i\sin(\alpha+i\beta)$$

$$=i[\sin\alpha\cos i\beta+\cos\alpha\sin i\beta]$$

$$=i[\sin\alpha\cos\beta+i\cos\alpha\sin\beta]$$

$$=i\sin\alpha\cos\beta-\cos\alpha\sinh\beta \qquad .....(3)$$

$$\therefore \quad \tan(\gamma-i\delta)=-i\sin\alpha\cosh\beta-\cos\alpha\sinh\beta \qquad .....(4)$$

$$\tan 2\,i\delta=\tan[(\gamma+i\delta)-(\gamma-i\delta)]$$

or
$$i\tanh 2\delta=\frac{\tan(\gamma+i\delta)-\tan(\gamma-i\delta)}{1+\tan(\gamma+i\delta)\tan(\gamma-i\delta)}$$

$$i\frac{2\tanh\delta}{1+\tan^2\delta}=\frac{i\sin\alpha\cosh\beta-\cos\alpha\sinh\beta+i\sin\alpha\cosh\beta+\cos\alpha\sinh\beta}{1+(i\sin\alpha\cosh\beta-\cos\alpha\sinh\beta)(-i\sin\alpha\cosh\beta-\cos\alpha\sinh\beta)}$$

[From (3) and (4)]

$$=\frac{2i\sin\alpha\cosh\beta}{1+(\sin^2\alpha\cosh^2\beta+\cos^2\alpha\sinh^2\beta)}$$

$$=\frac{2i\sin\alpha\cosh\alpha}{1+\sin^2\alpha\cosh^2\beta+(1-\sin^2\alpha)(\cosh^2\beta-1)}$$

$$=\frac{2i\sin\alpha\cosh\beta}{1+\sin^2\alpha\cosh^2\beta+\cosh^2\beta-1-\sin^2\alpha\cosh^2\beta+\sin^2\alpha}$$

$$=\frac{2i\sin\alpha\cosh\beta}{\sin^2\alpha+\cosh^2\beta}$$

$$\frac{\tanh\delta}{1+\tanh^2\delta}=\frac{\sin\alpha\cosh\beta}{\sin^2\alpha+\cosh^2\beta}$$

$$\frac{1+\tanh^2\delta}{2\tanh\delta}=\frac{\sin^2\alpha+\cosh^2\beta}{2\sin\alpha\cosh\beta}$$

By componendo and divideno

$$\frac{1+\tanh^2\delta+2\tanh\delta}{1+\tanh^2\delta-2\tanh\delta}=\frac{\sin^2\alpha+\cosh^2\beta+2\sin\alpha\cosh\beta}{\sin^2\alpha+\cosh^2\beta-2\sin\alpha\cosh\beta}$$

$$\left(\frac{1+\tanh\delta}{1-\tanh\delta}\right)^2=\left(\frac{\sin\alpha+\cosh\beta}{\cosh\beta-\sin\alpha}\right)^2$$

$$\frac{1+\tanh\delta}{1-\tanh\delta}=\frac{\sin\alpha+\cosh\alpha}{\cosh\beta-\sin\alpha}$$

By componendo and dividendo

$$\frac{2}{2\tanh\delta}=\frac{2\cosh\beta}{2\sin\alpha} \quad \text{or} \quad \tanh\delta=\frac{\sin\alpha}{\cosh\beta}$$

$$\tanh^2\delta\cosh^2\beta=\sin^2\alpha$$

***Example* 2.34:** If $\tan(x+iy)=\sin(u+iv)$, prove that $\frac{\sin 2x}{\sinh 2y}=\frac{\tan u}{\tanh v}$.

***Solution*:** Now $\tan(x+iy)=\sin(u+iv)$

Separating the real and imaginary parts of both sides, we have

$$\frac{\sin 2x}{\cos 2x+\cosh 2y}+\frac{i\sinh 2y}{\cos 2x+\cosh 2y}=\sin u\cosh v+i\cos u\sinh v$$

Equating real and imaginary parts, we get

$$\frac{\sin 2x}{\cos 2x+\cosh 2y}=\sin u\cosh v \qquad \text{.....(1)}$$

and

$$\frac{\sin 2y}{\cos 2x+\cosh 2y}=\cos u\sinh v \qquad \text{.....(2)}$$

Dividing (1) by (2), we obtain

$$\frac{\sin 2x}{\sinh 2y}=\frac{\sin u\cosh v}{\cos u\sinh v}$$

or

$$\frac{\sin 2x}{\sinh 2y}=\frac{\tan u}{\tanh v}$$

Proved.

***Example* 2.35:** If $\tan(x+iy)=i$, where x and y are real, prove that x is indeterminate and y is infinite.

***Solution*:** $\tan(x+iy)=i \Rightarrow \tan(x-iy)=-i$

$$\tan 2x=\tan(\overline{x+iy}+\overline{x-iy})=\frac{\tan(x+iy)+\tan(x-iy)}{1-\tan(x+iy)\tan(x-iy)}$$

$$=\frac{i-i}{1-i(-i)}=\frac{i-i}{1-1}=\frac{0}{0}$$

which is indeterminate.

Also, $$\tan 2iy = \tan[(x+iy)-(x-iy)] = \frac{\tan(x+iy)-\tan(x-iy)}{1+\tan(x+iy)\tan(x-iy)}$$

$$= \frac{i-(-i)}{1+i(-i)} = \frac{2i}{1+1} = i$$

$$i\tanh 2y = i \Rightarrow \tanh 2y = 1 \text{ or } 2y = \tanh^{-1}(1) = \frac{1}{2}\log\frac{1+1}{1-1} = \infty$$

$\therefore$ y is infinite. **Proved.**

Exercise 2.2

1. Separate the real and imaginary parts of (a) sech (x + iy) (b) $\tanh^{-1}$ (x + iy).

Ans. (a) $\dfrac{2a\cosh x\cos y}{\cosh 2x+\cos 2\beta} - i\dfrac{2\sinh x\sin y}{\cosh 2x+\cos 2\beta}$

(b) $\dfrac{1}{2}\tanh^{-1}\dfrac{2x}{1+x^2+y^2} + \dfrac{i}{2}\tan^{-1}\dfrac{2y}{1-x^2-y^2}$

2. If $\cosh(\alpha+\beta) = x+iy$, prove that

(a) $\dfrac{x^2}{\cosh^2\alpha} + \dfrac{y^2}{\sinh^2\alpha} = 1$ (b) $\dfrac{x^2}{\cos^2\beta} - \dfrac{y^2}{\sinh^2\beta} = 1$

3. Prove that $\sin^{-1}(\operatorname{cosec}\theta) = \dfrac{\pi}{2} + i\log\cot\dfrac{\theta}{2}$.

4. If $\tan(\theta+i\phi) = e^{i\alpha}$, show that

(a) $\theta = \left(n+\dfrac{1}{2}\right)\dfrac{\pi}{2}$ (b) $\phi = \dfrac{1}{2}\log\tan\left(\dfrac{\pi}{2}+\dfrac{\alpha}{2}\right)$

5. If $\tan(\alpha+i\beta) = \sin(x+iy)$, the prove that

(a) $x^2+y^2+2x\cot 2\alpha = 1$ (b) $x^2+y^2-2y\coth 2\beta = -1$

6. If $\tan(\theta+i\phi) = \sin(x+iy)$, then prove that coth y sinh 2ϕ = cot x sin 2θ.

7. If $\sin^{-1}(\cos\theta + i\sin\theta) = x+iy$, show that

(a) $x = \cos^{-1}\sqrt{\sin\theta}$ (b) $y = \log[\sqrt{\sin\theta}+\sqrt{1+\sin\theta}]$

8. Prove that $\tanh^{-1} x = \sinh^{-1}\dfrac{x}{\sqrt{1-x^2}}$

9. Prove that $\tan^{-1} i\dfrac{x-a}{x+a} = -\dfrac{i}{2}\log\dfrac{a}{x}$

10. If $\cos(\theta + i\phi) = R(\cos\alpha + i\sin\alpha)$, prove that $\phi = \dfrac{1}{2}\log_e\left[\dfrac{\sin(\theta-\alpha)}{\sin(\theta+\alpha)}\right]$

11. If $\tan(\theta + i\phi) = \tan\alpha + i\sec\alpha$, show that $2\theta = n\pi + \dfrac{\pi}{2} + \alpha$, $e^{2\phi} = \pm\cot\dfrac{\alpha}{2}$

12. If $\cos(\alpha + i\beta)\cos(\gamma + i\delta) = 1$, prove that

(i) $\tanh^2\delta\cosh^2\beta = \sin^2\alpha$ (ii) $\tanh^2\beta\,\cosh^2\delta = \sin^2\gamma$

13. If $\dfrac{u-1}{u+1} = \sin(x+iy)$, find u. **Ans.** $\tan^{-1}\dfrac{2\cos x\sinh y}{\cos^2 x - \sinh^2 y}$

14. If $x + iy = c\cot(u + iy)$, show that

$$\frac{x}{\sin 2u} = \frac{-y}{\sinh 2v} = \frac{c}{\cosh 2v - \cos 2u}$$

15. If $\tan y = \tan\alpha\tanh\beta$ and $\tan z = \cot\alpha\tanh\beta$, prove that $\tan\ (y+z) = \sinh 2\beta\,\mathrm{cosec}\,2\alpha$

16. If $\mathrm{cosec}\left(\dfrac{\pi}{4} + ix\right) = u + v$, show that $(u^2 + v^2) = 2(u^2 - v^2)$

17. If $x + iy = 2\cosh\left(\alpha + i\dfrac{\pi}{4}\right)$, prove that $x^2 - y^2 = 2$

18. If $a + ib = \tanh\left(\alpha + i\dfrac{\pi}{u}\right)$, prove that $a^2 + b^2 = 1$

19. Separate into real and imaginary parts $\sin^{-1}(e^{i\theta})$

Ans. $\cos^{-1}\sqrt{\sin\theta} + i\log[\sqrt{\sin\theta} + \sqrt{1+\sin\theta}]$

***Example* 2.36:** Separate $\tan^{-1}(\cos\theta + i\sin\theta)$ into real and imaginary parts.

Solution: Let $\tan^{-1}(\cos\theta + i\sin\theta) = x + iy$

$$\cos\theta + i\sin\theta = \tan(x+iy)$$

$$\cos\theta - i\sin\theta = \tan(x-iy)$$

$$\tan 2x = \tan[(x+iy)+(x-iy)] = \frac{\tan(x+iy)+\tan(x-iy)}{1-\tan(x+iy)\tan(x-iy)}$$

$$= \frac{(\cos\theta + i\sin\theta)+(\cos\theta - i\sin\theta)}{1-(\cos\theta + i\sin\theta)(\cos\theta - i\sin\theta)} = \frac{2\cos\theta}{1-(\cos^2\theta+\sin^2\theta)}$$

$$= \frac{2\cos\theta}{1-1} = \frac{2\cos\theta}{0} = \infty = \tan\frac{\pi}{2}$$

$$\tan 2x = \tan\left(n\pi + \frac{\pi}{2}\right)$$

$$2x = n\pi + \frac{\pi}{2} \quad \text{or} \quad x = \frac{n\pi}{2} + \frac{\pi}{4}$$

$$\tan 2iy = \tan[(x+iy)-(x-iy)] = \frac{\tan(x+iy)-\tan(x-iy)}{1+\tan(x+iy)\tan(x-iy)}$$

$$= \frac{(\cos\theta + i\sin\theta)-(\cos\theta - i\sin\theta)}{1+(\cos\theta + i\sin\theta)(\cos\theta - i\sin\theta)}$$

$$= \frac{2i\sin\theta}{1+(\cos^2\theta+\sin^2\theta)} = \frac{2i\sin\theta}{1+1} = i\sin\theta$$

$$i\tanh 2y = i\sin\theta$$

$$\frac{e^{2y}-e^{-2y}}{e^{2y}+e^{-2y}} = \frac{\sin\theta}{1}$$

By componendo and dividend, we have

$$\frac{2e^{2y}}{2e^{-2y}} = \frac{1+\sin\theta}{1-\sin\theta}$$

$$e^{4y} = \frac{1+\cos\left(\frac{\pi}{2}-\theta\right)}{1-\cos\left(\frac{\pi}{2}-\theta\right)} = \frac{1+2\cos^2\left(\frac{\pi}{4}-\frac{\theta}{2}\right)-1}{1-\phi\left[1-2\sin^2\left(\frac{\pi}{4}-\frac{\theta}{2}\right)\right]} = \frac{\cos^2\left(\frac{\pi}{4}-\frac{\theta}{2}\right)}{\sin^2\left(\frac{\pi}{4}-\frac{\theta}{2}\right)} = \cot^2\left(\frac{\pi}{4}-\frac{\theta}{2}\right)$$

$$e^{2y} = \cot\left(\frac{\pi}{4} - \frac{\theta}{2}\right)$$

$$2y = \log\cot\left(\frac{\pi}{4} - \frac{\theta}{2}\right) \quad \text{or} \quad y = \frac{1}{2}\log\cot\left(\frac{\pi}{4} - \frac{\theta}{2}\right)$$

Imaginary part $= \frac{1}{2}\log\cot\left(\frac{\pi}{4} - \frac{\theta}{2}\right)$

Real part $= \frac{n\pi}{2} + \frac{\pi}{4}$

$$\tan^{-1}(\cos\theta + i\sin\theta) = \frac{n\pi}{2} + \frac{\pi}{4} + \frac{i}{2}\log\cot\left(\frac{\pi}{4} - \frac{\theta}{2}\right)$$ **Ans.**

***Example* 2.37**: Separate $\sin^{-1}(\alpha + i\beta)$ into real and imaginary parts.

Solution: Let $\sin^{-1}(\alpha + i\beta) = x + iy$

$$\alpha + i\beta = \sin(x + iy)$$

or $$\alpha + i\beta = \sin x \cos iy + \cos x \sin iy$$

$$= \sin x \cosh y + i\cos x \sinh y$$

Equating real and imaginary parts, we have

$$\alpha = \sin x \cosh y \quad \text{and} \quad \beta = \cos x \sinh y$$

We know that $\cosh^2 y - \sinh^2 y = 1$

$$\left(\frac{\alpha}{\sin x}\right)^2 - \left(\frac{\beta}{\cos x}\right)^2 = 1$$

$\Rightarrow$ $$\alpha^2\cos^2 x - \beta^2\sin^2 x = \sin^2 x\cos^2 x$$

$$\alpha^2(1 - \sin^2 x) - \beta^2\sin^2 x = \sin^2 x(1 - \sin^2 x)$$

$$\sin^4 x - (\alpha^2 + \beta^2 + 1)\sin^2 x + \alpha^2 = 0$$

This is quadratic equation in $\sin^2 x$

$$\sin^2 x = \frac{(\alpha^2 + \beta^2 + 1) \pm \sqrt{(\alpha^2 + \beta^2 + 1)^2 - 4\alpha^2}}{2}$$

$$\sin x = \sqrt{\frac{1}{2}[(\alpha^2 + \beta^2 + 1) \pm \sqrt{(\alpha^2 + \beta^2 + 1)^2 + 4\alpha^2}]}$$

$$x = \sin^{-1}\sqrt{\frac{1}{2}[(\alpha^2+\beta^2+1) \pm \sqrt{(a^2+\beta^2+1)^2 - 4\alpha^2}]}$$

We know that $\sin^2 x + \cos^2 x = 1$

$$\left(\frac{\alpha}{\cosh\, y}\right)^2 + \left(\frac{\beta}{\sinh\, y}\right)^2 = 1$$

$$\alpha^2 \sinh^2 y + \beta^2 \cosh^2 y = \sinh^2 y \cosh^2 y$$

$$\alpha^2 \sinh^2 y + \beta^2 (1+\sinh^2 y) = \sinh^2 y(1+\sinh^2 y)$$

$$\sinh^4 y - (\alpha^2+\beta^2-1)\sinh^2 y - \beta^2 = 0$$

This is quadratic equation in $\sinh^2 y$.

$$\sinh^2 y = \frac{(\alpha^2+\beta^2-1) \pm \sqrt{(\alpha^2+\beta^2-1)^2 + 4\beta^2}}{2}$$

$$\sinh y = \sqrt{\frac{1}{2}[(\alpha^2+\beta^2-1) \pm \sqrt{(\alpha^2+\beta^2-1)^2 + 4\beta^2}]}$$

$$y = \sinh^{-1}\sqrt{\frac{1}{2}[(\alpha^2+\beta^2-1) \pm \sqrt{(a^2+\beta^2-1)^2 + 4\beta^2}]}$$

Real part $= \sin^{-1}\sqrt{\frac{1}{2}[(\alpha^2+\beta^2+1) \pm \sqrt{(a^2+\beta^2+1)^2 - 4\alpha^2}]}$

Imaginary part $= \sinh^{-1}\sqrt{\frac{1}{2}[(\alpha^2+\beta^2-1) \pm \sqrt{(\alpha^2+\beta^2-1)^2 + 4\beta^2}]}$ **Ans.**

***Example* 2.38:** If $\cos^{-1}(u+iv) = \alpha + i\beta$, prove that $\cos^2\alpha$ and $\cosh^2\beta$ are the roots of the equation.

$$x^2 = (1+u^2+v^2)x + u^2 = 0$$

***Solution*:** $\cos^{-1}(u+iv) = \alpha + i\beta$

$$u + iv = \cos(\alpha+i\beta) = \cos\alpha\cos i\beta - \sin\alpha\sin i\beta$$

$$= \cos\alpha\cosh\beta - i\sin\alpha\sinh\beta$$

Equating real and imaginary parts, we have

$$u = \cos\alpha\cosh\beta \qquad \text{.....(1)}$$

$$v = -\sin\alpha \sinh\beta$$

$$u^2 + v^2 = \cos^2\alpha \cosh^2\beta + \sin^2\alpha \sinh^2\beta$$

$$1 + u^2 + v^2 = 1 + \cos^2\alpha \cosh^2\beta + \sin^2\alpha \sinh^2\beta$$

$$= 1 + \cos^2\alpha \cosh^2\beta + (1 - \cos^2\alpha)(\cosh^2\beta - 1)$$

$$= 1 + \cos^2\alpha \cosh^2\beta + \cos^2\alpha + \cosh^2\beta - 1 - \cos^2\alpha \cosh^2\beta$$

$$= \cos^2\alpha + \cosh^2\beta \qquad \text{.....(2)}$$

The equation whose roots are $\cos^2\alpha$ and $\cos^2\beta$ is

$$x^2 - (\cos^2\alpha + \cosh^2\beta)x + \cos^2\alpha \cosh^2\beta = 0$$

or $\quad x^2 - (1 + u^2 + v^2)x + u^2 = 0 \quad$ From (1) and (2) **Proved.**

Exercise 2.3

Prove that

1. $\tanh^{-1}(\sin\theta) = \cosh^{-1}(\sec\theta)$

2. $\tan^{-1}(e^{i\theta}) = \dfrac{n\pi}{2} + \dfrac{\pi}{4} = \log\tan\left(\dfrac{\pi}{4} - \dfrac{\theta}{2}\right)$

3. $\cosh^{-1}\left(\dfrac{b + a\cos x}{a + b\cos x}\right) = \log\left[\dfrac{\sqrt{b+a} + \sqrt{b-a}\tan\dfrac{x}{2}}{\sqrt{b+a} - \sqrt{b-a}\tan\dfrac{x}{2}}\right]$

4. $\tan^{-1}\left(\dfrac{\tan 2\theta + \tan 2\phi}{\tan 2\theta - \tan 2\phi}\right) + \tan^{-1}\left(\dfrac{\tan\theta - \tan\phi}{\tan\theta + \tan\phi}\right) = \tan^{-1}(\cot\theta \coth\phi)$

5. If $\cosh x = \sec\theta$ prove that $\theta = \dfrac{\pi}{2} - 2\tan^{-1}(e^{-x})$ and $\tanh\dfrac{x}{2} = \pm\tan\theta/2$

6. If $u = \log\tan\left(\dfrac{\pi}{4} + \dfrac{x}{2}\right)$, prove that $\tanh\dfrac{u}{2} = \tan\dfrac{x}{2}$ and $\cosh u = \sec x$.

7. Prove that if $(1 + i\tan\alpha)^{1 + i\tan\beta}$ can have real values, one of them is $(\sec\alpha)^{\sec^2\beta}$.

8. If $\dfrac{(1+i)^{x+iy}}{(1-i)^{x-iy}} = \alpha + i\beta$, prove that the value of $\tan^{-1}\dfrac{\beta}{\alpha}$ is $\dfrac{\pi x}{2} + y\log 2$.

2.23 Gregory's Series

Introduction

In this section an infinite series will be introduced known as Gregory's series. Its utility is to express the value of π as the sum of an infinite series and to compute the sum of an infinite series.

In this context and in the summation of series in the next section it is essential to express e^{x+iy} and log (x + iy) as a complex quantity (A + iB). At the same time the knowledge of the expansion of log(1 + z), where z is a complex quantity, is expected from the readers.

Exponential series

If x be real then it can be proved with the help of Algebra that

$$e^x = 1 + \frac{x}{1!} + \frac{x^2}{2!} + \frac{x^3}{3!} + \ldots \text{ to } \infty$$

and $$e = 1 + \frac{1}{1!} + \frac{1}{2!} + \frac{1}{3!} + \ldots \text{ to } \infty$$

If x be a complex number that is a number of the form a + ib then the following series is called an exponential series:

$$1 + \frac{x}{1!} + \frac{x^2}{2!} + \frac{x^3}{3!} + \ldots + \frac{x^n}{n!} + \ldots \text{ to } \infty$$

Symbolically it is denoted by e^x. But in the case e does not refer to the series $1 + \frac{1}{1!} + \frac{1}{2!} + \frac{1}{3!} + \ldots$ to ∞. e^x is called the exponential function.

Product of two exponential functions

If x and y be two complex numbers then $e^x . e^y = e^{x+y}$

The reason behind it is that

$$e^x = 1 + \frac{x}{1!} + \frac{x^2}{2!} + \frac{x^3}{3!} + \ldots + \frac{x^{n-1}}{(n-1)!} + \frac{x^n}{n!} + \ldots \text{ to } \infty$$

and $$e^y = 1 + \frac{y}{1!} + \frac{y^2}{2!} + \frac{y^3}{3!} + \ldots + \frac{y^{n-1}}{(n-1)!} + \frac{y^n}{n!} + \ldots \text{ to } \infty$$

Multiplying these series, we find that

$$e^x . e^y = 1 + \frac{(x+y)}{1!} + \left(\frac{x^2}{2!} + xy + \frac{y^2}{2!}\right) + \ldots$$

$$+\left(\frac{x^n}{n!}+\frac{x^{n-1}}{(n-1)!}.\frac{y}{1!}+\frac{x^{n-2}}{(n-2)!}.\frac{y^2}{2!}+...+\frac{y^n}{n!}\right)+...\text{ to }\infty$$

$$=1+\frac{(x+y)}{1!}+\frac{1}{2!}\left(x^2+2xy+y^2\right)+...$$

$$+\frac{1}{n!}\left\{x^n+nx^{n-1}y+\frac{n(n-1)}{2!}x^{n-2}y^2+...+y^n\right\}+...\text{ to }\infty$$

$$=1+\frac{(x+y)}{1!}+\frac{(x+y)^2}{2!}+...+\frac{(x+y)^n}{n!}+...\text{ to }\infty$$

$= e^{x+y}$ (by definition) .....(1)

Similarly, $e^{x_1}.e^{x_2}.e^{x_3}...e^{x_n} = e^{x_1+x_2+x_3+...+x_n}$

If $x_1 = x_2 = x_3 = ... = x_n = x$

then $(e^x)^n = e^{nx}$, where n is an integer

Putting $y = -x$ in eq.(1), we get $e^x . e^{-x} = e^{x-x} = e^0 = 1$

Again $e^{x-y} . e^y = e^{x-y+y} = e^x \quad \therefore \quad e^x/e^y = e^{x-y}$

Exponential values of circular functions

From definition, we get

$$e^{ix} = 1 + ix + \frac{i^2x^2}{2!}+\frac{i^3x^3}{3!}+\frac{i^4x^4}{4!}+\frac{i^5x^5}{5!}+...\text{ to }\infty$$

$$=\left(1-\frac{x^2}{2!}+\frac{x^4}{4!}-...\text{to }\infty\right)+i\left(x-\frac{x^3}{3!}+\frac{x^5}{5!}-...\text{to }\infty\right)$$

If x be real then

$$1-\frac{x^2}{2!}+\frac{x^4}{4!}-...\text{ to }\infty = \cos x \text{ and } x-\frac{x^3}{3!}+\frac{x^5}{5!}-...\text{ to }\infty = \sin x$$

$\therefore \quad e^{-ix} = \cos x + i \sin x$ .....(1)

Similarly, $e^{-ix} = \cos x - i\sin x$ .....(2)

Adding and subtracting, we get

$$\cos x = \frac{1}{2}\left(e^{ix}+e^{-ix}\right) \qquad \sin x = \frac{1}{2i}\left(e^{ix}-e^{-ix}\right)$$

These are called Euler's exponential values of sin x and cos x.

Putting $x = 2n\pi$ in eq. (1), we find that

$$e^{2n\pi i} = \cos 2n\pi + i \sin 2n\pi = 1 \qquad \because e^{2npi} = 1$$

Generalised Circular Functions

We know the values of sin x and cos x, when x is real. At the same time we find that

$$\left.\begin{aligned} \cos x &= \frac{1}{2}\left(e^{ix}+e^{-ix}\right)=1-\frac{x^2}{2!}+\frac{x^4}{4!}-\ldots \text{ to } \infty \\ \sin x &= \frac{1}{2i}\left(e^{ix}-e^{-ix}\right)=x-\frac{x^3}{3!}+\frac{x^5}{5!}-\ldots \text{ to } \infty \end{aligned}\right\} \quad (A)$$

sin x and cos x are also defined by (A) when x is complex.

Hence, for all real and complex values of x we conclude that

$$\cos x = \frac{1}{2}\left(e^{ix}+e^{-ix}\right)=1-\frac{x^2}{2!}+\frac{x^4}{4!}-\ldots \text{ to } \infty$$

$$\sin x = \frac{1}{2i}\left(e^{ix}-e^{-ix}\right)=x-\frac{x^3}{3!}+\frac{x^5}{5!}-\ldots \text{ to } \infty$$

$$e^{ix} = \cos x + i\sin x \text{ and } e^{-ix} = \cos x - i\sin x$$

Note: **If x be real then the above results can be established. But when x is complex then the above results can be taken as definitions.**

The other trigonometrical functions are defined as follows:

$$\tan x = \frac{\sin x}{\cos x},\ \sec x = \frac{1}{\cos x},\ \operatorname{cosec} x = \frac{1}{\sin x},\ \cot x = \frac{1}{\tan x}$$

All the trigonometrical relations, which are true for real values of x, are also true for complex values of x. For examples

(i) $\sin(-x) = \frac{1}{2i}\left(e^{-ix}-e^{ix}\right) = -\frac{1}{2i}\left(e^{ix}-e^{-ix}\right) = -\sin x$

(ii) $\cos(-x) = \frac{1}{2}\left(e^{-ix}+e^{ix}\right) = \cos x$

(iii) $\sin^2 x + \cos^2 x = \left(\frac{e^{ix}-e^{-ix}}{2!}\right)^2 + \left(\frac{e^{ix}+e^{-ix}}{2}\right)^2$

$$= -\frac{1}{4}\left(e^{2ix}-2+e^{-2ix}\right)+\frac{1}{4}\left(e^{2ix}+2+e^{-2ix}\right)=1$$

(iv) $\sin x \cos y + \cos x \sin y$

$$= \frac{e^{ix}-e^{-ix}}{2i}\cdot\frac{e^{iy}+e^{-iy}}{2}+\frac{e^{ix}+e^{-ix}}{2}\cdot\frac{e^{iy}-e^{-iy}}{2i}$$

$$= \frac{1}{4i}\left(2e^{ix}.e^{iy}-2e^{-ix}.e^{-iy}\right)=\frac{1}{2i}\left\{e^{i(x+y)}-e^{-i(x+y)}\right\}$$

$$= \sin(x+y)$$

Logarithms of complex numbers

If $e^z = w$, where z and w are complex quantities then z is called the logarithm of w.

$\because \qquad e^{2n\pi i} = 1$

$\therefore \qquad e^{z+2n\pi i} = e^z \, . \, e^{2n\pi i} = e^z = w, \quad \text{where } n = 0, \pm 1, \pm 2, \ldots$

Hence, by definition, $z + 2\pi ni$ is also the logarithm of w.

Hence, $\log w = z + 2\pi ni$

If we put $n = 0$ in the above result then it is called the *principal value.* The general value of z is denoted by log w and the principal value by log w. If $e^z = w$ then

$\log w = z + 2\pi ni$ and $\log w = z.$

$\therefore \quad \log w = \log w + 2\pi ni$

Logarithm of $\alpha + i\beta$

(i) To prove that $\log_e(\alpha + i\beta) = \frac{1}{2}\log_e\left(\alpha^2 + \beta^2\right) + i\tan^{-1}\frac{\beta}{\alpha}$

Put $\alpha = r\cos\theta$ .....(

and $\beta = r\sin\theta$ .....(

Squaring and adding (1) and (2), we get

$$\alpha^2 + \beta^2 = r^2(\cos^2\theta + \sin^2\theta) = r^2 \qquad \therefore \qquad r = \sqrt{\alpha^2 + \beta^2}$$

Dividing (2) by (1), we get

$$\frac{\beta}{\alpha} = \frac{r\sin\theta}{r\cos\theta} = \tan\theta \qquad \therefore \qquad \theta = \tan^{-1}\frac{\beta}{\alpha}$$

The given expression $= \log_e(\alpha + i\beta)$

$$= \log_e(r\cos\theta + ir\sin\theta) = \log_e[r(\cos\theta + i\sin\theta)]$$

$$= \log_e(r.e^{i\theta}) = \log_e r + \log_e e^{i\theta}$$

$$= \log_e\left(\alpha^2 + \beta^2\right)^{\frac{1}{2}} + i\theta\log e$$

$$= \frac{1}{2}\log_e\left(\alpha^2 + \beta^2\right) + i\tan^{-1}\frac{\beta}{\alpha}$$

Hence, $\log_e(\alpha + i\beta) = \frac{1}{2}\log_e\left(\alpha^2 + \beta^2\right) + i\tan^{-1}\frac{\beta}{\alpha}$

Note: Putting $-i$ in place of i in the above result, we find that

$$\log_e(\alpha - i\beta) = \frac{1}{2}\log_e\left(\alpha^2 + \beta^2\right) - i\tan^{-1}\frac{\beta}{\alpha}$$

(ii) To find $\log_e(\alpha + i\beta)$, where α, β are real to prove that

$$\log_e(\alpha + i\beta) = \frac{1}{2}\log_e(\alpha^2 + \beta^2) + i\left(2n\pi + \tan^{-1}\frac{\beta}{\alpha}\right)$$

Put $\alpha = r\cos\theta$ .....(1)

and $\beta = r\sin\theta$ .....(2)

Squaring and adding (1) and (2), we get

$$\alpha^2 + \beta^2 = r^2(\cos^2\theta + \sin^2\theta) = r^2 \qquad \therefore r = \sqrt{\alpha^2 + \beta^2}$$

Dividing (2) by (1), we get

$$\frac{\beta}{\alpha} = \frac{r\sin\theta}{r\cos\theta} = \tan\theta \qquad \therefore \theta = \tan^{-1}\frac{\beta}{\alpha}$$

The given expression $= \log_e(\alpha + i\beta)$

$$= \log_e(r\cos\theta + i\sin\theta)]$$

$$= \log_e[r(\cos\theta + i\sin\theta)]$$

$$= \log_e(re^{i\theta}) = 2n\pi i + \log_e(re^{i\theta})$$

$$= 2n\pi i + \log_e r + \log_e e^{i\theta}$$

$$= 2n\pi i + \log_e(\alpha^2 + \beta^2)^{\frac{1}{2}} + i\theta\log_e e$$

$$= \frac{1}{2}\log_e(\alpha^2 + \beta^2) + i\left(2n\pi + \tan^{-1}\frac{\beta}{\alpha}\right)$$

Hence, $\log_e(\alpha + i\beta) = \frac{1}{2}\log_e(\alpha^2 + \beta^2) + i\left(2n\pi + \tan^{-1}\frac{\beta}{\alpha}\right)$

Note: Putting $-i$ in place of i in the above result, we find that

$$\log_e(\alpha - i\beta) = \frac{1}{2}\log_e(\alpha^2 + \beta^2) - i\left(2n\pi + \tan^{-1}\frac{\beta}{\alpha}\right)$$

Rules for Logarithm of complex numbers and expansions

The following rules of logarithm for real numbers are also true for complex numbers:

(i) $\log(x.y) = \log x + \log y$

(ii) $\log x^y = y\log x$

For real number x, we know that

$$\log(1+x) = x - \frac{1}{2}x^2 + \frac{1}{3}x^3 - \dots \text{ to } \infty \quad \text{when } -1 < x \le 1$$

This expansion is also true for the complex number x.

Example: $\log(1+\alpha+i\beta) = (\alpha+i\beta) - \frac{1}{2}(\alpha+i\beta)^2 + \frac{1}{3}(\alpha+i\beta)^3 - \dots$

where $|\alpha+i\beta| \le 1$

Gregory's series

Statement of Gregory's series

If $-\frac{1}{4}\pi \le \theta \le \frac{1}{4}\pi$ then

$$\theta = \tan\theta - \frac{1}{3}\tan^3\theta + \frac{1}{5}\tan^5\theta - \dots + (-1)^{n-1}\frac{\tan^{2n-1}\theta}{2n-1} + \dots \text{ to } \infty$$

Alternative form of the Gregory's series

Let $\tan\theta = x$ or $\theta = \tan^{-1}x$

Then from $\theta = \tan\theta - \frac{1}{3}\tan^3\theta + \frac{1}{5}\tan^5\theta - \dots$ to ∞

We get $\tan^{-1}x = x - \frac{1}{3}x^3 + \frac{1}{5}x^5 - \dots + (-1)^{n-1}\frac{x^{2n-1}}{2n-1} + \dots$ to ∞

where $-1 \le x \le 1$ as $-\frac{1}{4}\pi \le \theta \le \frac{1}{4}\pi$

A Deduction of Gregory's series

If $n\pi - \frac{\pi}{4} \le \theta \le n\pi + \frac{\pi}{4}$ then

$$\theta - n\pi = \tan\theta - \frac{1}{3}\tan^3\theta + \frac{1}{5}\tan^5\theta - \dots \text{ up to infinity}$$

Here let $\theta - n\pi = \varphi$, where $-\frac{\pi}{4} \le \varphi \le \frac{\pi}{4}$

$$1 + i\tan\theta = 1 + i\tan(n\pi + \varphi)$$

$$= 1 + i\tan\varphi = \sec\varphi(\cos\varphi + i\sin\varphi)$$

$$= \sec\varphi . e^{i\varphi}$$

Taking logarithm of both sides with base e, we get

$$\log_e \sec\varphi + i\varphi = \log_e(1 + i\tan\theta)$$

$$= i\tan\theta - \frac{1}{2}t^2\tan^2\theta + \frac{1}{3}t^3\tan^3\theta - \ldots \text{ to } \infty,$$

as $(i\tan\theta \mid \le 1)$

$$= i\tan\theta + \frac{1}{2}\tan^2\theta - \frac{1}{3}i\tan^3\theta - \ldots$$

Equating imaginary parts from both sides, we get

$$\varphi = \tan\theta - \frac{1}{3}\tan^3\theta + \frac{1}{5}\tan^5\theta - \ldots \text{ to } \infty$$

Hence, $\theta - n\pi = \tan\theta - \frac{1}{3}\tan^3\theta + \frac{1}{5}\tan^5\theta - \ldots \text{ to } \infty$

Note: If in the above result we put n = 0 then

$$\theta = \tan\theta - \frac{1}{3}\tan^3\theta + \frac{1}{5}\tan^5\theta - \ldots \text{ to } \infty$$

Example: If $\frac{3}{4}\pi \le \theta \le \frac{5}{4}\pi$, that is if $\pi - \frac{1}{4}\pi \le \theta \le \pi + \frac{1}{4}\pi$ then putting n = 1 in the above result, we get

$$\theta - \pi = \tan\theta - \frac{1}{3}\tan^3\theta + \frac{1}{5}\tan^5\theta - \ldots \text{ to } \infty$$

Evaluation of π

The value of π can be expressed as an sum of the infinite series by using the different forms of Gregory's series. The Gregory's series is as follows:

$$\theta = \tan\theta - \frac{1}{3}\tan^3\theta + \frac{1}{5}\tan^5\theta - \frac{1}{7}\tan^7\theta + \frac{1}{9}\tan^9\theta - \frac{1}{11}\tan^{11}\theta + \ldots \text{ to } \infty$$

Putting $\theta = \frac{\pi}{4}$ that is $\tan\theta = \tan\frac{\pi}{4} = 1$, we get

$$\frac{\pi}{4} = 1 - \frac{1}{3} + \frac{1}{5} - \frac{1}{7} + \frac{1}{9} - \frac{1}{11} + \ldots$$

$$= \frac{2}{1.3} + \frac{2}{5.7} + \frac{2}{9.11} + \ldots \text{ to } \infty$$

$$\therefore \quad \frac{\pi}{8} = \frac{1}{1.3} + \frac{1}{5.7} + \frac{1}{9.11} + \ldots \text{ to } \infty$$

It can easily be proved that $\frac{1}{4}\pi = \tan^{-1}\frac{1}{2} + \tan^{-1}\frac{1}{3}$, using Gregory's series.

Putting $x = \frac{1}{2}$ and $x = \frac{1}{3}$ in the expansion of $\tan^{-1} x$, we find

$$\frac{\pi}{4} = \left[\frac{1}{2} - \frac{1}{3}.\frac{1}{2^3} + \frac{1}{5}.\frac{1}{2^5} - \ldots \text{ to } \infty\right]$$

$$+\left[\frac{1}{3} - \frac{1}{3}.\frac{1}{3^3} + \frac{1}{5}.\frac{1}{3^5} - \ldots \text{ to}\right]$$

$$= \left(\frac{1}{2} + \frac{1}{3}\right) - \frac{1}{3}\left(\frac{1}{2^3} + \frac{1}{3^3}\right) + \frac{1}{5}\left(\frac{1}{2^5} + \frac{1}{3^5}\right) - \ldots \text{ to } \infty$$

Solved Examples

***Model*: On expressing $e^{i\theta}$ in the form A + iB**

***Example* 2.39:** Write $e^{i(2\pi+\alpha)}$ in the form A + iB

***Solution*:** $e^{i(2\pi+\alpha)} = \cos(2\pi + \alpha) + i\sin(2\pi + \alpha)$

$$= \cos\alpha + i\sin\alpha$$

***Model*: On establishing identities from the general definition of sin θ, cos θ**

***Example* 2.40:** Apply the exponential values of sine and cosine to prove that

(i) $\cos 2x = 2\cos^2 x - 1$

(ii) $\cos 3x = 4\cos^3 x - 3\cos x$

***Solution*:**

(i) $2\cos^2 x - 1 = 2\left(\frac{e^{ix} + e^{-ix}}{2}\right)^2 - 1 = \frac{e^{2ix} + e^{-2ix} + 2e^{ix}e^{-ix}}{2} - 1$

(ii) $\because$ $e^{3ix} = \cos 3x + i\sin 3x$ and $e^{-3ix} = \cos 3x - i\sin 3x$

$\therefore$ $2\cos 3x = e^{3ix} + e^{-3ix} = (e^{ix})^3 + (e^{-ix})^3$

$$= \left(e^{ix} + e^{-ix}\right)^3 - 3e^{ix}.e^{-ix}\left(e^{ix} + e^{-ix}\right)$$

$$= (2\cos x)^3 - 3 . e^0(2\cos x)$$

$$= 2(4\cos^3 x - 3\cos x)$$

$\therefore$ $\cos 3x = 4\cos^3 x - 3\cos x$

***Model*: On logarithm of complex numbers**

***Example* 2.41:** Express log i and log i as complex numbers

***Solution*:** We know that $\log(x + iy) = \frac{1}{2}\log(x^2 + y^2) + i\tan^{-1}\frac{y}{x}$

Here x = 0, y = 1

$$\therefore \quad \log(x+iy) = \frac{1}{2}\log\left(0^2+1^2\right)+i\tan^{-1}\frac{1}{0}$$

$$= \frac{1}{2}\log 1 + i\tan^{-1}\infty = \frac{1}{2}\times 0 + i\frac{\pi}{2} = i\frac{\pi}{2}$$

$$\log(x+iy) = \frac{1}{2}\log\left(x^2+y^2\right)+i\left(2n\pi+\tan^{-1}\frac{y}{x}\right)$$

$$= \frac{1}{2}\log 1 + i\left(2n\pi+\frac{\pi}{2}\right) = i\left(2n+\frac{1}{2}\right)\pi$$

***Example* 2.42:** Prove that

$$\log(1+\cos\alpha+i\sin\alpha) = \log\left(2\cos\frac{\alpha}{2}\right)+\frac{1}{2}i\alpha,$$

where $\quad -\pi < \alpha < \pi$

***Solution*:** We know that $\log(x + iy) = \frac{1}{2}\log\left(x^2+y^2\right)+i\tan^{-1}\frac{y}{x}$

Here $\quad x = 1 + \cos\alpha,\ y = \sin\alpha$

$$\therefore \quad \text{LHS} = \frac{1}{2}\log[(1+\cos\alpha)^2+\sin^2\alpha]+i\tan^{-1}\frac{\sin\alpha}{1+\cos\alpha}$$

$$= \frac{1}{2}\log\left(1+2\cos\alpha+\cos^2\alpha+\sin^2\alpha\right)i\tan^{-1}\frac{2\sin\frac{\alpha}{2}\cos\frac{\alpha}{2}}{2\cos^2\frac{\alpha}{2}}$$

$$= \frac{1}{2}\log(2+2\cos\alpha)+i\tan^{-1}\left(\tan\frac{\alpha}{2}\right)$$

$$= \frac{1}{2}\log\left(2.2\cos^2\frac{\alpha}{2}\right)+i\frac{\alpha}{2}$$

$$= \log\left(2\cos\frac{\alpha}{2}\right)+i\frac{\alpha}{2} = \text{RHS}$$

***Model*: On evaluation of $(x + iy)^{\alpha+i\beta}$**

***Example* 2.43:** Prove that $i^i = e^{-\frac{1}{2}(4n+1)\pi}, n \in Z$

***Solution*:** $i^i = e^{\log i^i} = e^{i\log i}$

$$= e^{i\left[\frac{1}{2}\log\left(0^2 + 1^2\right) + i\left(\tan^{-1}\frac{1}{0} + 2n\pi\right)\right]}$$

$$= e^{i.i\left(\frac{\pi}{2} + 2n\pi\right)} = e^{-(4n+1)\pi/2}, n \in Z$$

Important Formulae

1. $e^{ix} = \cos x + i\sin x$
2. $e^{-ix} = \cos x - i\sin x$
3. $\cos x = \frac{1}{2}\left(e^{ix} + e^{-ix}\right)$
4. $\sin x = \frac{1}{2i}\left(e^{ix} - e^{-ix}\right)$
5. $e^{2\pi ni} = 1$
6. $i = \cos\frac{\pi}{2} + i\sin\frac{\pi}{2} = e^{i\frac{\pi}{2}}$
7. $-i = \cos\frac{\pi}{2} - i\sin\frac{\pi}{2} = e^{-i\frac{\pi}{2}}$
8. $\log_e(\alpha + i\beta) = \log_e\sqrt{\alpha^2 + \beta^2} + i\tan^{-1}\frac{\beta}{\alpha}$
9. $\log_e(\alpha - i\beta) = \log_e\sqrt{\alpha^2 + \beta^2} + i\tan^{-1}\frac{\beta}{\alpha}$
10. If $-\frac{1}{4}\pi \le \theta \le \frac{1}{4}\pi$ then

    $\theta = \tan\theta - \frac{1}{3}\tan^3\theta + \frac{1}{5}\tan^5\theta - \ldots$ to ∞
11. If $-1 \le x \le 1$ then

    $\tan^{-1} x = x - \frac{1}{3}x^3 + \frac{1}{5}x^5 - \ldots + (-1)^{n-1}\frac{x^{2n-1}}{2n-1} + \ldots$ to ∞

Exercise 2.4

***Model*: On expressing $e^{i\theta}$ in the form A + iB**

1. (i) Express $e^{i(\alpha+i\beta)}$ in the form A + iB **Ans.** $e^{-\beta}\cos\alpha + ie^{-\beta}\sin\alpha$

   (ii) Find the value of $e^{i\frac{\pi}{2}}$ **Ans.** i

***Model*: On establishing identities from the general definition of sin θ, cos θ**

2. Apply the exponential values of sine and cosine to prove that
   (i) $\cos(x + y) = \cos x \cdot \cos y - \sin x \cdot \sin y$
   (ii) $\sin 2x = 2\sin x \cdot \cos x$
   (iii) $\sin 3x = 3\sin x - 4\sin^3 x$

***Model*: On logarithm of complex numbers**

3. Prove that
   (i) $\log(-2) = \log 2 + i(2n + 1)\pi$
   (ii) $\log(-3) = \log 3 + i\pi$
   (iii) $\log(-i) = \frac{1}{2}(4n-1)\pi i$

4. If $A + iB = \log(x + iy)$, prove that
$$A = \frac{1}{2}\log\left(x^2 + y^2\right) \text{ and } B = \tan^{-1}\frac{y}{x}$$

5. Prove that
   (i) $\log\left(1 + e^{2i\theta}\right) = \log\left(2\cos\theta\right) + i\theta$
   (ii) $\log\frac{1}{1-e^{i\theta}} = \log\left(\frac{1}{2}\operatorname{cosec}\frac{\theta}{2}\right) + i\left(\frac{\pi}{2} - \frac{\theta}{2}\right)$
   (iii) $\log\frac{a+ib}{a-ib} = 2i\tan^{-1}\frac{b}{a}$
   (iv) $i\log\frac{x-i}{x+i} = \pi - 2\tan^{-1}x$

6. If $(a_1 + ib_1)(a_2 + ib_2)\ldots(a_n + ib_n) = A + iB$ then prove that
   (i) $\left(a_1^2 + b_1^2\right)\left(a_2^2 + b_2^2\right)\ldots\left(a_n^2 + b_n^2\right) = A^2 + B^2$
   (ii) $\tan^{-1}\frac{b_1}{a_1} + \tan^{-1}\frac{b_2}{a_2} + \ldots + \tan^{-1}\frac{b_n}{a_n} = \tan^{-1}\frac{B}{A}$

***Model*: On evaluation of $(x + iy)^{\alpha+i\beta}$**

7. Prove that $(-i)^{-i} = e^{(4n-1)\pi/2}$, $n \in Z$

2.24 Summation of Series

Introduction

A number of methods for summing up the trigonometrical series are available to us. Here we will concentrate our attention particularly towards the method, namely

Summation with the aid of complex numbers or C + iS method.

C + iS method

By this method the summation of the following two types of series can be obtained:

$$a_0\cos\alpha + a_1\cos(\alpha+\beta) + a_2\cos(\alpha+2\beta) + \ldots \quad \text{.....(1)}$$

and $$a_0\sin\alpha + a_1\sin(\alpha+\beta) + a_2\sin(\alpha+2\beta) + \ldots \quad \text{.....(2)}$$

But the summation can be carried out provided the sum of the series $a_0 + a_1x + a_2x^2 + \ldots$ can be computed. The above two series may be finite or infinite.

Working rule

Step 1 The following notations are used for the series (1) and (2).

$$C = a_0\cos\alpha + a_1\cos(\alpha+\beta) + a_2\cos(\alpha+2\beta) + \ldots$$

$$S = a_0\sin\alpha + a_1\sin(\alpha+\beta) + a_2\sin(\alpha+2\beta) + \ldots$$

Step 2 Find C + iS

$$C + iS = a_0(\cos\alpha + i\sin\alpha) + a_1[\cos(\alpha+\beta) + i\sin(\alpha+\beta)] + \ldots$$

$$= a_0e^{i\alpha} + a_1e^{i(\alpha+\beta)} + a_2e^{i(\alpha+2\beta)} + \ldots$$

$$= e^{i\alpha}(a_0 + a_1e^{i\beta} + a_2e^{2i\beta} + \ldots)$$

$$= e^{i\alpha}(a_0 + a_1x + a_2x^2 + \ldots), \text{ where } e^{ib} = x$$

Step 3 Compute the sum of the series $a_0 + a_1x + a_2x^2 + \ldots$

Since $x = e^{i\beta}$ = a complex quantity, the sum of the series will be a complex quantity.

Let the sum be A + IB

Then $C + iS = e^{i\alpha} - (A + iB) = (\cos\alpha + i\sin\alpha) . (A + iB)$

$$= (A\cos\alpha - B\sin\alpha) + i(B\cos\alpha + A\sin\alpha)$$

Step 4 Now, equating real and imaginary parts separately, we get

$$C = A\cos\alpha - B\sin\alpha \text{ and } S = B\cos\alpha + A\sin\alpha$$

Hence, we get the values of C and S both

Note: After finding C + iS we can find C – iS by replacing I by –i.

Now, use $C = \frac{1}{2}[(C + iS) + (C - iS)]$

Remember the Results Carefully

(i) $a + ar + ar^2 + ... + ar^{n-1} = \dfrac{a\left(r^n - 1\right)}{r-1}$

(ii) $a + ar + ar^2 + ...$ to $\infty = \dfrac{a}{1-r}$, $\qquad (|r| < 1)$

(iii) $1 + x + \dfrac{x^2}{2!} + \dfrac{x^3}{3!} + ... = e^x$

(iv) $x - \dfrac{1}{2}x^2 + \dfrac{1}{3}x^3 - ... = \log(1+x)$

(v) $1 - \dfrac{x^2}{2!} + \dfrac{x^4}{3!} - ... = \cos x$

(vi) $x - \dfrac{x^3}{3!} + \dfrac{x^5}{5!} - ... = \sin x$

(vii) $x - \dfrac{1}{3}x^3 + \dfrac{1}{5}x^5 - ... = \tan^{-1} x$, $\qquad$ (Gregory's series)

(viii) $2\left[x + \dfrac{x^3}{3} + \dfrac{x^5}{5} + ... \text{ to } \infty\right] = \log\dfrac{1+x}{1-x}$

Solved Examples

***Model*: On summation of series based on logarithmic series**

***Example* 2.44: Find the sum of the series**

$$\cos\theta - \frac{1}{2}\cos 2\theta + \frac{1}{3}\cos 3\theta - ... \text{ to } \infty$$

Solution: Let $C = \cos\theta - \dfrac{1}{2}\cos 2\theta + \dfrac{1}{3}\cos 3\theta - ...$ to ∞

and $\qquad S = \sin\theta - \dfrac{1}{2}\sin 2\theta + \dfrac{1}{3}\sin 3\theta - ...$ to ∞

$$\therefore \quad C + iS = (\cos\theta + i\sin\theta) - \frac{1}{2}(\cos 2\theta + i\sin 2\theta)$$

$$+\frac{1}{3}(\cos 3\theta + i\sin 3\theta) - ... \text{ to } \infty$$

$$= e^{i\theta} - \frac{1}{2}e^{2i\theta} + \frac{1}{3}e^{3i\theta} - \ldots \text{ to } \infty$$

$$= \log(1+e^{i\theta}) = \log(1+\cos\theta + i\sin\theta)$$

$$= \log\left(2\cos^2\frac{1}{2}\theta + i2\sin\frac{1}{2}\theta\cos\frac{1}{2}\theta\right)$$

$$= \log\left[2\cos\frac{1}{2}\theta\left(\cos\frac{1}{2}\theta + i\sin\frac{1}{2}\theta\right)\right]$$

$$= \log\left[2\cos\frac{1}{2}\theta.e^{i\frac{1}{2}\theta}\right] = \log(2\cos\frac{1}{2}\theta) + \log e^{i\frac{1}{2}\theta}$$

$$= \log\left(2\cos\frac{1}{2}\theta\right) + i\frac{1}{2}\theta$$

Equating real parts from both sides, we get

$$C = \log\left(2\cos\frac{1}{2}\theta\right)$$

Note: Similarly, equating imaginary parts

$$S = \sin\theta - \frac{1}{2}\sin 2\theta + \frac{1}{3}\sin 3\theta - \ldots \text{ to } \infty = \frac{1}{2}\theta$$

***Example* 2.45:** Sum the series

$$c\cos\alpha + \frac{c^2}{2}\cos(\alpha+\beta) + \frac{c^3}{3}\cos(\alpha+2\beta) + \ldots \text{ to } \infty$$

and $$c\sin\alpha + \frac{c^2}{2}\sin(\alpha+\beta) + \frac{c^2}{3}\sin(\alpha+2\beta) + \ldots \text{ to } \infty$$

***Solution*:** Let $C = c\cos\alpha + \frac{c^2}{2}\cos(\alpha+\beta)\frac{c^3}{3}\cos(\alpha+2\beta) + \ldots \text{ to } \infty$

and $$S = c\sin\alpha + \frac{c^2}{2}\sin(\alpha+\beta) + \frac{c^3}{3}\sin(\alpha+2\beta) + \ldots \text{ to } \infty$$

$\therefore$ $$C + iS = c(\cos\alpha + i\sin\alpha) + \frac{c^2}{2}\left[\cos(\alpha+\beta) + i\sin(\alpha+\beta)\right]$$

$$+\frac{c^3}{3}\left[\cos(\alpha+2\beta)+i\sin(\alpha+2\beta)\right]+\ldots \text{ to } \infty$$

$$= ce^{i\alpha}+\frac{c^2}{2}e^{i(\alpha+\beta)}+\frac{c^3}{3}e^{i(\alpha+2\beta)}+\ldots \text{ to } \infty$$

$$= ce^{i\alpha}+\frac{c^2}{2}e^{i\alpha}.e^{i\beta}+\frac{c^3}{3}e^{i\alpha}.e^{2i\beta}\pm\ldots \text{ to } \infty$$

$$=\frac{e^{i\alpha}}{e^{i\beta}}\left(ce^{i\beta}+\frac{c^2}{2}e^{2i\beta}+\frac{c^3}{3}e^{3i\beta}+\ldots \text{ to } \infty\right)$$

$$= e^{i(\alpha-\beta)}\left[-\log\left(1-ce^{i\beta}\right)\right]$$

$$= -e^{i(\alpha-\beta)}\log\left[1-c(\cos\beta+i\sin\beta)\right]$$

$$= -e^{i(\alpha-\beta)}\log\left[(1-c\cos\beta)-i(c\sin\beta)\right]$$

$$= -\left[\cos(\alpha-\beta)+i\sin(\alpha-\beta)\right]$$

$$\times\left[\frac{1}{2}\log\left[(1-\cos\beta)^2+c^2\sin^2\beta\right]-i\tan^{-1}\frac{c\sin\beta}{1-c\cos\beta}\right]$$

Equating real and imaginary parts from both sides separately, we get

$$C = -\sin(\alpha-\beta)\tan^{-1}\frac{c\sin\beta}{1-c\cos\beta}$$

$$-\frac{1}{2}\cos(\alpha-\beta)\log\left(1-2\cos\beta+c^2\right)$$

and $$S = \cos(\alpha-\beta)\tan^{-1}\frac{c\sin\beta}{1-c\cos\beta}$$

$$-\frac{1}{2}\sin(\alpha-\beta)\log\left(1-2\cos\beta+c^2\right)$$

***Example* 2.46:** If $\theta-\alpha=\tan^2\frac{1}{2}\omega\sin 2\theta-\frac{1}{2}\tan^4\frac{1}{2}\omega\sin 4\theta$

$$+\frac{1}{3}\tan^6\frac{1}{2}\omega\sin 6\theta-\ldots \text{ to infinite terms,}$$

show that $\tan\alpha=\tan\theta\,.\,\cos\omega$

***Solution*:** Let $S=\tan^2\frac{1}{2}\omega\sin 2\theta-\frac{1}{2}\tan^4\frac{1}{2}\omega\sin 4\theta$

$$+\frac{1}{3}\tan^6\frac{1}{2}\omega\sin 6\theta - \dots \text{ to } \infty$$

and
$$C = \tan^2\frac{1}{2}\omega\cos 2\theta - \frac{1}{2}\tan^4\frac{1}{2}\omega\cos 4\theta$$
$$+\frac{1}{3}\tan^6\frac{1}{2}\omega\cos 6\theta - \dots \text{ to } \infty$$

$\therefore$
$$C + iS = \tan^2\frac{1}{2}\omega(\cos 2\theta + i\sin 2\theta) - \frac{1}{2}\tan^4\frac{1}{2}\omega(\cos 4\theta + i\sin 4\theta)$$
$$+\frac{1}{3}\tan^6\frac{1}{2}\omega(\cos 6\theta + i\sin 6\theta) - \dots \text{ to } \infty$$
$$= \tan^2\frac{1}{2}\omega.e^{2i\theta} - \frac{1}{2}\tan^4\frac{1}{2}\omega.e^{4i\theta} + \frac{1}{3}\tan^6\frac{1}{2}\omega.e^{6i\theta} - \dots \text{ to } \infty$$
$$= \log\left(1 + \tan^2\frac{1}{2}\omega.e^{2i\theta}\right)$$
$$= \log\left[1 + \tan^2\frac{1}{2}\omega(\cos 2\theta + i\sin 2\theta)\right]$$
$$= \log\left[\left(1 + \tan^2\frac{1}{2}\omega.\cos 2\theta\right) + i\tan^2\frac{1}{2}\omega.\sin 2\theta\right]$$
$$= \frac{1}{2}\log\left[(1 + \tan^2\frac{1}{2}\omega.\cos 2\theta)^2 + \tan^4\frac{1}{2}\omega.\sin^2 2\theta\right]$$
$$+ i\tan^{-1}\frac{\tan^2\frac{1}{2}\omega.\sin 2\theta}{1 + \tan^2\frac{1}{2}\omega.\cos 2\theta}$$

Equating imaginary parts from both sides, we get

$$S = \tan^{-1}\frac{\tan^2\frac{1}{2}\omega.\sin 2\theta}{1 + \tan^2\frac{1}{2}\omega.\cos 2\theta}$$

or
$$\theta - \alpha = \tan^{-1}\frac{\tan^2\frac{1}{2}\omega.\sin 2\theta}{1 + \tan^2\frac{1}{2}\omega.\cos 2\theta}$$

or $$\tan(\theta-\alpha)=\frac{\frac{\sin^2\frac{1}{2}\omega}{\cos^2\frac{1}{2}\omega}.\frac{2\tan\theta}{1+\tan^2\theta}}{1+\frac{\sin^2\frac{1}{2}\omega}{\cos^2\frac{1}{2}\omega}.\frac{1-\tan^2\theta}{1+\tan^2\theta}}$$

or $$\tan(\theta-\alpha)=\frac{\frac{1-\cos\omega}{1+\cos\omega}.\frac{2\tan\theta}{1+\tan^2\theta}}{1+\frac{1-\cos\omega}{1+\cos\omega}.\frac{1-\tan^2\theta}{1+\tan^2\theta}}$$

or $$\frac{\tan\theta-\tan\alpha}{1+\tan\theta.\tan\alpha}=\frac{2\tan\theta(1-\cos\omega)}{(1+\cos\omega)(1+\tan^2\theta)+(1-\cos\omega)(1-\tan^2\theta)}$$

or $$\frac{\tan\theta-\tan\alpha}{1+\tan\theta.\tan\alpha}=\frac{2\tan\theta(1-\cos\omega)}{1+\tan^2\theta+\cos\omega+\cos\omega.\tan^2\theta+1-\tan^2\theta}-\cos\omega+\cos\omega.\tan^2\theta$$

or $$\frac{\tan\theta-\tan\alpha}{1+\tan\theta.\tan\alpha}=\frac{2\tan\theta(1-\cos\omega)}{2(1+\cos\omega.\tan^2\theta)}$$

or $$(\tan\theta-\tan\alpha)(1+\cos\omega.\tan^2\omega)$$

$$=(\tan\theta-\tan\theta.\cos\omega)(1+\tan\theta\tan\alpha)$$

or $$\cos\omega.\tan\theta(1+\tan^2\theta)=\tan\alpha(1+\tan^2\theta)$$

or $$\tan\alpha=\tan\theta.\cos\omega$$

***Model*: On summation of series based on the Gregory's series**

***Example* 2.47:** Find the sum of the series

$$\sin\theta.\cos\theta-\frac{1}{3}\sin^3\theta.\cos 3\theta+\frac{1}{5}\sin^5\theta.\cos 5\theta-\ldots\text{ to }\infty$$

Solution: Let $C=\sin\theta.\cos\theta-\frac{1}{3}\sin^3\theta.\cos 3\theta+\frac{1}{5}\sin^5\theta\cos 5\theta-\ldots\text{ to }\infty$

and $$S=\sin\theta.\sin\theta-\frac{1}{3}\sin^3\theta.\sin 3\theta+\frac{1}{5}\sin^5\theta.\sin 5\theta-\ldots\text{ to }\infty$$

$\therefore$ $$C+iS=\sin\theta(\cos\theta+i\sin\theta)-\frac{1}{3}\sin^3\theta(\cos 3\theta+i\sin 3\theta)$$

$$+\frac{1}{5}\sin^5\theta.(\cos 5\theta+i\sin 5\theta)-\ldots\text{ to }\infty$$

$$= \sin\theta.e^{i\theta} - \frac{1}{3}\sin^3\theta.e^{3i\theta} + \frac{1}{5}\sin^5\theta.e^{5i\theta} - \dots \text{ to } \infty$$

$$= \tan^{-1}(\sin\theta.e^{i\theta}), \qquad \text{from the Gregory's series}$$

Similarly, $C - iS = \tan^{-1}(\sin\theta.e^{-i\theta})$

Adding, we get $2C = \tan^{-1}(\sin\theta.e^{i\theta}) + \tan^{-1}(\sin\theta.e^{-i\theta})$

$$= \tan^{-1}\frac{\sin\theta.e^{i\theta} + \sin\theta.e^{-i\theta}}{1 - \sin\theta.e^{i\theta}.\sin\theta.e^{-i\theta}}$$

$$= \tan^{-1}\frac{\sin\theta(e^{i\theta} + e^{-i\theta})}{1 - \sin^2\theta}$$

$$= \tan^{-1}\frac{2\sin\theta.\cos\theta}{\cos^2\theta} = \tan^{-1}(2\tan\theta)$$

$$\therefore \qquad C = \frac{1}{2}\tan^{-1}(2\tan\theta)$$

***Model*: On summation of series based on exponential series**

***Example* 2.48:** Sum up to infinite terms of the series

$$\cos\theta + \frac{a\cos 2\theta}{1!} + \frac{a^2\cos 3\theta}{2!} + \frac{a^3\cos 4\theta}{3!} + \dots, (\alpha < 1)$$

Solution: Let $\quad C = \cos\theta + \frac{a\cos 2\theta}{1!} + \frac{a^2\cos 3\theta}{2!} + \frac{a^3\cos 4\theta}{3!} + \dots, \text{ to } \infty$

and $\quad S = \sin\theta + \frac{a\sin 2\theta}{1!} + \frac{a^2\sin 3\theta}{2!} + \frac{a^3\sin 4\theta}{3!} + \dots \text{ to } \infty$

$$\therefore \qquad C + iS = (\cos\theta + i\sin\theta) + \frac{a}{1!}(\cos 2\theta + i\sin 2\theta)$$

$$+ \frac{a^2}{2!}(\cos 3\theta + i\sin 3\theta) + \dots \text{ to } \infty$$

$$= e^{i\theta} + \frac{a}{1!}e^{2i\theta} + \frac{a^2}{2!}e^{3i\theta} + \dots \text{ to } \infty$$

$$= e^{i\theta}\left[1 + \frac{ae^{i\theta}}{1!} + \frac{(ae^{i\theta})^2}{2!} + \dots \text{ to } \infty\right]$$

$$= e^{i\theta}.e^{ae^{i\theta}} = e^{i\theta}.e^{a(\cos\theta + i\sin\theta)}$$

$$= e^{a\cos\theta}.e^{i(\theta + a\sin\theta)}$$

$$= e^{a\cos\theta}[\cos(\theta + a\sin\theta) + i\sin(\theta + a\sin\theta)]$$

Equating real parts on both sides, we get

$$C = e^{a\cos\theta}.\cos(\theta + a\sin\theta)$$

***Example* 2.49:** Find the sum of the series

$$\cos\theta - \frac{\cos 2\theta}{2!} + \frac{\cos 3\theta}{3!} - \ldots \text{ to } \infty$$

Solution: Let $C = \cos\theta - \frac{\cos 2\theta}{2!} + \frac{\cos 3\theta}{3!} - \ldots \text{ to } \infty$

and $S = \sin\theta - \frac{\sin 2\theta}{2!} + \frac{\sin 3\theta}{3!} - \ldots \text{ to } \infty$

$\therefore$

$$C + iS = (\cos\theta + i\sin\theta) - \frac{1}{2!}(\cos 2\theta + i\sin 2\theta)$$

$$+ \frac{1}{3!}(\cos 3\theta + i\sin 3\theta) - \ldots \text{ to } \infty$$

$$= e^{i\theta} - \frac{1}{2!}e^{2i\theta} + \frac{1}{3!}e^{3i\theta} - \ldots \text{ to } \infty$$

$$= 1 - \left[1 - e^{i\theta} + \frac{(e^{i\theta})^2}{2!} - \frac{(e^{i\theta})^2}{3!} + \ldots \text{to } \infty\right]$$

$\because$

$$= 1 - e^{-e^{i\theta}}, \qquad \left[\therefore 1 - x + \frac{x^2}{2!} - \frac{x^3}{3!} + \ldots \text{to } \infty = e^{-x}\right]$$

$$= 1 - e^{-(\cos\theta + i\sin\theta)} = 1 - e^{-\cos\theta}.e^{-i\sin\theta}$$

$$= 1 - e^{-\cos\theta}[\cos(\sin\theta) - i\sin(\sin\theta)]$$

Equating real parts on both sides, we get

$$C = 1 - e^{-\cos\theta}.\cos(\sin\theta)$$

***Model*: On summation of series based on geometric progression**

***Example* 2.50:** Sum the infinite series

$$\sin\alpha + \frac{1}{2}\sin 2\alpha + \frac{1}{2^2}\sin 3\alpha + \ldots \text{ to } \infty$$

Solution: Let $S = \sin\alpha + \frac{1}{2}\sin 2\alpha + \frac{1}{2^2}\sin 3\alpha + \ldots \text{ to } \infty$

and $\quad C = \cos\alpha + \frac{1}{2}\cos 2\alpha + \frac{1}{2^2}\cos 3\alpha + \ldots \text{ to } \infty$

$$\therefore \quad C + iS = (\cos\alpha + i\sin\alpha) + \frac{1}{2}(\cos 2\alpha + i\sin 2\alpha) + \frac{1}{2^2}(\cos 3\alpha + i\sin 3\alpha) + \ldots \text{ to } \infty$$

$$= e^{i\alpha} + \frac{1}{2}e^{2i\alpha} + \frac{1}{2^2}e^{3i\alpha} + \ldots \text{ to } \infty$$

$$= \frac{e^{i\alpha}}{1 - \frac{1}{2}e^{i\alpha}}, \qquad \left(\because \text{ In GP, } |r| = \left|\frac{1}{2}e^{i\alpha}\right| = \frac{1}{2} < 1\right)$$

$$= \frac{2e^{i\alpha}}{2 - e^{i\alpha}} = \frac{2(\cos\alpha + i\sin\alpha)}{(2 - \cos\alpha) - i\sin\alpha}$$

$$= \frac{2(\cos\alpha + i\sin\alpha)\{(2 - \cos\alpha) + i\sin\alpha\}}{(2 - \cos\alpha)^2 + \sin^2\alpha}$$

$$= \frac{\{2\cos\alpha(2 - \cos\alpha) - 2\sin^2\alpha\} + i\{2\cos\alpha.\sin\alpha + 2\sin\alpha(2 - \cos\alpha)\}}{5 - 4\cos\alpha}$$

$$= \frac{(4\cos\alpha - 2) + 4i\sin\alpha}{5 - 4\cos\alpha}$$

Equating imaginary parts on both sides, we get

$$S = \frac{4\sin\alpha}{5 - 4\cos\alpha}$$

***Model*: On summation of series based on binomial series**

***Example* 2.51:** Find the sum of the series

$$\frac{1}{2}\cos\alpha.\sin\alpha+\frac{1.3}{2.4}\cos^2\alpha.\sin 2\alpha+\frac{1.3.5}{2.4.6}\cos^3\alpha.\sin 3\alpha+\ldots\text{ to }\infty$$

***Solution*:** Let $S=\frac{1}{2}\cos\alpha.\sin\alpha+\frac{1.3}{2.4}\cos^2\alpha.\sin 2\alpha$

$$+\frac{1.3.5}{2.4.6}\cos^3\alpha.\sin 3\alpha+\ldots\text{ to }\infty$$

and $$C=\frac{1}{2}\cos\alpha.\cos\alpha+\frac{1.3}{2.4}\cos^2\alpha.\cos 2\alpha$$

$$+\frac{1.3.5}{2.4.6}\cos^3\alpha+\cos 3\alpha+\ldots\text{ to }\infty$$

$$\therefore\quad C+iS=\frac{1}{2}\cos\alpha.e^{i\alpha}+\frac{1.3}{2.4}\cos^2\alpha e^{2i\alpha}+\frac{1.3.5}{2.4.6}\cos^3\alpha.e^{3i\alpha}+\ldots\text{ to }\infty$$

$$=\frac{1}{2}x+\frac{1.3}{2.4}x^2+\frac{1.3.5}{2.4.6}x^3+\ldots\text{ to }\infty$$

(putting $\cos\alpha\,.\,e^{i\alpha}=x$)

$$=(1-x)^{-\frac{1}{2}}-1,\qquad\text{by binomial theorem}$$

$$=(1-\cos\alpha.e^{i\alpha})^{-\frac{1}{2}}-1=\left[1-\cos\alpha(\cos\alpha+i\sin\alpha)\right]^{-\frac{1}{2}}-1$$

$$=\left[\sin\alpha(\sin\alpha-i\cos\alpha)\right]^{-\frac{1}{2}}-1$$

$$=\frac{1}{\sqrt{\sin\alpha}}\left\{\cos\left(\frac{\pi}{2}-\alpha\right)-i\sin\left(\frac{\pi}{2}-\alpha\right)\right\}^{-\frac{1}{2}}-1$$

$$=\frac{1}{\sqrt{\sin\alpha}}\left[\cos\left(\frac{\pi}{4}-\frac{\alpha}{2}\right)+i\sin\left(\frac{\pi}{4}-\frac{\alpha}{2}\right)\right]-1$$

Equating imaginary parts on both sides, we get

$$S=\frac{1}{\sqrt{\sin\alpha}}.\sin\left(\frac{\pi}{4}-\frac{\alpha}{2}\right)=\frac{1}{\sqrt{2\sin\alpha}}\left(\cos\frac{\alpha}{2}-\sin\frac{\alpha}{2}\right)$$

Exercise 2.5

***Model*: On summation of series based on logarithmic series**

1. Sum to infinity the following series.

(i) $\sin\theta - \frac{1}{2}\sin 2\theta + \frac{1}{3}\sin 3\theta - \dots$ to ∞

(ii) $c\cos\alpha - \frac{1}{2}c^2\cos 2\alpha + \frac{1}{3}c^3\cos 3\alpha - \dots$ to ∞

and $c\sin\alpha - \frac{1}{2}c^2\sin 2\alpha + \frac{1}{3}c^3\sin 3\alpha - \dots$ to ∞

(iii) $\cos\alpha\cos\alpha - \frac{1}{2}\cos^2\alpha\cos 2\alpha + \frac{1}{3}\cos^3\alpha\cos 3\alpha - \dots$ to ∞

(iv) $\sin\alpha\sin\alpha - \frac{1}{2}\sin^2\alpha\sin 2\alpha + \frac{1}{3}\sin^3\alpha\sin 3\alpha - \dots$ to ∞

and $\sin\alpha\cos\alpha - \frac{1}{2}\sin^2\alpha\sin 2\alpha + \frac{1}{3}\sin^3\alpha\cos 3\alpha - \dots$ to ∞

(v) $c\cos\alpha + \frac{1}{2}c^2\cos 2\alpha + \frac{1}{3}c^3\cos 3\alpha + \dots$ to ∞

and $c\sin\alpha + \frac{1}{2}c^2\sin 2\alpha + \frac{1}{3}c^3\sin 3\alpha + \dots$ to ∞, where $|c| < 1$

(vi) $\cos\alpha\sin\alpha + \frac{1}{2}\cos^2\alpha\sin 2\alpha + \frac{1}{3}\cos^3\alpha\sin 3\alpha + \dots$ to ∞

and $\cos\alpha\cos\alpha + \frac{1}{2}\cos^2\alpha\cos 2\alpha + \frac{1}{3}\cos^3\alpha\cos 3\alpha + \dots$ to ∞, where $0 < \alpha < \pi$

(vii) $\sin\theta + \frac{1}{2}\sin 2\theta + \frac{1}{3}\sin 3\theta + \dots$ to ∞

(viii) $\cos\frac{\pi}{3} + \frac{1}{2}\cos\frac{2\pi}{3} + \frac{1}{3}\cos\frac{3\pi}{3} + \dots$ to ∞

and $\sin\frac{\pi}{3} + \frac{1}{2}\sin\frac{2\pi}{3} + \frac{1}{3}\sin\frac{3\pi}{3} + \dots$ to ∞

***Model*: On summation of series based on the Gregory's series**

2. (i) If $C = \cos^2\theta - \frac{1}{3}\cos^3\theta\cos 3\theta + \frac{1}{5}\cos^5\theta\cos 5\theta - \dots$ to ∞,

then prove that $\tan 2C = 2\cot^2\theta$.

(ii) Sum the series

$$\cos\theta - \frac{1}{3}\cos 3\theta + \frac{1}{5}\cos 5\theta - \ldots \text{ to } \infty$$

(iii) Sum the series

$$\cos\frac{\pi}{3} + \frac{1}{3}\cos\frac{2\pi}{3} + \frac{1}{5}\cos\frac{3\pi}{3} + \ldots \text{ to } \infty,$$

Hint: $C + iS = e^{\frac{2\pi i}{6}} + \frac{1}{3}e^{\frac{4\pi i}{6}} + \frac{1}{5}e^{\frac{6\pi i}{6}} + \ldots \text{ to } \infty$

$$= e^{\frac{\pi i}{6}}\left(e^{\frac{\pi i}{6}} + \frac{1}{3}e^{\frac{3\pi i}{6}} + \frac{1}{5}e^{\frac{5\pi i}{6}} + \ldots \text{ to } \infty\right)$$

$$= \frac{1}{2}e^{\frac{\pi i}{6}}\log\frac{1 + e^{\frac{\pi i}{6}}}{1 - e^{\frac{\pi i}{6}}}$$

***Model*: On summation of series based on exponential series**

3. Sum the following series

(i) $\sin\theta + \frac{a\sin 2\theta}{1!} + \frac{a^2\sin 3\theta}{2!} + \frac{a^2\sin 4\theta}{3!} + \ldots \text{ to } \infty$

(ii) $\cos\theta + \frac{\cos\theta}{1!}\cos 2\theta + \frac{\cos^2\theta}{2!}\cos 3\theta + \ldots \text{ to } \infty$

and $\sin\theta + \frac{\cos\theta}{1!}\sin 2\theta + \frac{\cos^2\theta}{2!}\cos 3\theta + \ldots \text{ to } \infty$

(iii) $\sin\theta\cos\theta + \frac{\sin 2\theta\cos^2\theta}{2!} + \frac{\sin 3\theta\cos^3\theta}{3!} + \ldots \text{ to } \infty$

and $1 + \cos\theta.\cos\theta + \frac{\cos 2\theta\cos^2\theta}{2!} + \frac{\cos 3\theta\cos^3\theta}{3!} + \ldots \text{ to } \infty$

(iv) $\cos\theta + \frac{\operatorname{cosec}\theta}{1!}\cos 2\theta + \frac{\operatorname{cosec}^2\theta}{2!}\cos 3\theta + \ldots \text{ to } \infty$

(v) $\cos\theta + \frac{\sin\theta}{1!}\cos 2\theta + \frac{\sin^2\theta}{2!}\cos 3\theta + \ldots \text{ to } \infty$

and $\sin\theta + \frac{\sin\theta}{1!}\sin 2\theta + \frac{\sin^2\theta}{2!}\sin 3\theta + \ldots \text{ to } \infty$

(vi) $1+\cos\theta+\frac{\cos 2\theta}{2!}+\frac{\cos 3\theta}{3!}+\ldots$ to ∞

and $\frac{\sin\theta}{1!}+\frac{\sin 2\theta}{2!}+\frac{\sin 3\theta}{3!}+\ldots$ to ∞

(vii) $1-\cos\alpha\cos\beta+\frac{\cos^2\alpha}{2!}\cos 2\beta-\frac{\cos^3\alpha}{3!}\cos 3\beta+\ldots$ to ∞

and $\cos\alpha\sin\beta-\frac{\cos^2\alpha}{2!}\sin 2\beta+\frac{\cos^2\alpha}{3!}\sin 3\alpha-\ldots$ to ∞

(viii) $1+\cos\alpha\cos\beta+\frac{\cos^2\alpha}{2!}\cos 2\beta+\frac{\cos^3\alpha}{3!}\cos 3\beta+\ldots$ to ∞

4. Sum the series: $\sin\theta-\frac{\sin 2\theta}{2!}+\frac{\sin 3\theta}{3!}-\ldots$ to ∞

***Model*: On summation of series based on geometric progression**

5. Sum the following series.

(i) $\cos\alpha+\frac{1}{2}\cos 2\alpha+\frac{1}{2^2}\cos 3\alpha+\ldots$ to ∞

(ii) $\cos\theta.\cos\theta+\cos^2\theta\cos 2\theta+\cos^3\theta\cos 3\theta+\ldots$ to ∞

and $\cos\theta\sin\theta+\cos^2\theta\sin 2\theta+\cos^3\theta\sin 3\theta+\ldots$ to ∞

(iii) $1+x\cos\alpha+x^2\cos 2\alpha+\ldots$ to ∞, where $|x|<1$

***Model*: On summation of series based on binomial series**

6. Sum the following series.

(i) $1+\frac{1}{2}\cos 2\theta-\frac{1}{2.4}\cos 4\theta+\frac{1.3}{2.4.6}\cos 6\theta-\ \ldots$ to ∞,

Where $-\frac{\pi}{2}<\theta<\frac{\pi}{2}$

Hint: $C+iS=\left(1+e^{2i\theta}\right)^{\frac{1}{2}}$

(ii) $1+\frac{1}{2}\cos\alpha+\frac{1.3}{2.4}\cos 2\alpha+\frac{1.3.5}{2.4.6}\cos 3\alpha+\ldots$ to ∞

and $\frac{1}{2}\sin\alpha+\frac{1.3}{2.4}\sin 2\alpha+\frac{1.3.5}{2.4.6}\sin 3\alpha+\ldots$ to ∞

Hint: $C+iS+iS=\left(1-e^{i\theta}\right)^{-\frac{1}{2}}$

(iii) $1-\frac{1}{2}\cos\theta+\frac{1.3}{2.4}\sin 2\theta-\frac{1.3.5}{2.4.6}\cos 3\theta+\ldots$ to ∞

Answers

1. (i) $\frac{1}{2}\theta$

(ii) $\frac{1}{2}\log\left(1+2c\cos\alpha+c^2\right)$ and $\tan^{-1}\frac{c\sin\alpha}{1+c\cos\alpha}$

(iii) $\frac{1}{2}\log\left(1+3\cos^2\alpha\right)$ and $\tan^{-1}\frac{\sin\alpha\cos\alpha}{1+\cos^2\alpha}$

(iv) $\tan^{-1}\frac{\sin^2\alpha}{1+\sin\alpha\cos\alpha}$ and $\frac{1}{2}\log\left(1+\sin 2\alpha+\sin^2\alpha\right)$

(v) $-\frac{1}{2}\log\left(1-2c\cos\alpha+c^2\right)$ and $\tan^{-1}\frac{c\sin\alpha}{1-c\cos\alpha}$

(vi) $\frac{\pi}{2}-\alpha$ and $\log\operatorname{cosec}\alpha$

(vii) $\frac{\pi}{2}-\frac{\theta}{2}$

(viii) 0 and $\frac{\pi}{3}$

2. (ii) $\pm\frac{\pi}{4}$ (iii) $\frac{1}{8}\left[2\sqrt{3}\log\left(2+\sqrt{3}\right)-\pi\right]$

3. (i) $e^{a\cos\theta}.\sin(\theta+\alpha\sin\theta)$

(ii) $e^{\cos^2\theta}\cos(\theta+\cos\theta\sin\theta)$ and $e^{\cos^2\theta}\sin(\theta+\sin\theta\cos\theta)$

(iii) $e^{\cos^2\theta}\sin(\sin\theta\cos\theta)$ and $e^{\cos^2\theta}\cos(\sin\theta\cos\theta)$

(iv) $e^{\cos\theta}.\cos(1+\theta)$

(v) $e^{\sin\theta\cos\theta}\cos\left(\theta+\sin^2\theta\right)$ and $e^{\sin\theta\cos\theta}\sin\left(\theta+\sin^2\theta\right)$

(vi) $e^{\cos\theta}.\cos(\sin\theta)$ and $e^{\cos\theta}.\sin(\sin\theta)$

(vii) $e^{-\cos\alpha\cos\beta}.\cos(\cos\alpha.\sin\beta)$ and $e^{-\cos\alpha\cos\beta}\sin(\cos\alpha.\sin\beta)$

(viii) $e^{\cos\alpha\cos\beta}.\cos(\cos\alpha.\sin\beta)$

4. $e^{-\cos\theta}$. sin (sin θ)

5. (i) $\dfrac{4\cos\alpha-2}{5-4\cos\alpha}$ (ii) 0 and cot θ (iii) $\dfrac{1-x\cos\alpha}{1-2x\cos\alpha+x^2}$

6. (i) $\sqrt{2\cos\theta}.\cos\dfrac{1}{2}\theta$

(ii) $\left(2\sin\dfrac{\alpha}{2}\right)^{-\frac{1}{2}}\cos\dfrac{\pi-\alpha}{4}$ and $\left(2\sin\dfrac{\alpha}{2}\right)^{-\frac{1}{2}}\sin\dfrac{\pi-\alpha}{4}$

(iii) $\dfrac{\cos\frac{1}{4}\theta}{\sqrt{2\cos\frac{1}{2}\theta}}$

CHAPTER 3

Theory of Equations

3.1 Introduction

An expression of the form

$$f(x) = a_0x^n + a_1x^{n-1} + \ldots..+ a_{n-1}\,x + a_n$$

where a's are constants ($a_0 \neq 0$) and n is positive integer, is called a polynomial in x of degree n. The polynomial f(x) = 0 is called an algebraic equation of degree n. If f(x) contains some other functions such as trigonometric, logarithemic, exponential etc., then f(x) = 0 is called a transcendental equation.

The value α of x which satisfies f(x) = 0 .....(3.1)

is called a root of f(x) = 0. Geometrically, a root of (3.1) is that value of x where the graph of y = f(x) crosses the x-axis.

The process of finding the roots of an equation is known as the solution of that equation. This is a problem of great importance in applied mathematics.

If f(x) is a quadratic, cubic or a biquadratic expression, algebraic solutions of equation are available. But the need often arises to solve higher degree or transcendental equations for which no direct methods exist. Such equations can best be solved by approximate methods.

3.2 Theory of Equation

State and prove fundamental theorem of Algebra.

Statement: Every polynomial equation with real coefficients has at least one root.

Or, Every equation of nth degree has n roots and no more.

Proof: Let $f(x) = a_0x^n + a_1x^{n-1} + a_2x^{n-2} + \ldots + a_{n-1} + a_n = 0$ ....(A)

be a general equation of nth degree.

Then we have to prove that this equation has n roots.

In the first stage of the proof, let us assume that every equation has a root. This is the subject of complex analysis. So it is taken for granted here.

Let α_1 be one of the roots of the equation. Then it is clear that $(x - \alpha_1)$ is a factor of f(x). It means that when the division of f(x) is done by $(x - \alpha_1)$ then there is no remainder.

Let $\phi_1(x)$ be the quotient in this case. Thus we can write

$$f(x) = (x - \alpha_1)\,\phi_1(x) \qquad .....(3.2)$$

It is to be noted here that $\phi_1(x)$ is a polynominal of $(n-1)^{th}$ degree.

Then according to our elementary assumption this $\phi_1(x)$ will also have a root and let, α_2 be its root.

Then again $(x - \alpha_2)$ will be a factor of $\phi_1(x)$ and we can write

$$\phi_1(x) = (x - \alpha_2)\,\phi_2(x) \qquad .....(3.2A)$$

From (3.2) and (3.2A) we have

$$f(x) = (x - \alpha_1)(x - \alpha_2)\,\phi_2(x) \qquad .....(3.3)$$

It may be noted here that $\phi_2(x)$ is a polynomial of $(n-2)^{th}$ degree.

Again $\phi_2(x)$ has a root. Let α_3 be its root

Then $(x - \alpha_3)$ is a factor of $\phi_2(x)$

$$\phi_2(x) = (x - \alpha_3)\,\phi_2(x) \qquad .....(3.4)$$

From (3.3) and (3.4) we have

$$f(x) = (x - \alpha_1)(x - \alpha_2)(x - \alpha_3)\,\phi_3(x) \qquad .....(3.5)$$

where $\phi_3(x)$ is a polynomial of $(n-3)^{th}$ degree.

Proceeding in this way, we observe that after n^{th} factor degree of the quotient polynomial becomes zero, and so it cannot. Then we can write

$$f(x) = (x - \alpha_1)(x - \alpha_2)\,....(x - \alpha_n)K \qquad .....(B)$$

Now equating the coefficient of x^n from (A) & (B) we have $K = a_0$

Thus $f(x) = a_0 (x - \alpha_1)(x - \alpha_2)....(x - \alpha_n)$

If we have put $x = \alpha_1\ \alpha_2, ...\alpha_n$ successively in the above equation is satisfied. Thus there are n^{th} roots of the given equation.

It is also clear that no other quantity than these n can satisfy the above equation; having more than n^{th} roots doesn't arise.

Thus we conclude that every equation of nth degree has n roots and no more.

3.3 Establish Cardon's Method of Solving Cubic Equation

The general equation of a cubic is taken as

$$ax^3 + 3bx^2 + 3cx + d = 0$$

In order to obtain the solution of a cubic we first reduce it to the cubic in which the second degree term is absent.

For this we adopt the substitution $z = ax + b$

$$ax = z - b \quad \therefore \quad x = \frac{z-b}{a}$$

Substituting this value of x in the general equation of a cubic, we have

$$a\left(\frac{z-b}{a}\right)^3 + 3b\left(\frac{z-b}{a}\right)^2 + 3c\left(\frac{z-b}{a}\right) + d = 0$$

i.e.,
$$a\left(\frac{z-b}{a^3}\right)^3 + \frac{3b(z-b)^2}{a^2} + \frac{3c(z-b)}{a} + d = 0$$

i.e.,
$$(z-b)^3 + 3b(z-b)^2 + 3ac(z-b) + a^2d = 0$$

i.e.,
$$z^3 - 3z^2b + 3b^2z - b^3 + 3b(z^2 - 2bz + b^2)$$
$$+ 3\,acz - 3abc + a^2d = 0$$

i.e.,
$$z^3 - 3z^2b + 3zb^2 - b^3 + 3z^2b - 6b^2z + 3b^3 +$$
$$3acz - 3abc + a^2d = 0$$

i.e.,
$$z^3 - 3b^2z + 2b^3 + 3acz - 3abc + a^2d = 0$$

i.e.,
$$z^3 + z(3ac - 3b^2) + (2b^3 - 3abc + a^2d) = 0$$

i.e.,
$$z^3 + 3z\,(ac - b^2) + (2b^3 - 3abc + a^2d) = 0$$

i.e.,
$$z^3 + 3Hz + G = 0 \quad \text{.....(3.6)}$$

where $H = ac - b^2$ & $G = 2b^3 - 3abc + a^2d$.

In order to solve the above cubic (which is known as Cardon's reducing cubic) Cardon took.

$$Z = P^{1/3} + q^{1/3}$$

$$Z^3 = p + q + 3\,P^{1/3}\,q^{1/3}\,(p + q)$$

$$Z^3 = p + q + 3\,p^{1/3}\,q^{1/3}z$$

i.e., $$z^3 - 3p^{1/3}\,q^{1/3}z - (p + q) = 0 \qquad .....(3.7)$$

Then it is clear that this cubic must be identical with the cubic (3.6). Hence equating the coefficients in equation (3.6) & (3.7) we have

$H = -p^{1/3}\,q^{1/3}$ $\quad \therefore\ P^{1/2}\,q^{1/3} = -H$

$G = -(p + q)$ $\quad \therefore\ p + q = -G$

$pq = -H^3$

$$(p - q)^2 = (p + q)^2 - 4pq$$
$$= G^2 - 4(-H^3) = G^2 + 4H^3$$
$$p - q = \sqrt{G^2 + 4H^3}$$

Thus we have $p + q = -G$

$$p - q = \sqrt{G^2 + 4H^3}$$

$\therefore$ $$2p = -G + \sqrt{G^2 + 4H^3}$$

$\therefore$ $$p = \frac{-G + \sqrt{G^2 + 4H^3}}{2}$$

and $$2q = -G - \sqrt{G^2 + 4H^3}$$

$\therefore$ $$q = -G - \frac{\sqrt{G^2 + 4H^3}}{2}$$

Thus we have succeeded in finding out the separate values of p & q. It is clear that $p^{1/3}$ (cubic root of p) has three values p, wp, w^2p. Similarly $q^{1/3}$ has three values q, wq, w^2q. Thus the roots of reducing cubic will be

$$p^{1/3} + q^{1/3},\ wp^{1/3} + wq^{1/3},\ w^2p^{1/3} + w^2q^{1/3}$$

Thus these are three values of two which satisfies the given equation for each values of z so obtained we get the corresponding values of x which are the roots of the given cubic equation.

3.4 Obtain the Reducing Cubic of the Biquadratic Equation $ax^4 + 4bx^3 + 6cx^2 + 4dx + e = 0$

The general form of a biquadratic is $ax^4 + 4bx^3 + 6cx^2 + 4dx + e = 0$

In order to solve this biquadratic equation, we transform it into a biquadratic in which the third degree term is absent. For this we adopt the substitution

$$z = ax + b \qquad \therefore\ x = \frac{z-b}{a}$$

i.e., $$a\left(\frac{z-b}{a}\right)^4 + 4b\left(\frac{z-b}{a}\right)^3 + 6c\left(\frac{z-b}{a}\right)^2 + 4d\left(\frac{z-b}{a}\right) + e = 0$$

i.e., $$(z-b)^4 + 4b(z-b)^3 + 6ac(z-b)^2 + 4a^2d(z-b) + a^3e = 0$$

i.e., $$z^4 - 4z^3b + 6z^2b^2 - 4zb^3 + b^4 + 4bz^3 - 12z^2b^2 + 12zb^3$$
$$-4b^4 + 6acz^2 - 12abcz + 6ab^2c + 4a^2dz - 4a^2bd + a^3e = 0$$

i.e., $$z^4 - 6z^2b^2 + 6acz^2 + 8zb^3 - 12abcz - 3b^4 + a^2dz + 6ab^2c$$
$$-4a^2bd + a^3e + 3a^2c^2 - 3a^2c^2 = 0$$

i.e., $$z^4 + 6\left(ac - b^2\right)z^2 + 4c\left(2b^3 - 3abc + a^2d\right)z + a^2\left(ae - 4bd + 3c^2\right)$$
$$-3\left(a^2c^2 - 2ab^2c + b^4\right) = 0$$

i.e., $$z^4 + 6Hz^2 + 4Gz + a^2I - 3H^2 = 0 \qquad \text{.....(3.8)}$$

where $H = ac - b^2$, $\quad G = 12b^3 - 3abc + a^2 d$

$I = ac - 4bd + 3c^2$

In order to solve this biquadratic Euler assumed an expression for its root's as

$$z = p^{1/2} + q^{1/2} + r^{1/2}$$

$$z^2 = p + q + r + 2p^{1/2}\, q^{1/2} + 2q^{1/2}\, r^{1/2} + 2p^{1/2}\, r^{1/2}$$

$$z^2 - (p+q+r) = 2\left(p^{1/2}\, q^{1/2} + q^{1/2}\, r^{1/2} + p^{1/2}\, r^{1/2}\right)$$

Squaring both sides we get

$$\left(z^2 - (p+q+r)\right)^2 = 4\left(p^{1/2}\, q^{1/2} + q^{1/2}\, r^{1/2} + p^{1/2}\, r^{1/2}\right)^2$$

i.e., $$z^4 - 2(p+q+r)z^2 + (p+q+r)^2$$
$$= 4\left(pq + qr + rp + 2p^{1/2}\, q^{1/2}\, r + 2p^{1/2}\, r^{1/2}\, q + 2q^{1/2}\, r^{1/2}\, q\right)$$

i.e., $$z^4 - 2(p+q+r)z^2 + (p+q+r)^2$$
$$= 4(pq + qr + rp) + 8\left(p^{1/2}\, r^{1/2}\, q + q^{1/2}\, p^{1/2}\, r + q^{1/2}\, r^{1/2}\, p\right)$$

i.e., $z^4-2(p+q+r)z^2+(p+q+r)^2=4(pq+qr+rp)+8p^{1/2}q^{1/2}r^{1/2}z$

i.e., $z^4-2(p+q+r)z^2-8p^{1/2}q^{1/2}r^{1/2}z+(p+q+r)^2-4(pq+qr+rp)=0$ .....(3.9)

Comparing (3.8) and (3.9) we get

$$6H=-2(p+q+r)$$

$$4G=-8p^{1/2}q^{1/2}r^{1/2}$$

$$a^2I-3H^2=(p+q+r)^2-4(pq+qr+rp) \quad .....(3.10)$$

$$p+q+r=-3H$$

$$p^{1/2}q^{1/2}r^{1/2}=-G/2$$

$$pqr=G^2/4$$

From (3.8)

$$4(pq+qr+rp)=(p+q+r)^2-a^2I+3H^2$$

$$=9H^2-a^2I+3H^2=12H^2-a^2I$$

$$pq+qr+rp=3H^2-\frac{a^2I}{4}$$

Thus we see that the p, q & r be the roots of the equation.

$$y^3-(p+q+r)\,y^2+(pq+qr+rp)^4-pqr=0$$

$$y^3-(-3H)y^2+(3H^2-\frac{a^2I}{4})y-\frac{G^2}{4}=0$$

$$y^3+3Hy^2+\left(3H^2y-\frac{a^2I}{4}\right)y-\frac{G^2}{4}=60$$

$\therefore$ $$4y^3+12Hy^2+12H^2\,y-a^2Iy-G^2=0$$

$$4(y+H)^3-4H^3-a^2Iy-G^2=0$$

$\Rightarrow$ $$4(y+H)^3-a^2I(y+H)+a^2IH-4H^3-G^2=0$$

Now we put $y+H=a^2t$ & $a^2IH-4H^3-G^2=a^3j$

The equation reduces to

$$4\left(a^2t\right)^3-a^2I.a^2t+a^3j=0$$

$$4a^6t^3-a^4\ It+a^3j=0$$

i.e.,
$$4a^3t^3-aIt+j=0$$

Which is the third degree equation in a t & this cubic is known as Euler's reducing cubic.

This cubic equation is such that the second degree from is wanting in it. Thus it can be solved to give us three values of a^2t. Corresponding to each value of a^2t we shall be getting a value of y and those three values of y namely a^2t_1-H, a^2t_2-H & a^2t_3-H will be the values of p, q, r.

From the above equation we get eight values of z out of which only four will satisfy the condition $p^{1/2}\ q^{1/2}\ r^{1/2}=-G/2$ and these values will be accepted. Corresponding to these four values of x we get four values of z which is the complete solution of the biquadratic.

3.5 Find out the Relation between the Roots of the Biquadratic and Euler's Cubic

We have shown in the previous article that we get only four values of $z=\sqrt{p}+\sqrt{q}+\sqrt{r}$ due to the relation pqr = G/2

If G is negative and consequently in this case either all of $\sqrt{pqr}$ should be + ve or two – ve and one + ve.

If z_1, z_2, z_3, z_4 be the values of z and α, β, γ, δ be the roots of the given biquadratic then in this case we have

$$z_1=a_0\alpha+a_1=\sqrt{p}-\sqrt{q}-\sqrt{r}$$

$$z_2=a_0\beta+a_1=-\sqrt{p}+\sqrt{q}-\sqrt{r}$$

$$z_3=a_0\gamma+a_1=\sqrt{-p}-\sqrt{q}+\sqrt{r} \qquad \text{(A)}$$

$$z_4=a_0\delta+a_1=\sqrt{p}+\sqrt{q}+\sqrt{r}$$

From (A)
$$a_0\left(\beta+\gamma\right)+2a_1=2\sqrt{p}$$

and $$a_0(\alpha+\delta)+2a_1=2\sqrt{p}$$

Substracing

$$a_0(\beta+\gamma-\alpha-\delta)=-4\sqrt{p}$$

$\Rightarrow$ $$a_0^2(\beta+\gamma-\alpha-\delta)^2=16p$$

$\therefore$ $$p=\frac{a_0^2}{16}(\beta+\gamma-\alpha-\delta)^2$$

Similarly, $$q=\frac{a_0^2}{16}(\gamma+\alpha-\beta-\delta)^2 \qquad \text{(B)}$$

and $$r=\frac{a_0^2}{16}(\alpha+\beta-\gamma-\delta)^2$$

The equation (B) gives the necessary relation between the roots of the biquadratic and those of the Euler's cubic.

3.6 Establish a Relation between the Roots of the Biquadratic and the Reducing Cubic

From the relation (A) we get

$$a_0(\alpha-\beta)=2\left(\sqrt{p}-\sqrt{q}\right)$$

$$a_0(\gamma-\delta)=-2\left(\sqrt{p}+\sqrt{q}\right)$$

Multiplying these we get

$$a_0^2(\alpha-\beta)(\gamma-\delta)=-4(p-q)$$

i.e., $$4(p-q)=-a_0^2(\alpha-\beta)(\gamma-\delta)$$

$$4(q-\gamma)=-a_0^2(\beta-r)(\alpha-\delta) \qquad \text{(C)}$$

$$4(r-p)=-a_0^2(\gamma-\alpha)(\beta-\delta)$$

But $$p+H=a_0^2\delta_1,\ q+H=a_0^2\delta_2,\ r+H=a_0^2\delta_3$$

Hence from (C) we get

$$4(p-q)=4a_0^2(\delta_1-\delta_2)$$
$$=-a_0^2(\alpha-\beta)(\gamma-\delta)$$
$$4(q-r)=4a_0^2(\delta_2-\delta_3) \qquad \text{(D)}$$
$$=-a_0^2(\beta-\gamma)(\alpha-\delta)$$
$$4(r-p)=4a_0^2(\delta_3-\delta_1)=-a_0^2(\gamma-\alpha)(\beta-\delta)$$

Again from (D) we get

$$4a_0^2(\delta_3 - 2\delta_1 + \delta_2) = -a_0^2\left[(\gamma-\alpha)(\beta-\delta)-(\alpha-\beta)(\gamma-\delta)\right]$$

$$\Rightarrow \quad 4a_0^2(\delta_3 - 2\delta_1 + \delta_2) = -a_0^2\left[(\gamma-\alpha)(\beta-\delta)-(\alpha-\beta)(\gamma-\delta)\right]$$

But $\quad \delta_1 + \delta_2 + \delta_3 = 0 \; \delta_2 + \delta_3 = -\delta_1$

$$4a_0^2(+3\delta_1) = +a_0^2\left[(\gamma-\alpha)(\beta-\delta)-(\alpha-\beta)(\gamma-\delta)\right]$$

$$12\delta_1 = (\gamma-\alpha)(\beta-\delta)-(\alpha-\beta)(\gamma-\delta)$$

$$12\delta_2 = (\alpha-\beta)(\gamma-\delta)-(\beta-\gamma)(\gamma-\delta)$$

$$12\delta_3 = (\beta-\gamma)(\alpha-\delta)-(\gamma-\alpha)(\beta-\delta)$$

Synthetic Division: This is a process of dividing a polynomial by a binomial, that is $a_0x^n + a_1x^{n-1} + a_2x^{n-2} + \ldots\ldots + a_{n-1}x + a_n$ be a polynomial of nth degree. We propose to divide it by a binomial $x - \alpha$.

Let $b_1x^{n-1} + b_2x^{n-2} + \ldots\ldots + b_{n-1} + b_n$ be the quotient and R be the remainder. When the actual division of the polynomial by a binomial it is done. Then we can write

$$(x-\alpha)\left(b_1x^{n-1} + b_2x^{n-2} + \ldots\ldots + b_{n-1}x + b_n\right) + R$$

$$= a_0x^n + a_1x^{n-1} + a_2x^{n-2} + \ldots\ldots + a_{n-1}x + a_n$$

The above identity in x will hold for all values of x. Thus equating the coefficient of like powers of x from bothsides of the identity we have

$$b_1 = a_0 \qquad\qquad b_1 = a_0$$

$$b_2 - b_1\alpha = a_1 \qquad\qquad b_2 = a_1 + b_1\alpha$$

$$b_3 - b_2\alpha = a_2 \qquad\qquad b_3 = a_2 + b_2\alpha$$

....

....

$$R - b_n\alpha = a_n \qquad\qquad R = a_n + b_n\alpha$$

Thus we see that every coefficient in the quotient polynomial may be obtain in terms of α and the coefficients of the given polynomial as follows–

$$b_1 = a_0$$

$$b_2 = a_1 + b_1\alpha = a_1 + a_0\alpha$$

$$b_3 = a_2 + b_2\alpha = a_2 + a_1\alpha + b_1\alpha^2$$

....

Thus the whole process may be compared in a compact form as follows –

$$b_1 = a_0, b_2 = a_1 + b_1\alpha$$

$$b_3 = a_2 + b_2\alpha \; R = a_n + b_n\alpha$$

$$\begin{array}{c|l} \alpha & a_0 \; a_1 \; a_2 \; \ldots\ldots \; a_n \\ & b_1\alpha \; b_2\alpha \; \ldots. \; b_n\alpha \\ \hline & b_1 \; b_2 \; b_3 \qquad R \end{array}$$

3.7 State Descarte's Rule of Signs

The equation f(x) = 0 cannot have more positive roots than the no. of changes in the sign of f(x) from + ve to – ve or – ve to + ve in terms of its first member. Let us consider the equation

$$f(x) = x^3 + x^2 - x + 1 = 0$$

The sign of the terms in the above equation is given as follows –

$$+ + - +$$

We observe that there are two changes in the signs, one from + ve to – ve and another – ve to + ve. Thus above equation cannot have more than two + ve roots.

Second Rule: The equation f(x) = 0 cannot have more (–) ve roots than the changes of signs in f (–x). Let us consider the equation

$$f(x) = x^3 + x^2 - x + 1 = 0$$

$$f(-x) = -x^3 + x^2 + x + 1 = 0$$

Thus the changes of sign in f(–x) may be sign as follows –+++. We observe here that there is only one change of sign in f(–x). Thus equation f(x) = 0 cannot have more than one negative root.

***Note*:** We know that if a function f(x) is continuous in closed interval [a, b] such that it has opposite signs of both the ends of the interval. Then at both the ends of the interval there exist a point between a & b. Such that the function vanishes at that point. This property of continous function leads us to a conclusion that if f(a) & f(b) has opposite sign, then the equation f(x) = 0 has one of its roots between a and b.

3.8 State and Prove Factor Theorem

Statement: If α is a root of an equation nx, then (x – α) must be fracted.

Proof: Let $f(x) = a_0x^n + a_1x^{n-1} + a_2x^{n-2} + a_{n-1}x + a_n = 0$ be an equation of nth degree. Let be its roots. Then we have to prove that (x–) is a factor of f(x) = 0.

As α is a root of $f(x) = 0$

We have $f(x) = 0$ .....(3.11)

If $x - \alpha$ is to be a factor of $f(x) = 0$. Then if we divide $f(x)$ by $x - \alpha$, then there should be no remainder. The division is not exact. Let $\phi(x)$ be the quotient and R be the remainder.

$$f(x) = \phi(x)(x - \alpha) + R$$

Put $$x = \alpha,\ f(x) = \phi(\alpha)(\alpha - \alpha) + R$$

$$f(x) = 0 + R \qquad \text{(From} - 3.11)$$

This shows that the division of $f(x)$ by $x - \alpha$ is exact. Hence $(x - \alpha)$ is a factor.

3.9 The Imaginary Roots of an Equation with Real Coefficients always Occur in Conjugate Pair

Proof: Let $f(x) = a_0x^n + a_1x^{n-1} + ... + a_{n-1}\ x + a_n = 0$ be a general polynomial of nth degree.

Let $\alpha + i\beta$ be one of its roots. Then we have to prove that $\alpha + i\beta$ must be another root of this equation.

As $\alpha + i\beta$ is given to be roots of the equation $f(x) = 0$. We have

$$f(\alpha + i\beta) = 0 \qquad .....(3.12)$$

Also from factor theorem it is clear that $f\{x - (\alpha + i\beta)\}$ must be factor of the equation. If $(\alpha - i\beta)$ be a root of the equation, then $x - (\alpha - i\beta)$ should be another factor of the equation. Thus $\{x - (\alpha + i\beta)\}$ and $\{x - (\alpha - i\beta)\}$ should be a factor of the equation.

i.e., $(x - \alpha - i\beta)(x - \alpha + i\beta)$ should be a factor.

i.e., $\{(x - \alpha)^2 - i^2\beta^2\}$ should be a factor.

i.e., $\{(x - \alpha)^2 + \beta^2\}$ should be a factor.

The above expression will be a factor of the given equation. If we divide $f(x)$ by $\{(x - \alpha)^2 + \beta^2\}$ then there should be no remainder.

Let us suppose that when actual division is done $\phi(x)$ is the quotient and $px + \theta$ is the remainder.

Thus we have $f(x)=\{(x-\alpha)^2+\beta^2\}\phi(x)+px+\theta$

Now we put $x=\alpha+i\beta$ then

$$f(\alpha+i\beta)=\{(x+i\beta-\alpha)^2+\beta^2\}\phi(\alpha+i\beta)+p(\alpha+i\beta)+\theta$$

i.e., $$0=\{i^2\beta^2+\beta^2\}\phi(\alpha+i\beta)+p(\alpha+i\beta)+\theta$$

i.e., $$0=(-\beta^2+\beta^2)\phi(\alpha+i\beta)+p(\alpha+i\beta)+\theta$$

i.e., $$p(\alpha+\beta)+\theta=0$$

$$p\alpha+\theta+ip\beta=0$$

But we known that if a complex number is equal to zero. Then its real and imaginary part are separately equal to zero.

$$p\alpha+\theta=0 \quad p\beta=0 \quad \text{i.e., } p=0 \text{ as } \beta\neq 0$$

$\therefore$ $$0\alpha+\theta=0 \quad \Rightarrow \theta=0$$

As p and θ are separately equal to zero. It is obvious that the remainder $px+\theta=0$. Thus we observe here that the division of f(x) is exactly by the product of factors.

Hence $\alpha-i\beta$ is another root.

Problems on Descrete's Rule of Sign

3.10 Apply Descarte's Rule of Sign to Discuss the Nature of the Roots of the Equation

$$x^4+15x^2+7x-11=0$$

$$f(x)=x^4+15x^2+7x-11=0.$$

It has only one change of sign and it must have one (+) ve root.

$$f(-x)=x^4+15x^2+7x-11=0$$

There is only one change in sign and it must have one (–) ve root.

Thus the equation has two real roots in which one root is + ve and other is – ve and hence the other two roots must be imaginary.

3.11 Find the Minimum Number of Imaginary Roots which the Equation $2x^7 - x^4 + 4x^3 - 5 = 0$ must Possess

Solution: Since f(x) = 0 has three changes of signs and as such it must have three positive roots.

Also $$f(x) = -2x^7 - x^4 - 4x^2 - 5 = 0$$

$$2x^7 + x^4 + 4x^2 + 5 = 0$$

f(–x) = 0 has no changes of signs and as such it will have no (+) ve roots which means that f(n) = 0 has no (–) ve root. Hence the given equation can have three real roots i.e., (+) ve roots and it being to 7th degree and hence the minimum number of imaginary roots is 7 – 3 = 4.

Relation between the Roots and the Coefficients

We know that if α & β be the roots of a quadratic equation $ax^2 + bx + c = 0$ then

Sum of the roots $\alpha + \beta = -b/a$

Product of the roots $\alpha\beta = c/a$

Now we have to find out the relation between the roots and coefficients of an equation of nth degree.

Let the general equation of nth degree be

$$f(x) = a_0x^n + a_1x^{n-1} + a_2x^{n-2} + \ldots + a_{n-1}x + a_n = 0 \quad \ldots..(3.13)$$

and its roots be $\alpha_1, \alpha_2, \alpha_3 \ldots \alpha_n$ then clearly

$$f(x) = K(x - \alpha_1)(x - \alpha_2)\ldots(x - \alpha_n)$$

Comparing the coefficients of x^n in both sides of the above identity we get $a_0 = K$ and hence

$$f(x) = a_0(x - \alpha_1)(x - \alpha_2)\ldots(x - \alpha_n)$$

$$= a_0\left[x^n - x^{n-1}(\alpha_1 + \alpha_2 + \alpha_n) + x^{n-2}\right.$$

$$(\alpha_1\alpha_2 + \alpha_1\alpha_3 + \ldots) - x^{n-3}\ (\alpha_1\alpha_2\alpha_3 + \alpha_1\alpha_3\alpha_4)$$

$$+(-1)^n(\alpha_1\alpha_2\alpha_3 \ldots \alpha_n) \quad \ldots..(3.14)$$

Comparing the coefficients of like powers of x in both sides we get

$$-a_0\sum\alpha_1 = a_1 \quad \therefore \sum\alpha_1 = -\frac{a_1}{a_0} = (-1)\frac{a_1}{a_0}$$

$$a_0\sum\alpha_1\alpha_2 = a_2 \quad \therefore \sum\alpha_1\alpha_2 = \frac{a_2}{a_0} = (-1)^2\frac{a_2}{a_0}$$

$$-a_0\sum\alpha_1\alpha_2\alpha_3 = a_3 \quad \therefore \sum\alpha_1\alpha_2\alpha_3 = \frac{a_3}{a_0} = (-1)^3\frac{a_3}{a_0}$$

$$a_0(-1)^n(\alpha_1\alpha_2\alpha_3...\alpha_n) = a_n \therefore \alpha_1\alpha_2\alpha_3\alpha_4...\alpha_n = (-1)^n\frac{a_n}{a_0}$$

Particular cases: Relation between the roots and the coefficient of cubic and biquadratic equation.

Cubic: $a_0x^3 + a_1x^2 + a_2x + a_3 = 0$

Let α, β, γ be the roots of the given equation.

$$\therefore \quad \sum\alpha = \alpha+\beta+\gamma = \frac{a_1}{a_0}$$

$$\sum\alpha\beta = \alpha\beta+\beta\gamma+\alpha\gamma = \frac{a_2}{a_0}$$

$$\therefore \quad \sum\alpha\beta\gamma = \frac{-a_3}{a_0}$$

Biquadratic: $a_0x^4 + a_1x^3 + a_2x^2 + a_3x + a_4 = 0$

Let α, β, γ, δ be the roots of the above given biquadratic equation.

$$\sum\alpha = \alpha+\beta+\gamma+\delta = \frac{-a_1}{a_0}$$

$$\sum\alpha\beta = \alpha\beta+\alpha\gamma+\alpha\delta+\beta\delta+\gamma\delta = \frac{a_2}{a_0}$$

or, $$\sum\alpha\beta = (\alpha+\beta)(\gamma+\delta)+\alpha\beta+\gamma\delta = \frac{a_2}{a_0}$$

$$\sum\alpha\beta\gamma = \alpha\beta\gamma+\alpha\beta\delta+\gamma\alpha\delta+\gamma\beta\delta = -\frac{-a_3}{a_0}$$

or, $$\sum\alpha\beta\gamma = \alpha\beta\gamma(\gamma+\delta)+\gamma\delta(\alpha+\beta) = -\frac{-a_3}{a_0}$$

$$\sum\alpha\beta\gamma\delta = \alpha\beta\gamma\delta = \frac{a_4}{a_0}$$

Solved Examples

***Example* 3.1:** Solve the equation $x^3 - 7x^2 + 36 = 0$ given that one root is double to another

***Solution*:** Let the roots be α, 2α, & β

$$\sum\alpha = \alpha + 2\alpha + \beta = 7 \quad \therefore \ \beta = 7 - 3\alpha \qquad .....(3.15)$$

$$\sum\alpha\beta = \alpha - 2\alpha + \alpha\beta \ + 2\alpha\beta = 0$$

$$2\alpha^2 + 3\alpha(7 - 3\alpha) = 0$$

$$2\alpha^2 + 21\alpha - 9\alpha^2 = 0$$

$$21\alpha - 7\alpha^2 = 0$$

$$\alpha(21 - 7\alpha) = 0 \qquad \therefore \ \alpha = 0, \ 3$$

when $\alpha = 3, \quad \beta = -2$

$$\alpha = 0 \quad \beta = 7$$

$$\alpha\beta\gamma = \alpha \cdot 2\alpha \cdot \beta = -36$$

$$\alpha^2\beta = -18$$

$\Rightarrow \alpha\beta\gamma = 0$. Hence its roots satisfied clearly the first set of values of α & β satisfied it. Hence the roots are α, 2α, β i.e., 3, 6, – 2.

***Example* 3.2:** Find the condition that the roots of the equation $x^3 - px^2 + qx - r = 0$ may be in arithmetic progression and hence solve the equation $x^3 - 12x^2 + 39x - 28 = 0$.

Solution: Let the roots of the given equation be a – d, a, a + d being in AP.

$$\sum\alpha = a - d + a + a + d = 3a = p \qquad \therefore \ a = \frac{p}{3} \qquad .....(3.16)$$

But (a) is a root of the given equation and as such it will satisfy it.

$$\left(\frac{p}{3}\right)^3 - p\left(\frac{p}{3}\right)^2 + q\left(\frac{p}{3}\right) - r = 0$$

$$\frac{p^3}{27} - \frac{p^3}{9} + \frac{pq}{3} - r = 0$$

$$p^3 - 3p^3 + 9pq - 27r = 0$$

$$\Rightarrow \quad 2p^3 - 9pq + 27r = 0$$

This is the required condition.

From (3.16) $\frac{p}{3} = a \quad \therefore \quad \frac{12}{3} = 4$

So we divide it by 4

$$\begin{array}{c|ccc|} 4 & 1 & -12 & 39 & -28 \\ & & 4 & -32 & 28 \\ \hline & & -8 & 7 & 0 \end{array}$$

So the quadratic equation

$$x^2 - 8x + 7 = 0$$

$$x^2 - 7x - x + 7 = 0 \Rightarrow x(x-7) - (x-7) = 0$$

$$\Rightarrow \quad (x-1)(x-7) = 0 \qquad \therefore x = 1, 7$$

Clearly 1, 4, 7 are in AP.

***Example* 3.3:** Find the condition that the roots of the equation $x^3 - px^2 + qx - r = 0$ be in G.P. Hence solve the equation $x^3 - 19x^2 + 114x - 216 = 0$

Solution: Let the roots of the given equation be ax, a, a/x be in G.P.

$$\sum \alpha\beta\gamma = ax - a, \ a/x = a^3 = \gamma \qquad .....(3.17)$$

But 'a' is a root of the given sum as such $a^3 - pa^2 - qa - \gamma = 0 - pa^2 + qa = 0$

$\therefore$ pa = q is the required condition.

From (3.17) $a^3 = 216 = 6^3 \Rightarrow a = 6$

$$\begin{array}{c|ccc|} 6 & 1 & -19 & 114 & -216 \\ & & 6 & -78 & 216 \\ \hline & & -13 & 36 & 0 \end{array}$$

So the quadratic equation $= x^2 - 13x + 36 = 0$

$$x^2 - 9x - 4x + 36 = 0$$

$$x(x-9) - 4(x-9) = 0$$

$$(x-4)(x-9) = 0 \ \Rightarrow x = 4, 9$$

Clearly 4, 6, 9 are in GP.

***Example* 3.4:** Find the condition that the roots of the equation $x^3 - px^2 + qx - r = 0$ be in the harmonical progression. Hence solve the eqution $6x^3 - 11x^2 - 3x + 2 = 0$

Solution: Let the roots be α, β, γ which are in HP and hence $\frac{1}{\alpha}, \frac{1}{\beta}, \frac{1}{\gamma}$ are in AP.

$$\frac{1}{\alpha} + \frac{1}{\gamma} = \frac{2}{\beta}$$

$$\alpha\beta + \gamma\beta = 2\alpha\gamma$$

Adding αγ to bothsides we get,

$$\alpha\beta + \beta\gamma + \alpha\gamma = 3\alpha\gamma$$

$$q = 3\alpha\gamma \qquad \therefore\ \alpha\gamma = \frac{q}{3} \qquad .....(3.18)$$

Again $\qquad \alpha\beta\gamma = \gamma$

$$q\beta = 3r \qquad \therefore\ \beta = \frac{3r}{q}$$

But β is the root of the given equation and as such

$$\left(\frac{3r}{q}\right)^3 - p\left(\frac{3r}{q}\right)^2 + q\left(\frac{3r}{q}\right) - r = 0$$

$$\frac{27r^3}{q^3} - \frac{p9r^2}{q^2} + 3r - r = 0 \ \Rightarrow 27r^3 - 9pqr^2 + 3rq^3 = 0$$

$$\Rightarrow \qquad 27r^2 = 9pqr - 3q^3 = 0$$

This is required condition.

$$x^3 - \frac{11}{6}x^2 - \frac{1}{2}x + \frac{1}{3} = 0$$

From (3.18) $\quad \alpha\gamma = -\frac{3}{3-6} = -\frac{1}{6}\alpha\beta\gamma = -\frac{1}{3}\beta = 2$

So we divide the given equation by 2

$$\begin{array}{c|ccc} 2 & 1 \quad -\frac{11}{6} & -\frac{1}{2} & \frac{1}{3} \\ & \quad \frac{2}{} & 1/3 & -1/3 \\ \hline & 1/6 & -1/6 & 0 \end{array}$$

$$x^2 + \frac{x}{6} - \frac{1}{6} = 0$$

$$6x^2 + x - 1 = 0$$

$$x = \frac{-1 \pm \sqrt{1+24}}{12} = \frac{-1 \pm \sqrt{25}}{12} = \frac{-1 \pm 5}{12} = \frac{1}{2}, \frac{-1}{3}$$

Hence the other roots are $\frac{1}{2}, \frac{-1}{3}$

Clearly, $\frac{1}{2}$, 2, $\frac{-1}{3}$ are roots.

***Example* 3.5:** Find the condition that the roots α, β, γ, δ of equation $x^4 + px^3 + qx^2 + \gamma x + 5 = 0$ should be connected by the relative. (i) αβ = γδ (ii) ratio of two roots = ratio of two other roots.

Solution: Case (i) αβ = γδ given

But $\alpha\beta\,\gamma\delta = s$ $\quad \therefore \quad \alpha\beta = \gamma\delta = \pm\sqrt{s}$ .....(3.19)

$$\sum \alpha = \alpha + \beta + \gamma + \delta = -p$$

$$\sum \alpha\beta = (\alpha+\beta)(\gamma+\delta) + \alpha\beta + \gamma\delta = q$$

$$\sum \alpha\beta\gamma = \alpha\beta(\gamma+\delta) \pm \gamma\delta(\alpha+\beta) = -r$$

i.e., $\pm\sqrt{s}(\alpha+\beta+\gamma+\delta) = -r$

$\pm\sqrt{s} = \frac{r}{p}$ is the required condition

$P^2 s = r^2$ is the required condition

Case (ii) Let $\frac{\alpha}{\delta} = \frac{\gamma}{\beta}$ $\quad \alpha\beta = \gamma\delta$ & hence etc.

***Example* 3.6:** Solve the equation $x^4 + 15x^3 + 70x^2 + 120x + 64 = 0$ whose roots are in Geometrical progression.

Solution: Let the roots be a, ar, ar^2, ar^3 etc. Let us call them as α, β, γ, δ respectively.

We find $\quad \alpha\delta = a - ar^3 = a^2 - a^3$

$\beta\gamma = ar.ar^2 = a^2 - r^3 \quad \therefore \quad \alpha\delta = \beta\gamma$

But $\quad \alpha\,\beta\,\gamma\,\delta = 64 \quad \therefore \quad \alpha\gamma = \beta\delta = +8$ or -8

Let $\alpha+\beta=1 \;\&\; \gamma+\delta=x$

$$f(x)=(x^2-1x+8)(x^2-1x+8)$$

Comparing the coefficient we get

$$1+x=15$$

$$(\alpha+\beta)\ (\gamma+\delta)-\alpha\beta+\gamma\delta=70$$

$$lm=54$$

Since l and m are the roots of the equation.

$$r^2+15t+54=0$$

$$t=\frac{-15\pm\sqrt{225-216}}{2}=\frac{-15\pm\sqrt{9}}{2}$$

$$=\frac{-15\pm 3}{2}=-9,-6$$

$$f(x)=(x^2+9x+8)(x^2+6x+8)=0$$

$$(x^2+1)(x+8)(x+4)(x+2)=0$$

Roots of f (x) = – 1, – 2, – 4, – 8 which also in GP.

***Example* 3.7:** The roots of the equation $x^4-6x^3+18x^2-30x+25=0$ are of the form $\alpha+i\beta$ and $\beta+i\alpha$ where α and β are real. Find out all the roots.

Solution: If $\alpha+i\beta$ is a root then $\alpha-i\beta$ is also a root. Similarly $\beta+i\alpha$ is a root & $\beta-i\alpha$ is also a root.

$$\sum\alpha=\alpha+i\beta+\alpha-i\beta+\beta+i\alpha+\beta i\alpha=6$$

$$2(\alpha+\beta)=6 \quad \therefore\ \alpha+\beta=3$$

$$\text{Product of all roots }=(\alpha^2+\beta^2)(\beta^2+\alpha^2)=25$$

$$\alpha^2+\beta^2=\pm 5$$

$$2\,\alpha\beta=(\alpha+\beta)^2-(\alpha^2+\beta^2)$$

$$=9\ -5=4 \quad \therefore\ \alpha\beta=2$$

α and β are roots of a quadratic equation in t

$$t^2 - 3t + 2 = 0$$

$$t = \frac{3 \pm \sqrt{1}}{2} = \frac{3 \pm 1}{2} = 2,\ 1$$

$$\alpha = 2,\ \&\ \beta = 1$$

Hence the roots are $2 \pm i \quad 1 + 2i$

***Example* 3.8:** Find the condition that the sum of two roots of the equations $x^4 + px^3 + qx^2 + rx + 5 = 0$ be equal to the sum of the other two roots. Hence solve $x^4 - 8x^3 + 21x^2 - 20x + 5 = 0$

Solution: $\alpha + \beta = \gamma + \delta$ .....(3.20)

$$\sum \alpha = \alpha + \beta + \gamma + \delta = -p$$

$$\alpha + \beta = \gamma + \delta = -p/2 \quad .....(3.21)$$

$$\sum \alpha\beta = (\alpha + \beta)(\gamma + \delta) + \alpha\beta + \gamma\delta = q$$

$$\frac{p^2}{4} + \alpha\beta + \gamma\delta = q$$

$$\alpha\beta + \gamma\delta = \frac{4q - p^2}{4} \quad .....(3.22)$$

$$\sum \alpha\beta\gamma = \alpha\beta(\gamma + \delta) + \gamma\delta(\alpha + \beta) = -r$$

$$\alpha\beta\left(-\frac{p}{2}\right) + \gamma\delta\left(-\frac{p}{2}\right) = -r$$

$$\alpha\beta + \gamma\delta = \frac{2r}{p}$$

$$\frac{p}{2}\left(\frac{4q - p^2}{4}\right) = r \quad .....(3.23)$$

$$4pq - p^3 = 8r$$

In the given equation $\alpha + \beta = \gamma + \delta = -\frac{p}{2} = 4$

and $\alpha\beta = 1\ \&\ \gamma\delta = m\ \&$ hence

$$x^4 - 8x^3 + 21x^2 - 20x + 5 = 0(x^2 - 4x + 1)(x^2 - 4x + m)$$

Comparing the coefficient of like powers of x, we get,

$$4(1+m)=20 \quad 1+m=5 \quad lm=5$$

l and m are the roots of

$$t^2-5t+5=0$$

$$t=\frac{5\pm\sqrt{5}}{2}$$

Hence the two quadratic are

$$x^4-4x+\frac{5\pm\sqrt{5}}{2}=0$$ which on solving we get required root.

***Example* 3.9:** Find the condition that the equation $x^4 + px^3 + qx^2 + rx + 5$ should have two roots connected by the relation $\alpha + \beta = 0$ & determine in that case the two quadratic which shall have four roots (α, β), (γ, δ) and hence solve the equation $x^4-2x^3+4x^2+6x-21=0$

Solution: $\alpha+\beta=0$ (given)

$$\sum\alpha=\alpha+\beta+\gamma+\delta=-p \quad \therefore \gamma+\delta=-p \qquad .....(3.24)$$

$$\sum\alpha\beta=(\alpha+\beta)(\gamma+\delta)+\alpha\beta+\gamma\delta=q$$

$$\alpha\beta+\gamma\delta=q \qquad .....(3.25)$$

$$\sum\alpha\beta\gamma=\alpha\beta(\gamma+\delta)+\gamma\delta(\alpha+\beta)=-r$$

$$\alpha\beta(-p)=-r \quad \therefore \alpha\beta=\frac{r}{p}$$

$$\sum\beta\gamma\delta=\alpha\beta\gamma\delta=s\Rightarrow\left(\frac{\gamma}{p}\right)\gamma\delta=s$$

$$\gamma\delta=\frac{ps}{\gamma} \qquad .....(3.26)$$

Substituting the values of $\alpha\beta$ & $\gamma\delta$ in equation (3.25)

$$\frac{\gamma}{p}+\frac{ps}{\gamma}=q$$

$$\gamma^2+p^2s=pqr$$

$pqr - p^2s - r^2 = 0$. This is the required condition.

Case (1) Again since $\alpha + \beta = 0 \quad \alpha\beta = \frac{r}{p} \quad \alpha \,\&\, \beta$ are the roots of the quadratic.

$$x^2 - 0.x + \frac{r}{p} = 0$$

$$px^2 + r = 0$$

Case (2) Also $\quad \gamma + \delta = -p \,\&\, \gamma\delta = \frac{ps}{\gamma}$

$\therefore \quad \gamma$ and δ be the root of the quadratic

$$x^2 - (-px) + \frac{ps}{r} \Rightarrow rx^2 + prx + ps = 0$$

Now $\quad f(x) = x^4 - 2x^2 + 4x^2 + 6x - 21$

Since $\quad \alpha + \beta + \gamma + \delta = 2 \;\&\; \alpha + \beta = 0$

$\therefore \quad \gamma + \delta = 2$

Let $\quad \alpha\beta = 1 \,\&\, \gamma\delta = m$

$$f(n) = \{x^2 + (\alpha + \beta)x + \alpha\beta\} \times (x^2 + (\gamma + \delta)x + \gamma\delta)$$

$$f(n) = (x^2 + 1)(x^2 - 2x + m)$$

Comparing the coefficients we get

$$l + m = 4 \quad -2l = 0 \quad l = -3 \quad m = 7$$

$$f(n) = (x^2 - 3)(x^2 - 2x + 7) = 0$$

$$x^2 - 3 = 0 \Rightarrow x = \pm\sqrt{3}$$

$$x^2 - 2x + 7 = 0 \;\therefore\; x = \frac{2 \pm \sqrt{4 - 28}}{2} = 1 \pm i\sqrt{6}$$

Hence all the roots are $= \pm\sqrt{3x} \quad = 1 \pm \sqrt{6i}$

Other Important Problems

1. If α, β, γ are the roots of the equation $x^2 + px^2 + qx + r = 0$. Find the equation whose roots are $\beta + \gamma$, $\gamma + \alpha$, $\alpha + \beta$ and hence find the value of

(i) $\quad \sum(\beta + \gamma)(\gamma + \alpha)$ and

(ii) $(\beta+\gamma)(\gamma+\alpha)(\alpha+\beta)$

Solution: Let $y=\beta+\gamma=\alpha+\beta+\gamma-\alpha=-P-\alpha$

$\therefore \qquad \alpha=-(\gamma+p)$

But α is the root of given equation and therefore the required equation is $-(\gamma+P)^3+P(\gamma+P)^2-q(\gamma+P)+r=0$

or, $\gamma^3+2py^2+(P^2+q)y+(Pq-r)=0$ and let its roots $\beta+\gamma$ etc., be denoted by A, B and C.

$\therefore$ (i) $\sum AB=(P^2+q)$ and (ii) $ABC=r-Pq$

2. Find for the same cubic the equation whose roots are $\beta\gamma+\frac{1}{\alpha},\gamma\alpha+\frac{1}{\beta},\ \alpha\beta+\frac{1}{\gamma}$.

Solution: $\beta\gamma+\frac{1}{\alpha}=\frac{\alpha\beta\gamma+1}{\alpha}=\frac{1-r}{\alpha};\ \therefore\ \alpha=\frac{1-r}{\gamma}$ etc.

$$ry^3+q(1-r)y^2+P(1-r^2)y+(1-r)^3=0$$

3. Find for the same cubic the equation whose roots are $\alpha(\beta+\gamma),\beta(\gamma+\alpha),\gamma(\alpha+\beta)$

Solution: $\qquad y=\alpha\beta+\alpha\gamma=\alpha\beta+\beta\gamma+\gamma\alpha-\beta\gamma=q-\beta\gamma$

or, $\qquad y=q-\frac{\alpha\beta\gamma}{\alpha}=q+\frac{r}{\alpha};\quad \therefore\ \alpha=\frac{r}{y-q}$ etc.,

$$y^3-2qy^2+(Pr+q^2)y+(r^2-pqr)=0$$

4. If α, β, γ be the roots of $x^3-px^2+qx-r=0$. Find the equation whose roots are, $\beta+\gamma-\alpha,\ \gamma+\alpha-\beta,\ \alpha+\beta-\gamma$ and deduce that

$(\beta+\gamma-\alpha)\ (\gamma+\alpha-\beta)\ (\alpha+\beta-\gamma)=4pq-p^3-8r$

Solution: $\qquad y=\beta+\gamma-\alpha=\alpha+\beta+\gamma-2\alpha=P-2\alpha\ \ \therefore\ \alpha=\frac{(P-y)}{2}$

But α is roots of the given equation. Hence the transformed equation is

$$\left(\frac{P-y}{2}\right)^3-P\left(\frac{P-y}{2}\right)^2+q\left(\frac{P-y}{2}\right)-r=0$$

$$\left(P^3-3P^2y+3py^2-y^3\right)-2P\left(p^2-2Py+y^2\right)+4q(P-y)-8r=0$$

or, $$y^3-Py^2+y\left(4q-p^2\right)+\left(8r-4pq-p^3\right)=0$$

$\therefore$ $$(\beta+\gamma-\alpha)(\gamma+\alpha-\beta)(\alpha+\beta-\gamma)=\text{Products of all the roots of above}$$

$$=-\left(8r-4pq-p^3\right)$$

5. Find for the same cubic the equation whose roots are $\beta\gamma-\alpha^2,\gamma\alpha-\beta^2,\alpha\beta-\gamma^2$. Also find the value of

$$\sum\left(\beta\gamma-\alpha^2\right)\left(\gamma\alpha-\beta^2\right) \text{ and } \left(\beta\gamma-\alpha^2\right)\left(\gamma\alpha-\beta^2\right)\left(\alpha\beta-\gamma^2\right)$$

Solution: $y=\beta\gamma-\alpha^2=(\alpha\beta+\beta\gamma+\gamma\alpha)-\alpha\beta-\gamma\alpha-\alpha^2$

or, $$y=q-\alpha(\alpha+\beta+\gamma)=q+P\alpha,\ \therefore\ \alpha=\frac{y-q}{P}$$

$$y^3-y^2\left(3q^2-p^2\right)+y\left(3q^2-p^2q\right)+\left(rp^3-q^3\right)=0$$

(i) $3q^2-p^2q$ (ii) q^3-rp^3

6. Find for the same cubic the equation whose roots are

$$\alpha-\frac{1}{\beta\gamma},\ \beta-\frac{1}{\gamma\alpha},\ \gamma-\frac{1}{\alpha\beta}$$

Solution: Here, $y=\alpha-\dfrac{1}{\beta\gamma}=\dfrac{\alpha\beta\gamma-1}{\beta\gamma}=\dfrac{-r-1}{\beta\gamma}=-\dfrac{(r+1)\alpha}{\alpha\beta\gamma}$

or, $$y=-\frac{(r+1)}{-r}\alpha \quad \therefore \quad \alpha=\frac{ry}{r+1}$$

But α is a root of given equation

$\therefore$ $$r^3y^3+Pr^2y^2(r+1)+qry(r+1)^2+r(r+1)^3=0$$

or, $$r^2y^3+Pr(r+1)y^2+qy(r+1)^2+(r+1)^3=0$$

7. If α, β, γ be the roots of equation $x^2+qx+r=0$. Find the equation whose roots are

$$\frac{\beta}{\gamma}+\frac{\gamma}{\beta},\frac{\gamma}{\alpha}+\frac{\alpha}{\gamma},\frac{\alpha}{\beta}+\frac{\beta}{\alpha}$$

Solution: $$y=\frac{\beta}{\gamma}+\frac{\gamma}{\beta}=\frac{\beta^2+\gamma^2}{\beta\gamma}=\frac{(\beta+\gamma)^2-\beta\gamma}{\beta\gamma}+\frac{\alpha^2}{\beta\gamma}-2$$

$\therefore \quad \alpha+\beta+\gamma=0 \quad \therefore \beta+\gamma=-\alpha$

or, $\quad y+2=\dfrac{\alpha^3}{\alpha\beta\gamma}=\dfrac{\alpha^3}{-r} \quad \therefore \alpha^3+r(y+2)=0$

we have also $\alpha^3+q\alpha+r=0$

Subtracting we get $-q\alpha+r(y+1)=0$

or, $\quad \alpha=\dfrac{r(y+1)}{q}$

But α is a root of given equation

$$\therefore \quad \frac{r^3}{q^3}(y+1)^3+q.\frac{r(y+1)}{q}+r=0$$

or, $\quad r^2(y+1)^3+q^3(y+1)+q^3=0$

8. If $\alpha_1, \alpha_2, \ldots \alpha_n$ be the roots of the equation

$$x^n+P_1x^{n-1}+P_2x^{n-2}+\ldots+P_{n-1}x+P_n=0.$$

Find the values of the following symeteric functions:

(i) $\sum\dfrac{\alpha_1^2+\alpha_2^2}{\alpha_1\alpha_2}$ (ii) $\sum\dfrac{\alpha_1^2}{\alpha_2}$

(iii) $\sum(\alpha_1-\alpha_2)^2\alpha_3\alpha_4\ldots\alpha_n$ (iv) $\sum\dfrac{\alpha_1}{\alpha_2^2}$

(v) $\sum\alpha_1^2\alpha_2$ (vi) $\sum\alpha_1^2\alpha_2\alpha_3$

(vii) $\sum\alpha_1^2\alpha_2^2$ (viii) $\sum\dfrac{\alpha_1^2}{\alpha_2^2}$

(ix) $\sum\alpha_1^2\alpha_2\alpha_3\alpha_4$ (x) $\sum\alpha_1^4$

Solution:

(i)
$$\sum\frac{\alpha_1^2+\alpha_2^2}{\alpha_1\alpha_2}=\sum\left(\frac{\alpha_1}{\alpha_2}+\frac{\alpha_2}{\alpha_1}\right)$$

$$=\alpha_1\left(\frac{1}{\alpha_1}+\frac{1}{\alpha_2}+\ldots\frac{1}{\alpha_n}\right)-1+\alpha_2\left(\frac{1}{\alpha_1}+\frac{1}{\alpha_2}+\ldots+\frac{1}{\alpha_n}\right)-1$$

$$+\ldots+\alpha_n\left(\frac{1}{\alpha_1}+\frac{1}{\alpha_2}+\ldots+\frac{1}{\alpha_n}\right)-1$$

Adding, we get the values of given sigma as

$$\sum\alpha_1\sum\frac{1}{\alpha_1}-n=(-P_1)\left(\frac{-P_n-1}{P_n}\right)-n=\frac{P_1P_{n-1}}{P_n}-n$$

$\therefore$ $\frac{1}{\alpha_1},\frac{1}{\alpha_2}...\frac{1}{\alpha_n}$ are the roots of

$$P_nx^n+P_{n-1}x^{n-1}+...+P_1x+1=0$$

(ii) $\sum\frac{\alpha_1^{\,2}}{\alpha_2}$

$$\sum\frac{\alpha_1^{\,2}}{\alpha_2}=\frac{1}{\alpha_1}\left(\alpha_1^{\,2}+\alpha_2^{\,2}+...+\alpha_n^{\,2}\right)-\alpha_1+\frac{1}{\alpha_2}\left(\alpha_1^{\,2}+\alpha_2^{\,2}+...+\alpha_n^{\,2}\right)-\alpha_2$$

$$+...+=\sum\alpha_1^{\,2}.\sum\frac{1}{\alpha_1}-\sum\alpha_1=\left(P_1^2-2P_2\right)\left(-\frac{P_{n-1}}{P_n}\right)+P_1$$

$$=P_1-\frac{P_{n-1}\left(P_1^2-2P_2\right)}{P_n}$$

(iii) $\sum(\alpha_1-\alpha_2)^2\alpha_3\alpha_4...\alpha_n$

Now $\sum(\alpha_1-\alpha_2)^2\alpha_3\alpha_4...\alpha_n$

$$=\alpha_1\alpha_2\alpha_3\alpha_4...\alpha_n\sum\frac{(\alpha_1-\alpha_2)^2}{\alpha_1\alpha_2}$$

$$=(-1)^nP_n\left[\sum\left\{\frac{\left(a_1^{\,2}+\alpha_2^{\,2}\right)}{\alpha_1\alpha_2}-2\right\}\right]$$

Now we have n roots and we are taking two roots at a time i.e., there will be ${}^nC_2$ combinations and hence in the symmetric function 2 will occur ${}^nC_2$ times and the value of $\sum\frac{\alpha_1^{\,2}+\alpha_2^{\,2}}{\alpha_1\alpha_2}$

Hence $(-1)^nP_n\left[\frac{P_1\,P_{n-1}}{P_n}-n-2.\frac{n(n-1)}{2!}\right]$

$$=(-1)^n\left(P_1\,P_{n-1}-n^2\,P_n\right).$$

(iv) $\sum\frac{\alpha_1}{\alpha_2^{\,2}}$ for calculating $\sum\frac{\alpha_1}{\alpha_2^{\,2}}$ we multiply $\sum\alpha_1$ and $\sum\frac{1}{\alpha_1^{\,2}}$

$$\sum\alpha_1\sum\frac{1}{\alpha_1^{\,2}}=(\alpha_1+\alpha_2+....+\alpha_n)\left(\frac{1}{\alpha_1^{\,2}}+\frac{1}{\alpha_2^{\,2}}+...\frac{1}{\alpha_n^{\,2}}\right)$$

$$=\sum\frac{1}{\alpha_1}=\sum\frac{\alpha_1}{\alpha_2^2}$$

$$\therefore\qquad \sum\frac{\alpha_1}{\alpha_2^{\,2}}=\sum\alpha_1\sum\frac{1}{\alpha_1^{\,2}}-\sum\frac{1}{\alpha_1}$$

$$=\sum\alpha_1\left[(\sum\frac{1}{\alpha_1})^2-2\sum\frac{1}{\alpha_1\alpha_2}\right]-\sum\frac{1}{\alpha_1}$$

$\frac{1}{\alpha_1},\frac{1}{\alpha_2},\frac{1}{\alpha_3},\ldots\frac{1}{\alpha_n}$ are the roots of the equation

$$P_nx^n+P_{n-1}x^{n-1}+\ldots+P_1x+1=0$$

$$\therefore\ \sum\frac{\alpha_1}{\alpha_2^{\,2}},=-P_1\left(\frac{P^{2n-1}}{P_n^2}-2\frac{P_{n-2}}{P_n}\right)+\frac{P_{n-1}}{P_n}$$

$$=\frac{P_{n-1}P_n-P_1\,P_{n-1}^2+2P_1\,P_{n-2}P_n}{P_n^2}$$

(v) $\sum\alpha_1^{\,2}\,\alpha_2$

The sum of the powers in the above sigma is 2 + 1 = 3 and we know that in the case of a cubic

$$\sum\alpha^2\beta=\sum\alpha\,.\,\sum\beta-3\alpha\,\beta\gamma$$

or, $$\sum\alpha_1^{\,2}\alpha_2=\sum\alpha_1\,.\,\sum\alpha_1\alpha_2-3\alpha_1\,\alpha_2\alpha_3$$

Since the equation of nth degree we write

$$\sum\alpha_1^{\,2}\alpha_2=\sum\alpha_1\,.\,\sum\alpha_1\alpha_2-3\sum\alpha_1\,\alpha_2\alpha_3$$

$$=(-P_1)P_2-3(-P_3)=3P_3-P_1P_1$$

(vi) $$\sum \alpha_1^2 \alpha_2\alpha_3$$

The sum of the suffixes is 4 and we know that in a biquadratic

$$\sum \alpha^2\beta\gamma = \sum \alpha \,.\, \sum \alpha\beta\gamma - 4\alpha\beta\gamma\delta$$

or, $$\sum \alpha_1^2\, \alpha_2\, \alpha_3 = \sum \alpha_1 \sum \alpha_1\, \alpha_2\, \alpha_3 - 4\alpha_1\, \alpha_2\, \alpha_3\, \alpha_4$$

Since the equation of nth degree we write

$$\sum \alpha_1^2\alpha_2\alpha_3 = \sum \alpha_1 \sum \alpha_1\, \alpha_2\, \alpha_3 - 4\sum \alpha_1\, \alpha_2\, \alpha_3\, \alpha_4$$

$$= (-P_1)(-P_3) - 4P_4 = P_1P_3 - 4P_4$$

(vii) $$\sum \alpha_1^2\, \alpha_2^2$$

The sum of the suffixes is 4 and we know that in a biquadratic,

$$\sum \alpha^2\beta^2 = \left(\sum \alpha\beta\right)^2 - 2\sum \alpha^2\, \beta\gamma - 6\alpha\,\beta\,\gamma\delta$$

$$= \left(\sum \alpha\,\beta\right)^2 - 2\sum \alpha \,.\, \sum(\alpha\beta\gamma - 4\alpha\beta\gamma\delta) - 6\alpha\beta\gamma\delta$$

$$= \left(\sum \alpha\,\beta\right)^2 - 2\sum \alpha \,.\, \sum(\alpha\beta\gamma - 4\alpha\beta\gamma\delta) + 2\alpha\beta\gamma\delta$$

or, $$\sum \alpha_1^2\, \alpha_2^2 = \left(\sum \alpha_1\, \alpha_2\right)^2 - 2\sum \alpha_1 \sum \alpha_1\alpha_2\alpha_3 + 2\alpha_1\alpha_2\alpha_3\alpha_4$$

Since the equation is of nth degree we write the above as

$$\sum \alpha_1^2\alpha_2^2 = \left(\sum \alpha_1\, \alpha_2\right)^2 - 2\sum \alpha_1 \sum \alpha_1\alpha_2\alpha_3 + 2\sum \alpha_1\alpha_2\alpha_3\alpha_4$$

$$= (P_2)^2 - 2(-P_1)(-P_3) + 2P_4 \;\; = P_2^2 - 2P_1P_3 + 2P_4$$

(viii) $$\sum \frac{\alpha_1^2}{\alpha_2^2}$$

$$\sum \frac{\alpha_1^2}{\alpha_2^2} = \alpha_1^2\left(\frac{1}{\alpha_1^2} + \frac{1}{\alpha_2^2} + \ldots\ldots + \frac{1}{\alpha_n^2}\right) - 1$$

$$+\left(\alpha_2^2 \,.\, \sum \frac{1}{\alpha_1^2} - 1\right) + \left(\alpha_3^2 \,.\, \sum \frac{1}{\alpha_1^2} - 1\right) + \ldots\ldots$$

$$=\sum\frac{1}{\alpha_1^{\ 2}}\sum\alpha_1^{\ 2}-n$$

$$=\left\{\left(\sum\frac{1}{\alpha_1}\right)-2\sum\frac{1}{\alpha_1\alpha_2}\right\}\left\{\left(\sum\alpha_1\right)^2-2\sum\alpha_1\alpha_2\right\}-n$$

$$=\left(\frac{P^2_{\ n-1}}{P_n^2}-2\frac{P_{n-2}}{P_n}\right)\left(P_1^{\ 2}-2P_2\right)-n$$

$$=\frac{\left(P_1^2-2P_2\right)\left(P^2_{\ n-1}-2P_nP_{n-2}\right)}{P_n^2}-n$$

(ix) $\sum\alpha_1^{\ 2}\ \alpha_2\alpha_3\alpha_4$

Here sum of the powers is five and it can be easily verified that if the equation were of fifth degree then

$$\underset{5.5}{\sum\alpha_1\ .\ \sum\alpha_1\ \alpha_2\ \alpha_3\ \alpha_4}=\underset{5.4}{\sum\alpha_1^{\ 2}\ \alpha_2\ \alpha_3\ \alpha_4}+\underset{5.1}{\sum\alpha_1\ \alpha_2\ \alpha_3\ \alpha_4\ \alpha_5}$$

$\therefore$ In an equation of nth degree, we have

$$\sum\alpha_1^{\ 2}\alpha_2\ \alpha_3\ \alpha_4=\sum\alpha_1\sum\alpha_1\ \alpha_2\ \alpha_3\ \alpha_4-5\sum\alpha_1\ \alpha_2\ \alpha_3\ \alpha_4\ \alpha_5$$

$$=(-P_1)(P_4)-5(-P_5)=5\ P_5-P_1P_4$$

(x) $\sum\alpha_1^{\ 4}$

$$\sum\alpha_1^{\ 4}=\left(\sum\alpha_1^{\ 2}\right)^2-2\sum\alpha_1^{\ 2}\ \alpha_2^{\ 2}$$

$$=\left(P_1^{\ 2}-2\ P_2\right)^2-2\left(P_2^{\ 2}-2P_1\ P_3+2P_4\right)$$

$$=P_1^{\ 4}-4\ P_1^{\ 2}\ P_2+4P_2^{\ 2}-2P_2^{\ 2}+4P_1P_2-4P_4$$

$$=P_1^{\ 4}-4\ P_1^{\ 2}\ P_2+2\ P_2^{\ 2}-4P_1P_3-4P_4$$

Illustrative Examples

***Example* 3.10:** Solve the cubic by Cardon's method $x^3-18x-35=0$

***Solution*:** The given cubic is

$$x^3-18x-35=0 \qquad .....(3.27)$$

in which the term involving x^2 is missing

So assume that $x=p^{1/3}+q^{1/3}$, then on cubing we have

$$x^3 = p + q + 3\,p^{1/3}q^{1/3}\,(p^{1/3}+q^{1/3})$$

$$x^3 - 3p^{1/3}q^{1/3}x - (p + q) = 0 \qquad \text{.....(3.28)}$$

Comparing (3.27) and (3.28), we have

$$p + q = 35 \text{ and } p^{1/3}\,q^{1/3} = 6 \text{ or } pq = 216$$

Solving these equations, we get $q = 27$ and $q^{1/3} = 6/p^{1/3}$

$$\therefore \qquad p^{1/3} = 2,\ 2\omega \text{ or } 2\omega^2$$

The corresponding values of $q^{1/3}$ are given by $q^{1/3} = 3,\ 3\omega^2$ or 3ω

Hence the roots of (3.27) are 5, $2\omega + 3\omega^2$ and $2\omega^2 + 3\omega$, Substituting the value of

$$\omega = \frac{1}{2}\left(-1+i\sqrt{3}\right) \text{ and } \omega^2 = \frac{1}{2}\left(-1-i\sqrt{3}\right)$$

The required roots of the cubic (3.27) are 5 and $\frac{1}{2}\left(-5 \pm i\sqrt{3}\right)$

***Example* 3.11:** Solve the equation $x^3 - 15x - 126 = 0$, by Cardon's method.

***Solution*:** The given cubic is

$$x^3 - 15x - 126 = 0 \qquad \text{.....(3.29)}$$

in which the term involving x^2 is missing

$\therefore$ Let $x = p^{1/3} + q^{1/3}$, be a solution of the given cubic

Cubic both sides, we get

$$x^3 = p + q + 3p^{1/3}q^{1/3}\,(p^{1/3} + q^{1/3})$$

or

$$x^3 - 3p^{1/3}q^{1/3}\,x - (p + q) = 0$$

Comparing (3.28) and (3.29), we get

$$p^{1/3}q^{1/3} = 5 \text{ or } pq = 125 \text{ and } p + q = 126$$

Therefore, p and q are roots of the quadratic

$$t^2 - (p + q)\,t + pq = 0 \text{ or } t^2 - 126t + 125 = 0$$

or

$$(t - 1)\,(t - 125) = 0 \Rightarrow t = 1, 125$$

Let $p = 1$ and $q = 125$

$$\therefore \qquad p^{1/3} = 1 \text{ and } q^{1/3} = 5$$

Hence the roots of the given cubic (3.29) are

$$p^{1/3} + q^{1/3} = 1 + 5 = 6,$$

$$\omega p^{1/3}+\omega^2 q^{1/3}=\left(-\frac{1}{2}+i\frac{\sqrt{3}}{2}\right).1+\left(-\frac{1}{2}-i\frac{\sqrt{3}}{2}\right).5=-3-2\sqrt{3}i,$$

and $$\omega^2 p^{1/3}+\omega q^{1/3}=\left(-\frac{1}{2}-i\frac{\sqrt{3}}{2}\right).1+\left(-\frac{1}{2}+i\frac{\sqrt{3}}{2}\right).5=-3+2\sqrt{3}i,$$

***Example* 3.12:** Solve the equation $x^3 - 3x + 1 = 0$ by Cardon's method

Solution: The given cubic is

$$x^3 - 3x + 1 = 0 \qquad \text{.....(3.30)}$$

in which the term involving x^2 is missing

Let $x = p^{1/3} + q^{1/3}$ be a solution of (3.30)

Cubic both sides, we get

$$x^3 = p + q + 3\,p^{1/3}\,q^{1/3}\,(p^{1/3} + q^{1/3})$$

$$x^3 - 3p^{1/3}q^{1/3}x - (p + q) = 0 \qquad \text{.....(3.31)}$$

Comparing (3.30) and (3.31), we get

$$p^{1/3}\,q^{1/3} = 1 \text{ or } pq = 1 \text{ and } p + q = -1$$

$\therefore$ p and q are roots of the quadratic $t^2 + t + 1 = 0$

$$\therefore \qquad t=\frac{-1\pm i\sqrt{3}}{2}$$

Let $$p=-\frac{1}{2}+i\frac{\sqrt{3}}{2} \text{ and } q=-\frac{1}{2}-i\frac{\sqrt{3}}{2}$$

$$\therefore \qquad x=\left(-\frac{1}{2}+i\frac{\sqrt{3}}{2}\right)^{1/3}+\left(-\frac{1}{2}-i\frac{\sqrt{3}}{2}\right)^{1/3}$$

Put $$-\frac{1}{2}=r\cos\theta \text{ and } \frac{\sqrt{3}}{2}=r\sin\theta \Rightarrow r^2=1 \text{ or } r=1 \text{ and } \tan\theta=-\sqrt{3} \text{ or } \theta=\frac{2\pi}{3}$$

$$x = r^{1/3}\,[(\cos\theta + i\sin\theta)^{1/3} + (\cos\theta - i\sin\theta)^{1/3}]$$

$$=\left[\left\{\cos\left(\frac{2n\pi+\theta}{3}\right)+i\sin\left(\frac{2n\pi+\theta}{3}\right)\right\}+\left\{\cos\left(\frac{2n\pi+\theta}{3}\right)-i\sin\left(\frac{2np+\theta}{3}\right)\right\}\right]$$

$$=2\cos\left(\frac{2n\pi}{3}+\frac{2\pi}{9}\right), \qquad n=0,\ 1,\ 2$$

Hence the roots of given cubic are $2\cos\frac{2\pi}{9}$, $2\cos\frac{8\pi}{9}$ and $2\cos\frac{14\pi}{9}$

***Example* 3.13:** Solve the cubic $x^3 + 6x^2 - 12x + 32 = 0$, by Cardon's method.

***Solution*:** The given cubic is

$$x^3 + 6x^2 - 12x + 32 = 0 \qquad .....(3.32)$$

First we remove the second term of the equation (3.32) by diminishing its roots by

$$h = -\frac{b}{na} = -\frac{6}{3.1} = -2$$

The procedure is as given below:

$$\begin{array}{r|rrrr}
-2 & 1 & 6 & -12 & 32 \\
 & & -2 & -8 & 40 \\
\hline
 & & 4 & -20 & 72 \\
 & & -2 & -4 & \\
\hline
 & & 2 & -24 & \\
 & & -2 & & \\
\hline
 & & 0 & &
\end{array}$$

∴ The transformed equation is

$$z^3 - 24z + 72 = 0$$

where $\quad z = x - h = x - (-2) = x + 2 \qquad .....(3.33)$

Let $z = p^{1/3} + q^{1/3}$ be a solution of the cubic (3.33)

Cubing both sides, we get

$$z^3 = p + q + 3p^{1/3} q^{1/3} (p^{1/3} + q^{1/3})$$

or $\quad z^3 - 3p^{1/3} q^{1/3} z - (p + q) = 0 \qquad .....(3.34)$

Comparing (3.33) and (3.34), we get

$$p^{1/3} q^{1/3} = 8 \text{ or } pq = 512 \text{ and } p + q = -72$$

∴ p and q are the roots of the quadratic

$$t^2 - (p + q)\, t + pq = 0 \text{ or } t^3 + 72t + 512 = 0$$

or $\quad (t + 8)(t + 64) = 0$; giving $t = -8, -64$

Let $\quad p = -8$ and $q = -64$

∴ $\quad p^{1/3} = -2$ and $q^{1/3} = -4$

∴ Roots of the cubic (3.33) are $p^{1/3} + q^{1/3} = -2 - 4 = -6$

$$\omega p^{1/3} + \omega^2 q^{1/3} = \left(-\frac{1}{2} + i\frac{\sqrt{3}}{2}\right).(-2) + \left(-\frac{1}{2} - i\frac{\sqrt{3}}{2}\right).(-4) = 3 + i\sqrt{3}$$

and
$$\omega^2 p^{1/3} + \omega q^{1/3} = \left(-\frac{1}{2} - i\frac{\sqrt{3}}{2}\right).(-2) + \left(-\frac{1}{2} + i\frac{\sqrt{3}}{2}\right).(-4) = 3 - i\sqrt{3}$$

But $x = z - 2$. Hence the roots of the given cubic (3.32) are $-6-2$, $3 + i\sqrt{3} - 2$ and $3 - i\sqrt{3} - 2 - 8$, $1 + i\sqrt{3}$ and $1 - i\sqrt{3}$

Example **3.14:** Solve the Cardon's method, the equation $27x^3 + 54x^2 + 198x - 73 = 0$.

Solution: To remove the second term, the roots of the given equation is to be diminished by $h = -\frac{\alpha_1}{na_0} = -\frac{54}{3\times 27} = -\frac{2}{3}$, so we apply synthetic division method:

$$\begin{array}{r|rrrr}
-\frac{2}{3} & 27 & 54 & 198 & -73 \\
 & & -18 & -24 & -116 \\
\hline
 & 27 & 36 & 174 & -189 \\
 & & -18 & -12 & \\
\hline
 & 27 & 18 & 162 & \\
 & & -18 & & \\
\hline
 & 27 & 0 & & \\
\hline
 & 27 & & &
\end{array}$$

Hence the transformed equation is

$$27y^3 + oy^2 + 162y - 189 = 0$$

i.e., $$27\,y^3 + 162\,y - 189 = 0$$

i.e., $$y^3 + 6y - 7 = 0 \qquad \text{.....(3.35)}$$

in which the term involving y^2 is missing

So assume that $y = p^{1/3} + q^{1/3}$, then on cubing, we have

$$y^3 = p + q + 3p^{1/3}\,q^{1/3}\,(p^{1/3} + q^{1/3})$$

or $$y^3 - 3p^{1/3}\,q^{1/3}\,y - (p + q) = 0 \qquad \text{.....(3.36)}$$

Comparing (3.35) and (3.36), we have

$$p + q = 7 \text{ and } p^{1/3}\,q^{1/3} = -2$$

or $$p + q = 7 \text{ and } pq = -8$$

$\therefore$ p and q are roots of the quadratic $t^2 - 7t - 8 = 0$

$\therefore$ $(t - 8)(t + 1) = 0 \Rightarrow t = 8, -1$

Let $p = 8$ and $q = -1$. Then $p^{1/3} = 2$ and $q^{1/3} = -1$

$\therefore$ Roots of the cubic (3.35) are $p^{1/3} + q^{1/3} = 2 - 1 = 1$,

$$\omega p^{1/3} + \omega^2 q^{1/3} = \left(-\frac{1}{2} + i\frac{\sqrt{3}}{2}\right).(2) + \left(-\frac{1}{2} - i\frac{\sqrt{3}}{2}\right).(-1) = -\frac{1}{2} + \frac{3\sqrt{3}}{2}i$$

and
$$\omega^2 p^{1/3} + \omega q^{1/3} = \left(-\frac{1}{2} - i\frac{\sqrt{3}}{2}\right).(2) + \left(-\frac{1}{2} + i\frac{\sqrt{3}}{2}\right).(-1) = -\frac{1}{2} - \frac{3\sqrt{3}}{2}i$$

But $x = y - \frac{2}{3}$. Hence the roots of the given cubic are $\frac{1}{3}, -\frac{7}{6} + \frac{3\sqrt{3}}{2}i$ and $-\frac{7}{6} - \frac{3\sqrt{3}}{2}i$

***Example* 3.15:** Solve by Cardon's method $x^3 - 3x^2 + 3 = 0$.

***Solution*:** The given cubic equation is

$$x^3 - 3x^2 + 3 = 0 \qquad .....(3.37)$$

We can remove the second term of the eq. (3.37) by diminishing its roots by

$$h = -\frac{b}{na} = -\frac{-3}{3.1} = 1$$

The procedure is as follows:

1	1	−3	0	3
		1	−2	−2
		−2	−2	1
		1	−1	
		−1	−3	
		1		
		0		

$\therefore$ The transformed equation is

$$z^3 - 3z + 1 = 0$$

where $z = x - h = x - 1$ .....(3.38)

Let $z = p^{1/3} + q^{1/3}$ be a solution of the cubic (3.38)

On cubing both sides, we get

$$z^3 = p + q + 3\,p^{1/3}q^{1/3}\,(p^{1/3} + q^{1/3})$$

or $$z^3 - 3p^{1/3}\,q^{1/3}\,z - (p + q) = 0 \qquad .....(3.39)$$

Comparing (3.38) and (3.39), we get

$$p^{1/3}\,q^{1/3} = 1 \text{ or } pq = 1 \text{ and } p + q = -1$$

∴ p and q are roots of the quadratic

$$t^2 - (p + q)\,t + pq = 0 \text{ or } t^3 + t + 1 = 0 \text{ or } t = \frac{-1 \pm i\sqrt{3}}{2}$$

Let $$p = \frac{-1 \pm i\sqrt{3}}{2} \text{ and } q = \frac{-1 - i\sqrt{3}}{2}$$

∴ Roots of the cubic (3.38), are

$$p^{1/3} + q^{1/3} = \left(-\frac{1}{2} + i\frac{\sqrt{3}}{2}\right)^{1/3} + \left(-\frac{1}{2} - i\frac{\sqrt{3}}{2}\right)^{1/3}$$

$$= \left(\cos\frac{2\pi}{3} + i\sin\frac{2\pi}{3}\right)^{1/3} + \left(\cos\frac{2\pi}{3} - i\sin\frac{2\pi}{3}\right)^{1/3}$$

$$\left[\text{Put } -\frac{1}{2} = r\cos\theta \text{ and } \frac{\sqrt{3}}{2} = r\sin\theta \text{ so that } r = 1, \theta = \frac{2\pi}{3}\right]$$

$$= \left(\cos\frac{2\pi}{9} + i\sin\frac{2\pi}{9}\right) + \left(\cos\frac{2\pi}{9} - i\sin\frac{2\pi}{9}\right) = 2\cos\frac{2\pi}{9};$$

$$\omega p^{1/3} + \omega^2 q^{1/3} = \left(-\frac{1}{2} + i\frac{\sqrt{3}}{2}\right)\left(-\frac{1}{2} + i\frac{\sqrt{3}}{2}\right)^{1/3} + \left(-\frac{1}{2} - i\frac{\sqrt{3}}{2}\right)\left(-\frac{1}{2} - i\frac{\sqrt{3}}{2}\right)^{1/3}$$

$$= \left(-\frac{1}{2} + i\frac{\sqrt{3}}{2}\right)^{4/3} + \left(-\frac{1}{2} - i\frac{\sqrt{3}}{2}\right)^{4/3} = \left(\cos\frac{2\pi}{3} + i\sin\frac{2\pi}{3}\right)^{4/3} + \left(\cos\frac{2\pi}{3} - i\sin\frac{2\pi}{3}\right)^{4/3}$$

$$= \left(\cos\frac{8\pi}{9} + i\sin\frac{8\pi}{9}\right) + \left(\cos\frac{8\pi}{9} - i\sin\frac{8\pi}{9}\right) = 2\cos\frac{8\pi}{9}$$

and $$\omega^2 p^{1/3} + \omega q^{1/3} = \left(-\frac{1}{2} - i\frac{\sqrt{3}}{2}\right)\left(-\frac{1}{2} + i\frac{\sqrt{3}}{2}\right)^{1/3} + \left(-\frac{1}{2} + i\frac{\sqrt{3}}{2}\right)\left(-\frac{1}{2} - i\frac{\sqrt{3}}{2}\right)^{1/3}$$

$$= \left(\cos\frac{2\pi}{3} - i\sin\frac{2\pi}{3}\right)\left(\cos\frac{2\pi}{3} + i\sin\frac{2\pi}{3}\right)^{1/3} + \left(\cos\frac{2\pi}{3} + i\sin\frac{2\pi}{3}\right)\left(\cos\frac{2\pi}{3} - i\sin\frac{2\pi}{3}\right)^{1/3}$$

$$= \left(\cos\frac{2\pi}{3} - i\sin\frac{2\pi}{3}\right)\left(\cos\frac{2\pi}{9} + i\sin\frac{2\pi}{9}\right) + \left(\cos\frac{2\pi}{3} + i\sin\frac{2\pi}{3}\right)\left(\cos\frac{2\pi}{9} - i\sin\frac{2\pi}{9}\right)$$

$$= 2\left[\cos\frac{2\pi}{3}\cos\frac{2\pi}{9} + \sin\frac{2\pi}{3}\sin\frac{2\pi}{9}\right] = 2\cos\left(\frac{2\pi}{3} - \frac{2\pi}{9}\right) = 2\cos\frac{4\pi}{9} = 2\cos\left(2\pi - \frac{4\pi}{9}\right)$$

$$= 2\cos\frac{14\pi}{9}$$

But x = 1 + z, Hence the roots of the given cubic (3.37) are

$$1+2\cos\frac{2\pi}{9},\ 1+2\cos\frac{8\pi}{9},\ 1+2\cos\frac{14\pi}{9}.$$

Exercise 3.1

Solve the following cubic equations by Cardon's method:

1. $x^3 - 21x - 344 = 0$
2. $x^3 - 6x - 9 = 0$
3. $x^3 - 6x - 4 = 0$
4. $8x^3 + 24x - 63 = 0$
5. $9x^3 - 6x^2 + 1 = 0$
6. $28x^3 - 9x^2 + 1 = 0$
7. $x^3 - 12x^2 - 6x - 10 = 0$
8. $x^3 + 3x^2 + 27x + 104 = 0$
9. $x^3 + 6x^2 + 9x + 4 = 0$
10. $x^3 - 15x^2 - 357x + 5491 = 0$
11. $x^3 - 15x^2 - 33x + 847 = 0$
12. $x^3 + x^2 - 16x + 20 = 0$
13. $x^3 - 6x^2 + 6x - 5 = 0$
14. $x^3 - 3x^2 + 3 = 0$
15. $2x^3 + 5x^2 + x - 2 = 0$
16. Show by Cardon's method that the roots of the cubic equation $x^3 - 3a^2x - 2a^3 \cos 3A = 0$ are $2a \cos A$, $2a \cos\left(\frac{2\pi}{3} \pm A\right)$
17. Show that the roots of the cube $x^3 + 3ax^2 + 3(a^2 - bc)x + a^3 + b^3 + c^3 - 3abc = 0$ are $-(a + b + c)$, $-(a + bw + cw^2)$, $-(a + bw^2 + cw)$
18. Find the limits to the values of c such that the cubic $x^3 - 3x + c = 0$ may have all its roots real.

Answers

1. $8, -4 \pm i3\sqrt{3}$

2. $3, -\frac{3}{2} \pm \frac{i\sqrt{3}}{2}$

3. $2\sqrt{2}\cos 15°, 2\sqrt{2}\cos 135°, 2\sqrt{2}\cos 255°$

4. $\frac{3}{2}, \frac{1}{4}\left(-3 \pm i5\sqrt{3}\right)$

5. $\frac{1}{3}, \frac{1}{6}\left(3 \pm i\sqrt{3}\right)$

6. $-\frac{1}{4}, \frac{1}{7}\left(2 \pm i\sqrt{3}\right)$

7. $4 + p^{1/3} + q^{1/3}, 4 + \omega p^{1/3} + \omega^2 q^{1/3}, 4 + \omega^2 p^{1/3} + \omega q^{1/3}$,
where $p^{1/3} = 3\ (2)^{1/3}$ and $q^{1/3} = 3.\ (4)^{1/3}$

8. $-8, \frac{1}{2}\left(5 \pm i3\sqrt{3}\right)$

9. $-4, -1, -1$

10. $-19, 17, 17$

11. $-7, 11, 11$

12. $-5, 2, 2$

13. $5, \frac{1}{2}\left(1 \pm \sqrt{3}\right)$

14. $1 + 2\cos\frac{2\pi}{9}, 1 + 2\cos\frac{8\pi}{9}, 1 + 2\cos\frac{14\pi}{9}$

15. $-1, -2, \frac{1}{2}$

18. $-2 \le c \le 2$

Illustrative Examples

***Example* 3.16:** Solve the equation by Ferrari's method $x^4 - 4x^3 - x^2 + 16x - 12 = 0$.

Solution: Assume that

$$f(x) = x^4 - 4x^3 - x^2 + 16x - 12 = (x^2 - 2x + \lambda)^2 - (mx + n)^2$$

Comparing coefficients of x^2, x and absolute terms from both sides, we get

$$\left.\begin{aligned} 2\lambda + 4 - m^2 &= -1 \text{ or } m^2 = 2\lambda + 5, \\ -4\lambda - 2mn &= 16 \quad \text{or } mn = -2\lambda - 8, \\ \lambda^2 - n^2 &= -12 \quad \text{or } n^2 = \lambda^2 + 12 \end{aligned}\right\} \qquad \text{.....(3.40)}$$

Eliminating m and n from the above relations, we have

$$(2\lambda + 5)(\lambda^2 + 12) = (-2\lambda - 8)^2$$

or $\quad 2\lambda^3 + \lambda^2 - 8\lambda - 4 = 0$ or $(\lambda^2 - 4)(2\lambda + 1) = 0$ or $(\lambda - 2)(\lambda + 2)(2\lambda + 1) = 0$

Hence $\lambda = 2, \ -2, \ -\dfrac{1}{2}$

Taking $\lambda = 2$, we obtain from (3.40):

$$m^2 = 9, \ mn = -12 \text{ and } n^2 = 16$$

Hence m = 3 and n = –4

$$\therefore \quad f(x) = (x^2 - 2x + 2)^2 - (3x - 4)^2$$

$$= (x^2 + x - 2)(x^2 - 5x + 6) = (x + 2)(x - 1)(x - 2)(x - 3)$$

Now $\quad f(x) = 0 \Rightarrow (x + 2)(x - 1)(x - 2)(x - 3) = 0$

$\Rightarrow \quad x = -2, 1, 2, 3$

Hence the required roots are –2, 1, 2, 3

***Example* 3.17:** Solve the equation by Ferrari's method: $x^4 - 4x^3 - 4x^2 - 24x + 15 = 0$

Solution: Assume that

$$f(x) = x^4 - 4x^3 - 4x^2 - 24x + 15 = (x^2 - 2x + \lambda)^2 - (mx + n)^2$$

Comparing coefficients of x^2, x and absolute terms on both sides we obtain

$$\left.\begin{aligned} 2\lambda + 4 - m^2 = -4 \text{ or } m^2 &= 2\lambda + 8, \\ -4\lambda - 2mn = 24 \text{ or } mn &= -12\lambda - 8 \\ \lambda^2 - n^2 = -15 \text{ or } n^2 &= \lambda^2 + 15 \end{aligned}\right\} \quad .....(3.41)$$

Eliminating m and n from relations (3.41), we get

$$(2\lambda + 8)(\lambda^2 - 15) = (-2\lambda + 12)^2$$

or $\quad \lambda^3 + 2\lambda^2 + 9\lambda - 132 = 0$

or $\quad (\lambda - 4)(\lambda^2 + 6\lambda + 33) = 0$

A root of this equation is $\lambda = 4$. This value gives m = 4; n = 1

$$\therefore \quad f(x) = (x^2 - 2x + 4)^2 - (4x + 1)^2 = (x^2 - 6x + 3)(x^2 + 2x + 5)$$

Now $\quad f(x) = 0 \Rightarrow (x^2 - 6x + 3)(x^2 + 2x + 5) = 0$

$$\Rightarrow x^2 - 6x + 3 = 0, x^2 + 2x + 5 = 0$$

$$\Rightarrow x = 3 \pm \sqrt{6}, -1 \pm 2i$$

Hence the required roots are $3 + \sqrt{6}, \ 3 - \sqrt{6}, \ -1 + 2i$ and $-1 - 2i$

***Example* 3.18:** Solve the equation by Ferrari's method $x^4 - 12x^3 + 41x^2 - 18x - 72 = 0$.

***Solution*:** The given biquadratic equation is

$$x^4 - 12x^3 + 41x^2 - 18x - 72 = 0. \qquad .....(3.42)$$

Combining x^4 and x^3 terms into a perfect square, the given eq. (3.42) can be written as

$$(x^2 - 6x + \lambda)^2 + 41x^2 - 18x - 72 - 36x^2 - 2\lambda x^2 + 12\lambda x - \lambda^2 = 0$$

or $$(x^2 - 6x + \lambda)^2 + (5 - 2\lambda)\,x^2 + (12\lambda - 18)\,x - (\lambda^2 + 72) = 0$$

or $$(x^2 - 6x + \lambda)^2 - [(2\lambda - 5)\,x^2 + (18 - 12\lambda)\,x + (\lambda^2 + 72)] = 0 \qquad .....(3.43)$$

This equation can be factorized if

$$(2\lambda - 5)\,x^2 + (18 - 12\lambda)\,x + \lambda^2 + 72$$

is a perfect square, i.e., if $B^2 - 4AC = 0$

i.e., if $(18 - 12\lambda)^2 - 4(2\lambda - 5)(\lambda^2 + 72) = 0$

i.e., if $2\lambda^3 - 41\lambda^2 + 252\lambda - 441 = 0$

i.e., if $2\lambda^3 - 6\lambda^2 - 35\lambda^2 + 105\lambda + 147\lambda - 441 = 0$

i.e., if $2\lambda^2(\lambda - 3) - 35\lambda(\lambda - 3) + 147(\lambda - 3) = 0$

i.e., if $(\lambda - 3)(2\lambda^2 - 35\lambda + 147) = 0$

i.e., if $\lambda = 3,\ 7,\ \dfrac{21}{2}$

Taking $\lambda = 3$, the eq. (3.42) reduces to

$$(x^2 - 6x + 3) - (x - 9)^2 = 0$$

or $$(x^2 - 5x - 6)(x^2 - 7x + 12) = 0$$

or $$(x + 1)(x - 6)(x - 3)(x - 4) = 0$$

or $$x = -1, 3, 4, 6$$

Hence, the roots of the given equations are –1, 3, 4 and 6

Exercise 3.2

Solve the following biquadratic by Ferrari's method:

1. $x^4 - 8x^2 - 24x + 7 = 0$
2. $x^4 - 5x^2 - 6x - 5 = 0$
3. $x^4 + 12x - 5 = 0$
4. $x^4 - 12x + 3 = 0$
5. $x^4 - 2x^2 + 8x - 3 = 0$
6. $x^4 - 6x^3 - 3x^2 + 22x - 6 = 0$

7. $x^4 - 10x^2 - 20x - 16 = 0$
8. $x^4 - 12x - 5 = 0$
9. $x^4 - 3x^2 - 6x - 2 = 0$
10. $x^4 + 8x^3 + 9x^2 - 8x - 10 = 0$
11. $x^4 + 2x^3 - 7x^2 - 8x + 12 = 0$
12. $x^4 - 3x^2 - 42x - 40 = 0$
13. $x^4 + 6x^2 + 8x + 21 = 0$
14. $x^4 + 6x^2 + 10x + 50 = 0$
15. $x^4 + 4x^3 + 8x^2 + 7x + 4 = 0$
16. $x^4 - 11x^3 + 12x^2 - 20x - 12 = 0$
17. $x^4 - 8x^3 - 12x^2 + 60x + 63 = 0$
18. $x^4 - 10x^3 + 44x^2 - 104x + 96 = 0$
19. $x^4 - 10x^3 + 35x^2 - 50x + 24 = 0$
20. $x^4 + 4x^3 + 12x^2 - 8x + 95 = 0$
21. $x^4 - 2x^3 - 5x^2 + 10x - 3 = 0$
22. $2x^4 + 6x^3 - 3x^2 + 2 = 0,$
23. $x^4 + 12x - 5 = 0$
24. $x^4 - 12x^3 + 41x^2 - 18x - 72 = 0$

Answers

1. $-2 \pm i\sqrt{3}, 2 \pm \sqrt{3}$
2. $\frac{1}{2}\left(-1 \pm i\sqrt{3}\right), \frac{1}{2}\left(1 \pm \sqrt{21}\right)$
3. $-1 \pm \sqrt{2}, 1 \pm 2i$
4. $\frac{1}{2}\left[-\sqrt{6} \pm \sqrt{4\sqrt{6} - 6}\right], \frac{1}{2}\left[\sqrt{6} \pm \sqrt{-6 - 4\sqrt{6}}\right]$
5. $1 \pm i\sqrt{2}, -1 \pm \sqrt{2}$
6. $1 \pm \sqrt{7}, 2 \pm \sqrt{3}$
7. $4, -2, -1 \pm i$
8. $-1 \pm 2i, 1 \pm \sqrt{2}$
9. $-1 \pm i, 1 \pm \sqrt{2}$
10. $\pm 1, -4 \pm \sqrt{2}$
11. $1, 2, -2, -3$
12. $4, -1, -\frac{1}{2}\left(3 \pm i\sqrt{31}\right)$
13. $1 \pm i\sqrt{6}, -1 \pm i\sqrt{2}$
14. $1 \pm 3i, -1 \pm 2i$
15. $-\frac{1}{2}\left(-1 \pm i\sqrt{3}\right), \frac{1}{2}\left(-3 \pm i\sqrt{7}\right)$
16. $2 \pm \sqrt{6}, 1 \pm i\sqrt{5}$
17. $3 \pm \sqrt{30}, 3, -1$
18. $2, 4, 2\,2 \pm i2\sqrt{2}$
19. $1, 2, 3, 4$
20. $-3 \pm i\sqrt{10}, 1 \pm 2i$
21. $\frac{1}{2}\left(3 \pm \sqrt{5}\right), \frac{1}{2}\left(-1 \pm \sqrt{13}\right)$
22. $-2 \pm \sqrt{2}, \frac{1}{2}(1 \pm i)$
23. $-1 - \sqrt{2}, -1 \pm 2i$
24. $-1, 3, 4$ and 6

3.12 Empirical Law

A law which connects the two variables of a given data is known as empirical law. Several equations of different types can be obtained to express the given data approximately.

3.13 Curve Fitting

To find the empirical law, a curve of "best fit" which can pass through most of the points of the given data or nearest to them is drawn. The process of finding such an equation of "best fit" is known as curve fitting. The equation of the curve (best fit) is used to predict the unknown values. To obtain an equation representing the data, we can apply Graphical method or Method of least-squares.

3.14 Graphical Method

If we are required to fit (a st. line) a linear law $y = a + bx$ to the given data with the help of graph, we proceed as follows:

(i) Plot the given points.

(ii) Draw the straight line of best fit such that, it may pass through most of the points or nearest to them.

(iii) Substitutute the co-ordinates of two suitable points in the equation $y = a + bx$. Two simultaneous equations are obtained. On solving these equations, we get, the values of a and b, so the law $y = a + bx$ is determined.

If a straight line is not suitable to the points of the data, a smooth curve is drawn through them. From the shape of the curve, we can infer the approximate equation of the curve. The equation of the curve can be transformed to the form $y = a + bx$.

3.15 Determination of other Empirical Laws Reducible to Linear Form

By simple transformation we can reduce the non-linear to linear form

A list of transformation is given below:

S.No	Equation of the curve	Substitution	Reduced linear equation
1.	$y = a + bx^2$	$y = y, x^2 = X$	$Y = a + bX$
2.	$xy = ax + by$ or $y = a + b\dfrac{y}{x}$	$y = Y, \dfrac{y}{x} = X$	$Y = a + bX$
3.	$y = a + x + bx^2$ or $y - x = a + bx^2$	$y - x = Y, x^2 = X$	$Y = a + bx$
4.	$y = Ax^n$ $\log y = \log A + n \log x$	$\log y = Y, \log A = a$ $\log x = X$	$Y = a + nX$
5.	$y x^n = m$ $\log y + n \log x = \log m$	$Y = \log y, X = \log x$ $a = \log m$	$Y = a - nX$
6.	$y = A x^b$ $\log y = \log A + b \log x$	$Y = \log y, a = \log A$ $X = \log x$	$Y = a + bX$

***Example* 3.19:** In the following table some observed values of x and y are given:

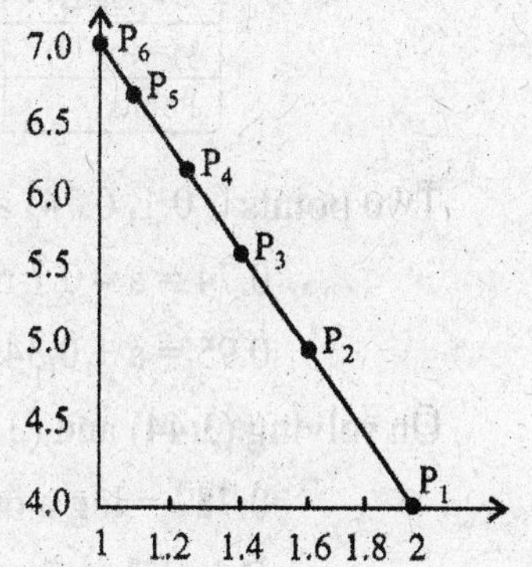

x	2	3	4	5	6	7
y	4	5	5.71	6.25	6.67	7

The law connecting x and y is xy = ax + by. Find the best values of a and b

Solution: $xy = ax + by$ or $y = a + b\dfrac{y}{x}$

Put $y = Y$ and $\dfrac{y}{x} = X$. The transformed equation is $Y = a + bX$.

The following tables gives the value of X and Y.

$X = \dfrac{y}{x}$	2	1.66	1.43	1.25	1.11	1
Y = y	4	5	5.71	6.25	6.67	7
Points	P_1	P_2	P_3	P_4	P_5	P_6

A best fit line Y = a + bX is drawn. Two points (2, 4) and (1, 7) of this line satisfy the equation Y = a + bX.

$$4 = a + 2b,\ 7 = a + b$$

On solving $a = 10,\ b = -3$

$$Y = 10 - 3X$$

$$y = 10 - 3\frac{y}{x} \text{ or } xy = 10x - 3y$$

***Example* 3.20:** The following figures were obtained in a calibration test of the discharge of water through an orifice.

Head H	2.2	1.8	1.4	1.1	0.8	0.6
Quantity Q	8.9	8.03	7.23	6.4	5.5	4.85

The law connecting H and Q has the form $Q = AH^n$. Find this law.

Solution: $Q = AH^n$

$$\log_{10} Q = \log_{10} A + n \log_{10} H$$

Put $\log_{10} Q = y,\ \log_{10} A = a,\ \log_{10} H = x$

The equation becomes $y = a + nx$

x(= log H)	0.34	0.26	0.15	0.04	–0.10	–0.22
y(= log Q)	0.95	0.91	0.86	0.81	0.74	0.69
Points	P_1	P_2	P_3	P_4	P_5	P_6

Two points (–0.1, 0.74) and (0.34, 0.95) are taken on this line

$$0.74 = a - 0.1\,n \quad \text{.....(3.44)}$$

$$0.95 = a + 0.34\,n \quad \text{.....(3.45)}$$

On solving (3.44) and (3.45), we get a = 0.787, n = 0.47

$$0.787 = \log a \text{ or } a = 6.13$$

$$Q = aH^n \text{ or } Q = 6.13\,H^{0.47}$$

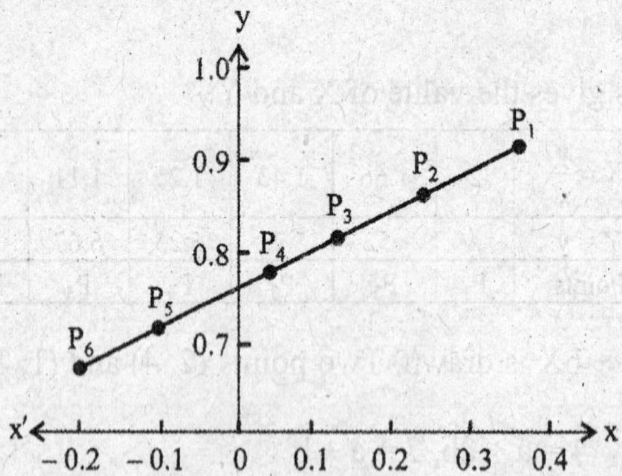

Exercise 3.3

1. Fit a straight line to the following data y as the dependent variable

x	1	2	3	4	5
y	5	7	9	10	11

Ans. y = 1.5, x – 0.3

2. The following values of x and y obey the law $y = ae^{bx}$. Find the best values a and b.

x	2.70	2.87	3.26	3.68	3.89
y	3.86	4.2	5.1	6.3	7

Ans. a = 0.91, b = 0.53

3. The following readings are recorded in an experiment designed to verify Boyle's law.

P	15	17	20	25	30	34	40
v	5.0	4.83	3.60	2.76	2.20	1.87	1.50

Find a law of the form $V = \dfrac{a}{P} + b$ **Ans.** a = 84.5, b = –0.6

4. The tension T in a belt just slipping over a pulley satisfies the law $T = T_0 e^{\mu Q}$ where μ is the coefficient of friction and θ the angle of contact between the pulley and the belt in radian. If T = 5 kg for all values of θ, find the best value of μ from the experimental results given below:

θ	$\frac{\pi}{8}$	$\frac{\pi}{4}$	$\frac{3\pi}{8}$	$\frac{\pi}{2}$	$\frac{5\pi}{8}$	$\frac{3\pi}{4}$	$\frac{7\pi}{8}$	π
T	5.57	6.14	6.84	7.61	8.47	9.34	10.32	11.45

Ans. $\mu = 0.27$

5. R is the resistance to maintain a train at speed V, find a law of the type $R = a + bv^2$ connecting R and V, using the following data:

V(miles / hour)	10	20	30	40	50
R(lb / ton)	8	10	15	21	30

Ans. a = 7.35, b = 0.0085

3.16 Principle of Least Squares

The graphical method has the obvious drawback in that the straight line drawn may not be unique. The method of least squares is probably the most systematic procedure to fit a unique curve through the given data points.

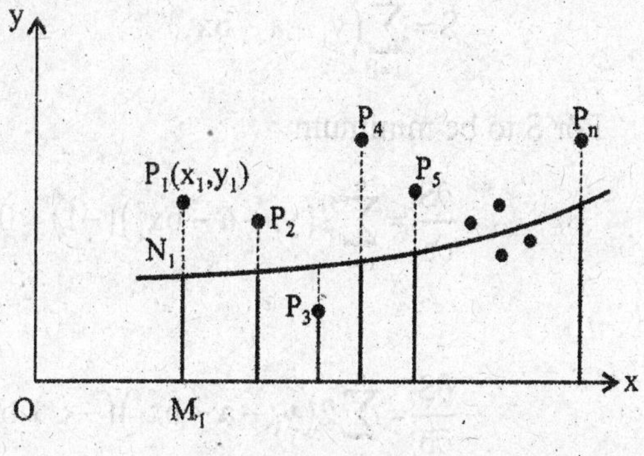

Let y = f(x) be the equation of curve to be fitted to the given data points (x_1, y_1), (x_2, y_2)... (x_n, y_n). At $x = x_i$, the observed (or experimental) value of the ordinate PM is y_i and the corresponding value on the fitting curve is NM, i.e., $[f(x_1)]$. The difference of the observed and the expected (theoretical) value is $PN = P_1 M_1 - N_1 M_1 = e_1$. This difference is called the error.

$$e_1 = y_1 - f(x_1)$$

Similarly, $e_2 = y_2 - f(x_2)$

$$e_3 = y_3 - f(x_3)$$

$$e_n = y_n - f(x_n)$$

Some of the errors e_1, e_2, e_3, ... e_n will be positive and others negative. To make all errors positive, we square each of the errors.

$$S = e_1^2 + e_2^2 + e_3^2 + ... + e_n^2$$

The curve of the best fit is that for which the sum of the squares of errors (S) is minimum. This is called the principle of least squares.

3.17 Method of Least Squares

Let $\quad y = a + bx \quad$ .....(3.46)

be the straight line to be fitted to the given data points (x_1, y_1), (x_2, y_2) ... (x_n, y_n).

Let y_{t_1} be the theoretical value for x_1.

Then $\quad e_1 = y_1 - y_{t_1}$

$$e_1 = y_1 - (a + bx_1)$$

$$e_1^2 = (y_1 - a - bx_1)^2$$

$$S = e_1^2 + e_2^2 + e_3^2 + ... + e_n^2 = \Sigma e_1^2$$

$$S = \sum_{i=1}^{n} (y_i - a - bx_i)^2$$

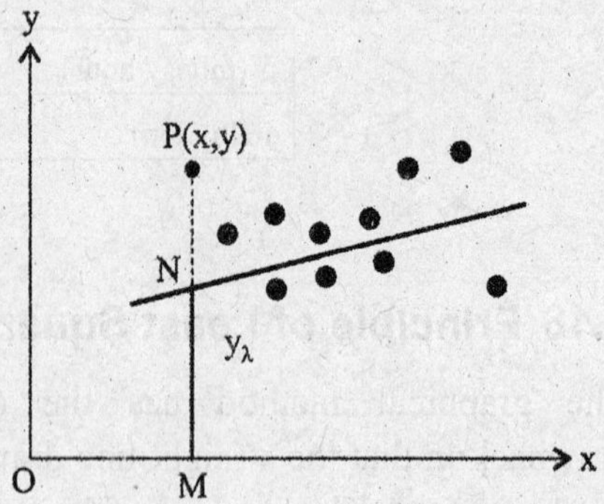

For S to be minimum

$$\frac{\partial S}{\partial a} = \sum_{i=1}^{n} 2(y_i - a - bx_i)(-1) = 0 \text{ or } \Sigma(y - a - bx) = 0 \qquad .....(3.46(a))$$

[To generalize y_i, y_1 is written as y]

$$\frac{\partial S}{\partial b} = \sum_{i=1}^{n} 2(y_i - a - bx_i)(-x_i) \text{ or } \Sigma(xy - ax - bx^2) = 0 \qquad .....(3.47)$$

On simplification eq. (3.46(a)) and (3.47) becomes

$$\Sigma y = na + b\Sigma x \qquad .....(3.48)$$

$$\Sigma xy = a\Sigma x + b\Sigma x^2 \qquad .....(3.49)$$

The equation (3.46(a)) and (3.47) are known as Normal equations.

On solving eq.s (3.46(a)) and (3.47), we get the value of a and b

(b) To fit the parabola:

$$y = a + bx + cx^2 \qquad .....(3.50)$$

The normal equations are

$$\Sigma y = na + b\Sigma x + c\Sigma x^2 \quad .....(3.51)$$

$$\Sigma xy = a\Sigma x + b\Sigma x^2 + c\Sigma x^3 \quad .....(3.52)$$

$$\Sigma x^2 y = a\Sigma x^2 + b\Sigma x^3 + c\Sigma x^4 \quad .....(3.53)$$

On solving three normal equations, we get the values of a, b and c

Note: 1. The normal eq. (3.47) has been obtained by putting Σ on both sides of eq. (3.46), eq. (3.48) is obtained by multiplying Σx on both sides of (3.46).

2. The normal eq. (3.51), (3.52), (3.53) are obtained by multiplying by Σ, Σx and Σx^2 on both sides of eq. (3.50).

***Example* 3.21:** Find the best values of a and b so that y = a + bx fits the data given in the table.

x	0	1	2	3	4
y	1.0	2.9	4.8	6.7	8.6

***Solution*:** y = a + bx .....(3.54)

x	y	xy	x^2
0	1.0	0	0
1	2.9	2.9	1
2	4.8	9.6	4
3	6.7	20.1	9
4	8.6	13.4	16
Σx = 10	Σy = 24.0	Σxy = 67.0	$\Sigma x^2 = 30$

Normal equations $\Sigma y = na + b\Sigma x$,....(3.55)

$$\Sigma xy = a\Sigma x + b\Sigma x^2 \quad .....(3.56)$$

On putting the value of $\Sigma x, \Sigma y, \Sigma xy, \Sigma x^2$ in (3.55) and (3.56), we have

$$24 = 5a + 10b \quad .....(3.57)$$

$$67 = 10a + 30b \quad .....(3.58)$$

On solving (3.57) and (3.58), we get

$$a = 1, b = 1.9$$

On substituting the values of a and b in (3.54), we get

$$y = 1 + 1.9\,x$$

Ans.

***Example* 3.22:** By the method of least squares, find the straight line that best fits the following data:

x	1	2	3	4	5
y	14	27	40	55	68

Solution: Let the equation of the straight line best fit be $y = a + bx$ .....(3.59)

x	**y**	**xy**	x^2
1	14	14	1
2	27	54	4
3	40	120	9
4	55	220	16
5	68	340	25
$\Sigma x = 15$	$\Sigma y = 204$	$\Sigma xy = 748$	$\Sigma x^2 = 55$

Normal equations are $\Sigma y = na + b\Sigma x$ .....(3.60)

$$\Sigma\, xy = a\Sigma x + b\Sigma x \quad .....(3.61)$$

On putting the values of Σx, Σy, Σxy and Σx^2 in (3.60) and (3.61), we have

$$204 = 5a + 15b \quad .....(3.62)$$

$$748 = 15a + 55b \quad .....(3.63)$$

On solving eq. (3.62) and (3.63), we get

$$a = 0,\ b = 13.6$$

On substituting the values of a and b in (3.59), we get

$$y = 13.6x$$

***Example* 3.23:** Find least squares polynomial approximation of degree two to the data.

x	0	1	2	3	4
y	–4	–1	4	11	20

Also compute the least error

***Solution*:** Let the equation of the polynomial be

$$y = a + bx + cx^2 \quad .....(3.64)$$

x	y	xy	x^2	x^2y	x^3	x^4
0	– 4	0	0	0	0	0
1	– 1	–1	1	–1	1	1
2	4	8	4	16	8	16
3	11	33	9	99	27	81
4	20	80	16	320	64	256
$\Sigma x = 10$	$\Sigma y = 30$	$\Sigma xy = 120$	$\Sigma x^2 = 30$	$\Sigma x^2y = 434$	$\Sigma x^3 = 100$	$\Sigma x^4 = 354$

Normal equations are $\Sigma y = 5a + b\,\Sigma x + c\,\Sigma\,x^2$ .....(3.65)

$$\Sigma\,xy = a\,\Sigma\,x + b\Sigma x^2 + c\,\Sigma x^3 \quad .....(3.66)$$

$$434 = 30a + 100b + 354c \quad .....(3.67)$$

On putting the values of Σx, Σy, $\Sigma\,xy$, Σx^2, Σx^2y, Σx^3, Σx^4 in eq.s (3.65), (3.66), (3.67), we obtain

$$30 = 5a + 10b + 30c \quad .....(3.68)$$

$$120 = 10a + 30b + 100c \quad .....(3.69)$$

$$434 = 30a + 100b + 354c \quad .....(3.70)$$

On solving these equations, we get $a = -4$, $b = 2$, $c = 1$

The required polynomial is

$$y = -4 + 2x + x^2, \quad \text{Error} = 0$$

Ans.

***Example* 3.24:** Find the least squares approximation of second degree for the discrete data:

x	–2	–1	0	1	2
y	15	1	1	3	19

***Solution*:** Let the equation of second degree polynomial be $y = a + bx + cx^2$ ... (3.71)

x	y	xy	x^2	x^2y	x^3	x^4
–2	15	– 30	4	60	– 8	16
– 1	1	– 1	1	1	– 1	1
0	1	0	0	0	0	0
1	3	3	1	3	1	1
2	19	38	4	76	8	16
$\Sigma x = 0$	$\Sigma y = 39$	$\Sigma xy = 10$	$\Sigma x^2 = 10$	$\Sigma x^2y = 140$	$\Sigma x^3 = 0$	$\Sigma x^4 = 34$

Normal equations are $\Sigma y = na + b\Sigma x + c\Sigma x^2$ ...(3.72)

$$\Sigma xy = a\Sigma x + b\Sigma x^2 + c\Sigma x^3 \quad .....(3.73)$$

$$\Sigma x^2y = a\Sigma x^2 + b\Sigma x^3 + c\Sigma x^4 \quad .....(3.74)$$

On putting the values of Σx, Σy, Σxy, Σx^2, Σx^2y, Σx^3, Σx^4 in eq.s. (3.72), (3.73), (3.74), we have

$$39 = 5a + 0b + 10c \quad .....(3.75)$$

$$10 = 0 + 10b + 0.c \quad .....(3.76)$$

$$140 = 10a + 0b + 34c \quad .....(3.77)$$

On solving (3.75), (3.76), (3.77), we get: $a = -\frac{37}{35}$, $b = 1$, $c = \frac{31}{7}$

The required polynomial of second degree is

$$y = -\frac{37}{35} + x + \frac{31}{7}x^2$$ **Ans.**

3.18 Change of Scale

If the data is equal interval in large numbers then we change the scale as $u = \frac{x - x_0}{h}$.

***Example* 3.25:** Fit a second degree parabola to the following data by least squares method.

x	1929	1930	1931	1932	1933	1934	1935	1936	1937
y	352	356	357	358	360	361	361	360	359

Solution: Taking $x_0 = 1933$, $y_0 = 357$

Taking $u = x - x_0$, $v = y - y_0$

$$u = x - 1933,\ v = y - 357$$

The equation $y = a + bx + cx^2$ is transformed to $v = A + Bu + Cu^2$ .....(3.78)

x	u = x = 1933	y	v = y = 357	uv	u^2	u^2v	u^3	u^4
1929	–4	352	–5	20	16	–80	–64	256
1930	–3	360	–1	3	9	–9	–27	81
1931	–2	357	0	0	4	0	–8	16
1932	–1	358	1	–1	1	1	–1	1
1933	0	360	3	0	0	0	0	0
1934	1	361	4	4	1	4	1	1
1935	2	361	4	8	4	16	8	16
1936	3	360	3	9	9	27	27	81
1937	4	359	2	8	16	32	64	256
Total	$\Sigma u = 0$		$\Sigma v = 11$	$\Sigma uv = 51$	$\Sigma u^2 = 60$	$\Sigma u^2v = -9$	$\Sigma u^3 = 0$	$\Sigma u^4 = 708$

Normal equations are

$\Sigma v = 9A + B\Sigma u + C\Sigma u^2$ or $11 = 9A + 0B + 60\,C$ or $11 = 9A + 60\,C$

$\Sigma uv = A\Sigma u + B\Sigma u^2 + C\Sigma u^3$ or $51 = 0A + 60B + 0C$ or $51 = 60\,B$ or $B = \frac{17}{20}$

$\Sigma u^2 v = A\Sigma u^2 + B\Sigma u^3 + C\Sigma u^4$ or $-9 = 60A + 0B + 708C$ or $-9 = 60A + 708\,C$

On solving these equations, we get $A = \frac{694}{231}$, $B = \frac{17}{20}$, $C = \frac{247}{924}$

$$V = \frac{694}{231} + \frac{17}{20}u - \frac{247}{924}u^2$$

$$y - 357 = \frac{694}{231} + \frac{17}{20}(x - 1933) - \frac{247}{924}(x - 1933)^2$$

$$= \frac{694}{231} + \frac{17x}{20} - \frac{32861}{20} - \frac{247x^2}{924} \frac{-247}{-924}(-3866x) - \frac{247}{924} + (1933)^2$$

$$y = \frac{694}{231} + -\frac{32861}{20} - \frac{247}{924}(1933)^2 + \frac{17}{20}x + \frac{247 \times 3866}{924}x - \frac{247}{924}x^2$$

$$y = 3 - 1643.05 - 998823\,.\,36 + 357 + .85\,x + 1033\,.\,44x - 267x^2$$

$$y = -1000106.41 + 1034.29x - 0.267\,x^2$$ **Ans.**

***Example* 3.26:** Find the least squares fit of the form $y = a + bx^2$ to the following data:

x	–1	0	1	2
y	2	5	3	0

***Solution*:** $y = a + bx^2$ .....(3.79)

Let $x^2 = z,\ y = a + bz$ .....(3.80)

x	y	$Z = x^2$	yz	z^2
–1	2	1	2	1
0	5	0	0	0
1	3	1	3	1
2	0	4	0	16
	$\Sigma y = 10$	$\Sigma z = 6$	$\Sigma yz = 5$	$\Sigma z^2 = 18$

Normal equations are $\Sigma y = na + b\Sigma z$ .....(3.81)

$\Sigma yz = a\Sigma z + b\Sigma z^2$ .....(3.82)

On putting the values of Σy, Σz, Σyz, Σz^2 in (3.81) and (3.82), we have

$$10a = 4a + 6b \quad .....(3.83)$$

$$5 = 6a + 18b \quad .....(3.84)$$

On solving eq. (3.83) and (3.84), we get, $a = \frac{25}{6}, b = \frac{10}{9}$

On substituting the values of a, b in (3.80), we obtain

$$y = \frac{25}{6} + \frac{10}{9}z$$

or $$y = \frac{25}{6} + \frac{10}{9}x^2 \text{ or } 18y = 75 - 20x^2$$ **Ans.**

***Example* 3.27:** Use least-squares method to fit a curve of the form $y = ae^{bx}$ to the data.

x	1	2	3	4	5	6
y	7.209	5.265	3.846	2.809	2.052	1.499

Solution: $$y = ae^{bx} \quad .....(3.85)$$

$$\log_e y = \log_e a + bx \quad .....(3.86)$$

On putting $\log_e y = Y$, $\log_e a = c$ in (3.86), it becomes

$$Y = c + bx \quad .....(3.87)$$

x	y	$Y = \log_e y$	xY	x^2
1	7.209	1.97533	1.97533	1
2	5.265	1.66108	3.32216	4
3	3.846	1.34703	4.04109	9
4	2.809	1.03283	4.13132	16
5	2.052	0.71881	3.59405	25
6	1.499	0.40480	2.4288	36
$\Sigma x = 21$		$\Sigma y = 7.13988$	$\Sigma xY = 19.49275$	$\Sigma x^2 = 91$

Normal equations are $\Sigma Y = nc + b\Sigma x$ .....(3.88)

$$\Sigma xY = c\Sigma x + b\Sigma x^2 \quad .....(3.89)$$

On putting the value of Σx, ΣxY and Σx^2 in equations (3.88) and (3.89), we get

$$7.13988 = 6c + 21b \quad .....(3.90)$$

$$19.49275 = 21c + 91b \quad .....(3.91)$$

On solving (3.90) and (3.91), we obtain b = –0.3141, c = 2.28933

$$c = \log_e a = 2.28933 \Rightarrow a = 9.86832$$

On putting the value of a and b in (3.85), we obtain

$$y = 9.86832\ e^{-0.3141}$$ **Ans.**

Exercise 3.4

1. Find the linear least square polynomials based on data

x	–2	–1	0	1
y	6	3	2	2

is given. Find the least square straight line approximation to the data.

2. The data

x	1	2	3	4
y	3	7	13	21

is given. Find square straight line approximation to the data.

Ans. $y = -4 + 6x$

3. Use the least-square method to obtain a parabola that approximates the data

x	1.0	1.2	1.4	1.6	1.8	2
y	2.345	2.419	2.592	2.863	3.233	3.702

Ans. $y = 3.124 - 2.477 + 1.458\ x^2$

4. Find the values of a, b, c so that $y = a + bx + cx^2$ is the best fit to the data:

x	0	1	2	3	4
y	1	0	3	10	21

Ans. a = 1, b = –3, c = 2

5. Find the least squares approximation of the form $y = a + bx^2$, for the data.

x	0	0.1	0.2	0.3	0.4	0.5
y	1	1.01	0.99	0.85	0.81	0.75

Ans. $y = 1.0032 - 1.1081x^2$

6. If V (km/hr) and R (kg/ton) are related by a relation of type $R = a + bl^2$, find by the method of least squares, a and b with the help of the following table.

V	10	20	30	40	50
R	8	10	15	21	30

Ans. a = 632, b = 0.0095

7. Determine the constants a and b by the least-squares method such that $y = ae^{bx}$ fits the following data:

x	1.0	1.2	1.4	1.6
Y	40.170	73.196	133.372	243.02

Ans. a = 2, b = 3

8. The pressure and volume of a gas related by the equation $PV^r = k$, r and k being constants. Fit this equation to the following set of observations:

P(kg/cm^2)	0.5	1.0	1.5	2.0	2.5	3.0
V (litres)	1.62	1.00	0.75	0.62	0.52	0.46

Ans. $PV^{1.276} = 1.039$

9. Fit a least-square geometric curve $y = ax^b$ to the following data:

x	1	2	3	4	5
y	0.5	2	4.5	8	12.5

Ans. a = 0.5012, b = 1.9977

10. Using method of least-squares, fit a relation of the form $y = ab^x$ to the following data:

x	2	3	4	5	6
Y	144	172.8	207.4	248.8	298.5

Ans. a = 99.86, b = 1.2

11. The following table gives the results of the measurements of train resistances; V is the velocity in miles per hour, R is the resistance in pounds per ton:

V	20	40	60	80	100	120
R	5.5	9.1	14.9	22.8	33.3	46.0

If R is related to V by the relation $R = a + bV + cv^2$, find a, b and c by using the method of least squares.

[**Hint:** $131.6 = 6a + 420b + 36400\,c$

$$12042 = 420\,a + 36400\,b + 3528000\,c$$

$$1219720 = 36400\,a + 3528000\,b + 364000000\,c]$$

Ans. $R = 4.2247 + 0.0016\,V + 002877\,V$

12. Employ the method of least square to fit a parabola $y = a + bx + cx^2$ in the following data.

(x, y) : (–1, 2), (0, 0) (0, 1) (1, 2)

[**Hint:** $5 = 4a + 0 + 2c$

$$0 = 0 + 2b + 0$$

$$4 = 2a + 0 + 2c$$

On solving, $a = \frac{1}{2}$, $c = \frac{3}{2}$] **Ans.** $y = \frac{1}{2} + \frac{3}{2}x^3$

13. If F is pull required to lift a load W by means of a pulley, fit a linear law F = a – bw connecting F and W against the following data:

W	50	70	100	120
F	12	15	21	25

Ans. $F = 2.2785 - 0.1879\,W$

Part II

CHAPTER 4

Successive Differentiation

4.1 Definitions and Notations

We know that the derivative of a function of x is in general also a function of x. This new function may also be differentiable, in which case the derivative of the *first derivative* is called the *second derivative* of the original function. Similarly, the derivative of the second derivative is called the *third derivative*; and so on to the nth *derivative*.

If y = f(x) be a function of x, the derivative (i.e., differential co-efficient) of y w.r.t. x is denoted by one or other of the following:

$$\frac{dy}{dx}y_1,\ y_1',\ f'(x) \text{ or } Dy \text{ where } D \equiv \frac{d}{dx}.$$

This derivative is also called the first derivative of y. Clearly $\frac{dy}{dx}$ itself is a function of x.

If this derived function is again derivable then the differential co-efficient of $\frac{dy}{dx}$ w.r.t x is called the *second derivative* of y w.r.t. x (or the derivative of the second order) and it is denoted by one or other of the following:

$$\frac{d}{dx}\left(\frac{dy}{dx}\right),\ \frac{d^2y}{dx^2},\ y_2,\ y'',\ f''(x) \text{ or } D^2y$$

By the definition of derivative we find that

$$\frac{d^2y}{dx^2} \text{ i.e., } \boxed{f''(x) = \lim_{\delta x \to 0} \frac{f'(x+\delta x) - f'(x)}{\delta x}}$$

Note: For the geometrical interpretation of the second derivative and significance of its sign, see Appendix I.

The second derivative is also a function of x. So this derived function may again be derivable (that is, differentiable) and the derivative of $\frac{d^2y}{dx^2}$, w.r.t. x can be obtained. This derivative will be called the *third derivative* of y w.r.t. x and it is denoted by one or other of the following:

$$\frac{d}{dx}\left(\frac{d^2y}{dx^2}\right) \text{ or } \frac{d^3y}{dx^3} \text{ or } y_3 \text{ or } f'''(x) \text{ or } D^3y$$

Clearly the second derivative of y w.r.t. x is the first derivative of $\frac{dy}{dx}$ w.r.t. x and the third derivative of y w.r.t. x is the first derivative of the second derivative of y w.r.t. x. Thus

$$\frac{d^2y}{dx^2} = \frac{d}{dx}\left(\frac{dy}{dx}\right); \frac{d^3y}{dx^3} = \frac{d}{dx}\left(\frac{d^2y}{dx^2}\right)$$

In this way we can go on finding the fourth, fifth... derivatives of y w.r.t. x successively. The successive derivatives of $y = f(x)$ w.r.t. x are denoted as follows:

	First derivative	Second derivative	Third derivative	...	n^{th} derivative	...
	y_1	y_2	y_3	...	y_n	...
or	y'	y''	y'''	...	$y^{(n)}$	...
or	$f'(x)$	$f''(x)$	$f'''(x)$	...	$f^n(x)$	...
or	Dy	D^2y	D^3y	...	$D^n(y)$	...

where $D \equiv \frac{d}{dx}$

If the independent variable is the time t then usually the successive derivatives are denoted as

$$\dot{y}, \ddot{y}, \dddot{y}, \ldots$$

Illustration: If $y = x^n$, then

$\frac{dy}{dx} = nx^{n-1}$, which is a function of x

$\frac{d^2y}{dx^2} = \frac{d}{dx}\left(\frac{dy}{dx}\right) = \frac{d}{dx}(nx^{n-1}) = n(n-1)x^{n-2}$, which is a function of x.

$$\frac{d^3y}{dx^3} = \frac{d}{dx}\left(\frac{d^2y}{dx^2}\right) = \frac{d}{dx}\{n(n-1)x^{n-2}\} = n(n-1)(n-2)x^{n-3}$$

$$\frac{d^4y}{dx^4} = \frac{d}{dx}\left(\frac{d^3y}{dx^3}\right) = \frac{d}{dx}\{n(n-1)(n-2)x^{n-3}\}$$

$$= n(n-1)(n-2)(n-3)x^{n-4} \text{ etc.}$$

Also remember the following facts.

$$\frac{d^ny}{dx^n} = \frac{d}{dx}\left(\frac{d^{n-1}y}{dx^{n-1}}\right) = \frac{d^{n-1}}{dx^{n-1}}\left(\frac{dy}{dx}\right) = \frac{d^{n-2}}{dx^{n-2}}\left(\frac{d^2y}{dx^2}\right) \text{ etc.}$$

i.e., the nth derivative

= the first derivative of the $(n-1)^{th}$ derivative

= $(n-1)^{th}$ derivative of the first derivative

= $(n-2)^{th}$ derivative of the second derivative etc.

Work Out Examples

(On finding a Derivative of Higher Order)

1. (i) Find y_3 if $y = x^2 \log x$

(ii) Find y_2 if $y = \tan(x + y)$

(iii) Find $\frac{d^2y}{dx^2}$ if $x = a\cos\theta$, $y = a\sin\theta$

Solution:

(i) Here $y = x^2 \log x$

Differentiating w.r.t. x,

$$\frac{dy}{dx} = y_1 = 2x\log x + x^2.\frac{1}{x} = x + 2x\log x$$

Again differentiating w.r.t. x,

$$\frac{d^2y}{dx^2} = y_2 = 1 + 2\left(1.\log x + x.\frac{1}{x}\right)$$

$$= 2\log x + 3$$

Again differentiating w.r.t. x,

$$\frac{d^3y}{dx^3} = y_3 = \frac{2}{x}$$

(ii) Here $y = \tan(x+y)$

Differentiating w.r.t. x,

$$y_1 = \sec^2(x+y).(1+y_1)$$

or $\{\sec^2(x+y) - 1\}y_1 = -\sec^2(x+y)$

or $\tan^2(x+y).y_1 = -\sec^2(x+y)$

or $y_1 = -\operatorname{cosec}^2(x+y)$

Differentiating again w.r.t. x,

$$y_2 = -2\operatorname{cosec}(x+y).\{-\operatorname{cosec}(x+y).\cot(x+y)\}(1+y_1)$$

$$= 2\operatorname{cosec}^2(x+y).\cot(x+y).\{1-\operatorname{cosec}^2(x+y)\}$$

$$= -2\operatorname{cosec}^2(x+y)\cot(x+y)\cot^2(x+y)$$

$$= -2.\{1+\cot^2(x+y)\}.\cot^3(x+y)$$

$$= -2.\left\{1+\frac{1}{y^2}\right\}.\frac{1}{y^3} = \frac{-2(1+y^2)}{y^5}$$

(iii) Here $\frac{dx}{d\theta} = -a\sin\theta$; $\frac{dy}{d\theta} = a\cos\theta$

$\therefore$ $$\frac{dy}{dx} = \frac{\frac{dy}{d\theta}}{\frac{dx}{d\theta}} = \frac{a\cos\theta}{-a\sin\theta} = -\cot\theta \quad \therefore \quad \frac{d}{d\theta}\left(\frac{dy}{dx}\right) = \operatorname{cosec}^2\theta$$

Now $$\frac{d^2y}{dx^2} = \frac{d}{dx}\left(\frac{dy}{dx}\right) = \frac{\frac{d}{d\theta}\left(\frac{dy}{dx}\right)}{\frac{dx}{d\theta}} = \frac{\operatorname{cosec}^2\theta}{-a\sin\theta} = -\frac{1}{a}.\operatorname{cosec}^3\theta$$

2. (a) If $ax^2 + 2hxy + by^2 = 1$, show that

$$\frac{d^2y}{dx^2} = \frac{h^2 - ab}{(hx+by)^3}$$

Solution: We have $ax^2 + 2hxy + by^2 = 1$ .....(1)

Differentiating both sides with respect to x, we get

$$2ax + 2h(1.y + x.y_1) + 2byy_1 = 0$$

$\therefore$ $$ax + hy + hxy_1 + byy_1 = 0$$

or $$y_1(hx+by) = -(ax+hy)$$

or $$y_1 = -\frac{ax+hy}{hx+by} \qquad \text{.....(2)}$$

Again differentiating both sides with respect to x, we get

$$y_2 = -\frac{(hx+by)(a+hy_1)-(ax+hy)(h+by_1)}{(hx+by)^2}$$

$$= -\frac{ahx+aby+h^2xy_1+bhyy_1-ahx-abxy_1-h^2y-bhyy_1}{(hx+by)^2}$$

$$= -\frac{ab(y-xy_1)-h^2(y-xy_1)}{(hx+by)^2} = \frac{(ab-h^2)(y-xy_1)}{(hx+by)^2} \qquad \text{.....(3)}$$

But $$y - xy_1 = y + x.\frac{ax+hy}{hx+by}, \qquad \text{from (2)}$$

$$= \frac{hxy+by^2+ax^2+hxy}{hx+by}$$

$$= \frac{ax^2+2hxy+by^2}{hx+by} = \frac{1}{hx+by}, \qquad \text{from (1)}$$

Putting this value in (3), we get

$$y_2 = -\frac{ab-h^2}{(hx-by)^3}. \text{ Hence } \frac{d^2y}{dx^2} = \frac{h^2-ab}{(hx+by)^3}$$

2. (b) If $\sqrt{x+y}+\sqrt{y-x}=c$

Show that $y_2 = \frac{2}{c^2}$.

Solution: We have $\sqrt{x+y}+\sqrt{y-x}=c$.

Squaring both sides, we get

$$x+y+y-x+2\sqrt{y^2-x^2}=c^2$$

or $$2\sqrt{y^2-x^2}=c^2-2y$$

Again squaring both sides, we get

$$4(y^2 - x^2) = c^4 - 4c^2y + 4y^2$$

or $$4x^2 - 4c^2y + c^4 = 0$$

Differentiating both sides with respect to x, we get

$$8x - 4c^2y_1 = 0$$

Again differentiating both sides with respect to x, we get

$$2 - c^2y_2 = 0$$

Hence $$y_2 = \frac{2}{c^2}$$

2. (c) If $p = a^2\cos^2\theta + b^2\sin^2\theta$, show that

$$p + \frac{d^2p}{d\theta^2} = 2a^2 + 2b^2 - 3p$$

Solution: We have $p = a^2\cos^2\theta + b^2\sin^2\theta$ .....(1)

Differentiating both sides with respect to θ, we get

$$\frac{dp}{d\theta} = -2a^2\cos\theta\sin\theta + 2b^2\sin\theta\cos\theta$$

$$= -a^2\sin 2\theta + b^2\sin 2\theta = (b^2 - a^2)\sin 2\theta \quad .....(2)$$

Again differentiating both sides with respect to θ, we get

$$\frac{d^2p}{d\theta^2} = 2(b^2 - a^2)\cos 2\theta \quad .....(3)$$

Adding (1) and (3), we get

$$p + \frac{d^2p}{d\theta^2} = 2(b^2 - a^2)(\cos^2\theta - \sin^2\theta) + a^2\cos^2\theta + b^2\sin^2\theta$$

$$= \cos^2\theta(2b^2 - 2a^2 + a^2) + \sin^2\theta(2a^2 - 2b^2 + b^2)$$

$$= \cos^2\theta(2b^2 - 3a^2 + 2a^2) + \sin^2\theta(2a^2 - 3b^2 + 2b^2)$$

$$= (2a^2 + 2b^2)(\cos^2\theta + \sin^2\theta) - 3(a^2\cos^2\theta + b^2\sin^2\theta)$$

$$= 2a^2 + 2b^2 - 3p$$

(On Relation between Derivatives of Different Order)

3. (a) If $y^{\frac{1}{m}} + y^{-\frac{1}{m}} = 2x$, prove that

$$(x^2 - 1)y_2 + xy_1 - m^2y = 0$$

Solution: We have $y^{\frac{1}{m}} + y^{-\frac{1}{m}} = 2x$ .....(1)

Differentiating with respect to x, we get

$$\frac{1}{m}y^{\frac{1}{m}-1}y_1 + \left(-\frac{1}{m}\right)y^{-\frac{1}{m}-1}y_1 = 2$$

or $$\frac{1}{m}y^{-1}y_1\left(y^{\frac{1}{m}} - y^{-\frac{1}{m}}\right) = 2$$

or $$y_1(y^{\frac{1}{m}} - y^{-\frac{1}{m}}) = 2my$$

Squaring both sides, we get

$$y_1^2(y^{\frac{1}{m}} - y^{-\frac{1}{m}})^2 = 4m^2y^2$$

or $$y_1^2[(y^{\frac{1}{m}} + y^{-\frac{1}{m}})^2 - 4y^{\frac{1}{m}}y^{-\frac{1}{m}}] = 4m^2y^2$$

or $$y_1^2[(2x)^2 - 4] = 4m^2y^2, \quad \text{from (1)}$$

or $$y_1^2(x^2-1) = m^2y^2$$

Again differentiating with respect to x, we get

$$(x^2-1).\frac{d}{dx}(y_1^2) + y_1^2.\frac{d}{dx}(x^2-1) = m^2\frac{d}{dx}(y^2)$$

or $$(x^2-1).2y_1y_2 + y_1^2(2x) = m^2.2yy_1$$

Dividing both sides by $2y_1$, we get

$$(x^2-1)y_2 + xy_1 - m^2y = 0$$

3. (b) If $y = \frac{ax+b}{a+bx}$, show that

$$2y_1y_3 = 3y_2^2$$

Solution: Dividing $ax+b$ by $a+bx$, we get

$$\therefore \quad y = \frac{a}{b} + \left(b - \frac{a^2}{b}\right).\frac{1}{bx+a}$$

Differentiating both sides with respect to x, we get

$$y_1 = 0 + \left(b - \frac{a^2}{b}\right) \cdot \frac{-b}{(bx+a)^2} = \frac{-(b^2 - a^2)}{(bx+a)^2}$$

Differentiating the above with respect to x, we get

$$y_2 = \frac{2b(b^2 - a^2)}{(bx+a)^3} \quad \text{Also} \quad y_3 = \frac{-6b^2(b^2 - a^2)}{(bx+a)^4}$$

$$\text{Now L.H.S} = 2y_1 2y_3 = 2\frac{-(b^2 - a^2)}{(bx+a)^2} \times \frac{-6b^2(b^2 - a^2)}{(bx+a)^4}$$

$$3\left[\frac{2b(b^2 - a^2)}{(bx+a)^3}\right]^2 = 3y_2^2 = \text{R.H.S}$$

3. (c) If $y = \dfrac{ax^2 + bx + c}{1 - x}$, show that $(1 - x)y_3 = 3y_2$

Solution: We have $y(1 - x) = ax^2 + bx + c$

Differentiating both sides with respect to x, we get

$$y_1(1 - x) - y = 2ax + b$$

Again differentiating both sides with respect to x, we get

$$y_2(1 - x) + y_1(-1) - y_1 = 2a$$

Again differentiating both sides with respect to x, we get

$$y_3(1 - x) - y_2 - 2y_2 = 0$$

Hence $\quad (1 - x)y_3 = 3y_2$

3. (d) If $y = \left(x + \sqrt{1 + x^2}\right)^m$, prove that

$$\left(1 + x^2\right)\frac{d^2y}{dx^2} + x\frac{dy}{dx} - m^2y = 0$$

Solution: We have $y = \left(x + \sqrt{1 + x^2}\right)^m$ .....(1)

Differentiating both sides with respect to x, we get

$$\frac{dy}{dx}=m\left(x+\sqrt{1+x^2}\right)^{m-1}\left(1+\frac{1}{2\sqrt{1+x^2}}.2x\right)$$

$$=m(x+\sqrt{1+x^2})^{m-1}\left(\frac{x+\sqrt{1+x^2}}{\sqrt{1+x^2}}\right)$$

$$=\frac{m(x+\sqrt{1+x^2})^m}{\sqrt{1+x^2}}=\frac{my}{\sqrt{1+x^2}}, \qquad \text{from (1)}$$

or $$\sqrt{1+x^2}\frac{dy}{dx}=my \qquad .....(2)$$

Again differentiating both sides with respect to x, we get

$$\sqrt{1+x^2}\frac{d^2y}{dx^2}+\frac{1}{2\sqrt{1+x^2}}.2x\frac{dy}{dx}=m\frac{dy}{dx}$$

or $$\left(1+x^2\right)\frac{d^2y}{dx^2}+x\frac{dy}{dx}=m.\sqrt{1+x^2}\frac{dy}{dx}$$

Hence $$\left(1+x^2\right)\frac{d^2y}{dx^2}+x\frac{dy}{dx}-m^2y=0 \qquad \text{from (2)}$$

3. (e) If $y=A\left(x+\sqrt{x^2-1}\right)^n+B\left(x-\sqrt{x^2-1}\right)^n$, prove that

$$(x^2-1)y_2+xy_1-n^2y=0$$

Solution: We have $y=A(x+\sqrt{x^2-1})^n+B(x-\sqrt{x^2-1})^n$

Differentiating both sides with respect to x, we get

$$y_1=nA\left(x+\sqrt{x^2-1}\right)^{n-1}\left(1+\frac{1}{2\sqrt{x^2-1}}.2x\right)$$

$$+nB\left(x-\sqrt{x^2-1}\right)^{n-1}\left(1-\frac{1}{2\sqrt{x^2-1}}.2x\right)$$

$\therefore$ $$y_1=nA\left(x+\sqrt{x^2-1}\right)^{n-1}\left(\frac{x+\sqrt{x^2-1}}{\sqrt{x^2-1}}\right)$$

$$-nB\left(x-\sqrt{x^2-1}\right)^{n-1}\left(\frac{x-\sqrt{x^2-1}}{\sqrt{x^2-1}}\right)$$

or $\sqrt{x^2-1}\, y_1 = nA(x+\sqrt{x^2-1})^n - nB(x-\sqrt{x^2-1})^n$

Again differentiating both sides with respect to x, we get

$$\sqrt{x^2-1}\, y_2 + \frac{1}{2\sqrt{x^2-1}}.2xy_1$$

$$= n^2A(x+\sqrt{x^2-1})^{n-1}\left(1+\frac{1}{2\sqrt{x^2-1}}.2x\right)$$

$$-n^2B(x-\sqrt{x^2-1})^{n-1}\times\left(1-\frac{1}{2\sqrt{x^2-1}}.2x\right)$$

$$(x^2-1)y_2 + xy_1 = n^2A(x+\sqrt{x^2-1})^n + n^2B(x-\sqrt{x^2-1})^n$$

or $(x^2-1)y_2 + xy_1 = n^2[A(x+\sqrt{x^2-1})^n + B(x-\sqrt{x^2-1})^n]$

Hence $(x^2-1)y_2 + xy_1 - n^2y = 0$

4. If $y = (a+bx)\cos kx + (c+dx)\sin kx$, prove that

$$y_4 + 2k^2y_2 + k^4y = 0$$

where a, b, c d and k are all constants.

Solution: We have

$$y = (a + bx) \cos kx + (c + dx) \sin kx \quad .....(1)$$

Differentiating with respect to x, we get

$$y_1 = \cos kx \frac{d}{dx}(a+bx) + (a+bx)\frac{d}{dx}(\cos kx)$$

$$+\sin kx \frac{d}{dx}(c+dx) + (c+dx)\frac{d}{dx}(\sin kx)$$

$$= \cos kx.b - k(a+bx)\sin kx + \sin kx.d + k(c+dx)\cos kx$$

Again differentiating with respect to x, we get

$$y_2 = bk\sin kx - kb\sin kx - k^2(a+bx)\cos kx$$

$$+dx\cos kx + kd\cos kx - k^2(x+dx)\sin kx$$

$$= 2dk\cos kx - 2bk\sin kx - k^2[(a+bx)\cos kx + (c+dx)\sin kx] \quad .....(2)$$

$$= 2dk\cos kx - 2bk\sin kx - k^2y \quad .....(3)$$

or $$k^2y_2 = 2dk^3 \cos kx - 2bk^3 \sin kx - k^4 y$$

Again differentiating (2) with respect to x, we get

$$y_3 = -2dk^2 \sin kx - 2bk^2 \cos kx - k^2 y_1$$

Again differentiating with respect to x, we get

$$y_4 = -2dk^3 \cos kx - 2bk^3 \cos kx - k^2 y_2 \qquad .....(4)$$

Adding (3) and (4), we get

$$y_4 + 2k^2 y_2 + k^4 y = 0$$

5. (a) If $y = \sin(\sin x)$, prove that

$$y_2 + y_1 \tan x + y \cos^2 x = 0$$

Solution: We have $y = \sin(\sin x)$ .....(1)

Differentiating both sides with respect to x, we get

$$y_1 = \cos(\sin x).\cos x. \qquad .....(2)$$

Again differentiating both sides with respect to x, we get

$$y_2 = -\sin(\sin x)\cos^2 x - \cos(\sin x)\sin x \qquad .....(3)$$

Now L.H.S = $y_2 + y_1 \tan x + y \cos^2 x$

$$= -\sin(\sin x)\cos^2 x - \cos(\sin x)\sin x + \cos(\sin x)\sin x$$

$$+\sin(\sin x)\cos^2 x, \quad \text{from (1), (2) and (3)}$$

$$= 0$$

5. (b) If $y = \sin^{-1} x$, prove that

$$(1 - x^2)y_2 - xy_1 = 0$$

Solution: We have $y = \sin^{-1} x$

Differentiating both sides with respect to x, we get

$$y_1 = \frac{1}{\sqrt{1 - x^2}}$$

or $\sqrt{1-x^2}\,y_1 = 1$

Again differentiating both sides with respect to x, we get

$$\sqrt{1-x^2}\,y_2 + \frac{1}{2\sqrt{1-x^2}}(-2x)y_1 = 0$$

or $(1-x^2)y_2 - xy_1 = 0$

6. If $y = e^{\cos^{-1}x}$, prove that

$$(1-x^2)y_2 - xy_1 - y = 0$$

Solution: We have $y = e^{\cos^{-1}x}$ .....(1)

Differentiating both sides with respect to x, we get

$$y_1 = e^{\cos^{-1}x}.\left(-\frac{1}{\sqrt{1-x^2}}\right)$$

or $\sqrt{1-x^2}\,y_1 = -y$ from (1)

Again differentiating both sides with respect to x, we get

$$\sqrt{1-x^2}\,y_2 + \frac{1}{2\sqrt{1-x^2}}(-2x)y_1 = -y_1$$

or $(1-x^2)y_2 - xy_1 = -y_1\sqrt{1-x^2}$

or $(1-x^2)y_2 - xy_1 = y$

Hence $(1-x^2)y_2 - xy_1 = y = 0$

7. If $x = \cosh\left(\frac{\log y}{m}\right)$, prove that $(x^2-1)y_2 + xy_1 - m^2y = 0$

Solution: We have $x = \cosh\left(\frac{\log y}{m}\right)$; $\cosh^{-1}x = \frac{1}{m}\log y$

Differentiating both sides with respect to x, we get

$$\frac{1}{\sqrt{x^2-1}} = \frac{1}{my}.y_1 \quad \text{or} \quad y_1\sqrt{x^2-1} = my \qquad .....(1)$$

Again differentiating both sides with respect to x, we get

$$\sqrt{x^2-1}.y_2 + y_1.\frac{1}{2\sqrt{x^2-1}}.2x = my_1$$

or $$(x^2-1)y_2 + xy_1 = my_1\sqrt{x^2-1}$$

or $$(x^2-1)y_2 + xy_1 = m^2y \qquad \text{from (1)}$$

Hence $$(x^2-1)y_2 + xy_1 = m^2y = 0$$

8. If $y = \log\left(\frac{x}{a+bx}\right)^x$, prove that

$$x^3y_2 = (y - xy_1)^2$$

Solution: We have $y = \log\left(\frac{x}{a+bx}\right)^x$

or $$y = x\log\frac{x}{a+bx} \qquad .....(1)$$

or $$y = x[\log x - \log(a+bx)]$$

or $$y = x\log x - x\log(a+bx)$$

Differentiating both sides with respect to x, we get

$$y_1 = \log x + x.\frac{1}{x} - 1.\log(a+bx) - \frac{bx}{a+bx}$$

$$= [\log x - \log(a+bx)] + \left[1 - \frac{bx}{a+bx}\right]$$

$$= \log\frac{x}{a+bx} + \frac{a+bx-bx}{a+bx}$$

or $$y_1 = \frac{y}{x} + \frac{a}{a+bx} \qquad ......(2)$$

from (1) Again differentiating both sides with respect to x, we get

$$y_2 = \frac{xy_1 - y}{x^2} - \frac{ab}{(a+bx)^2}$$

Multiplying both sides by x^3, we get

$$x^3y_2 = x(xy_1 - y) - \frac{abx^3}{(a+bx)^2} \qquad .....(3)$$

From (2) $\quad y_1 - \frac{y}{x} = \frac{a}{a+bx} \quad$ or $\quad xy_1 - y = \frac{ax}{a+bx}$

Putting this values in (3), we get

$$x^3y_2 = \frac{ax^2}{a+bx} - \frac{abx^3}{(a+bx)^2}$$

$$= \frac{ax^2(a+bx-bx)}{(a+bx)^2} = \frac{a^2x^2}{(a+bx)^2}$$

$$= \left(\frac{ax}{a+bx}\right)^2 = (xy_1 - y)^2 = (y - xy_1)^2$$

9. If $y = (a+bx)e^{-x}$, prove that

$$ya + 2y_1 + y = 0.$$

Solution: We have $y = (a+bx)e^{-x}$ .....(1)

Differentiating both sides with respect to x, we get

$$y_1 = be^{-x} - (a+bx)e^{-x}$$

$$= be^{-x} - y, \qquad \text{from (1)}$$

or $\quad y_1 + y = be^{-x}$. .....(2)

Differentiating both sides of (2) with respect to x, we get

$$y_2 + y_1 = -be^{-x}$$

or $\quad y_2 + y_1 + be^{-x} = 0$

or $\quad y_2 + y_1 + y_1 + y = 0, \qquad$ from (2)

or $\quad y_2 + 2y_1 + y = 0.$

10. If $y = Ae^{-kt}\cos(pt+c)$, prove that

$$\frac{d^2y}{dt^2} + 2k\frac{dy}{dt} + (p^2+k^2)y = 0.$$

Solution: We have $y = Ae^{-kt}\cos(pt + c)$ .....(1)

Differentiating both sides with respect to t, we get

$$\frac{dy}{dt} = -Ake^{-kt}\cos(pt+c) - Ape^{-kt}\sin(pt+c)$$

or $$\frac{dy}{dt} = -ky - Ape^{-kt}\sin(pt+c).$$ .....(2)

Differentiating both sides of (2) with respect to t, we get

$$\frac{d^2y}{dt^2} = -k\frac{dy}{dt} + Ape^{-kt}\sin(pt+c) - Ap^2e^{-kt}\cos(pt+c)$$

or $$\frac{d^2y}{dt^2} + k\frac{dy}{dt} = k\left(-ky - \frac{dy}{dt}\right) - p^2y,$$ from (2) and (1)

or $$\frac{d^2y}{dt^2} + 2k\frac{dy}{dt} + (p^2 + k^2)y = 0.$$

Exercise 4.1

1. (a) (i) Find y_2, if $y = e^{-1/x}$ (ii) Find y_2, if $y = \cos^{-1} x$

(iii) Find y_a, if $y = \frac{1}{a + bx}$ (iv) Find y_4, if $y = x^3 \log x$

(b) (i) Find y_2, if $x^2 + y^2 = (x+y)^2$

(ii) Find y_2, if $xy = \sin(xy)$

(c) (i) Find $\frac{d^2y}{dx^2}$, if $x = a(\varphi + \sin\varphi)$, $y = a(1 - \cos\varphi)$.

(ii) Find $\frac{d^3y}{dx^3}$, if $x = a\cos\theta$, $y = b\sin\theta$.

(d) If $x = F(t)$, $y = f(t)$, then prove that

$$\frac{d^2y}{dx^2} = \frac{\dot{x}\,\ddot{y} - \dot{y}\,\ddot{x}}{(\dot{x})^3},$$

where $\dot{x} = \frac{dx}{dt}$, $\ddot{x} = \frac{d^2y}{dx^2}$ etc.

2. (i) If $\frac{x^2}{a^2}+\frac{y^2}{b^2}=1$, show that $a^2y^3\frac{d^2y}{dx^2}+b^4=0$.

(ii) If $y = A \sin mx + B \cos mx$, prove that

$$y_2 + m^2y = 0$$

(iii) If $y = ae^{nx} + be^{-nx}$, prove that $y_2 = n^2y$

3. (i) If $y = (x^2 - 1)^n$, prove that

$$(x^2 - 1)\, y_2 + 2\,(1 - n)\, xy_1 - 2ny = 0$$

(ii) If $y=\left(x+\sqrt{1+x^2}\right)^4$, prove that

$$\left(1+x^2\right)\frac{d^2y}{dx^2}+x\frac{dy}{dx}-16y=0$$

(iii) If $y=A\left(x+\sqrt{x^2-1}\right)^2+B\left(x-\sqrt{x^2-1}\right)^2$, prove that

$$(x^2 - 1)\, y_2 + xy_1 - 4y = 0$$

4. If $y = x \sin x$, prove that

$$\frac{d^5y}{dx^5}-\frac{dy}{dx}=4\frac{y}{x}$$

5. (i) If $y = \sin^{-1} x$, show that $\frac{y_1}{y_2}=\frac{1}{x}-x$

(ii) If $y = \cos(m\sin^{-1} x)$, prove that

$$(1 - x^2)\, y_2 = xy_1 - m^2y$$

(iii) If $y = \sin(m\sin^{-1} x)$, prove that

$$(1 - x^2)\, y_2 - xy_1 + m^2y = 0$$

(iv) If $y = (\tan^{-1} x)^2$, prove that

$$\left(x^2+1\right)^2\frac{d^2y}{dx^2}+2x\left(x^2+1\right)\frac{dy}{dx}=2$$

(v) If $y = (\sin^{-1} x)^2$, prove that

$$(1 - x^2)\, y_2 - xy_1 = 2$$

6. (i) If $y = e^{a\sin^{-1} x}$, prove that

$$\left(1-x^2\right)\frac{d^2y}{dx^2}-x\frac{dy}{dx}=a^2y$$

(ii) If $y = e^{\sin^{-1}x}$, prove that $(1 - x^2)\, y_2 - xy_1 - y = 0$

7. (i) If $x = \sinh\left(\frac{\log y}{m}\right)$, prove that

$$\left(x^2+1\right)\frac{d^2y}{dx^2}+x\frac{dy}{dx}=m^2y$$

(ii) If x = cosh (logy), prove that $(x^2 - 1)\, y_2 + xy_1 = y$

8. (i) If x = cos (logy), show that $(1 - x^2)\, y_2 - xy_1 = y$

(ii) If y = a cos (logx) + b sin(logx), prove that

$$x^2y_2 + xy_1 + y = 0$$

(iii) If x = sin (logy), show that $(1 - x^2)\, y_2 - xy_1 = y$

(iv) If y = a cos(logx), prove that $x^2y_2 + xy_1 + y = 0$

(v) If y = a sin(logx), prove that $x^2y_2 + xy_1 + y = 0$

(vi) If $y=\left[\log\left(x+\sqrt{x^2+1}\right)\right]^2$, prove that

$$\left(1+x^2\right)\frac{d^2y}{dx^2}+x\frac{dy}{dx}=2$$

9. If $y = (a + bx)e^{-nx}$, prove that

$$y_2 + 2ny_1 + n^2y = 0$$

10. If $y = e^{ax}\cos bx$, prove that

$$y_2 - 2ay_1 + (a^2 + b^2)\, y = 0$$

4.2 Formulae for n^th Derivative of some Standard Functions

It is clear that the work of finding derivatives of higher order like y_6, y_{10}, y_{20} will be a tedious job. Fortunately the nth derivative of some standard functions take nice forms and they can be used to find n^{th} derivatives of other functions.

(i) To find y_n when $y = (ax + b)^m$

Here $y = (ax + b)^m$

Differentiating both sides with respect to x successively, we get

$$y_1 = m(ax + b)^{m-1}.a = ma(ax + b)^{m-1}$$

$$y_2 = ma(m - 1)\,(ax + b)^{m-2}.a = m(m - 1)a^2.(ax + b)^{m-2}$$

$$y_3 = m(m - 1)\,(m - 2)a^3.\,(ax + b)^{m-3}$$

....

$\therefore$ Intuitively,

$$y_n = \{m(m-1)(m-2) \ldots \text{to n factors}\} a^n (ax+b)^{m-n}$$

i.e.,
$$\boxed{y_n = m(m-1)(m-2) \ldots (m-n+1)a^n (ax+b)^{m-n}}$$

where m is any number.

If m is a positive integer $\geq n$ then

$$y_n = \frac{m(m-1)(m-2) \ldots (m-n+1)\lfloor m-n}{\lfloor m-n} a^n (ax+b)^{m-n}.$$

$$\therefore \quad \boxed{D^n \left\{(ax+b)^m\right\} = \frac{\lfloor m}{\lfloor m-n} a^n (ax+b)^{m-n}}, \qquad \ldots\ldots(4.1)$$

when m is a positive integer and $m \geq n$.

Note: The rigorous proof can be provided by the "Principle of Mathematical Induction".

Particular Cases

(A) If $y = x^m$, comparing with the above result (4.1), $a = 1, b = 0$.

$$\therefore \quad \boxed{D^n \left(x^m\right) = \frac{\lfloor m}{\lfloor m-n} x^{m-n}},$$

where m is a positive integer and $m \geq n$.

(B) If $y = x^n$, comparing with the above result (4.1), $a = 1, b = 0, m = n$.

$$\therefore \quad \boxed{D^n \left(x^m\right) = \lfloor n},$$

Note: If m is a positive integer less than n then

$$D^n \left(x^m\right) = 0 \text{ because } D^m \left(x^m\right) = \lfloor m = \text{constant}$$

and so further differentiation will make the derivative zero.

(ii) To find y_n when $y = \frac{1}{(ax+b)^m}$, where m is a positive integer.

Here $\quad y = (ax+b)^{-m}$

Differentiating successively w.r.t. x we get

$$y_1 = -m(ax+b)^{-m-1}.a = \frac{(-1)ma}{(ax+b)^{m+1}} = (-1)^1.\frac{\lfloor m}{\lfloor m-1}.\frac{a}{(ax+b)^{m+1}}$$

$$y_2 = \frac{d}{dx}\left\{(-1)^1.\frac{\lfloor m}{\lfloor m-1}a.(ax+b)^{-m-1}\right\}$$

$$= (-1)^1\frac{\lfloor m}{\lfloor m-1}.a.(-m-1)(ax+b)^{-m-2}.a$$

$$= (-1)^2.\frac{\lfloor m(m-1)}{\lfloor m-1}.\frac{a^2}{(ax+b)^{m+2}} = (-1)^2.\frac{\lfloor m+1}{\lfloor m-1}.\frac{a^2}{(ax+b)^{m+2}},$$

$$y_3 = \frac{d}{dx}\left\{(-1)^2.\frac{\lfloor m+1}{\lfloor m-1}.a^2.(ax+b)^{-m-2}\right\}$$

$$= (-1)^2.\frac{\lfloor m+1}{\lfloor m-1}.a^2.(-m-2)(ax+b)^{-m-3}.a$$

$$= (-1)^3.\frac{\lfloor m+1(m+2)}{\lfloor m-1}.\frac{a^3}{(ax+b)^{m+3}}$$

$$= (-1)^3.\frac{\lfloor m+2}{\lfloor m-1}.\frac{a^3}{(ax+b)^{m+3}}$$

Intuitively we get,

$$y_n = (-1)^n.\frac{\lfloor m+(n-1)}{\lfloor m-1}.\frac{a^n}{(ax+b)^{m+n}}$$

$$\therefore \quad \boxed{D^n\left\{\frac{1}{(ax+b)^m}\right\} = \frac{(-1)^n.\lfloor m+n-1}{\lfloor m-1}.\frac{a^n}{(ax+b)^{m+n}}} \qquad .....(4.2)$$

Particular Cases

(A) If $y = x^{-m}$ where m is a positive integer, comparing with the above result (4.2), $a = 1, b = 0$ we get

$$D^n\left(x^{-m}\right) = \frac{(-1)^n\lfloor m+n-1}{\lfloor m-1}.\frac{1}{x^{m+n}}$$

(B) If $y = \frac{1}{x+a}$, comparing with the above result (4.2), a = 1, b is a and m = 1 we get

$$\boxed{D^n\left(\frac{1}{x+a}\right)=\frac{(-1)^n \lfloor n}{(x+a)^{n+1}}}$$

(C) If $y=\dfrac{1}{ax+b}$, comparing with the above result (4.2), m = 1 and we get

$$\boxed{D^n\left(\frac{1}{ax+b}\right)=\frac{(-1)^n \lfloor n.\, a^n}{(ax+b)^{n+1}}}$$

(iii) To find y_n when $y = e^{mx}$

Here $y = e^{mx}$. Differentiating successively with respect to x,

$$y_1 = e^{mx}.m = me^{mx}$$

$$y_2 = m.e^{mx}.m = m^2e^{mx}$$

$$y_3 = m^2.e^{mx}.m = m^3e^{mx}$$

....

Intuitively we get $y_n = m^n.e^{mx}$

$$\therefore \quad \boxed{D^n\left(e^{mx}\right)=m^n.e^{mx}} \qquad .....(4.3)$$

Particular Case

If $y = e^x$, comparing with the above result (4.3), m = 1 and we get

$$\boxed{D^n\left(e^x\right)=e^x}$$

(iv) To find y_n when $y = a^{mx}$

Here $y = a^{mx}$. Differentiating successively w.r.t. x,

$$y_1 = \log_e a.a^{mx}.m = m\log_e a.a^{mx}$$

$$y_2 = m\log_e a.\left\{\log_e a.a^{mx}.m\right\}=\left(m\log_e a\right)^2.a^{mx}$$

$$y_3 = \left(m\log_e a\right)^2\left\{\log_e a.a^{mx}.m\right\}=\left(m\log_e a\right)^3.a^{mx}$$

....

Intuitively we get $y_n = \left(m\log_e a\right)^n.a^{mx}$

$$\therefore \quad \boxed{D^n\left(a^{mx}\right)=\left(m\log_e a\right)^n.a^{mx}} \qquad .....(4.4)$$

Particular Case

If $y = a^x$, comparing with the above result (4.4), m = 1 and we get

$$\boxed{D^n\left(a^x\right) = \left(\log_e a\right)^n . a^x}$$

(v) To find y_n, when $y = \log_e(ax + b)$

Here $y = \log_e(ax + b)$

Differentiating both sides successively with respect to x, we get

$$y_1 = \frac{a}{ax+b}$$

$$y_2 = \frac{(-1)a^2}{(ax+b)^2}$$

$$y_3 = \frac{(-1)(-2)a^3}{(ax+b)^3} = \frac{(-1)^2 \lfloor 2 \, a^3}{(ax+b)^3}$$

$$y_4 = \frac{(-1)(-2)(-3)a^4}{(ax+b)^4} = \frac{(-1)^3 \lfloor 3 \, a^4}{(ax+b)^4}$$

....

$$y_n = \frac{(-1)^{n-1} \lfloor n-1 \, a^n}{(ax+b)^n}$$

That is, $$\left| D^n\left[\log_e(ax+b)\right] = \frac{(-1)^{n-1} \lfloor n-1 \, a^n}{(ax+b)^n} \right| \qquad .....(4.5)$$

Particular Case

If $y = \log_e x$, comparing with the above result (4.5), a = 1, b = 0 and we get

$$\boxed{D^n\left(\log_e x\right) = \frac{(-1)^{n-1} \lfloor n-1}{x^n}}.$$

(vi) To find y_n, when $y = \sin(ax + b)$.

We have $y = \sin(ax + b)$

Differentiating both sides successively with respect to x, we get

$$y_1 = a\cos(ax+b) = a\sin\left(\frac{\pi}{2} + ax + b\right)$$

$$y_2 = -a^2 \sin(ax+b) = a^2 \sin(\pi + ax + b) = a^2 \sin\left(\frac{2\pi}{2} + ax + b\right)$$

$$y_3 = -a^3 \cos(ax+b) = a^3 \sin\left(\frac{3\pi}{2} + ax + b\right)$$

....

Intuitively we get $y_n = a^n \sin\left(\frac{n\pi}{2} + ax + b\right)$

That is, $$\boxed{D^n\left[\sin(ax+b)\right] = a^n \sin\left(\frac{n\pi}{2} + ax + b\right)} \quad .....(4.6)$$

Particular Cases

(A) If y = sinax, comparing with the above result (4.6), b = 0 we get

$$\boxed{D^n(\sin ax) = a^n \sin\left(\frac{n\pi}{2} + ax\right)}$$

(B) If y = sinx, comparing with the above result (4.6), a = 1, b = 0 we get

$$\boxed{D^n(\sin x) = \sin\left(\frac{n\pi}{2} + x\right)}$$

(vii) To find y_n when y = cos(ax + b)

Here y = cos(ax + b)

Differentiating both sides successively with respect to x, we get

$$y_1 = -a \sin(ax+b) = a \cos\left(\frac{\pi}{2} + ax + b\right)$$

$$y_2 = -a^2 \cos(ax+b) = a^2 \cos(\pi + ax + b)$$

$$= a^2 \cos\left(\frac{2\pi}{2} + ax + b\right)$$

$$y_3 = a^3 \sin(ax+b) = a^3 \cos\left(\frac{3\pi}{2} + ax + b\right)$$

....

Intuitively we have $y_n = a^n.\cos\left(\frac{n\pi}{2} + ax + b\right)$

$\therefore$ $$\boxed{D^n\left[\cos(ax+b)\right] = a^n \cos\left(\frac{n\pi}{2} + ax + b\right)} \quad .....(4.7)$$

Particular Cases

(A) If $y = \cos ax$, comparing with the above result (4.7), $b = 0$ and we get

$$\boxed{D^n(\cos ax) = a^n \cos\left(\frac{n\pi}{2} + ax\right)}$$

(B) If $y = \cos x$, comparing with the above result (4.7), $b = 0$, $a = 1$ and we get

$$\boxed{D^n(\cos x) = \cos\left(\frac{n\pi}{2} + x\right)}$$

(viii) To find y_n when $y = e^{ax}\sin bx$.

Here $\qquad y = e^{ax}\sin bx$.

Differentiating both sides with respect to x, we get

$$y_1 = ae^{ax}\sin bx + be^{ax}\cos bx$$

$$= e^{ax}(a\sin bx + b\cos bx)$$

Put $\qquad a = r\cos\theta$ and $b = r\sin\theta$.

$$\therefore \quad a^2 + b^2 = r^2\cos^2\theta + b^2\sin^2\theta = r^2(\cos^2\theta + \sin^2\theta) = r^2$$

and $\qquad \dfrac{b}{a} = \dfrac{r\sin\theta}{r\cos\theta} = \tan\theta; \ \therefore \ \theta = \tan^{-1}\dfrac{b}{a}$

$$\therefore \quad y_1 = e^{ax}(r\cos\theta\sin bx + r\sin\theta\cos bx)$$

$$= re^{ax}\sin(bx + \theta)$$

Similarly $\qquad y_2 = r^2e^{ax}\sin(bx + 2\theta)$

$$y_3 = r^3e^{ax}\sin(bx + 3\theta)$$

....

Intuitively we get $y_n = r^n e^{ax}\sin(bx + n\theta)$

$$= (a^2 + b^2)^{\frac{n}{2}} e^{ax}\sin\left(bx + n\tan^{-1}\frac{b}{a}\right)$$

$$\therefore \quad \boxed{D^n(e^{ax}\sin bx) = (a^2 + b^2)^{\frac{n}{2}} e^{ax}\sin\left(bx + n\tan^{-1}\frac{b}{a}\right)} \quad .....(4.8)$$

Particular Case

If $y = e^x \sin x$, comparing with the above result (4.8), a = 1, b = 1 and we get

$$\boxed{D^n\left(e^x \sin x\right) = 2^{\frac{n}{2}} e^x \sin\left(x + \frac{n\pi}{4a}\right)}$$

(ix) To find y_n, when $y = e^{ax} \sin (bx + c)$

We have $y = e^{ax} \sin (bx + c)$

Differentiating both sides with respect to x, we get

$$y_1 = ae^{ax} \sin(bx + c) + be^{ax} \cos(bx + c)$$

or $$y_1 = e^{ax}\left[a\sin(bx + c) + b\cos(bx + c)\right]$$

Put $a = r\cos\theta$ and $b = r\sin\theta$.

Squaring and adding these two, we get

$$a^2 + b^2 = r^2\cos^2\theta + r^2\sin^2\theta$$

$$= r^2\left(\cos^2\theta + \sin^2\theta\right) = r^2.1 = r^2$$

By division, we get

$$\frac{b}{a} = \frac{r\sin\theta}{r\cos\theta} = \tan\theta$$

$$\therefore \quad \theta = \tan^{-1}\frac{b}{a}$$

$$\therefore \quad y_1 = e^{ax}\left[r\cos\theta\sin(bx + c) + r\sin\theta\cos(bx + c)\right]$$

or $$y_1 = re^{ax}\left[\sin(bx + c)\cos\theta + \cos(bx + c)\sin\theta\right]$$

or $$y_1 = re^{ax}\sin(bx + c + \theta)$$

Similarly $y_2 = r^2e^{ax}\sin(bx + c + 2\theta)$

$$y_3 = r^3e^{ax}\sin(bx + c + 3\theta)$$

....

Intuitively we get $y_n = r^ne^{ax}\sin(bx + c + n\theta)$

Hence $y_n = \left(a^2+b^2\right)^{\frac{n}{2}} e^{ax} \sin\left(bx + c + n\tan^{-1}\frac{b}{a}\right)$

That is, $$\boxed{D^n\left[e^{ax}\sin(bx+c)\right] = \left(a^2+b^2\right)^{\frac{n}{2}} e^{ax}\sin\left(bx+c+n\tan^{-1}\frac{b}{a}\right)}$$ .....(4.9)

(x) To find y_n when $y = e^{ax}\cos bx$

Here $y = e^{ax}\cos bx$

Differentiating both sides with respect to x, we get

$$y_1 = ae^{ax}\cos bx - be^{ax}\sin bx$$

$$= e^{ax}(a\cos bx - b\sin bx)$$

Put $a = r\cos\theta$ and $b = r\sin\theta$.

$$\therefore \quad a^2+b^2 = r^2\cos^2\theta + r^2\sin^2\theta$$

$$= r^2\left(\cos^2\theta+\sin^2\theta\right) = r^2.1 = r^2$$

and $\dfrac{b}{a} = \dfrac{r\sin\theta}{r\cos\theta} = \tan\theta; \quad \therefore \ \theta = \tan^{-1}\dfrac{b}{a}$

$$\therefore \quad y_1 = e^{ax}(r\cos\theta\cos bx - r\sin\theta\sin bx)$$

or $y_1 = re^{ax}(\cos bx\cos\theta - \sin bx\sin\theta)$

or $y_1 = re^{ax}\cos(bx+\theta)$

Similarly $y_2 = r^2e^{ax}\cos(bx+2\theta)$

$$y_3 = r^3e^{ax}\cos(bx+3\theta)$$

....

Intuitively, $y_n = r^ne^{ax}\cos(bx+n\theta)$

$$= \left(a^2+b^2\right)^{\frac{n}{2}} e^{ax}\cos\left(bx+n\tan^{-1}\frac{b}{a}\right)$$

That is, $$\boxed{D^n\left(e^{ax}\cos bx\right) = \left(a^2+b^2\right)^{\frac{n}{2}} e^{ax}\cos\left(bx+n\tan^{-1}\frac{b}{a}\right)}$$ .....(4.10)

Particular Case

If $y = e^x \cos x$, comparing with the above result (4.10), $a = 1$ and $b = 1$ and we get

$$\boxed{D^n\left(e^x \cos x\right) = 2^{\frac{n}{2}}\, e^x \cos\left(x + \frac{n\pi}{4}\right)}.$$

(xi) To find y_n when $y = e^{ax}\cos(bx + c)$

Here $\quad y = e^{ax}\cos(bx + c)$

Differentiating both sides with respect to x, we get

$$y_1 = ae^{ax}\cos(bx + c) - be^{ax}\sin(bx + c)$$
$$= e^{ax}\left[a\cos(bx + c) - b\sin(bx + c)\right]$$

Put $\quad a = r\cos\varphi$ and $b = r\sin\varphi$

$$\therefore \quad a^2 + b^2 = r^2\cos^2\varphi + r^2\sin^2\varphi$$
$$= r^2\left(\cos^2\varphi + \sin^2\varphi\right) = r^2.1 = r^2$$

and $\quad \dfrac{b}{a} = \dfrac{r\sin\varphi}{r\cos\varphi} = \tan\varphi; \quad \therefore \ \varphi = \tan^{-1}\dfrac{b}{a}$

$$\therefore \quad y_1 = e^{ax}\left[r\cos\varphi\cos(bx + c) - r\sin\varphi\sin(bx + c)\right]$$
$$= re^{ax}\left[\cos(bx + c)\cos\varphi - \sin(bx + c)\sin\varphi\right]$$
$$= re^{ax}\cos(bx + c + \varphi)$$

Similarly $y_2 = r^2e^{ax}\cos(bx + c + 2\varphi)$

$$y_3 = r^3e^{ax}\cos(bx + c + 3\varphi)$$

....

$$y_n = r^ne^{ax}\cos(bx + c + n\varphi)$$

Hence $\quad y_n = \left(a^2 + b^2\right)^{\frac{n}{2}} e^{ax}\cos\left(bx + c + n\tan^{-1}\dfrac{b}{a}\right)$

That is,

$$\boxed{D^n\left[e^{ax}\cos(bx + c)\right] = \left(a^2 + b^2\right)^{\frac{n}{2}} e^{ax}\cos\left(bx + c + n\tan^{-1}\frac{b}{a}\right)}$$

4.3 Rules for finding n^{th} Derivatives

If a given function y = f(x) can be written as cu or $u \pm v$ (where c is a constant and u, v are standard functions whose n^{th} derivatives are known) then y_n can be obtained by the following rules, which can be established very easily:

(i) $D^n(cu) = cD^n(u)$ (ii) $D^n(u \pm v) = D^n(u) \pm D^n(v)$

If y = u . v where u, v are functions of x then the nth derivative y_n w.r.t. x can be obtained by the theorem which we shall discuss below. The credit for the method goes to a German mathematician G. W. Leibnitz.

Leibnitz's Theorem

Statement: If u and v be two functions of x, possessing derivatives of the nth order, then

$$(uv)_n = {}^nC_0 u_n v + {}^nC_1 u_{n-1} v_1 + {}^nC_2 u_{n-2} v_2 + \ldots + {}^nC_r u_{n-r} v_r + \ldots + {}^nC_n u v_n$$

Proof: Let y = uv. Differentiating both sides with respect to x,

$$y_1 = u_1 v + u v_1 = {}^1C_0 u_1 v + {}^1C_1 u v_1 \quad \ldots..(1)$$

Differentiating again w.r.t. x,

$$y_2 = u_2 v + u_1 v_1 + u v_2 + u_1 v_1 = u_2 v + 2u_1 v_1 + u v_2$$

$$= {}^2C_0 u_2 v + {}^2C_1 u_1 v_1 + {}^2C_1 u v_2 \quad \ldots..(2)$$

Differentiating again w.r.t. x,

$$y_3 = u_3 v_1 + u_2 v_1 + 2(u_2 v_1 + u_1 v_2) + u_1 v_2 + u v_3$$

$$= u_3 v_1 + 3u_2 v_1 + 3u_1 v_2 + u v_3$$

$$= {}^3C_0 u_3 v_1 + {}^3C_1 u_2 v_1 + {}^3C_2 u_1 v_2 + {}^3C_3 u v_3$$

We find that the theorem is true for n = 1, 2, 3. Let us suppose that the above theorem is true for the particular value, say m, of n.

$$\therefore \quad y_m = {}^mC_0 u_m v + {}^mC_1 u_{m-1} v_1 + {}^mC_2 u_{m-2} v_2 + \ldots..$$
$$+ {}^mC_{r-1} u_{m-r+1} v_{r-1} + {}^mC_r u_{m-r} v_r + \ldots.. + {}^mC_m u v_m$$

Differentiating both sides of this with respect to x, we get

$$y_{m+1} = {}^mC_0 u_{m+1} v + {}^mC_0 u_m v_1 + {}^mC_1 u_m v_1 + {}^mC_1 u_{m-1} v_2$$
$$+ {}^mC_2 u_{m-1} v_2 + {}^mC_2 u_{m-2} v_3 + \ldots.$$
$$+ {}^mC_{r-1} u_{m-r+2} v_{r-1} + {}^mC_{r-1} u_{m-r+1} + v_r$$
$$+ {}^mC_r u_{m-r+1} v_r + {}^mC_r u_{m-r} v_{r+1} + \ldots.$$
$$+ {}^mC_m u_1 v_m + {}^mC_m u v_{m+1}$$

or $\quad y_{m+1} = {}^mC_0u_{m+1}v + ({}^mC_0+{}^mC_1)u_mv_1 + ({}^mC_1 + {}^mC_2)u_{m-1}v_2+...$

$$+ ({}^mC_{r-1}+{}^mC_r)\, u_{m-r+1}v_r+....+{}^mC_muv_{m+1}.... \quad .....(3)$$

But we know that

$${}^mC_{r-1} + {}^mC_r = {}^{m+1}C_r$$

Putting $r = 1, 2, ...$successively, we get

$${}^mC_0 + {}^mC_1 = {}^{m+1}C_1,\ {}^mC_1 + {}^mC_2 = {}^{m+1}C_2, \text{ and so on}$$

Also, $\quad {}^mC_m = {}^{m+1}C_{m+1}$ and ${}^mC_0 = {}^{m+1}C_0$

Putting these values in (3), we get

$$y_{m+1} = {}^{m+1}C_0u_{m+1}v+{}^{m+1}C_1u_mv_1 +{}^{m+1}C_2u_{m-1}v_2+....$$

$$+ {}^{m+1}C_ru_{m-r+1}v_r+...+{}^{m+1}C_{m+1}uv_{m+1}$$

Therefore, if Leibnitz's theorem holds for $n = m$, it is also holds good for $n = m +1$. But this theorm also holds good for $n = 1, 2, 3$; so it holds for $n = 4$. Again as it holds for $n = 4$, it holds for $n = 5$ and so on.

Hence Leibinitz's theorem holds for all positive integral values of n.

Thus,

$$(uv)_n = {}^nC_0u_nv + {}^nC_1u_{n-1}v_1 + {}^nC_2u_{n-2}v_2$$

$$+....+{}^nC_ru_{n-r}v_r +....+{}^nC_n\, uv_n$$

Note: The method adopted in the proof of the above theorem is known as "Principle of Mathematical Induction" which states–

If P(n) is a statement for which

(i) P(1) is true, (ii) P(m) is true $\Rightarrow$ P(m+1) is true then P(n) is true for all positive integral values of n. In the above proof we could have dispensed with the verification for $n = 2, 3$.

Illustration: If $y = x^2e^{az}$, find y_n.

Solution: We have $y = x^2e^{ax} = (e^{ax})\,(x^2)$.

Let $u = e^{ax}$

$\therefore \quad u_n = a^ne^{ax},\ u_{n-1} = a^{n-1}e^{ax},$

$u_{n-2} = a^{n-2}e^{ax}$, etc.

Again, let $v = x^2$, $\therefore\ v_1 = 2x,\ v_2 = 2,\ v_3 = 0,\ v_4 = 0$, etc.

$\therefore \quad y = uv.$

Differentiating both sides n times with respect to x by Leibnitz's theorem, we get

$$y_n = (uv)_n$$

or $$y_n = {}^nC_0u_nv + {}^nC_1u_{n-1}v_1 + {}^nC_2u_{n-2}v_2 + \ldots + {}^nC_nuv_n$$

$$y_n = a^ne^{ax}x^2 + n.a^{n-1}e^{ax}.(2x) + \frac{n(n-1)}{1.2}a^{n-2}e^{ax}.(2)$$

+ other terms which vanish.

Hence $$y_n = e^{ax}a^{n-2}\,[x^2a^2 + 2nax + n(n-1)]$$

4.4 Method of Determination of the nth Derivative of an Algebraic Rational Function

If the denominator of the given algebraic rational function contains real linear factors, then resolve the fraction into partial fractions. Let us first learn how to resolve a given Algebraic Rational Function into partial fractions.

Proper Fraction

If the numerator be of a lower degree than that of the denominator, then the fraction is called a proper fraction.

For example, $\frac{27x+10}{3x^2+5x+8}$ is a proper fraction.

If the degree of the numerator be equal to or greater than that of the denominator, then the fraction may be reduced to a mixed quantity by dividing the numerator by the denominator until the degree of the remainder is less than that of the denominator.

For example, $$\frac{12x^3+4x^2+8}{6x^2+11x+3} = 2x-3+\frac{27x+17}{6x^2+11x+3}$$

Partial Fractions

An algebraical fraction may be sometimes separated into a group of simple fractions; for example, $$\frac{x-5}{(2-x)(x-3)} = \frac{3}{2-x} + \frac{2}{x-3}$$

These simpler fractions into which a given fraction is decomposed are called partial fractions.

To resolve a given proper fraction into partial fractions, the denominator of the fraction should be factorised into real prime factors. Generally these are linear non-repeated, linear repeated, quadratic non-repeated or quadratic repeated factors.

First Form

When the denominator of the given fraction is resolvable into real linear factors, non-repeated, to every non-repeated linear factor, such as x–a occurring in the denominator, there corresponds a partial fraction of the form $\frac{A}{x-a}$, where A is a constant to be determined.

The given fraction can be expressed as a sum of fractions of this form.

Second Form

When the factors of the denominator are all linear and some repeated.

To every n-fold linear factor, such as $(x-a)^n$, there will correspond the sum of n partial fractions,

$$\frac{A}{(x-a)}+\frac{B}{(x-a)^2}+....+\frac{L}{(x-a)^n},$$

in which A, B,.....L are constants to be determined.

If the denominator of the given algebraic rational function contains a quadratic factor, then it would be convenient to use De Moivre's theorem for finding successive derivatives.

Illustration: Find y_m, when $y=\frac{1}{x^2-x-2}$

Solution: We have $y=\frac{1}{x^2-x-2}=\frac{1}{x^2-2x+x-2}$

$$=\frac{1}{x(x-2)+1(x-2)}$$

Let $=\frac{1}{(x+1)(x-2)}=\frac{A}{x+1}+\frac{B}{x-2}$

or $1 = A(x-2) + B(x+1)$

putting x = 2, we get $1 = 3B$, or $B=\frac{1}{3}$

Putting x = –1, we get $1 = -3A$; or $A=-\frac{1}{3}$

$$\therefore \quad y=\frac{1}{(x+1)(x-2)}=-\frac{1}{3}\frac{1}{x+1}+\frac{1}{3}.\frac{1}{x-2}$$

$$=\frac{1}{3}\left[\frac{1}{x-2}-\frac{1}{x+1}\right].$$

Differentiating both sides n times with respect to x, we get

$$y_n = \frac{1}{3}\left[\frac{(-1)^n \lfloor n}{(x-2)^{n+1}} - \frac{(-1)^n \lfloor n}{(x+1)^{n+1}}\right]$$

Hence $$y_n = \frac{(-1)^n \lfloor n}{3}\left[\frac{1}{(x-2)^{n+1}} - \frac{1}{(x+1)^{n+1}}\right]$$

4.5 Some Important Results

(i) $D^T[e^{nx} f(x)] = e^{nx} (D + n)^r f(x)$

Let us consider the function $e^{nx} f(x)$

Let $D \equiv \frac{d}{dx}$

Then $D\left[e^{nx} f(x)\right] = ne^{nx} f(x) + e^{nx} D\left[f(x)\right]$

$$= e^{nx}(D+n)f(x)$$

$$D^2[e^{nx}f(x)] = D[e^{nx} (D+n) f(x)]$$

$$= ne^{nx} (D + n) f(x) + e^{nx} D(D+n) f(x)$$

$$= e^{nx} (nD + n^2 + D^2 + nD) f(x)$$

$$= enx (D + n)^2 f(x)$$

$$D^r [e^{nx} f(x) = e^{nx} (D + n)^r f(x)$$

Hence, in general, if $\varphi(D)$ be a polynomial in D,

$$\boxed{\varphi(D)\left[e^{nx} f(x)\right] = e^{nx}\, \varphi(D+n)\, f(x)}$$

(ii) To prove that

$$\frac{d^3y}{dx^3} = -\frac{\frac{d^3x}{dy^3}\cdot\frac{dx}{dy} - 3\left(\frac{d^2x}{dy^2}\right)^2}{\left(\frac{dx}{dy}\right)^5}$$

We have $\frac{dy}{dx} = \frac{1}{\frac{dx}{dy}}$

or $$\frac{d^2y}{dx^2}=\frac{d}{dx}\left(\frac{1}{\frac{dx}{dy}}\right)=\frac{d}{dy}\left(\frac{1}{\frac{dx}{dy}}\right)\frac{dy}{dx}$$

$$=-\frac{1}{\left(\frac{dx}{dy}\right)^2}.\frac{d^2x}{dy^2}.\frac{1}{\frac{dx}{dy}}=-\frac{\frac{d^2x}{dy^2}}{\left(\frac{dx}{dy}\right)^3}$$

or $$\frac{d^3y}{dx^3}=-\frac{d}{dx}\left[\frac{\frac{d^2x}{dy^2}}{\left(\frac{dx}{dy}\right)^2}\right]=-\frac{d}{dy}\left[\frac{\frac{d^2x}{dy^2}}{\left(\frac{dx}{dy}\right)^3}\right].\frac{dy}{dx}$$

$$=-\left[\frac{\left(\frac{dx}{dy}\right)^3.\frac{d^3x}{dy^3}-3\left(\frac{dx}{dy}\right)^2\left(\frac{d^2x}{dy^2}\right)^2}{\left(\frac{dx}{dy}\right)^6.\frac{dx}{dy}}\right]$$

$\therefore$ $$\frac{d^3y}{dx^3}=-\frac{\frac{d^3x}{dy^3}.\frac{dx}{dy}-3\left(\frac{d^2x}{dy^2}\right)^2}{\left(\frac{dx}{dy}\right)^5}$$

Another form

To prove that

$$\frac{d^3x}{dy^3}=-\frac{\frac{d^3y}{dx^3}.\frac{dy}{dx}-3\left(\frac{d^2y}{dx^2}\right)^2}{\left(\frac{dy}{dx}\right)^5}$$

We have $\frac{dx}{dy}=\frac{1}{\frac{dy}{dx}}$

or $$\frac{d^2x}{dy^2} = -\frac{d}{dy}\left(\frac{1}{\frac{dy}{dx}}\right) = \frac{d}{dx}\left(\frac{1}{\frac{dy}{dx}}\right)\frac{dx}{dy}$$

$$= -\frac{1}{\left(\frac{dy}{dx}\right)^2}.\frac{d^2y}{dx^2}.\frac{1}{\frac{dy}{dx}} = -\frac{\frac{d^2y}{dx^2}}{\left(\frac{dy}{dx}\right)^3}$$

or $$\frac{d^3x}{dy^3} = -\frac{d}{dy}\left[\frac{\frac{d^2y}{dx^2}}{\left(\frac{dy}{dx}\right)^3}\right] = -\frac{d}{dx}\left[\frac{\frac{d^2y}{dx^2}}{\left(\frac{dy}{dx}\right)^3}\right].\frac{dx}{dy}$$

$$= -\frac{\left(\frac{dy}{dx}\right)^3.\frac{d^3y}{dx^3} - 3\left(\frac{dy}{dx}\right)^2\left(\frac{d^2y}{dx^2}\right)^2}{\left(\frac{dy}{dx}\right)^6.\frac{dy}{dx}}$$

$$= -\frac{\frac{d^3y}{dx^3}.\frac{dy}{dx} - 3\left(\frac{d^2y}{dx^2}\right)^2}{\left(\frac{dy}{dx}\right)^5}$$

Work Out Examples

1. (a) If y = sinx + cosx, prove that $y_n = \left[1+(-1)^n \sin 2x\right]^{\frac{1}{2}}$.

Solution: We have y = sinx + cosx

Differentiating n times with respect to x, we get

$$y_n = \sin\left(\frac{n\pi}{2}+x\right)+\cos\left(\frac{n\pi}{2}+x\right)$$

$$= \left[\left\{\sin\left(\frac{n\pi}{2}+x\right)+\cos\left(\frac{n\pi}{2}+x\right)\right\}^2\right]^{\frac{1}{2}}$$

$$= \left[\sin^2\left(\frac{n\pi}{2}+x\right)+\cos^2\left(\frac{n\pi}{2}+x\right)+2\sin\left(\frac{n\pi}{2}+x\right)\cos\left(\frac{n\pi}{2}+x\right)\right]^{\frac{1}{2}}$$

$$= \left[1+\sin 2\left(\frac{n\pi}{2}+x\right)\right]^{\frac{1}{2}}$$

$$= \left[1+\sin(n\pi+2x)\right]^{\frac{1}{2}}$$

$$= \left[1\pm\sin 2x\right]^{\frac{1}{2}}, \text{ according as n is even or odd.}$$

Hence $y_n = \left[1+(-1)^n \sin 2x\right]^{\frac{1}{2}}$

1. (b) Find y_n, if $y = \sin^2 x.\cos^2 x$

Solution: Here $y = \frac{1}{4}\left(4\sin^2 x\cos^2 x\right) = \frac{\sin^2 2x}{4} = \frac{1}{8}\left(2\sin^2 2x\right)$

$$\therefore \quad y = \frac{1}{8}(1-\cos 4x) = \frac{1}{8} - \frac{1}{8}\cos 4x$$

$\therefore$ From formula,

$$y_n = -\frac{1}{8}.4^n.\cos\left(\frac{n\pi}{2}+4x\right) = -2^{2\pi-3}\cos\left(\frac{n\pi}{2}+4x\right)$$

1. (c) If $y = e^x\sin^2 x$, find y_n

Solution: Here $y = \frac{1}{2}e^x.\left(2\sin^2 x\right) = \frac{1}{2}e^x(1-\cos 2x),$

$$\therefore \quad y = \frac{1}{2}e^x - \frac{1}{2}e^x\cos 2x.$$

$\therefore$ By formula, $y_n = \frac{1}{2}e^x - \frac{1}{2}.\left(1^2+2^2\right)^{\frac{n}{2}}.e^x\cos\left(2x+n\tan^{-1}2\right)$

2. Find y_n, when $y = \frac{1}{(x-1)^3(x-2)}$

Solution: We have $y = \frac{1}{(x-1)^3(x-2)}$

Let $\frac{1}{(x-1)^3(x-2)} = \frac{A}{x-1} + \frac{B}{(x-1)^2} + \frac{C}{(x-1)^3} + \frac{D}{x-2}$

or $1 = A(x-1)^2(x-2) + B(x-1)(x-2) + C(x-2) + D(x-1)^3$ .....(1)

or $1 = A(x^3 - 2x + 1)(x-2) + B(x^2 - 3x + 2) + C(x-2)$
$+D(x^3 - 3x^2 + 3x - 1)$ .....(2)

putting x = 1 in (1), we get

$1 = -C$; or $C = -1$

Putting x = 2 in (1), we get 1 = D

Equating the co-efficients of x^3 on both sides in (2), we get

$0 = A + D = A + 1$;

or $A = -1$

Putting x = 0 in (2), we get

$1 = -2A + 2B - 2C - D$

$= 2 + 2B + 2 - 1$

or $2B = -2$; or $B = -1$.

$$\therefore\ y = \frac{-1}{x-1} - \frac{1}{(x-1)^2} - \frac{1}{(x-1)^3} + \frac{1}{x-2}$$

Differentiating n times with respect to x, we get

$$y_n = \frac{(-1)^n n!}{(x-1)^{n+1}} - \frac{(-1)^n (n+1)!}{1!\ (x-1)^{n+2}} - \frac{(-1)^n (n+2)!}{2!\ (x-1)^{n+3}} + \frac{(-1)^n n!}{(x-2)^{n+1}}$$

3. (a) If $y = \dfrac{1}{x^2 + a^2}$, find y_n,

Solution: We have $y = \dfrac{1}{x^2 + a^2} = \dfrac{1}{x^2 - (-1)a^2}$

$$= \frac{1}{x^2 - i^2a^2} = \frac{1}{(x-ia)(x+ia)} = \frac{1}{2ia}\left[\frac{(x+ia)-(x-ia)}{(x-ia)(x+ia)}\right]$$

$$= \frac{1}{2ia}\left[\frac{(x+ia)}{(x-ia)(x+ia)} - \frac{(x-ia)}{(x-ia)(x+ia)}\right]$$

$$= \frac{1}{2ia}\left[\frac{1}{x-ia} - \frac{1}{x+ia}\right]$$

Differentiating both sides n times with respect to x, we get

$$y_n = \frac{1}{2ia}\left[\frac{(-1)^n \lfloor n}{(x-ia)^{n+1}} - \frac{(-1)^n \lfloor n}{(x+ia)^{n+1}}\right]$$

or $$y_n = \frac{(-1)^n \lfloor n}{2ia}\left[\frac{1}{(x-ia)^{n+1}} - \frac{1}{(x+ia)^{n+1}}\right]$$

Put $x = r\cos\theta$ and $a = r\sin\theta$.

Squaring and adding these two, we get

$$x^2 + a^2 = r^2\cos^2\theta + r^2\sin^2\theta = r^2(\cos^2\theta + \sin^2\theta) = r^2.\ 1 = r^2,$$

and by division, we get

$$\frac{x}{a} = \frac{r\cos\theta}{r\sin\theta} = \cot\theta;\ \therefore\ \theta = \cot^{-1}\frac{x}{a}$$

$$\therefore\quad y_n = \frac{(-1)^n \lfloor n}{2ia}\left[\frac{1}{(r\cos\theta - ir\sin\theta)^{n+1}} - \frac{1}{(r\cos\theta + ir\sin\theta)^{n+1}}\right]$$

$$= \frac{(-1)^n \lfloor n}{2ia}\left[\frac{1}{r^{n+1}(\cos\theta - i\sin\theta)^{n+1}} - \frac{1}{r^{n+1}(\cos\theta + i\sin\theta)^{n+1}}\right]$$

$$= \frac{(-1)^n \lfloor n}{2ia.r^{n+1}}\left[(\cos\theta - i\sin\theta)^{-n+1} - (\cos\theta + i\sin\theta)^{-(n+1)}\right]$$

$$= \frac{(-1)^n \lfloor n}{2ia.r^{n+1}}\left[\cos(n+1)\theta + i\sin(n+1)\theta - \cos(n+1)\theta + i\sin(n+1)\theta\right]$$

$$= \frac{(-1)^n \lfloor n}{2ia.r^{n+1}}.2i\sin(n+1)\theta$$

or $$y_n = \frac{(-1)^n \lfloor n}{a.\left(\dfrac{a}{\sin\theta}\right)^{n+1}}\sin(n+1)\theta$$

Hence, $y_n = \dfrac{(-1)^n \lfloor n}{a^{n+2}}\sin^{n+1}\theta\sin(n+1)\theta$, where $\theta = \cot^{-1}\dfrac{x}{a}$

3. (b) Find y_n, when $y = \dfrac{x}{x^2 + a^2}$

Solution: We have $y = \dfrac{x}{x^2 + a^2}$

or $$y=\frac{x}{x^2-(-1)a^2}=\frac{x}{x^2-i^2a^2}$$

$$=\frac{x}{(x+ia)(x-ia)}=\frac{1}{2}\cdot\frac{(x+ia)+(x-ia)}{(x-ia)(x+ia)}$$

$$=\frac{1}{2}\left[\frac{x+ia}{(x-ia)(x+ia)}+\frac{x-ia}{(x-ia)(x+ia)}\right]$$

$$=\frac{1}{2}\left[\frac{1}{x-ia}+\frac{1}{x+ia}\right]$$

Differentiating both sides n times with respect to x, we get

$$y_n=\frac{1}{2}\left[\frac{(-1)^n\lfloor n}{(x-ia)^{n+1}}+\frac{(-1)^n\lfloor n}{(x+ia)^{n+1}}\right]$$

or $$y_n=\frac{(-1)^n\lfloor n}{2}\left[\frac{1}{(x-ia)^{n+1}}+\frac{1}{(x+ia)^{n+1}}\right]$$

put $x = r\cos\theta$ and $a = r\sin\theta$.

Squaring and adding these two, we get

$$x^2+a^2=r^2\cos^2\theta+r^2\sin^2\theta=r^2\left(\cos^2\theta+\sin^2\theta\right)$$

$$=r^2.1=r^2$$

and by division, we get

$$\frac{a}{x}=\frac{r\sin\theta}{r\cos\theta}=\tan\theta;\ \therefore\ \theta=\tan^{-1}\frac{a}{x}$$

$\therefore$ $$y_n=\frac{1}{2}(-1)^n\lfloor n\left[\frac{1}{(r\cos\theta-ir\sin\theta)^{n+1}}+\frac{1}{(r\cos\theta+ir\sin\theta)^{n+1}}\right]$$

$$=\frac{1}{2}(-1)^n\lfloor n\left[\frac{1}{r^{n+1}(\cos\theta-i\sin\theta)^{n+1}}+\frac{1}{r^{n+1}(\cos\theta+i\sin\theta)^{n+1}}\right]$$

$$=\frac{1}{2}\frac{(-1)^n\lfloor n}{r^{n+1}}\left[(\cos\theta-i\sin\theta)^{-(n+1)}+(\cos\theta+i\sin\theta)^{-(n+1)}\right]$$

$$=\frac{1}{2}\frac{(-1)^n\lfloor n}{r^{n+1}}\left[\cos(n+1)\theta+i\sin(n+1)\theta+\cos(n+1)\theta-i\sin(n+1)\theta\right]$$

$$= \frac{1}{2}\frac{(-1)^n \lfloor n}{r^{n+1}}.2\cos(n+1)\theta$$

Hence $$y_n = \frac{(-1)^n \lfloor n}{a^{n+1}}\sin^{n+1}\theta\cos(n+1)\theta, \text{ where } \theta = \tan^{-1}\frac{a}{x}$$

4. (a) If $y = x^2 \log x$, find y_n

Solution: We have $y = x^2 \log x$

Differentiating both sides successively with respect to x, we get

$$y_1 = 2x\log x + x^2.\frac{1}{x} = 2x\log x + x$$

$$y_2 = 2\log x + 2x.\frac{1}{x} + 1 = 2\log x + 3$$

$$y_3 = \frac{2}{x}$$

$$y_4 = \frac{2(-1)}{x^2} = \frac{2(-1)\lfloor 1}{x^2}$$

$$y_5 = \frac{2(-1)(-2)}{x^3} = \frac{2(-1)^2\lfloor 2}{x^3}$$

$$y_6 = \frac{2(-1)(-2)(-3)}{x^4} = \frac{2(-1)^3\lfloor 3}{x^4}$$

....

$\therefore$ Intuitively, $$y_n = \frac{2(-1)^{n-3}\lfloor n-3}{x^{n-2}}$$

Another Method

Here $y = uv$ where $u = \log x$ and $v = x^2$

By Leibnitz's theorem,

$$y_n = u_n v + {}^nC_1 u_{n-1} v_1 + {}^nC_2 u_{n-2} v_2 + ... + {}^nC_n u v_n$$

Here $u_n = \frac{(-1)^n \lfloor n-1}{x^n}$ and $v_1 = 2x$, $v_2 = 2$, $v_3 = v_4 = ... = 0$

$$\therefore \quad y_n = \frac{(-1)^n \lfloor n-1}{x^n} x^2 + {}^nC_1 . \frac{(-1)^{n-1} \lfloor n-2}{x^{n-1}} .2x + {}^nC_2 . \frac{(-1)^{n-2} \lfloor n-3}{x^{n-2}} .2$$

$$= \frac{(-1)^{n-2} \lfloor n-3}{x^{n-2}} \{(n-1)(n-2) - 2n(n-2) + n(n-1)\}$$

$$= \frac{(-1)^{n-2} \lfloor n-3}{x^{n-2}} \{n^2 - 3n + 2 - 2n^2 + 4n + n^2 - n\}$$

$$= \frac{2(-1)^{n-2} \lfloor n-2}{x^{n-2}}$$

4. (b) Find y_n, when $y = x^2 \sin x$

Solution: We have $y = x^2 \sin x$

Let $u = \sin x$ and $v = x^2$

$$\therefore \quad u_n = \sin\left(x + \frac{n\pi}{2}\right),\ u_{n-1} = \sin\left[x + \frac{(n-1)\pi}{2}\right],$$

$$u_{n-2} = \sin\left[x + \frac{(n-2)\pi}{2}\right], \text{etc.}$$

$$v_1 = 2x,\ v_2 = 2,\ v_3 = 0,\ v_4 = 0, \text{ etc.}$$

$$y = x^2 \sin x = (\sin x).(x^2) = uv$$

Differentiating both sides n times with respect to x by Leibnitz's theorem, we get

$$y_n = (uv)_n = u_n v + {}^nC_1 u_{n-1} v_1 + {}^nC_2 u_{n-2} v_2 + {}^nC_3 u_{n-3} v_3 + \ldots + uv_n$$

$$= \sin\left(x + \frac{n\pi}{2}\right).x^2 + n\sin\left[x + \frac{(n-1)\pi}{2}\right]2x$$

$$+ \frac{n(n-1)}{1.2}.\sin\left[x + \frac{(n-2)\pi}{2}\right].2 + \text{other terms vanish}$$

$$= x^2 \sin\left(x + \frac{n\pi}{2}\right) + 2nx\sin\left[x + \frac{(n-1)\pi}{2}\right]$$

$$+ n(n-1)\sin\left[x + \frac{(n-2)\pi}{2}\right]$$

4. (c) If $y = x^2 e^{2x}\sin x$, find $\dfrac{d^n y}{dx^n}$

Solution: We have $y = x^2 e^{2x}\sin x$

or $\quad y = (e^{2x}\sin x)(x^2)$

Let $\quad u_n = e^{2x}\sin x$

$$\therefore \quad u_n = \left(2^2+1^2\right)^{\frac{n}{2}} e^{2x}\sin\left(x + n\tan^{-1}\frac{1}{2}\right)$$

$$= 5^{\frac{n}{2}} e^{2x}\sin\left(x + n\tan^{-1}\frac{1}{2}\right),$$

$$u_{n-1} = 5^{\frac{n-1}{2}} e^{2x}\sin\left[x + (n-1)\tan^{-1}\frac{1}{2}\right],$$

$$u_{n-2} = 5^{\frac{n-1}{2}} e^{2x}\sin\left[x + (n-2)\tan^{-1}\frac{1}{2}\right], \text{etc.}$$

Let $v = x^2$, $\therefore v_1 = 2x$, $v_2 = 2$, $v_3 = 0$, etc.

$\therefore \quad y = uv.$

Differentiating both sides n times with respect to x by Leibnitz' theorem, we get

$$y_n = (uv)_n$$

$$= u_n v + {}^nC_1 u_{n-1} v_1 + {}^nC_2 u_{n-2} v_2 + {}^nC_3 u_{n-3} v_3 + \ldots + uv_n$$

$$= 5^{\frac{n}{2}} e^{2x}\sin\left(x + n\tan^{-1}\frac{1}{2}\right).x^2 + n.5^{\frac{n-1}{2}} e^{2x}\sin\left[x + (n-1)\tan^{-1}\frac{1}{2}\right].2x$$

$$+\frac{n(n-1)}{1.2}5^{\frac{n-2}{2}} e^{2x}\sin\left[x + (n-2)\tan^{-1}\frac{1}{2}\right].2 + \text{all other terms vanish.}$$

Hence $$y_n = 5^{\frac{n}{2}} x^2 e^{2x}\sin\left(x + n\tan^{-1}\frac{1}{2}\right)$$

$$+2nx.5^{\frac{n-1}{2}} e^{2x}\sin\left[x + (n-1)\tan^{-1}\frac{1}{2}\right]$$

$$+n(n-1).5^{\frac{n-2}{2}} e^{2x}\sin\left[x + (n-2)\tan^{-1}\frac{1}{2}\right]$$

[On Establishing Relation between Derivatives of Higher Order]

5. (a) If $y = e^{x^2}$, prove that

$$y_{n+1} - 2xy_n - 2ny_{n-1} = 0.$$

Solution: We have $y_1 = e^{x^2}$ .....(1)

Differentiating both sides with respect to x, we get

$$y_1 = e^{x^2}.2x = 2xy, \qquad \text{from (1).}$$

Differentiating both sides n times with respect to x by Leibnitz's theorem, we get

$$y_{n+1} = 2(yx)_n$$

$$= 2\left[y_n.x + {}^nC_1\, y_{n-1}.1 + \text{all other terms vanish}\right]$$

$$= 2\left[xy_n + ny_{n-1}\right].$$

Hence $\qquad y_{n+1} - 2xy_n - 2ny_{n-1} = 0.$

5. (b) If $y = \sin(m\sin^{-1}x)$, prove that [CSVTU 2005]

(i) $\left(1-x^2\right)y_{n+2} - (2n+1)xy_{n+1} + \left(m^2 - n^2\right)y_m = 0.$

(ii) $\lim\limits_{x\to 0} \dfrac{y_{n+2}}{y_n} = n^2 - m^2$

(iii) Find $(y_n)_0$

Solution: We have $\qquad y = \sin(m\sin^{-1}x)$

or $\qquad \sin^{-1}y = m\sin^{-1}x$

Differentiating both sides with respect to x, we get

$$\frac{1}{\sqrt{1-y^2}}.y_1 = m.\frac{1}{\sqrt{1-x^2}}$$

or $\qquad y_1\sqrt{1-x^2} = m\sqrt{1-y^2}\,.$ .....(1)

Squaring both sides, we get

$$y_1^2\left(1-x^2\right) = m^2\left(1-y^2\right).$$

Differentiating both sides with respect to x, we get

$$2y_1y_2\left(1-x^2\right) + y_1^2(-2x) = m^2(-2yy_1)$$

or $\quad 2y_1\left[y_2\left(1-x^2\right)-y_1x\right]=2y_1\left(-m^2y\right)$

or $\quad y_2\left(1-x^2\right)-y_1x=-m^2y \qquad \{\because y_1 \neq 0 \text{ in general}\}$

or $\quad y_2\left(1-x^2\right)-y_1x+m^2y=0.$

Differentiating both sides n times with respect to x by Leibnitz's theorem, we get

$$\left[y_2\left(1-x^2\right)\right]_n-\left[y_1x\right]_n+m^2\left[y\right]_n=0.$$

or $\quad \left[y_{n+2}\left(1-x^2\right)+{}^nC_1y_{n+1}(-2x)+{}^nC_2y_n(-2)\right]$

$$+ \text{all other terms vanish}]$$

$$-\left[y_{n+1}x+{}^nC_1y_n.1+\text{all other terms vanish}\right]+m^2y_n=0$$

or $\quad y_{n+2}\left(1-x^2\right)+ny_{n+1}(-2x)+\dfrac{n(n-1)}{1.2}y_n(-2)-y_{n+1}x-ny_n+m^2y_n=0$

or $\quad \left(1-x^2\right)y_{n+2}-(2n+1)xy_{n+1}+\left(-n^2+n-n+m^2\right)y_n=0$

hence $\quad \left(1-x^2\right)y_{n+2}-(2n+1)xy_{n+1}+\left(m^2-n^2\right)y_n=0 \qquad .....(2)$

This proves the part (i).

Let $x \to 0$ in (2);

Then $\lim\limits_{x\to0}(y_{n+2})-0+(m^2-n^2)\lim\limits_{x\to0}(y_n)=0$

or $\lim\limits_{x\to0}\dfrac{y_{n+2}}{y_2}=n^2-m^2$. This proves the part (ii).

(iii) Now putting $x=0$ in (2),

$$(y_{n+2})_0-0+\left(m^2-n^2\right)(y_n)_0=0, \qquad \left\{\because (y_{n+1})_0 \neq \infty \text{ at } x=0\right\}.$$

$$\therefore\ (y_{n+2})_0=\left(n^2-m^2\right)(y_n)_0;\ \therefore\ (y_n)_0=\left\{(n-2)^2-m^2\right\}(y_{n-2})_0$$

Using this result for values n – 2, n – 4, n – 6,…we get

$$(y_n)_0 = \{(n-2)^2 - m^2\}\{(n-4)^2 - m^2\}\{(n-6)^2 - m^2\}....(y)_0 \quad \text{if n is even.}$$

or $\{(n-2)^2 - m^2)\ \{(n-4)^2 - m^2\}\ \{(n-6)^2 - m^2\}...(y_1)_0$ if n is even

But $(y)_0 = \sin(m\sin^{-1}0) = 0;\ (y_1)_0 = m\sqrt{1-0^2} = m,$ by (1)

$\therefore$ $(y_n)_0 = 0$ if n is even and

$$(y_n)_0 = \{(n-2)^2 - m^2\}\{(n-4)^2 - m^2\}\{(n-6)^2 - m^2\}...(1-m^2)m$$

when n is odd.

6. (a) If $u_n = \frac{d^n}{dx^n}(x^n \log x)$, show that $u_n = nu_{n-1} + \underline{|n-1}$ and hence deduce that,

$$u_n = \underline{|n}\left(\log x + 1 + \frac{1}{2} + \frac{1}{3} + ... + \frac{1}{n}\right).$$

Solution: We have $u_n = \frac{d^n}{dx^n}(x^n \log x)$ .....(1)

or $$u_n = \frac{d^{n-1}}{dx^{n-1}}\left[\frac{d}{dx}(x^n \log x)\right] = \frac{d^{n-1}}{dx^{n-1}}\left(x^n.\frac{1}{x} + nx^{n-1}\log x\right)$$

$$= \frac{d^{n-1}}{dx^{n-1}}\left[n(x^{n-1}\log x) + x^{n-1}\right]$$

$$= n\frac{d^{n-1}}{dx^{n-1}}(x^{n-1}\log x) + \frac{d^{n-1}}{dx^{n-1}}(x^{n-1})$$

or $u_n = nu_{n-1} + \underline{|n-1}$ .....(2)

Hence the first part is proved.

Replacing n by n – 1 in (1), we get

$$u_{n-1} = (n-1)u_{n-2} + \underline{|n-2}.$$

Putting this value in (2), we get

$$u_n = n\{(n-1)u_{n-2} + \underline{|n-2}\} + \underline{|n-1}$$

or $u_n = n(n-1)u_{n-2} + n\underline{|n-2} + \underline{|n-1}$ .....(3)

or $$u_n = \frac{\underline{|n}}{\underline{|n-2}}u_{n-2} + \frac{\underline{|n}}{n-1} + \frac{\underline{|n}}{n}$$ .....(4)

Again replacing n by n – 2 in (2), we get

$$u_{n-2} = (n-2)u_{n-3} + \lfloor n-3$$

Putting this value in (3), we get

$$u_n = n(n-1)\{(n-2)u_{n-3} + \lfloor n-3\} + n\lfloor n-2 + \lfloor n-1$$

or $$u_n = n(n-1)(n-2)u_{n-3} + n(n-1)\lfloor n-3 + n\lfloor n-2 + \lfloor n-1$$

$$u_n = \frac{\lfloor n}{\lfloor n-3}u_{n-3} + \frac{\lfloor n}{(n-2)} + \frac{\lfloor n}{(n-1)} + \frac{\lfloor n}{n}$$

....

$$u_n = \frac{\lfloor n}{\lfloor 1}u_1 + \frac{\lfloor n}{2} + \frac{\lfloor n}{3} + ... + \frac{\lfloor n}{n-1} + \frac{\lfloor n}{n} \quad .....(5)$$

Putting n = 1 in (1), we get

$$u_1 = \frac{d}{dx}(x\log x) = \log x + x.\frac{1}{x} = \log x + 1.$$

Putting this value in (5), we get

$$u_n = \lfloor n(\log x + 1) + \frac{\lfloor n}{2} + \frac{\lfloor n}{3} + ... + \frac{\lfloor n}{n}$$

$$u_n = \lfloor n\left(\log x + 1 + \frac{1}{2} + \frac{1}{3} + ... + \frac{1}{n}\right).$$

Hence the second part is proved.

7. If $y = x^2e^x$, prove that

$$\frac{d^ny}{dx^n} = \frac{1}{2}n(n-1)\frac{d^2y}{dx^2} - n(n-2)\frac{dy}{dx} + \frac{1}{2}(n-1)(n-2)y$$

Solution: We have $y = x^2e^x$ .....(1)

Differentiating both sides with respect to x, we get

$$\frac{dy}{dx} = x^2e^x + 2xe^x \quad .....(2)$$

Again differentiating both sides of (2) with respect to x, we get

$$\frac{d^2y}{dx^2} = x^2e^x + 2xe^x + 2xe^x + 2e^x$$

or $$\frac{d^2y}{dx^2} = x^2e^x + 4xe^x + 2e^x \quad .....(3)$$

from (2), $\frac{dy}{dx} = y + 2xe^x$, by virtue of (1)

or $$2xe^x = \frac{dy}{dx} - y \quad .....(4)$$

From (3), $\frac{d^2y}{dx^2} = y + 2\left(\frac{dy}{dx} - y\right) + 2e^x$, by virtue of (1) and (4)

or $$2e^x = \frac{d^2y}{dx^2} - 2\frac{dy}{dx} + 2y. \quad .....(5)$$

Differentiating both sides of (1) n times with respect to x by Leibnitz's theorem, we get

$$y_n = \left(x^2e^x\right)_n = \left(e^x x^2\right)_n$$

or $$y_n = e^x.x^2 + {}^nC_1e^x(2x) + {}^nC_2e^x(2)$$

+ other terms which vanish

or $$y_n = y + n\left(\frac{dy}{dx} - y\right) + \frac{n(n-1)}{2}\left(\frac{d^2y}{dx^2} - 2\frac{dy}{dx} + y\right),$$

by virtue of (1), (4) and (5)

or $$\frac{d^ny}{dx^n} = \frac{1}{2}n(n-1)\frac{d^2y}{dx^2} - n(n-2)\frac{dy}{dx} + \frac{1}{2}\left[n(n-1) - 2(n-1)\right]y$$

Hence $$\frac{d^ny}{dx^n} = \frac{1}{2}n(n-1)\frac{d^2y}{dx^2} - n(n-2)\frac{dy}{dx} + \frac{1}{2}(n-1)(n-2)y$$

Formulae Gallery

1. $D^n(ax+b)^m = \frac{\lfloor m}{\lfloor m-n} a^n (ax+b)^{m-n}$, where m is a positive integer > n.

2. $D^n(ax+b)^n = \lfloor n.a^n$

3. $D^n(ax+b)^{-1} = (-1)^n \lfloor n \, a^n (ax+b)^{-n-1}$

4. $D^n \log(ax+b) = \dfrac{(-1)^{n-1} \lfloor n-1 \, a^n}{(ax+b)^n}$

5. $D^n(e^{mx}) = m^n e^{mx}$; $D^n(e^x) = e^x$

6. $D^n(a^x) = a^x (\log_e a)^n$

7. $D^n \sin(ax+b) = a^n \sin\left(ax+b+\dfrac{n\pi}{2}\right)$

8. $D^n \cos(ax+b) = a^n \cos\left(ax+b+\dfrac{n\pi}{2}\right)$

9. $D^n \{e^{ax} \sin(bx+c)\} = (a^2+b^2)^{\frac{n}{2}} e^{ax} \sin\left(bx+c+n\tan^{-1}\dfrac{b}{a}\right)$

10. $D^n \{e^{ax} \cos(bx+c)\} = (a^2+b^2)^{\frac{n}{2}} e^{ax} \cos\left(bx+c+n\tan^{-1}\dfrac{b}{a}\right)$

11. $(uv)_n = {}^nC_0 u_n v + {}^nC_1 u_{n-1} v_1 + {}^nC_2 u_{n-2} v_2 + \ldots + {}^nC_n u v_n$ (Leibnitz's Theorem)

Exercise 4.2

1. (i) If y = sinmx + cosmx, prove that

$$y_n = m^n\left[1+(-1)^n \sin 2mx\right]^{\frac{1}{2}}.$$

(ii) If $y = \cos^3 x$, find y_n.

(iii) Find the nth derivative of cosx.cos2x.cos3x.

(iv) Find y_n if $y = e^x$. sinx.sin2x.

2. (i) If $y = \dfrac{1}{6x^2+11x+3}$, find y_n.

(ii) $y = \dfrac{x^2}{(x-1)^2(x+2)}$, find y_n.

(iii) If $y = \dfrac{x^4}{(x-1)^2(x+2)}$, find y_n. when $n \geq 3$

[Hint. $y = x^2+3x+7-\dfrac{1}{x-1}+\dfrac{16}{x-2}$

$$\therefore \qquad y_3 = -\frac{d^3}{dx^3}\left(\frac{1}{x-1}\right)+16\frac{d^3}{dx^3}\left(\frac{1}{x-2}\right);$$

when $n \geq 3,\ y_n = -\frac{d^n}{dx^n}\left(\frac{1}{x-1}\right)+16\frac{d^n}{dx^n}\left(\frac{1}{x-2}\right)\Big]$

(iv) If $y=\frac{x^3}{x^3-1}$, find y_n

(v) if $y=\frac{x^3}{(x-1)(x-2)(x-3)}$, find y_n

3. (i) Find y_n, when $y=\frac{1}{x^2+1}$

(ii) Find y_n, when $y = \tan^{-1} x$

$\left[\textbf{Hint.}\ y_1 = \frac{1}{x^2+1}\right]$

(iii) Find y_n, when $y = \tan^{-1}\frac{2x}{1-x^2}$

(iv) Find y_n, when $y=\frac{1}{(x+1)(x^2+1)}$

(v) If $\tan A = \frac{1+x}{1-x}$, find $\frac{d^n A}{dx^n}$

Hint. $A = \tan^{-1}\frac{1+x}{1-x}, = \tan^{-1}1+\tan^{-1}x = \frac{\pi}{4}+\tan^{-1}x;$

$$\therefore \qquad \frac{d^n A}{dx^n} = \frac{d^n}{dx^n}\left(\tan^1 x\right)]$$

4. (i) Find y_n, when $y=\frac{x^n}{1+x}$

$\left[\textbf{Hint.}\ \text{Let}\ u = x^n \text{and}\ v = \frac{1}{1+x}\right]$

(ii) If $y = x^{n-1}\log x$, prove that $y_n = \frac{(n-1)!}{x}$

(iii) Find y_n, when $y = e^{ax}\{a^2x^2 - 2nax + n(n+1)\}$

(iv) if $y = x \sin x$, prove that

$$y_n = x \sin\left(x + \frac{nx}{2}\right) - n\cos\left(x + \frac{n\pi}{2}\right)$$

5. (a) If $(1-x^2)y_2 - xy_1 + a^2y = 0$, show that

$$(1-x^2)y_{n+2} - (2n+1)xy_{n+1} - (n^2 - a^2)y_n = 0$$

(b) (i) If $y = \tan^{-1} x$, prove that

$$(1+x^2)y_{n+1} + 2nxy_n + n(n-1)y_{n-1} = 0$$

(ii) If $y = (\sin^{-1} x)^2$, prove that

$$(1+x^2)y_{n+2} - (2n+1)xy_{n+1} - n^2y_n = 0$$

Also find the value of y_n at $x = 0$

(c) (i) If $y = (x^2 - 1)^n$, prove that

$$(x^2-1)y_{n+2} + 2xy_{n+1} - n(n+1)y_n = 0$$

(ii) If $y^{\frac{1}{m}} + y^{-\frac{1}{m}} = 2x$, prove that

$$(x^2-1)y_{n+2} + (2n+1)xy_{n+1} + (n^2 - m^2)y_n = 0$$

(d) (i) If $y = e^{a \sin^{-1} x}$, prove that

$$(1-x^2)y_{n+2} - (2n+1)xy_{n+1} - (n^2 + a^2)y_n = 0$$

Hence, find the value of $(y_n)_0$

(ii) If $\log y = \tan^{-1}x$, show that

(a) $(1-x^2)y_2 + (2x-1)y_1 = 0$

(b) $(1+x^2)y_{n+2} + (2nx + 2x - 1)y_{n+1} + n(n+1)y_n = 0$

(e) (i) If $y = a\cos(\log x) + b\sin(\log x)$, prove that

$$x^2y_{n+2} + (2n+1)xy_{n+1} + (n^2+1)y_n = 0$$

(ii) If $\cos^{-1}\left(\frac{y}{b}\right) = \log\left(\frac{x}{a}\right)^n$, prove that

$$x^2y_{n+2} + (2n+1)xy_{n+1} + 2n^2y_n = 0$$

(f) (i) If $y = \cosh(\sin^{-1}x)$, prove that

$(1-x^2)y_2 - xy_1 - y = 0$ and hence deduce that

$$(1-x^2)y_{n+2} - (2n+1)xy_{n+1} = (n^2+1)y_n$$

(ii) If $x = x = \cosh\left(\frac{1}{m}\log y\right)$, prove that

$$(x^2-1)y_{n+2} + (2n+1)xy_{n+1} + (n^2-m^2)y_n = 0$$

(g) If $y = \cos(m\cos^{-1}x)$, prove that

$$(1-x^2)y_{n+2} - (2n+1)xy_{n+1} + (m^2-n^2)y_n = 0$$

6. Show that $\frac{d^n}{dx^n}\left(\frac{\log x}{x}\right) = \frac{(-1)^n \lfloor n}{x^{n+1}}\left(\log x - 1 - \frac{1}{2} - \frac{1}{3} - \ldots - \frac{1}{n}\right)$

[**Hint.** Let $u = \frac{1}{x}$, then $u_n = \frac{(-1)^n \lfloor n}{x^{n+1}}$

$$u_{n-1} = \frac{(-1)^{n-1}\lfloor n-1}{x^n} = -\frac{(-1)^n \lfloor n-1}{x^n}, \ldots\ldots$$

Let $v = \log x$, then

$$v_1 = \frac{1}{x},\ v_2 = \frac{1}{x^2},\ v_3 = \frac{\lfloor 2}{x^3}, \ldots,\ v_n = \frac{(-1)^{n-1}\lfloor n-1}{x^n}$$

Now apply Leibnitz's theorem.]

7. If $y = \frac{1}{1+x+x^2+x^3}$, find y_n and show that $(y_n)_0 = 0$ when n is of the form $4p+2$ or $4p+3$, $= \lfloor n$ or $-\lfloor n$ according as n is the form $4p$ or $4p+1$.

$$\left[\textbf{Hint. } y=\frac{1}{(1+x)(1+x^2)}=\frac{1}{2}\cdot\frac{(1+x^3)+(1-x^2)}{(1+x)(1+x^2)}=\frac{1}{2}\left(\frac{1}{1+x}+\frac{1-x}{1+x^3}\right)\right]$$

8. Find the nth derivation of $\tan^{-1}\dfrac{2x}{1-x^2}$ in terms of r and θ.

4.6 Power Series

A series of the form

$$a_0+a_1x+a_2x^2+a_3x^3+\ldots+a_nx^n+\ldots..,$$

in which a_ns are independent of x and successive terms have ascending positive integral powers of the independent variable x, is called a *power series* in *x*. Infinite power series are not convergent for all values of x in general. The set of values of x for which the infinite power series is convergent is called the *interval of convergence.*

Illustration

(i) $1+2x+3x^2+4x^3+\ldots+(n+1)x^n+\ldots$ to ∞ is an infinite power series whose interval of convergence is

$$-1<x<1$$

(ii) $1+x^2+x^4+x^6+\ldots$ to ∞ is an infinite power series whose interval of convergence is

$$-1<x<1$$

(iii) ${}^nC_0+{}^nC_1x+{}^nC_2x^n+\ldots+{}^nC_nx^n$ is a finite power series which is the expansion of $(1+x)^n$.

In calculus, we are required to expand a function in a power series in many cases. Among the different methods, Taylor's and Maclaurin's series are of vital importance for expanding a function in power series. In this section we are primarily concerned with the elementary knowledge of the two series for expanding a function in infinite power series.

4.7 Taylor's (Infinite) Series [Elementary Treatment]

Statement: (Under certain circumstances) If the function f(x + h) can be expanded in a series of powers of h, then

$$f(x+h)=f(x)+\frac{h}{\underline{|1}}f'(x)+\frac{h^2}{\underline{|2}}f''(x)+\frac{h^3}{\underline{|3}}f'''(x)+\ldots$$

$$+\frac{h^n}{\underline{|n}}f^n(x)+\ldots \text{ to } \infty,$$

Proof: By assumption $f(x+h)$ can be expanded as a power series in h. Hence we can write

$$f(x+h) = A_0 + A_1\frac{h}{\underline{|1}} + A_2\frac{h^2}{\underline{|2}} + A_3\frac{h^3}{\underline{|3}} + A_4\frac{h^4}{\underline{|4}} + \ldots\ \infty \qquad \ldots..(1)$$

where $A_0, A_1, A_2, A_3, A_4, \ldots..$ are functions of x only (i.e., independent of h) and are to be determined.

Now $\quad \dfrac{d}{dh}f(x+h) = \dfrac{df(x+h)}{d(x+h)}.\dfrac{d(x+h)}{dh} = f'(x+h).1 = f'(x+h)$

as x and h are independent quantities and so x may be considered constant in differentiating with respect to h.

i.e., $\quad \dfrac{d}{dh}f(x+h) = 1$

Similarly $\dfrac{d^2}{dh^2}f(x+h) = f''(x+h)$ and so on.

Differentiating both sides of (1) with respect to h successively, we get

$$\frac{d}{dh}f(x+h) \text{ i.e., } f'(x+h) = 0 + A_1.1 + A_2h + A_3.\frac{h^2}{\underline{|2}} + A_4.\frac{h^3}{\underline{|3}} + \ldots \qquad \ldots.(2)$$

$$\frac{d^2}{dh^2}f(x+h), \text{ i.e., } f''(x+h) = A_2 + A_3h + A_4.\frac{h^2}{\underline{|2}} + \ldots \qquad \ldots..(3)$$

$$\frac{d^3}{dh^3}f(x+h), \text{ i.e., } f'''(x+h) = A_3 + A_4h + \ldots \qquad \ldots..(4)$$

....

....

Putting $h = 0$ in (1), (2), (3) (4),.. we get

$A_0 = f(x),\ A_1 = f'(x),\ A_2 = f''(x),\ A_3 = f'''(x),$

....

Substituting these values in (1), we obtain

$$f(x+h) = f(x) + \frac{h}{\underline{|1}}f'(x) + \frac{h^2}{\underline{|2}}f''(x) + \frac{h^3}{\underline{|3}}f'''(x),$$

$$+\ldots.. + \frac{h^n}{\underline{|n}}f^n(x) + \ldots. \text{ to infinity.}$$

Note:

(i) This series is named after the English mathematician Brook Taylor (1685-1731)

(ii) Rigorous treatment of Taylor's series follow from Taylor's theorem (Generalised Mean Value Theorem) which discussed in previous chapter.

Illustration: Expand the function $\sin(x+h)$ in powers of h with the help of Taylor's series.

Solution: Let $f(x+h)=\sin(x+h)$

Putting $h=0$, we get $f(x)=\sin x$

Differentiating both sides successively, we get

$$f'(x)=\cos x,\ f''(x)=-\sin x,\ f'''(x)=-\cos x,\ \text{etc}$$

The Taylor's series is

$$f(x+h)=f(x)+hf'(x)+\frac{h^2}{2!}f''(x)+\frac{h^3}{3!}f'''(x)+\ldots \text{ to } \infty$$

Hence $$\sin(x+h)=\sin c+h\cos x-\frac{h^2}{2!}\sin x-\frac{h^3}{3!}\cos x+\ldots \text{ to } \infty.$$

4.8 Maclaurin's Series [Elementary Discussion]

Statement: (Under certain circumstances) If the function f(x) can be expanded in powers of x, then

$$f(x)=f(0)+xf'(0)+\frac{x^2}{2!}f''(0)+\frac{x^3}{3!}f'''(0)+\ldots \text{ to } \infty$$

Proof: Let us assume the possibility of expanding f(x) in a convergent series of positive integral powers of x, so that

$$f(x)=A_0+A_1x+A_2\frac{x^2}{2!}+A_3\frac{x^3}{3!}+A_4\frac{x^4}{4!}+\ldots \text{ to } \infty \qquad \ldots\ldots(1)$$

where $A_0, A_1, A_2, \ldots\ldots$ are constants to be determined, not containing x.

Differentiating (1) with respect to x, we get

$$f'(x)=A_1+A_2\frac{2x}{2!}+A_3\frac{3x^2}{3!}+A_4\frac{4x^3}{4!}+\ldots \text{ to } \infty$$

or $$f'(x)=A_1+A_2x+A_3\frac{x^2}{2!}+A_4\frac{x^3}{3!}+\ldots \text{ to } \infty \qquad \ldots\ldots(2)$$

Differentiating (2) with respect to x, we get

$$f''(x) = A_2 + A_3\frac{2x}{2!} + A_4\frac{3x^2}{3!} + \ldots \text{ to } \infty$$

or $$f''(x) = A_2 + A_3x + A_4\frac{x^2}{2!} + \ldots \text{ to } \infty \qquad \ldots\ldots(3)$$

Differentiating (3) with respect to x, we get

$$f'''(x) = A_3 + A_4\frac{2x}{2!} + \ldots \text{ to } \infty$$

or $$f'''(x) = A_3 + A_4x + \ldots \text{ to } \infty \qquad \ldots\ldots(4)$$

....

....

....

Putting x = 0 in (1), (2), (3), (4),... we get

$$f(0) = A_0,\ f'(0) = A_1,\ f''(0) = A_2,\ f'''(0) = A_3, \ldots\ldots.$$

Substituting the values of these constants in (1), we get

$$\boxed{f(x) = f(0) + xf'(0) + \frac{x^2}{2!}f''(0) + \frac{x^3}{3!}f'''(0) + \ldots \text{ to } \infty}$$

Alternative Proof of Maclaurin's Series

Maclaurin's series can be deduced from Tayor's series.
Taylor's series is

$$f(x + h) = f(x) + hf'(x) + \frac{h^2}{2!}f''(x) + \frac{h^3}{3!}f'''(x) + \ldots\text{.to } \infty$$

In this series writing 0 for x and x for h, we arrive at the result

$$f(x) = f(0) + xf'(0) + \frac{x^2}{2!}f''(0) + \frac{x^3}{3!}f'''(0) + \ldots\text{to } \infty$$

This series is called Maclaurin's series or the power series expansion of f(x) about x = 0.

It is named after the Scottish mathematician Colin Maclaurin (1698 – 1746).

Illustration: Obtain the Maclaurin's series for e^x.

Solution: Let $f(x) = e^x$

Differentiating successively with respect to x, we get

$$f'(x)=e^x, f''(x)=e^x, f'''(x)=e^x, \text{ and so on.}$$

Thus, when x = 0, we find

$$f(0)=e^0=1,\ f'(0)=1,\ f''(0)=1,\ f'''(0)=1, \text{ and so on.}$$

Maclaurin's series is

$$f(x)=f(0)+xf'(0)+\frac{x^2}{2\,!}f''(0)+\frac{x^3}{3\,!}f'''(0)+\ldots\text{to } \infty$$

$$\therefore \qquad e^x=1+x+\frac{x^2}{2\,!}+\frac{x^3}{3\,!}+\ldots \text{ to } \infty$$

4.9 Expansions based on Algebraic and Trigonometrical Series

There are some functions which can be easily expanded with the help of the following formulae:

$$e^x=1+\frac{x}{1!}+\frac{x^2}{2!}+\frac{x^3}{3!}+\ldots\text{ to } \infty$$

$$\log(1+x)=x-\frac{1}{2}x^2+\frac{1}{3}x^3-\ldots\text{ to } \infty$$

$$(1+x^n)=1+nx+\frac{n(n-1)}{1.2}x^2+\ldots\text{ to } \infty$$

$$\sin x=x-\frac{x^3}{3!}+\frac{x^5}{5!}\ldots\text{ to } \infty$$

$$\cos x=1-\frac{x^2}{2!}+\frac{x^4}{4!}\ldots\text{ to } \infty$$

$$\tan x=x+\frac{1}{3}x^3+\frac{2}{15}x^5+\ldots\text{ to } \infty$$

4.10 Differentiation and Integration of Power Series

(a) **Differentiation of power series:** A power series representing the function f(x) may be differentiated term by term for all values of x within its interval of convergence.

Illustration: Differentiating the series

$$\sin x=x-\frac{x^3}{3!}+\frac{x^5}{5!}-\ldots\text{ to } \infty$$

We obtain

$$\cos x = 1 - \frac{x^2}{2!} + \frac{x^4}{4!} - .... \text{ to } \infty$$

Both series converge for all values of x.

(b) Integration of power series: A power series representing the function f(x) may be integrated term by term between any limits lying I its interval of convergence.

Illustration: Integrating the series

$$\frac{1}{1+x} = 1 - x + x^2 - ..... \text{ to } \infty, |x| < 1,$$

from 0 to x, we obtain

$$\log_e(1+x) = x - \frac{1}{2}x^2 + \frac{1}{3}x^3 - .... \text{ to } \infty, \quad |x|<1$$

4.11 Working Rule for Selection of Method for Expansion

(i) If the function be y = f(x) whose nth derivative can be found as an explicit function of x easily, then in order to expand f(x) as a power series in x use the Maclaurin's series.

(ii) If the function be y = f(x + h) and it is to be expanded in a power series of h, use Taylor's series. In case it is y = f(x) but it is to be expanded in powers of (x – a) then write y = f(x) = f(a + x – a) and expand by Taylor's series.

(iii) If the function be y = f(x) whose major factor can be expanded by binomial expansion or exponential theorem or logarithmic series or by any trigonometrical expansion, then they can be directly expanded by using the relevant expansion.

(iv) In case (iii) if it is true for $f'(x)$ then the expansion for y can be obtained by integrating $f'(x)$ between 0 and x. In case (iii) if it is true for $\int_0^x f(x)dx$ then the expansion for y can be obtained by differentiating $\int_0^x f(x)dx$.

(v) If the function be $y = f(x)$ whose nth derivative cannot be found conveniently but relation between two or three of y_{n+2}, y_n, y_{n-1} can be obtained easily by methods of successive differentiation, then find $(y_n)_0$ from the relation and use Maclaurin's series.

Work Out Examples

1. (a) Apply Taylor's series to prove that following expansion

$$\log\sin(x+h) = \log\sin x + h\cot x - \frac{1}{2}h^2 \operatorname{cosec}^2 x + \frac{1}{3}h^3 \cot x \operatorname{cosec}^2 x + ... \text{ to } \infty$$

Solution: Let $$f(x+h) = \log\sin(x+h)$$

Putting h = 0, we get $f(x) = \log\sin x$

Differentiating successively with respect to x, we get

$$f'(x) = \frac{1}{\sin x}.\cos x = \cot x$$

$$f''(x) = -\operatorname{cosec}^2 x$$

$$f'''(x) = -2\operatorname{cosec} x(-\operatorname{cosec} x\cot x)$$

....

Taylor's series is

$$f(x+h) = f(x) + hf'(x) + \frac{h^2}{2!}f''(x) + \frac{h^3}{3!}f'''(x) + \text{....to } \infty$$

Hence

$$\log\sin(x+h) = \log\sin x + h\cot x - \frac{1}{2}h^2$$

$$\operatorname{cosec}^2 x + \frac{h^3}{3}\cot x\operatorname{cosec}^2 x + \dots \text{ to } \infty$$

This is the required expansion

(b) Prove, by Taylor's series, that

$$\tan^{-1}(x+h) = \tan^{-1}x + h\sin\alpha\ \frac{\sin\alpha}{1} - (h\sin\alpha)^2\frac{\sin 2\alpha}{2} +$$

$$(h\sin\alpha)^3\frac{\sin 3a}{3} + \text{...to } \infty,$$

where $x = \cot\alpha$.

Solution: Let $$f(x+h) = \tan^{-1}(x+h)$$

Putting h = 0, we get

$$f(x) = \tan^{-1} x$$

Differentiating successively with respect to x, we get

$$f'(x) = \frac{1}{1+x^2} = \frac{1}{1+\cot^2\alpha} = \frac{1}{\operatorname{cosec}^2\alpha} = \sin^2\alpha, \text{ as } x = \cot\alpha$$

$$f''(x) = -\frac{2x}{(1+x^2)^2} = -\frac{2\cot\alpha}{\operatorname{cosec}^4\alpha} = -2\cos\alpha.\sin\alpha.\sin^2\alpha$$

$$= -(\sin a)2\sin 2a$$

$$f'''(x) = -\frac{2}{(1+x^2)^2} + \frac{8x^2}{(1+x^2)^3}$$

$$= -2\sin^4\alpha + 8\cot^2\alpha\sin^6\alpha$$

$$= -2\sin^4\alpha(1-4\cos^2\alpha)$$

$$= -2\sin^4\alpha(1-4+4\sin^2\alpha)$$

$$= 2\sin^3\alpha(3\sin\alpha - 4\sin^3\alpha)$$

$$= 2\sin^3\alpha.\sin 3\alpha$$

....

....

Taylor's series is

$$f(x+h) = f(x) + hf'(x) + \frac{h^2}{2!}f''(x) + \frac{h^3}{3!}f'''(x) + \text{...to } \infty$$

$$\therefore \quad \tan^{-1}(x+h) = \tan^{-1}x + h\sin\alpha\frac{\sin\alpha}{1} - (h\sin\alpha)^2\frac{\sin 2\alpha}{2}$$

$$+(h\sin\alpha)^3\frac{\sin 3\alpha}{3} + \text{... to } \infty$$

This is the required expansion

2. (a) Apply Maclaurin's series to prove $\sin x = x - x - \frac{x^3}{\lfloor 3} + \frac{x^5}{\lfloor 5} - \text{... to } \infty$.

Solution: Let $f(x) = \sin x$

Differentiating successively w.r.to x,

$$f'(x) = \cos x,$$

$$f''(x) = -\sin x,\ f'''(x) = -\cos x,\ f^{iv}(x) = \sin x$$

$$f^{v}(x) = \cos x, \text{ and so on}$$

At $x = 0$, $f(0) = 0$, $f'(0) = 1$, $f''(0) = 0$, $f'''(0) = -1$,

$$f^{iv}(0) = 0,\ f^{v}(0) = 1 \text{ and so on.}$$

By Maclaurin's series

$$f(x) = f(0) + xf'(0) + \frac{x^2}{\lfloor 2}f''(0) + \frac{x^3}{\lfloor 3}f'''(0)$$

$$+\frac{x^4}{\lfloor 4}f^{iv}(0) + \frac{x^5}{\lfloor 5}f^{v}(0) + \text{...to } \infty$$

Hence $\sin x = x - \frac{x^3}{\lfloor 3} + \frac{x^5}{\lfloor 5} - \ldots \text{to } \infty$

(b) Verify the following expansion:

$$\tan\left(\frac{\pi}{4}+x\right) = 1 + 2x + 2x^2 + \frac{8}{3}x^3 + \frac{10}{3}x^4 + \ldots$$

Solution: Let $f(x) = \tan\left(\frac{\pi}{4}+x\right)$

$$\therefore \quad f'(x) = \sec^2\left(\frac{\pi}{4}+x\right)$$

$$f''(x) = 2\sec^2\left(\frac{\pi}{4}+x\right).\tan\left(\frac{\pi}{4}+x\right)$$

$$f'''(x) = 4\sec^2\left(\frac{\pi}{4}+x\right).\tan^2\left(\frac{\pi}{4}+x\right) + 2\sec^4\left(\frac{\pi}{4}+x\right)$$

$$f^{iv}(x) = 8\sec^2\left(\frac{\pi}{4}+x\right).\tan^3\left(\frac{\pi}{4}+x\right) + 8\sec^4\left(\frac{\pi}{4}+x\right)$$

$$\times\tan\left(\frac{\pi}{4}+x\right) + 8\sec^4\left(\frac{\pi}{4}+x\right).\tan\left(\frac{\pi}{4}+x\right)$$

$$\therefore \quad f(0) = 1;\ f'(0) = 2;\ f''(0) = 4;$$

$$f'''(0) = 8 + 8 = 16;\ f^{iv}(0) = 16 + 32 + 32 = 80$$

$\therefore$ By Maclaurin's series

$$\tan\left(\frac{\pi}{4}+x\right) = 1 + 2x + \frac{4x^2}{\lfloor 2} + \frac{16x^3}{\lfloor 3} + \frac{80x^4}{\lfloor 4} + \ldots$$

$$= 1 + 2x + 2x^2 + \frac{8x^3}{3} + \frac{10}{3}x^4 + \ldots$$

3. Apply Maclaurin's series to prove that

$$\log_e(1+x) = x - \frac{1}{2}x^2 + \frac{1}{3}x^3 - \frac{1}{4}x^4 + \ldots + (-1)^{n-1}\frac{x^n}{n} + \ldots$$

Solution: We have $f(x) = \log_e(1 + x)$

Differentiating successively w.r.t. x, $f'(x) = \frac{1}{1+x}$

$$f''(x) = -\frac{1}{(1+x)^2}$$

$$f'''(x) = \frac{(-1)(-2)}{(1+x)^3} = (-1)^2 \frac{\lfloor 2}{(1+x)^3}$$

$$f^{iv}(x) = \frac{(-1)(-2)(-3)}{(1+x)^4} = (-1)^3 \frac{\lfloor 3}{(1+x)^4}$$

....

....

$$f^n(x) = (-1)^{n-1} \frac{\lfloor n-1}{(1+x)^n} \text{ and so on}$$

By Maclaurin's series

$$f(x) = f(0) + xf'(0) + \frac{x^2}{\lfloor 2} f''(0) + \frac{x^3}{\lfloor 3} f^{iv}(0) + \frac{x^4}{\lfloor 4} f^{iv}(0) +$$

$$\ldots + \frac{x^n}{\lfloor n} f^n(0) + \ldots\text{to } \infty$$

Hence $$\log_e(1+x) = x - \frac{x^2}{2} + \frac{x^3}{3} - \frac{x^4}{4} + \ldots + (-1)^{n-1}\frac{x^n}{n} + \ldots\text{to } \infty$$

4. (a) Apply Maclaurin's series to find the expansion of $\frac{e^x}{e^x+1}$ as far as the term x^3.

Solution: Let $$f(x) = \frac{e^x}{e^x+1} = \frac{(e^x+1)-1}{e^x+1} = 1 - \frac{1}{e^x+1}$$

Differentiating successively w.r.t. x,

$$f'(x) = \frac{e^x}{(e^x+1)^2}$$

$$f''(x) = \frac{e^x(e^x+1)^2 - 2e^x(e^x+1).e^x}{(e^x+1)^4}$$

$$= \frac{e^x(e^x+1)(e^x+1-2e^x).e^x}{(e^x+1)^4} = \frac{e^x - e^{2x}}{(e^x+1)^3}$$

$$f'''(x)=\frac{\left(e^x-2e^{2x}\right)\left(e^x+1\right)^3-3\left(e^x+1\right)^2e^x\left(ex-e^{2x}\right)}{\left(e^x+1\right)^6}$$

and so on,

At $x=0,\ f(0)=\frac{1}{2},\ f'(0)=\frac{1}{4},\ f''(0)=0,\ f'''(0)=\frac{-8}{64}=-\frac{1}{8}$ and so on

By Maclaurin's series,

$$f(x)=f(0)+xf'(0)+\frac{x^2}{\underline{|2}}f''(0)+\frac{x^3}{\underline{|3}}f'''(0)+\ldots\text{to }\infty$$

Hence $\frac{e^x}{e^x+1}=\frac{1}{2}+\frac{1}{4}x-\frac{1}{48}x^3+\ldots\text{to }\infty$

(b) Apply Maclaurin's series to expand $e^{x\sec x}$ as far as the term containing x^3.

Solution: Let $f(x) = e^{x\sec x}$

Differentiating successively w.r.t. x,

$$f'(x) = e^{x\sec x}(\sec x + x\sec x\tan x)$$

$$= f(x).\sec x(1 + x\tan x)$$

$$f''(x)=f'(x)\sec x(1+x\tan x)+f(x).\sec x\tan x(1+x\tan x)$$

$$+ f(x).\sec x(\tan x + x\sec^2 x)$$

$$= f'(x)\sec x(1+x\tan x)+2f(x).\sec x\ \tan x+xf(x)$$

$$\sec x\tan^2 x + xf(x)\sec x(1+\tan^2 x)$$

$$= f'(x)\sec x(1+x\tan x)+2f(x).\sec x\ \tan x+2xf(x)$$

$$\sec x\tan^2 x + x f(x)\sec x$$

$$= f'(x)\sec x(1+x\tan x)+2f(x).\sec x(1+x\tan x)$$

$$\tan x + x f(x)\sec x$$

$$=f'(x)\sec x(1+x\tan x)+2f'(x)\tan x+xf(x)\sec x$$

$$\therefore\quad f'''(x)=f''(x)\sec x(1+x\tan x)+f'(x)\sec x\tan x(1+x\tan x)$$

$$+f'(x)\sec x\left(\tan x+x\sec^2 x\right)+2f''(x)\tan x$$

$$+2f'(x)\sec^2 x+f(x)\sec x+xf'(x)\sec x$$

$$+ xf(x)\sec x\tan x$$

At x = 0, f(0) = 1, $f'(0)=1$, $f''(0)=1$,

$f'''(0)=4$ and so on

By Maclaurin's series

$$f(x) = f(0) + x f'(0) + \frac{x^2}{\lfloor 2} f''(0) + \frac{x^3}{\lfloor 3} f'''(0) + ...$$

Hence $e^{x\sec x} = 1 + x + \frac{x^2}{2} + \frac{2}{3}x^3 + ...$ to ∞

5. Find, by Maclaurin's series, the expansion of cos x cos hx in powers of x up to the eighth power.

***Solution*:** Let y = cos x cos hx

$$= \cos x\left(\frac{e^x + e^{-x}}{2}\right) = \frac{1}{2}\left(e^x \cos x + e^{-x} \cos x\right)$$

We know that the nth differential of e^{ax} cos bx

$$= \left(a^2 + b^2\right)^{\frac{n}{2}}.e^{ax}.\cos\left(bx + n\tan^{-1}\frac{b}{a}\right)$$

∴ The nth differential of e^x cosx

$$= 2^{\frac{n}{2}}.e^x.\cos\left(x + \frac{n\pi}{4}\right) = 2^{\frac{n}{2}}.e^x.\cos\left(x + n\tan^{-1}1\right)$$

and the nth differential of e^{-x} cos x

$$= 2^{\frac{n}{2}}.e^{-x}.\cos\left(x + \frac{3n\pi}{4}\right)$$

$$\therefore \quad y_n = \frac{1}{2}.2^{\frac{n}{2}}\left[e^x \cos\left(x + \frac{n\pi}{4}\right) + e^{-x} \cos\left(x + \frac{3n\pi}{4}\right)\right]$$

Putting x = 0, we get

$$\left(y_n\right)_0 = \frac{1}{2}.2^{\frac{n}{2}}\left[\cos\frac{n\pi}{4} + \cos\frac{3n\pi}{4}\right]$$

Putting n = 1, 2, 3, 4, 5, 6, 7, 8, ..., we get

$(y_1)_0 = 0$, $(y_2)_0 = 0$, $(y_2)_0 = 0$,

$(y_4)_0 = -4$, $(y_5)_0 = 0 = (y_6)_0 = (y_7)_0$,

$(y_8)_0 = 2^4$, etc

Also, $(y)_0 = \cos 0 \,.\, \cos h0 = 1$

Now, by Maclaurin's series,

$$y = (y)_0 + x(y_1)_0 + \frac{x^2}{2!}(y_2)_0 + \frac{x^3}{3!}(y_3)_0 + \frac{x^4}{4!}(y_4)_0$$

$$+\frac{x^5}{5!}(y_5)_0 + \frac{x^4}{6!}(y_6)_0 + \frac{x^7}{7!}(y_7)_0 + \frac{x^8}{8!}(y_8)_0 + \ldots \text{ to } \infty$$

$$\therefore \quad \cos x \cos hx = 1 - \frac{2^2 x^4}{4!} + \frac{2^4 x^8}{8!} - \ldots \text{to } \infty$$

This is the required expansion

6. Expand cosx in powers of $x - \frac{1}{4}\pi$

Expand cos x by Taylor's series about the point $\frac{\pi}{4}$

Solution: Here $\cos x = \cos\left(x - \frac{\pi}{4} + \frac{\pi}{4}\right)$

$$= \cos\left(y + \frac{\pi}{4}\right) = f\left(y + \frac{\pi}{4}\right), \qquad \text{where } y = x - \frac{\pi}{4}$$

By Taylor's series, we have

$$f\left(y + \frac{\pi}{4}\right) = f\left(\frac{\pi}{4}\right) + y\, f'\left(\frac{\pi}{4}\right) + \frac{y^2}{2!} f''\left(\frac{\pi}{4}\right) + \ldots\text{to } \infty \qquad \ldots..(1)$$

Now $f\left(y + \frac{\pi}{4}\right) = \cos\left(y + \frac{\pi}{4}\right)$

or $f(z) = \cos z$

$\therefore \quad f^n(z) = \cos\left(\frac{n\pi}{2} + z\right)$

or $f^n\left(\frac{\pi}{4}\right) = \cos\left(\frac{n\pi}{2} + \frac{\pi}{4}\right)$

$\therefore \quad f'\left(\frac{\pi}{4}\right) = \cos\left(\frac{\pi}{2} + \frac{\pi}{4}\right) = -\sin\frac{\pi}{4} = -\frac{1}{\sqrt{2}}$

$f''\left(\frac{\pi}{4}\right) = \cos\left(\pi + \frac{\pi}{4}\right) = -\cos\frac{\pi}{4} = -\frac{1}{\sqrt{2}}$

....

....

Also, $f\left(\frac{\pi}{4}\right)=\cos\frac{\pi}{4}=\frac{1}{\sqrt{2}}$

$\therefore$ From (1), we get

$$\cos\left(y+\frac{\pi}{4}\right)=\frac{1}{\sqrt{2}}+\left(x-\frac{\pi}{4}\right)\left(-\frac{1}{\sqrt{2}}\right)+\frac{\left(x-\frac{\pi}{4}\right)^2}{2!}\left(-\frac{1}{\sqrt{2}}\right)+\ldots \text{ to } \infty$$

or $$\cos x=\frac{1}{2}\sqrt{2}\left[1-\left(x-\frac{\pi}{4}\right)-\frac{\left(x-\frac{\pi}{4}\right)^2}{2!}+\ldots \text{ to } \infty\right]$$

This is the required expansion

Use of Algebraic or Trigonometrical Series

7. Prove that $x \operatorname{cosec} x = 1+\frac{1}{6}x^2+\frac{7}{360}x^4+\ldots \text{ to } \infty$

***Solution*:** We know, $\sin x = x-\frac{x^2}{\lfloor 3}+\frac{x^5}{\lfloor 5}-\frac{x^7}{\lfloor 7}+\ldots$

$$\therefore\ x \operatorname{cosec} x=\frac{x}{\sin x}=\frac{x}{x-\frac{x^2}{\lfloor 3}+\frac{x^5}{\lfloor 5}-\frac{x^7}{\lfloor 7}+\ldots\text{to } \infty}$$

$$=\frac{1}{1-\frac{x^2}{\lfloor 3}+\frac{x^4}{\lfloor 5}-\ldots\text{to } \infty}$$

$$=\left\{1-\left(\frac{x^2}{\lfloor 3}-\frac{x^4}{\lfloor 5}+\ldots\text{to } \infty\right)\right\}^{-1}$$

$$=1+\left(\frac{x^2}{\lfloor 3}-\frac{x^4}{\lfloor 5}+\ldots\right)+\left(\frac{x^2}{\lfloor 3}-\frac{x^4}{\lfloor 5}+\ldots\right)^2+\ldots \text{ to } \infty$$

$$=1+\frac{1}{6}x^2+\left(\frac{1}{36}-\frac{1}{120}\right)x^4+\ldots \text{ to } \infty$$

$$=1+\frac{1}{6}x^2+\frac{7}{360}x^4+\ldots \text{ to } \infty$$

(On Expansion by using Relation between Derivatives of Different Order)

8. Prove that $\sin^{-1} x = x + \frac{1}{2}.\frac{x^3}{3} + \frac{1}{2}.\frac{3}{4}\frac{x^5}{5} + \frac{1}{2}.\frac{2}{4}.\frac{5}{6}\frac{x^7}{7} + ...$

Solution: Let $y = \sin^{-1} x$, then $(y)_0 = 0$,

$$y_1 = \frac{1}{\sqrt{(1-x^2)}} \quad \because \quad (y_1)_0 = 1,$$

i.e., $(1 - x^2)\, y_1^2 = 1$

Differentiating, $(1 - x^2)\, 2y_1 y_2 - 2x\, y_1^2 = 0$

or $\quad (1 - x^2)y_2 - xy_1 = 0 \qquad (y_2)_0 = 0$

Differentiating n times by Leibnitz's theorem

$$(1 - x^2)\ y_{n+2} + n.y_{n+1}(-2x) + \frac{n(n-1)}{2} y_n(-2) - y_{n+1}.x - n\ .y_n.1 = 0$$

or $\quad (1 - x^2)\ y_{n+0} - (2n + 1)\ xy_{n+1} - n^2\ y_n = 0$

Putting x = 0, we have $(y_{n+2})_0 = n^2 . (y_n)_0$

Now putting n = 1, 2, 3, ... we get

$(y_3)_0 = 1^2(y_1)_0 = 1^2$, $(y_4)_0 = 2^2\,(y_2)_0 = 0$,

$(y_5)_0 = 3^2(y_3)_0 = 3^2 . 1^2$, $(y_6)_0 = 4^2 . (y_4)_0 = 0$ and so on

Thus all even differential co-efficients are zero

$(y_7)_0 = 5^2(y_5)_0 = 5^2 . 3^2 . 1^2$ etc

Hence by Maclaurin's series, we have

$$\sin^{-1} x = x + 1^2 . \frac{x^3}{\lfloor 3} + 3^2.\frac{x^5}{\lfloor 5} + 3^2.5^2.\frac{x^7}{\lfloor 7} + ...$$

$$= x + \frac{1}{2}.\frac{x^3}{3} + \frac{1}{2}.\frac{3}{4}.\frac{x^5}{5} + \frac{1}{2}.\frac{3}{4}.\frac{5}{6}.\frac{x^7}{7} + ...$$

Alternative Method

By the binomial theorem, we have

$$\frac{1}{\sqrt{1-x^2}} = (1 - x^2)^{-\frac{1}{2}} = 1 + \frac{1}{2}x^2 + \frac{1.3}{2.4}x^4 + \frac{1.3.5}{2.4.6}x^6 + ... \text{ to } \infty$$

Integrating from 0 to x, we obtain

$$\sin^{-1}x = \int_0^x \frac{dx}{\sqrt{1-x^2}} = x + \frac{1}{2}.\frac{x_3}{3} + \frac{1.3}{2.4}.\frac{x^5}{5} + ... + \text{ to } \infty$$

(Both series converge for |x| < 1)

9. Expand $e^{a\sin^{-1}x}$ in powers of x.

or

If $y = e^{a\sin^{-1}x}$, show that

$$(1-x^2)y_{n+2} - (2n+1)xy_{n+1} - (n^2+a^2).y_n = 0$$

Hence expand $e^{a\sin^{-1}x}$ in powers of x

Solution: Let $y = e^{a\sin^{-1}x}$ .....(1)

Differentiating both sides with respect to x, we get

$$y_1 = \frac{ae^{a\sin^{-1}x}}{\sqrt{1-x^2}} = \frac{ay}{\sqrt{1-x^2}} \text{ by (1)} \quad .....(2)$$

or $(1-x^2)\ y_1^2 = a^2y^2$

Differentiating this with respect to x, we get

$$(1-x^2).2y_1y_2 + (-2x)y_1^2 = 2a^2.yy_1$$

or $(1-x^2)y_2 - xy_1 - a^2\ y = 0$ .....(3)

Differentiating n times with respect to x by Leibnitz's theorem, we get

$$(1-x^2)y_{n+2} + {}^nC_1(-2x)y_{n+1} + {}^nC_2(-2)y_n - y_{n+1}x - {}^nC_1y_n.1 - a^2y_n = 0$$

or $(1-x^2)y_{n+2} - (2n+1)xy_{n+1} - (n^2+a^2)\ y_n = 0,$

By putting x = 0, we get

$$(y_{n+2})_0 = (n^2+a^2)\ (y_n)_0$$

Putting n = 1, we get

$$(y_3)_0 = (1+a^2)\ (y_1)_0$$

From (1), $(y)_0 = e^0 = 1$

From (2), $(y_1)_0 = a$

From (3), $(y_2)_0 = a^2$

$\therefore$ $(y_3)_0 = a(1^2 + a^2)$

Again $(y_4)_0 = (2^2 + a^2)(y_2)_0 = a^2(2^2 + a^2)$

$(y_5)_0 = (3^2 + a^2)(y_3)_0 = a(1^2 + a^2)(3^2 + a^2)$

$(y_6)_0 = (4^2 + a^2)(y_4)_0 = a^2(2^2 + a^2)(4^2 + a^2)$

....

By Maclaurin's series,

$$y = (y)_0 + x(y_1)_0 + \frac{x^2}{2!}(y_2)_0 + \frac{x^3}{3!}(y_3)_0 + \frac{x^4}{4!}(y_4)_0 + \frac{x^5}{5!}(y_5)_0 + \frac{x^6}{6!}(y_6)_0 + \ldots \text{ to } \infty$$

Hence

$$e^{a\sin^{-1}x} = 1 + ax + a^2\frac{x^2}{2!} + a(1^2 + a^2)\frac{x^3}{3!} + a^2(2^2 + a^2)\frac{x^4}{4!} + a(1^2 + a^2)(3^2 + a^2)\frac{x^5}{5!} + a^2(2^2 + a^2)(4^2 + a^2)\frac{x^5}{6!} \ldots \text{ to } \infty$$

This is the required expansion.

Formulae Gallery

1. Taylor's series:

$$f(x+h) = f(x) + hf'(x) + \frac{h^2}{2!}f''(x) + \frac{h^3}{3!}f'''(x) + \ldots \text{ to } \infty$$

2. Maclaurin's series

$$f(x) = f(0) + xf'(0) + \frac{x^2}{2!}f''(0) + \frac{x^3}{3!}f'''(0) + \ldots \text{ to } \infty$$

Exercise 4.3

1. Prove the following expansions with the help of Taylor's series:

(i) $e^{x+h} = e^x\left(1 + h + \frac{h^2}{2!} + \frac{h^3}{3!} + \ldots \text{ to } \infty\right)$

(ii) $\sin^{-1}(x+h) = \sin^{-1} x + \dfrac{h}{\sqrt{1-x^2}} + \dfrac{x}{(1-x^2)^{\frac{3}{2}}} \cdot \dfrac{h^2}{2!} + \ldots \text{ to } \infty$

Apply Maclaurin's series to prove the following expansions.

2. (i) $\sec x = 1 + \dfrac{x^2}{2!} + \dfrac{5x^4}{4!} + \ldots \text{ to } \infty$

(ii) $\cos x = 1 - \dfrac{x^2}{2!} + \dfrac{x^4}{4!} - \ldots \text{ to } \infty$

(iii) $\tan x = x + \dfrac{x^3}{3} + \dfrac{2x^5}{15} + \ldots \text{ to } \infty$

3. (i) $\log(1+\sin x) = x - \dfrac{x^2}{2} + \dfrac{x^3}{6} - \dfrac{x^4}{12} + \ldots \text{ to } \infty$

(ii) $\log \cos x = -\dfrac{1}{2}x^2 - \dfrac{1}{12}x^4 - \dfrac{1}{45}x^6 - \ldots\ldots \text{ to } \infty$

(iii) $\log(1+\tan x) = x - \dfrac{x^2}{2!} + \dfrac{4x^3}{3!} - \ldots \text{ to } \infty$

4. (i) $e^{\sin x} = 1 + x + \dfrac{1}{2}x^2 - \dfrac{1}{2}x^2 - \dfrac{1}{8}x^4 - \ldots \text{ to } \infty$

(ii) $e^{x\cos x} = 1 + x + \dfrac{1}{2}x^2 - \dfrac{1}{3}x^3 - \dfrac{1}{2}\dfrac{1}{4}x^4 - \ldots \text{ to } \infty$

(iii) $\dfrac{e^x}{\cos x} = 1 + x + x^2 + \dfrac{2}{3}x^3 + \ldots \text{ to } \infty$

(iv) $e^{ax}\cos bx = 1 + ax + (a^2 - b^2)\dfrac{x^2}{\underline{|2}} + a(a^2 - 3b^2)\dfrac{x^3}{\underline{|3}} + \ldots \text{ to } \infty$

(v) $e^x \sin x = x + \dfrac{2x^2}{2!} + \dfrac{2x^3}{3!} - \dfrac{2^2}{5!}x^5 - \ldots$

$$+\sin\left(\frac{1}{4}n\pi\right)\frac{2^{\frac{n}{2}}}{n!}x^n + \ldots \text{ to } \infty$$

(vi) $e^{\sec x} = e\left[1 + \dfrac{x^2}{\underline{|2}} + \dfrac{6x^4}{\underline{|4}} + \ldots \text{ to } \infty\right]$

(vii) $\sin(e^x - 1) = x + \dfrac{x^2}{2} - \dfrac{5}{24}x^4 + \ldots \text{ to } \infty$

(viii) $\log_e\left(1+e^x\right)=\log_e 2+\frac{1}{2}x+\frac{1}{4}.\frac{x^2}{2!}-\frac{1}{8}.\frac{x^4}{4!}+\ldots$ to ∞

(ix) Find, by Maclaurin's series, the expansion of $e^x\sin^2 x$ in powers of x up to the fourth power.

5. (i) Find by Maclaurin's series the expansion of sin x cos hx in powers of x up to the fifth power.

(ii) Obtain the expansion of e^{ax} cos hx in ascending powers of x

6. (i) Expand e^x in ascending powers of (x – 1)

(ii) Expand sin x in powers of $\left(x-\frac{1}{4}\pi\right)$

(iii) Find the Taylor's series expansion for the expansion of the function f(x) = log cos x about the point $\frac{\pi}{3}$.

[*Hint:* Expand log cos x = log cos $\left(\frac{\pi}{3}+x-\frac{\pi}{3}\right)$ by Taylor's series]

7. (i) Prove that $x\cot x=1-\frac{1}{3}x^2-\frac{1}{45}x^4+\ldots$ to ∞

[*Hint:* $\frac{x\cos x}{\sin x}=\frac{x\left(1-\frac{x^2}{2!}+\frac{x^4}{4!}-\ldots \text{ to } \infty\right)}{x-\frac{x^3}{3!}+\frac{x^5}{5!}-\ldots \text{ to } \infty}$]

(ii) Prove that loge (x cot x) = $-\frac{1}{2}x^2-\frac{7}{90}x^4+\ldots$ to ∞

(iii) Expand $\log\left(\frac{\tan x}{x}\right)$ in powers of x

(iv) Expand $\frac{x}{e^x-1}$ as far as x^4

(v) Expand x cothx in a series in ascending powers of x up to the fourth power.

(*Hint:* $x\frac{\cosh x}{\sinh x}=x.\frac{e^x+e^{-x}}{e^x-e^{-x}}=x\frac{\left(e^{2x}-1\right)+2}{e^{2x}-1}=x+2x\,(e^{2x}-1)^{-1}$]

8. (i) Prove that $\tan^{-1}x=x-\frac{1}{3}x^3+\frac{1}{5}x^5-\ldots$ to ∞

Also state restrictions on the values of x

[*Hint:* Let $y = \tan^{-1} x$

Establish $(1 + x^2)y_{n+2} + 2x(n + 1)y_{n+1} + n(n + 1)y_n = 0$

Put $x = 0$, $(y_{n+2})_0 = -n(n + 1)(y_n)_0$

Put $n = 1$, $(y_3)_0 = -1 . (1 + 1)(y_1)_0 = -\lfloor 2 \rfloor$]

(ii) Expand $\tan^{-1}\frac{x}{\sqrt{1-x^2}}$ in ascending integral powers of x

(*Hint*: Put $x = \sin\theta$]

9. (i) Prove that

$$\sin(m\sin^{-1} x) = mx + m(m^2 - 1^2)\frac{x^3}{3!} + m(m^2 - 1^2)(m^2 - 3^2).\frac{x^5}{5!} + \ldots \text{ to } \infty$$

Hence expand $\sin m\theta$ in powers of θ

(ii) Prove that

$$e^{\sin^{-1}x} = 1 + x + \frac{x^2}{2!} + \frac{2}{3!}x^3 + \frac{5}{4!}x^4 + \frac{20}{5!}x^5 + \ldots \text{ to } \infty$$

(iii) If $y = e^{\tan^{-1}x} = a_0 + a_1x + a_2x^2 + \ldots$ to ∞ prove that $(n + 2)\, a_{n+2} + na_n = a_{n+1}$

(iv) If $y = \left[\log_e\left(x + \sqrt{a^2 + x^2}\right)^2\right]$, prove that

$(x^2 + a^2)\, y_{n+2} + (2n + 1)\, xy_{n+1} + n^2 y_n = 0$

Hence write down the first four terms of the series for y when expanded in ascending powers of x.

10. Expand sin x in powers of $x - \frac{\pi}{2}$. Hence find the value of sin 99° correct to four decimal places.

11. Expand $\tan^{-1} x$ in powers of $x - 1$.

12. Apply Maclaurin's theorem to prove that $\log_e(1 + e^x) = \log_e 2 + \frac{1}{2}x + \frac{1}{8}x^2 \ldots \infty$.

CHAPTER 5

Partial Differentiation

5.1 Functions of Two Real Variables

If the value of a real variable z depends upon the values of two independent real variables x and y then z is said to be a function of two real variables x and y, represented symbolically by

$$z = f(x, y)$$

In set theoretic language, a mapping f: R × R → R is a real function of two real variables.

5.2 Partial Derivatives

Let u = f(x, y) be a function of two independent variables x and y. Let x change from x to x + δx while y remains constant then the limit of

$$\frac{f(x+\delta x, y) - f(x, y)}{\delta x}$$

as $\delta x \to 0$ is called the partial derivative of f(x, y) w.r.t. x and it is denoted by any one of the following.

$$\frac{\partial u}{\partial x}, \frac{\partial f}{\partial x}, \frac{\partial f(x,y)}{\partial x}, u_x, u_x(x,y), f_x, f_x(x, y)$$

Thus by definition, if u = f(x, y),

$$\boxed{\frac{\partial u}{\partial x} = \lim_{\delta x \to 0} \frac{f(x+\delta x, y) - f(x, y)}{\delta x}}$$

provided the limit exists uniquely as $\delta x \to 0$ in any manner i.e., $\delta x \to 0$ such that $x + \delta x$ always remains in the domain of definition of the function.

Again the partial derivation of $u = f(x,y)$ w.r.t. x at the point (a, b) is denoted by $\left(\frac{\partial u}{\partial x}\right)_{(a,b)}$ or u_x (a, b) and is defined as

$$\left(\frac{\partial u}{\partial x}\right)_{(a,b)} = \lim_{h\to 0}\frac{f(a+h,b)-f(a,b)}{h}$$

Next let, for the function u = f(x, y), y changes from y to $y+\delta y$ but x remains constant. Then the partial derivative of u w.r.t. y, denoted by any one of following,

$$\frac{\partial u}{\partial y}, u_y, u_y(x,y),\ f_y,\ f_y(x,y),\ \frac{\partial f}{\partial y},\ \frac{\partial f(x,y)}{\partial y}$$

is defined as below–

$$\frac{\partial u}{\partial y} = \lim_{\delta y\to 0}\frac{f(x,y+\delta y)-f(x,y)}{\delta y}$$

provided the limit exists uniquely as $\delta y \to 0$ is any manner so that $y + \delta y$ always remains in the domain of definition of the function.

Again the partial derivative of u = f(x, y) w.r.t. y at (a, b) is defined as follows

$$\left(\frac{\partial u}{\partial y}\right)_{(a,b)} = \lim_{k\to 0}\frac{f(a,b+k)-f(a,b)}{k}$$

Working Rule for Finding Partial Derivatives: Let u = f(x, y), a function of two independent variables x and y.

(i) In order to find $\frac{\partial u}{\partial y}$ i.e., the partial derivatives of u w.r.t. x, treat y as a constant and differentiate the function w.r.t. x by using the formulae and rules for ordinary differentiation of functions of one variable x.

To find $\left(\frac{\partial u}{\partial y}\right)_{(a,b)}$ put x = a and y = b in the value of $\frac{\partial u}{\partial x}$.

(ii) In order to find $\frac{\partial u}{\partial y}$ i.e., the partial derivative of u w.r.t. y, treat x as a constant and differentiate the function w.r.t. y by using the methods of ordinary differentiation of functions of one variable.

To find $\left(\frac{\partial u}{\partial y}\right)_{(a,b)}$ put x = a, y = b in the value of $\frac{\partial u}{\partial y}$.

Illustration: If $u = x^3 + 2x^2y + 3xy^2 - y^3$, find $\frac{\partial u}{\partial x}$ and $\frac{\partial u}{\partial y}$

Solution: Here $u = x^3 + 2x^2y + 3xy^2 - y^3$ .....(1)

Differentiating partially with respect to x, keeping y constant, we get

$$\frac{\partial u}{\partial x} = 3x^2 + 4xy + 3y^2$$

Again differentiating (1) partially with respect to y, treating x as constant, we get

$$\frac{\partial u}{\partial y} = 2x^2 + 6xy - 3y^2$$

Similarly if u = (x, y, z), a function of three independent variables, the partial derivative of u w.r.t. x is

$$\frac{\partial u}{\partial x} = \lim_{\delta x \to 0} \frac{f(x+\delta x, y, z) - f(x, y, z)}{\delta x}$$

i.e., the differential co-efficient of u w.r.t. x treating y, z as constants.

The partial derivative of u w.r.t. y is

$$\frac{\partial u}{\partial y} = \lim_{\delta y \to 0} \frac{f(x, y+\delta y, z) - f(x, y, z)}{\delta y}$$

i.e., the differential co-efficient of u w.r.t. y treating z, x as constants.

The partial derivative of u w.r.t. z is

$$\frac{\partial u}{\partial z} = \lim_{\delta z \to 0} \frac{f(x, y, z+\delta z) - f(x, y, z)}{\delta z}$$

i.e., the differential co-efficient of u w.r.t. z treating x, y as constants.

Similarly for the functions of four or more independent variables.

5.3 Partial Derivatives of Second and Higher Orders

Since the partial derivatives of a function u = f(x, y) are themselves function of x and y, they may in turn be differentiated. Their partial derivatives, if they exist, are called the second partial derivatives of u, and each is denoted by one of the following symbols.

$$\frac{\partial}{\partial x}\left(\frac{\partial u}{\partial x}\right), \frac{\partial^2 u}{\partial x^2}, u_{xx}, \frac{\partial^2 f}{\partial x^2}, f_{xx},$$

(second partial derivative w.r.t. x)

$$\frac{\partial}{\partial x}\left(\frac{\partial u}{\partial y}\right), \frac{\partial^2 u}{\partial x \partial y}, u_{yx}, \frac{\partial^2 f}{\partial x \partial y}, f_{yx},$$

(second partial derivative w.r.t. y and x)

$$\frac{\partial}{\partial y}\left(\frac{\partial u}{\partial x}\right), \frac{\partial^2 u}{\partial y \partial x}, u_{xy}, \frac{\partial^2 f}{\partial y \partial x}, f_{xy},$$

(second partial derivative w.r.t. x and y)

$$\frac{\partial}{\partial y}\left(\frac{\partial u}{\partial y}\right), \frac{\partial^2 u}{\partial y^2}, u_{yy}, \frac{\partial^2 f}{\partial y^2}, f_{yy},$$

(second partial derivative w.r.t. y)

It will be noticed that there is primarily a distinction of meaning between the symbols u_{yx} and u_{xy}, the operations indicated being performed in inverse orders in the two cases.

It can be shown that under certain conditions, which are generally satisfied in our cases, the results are identical i.e., $u_{yx} = u_{xy}$ etc.

***Illustration*:** If $u = x^m y^n$, find $\frac{\partial^2 u}{\partial x^2}, \frac{\partial^2 u}{\partial y^2}, \frac{\partial^2 u}{\partial x \partial y}, \frac{\partial^2 u}{\partial x \partial y}$ and show that $\frac{\partial^2 u}{\partial x \partial y} = \frac{\partial^2 u}{\partial y \partial x}$

***Solution*:** Here $u = x^m y^n$

$$\therefore \quad \frac{\partial u}{\partial x} = mx^{m-1}y^n, \quad \frac{\partial u}{\partial y} = nx^m y^{n-1},$$

$$\frac{\partial^2 u}{\partial x^2} = m(m-1)x^{m-2}y^n, \quad \frac{\partial^2 u}{\partial y^2} = n(n-1)x^m y^{n-2},$$

$$\frac{\partial^2 u}{\partial x \partial y} = \frac{\partial}{\partial x}\left(\frac{\partial u}{\partial y}\right) = \frac{\partial}{\partial x}\left(nx^m y^{n-1}\right) = mnx^{m-1}y^{n-1},$$

$$\frac{\partial^2 u}{\partial y \partial x} = \frac{\partial}{\delta y}\left(\frac{\partial u}{\partial x}\right) = \frac{\partial}{\partial y}\left(mx^{m-1}y^n\right) = mnx^{m-1}y^{n-1}.$$

Obviously $\dfrac{\partial^2 u}{\partial x \partial y} = \dfrac{\partial^2 u}{\partial y \partial x}$.

Similarly the partial derivatives of third order are

$$\frac{\partial^2 u}{\partial x^3}, \frac{\partial^2 u}{\partial x^2 \partial y}, \frac{\partial^3 u}{\partial x \partial y^2}, \frac{\partial^3 u}{\partial y^3}.$$

The partial derivatives of higher order can be formed as above.

Worked out Examples

1. (a) Find $\dfrac{\partial u}{\partial x}$ from the first principle if $u = \log\left(x^2 + y^2\right)$.

(b) Find $\dfrac{\partial u}{\partial y}$ at (1, 1) from the first principle if $u = \dfrac{1}{\sqrt{x^2 + y^2}}$

(c) Find f_x, f_y if $f(x, y) = \tan^{-1}\dfrac{y}{x}$; also find f_x (2, 1)

Solution:

(a) If $u = f(x, y)$ by definition

$$\frac{\partial u}{\partial x} = \lim_{\delta x \to 0} \frac{f(x + \delta x, y) - f(x, y)}{\delta x}$$

$$\therefore \qquad \frac{\partial u}{\partial x} = \lim_{\delta x \to 0} \frac{\log\left\{(x + \delta x)^2 + y^2\right\} - \log\left(x^2 + y^2\right)}{\delta x}$$

$$\left\{\because f(x, y) = \log\left(x^2 + y^2\right)\right\}$$

$$= \lim_{\delta x \to 0} \frac{\log \dfrac{x^2 + 2x.\delta x + (\delta x)^2 + y^2}{x^2 + y^2}}{\delta x}$$

$$= \lim_{\delta x \to 0} \frac{\log\left\{1 + \dfrac{2x + \delta x}{x^2 + y^2}.\delta x\right\}}{\delta x}$$

$$= \lim_{\delta x \to 0} \frac{1}{\delta x}\left\{\frac{2x+\delta x}{x^2+y^2}.\delta x - \frac{1}{2}.\left(\frac{2x+\delta x}{x^2+y^2}\right)^2.(\delta x)^2 + \ldots \text{ to } \infty\right\}$$

$$= \lim_{\delta x \to 0}\left[\frac{2x+\delta x}{x^2+y^2} - \frac{1}{2}\left(\frac{2x+\delta x}{x^2+y^2}\right)^2.\delta x + \ldots \text{ to } \infty\right]$$

$$= \frac{2x}{x^2+y^2}$$

(b) If u = f(x, y), by definition

$$\left(\frac{\partial u}{\partial y}\right)_{(1,1)} = \lim_{k \to 0} \frac{f(1,1+k) - f(1,1)}{k}$$

$$= \lim_{k\to 0}\frac{\dfrac{1}{\sqrt{1+(1+k)^2}} - \dfrac{1}{\sqrt{1+1}}}{k}, \quad \left\{\because f(x,y) = \frac{1}{\sqrt{x^2+y^2}}\right\}$$

$$= \lim_{k\to 0}\frac{\sqrt{2} - \sqrt{1+(1+k)^2}}{\sqrt{2}.k.\sqrt{1+(1+k)^2}}$$

$$= \lim_{k\to 0}\frac{2 - \left\{1+(1+k)^2\right\}}{\sqrt{2}.k.\sqrt{1+(1+k)^2}.\left\{\sqrt{2}+\sqrt{1+(1+k)^2}\right\}}$$

$$= \lim_{k\to 0}\frac{-2-k}{\sqrt{2}\sqrt{1+(1+k)^2}.\left\{\sqrt{2}+\sqrt{1+(1+k)^2}\right\}}$$

$$= \frac{-2}{\sqrt{2}.\sqrt{2}.2\sqrt{2}} = -\frac{1}{2\sqrt{2}}$$

(c) Here $f(x, y) = \tan^{-1}\dfrac{y}{x}$

Differentiating w.r.t. x, treating y as constant

$$f_x = \frac{1}{1+\left(\dfrac{y}{x}\right)^2}.\left(\frac{-y}{x^2}\right) = \frac{-y}{x^2+y^2}; \quad \therefore \quad f_x(2,1) = \frac{-1}{2^2+1^2} = \frac{-1}{5}$$

Differentiating w.r.t. y, treating x as constant

$$f_y = \frac{1}{1+(y/x)^2}\cdot\left(\frac{1}{x}\right) = \frac{x}{x^2+y^2}$$

2. If $u = \frac{x^2y^2}{x+y}$, show that $x\frac{\partial u}{\partial x} + y\frac{\partial u}{\partial y} = 3u$.

***Solution*:** We have $u = \frac{x^2y^2}{x+y}$ .....(1)

Differentiating eq.(1) partially with respect to x, (treating y as constant) we get

$$\frac{\partial u}{\partial x} = y^2\left[\frac{2x}{x+y} - \frac{x^2}{(x+y)^2}\right]$$

Multiplying both sides by x, we get

$$x\frac{\partial u}{\partial x} = \frac{2x^2y^2}{x+y} - \frac{x^3y^2}{(x+y)^2} \quad .....(2)$$

Again differentiating eq.(1) partially with respect to y (treating x as constant), we get

$$\frac{\partial u}{\partial y} = x^2\left[\frac{2y}{x+y} - \frac{y^2}{(x+y)^2}\right]$$

Multiplying both sides by y, we get

$$y\frac{\partial u}{\partial y} = \frac{2x^2y^2}{x+y} - \frac{x^2y^3}{(x+y)^2} \quad .....(3)$$

Adding eq.(2) and eq.(3), we get

$$x\frac{\partial u}{\partial x} + y\frac{\partial u}{\partial y} = \frac{4x^2y^2}{x+y} - \frac{x^2y^2}{(x+y)^2}(x+y)$$

$$= \frac{4x^2y^2}{x+y} - \frac{x^2y^2}{x+y} = \frac{3x^2y^2}{x+y}$$

Hence $x\frac{\partial u}{\partial x} + y\frac{\partial u}{\partial y} = 3u$, by virtue of eq.(1)

3. If $u = x\sin(y-x)$, prove that

$$\frac{\partial u}{\partial x} + \frac{\partial u}{\partial y} = \frac{u}{x}$$

Solution: We have $u = x\sin(y-x)$ .....(1)

Differentiating (1) partially with respect to x (keeping y constant), we get

$$\frac{\partial u}{\partial x} = \sin(y-x) - x\cos(y-x) \quad .....(2)$$

Differentiating (1) partially with respect to y (keeping x constant), we get

$$\frac{\partial u}{\partial y} = x\cos(y-x) \quad .....(3)$$

Adding eq.(2) and (3), we get

$$\frac{\partial u}{\partial x} + \frac{\partial u}{\partial y} = \sin(y-x) = \frac{u}{x}, \qquad \text{by virtue of (1)}$$

4. If $v = \cot^{-1}\frac{y}{x}$, prove that $v_x^{\ 2} + v_y^{\ 2} = \frac{1}{x^2+y^2}$.

Solution: Here $v = \cot^{-1}\frac{y}{x}$ .....(1)

Differentiating partially with respect to x, (treating y as constant), we get

$$v_x = -\frac{1}{1+\left(\frac{y}{x}\right)^2}.y.\left(-\frac{1}{x^2}\right) = \frac{y}{x^2+y^2}.$$

Again differentiating (1) partially with respect to y, (keeping x constant), we get

$$v_y = -\frac{1}{1+\left(\frac{y}{x}\right)^2}.\frac{1}{x} = -\frac{x}{x^2+y^2}$$

$$\therefore \qquad v_x^{\ 2} + v_y^{\ 2} = \frac{y^2}{\left(x^2+y^2\right)^2} + \frac{x^2}{\left(x^2+y^2\right)^2}$$

$$= \frac{x^2+y^2}{\left(x^2+y^2\right)^2} = \frac{1}{x^2+y^2}.$$

5. If $u = xf\left(\frac{y}{x}\right)$, show that $xu_x + yu_y = u$.

Solution: Here $u = xf\left(\frac{y}{x}\right)$ .....(1)

Differentiating partially w.r.t. x,

$$\frac{\partial u}{\partial x} = 1.f\left(\frac{y}{x}\right) + x.f'\left(\frac{y}{x}\right).\left(\frac{-y}{x^2}\right);$$

$$\therefore \quad u_x = f\left(\frac{y}{x}\right) - \frac{y}{x}f'\left(\frac{y}{x}\right) \quad .....(2)$$

Differentiating (1) partially w.r.t. y,

$$\frac{\partial u}{\partial y} = x.f'\left(\frac{y}{x}\right).\frac{1}{x};$$

$$\therefore \quad u_y = f'\left(\frac{y}{x}\right) \quad .....(3)$$

From (2) and (3),

$$xu_x + yu_y = x\left[f\left(\frac{y}{x}\right) - \frac{y}{x}f'\left(\frac{y}{x}\right)\right] + yf'\left(\frac{y}{x}\right)$$

$$= xf\left(\frac{y}{x}\right) = u.$$

6. If $f(x, y) = x\cos y + y\cos x$,

(i) find f_{yy}

(ii) prove that $\frac{\partial^2 f}{\partial x \partial y} = \frac{\partial^2 f}{\partial y \partial x}$

Solution: Here $f(x, y) = x\cos y + y\cos x$.

(i) Differentiating partially w.r.t. y,

$$f_y = x(-\sin y) + \cos x.$$

Again differentiating partially w.r.t. y,

$$f_{yy} = x(-\cos y) + 0 = -x\cos y.$$

(ii) Next $\frac{\partial f}{\partial y} = -x\sin y + \cos x.$

Now differentiating w.r.t. x partially

$$\frac{\partial}{\partial x}\left(\frac{\partial f}{\partial y}\right) = \frac{\partial^2 f}{\partial x \partial y} = -\sin y - \sin x \qquad .....(1)$$

Again $\frac{\partial f}{\partial x} = \cos y + y(-\sin x)$

$$\therefore \quad \frac{\partial}{\partial y}\left(\frac{\partial f}{\partial x}\right) = \frac{\partial^2 f}{\partial y \partial x} = -\sin y - \sin x \qquad .....(2)$$

$\therefore$ From eq.(1) and (2), $\frac{\partial^2 f}{\partial x \partial y} = \frac{\partial^2 f}{\partial y \partial x}.$

7. If ∇^2 stands for the operator $\frac{\partial^2}{\partial x^2} + \frac{\partial^2}{\partial y^2} + \frac{\partial^2}{\partial z^2}$,

prove that $\nabla^2 r = 0$, where $\frac{1}{r^2} = x^2 + y^2 + z^2$.

Or

If $u = \frac{1}{\sqrt{x^2 + y^2 + z^2}}$, prove that

$$\frac{\partial^2 u}{\partial x^2} + \frac{\partial^2 u}{\partial y^2} + \frac{\partial^2 u}{\partial z^2} = 0$$

Solution: We have $\frac{1}{r^2} = x^2 + y^2 + z^2 \qquad .....(1)$

Differentiating (1) partially with respect to x (keeping y and z constant), we get

$$-\frac{2}{r^3}.\frac{\partial r}{\partial x} = 2x \quad \text{or} \quad \frac{1}{r^3}.\frac{\partial r}{\partial x} = -x. \qquad .....(2)$$

Differentiating this partially with respect to x (keeping y and z constant), we get

$$\frac{3}{r^4}\left(\frac{\partial r}{\partial x}\right)^2 + \frac{1}{r^3}.\frac{\partial^2 r}{\partial x^2} = -1,$$

or $$\frac{3}{r^4}.\left(-xr^3\right)^2 + \frac{1}{r^3}.\frac{\partial^2 r}{\partial x^2} = -1, \qquad \text{by virtue of (2)}$$

or $$-3x^2r^2+\frac{1}{r^3}.\frac{\partial^2 r}{\partial x^2}=-1 \qquad .....(3)$$

Similarly $$-3y^2r^2+\frac{1}{r^3}.\frac{\partial^2 r}{\partial y^2}=-1 \qquad .....(4)$$

and $$-3z^2r^2+\frac{1}{r^3}.\frac{\partial^2 r}{\partial z^2}=-1 \qquad .....(5)$$

Adding eq.(3), (4) and (5), we get

$$-3r^2\left(x^2+y^2+z^2\right)+\frac{1}{r^3}\left(\frac{\partial^2 r}{\partial x^2}+\frac{\partial^2 r}{\partial y^2}+\frac{\partial^2 r}{\partial z^2}\right)=-3$$

or $$-3r^2.\frac{1}{r^2}+\frac{1}{r^3}\left(\frac{\partial^2}{\partial x^2}+\frac{\partial^2}{\partial y^2}+\frac{\partial^2}{\partial z^2}\right)r=-3, \qquad \text{by virtue of (1),}$$

or $$\frac{1}{r^3}\nabla^2 r=0. \qquad \text{Hence } \nabla^2 r=0.$$

8. If $u=t^{-\frac{3}{2}}.e^{-\frac{r^2}{4kt}}$, prove that

$$\frac{\partial u}{\partial t}=k\left(\frac{\partial^2 u}{\partial r^2}+\frac{2}{r}.\frac{\partial u}{\partial r}\right)$$

***Solution*:** We have $u=t^{-\frac{3}{2}}.e^{-\frac{r^2}{4kt}}$. .....(

Differentiating eq.(1) partially with respect to t (keeping r constant), we get

$$\frac{\partial u}{\partial t}=-\frac{3}{2}t^{-\frac{5}{2}}e^{-\frac{r^2}{4kt}}+t^{-\frac{3}{2}}e^{-\frac{r^2}{4kt}}\left(-\frac{r^2}{4k}\right)\left(-\frac{1}{t^2}\right)$$

$$\frac{\partial u}{\partial t}=-\frac{3}{2}t^{-\frac{5}{2}}e^{-\frac{r^2}{4kt}}+\frac{r^2}{4k}t^{-\frac{7}{2}}e^{-\frac{r^2}{4kt}} \qquad .....($$

Differentiating (1) partially with respect to r (treating t as constant), we get

$$\frac{\partial u}{\partial r}=t^{-\frac{3}{2}}e^{-\frac{r^2}{4kt}}\left(-\frac{2r}{4kt}\right)$$

or $$\frac{\partial u}{\partial r} = -\frac{t^{-\frac{5}{2}}}{2k}\left(re^{-\frac{r^2}{4kt}}\right) \qquad .....(3)$$

Differentiating eq.(3) partially with respect to r (keeping t constant), we get

$$\frac{\partial^2 u}{\partial r^2} = -\frac{t^{-\frac{5}{2}}}{2k}\left(e^{-\frac{r^2}{4kt}} - \frac{2r^2}{4kt}e^{-\frac{r^2}{4kt}}\right) \qquad .....(4)$$

$$\text{R.H.S.} = k\left(\frac{\partial^2 u}{\partial r^2} + \frac{2}{r}.\frac{\partial u}{\partial r}\right)$$

$$= -\frac{1}{2}t^{-\frac{5}{2}}e^{-\frac{r^2}{4kt}} + \frac{r^2}{4k}t^{-\frac{7}{2}}.e^{-\frac{r^2}{4kt}} - t^{-\frac{5}{2}}e^{-\frac{r^2}{4kt}}$$

$$= -\frac{3}{2}t^{-\frac{5}{2}}.e^{-\frac{r^2}{4kt}} + \frac{r^2}{4k}t^{-\frac{7}{2}}.e^{-\frac{r^2}{4kt}} = \frac{\partial u}{\partial t} = \text{L.H.S.}$$

9. If $z = \tan^{-1}\frac{2xy}{x^2 - y^2}$, prove that $z_{xx} + z_{yy} = 0$.

Solution: We have $z = \tan^{-1}\frac{2xy}{x^2 - y^2}$ or $z = \tan^{-1}\frac{2\frac{y}{x}}{1-\left(\frac{y}{x}\right)^2}$

or $$z = 2\tan^{-1}\frac{y}{x} \qquad .....(1)$$

Differentiating eq.(1) partially with respect to x (keeping y constant), we get

$$z_x = 2.\frac{1}{1+\frac{y^2}{x^2}}.y\left(-\frac{1}{x^2}\right) = -\frac{2y}{x^2+y^2}$$

Differentiating this partially with respect to x (keeping y constant), we get

$$z_{xx} = \frac{4xy}{\left(x^2+y^2\right)^2} \qquad .....(2)$$

Differentiating eq.(1) partially with respect to y (treating x as constant), we get

$$z_y = 2\frac{1}{1+\frac{y^2}{x^2}}.\frac{1}{x} \quad \text{or} \quad z_y = \frac{2x}{x^2+y^2}$$

Differentiating this partially with respect to y (keeping x as constant), we get

$$z_{yy} = -\frac{4xy}{\left(x^2+y^2\right)^2} = -z_{xx}, \qquad \text{from (2)}$$

Hence $z_{xx} + z_{yy} = 0$.

10. If $u = \log\left(x^3+y^3+z^3-3xyz\right)$, show that

$$\frac{\partial^2 u}{\partial x^2}+\frac{\partial^2 u}{\partial y^2}+\frac{\partial^2 u}{\partial z^2} = -\frac{3}{(x+y+z)^2}.$$

Solution: We have $u = \log\left(x^3+y^3+z^3-3xyz\right)$

or $u = \log\left[(x+y+z)\left(x^2+y^2+z^2-xy-yz-zx\right)\right]$

$$= \log(x+y+z) + \log\left(x^2+y^2+z^2-xy-yz-zx\right)$$

Differentiating partially with respect to x (keeping y and z as constant), we get

$$\frac{\partial u}{\partial x} = \frac{1}{x+y+z} + \frac{2x-y-z}{x^2+y^2+z^2-xy-yz-zx}.$$

Differentiating this partially with respect to x (treating y and z as constant), we get

$$\frac{\partial^2 u}{\partial x^2} = -\frac{1}{(x+y+z)^2} + \frac{2}{\left(x^2+y^2+z^2-xy-yz-zx\right)} - \frac{(2x-y-z)(2x-y-z)}{\left(x^2+y^2+z^2-xy-yz-zx\right)^2}$$

$$= -\frac{1}{(x+y+z)^2} + \frac{2\left(x^2+y^2+z^2-xy-yz-zx\right)-\left(4x^2+y^2+z^2-4xy-2yz-4xz\right)}{\left(x^2+y^2+z^2-xy-yz-zx\right)^2}$$

$$= -\frac{1}{(x+y+z)^2} + \frac{y^2+z^2+2xy+2xz-2x^2-4yz}{\left(x^2+y^2+z^2-xy-yz-zx\right)^2}$$

Similarly $\displaystyle \frac{\partial^2 u}{\partial y^2} = -\frac{1}{(x+y+z)^2} + \frac{z^2+x^2+2yz+2yx-2y^2-4zx}{\left(x^2+y^2+z^2-xy-yz-zx\right)^2}$

$$\frac{\partial^2 u}{\partial z^2} = -\frac{1}{(x+y+z)^2} + \frac{x^2+y^2+2zx+2zy-2z^2-4xy}{\left(x^2+y^2+z^2-xy-yz-zx\right)^2}$$

Hence $$\frac{\partial^2 u}{\partial x^2}+\frac{\partial^2 u}{\partial y^2}+\frac{\partial^2 u}{\partial z^2}=-\frac{3}{(x+y+z)^2}.$$

11. If $u=e^{xyz}$, prove that

$$\frac{\partial^3 u}{\partial x\,\partial y\,\partial z}=\left(1+3xyz+x^2y^2z^2\right)e^{xyz}$$

Solution: We have $u=e^{xyz}$

Differentiating partially with respect to z (keeping x and y constants), we get

$$\frac{\partial u}{\partial z}=(xy)e^{xyz}$$

Differentiating this partially with respect to y (keeping x and z constants), we get

$$\frac{\partial^2 u}{\partial y\partial z}=(x.1)e^{xyz}+(xy)(xz)e^{xyz}$$

Differentiating this partially with respect to x (keeping y and z constants), we get

$$\frac{\partial^3 u}{\partial x\,\partial y\,\partial z}=1.e^{xyz}+(x)(yz)e^{xyz}+\left(x^2yz\right)(yz)e^{xyz}+(2xyz)e^{xyz}$$

Hence $$\frac{\partial^3 u}{\partial x\,\partial y\,\partial z}=\left(1+3xyz+x^2y^2z^2\right)e^{xyz}$$

Exercise 5.1

1. (a) (i) Find $\frac{\partial u}{\partial x}$ from the first principle, if $u=x^3+y^3$.

(ii) Find u_y from the definition, if $u=\sin(x+y)$.

(b) Find $\frac{\partial u}{\partial x}$ at (2, 1) from the first principle, if $u=e^{x-y}$.

(c) (i) If $f(x,y)=\frac{x^2}{a^2}-\frac{y^2}{b^2}-1$, find f_x and f_y.

(ii) If $x^3+y^3-x^3y^2z=0$, find $\frac{\partial z}{\partial x}$ and $\frac{\partial z}{\partial y}$ when $x=y=1$.

2. (i) If $u = \dfrac{x+y}{x-y}$, find u_x, u_y and prove that $xu_x + yu_y = 0$.

(ii) If $u = x^2 + y^2 + z^2$, prove that

$$x.\frac{\partial u}{\partial x} + y.\frac{\partial u}{\partial y} + z.\frac{\partial u}{\partial z} = 2u.$$

(iii) If $u = \left(x^2 + y^2 + z^2\right)^{-\frac{1}{2}}$, prove that

$$x\frac{\partial u}{\partial x} + y\frac{\partial u}{\partial y} + z\frac{\partial u}{\partial z} + u = 0.$$

(iv) If $u = x^2y + y^2z + z^2x$, show that

$$\frac{\partial u}{\partial x} + \frac{\partial u}{\partial y} + \frac{\partial u}{\partial z} = (x+y+z)^2.$$

(v) If $u = \dfrac{x^2 + y^2}{x+y}$, prove that

$$\left(\frac{\partial u}{\partial x} - \frac{\partial u}{\partial y}\right)^2 = 4\left(1 - \frac{\partial u}{\partial x} - \frac{\partial u}{\partial y}\right)$$

(vi) If $u = \begin{vmatrix} x^2 & y^2 & z^2 \\ x & y & z \\ 1 & 1 & 1 \end{vmatrix}$,

prove that $u_x + u_y + u_z = 0$.

Hint: Break the determinant and then differentiate, or use the following

$$\text{if } u = \begin{vmatrix} f_1(x) & F_1(y) & \varphi_1(z) \\ f_2(x) & F_2(y) & \varphi_2(z) \\ f_3(x) & F_3(y) & \varphi_3(z) \end{vmatrix} \text{ then}$$

$$\frac{\partial u}{\partial x}=\begin{vmatrix}\frac{df_1}{dx} & F_1(y) & \varphi_1(z)\\ \frac{df_2}{dx} & F_2(y) & \varphi_2(z)\\ \frac{df_3}{dx} & F_3(y) & \varphi_3(z)\end{vmatrix},\ \frac{\partial u}{\partial y}=\begin{vmatrix}f_1(x) & \frac{dF_1}{dy} & \varphi_1(z)\\ f_2(x) & \frac{dF_2}{dy} & \varphi_2(z)\\ f_3(x) & \frac{dF_3}{dy} & \varphi_3(z)\end{vmatrix}\text{ etc.]}$$

3. (i) If $u=a\sin\frac{x}{y}+b\cos\frac{y}{x}$, prove that

$$x\frac{\partial u}{\partial x}+y\frac{\partial u}{\partial y}=0.$$

(ii) If $u=\sin\left(\sin\frac{y}{x}\right)$, find the value of

$$x\frac{\partial u}{\partial x}+y\frac{\partial u}{\partial y}.$$

4. (i) If $u=\sin^{-1}\frac{x}{y}$, prove that $x\frac{\partial u}{\partial x}+y\frac{\partial u}{\partial y}=0.$

(ii) If $u=\sin^{-1}\frac{x}{y}+\tan^{-1}\frac{y}{x}$, show that $x\frac{\partial u}{\partial x}+y\frac{\partial u}{\partial y}=0.$

5. (a)

(i) If $u=\log\frac{x^2+y^2}{x+y}$, prove that

$$xu_x+yu_y=1.$$

(ii) If $u=\log(\tan x+\tan y+\tan z)$, prove that

$$\sin 2x.\frac{\partial u}{\partial x}+\sin 2y.\frac{\partial u}{\partial y}+\sin 2z.\frac{\partial u}{\partial z}=2.$$

(b) **(i)** If $V=e^{\frac{x}{y}}+e^{\frac{y}{z}}+e^{\frac{z}{x}}$, show that

$$xV_x+yV_y+zV_z=0$$

(ii) If $u = \dfrac{e^{x+y}}{e^x + e^y}$, show that $\dfrac{\partial u}{\partial x} + \dfrac{\partial u}{\partial y} = u$.

(c) (i) If $u = F\left(\dfrac{y}{x}\right)$, show that $xu_x + yu_y = 0$.

(ii) If $z = e^{ax+by} f(ax - by)$, prove that

$$b\frac{\partial z}{\partial x} + a\frac{\partial z}{\partial y} = 2abz.$$

6. If $f(x, y) = \log(x^2 y + xy^2)$, find f_{xx} and f_{yx}.

7. (i) If $u = (x^2 + y^2)^{-\frac{1}{2}}$, prove that $\dfrac{\partial^2 u}{\partial x^2} + \dfrac{\partial^2 u}{\partial y^2} = u^3$.

(ii) If $u = \sqrt{x^2 + y^2 + z^2}$, prove that

$$u_{xx} + u_{yy} + u_{zz} = \frac{2}{u} \text{ (i.e., a function of u only)}$$

(iii) If $u = 2(px + qy)^2 - (x^2 + y^2)$ and $P^2 + q^2 = 1$, prove that

$$\frac{\partial^2 u}{\partial x^2} + \frac{\partial^2 u}{\partial y^2} = 0.$$

(iv) Show that, at the point of the surface $x^x y^y z^z = c$

where $x = y = z$, $\dfrac{\partial^2 z}{\partial x \partial y} = -\dfrac{1}{x \log_e ex}$

(v) If ∇^2 stands for the operator $\dfrac{\partial^2}{\partial x^2} + \dfrac{\partial^2}{\partial y^2} + \dfrac{\partial^2}{\partial z^2}$,

prove that $\nabla^2 r = \dfrac{2}{r}$ and $\nabla^2\left(\dfrac{1}{r}\right) = 0$

where $r^2 = x^2 + y^2 + z^2$

8. (i) If $u = e^{-mx} \sin(nt - mx)$, prove that

$$\frac{\partial u}{\partial t} = \frac{n}{2m^2} \cdot \frac{\partial^2 u}{\partial x^2}$$

(ii) If $u = \frac{\sin(ct - x)}{x}$, prove that

$$\frac{\partial^2 u}{\partial t^2} = c^2\left(\frac{\partial^2 u}{\partial x^2} + \frac{2}{x}.\frac{\partial u}{\partial x}\right).$$

(iii) Prove that if $\varphi = \left(Ax^n + Bx^{-n}\right)\cos n(y - \alpha)$, where A, B, n, α are constants, then

$$\frac{\partial^2 \varphi}{\partial x^2} + \frac{1}{x}.\frac{\partial \varphi}{\partial x} + \frac{1}{x^2}.\frac{\partial^2 \varphi}{\partial y^2} = 0$$

(iv) If $\theta = t^n e^{-r^2/4t}$, find what value of n will make

$$\frac{1}{r^2}.\frac{\partial}{\partial r}\left(r^2\frac{\partial \theta}{\partial r}\right) = \frac{\partial \theta}{\partial t}$$

9. (i) If $u = x^2 \tan^{-1}\frac{y}{x} - y^2 \tan^{-1}\frac{x}{y}$, prove that

$$\frac{\partial^2 u}{\partial x \partial y} = \frac{x^2 - y^2}{x^2 + y^2}$$

(ii) If $z = \sin^{-1}\frac{y}{x}$, prove that $\frac{\partial^2 z}{\partial x \partial y} = \frac{\partial^2 z}{\partial y \partial x}$

(iii) Prove that $\frac{\partial^2 z}{\partial x^2} + \frac{\partial^2 z}{\partial y^2} = 0,$

where $z = \tan^{-1}\frac{y}{x}$

10. (i) If $u = \log\left(x^2 + y^2\right)$, show that $\frac{\partial^2 u}{\partial x^2} + \frac{\partial^2 u}{\partial y^2} = 0.$

(ii) If $u = \log\left(x^2 + y^2 + z^2\right)$, prove that

$$x\frac{\partial^2 u}{\partial y \partial z} = y\frac{\partial^2 u}{\partial z \partial x} = z\frac{\partial^2 u}{\partial x \partial y}$$

(iii) If $u = \log r$, where $r^2 = (x-a)^2 + (y-b)^2 + (z-c)^2$,

show that $\frac{\partial^2 u}{\partial x^2} + \frac{\partial^2 u}{\partial y^2} + \frac{\partial^2 u}{\partial z^2} = \frac{1}{r^2}$

(iv) If $u = e^x(x\cos y - y\sin y)$, prove that

$$u_{xx} + u_{yy} = 0.$$

(v) If $r = ax^2 + 2hxy + by^2$ then find the value of

$$\frac{\partial^2 r}{\partial x^2} + \frac{\partial^2 r}{\partial y^2}.$$

11. If $V = x^2 + y^2 + z^2$, prove that

$$V_{xxx} = V_{yyy} = V_{zzz} = 0.$$

12. If $u = \log(x^3 + y^3 + z^3 - 3xyz)$, prove that

$$\left(\frac{\partial}{\partial x} + \frac{\partial}{\partial y} + \frac{\partial}{\partial z}\right)^2 u = -\frac{9}{(x+y+z)^2}$$

[*Hint*: If $\left(\frac{\partial}{\partial x} + \frac{\partial}{\partial y} + \frac{\partial}{\partial z}\right)u = V$

i.e., $\frac{\partial u}{\partial x} + \frac{\partial u}{\partial y} + \frac{\partial u}{\partial z} = V$ then L.H.S. $= \left(\frac{\partial}{\partial x} + \frac{\partial}{\partial y} + \frac{\partial}{\partial z}\right)V$.]

13. If $x = r\cos\theta, y = r\sin\theta$, show that

$$\frac{\partial x}{\partial r} = \frac{\partial r}{\partial x}$$

and $$\frac{1}{r}.\frac{\partial x}{\partial \theta} = r\frac{\partial \theta}{\partial x}.$$

[*Hint*: Use $x^2 + y^2 = r^2$ to calculate $\frac{\partial r}{\partial x}$; similarly for $\frac{\partial \theta}{\partial x}$ use $\theta = \tan^{-1}\frac{y}{x}$.]

14. If $y = f(x+ct) + \varphi(x-ct)$, show that

$$\frac{\partial^2 y}{\partial t^2} = c^2\frac{\partial^2 y}{\partial x^2}$$

[*Hint*: $\frac{\partial y}{\partial t} = f'(x+ct).(0+c) + \varphi'(x-ct).(0-c)$

$\frac{\partial y}{\partial x} = f'(x+ct).(1+0) + \varphi'(x-ct).(1-0)$ etc.]

5.4 Total Differential

Let $u = f(x, y)$ be a function of two independent variables x and y and let $\frac{\partial f}{\partial x}, \frac{\partial f}{\partial y}$ be continuous in the neighbourhood of the point (x, y).

[*Note*: A function $f(x, y)$ of two independent variables x, y is said to be continuous at (a, b) if given $\in > 0$ these exists a positive number $\delta(\in)$ such that $|f(x,y) - f(a,b)| < \in$ whenever

$$|x-a| \le \delta, \ |y-b| \le \delta.]$$

If x and y are assigned arbitrary increments δx and δy respectively, then let the given function become $u + \delta u$.

$$\therefore \ u + \delta u = f(x+\delta x, y+\delta y).$$

$$\therefore \ u + \delta u - u = f(x+\delta x, y+\delta y) - f(x, y)$$

or $$\delta u = \frac{f(x+\delta x, y+\delta y) - f(x, y+\delta y)}{\delta x}.\delta x + \frac{f(x, y+\delta y) - f(x, y)}{\delta y}.\delta y \qquad .....(5.1)$$

by Mean Value Theorem, we have

$$\frac{f(x+\delta x, y+\delta y) - f(x, y+\delta y)}{\delta x} = \frac{\partial}{\partial x} f(x+\theta_1 \delta x, y+\delta y),$$

and $$\frac{f(x, y+\delta y) - f(x, y)}{\delta y}$$

$$= \frac{\partial}{\partial y} f(x, y+\theta_2 \delta y),$$

where $0 < \theta_1 < 1$ and $0 < \theta_2 < 1$.

Substituting these two values in eq.(5.1), we get

$$\delta u = \frac{\partial}{\partial x} f(x+\theta_1 \delta x, y+\delta y).\delta x + \frac{\partial}{\partial y} f(x, y+\theta_2 \delta y).\delta y$$

$$= f_x(x+\theta_1 \delta x, y+\delta y).\delta x + f_y(x, y+\theta_2 \delta y).\delta y \qquad .....(5.2)$$

Since $\frac{\partial f}{\partial x}$ and $\frac{\partial f}{\partial y}$ are continuous near (x, y), therefore, we may write $$f_x\left(x+\theta_1\,\delta x, y+\delta y\right)=f_x\left(x,y\right)+\epsilon_1,$$

and $$f_y\left(x,y+\theta_2\delta y\right)=f_y\left(x,y\right)+\epsilon_2,$$

where ϵ_1 and ϵ_2 are quantities such that

$$\delta x,\lim_{\delta y\to 0}\epsilon 1=0 \text{ and } \lim_{\delta y\to 0}\epsilon 2=0$$

Substituting in (5.2), we get

$$\delta u=f_x\left(x,y\right).\delta x+\epsilon_1.\delta_x+f_y\left(x,y\right).\delta y+\epsilon_2\,\delta y \qquad .....(5.3)$$

For sufficiently small values of δx and δy, we find that δu is given to any desired accuracy by the sum of the first two terms in the right side of eq.(5.3). This sum is called the total differential or differential of u and is denoted by the symbol du.

Hence, by definition,

$$du=f_x\delta x+f_y\delta y.$$

or $$du=\frac{\partial u}{\partial x}\delta x+\frac{\partial u}{\partial y}\delta y \qquad .....(5.4)$$

If we write $u=x$ and $u=y$, it follows that

$$dx=\delta x \text{ and } dy=\delta y,$$

as the differential of each independent variable is defined to be the same as its small increment.

The expression for du from eq.(5.4) is seen to be

$$\boxed{du=\frac{\partial u}{\partial x}\delta x+\frac{\partial u}{\partial y}\delta y}.$$

This is a fundamental formula in the theory of differentiability.

The formula can be extended to a function of three or of any number of variables.

In case of three variables, let

$$u=f\left(x,y,z\right),$$

then $$\boxed{du=\frac{\partial u}{\partial x}\delta x+\frac{\partial u}{\partial y}\delta y+\frac{\partial u}{\partial z}\delta z}$$

if $u=f(x_1,x_2,x_3,...,x_n)$, then we have

$$du=\frac{\partial u}{\partial x_1}dx_1+\frac{\partial u}{\partial x_2}dx_2+\frac{\partial u}{\partial x_3}dx_3+...+\frac{\partial u}{\partial x_n}dx_n$$

5.5 Total Differential Co-efficient

If $u=f(x_1,x_2)$, where x_1 and x_2 are known functions of a single variable x, we have

$$du=\frac{\partial u}{\partial x_1}dx_1+\frac{\partial u}{\partial x_2}dx_2,$$

and remembering that

$$du=\frac{\partial u}{\partial x}dx,\quad dx_1=\frac{dx_1}{dx}dx,$$

$$dx_2=\frac{dx_2}{dx}dx,$$

we obtain

$$\frac{du}{dx}=\frac{\partial u}{\partial x_1}.\frac{dx_1}{dx}+\frac{\partial u}{\partial x_2}.\frac{dx_2}{dx}$$

Similarly, if $u=f(x_1,x_2,x_3,...,x_n)$

where $x_1,x_2,x_3,...,x_n$ are known functions of x, we have

$$\frac{du}{dx}=\frac{\partial u}{\partial x_1}.\frac{dx_1}{dx}+\frac{\partial u}{\partial x_2}.\frac{dx_2}{dx}+\frac{\partial u}{\partial x_3}.\frac{dx_3}{dx}+...+\frac{\partial u}{\partial x_n}.\frac{dx_n}{dx}$$

Again, if $x_1,x_2,x_3,...,x_n$ be each known functions of several variables $x,y,z,...,$ we shall have in the same way the series of relations

$$\frac{\partial u}{\partial x}=\frac{\partial u}{\partial x_1}.\frac{\partial x_1}{\partial x}+\frac{\partial u}{\partial x_2}.\frac{\partial x_2}{\partial x}+...+\frac{\partial u}{\partial x_n}.\frac{\partial x_n}{\partial x},$$

$$\frac{\partial u}{\partial y}=\frac{\partial u}{\partial x_1}.\frac{\partial x_1}{\partial y}+\frac{\partial u}{\partial x_2}.\frac{\partial x_2}{\partial y}+...+\frac{\partial u}{\partial x_n}.\frac{\partial x_n}{\partial y}.$$

Illustration: If $u=x\log y$, $x=t^2$, $y=e^t$, then find $\frac{du}{dt}$.

Solution: Here $u = x \log y.$

$$\therefore \quad \frac{\partial u}{\partial x} = \log y; \quad \frac{\partial u}{\partial y} = \frac{x}{y}$$

$$\therefore \quad x = t^2, \quad \therefore \quad \frac{dx}{dt} = 2t. \qquad \because \quad y = e^t,$$

$$\therefore \quad \frac{dy}{dt} = e^t.$$

$$\therefore \quad \frac{du}{dt} = \frac{\partial u}{\partial x}.\frac{dx}{dt} + \frac{\partial u}{\partial y}.\frac{dy}{dt}$$

$$= \log y.2t + \frac{x}{y}.e^t \qquad = \log e^t.2t + \frac{t^2}{e^t}.e^t$$

$$= 2t^2 + t^2 = 3t^2$$

5.6 An Important Result

Let $u = f(x, y)$, where y is a function of x.

Then $$du = \frac{\partial u}{\partial x}.dx + \frac{\partial u}{\partial y}.dy$$

or $$\boxed{\frac{du}{dx} = \frac{\partial u}{\partial x} + \frac{\partial u}{\partial y}.\frac{dy}{dx}}, \quad \text{as} \quad \frac{dx}{dx} = 1.$$

5.7 Differentiation of Implicit Functions

Let $u = f(x, y)$, where y is a function of x.

Then, we have,

$$\frac{du}{dx} = \frac{\partial f}{\partial x} + \frac{\partial f}{\partial y}.\frac{dy}{dx}. \qquad \text{.....(5.5)}$$

If y is defined as an implicit function of x by the equation

$$f(x, y) = 0,$$

i.e., $u = 0$, then $\frac{du}{dx} = 0.$

Hence, eq.(5.5) becomes,

$$\frac{\partial f}{\partial x} + \frac{\partial f}{\partial y}.\frac{dy}{dx} = 0.$$

or $\boxed{\dfrac{dy}{dx} = -\dfrac{\frac{\partial f}{\partial x}}{\frac{\partial f}{\partial y}}}$, provided $\dfrac{\partial f}{\partial y} \neq 0.$

Illustration: If $2x^2 - 3xy + y^2 = 0$, find $\dfrac{dy}{dx}$ by partial differentiation.

Solution: Let $f(x, y) = 2x^2 - 3xy + y^2$.

$$\therefore \quad \frac{\partial f}{\partial x} = 4x - 3y, \quad \text{and} \quad \frac{\partial f}{\partial y} = 2y - 3x.$$

$$\text{Hence } \frac{dy}{dx} = -\frac{\frac{\partial f}{\partial x}}{\frac{\partial f}{\partial y}} = -\frac{4x - 3y}{2y - 3x}$$

5.8 Homogeneous Function

A function is said to be a homogeneous function of the degree n in the variables x, y, z,..., if it can be put in the form

$$x^n F\left(\frac{y}{x}, \frac{z}{x}, \ldots\right).$$

Illustration: Let the given function be $x^2 + 2xy + y^2$.

This can be written as

$$x^2\left[\left(\frac{y}{x}\right)^0 + 2\left(\frac{y}{x}\right) + \left(\frac{y}{x}\right)^2\right],$$

which is in the form of $x^2 F\left(\frac{y}{x}, \right)$.

Hence the given function is a homogeneous function of degree two.

Again, let us consider the function

$$A_0x^n + A_1x^{n-1}y + A_2x^{n-2}y^2 + \ldots + A_{n-1}xy^{n-1} + A_ny^n,$$

where $A_0, A_1, A_2, \ldots, A_n$ are constants.

Rewriting this function in the following form

$$x^n\left[A_0\left(\frac{y}{x}\right)^0+A_1\left(\frac{y}{x}\right)+A_2\left(\frac{y}{x}\right)^2+\ldots+A_{n-1}\left(\frac{y}{x}\right)^{n-1}+A_n\left(\frac{y}{x}\right)^n\right],$$

which is in the form of $\qquad x^n F\left(\frac{y}{x}\right)$.

Hence the function under consideration is a homogeneous function of degree n.

5.9 Euler's Theorem on Homogeneous Functions of Two Independent Variables

Statement: If $u=f(x,y)$ be a homogeneous function of two independent variables x and y and of the degree n, then

$$x\frac{\partial u}{\partial x}+y\frac{\partial u}{\partial y}=nu$$

Proof: Let $\quad u=A_1x^{\alpha_1}y^{\beta_1}+A_2x^{\alpha_2}y^{\beta_2}+A_3x^{\alpha_3}y^{\beta_3}+\ldots \qquad \ldots..(5.6)$

where $\quad n=\alpha_1+\beta_1=\alpha_2+\beta_2=\alpha_3+\beta_3=\ldots \qquad \ldots..(5.7)$

Differentiating (5.6) partially with respect to x (keeping y constant), we get

$$\frac{\partial u}{\partial x}=A_1\alpha_1x^{\alpha_1-1}y^{\beta_1}+A_2\alpha_2x^{\alpha_2-1}y^{\beta_2}+A_3\alpha_3x^{\alpha_3-1}y^{\beta_3}+\ldots.$$

Multiplying both sides by x, we get

$$x\frac{\partial u}{\partial x}=A_1\alpha_1x^{\alpha_1-1}y^{\beta_1}+A_2\alpha_2x^{\alpha_2}y^{\beta_2}+A_3\alpha_3x^{\alpha_3}y^{\beta_3}+\ldots.. \qquad \ldots..(5.8)$$

Again differentiating (5.6) partially with respect with respect to y, (treating x as constant), we get

$$\frac{\partial u}{\partial y}=A_1\beta_1x^{\alpha_1}y^{\beta_1-1}+A_2\beta_2x^{\alpha_2}y^{\beta_2-1}+A_3\beta_3x^{\alpha_3}y^{\beta_3-1}+\ldots..$$

Multiplying both sides by y, we get

$$y\frac{\partial u}{\partial y}=A_1\beta_1x^{\alpha_1}y^{\beta_1}+A_2\beta_2x^{\alpha_2}y^{\beta_2}+A_3\beta_3x^{\alpha_3}y^{\beta_3}+\ldots.. \qquad \ldots..(5.9)$$

Adding eq.(5.8) and (5.9), we get

$$x\frac{\partial u}{\partial x}+y\frac{\partial u}{\partial y}=A_1(\alpha_1+\beta_1)x^{\alpha_1}y^{\beta_1}+A_2(\alpha_2+\beta_2)x^{\alpha_2}y^{\beta_2}$$

$$+A_3(\alpha_2+\beta_2)x^{\alpha_3}y^{\beta_3}+.....$$

$$=A_1nx^{\alpha_1}y^{\beta_1}+A_2nx^{\alpha_2}y^{\beta_2}+A_3nx^{\alpha_3}y^{\beta_3}+......,\qquad \text{from eq(5.7)}$$

$$=n\left(A_1x^{\alpha_1}y^{\beta_1}+A_2x^{\alpha_2}y^{\beta_2}+A_3x^{\alpha_3}y^{\beta_3}+......\right).$$

Hence $\boxed{x\frac{\partial u}{\partial x}+y\frac{\partial u}{\partial y}=nu}$, by virtue of eq.(5.6).

Alternative method

Let u be a homogeneous function of the nth degree in x and y. Then by definition of homogeneous function,

$$u=x^n.f\left(\frac{y}{x}\right). \qquad .....(5.10)$$

Differentiating partially w.r.t. x,

$$\frac{\partial u}{\partial x}=nx^{n-1}.f\left(\frac{y}{x}\right)-x^n.f'\left(\frac{y}{x}\right).\left(\frac{-y}{x^2}\right),$$

$$\therefore \quad x\frac{\partial u}{\partial x}=nx^{n-1}.f\left(\frac{y}{x}\right)-x^{n-1}y.f'\left(\frac{y}{x}\right). \qquad .....(5.11)$$

Again differentiating eq.(5.10) w.r.t. y partially,

$$\frac{\partial u}{\partial y}=x^n.f'\left(\frac{y}{x}\right).\frac{1}{x}=x^{n-1}.f'\left(\frac{y}{x}\right).$$

$$\therefore\ y\frac{\partial u}{\partial y}=x^{n-1}y.f'\left(\frac{y}{x}\right) \qquad .....(5.12)$$

Adding eq.(5.11) and (5.12),

$$x\frac{\partial u}{\partial x}=y\frac{\partial u}{\partial y}=nx^n f\left(\frac{y}{x}\right)=nu.$$

***Illustration*:** If $u=x^2+2xy+y^2$, prove that

$$x\frac{\partial u}{\partial x}+y\frac{\partial u}{\partial y}=2u$$

Solution: Here $$u=x^2\left[\left(\frac{y}{x}\right)^0+2\left(\frac{y}{x}\right)+\left(\frac{y}{x}\right)^2\right],$$

which is evidently a homogeneous function of x and y and of the 2nd degree.

Hence, by Euler's theorem, we have

$$x\frac{\partial u}{\partial x}+y\frac{\partial u}{\partial y}=nu=2u.$$

5.10 Euler's Theorem on Homogeneous Function of Three Independent Variables

Statement: If u be a homogeneous function of three independent variables x, y, z and of the nth degree,

$$x\frac{\partial u}{\partial x}+y\frac{\partial u}{\partial y}+z\frac{\partial u}{\partial z}=nu.$$

Proof: Let $u= A_1x^{\alpha_1}y^{\beta_1}z^{\gamma_1}+A_2x^{\alpha_2}y^{\beta_2}z^{\gamma_2}$

$$+A_3x^{\alpha_3}y^{\beta_3}z^{\gamma_3}+......, \qquad .....(5.13)$$

where $\alpha_1+\beta_1+\gamma_1=n=\alpha_2+\beta_2+\gamma_2=\alpha_3+\beta_3+\gamma_3=.....$ .....(5.14)

Differentiating eq.(5.13) partially with respect to x, keeping y and z constants, we get

$$\frac{\partial u}{\partial x}=A_1\alpha_1.x^{\alpha_1-1}y^{\beta_1}z^{\gamma_1}+A_2\alpha_2x^{\alpha_2-1}y^{\beta_2}z^{\gamma_2}+A_3\alpha_3x^{\alpha_3-1}y^{\beta_3}z^{\gamma_3}+.....$$

Multiplying both sides by x, we get

$$x\frac{\partial u}{\partial x}=A_1\alpha_1x^{\alpha_1}y^{\beta_1}z^{\gamma_1}+A_2\alpha_2x^{\alpha_2}y^{\beta_2}z^{\gamma_2}$$

$$+A_3\alpha_3x^{\alpha_3}y^{\beta_3}z^{\gamma_3}+..... \qquad .....(5.15)$$

Again differentiating (5.13) partially with respect to y, treating x and z as constants, we get

$$\frac{\partial u}{\partial y}=A_1\beta_1x^{\alpha_1}y^{\beta_1-1}z^{\gamma_1}+A_2\beta_2x^{\alpha_2}y^{\beta_2-1}z^{\gamma_2}+A_3\beta_3x^{\alpha_3}y^{\beta_3-1}z^{\gamma_3}+.....$$

Multiplying both sides by y, we get

$$y\frac{\partial u}{\partial y} = A_1\beta_1 x^{\alpha_1} y^{\beta_1} z^{\gamma_1} + A_2\beta_2 x^{\alpha_2} y^{\beta_2} z^{\gamma_2}$$

$$+A_3\beta_3 x^{\alpha_3} y^{\beta_3} z^{\gamma_3} + ... \qquad .....(5.16)$$

Again differentiating (5.13) partially with respect to z, taking x and y constants, we get

$$\frac{\partial u}{\partial z} = A_1\gamma_1 x^{\alpha_1} y^{\beta_1} z^{\gamma_1 - 1} + A_2\gamma_2 x^{\alpha_2} y^{\beta_2} z^{\gamma_2 - 1} + ...A_3\gamma_3 x^{\alpha_3} y^{\beta_3} z^{\gamma_3 - 1} + ...$$

Multiplying both sides by z, we get

$$z\frac{\partial u}{\partial z} = A_1\gamma_1 x^{\alpha_1} y^{\beta_1} z^{\gamma_1} + A_2\gamma_2 x^{\alpha_2} y^{\beta_2} z^{\gamma_2}$$

$$+A_3\gamma_3 x^{\alpha_3} y^{\beta_3} z^{\gamma_3} + ... \qquad .....(5.17)$$

Adding (5.15), (5.16) and (5.17), we get

$$x\frac{\partial u}{\partial x} + y\frac{\partial u}{\partial y} + z\frac{\partial u}{\partial z} = A_1 x^{\alpha_1} y^{\beta_1} z^{\gamma_1}(\alpha_1 + \beta_1 + \gamma_1)$$

$$+A_2 x^{\alpha_2} y^{\beta_2} z^{\gamma_2}(\alpha_2 + \beta_2 + \gamma_2) + A_3 x^{\alpha_3} y^{\beta_3} z^{\gamma_3}(\alpha_3 + \beta_3 + \gamma_3) + ...$$

$$= n\left(A_1 x^{\alpha_1} y^{\beta_1} z^{\gamma_1} + A_2 x^{\alpha_2} y^{\beta_2} z^{\gamma_2} + A_3 x^{\alpha_3} y^{\beta_3} z^{\gamma_3} + ...\right), \quad \text{from eq.(5.14)}$$

$$= nu, \qquad \text{from eq.(5.13).}$$

Alternative Proof

As u is a homogeneous function of the nth degree in x, y and z we can write

$$u = x^n f\left(\frac{y}{x}, \frac{z}{x}\right).$$

Now, let $\frac{y}{x} = Y$, $\frac{z}{x} = Z$.

Then $\frac{\partial Y}{\partial x} = -\frac{y}{x^2}$, $\frac{\partial Y}{\partial y} = \frac{1}{x}$, $\frac{\partial Y}{\partial z} = 0$

and $\frac{\partial Z}{\partial x} = -\frac{z}{x^2}$, $\frac{\partial Z}{\partial z} = \frac{1}{x}$, $\frac{\partial Z}{\partial y} = 0$

Now $u = x^n f(Y, Z)$ where Y, Z are functions of x, y, z.

$$\therefore \quad \frac{\partial u}{\partial x} = nx^{n-1}f + x^n.\left[\frac{\partial f}{\partial Y}.\frac{\partial Y}{\partial x} + \frac{\partial f}{\partial Z}.\frac{\partial Z}{\partial x}\right]$$

$$= nx^{n-1}f + x^n.\left[\frac{\partial f}{\partial Y}.\left(\frac{-y}{x^2}\right) + \frac{\partial f}{\partial Z}.\left(-\frac{z}{x^2}\right)\right]$$

$$= nx^{n-1}f - x^{n-2}\left[y\frac{\partial f}{\partial Y} + z\frac{\partial f}{\partial Z}\right]$$

Again differentiating $u = x^n f(Y,Z)$ partially w.r.t. y,

$$\frac{\partial u}{\partial y} = x^n\left[\frac{\partial f}{\partial Y}.\frac{\partial Y}{\partial y} + \frac{\partial f}{\partial Z}.\frac{\partial Z}{\partial y}\right]$$

$$= x^n\left[\frac{\partial f}{\partial Y}.\frac{1}{x} + \frac{\partial f}{\partial Z}.0\right] = x^{n-1}\frac{\partial f}{\partial Y}.$$

Similarly, $$\frac{\partial u}{\partial z} = x^n\left[\frac{\partial f}{\partial Y}.\frac{\partial Y}{\partial z} + \frac{\partial f}{\partial Z}.\frac{\partial Z}{\partial z}\right]$$

$$= x^n\left[\frac{\partial f}{\partial Y}.0 + \frac{\partial f}{\partial Z}.\frac{1}{x}\right] = x^{n-1}\frac{\partial f}{\partial Z}.$$

$$\therefore \quad x\frac{\partial u}{\partial x} + y\frac{\partial u}{\partial y} + z\frac{\partial u}{\partial z}$$

$$= x\left[nx^{n-1}f - x^{n-2}\left\{y.\frac{\partial f}{\partial Y} + z.\frac{\partial f}{\partial Z}\right\}\right] + x^{n-1}y\frac{\partial f}{\partial Y} + x^{n-1}z\frac{\partial f}{\partial Z}$$

$$= nx^n f = nu$$

Illustration: If $u = x^3 + 3x^2y + 3xy^2 + y^3$, prove that

$$x\frac{\partial u}{\partial x} + y\frac{\partial u}{\partial y} + z\frac{\partial u}{\partial z} = 3u.$$

Solution: Here $u = x^3\left[\left(\frac{y}{x}\right)^0 + 3\left(\frac{y}{x}\right) + 3\left(\frac{y}{x}\right)^2 + \left(\frac{y}{x}\right)^3\right]$

which is clearly a homogeneous function in x, y, z and of the 3rd degree.

Hence, by Euler's Theorem, we have

$$x\frac{\partial u}{\partial x} + y\frac{\partial u}{\partial y} + z\frac{\partial u}{\partial z} = nu = 3u.$$

5.11 Extension of Eluer's Theorem in the Cases of Any Number of Independent Variables

Proof of Euler's Theorem in case of n Variables:

Let $$u = x^n F\left(\frac{y}{x}, \frac{z}{x}, ...\right). \quad .....(5.18)$$

Put $\frac{y}{x} = Y$, $\frac{z}{x} = Z$, and so on.

$\therefore$ $\frac{\partial Y}{\partial x} = -\frac{y}{x^2}$, $\frac{\partial Z}{\partial x} = -\frac{z}{x^2}$, and so on.

$\therefore$ From eq.(5.18), we have $u = x^n F(Y, Z, ...)$.

$$\therefore \quad \frac{\partial u}{\partial x} = nx^{n-1}F(Y,Z,...)$$

$$+x^n\left\{\frac{\partial F}{\partial Y}.\frac{\partial Y}{\partial x} + \frac{\partial F}{\partial Z}.\frac{\partial Z}{\partial x} + ...\right\}$$

$$= nx^{n-1}F(Y,Z,...) + x^n\left\{\frac{\partial F}{\partial Y}.\left(-\frac{y}{x^2}\right)\right.$$

$$\left.+\frac{\partial F}{\partial Z}.\left(-\frac{z}{x^2}\right) + ...\right\}$$

$$= nx^{n-1}F(Y,Z,...) - x^{n-2}\left\{y\frac{\partial F}{\partial Y} + z\frac{\partial F}{\partial Z} + ...\right\}.$$

Again $$\frac{\partial u}{\partial y} = x^n\frac{\partial F}{\partial Y}.\frac{\partial Y}{\partial y} = x^{n-1}\frac{\partial F}{\partial Y},$$

$$\frac{\partial u}{\partial z} = x^n\frac{\partial F}{\partial Z}.\frac{\partial Z}{\partial z} = x^{n-1}\frac{\partial F}{\partial Z},$$

....

$$\therefore \quad x\frac{\partial u}{\partial x} + y\frac{\partial u}{\partial y} + z\frac{\partial u}{\partial z} + ... = nx^n F(Y,Z,...)$$

$$= nu.$$

5.12 To obtain the following Results if u be a Homogenous Function of Two Independent Variables x and y and of Degree n

Statement:

(i) $x\frac{\partial^2 u}{\partial x^2} + y\frac{\partial^2 u}{\partial x \partial y} = (n-1)\frac{\partial u}{\partial x}$.

(ii) $x\frac{\partial^2 u}{\partial x \partial y}+y\frac{\partial^2 u}{\partial y^2}=(n-1)\frac{\partial u}{\partial y}$.

(iii) $x^2\frac{\partial^2 u}{\partial x^2}+2xy\frac{\partial^2 u}{\partial x \partial y}+y^2\frac{\partial^2 u}{\partial y^2}=n(n-1)u$.

Proof: Since u is a homogeneous function of two independent variables x and y and of degree n, therefore by Euler's Theorem, we have

$$x\frac{\partial u}{\partial x}+y\frac{\partial u}{\partial y}=nu. \qquad .....(5.19)$$

Differentiating (5.19) partially with respect to x, (keeping y constant), we get

$$x\frac{\partial^2 u}{\partial x^2}+\frac{\partial u}{\partial x}+y\frac{\partial^2 u}{\partial x \partial y}=n\frac{\partial u}{\partial y}$$

or
$$x\frac{\partial^2 u}{\partial x^2}+y\frac{\partial^2 u}{\partial x \partial y}=(n-1)\frac{\partial u}{\partial x}. \qquad .....(5.20)$$

Hence the first part is proved.

Again differentiating (5.19) partially with respect to y, (treating x as constant), we get

$$x\frac{\partial^2 u}{\partial x \partial y}+y\frac{\partial^2 u}{\partial y^2}+\frac{\partial u}{\partial y}=n\frac{\partial u}{\partial y}$$

or
$$x\frac{\partial^2 u}{\partial x \partial y}+y\frac{\partial^2 u}{\partial y^2}=(n-1)\frac{\partial u}{\partial y}. \qquad .....(5.21)$$

This proves the second part.

Multiplying eq.(5.20) and (5.21) by x and y respectively and adding, we get

$$=(n-1)\left(x\frac{\partial u}{\partial x}+y\frac{\partial u}{\partial y}\right)$$

$$=(n-1).nu, \qquad \text{by eq. (5.19)}$$

Hence
$$\boxed{x^2\frac{\partial^2 u}{\partial x^2}+2xy\frac{\partial^2 u}{\partial x \partial y}+y^2\frac{\partial^2 u}{\partial y^2}=n(n-1)u}$$

Hence the desired result (iii) is obtained.

5.13 Laplace's Equation

The operator ∇^2, viz.,

$$\frac{\partial^2}{\partial x^2}+\frac{\partial^2}{\partial y^2}+\frac{\partial^2}{\partial z^2}$$

plays an important role in Applied Mathematics.

The equation $\nabla^2\ V=0$ is called Laplace's Equation.

5.14 An Interesting Observation

It is evident that the expressions

$$h^2\frac{\partial^2 u}{\partial x^2}+2hk\frac{\partial^2 u}{\partial x\partial y}+k^2\frac{\partial^2 u}{\partial y^2} \text{ and } \left(h\frac{\partial}{\partial x}+k\frac{\partial}{\partial y}\right)^2 u$$

are identical, where h, k are constants.

But it is not so if h, k are functions of x and y.

One must take care to distinguish between

$$x^2\frac{\partial^2 u}{\partial x^2}+2xy\frac{\partial^2 u}{\partial x\partial y}+y^2\frac{\partial^2 u}{\partial y^2} \text{ and } \left(x\frac{\partial}{\partial x}+k\frac{\partial}{\partial y}\right)^2 u.$$

For, $\left(x\frac{\partial}{\partial x}+y\frac{\partial}{\partial y}\right)^2 u$

$$=\left(x\frac{\partial}{\partial x}+y\frac{\partial}{\partial y}\right)\left(x\frac{\partial}{\partial x}+y\frac{\partial}{\partial y}\right)u$$

$$=\left(x\frac{\partial}{\partial x}+y\frac{\partial}{\partial y}\right)\left(x\frac{\partial u}{\partial x}+y\frac{\partial u}{\partial y}\right)$$

$$=\left(x\frac{\partial}{\partial x}+y\frac{\partial}{\partial y}\right)\left(x\frac{\partial u}{\partial x}\right)+\left(x\frac{\partial}{\partial x}+y\frac{\partial}{\partial y}\right)\left(y\frac{\partial u}{\partial y}\right)$$

$$=x\frac{\partial}{\partial x}\left(x\frac{\partial u}{\partial x}\right)+y\frac{\partial}{\partial y}\left(x\frac{\partial u}{\partial x}\right)+x\frac{\partial}{\partial x}\left(y\frac{\partial u}{\partial y}\right)+y\frac{\partial}{\partial y}\left(y\frac{\partial u}{\partial y}\right)$$

$$=x\left(1.\frac{\partial u}{\partial x}+x\frac{\partial^2 u}{\partial x^2}\right)+y\left(x\frac{\partial^2 u}{\partial y\partial x}\right)+x\left(y\frac{\partial^2 u}{\partial x\partial y}\right)+y\left(1.\frac{\partial u}{\partial y}+y\frac{\partial^2 u}{\partial y^2}\right)$$

$$= x^2 \frac{\partial^2 u}{\partial x^2} + 2xy \frac{\partial^2 u}{\partial x \partial y} + y^2 \frac{\partial^2 u}{\partial y^2} + x \frac{\partial u}{\partial x} + y \frac{\partial u}{\partial y},$$

which differs from the former expression

$$x^2 \frac{\partial^2 u}{\partial x^2} + 2xy \frac{\partial^2 u}{\partial x \partial y} + y^2 \frac{\partial^2 u}{\partial y^2}$$

by the addition of the two terms

$$x \frac{\partial u}{\partial x} \text{ and } y \frac{\partial u}{\partial y},$$

$$\text{Hence } x^2 \frac{\partial^2 u}{\partial x^2} + 2xy \frac{\partial^2 u}{\partial x \partial y} + y^2 \frac{\partial^2 u}{\partial y^2} \neq \left(x \frac{\partial}{\partial x} + y \frac{\partial}{\partial y} \right)^2 u$$

5.15 Jacobian

If u, v are two functions of two independent variables x, y then the expression

$$\begin{vmatrix} \frac{\partial u}{\partial x} & \frac{\partial u}{\partial y} \\ \frac{\partial v}{\partial x} & \frac{\partial v}{\partial y} \end{vmatrix}$$

is called the Jacobian or functional determinant of u, v with respect to x, y and it is denoted by

$$\frac{\partial(u, v)}{\partial(x, y)} \text{ or } J\left(\frac{u, v}{x, y}\right)$$

$$\text{Thus } \frac{\partial(u, v)}{\partial(x, y)} = J\left(\frac{u, v}{x, y}\right) = \begin{vmatrix} \frac{\partial u}{\partial x} & \frac{\partial u}{\partial y} \\ \frac{\partial v}{\partial x} & \frac{\partial v}{\partial y} \end{vmatrix}$$

Similarly the Jacobian of u, v, w with respect to x, y, z is

$$\frac{\partial(u,v,w)}{\partial(x,y,z)}=\begin{vmatrix}\frac{\partial u}{\partial x} & \frac{\partial u}{\partial y} & \frac{\partial u}{\partial z}\\ \frac{\partial v}{\partial x} & \frac{\partial v}{\partial y} & \frac{\partial v}{\partial z}\\ \frac{\partial w}{\partial x} & \frac{\partial w}{\partial y} & \frac{\partial w}{\partial z}\end{vmatrix}$$

Illustration: If $x = r\cos\theta$, $y = r\sin\theta$, find $\dfrac{\partial(x,y)}{\partial(r,\theta)}$

Here $\qquad \dfrac{\partial x}{\partial r}=\cos\theta,\quad \dfrac{\partial x}{\partial\theta}=-r\sin\theta,$

$$\frac{\partial y}{\partial r}=\sin\theta,\quad \frac{\partial y}{\partial\theta}=r\cos\theta.$$

$$\therefore \qquad \frac{\partial(x,y)}{\partial(r,\theta)}=\begin{vmatrix}\frac{\partial x}{\partial r} & \frac{\partial x}{\partial\theta}\\ \frac{\partial y}{\partial r} & \frac{\partial y}{\partial\theta}\end{vmatrix}=\begin{vmatrix}\cos\theta & -r\sin\theta\\ \sin\theta & r\cos\theta\end{vmatrix}$$

$$=r\left(\cos^2\theta+\sin^2\theta\right)=r.$$

Worked Out Examples

1. Verify Euler's Theorem for the function

$$u=\frac{x^{\frac{1}{4}}+y^{\frac{1}{4}}}{x^{\frac{1}{5}}+y^{\frac{1}{5}}}$$

Solution: Euler's theorem states that if u is a homogeneous function of the nth degree in x and y then

$$x\frac{\partial u}{\partial x}+y\frac{\partial u}{\partial y}=nu$$

In the problem, $u=\dfrac{x^{\frac{1}{4}}+y^{\frac{1}{4}}}{x^{\frac{1}{5}}+y^{\frac{1}{5}}}=\dfrac{x^{\frac{1}{4}}\left[1+\left(\frac{y}{x}\right)^{\frac{1}{4}}\right]}{x^{\frac{1}{5}}\left[1+\left(\frac{y}{x}\right)^{\frac{1}{5}}\right]}=x^{\frac{1}{4}-\frac{1}{5}}.F\left(\dfrac{y}{x}\right)$

where $F\left(\frac{y}{x}\right)$ denotes some function of $\frac{y}{x}$

$$= x^{\frac{1}{20}}.F\left(\frac{y}{x}\right)$$

Hence u is a homogeneous function of the $\frac{1}{20}$ degree in x and y. So the Euler's theorem will be verified if we can prove by direct differentiation that

$$x\frac{\partial u}{\partial x} + y\frac{\partial u}{\partial y} = \frac{1}{20}.u \quad .....(1)$$

Differentiating $u = \dfrac{x^{\frac{1}{4}} + y^{\frac{1}{4}}}{x^{\frac{1}{5}} + y^{\frac{1}{5}}}$ ......(2)

Dirrerentiating partially w.r.t. x, we get

$$\frac{\partial u}{\partial x} = \frac{\frac{1}{4}x^{-\frac{3}{4}}.\left(x^{\frac{1}{5}} + y^{\frac{1}{5}}\right) - \left(x^{\frac{1}{4}} + y^{\frac{1}{4}}\right).\frac{1}{5}x^{-\frac{4}{5}}}{\left(x^{\frac{1}{5}} + y^{\frac{1}{5}}\right)^2}$$

$$\therefore \quad \frac{\partial u}{\partial x} = \frac{1}{4}\frac{x^{-\frac{3}{4}}}{x^{\frac{1}{5}} + y^{\frac{1}{5}}} - \frac{1}{5}\frac{x^{\frac{1}{4}} + y^{\frac{1}{4}}}{\left(x^{\frac{1}{5}} + y^{\frac{1}{5}}\right)^2}.x^{-\frac{4}{5}}$$

Multiplying both sides by x, we get

$$x\frac{\partial u}{\partial x} = \frac{1}{4}\frac{x^{-\frac{1}{4}}}{x^{\frac{1}{5}} + y^{\frac{1}{5}}} - \frac{1}{5}\frac{x^{\frac{1}{4}} + y^{\frac{1}{4}}}{\left(x^{\frac{1}{5}} + y^{\frac{1}{5}}\right)^2}.x^{-\frac{1}{5}} \quad .....(3)$$

Again differentiating (2) partially with respect to y, we get

$$\frac{\partial u}{\partial y} = \frac{\frac{1}{4}y^{-\frac{3}{4}}\left(x^{\frac{1}{5}} + y^{\frac{1}{5}}\right) - \left(x^{\frac{1}{4}} + y^{\frac{1}{4}}\right)\frac{1}{5}y^{-\frac{4}{5}}}{\left(x^{\frac{1}{5}} + y^{\frac{1}{5}}\right)^2}$$

$$\therefore \quad \frac{\partial u}{\partial y}=\frac{1}{4}\frac{y^{-\frac{3}{4}}}{x^{\frac{1}{5}}+y^{\frac{1}{5}}}-\frac{1}{5}\frac{x^{\frac{1}{4}}+y^{\frac{1}{4}}}{\left(x^{\frac{1}{5}}+y^{\frac{1}{5}}\right)^2}.y^{-\frac{4}{5}}$$

Multiplying both sides by y, we get

$$y\frac{\partial u}{\partial y}=\frac{1}{4}\frac{y^{\frac{1}{4}}}{x^{\frac{1}{5}}+y^{\frac{1}{5}}}-\frac{1}{5}\frac{x^{\frac{1}{4}}+y^{\frac{1}{4}}}{\left(x^{\frac{1}{5}}+y^{\frac{1}{5}}\right)^2}.y^{\frac{1}{5}} \quad .....(4)$$

Adding (3) and (4), we get

$$x\frac{\partial u}{\partial x}+y\frac{\partial u}{\partial y}=\frac{1}{4}\frac{x^{\frac{1}{4}}+y^{\frac{1}{4}}}{x^{\frac{1}{5}}+y^{\frac{1}{5}}}-\frac{1}{5}\frac{x^{\frac{1}{4}}+y^{\frac{1}{4}}}{\left(x^{\frac{1}{5}}+y^{\frac{1}{5}}\right)^2}\left(x^{\frac{1}{5}}+y^{\frac{1}{5}}\right)$$

$$=\frac{1}{4}.u-\frac{1}{5}\frac{x^{\frac{1}{4}}+y^{\frac{1}{4}}}{x^{\frac{1}{5}}+y^{\frac{1}{5}}}=\frac{1}{4}u-\frac{1}{5}u=\frac{5u-4u}{20}$$

$$=\frac{1}{20}u.$$

$\therefore$ (1) and hence Euler's Theorem is verified.

2. If $u=\dfrac{xy}{x+y}$, prove that $x\dfrac{\partial u}{\partial x}+y\dfrac{\partial u}{\partial y}=u.$

Solution: We have $u=\dfrac{xy}{x+y}$ or $u=x\left(\dfrac{\frac{y}{x}}{1+\frac{y}{x}}\right)$

which is a function of the form $x^1 f\left(\dfrac{y}{x}\right)$, that is, a homogeneous function of the first degree in x and y.

Hence, by Euler's Theorem, we have

$$x\frac{\partial u}{\partial x}+y\frac{\partial u}{\partial y}=nu=1.u=u.$$

3. If $u = \sin^{-1}\dfrac{\sqrt{x}-\sqrt{y}}{\sqrt{x}+\sqrt{y}}$, show that $x\dfrac{\partial u}{\partial x}+y\dfrac{\partial u}{\partial y}=0.$

Solution: We have $u = \sin^{-1}\dfrac{\sqrt{x}-\sqrt{y}}{\sqrt{x}+\sqrt{y}}$

or $$\sin u = \frac{\sqrt{x}-\sqrt{y}}{\sqrt{x}+\sqrt{y}} = \frac{\sqrt{x}\left(1-\sqrt{\frac{y}{x}}\right)}{\sqrt{x}\left(1+\sqrt{\frac{y}{x}}\right)}$$

or $$\sin u = x^0\left(\frac{1-\sqrt{\frac{y}{x}}}{1+\sqrt{\frac{y}{x}}}\right) = V(\text{say}) \qquad .....(1)$$

We find that V is a homogeneous function of two independent variables x and y and of degree zero. Hence applying Euler's theorem for the function V, we get

$$x\frac{\partial V}{\partial x}+y\frac{\partial V}{\partial y} = nV = 0.V = 0 \qquad .....(2)$$

From (1), $V = \sin u$.

$$\therefore \qquad \frac{\partial V}{\partial x} = \cos u\frac{\partial u}{\partial x} \text{ and } \frac{\partial V}{\partial y} = \cos u\frac{\partial u}{\partial y}$$

Substituting these values in (2), we get

$$x\cos u\frac{\partial u}{\partial x}+y\cos u\frac{\partial u}{\partial y} = 0 \text{ or } \cos u\left(x\frac{\partial u}{\partial x}+y\frac{\partial u}{\partial y}\right)=0$$

But $$\cos u \neq 0 \text{ as } u \neq \frac{\pi}{2}$$

Hence $$x\frac{\partial u}{\partial x}+y\frac{\partial u}{\partial y}=0$$

4. (a) If $u = \cos^{-1}\dfrac{x+y}{\sqrt{x}+\sqrt{y}}$, show that

$$x\frac{\partial u}{\partial x}+y\frac{\partial u}{\partial y}+\frac{1}{2}\cot u=0$$

***Solution*:** We have $u=\cos^{-1}\dfrac{x+y}{\sqrt{x}+\sqrt{y}}$

or $$\cos u=\frac{x+y}{\sqrt{x}+\sqrt{y}} \quad \text{or} \quad \cos u=\frac{x\left(1+\frac{y}{x}\right)}{\sqrt{x}\left(1+\sqrt{\frac{y}{x}}\right)}$$

or $$\cos u=\frac{x^{\frac{1}{2}}\left(1+\frac{y}{x}\right)}{1+\sqrt{\frac{y}{x}}}=V(\text{say}) \qquad .....(1)$$

We find that V is a homogeneous function of two independent variables x and y and of degree $\frac{1}{2}$. Hence applying Euler's theorem for the function V, we have

$$x\frac{\partial V}{\partial x}+y\frac{\partial V}{\partial y}=nV=\frac{1}{2}V \qquad .....(2)$$

From (1), $V=\cos u.$

$\therefore$ $$\frac{\partial V}{\partial x}=-\sin u\frac{\partial u}{\partial x} \quad \text{and} \quad \frac{\partial V}{\partial y}=-\sin u\frac{\partial u}{\partial y}$$

Substituting these two values in (2), we get

$$-x\sin u\frac{\partial u}{\partial x}-y\sin u\frac{\partial u}{\partial y}=\frac{1}{2}\cos u$$

or $$-\sin u\left(x\frac{\partial u}{\partial x}+y\frac{\partial u}{\partial y}\right)=\frac{1}{2}\cos u$$

or $$x\frac{\partial u}{\partial x}+y\frac{\partial u}{\partial y}=\frac{1}{2}\cot u$$

or $$x\frac{\partial u}{\partial x}+y\frac{\partial u}{\partial y}+\frac{1}{2}\cot u=0$$

4. (b) If $u = \sec^{-1}\dfrac{x^2+y^2}{x-y}$, prove that

$$x\frac{\partial u}{\partial x} + y\frac{\partial u}{\partial y} = \cot u$$

Solution: We have $u = \sec^{-1}\dfrac{x^2+y^2}{x-y}$

or $$\sec u = \frac{x^2+y^2}{x-y} \quad \text{or} \quad \sec u = \frac{x^2\left[1+\left(\frac{y}{x}\right)^2\right]}{x\left(1-\frac{y}{x}\right)}$$

or $$\sec u = \frac{x\left[1+\left(\frac{y}{x}\right)^2\right]}{1-\frac{y}{x}} = V(\text{say}). \qquad .....(1)$$

We find that V is a homogeneous function of two independent variables x and y and of degree 1. Hence applying Euler's theorem for the function V, we get

$$x\frac{\partial u}{\partial x} + y\frac{\partial u}{\partial y} = nv \;\; =1.\ V=V. \qquad .....(2)$$

From (1), $V = \sec u$.

$$\therefore \quad \frac{\partial V}{\partial x} = \sec u \tan u \frac{\partial u}{\partial x} \quad \text{and} \quad \frac{\partial V}{\partial y} = \sec u \tan u \frac{\partial u}{\partial y}$$

Substituting these values in (2), we get

$$x \sec u \tan u \frac{\partial u}{\partial x} + y \sec u \tan u \frac{\partial u}{\partial y} = \sec u$$

or $$\sec u \tan u\left(x\frac{\partial u}{\partial x} + y\frac{\partial u}{\partial y}\right) = \sec u$$

or $$x\frac{\partial u}{\partial x} + y\frac{\partial u}{\partial y} = \cot u.$$

4. (c) If $V = \tan^{-1}\dfrac{x^3+y^3}{x-y}$, then show that [CSVTU, 2007]

$$x\frac{\partial V}{\partial x}+y\frac{\partial V}{\partial y}=\sin 2V$$

Solution: We have $V=\tan^{-1}\dfrac{x^3+y^3}{x-y}$

or $$\tan V=\frac{x^3+y^3}{x-y}=\frac{x^2\left[1+\left(\frac{y}{x}\right)^3\right]}{x\left(1-\frac{y}{x}\right)}$$

or $$\tan V=\frac{x^2\left[1+\left(\frac{y}{x}\right)^3\right]}{1-\frac{y}{x}}=u(\text{say}). \quad .....(1)$$

We find that u is a homogeneous function of two independent variables x and y and of degree two. Hence applying Euler's theorem for the function u, we have

$$x\frac{\partial u}{\partial x}+y\frac{\partial u}{\partial y}=nu=2u. \quad .....(2)$$

From (1), $u=\tan V$.

$$\therefore \quad \frac{\partial u}{\partial x}=\sec^2 V\frac{\partial V}{\partial x} \text{ and } \frac{\partial u}{\partial y}=\sec^2 V\frac{\partial V}{\partial y}$$

Substituting these two values in (2), we get

$$x\sec^2 V\frac{\partial V}{\partial x}+y\sec^2 V\frac{\partial V}{\partial y}=2\tan V$$

or $$\sec^2 V\left(x\frac{\partial V}{\partial x}+y\frac{\partial V}{\partial y}\right)=2\frac{\sin}{\cos}V$$

or $$x\frac{\partial V}{\partial x}+y\frac{\partial V}{\partial y}=2\frac{\sin V}{\cos V}.\cos^2 V=\sin 2V$$

5. If $u=\sin\left(\sqrt{x}+\sqrt{y}\right)$, prove that

$$x\frac{\partial u}{\partial x}+y\frac{\partial u}{\partial y}=\frac{1}{2}\left(\sqrt{x}+\sqrt{y}\right)\cos\left(\sqrt{x}+\sqrt{y}\right)$$

Solution: Since $u = \sin\left(\sqrt{x} + \sqrt{y}\right)$

or $$\sin^{-1} u = \sqrt{x} + \sqrt{y}$$

or $$\sin^{-1} u = \sqrt{x}\left(1 + \sqrt{\frac{y}{x}}\right) = V(\text{say}). \qquad .....(1)$$

We find that V is a homogeneous function of two independent variables x and y and of degree $\frac{1}{2}$. Hence applying Euler's theorem for the function V, we have

$$x\frac{\partial V}{\partial x} + y\frac{\partial V}{\partial y} = nV = \frac{1}{2}V. \qquad .....(2)$$

From (1), $V = \sin^{-1} u.$

$$\therefore \quad \frac{\partial V}{\partial x} = \frac{1}{\sqrt{1-u^2}}\frac{\partial u}{\partial x}$$

and $$\frac{\partial V}{\partial y} = \frac{1}{\sqrt{1-u^2}}\frac{\partial u}{\partial y}.$$

Substituting these two values in (2), we get

$$\frac{x}{\sqrt{1-u^2}}\frac{\partial u}{\partial x} + \frac{y}{\sqrt{1-u^2}}\frac{\partial u}{\partial y} = \frac{1}{2}\sin^{-1} u$$

or $$\frac{1}{\sqrt{1-u^2}}\left(x\frac{\partial u}{\partial x} + y\frac{\partial u}{\partial y}\right) = \frac{1}{2}\left(\sqrt{x} + \sqrt{y}\right)$$

or $$x\frac{\partial u}{\partial x} + y\frac{\partial u}{\partial y} = \frac{1}{2}\left(\sqrt{x} + \sqrt{y}\right)\sqrt{1-u^2}$$

$$= \frac{1}{2}\left(\sqrt{x} + \sqrt{y}\right)\sqrt{1 - \sin^2\left(\sqrt{x} + \sqrt{y}\right)}.$$

Hence $$x\frac{\partial u}{\partial x} + y\frac{\partial u}{\partial y} = \frac{1}{2}\left(\sqrt{x} + \sqrt{y}\right)\cos\left(\sqrt{x} + \sqrt{y}\right).$$

6. If $u = x^2 \sin^{-1}\frac{y}{x} + e^{-\frac{y}{x}}$, prove that

$$x^2u_{xx} + 2xy\,u_{xy} + y^2\,u_{yy} = 2\left(u - e^{-\frac{y}{x}}\right)$$

Solution: Let $X = x^2 \sin^{-1}\dfrac{y}{x}$ and $Y = e^{-\frac{y}{x}}$ so that $u = X + Y$.

As $x^2 \sin^{-1}\dfrac{y}{x}$ is of the form $x^2 f\left(\dfrac{y}{x}\right)$, it is homogeneous of the second degree. So by Euler's theorem

$$x^2\frac{\partial^2 X}{\partial x^2} + 2xy\frac{\partial^2 X}{\partial x \partial y} + y^2\frac{\partial^2 X}{\partial y^2} = 2(2-1)X. \qquad .....(1)$$

Again $e^{-\frac{y}{x}}$ is of the form $x^0 F\left(\dfrac{y}{x}\right)$, so it is homogeneous of the 0 degree. So by Euler's theorem,

$$x^2\frac{\partial^2 Y}{\partial x^2} + 2xy\frac{\partial^2 Y}{\partial x \partial y} + y^2\frac{\partial^2 Y}{\partial y^2} = 0(0-1)Y. \qquad .....(2)$$

Adding (1) and (2),

$$x^2\frac{\partial^2}{\partial x^2}(X+Y) + 2xy\frac{\partial^2}{\partial x \partial y}(X+Y) + y^2\frac{\partial^2}{\partial y^2}(X+Y) = 2X$$

or $$x^2\frac{\partial^2 u}{\partial x^2} + 2xy\frac{\partial^2 u}{\partial x \partial y} + y^2\frac{\partial^2 u}{\partial y^2} = 2(u - Y), \qquad \{\therefore\ X + Y = u\}$$

$\therefore$ $$x^2u_{xx} + 2xy\,u_{xy} + y^2\,u_{yy} = 2\left(u - e^{-\frac{y}{x}}\right)$$

7. Find $\dfrac{dy}{dx}$ by partial differentiation, if $x^y + y^x = a^b$

Solution: Let $f(x, y) \equiv x^y + y^x - a^b = 0$.

$\therefore$ $$\frac{\partial f}{\partial x} = yx^{y-1} + \log_e y.y^x$$

$$\frac{\partial f}{\partial y} = \log_e x.x^y + xy^{x-1}$$

$$\therefore \quad \frac{dy}{dx} = -\frac{\frac{\partial f}{\partial x}}{\frac{\partial f}{\partial y}} = -\frac{yx^{y-1} + \log_e y.y^x}{\log_e x.x^y + xy^{x-1}}$$

8. (a) If $r = \sqrt{x^2 + y^2}$ and $z = \varphi(r)$, show that

$$\frac{\partial^2 z}{\partial x^2} + \frac{\partial^2 z}{\partial y^2} = \frac{1}{r}\varphi'(r) + \varphi''(r).$$

***Solution*:** Here z is a function of r where r is a function of x and y.

$$\therefore \quad \frac{\partial z}{\partial x} = \frac{\partial z}{\partial r}.\frac{\partial r}{\partial x}$$

$$= \varphi'(r)\frac{\partial r}{\partial x}, \qquad \{\because z = \varphi(r)\}. \qquad .....(1)$$

Again $r^2 = x^2 + y^2$;

$$\therefore \quad 2r\frac{\partial r}{\partial x} = 2x \quad \text{and} \quad 2r\frac{\partial r}{\partial y} = 2y.$$

$$\therefore \quad \frac{\partial r}{\partial x} = \frac{x}{r}$$

and $$\frac{\partial r}{\partial y} = \frac{y}{r} \qquad .....(2)$$

$\therefore$ From (1) and (2),

$$\frac{\partial z}{\partial x} = \varphi'(r).\frac{x}{r}.$$

$$\therefore \quad \frac{\partial^2 z}{\partial x^2} = \frac{\partial}{\partial x}\left(\frac{\partial z}{\partial x}\right) = \frac{\partial}{\partial x}\left\{\varphi'(r).\frac{x}{r}\right\} = \frac{\partial}{\partial x}\left\{\frac{\varphi'(r)}{r}.x\right\}$$

$$= \frac{\partial}{\partial x}\left\{\frac{\varphi'(r)}{r}\right\}.x + \frac{\varphi'(r)}{r}.1 = \frac{\partial}{\partial r}\left\{\frac{\varphi'(r)}{r}\right\}\frac{\partial r}{\partial x}.x + \frac{\varphi'(r)}{r}$$

$$= \frac{\varphi''(r).r - \varphi'(r).1}{r^2}.\frac{x}{r}.x + \frac{\varphi'(r)}{r}.$$

$$\therefore \quad \frac{\partial^2 z}{\partial x^2} = \frac{x^2}{r^2}\varphi''(r) - \frac{x^2}{r^2}\varphi'(r) + \frac{\varphi'(r)}{r}.$$

Similarly we get

$$\frac{\partial^2 z}{\partial y^2} = \frac{y^2}{r^2}\varphi''(r) - \frac{y^2}{r^3}\varphi'(r) + \frac{\varphi'(r)}{r}.$$

$$\therefore \quad \frac{\partial^2 z}{\partial x^2} + \frac{\partial^2 z}{\partial y^2} = \frac{x^2+y^2}{r^2}\varphi''(r) - \frac{x^2+y^2}{r^3}\varphi'(r) + 2\frac{\varphi'(r)}{r}$$

$$= \varphi''(r) + \frac{1}{r}\varphi'(r), \qquad \{\because x^2+y^2 = r^2\}.$$

8. (b) If w = f(x, y), x = r cos θ, y = r sin θ show that

$$\left(\frac{\partial w}{\partial r}\right)^2 + \frac{1}{r^2}\left(\frac{\partial w}{\partial \theta}\right)^2 = \left(\frac{\partial f}{\partial x}\right)^2 + \left(\frac{\partial f}{\partial y}\right)^2$$

Solution: As w = f(x, y) where x, y are functions of r, θ we get

$$\frac{\partial w}{\partial r} = \frac{\partial f}{\partial x}\cdot\frac{\partial x}{\partial r} + \frac{\partial f}{\partial y}\cdot\frac{\partial y}{\partial r}, \qquad .....(1)$$

and
$$\frac{\partial w}{\partial \theta} = \frac{\partial f}{\partial x}\cdot\frac{\partial x}{\partial \theta} + \frac{\partial f}{\partial y}\cdot\frac{\partial y}{\partial \theta}. \qquad .....(2)$$

But x = r cos θ, y = r sin θ

$$\left.\begin{aligned} \therefore \quad & \frac{\partial x}{\partial r} = \cos\theta, \quad \frac{\partial r}{\partial \theta} = -r\sin\theta \\ \text{and} \quad & \frac{\partial y}{\partial r} = \sin\theta, \quad \frac{\partial y}{\partial \theta} = r\cos\theta. \end{aligned}\right\} \qquad .....(3)$$

From (1), (2), (3) we get

$$\frac{\partial w}{\partial r} = \frac{\partial f}{\partial x}.\cos\theta + \frac{\partial f}{\partial y}\sin\theta$$

and
$$\frac{\partial w}{\partial \theta} = \frac{\partial f}{\partial x}.(-r\sin\theta) + \frac{\partial f}{\partial y}(r\cos\theta)$$

$$\therefore \quad \text{L.H.S} = \left(\frac{\partial f}{\partial x}\cos\theta + \frac{\partial f}{\partial y}\sin\theta\right)^2 + \frac{1}{r^2}.r^2\left(-\frac{\partial f}{\partial x}\sin\theta + \frac{\partial f}{\partial y}\cos\theta\right)^2$$

$$= \left(\frac{\partial f}{\partial x}\right)^2\left(\cos^2\theta + \sin^2\theta\right) + \left(\frac{\partial f}{\partial y}\right)^2\left(\sin^2\theta + \cos^2\theta\right)$$

$$+2\frac{\partial f}{\partial x}.\frac{\partial f}{\partial y}\cos\theta.\sin\theta-2\frac{\partial f}{\partial x}.\frac{\partial f}{\partial y}\cos\theta\sin\theta$$

$$=\left(\frac{\partial f}{\partial x}\right)^2+\left(\frac{\partial f}{\partial y}\right)^2$$

$$= \text{R.H.S}$$

8. (c) If $u = F(x-y, y-z, z-x)$

and $\frac{\partial u}{\partial x},\frac{\partial u}{\partial y},\frac{\partial u}{\partial z}$ all exist, prove that $\frac{\partial u}{\partial x}+\frac{\partial u}{\partial y}+\frac{\partial u}{\partial z}=0.$

Solution: We have $u = F(x-y, y-z, z-x)$.

Let $X = x-y,\ Y = y-z,\quad \text{and}\quad Z = z-x.$

$$\therefore\quad \frac{\partial X}{\partial x}=1,\ \frac{\partial X}{\partial y}=-1,\ \frac{\partial X}{\partial z}=0.$$

$$\frac{\partial Y}{\partial x}=0,\ \frac{\partial Y}{\partial y}=1,\ \frac{\partial Y}{\partial z}=-1,$$

$$\frac{\partial Z}{\partial x}=-1,\ \frac{\partial Z}{\partial y}=0,\ \frac{\partial Z}{\partial z}=1. \qquad .....(1)$$

transforms into

$$u = F(X, Y, Z)$$

$$\therefore\quad \frac{\partial u}{\partial x}=\frac{\partial F}{\partial X}.\frac{\partial X}{\partial x}+\frac{\partial F}{\partial Y}.\frac{\partial Y}{\partial x}+\frac{\partial F}{\partial Z}.\frac{\partial Z}{\partial x}$$

$$=\frac{\partial F}{\partial X}.1+\frac{\partial F}{\partial Y}.0+\frac{\partial F}{\partial Z}.(-1)$$

$$=\frac{\partial F}{\partial X}-\frac{\partial F}{\partial Z}. \qquad .....(2)$$

$$\text{Again,}\quad \frac{\partial u}{\partial y}=\frac{\partial F}{\partial X}.\frac{\partial X}{\partial y}+\frac{\partial F}{\partial Y}.\frac{\partial Y}{\partial y}+\frac{\partial F}{\partial Z}.\frac{\partial Z}{\partial y}$$

$$=\frac{\partial F}{\partial X}.(-1)+\frac{\partial F}{\partial Y}.1+\frac{\partial F}{\partial Z}.0$$

$$= \frac{\partial F}{\partial Y} - \frac{\partial F}{\partial X}. \qquad .....(3)$$

Also, $\frac{\partial}{\partial z} = \frac{\partial F}{\partial X}.\frac{\partial X}{\partial z} + \frac{\partial F}{\partial Y}.\frac{\partial Y}{\partial z} + \frac{\partial F}{\partial Z}.\frac{\partial Z}{\partial z}$

$$= \frac{\partial F}{\partial X}.0 + \frac{\partial F}{\partial Y}.(-1) + \frac{\partial F}{\partial Z}.1 = \frac{\partial F}{\partial Z} - \frac{\partial F}{\partial Y}. \qquad .....(4)$$

Adding (2), (3) and (4), we get

$$\frac{\partial u}{\partial x} + \frac{\partial u}{\partial y} + \frac{\partial u}{\partial z} = 0.$$

9. If u be a homogeneous function of the nth degree in x, y, z and if u = f (X, Y, Z), where X, Y, Z are the first differential co-efficients of u with regard to x, y, z respectively, prove that

$$X\frac{\partial f}{\partial X} + Y\frac{\partial f}{\partial Y} + Z\frac{\partial f}{\partial Z} = \frac{n}{n-1}u.$$

Solution: Since u is a homogeneous function of the nth degree in three independent variables x, y and z, therefore, by Euler's Theorem,

we have $x.\frac{\partial u}{\partial x} + y.\frac{\partial u}{\partial y} + z.\frac{\partial u}{\partial z} = nu. \qquad .....(1)$

Also $u = f(X, Y, Z)$,

where $X = \frac{\partial u}{\partial x}, Y = \frac{\partial u}{\partial y}$ and $Z = \frac{\partial u}{\partial z}$.

$$\therefore \quad \frac{\partial u}{\partial x} = \frac{\partial f}{\partial X}.\frac{\partial X}{\partial x} + \frac{\partial f}{\partial Y}.\frac{\partial Y}{\partial x} + \frac{\partial f}{\partial Z}.\frac{\partial Z}{\partial x}. \qquad .....(2)$$

Again we have

$$\frac{\partial u}{\partial y} = \frac{\partial f}{\partial X}.\frac{\partial X}{\partial y} + \frac{\partial f}{\partial Y}.\frac{\partial Y}{\partial y} + \frac{\partial f}{\partial Z}.\frac{\partial Z}{\partial y}. \qquad .....(3)$$

Also $\frac{\partial u}{\partial z} = \frac{\partial f}{\partial X}.\frac{\partial X}{\partial z} + \frac{\partial f}{\partial Y}.\frac{\partial Y}{\partial z} + \frac{\partial f}{\partial Z}.\frac{\partial Z}{\partial z}. \qquad .....(4)$

Multiplying (2), (3) and (4) by x, y and z respectively and adding, we get

$$x\frac{\partial u}{\partial x}+y\frac{\partial u}{\partial y}+z\frac{\partial u}{\partial z}$$

$$=\frac{\partial f}{\partial X}\left[x\frac{\partial}{\partial x}\left(\frac{\partial u}{\partial x}\right)+y\frac{\partial}{\partial y}\left(\frac{\partial u}{\partial x}\right)+z\frac{\partial}{\partial z}\left(\frac{\partial u}{\partial x}\right)\right]$$

$$+\frac{\partial f}{\partial Y}\left[x\frac{\partial}{\partial x}\left(\frac{\partial u}{\partial y}\right)+y\frac{\partial}{\partial y}\left(\frac{\partial u}{\partial y}\right)+z\frac{\partial}{\partial z}\left(\frac{\partial u}{\partial y}\right)\right]$$

$$+\frac{\partial f}{\partial Z}\left[x\frac{\partial}{\partial x}\left(\frac{\partial u}{\partial z}\right)+y\frac{\partial}{\partial y}\left(\frac{\partial u}{\partial z}\right)+z\frac{\partial}{\partial z}\left(\frac{\partial u}{\partial z}\right)\right]$$

or $$nu=\frac{\partial f}{\partial X}\left(x\frac{\partial^2 u}{\partial x^2}+y\frac{\partial^2 u}{\partial x\partial y}+z\frac{\partial^2 u}{\partial x\partial z}\right)$$

$$+\frac{\partial f}{\partial Y}\left(x\frac{\partial^2 u}{\partial y\partial x}+y\frac{\partial^2 u}{\partial y^2}+z\frac{\partial^2 u}{\partial y\partial z}\right)$$

$$+\frac{\partial f}{\partial Z}\left(x\frac{\partial^2 u}{\partial z\partial x}+y\frac{\partial^2 u}{\partial z\partial y}+z\frac{\partial^2 u}{\partial z^2}\right). \quad .....(5)$$

Differentiating (1) partially with respect to x, (keeping y and z constant), we get

$$x\frac{\partial^2 u}{\partial x^2}+\frac{\partial u}{\partial x}+y\frac{\partial^2 u}{\partial x\partial y}+z\frac{\partial^2 u}{\partial x\partial z}=n\frac{\partial u}{\partial x}$$

or $$x\frac{\partial^2 u}{\partial x^2}+y\frac{\partial^2 u}{\partial x\partial y}+z\frac{\partial^2 u}{\partial x\partial z}=(n-1)\frac{\partial u}{\partial x}$$

$$=(n-1)X.$$

Similarly differentiating (1) partially with respect to y and z separately, we get

$$x\frac{\partial^2 u}{\partial y\partial x}+y\frac{\partial^2 u}{\partial y^2}+z\frac{\partial^2 u}{\partial y\partial z}=(n-1)Y$$

and $$x\frac{\partial^2 u}{\partial z\partial x}+y\frac{\partial^2 u}{\partial z\partial y}+z\frac{\partial^2 u}{\partial z^2}=(n-1)Z.$$

Substituting the above values in (5), we get

$$nu=(n-1)\left(X\frac{\partial f}{\partial X}+Y\frac{\partial f}{\partial Y}+Z\frac{\partial f}{\partial Z}\right).$$

Hence $$X\frac{\partial f}{\partial X}+Y\frac{\partial f}{\partial Y}+Z\frac{\partial f}{\partial Z}=\frac{n}{n-1}u.$$

10. If A, B, C be the angles of a triangle such that

$$\sin^2 A+\sin^2 B+\sin^2 C=\text{constant},$$

prove that

$$\frac{dA}{dB}=\frac{\tan C-\tan B}{\tan A-\tan C}.$$

Solution: We know that in a triangle,

$$A+B+C=\pi.$$

Differentiating with respect to B, we get

$$\frac{dA}{dB}+1+\frac{dC}{dB}=0$$

or $$\frac{dC}{dB}=-\left(1+\frac{dA}{dB}\right). \qquad .....(1)$$

By question, $\sin^2 A+\sin^2 B+\sin^2 C=\text{constant}.$

Differentiating with respect to B, we get

$$\sin 2A.\frac{dA}{dB}+\sin 2B+\sin 2C.\frac{dC}{dB}=0.$$

Using (1), we get

$$\sin 2A.\frac{dA}{dB}+\sin 2B-\sin 2C\left(1+\frac{dA}{dB}\right)=0$$

or $$\frac{dA}{dB}(\sin 2A-\sin 2C)=\sin 2C-\sin 2B$$

or $$\frac{dA}{dB}\left[2\cos\frac{2(A+C)}{2}\sin\frac{2(A-C)}{2}\right]$$

$$=2\cos\frac{2(C+B)}{2}\sin\frac{2(C-B)}{2}$$

or $$\frac{dA}{dB}.\cos(A+C)\sin(A-C)=\cos(C+B)\sin(C-B)$$

or $$\frac{dA}{dB}.\cos(\pi-B)(\sin A\cos C-\cos A\sin C)$$
$$=\cos(\pi-A)(\sin C\cos B-\cos C\sin B)$$

or $$-\frac{dA}{dB}\cos B(\sin A\cos C-\cos A\sin C)$$
$$=-\cos A(\sin C\cos B-\cos C\sin B)$$

Dividing both sides by cos A cos B cos C, we get

$$\frac{dA}{dB}(\tan A-\tan C)=\tan C-\tan B.$$

Hence $$\frac{dA}{dB}=\frac{\tan C-\tan B}{\tan A-\tan C}.$$

Formulae at a Glance

1. $$du=\frac{\partial u}{\partial x}dx+\frac{\partial u}{\partial y}dy.$$

2. $$du=\frac{\partial u}{\partial x}dx+\frac{\partial u}{\partial y}dy+\frac{\partial u}{\partial z}dz.$$

3. $$du=\frac{\partial u}{\partial x_1}dx_1+\frac{\partial u}{\partial x_2}dx_2+\frac{\partial u}{\partial x_3}dx_3+\ldots+\frac{\partial u}{\partial x_n}dx_n.$$

4. $$\frac{du}{dx}=\frac{\partial u}{\partial x_1}.\frac{dx_1}{dx}+\frac{\partial u}{\partial x_2}.\frac{dx_2}{dx}.$$

5. $$\frac{du}{dx}=\frac{\partial u}{\partial x_1}.\frac{dx_1}{dx}+\frac{\partial u}{\partial x_2}.\frac{dx_2}{dx}+\frac{\partial u}{\partial x_3}.\frac{dx_3}{dx}+\ldots+\frac{\partial u}{\partial x_n}.\frac{dx_n}{dx}$$

6. $$\frac{\partial u}{\partial x}=\frac{\partial u}{\partial x_1}.\frac{\partial x_1}{\partial x}+\frac{\partial u}{\partial x_2}.\frac{\partial x_2}{\partial x}+\ldots+\frac{\partial u}{\partial x_n}.\frac{\partial x_n}{\partial x}.$$

7. $$\frac{\partial u}{\partial y}=\frac{\partial u}{\partial x_1}.\frac{\partial x_1}{\partial y}+\frac{\partial u}{\partial x_2}.\frac{\partial x_2}{\partial y}+\ldots+\frac{\partial u}{\partial x_n}.\frac{\partial x_n}{\partial y}.$$

8. If y be a function of x, then $\dfrac{du}{dx}=\dfrac{\partial u}{\partial x}+\dfrac{\partial u}{\partial y}.\dfrac{dy}{dx}.$

9. $$\frac{dy}{dx} = -\frac{\frac{\partial f}{\partial x}}{\frac{\partial f}{\partial y}}$$

10. Euler's Theorem: $$x\frac{\partial u}{\partial x} + y\frac{\partial u}{\partial y} = nu.$$

$$x\frac{\partial u}{\partial x} + y\frac{\partial u}{\partial y} + z\frac{\partial u}{\partial z} = nu.$$

11. $$x^2\frac{\partial^2 u}{\partial x^2} + 2xy\frac{\partial^2 u}{\partial x \partial y} + y^3\frac{\partial^2 u}{\partial y^2} = n(n-1)u.$$

12. $$\nabla^2 \equiv \frac{\partial^2}{\partial x^2} + \frac{\partial^2}{\partial y^2} + \frac{\partial^2}{\partial z^2}.$$

Exercise 5.2

1. (a) Verify Euler's Theorem in the following cases –

(i) $f(x,y) = ax^2 + 2hxy + by^2$.

(ii) $u = x^2 \cos^{-1}\left(\frac{y}{2x}\right)$.

(iii) $u = x^n \sin\frac{y}{x}$.

(iv) $u = \frac{x(x^3 - y^3)}{x^3 + y^3}$.

(v) $u = x^3.\log\frac{y}{x}$.

(vi) $u = \sin^{-1}\left(\frac{x}{y}\right) + \tan^{-1}\left(\frac{y}{x}\right)$.

(b) If $u = \sin\sqrt{\frac{x-y}{x+y}}$, prove that the Euler's formula

$$x\frac{\partial u}{\partial x} + y\frac{\partial u}{\partial y} = nu \text{ holds good.}$$

2. (i) If $u = \frac{x-y}{x+y}$, prove that $x\frac{\partial u}{\partial x} + y\frac{\partial u}{\partial y} = 0.$

(ii) If $u = x^3 + y^3 + z^3 - 3xyz$, show that $x\frac{\partial u}{\partial x} + y\frac{\partial u}{\partial y} + z\frac{\partial u}{\partial z} = 3u$.

(iii) If $f(x, y, z) = 3x^2yz + 5xy^2z + 4z^4$, prove that

$$x\frac{\partial f}{\partial x} + y\frac{\partial f}{\partial y} + z\frac{\partial f}{\partial z} = 4f(x, y, z).$$

3. (i) If $u = \sin^{-1}\frac{x - y}{x + y}$, show that $x\frac{\partial u}{\partial x} + y\frac{\partial u}{\partial y} = 0$.

(ii) If $u = \cos^{-1}\frac{x - y}{x + y}$, show that $x.u_x + y.u_y = 0$.

4. (i) Prove that $x\frac{\partial u}{\partial x} + y\frac{\partial u}{\partial y} = \frac{1}{2}\tan u$,

when $u = \sin^{-1}\left(\frac{x + y}{\sqrt{x} + \sqrt{y}}\right)$.

(ii) If $V = \tan^{-1}\frac{x^2 + y^2}{x - y}$, prove that $x\frac{\partial V}{\partial x} + y\frac{\partial V}{\partial y} = \frac{1}{2}\sin 2V$.

(iii) If $u = \tan^{-1}\frac{x^2 + y^2}{x + y}$, show that $x\frac{\partial u}{\partial x} + y\frac{\partial u}{\partial y} = \frac{1}{2}\sin 2u$.

(iv) If $u = \sin^{-1}\frac{x^2 + y^2}{x + y}$, show that $x\frac{\partial u}{\partial x} + y\frac{\partial u}{\partial y} = \tan u$.

5. (i) If $\sin u = \frac{x^3 + y^3}{x - y}$, show that $x\frac{\partial u}{\partial x} + y\frac{\partial u}{\partial y} = 2\tan u$.

(ii) If $\sec u = \frac{x^3 + y^3}{x + y}$, show that $x\frac{\partial u}{\partial x} + y\frac{\partial u}{\partial y} = 2\cot u$.

6. If $u = x\,\varphi\left(\frac{y}{x}\right) + \psi\left(\frac{y}{x}\right)$, prove that

$$x^2\frac{\partial^2 u}{\partial x^2} + 2xy\frac{\partial^2 u}{\partial x \partial y} + y^2\frac{\partial^2 u}{\partial y^2} = 0.$$

7. Find $\frac{dy}{dx}$ using partial differentiation of

(i) $x^2 + y^2 = 3xy$, **(ii)** $(\cos y)^x = (\sin x)^y$.

8. If $x = r\cos\theta$, $y = r\sin\theta$, prove that

$$\left(\frac{dx}{dr}\right)^2 + \left(\frac{dy}{dr}\right)^2 = 1 + r^2\left(\frac{d\theta}{dr}\right)^2,$$

$$\left\{\text{or } (dx)^2 + (dy)^2 = (dr)^2 + (r\,d\theta)^2.\right\}$$

9. If $u = f(z)$, z being a homogeneous function of degree n in x, y, prove that

$$x\frac{\partial u}{\partial x} + y\frac{\partial u}{\partial y} = nu\frac{du}{dz}.$$

10. (i) If the sides of a plane triangle ABC vary in such a way that its circum-radius remains constant, prove that

$$\frac{\delta a}{\cos A} + \frac{\delta b}{\cos B} + \frac{\delta c}{\cos C} = 0.$$

[*Hint.* $\frac{a}{\sin A} = 2R.$ $\because$ $\delta a = 2R\cos A\,\delta A.$

or $\frac{\delta a}{\cos A} = 2R\delta A.$

Remember $A + B + C = 180^\circ$

or $\delta A + \delta B + \delta C = 0.$]

(ii) In a plane triangle ABC, if the sides a, b be constant, prove that the variations of its base angles are connected by the equation

$$\frac{dA}{\sqrt{a^2 - b^2\sin^2 A}} = \frac{dB}{\sqrt{b^2 - a^2\sin^2 B}}$$

[*Hint.* $\because$ $\frac{\sin A}{a} = \frac{\sin B}{b}$

$$\because \qquad \frac{\cos A dA}{a} = \frac{\cos B dB}{b}, \qquad \text{as a, b are constants]}$$

(iii) The side c of a triangle ABC is determined by measuring the sides a, b and the angle C.

If the possible errors of measurement are δa, δb and δC respectively, show that

$$\delta c = \cos B.\ \delta a + \cos A.\ \delta b + a \sin B.\ \delta C.$$

[*Hint.* $c^2 = a^2 + b^2 - 2ab \cos C$.

$$\because \qquad 2c\delta c = 2a\delta a + 2b\delta b - 2b \cos C\ \delta a - 2a \cos C\ \delta b + 2ab \sin C\ \delta C.$$

or $\qquad c\delta c = (a - b\cos C) + (b - a\cos C) + ab\sin C\ \delta C.$

Use $\qquad a = b\cos C + c\cos B,\ \ b = a\cos C + c\cos A$

and $\qquad \frac{b}{c} = \frac{\sin B}{\sin C}$.]

(iv) If the three sides a, b, c of a triangle ABC are measured, prove that the error in the angle A, due to small errors in the sides, is given by

$$\delta A = \frac{\sin A}{\sin B \sin C}.\frac{\delta a}{a} - \cot C\frac{\delta b}{b} - \cot B\frac{\delta c}{c}.$$

[*Hint.* Differentiate

$$\cos A = \frac{b^2 + c^2 - a^2}{2bc}$$

$$-2 \sin A\delta A = \frac{2(b\delta b + c\delta c - a\delta a)(bc) - (b^2 + c^2 - a^2)(b\delta c + c\delta b)}{b^2c^2}$$

$$= \frac{2(b\delta b + c\delta c - a\delta a)}{bc} - \frac{2bc\cos A(b\delta c + c\delta b)}{b^2c^2}$$

Use $b = a \cos C + c \cos A$, $c = b \cos A + a \cos B$.]

5.16 Exact Differential

Definition. If an expression of the type M dx + N dy, (M and N are functions of x and y or constants), may be reduced to du, where u is a function of x and y, then M dx + N dy is said to be an exact differential.

5.17 An Important Theorem

To prove that the necessary and sufficient condition that the expression M dx + N dy should be an exact differential is

$$\frac{\partial M}{\partial y} = \frac{\partial N}{\partial x}.$$

Proof: The given expression is

$$M\,dx + N\,dy. \qquad .....(5.22)$$

Let this expression be an exact differential and let it be equal to du.

Then $$du = \frac{\partial u}{\partial x}dx + \frac{\partial u}{\partial y}dy.$$

$\because$ $$M\,dx + N\,dy = \frac{\partial u}{\partial x}dx + \frac{\partial u}{\partial y}dy.$$

Equating the coefficients of dx and dy separately from both sides, we get

$$M = \frac{\partial u}{\partial x} \quad \text{and} \quad N = \frac{\partial u}{\partial y}.$$

$\because$ $$\frac{\partial M}{\partial y} = \frac{\partial^2 u}{\partial y \partial x} \quad \text{and} \quad \frac{\partial N}{\partial x} = \frac{\partial^2 u}{\partial x \partial y}$$

Assuming that, under certain conditions,

$$\frac{\partial^2 u}{\partial y \partial x} = \frac{\partial^2 u}{\partial x \partial y} \quad \text{or} \quad \frac{\partial M}{\partial y} = \frac{\partial N}{\partial x} \qquad .....(5.23)$$

This is the necessary condition.

Conversely, we can show that, if the condition (5.23) holds, the expression eq.(5.22) will be an exact differential.

Let y = constant, then dy = 0.

$\because$ eq.(5.22) becomes Mdx.

Let $$v = \int M\,dx. \quad \text{or} \quad \frac{\partial v}{\partial x} = M. \qquad .....(5.24)$$

$\therefore$ From (5.23), $$\frac{\partial N}{\partial x}-\frac{\partial M}{\partial y}=\frac{\partial^2 y}{\partial x \partial y}$$

or $$\frac{\partial N}{\partial x}-\frac{\partial^2 v}{\partial x \partial y}=0 \quad \text{or} \quad \frac{\partial}{\partial x}\left(N-\frac{\partial v}{\partial y}\right)=0$$

This shows that the function $N-\frac{\partial v}{\partial y}$ is constant so far as x is concerned and is therefore a function of y only. Denoting its value by f (y), we have

$$N-\frac{\partial v}{\partial y}=f'(y) \quad \text{or} \quad N=\frac{\partial v}{\partial y}+f'(y) \qquad .....(5.25)$$

Hence, if we write $u = v + f(y)$,

$\because$ $$\frac{\partial u}{\partial x}=\frac{\partial v}{\partial x} \quad \text{and} \quad \frac{\partial u}{\partial y}=\frac{\partial v}{\partial y}+f'(y)$$

then, by eq.(5.24) and (5.25), we get

$$\frac{\partial u}{\partial x}=\frac{\partial v}{\partial x}=M \quad \text{and} \quad \frac{\partial u}{\partial y}=\frac{\partial v}{\partial y}+f'(y)=N.$$

$\because$ $$M\,dx+N\,dy=\frac{\partial u}{\partial x}dx+\frac{\partial u}{\partial y}dy$$

or $$M\,dx+N\,dy=du,$$

which is an exact differential.

Hence the sufficient condition is also proved.

Worked Out Examples

1. (a) Prove that the expression

$$(2x+y-3)dx+(x-4y+1)dy$$

is an exact differential.

Solution: The given expression

$$=(2x+y-3)dx+(x-4y+1)dy$$

Here $$M=2x+y-3 \qquad .....(1)$$

and $\quad N = x - 4y + 1.$ .....(2)

Differentiating (1) partially with respect to y, (keeping x constant), we get.

$$\frac{\partial M}{\partial y} = 1. \qquad .....(3)$$

Now differentiating (2) partially with respect to x, (keeping y constant), we get

$$\frac{\partial N}{\partial x} = 1. \qquad .....(4)$$

From (3) and (4), we observe that $\frac{\partial M}{\partial y} = \frac{\partial N}{\partial x}$,

which is the necessary and sufficient condition for the given expression to be an exact differential. Hence the given expression is an exact differential.

(b) Prove that $\frac{\sin hy\, dx - \sin x\, dy}{\cos hy - \cos x}$ is an exact differential.

***Solution*:** The given expression

$$= \frac{\sinh y\, dx - \sin x\, dy}{\cos hy - \cos x} = \frac{\sin hy}{\cos hy - \cos x} dx + \frac{-\sin x}{\cos hy - \cos x} dy$$

Hence $\quad M = \frac{\sin hy}{\cos hy - \cos x},$ .....(1)

and $\quad N = \frac{-\sin x}{\cos hy - \cos x}.$ .....(2)

Differentiating (1) partially with respect to y, (keeping x constant), we get

$$\frac{\partial M}{\partial y} = \frac{\cos hy\left(\cos hy - \cos x\right) - \sin hy.\, \sin hy}{\left(\cos hy - \cos x\right)^2}$$

$$= \frac{\cos h^2 y - \sin h^2 y - \cos hy \cos x}{\left(\cos hy - \cos x\right)^2} = \frac{1 - \cos hy \cos x}{\left(\cos hy - \cos x\right)^2}. \qquad .....(3)$$

Differentiating (2) partially with respect to x (keeping y constant), we get

$$\frac{\partial N}{\partial x} = \frac{-\cos x\left(\cos hy - \cos x\right) + \sin x.\, \sin x}{\left(\cos hy - \cos x\right)^2}$$

$$= \frac{-\cos x \cos hy + \cos^2 x + \sin^2 x}{(\cos hy - \cos x)^2}$$

$$= \frac{1 - \cos hy \cos x}{(\cos hy - \cos x)^2}. \qquad .....(4)$$

From (3) and (4), we get that $\frac{\partial M}{\partial y} = \frac{\partial N}{\partial x}$

which is the necessary and sufficient condition for the given expression to be an exact differential.

Hence the given expression $\frac{\sin hy\, dx - \sin x\, dy}{\cos hy - \cos x}$

is an exact differential

2. Prove that $y(1+xy)dx - x\,dy$ is not an exact differential but

$\frac{1}{y^2}[y(1+xy)dx - x\,dy]$ is an exact differential.

Solution: For $y(1+xy)dx - x\,dy$,

$M = y(1+xy),\ N = -x.$

Now $\frac{\partial M}{\partial y} = 1 + xy + y(0+x) = 1 + 2xy$

$\frac{\partial N}{\partial x} = -1;\quad \therefore \quad \frac{\partial M}{\partial y} \neq \frac{\partial N}{\partial x}$, so it is not exact differential

For $\frac{1}{y^2}[y(1+xy)dx - x\,dy]$,

$M \equiv \frac{1+xy}{y},\ N = -\frac{x}{y^2}.\quad \therefore \quad \frac{\partial M}{\partial y} = \frac{xy - (1+xy).1}{y^2} = \frac{-1}{y^2}.$

$\frac{\partial N}{\partial x} = -\frac{1}{y^2};\quad \therefore \quad \frac{\partial M}{\partial y} = \frac{\partial N}{\partial x}.$

Hence $\frac{1}{y^2}[y(1+xy)dx - x\,dy]$ is an exact differential.

Formulae at a Glance

1. The necessary and sufficient condition that the expression

$$M\,dx + N\,dy$$

should be an exact differential is

$$\frac{\partial M}{\partial y} = \frac{\partial N}{\partial x}.$$

Exercise 5.3

1. Prove that the following are exact (that is prefect differential)
   (i) $y\,dx + x\,dy$.
   (ii) $(2x+3y)dx + (3x-4y)dy$.
   (iii) $\sin y \cos x\; dx + \sin x \cos y\; dy$.
   (iv) $\sin x \sin y\; dx - \cos x \cos y\; dy$.
   (v) $ye^x dx + e^x dy$.
   (vi) $(2x+3y-5)dx + (3x-2y-1)dy$.
   (vii) $(ax+hy+g)dx + (hx+by+f)dy$.
2. Show that, though $y\,dx - x\,dy$ is not a perfect differential, $\frac{1}{y^2}(y\,dx - x\,dy)$ is a perfect differential.
3. Show that $(x^2+y^2)dx - 2xy\,dy$ is not exact differential in x, y but substituting $y = vx$ it can be put in the form $x^2 d\varphi$ where φ is a function of x and v.
   [*Hint.* $dy = x\,dv + v\,dx$; put for dy in the given expression and then verify.]
4. If $Pdx + Qdy + Rdz$ can be made a perfect differential of some function of x, y, z by multiplying each term by a common factor, prove that

$$P\left(\frac{\partial Q}{\partial z} - \frac{\partial R}{\partial y}\right) + Q\left(\frac{\partial R}{\partial x} - \frac{\partial P}{\partial z}\right) + R\left(\frac{\partial P}{\partial y} - \frac{\partial Q}{\partial x}\right) = 0.$$

[*Hint.* Let $du = X(Pdx + Qdy + Rdz)$

$$= \frac{\partial u}{\partial x}dx + \frac{\partial u}{\partial y}dy + \frac{\partial u}{\partial z}dz.$$

$$\therefore \quad XP = \frac{\partial u}{\partial x};\; XQ = \frac{\partial u}{\partial y};\; XR = \frac{\partial u}{\partial z},$$

$$\therefore \quad P\frac{\partial X}{\partial y} + X\frac{\partial P}{\partial y} = \frac{\partial^2 u}{\partial x \partial y} = Q\frac{\partial X}{\partial x} + X\frac{\partial Q}{\partial x}.]$$

5.18 Maxima and Minima of Functions of Two Variables

The function F(x, y) is maximum at (x, y) if for all small positive or negative values of h and k; we have

$$f(x+h, y+k) - F(x, y) < 0 \qquad ..... (5.26)$$

Similarly, F(x, y) is minimum at (x, y) if for all small positive or negative values of h and k; we have

$$F(x+h, y+k) - F(x, y) > 0 \qquad .....(5.27)$$

Thus if F(x, y) has a minimum or maximum value then [F(x + h, y + k) –F(x, y)] keeps the same sign for all small positive or negative values of h and k.

Now using Taylor's expansion, we get

$$F(x+h, y+k) - F(x, y) = \left(h\frac{\partial F}{\partial x} + k\frac{\partial F}{\partial y}\right)$$

$$+\frac{1}{2!}\left(h^2\frac{\partial^2 F}{\partial x^2} + 2hk\frac{\partial^2 F}{\partial x\partial y} + k^2\frac{\partial^2 F}{\partial y^2}\right) \qquad .....(5.28)$$

Assuming h, k to be sufficiently small; the sign of the expression on the left can be made to depend on that of $h\frac{\partial F}{\partial x} + k\frac{\partial F}{\partial y}$

Hence the necessary condition for f(x, y) to be a maximum or minimum is that

$$h\frac{\partial F}{\partial x} + k\frac{\partial F}{\partial y} = 0 \quad \forall \text{ h and k}$$

$$\therefore \quad \frac{\partial F}{\partial x} = 0 \text{ and } \frac{\partial F}{\partial y} = 0$$

i.e., $p = 0$ and $q = 0$ where $p = \frac{\partial F}{\partial x}$, $q = \frac{\partial F}{\partial y}$

then (5.28) $\Rightarrow$ $$f(x+h, y+k) - f(x, y) = \frac{1}{2!}\left[h^2\frac{\partial^2 F}{\partial x^2} + 2hk\frac{\partial^2 F}{\partial x\partial y} + k^2\frac{\partial^2 F}{\partial y^2}\right] + ...$$

$$= \frac{1}{2!}[rh^2 + 2shk + tk^2]$$

$$\left(\text{where } r = \frac{\partial^2 F}{\partial x^2}, \quad s = \frac{\partial^2 F}{\partial x\partial y}, \quad t = \frac{\partial^2 F}{\partial y^2}\right)$$

$$=\frac{r}{2!}\left[h^2+2hk\frac{s}{r}+\frac{t}{r}k^2\right]$$

$$=\frac{r}{2!}\left[\left(h+\frac{s}{r}k\right)^2+\left(\frac{-s^2}{r^2}k^2+t\frac{k^2}{r}\right)\right]$$

$$=\frac{r}{2!}\left[\left(h+\frac{s}{r}k\right)^2+\left(rt-s^2\right)\frac{k^2}{r^2}\right]$$

It has the same sign for all h, k if and only if $(rt - s^2)$ is positive.

When (i) $(rt - s^2)$ positive, f(x, y) is maximum for negative 'r' and minimum for positive 'r'.

(ii) $(rt - s^2)$ is negative, we have neither a maximum nor minimum and such stationary points are called *saddle points*.

Solving p = 0, q = 0, we get the extreme points

Extreme value :

f(a, b) is said to be an extreme value of f(x, y) if it is either maximum or minimum.

Solved Examples

***Example* 5.1:** Find the maximum and minimum values of $x^3 + 3xy^2 - 3y^2 + 4$.

***Solution*:** We have $f(x, y) = x^3 + 3xy^2 - 3x^2 - 3y^2 + 4$

$f_x = 3x^2 + 3y^2 - 6x,\quad f_y = 6xy - 6y,\quad f_{yy} = 6x - 6,\quad f_{xy} = 6y$

$f_{xx} = 6x - 6$

We now solve

$f_x = 0, f_y = 0$ simultaneously

i.e., $x^2 + y^2 - 2x = 0$ ..... (1)

and $xy - y = 0$ ..... (2)

$\therefore$ $xy - y = 0 \Rightarrow y = 0$ or $x = 1$

From $x^2 + y^2 - 2x = 0$ we get

when $y = 0, x^2 - 2x = 0$

$\therefore$ $x = 2$ or $x = 0$

when $x = 1, y^2 - 1 = 0$ i.e., $y = 1$ or -1

Thus the points are (0, 0), (2, 0), (1, 1), (1, –1)

(a) For $x = 0, y = 0, r = f_{xx} = -6, s = 0, t = -6$

$\therefore$ $rt - s^2 = 36 - 0 > 0$

$\therefore$ f(x, y) is stationary at $x = 0, y = 0$

But $r = f_{xx} = -6 < 0$

$\therefore$ f(x, y) is maximum at $x = 0, y = 0$ and the maximum value $f(0, 0) = 4$

(b) For $x = 2, y = 0, r = f_{xx} = 6, s = f_{xy} = 0, t = f_{yy} = 6$

Again $rt - s^2 > 0$

$\therefore$ f(x, y) is stationary at $x = 2, y = 0$

But $r = f_{xx} = 6 > 0$

$\therefore$ f(x, y) is minimum at $x = 2, y = 0$ and minimum value $f(2, 0) = 0$

(c) At $x = 1, y = 1$ or at $x = 1, y = -1$

$rt - s^2 > 0$ and we can reject these points as they are not stationary points.

***Example* 5.2:** Discuss the maximum and minimum of

$$f(x,y) = x^2 - y^2 + 6x - 12$$

$$p = 2x + 6,\ q = -2y,\ r = 2,\ s = 0,\ t = -2$$

$$p = 0 \text{ gives } x = -3,\ q = 0 \rightarrow y = 0$$

Stationary point is (–3, 0); and here $r = 2 > 0$; and

$$rt - s^2 = 4 - 0 = 4 > 0$$

$x^2 - y^2 + 6x - 12$ is minimum at (–3,0) the minimum value is

$f(-3, 0) = -21$.

Exercise 5.4

1. Find the maximum and minimum values of:

1. $x^3 + 3xy^2 - 3y^2 + 4$ [**Ans:** (0,0) max, its value 4,(2,0) min. its value '0']

2. $x^3+3xy^2-3x^2-3y^2+7$ [**Ans:** max. value 7 at (0,0) min. value '3' at (2,0)]

3. $x^2+y^2+6x+12$ [**Ans:** min. value '3' at (–3,0)]

4. $x^3+xy^2+21x-12x^2-2y^2$ [**Ans:** max. value 10 at (1,0) min. value –98 at (7,0)]

5. $x^4+xy^2+y^2$ [**Ans:** minimum value '0' at (0,0)]

6. $x^4+y^4-x^2-y^2+1$

[**Ans:** max.value 1 at (0, 0) min.value $\frac{1}{2}$ at four points $\left(\pm\frac{1}{\sqrt{2}},\pm\frac{1}{\sqrt{2}}\right)$]

7. $x^2y^2-x^3y^2-x^2y^3$ [**Ans:** max.value $\frac{1}{324}$ at $\left(\frac{1}{2},\frac{1}{2}\right)$]

8. Discuss the statoinary values of $u=x^2y-3x^2-2y^2-4y+3$ [**Ans:** u is a maximum at (0,–1). The maximum value is 5, u is neither maximum nor minimum at $(\pm4,3)$]

9. Find the maximum and minimum values of the following functions.

(a) $x^3y^2(12-3x-4y)$ [**Ans:** maximum value at (2,1)]

(b) $x^3+y^3-3axy\ (a>0)$ [**Ans:** minimum value is $-a^3$ at (a,a)]

(c) $xy(a-x-y)\ (a>0)$ [**Ans:** maximum value is $\frac{a^3}{27}$ at (a/3,a/3)]

10. Determine the values of x, y for which the following functions are maximum or minimum.

(a) $x^3y^2(1-x-y)$ [**Ans:** maximum for x = 1/2, y = 1/3]

(b) $(x^2+y^2)^2-2a^2(x^2-y^2)$ [**Ans:** minimum for $x=\pm a, y=0$]

(c) $\sin x\sin y\sin(x+y)$ [**Ans:** maximum for $x=y=\pi/3$]

(d) $a[\sin x+\sin y+\sin(x+y)]$ [**Ans:** maximum for $x=y=\pi/3$]

(e) $xy^2(3x+6y-2)$ [**Ans:** minimum at x = y = 1/6]

5.18.1 Lagrange's Method of Undetermined Multipliers

Theorem: Find the maximum and minimum values of $f(x_1, x_2, x_3, \ldots\ldots x_n)$ where $x_1, x_2, x_3, \ldots\ldots x_n$ are connected by the following m equations.

$$\phi_1(x_1, x_2, \ldots\ldots x_n) = 0$$

$$\phi_2(x_1, x_2, \ldots\ldots x_n) = 0$$

$$\cdots\cdots\cdots\cdots$$

$$\cdots\cdots\cdots\cdots$$

$$\phi_m(x_1, x_2, \ldots\ldots x_n) = 0$$

Proof: Let $u = f(x_1, x_2, \ldots\ldots x_n)$. Since n variables are connected by the m given relations, only (n-m) of the variables are independent.

The maximum and minimum values of u can be found by the method of Lagrange's method of undetermined Multipliers. For u to be max or min du = 0

i.e. $$\frac{\partial \phi_1}{\partial x_1}dx_1 + \frac{\partial u}{\partial x_2}dx_2 + \ldots\ldots + \frac{\partial u}{\partial x_n}dx_n = 0 \qquad \ldots\ldots(5.29)$$

Also from the m given relations

$$\left.\begin{array}{l} \dfrac{\partial \phi_1}{\partial x_1}dx_1 + \dfrac{\partial \phi_1}{\partial x_2}dx_2 + \ldots\ldots + \dfrac{\partial \phi_1}{\partial x_n}dx_n = 0 \\ \dfrac{\partial \phi_1}{\partial x_2}dx_1 + \dfrac{\partial \phi_2}{\partial x_2}dx_2 + \ldots\ldots + \dfrac{\partial \phi_2}{\partial x_n}dx_n = 0 \\ - \; - \; - \; - \; - \; - \; - \; - \; - \\ - \; - \; - \; - \; - \; - \; - \; - \; - \\ \dfrac{\partial \phi_m}{\partial x_1}dx_1 + \dfrac{\partial \phi_m}{\partial x_2}dx_2 + \ldots\ldots + \dfrac{\partial \phi_m}{\partial x_n}dx_n = 0 \end{array}\right\} \qquad \ldots\ldots(5.30)$$

multiplying (5.29) by 1 and (5.30) by $\lambda_1, \lambda_2, \ldots\ldots \lambda_m$ respectively and adding, we get

$$\left(\frac{\partial u}{\partial x_1}+\lambda_1\frac{\partial \phi_1}{\partial x_1}+\lambda_2\frac{\partial \phi_2}{\partial x_1}+......+\lambda_m\frac{\partial \phi_m}{\partial x_1}\right)dx_1+\left(\frac{\partial u}{\partial x_2}+\lambda_1\frac{\partial \phi_1}{\partial x_2}+\lambda_2\frac{\partial \phi_2}{\partial x_2}+......+\lambda_m\frac{\partial \phi_m}{\partial x_2}\right)dx_2$$

$$+...$$

$$+\left(\frac{\partial u}{\partial x_n}+\lambda_1\frac{\partial \phi_1}{\partial x_n}+\lambda_2\frac{\partial \phi_2}{\partial x_n}+......+\lambda_m\frac{\partial \phi_m}{\partial x_n}\right)dx_n=0. \qquad .....(5.31)$$

The values $\lambda_1, \lambda_2, ...\lambda_m$ are at our choice. We can, therefore choose them so as to satisfy 'm' linear equations .....(5.32)

$$\left.\begin{aligned}
&\frac{\partial u}{\partial x_1}+\lambda_1\frac{\partial \phi_1}{\partial x_1}+\lambda_2\frac{\partial \phi_2}{\partial x_1}+......+\lambda_m\frac{\partial \phi_m}{\partial x_1}=0\\
&\frac{\partial u}{\partial x_2}+\lambda_1\frac{\partial \phi_1}{\partial x_2}+\lambda_2\frac{\partial \phi_2}{\partial x_2}+......+\lambda_m\frac{\partial \phi_m}{\partial x_2}=0\\
&- \quad - \quad - \quad - \quad -\\
&- \quad - \quad - \quad - \quad -\\
&\frac{\partial u}{\partial x_m}+\lambda_1\frac{\partial \phi_1}{\partial x_m}+\lambda_2\frac{\partial \phi_2}{\partial x_m}+......+\lambda_m\frac{\partial \phi_m}{\partial x_m}=0
\end{aligned}\right\} \qquad .....(5.33)$$

The equation (5.31) now reduces to

$$\left(\frac{\partial u}{\partial x_{m+1}}+\lambda_1\frac{\partial \phi_1}{\partial x_{m+1}}+\lambda_2\frac{\partial \phi_2}{\partial x_{m+1}}+......+\lambda_m\frac{\partial \phi_m}{\partial x_{m+1}}\right)dx_{m+1}$$

$$+\left(\frac{\partial u}{\partial x_{m+2}}+\lambda_1\frac{\partial \phi_1}{\partial x_{m+2}}+\lambda_2\frac{\partial \phi_2}{\partial x_{m+2}}+......+\lambda_m\frac{\partial \phi_m}{\partial x_{m+2}}\right)dx_{m+2}$$

$$+\left(\frac{\partial u}{\partial x_n}+\lambda_1\frac{\partial \phi_1}{\partial x_n}+\lambda_2\frac{\partial \phi_2}{\partial x_n}+......+\lambda_m\frac{\partial \phi_m}{\partial x_n}\right)dx_n=0 \qquad .....(5.34)$$

It is immaterial which n – m of the n variable are regarded as independent.

Let them be $x_{m+1,} x_{m+2,.....} x_n$. Then since the n – m quantities $dx_{m+1,} dx_{m+2,.....} dx_n$ are all independent, their coefficient must be separately zero. This gives additional equations.

$$\left.\begin{aligned}
&\frac{\partial u}{\partial x_{m+1}}+\lambda_1\frac{\partial \phi_1}{\partial x_{m+1}}+\lambda_2\frac{\partial \phi_2}{\partial x_{m+1}}+.....+\lambda_m\frac{\partial \phi_m}{\partial x_{m+1}}=0\\
&\frac{\partial u}{\partial x_{m+2}}+\lambda_1\frac{\partial \phi_1}{\partial x_{m+2}}+\lambda_2\frac{\partial \phi_2}{\partial x_{m+2}}+.....+\lambda_m\frac{\partial \phi_m}{\partial x_{m+2}}=0\\
&\dots\dots\dots\dots\dots\dots\dots\dots\\
&\frac{\partial u}{\partial x_n}+\lambda_1\frac{\partial \phi_1}{\partial x_n}+\lambda_2\frac{\partial \phi_2}{\partial x_n}+.....+\lambda_m\frac{\partial \phi_m}{\partial x_n}=0
\end{aligned}\right\} \quad .....(5.35)$$

Now we have m + n equations in all and given relations $\phi_1 = 0, \phi_2 = 0, ... \phi_m = 0$
The equations (5.32) and n–m equations (5.35) and $\phi_1 = 0, \phi_2 = 0, ... \phi_m = 0$ are sufficient to determine m + n quantities

$\lambda_1 + \lambda_2, ... \lambda_m$ and $x_1, x_2, ... x_m$ for which maximum and minimum values of u are possible. The m multipliers $\lambda_1 + \lambda_2, ... \lambda_m$ are called Lagrange's Multipliers.

Theorem: If $u = \phi(x, y, z)$, $f_1(x, y, z) = 0.$, $f_2(x, y, z) = 0$ discuss the maximum and minimum values of u

Proof: For a maximum or minimum, we have du = 0

$$\text{i.e., } \frac{\partial \phi}{\partial x}dx+\frac{\partial \phi}{\partial y}dy+\frac{\partial \phi}{\partial z}dz=0$$

$$\left.\begin{aligned}
&\text{Also form } f_1 = 0 \text{ and } f_2 = 0\\
&\frac{\partial f_1}{\partial x}dx+\frac{\partial f_1}{\partial y}dy+\frac{\partial f_1}{\partial z}dz=0\\
&\frac{\partial f_2}{\partial x}dx+\frac{\partial f_2}{\partial y}dy+\frac{\partial f_2}{\partial z}dz=0
\end{aligned}\right\} \quad .....(5.36)$$

Multiplying these equations by λ_1, λ_2 respectively and adding

$$\left(\frac{\partial \phi}{\partial x}+\lambda_1\frac{\partial f_1}{\partial x}+\lambda_2\frac{\partial f_2}{\partial x}\right)dx+\left(\frac{\partial \phi}{\partial y}+\lambda_1\frac{\partial f_1}{\partial y}+\lambda_2\frac{\partial f_2}{\partial y}\right)dy+\left(\frac{\partial \phi}{\partial z}+\lambda_1\frac{\partial f_1}{\partial z}+\lambda_2\frac{\partial f_2}{\partial z}\right)dz \quad .....(5.37)$$

The Lagrange's multipliers λ_1, λ_2 are chosen such that

$$\frac{\partial \phi}{\partial x}dx + \frac{\partial \phi}{\partial y}dy + \frac{\partial \phi}{\partial z}dz = 0 \quad .....(5.38)$$

$$\frac{\partial \phi}{\partial y} + \lambda_1 \frac{\partial f_1}{\partial y} + \lambda_2 \frac{\partial f_2}{\partial y} = 0 \quad .....(5.39)$$

then equation (5.37) becomes

$$\frac{\partial \phi}{\partial z} + \lambda_1 \frac{\partial f_1}{\partial z} + \lambda_2 \frac{\partial f_2}{\partial z} = 0 \quad .....(5.40)$$

These equations (5.38), (5.39), (5.40) and given $f_1(x, y, z) = 0$, $f_2(x, y, z) = 0$ are sufficient to find λ_1, λ_2, x, y, z which give maximum and minimum values of u.

Theorem: If $u = \phi(x, y, z)$ and $f(x, y, z) = 0$ find the maximum or minimum values of u.

Proof: For a maximum or a minimum u, we have du = 0

i.e.,
$$\frac{\partial \phi}{\partial x}dx + \frac{\partial \phi}{\partial y}dy + \frac{\partial \phi}{\partial z}dz = 0 \quad .....(5.41)$$

and

from given $f(x, y, z) = 0$

$$\frac{\partial f}{\partial x}.dx + \frac{\partial f}{\partial y}.dy + \frac{\partial f}{\partial z}.dz = 0 \quad .....(5.42)$$

Multiplying (5.41) by (5.41) and (5.42) by λ and adding

$$\left(\frac{\partial \phi}{\partial x} + \lambda \frac{\partial f}{\partial x}\right)dx + \left(\frac{\partial \phi}{\partial y} + \lambda \frac{\partial f}{\partial y}\right)dy + \left(\frac{\partial \phi}{\partial z} + \lambda \frac{\partial f}{\partial z}\right)dz = 0 \quad .....(5.43)$$

Lagrange's multiplier λ is choosen such that

$$\frac{\partial \phi}{\partial x} + \lambda \frac{\partial f}{\partial x} = 0 \quad .....(5.44)$$

and
$$\frac{\partial \phi}{\partial y} + \lambda \frac{\partial f}{\partial y} = 0 \quad .....(5.45)$$

The equation (5.43) gives

$$\frac{\partial \phi}{\partial z} + \lambda \frac{\partial f}{\partial z} = 0 \quad .....(5.46)$$

Solved Examples

***Example* 1:** Find the point upon the plane ax + by + cz = p at which the function $\phi = x^2 + y^2 + z^2$ has a minimum value and find this minimum ϕ.

Solution:

Given $\quad ax + by + cz = P$

$$ax + by + cz = p \Rightarrow z = \frac{p - ax - by}{c} \qquad ..... (1)$$

Again $\phi = x^2 + y^2 + z^2 = x^2 + y^2 + \left(\frac{p - ax - by}{c}\right)^2$

Then for min./max., we have

$$\frac{\partial\phi}{\partial x} = 2x - \frac{2a}{c^2}(p - ax - by) = 0 \qquad .....(2)$$

$$\frac{\partial\phi}{\partial y} = 2y - \frac{2b}{c^2}(p - ax - by) = 0 \qquad .....(3)$$

$$\Rightarrow \frac{x}{a} = \frac{p - ax - by}{c^2}$$

$$\Rightarrow \frac{y}{b} = \frac{p - ax - by}{c^2}$$

$$\Rightarrow \frac{x}{a} = \frac{y}{b} \Rightarrow y = \frac{bx}{a}$$

Now, substituting this value of y in (2) we obtain

$$x - \frac{a}{c^2}\left(p - ax - \frac{b^2x}{a}\right) = 0$$

$$\Rightarrow \qquad c^2x - ap + a^2x + b^2x = 0$$

$$\Rightarrow \qquad x(a^2 + b^2 + c^2) = ap$$

$$\Rightarrow \qquad x = \frac{ap}{a^2 + b^2 + c^2} \text{ and } y = \frac{bx}{a} = \frac{bp}{a^2 + b^2 + c^2}$$

Then, $\quad r = \dfrac{\partial^2\phi}{\partial x^2} = 2 + \dfrac{2a^2}{c^2}; \dfrac{\partial^2\phi}{\partial x \partial y} = \dfrac{2ab}{c^2} = s; \qquad t = \dfrac{\partial^2\phi}{\partial y^2} = 2 + \dfrac{2b^2}{c^2}$

and $\quad rt - s^2 = 4\left(1 + \dfrac{a^2}{c^2} + \dfrac{b^2}{c^2}\right) > 0$ and r is positive

Hence $\exists$ a min, when

$$\frac{x}{a}=\frac{y}{b}=\frac{p}{a^2+b^2+c^2}$$

Also $$z=\frac{p-ax-by}{c}=\frac{1}{c}\left[p-\frac{a^2p}{a^2+b^2+c^2}-\frac{b^2p}{a^2+b^2+c^2}\right]$$

$$=\frac{cp}{a^2+b^2+c^2}$$

i.e., $$\frac{z}{c}=\frac{p}{a^2+b^2+c^2}$$

Hence $\exists$ a min. at

$$\frac{x}{a}=\frac{y}{b}=\frac{z}{c}=\frac{p}{a^2+b^2+c^2}$$

$$\phi_{min}=x^2+y^2+z^2=\frac{a^2p^2+b^2p^2+c^2p^2}{\left(a^2+b^2+c^2\right)^2}=\frac{p^2}{a^2+b^2+c^2}$$

***Example* 2:** Divide 24 into the three parts such that the continued product of the first, square of the second and cube of third is maximum.

Solution:

Let 24 be divided into parts x, y, z then $x+y+z=24$ .....(1)

Given $\phi(x, y, z) = x^3 y^2 z$

then $\phi(x, y) = x^3y^2 (24 - x - y)$ from (1)

or $\phi(x, y) = 24x^3y^2 - x^4y^2 - x^3y^3$

Differentiating, we get

$$p=\frac{\partial\phi}{\partial x}=72x^2y^2-4x^3y^2-3x^2y^3$$

$$r=\frac{\partial^2\phi}{\partial x^2}=144xy^2-12x^2y^2-6xy^3$$

$$q=\frac{\partial\phi}{\partial y}=48x^3y-2x^4y-3x^3y^2$$

$$t=\frac{\partial^2\phi}{\partial y^2}=48x^3-2x^4-6x^3y$$

and $$s = \frac{\partial^2 \phi}{\partial x \partial y} = \frac{\partial}{\partial x}\left(\frac{\partial \phi}{\partial y}\right) = 144x^2y - 8x^3y - 9x^2y^2$$

For a maximum value of $\phi(x, y)$we have

$$\frac{\partial \phi}{\partial x} = 0 \Rightarrow 72x^2y^2 - 4x^3y^2 - 3x^2y^3 = 0$$

$$\Rightarrow x^2y^2 (72 - 4x - 3y) = 0$$

$$\Rightarrow 72 - 4x - 3y = 0, \quad x = 0, y = 0 \qquad \text{..... (2)}$$

and $$\frac{\partial \phi}{\partial y} = 0 \Rightarrow 48x^3y - 2x^4y - 3x^3y^2 = 0$$

$$\Rightarrow x^3y (48 - 2x - 3y) = 0$$

$$\Rightarrow 48 - 2x - 3y = 0, \quad x = 0, y = 0 \qquad \text{..... (3)}$$

Solving (2) and (3) we get $x = 12$ and $y = 8$ etc

At (12, 8), we have

$$r = 144, 12(8)^2 - 12(12)^2(8)^2 - 6(12)(8)^3$$

$$= 12(8)^2 (144 - 144 - 48) = 12(8)^2 (-48)$$

$$t = 48(12)^3 - 2(12)^4 - 6(12)^3 .8$$

$$= (12)^3 (48 - 24 - 48)$$

$$s = 144(12)^2 . 8 - 8(12)^3 . 8 - 9(12)^2 (8)^2$$

$$= (12)^2 . 8 (144 - 96 - 72) = (12)^2 . 8(-24)$$

$$\therefore \quad rt - s^2 = 12 . (8)^2 (-48) . (12)^3 (-24) - [(12)^2 . 8. (-24)]^2$$

$$= (12)^4 . 8^2 . 24 (48 - 24)$$

$$= (12)^4 . 8^2 . 24^2 > 0 ;$$

Since $rt - s^2 > 0$ and $r < 0$, therefore $\phi (x, y)$ is maximum at (12, 8).

Putting $x = 12$, and $y = 8$ in (i), we get $z = 4$

The values of x, y, z are 12, 8, 4 respectively. This is the division of 24 for maximum $\phi(x, y, z)$.

***Example* 3:** A rectangular box, open at the top, is to have a volume of 32 c.c. Find the dimensions of the box requiring least material for its construction.

Solution:

Let l, b and h be the length, breadth and height of the box respectively. Then, we have

$$v = lbh = 32\ ;\ \text{surface} = 2(l + b)\,h + lb = S\ (\text{say})$$

$$\therefore \qquad S = 2(l + b)h + lb \text{ and } b = \frac{32}{lh} \qquad .....(1)$$

$$\Rightarrow \qquad s = 2\left(l + \frac{32}{lh}\right)h + l\left(\frac{32}{lh}\right) = 2lh + \frac{64}{l} + \frac{32}{h}$$

Now $$\frac{\partial s}{\partial l} = 2h - \frac{64}{l^2}; \qquad \frac{\partial s}{\partial h} = 2l - \left(\frac{32}{h^2}\right)$$

$$\Rightarrow \qquad \frac{\partial^2 s}{\partial l^2} = \frac{128}{l^3}, \qquad \frac{\partial^2 s}{\partial l \partial h} = 2, \frac{\partial^2 s}{\partial h^2} = \frac{64}{h^3}$$

For maximum and minimum of S, we get

$$\frac{\partial s}{\partial l} = 0 = \frac{\partial s}{\partial h}$$

$$\Rightarrow \qquad \frac{\partial s}{\partial l} = 0 \Rightarrow 2h - \frac{64}{l^2} = 0, h = \frac{32}{l^2}$$

and $$\frac{\partial s}{\partial h} = 0 \Rightarrow 2l - \frac{32}{h^2} = 0$$

or $$l = \frac{16}{h^2}$$

These give $l = 4$, h = 2 and b = 4

Now $$\frac{\partial^2 s}{\partial l^2} = \frac{128}{l^3} = \frac{128}{64} = 2 \frac{\partial^2 s}{\partial l \partial h} = 2$$

$$\frac{\partial^2 s}{\partial h^2} = \frac{64}{h^3} = \frac{64}{8} = 8$$

$$\Rightarrow \frac{\partial^2 s}{\partial l^2} \cdot \frac{\partial^2 s}{\partial h^2} - \left(\frac{\partial^2 s}{\partial l \partial h}\right)^2 = 2.8 - (2)^2 = 12 = +\text{ve}.$$

and $\dfrac{\partial^2 s}{\partial h^2} = +2$ so 'S' is minimum for $l = 4$, b = 4, h = 2.

***Example* 4:** If $u = a^3 x^2 + b^3 y^2 + c^3 z^2$ where $\frac{1}{x}+\frac{1}{y}+\frac{1}{z}=1$ show that the stationary point of u is given by $x=\frac{a+b+c}{a}$, $y=\frac{a+b+c}{b}$, $z=\frac{a+b+c}{c}$

***Solution*:**

For a stationary value of 'u', du = 0

i.e., $a^3 2xdx + b^3 2ydy + c^3 2zdz = 0$ .....(1)

Also we have the given relation $\frac{1}{x}+\frac{1}{y}+\frac{1}{z}=1$

$$\frac{1}{x^2}dx+\frac{1}{y^2}dy+\frac{1}{Z^2}dz=0 \quad .....(2)$$

Equating the coefficients of dx, dy, dz from (1) + λ (2) separately to zero

We get $a^3x+\frac{\lambda}{x^2}=0,\ b^3y+\frac{\lambda}{y^2}=0,\ c^3z+\frac{\lambda}{z^2}=0$

$\therefore \quad a^3x^3 = b^3y^3 = c^3z^3 = (-\lambda)$

or $ax = by = cz = (-\lambda)^{1/3} = k(\text{say})$ .....(3)

$x = k/a$; $y = k/b$; $z = k/c$

Subsituting for x, y, z in $\frac{1}{x}+\frac{1}{y}+\frac{1}{z}=1$

$$\frac{a}{k}+\frac{b}{k}+\frac{c}{k}=1$$

$\therefore \quad k = a + b + c$

Substituting in (3) we get

$$x=\frac{a+b+c}{a}, y=\frac{a+b+c}{b}, z=\frac{a+b+c}{c}$$ for which u is stationary.

Exercise 5.5

1. Find the maximum and minimum distances from the origin to the curve $5x^2 + 6xy + 5y^2 - 8 = 0$. **[Ans :** max. 4, min. 16,6,3]
2. Find the dimensions of the rectangular box, with out a top of maximum capacity whose surface is 108 sq.inches. **[Ans:** 6",6",3"]

3. If $f = u^2 + v^2 + w^2$ where $u + v + w = 1$, show that f is stationary when $u = v = w = \frac{1}{3}$

4. Use Lagrange's method of multiplies to determine the minimum distance from the origin to the plane $x + 2y + 3z = 14$. **[Ans:** $\sqrt{14}$]

5. Find the maximum value of $f = xyz$ when $xy + yz + zx = k$. **[Ans:** $(k/3)^{3/2}$]

6. If the distance from the origin of any point p(x,y,z) on the plane $3x + 2y + z - 12 = 0$ is $p = \sqrt{x^2 + y^2 + z^2}$, find the minimum value of p (use the Langrange's method).

7. Find the maximum volume of a parallelepiped inscribed in a sphere $x^2 + y^2 + z^2 = r^2$. **[Ans:** $\frac{8r^2}{3\sqrt{3}}$]

8. Find the volume of the largest rectangular parallelepiped that can be inscribed in the ellipsoid $\frac{x^2}{a^2} + \frac{y^2}{b^2} + \frac{z^2}{c^2}$ **[Ans:** $\frac{8abc}{3\sqrt{3}}$]

9. Given $x + y + z = k$, find the maximum value of $x^{\alpha}, y^{\beta}, z^{\gamma}$

[Ans: $\left[\frac{\alpha^{\alpha}.\beta^{\beta}.\gamma^{\gamma}.k^{(\alpha+\beta+\gamma)}}{(\alpha+\beta+\gamma)^{\alpha+\beta+\gamma}}\right]$]

10. If $r^2 = x^2 + y^2 + z^2$ and $x+y+z = 30$ find the values of x, y, z for which r is a minimum **[Ans:** $x = y = z = 10$]

11. Find the dimensions of a rectangular box without a top of maximum capacity whose surface area is 108 square inches. **[Ans:** 6,6,3 inches]

12. Find the dimensions of a rectangular box with open top, so that the total surface area S of the box is a minimum, given that the volume 'V' of the box is constant. **[Ans:** $x = y = 2z = (2v)^{1/3}$]

13. Show that the rectangular solid of maximum volume of that can be inscribed in a sphere is a cube

14. If the total surface area of a closed rectangular box is 108 sq.cm. Find the dimensions of the box having maximum capacity **[Ans:** $\sqrt{18}, \sqrt{18}, \sqrt{18}$]

15. Perimeter of a triangle is constant, prove that the area of this triangle is maximum when the triangle is equilateral.

16. If $u = \phi(x,y)$ where $f(x,y) = 0$, find the maximum and minimum values of u using Lagrange's multipliers Method

CHAPTER 6

Reduction Formulae

We have come across so many problems whose integration are neither immediately reduced to some standard form nor are directly obtained. In such cases we find out a formula, called reduction formula, which connects one integral with another in which the integral is of same type but of lower degree.

Repeated application of reduction formula makes it easy to evaluate the original integral completely. In most cases the reduction formula is obtained by the process of integration by parts but in some cases the method of differentiation or other devices are adopted.

Some Important Reduction Formulae

1. Obtain a reduction formula for

(i) $\int \sin^n x\,dx$, (ii) $\int \cos^n x\,dx$.

(i) Let $I_n = \int \sin^n x\,dx$. We have $\int \sin^n x\,dx = \int \sin^{n-1} x \sin x\,dx$

Integrating by parts, we get

$$I_n = \sin^{n-1} x(-\cos x) - \int (-\cos x)(n-1)\sin^{n-2} x.\cos x\,dx$$

$$= -\sin^{n-1} x\cos x + (n-1)\int \sin^{n-2} x.\cos^2 x\,dx$$

$$= -\sin^{n-1} x\cos x + (n-1)\int \sin^{n-2} x.\left(1-\sin^2 x\right)dx$$

$$= -\sin^{n-1} x\cos x + (n-1)\int \sin^{n-2} x\,dx - (n-1)\int \sin^n x\,dx$$

or $n\int \sin^n x\,dx = -\sin^{n-1} x\cos x + (n-1)\int \sin^n x\,dx$

$$\therefore\ \int \sin^n x\,dx = \frac{\sin^{n-1} x\cos x}{n} + \frac{n-1}{n}\int \sin^{n-2} x\,dx$$

which is the reduction formula.

(ii) We have

$$I_n = \int \cos^n x\,dx = \int \cos^{n-1} x.\cos x\,dx.$$

Integrating by parts, we get

$$I_n = \cos^{n-1} x.\sin x - (n-1)\int \cos^{n-2} x.(-\sin x).\sin x\,dx$$

$$= \cos^{n-1} x.\sin x + (n-1)\int \cos^{n-2} x.\sin^2 x\,dx$$

$$= \cos^{n-1} x.\sin x + (n-1)\int \cos^{n-2} x.\left(1-\cos^2 x\right)dx$$

$$= \cos^{n-1} x.\sin x + (n-1)\int \cos^{n-2} x.dx - (n-1)\int \cos^n x\,dx$$

or $\quad n\int \cos^n x\,dx = \cos^{n-1} x \sin x + (n-1)\int \cos^{n-2} x\,dx$

$\therefore \quad \int \cos^n x\ dx = \dfrac{\cos^{n-1} x \sin x}{n} + \dfrac{n-1}{n}\int \cos^{n-2} x\ dx$

which is the reduction formula.

2. Obtain a reduction formula for

(i) $\int \tan^n x\,dx$, (ii) $\int \cot^n x\,dx$.

(i) We have

$$\int \tan^n x\,dx = \int \tan^{n-2} x.\tan^2 x\,dx = \int \tan^{n-2} x.\left(\sec^2 x - 1\right)dx$$

$$= \int \tan^{n-1} x.\sec^2 n\,dx - \int \tan^{n-2} x.dx$$

$$= \frac{\tan^{n-1} x}{n-1}\int \tan^{n-2} x\,dx\,.$$

If $I_n = \int \tan^n x\ dx$, the above reduction formula can be written as

$$I_n = \frac{\tan^{n-1} x}{n-1} - I_{n-2}$$

(ii) We have

$$\int \cot^n x\ dx = \int \cot^{n-2} x \cot^2 x\ dx = \int \cot^{n-2} x\left(\text{cosec}^2 x - 1\right)\ dx$$

$$= \int \cot^{n-2} x \operatorname{cosec}^2 x \, dx - \int \cot^{n-2} x \, dx$$

$$= \frac{\cot^{n-1} x}{n-1} - \int \cot^{n-2} x \, dx.$$

If $I_n = \int \cot^n x \, dx$, the required reduction formula takes the form

$$I_n = \frac{\cot^{n-1} x}{n-1} - I_{n-2}$$

3. Obtain a reduction formula for

(i) $\int \sec^n x \, dx$, (ii) $\int \operatorname{cosec}^n x \, dx$.

(i) We have

$$\int \sec^n x \, dx = \int \sec^{n-2} x \sec^2 x \, dx.$$

Integrating by parts, we get

$$\int \sec^n x \, dx = \sec^{n-2} x \tan x - (n-2) \int \sec^{n-3} x \sec x \tan x \tan x \, dx$$

$$= \sec^{n-2} x \tan x - (n-2) \int \sec^{n-2} x \left(\sec^2 x - 1\right) dx$$

$$= \sec^{n-2} x \tan x + (n-2) \int \sec^{n-2} x \, dx - (n-2) \int \sec^n x \, dx.$$

If we write $I_n = \int \sec^n x \, dx$ then

$$I_n = \sec^{n-2} x \tan x + (n-2) I_{n-2} - (n-2) I_n$$

or $\quad (1+n-2) I_n = \sec^{n-2} x \tan x + (n-2) I_{n-2}$

$\therefore \quad I_n = \dfrac{\sec^{n-2} x \tan x}{n-1} + \dfrac{n-2}{n-1} I_{n-2}$

(ii) We have $\int \operatorname{cosec}^n x \, dx = \int \operatorname{cosec}^{n-2} x \operatorname{cosec}^2 x \, dx.$

Integrating by parts, we get

$$I = -\operatorname{cosec}^{n-2} x \cot x - (n-2) \int \cot . x . \operatorname{cosec}^{n-3} x . \operatorname{cosec} x \cot x \, dx$$

$$= -\operatorname{cosec}^{n-2} x \cot x - (n-2) \int \operatorname{cosec}^{n-2} x \left(\operatorname{cosec}^2 x - 1\right) dx$$

$$= -\operatorname{cosec}^{n-2} \cot x + (n-2) \int \operatorname{cosec}^{n-2} x \, dx - (n-2) \int \operatorname{cosec}^n x \, dx.$$

If we write $I_n = \int \operatorname{cosec}^n x\,dx$ then

$$I_n + (n-2)I_n = -\operatorname{cosec}^{n-2} x \cot x + (n-2)I_{n-2}$$

$$\therefore \quad I_n = -\frac{\operatorname{cosec}^{n-2} x \cot x}{n-1} + \frac{n-2}{n-1} I_{n-2}$$

4. Reduction formula for $\int \sin^m x \cos^n x\,dx$

We take $P = \sin^{\lambda+1} x \cos^{\mu+1} x$, where λ and μ are smaller indices of sin x and cos x respectively in the two integrals to be connected. Then we find $\frac{dp}{dx}$ and rearrange it as a linear function of the two integrals to be connected. Then we integrate both sides and solve for the required integrals by transposition.

(i) Obtain a reduction formula for $\int \sin^m x \cos^n x\,dx$ by connecting it with $\int \sin^{m-2} x \cos^n x\,dx$

In other words, if $I_{m,n} = \int \sin^m x \cos^n x\,dx$, establish

$$I_{m,n} = \frac{m-1}{m+n} I_{m-2,\,n} - \frac{\sin^{m-1} x \cos^{m+1} x}{m+n}$$

In the two integrals, the smaller index of sin x is m – 2 and that of cos x is n.

$$\therefore \quad \lambda = m-2, \ \mu = n.$$

So, here $P = \sin^{\lambda+1} x \cos^{\mu+1} x = \sin^{m-1} x \cos^{n+1} x$

$$\therefore \quad \frac{dp}{dx} = (m-1)\sin^{m-2} x \cos x \cos^{n+1} x + \sin^{m-1} x (n+1)\cos^n x(-\sin x)$$

$$= (m-1)\sin^{m-2} x \cos^n x \cos^2 x - \sin^m x (n+1)\cos^n x$$

$$= (m-1)\sin^{m-2} x \cos^n x (1-\sin^2 x) - (n+1)\sin^m x \cos^n x$$

$$= (m-1)\sin^{m-2} x \cos^n x - (m-1)\sin^m x \cos^n x - (n+1)\sin^m x \cos^n x$$

$$= (m-1)\sin^{m-2} x \cos^n x - [(m-1)+(n+1)]\sin^m x \cos^n x$$

$$= (m-1)\sin^{m-2} x \cos^n x - (m-1)\sin^m x \cos^n x.$$

Integrating we get

$$P=(m-1)\int \sin^{m-2} x \cos^n x\, dx-(m+n)\int \sin^m x \cos^n x\, dx$$

$$\therefore \quad \int \sin^m x \cos^n x\, dx=-\frac{P}{m+n}+\frac{m-1}{m+n}\int \sin^{m-2} \cos^n x\, dx$$

$$=-\frac{\sin^{m-1} x \cos^{n+1} x}{m+n}+\frac{m-1}{m+n}+\frac{m-1}{m+n}\sin^{m-2} x \cos^n x\, dx,$$

i.e., $$I_{m,n}=\frac{m-1}{m+n}I_{m-2,n}-\frac{\sin^{m-1} x \cos^{n+1} x}{m+n}$$

or $$(m+n)I_{m,n}=(m-1)I_{m-2,n}-\sin^{m-1} x \cos^{n+1} x.$$

(ii) Obtain a reduction formula for $\int \sin^m x \cos^n x\, dx$ by connecting it with $\int \sin^m x \cos^{n-2} x\, dx$.

In other words, if $I_{m,n}=\int \sin^m x \cos^n x\, dx$, establish

$$I_{m,n}=\frac{\cos^{n-1} x \sin^{m+1} x}{m+n}+\frac{n-1}{m+n}I_{m,n-2}$$

i.e., $$(m+n)I_{m,n}=\sin^{m+1} x \cos^{n-1} x+(n-1)I_{m,\,n-2}$$

Here the smaller indices are m and n − 2, so we take

$$P=\sin^{\lambda+1} x \cos^{\mu+1} x=\sin^{m+1} x \cos^{(n-2)+1} x=\sin^{m+1} x \cos^{n-1} x$$

$$\therefore \quad \frac{dP}{dx}=(m+1)\sin^m x \cos x \cos^{n-1} x+\sin^{m+1}(n-1)\cos^{n-2} x(-\sin x)$$

$$=(m+1)\sin^m x \cos^n x-(n-1)\sin^{m+2} x \cos^{n-2} x$$

$$=(m+1)\sin^m x \cos^n x-(n-1)\sin^m x \cos^{n-2} x\left(1-\cos^2 x\right)$$

$$=\{(m+1)+(n-1)\}\sin^m x \cos^n x-(n-1)\sin^m x \cos^{n-2} x.$$

Integrating, we get

$$P=(m+n)\int \sin^m x \cos^n x\, dx-(n-1)\int \sin^m x \cos^{n-2} x\, dx$$

$$\therefore \quad \int \sin^m x \cos^n x\, dx=\frac{P}{m+n}+\frac{n-1}{m+n}\int \sin^m x \cos^{n-2} x\, dx,$$

i.e., $$I_{m,n}=\frac{\sin^{m+1} x \cos^{n-1} x}{m+n}+\frac{n-1}{m+n}I_{m,\,n-2}$$

i.e., $$(m+n)I_{m,n} = \sin^{m+1} x \cos^{n-1} x + (n-1)I_{m,\,n-2}$$

[*Note*:1. Similarly we may connect $\int \sin^m x \cos^n x\, dx = I_{m,n}$ with either of $I_{m+2,n}$, $I_{m,\,n+2}$, $I_{m-2,n+2}$, $I_{m+2,n-2}$

2. Results (i) and (ii) can be obtained directly by integrating $\int \sin^m x \cos^n x\, dx$ by parts.]

(iii) If $I_{m,n} = \int \cos^m x \sin^n x\, dx$ then show that

$$(m+n)I_{m,n} = \cos^{m-1} x \sin^{n+1} x + (m-1)I_{m-2,\,n}$$

Here we have to connect the integrals

$$\int \cos^m x \sin^n x\, dx \text{ and } \int \cos^{m-2} x \sin^n x\, dx.$$

Here $\lambda = n,\ \mu = m-2,$ so we take

$$P = \sin^{\lambda+1} x \cos^{\mu+1} x = \sin^{n+1} x \cos^{(m-2)+1} x = \sin^{n+1} x \cos^{m-1} x$$

$$\therefore \quad \frac{dP}{dx} = (n+1)\sin^n x \cos x \cos^{m-1} x + \sin^{n+1} x (m-1)\cos^{m-2} x(-\sin x)$$

$$= (n+1)\sin^n x \cos^m x - (m-1)\cos^{m-2} x \sin^{n+2} x$$

$$= (n+1)\sin^n x \cos^m x - (m-1)\cos^{m-2} x \sin^n x\left(1-\cos^2 x\right)$$

$$= \{(n+1)+(m-1)\}\sin^n x \cos^m x - (m-1)\cos^{m-2} x \sin^n x.$$

Integrating, we get

$$P = (m+n)\int \cos^m x \sin^n x\, dx - (m-1)\int \cos^{m-2} x \sin^n x\, dx$$

or $$\sin^{n+1} x \cos^{m-1} x = (m+n)I_{m,n} - (m-1)I_{m-2,\,n}$$

$$\therefore \quad (m+n)I_{m,n} = (m-1)I_{m-2,\,n} + \cos^{m-1} x \sin^{n+1} x.$$

5. Obtain a reduction formula for $\int \cos^m x \sin nx\, dx$.

Let $$I_{m,n} = \int \cos^m x \sin nx\, dx$$

$$= -\frac{\cos^m x \cos nx}{n} - \frac{m}{n}\int \cos^{m-1} x \cos nx \sin x \, dx$$

(integrating by parts)

But, $\sin(n-1)x = \sin nx \cos x - \cos nx \sin x$

$\therefore \quad \cos nx \sin x = \sin nx \cos x - \sin(n-1)x$

$$\therefore \quad I_{m,n} = -\frac{\cos^m x \cos nx}{n} - \frac{m}{n}\int \cos^{m-1} x \{\sin nx \cos x - \sin(n-1)x\} dx$$

$$= -\frac{\cos^m x \cos nx}{n} - \frac{m}{n}\int \cos^m x \sin nx \, dx + \frac{m}{n}\int \cos^{m-1} x \sin(n-1)x \, dx$$

$$= -\frac{\cos^m x \cos nx}{n} - \frac{m}{n} I_{m,n} + \frac{m}{n} I_{m-1,n-1}$$

or $$\left(1+\frac{m}{n}\right) I_{m,n} = -\frac{\cos^m x \cos nx}{n} + \frac{m}{n} I_{m-1,n-1}$$

$$\therefore \quad I_{m,n} = -\frac{\cos^m x \cos nx}{m+n} + \frac{m}{m+n} I_{m-1,n-1}$$

6. Obtain a reduction formula for $\int \cos^m x \cos nx \, dx$

Let $I_{m,n} = \int \cos^m x \cos nx \, dx$.

Integrating by parts, we get

$$I_{m,n} = \frac{\cos^m x \sin nx}{n} + \frac{m}{n}\int \cos^{m-1} x \sin x \sin nx \, dx$$

But, $\cos(n-1)x = \cos nx \cos x + \sin nx \sin x$

$\therefore \quad \sin nx \sin x = \cos(n-1)x - \cos nx \cos x$

$$\therefore \quad I_{m,n} = \frac{\cos^m x \sin nx}{n} + \frac{m}{n}\int \cos^{m-1} x \{\cos(n-1)x - \cos nx \cos x\} dx$$

$$= \frac{\cos^m x \sin nx}{n} + \frac{m}{n}\int \cos^{m-1} x \cos(n-1)x \, dx - \frac{m}{n}\int \cos^m x \cos nx \, dx$$

$$= \frac{\cos^m x \sin nx}{n} + \frac{m}{n} I_{m-1,n-1} - \frac{m}{n} I_{m,n}$$

or $\left(1+\frac{m}{n}\right)I_{m,n}=\frac{\cos^m x\sin nx}{n}+\frac{m}{n}I_{m-1,n-1}$

or $I_{m,n}=\frac{\cos^m x\sin nx}{m+n}+\frac{m}{m+n}I_{m-1,n-1}$

7. Obtain a reduction formula for $\int \sin^m x\sin nx\, dx$.

Let $I_{m,n}=\int \sin^m x\sin nx\, dx.$

Integrating by parts, we get

$$I_{m,n}=\sin^m x\left(-\frac{\cos nx}{n}\right)+\frac{m}{n}\int \sin^{m-1} x(\cos x)\cos nx\, dx$$

$$=-\frac{\sin^m x\cos nx}{n}+\frac{m}{n}\int\left(\sin^{m-1} x\cos x\right)\cos nx\, dx \quad .....(1)$$

But, $\int\left(\sin^{m-1} x\cos x\right)\cos nx\, dx$

$$=\sin^{m-1} x\cos x\frac{\sin nx}{n}-\int\frac{\sin nx}{n}\frac{d}{dx}\left(\sin^{m-1} x\cos x\right)dx \quad .....(2)$$

Now, $\frac{d}{dx}\left(\sin^{m-1} x\cos x\right)$

$$=\sin^{m-1} x(-\sin x)+(m-1)\sin^{m-2} x\cos^2 x$$

$$=-\sin^m x+(m-1)\sin^{m-2} x\left(1-\sin^2 x\right)$$

$$=-\sin^m x+(m-1)\sin^{m-2} x-(m-1)\sin^m x$$

$$=-m\sin^m x+(m-1)\sin^{m-2} x$$

$\therefore$ eq. (2) becomes $\int\left(\sin^{m-1} x\cos x\right)\cos nx\, dx$

$$=\frac{\sin^{m-1} x\cos x\sin nx}{n}\int\left\{-m\sin^m x+(m-1)\sin^{m-2} x\right\}\frac{\sin nx}{n}dx$$

$$=\frac{\sin^{m-1} x\cos x\sin nx}{n}+\frac{m}{n}\int \sin^m x\sin nx\, dx-\frac{m-1}{n}\int \sin^{m-2}x\sin nx\, dx$$

$$=\frac{\sin^{m-1} x\cos x\sin nx}{n}+\frac{m}{n}I_{m,n}-\frac{m-1}{n}I_{m-2,n}$$

$\therefore$ eq. (1) becomes

$$I_{m,n} = -\frac{\sin^m x \cos nx}{n} + \frac{m \sin^{m-1} x.\cos x \sin nx}{n^2} + \frac{m^2}{n^2} I_{m,n} - \frac{m(m-1)}{n^2} I_{m-2,n}$$

or
$$\left(1 - \frac{m^2}{n^2}\right) I_{m,n} = -\frac{\sin^m x \cos nx}{n} + \frac{m \sin^{m-1} x \cos x \sin nx}{n^2} - \frac{m(m-1)}{n^2} I_{m-2,n}$$

or
$$I_{m,n} = \frac{n \sin^m x \cos nx}{m^2 - n^2} - \frac{m \sin^{m-1} x \cos x \sin nx}{m^2 - n^2} + \frac{m(m-1)}{m^2 - n^2} I_{m-2}$$

[*Note:* In a similar way

$$\int \sin^m x \cos nx \, dx = \frac{-m \cos x \cos nx + n \sin x \sin nx}{m^2 - n^2} \sin^m x + \frac{m(m-1)}{m^2 - n^2} \int \sin^{m-2} x \cos nx dx]$$

Beta and Gamma Functions

Beta function: The beta function or the first Eulerian integral is defined by the definite integral $\int_0^1 x^{m-1} (1-x)^{n-1} dx$, where m and n are supposed positive in all cases. It is denoted by $B(m,n)$.

Gamma function: The definite integral $\int_0^\infty e^{-x} x^{x-1} dx$ is known as the gamma function or the second Eulerian integral denoted by $\Gamma(n)$ and read as gamma n. Thus

$$\Gamma(n) = \int_0^\infty e^{-x} x^{n-1} dx$$

where n is a positive number, integer or fraction.

We shall be chiefly concerned with the following properties of gamma function:

$$\Gamma(1) = 1,\ \Gamma(0) = \infty,\ \Gamma(-n) = \infty,\ \Gamma(n+1) = n\,\Gamma(n),\ \Gamma\left(\frac{1}{2}\right) = \sqrt{\pi}$$

$\Gamma(n) = (n-1)!$, i.e., $\Gamma(n) = (n-1)(n-2)...2.\,1\,\Gamma(1)$, n being a positive integer.

$$\Gamma\left(\frac{7}{2}\right) = \Gamma\left(\frac{5}{2}+1\right) = \frac{5}{2}\Gamma\left(\frac{5}{2}\right) \qquad [\text{using } \Gamma(n+1) = n\,\Gamma(n)]$$

$$= \frac{5}{2}.\Gamma\left(\frac{3}{2}+1\right) = \frac{5}{2}.\frac{3}{2}\Gamma\left(\frac{3}{2}\right) = \frac{5}{2}.\frac{3}{2}\Gamma\left(\frac{1}{2}+1\right) = \frac{5}{2}.\frac{3}{2}.\frac{1}{2}\Gamma\left(\frac{1}{2}\right)$$

$$= \frac{15}{8}\sqrt{\pi}.$$

Similarly for gamma of half of any odd number.

Further, $B(m,n)=\dfrac{\Gamma(m)\Gamma(n)}{\Gamma(m+n)}$.

8. Evaluate $\int_0^{\pi/2}\sin^m x\cos^n x\,dx$

or to prove that $\int_0^{\pi/2}\sin^m x\cos^n x\,dx=\dfrac{\Gamma\left(\dfrac{m+1}{2}\right)\Gamma\left(\dfrac{n+1}{2}\right)}{2\Gamma\dfrac{(m+n+2)}{2}}$

We know that

$$\int\sin^m x\cos^n x\,dx=\frac{\sin^{m+1}x\cos^{n-1}x}{m+n}+\frac{n-1}{m+n}\int\sin^m x\cos^{n-2}x\,dx$$

$$\therefore\quad \int_0^{\pi/2}\sin^m x\cos^n x\,dx=\left[\frac{\sin^{m+1}x\cos^{n-1}x}{m+n}\right]_0^{\pi/2}+\frac{n-1}{m+n}\int_0^{\pi/2}\sin^m x\cos^{n-1}x\,dx$$

$$=\frac{n-1}{m+n}\int_0^{\pi/2}\sin^m x\cos^{n-1}x\,dx \qquad .....(1)$$

$$\therefore\quad \left(\left[\frac{\sin^{m+1}x\cos^{n-1}x}{m+n}\right]_0^{\pi/2}=0.\right)$$

Now four cases arise.

Case I: When m and n are both even

Repeating the above formula (1) till the power of cos x is zero, we get

$$\int_0^{\pi/2}\sin^m x\cos^n x\,dx=\frac{n-1}{m+n}.\frac{n-3}{m+n-2}.\frac{n-5}{m+n-4}\cdots\frac{n-(n-1)}{(m+n)-(n-2)}\int_0^{\pi/2}\sin^m x\,dx \qquad .....(2)$$

But we know $\int_0^{\pi/2}\sin^m x\,dx=\dfrac{m-1}{m}\int_0^{\pi/2}\sin^{m-2}x\,dx$ .....(3)

Since m is even, repeating the above formula (3),

$$\int_0^{\pi/2} \sin^m x\, dx = \frac{m-1}{m}.\frac{m-3}{m-2}.\frac{m-5}{m-4} \cdots \frac{1}{2}.\frac{\pi}{2}$$

$\therefore$ (2) becomes

$$\int_0^{\pi/2} \sin^m x \cos^n x\, dx = \frac{n-1}{m+n}.\frac{n-3}{m+n-2} \cdots \frac{1}{m+2} \times \frac{m-1}{m}.\frac{m-3}{m-2}.\frac{m-5}{m-4} \cdots \frac{1}{2}.\frac{\pi}{2}$$

$$= \frac{(n-1)(n-3)(n-3) \dots 1}{(m+n)(m+n-2) \dots (m+2)} \times \frac{(m-1)(m-3) \dots 1}{m(m-2) \dots 2}.\frac{\pi}{2}$$

$$= \frac{\left\{\left(\frac{n-1}{2}\right)\left(\frac{n-3}{2}\right) \cdots \frac{1}{2}\right\}\left\{\left(\frac{m-1}{2}\right)\left(\frac{m-3}{2}\right) \cdots \frac{1}{2}\right\}}{\left\{\frac{m+n}{2}.\frac{m+n-2}{2} \cdots 1\right\}}.\frac{\pi}{2}$$

$$= \frac{\Gamma\left(\frac{n+1}{2}\right).\Gamma\left(\frac{m+1}{2}\right)}{2\,\Gamma\left(\frac{m+n+2}{2}\right)}$$

Case II: When n is even and m is odd

In this case $\int_0^{\pi/2} \sin^m x\, dx = \frac{m-1}{m}.\frac{m-3}{m-2} \cdots \frac{2}{3}.1$

$\therefore \quad \int_0^{\pi/2} \sin^m x \cos^n x\, dx$

$$= \frac{\left\{\left(\frac{n-1}{2}\right)\left(\frac{n-3}{2}\right) \cdots \frac{1}{2}\right\}\left\{\left(\frac{m-1}{2}\right)\left(\frac{m-3}{2}\right) \cdots 1\right\}}{\left\{\left(\frac{m+n}{2}\right)\left(\frac{m+n-2}{2}\right) \cdots \frac{3}{2}\right\}}$$

$$= \frac{\Gamma\left(\frac{m+1}{2}\right).\Gamma\left(\frac{n+1}{2}\right)}{2\,\Gamma\left(\frac{m+n+2}{2}\right)}.$$

Case III: When n is odd and m is even

Putting $x = \frac{\pi}{2} + x$, it reduces to case II. So, the formula is true in this case also.

Case IV: When m and n are both odd

Repeating the formula (1) above till the power of cos x is unity, we get

$$\int_0^{\pi/2} \sin^m x \cos^n x\, dx = \frac{n-1}{m+n}\cdot\frac{n-3}{m+n-2} \cdots \frac{2}{m+3}\int_0^{\pi/2} \sin^m x \cos x\, dx$$

$$= \frac{(n-1)(n-3)\ldots 2}{(m+n)(m+n-2)\ldots(m+3)}\cdot\frac{m-1}{m+1}\cdot\frac{m-3}{n-1}\cdots\frac{2}{4}\times\int_0^{\pi/2} \sin x \cos x\, dx$$

$$= \frac{\left\{\left(\frac{n-1}{2}\right)\left(\frac{n-3}{2}\right)\ldots 1\right\}\left\{\left(\frac{m-1}{2}\right)\left(\frac{m-3}{2}\right)\ldots 1\right\}}{\left\{\left(\frac{m+n}{2}\right)\left(\frac{m+n-2}{2}\right)\ldots 2\right\}}\cdot\frac{1}{2}$$

$$= \frac{\Gamma\left(\frac{n+1}{2}\right)\Gamma\left(\frac{m+1}{2}\right)}{2\,\Gamma\left(\frac{m+n+2}{2}\right)}$$

9. Obtain reduction formula for

(i) $\int x^n \sin mx\, dx$, (ii) $\int x^n \cos mx\, dx$.

(i) Integrating by parts,

$$I_n = \int x^n \sin mx\, dx = x^n\left(\frac{-\cos mx}{m}\right) - \int nx^{n-1}\left(\frac{-\cos mx}{m}\right)dx$$

$$= -\frac{x^n \cos mx}{m} + \frac{n}{m}\int x^{n-1} \cos mx\, dx.$$

Integrating by parts again,

$$I_n = \frac{-x^n \cos mx}{m} + \frac{n}{m}\left[\frac{x^{n-1}\sin mx}{m} - \frac{n-1}{m}\int x^{n-2}\sin mx\, dx\right]$$

$$= \frac{-x^n \cos mx}{m} + n\frac{x^{n-1}\sin mx}{m^2} - \frac{n(n-1)}{m^2}\int x^{n-2}\sin mx\, dx,$$

which is the required reduction formula.

(ii) Do yourself just as above.

Required formula is

$$\frac{x^n \sin mx}{m} + \frac{n\,x^{n-1}\cos mx}{m^2} - \frac{n(n-1)}{m^2}\int x^{n-2}\cos mx\, dx.$$

10. Obtain reduction formula for

(i) $\int x\sin^n x\,dx$, (ii) $\int x\cos^n x\,dx$.

(i) Let $I_n = \int x\sin^n x\,dx = \int \left(x\sin^{n-1} x\right)\sin x\,dx$

Integrating by parts,

$$I_n = \left(x\sin^{n-1} x\right)\left(-\cos x\right) + \int \cos x\left\{\sin^{n-1} x + x\left(n-1\right)\sin^{n-2} x\cos x\right\}dx$$

$$= -x\cos x\sin^{n-1} x + \int \sin^{n-1} x\cos x\,dx + \left(n-1\right)\int x\sin^{n-2} x\cos^2 x\,dx$$

$$= -x\cos x\sin^{n-1} x + \frac{\sin^n x}{n} + \left(n-1\right)\int x\sin^{n-2} x\left(1-\sin^2 x\right)dx$$

$\therefore$ $$\left\{\because \int \sin^{n-1} x\cos x\,dx = \int \sin^{n-1} x\,d\left(\sin x\right)\right\}$$

$$= -x\cos x\sin^{n-1} x + \frac{\sin^n x}{n} + \left(n-1\right)\int x\sin^{n-2} x\,dx - \left(n-1\right)\int x\sin^n x\,dx$$

$$= -x\cos x\sin^{n-1} x + \frac{\sin^n x}{n} + \left(n-1\right)I_{n-2} - \left(n-1\right)I_n$$

or $$I_n\left(1+n-1\right) = -x\cos x\sin^{n-1} x + \frac{\sin^n x}{n} + \left(n-1\right)I_{n-2}$$

i.e., $$\int x\sin^n x\,dx = \frac{-x\cos x\sin^{n-1} x}{n} + \frac{\sin^n x}{n^2} + \left(\frac{n-1}{n}\right)\int x\sin^{n-2} x\,dx.$$

(ii) Do yourself as above.

11. Find a reduction formula for the integral $\int \frac{\sin nx}{\sin x}dx$ and show that

$\int_0^{\pi} \frac{\sin nx}{\sin x}dx = \pi$ or 0 according as n is odd or even.

We know from trigonometry,

$$\sin nx - \sin\left(n-2\right)x = 2\cos\left(n-1\right)x\sin x$$

or $$\frac{\sin nx - \sin\left(n-2\right)x}{\sin x} = 2\cos\left(n-1\right)x$$

or $$\frac{\sin nx}{\sin x} = 2\cos(n-1)x + \sin(n-2)x$$

$\therefore$ $$\int \frac{\sin nx}{\sin x}\,dx = \frac{2\sin(n-1)x}{n-1} + \int \frac{\sin(n-2)x\,dx}{\sin x}$$

If $I_n = \int \frac{\sin nx}{\sin x}\,dx$ then

$$I_n = \frac{2\sin(n-1)x}{n-1} + I_{n-2}$$ is the required reduction formula.

2nd part Let $u_n = \int_0^{\pi} \frac{\sin nx}{\sin x}\,dx$

$\therefore$ $$u_n = \left[\frac{2\sin(n-1)x}{n-1}\right]_0^{\pi} + u_{n-2} \qquad \text{(as above)}$$

$$= 0 + u_{n-2} = u_{n-2}$$

Repeating this process,

$$u_n = u_{n-2} = u_{n-4} = \cdots.$$

When n is even,

$$u_2 = \int_0^{\pi} \frac{\sin 2x}{\sin x} \qquad (\because u_n = u_{n-2} = u_{n-4} = u_{n-6} = \cdots = u_2)$$

$$= 2\int_0^{\pi} \cos x\,dx = 2[\sin x]_0^{\pi} = 0.$$

When n is odd,

$$u_1 = \int_0^{\pi} \frac{\sin x}{\sin x}\,dx \qquad (\because u_n = u_{n-2} = u_{n-4} = u_{n-6} = \cdots u_1)$$

$$= [x]_0^{\pi} = \pi.$$

12. Prove the reduction formula

$$\int (a^2+x^2)^{n/2}\,dx = \frac{x(a^2+x^2)^{n/2}}{n+1} + \frac{na^2}{n+1}\int (a^2+x^2)^{(n-1)/2}\,dx$$

We have

$$\int 1(a^2+x^2)^{n/2}\,dx$$

$$= (a^2+x^2)^{n/2}\,x - \int \frac{n}{2}(a^2+x^2)^{(n/2)-1}\,2x\cdot x\,dx \text{ (integrating by parts)}$$

$$= x(a^2+x^2)^{n/2} - n\int (a^2+x^2)^{(n/2)-1}(a^2+x^2-a^2)\,dx$$

$$= x(a^2+x^2)^{n/2} - n\int (a^2+x^2)^{n/2}\,dx + na^2\int (a^2+x^2)^{(n/2)-1}\,dx$$

or $$(1+n)\int\left(a^2+x^2\right)^{n/2}dx = x\left(a^2+x^2\right)^{n/2} + na^2\int\left(a^2+x^2\right)^{(n/2)-1}dx$$

$$\therefore\ \int\left(a^2+x^2\right)^{n/2}dx = \frac{x\left(a^2+x^2\right)^{n/2}}{n+1} + \frac{na^2}{n+1}\int\left(a^2+x^2\right)^{(n/2)-1}dx.$$

13. Obtain a reduction formula for $\int x^m\sqrt{2ax - x^2}\,dx$, m being a positive integer.

We have $I_n = \int x^m\sqrt{2ax-x^2}\,dx = \int x^{(2m+1)/2}(2a-x)^{1/2}\,dx$

Integrating by parts, we get

$$I_n = x^{(2m+1)/2}\left[-\frac{2}{3}(2a-x)^{3/2}\right] + \frac{2}{3}\left(m+\frac{1}{2}\right)\int x^{(2m-1)/2}(2a-x)^{3/2}\,dx$$

$$= -\frac{2}{3}x^{(2m+1)/2}(2a-x)^{3/2} + \frac{2}{3}\left(m+\frac{1}{2}\right)\int x^{(2m-1)/2}(2a-x)^{1/2}\,(2a-x)dx$$

$$= -\frac{2}{3}x^{(2m+1)/2}(2a-x)^{3/2} + \frac{2}{3}\left(m+\frac{1}{2}\right)\cdot 2a\int x^{(2m-1)/2}(2a-x)^{1/2}\,dx$$

$$-\frac{2}{3}\left(m+\frac{1}{2}\right)\int x^{(2m+1)/2}(2a-x)^{1/2}\,dx$$

or $$\left(1+\frac{2}{3}m+\frac{1}{3}\right)\int x^m\sqrt{2ax-x^2}\,dx$$

$$= -\frac{2}{3}x^{m+\frac{1}{2}}(2a-x)^{3/2} + \frac{2m+1}{3}2a\int x^{m-\frac{1}{2}}(2a-x)^{\frac{1}{2}}dx$$

$$\therefore \int x^m\sqrt{2ax-x^2}\,dx$$

$$= -\frac{x^{m-1}\left(2a-x^2\right)^{\frac{3}{2}}}{m+2} + \frac{2m+1}{m+2}\int x^{m-1}\left(2ax-x^2\right)^{\frac{1}{2}}\,dx,$$

which is the required reduction formula.

14. Find a reduction formula for $\int \frac{x^{2n+1}}{\sqrt{1-x^2}}dx$, where n is a positive integer.

Let $I = \int \frac{x^{2n+1}}{\sqrt{1-x^2}}dx$

We have $I=\int \frac{x^{2n+1}}{\sqrt{1-x^2}}dx=-\int \frac{x^{2n-1}\left(-x^2\right)dx}{\sqrt{1-x^2}}$

$$=-\int \frac{x^{2n-1}\left(1-x^2\right)}{\sqrt{1-x^2}}dx+\frac{x^{2n-1}}{\sqrt{1-x^2}}dx$$

$$=-\int x^{2n-1}\left(1-x^2\right)^{1/2}dx+\int \frac{x^{2n-1}}{\sqrt{1-x^2}}dx.$$

Integrating by parts the first integral, we get

$$I=-\left(1-x^2\right)^{1/2}\frac{x^{2n}}{2n}+\int \frac{1}{2\sqrt{1-x^2}}(-2x)\frac{x^{2n}}{2n}dx+\int \frac{x^{2n-1}}{\sqrt{1-x^2}}dx$$

$\therefore \quad 2nI=-\sqrt{1-x^2}\,x^{2n}-I+2n\int \frac{x^{2n-1}}{\sqrt{1-x^2}}dx$

or $\quad (2n+1)I=-\left(1-x^2\right)^{\frac{1}{2}}x^{2n}+2n\int \frac{x^{2n-1}}{\sqrt{1-x^2}}dx,$

which is the required reduction formula.

15. If $I_n=\int x^n\left(a^2-x^2\right)^{1/2}dx$, prove that $I_n=\frac{x^{n-1}\left(a^2-x^2\right)^{3/2}}{n+2}+\frac{n-1}{n+2}a^2 I_{n-2}$

We have $I_n=\int x^n\left(a^2-x^2\right)^{1/2}dx=\frac{1}{2}\int x^{n-1}2x\left(a^2-x^2\right)^{1/2}dx.$

Integrating by parts, taking x^{n-1} as first function,

$$I_n=\frac{1}{2}\left[x^{n-1}\times\left\{-\frac{\left(a^2-x^2\right)^{3/2}}{3/2}\right\}+\frac{2}{3}\int\left(a^2-x^2\right)^{3/2}\cdot(n-1)x^{n-2}dx\right]$$

$$=-\frac{1}{3}x^{n-1}\left(a^2-x^2\right)^{3/2}+\frac{n-1}{3}\int x^{n-2}\left(a^2-x^2\right)^{3/2}dx$$

$$=-\frac{1}{3}x^{n-1}\left(a^2-x^2\right)^{3/2}+\frac{n-1}{3}\int x^{n-2}\left(a^2-x^2\right)^{1/2}\left(a^2-x^2\right)dx$$

$$=-\frac{1}{3}x^{n-1}\left(a^2-x^2\right)^{3/2}+\frac{n-1}{3}a^2$$

$$\int x^{n-2}\left(a^2-x^2\right)^{1/2} dx - \frac{n-1}{3}\int x^2\left(a^2-x^2\right)^{1/2} dx$$

$$= -\frac{1}{3}x^{n-1}\left(a^2-x^2\right)^{3/2} + \frac{n-1}{3}a^2\, I_{n-2} - \frac{n-1}{3}I_n$$

or $$I_n\left(1+\frac{n-1}{3}\right) = -\frac{1}{3}x^{n-1}\left(a^2-x^2\right)^{3/2} + \frac{n-1}{3}a^2 I_{n-2}$$

$$\therefore \quad I_n = -\frac{x^{n-1}\left(a^2-x^2\right)^{3/2}}{n+2} + \frac{n-1}{n+2}a^2 I_{n-2}$$

16. If $I_n = \int_0^1 x^p\left(1-x^q\right)^n dx$, where p, q and n are positive, prove that $(qn+p+1)I_n = q^n I_{n-1}$

Here we have to connect

$$\int_0^1 x^p\left(1-x^q\right)^n dx \quad \text{and} \quad \int_0^1 x^p\left(1-x^q\right)^{n-1} dx.$$

Clearly λ = smaller index of $x = p$,

μ = smaller index of $\left(1-x^q\right) = n-1$

$$\therefore \quad P = x^{\lambda+1}\left(1-x^q\right)^{\mu+1} = x^{p+1}\left(1-x^q\right)^{(n-1)+1} = x^{p+1}\left(1-x^q\right)^n$$

$$\therefore \quad \frac{dP}{dx} = (p+1)x^p\left(1-x^q\right)^n + n\left(1-x^q\right)^{n-1}\cdot\left(-q\,x^{q-1}\right)\cdot x^{p+1}$$

$$= (p+1)x^p\left(1-x^q\right)^n + n\,q\,x^p\left(1-x^q\right)^{n-1}\left(-x^q\right)$$

$$= (p+1)x^p\left(1-x^q\right)^n + n\,q\,x^p\left(1-x^q\right)^{n-1}\left(1-x^q-1\right)$$

$$= (p+1)x^p\left(1-x^q\right)^n + n\,q\,x^p\left(1-x^q\right)^{n-1}\left(1-x^q\right) - n\,q\,x^p\left(1-x^q\right)^{n-1}$$

$$= (p+1+nq)x^p\left(1-x^q\right)^n - n\,q\,x^p\left(1-x^q\right)^{n-1}$$

Integrating between 0 and 1,

$$\left[x^{p+1}\left(1-x^q\right)^n\right]_0^1 = (p+1+nq)I_n - nq\,I_{n-1} \qquad \left(\because P = x^{p+1}\left(1-x^q\right)^n\right)$$

i.e., $$0=(nq+p+1)I_n - nq\, I_{n-1}$$

$\therefore$ $$(nq+p+1)I_n = nq\, I_{n-1}$$

17. Obtain a reduction formula for $\int x^m (\log x)^n\, dx$.

Integrating by parts, we have

$$\int x^m (\log x)^n\, dx = (\log x)^n \frac{x^{m+1}}{m+1} - \int \frac{x^{m+1}}{m+1} n(\log x)^{n-1}\frac{1}{x}dx$$

$$= \frac{x^{m+1}(\log x)^n}{m+1} - \frac{n}{m+1}\int x^m (\log x)^{n-1}\, dx,$$

which is the required reduction formula.

18. Find the reduction formula for $\int \frac{x^m}{(\log x)^n} dx$.

$$\int \frac{x^m}{(\log x)^n} dx = \int x^{m+1}\left[\frac{1}{(\log x)^n}\cdot\frac{1}{x}\right] dx$$

$$= \int x^{m+1}\left[(\log x)^{-n}\cdot\frac{1}{x}\right] dx$$

$$= x^{m+1}\frac{(\log x)^{-n+1}}{-n+1} - \int (m+n)x^m \frac{(\log x)^{-n+1}}{-n+1} dx$$

$$= -\frac{x^{m+1}}{(n-1)(\log x)^{n-1}} + \frac{m+n}{n-1}\int \frac{x^m}{(\log x)^{n-1}} dx,$$

which is the required reduction formula.

Solved Examples

1. Evaluate (i) $\int_0^{\pi/2} \sin^n x\, dx$, (ii) $\int_0^{\pi/2} \cos^n x\, dx$.

(i) Let

$$I_n = \int_0^{\pi/2} \sin^n x\, dx = \int_0^{\pi/2} \sin^{n-1} x \sin x\, dx = -\int_0^{\pi/2} \sin^{n-1} x\, d(\cos x)$$

$$= -\left[\sin^{n-1} x \cos x\right]_0^{\pi/2} + \int_0^{\pi/2} \cos x (n-1)\sin^{n-2} x \cdot \cos x \, dx$$

$$= 0 + (n-1)\int_0^{\pi/2} \sin^{n-2} x\left(1-\sin^2 x\right)dx$$

$$= (n-1)\int_0^{\pi/2} \sin^{n-2} x \, dx - (n-1)\int_0^{\pi/2} \sin^n x \, dx$$

$$= (n-1)I_{n-2} - (n-1)I_n$$

or $$(1+n-1)I_n = (n-1)I_{n-2}$$

or $$I_n = \frac{n-1}{n} I_{n-2} \qquad \text{.....(1)}$$

Similarly, $$I_{n-2} = \int_0^{\pi/2} \sin^{n-2} x \, dx = \frac{n-3}{n-2} I_{n-4}$$

$$I_{n-4} = \frac{n-5}{n-4} I_{n-6} \text{ and so on}$$

(1) becomes $I_n = \dfrac{n-1}{n}\dfrac{n-3}{n-2}\dfrac{n-5}{n-4} I_{n-6}$ and so on.

Now two cases arise.

Case I: When n is odd

$$I_n = \frac{n-1}{n}\frac{n-3}{n-2}\frac{n-5}{n-4} \cdots \frac{4}{5}\frac{2}{3}\int_0^{\pi/2} \sin x \, dx,$$

i.e., $$\int_0^{\pi/2} \sin^n x \, dx = \frac{n-1}{n}\frac{n-3}{n-2}\frac{n-5}{n-4} \cdots \frac{4}{5}\frac{2}{3}.1$$

Case II: When n is even

$$I_n = \frac{n-1}{n}\frac{n-3}{n-2}\frac{n-5}{n-4} \cdots \frac{3}{4}\cdot\frac{1}{2}\int_0^{\pi/2} (\sin x)^0 \, dx,$$

i.e., $$\int_0^{\pi/2} \sin^n x \, dx = \frac{n-1}{n}\cdot\frac{n-3}{n-2}\cdot\frac{n-5}{n-4} \cdots \frac{3}{4}\frac{1}{2}.\frac{\pi}{2}.$$

(ii) Proceeding as above, we can see that

$$\int_0^{\pi/2} \sin^n x \, dx = \int_0^{\pi/2} \cos^n x \, dx.$$

The above formula is called Wallis's formula.

It may be observed that in the numerator we go on writing $(n-1)\ (n-3)\ (n-5)$ till we get 2 or 1 and in the denominator. We go on writing $n(n-2)\ (n-4)$ till we get 2 or 1 and if n is even, multiply by $\pi/2$.

2. Evaluate (i) $\int_0^{\pi/2} \sin^6 x\, dx$, (ii) $\int_0^{\pi/2} \cos^9 x\, dx$

(i) Here n is even. So,

$$\int_0^{\pi/2} \sin^6 x\, dx = \frac{(6-1)(6-3)(6-5)}{6\cdot(6-2)(6-4)}\cdot\frac{\pi}{2} = \frac{5}{6}\cdot\frac{3}{4}\cdot\frac{1}{2}\cdot\frac{\pi}{2} = \frac{5\pi}{32}$$

(ii) Here n is odd. So,

$$\int_0^{\pi/2} \cos^9 x\, dx = \frac{(9-1)(9-3)(9-5)(9-7)}{9\cdot(9-2)(9-4)(9-6)(9-8)} = \frac{8\cdot 6\cdot 4\cdot 2}{9\cdot 7\cdot 5\cdot 3\cdot 1}.$$

3. Prove that $\int_0^{\pi/2} \sin^{2m}\theta\, d\theta = \frac{(2m)!}{\{2^m m!\}^2}\cdot\frac{\pi}{2}$

Since here $n = 2m$ (even), so

$$\int_0^{\pi/2} \sin^{2m}\theta\, d\theta = \frac{(2m-1)(2m-3)\ \cdots\ 3\cdot 1}{2m\cdot(2m-2)\ \cdots\ 4\cdot 2}\cdot\frac{\pi}{2}$$

$$= \frac{(2m-1)(2m-3)\ \cdots\ 3\cdot 1}{2m\cdot(2m-2)\ \cdots\ 4\cdot 2}\times\frac{2m(2m-2)\ \cdots\ 4\cdot 2}{2m(2m-2)\ \cdots\ 4\cdot 2}\times\frac{\pi}{2}$$

$$= \frac{(2m)!}{\{2m(2m-2)\ \cdots\ 4\cdot 2\}^2}\times\frac{\pi}{2}$$

$$= \frac{(2m)!}{\{2^m\cdot m(m-1)\ \cdots\ 2\cdot 1\}^2}\times\frac{\pi}{2} = \frac{(2m)!}{\{2^m m!\}^2}\cdot\frac{\pi}{2}$$

4. If $I_n = \int_0^{\pi/4} \tan^n x\, dx$, prove that $I_n + I_{n-2} = \frac{1}{n-1}$

We have

$$\int \tan^n x\, dx = \int \tan^{n-2} x \tan^2 x\, dx = \int \tan^{n-2} x\left(\sec^2 x - 1\right)dx$$

$$= \int \tan^{n-2} x \sec^2 x\, dx - \int \tan^{n-2} x\, dx$$

$$= \frac{\tan^{n-1} x}{n-1} - \int \tan^{n-2} x\, dx$$

$$\therefore \quad I_n = \int_0^{\pi/4} \tan^n x\, dx = \left[\frac{\tan^{n-1} x}{n-1}\right]_0^{\pi/4} - \int_0^{\pi/4} \tan^{n-2} x\, dx$$

$$= \frac{1}{n-1} - I_{n-2}$$

$$\therefore \quad I_n + I_{n-2} = \frac{1}{n-1}$$

5. Evaluate $\int_0^{\pi/4} \tan^5 x\, dx$

We have

$$\int \tan^n x\, dx = \frac{\tan^{n-1} x}{n-1} - \int \tan^{n-2} x\, dx \qquad \text{(derive it as in example 4)}$$

$$\therefore \quad \int_0^{\pi/4} \tan^5 x\, dx = \left[\frac{\tan^{5-1} x}{5-1}\right]_0^{\pi/4} - \int_0^{\pi/4} \tan^{5-2} x\, dx$$

$$= \frac{1}{4} - \int_0^{\pi/4} \tan^3 x\, dx = \frac{1}{4} - \int_0^{\pi/4} \tan x\left(\sec^2 x - 1\right) dx$$

$$= \frac{1}{4} - \int_0^{\pi/4} \tan x \sec^2 x\, dx + \int_0^{\pi/4} \tan x\, dx$$

$$= \frac{1}{4} - \frac{1}{2}\left[\tan^2 x\right]_0^{\pi/4} + \left[\log \sec x\right]_0^{\pi/4}$$

6. $\int_0^{\pi/4} \tan^7 x\, dx$

We have $= \int \tan^n x\, dx = \dfrac{\tan^{n-1} x}{n-1} - \int \tan^{n-2} x\, dx$

(derive as in example 4)

$$\therefore \quad \int_0^{\pi/4} \tan^7 x\, dx = \left[\frac{\tan^6 x}{6}\right]_0^{\pi/4} - \int_0^{\pi/4} \tan^5 x\, dx$$

$$= \frac{1}{6} - \int_0^{\pi/4} \tan^5 x\, dx \qquad \text{.....(1)}$$

Now $\quad \int_0^{\pi/4} \tan^5 x\, dx = \left[\frac{\tan^4 x}{4}\right]_0^{\pi/4} - \int_0^{\pi/4} \tan^5 x\, dx$

$$= \frac{1}{4} - \int_0^{\pi/4} \tan x \left(\sec^2 x - 1\right) dx$$

$$= \frac{1}{4} - \int_0^{\pi/4} \tan x \ \sec^2 x \ dx + \int_0^{\pi/4} \tan x \ dx$$

$$= \frac{1}{4} - \frac{1}{2}\left[\tan^2 x\right]_0^{\pi/4} + \left[\log \sec x\right]_0^{\pi/4}$$

$$= \frac{1}{4} - \frac{1}{2} + \log\sqrt{2} = \frac{1}{2}\left(\log 2 - \frac{1}{2}\right)$$

$\therefore$ from (1),

$$\int_0^{\pi/4} \tan^7 x \ dx = \frac{1}{6} - \frac{1}{2}\left(\log 2 - \frac{1}{2}\right) = \frac{1}{6} + \frac{1}{4} - \frac{1}{2}\log 2 = \frac{5}{12} - \frac{1}{2}\log 2$$

7. Evaluate $\int_0^{\pi/2} \cos^n x \cos nx \, dx$

Let $\qquad I_n = \int_0^{\pi/2} \cos^n x \ \cos \ nx \ dx$

Integrating by parts,

$$I_n = \left[\frac{\cos^n x \sin nx}{n}\right]_0^{\pi/2} - \int_0^{\pi/2} \frac{\sin nx}{x} n \cos^{n-1} x(-\sin x) dx$$

$$= 0 + \int_0^{\pi/2} \cos^{n-1} x \sin nx \sin x \ dx$$

But, $\qquad \cos(n-1)x = \cos nx . \cos x + \sin nx \sin x$

or $\qquad \sin nx \sin x = \cos(n-1)x - \cos nx \cos x$

$\therefore \qquad I_n = \int_0^{\pi/2} \cos^{n-1} x\left\{\cos(n-1)x - \cos nx \cos x\right\} dx$

$$= \int_0^{\pi/2} \cos^{n-1} x \cos(n-1)x \ dx - \int_0^{\pi/2} \cos^n x \cos nx dx$$

$$= I_{n-1} - I_n$$

or $\qquad 2I_n = I_{n-1}; \qquad \therefore \ I_n = \frac{1}{2} I_{n-1}$

Similarly $I_{n-1} = \frac{1}{2} I_{n-2}$

$\therefore \quad I_n = \frac{1}{2} \cdot \frac{1}{2} I_{n-2}$ and so on.

$$\therefore \quad I_n = \frac{1}{2^n} I_{n-n} = \frac{1}{2^n} I_0 = \frac{1}{2^n} \int_0^{\pi/2} \cos^0 x \cos 0 \, dx$$

$$= \frac{1}{2^n} \int_0^{\pi/2} dx = \frac{1}{2} \frac{\pi}{2^n} = \frac{\pi}{2^{n+1}}$$

8. If $I_{m,n} = \int_0^{\pi/2} \cos^m x \sin nx \, dx$, prove that

$$(m+n) I_{m,n} = 1 + m I_{m-1, n-1}$$

or $$I_{m,n} = \frac{1}{m+n} + \frac{m}{m+n} - I_{m-1, n-1}$$

We have $$I_{m,n} = \int_0^{\pi/2} \cos^m x \sin nx \, dx$$

$$= [-\frac{\cos^m x \cos nx}{n}]_0^{\pi/2} - \frac{m}{n} \int_0^{\pi/2} \cos^{m-1} x \cos nx \sin x \, dx$$

(integrating by parts)

But, $\quad \sin(n-1)x = \sin nx \cos x - \cos nx \sin x$

$\therefore \quad \cos nx \sin x = \sin nx \cos x - \sin(n-1)x$

$$\therefore \quad \int_0^{\pi/2} \cos^m x \sin nx \, dx = \left[-\frac{\cos^m x \cos nx}{n} \right]_0^{\pi/2}$$

$$-\frac{m}{n} \int_0^{\pi/2} \cos^{m-1} x \{ \sin nx \cos x - \sin (n-1)x \} dx,$$

i.e., $$I_{m,n} = \left[-\frac{\cos^m x \cos nx}{n} \right]_0^{\pi/2} - \frac{m}{n} I_{m,n} + \frac{m}{n} I_{m-1, n-1}$$

$$\therefore \quad \left(1 + \frac{m}{n}\right) I_{m,n} = \frac{1}{n} + \frac{m}{n} I_{m-1, n-1}$$

or $$I_{m,n} = \frac{1}{m+n} + \frac{m}{m+n} I_{m-1, n-1}$$

9. Prove that $\int_0^{\pi/2} \cos^m x \sin nx\, dx = \frac{1}{2^{m+1}}\left(2+\frac{2^2}{2}+\frac{2^3}{3}+\cdots+\frac{2^m}{m}\right)$

(Write down the solution of the above example)

We have

$$I_{m,n} = \int_0^{\pi/2} \cos^m x \sin nx\, dx$$

$$= \frac{1}{m+n} + \frac{m}{m+n} - I_{m-1,n-1}$$

Putting $n = m$,

$$I_{m,n} = \int_0^{\pi/2} \cos^m x \sin nx\, dx$$

$$= \frac{1}{2m} + \frac{1}{2} I_{m-1,\, m-1}$$

$$= \frac{1}{2m} + \frac{1}{2}\left\{\frac{1}{2(m-1)} + \frac{1}{2} I_{m-2,\, m-2}\right\}$$

$$= \frac{1}{2m} + \frac{1}{2^2(m-1)} + \frac{1}{2^2} I_{m-2,\, m-2}$$

$$= \cdots\cdots\cdots\cdots\cdots\cdots\cdots\cdots\cdots\cdots$$

$$= \frac{1}{2m} + \frac{1}{2^2(m-1)} + \frac{1}{2^3(m-2)} + \cdots \text{ to m terms} + \frac{1}{2^m} I_{m-m,\, m-m}$$

$$= \frac{1}{2m} + \frac{1}{2^2(m-1)} + \frac{1}{2^3(m-2)} + \cdots + \frac{1}{2^m} + 0$$

$$= \frac{1}{2m} + \frac{1}{2^2(m-1)} + \frac{1}{2^3(m-2)} + \cdots + \frac{1}{2^m}$$

$$= \frac{1}{2^{m+1}}\left\{\frac{2^m}{m} + \frac{2^{m-1}}{m-1} + \cdots + \frac{2^2}{2} + \frac{2^1}{1}\right\}$$

$$= \frac{1}{2^{m+1}}\left\{2 + \frac{2^2}{2} + \frac{2^3}{3} + \cdots + \frac{2^m}{m}\right\}$$

10. If $I_{m,n} = \int_0^{\pi/2} \cos^m x \cos nx\, dx$, prove that

$$I_{m,n} = \frac{m(m-1)}{m^2 - n^2} I_{m-2,n}$$

Integrating by parts, we get

$$I_{m,n} = \left[\frac{\cos^m x \sin nx}{n}\right]_0^{\pi/2} + \frac{m}{n}\int_0^{\pi/2} \cos^{m-1} x \sin x \sin nx \, dx.$$

Again integrating by parts, we get

$$I_{m,n} = 0 + \frac{m}{n}\left[-\cos^{m-1} x \sin x \frac{\cos nx}{n}\right]_0^{\pi/2}$$

$$+\frac{m}{n}\int_0^{\pi/2} \left\{(m-1)\cos^{m-2} x(-\sin x)\sin x + \cos^m x\right\}\frac{\cos nx}{n} dx$$

$$= 0 + 0 + \frac{m}{n^2}\int_0^{\pi/2} \left\{-(m-1)\cos^{m-2} x\left(1-\cos^2 x\right)+\cos^m x\right\}\cos nx \, dx$$

$$= -\frac{m(m-1)}{n^2} I_{m-2,n} + \frac{m(1+m-1)}{n^2} I_{m,n}$$

$$\therefore \qquad \left(1+\frac{m^2}{n^2}\right) I_{m,n} = -\frac{m(m-1)}{n^2} I_{m-2,n}$$

or $$I_{m,n} = \frac{m(m-1)}{m^2 - n^2} I_{m-2,n}$$

11. If $I_{m,n} = \int_0^{\pi/2} \cos^m x \cos nx \, dx$, prove that

$$(m+n) I_{m,n} = m \, I_{m-1,n-1}$$

We have

$$I_{m,n} = \int_0^{\pi/2} \cos^m x \cos nx \, dx$$

Integrating by parts, we get

$$I_{m,n} = \left[\frac{\cos^m x \sin nx}{n}\right]_0^{\pi/2} + \frac{m}{n}\int_0^{\pi/2} \cos^{m-1} x \sin nx \, dx$$

But, $\cos(n-1)x = \cos nx \cos x + \sin nx \sin x$

$\therefore$ $\sin nx \sin x = \cos(n-1)x - \cos nx \cos x$

$\therefore$
$$I_{m,n} = 0 + \frac{m}{n}\int_0^{\pi/2} \cos^{m-1} x\{\cos(n-1)x - \cos nx \cos x\}dx$$

$$= \frac{m}{n}\int_0^{\pi/2} \cos^{m-1} x \cos(n-1)x\ dx - \frac{m}{n}\int_0^{\pi/2} \cos^m x \cos nx\ dx$$

$$= \frac{m}{n} I_{m-1,\,n-1} - \frac{m}{n} I_{m,\,n}$$

or
$$\left(1+\frac{m}{n}\right) I_{m,\,n} = \frac{m}{n} I_{m-1,\,n-1}$$

$\therefore$
$$(m+n) I_{m,\,n} = m I_{m-1,\,n-1}$$

12. If n be a positive integer greater than one then prove that

$$\int_0^{\pi/2} \cos^{n-2} x \sin nx\ dx = \frac{1}{n-1}$$

We have

$$I_{m,\,n} = \int_0^{\pi/2} \cos^m x \sin nx\ dx$$

$$= \left[-\frac{\cos^m x \cdot \cos nx}{n}\right]_0^{\pi/2} - \frac{m}{n}\int_0^{\pi/2} \cos^{m-1} x \cos nx\ \sin x\ dx$$

(integrating by parts)

$$= \frac{1}{n} - \frac{m}{n}\int_0^{\pi/2} \cos^{m-1} x\{\sin nx \cos x - \sin(n-1)x\}dx$$

$\therefore$ $(\because \sin(n-1)x = \sin nx \cos x - \cos nx \sin x)$

$$= \frac{1}{n} - \frac{m}{n}\int_0^{\pi/2} \cos^m x \sin nx\ dx$$

$$+ \frac{m}{n}\int_0^{\pi/2} \cos^{m-1} x \sin(n-1)x\ dx$$

$$= \frac{1}{n} - \frac{m}{n} I_{m,\,n} + \frac{m}{n} I_{m-1,\,n-1}$$

$$\therefore \qquad \left(1+\frac{m}{n}\right)I_{m,\,n} = \frac{1}{n}+\frac{m}{n}I_{m-1,\,n-1}$$

or $$I_{m,\,n} = \frac{1}{m+n}+\frac{m}{m+n}I_{m-1,\,n-1}$$

Putting $m = n-2$,

$$I_{n-2,\,n} = \frac{1}{2(n-1)}+\frac{n-2}{2(n-1)}I_{n-3,\,n-1}$$

$$= \frac{1}{2(n-1)}+\frac{n-2}{2(n-1)}\left\{\frac{1}{2(n-2)}+\frac{n-3}{2(n-2)}I_{n-4,\,n-2}\right\}$$

$$= \frac{1}{2(n-1)}+\frac{1}{2^2(n-1)}+\frac{n-3}{2^2(n-1)}I_{n-4,\,n-2}$$

$$= \ldots\ldots\ldots\ldots\ldots\ldots\ldots\ldots\ldots\ldots\ldots\ldots$$

$$= \frac{1}{2(n-1)}+\frac{1}{2^2(n-1)}+\frac{1}{2^3(n-1)}+\cdots \text{ to}(n-2) \text{ terms}$$

$$+\frac{1}{2^{n-2}(n-1)}I_{0,\,2}$$

But, $$I_{0,\,2} = \int_0^{\pi/2}\cos^0 x\sin 2x\,dx = \int_0^{\pi/2}\sin 2x\,dx$$

$$= \left[-\frac{\cos 2x}{2}\right]_0^{\pi/2} = 1$$

$$\therefore \qquad I_{n-2,\,n} = \frac{1}{2(n-1)}+\frac{1}{2^2(n-1)}+\frac{1}{2^3(n-1)}+\cdots$$

$$\frac{1}{2^{n-2}(n-1)}+\frac{1}{2^{n-2}(n-1)}$$

$$= \frac{1}{n-1}\left\{\frac{1}{2}+\frac{1}{2^2}+\frac{1}{2^3}+\cdots+\frac{1}{2^{n-2}(n-1)}+\frac{1}{2^{n-2}}\right\}$$

$$= \frac{1}{n-1}\left\{\frac{\frac{1}{2}\left(1-\frac{1}{2^{n-2}}\right)}{1-\frac{1}{2}}+\frac{1}{2^{n-2}}\right\}$$

$$= \frac{1}{n-1}\left\{\frac{2^{n-2}-1}{2^{n-2}}+\frac{1}{2^{n-2}}\right\} = \frac{1}{n-1}$$

13. If $I_m = \int_0^{\infty} e^{-x} \sin^m x \, dx \, (m > 2)$, prove that

$$\left(1+m^2\right)I_m = m(m-1)I_{m-2}$$

Integrating by parts, we get

$$I_m = \left[\sin^m x\left(-e^{-x}\right)\right]_0^{\infty} + \int_0^{\infty} m \sin^{m-1} x \cos x \; e^{-x} dx$$

$$= 0 + m \int_0^{\infty} \left(\sin^{m-1} x \cos x\right) e^{-x} dx$$

Again integrating by parts, we get

$$m\left[\sin^{m-1} x \cos x \left(-e^{-x}\right)\right]_0^{\infty}$$

$$- m \int_0^{\infty} \left(-e^{-x}\right) \cdot \frac{d}{dx}\left(\sin^{m-1} x \cos x\right) dx$$

$$= 0 + m\int_0^{\infty} e^{-x} \left\{(m-1)\sin^{m-2} x \cdot \cos x \cdot \cos x + \sin^{m-1} x(-\sin x)\right\} dx$$

$$= m\int_0^{\infty} e^{-x} \left\{(m-1)\sin^{m-2} x \cos^2 x - \sin^m x\right\} dx$$

$$= m\int_0^{\infty} e^{-x} \left\{(m-1)\sin^{m-2} x\left(1-\sin^2 x\right) - \sin^m x\right\} dx$$

$$= m\int_0^{\infty} e^{-x} \left\{(m-1)\sin^{m-2} x - m \sin^m x\right\} dx$$

$$= m(m-1)I_{m-2} - m^2 I_m$$

$$\therefore \quad \left(1+m^2\right)I_m = m(m-1)I_{m-2}$$

14. If $I_n = \int_0^{\pi/2} x^n \sin x \, dx$ and $n > 1$ then show that

$$I_n + n(n-1)I_{n-2} = n\left(\frac{1}{2}\pi\right)^{n-1}$$

Hence evaluate $\int_0^{\pi/2} x^5 \sin x \, dx$

$$I_n = \int_0^{\pi/2} x^n \sin x \, dx.$$

Integrating by parts,

$$I_n = \left[x^n(-\cos x)\right]_0^{\pi/2} - \int_0^{\pi/2} nx^{n-1}(-\cos x)\, dx$$

$$= 0 + n\int_0^{\pi/2} x^{n-1}\cos x\, dx.$$

Again integrating by parts,

$$I_n = n\left[x^{n-1}\sin x\right]_0^{\pi/2} - n\int_0^{\pi/2}(n-1)x^{n-2}\sin x\, dx$$

$$= n\left(\frac{1}{2}\pi\right)^{n-1} - n(n-1)\int_0^{\pi/2} x^{n-2}\sin x\, dx,$$

i.e., $$I_n = n\left(\frac{1}{2}\pi\right)^{n-1} - n(n-1)I_{n-2}$$

$\therefore$ $$I_n + n(n-1)I_{n-2} = n\left(\frac{1}{2}\pi\right)^{n-1}$$

2nd part

$$I_5 = 5\left(\frac{1}{2}\pi\right)^{5-1} - 5(5-1)I_3$$

$$= 5\left(\frac{1}{2}\pi\right)^4 - 20\left\{3\left(\frac{1}{2}\pi\right)^{3-1} - 3(3-1)I_1\right\}$$

$$= \frac{5}{16}\pi^4 - 15\pi^2 + 120\, I_1$$

$$= \frac{5}{16}\pi^4 - 15\pi^2 + 120\int_0^{\pi/2} x\sin x\, dx$$

$$= \frac{5}{16}\pi^4 - 15\pi^2 + 120.1$$

$$= \frac{5}{16}\pi^4 - 15\pi^2 + 120$$

15. If $I_n = \int_0^{\pi/2} x^n \sin\,(2p+1)x\, dx$ then prove that

$$I_n + \frac{n(n-1)}{(2p+1)^2}I_{n-2} = (-1)^p \cdot \frac{n}{(2p+1)^2}\left(\frac{\pi}{2}\right)^{n-1}$$

where n and p are positive integers.

Hence deduce that $\int_0^{\pi/2} x^3 \sin 3x\,dx = \frac{2}{27} - \frac{\pi^2}{12}$

Integrating by parts, the given integral

$$I_n = \left[-x^n \frac{\cos(2p+1)x}{2p+1}\right]_0^{\pi/2} + \frac{n}{2p+1}\int_0^{\pi/2} x^{n-1}\cos(2p+1)x\,dx$$

$$= 0 + \frac{n}{2p+1}\left[x^{n-1}\frac{\sin(2p+1)x}{2p+1}\right]_0^{\pi/2} - \frac{n(n-1)}{(2p+1)^2} I_{n-2}$$

Integrating by parts again,

$$I_n = \frac{n}{(2p+1)^2}\left(\frac{\pi}{2}\right)^{n-1} \sin(2p+1)\frac{\pi}{2} - \frac{n(n-1)}{(2p+1)^2} I_{n-2}$$

$$\therefore \quad I_n + \frac{n(n-1)}{(2p+1)^2} I_{n-2} = (-1)^n \frac{n}{(2p+1)^2}\left(\frac{\pi}{2}\right)^{n-1}$$

2nd Part

$$I_3 = \int_0^{\pi/2} x^3 \sin 3x\,dx$$

$$= (-1)\frac{3}{3^2}\left(\frac{\pi}{2}\right)^{3-1} - \frac{3.2}{3^2} I_1 \qquad \text{(here } n = 3,\ p = 1\text{)}$$

$$= -\frac{\pi^2}{12} - \frac{2}{3}\int_0^{\pi/2} x \sin 3x\,dx$$

$$= -\frac{\pi^2}{12} - \frac{2}{3}\left[-x\frac{\cos 3x}{3}\right]_0^{\pi/2} - \frac{2}{3^2}\int_0^{\pi/2} \cos 3x\,dx$$

$$= -\frac{\pi^2}{12} - 0 - \frac{2}{3^3}[\sin 3x]_0^{\pi/2}$$

$$= -\frac{\pi^2}{12} + \frac{2}{27} = \frac{2}{27} - \frac{\pi^2}{12}$$

16. Show that $\int_0^a \frac{x^4}{\sqrt{a^2 - x^2}}\,dx = \frac{3a^4\pi}{16}$

Putting $x = a\sin\theta, dx = a\cos\theta\,d\theta$

When $x = 0,\ \theta = 0$; when $x = a,\ \theta = \pi/2$

$\therefore$ given integral

$$= \int_0^{\pi/2} \frac{a^4 \sin^4 \theta}{a \cos\theta} a \cos\theta \, d\theta = a^4 \int_0^{\pi/2} \sin^4\theta \cos^0 \theta \, d\theta$$

$$= a^4 \cdot \frac{\Gamma\left(\frac{5}{2}\right)\Gamma\left(\frac{1}{2}\right)}{2\Gamma(3)} = \frac{a^4 \cdot \frac{3}{2} \cdot \frac{1}{2}\sqrt{\pi}\sqrt{\pi}}{2 \cdot 2 \cdot 1}$$

$$= \frac{3\pi}{16} a^4$$

17. Evaluate $\int_0^1 x^m (1-x)^n \, dx$

Putting $x = \sin^2 \theta$, $dx = 2\sin\theta \cos \theta \, d\theta$.

When $x = 0$, $\theta = 0$; when $x = 1$, $\theta = \pi/2$

$$\therefore \text{ integral } = \int_0^{\pi/2} \sin^{2m}\theta \cdot \cos^{2n} \theta \cdot 2\sin \theta \cos \theta \, d\theta$$

$$= 2\int_0^{\pi/2} \sin^{2m+1}\theta \cos^{2n+1} \theta \, d\theta$$

$$= 2\frac{\Gamma(m+1)\Gamma(n+1)}{2\Gamma(m+n+2)} = \frac{\Gamma(m+1)\Gamma(n+1)}{2\Gamma(m+n+2)}.$$

18. Evaluate $\int_0^{2a} x^m \sqrt{2ax - x^2} \, dx$, m being a positive integer

Given integral = $\int_0^{2a} x^{m+\frac{1}{2}} \sqrt{2a - x} \, dx$

Putting $x = 2a\sin^2 \theta$, $dx = 4a\sin \theta \cos \theta \, d\theta$

When $x = 0$, $\theta = 0$; when $x = 2a$, $\theta = \pi/2$.

$$\therefore \text{ integral } = \int_0^{\pi/2} \left(2a \sin^2 \theta\right)^{m+\frac{1}{2}} \sqrt{2a - 2a \sin^2 \theta}.4a \sin\theta \cos\theta \, d\theta$$

$$= \int_0^{\pi/2} 2^{\frac{2m+1}{2}} a^{\frac{2m+1}{2}} \sin^{2m+1} \theta . 2^{1/2} a^{1/2} \cos\theta . 2^2 a$$

$$\times \sin \theta \cos \theta \, d\theta$$

$$= \int_0^{\pi/2} 2^{m+3} a^{m+2} \sin^{2m+2} \cos^2 \theta \, d\theta$$

$$= 2^{m+3} a^{m+2} \frac{\Gamma\left(\frac{2m+3}{2}\right)\Gamma\left(\frac{3}{2}\right)}{2\Gamma\left(\frac{2m+2+2}{2}\right)}$$

$$= 2^{m+2} a^{m+2} \frac{\frac{2m+1}{2} \cdot \frac{2m-1}{2} \ldots \frac{3}{2} \cdot \frac{1}{2}\sqrt{\pi} \cdot \frac{1}{2}\sqrt{\pi}}{(m+2)(m+1)m(m-1) \ldots 2.1}$$

$$= a^{m+2} \frac{(2m+1)(2m-1) \ldots 3.1}{(m+2)(m+1)m(m-1) \ldots 2.1} \pi.$$

19. If $u_n = \int x^n \left(a^2 - x^2\right)^{1/2} dx$, prove that

$$u_n = \frac{x^{n-1}\left(a^2 - x^2\right)^{3/2}}{n+2} + \frac{n-1}{n+2} a^2 u_{n-2}$$

Hence evaluate $\int_0^a x^4 \sqrt{a^2 - x^2}\ dx$

We have

$$u_n = \int x^n \left(a^2 - x^2\right)^{1/2} dx = \frac{1}{2}\int x^{n-1} 2x \left(a^2 - x^2\right)^{1/2} dx$$

Integrating by parts, we get

$$u_n = \frac{1}{2} x^{n-1} \left\{-\frac{2}{3}\left(a^2 - x^2\right)^{3/2}\right\} + \frac{1}{3}(n-1)\int x^{n-2} \left(a^2 - x^2\right)^{3/2} dx$$

$$= -\frac{1}{3} x^{n-1} \left(a^2 - x^2\right)^{3/2} + \frac{1}{3}(n-1)\int x^{n-2} \left(a^2 - x^2\right)^{1/2} \left(a^2 - x^2\right) dx$$

$$= -\frac{1}{3} x^{n-1} \left(a^2 - x^2\right)^{3/2} + \frac{1}{3}(n-1) a^2 \int x^{n-2} \left(a^2 - x^2\right)^{1/2} dx$$

$$- \frac{1}{3}(n-1)\int x^n \left(a^2 - x^2\right)^{1/2} dx$$

$$= -\frac{1}{3} x^{n-1} \left(a^2 - x^2\right)^{3/2} + \frac{1}{3}(n-1) a^2 u_{n-2} - \frac{1}{3}(n-1) u_n$$

$$\therefore\ u_n \left[1 + \frac{1}{3}(n-1)\right] = -\frac{1}{3} x^{n-1} \left(a^2 - x^2\right)^{3/2} + \frac{1}{3}(n-1) a^2 u_{n-2}$$

or $$u_n = \frac{x^{n-1}\left(a^2 - x^2\right)^{3/2}}{n+2} + \frac{n-1}{n+2}a^2 u_{n-2}$$

2nd part

$$\int_0^a x^n \left(a^2 - x^2\right)^{1/2} dx = \left[\frac{-x^{n-1}\left(a^2 - x^2\right)^{3/2}}{n+2}\right]_0^a + \frac{n-1}{n+2}a^2 \int_0^a u_{n-2} dx$$

$$= 0 + \frac{n-1}{n+2}a^2 \int_0^a x^{n-2}\left(a^2 - x^2\right)^{1/2} dx$$

$$\therefore \quad \int_0^a x^4 \left(a^2 - x^2\right)^{1/2} dx = \frac{4-1}{4+2}a^2 \int_0^a x^2 \left(a^2 - x^2\right)^{1/2} dx$$

$$= \frac{3}{6}a^2 . \frac{2-1}{2+2}a^2 \int_0^a \left(a^2 - x^2\right)^{1/2} dx$$

(repeating the same process)

$$= \frac{1}{8}a^4 \left\{\frac{1}{2}x\sqrt{a^2 - x^2} + \frac{1}{2}a^2 \sin^{-1}\frac{x}{a}\right\}_0^a$$

$$= \frac{1}{8}a^4 . \frac{1}{2}a^2 . \frac{\pi}{2} = \frac{1}{32}\pi a^6$$

2nd Method

Putting $x = a\sin\theta$, $dx = a\cos\theta\, d\theta$. When $x = 0$, $\theta = 0$; when $x = a$, $\theta = \pi/2$.

$$\therefore \quad \text{integral} = \int_0^{\pi/2} a^4 \sin^4\theta \sqrt{a^2 - a^2\sin^2\theta}.a\cos\theta\, d\theta$$

$$= a^6 \int_0^{\pi/2} \sin^4\theta . \cos^2\theta\, d\theta = \frac{a^6 \Gamma\left(\frac{5}{2}\right)\Gamma\left(\frac{3}{2}\right)}{2\Gamma(4)}$$

$$= a^6 . \frac{\frac{3}{2}.\frac{1}{2}\sqrt{\pi}.\frac{1}{2}\sqrt{\pi}}{2.3.2.1} = \frac{\pi a^6}{32}$$

20. Show that $\int_0^1 x^2\left(1-x^2\right)^{3/2} dx = \frac{\pi}{32}$

Putting x = sinθ, dx = cosθ dθ. When x = 0, θ = 0; when x = 1, θ = π/2

$$\therefore \quad \text{integral} = \int_0^{\pi/2} \sin^2\theta.\cos^4\theta\, d\theta = \frac{\Gamma\left(\frac{2+1}{2}\right)\Gamma\left(\frac{4+1}{2}\right)}{\Gamma\left(\frac{2+4+2}{2}\right)} = \frac{\Gamma\left(\frac{3}{2}\right)\Gamma\left(\frac{5}{2}\right)}{2\Gamma(4)}$$

$$= \frac{\frac{1}{2}\sqrt{\pi}.\frac{3}{2}.\frac{1}{2}\sqrt{\pi}}{2.3.2.1} = \frac{\pi}{32}$$

21. Show that $\int_0^{2a} x^m\sqrt{2ax-x^2}\, dx = \pi\frac{a^{m+2}\times(2m+1)!}{2^m\times m!(m+2)!}$

Let $\quad I_m = \int_0^{2a} x^m\left(2ax-x^2\right)^{1/2} dx$, then as we know (derive reduction formula 13)

$$I_m = \left[-\frac{x^{m-1}\left(2ax-x^2\right)^{3/2}}{m+2}\right]_0^{2a} + \frac{2m+1}{m+2}a\int_0^a x^{m-1}\sqrt{2ax-x^2}\,dx,$$

i.e., $$I_m = \frac{2m+1}{m+2}aI_{m-1}$$

Repeating this formula,

$$I_m = \frac{2m+1}{m+2}a\frac{2m-1}{m+1}aI_{m-2}$$

$$= \frac{(2m+1)(2m-1)\text{ ... to m factors}}{(m+2)(m+1)\text{ ... to n factors}}a^m I_0 \qquad \text{.....(1)}$$

But $$I_0 = \int_0^{2a}\sqrt{2ax-x^2}\,dx$$

$$= \int_0^{2a}\sqrt{a^2-(x-a)^2}\,dx$$

$$= \left[\frac{1}{2}(x-a)\sqrt{a^2-(x-a)^2} + \frac{1}{2}a^2\sin^{-1}\frac{x-a}{a}\right]_0^{2a}$$

$$\left(\because \int \sqrt{a^2 - x^2}\ dx = \frac{x\sqrt{a^2 - x^2}}{2} + \frac{a^2}{2}\sin^{-1}\frac{x}{a}\right)$$

$$= \frac{1}{2}a^2 \sin^{-1} 1 - \frac{a^2}{2}\sin^{-1}(-1)$$

$$= \frac{a^2}{2}\left\{\frac{\pi}{2} + \frac{\pi}{2}\right\} = \frac{a^2}{2}\pi.$$

$$\therefore \text{ from (1), } I_m = \frac{(2m+1)(2m-1)\ \ldots\ 5.3}{(m+2)(m+1)\ \ldots\ 4.3} a^m \frac{a^2}{2}\pi$$

$$= \frac{1}{2^m}.\frac{(2m+1)!}{m!(m+2)!} a^{m+2}\pi$$

$$= \pi\frac{a^{m+2}.(2m+1)!}{2^m.m!(m+2)!}$$

22. Evaluate $\int_0^a x^4\sqrt{a^2 - x^2}\,dx$

We know (derive reduction formula 15)

$$\int_0^a x^n\sqrt{a^2 - x^2}\,dx = \left[-\frac{x^{n-1}\left(a^2 - x^2\right)^{3/2}}{n+2}\right]_0^a + \frac{n-1}{n+2}a^2 \times \int_0^a x^{n-2}\sqrt{a^2 - x^2}\,dx$$

$$= 0 + \frac{n-1}{n+2}a^2\int_0^a x^{n-2}\sqrt{a^2 - x^2}\,dx$$

$$\therefore \int_0^a x^4\sqrt{a^2 - x^2}\,dx = 0 + \frac{4-1}{4+2}a^2\int_0^a x^2\left(a^2 - x^2\right)^{1/2}dx$$

$$= \frac{3}{6}a^2\frac{2-1}{2+2}a^2\int_0^a \left(a^2 - x^2\right)^{1/2}dx$$

$$= \frac{1}{8}a^4\left[\frac{1}{2}x\sqrt{a^2 - x^2} + \frac{1}{2}a^2\sin^{-1}\frac{x}{a}\right]_0^a$$

$$= \frac{1}{8}a^4\left[\frac{1}{2}a^2\sin^{-1}1\right] = \frac{1}{16}a^6.\frac{\pi}{2}$$

$$= \frac{1}{32}\pi a^6$$

23. Evaluate $\int_0^1 x^m (\log x)^n dx$, where $m \geq 0$ and n is a positive integer

We know that (derive reduction formula 17)

$$\int_0^1 x^m (\log x)^n dx = \frac{x^{m+1}(\log x)^n}{m+1} - \frac{n}{m+1}\int x^m (\log x)^{n-1} dx$$

$$\therefore \quad \int_0^1 x^m (\log x)^n dx = \left[\frac{x^{m+1}(\log x)^n}{m+n}\right]_0^1 - \frac{n}{m+1}\int_0^1 x^m (\log x)^{n-1} dx$$

$$= -\frac{n}{m+1}\int_0^1 x^m (\log x)^{n-1} dx,$$

i.e.,

$$I_{m,n} = -\frac{n}{m+1} I_{m,\,n-1}$$

repeating it, we get

$$I_{m,n} = \left(-\frac{n}{m+1}\right)\left(-\frac{n-1}{m+1}\right) I_{m,\,n-2}$$

$$= \dots\dots\dots\dots\dots\dots\dots\dots$$

$$= (-1)^n \frac{n(n-1)(n-2)\ \dots \text{ to } n \text{ factors}}{(m+1)^n} I_{m,0}$$

But,

$$I_{m,0} = \int_0^1 x^m dx = \left[\frac{x^{m+1}}{m+1}\right]_0^1 = \frac{1}{m+1}$$

$$\therefore \quad I_{m,0} = (-1)^n \frac{n(n-1)(n-2)\ \dots\ 2.1}{(m+1)^n} \cdot \frac{1}{m+1}$$

$$= (-1)^n \frac{n!}{(m+1)^{n+1}}$$

24. Show that $\int_0^\infty e^{-x} x^n dx = \Gamma n$

We have $\quad I_n = \int_0^\infty e^{-x} x^n dx$

Integrating by parts, we get

$$= \left[-x^n e^{-x}\right]_0^\infty + n\int_0^\infty e^{-x} x^{n-1} dx$$

But, $$\underset{x\to\infty}{\text{Lt}}\, x^n e^{-x} = \underset{x\to\infty}{\text{Lt}}\, \frac{x^n}{e^x} = \underset{x\to\infty}{\text{Lt}}\, \frac{nx^{n-1}}{e^x}$$

$$= \underset{x\to\infty}{\text{Lt}} \left\{ \frac{n(n-1)(n-2)\ldots 1}{e^x} x^0 \right\}$$

$$= \frac{n(n-1)(n-2)\ldots 1}{\infty} = 0$$

$\therefore$ $$I_n = nI_{n-1}$$

$$= n(n-1)I_{n-2} \text{ (similarly)}$$

$$= \ldots\ldots\ldots$$

$$= n(n-1)(n-2)\ldots 1.I_0$$

But, $$I_0 = \int_0^\infty e^{-x} x^0 dx = \left[-e^x\right]_0^\infty = 1$$

$\therefore$ $$I_n = n(n-1)(n-2)\ldots 1 = n!$$

Exercise 6.1

1. Evaluate $\int_0^{\pi/2} \sin^{10} x dx$ or $\int_0^{\pi/2} \cos^{10} x dx$

2. If $I_n = \int_0^{\pi/2} \tan^n x dx$, prove that $n(I_{n-1} + I_{n+1}) = 1$. [*Hint*: See example 14.]

3. Evaluate $\int_0^{\pi/2} \sin^4 \theta \cos^2 \theta d\theta$. [*Hint*: $I = \dfrac{\Gamma\left(\frac{4+1}{2}\right)\Gamma\left(\frac{2+1}{2}\right)}{2\Gamma\left(\frac{4+2+2}{2}\right)}$]

4. Evaluate $\int_0^{\pi/2} \sin^4 \theta \cos^6 \theta d\theta$ [*Hint*: $I = \dfrac{\Gamma\left(\frac{4+1}{2}\right)\Gamma\left(\frac{2+1}{2}\right)}{2\Gamma\left(\frac{4+2+2}{2}\right)}$]

5. Evaluate $\int_0^{\pi/2} \sin^3 \theta \cos^3 \theta d\theta$ [*Hint*: $I = \dfrac{\Gamma\left(\frac{3+1}{2}\right)\Gamma\left(\frac{3+1}{2}\right)}{2\Gamma\left(\frac{3+3+2}{2}\right)}$]

6. Evaluate $\int_0^1 x^4(1-x)^{5/2}dx$

[*Hint*: Put $x = \sin\theta, I = \int_0^{\pi/2} \sin^4\theta\cos^6\theta d\theta = \dfrac{\Gamma(5/2)\Gamma(1/2)}{2\Gamma(6)}$]

7. Evaluate $\int_0^3 \sqrt{\dfrac{x^3}{3-x}}dx$

[*Hint*: Put $x = 3\sin^2\theta,\ I = 18\int_0^{\pi/2} \sin^4\theta d\theta = 18\dfrac{\Gamma\left(\frac{5}{2}\right)\Gamma\left(\frac{1}{2}\right)}{2\Gamma(3)}$]

8. Evaluate $\int_0^{2a} x^{9/2}(2a-x)^{-1/2}dx$

[*Hint*: Put $x = 2a\sin^2\theta, I = 2^6a^5\int_0^{\pi/2} \sin^{10}\theta d\theta = 2^6a^5\dfrac{\Gamma\left(\frac{11}{2}\right)\Gamma\left(\frac{1}{2}\right)}{2\Gamma(6)}$]

9. Evaluate $\int_0^{\pi/2} \dfrac{\sin 7x}{\sin x}dx$ [*Hint*: See reduction formula 11]

10. Evaluate $\int (a^2+x^2)^{5/2}dx$ [*Hint*: See reduction formula 12]

11. Evaluate $\int_0^{2a} x^3\sqrt{2ax-x^2}dx$

[*Hint*: Put $x = 2a\sin^2\theta, I = 64a^5\int_0^{\pi/2} \sin^8\theta\cos^2\theta d\theta$

$$= 64a^5\frac{\Gamma\left(\frac{9}{2}\right)\Gamma\left(\frac{3}{2}\right)}{2\Gamma(6)}$$

12. Evaluate $\int_0^a x^2\sqrt{ax-x^2}dx$

[*Hint*: Put $x = a\sin^2\theta,\ I = a^4\int_0^{\pi/2} \sin^6\theta\cos^2\theta d\theta = a^4\dfrac{\Gamma\left(\frac{7}{2}\right)\Gamma\left(\frac{3}{2}\right)}{2\Gamma(5)}$]

Answers

1. $\dfrac{63\pi}{512}$ 3. $\dfrac{\pi}{32}$ 4. $\dfrac{3\pi}{512}$ 5. $\dfrac{1}{12}$

6. $\dfrac{3\pi}{512}$ 7. $\dfrac{27\pi}{8}$ 8. $\dfrac{63}{8}\pi a^5$ 9. π

10. $\dfrac{x(a^2+x^2)^{5/2}}{6}+\dfrac{5a^2}{24}x(a^2+x^2)^{3/2}+\dfrac{5a^2}{16}\left[x\sqrt{a^2+x^2}+a^2\sinh^{-1}\dfrac{x}{a}\right]$

11. $\dfrac{7}{8}\pi a^5$ 12. $\dfrac{5\pi a^5}{256}$

CHAPTER 7

Curve Tracing

The process of determining the shape, location, singular points and other peculiarities of a curve from its equations is called curve tracing.

This will help us to know the area, the length, the volume, centre of gravity etc., of plane curves.

Rules for Tracing of Cartesian Curves

I ***Detect symmetry with the help of following rules***

(i) A curve is symmetrical about x-axis if its equation has only even powers of y (i.e., if by putting – y for y the equation remains unchanged) e.g., $y^2 = 4ax$ is symmetrical about x-axis.

(ii) A curve is symmetrical about y-axis if its equation contains only even powers of x, e.g., $x^2 = 4ay$ is symmetrical about y-axis.

(iii) If the equation of the curve contains only even powers of x and y, it is symmetrical about both axes, e.g., $x^2 + y^2 = a^2$ is symmetrical about both axes.

(iv) If on interchanging x and y the equation of the curve remains unchanged, it is symmetrical about the line y = x, e.g., $x^3 + y^3 = 3axy$ is symmetrical about the line y = x.

II ***Check whether the curve passes through the origin***

For this put x = 0, y = 0 in the equation of the curve; if the equation is satisfied, the curve passes through the origin, e.g., $x^2 = 4ay$ passes through the origin.

III ***Find the equation of the tangent or tangents at origin***

If the curve passes through the origin, find the equation of the tangent or tangents at origin. For this, equate the lowest-degree term in the equation to zero, e.g., in the curve $x^2 = 4ay$, the lowest-degree term is 4ay. Equating it to zero, we get $y = 0$, i.e., x-axis is the tangent at origin.

IV ***Find the point of intersection of the curve and the axes***

To know the point of intersection of the curve and the x-axis, put $y = 0$ in the equation of the curve. To know the point of intersection of the curve with y-axis, put $x = 0$ in the equation of the curve.

V ***Find the regions of the curve***

For this, solve the equation for y. Find such values of x for which y becomes imaginary. For example, if the curve is

$$y^2(2a - x) = x^3 \text{ then } y^2 = \frac{x^3}{2a - x}$$

If $x < 0$, y is imaginary and if $x > 2a$, y is again imaginary. So, no part of the curve is either in the left of the origin or in the right of $x = 2a$.

See the behavior of y (or x) for different values of x (or y) giving particular importance to those values for which y (or x) tends to infinity or zero.

VI ***Find $\frac{dy}{dx}$ and the points where tangents are parallel to axes.***

VII ***Find the asymptote or asymtotes, if any***

Equating to zero the coefficient of highest power of y we get asymptote parallel to y-axis and equating the coefficient of highest power of x to zero we get asymptote parallel to x-axis.

For example, if the curve be of nth degree and the term containing y^n be absent then the coefficient of y^{n-1} equated to zero will give asymptote parallel to y-axis.

If the equation of the curve be of nth degree and the coefficient of x^n is not zero then there will be no asymptote parallel to x-axis. Similarly, if the coefficient of y^n is not zero then there will be no asymptote parallel to y-axis. For example, $x^3 + y^3 = 3axy$ will have no asymptote parallel either to x-axis or y-axis.

Tracing of Polar Curves

(i) ***Symmetry***

(a) Replace θ by – θ; if the equation of the curve remains unchanged, the curve is symmetrical about the initial line.

(b) Replace r by – r; if the equation of the curve is unchanged, i.e., if only even powers of r occur, the curve is symmetrical about the pole.

(c) Replace θ by $\pi - \theta$; if the equation of the curve is unchanged, the curve is symmetrical about the line $\theta = \frac{\pi}{2}$ (i.e., symmetrical about OX).

(d) Replace θ by $\frac{\pi}{2} - \theta$; if the equation remains unchanged, the curve is symmetrical about the line $\theta = \frac{\pi}{4}$.

(ii) ***Whether the curve passes through the pole***

If $r = 0$ for some real value of θ, the curve passes through the pole.

If the curve passes through the pole, the value of θ for which $r = 0$ gives the equation of the tangent at pole.

If $r = 0$ for more than one value of θ, the curve is said to have a loop.

(iii) ***Intersection of the curve with the initial line OX and also with OY***

For this, we know r by putting $\theta = 0$ and $\theta = \frac{\pi}{2}$ in the equation of the curve.

(iv) ***Region of the curve***

To find the region in which no portion of the curve lies, we determine those values of θ for which r is imaginary.

(v) ***Asymptotes, if any***

If the curve possesses never-ending branch, find its asymptote. This is sometimes best accomplished by transforming the equation to the Cartesian form.

(vi) ***Some more points***

Some specific points on the curve should be plotted. For this we determine corresponding values of r for some special values of θ, e.g., $\theta = 0, \frac{\pi}{6}, \frac{\pi}{4}, \frac{\pi}{3}, \frac{\pi}{2}, \ldots$ and plot them.

(vii) Sometimes tracing of the curve becomes easy by transforming its equation into cartesian form.

Examples

1. Trace the curve $y^2 (2a - x) = x^3$

(i) Since only even power of y occurs, so the curve is symmetrical about x-axis.

(ii) Since $x = 0, y = 0$ satisfies the equation, so the curve passes through origin.

(iii) Equating the lowest-degree term to zero, the equation of tangent at origin is $2ay^2 = 0$, i.e., $y^2 = 0$, i.e., two coincident tangents at origin. So, origin is a cusp.

(iv) Putting $x = 0$, we get $y = 0$ and putting $y = 0$ we get $x = 0$. So, the curve does not cut the axes at any point other than the origin.

(v) Solving the equation of the curve,

$$y^2 = \frac{x^3}{2a - x}$$

Clearly, y^2 is –ve, i.e., y is imaginary if $x > 2a$

So, no portion of curve is beyond the line $x = 2a$

Also y is imaginary if x is –ve. So, no portion of the curve is in the left of $x = 0$, (i.e., y-axis).

At $x = 2a$, $y = \pm\infty$, so the curve goes to infinity

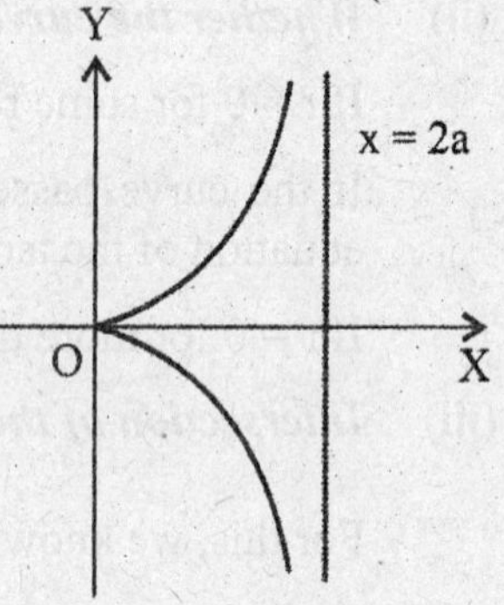

(vi) Equating to zero the coefficient of the highest power of y we get $2a - x = 0$, i.e., $x = 2a$. So the asymptote parallel to y-axis is given by $x = 2a$.

Hence the curve is as shown in the figure. The curve is called a cissoid.

2. Trace the curve $y^2 (a - x) = x^2 (a + x)$

(i) The curve is symmetrical about x-axis since only even power of y occurs.

(ii) Putting $x = 0$, $y = 0$, the equation is satisfied. So, it passes through the origin.

(iii) Equating lowest-degree term to zero, we get $y^2 - x^2 = 0$, i.e., $y = \pm x$. So, $y = \pm x$ are two real and distinct tangents at origin. So, origin is a node.

(iv) Putting $y = 0$ in the equation of the curve we get $x^2(a + x) = 0$; $\therefore$ $x = 0$ and $x = -a$. So, the curve cuts x-axis at $(-a, 0)$ and $(0, 0)$. Again putting $x = 0$, we get $y = 0$. So, the curve cuts y-axis at $(0, 0)$

(v) Solving for y, we get

$$y^2 = \frac{x^2(a + x)}{a - x};$$

$$\therefore \quad y = x\sqrt{\frac{a + x}{a - x}}$$

Clearly y is imaginary when $x < -a$ and when $x > a$. So, no portion of the curve is in the left of the line $x = -a$, and in the right of $x = a$. Also $y = 0$

when $x = 0$ and $x = -a$, and $y^2 > 0$ if $-a < x < 0$, so there is a loop between $x = 0$ and $x = -a$.

(vi) Equating the coefficient of the highest power of y to zero, $a - x = 0$, $x = a$ is the equation of the asymptote parallel to y-axis.

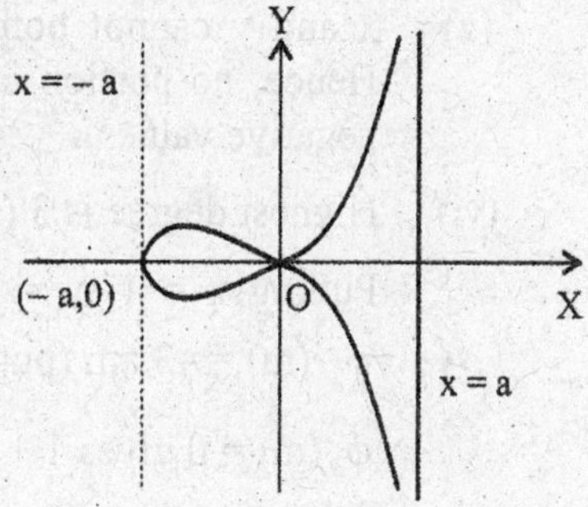

(vii) At $(-a, 0)$, $\frac{dy}{dx} = \infty$, i.e., tangent at $(-a, 0)$ is at right angles to the axis of x.

Hence the curve is as shown.

3. Trace the curve $y^2 (a + x) = x^2 (3a - x)$

(i) Since only even powers of y occurs, the curve is symmetrical about x-axis.

(ii) The curve passes through origin and tangent at origin is given by $y^2 = 3x^2$, which gives two real and distinct tangents, their equations being

$$y = \pm\sqrt{3}x.$$

(iii) Putting $y = 0$, $x^2(3a - x) = 0$; $\therefore$ $x = 0, 3a$. So, the curve cuts x-axis at (0, 0) and (3a, 0). Putting $x = 0$, we get $y = 0$, so the curve cuts y-axis only at (0, 0).

(iv) Equating the coefficient of the highest power of y to zero, the asymptote parallel to y-axis is $x = -a$.

(v) Solving for y, $y = \pm x\sqrt{\frac{3a - x}{a + x}}$.

If $x > 3a$ or $x < -a$, y is imaginary. So, no portion of the curve is in the right of $x = 3a$ or in the left of $x = -a$.

Hence the curve is as shown.

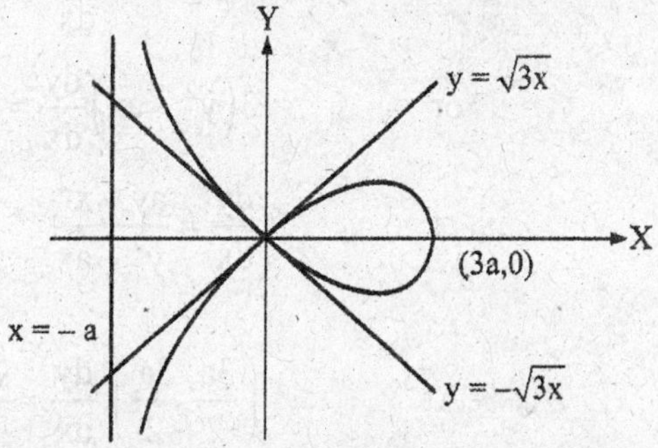

4. Trace the curve $x^3 + y^3 = 3axy$ (Folium of Descartes)

(i) Interchanging x and y, the equation remains unaltered. So, the curve is symmetrical about the line $y = x$.

(ii) There is no constant term in the equation of the curve, so it passes through the origin.

(iii) Equating the lowest-degree term to zero we get $3axy = 0$; i.e., $x = 0$, $y = 0$. So, at origin both axes are tangent. So origin is node.

(iv) Putting $y = 0$ in the equation of the curve, we get $x = 0$, and putting $x = 0$ we get $y = 0$. So, the curve cuts the axes at origin only.

(v) x and y cannot both be negative since then LHS = –ve and RHS = +ve. Hence, no portion of the curve is in the 3rd quadrant (where x and y have negative values).

(vi) Highest degree is 3 (i.e., $n = 3$)

Putting $x = 1$, $y = m$ in third-degree terms, we have $\varphi_3(m) = 1 + m^3$, $\varphi_2(m) = -3am$ (putting $x = 1$, $y = m$ in 2nd degree terms).

$\varphi_3(m) = 0$ gives $1 + m^2 = 0$, i.e., $(1 + m)(1 - m + m^2) = 0$

$\therefore \quad m = -1$, is the only real value of m

$$\text{Now } c = \frac{\varphi_{n-1}(m)}{\varphi_n'(m)} = -\frac{\varphi_2(m)}{\varphi_3'(m)} = -\frac{-3am}{3m^2} = \frac{a}{m} = -a, \text{ for } m = -1$$

$\therefore$ the only asymptote is $y = -x - a$, i.e., $x + y + a = 0$

(vii) Putting $x = y$ in the equation of the curve we find that the line $y = x$ cuts the curve at the point $\left(\frac{3a}{2}, \frac{3a}{2}\right)$

(viii) Differentiating the equation of the curve,

$$3x^2 + 3y^2\frac{dy}{dx} = 3a\left(y + x\frac{dy}{dx}\right)$$

or

$$3\left(y^2 - ax\right)\frac{dy}{dx} = 3ay - 3x^2$$

$$\therefore \quad \frac{dy}{dx} = \frac{ay - x^2}{y^2 - ax}$$

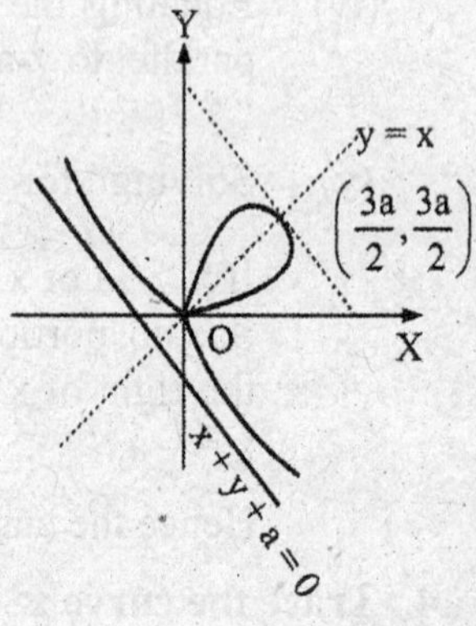

At

$$\left(\frac{3a}{2}, \frac{3a}{2}\right), \frac{dy}{dx} = \frac{a.\frac{3a}{2} - \frac{9a^2}{4}}{\frac{9a^2}{4} - a\frac{3a}{2}}$$

$$= \frac{\frac{3a^2}{2} - \frac{9a^2}{4}}{\frac{9a^2}{4} - \frac{3a^2}{2}} = -1$$

So, the tangent at this point makes an angle 135° with the positive direction of x-axis.

Hence the curve is as shown.

5. Trace the curve $x^4 + y^4 = a^2 (x^2 - y^2)$

(i) Since only even powers of x and y occur, the curve is symmetrical about both axes.

(ii) $x = 0, y = 0$ satisfy the equation of the curve, so the curve passes through the origin.

(iii) Equating the term of lowest degree to zero we get $x^2 - y^2 = 0$, i.e., $y = \pm x$.

So the equations of tangent at origin are $y = x, y = -x$.

(iv) Putting $y = 0$, $x^4 - a^2 x^2 = 0$, i.e., $x^2(x^2 - a^2) = 0$, i.e., $x = 0, + a$.

So, the curve cuts x-axis at three points (0, 0), (a, 0), (0, –a).

Again by putting $x = 0$, $y^4 = -a^2 y^2$ or $y^4 + a^2 y^2 = 0$

$\therefore$ $y = 0$, i.e., the curve cuts y-axis at the origin only

(v) The curve has no asymptote.

(vi) Solving the equation for y,

$$y^4 + a^2 y^2 = a^2 x^2 - x^4 = x^2(a^2 - x^2)$$

If $x > a$ or $x < -a$, y is imaginary

So, no portion of the curve is in the right of $x = a$ or in the left of $x = -a$.

(vii) Differentiating the equation of the curve,

$$4x^3 + 4y^3 \frac{dy}{dx} = a^2 \left(2x - 2y\frac{dy}{dx}\right)$$

$$\therefore \frac{dy}{dx} = \frac{2a^2x - 4x^3}{4y^3 + 2a^2y}$$

$\therefore$ at (a, 0), $\frac{dy}{dx} = \infty$ and

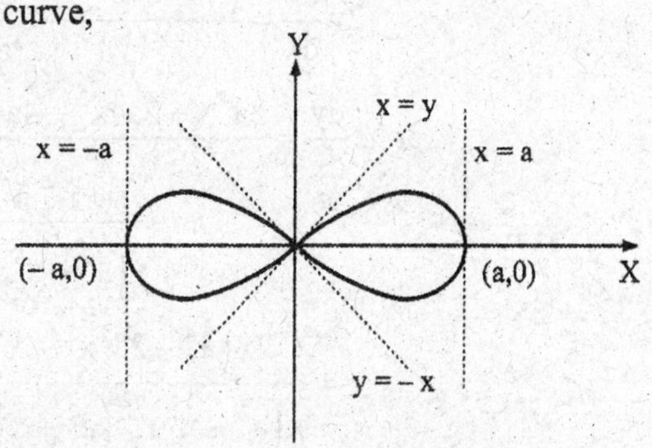

also at (–a, 0), $\frac{dy}{dx} = \infty$.

So, at (a, 0) and at (– a, 0) tangent is perpendicular to x-axis. The curve is as shown.

6. Trace the curve $y^2 (a^2 + x^2) = x^2 (a^2 - x^2)$

(i) The curve is symmetrical about both the axes, since there occur even powers of x and y both. On changing signs of x and y both the equation remains unaltered, so there is symmetry in opposite quadrants.

(ii) The curve passes through the origin as by putting x = 0 and y = 0, the equation is satisfied.

(iii) Equating the lowest-degree term to zero, the tangents at origin are given by $y^2 - x^2 = 0$, i.e., $y = \pm x$.

Hence the origin is a node.

(iv) Putting y = 0, we have $x^2(a^2 - x^2) = 0$, i.e., x = 0, $x = \pm a$. So, the curve cuts the x-axis at three points (0, 0), (a, 0) and (– a, 0).

The curve does not cut y-axis at any point other than (0, 0) since by putting x = 0, we get y = 0.

(v) Equating the coefficient of highest power of y to zero. $a^2 + x^2 = 0$, giving asymptotes parallel to y-axis which are clearly imaginary. There is also no other real asymptote.

(vi) Solving for y, we get

$$y^2 = \frac{x^2(a^2 - x^2)}{a^2 + x^2}$$

If x > a, y is imaginary. Also if x < – a, y is imaginary. So, no portion of the curve lies beyond x = a and x = – a.

(vii) Differentiating the equation of the curve w.r.t. x,

$$2y\frac{dy}{dx} = \frac{(2a^2x - 4x^3)(a^2 + x^2) - 2x.x^2(a^2 - x^2)}{(a^2 + x^2)^2}$$

$$\text{or } 2\frac{dy}{dx} = \frac{2a^4x + 2a^2x^3 - 4a^2x^3 - 4x^5 - 2x^3a^2 + 2x^5}{\dfrac{x(a^2 - x^2)^{1/2}}{(a^2 + x^2)^{1/2}}.(a^2 + x^2)^2}$$

$$= \frac{2x(a^4 - 2a^2x^2 - x^4)}{x(a^2 - x^2)^{1/2}(a^2 + x^2)^{3/2}}$$

$$\therefore \quad \frac{dy}{dx} = \frac{a^4 - 2a^2x^2 - x^4}{(a^2 - x^2)^{1/2}(a^2 + x^2)^{3/2}}$$

$$= \infty \text{ when } x = \pm a$$

Thus at the points (a, 0) and (–a, 0), tangents are perpendicular to x-axis. Hence, the curve is as shown.

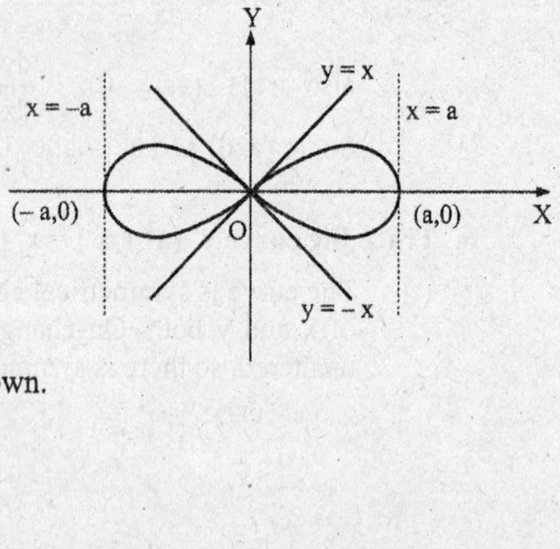

7. Trace the curve $\frac{a^2}{x^2} - \frac{b^2}{y^2} = 1$

The curve is $a^2 y^2 - b^2 x^2 = x^2 y^2$

(i) Since even powers of x and y occur, so the curve is symmetrical about both axes.

(ii) Putting $x = 0$, $y = 0$, the equation of the curve is satisfied, so the curve passes through the origin.

(iii) Equating the lowest-degree term to zero, we get the equation of tangent at the origin as

$$a^2y^2 - b^2x^2 = 0, \text{ i.e., } ay = \pm bx$$

So, origin is a node.

(iv) Equating the coefficient of highest power of y, the asymptotes parallel to y-axis are given by $x^2 - a^2 = 0$, i.e., $x = \pm a$.

Equating the coefficient of highest power of x, $b^2 + y^2 = 0$, i.e., $y^2 = -b^2$, so asymptotes parallel to x-axis are imaginary.

(v) The curve does not intersect the axes at any point other than (0, 0).

(vi) Solving for y, $y^2 (a^2 - x^2) = b^2 x^2$ or $y^2 = \frac{b^2x^2}{a^2 - x^2}$

$$\therefore \quad y = \pm\frac{bx}{\sqrt{a^2 - x^2}}$$

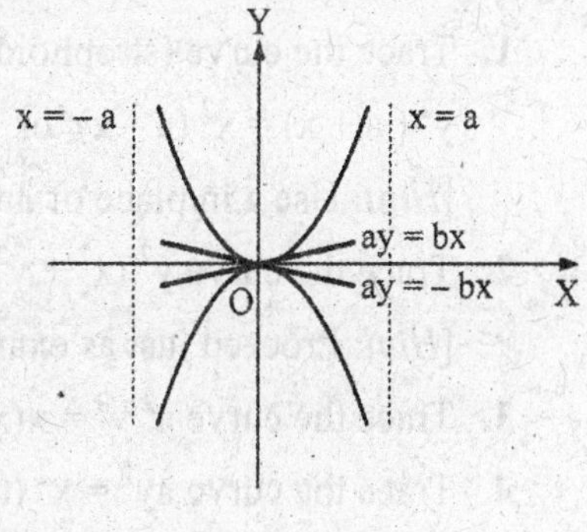

So, y is real only when x lies between a and – a. So, the whole curve lies between x = a and x = – a.

As $x \to \pm a$, $y \to \pm\infty$. The origin is the point of inflexion as $\frac{d^2y}{dx^2} = 0$ at origin and $\frac{d^3y}{dx^3}$ is not zero at origin. So, the curve is as shown.

8. Trace the curve $xy^2 + (x + a)^2 (x + 2a) = 0$

(i) The curve is symmetric about x-axis as only even powers of y occur.

(ii) The curve does not pass through the origin as $x = 0, y = 0$ do not satisfy the equation.

(iii) Putting $x = 0$, we get $2a^3 = 0$, which is impossible. So the curve never cuts the y-axis. Putting $y = 0$, we get $(x + a)^2 (x + 2a) = 0$.

$\therefore$ $x = -a$ and $x = -2a$. So the curve cuts x-axis at $(-a, 0)$ and $(-2a, 0)$.

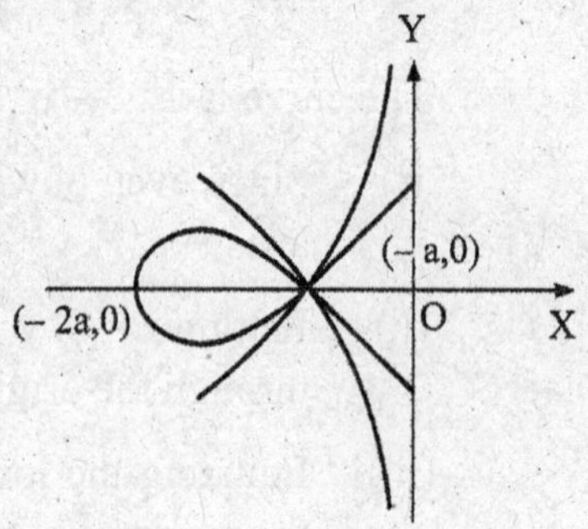

(iv) Equating to zero the coefficient of highest power of y, i.e., of y^2, we get $x = 0$. So, the asymptote parallel to y-axis is $x = 0$.

(v) Solving the equation of the curve for y,

$$y^2 = \frac{(x+a)^2 (x+2a)}{x}$$

Clearly, y is imaginary if $x > 0$. Hence the curve lies in the left of $x = 0$, i.e., the y-axis. Also y is imaginary if $x < -2a$. So no portion of the curve lies in the left of $x = -2a$.

Thus using (iii), we find that there is a loop between $x = -2a$ and $x = -a$.

Tangents at $(-a, 0)$ are parallel to $y = \pm x$. The curve is as shown.

Exercise 7.1

1. Trace the curve (strophoid)

$y^2 (a + x) = x^2 (a - x)$ or $(x^2 + y^2) x - a (x^2 - y^2) = 0$

[*Hint*: Use a in place of 3a in example 3]

2. Trace the curve $y^2 (x^2 + y^2) + a^2 (x^2 - a^2) = 0$

[*Hint*: Proceed just as example 6]

3. Trace the curve $a^2 y^2 = x(x - a)^2$

4. Trace the curve $ay^2 = x^2 (a - x)$

5. Trace the cuve $a^2 y^2 = x^2 (a^2 - x^2)$

6. Trace the semicubical parabola $y^2 = x^3$

7. Trace the curve $(a - x) y^2 = a^2 x$

Note: Tracing itself contains hint. You have to only justify it.

Answers

1.

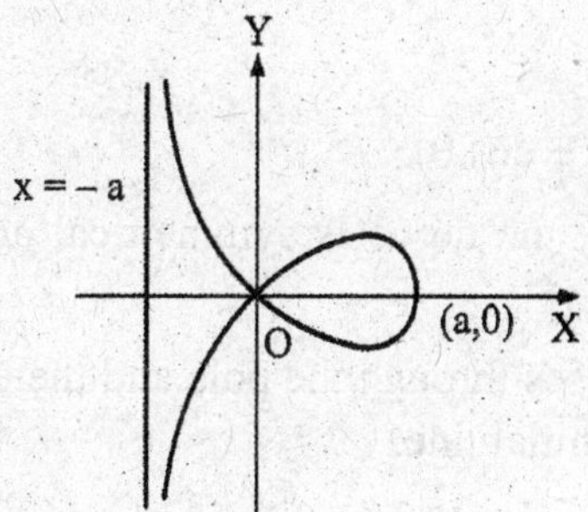

2.

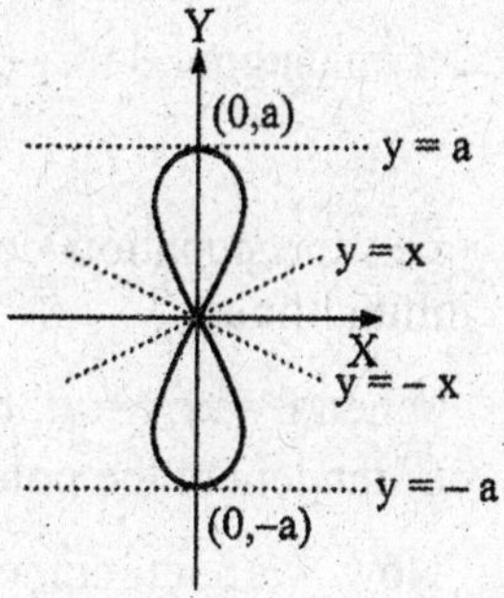

3.

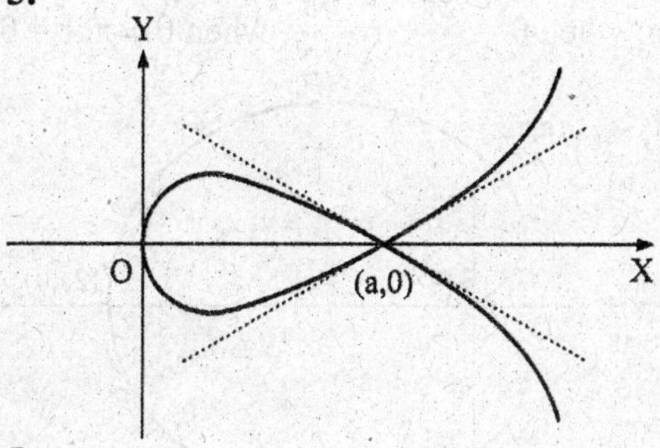

4.

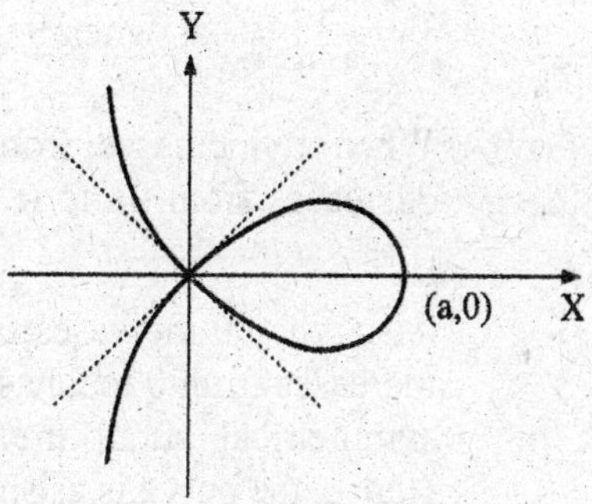

5.

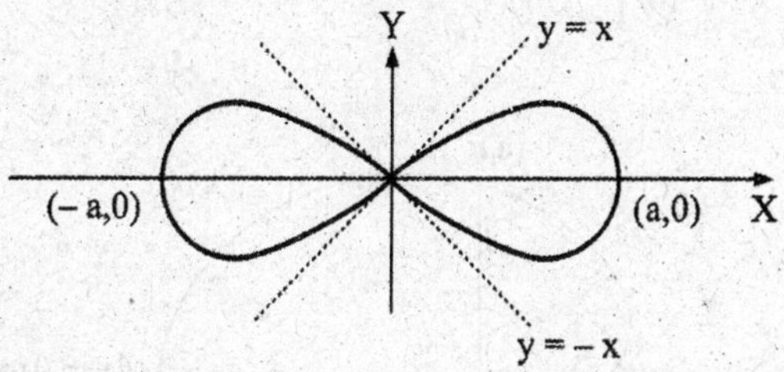

6.

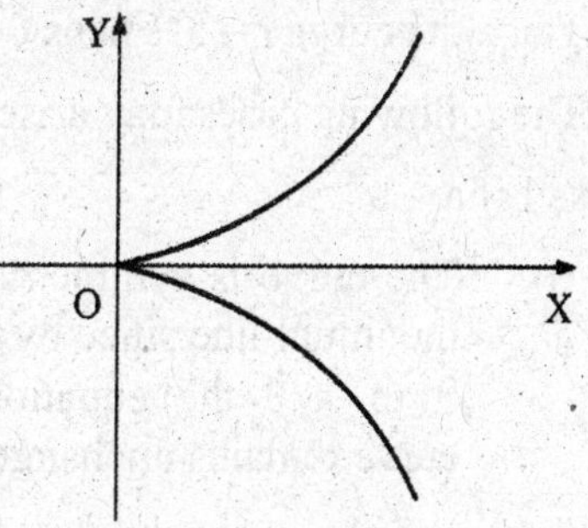

7.

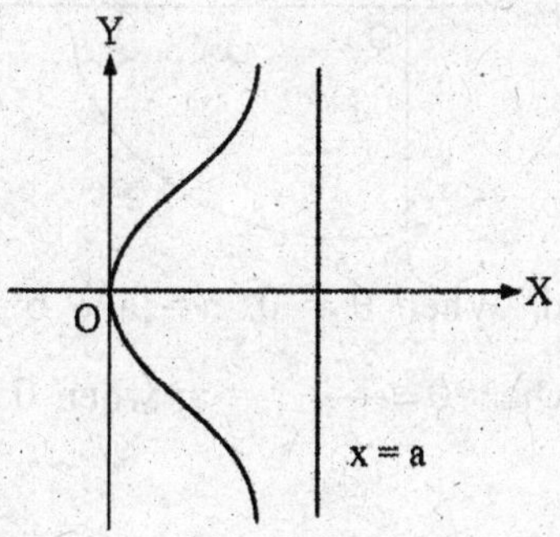

Solved Examples

1. Trace the cardiod $r = a(1 + \cos\theta)$

(i) By changing θ to $-\theta$, we get

$$r = a[1 + \cos(-\theta)] = a(1 + \cos\theta),$$

i.e., the equation does not change. So, the curve is symmetrical about the initial line.

(ii) When $\theta = \pi$, $r = 0$, hence the curve passes through the pole and the equation of tangent at the pole is $\theta = \pi$, i.e., the initial line.

(iii) Now we put some points on the curve. When $\theta = 0$, $r = 2a$; when $\theta = \frac{\pi}{3}$, $r = \frac{3}{2}a$; when $\theta = \frac{\pi}{2}$, $r = a$; when $\theta = \frac{2\pi}{3}$, $r = \frac{a}{2}$, when $\theta = \pi$, $r = 0$.

(iv) When θ increases from 0 to π, r decreases from 2a to 0 as shown in (iii)

Also when θ increases from π to 2π, increases from 0 to 2a as the curve is symmetrical about the initial line. Hence, the curve is as shown.

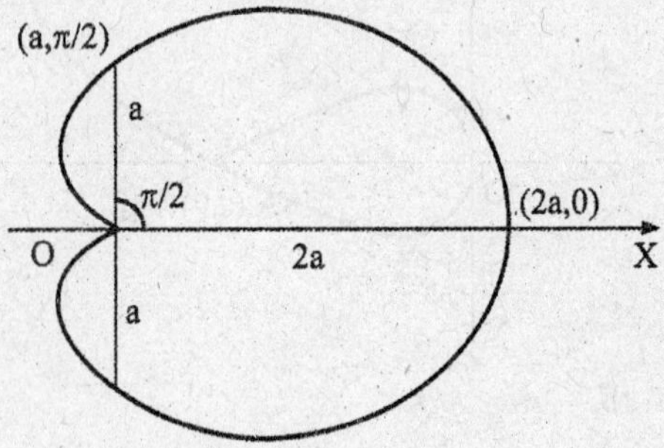

2. Trace the curve $r = a + b\cos\theta$ (limacon)

The following cases may arise:

Case I if $a > b$

(i) The curve is symmetrical about the initial line since by changing θ to $-\theta$ the equation of the curve remains unchanged.

(ii) r is always positive and is never greater than $a + b$.

(iii) Since $a > b$, r can never be zero. So, the curve does not pass through the pole.

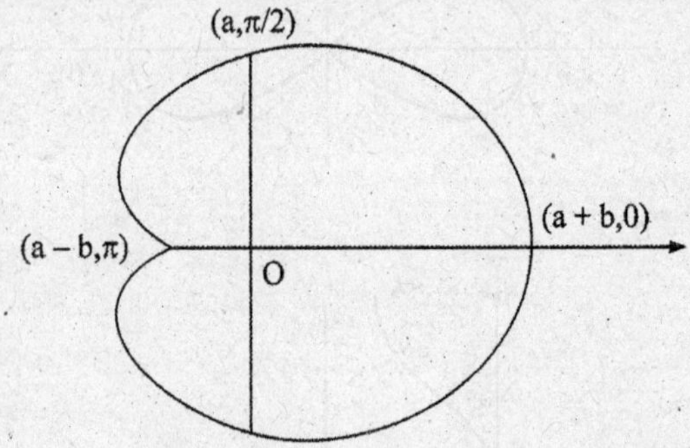

(iv) Some points on the curve are evaluated. When $\theta = 0$, $r = a + b$; when $\theta = \frac{\pi}{2}$, $r = a$; when $\theta = p$, $r = a - b$; when $\theta = \frac{3\pi}{2}$, $r = a$; when $\theta = 2\pi$, $r = a + b$.

(v) It is clear from (iv) that when θ increases from 0 to π, r decreases from a + b to a – b. Also when θ increases from π to 2π, r increases from a – b to a + b. Hence the curve is as shown.

Case II if a = b

In this case the curve becomes a cardoid which has been traced in the above sum.

Case III if a < b

(i) The curve is symmetrical about the initial line.

(ii) r is never greater than a + b.

(iii) r = 0 when $\theta = \cos^{-1}\left(-\frac{a}{b}\right)$. So, the curve passes through the pole.

But $\cos^{-1}\left(-\frac{a}{b}\right)$ has two values in the region $\theta = 0$ to $\theta = 2\pi$, one value being in the second and the other in the third quadrant. If these values be θ_1, and θ_2 then the equation of tangents at the pole are $\theta = \theta_1$ and $\theta = \theta_2$. Here is a loop between $\theta = \theta_1$ and $\theta = \theta_2$.

(iv) Some points on the curve are taken below.

When $\theta = 0$, r = a + b;

when $\theta = \pi$, r = – (b – a);

when $\theta = \frac{3\pi}{2}$, r = a;

when $\theta = 2\pi$, r = a + b

The curve is as shown

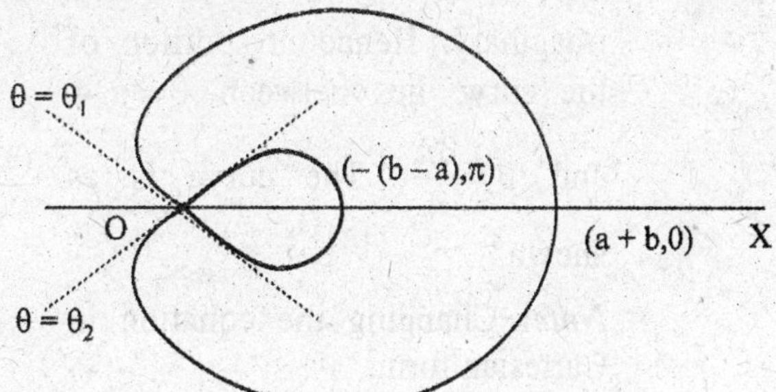

3. Trace the curve $r^2 = a^2 \cos 2\theta$ (lemnislate of Bernoulli)

(i) (a) Replacing θ by – θ, the equation of the curve remains unchanged, so the curve is symmetrical about the initial line.

(b) Replacing r by – r, the equation of the curve remains unchanged, so the curve is symmetrical about the pole.

(c) Putting $\pi - \theta$ for θ, we get

$$r^2 = a^2 \cos^2 (\pi - \theta) = a^2 \cos 2\theta,$$

i.e., the equation of the curve remains unchanged, so the curve is symmetrical about the line $\theta = \frac{\pi}{2}$

(ii) $r = 0$ when $\cos 2\theta = 0$, i.e., $2\theta = \frac{\pi}{2}, \frac{3\pi}{2}$

$$\therefore \quad \theta = \frac{\pi}{4}, \frac{3\pi}{4}$$

So, the curve passes through the pole and the equation of tangents at the pole are

$$\theta = \frac{\pi}{4}, \quad \theta = \frac{3\pi}{4}$$

(iii) When $\theta = 0$, $r^2 = a^2$, $\therefore$ $r = \pm\, a$. So, the points (a, 0) and (– a, 0) lie on the curve.

(iv) When θ increases from 0 to $\frac{\pi}{4}$, r^2 decreases from a^2 to 0. Again when θ increases from $\frac{3\pi}{4}$ to π, r^2 increases from 0 to a^2. But when θ lies between $\frac{\pi}{4}$ and $\frac{3\pi}{4}$ then r^2 = –ve. So, r is imaginary. Hence no portion of the curve lies between $\theta = \pi/4$ and $\theta = \frac{3\pi}{4}$. The curve is as shown.

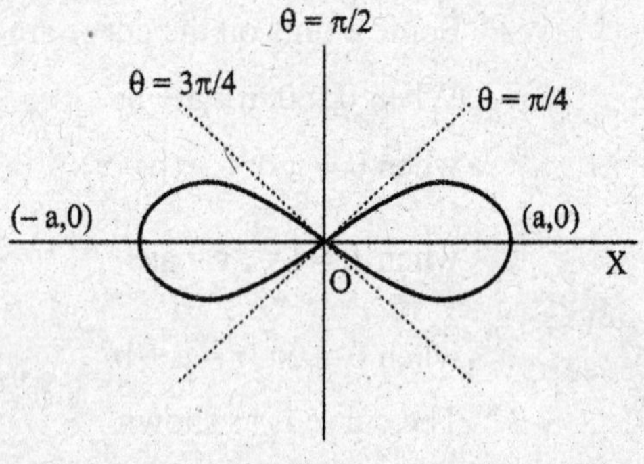

Note: Changing the equation in Cartesian form,

$r^2 = a^2 (\cos^2\theta - \sin^2\theta)$ or $r^4 = a^2 (r^2\cos^2\theta - r^2\sin^2\theta)$ i.e., $(x^2 + y^2)^2 = a^2 (x^2 - y^2)$. Now the curve can be traced as Cartesian curve.

4. Trace the curve $r = ae^{m\theta}$ (equiangular spiral)

(i) r is always positive and increases with θ and when $\theta \to \infty$, $r \to \infty$.

(ii) r is never zero for any finite value of θ, hence pole does not lie on the curve.

(iii) Some points on the curve are as follows: When $\theta = 0$, $r = a$;

when $\theta = \frac{\pi}{2}$, $r = ae^{\frac{m\pi}{2}}$; when $\theta = \pi$, $r = ae^{m\pi}$

Thus as θ increases from 0 to ∞, r increases from a to ∞.

(iv) Differentiating the equation of the curve w.r.t. θ, we get

$$\frac{dr}{d\theta} = ae^{m\theta}m = mr$$

$$\therefore \tan\varphi = r\frac{d\theta}{dr} = \frac{r}{mr} = \frac{1}{m}$$

$$\therefore \varphi = \tan^{-1}\left(\frac{1}{m}\right) = \text{constant}$$

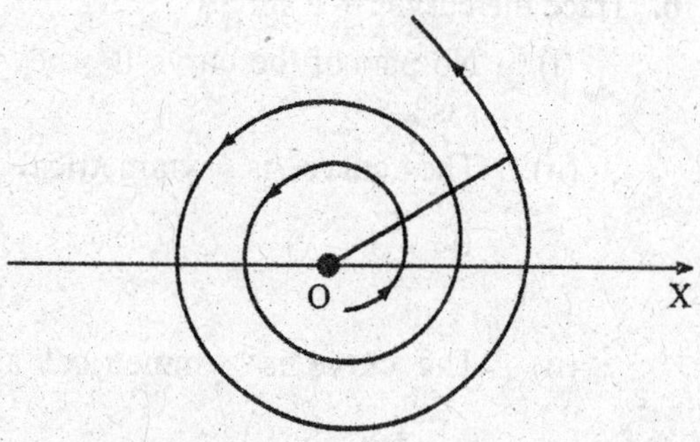

So, at every point the angle between the radius vector and the tangent is constant. This is way the curve is called equiangular spiral. The shape of the curve is as shown.

5. Trace the curve $r = a \sin 3\theta$.

(i) The curve is symmetrical about the line $\theta = \pi/2$ since the equation remains unchanged when θ is replaced by $\pi - \theta$.

(ii) $r = 0$ when $\sin 3\theta = 0$, i.e., when $3\theta = n\pi$

$$\therefore\ \theta = \frac{n\pi}{3}, \text{ i.e., } r = 0 \text{ when } \theta = 0, \frac{\pi}{3}, \frac{2\pi}{3}, \pi, \ldots$$

So, the curve passes through the pole and the equations of tangents at pole are

$$\theta = 0,\ \theta = \pi/3,\ \theta = \frac{2\pi}{3}$$

Also when $\theta = \pi/6$, $r = a$; when $\theta = \pi/2$, $r = -a$; when $\theta = \frac{5\pi}{6}$, $r = a$.

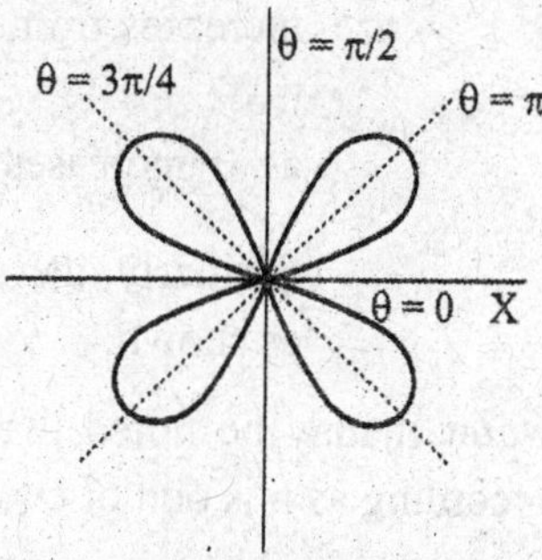

(a) We find that as θ increases from 0 to $\pi/6$, r increases from 0 to a. As θ increases from $\pi/6$ to $\pi/3$, r decreases from a to 0. Thus there is a loop between $\theta = 0$ and $\theta = \pi/3$.

(b) when θ increases from $\pi/3$ to $\pi/2$, r becomes – a from 0. When θ increases from $\pi/2$ to $2\pi/3$, r becomes 0 from – a. Thus, there is second loop between $\theta = \pi/3$ and $\theta = 2\pi/3$.

(c) Similarly, there is third loop between $\theta = 2\pi/3$ and $\theta = \pi$. Thus, the curve is as shown.

6. Trace the curve $r = a \sin 2\theta$

(i) No part of the curve lies beyond the region r = a as the maximum value of r is a.

(ii) The curve is symmetrical about the line $\theta = \frac{\pi}{2}$.

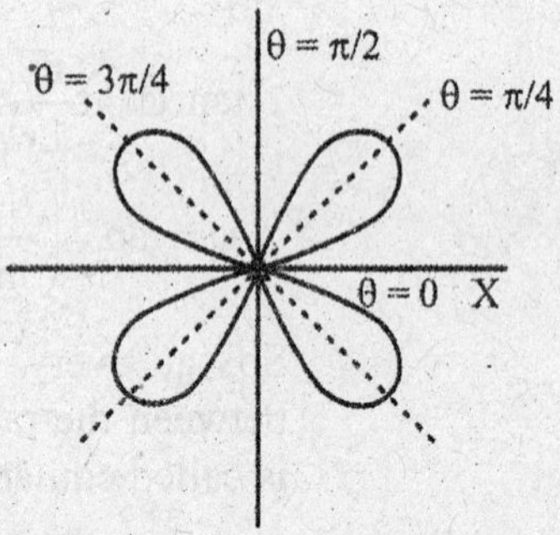

(ii) The curve is symmetrical about $\theta = \frac{\pi}{4}$ and $\theta = \frac{3\pi}{4}$.

(iv) $r = 0$ when $\theta = 0, \frac{\pi}{2}, \pi, \ldots$

The curve passes through the pole and tangents at pole are $\theta = 0$ and the $\theta = \frac{\pi}{2}$, the other values of θ giving the same tangents.

(v) When $\theta = \frac{\pi}{4}$, r = a; when $\theta = \frac{3\pi}{4}$, r = – a; when $\theta = \frac{5\pi}{4}$, r = a. Thus r increases from 0 to a as θ increases from 0 to $\frac{\pi}{4}$ and r decreases from a to 0 as θ increases from $\frac{\pi}{4}$ to $\frac{\pi}{2}$. Thus there is a loop in the first quadrant. Similarly, there are loops each in 2nd, 3rd and 4th quadrant. Hence the curve is as shown.

***Note*:** It may be noted that for $r = a \sin n\theta$ or $r = a \cos n\theta$, there are n or 2n loops according as n is odd or even.

Exercise 7.2

1. Trace the curve $r = a(1 - \cos \theta)$.

(*Hint*: Proceed as example 1]

2. Trace the curve $r^2 = 4 \cos 2\theta$

[*Hint*: Put a = 2 in example 3, curve shape is as in example 3]

3. Trace the curve $r = a \cos 2\theta$

[*Hint*: Proceed as example 6]

4. Trace the curve $r = a \cos 3\theta$

[*Hint*: Proceed as example 5.]

Answers

1.

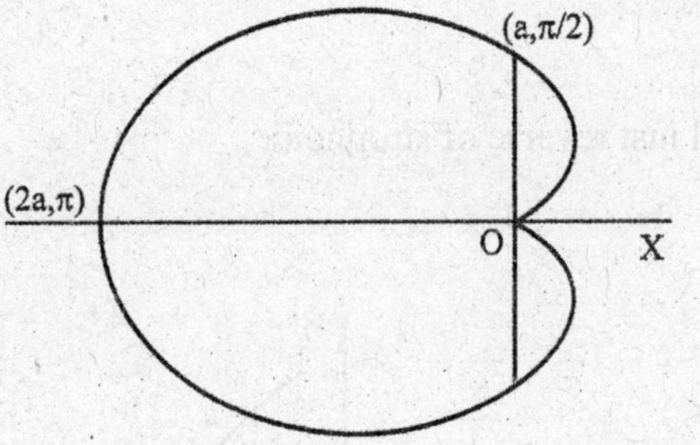

3.

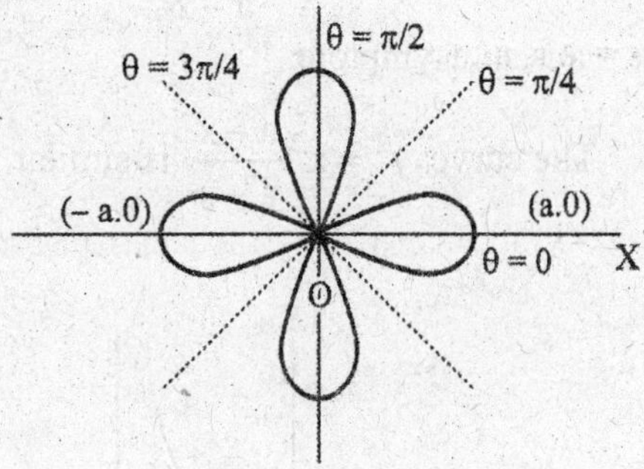

4.

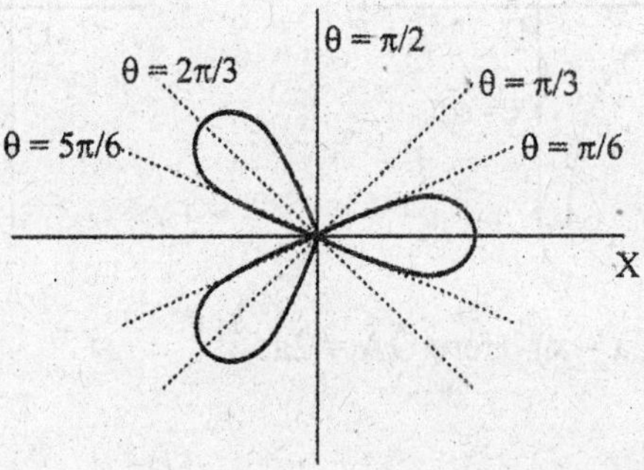

Some well-known Curves

The following curves should always be kept in mind.

Folium of Descartes

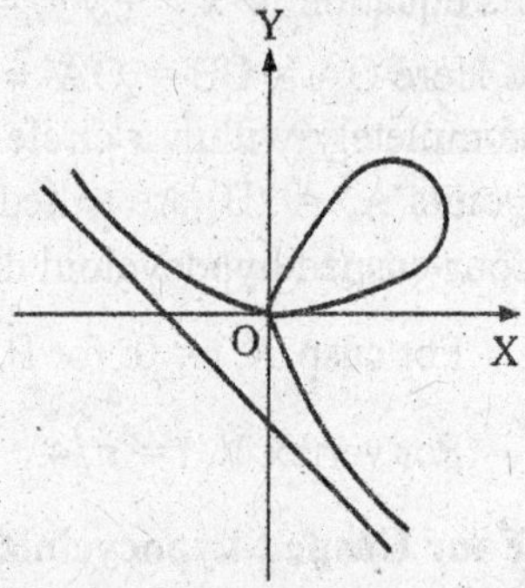

Its equation is $x^3 + y^3 = 3axy$. It is symmetrical about the line $y = x$. The axes are tangents at the origin and there is a loop in the first quadrant. It has an asymptote $x + y + a = 0$.

Cissoid of Diocles

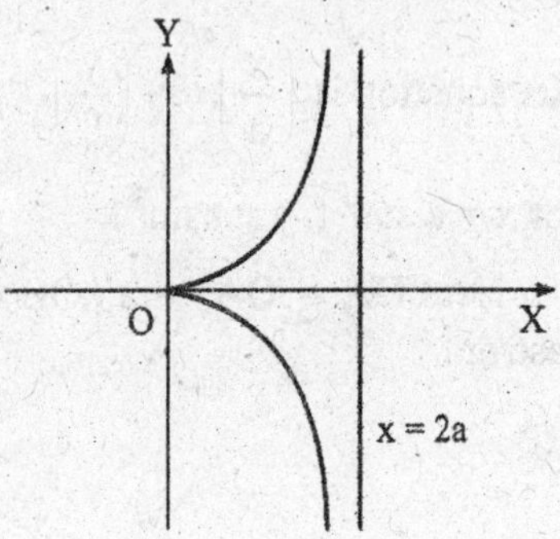

Its equation is $y^2(2a - x) = x^3$ or $y^3 = \dfrac{x^3}{2a - x}$; $x = 2a$ is an asymptote.

Strophoid

Its equation is $y^2 = x^2 \dfrac{a+x}{a-x}$ or $y(a-x) = x^2(a+x)$. OABCO is a loop OD = OB = a. $x = a$ is an asymptote.

The curve $y^2 = x^2 \dfrac{a-x}{a+x}$ is similar and just reverse of strophoid.

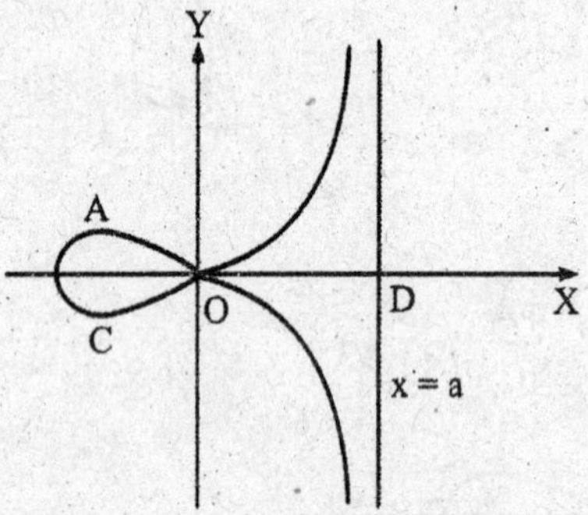

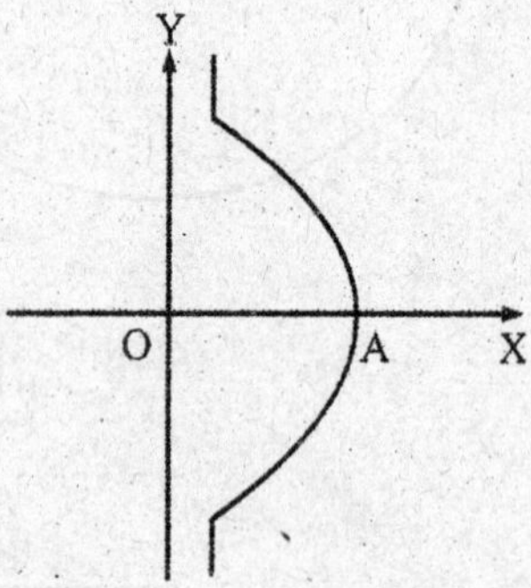

Witch of Agnesi

Its equation is $xy^2 = 4a^2(2a - x)$. Here OA = 2a.

Astroid

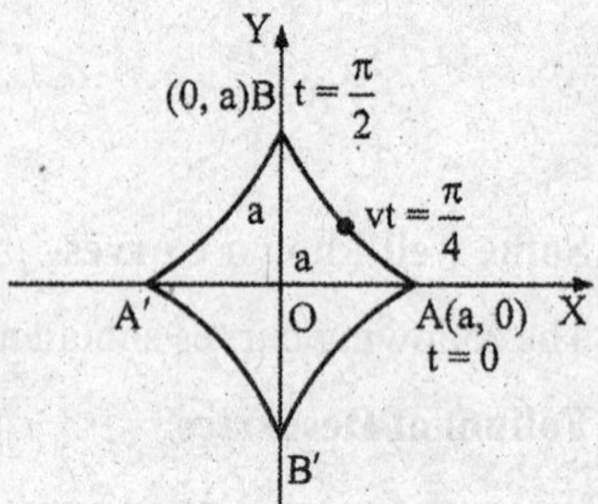

Its equation is $x^{2/3} + y^{2/3} = a^{2/3}$ or $x = a\cos^3 t$, $y = a\sin^3 t$. Here OA = OB = OA′ = OB′ = a. The whole figure lies completely within a circle of radius a and centre O. The points A, A′, B′ are called cusps. It is a special case of a four-cusped hypocycloid described below.

For cusp A, $t = 0$; for B, $t = \pi/2$;

For vertex V, $t = \pi/4$.

Four Cusped hypocycloid

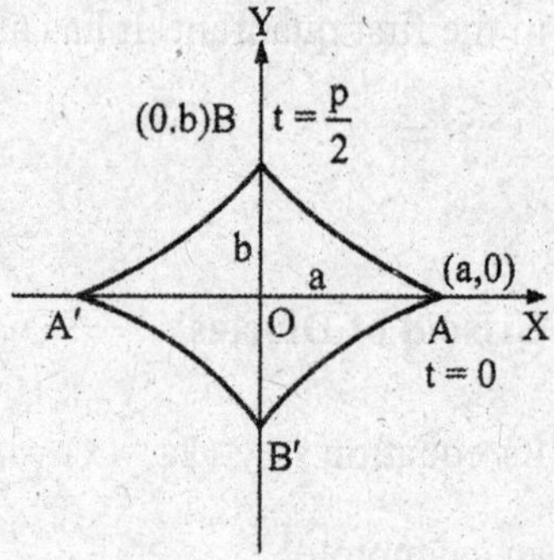

Its equation is $\left(\dfrac{x}{a}\right)^{2/3} + \left(\dfrac{y}{b}\right)^{2/3} = 1$

or $x = a\cos^2 t$, $y = b\sin^3 t$.

Here OA = OA′ = a; OB = OB′ = b. When a = b, it is astroid.

Cycloid

The cycloid is the curve traced out by a point on the circumference of a circle which rolls (without sliding) on a straight line. The parametric equations of the cycloid with the starting point as the origin and the line on which the circle rolls, called the base, as x-axis are

$$x = a(\theta - \sin\theta), \quad y = a(1 - \cos\theta)$$

For O and O′, y = 0, $\therefore$ cos θ = 1, $\therefore$ θ = 0 and 2π and x = 0 and 2aπ.

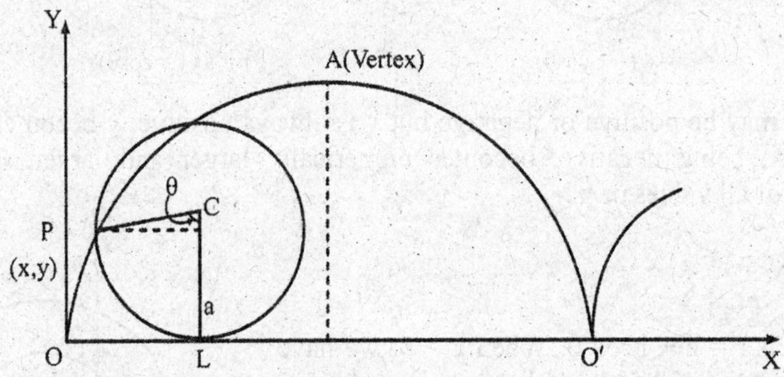

When the circle rolls on, arches like OAO′ are generated over and over again and any single arch is called a cycloid.

The equation of the cycloid with the vertex as origin and tangent at the vertex as x-axis is

$$x = a(\theta + \sin\theta), y = a(1 - \cos\theta)$$

For vertex A, θ = 0; for O. θ = π and for O′, θ = −π

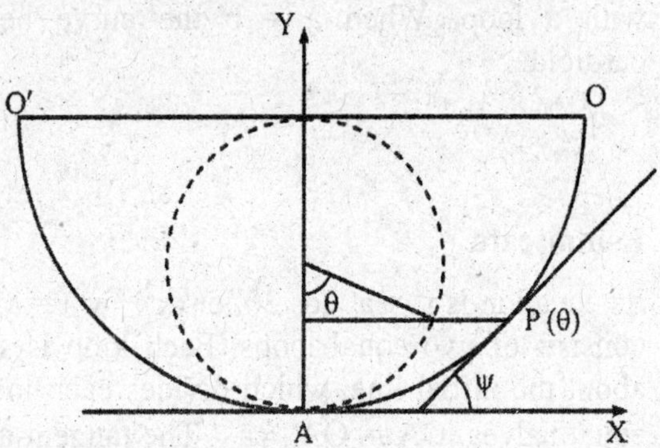

Catenary

The catenary is the curve in which a uniform heavy string hangs under the action of gravity when suspended freely from two points.

Its equation is $y = c\cosh\frac{x}{c} = \frac{c}{2}\left(e^{x/c} + e^{-x/c}\right)$.

C is called vertex, OC = c, OX is called directrix.

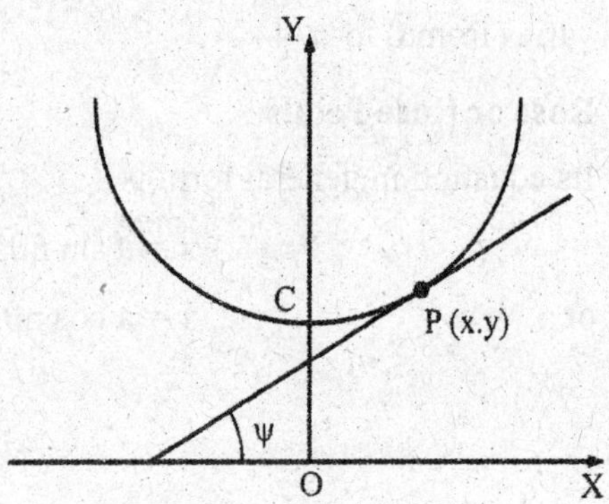

Logarithmic and Exponential Curves

In $y = \log x$, x is always positive; $y = 0$ when $x = 1$ and as x becomes smaller and smaller, y, being negative, becomes numerically larger and larger. For $x > 0$ the curve is continuous.

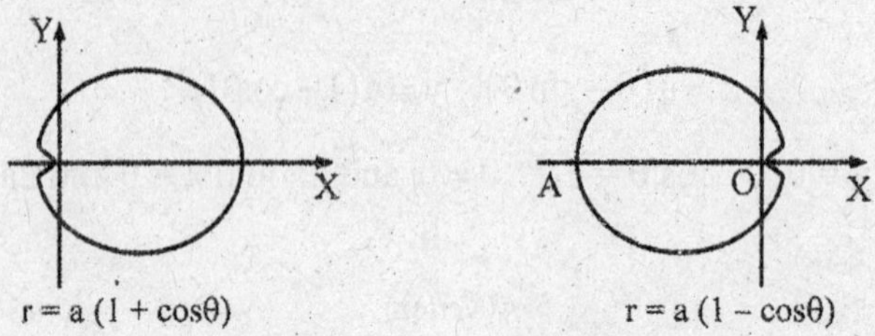

$y = e^x$, x may be positive or negative but y is always positive. y becomes smaller and smaller as x, being negative, becomes numerically larger and larger. The curve is continuous for all values of x.

Limacon

Its equation is $r = a + b\cos\theta$. When $a > b$, we have the outer curve when $a < b$ we have the inner curve with a loop. When $a = b$ the curve becomes a cardioide.

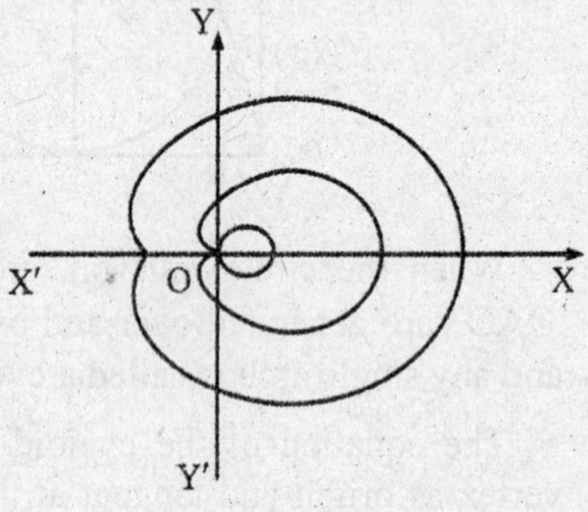

Lemniscate

Its equation is $r^2 = a^2 \cos 2\theta$, or $(x^2 + y^2) = a^2(x^2 - y^2)$. It consists of two equal loops. Each loop is symmetrical about the initial line, which divides each loop into two equal halves. $OA = OA' = a$. The tangent at origin are $y = \pm x$. For the upper half of the right-hand loop, θ varies from 0 to $\pi/4$.

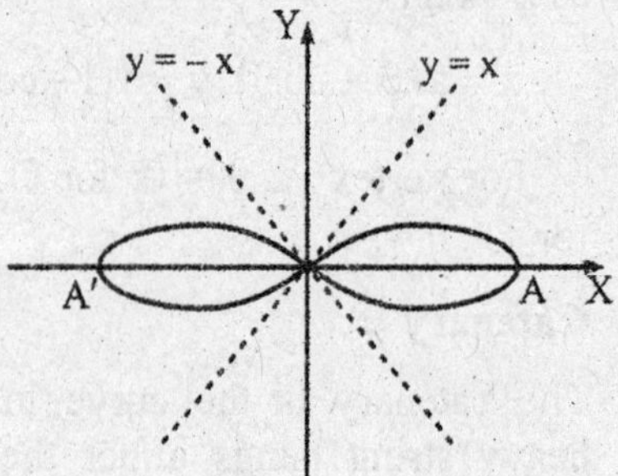

Rose or Rose Petals

Its equation in general form is

$$r = a \sin n\theta,$$

or

$$r = a \cos n\theta.$$

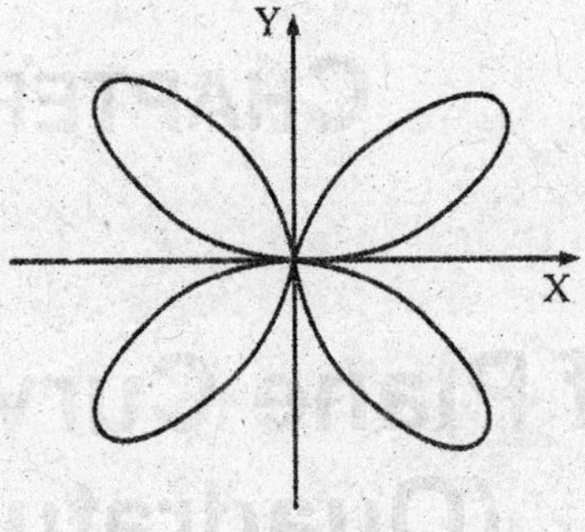

r = asin 2θ

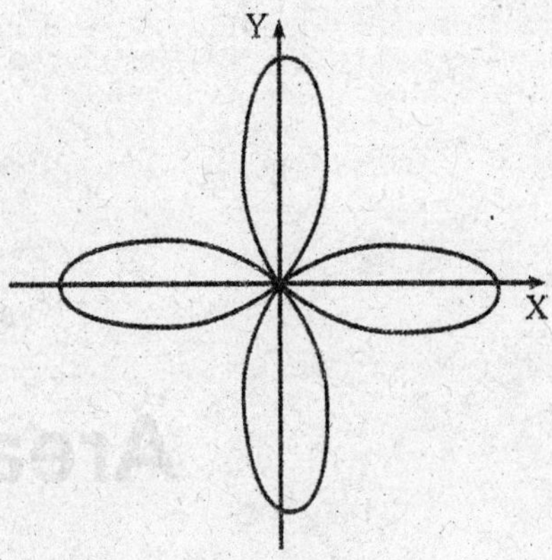

r = acos 2θ

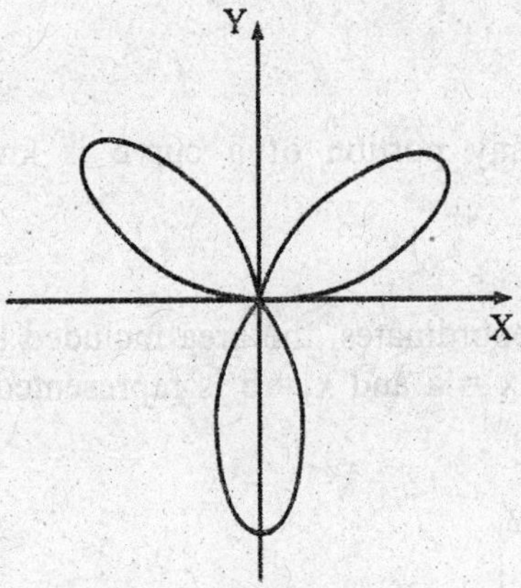

r = asin 3θ

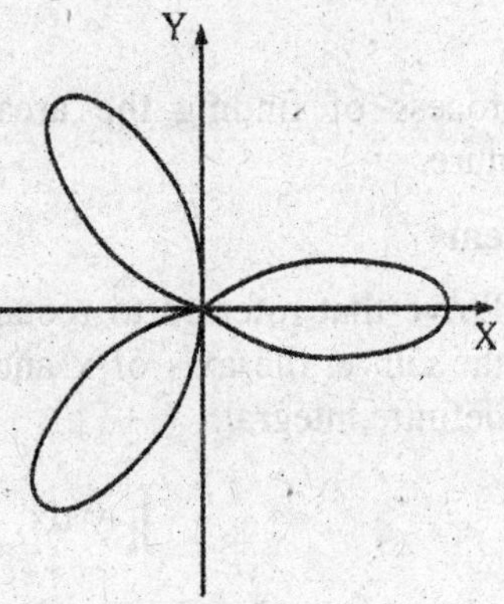

r = acos 3θ

If n is odd, the number of loops is n and if n is even, the number of loops is 2n.

CHAPTER 8

Areas of Plane Curves (Quadrature)

The process of finding the area bounded by any portion of a curve is known as quadrature.

Theorems

I. Prove that referred to rectangular axes of coordinates, the area included between the curve, the axis of x and two ordinates x = a and x = b is represented by the definite integral.

$$\int_a^b y\, dx,$$

where $y = f(x)$ is the equation of the curve.

Let AB be the curve $y = f(x)$, continuous in $[a, b]$. Let AL and BM be the two ordinates at $x = a$ and $x = b$ respectively where $b > a$. Let the ordinates, i.e., the values of $f(x)$, are all positive throughout the range. To find the area ALMBA.

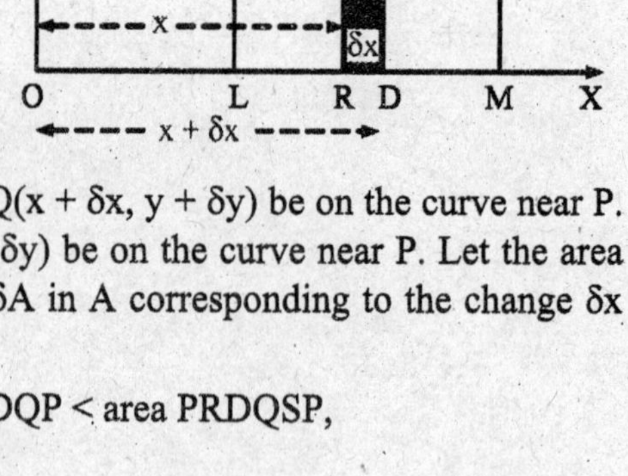

Let any point on the curve be $P(x, y)$. Let $Q(x + \delta x, y + \delta y)$ be on the curve near P. Let the area be $P(x, y)$. Let $Q(x + \delta x, y + \delta y)$ be on the curve near P. Let the area LRPAL be denoted by A. The increment δA in A corresponding to the change δx in x is the area PRDQP. Now,

$$\text{area PRDTP} < \text{area PRDQP} < \text{area PRDQSP},$$

i.e., $y\,\delta x < \delta A < (y + \delta y)\,\delta x$

or $y < \dfrac{\delta A}{\delta x} < y + \delta y$

$\therefore$ as $Q \to P, \lim\limits_{\delta x \to 0} y < \lim\limits_{\delta x \to 0} \dfrac{\delta A}{\delta x} < \lim\limits_{\delta y \to 0}(y + \delta y),$

i.e., $y < \dfrac{dA}{dx} < y$ $\quad (\because \delta x \to 0, \delta y \to 0 \text{ as } Q \to P)$

$\Rightarrow$ $\dfrac{dA}{dx} = y,$

i.e., $\dfrac{dA}{dx} = f(x);$ $\quad \because\ dA = f(x)\,dx$

$\therefore$ $\int dA = \int f(x)\,dx;$ $\quad$ i.e., $A = f(x) + C,$

where f(x) is an indefinite integral of f(x)

Now when x = a, A = 0

$\therefore$ $0 = F(a) + C$

When x = b, A = required area

Required area = F(b) + C

$\therefore$ required area $= F(b) - F(a) = \int_a^b f(x)\,dx$

Similarly it can be proved that the area bounded by the curve x = f(y), the axis of y and the abscissae y = c, y = d is given by

$$\int_c^d x\,dy; \quad \text{i.e.,} \quad \int_c^d f(y)\,dy$$

II. Prove that if r = f(θ) be the equation of the curve, then the area of the sector enclosed by the curve and two radii vectors θ = α, θ = β is

$$\frac{1}{2}\int_\alpha^\beta r^2 d\theta,$$

i.e., $\dfrac{1}{2}\int_\alpha^\beta \int [f(\theta)]^2 d\theta$

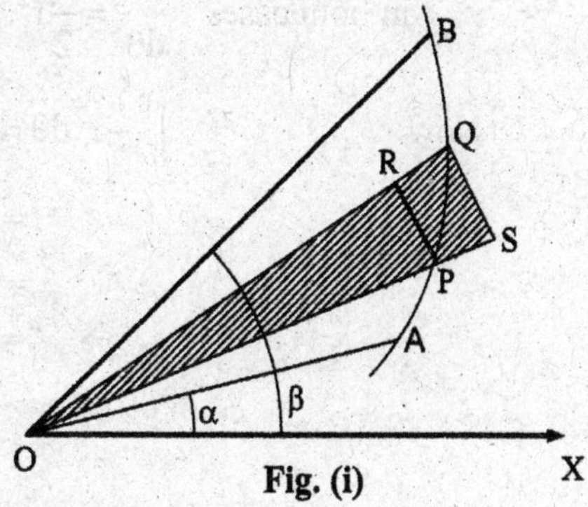

Fig. (i)

Let P(r, θ) be any point on the curve. Let (r + δr, θ + δθ) be any other point on the curve near P so that as a point moves on the curve from P to Q, the radius vector either constantly increases [Fig. (i)] or constantly decreases [Fig. (ii)].

Let the area of the sector AOP be denoted by A and the increment δA in A is the area POQ. If we draw arcs of circles with O as centre and OP, OQ as radii cutting OP and OQ at S and R respectively then from Fig. (i), area of circular sector OPR < δA < area of circular sector OQS,

i.e.,

$$\frac{1}{2}r^2\delta\theta < \delta A < \frac{1}{2}(r+\delta r)^2\,\delta\theta$$

or

$$\frac{1}{2}r^2 < \frac{\delta A}{\delta\theta} < \frac{1}{2}(r+\delta r)^2$$

As Q $Q \to P$,

$$\lim_{\delta r\to 0}\frac{1}{2}r^2 < \lim_{\delta\theta\to 0}\frac{\delta A}{\delta\theta}$$

$$< \lim_{\delta r\to 0}\frac{1}{2}(r+\delta r)^2$$

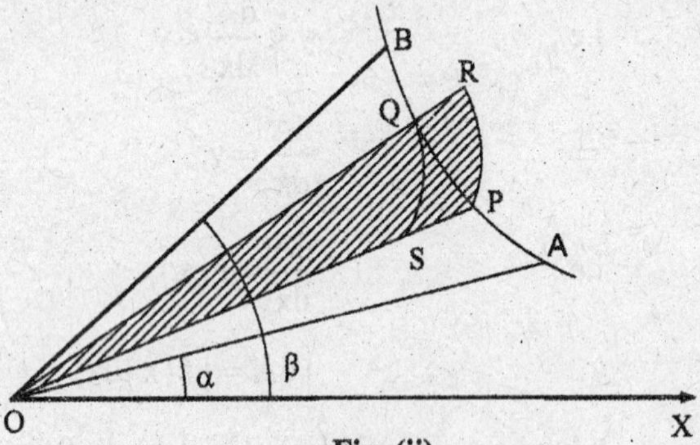

Fig. (ii)

$$(\because \text{as } Q \to P, \delta r \to 0, \delta\theta \to 0)$$

i.e., $$\frac{1}{2}r^2 < \frac{dA}{d\theta} < \frac{1}{2}r^2$$

$$\Rightarrow \quad \frac{dA}{d\theta} = \frac{1}{2}r^2$$

From Fig. (ii), $\frac{1}{2}(r+\delta r)^2\,\delta\theta < \delta A < \frac{1}{2}r^2\delta\theta$

$$\Rightarrow \quad \frac{dA}{d\theta} = \frac{1}{2}r^2 \quad \text{(as above)}$$

$\therefore$ in both cases $\frac{dA}{d\theta} = \frac{1}{2}r^2$

Now $$\int_\alpha^\beta \frac{1}{2}r^2 d\theta = \int_\alpha^\beta \frac{dA}{d\theta}.\,d\theta = [A]_\alpha^\beta$$

$$= (A \text{ when } \theta = \beta) - (A \text{ when } \theta = \alpha)$$

$$= \text{sectorial area AOB} - 0$$

$$= \text{sectorial area AOB}$$

III. Explain Simpson's rule for approximate evaluation of areas (or definite integrals)

Statement: The definite integral $\int_a^b y\,dx$ is approximately equal to $\frac{h}{3}\{(y_1+y_{2n+1})+2(y_3+y_5+...+y_{2n-1})+4(y_2+y_4+...y_{2n})\}$

where $h=\frac{\beta-\alpha}{2n}$ and $y_1, y_2, y_3, \ldots y_{2n+1}$ are the values of y when x = a, a + h, a + 2h, ..., a + 2nh respectively.

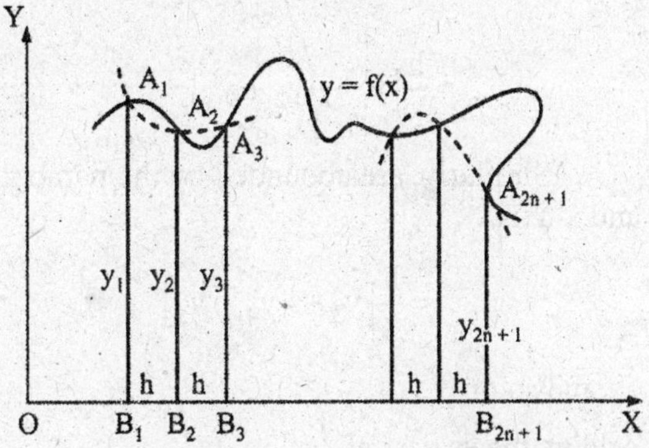

We consider the region $A_1 B_1 B_{2n+1} A_{2n+1}$ bounded by any curve y = f(x), x – axis and two ordinates $A_1 B_1$ and $A_{2n+1} B_{2n+1}$. We divide the interval from $x = a = OB_1$ to $x = b = OB_{2n+1}$ into 2n equal parts.

$$B_1 B_2 = B_2 B_3 = \ldots = h = \frac{b-a}{2}n.$$

Let $A_1, A_2, \ldots A_{2n+1}$ be the points on the curve corresponding to the points of division $B_1, B_2, \ldots, B_{2n+1}$. Through each successive set of three points $A_1 A_2 A_3$, $A_3, A_4 A_5,\ldots$ $A_{2n-1} A_{2n} A_{2n+1}$ we draw arcs of parabolas with axis parallel to y-axis.

The equation of the parabola drawn through A_1, A_2, A_3 referred to (a + h, 0) as origin is of the form

$$y = Ax^2 + Bx + C \quad \text{.....(I)}$$

Then the area bounded by the parabolic arc $A_1 A_2 A_3$, the ordinates $A_1 B_1$, $A_3 B_3$ and x-axis, i.e., the area $B_1B_3A_3A_1$

$$= \int_{-h}^{h}\left(Ax^2+bx+C\right)dx = \frac{2}{3}Ah^2+2Ch$$

$$= 2h\left(C+\frac{Ah^2}{3}\right) \quad \text{.....(II)}$$

$\because$ $A_1(-h, y_1)$, $A_2(0, y_2)$, $A_3(h, y_3)$ are on the parabola (I),

$\therefore$ $y_1 = A(-h)^2 + B(-h) + C = Ah^2 - Bh + C$

$y_2 = C$; $y_3 = Ah^2 + Bh + C$

$$\therefore \quad \frac{y_1+y_3}{2} = Ah^2 + C = Ah^2 + y^2 \qquad (\because C = y_2)$$

$$\therefore \quad Ah^2 = \frac{y_1 + y_3}{2} - y_2 = \frac{y_1 - 2y_2 + y_3}{2}$$

$\therefore$ from (II), the area under consideration

$$= 2h\left(y_2 + \frac{1}{3}.\frac{y_1 - 2y_2 + y_3}{2}\right) = 2h\left(\frac{6y_2 + y_1 - 2y_2 + y_3}{6}\right)$$

$$= \frac{h}{3}[y_1 + 4y_2 + y_3]$$

Similarly, area bounded by the parabolic are A_3 A_4 A_5, the ordinates A_3 B_3, A_5 B_5 and x-axis.

$$= \frac{h}{3}[y_3 + 4y_4 + y_5]$$

and so on

Adding these, we get

Approximate area = approximate value $\int_a^b ydx$

$$= \frac{h}{3}\left[\{y_1 + 4y_2 + y_3\} + \{y_3 + 4y_4 + y\overline{5}\} + \ldots\right.$$

$$\left. + \{y_{2n-1} + 4y_{2n} + y_{2n+1}\}\right]$$

$$= \frac{h}{3}\left[(y_1 + y_{2n+1}) + 2(y_3 + y_5 + \ldots y_{2n-1}) + 4(y_2 + y_4 + \ldots + y_{2n})\right]$$

Note: Simpson's rule is also stated as follows:

To obtain an approximate value of an area (or of a definite integral) divide it into an even number of strips by drawing equidistant ordinates and multiply one third of the distance between consecutive ordinates by the sum of

(i) the first and the last ordinate

(ii) twice all the other ordinates and

(iii) four times all the even ordinates

Remember

1. Area of Cartesian curve

$$= \int_a^b ydx \text{ or } \int_a^b x\,dy$$

2. Area of a polar curve

$$=\frac{1}{2}\int_{\alpha}^{\beta} r^2 d\theta$$

3. Area included between two curves

$$=\int_a^b [\varphi(x)-\psi(x)dx] \quad \text{or} \quad \int_a^b [\varphi(y)-\psi(y)]\, dy$$

4. Area of a closed curve in parametric form

$$=\frac{1}{2}\int\left(x\frac{dy}{dt}-y\frac{dx}{dt}\right)dt$$

Examples

1. Find the area of the ellipse $\frac{x^2}{a^2}+\frac{y^2}{b^2}=1$. The ellipse is symmetrical about both the axes (as its equation contains even powers of x and y).

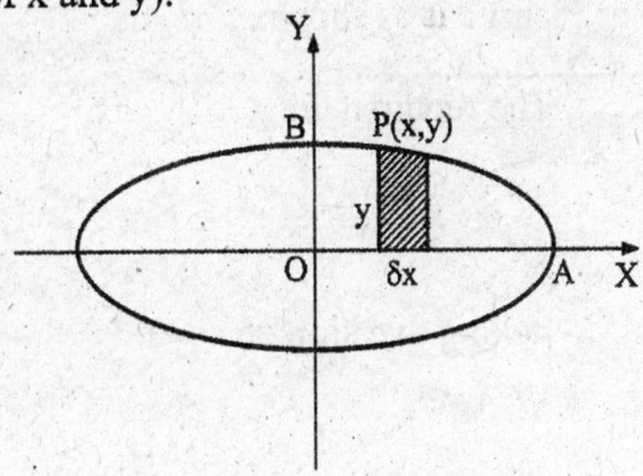

Hence the area of the ellipse is divided into four equal parts. We shall determine the area of the portion OABO lying in the first quadrant and the whole area of the ellipse is 4 times the area of the portion OABO.

Now area OABO

$$=\int_0^a ydx=\frac{b}{a}\int_0^a \sqrt{a^2-x^2}dx$$

$$\left[\because \frac{x^2}{a^2}+\frac{y^2}{b^2}=1, \quad \therefore y=\frac{b}{a}\sqrt{(a^2-x^2)}\right]$$

Putting $x = a\sin\theta$, $dx = a\cos\theta\, d\theta$.

Also when $x = 0$, $\theta = 0$; when $x = a$, $\theta = \pi/2$

$$\therefore \text{ area OABO } =\frac{b}{a}\int_0^{\pi/2}\sqrt{a^2-a^2\sin^2\theta}\; a\cos\theta\; d\theta$$

$$=\frac{b}{a}\int_0^{\pi/2} a^2\cos^2\theta\; d\theta = ab\int_0^{\pi/2}\frac{1+\cos 2\theta}{2}d\theta$$

$$=\frac{ab}{2}\left[\theta+\frac{\sin 2\theta}{2}\right]_0^{\pi/2}=\frac{ab}{2}.\frac{\pi}{2}=\frac{1}{4}\pi ab$$

$$\therefore \text{ area of the ellipse } =4\times\frac{1}{4}\pi ab=\pi ab.$$

2. Find the area of the loop of the curve $y^2 = x(x-1)^2$

The curve is symmetrical about x-axis (as its equation contains even power of y). It passes through the origin (as its equation is satisfied by x = 0, y = 0,)

x = 0, i.e., y-axis is the tangent at origin (by equating the lowest degree term to zero).

Putting x = 0, we get y = 0 and putting y = 0 we get x = 0, 1. Hence the curve cuts the y-axis at origin only while it cuts the x-axis at the origin and at another point whose abscissa is 1.

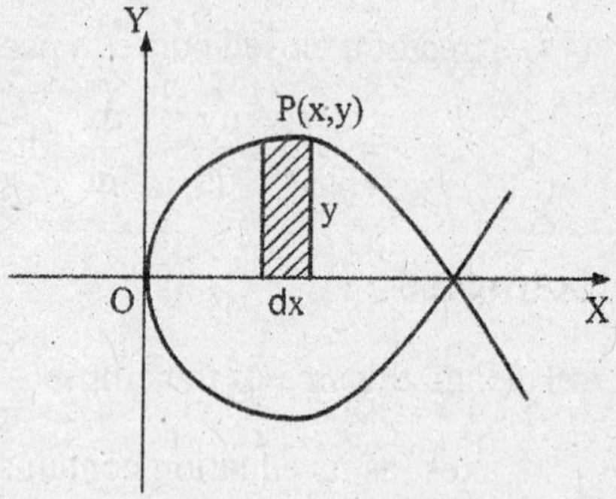

When x is negative, y is imaginary, so no part of the curve lies to the left of y-axis and the curve is as shown.

The required area $= 2\int_0^1 y\,dx$

$$= -2\int_0^1 \sqrt{x}(x-1)\,dx$$

[taking –ve sign of $y = \pm\sqrt{x}(x-1)$ to keep y + ve as x has values less than 1]

$$= -2\left\{\int_0^1 \left(x^{3/2} - x^{1/2}dx\right)\right\} = -2\left[\frac{2}{5}x^{5/2} - \frac{2}{3}x^{3/2}\right]_0^1$$

$$= -2\left[\frac{2}{5} - \frac{2}{3}\right] = \frac{8}{15}$$

3. Find the area of the loop of the curve $ay^2 = x^2(a-x)$

The curve is symmetrical about x-axis as the equation contains even power of y.

The curve passes through origin and the tangents at origin are $ay^2 - ax^2 = 0$ (equating lowest-degree term to zero) i.e., $y = \pm x$.

Putting y = 0, we get x = 0, x = a.

So, the points of intersection of the curve and the x-axis are (0, 0) and (a, 0) and there is a loop lying between x = 0 and x = a.

For x > a, y becomes imaginary, so no part of the curve is in the right of (a, 0).

But y is real if $x < 0$, so the curve extends in the left of O and the curve is as shown in the figure.

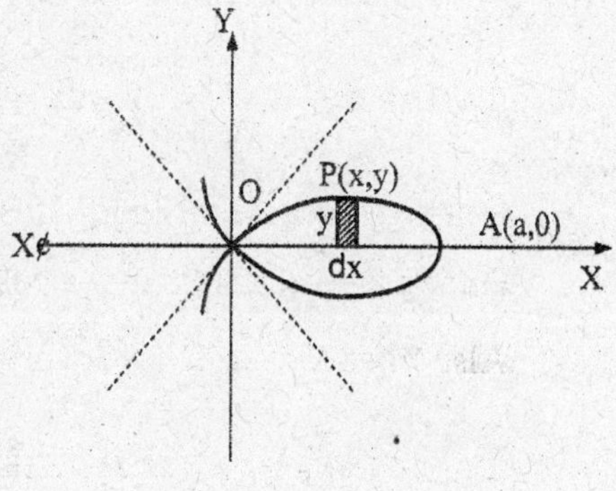

$\therefore$ the required area

$$= 2.\ \text{Area OAPO}$$

$$= 2\int_0^a y\,dx$$

$$= 2\int_0^a \frac{x\sqrt{a-x}}{\sqrt{a}}\,dx$$

Putting $a - x = z$, $-dx = dz$; also when $x = 0$, $z = a$ and when $x = a$, $z = 0$

$$\therefore \quad \text{required area} = \frac{-2}{\sqrt{a}}\int_a^0 (a-z)\sqrt{z}\,dz = \frac{2}{\sqrt{a}}\int_a^0 \left(az^{1/2} - z^{3/2}\right)dz$$

$$= \frac{2}{\sqrt{a}}\left[a\frac{2}{3}z^{3/2} - \frac{2}{5}z^{5/2}\right]_a^0 = \frac{2}{\sqrt{a}}\left[\frac{2a}{3}a^{3/2} - \frac{2}{5}a^{5/2}\right]$$

$$= \frac{2}{\sqrt{a}}2a^{5/2}\left(\frac{1}{3} - \frac{1}{5}\right) = \frac{8}{15}a^2$$

4. Find the area between the curve $x(x^2 + y^2) = a(x^2 - y^2)$ and its asymptote. Also find the area of its loop.

The curve is symmetrical about x-axis as only even powers of y occur. It passes through origin and $y = + x$ (by equating $x^2 - y^2$ to zero).

Equating to zero the coefficient of the highest power of y, $a + x = 0$; so $x = -a$ is the asymptote of the curve.

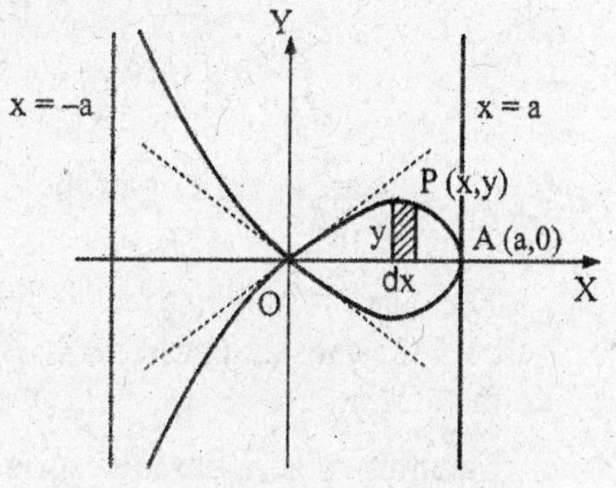

When $x = 0$, $y = 0$; when $x = a$, $y = 0$, so the curve cuts x-axis at (0, 0) and (a, 0)

The curve is $x(x^2 + y^2) = a(x^2 - y^2)$ or $y^2(x + a) = (a - x)x^2$

$$\therefore \quad y = \pm x\sqrt{\frac{a-x}{a+x}}$$

If $x > a$, y becomes imaginary, so no part of the curve is in the right of (a, 0).

The loop is situated between $x = 0$ and $x = a$ and the curve is as shown in the figure.

$\therefore$ the area between the curve and its asymptote

$$= 2\int_{a}^{0} y\,dx = 2\int_{-a}^{0} -x\sqrt{\frac{a-x}{a+x}}\,dx = -2\int_{-a}^{0} \frac{x(a-x)}{\sqrt{a^2-x^2}}\,dx$$

(–ve sign because this portion is in the left of O).

Putting $x = -a \sin\theta$, $dx = -a\cos\theta\, d\theta$.

Also when $x = -a$, $\theta = \frac{\pi}{2}$; when $x = 0$, $\theta = 0$

$$\therefore \quad \text{the area} = -2\int_{\pi/2}^{0} \frac{-a\sin\theta(a + a\sin\theta)\times(-a\cos\theta)\,d\theta}{\sqrt{a^2 - a^2\sin^2\theta}}$$

$$= -2a^2\int_{\pi/2}^{0}\left(\sin\theta + \sin^2\theta\right)d\theta$$

$$= 2a^2\left\{\int_{\pi/2}^{0} \sin\theta\, d\theta + \int_{\pi/2}^{0} \sin^2\theta\, d\theta\right\} = \frac{4+\pi}{2}a^2$$

The area of the loop

$$= 2\int_{0}^{a} y\,dx = 2\int_{0}^{a} x\sqrt{\frac{a-x}{a+x}}\,dx$$

$$= 2\int_{0}^{a} \frac{x(a-x)}{\sqrt{a^2-x^2}}\,dx$$

Putting $x = a\sin\theta$, $dx = a\cos\theta\, d\theta$

Also when $x = 0$, $\theta = 0$; when $x = a$, $\theta = \pi/2$

$$\therefore \ \text{area} = 2\int_{0}^{\pi/2} a^2\left(\sin\theta - \sin^2\theta\right)d\theta = 2a^2\left(1 - \frac{1}{2}.\frac{\pi}{2}\right) = \frac{4-\pi}{2}a^2$$

5. Find the area of a loop of the curve $y^2(a^2 + x^2) = x^2(a^2 - x^2)$

The curve is symmetrical about both the axes and passes through the origin.

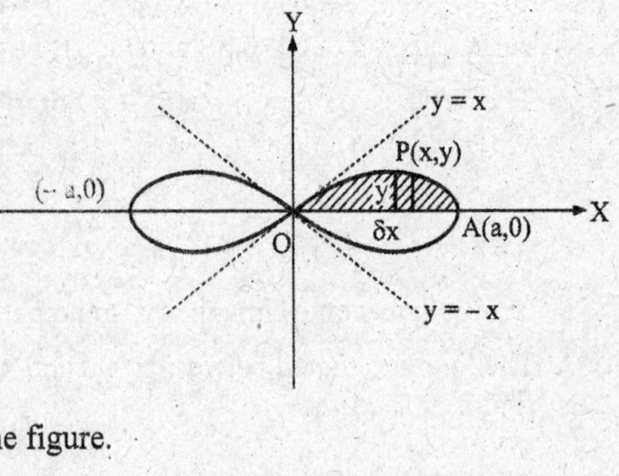

Equating to zero the lowest-degree term, $x^2 - y^2 = 0$, i.e., $y = \pm x$ are the tangents at origin.

Putting in the equation of the curve $y = 0$ we get $x = 0$, $x = \pm a$.

So there are two loops as shown in the figure.

∴ the required area (i.e., area of one loop)

$$= 2 \times \text{area OPAO}$$

$$= 2\int_0^a y\,dx = 2\int_0^a x\sqrt{\frac{a^2 - x^2}{a^2 + x^2}}\,dx$$

Putting $z^2 = a^2 + x^2$, $z^2 - a^2 = x^2$; $2z\,dz = 2x\,dx$,

∴ $a^2 - x^2 = a^2 - z^2 + a^2 = 2a^2 - z^2$

When $x = 0$, $z = a$; when $x = a$, $z = \sqrt{2}a$

$$\therefore \quad \text{area} = 2\int_0^{\sqrt{2}a} z.\frac{\sqrt{2a^2 - z^2}}{z}dz = 2\int_a^{\sqrt{2}a} \sqrt{2a^2 - z^2}\,dz$$

$$= 2\left[\frac{1}{2}z\sqrt{2a^2 - z^2} + \frac{1}{2}.2a^2 \sin^{-1}\left(\frac{z}{a\sqrt{2}}\right)\right]_a^{a\sqrt{2}}$$

$$\left(\because \int \sqrt{a^2 - x^2}\,dx = \frac{1}{2}x\sqrt{a^2 - x^2} + \frac{1}{2}a^2 \sin^{-1}\frac{x}{a}\right)$$

$$= 2\left[0 - \frac{1}{2}a.a + \frac{1}{2}a^2 \sin^{-1}(1) - a^2 \sin^{-1}\left(\frac{1}{\sqrt{2}}\right)\right]$$

$$= 2\left(a^2\frac{\pi}{2} - \frac{1}{2}a^2 - a^2\frac{\pi}{4}\right)$$

$$= 2a^2\left(\frac{\pi}{2} - \frac{1}{2} - \frac{\pi}{4}\right) = 2a^2\left(\frac{\pi - 2}{4}\right) = \frac{1}{2}a^2(\pi - 2)$$

***Note*:** The whole area enclosed by the curve

$$= 2 \times \frac{1}{2}a^2(\pi - 2) = a^2(\pi - 2)$$

6. Find the whole area of the curve $a^2 y^2 = x^2 (a^2 - x^2)$

The curve is symmetrical about both axes as only even powers of x and y occur.

It passes through the origin and equations of tangents are (equating the lowest-degree term to zero, i.e., $y^2 - x^2 = 0$) $y = \pm x$. Putting $y = 0$, in the equation of the curve, we get $x = 0$, $x = \pm a$. So, the curve cuts the x-axis at (0, 0), (a, 0) and (–a, 0) and there are two loops, as shown in the figure.

∴ the required area

$$= 4 \times \text{area OPAO} = 4\int_0^a y\,dx$$

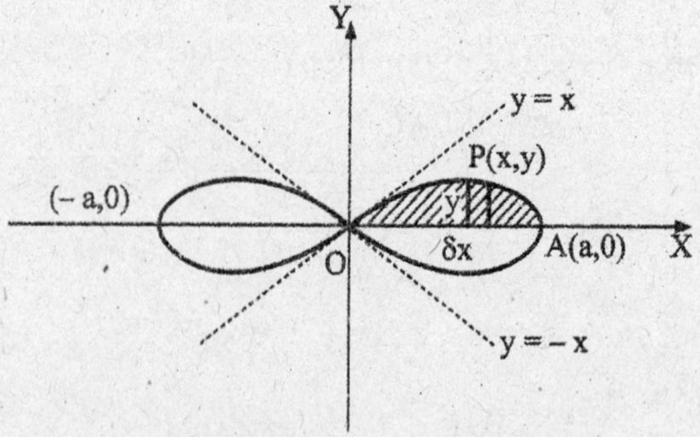

$$= 4\int_0^a \frac{x\sqrt{a^2 - x^2}}{a}\, dx$$

Putting $x = a \sin \theta$, $dx = a \cos \theta\, d\theta$. When $x = 0$, $\theta = 0$; when $x = a$, $\theta = \pi/2$

$$\therefore \quad \text{area} = \frac{4}{a}\int_0^{\pi/2} a \sin\theta\sqrt{a^2 - a^2 \sin^2\theta}\; a\cos\theta\, d\theta$$

$$= \frac{4}{a} a^3 \int_0^{\pi/2} \sin\theta\cos^2\theta\, d\theta = -4a^2\int_0^{\pi/2} \cos^2\theta d(\cos\theta)$$

$$= -4a^2\left[\frac{\cos^3\theta}{3}\right]_0^{\pi/2} = -4a^2\left[0 - \frac{1}{3}\right] = \frac{4}{3}a^2.$$

7.

(a) Find the whole area of the curve $a^2 y^2 = x^3 (2a - x)$.

(b) Show that the whole area of the curve $a^4 y^2 = x^5 (2a - x)$ is to that of the circle whose radius is a is 5 : 4.

(a) The curve is $a^2 y^2 = x^3 (2a - x)$. Since even power of y occurs, so the curve is symmetrical about x-axis. It passes through origin (0, 0) and the point (2a, 0).

$\therefore$ the required area = 2 × area OABO

$$= 2\int_0^{2a} y\, dx$$

$$= 2\int_0^{2a} \frac{x^{3/2}(2a - x)^{1/2}}{a} dx$$

(from the equation of the curve $y = \dfrac{x^{3/2}(2a - x)^{1/2}}{a}$).

Putting $x = 2a \sin^2 \theta$, $dx = 4a \sin\theta \cos\theta \, d\theta$. When $x = 0$, $\theta = 0$; when $x = 2a$, $\theta = \pi/2$

$$\therefore \quad \text{area} = \frac{2}{a}\int_0^{\pi/2} (2a)^{3/2} \sin^3\theta . (2a)^{1/2} \cos\theta \, . \, 4a \sin\theta \cos\theta \, d\theta$$

$$= 32a^2 \int_0^{\pi/2} \sin^4\theta \cos^2\theta \, d\theta$$

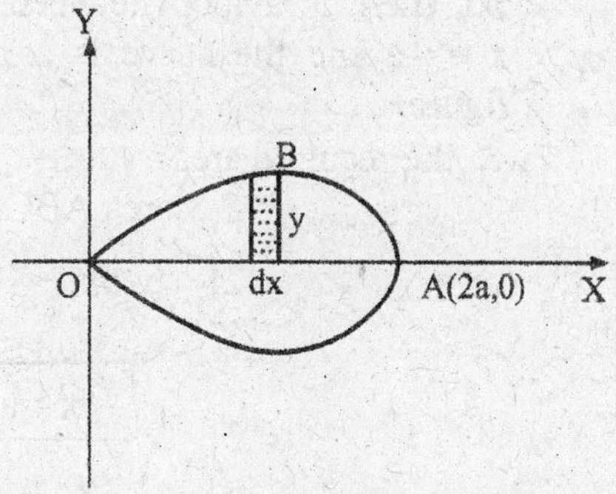

$$= 32a^2 . \frac{\Gamma\left(\frac{5}{2}\right)\Gamma\left(\frac{3}{2}\right)}{2\Gamma(4)}$$

$$= 32a^2 . \frac{\frac{3}{2}.\frac{1}{2}\sqrt{\pi}\frac{1}{2}\sqrt{\pi}}{2.3.2.1} = \pi a^2$$

(b) Draw the curve as in (a).

$$\text{Area} = 2\int_0^{2a} y dx = 2\int_0^{2a} \frac{x^{5/2}(2a - x)^{1/2}}{a^2} dx$$

$$= 2\int_0^{\pi/2} \frac{(2a)^{5/2} \sin^5\theta (2a)^{1/2} \cos\theta . 4a \sin\theta \cos\theta d\theta}{a^2}$$

Putting $x = 2a\sin^2\theta$ (as above)

$$= 64a^2 \int_0^{\pi/2} \sin^6\theta \cos^2\theta \, d\theta$$

$$= \frac{64a^2\Gamma\left(\frac{7}{2}\right)\Gamma\left(\frac{3}{2}\right)}{2\Gamma(5)} = \frac{64a^2 . \frac{5}{2}.\frac{3}{2}.\frac{1}{2}\sqrt{\pi}.\frac{1}{2}\sqrt{\pi}}{2.4.3.2.1}$$

$$= \frac{5}{4}\pi a^2$$

$$\therefore \quad \frac{\text{area of the curve}}{\text{area of circle}} = \frac{\frac{5}{4}\pi a^2}{\pi a^2} = \frac{5}{4}.$$

8. Find the area of the loop of the curve

$$xy^2 + (x + a)^2 (x + 2a) = 0$$

Since the equation contains even power of y only, so the curve is symmetrical about x-axis.

Putting y = 0, we get x = –a, x = –2a

So, the curve passes through (–a, 0) and (–2a, 0). Equating to zero the coefficient of the highest power of y, we get the asymptote as x = 0, i.e., y-axis is the asymptote of the curve.

So, there is a loop between x = –2a and x = –a and the curve is as shown in the figure.

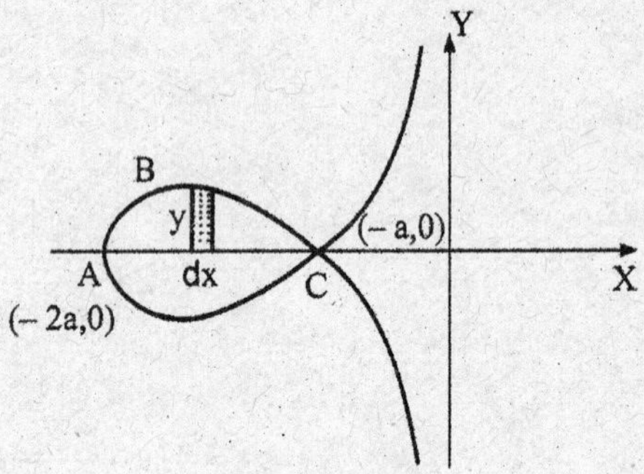

∴ the required area

$$= 2 \times \text{area ABCA}$$

$$= 2\int_{-2a}^{-a} y\,dx$$

$$= 2\int_{-2a}^{-a} \sqrt{\frac{-(x+a)^2(x+2a)}{x}}\,dx$$

(from the equation of the curve)

Put x + 2a = z. ∴ dx = dz; x + a = (z – 2a) + a = z – a

When x = –2a, z = 0; when x = –a, z = a

$$\therefore \qquad \text{area} = 2\int_0^a \sqrt{\frac{-(z-a)^2 .z}{z-2a}}\,.\,dz = 2\int_0^a (a-z)\sqrt{\frac{z}{2a-z}}\,dz$$

Putting a – z = a sin φ, – dz = a cos φ dφ

When z = 0, φ = π/2; when z = a, φ = 0

$$\therefore \qquad \text{area} = 2\int_{\pi/2}^{0} a\sin\varphi \sqrt{\frac{a-a\sin\varphi}{a+a\sin\varphi}} - a\cos\varphi\,d\varphi$$

$$= 2a^2\int_0^{\pi/2} \sin\theta.\frac{1-\sin\theta}{\cos\theta}.\cos\theta\,d\theta$$

$$\left[\because \sqrt{\frac{1-\sin\theta}{1+\sin\theta}} = \sqrt{\left(\frac{1-\sin\theta}{1+\sin\theta}\times\frac{1-\sin\theta}{1-\sin\theta}\right)}\right.$$

$$\left. = \sqrt{\frac{(1-\sin\theta)^2}{1-\sin^2\theta}} = \frac{1-\sin\theta}{\cos\theta}\right]$$

$$= a^2\int_0^{\pi/2} \left(2\sin\theta - 2\sin^2\theta\right)d\theta$$

$$= a^2 \int_0^{\pi/2} \{2 \sin\theta - (1 - \cos 2\theta)\} d\theta$$

$$= a^2 \left[-2\cos\theta - \theta + \frac{\sin^2\theta}{2} \right]_0^{\pi/2}$$

$$= a^2 \left(0 + 2 - \frac{\pi}{2} + 0 \right) = \frac{a^2}{2}(4 - \pi)$$

9. Find the area included between the curve $y^2 = x^2 \frac{a+x}{a-x'}$, i.e., $y^2 (a - x) = x^2 (a + x)$ and its asymptote excluding the loop. Also find the area of the loop.

The curve is symmetrical about x-axis, as the equation contains even power of y. It passes through (0, 0) and (–a, 0) since y = 0 when x = 0, x = –a. Thus a loop is formed. By equating to zero the co-efficient of the highest power of y, a – x = 0, i.e., x = a. So x = a is the asymptote parallel to y-axis. If x < –a or x > a, y is imaginary, so no part of the curve is in the left of A.

At $(-a, 0), \frac{dy}{dx} = \infty,$

i.e., tangent of (a, 0) is perpendicular to x-axis and the curve is as shown in the figure.

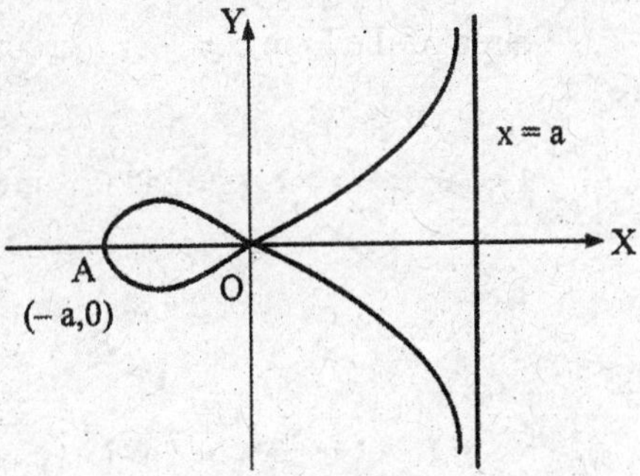

∴ the required area

$$= 2\int_0^a y\, dx = 2\int_0^a x\sqrt{\frac{a+x}{a-x}} dx$$

Putting $x = a \cos\theta$, $dx = -a \sin\theta\, d\theta$; also when $x = 0$, $\theta = \pi/2$, when $x = a$, $\theta = 0$

$$\therefore \quad \text{area} = 2\int_{\pi/2}^0 a\cos\theta \sqrt{\frac{a + a\cos\theta}{a - \cos\theta}} (-a \sin\theta) d\theta$$

$$= 2a^2 \int_0^{\pi/2} \cos\theta \frac{\cos\theta/2}{\sin\theta/2} 2 \sin\frac{\theta}{2} \cos\frac{\theta}{2}\, d\theta$$

$$= 2a^2 \int_0^{\pi/2} \cos\theta . 2\cos^2\frac{\theta}{2} = 2a^2 \int_0^{\pi/2} \cos\theta(1 + \cos\theta) d\theta$$

$$= 2a^2 \int_0^{\pi/2} (\cos\theta + \cos^2\theta)\, d\theta$$

$$= 2a^2\left\{\int_0^{\pi/2}\cos\theta\, d\theta + \int_0^{\pi/2}\cos^2\theta\, d\theta\right\}$$

$$= 2a^2.[\sin\theta]_0^{\pi/2} + a^2.\int_0^{\pi/2}(1+\cos 2\theta)d\theta$$

$$= 2a^2.1 + a^2.\left[[\theta]_0^{\pi/2} + \left[\frac{\sin 2\theta}{2}\right]_0^{\pi/2}\right]$$

$$= 2a^2 + a^2.\frac{\pi}{2} + 0 = \frac{1}{2}(4+\pi)a^2$$

Area of the loop $= 2\int_{-a}^{0} y\, dx = -2\int_{-a}^{0} x\sqrt{\frac{a+x}{a-x}}dx$

(since in this case x and hence y is negative)

Putting $x = a\cos\theta$, $dx = -a\sin\theta\, d\theta$

When $x = -a$, $\theta = \pi$; when $x = 0$, $\theta = \pi/2$

$\therefore$ area of the loop $= -2\int_{\pi}^{\pi/2} a\cos\theta\sqrt{\frac{a+a\cos\theta}{a-a\cos\theta}}.(-a\sin\theta\, d\theta)$

$$= 2a^2\int_{\pi}^{\pi/2}\cos\theta\cot\frac{\theta}{2}\sin\theta\, d\theta$$

$$= 2a^2\int_{\pi}^{\pi/2}\cos\theta.\cot\frac{\theta}{2}.2\sin\frac{\theta}{2}\cos\frac{\theta}{2}\, d\theta$$

$$= 2a^2\int_{\pi}^{\pi/2}\cos\theta.2\cos^2\frac{\theta}{2}d\theta$$

$$= 2a^2\int_{\pi}^{\pi/2}\cos\theta(1+\cos\theta)d\theta$$

$$= 2a^2\int_{\pi}^{\pi/2}\left(\cos\theta+\cos^2\theta\right)d\theta$$

$$= 2a^2\int_{\pi}^{\pi/2}\cos\theta\, d\theta + a^2\int_{\pi}^{\pi/2}(1+\cos 2\theta)d\theta$$

$$= 2a^2[\sin\theta]_{\pi}^{\pi/2} + a^2\left[\theta + \frac{1}{2}\sin 2\theta\right]_{\pi}^{\pi/2}$$

$$= 2a^2 + a^2\left(\frac{\pi}{2}-\pi\right) = 2a^2\left(1-\frac{1}{4}\pi\right).$$

10. Find the whole area between the curve $x^2 y^2 = a^2 (y^2 - x^2)$ and its asymptote.

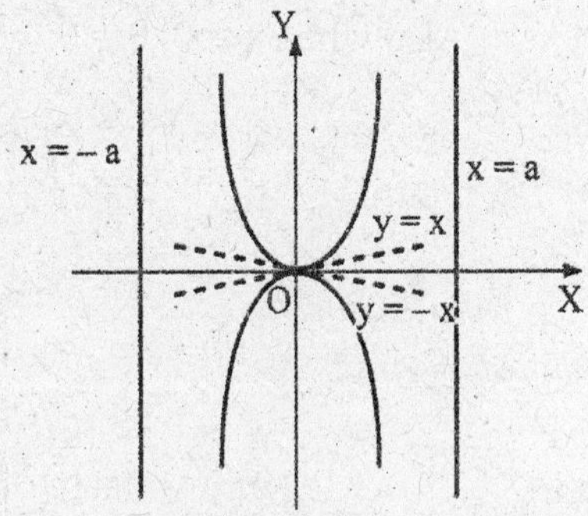

The curve is symmetrical about both axes as its equation contains even powers of x and y.

It passes through origin as it is satisfied by $x = 0$, $y = 0$.

The equations of tangents at origin are (equating lowest term to zero) $y^2 - x^2 = 0$ i.e., $y = \pm x$.

The curve is $x^2 y^2 = a^2 (y^2 - x^2)$

or $\quad x^2 y^2 = a^2 y^2 - a^2 x^2$

or $\quad a^2 y^2 - x^2 y^2 - a^2 x^2;$

$$\therefore \qquad y^2 = \frac{a^2x^2}{a^2 - x^2}; \qquad \therefore\ y = \pm\frac{ax}{\sqrt{a^2 - x^2}}$$

Equating to zero the coefficient of the highest power of y, we get $x^2 - a^2 = 0$;

$\therefore\ x = \pm a$. So $x = a$, $x = -a$ are its asymptotes parallel to y-axis.

y is real when x lies between $-a$ and a. Thus the curve is as shown here.

$$\therefore \text{ the required area } = 4\int_0^a y\, dx = 4\int_0^a \frac{ax}{\sqrt{a^2 - x^2}} dx$$

Putting $x = a \sin\theta$, $dx = a\cos\theta\, d\theta$

Also when $x = 0$, $\theta = 0$, when $x = a$, $\theta = \pi/2$

$$\therefore \qquad \text{area } = 4a\int_0^{\pi/2} \frac{a \sin\theta . a\cos\theta\, d\theta}{a\cos\theta}$$

$$= 4a\int_0^{\pi/2} \sin\theta\, d\theta = 4a^2\left[-\cos\theta\right]_0^{\pi/2} = -4a^2(0-1)$$

$$= 4a^2.$$

11. Find the whole area included between the curve $x^2 (x^2 + y^2) = a^2 (y^2 - x^2)$ and its asymptotes.

The curve is symmetrical about both the axes. It passes through origin. Equating coefficient of highest power of y to zero, $a^2 - x^2 = 0$, i.e., $x = \pm a$, i.e., $x = \pm a$ are the asymptotes parallel to y-axis. (Figure is similar as above)

$$\therefore \qquad \text{area } = 4\int_0^a y dx = 4\int_0^a \frac{\sqrt{a^2 + x^2}}{a^2 - x^2} dx$$

[from the equation of the curve, $y^2 (a^2 - x^2) = x^2 (a^2 + x^2)$, so $y^2 = x^2 \dfrac{a^2 + x^2}{a^2 - x^2}$]

Putting $x^2 = a^2 \sin\theta$, $2x\,dx = a^2\cos\theta\,d\theta$

Also when $x = 0$, $\theta = 0$; when $x = a$, $\theta = \pi/2$

$$\therefore \quad \text{area} = 2\int_0^{\pi/2} \frac{\sqrt{a^2 + a^2\sin\theta}}{\sqrt{a^2 - a^2\sin\theta}}\, a^2\cos\theta\, d\theta$$

$$= 2a^2\int_0^{\pi/2} \frac{\sqrt{1+\sin\theta}}{\sqrt{1-\sin\theta}}\cos\theta\, d\theta$$

$$= 2a^2\int_0^{\pi/2} \frac{1+\sin\theta}{\sqrt{1-\sin^2\theta}}\cos\theta\, d\theta$$

$$= 2a^2\int_0^{\pi/2} (1+\sin\theta)\,d\theta = 2a^2\left[\theta - \cos\theta\right]_0^{\pi/2}$$

$$= 2a^2\left(\frac{\pi}{2}+1\right).$$

12. Find the area between the curve $y^2(a+x) = (a-x)^3$ and its asymptote.

Only even power of y occur, so the curve is symmetrical about x-axis.

Putting $y = 0$, we get $x = a$, so the curve passes through $(a, 0)$

Putting $x = 0$, we get $y = \pm a$, so the curve passes through $(0, \pm a)$

Equating the coefficient of highest power of y to zero, the asymptote parallel to y-axis is $x + a = 0$, i.e., $x = -a$ is the asymptote.

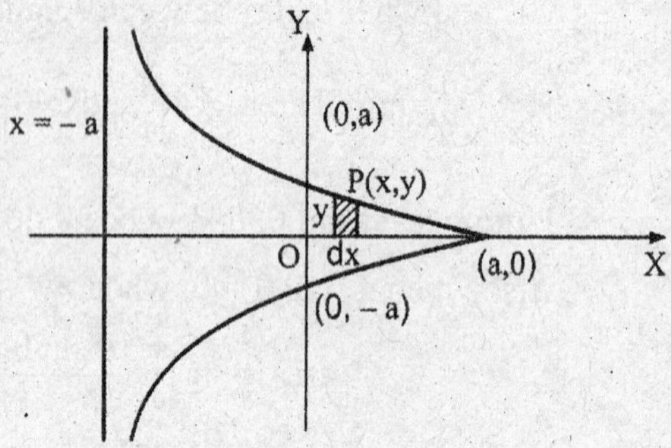

The curve is as shown in the figure

$\therefore$ the required area $= 2 \times$ area of the upper half

$$= 2\int_{-a}^{a} y\,dx$$

$$= 2\int_{-a}^{a} \frac{(a-x)^{3/2}}{\sqrt{a+x}}\,dx$$

$$= 2\int_{-a}^{a} (a-x)\sqrt{\frac{(a-x)}{a+x}}\,dx$$

Putting x = a sin θ, dx = a cos θ dθ

When x = –a, θ = π/2 ; when x = a, θ = π/2

$$\therefore \qquad \text{area} = 2\int_{-\pi/2}^{\pi/2}(a - a\sin\theta)\sqrt{\frac{a - a\sin\theta}{a + a\sin\theta}}\ a\cos\theta\, d\theta$$

$$= 2a^2\int_{-\pi/2}^{\pi/2}(1-\sin\theta)\sqrt{\frac{1-\sin\theta}{1+\sin\theta}\times\frac{1-\sin\theta}{1-\sin\theta}}\ \cos\theta\, d\theta$$

$$= 2a^2\int_{-\pi/2}^{\pi/2}\frac{(1-\sin\theta)^2}{\sqrt{1-\sin^2\theta}}\cos\theta\, d\theta$$

$$= 2a^2\int_{-\pi/2}^{\pi/2}(1-\sin\theta)^2 d\theta$$

$$= 2a^2\int_{-\pi/2}^{\pi/2}\left(1-2\sin\theta+\sin^2\theta\right) d\theta$$

$$= 2a^2\left\{\int_{-\pi/2}^{\pi/2} d\theta - 2\int_{-\pi/2}^{\pi/2}\sin\theta\, d\theta + \int_{-\pi/2}^{\pi/2}\sin^2 d\theta\right\}$$

$$= 2a^2\left[\theta + 2\cos\theta + \frac{\theta}{2} - \frac{\sin 2\theta}{4}\right]_{-\pi/2}^{\pi/2}$$

$$= 3\pi a^2$$

13. Find the area included between the curve $y^2 (a - x) = x^3$ and its asymptote.

The curve is symmetrical about x-axis since even power of y occurs in its equation. The curve passes through the origin and equation of tangents at origin are $y^2 = 0$, i.e., y = 0, y = 0, i.e., two coincident tangents, and so origin is a cusp. The curve does not meet the axes at any other point since by putting x = 0, we get y = 0 and by putting y = 0, we get x = 0.

From the equation of the curve,

$$y = \pm\sqrt{\frac{x^3}{a-x}} = \pm x\sqrt{\frac{x}{a-x}}$$

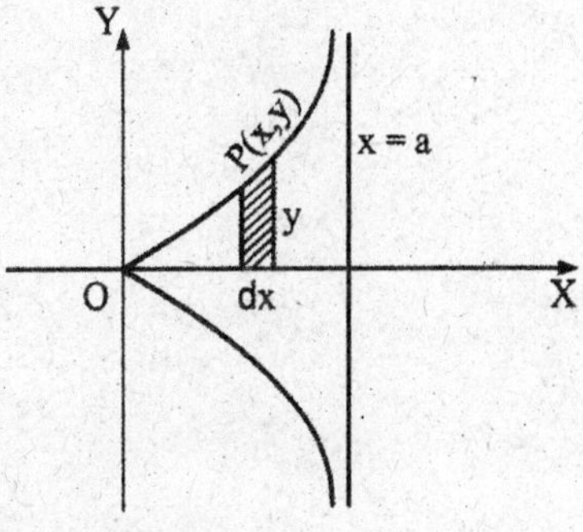

When x > a, y is imaginary, so x cannot be greater than a.

Equating coefficients of highest power of y, x – a = 0, asymptote parallel to y-axis is x = a.

Thus the curve wholly lies between the line x = 0 (y-axis) and the asymptote x = a and is as shown in the figure.

$\therefore$ the required area $= 2\int_0^a y dx = 2\int_0^a \frac{x\sqrt{x}}{\sqrt{a-x}} dx$

Putting $x = a \sin^2\theta$, $dx = 2 a \sin\theta \cos\theta \, d\theta$

When $x = 0, \theta = 0$; when $x = a, \theta = \pi/2$

$$\therefore \quad \text{area} = 4a^2 \int_0^{\pi/2} \frac{a \sin^2\theta . \sqrt{a \sin^2\theta}. 2a \sin\theta \cos\theta \, d\theta}{\sqrt{a - a\sin^2\theta}}$$

$$= 4a^2 \int_0^{\pi/2} \sin^4\theta \, d\theta = 4a^2 . \frac{3}{2} . \frac{1}{2} . \frac{\pi}{2}$$

$$= \frac{3}{2}\pi a^2 .$$

14. Find the area bounded by the semicubical parabola $y^2 = ax^3$ and a double ordinate.

The curve is symmetrical about x-axis as only even power of y occurs. The curve passes through origin. The equation of tangent at origin is $y^2 = 0$, i.e., $y = 0$, $y = 0$, i.e., two coincident tangents. So, the origin is a cusp. The curve does not meet the axes in any points since by putting $x = 0$, we get $y = 0$, and by putting $y = 0$, we get $x = 0$. From the equation of the curve,

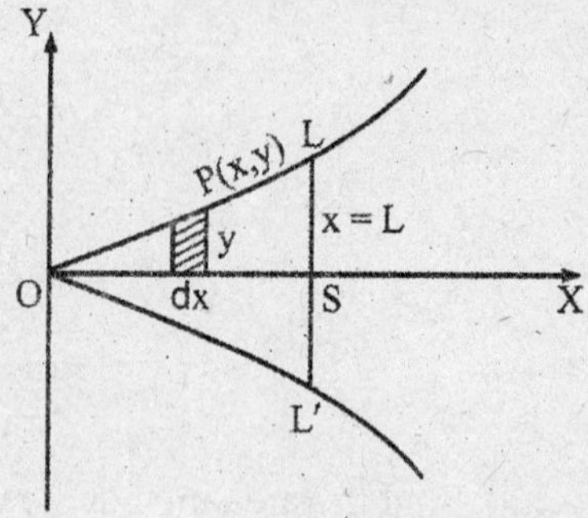

$$y = \pm x\sqrt{ax}$$

So, x cannot be negative, since then y becomes imaginary. Thus we get a curve shown in the figure. Let $x = b$, be any double ordinate.

The area bounded by the double ordinate and the curve

$$= 2 \times \text{area LSO} = 2\int_0^1 y \, dx$$

$$= 2\int_0^1 x\sqrt{ax} \, dx = 2\sqrt{a}\left[\frac{x^{5/2}}{\frac{5}{2}}\right]_0^1$$

$$= \frac{4}{5}\sqrt{a}\, 1^{5/2} \qquad .....(1)$$

Again since L lies on the curve, if $(1, y_1)$ be its coordinates, $y_1^2 = al^3$, $\therefore y_1 = \sqrt{a}\,l^{3/2}$

Also $$LL' = 2y_1 = 2\sqrt{a}\,l^{3/2}$$

$\therefore$ area of the rectangle formed by this double ordinate LL' and the abscissa OS

$$= LL' \times OS = 2\sqrt{a}\,l^{3/2} \times l = 2\sqrt{a}\,l^{5/2} \quad .....(2)$$

From (1) and (2), the required area $= \frac{2}{5}$ of the rectangle formed by the double ordinate and the abscissa.

15. Find the area bounded by the 'Witch of Agensi' $y^2 = \frac{4a^2(2a-x)}{x}$ and its asymptotes.

The curve is $xy^2 = 4a^2(2a - x)$. It passes through $(2a, 0)$. Equating to zero the coefficient of highest power of y, the asymptote parallel to y-axis is $x = 0$ (i.e., y-axis)

The curve is as shown in the figure

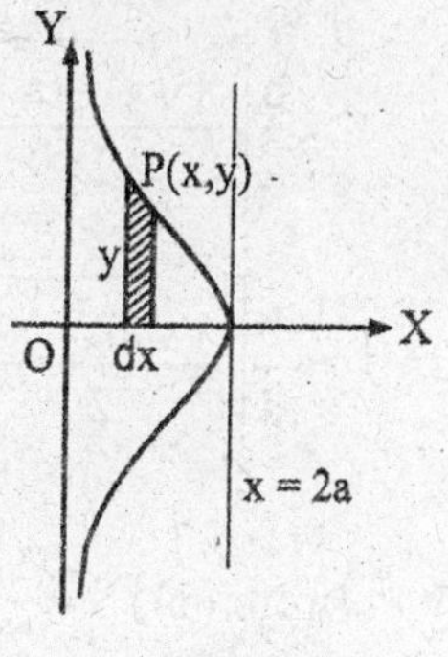

$$\text{Area} = 2\int_0^{2a} y\,dx$$

$$= 2\int_0^{2a} 2a\sqrt{\frac{2a-x}{x}}\,dx$$

Putting $x = 2a\sin^2\theta$, $dx = 4a\sin\theta\cos d\theta$

When $x = 0, \theta = 0$; when $x = 2a, \theta = \pi/2$

$$\text{Area} = 4a\int_0^{\pi/2}\sqrt{\frac{2a - 2a\sin^2\theta}{2a\sin^2\theta}}.4a\sin\theta\cos\theta\,d\theta$$

$$= 16a^2\int_0^{\pi/2}\frac{\cos\theta}{\sin\theta}\times\sin\theta\cos\theta\,d\theta$$

$$= 16a^2\int_0^{\pi/2}\cos^2\theta\,d\theta = 8a^2\int_0^{\pi/2}(1+\cos 2\theta)d\theta$$

$$= 8a^2\left[\theta + \frac{\sin 2\theta}{2}\right]_0^{\pi/2} = 8a^2\left[\left(\frac{\pi}{2} - 0\right) + \frac{1}{2}(\sin\pi - \sin\theta)\right]$$

$$= 8a^2.\frac{\pi}{2} = 4a^2\pi$$

16. If A is the vertex, O the centre and P(x, y) a point on the hyperbola $\frac{x^2}{a^2}-\frac{y^2}{b^2}=1$;

prove that $x = a\cos h\frac{25}{ab'}$ $y = b\sin\ h\frac{25}{ab}$ when S is the sectorial area OPA.

S = area OAP

= area ONP – area ANP

$=\frac{1}{2}xy$ – area ANP, ...(1)

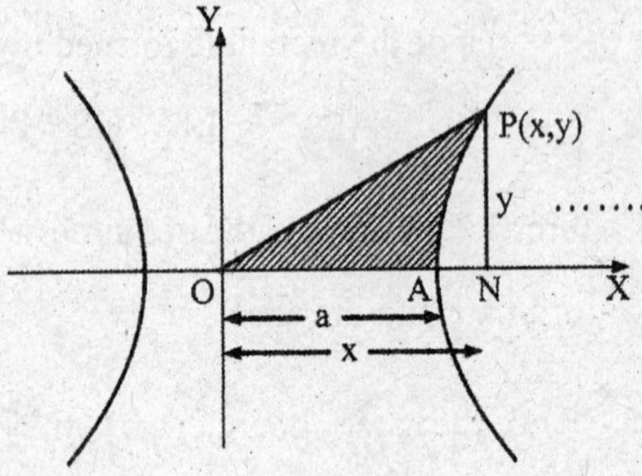

........(8.3)

where PN ⊥ OX

From the equation of the hyperbola

$$\frac{y^2}{b^2}=\frac{x^2}{a^2}-1=\frac{x^2-a^2}{a^2};\ \therefore y=\frac{b}{a}\sqrt{x^2-a^2}$$

$$\therefore\ \text{area ANP} = \int_a^x y\,dx=\frac{b}{a}\int_a^x\sqrt{x^2-a^2}\,dx$$

$$=\frac{b}{a}\left[\frac{x\sqrt{x^2-a^2}}{2}-\frac{a^2}{2}\cosh^{-1}\frac{x}{a}\right]_a^x$$

$$=\frac{b}{a}\left[\frac{x\sqrt{x^2-a^2}}{2}-\frac{a^2}{2}\cosh^{-1}\frac{x}{a}\right]$$

$$\therefore\ \text{from (1), } S=\frac{1}{2}x\frac{b}{a}\sqrt{x^2-a^2}-\frac{b}{a}\left[\frac{x\sqrt{x^2-a^2}}{2}-\frac{a^2}{2}\cosh^{-1}\frac{x}{a}\right]$$

$$=\frac{ab}{2}\cosh^{-1}\frac{x}{a};\quad\text{or}\quad\frac{25}{ab}=\cosh^{-1}\frac{x}{a}$$

$$\text{or}\quad x=a\cos h\frac{2s}{ab}$$

$$\therefore\quad y=\frac{b}{a}\sqrt{x^2-a^2}=\frac{b}{a}\sqrt{a^2\cosh^2\left(\frac{2s}{ab}\right)-a^2}$$

$$=b\sin h\frac{2s}{ab}$$

17. Trace and find the loop of the curve $x^3 + y^3 = 3axy$.

The curve is symmetrical about the line $y = x$ as by putting x for y and y for x the equation of the curve remains unchanged. The curve passes through the origin.

The equation of the tangent is $x = 0$ and $y = 0$

(by equating the lowest-degree term 3axy to zero)

i.e., x-axis and y-axis are tangents to the curve at origin

To find asymptote we see that in the highest-degree term putting $x = 1$, $y = m$ and equating to zero, we get

$$\varphi_3(m) = 1 + m^3 = 0; \quad \text{or } (1 + m)(m^2 - m + 1) = 0$$

$$\therefore \quad m = -1 \text{ or } m = \frac{1 \pm i\sqrt{3}}{2}$$

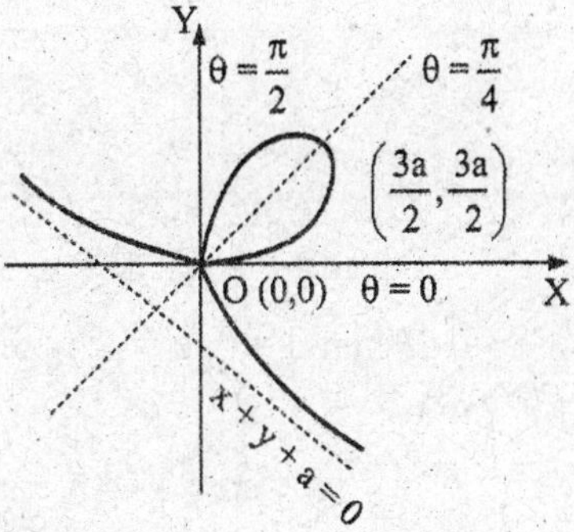

Again $\varphi_2(m) = -3am$

$$\therefore \quad C = -\frac{\varphi_2(m)}{\varphi_3(m)} = -\frac{-3am}{3m^2} = \frac{a}{m}$$

If $m = -1$ then $C = -a$

$\therefore$ from $y = mx + C$, $y = -1\,x - a = -x - a$,

i.e., $x + y + a = 0$ is the only real asymptote

Putting $y = x$ in $x^3 + y^3 = 3axy$, we get

$$2x^3 = 3ax^2 \text{ or } x^2(2x - 3a) = 0$$

$$\therefore \quad x = 0, x = \frac{3a}{2}$$

So the curve cuts the line $y = x$ in two points $(0, 0)$, $\left(\frac{3a}{2}, \frac{3a}{2}\right)$ and the curve is as shown in the figure.

Putting $x = r\cos\theta$, $y = r\sin\theta$ in the equation of the curve

$$r^3\cos^3\theta + r^3\sin^3\theta = 3ar\cos\theta \,.\, r\sin\theta,$$

i.e., $r^3(\cos^3\theta + \sin^3\theta) = 3ar^2\sin\theta\cos\theta$,

$$\therefore \quad r = \frac{3a\sin\theta\cos\theta}{\cos^3\theta + \sin^3\theta}$$

Which is the polar equation of the curve

Now $r = 0$ when $\theta = 0$ and $\theta = \pi/2$

$$r = \frac{3a}{\sqrt{2}} \text{ when } \theta = \pi/4$$

Thus as θ increases from 0 to $\pi/2$, r increases from 0 to $\frac{3a}{\sqrt{2}}$ having its maximum value at $\theta = \pi/4$ and again diminishes to 0. So it forms a loop in the first quadrant.

The area of the loop

$$= \frac{1}{2}\int_0^{\pi/2} r^2 d\theta = \frac{1}{2}\int_a^{\pi/2} \frac{9a^2 \sin^2\theta \cos^2\theta}{(\cos^3\theta + \sin^3\theta)^2} d\theta$$

$$= \frac{9}{2} a^2 \int_0^{\pi/2} \frac{\tan^2\theta \sec^2\theta}{(1+\tan^3\theta)^2} d\theta$$

(dividing numerator and denominator by $\cos^6\theta$)

Putting $1 + \tan^3\theta = z$, $3\tan^2\theta \sec^2\theta \, d\theta = dz$

When $\theta = 0$, $z = 1$; when $\theta = \pi/2$, $z = \infty$

$$\therefore \quad \text{area} = \frac{9}{2 \times 3} a^2 \int_1^{\infty} \frac{dz}{z^2} = \frac{3a^2}{2}\left[-\frac{1}{z}\right]_1^{\infty} = \frac{3a^2}{2}$$

18. Find the area of the loop of the curve $x^4 + y^4 = 2a^2 xy$

Putting $x = r\cos\theta$, $y = r\sin\theta$

$$r^2 \cos^4\theta + r^4 \sin^4\theta = 2a^2 r\cos\theta . r\sin\theta$$

$$\therefore \quad r^2 = \frac{2a^2 \sin\theta\cos\theta}{\cos^4\theta + \sin^4\theta}$$

which is polar equation of the curve

when $\theta = 0$ and $\theta = \pi/2$, $r = 0$

So, a loop is formed in the first quadrant and its area

$$= \frac{1}{2}\int_0^{\pi/2} r^2 d\theta$$

$$= \frac{1}{2}\int_0^{\pi/2} \frac{2a^2 \sin\theta\cos\theta}{\cos^4\theta + \sin^4\theta} d\theta = a^2 \int_0^{\pi/2} \frac{\tan\theta \sec^2\theta \, d\theta}{1+\tan^4\theta}$$

Putting $z = \tan^2 \theta$, $dz = 2 \tan \theta \sec^2 \theta \, d\theta$

When $\theta = 0, z = 0$; when $\theta = \pi/2, z = \infty$

$$\therefore \quad \text{area} = \frac{a^2}{2}\int_0^{\infty} \frac{dz}{1+z^2} = \frac{a^2}{2}\left[\tan^{-1} z\right]_0^{\infty}$$

$$= \frac{1}{2}a^2\left[\tan^{-1}\infty - \tan^{-1} 0\right] = \frac{1}{2}a^2 . \frac{\pi}{2}$$

$$= \frac{1}{4}\pi a^2.$$

19. Find the area of the curve $x^{2/3} + y^{2/3} = a^{2/3}$

The curve (astroid) is symmetrical about both the axes. It does not pass through the origin.

Putting $y = 0$ we get $x = \pm a$ and putting $x = 0$ we get $y = \pm a$

So, the curve cuts x-axis at (a, 0) and (–a, 0) and y-axis at (0, a) and (0, –a). For $x > a$ or $x < -a$, y becomes imaginary. So the curve has no part in the right of $x = a$ and in the left of $x = -a$. Similarly, the curve has no part above $y = a$ and below $y = -a$.

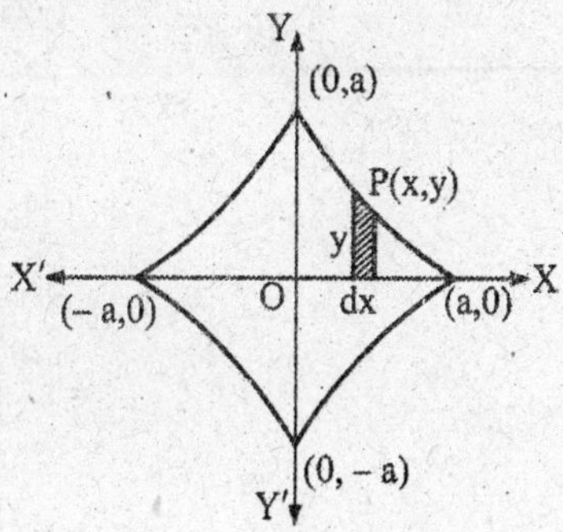

Thus the curve is as shown in the figure.

$\therefore$ the required area

$$= 4 \times \text{area in the first quadrant}$$

$$= 4\int_0^a y dx = 4\int_0^a \left(a^{2/3} - x^{3/2}\right)^{3/2} dx$$

Putting $x = a \sin^2 \theta$, $dx = 3a \sin^2 \theta \cos \theta \, d\theta$

When $x = 0, \theta = 0$; when $x = a, \theta = \pi/2$

$$\therefore \text{ the required area } = 4\int_0^{\pi/2} \left(a^{2/3} - a^{2/3} \sin^2 \theta\right)^{3/2} 3a \sin^2 \theta \cos\theta \, d\theta$$

$$= 12a^2 \int_0^{\pi/2} \cos^3 \theta \sin^2\theta \cos\theta \, d\theta$$

$$= 12a^2 \int_0^{\pi/2} \cos^4 \theta \sin^2\theta \, d\theta$$

$$= 12a^2 \int_0^{\pi/2} \left(1 - \cos^2 \theta\right) \cos^4 \theta \, d\theta$$

$$= 12a^2 \left\{ \int_0^{\pi/2} \cos^4 \theta \, d\theta - \int_0^{\pi/2} \cos^6 \theta \, d\theta \right\}$$

$$= \frac{3\pi a^2}{8}.$$

20. Find the area of the cycloid x = a(θ – sin θ), y = a(1 – cos θ) and its base.

From the equation of the curve we find that when θ = 0 or 2π, y = 0; when θ = 0, x = 0; when θ = 2π, x = 2πa.

So, in this case θ varies from 0 to 2π and the coordinates of A are (2πa, 0)

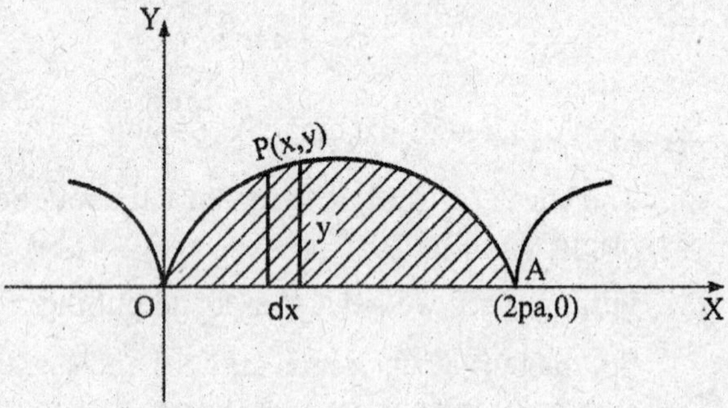

∴ the required area

$$= \int_0^{2\pi} y \, dx$$

$$= \int_0^{2\pi} a(1-\cos\theta).a(1-\cos\theta)d\theta$$

[∵ x = a(θ – sin θ), ∴ dx = a (1 – cos θ) dθ]

$$= a^2 \int_0^{2\pi} (1 - 2\cos\theta + \cos^2\theta) d\theta$$

$$= a^2 \int_0^{2\pi} \left(1 - 2\cos\theta + \frac{1+\cos 2\theta}{2}\right) d\theta$$

$$= a^2 \left[\theta - 2\sin\theta + \frac{\theta}{2} + \frac{\sin 2\theta}{4}\right]_0^{2\pi}$$

$$= 3a^2 \pi.$$

21. Find the area of the curve (lemniscate) $r^2 = a^2 \cos 2\theta$

The curve is symmetrical about the initial line since by putting – θ for θ, its equation remains unchanged.

The curve passes through the pole.

The straight lines $\theta = \pm\pi/4$ are tangents at pole since by putting r = 0, we get cos 2θ = 0, so $\theta = \pm\pi/4$.

For different values of θ the corresponding values of r are given as follows.

θ	0	$\frac{\pi}{6}$	$\frac{\pi}{4}$	$\frac{\pi}{2}$	$\frac{3\pi}{4}$	$\frac{5\pi}{6}$	π
r^2	a^2	$\frac{a^2}{2}$	0	$-a^2$	0	$\frac{a^2}{2}$	a^2
r	$\pm a$	$\pm\frac{a}{\sqrt{2}}$	0	Imaginary	0	$\pm\frac{a}{\sqrt{2}}$	$\pm a$

It is clear from the above table that no part of the curve lies between $\theta = \pi/4$ and $\theta = 3\pi/4$.

Combining the above facts, the curve can be traced as shown in the figure.

$\therefore$ the required area $= 4 \times$ area OABO

$$= 4 \times \frac{1}{2}\int_0^{\pi/4} r^2 d\theta$$

$$= 2\int_0^{\pi/4} a^2 \cos 2\theta d\theta = 2a^2\left[\frac{\sin 2\theta}{2}\right]_0^{\pi/4}$$

$$= 2a^2 . \frac{1}{2} = a^2 .$$

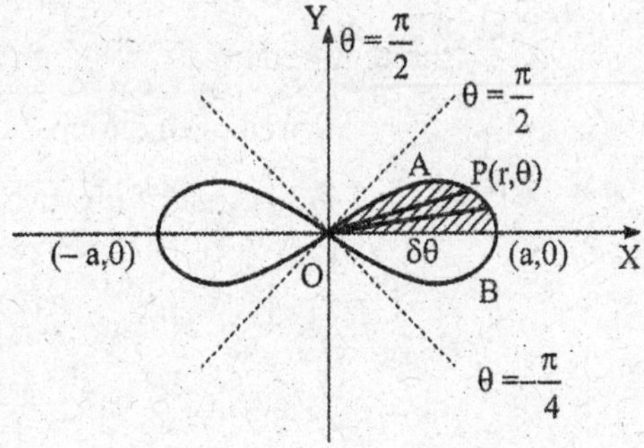

22. Find the entire area enclosed by the curve r = a sin 3θ.

The equation of the curve is r = a sin 3θ. When $\theta = 0, \frac{\pi}{3}, \frac{2\pi}{3}, \pi$, we get r = 0. So, the curve has three equal loops as shown.

$\therefore$ the required area

$$= 3 \times \text{area of one loop}$$

$$= 3 \times \frac{1}{2}\int_0^{\pi/2} r^2 d\theta$$

$$= \frac{3}{4}a^2\int_0^{\pi/3}(1 - \cos 6\theta)d\theta$$

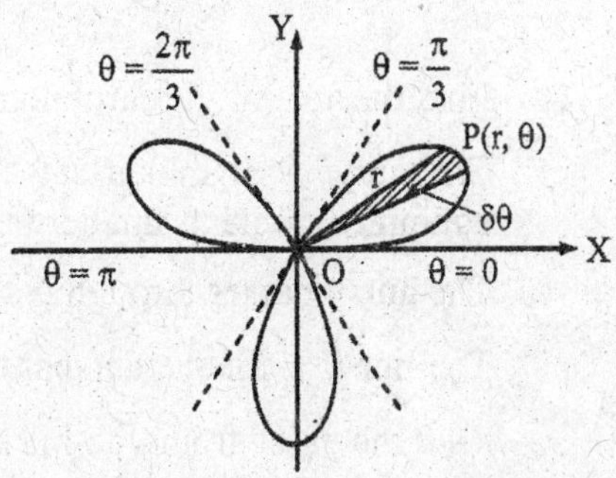

$$= \frac{3}{4}a^2\left[\theta - \frac{\sin 6\theta}{6}\right]_0^{\pi/3} = \frac{3}{4}a^2\left[\frac{\pi}{3} - 0\right] = \frac{1}{4}\pi a^2$$

23. Find the entire area enclosed by the curve r = a cos 2θ.

The curve is symmetrical about the initial line since by taking – θ for θ, the equation remains unchanged.

The corresponding values of r for different values of θ are as follows.

θ	0	$\frac{\pi}{4}$	$\frac{\pi}{2}$	$\frac{3\pi}{4}$	π	$\frac{5\pi}{4}$	$\frac{3\pi}{2}$	$\frac{7\pi}{4}$	2
2θ	0	$\frac{\pi}{2}$	π	$\frac{3\pi}{2}$	2π	$\frac{5\pi}{2}$	3π	$\frac{7\pi}{2}$	4π
r	a	0	–a	0	a	0	–a	0	a

Thus by plotting these points, we get the curve as shown in the figure.

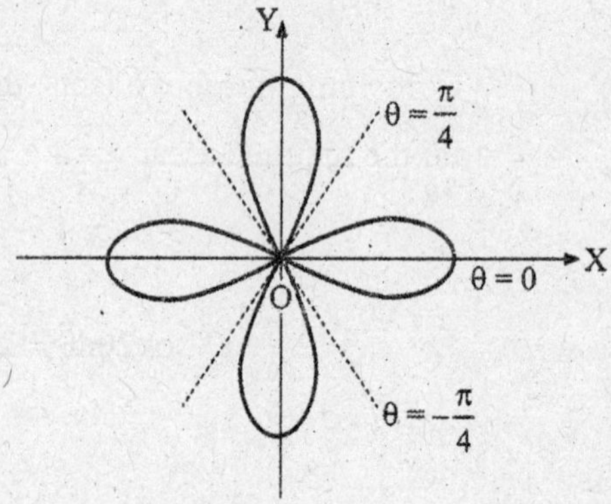

The curve has four equal loops.

∴ the required area

= 4 × area of one loop

= 4 × 2 × area of upper half of one loop

$$= 4 \times 2\int_0^{\pi/4} \frac{1}{2} r^2 d\theta$$

$$= 4\int_0^{\pi/4} a^2 \cos^2 2\theta \, d\theta = 2a^2 \int_0^{\pi/4} (1 + \cos 4\theta) d\theta$$

$$= 2a^2 \left[\theta + \frac{\sin 4\theta}{4}\right]_0^{\pi/4} = 2a^2 [\pi/4 + 0] = \frac{\pi a^2}{2}$$

24. Find the area of the cardioid r = a (1 + cos θ)

The curve is symmetrical about the initial line since by putting – θ for θ the equipment remains unaltered.

The curve passes through the pole

The line $\theta = \pi$ is the tangent to the curve at pole for by putting $r = 0$ we get $\theta = \pi$.

As θ increases from 0 to π, r decreases from 2a to 0 as follows.

θ	0	$\frac{\pi}{3}$	$\frac{\pi}{2}$	π
r	2a	$\frac{3a}{2}$	a	0

Hence the curve is as shown in the figure

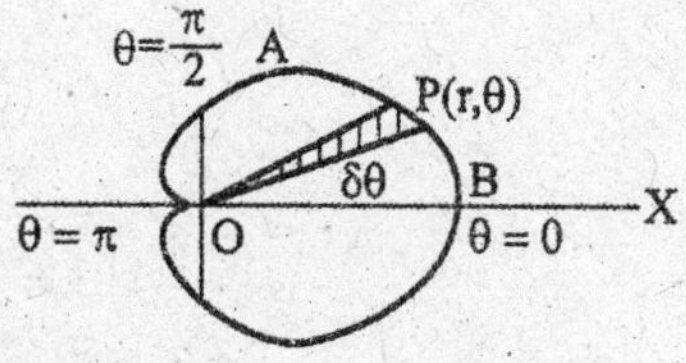

The required area

$= 2 \times \text{area OABO}$

$= 2 \times \frac{1}{2}\int_0^{\pi} r^2 d\theta$

$= \int_0^{\pi} a^2 (1+\cos\theta)^2 \, d\theta$

$= a^2 \int_0^{\pi} \left(2\cos^2\frac{\theta}{2}\right)^2 d\theta$

$= 4a^2 \int_0^{\pi} \cos^4\frac{\theta}{2} d\theta$

Putting $\frac{\theta}{2} = \alpha$, $d\theta = 2d\alpha$. When $\theta = 0, \alpha = 0$; when $\theta = \pi, \alpha = \pi/2$

$\therefore$ area $= 4a^2 \int_0^{\pi/2} \cos^4\alpha . 2d\alpha = 8a^2 . \frac{3}{4} . \frac{1}{2} \frac{\pi}{2} = \frac{3}{2}\pi a^2$

Note: If the cardioid be $r = a(1 - \cos\theta)$, it is shown as in the figure and

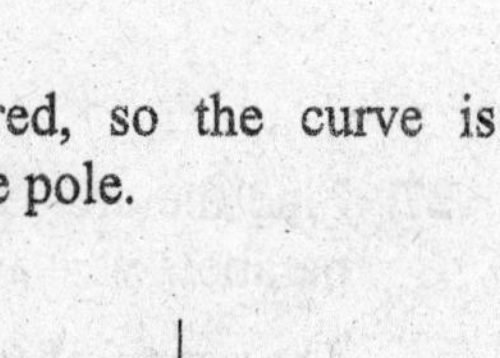

$\text{area} = 2\int_0^{\pi} \frac{1}{2} r^2 d\theta$

$= a^2 \int_0^{\pi} (1-\cos\theta)^2 \, d\theta$

$= \frac{3a^2\pi}{2}$ (as above)

25. Find the area of the curve $r = a\,(1 + \sin\theta)$

By replacing θ by $\pi - \theta$, the equation remains unaltered, so the curve is symmetrical about a line perpendicular to the initial line at the pole.

When $\theta = -\pi/2$, $r = 0$ and when $\theta = \pi/2$, $r = 2a$

The curve is as shown in the figure

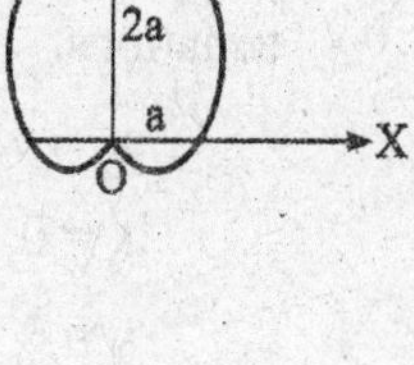

$\therefore$ the required area $= 2\int_{-\pi/2}^{\pi/2} \frac{1}{2} r^2 d\theta = \int_{-\pi/2}^{\pi/2} a^2 (1+\sin\theta)^2 \, d\theta$

$= \int_{-\pi/2}^{\pi/2} \left(1 + 2\sin\theta + \sin^2\theta\right) d\theta$

$$= a^2 \left\{ \int_{-\pi/2}^{\pi/2} d\theta + 2\int_{-\pi/2}^{\pi/2} \sin\theta \, d\theta + \int_{-\pi/2}^{\pi/2} \frac{1-\cos 2\theta}{2} d\theta \right\}$$

$$= a^2 \left[\theta - 2\cos\theta + \frac{\theta}{2} - \frac{\sin 2\theta}{4} \right]_{-\pi/2}^{\pi/2}$$

$$= \frac{3}{2}\pi a^2$$

26. Find the area of the curve $r = 3 + 2 \cos \theta$

The curve is symmetrical about the initial line

When $\theta = 0$, $r = 5$; when $\theta = \pi/2$, $r = 3$; when $\theta = \pi$, $r = 1$; when $\theta = \frac{3\pi}{2}$, $r = 3$; when $\theta = 2\pi$, $r = 5$.

The curve is as shown in the figure.

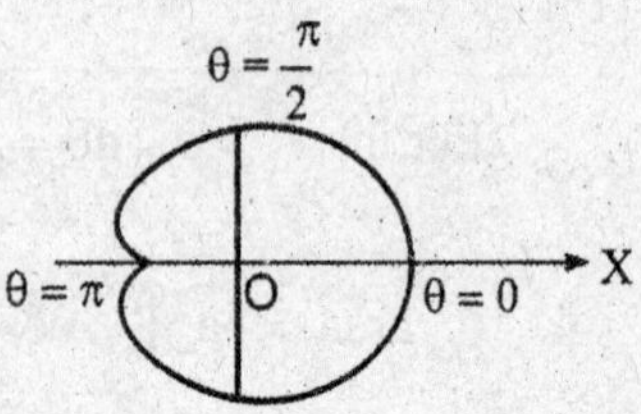

The required area $= 2\int_0^{\pi} \frac{1}{2} r^2 d\theta$

$$= \int_0^{\pi} (3 + 2\cos\theta)^2 \, d\theta$$

$$= \int_0^{\pi} \left(9 + 12\cos\theta + 4\cos^2\theta\right) d\theta$$

$$= \int_0^{\pi} (9 + 12\cos\theta + 2 + 2\cos 2\theta) d\theta$$

$$= \int_0^{\pi} (11 + 12\cos\theta + 2\cos 2\theta) d\theta$$

$$= [11\theta + 12\sin\theta + \sin 2\theta]_0^{\pi}$$

$$= 11\,\pi.$$

27. Find the area above the x-axis included between the circle $x^2 + y^2 = 2ax$ and the parabola $y^2 = ax$.

The centre of the given circle is (a, 0) and radius a. It passes through the origin. Solving the given equations,

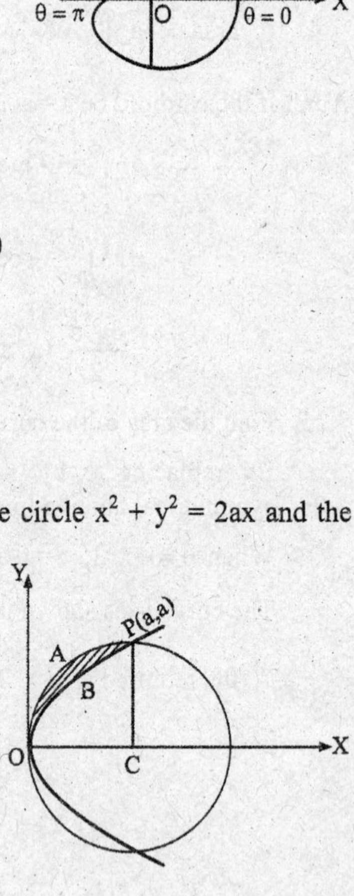

$$x^2 + ax = 2ax \text{ or } x(x - a) = 0$$

$\therefore \quad x = 0, x = a$

$\therefore \quad y = 0, y = \pm a$

Thus the points of intersection of the circle and the parabola are (0, 0), (a, a) and (a, – a). We have to find the area OAPBO, where P(a, a) is point of intersection of the curves.

Now the required area = area OAPCO – area OBPCO

$$= \int_0^a \sqrt{2ax - x^2}\,dx - \int_0^a \sqrt{ax}\,dx$$

$$= \int_0^a \sqrt{a^2 - (a-x)^2}\,dx - \int_0^a \sqrt{ax}\,dx$$

In the first integral, putting a – x = sin θ,
dx = a cos θ dθ

When x = 0, θ = π/2; when x = a, θ = 0

$$\therefore \text{ area } = \int_{\pi/2}^{0} a\cos\theta.-a\cos\theta d\theta - \sqrt{a}\int_0^a \sqrt{x}\,dx$$

$$= a^2\int_{\pi/2}^{\pi/2} \cos^2\theta d\theta - \sqrt{a}\frac{2}{3}\left[x^{3/2}\right]_0^a$$

$$= a^2.\frac{1}{2}.\frac{\pi}{2} - \frac{2}{3}a^2$$

$$= a^2\left\{\frac{\pi}{4} - \frac{2}{3}\right\}.$$

28. Find the area common to the circle $x^2 + y^2 = 25$ and the parabola $3x^2 = 16y$.

The centre of the circle is (0, 0) and its radius is 5; $3x^2 = 16y$ is a parabola whose vertex is (0, 0) and axis is the y-axis.

To find the points of intersection of the given curves we have

$$x^2 + y^2 = 25; \quad \therefore x^2 = 25\ y^2$$

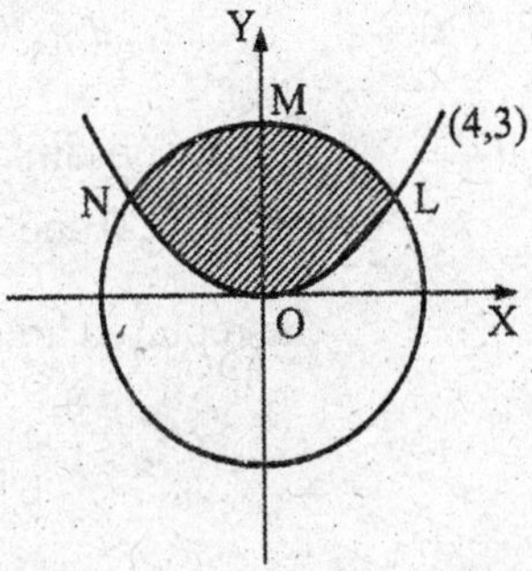

∴ from $3x^2 = 16y$, $3(25 - y^2) = 16y$

or $75 - 3y^2 = 16y$

or $3y^2 + 16y - 75 = 0$ or $(y - 3)(3y + 25) = 0$

∴ $y = 3$ or $y = -25/3$

Rejecting –ve value (to keep x real), y = 3

∴ $x^2 = 25 - 9 = 16$; ∴ $x = \pm 4$

So, x-coordinate of L is 4

∴ the required area = 2 × area OLMO

$$= 2\int_0^4 [\varphi(x) - \psi(x)]dx$$

$$= 2\int_0^4 \left(\sqrt{25 - x^2} - \frac{3}{16}x^2\right)dx \qquad \left\{\because \text{here } \varphi(x) = \sqrt{25 - x^2},\ \psi(x) = \frac{3}{16}x^2\right\}$$

$$= 2\left\{\int_0^4 \sqrt{5^2 - x^2}\,dx - \frac{3}{16}\int_0^4 x^2 dx\right\}$$

$$= 2\left[\left\{\frac{25}{2}\sin^{-1}\frac{x}{5} + \frac{x\sqrt{25 - x^2}}{2}\right\} - \frac{3}{16}\frac{x^3}{3}\right]_0^4$$

$$\left[\because \int \sqrt{a^2 - x^2}\,dx = \frac{a^2}{2}\sin^{-1}\frac{x}{a} + \frac{\sqrt{a^2 - x^2}}{2}\right]$$

$$= 2\left[\frac{25}{2}\sin^{-1}\frac{4}{5} + \frac{1}{2}4\sqrt{25 - 16} - \frac{1}{16}4^3\right]$$

$$= 2\left[\frac{25}{2}\sin^{-1}\frac{4}{5} + 6 - 4\right] = 25\sin^{-1}\frac{4}{5} + 4.$$

29. Find the area enclosed by the parabolas $y^2 = 4ax$ and $x^2 = 4by$.

The vertex of $y^2 = 4ax$ and $x^2 = 4by$ is the origin and their axes are x-axis and y-axis respectively.

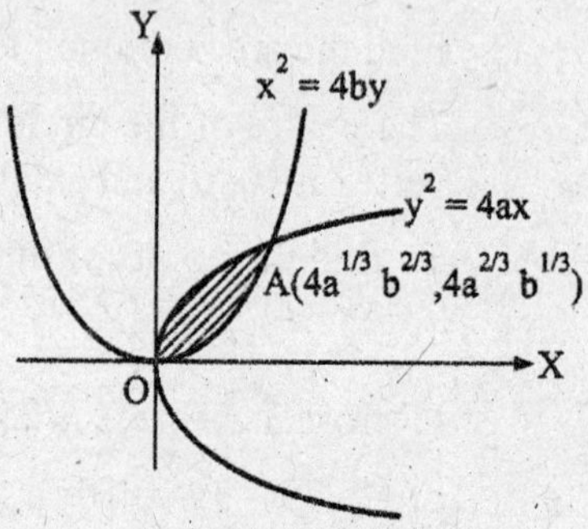

Solving $y^2 = 4ax$ and $x^2 = 4by$,

we get

$$y = 0 \text{ or } 4a^{2/3}\ b^{1/3}$$

$$\therefore \quad x = 0 \text{ or } 4a^{1/3}\ b^{2/3}$$

So, the x-coordinates of the points of intersection O and A are 0 and $4a^{1/3}\ b^{2/3}$ respectively.

$\therefore$ the required area

$$= \int_0^{4a^{1/3}b^{2/3}} [\varphi(x) - \psi(x)]dx, \quad \text{where } \varphi(x) = \sqrt{4ax}, \psi(x) = \frac{x^2}{4b}$$

$$=\int_0^{4a^{1/3}b^{2/3}}\left(\sqrt{4ax}-\frac{x^2}{4b}\right)dx=\left[2\sqrt{a}\,\frac{x^{3/2}}{3/2}-\frac{1}{4b}\cdot\frac{x^3}{3}\right]_0^{4a^{1/3}b^{2/3}}$$

$$=\left[\frac{4}{3}\sqrt{a}\left(4a^{1/3}b^{2/3}\right)^{3/2}-\frac{1}{4b}\cdot\frac{\left(4a^{1/3}b^{2/3}\right)}{3}\right]$$

$$=\frac{32}{3}ab-\frac{16}{3}ab=\frac{16}{3}ab.$$

30. Find the area outside the circle $r = 2a\cos\theta$ and inside the cardioide $r = a(1+\cos\theta)$.

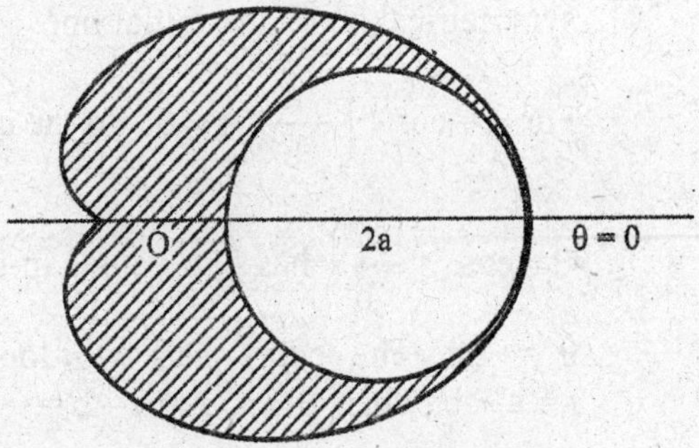

Here $r = 2a\cos\theta$ represents a circle through pole and initial line as diameter 2a. At common point of the two curves, we have $2a\cos\theta = a(1+\cos\theta)$ i.e., $\theta = 0$.

Required area = area of the cardiod – area of the circle .....(1)

Area of the cardioids $=2\int_0^{\pi}\frac{1}{2}r^2d\theta$

$$=\int_0^{\pi}a^2(1+\cos\theta)^2\,d\theta=a^2\int_0^{\pi}4\cos^4\frac{\theta}{2}d\theta$$

Putting $\frac{\theta}{2}=\varphi,\ d\theta=2d\varphi$, when $\theta=0,\varphi=0$; when $\theta=\pi,\varphi=\frac{\pi}{2}$;

$\therefore$ area of the cardioid $=8a^2\int_0^{\pi/2}\cos^4\varphi\,d\varphi$

$$=8a^2\int_0^{\pi/2}\cos^4\varphi\sin^0\varphi\,d\varphi$$

$$=8a^2\frac{\Gamma\left(\frac{5}{2}\right)\Gamma\left(\frac{1}{2}\right)}{2\Gamma(3)}=8a^2\frac{\frac{3}{2}\frac{1}{2}\sqrt{\pi}\sqrt{x}}{2.2.1}=\frac{3}{2}\pi a^2$$

Area of the circle $=2\int_0^{\pi/2}\frac{1}{2}r^2d\theta$

$$=\int_0^{\pi/2}(2a\cos\theta)^2\,d\theta=4a^2\int_0^{\pi/2}\cos^2\theta\,d\theta$$

$$= 2a^2 \int_0^{\pi/2} (1 + \cos 2\theta)\, d\theta$$

$$= 2a^2 \left[\theta + \frac{\sin 2\theta}{2} \right]_0^{\pi/2} = \pi a^2$$

$\therefore$ the required area $= \frac{3}{2}\pi a^2 - \pi a^2 = \frac{1}{2}\pi a^2$.

31. Prove that the sum of the areas of two loops of the limacon $r = a + b\cos\theta$, $b > a$ is equal to $\frac{1}{2}\pi(2a^2 + b^2)$

Changing θ to $-\theta$, the equation of the curve remains unchanged, so the curve is symmetrical about the initial line.

For $\theta = \cos^{-1}\left(-\frac{a}{b}\right)$, $r = 0$. So the curve passes through the pole

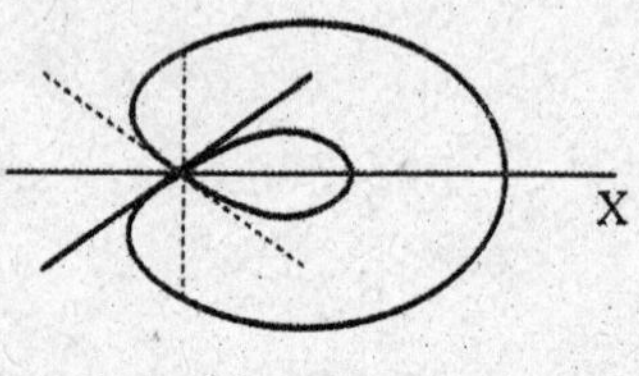

$\theta = \cos^{-1}\left(-\frac{a}{b}\right)$ has two values in the region. $\theta = 0$ to $\theta = 2\pi$. The curve has two loops. When $\theta = 0$, $r = a + b$; when $\theta = \pi$, $r = a - b = -(b - a)$

Thus the curve is as shown in the figure

The sum of the areas of both the loops

$$= 2\int_0^{\pi} \frac{1}{2} r^2 d\theta$$

$$= \int_0^{\pi} (a + b\cos\theta)^2\, d\theta = \int_0^{\pi} \left(a^2 + 2ab\cos\theta + b^2\cos^2\theta\right) d\theta$$

$$= a^2[\theta]_0^{\pi} + 2ab[\sin\theta]_0^{\pi} + \frac{b^2}{2}\int_0^{\pi}(1 + \cos 2\theta)\, d\theta$$

$$= a^2\pi + 0 + \frac{b^2}{2}\pi + \frac{b^2}{4}[\sin 2\theta]_0^{\pi}$$

$$= a^2\pi + \frac{b^2}{2}\pi + 0$$

$$= \frac{1}{2}\pi\left(2a^2 + b^2\right)$$

32. Find the area of the portion included between the cardioides $r = a(1 + \cos\theta)$ and $r = a(1 - \cos\theta)$.

Required area = 4 × area OABO

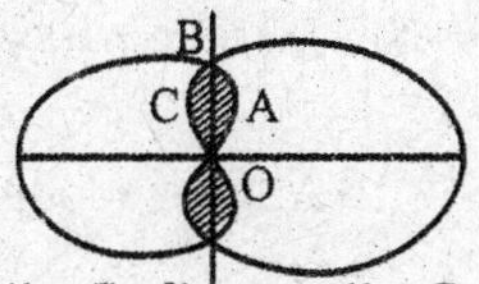

$$= 4\int_0^{\pi/2} \frac{1}{2} r^2 d\theta$$

$$= 2\int_0^{\pi/2} a^2 (1-\cos\theta)^2 \, d\theta$$

[∵ OABO is the area of $r = a(1 - \cos\theta)$)

$$= 2a^2 \int_0^{\pi/2} \left(1 - 2\cos\theta + \cos^2\theta\right) d\theta$$

$$= 2a^2 \int_0^{\pi/2} \left[1 - 2\cos\theta + \frac{1}{2}(1+\cos 2\theta)\right] d\theta$$

$$= 2a^2 \left[\frac{3\theta}{2} - 2\sin\theta + \frac{\sin 2\theta}{4}\right]_0^{\pi/2}$$

$$= 2a^2\left[\frac{3}{2}\times\frac{\pi}{2} - 2\right] = 2a^2\left[\frac{3}{4}\pi - 2\right] = 2a^2\left[\frac{3\pi - 8}{4}\right]$$

$$= \frac{a^2}{2}[3\pi - 8].$$

33. Prove that the area of the loop of the curve $x^3 + y^3 = 3axy$ is three times the area of a loop of the curve $r^2 = a^2 \cos 2\theta$ (lemniscate).

Area of a loop of $r^2 = a^2 \cos 2\theta = \frac{1}{2} a^2$ (Prove as in example 21)

Area of the loop of $x^3 + y^3 = 3axy = \frac{3}{2} a^2$ (prove as in example 17)

= 3 (area of a loop of $r^2 = a^2 \cos 2\theta$).

34. Prove that the area of a loop of the curve $x^5 + y^5 = 5ax^2y^2$ is five times the area of one loop of the curve $r^2 = a^2 \cos 2\theta$.

Putting $x = r\cos\theta$, $y = r\sin\theta$ in $x^5 + y^5 = 5ax^2y^2$, its polar equation is $r^5\cos^5\theta + r^5\sin^5\theta = 5ar^2\cos^2\theta \,.\, r^2\sin^2\theta$

i.e., $$r = \frac{5a\sin^2\theta\cos^2\theta}{\sin^5\theta + \cos^5\theta}$$

Obviously r = 0 when θ = 0 and $\theta = \dfrac{\pi}{2}$

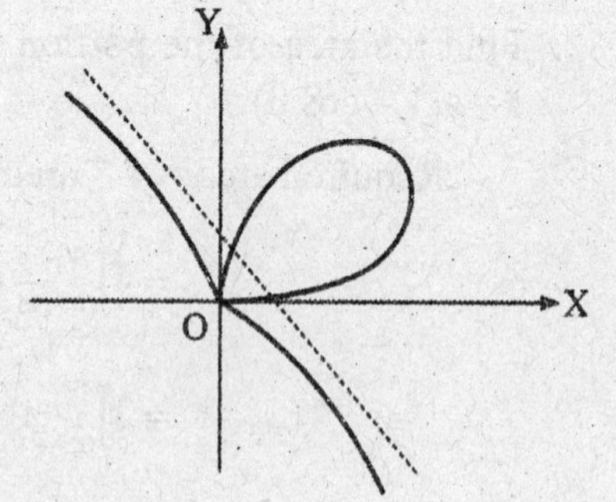

So, for a loop, θ varies from 0 to π/2

$$\text{Area of a loop} = \int_0^{\pi/2} \frac{1}{2} r^2 d\theta$$

$$= \frac{1}{2}\int_0^{\pi/2} \frac{25a^2 \sin^2\theta \cos^4\theta}{\left(\sin^5\theta + \cos^5\theta\right)^2} d\theta$$

$$= \frac{25a^2}{2}\int_0^{\pi/2} \frac{\tan^4\theta . \sec^2\theta}{\left(\tan^5\theta + 1\right)^2} d$$

dividing the number and denominator by $\cos^{10}\theta$

Putting $1 + \tan^5\theta = z$, $5\tan^4\theta \sec^2\theta \, d\theta = dz$

Also when θ = 0, z = 1; when θ = π/2, z = ∞

$$\therefore \text{ area of a loop} = \frac{25a^2}{2\times 5}\int_1^{\infty} \frac{dz}{z^2}$$

$$= \frac{5a^2}{2}\left[-\frac{1}{2}\right]_1^{\infty} = \frac{5}{2}a^2 \qquad .....(1)$$

Also we have the area of a loop of the lemniscate

$$r^2 = a^2 \cos 2\theta \text{ as } \frac{1}{2}a^2 \qquad .....(2)$$

(Prove as in example 21)

∴ from (1) and (2),

area of a loop of $x^5 + y^5 = 5ax^2y^2$ = 5(area of a loop of $r^2 = a^2 \cos 2\theta$).

35. Show that the area common to ellipses $a^2x^2 + b^2y^2 = 1$, $b^2x^2 + a^2y^2 = 1$ when $0 < a, < b$ is $\dfrac{4}{ab}\tan^{-1}\left(\dfrac{a}{b}\right)$

Rewriting the given equations as

$$\frac{x^2}{1/a^2} + \frac{y^2}{1/b^2} = 1 \qquad .....(1)$$

$$\frac{x^2}{1/b^2}+\frac{y^2}{1/a^2}=1 \qquad .....(2)$$

Solving (1) and (2), a point common to both the ellipses is given by

$$a^2x^2=1-b^2y^2 \quad \text{or} \quad x^2=\frac{1-b^2y^2}{a^2}$$

Also $b^2x^2=1-a^2b^2$ or $x^2=\dfrac{1-a^2y^2}{b^2}$

$$\therefore \quad \frac{1-b^2y^2}{a^2}=\frac{1-a^2y^2}{b^2} \text{ or } b^2-b^4y^2=a^2-a^4y^2$$

or $\quad y^2\left(a^4-b^4\right)=a^2-b^2,\quad$ or $\quad y^2=\dfrac{a^2-b^2}{a^4-b^4}$

or $\quad y^2=\dfrac{a^2-b^2}{\left(a^2-b^2\right)\left(a^2+b^2\right)}=\dfrac{1}{a^2+b^2} \quad \therefore \quad y=\dfrac{1}{\sqrt{a^2+b^2}}$

$$\therefore \quad x^2=\frac{1-\dfrac{a^2}{a^2+b^2}}{b^2}=\frac{a^2+b^2-a^2}{b^2\left(a^2+b^2\right)}=\frac{1}{a^2+b^2}$$

$$\therefore \quad x=\frac{1}{\sqrt{a^2+b^2}}$$

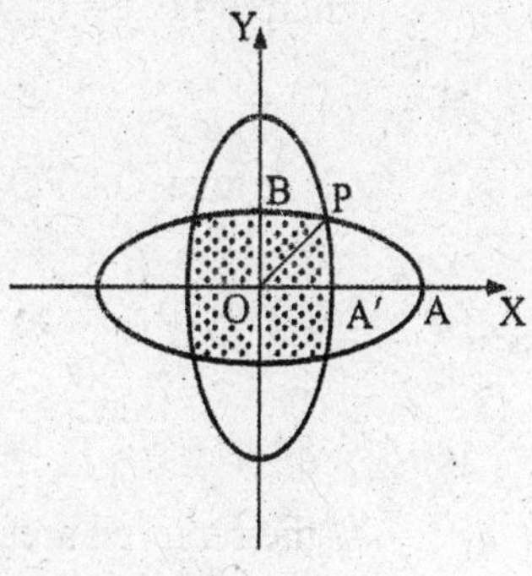

So, if P is a point of intersection, then

$$P\left(\frac{1}{\sqrt{a^2+b^2}},\frac{1}{\sqrt{a^2+b^2}}\right)$$

Thus $\angle XOP=\pi/4$ and the equation of OP is $y=x$

Interchanging x and y we find that one ellipse is transformed into other and accordingly the area of the region $OA'PBO$ is twice that of $OA'PO$.

For sectorial area $OA'PO$, θ varies from 0 to $\pi/4$

Putting $x=r\cos\theta$, $y=r\sin\theta$, the popular equation of the ellipse (2) of which are $A'P$ is a part is

$$\frac{r^2\cos^2\theta}{1/b^2}+\frac{r^2\sin^2\theta}{1/a^2}=1$$

or $\quad b^2r^2\cos^2\theta+a^2r^2\sin^2\theta=1$

or $\quad r^2 (b^2 \cos^2 \theta + a^2 \sin^2 \theta) = 1$

or $$r^2 = \frac{1}{b^2 \cos^2 \theta + a^2 \sin^2 \theta}$$

$\therefore$ $$\text{area } OA'PO = \frac{1}{2}\int_0^{\pi/4} \frac{1}{b^2 \cos^2 \theta + a^2 \sin^2 \theta} d\theta$$

$$= \frac{1}{2}\int_0^{\pi/4} \frac{\sec^2 \theta \, d\theta}{b^2 + a^2 \tan^2 \theta}$$

$$= \frac{1}{2} a^2 \int_0^{\pi/4} \frac{\sec^2 \theta \, d\theta}{\tan^2 \theta + \frac{b^2}{a^2}}$$

$$= \frac{1}{2a^2}\int_0^1 \frac{dt}{t^2 + \frac{b^2}{a^2}} \quad \text{where } t = \tan\theta \text{ and } dt = \sec^2 \theta \, d\theta$$

When $\theta = 0, t = 0$; when $\theta = \frac{\pi}{4}, t = 1$

$\therefore$ $$\text{area } OA'PO = \frac{1}{2a^2}\frac{a}{b}\left[\tan^{-1}\left(\frac{t}{b/a}\right)\right]_0^1$$

$$= \frac{1}{2ab}\tan^{-1}\left(\frac{a}{b}\right)$$

$\therefore$ the required area $= 8 \times \frac{1}{2ab}\tan^{-1}\left(\frac{a}{b}\right)$

$$= \frac{4}{ab}\tan^{-1}\left(\frac{a}{b}\right)$$

36. If a curve be referred to its radius vector r and the perpendicular p on the tangent, then prove that its area is given by

$$\frac{1}{2}\int \frac{prdr}{\sqrt{r^2 - p^2}}$$

We know that with usual symbols $\tan\varphi = \frac{r\, d\theta}{dr}$ and $p = r \sin \varphi$.

$$\therefore \qquad \text{area} = \frac{1}{2}\int r^2 d\theta = \frac{1}{2}\int r.\frac{r\,d\theta}{dr}dr$$

$$= \frac{1}{2}\int r\tan\varphi dr = \frac{1}{2}\int \frac{r\,\sin\varphi}{\cos\varphi}dr = \frac{1}{2}\int \frac{r\sin\varphi}{\sqrt{r^2-p^2}}dr$$

$$= \frac{1}{2}\int \frac{rp}{\sqrt{r^2-p^2}}dr.$$

37. Obtain an approximate value of log 2 by calculating the definite integral $\int_1^2 \frac{dx}{x}$ by Simpson's rule using eleven ordinates.

We have $\qquad \int_1^2 \frac{dx}{x} = \log 2$

We divide the interval [1, 2] into 10 equal parts

Let $\qquad f(x) = \frac{1}{x}$; here $h = \frac{2-1}{10} = \frac{1}{10}$

We have

$y_1 = f(1, 0) = 1.00000$, $\qquad y_2 = f(1.1) = 0.90909$

$y_3 = f(1, 2) = 1.83333$, $\qquad y_4 = f(1.3) = 0.76923$

$y_5 = f(1, 4) = 0.71429$, $\qquad y_6 = f(1.5) = 0.66667$

$y_7 = f(1.6) = 0.62500$, $\qquad y_8 = f(1.7) = 0.58824$

$y_9 = f(1.8) = 0.55556$, $\qquad y_{10} = f(1.9) = 0.52632$

$y_n = f(2.0) = 0.50000$

$\therefore\ y_1 + y_{11}$ = sum of first and last ordinates

$= 1.00000 + 0.50000 = 1.50000$

Sum of remaining odd ordinates = 2.72818;

Sum of the even ordinates = 3.45955

$$\therefore \qquad \log 2 = \frac{1}{3}.\frac{1}{10}\ (1.50000 + 2\times 2.72818 + 4\times 3.45955)$$

$$= \frac{1}{30}\times 20.79456 = 0.69315$$

38. Find $\int_2^{10} \frac{dx}{1+x}$ by Simpson's rule using nine ordinates

Here $f(x) = \frac{1}{1+x}$ and $h = \frac{10-2}{8} = 1$

We have

$$y_1 = f(2) = \frac{1}{3}, \quad y_2 = f(3) = \frac{1}{4}, \quad y_3 = f(4) = \frac{1}{5}, \quad y_4 = f(5) = \frac{1}{6},$$

$$y_5 = f(6) = \frac{1}{7}, \quad y_6 = f(7) = \frac{1}{8}, \quad y_7 = f(8) = \frac{1}{9}, \quad y_8 = f(9) = \frac{1}{10},$$

$$y_9 = f(10) = \frac{1}{11},$$

$$\therefore \quad \int_2^{10} \frac{dx}{1+x} = \frac{1}{3}\left[\left(\frac{1}{3}+\frac{1}{11}\right) + 2\left(\frac{1}{5}+\frac{1}{7}+\frac{1}{9}\right) + 4\left(\frac{1}{4}+\frac{1}{6}+\frac{1}{8}+\frac{1}{10}\right)\right]$$

$$= \frac{1}{3}\,[\,(0.3333 + 0.0909) + 2\,(0.2 + 0.1428 + 0.1111)$$

$$+ 4\,(0.25 + 0.1666 + 0.125 + 0.1)]$$

$$= \frac{1}{2}(3.8984) = 1.2995$$

39. Evaluate $\int_0^1 \frac{dx}{1+x^2}$ by Simpson's rule. Hence calculate the value of π.

Let the points of division of [0, 1] in four equal parts be y_1, y_2, y_3, y_4, y_5.

Now $h = \frac{1-0}{4} = 0.25$; $f(x) = \frac{1}{1+x^2}$

$$y_1 = f(0) = \frac{1}{1} = 1$$

$$y_2 = f(0.25) = \frac{1}{1+(0.25)^2} = \frac{1}{1+\left(\frac{25}{100}\right)^2} = \frac{1}{1+\left(\frac{1}{4}\right)^2} = \frac{16}{17} = 0.94117$$

$$y_3 = f(0.50) = \frac{1}{1+(0.50)^2} = 0.8000, \quad y_4 = f(0.75) = 0.64000$$

$$y_5 = f(1) = 0.50000$$

By Simpson's rule $\int_0^1 \frac{dx}{1+x^2}$

$$= \frac{h}{3}[(y_1 + y_5) + 2y_3 + 4(y_2 + y_4)]$$

$$= \frac{0.25}{3}[(1 + 0.50000) + 2 \times 0.80000 + 4 \times (0.94117 + 0.64000)]$$

$$= \frac{25}{300}[1.50000 + 1.60000 + 4 \times 1.58117]$$

$$= \frac{1}{12} \times 9.42468 = 0.78539 \qquad .....(1)$$

Also by actual integration,

$$\int_0^1 \frac{dx}{1+x^2} = \left[\tan^{-1} x\right]_0^1 = \frac{\pi}{4} \qquad .....(2)$$

From (1) and (2)

$$\frac{\pi}{4} = 0.78539; \qquad \because \pi = 3.14156$$

Exercise 8.1

1. Find the area of the curve $4y^2 = x^2(4 - x^2)$

   [*Hint*: Put a = 2 in example 6]

2. Prove that the area of the curve $a^2x^2 = y^2(2a - y)$ is equal to that of the circle whose radius is a

   [*Hint*: A = 2 × area of AOBA

   $$= 2\int_0^{2a} x\,dy$$

   $$= \pi a^2 \text{ (as in example 7)}]$$

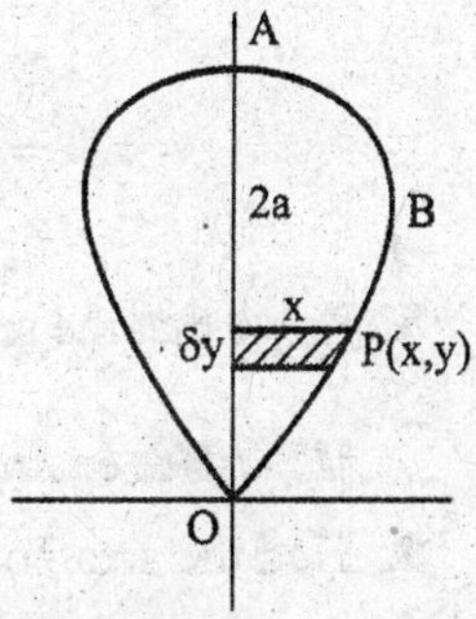

3. Find the area enclosed by the curve $xy^2 = a^2(a - x)$ and y-axis.

   [*Hint*: The curve passes through origin. y-axis is asymptote and the curve is symmetrical about x-axis.

   $$\text{Area} = 2\int_0^{\infty} x\,dy = 2\int_0^{\infty} \frac{a^3}{a^2 + y^2} dy = 2a^3\left[\frac{1}{a}\tan^{-1}\left(\frac{y}{a}\right)\right]_0^{\infty}$$

4. Find the area between the curve $y^2(2a - x) = x^3$ and its asymptote.

[*Hint*: See example 13]

5. Show that the area bounded by the curve $x = a\sin^2 t$, $y = a\dfrac{\sin^3 t}{\cos t}$ and its asymptote is $\dfrac{3}{4}\pi a^2$.

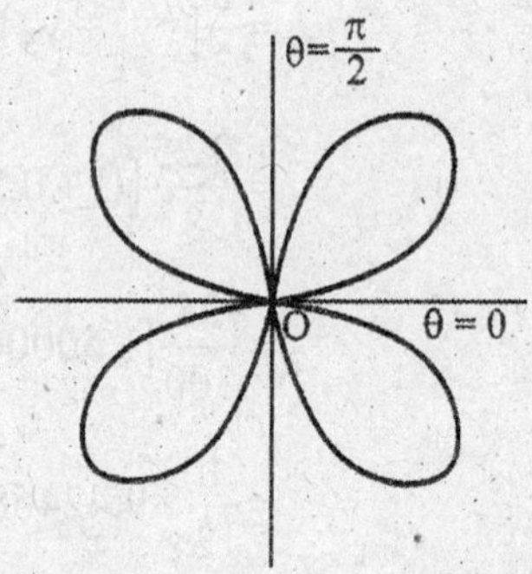

[*Hint*: $x^3 = a^3 \sin^6 t$, $y^2 = \dfrac{a^2 \sin^6 t}{\cos^2 t} = \dfrac{x^2/a}{1 - x/a} = \dfrac{x^3}{a - x}$

i.e., $y^2(a - x) = x^3$ [Now this is example 13].

6. Find the whole area of the curve $r = a \sin 2\theta$

[*Hint*: Area $= 4\int_0^{\pi/2} \dfrac{1}{2} r^2 d\theta$

$$= 2\int_0^{\pi/2} a^2 \sin^2 2\theta \, d\theta \,].$$

7. Find the total area inside $r = \sin\theta$ and outside $r = 1 - \cos\theta$

[*Hint*: Area

$$= \frac{1}{2}\int_0^{\pi/2} \left\{(\sin\theta)^2 - (1 - \cos\theta)^2\right\} d\theta$$

$$= \frac{1}{2}\int_0^{\pi/2} \left(2\cos\theta - 2\cos^2\theta\right) d\theta$$

$$= \int_0^{\pi/2} \left(\cos\theta - \cos^2\theta\right) d\theta$$

$$= \left[\sin\theta\right]_0^{\pi/2} - \frac{1}{2}\int_0^{\pi/2} (1 + \cos 2\theta)\, d\theta \,].$$

8. Prove that the area of a loop $x^4 + y^4 = 4a^2xy$ is $\dfrac{1}{2}\pi a^2$.

[*Hint*: See example 18].

9. Find the area bounded by the curve $y^2 = x^3$ and the line $y = 2x$

[*Hint*: For point of intersection $x = 0$ or $x = 4$

$$\therefore \text{ area} = \int_0^4 \left(2x - x^{\frac{1}{3}}\right) dx \,].$$

Answers

1. $\frac{16}{3}$ 3. πa^2 4. $3\pi a^2$

6. $\frac{1}{2}\pi a^2$ 7. $1-\frac{\pi}{4}$ 9. $\frac{16}{5}$

CHAPTER 9

Lengths of Plane Curves (Rectification)

The Process of determining the length of an arc between two points on a curve is known as rectification.

When the Curve is in Cartesian Form: The length of the arc of the curve $y = f(x)$ included between two points whose abscissae are a and b is

$$\int_a^b \sqrt{1+(dy/dx)^2}\,dx$$

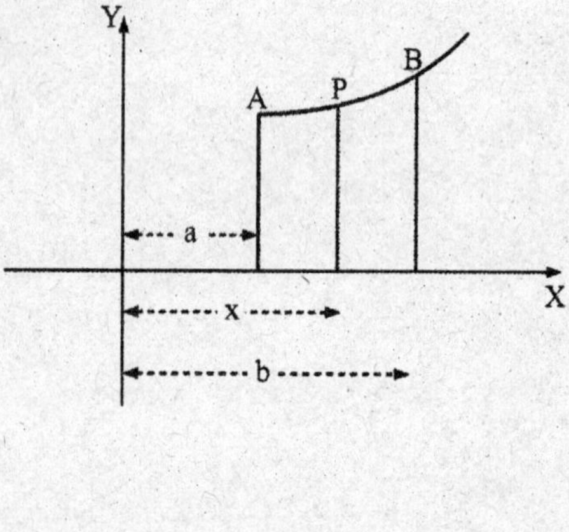

Let A, B be two points with abscissae a and b on the curve $y = f(x)$.

If S denotes the length of the arc AP where A is some fixed point on the curve and P is a variable point whose abscissa is a variable point whose abscissa is x (a function of x) then from differential calculus,

$$\frac{ds}{dx} = \sqrt{1+\left(\frac{dy}{dx}\right)^2}$$

$$\therefore \quad \int_a^b \sqrt{1+(dy/dx)^2}\,dx = \int_a^b \frac{ds}{dx}\,dx = [s]_a^b$$

$$= (\text{value of s when } x = b) - (\text{value of s when } x = a)$$

Similarly, for the curve $x = f(y)$, length included between two points whose ordinates are c, d is

$$\int_c^b \sqrt{1+(dx/dy)^2}\;dy$$

When the Curve is in Polar Form: The length of the arc of the curve $r = f(\theta)$ included between two points whose vectorial angles are α, β is

$$\int_{\alpha}^{\beta} \sqrt{r^2 + \left(\frac{dr}{d\theta}\right)^2}\, d\theta$$

We have from differential calculus,

$$(ds)^2 = (dr)^2 + (rd\theta)^2 ; \quad \left(\frac{ds}{d\theta}\right)^2 = r^2 + \left(\frac{dr}{d\theta}\right)^2$$

$$\therefore \quad \frac{ds}{d\theta} = \sqrt{r^2 + \left(\frac{dr}{d\theta}\right)^2} ; \qquad \therefore\ s = \int_{\alpha}^{\beta} \sqrt{r^2 + \left(\frac{dr}{d\theta}\right)^2}\, d\theta .$$

When the Curve is in Parametric Form: The length of the arc of the curve $x = f(t)$, $y = p(t)$ included between two points whose parametric values are α, β is

$$\int_{\alpha}^{\beta} \sqrt{\left(\frac{dx}{dt}\right)^2 + \left(\frac{dy}{dt}\right)^2}\, dt$$

When the Curve is in Pedal Form (p – r form):

If the curve be $p = f(r)$, the length of the curve between the two points whose radius vectors are r_1 and r_2 is

$$s = \int_{r_1}^{r_2} \frac{r\, dr}{\sqrt{r^2 - p^2}}$$

The arc length between radius vectors r_1 and r_2 is

$$\int_{r_1}^{r_2} ds = \int_{r_1}^{r_2} \frac{ds}{dr} dr = \int_{r_1}^{r_2} \sec\varphi\, dr \qquad \left(\because \frac{dr}{ds} = \cos\varphi\right)$$

$$= \int_{r_1}^{r_2} \frac{1}{\sqrt{1-\sin^2\varphi}} dr = \int_{r_1}^{r_2} \frac{1}{\sqrt{1-\left(\frac{p}{r}\right)^2}} dr \qquad (\because p = r\sin\varphi)$$

$$s = \int_{r_1}^{r_2} \frac{r\, dr}{\sqrt{r^2 - p^2}}$$

Intrinsic* Equation of a Curve

A relation between s and ψ is called an intrinsic equation, where s denotes the length of the arc of a curve measured from some fixed point to a variable point p and ψ denotes the angle that the tangent at p makes with x-axis.

---

*The word 'intrinsic' means inherent. s and ψ for a point depend only upon the form of the curve and not on its position in the plane, so that they are inherently associated with the curve.

Intrinsic Equation from Cartesian Equation: Elimination of x from

$$\frac{dy}{dx} = \tan\psi \text{ and } s = \int_{\alpha}^{x} \sqrt{1 + (dy/dx)^2}\, dx$$

gives the intrinsic equation, where a is the abscissa of some fixed point A on the curve, P (x, y) a variable point on the curve and ψ the angle that tangent at P makes with x-axis. Arc AP = s.

Intrinsic Equation from Polar Equation: Elimination of θ and φ from

$$\frac{dy}{dx} = \tan\psi, \psi = 0 + \varphi \text{ and } s = \int_{\alpha}^{\theta} \sqrt{r^2 + \left(\frac{dr}{d\theta}\right)^2}\, d\theta$$

gives the intrinsic equation, α being vectorial angle of the fixed point A.

Intrinsic Equation from Parametric Form:

If x = f(t), y = φ(t) then

$$\tan\psi = \frac{dy}{dx} = \frac{dy}{dt}\bigg/\frac{dx}{dt} = \frac{\varphi'(t)}{f'(t)} \qquad \text{...(I)}$$

and

$$s = \int_{t_1}^{t} \sqrt{\left(\frac{dx}{dt}\right)^2 + \left(\frac{dy}{dt}\right)^2}\, dt \qquad \text{...(II)}$$

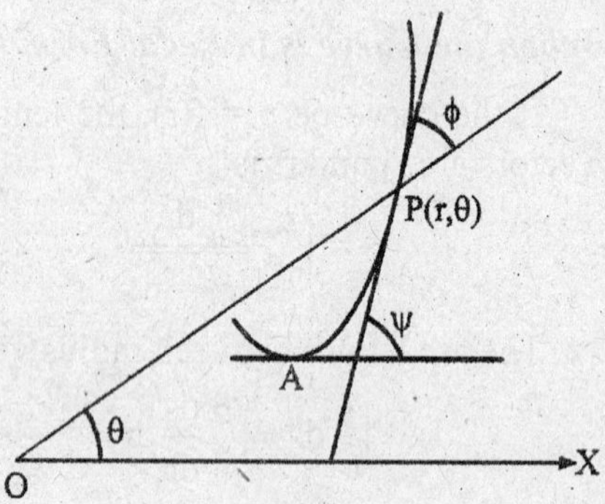

where t_1 and t correspond to the fixed and variable points respectively.

Eliminating t between (I) and (II), we get the required intrinsic equation.

Intrinsic from Pedal

$$\text{If } p = f(r), \quad \text{then } \quad s = \int_a^r \frac{r\,dr}{\sqrt{r^2 - p^2}} = \int_a^r \frac{r\,dr}{\sqrt{r^2 - \{f(r)\}^2}} \qquad \text{.....(1)}$$

Also from differential calculus,

$$\frac{ds}{d\psi} = \rho = r\frac{dr}{dp} = \frac{r}{f'(r)} \qquad \text{.....(2)}$$

Eliminating r between (1) and (2), we get a relation of the form $\frac{ds}{d\psi} = \varphi(s)$, which on integration gives the intrinsic equation.

Remember:

1. For Cartesian curve, $s = \int \sqrt{1+\left(\frac{dy}{dx}\right)^2}\, dx$ or

$$= \int \sqrt{1+\left(\frac{dx}{dy}\right)^2}\, dy$$

2. For polar curve, $s = \int \sqrt{r^2 + \left(\frac{dr}{d\theta}\right)^2}\, d\theta$

3. For parametric form, $s = \int \sqrt{\left(\frac{dx}{dt}\right)^2 + \left(\frac{dy}{dt}\right)^2}\, dt$

4. For pedal form, $s = \int \frac{r\,dr}{\sqrt{r^2 - p^2}}$

where symbols have their usual meanings and range of integration to be determined according to the problem

Solved Examples

1. Find the perimeter of the circle $x^2 + y^2 = a^2$.

The curve is symmetrical about both axes.

The required perimeter, s = 4 × length of the arc AB

$$= \int_0^a \sqrt{1 + (dy/dx)^2}\, dx,$$

Since for B, x = 0 and for A, x = a.

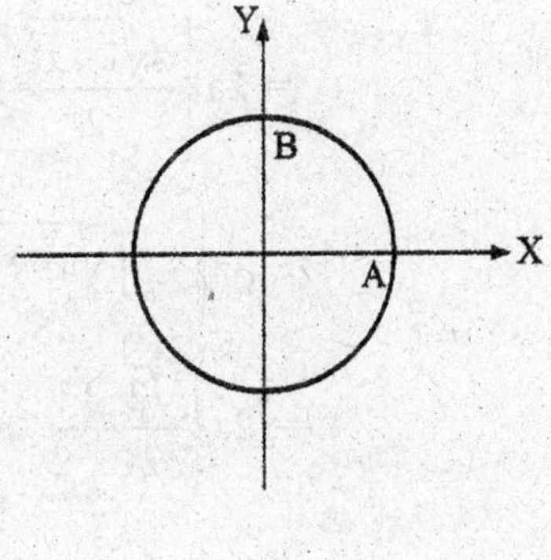

Here $x^2 + y^2 = a^2$

$\therefore$ $2x + 2y\frac{dy}{dx} = 0$

$\therefore$ $\frac{dy}{dx} = -\frac{x}{y}$

$$\therefore \quad 1+\left(\frac{dy}{dx}\right)^2 = 1+\frac{x^2}{y^2} = \frac{x^2+y^2}{y^2} = \frac{a^2}{y^2}$$

$$\therefore \quad s = 4\int_0^a \frac{a}{y}dx = 4a\int_0^a \frac{dx}{\sqrt{a^2-x^2}}$$

$$= 4a\left[\sin^{-1}\frac{x}{a}\right]_0^a = 4a\frac{\pi}{2} = 2\pi a.$$

2. Find the length of the arc of the parabola $y^2 = 4ax$ cut off by the line $y = 2x$.

Solving the given equations $x = 0$, $x = a$, the corresponding values of y are $y = 0$, $y = 2a$. So, the points of intersection of the parabola and the line are (0, 0) and (a, 2a).

Thus for the arc APB, x varies from 0 to a.

Differentiating the equation of the parabola,

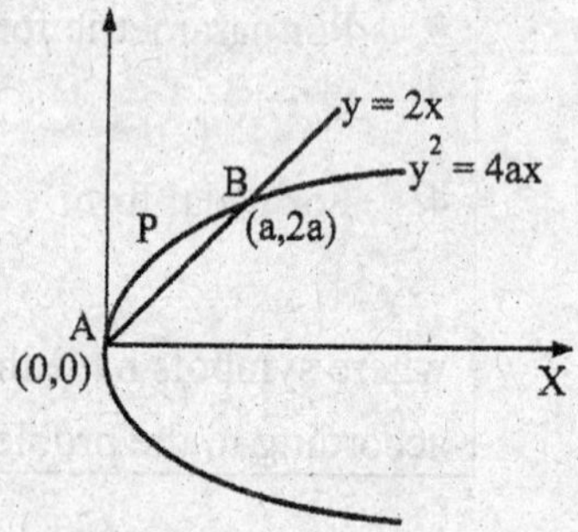

$$2y\frac{dy}{dx} = 4a$$

or $$\frac{dy}{dx} = \frac{2a}{y} = \frac{2a}{2\sqrt{ax}} = \sqrt{\frac{a}{x}}.$$

∴ the required length

$$= \int_0^a \sqrt{1+(dy/dx)^2}\ dx = \int_0^a \sqrt{\left(1+\frac{a}{x}\right)}$$

$$dx = \int_0^a \sqrt{\frac{x+a}{x}}\ dx.$$

Putting $x = az^2$, $dx = 2az\ dz$; when $x = 0$, $z = 0$; when $x = a$, $z = 1$.

$$\therefore \text{ length} = \int_0^1 \sqrt{\frac{az^2+a}{az^2}}\ .\ 2az\ dz = \int_0^a \sqrt{1+z^2}\ dz$$

$$= 2a\left\{\frac{z\sqrt{1+z^2}}{2} + \frac{1^2}{z}\log\frac{z+\sqrt{1+z^2}}{1}\right\}_0^1$$

$$\left(\because \int\sqrt{a^2+x^2}dx = \frac{x\sqrt{a^2+x^2}}{2} + \frac{a^2}{2}\log\frac{x+\sqrt{x^2+a^2}}{a}\right)$$

$$= 2a\left\{\frac{\sqrt{2}}{2} + \frac{1}{2}\ .\ \log\left(1+\sqrt{2}\right)\right\} = a\left\{\sqrt{2} + \log\ (1+\sqrt{2})\right\}.$$

3. Find the length of the arc of the parabola $y^2 = 4ax$ cut off by the line $3y = 8x$.

Solving the given equations, $\left(\dfrac{8x}{3}\right)^2 = 4ax$

or $\quad x^2 = \dfrac{9}{64}4ax = \dfrac{9}{16}ax$ or $\left(x - \dfrac{9}{16}a\right) = 0;\quad \therefore\ x = 0,\ x = \dfrac{9}{16}a.$

Now $\quad 2y\dfrac{dy}{dx} = 4a;\quad \therefore \dfrac{dy}{dx} = \dfrac{2a}{y};\quad \therefore \left(\dfrac{dy}{dx}\right)^2 = \dfrac{4a^2}{y^2} = \dfrac{4a^2}{4ax} = \dfrac{a}{x}.$

$\therefore$ required length

$$= \int_0^{\frac{9}{16}a} \sqrt{1 + \left(\frac{dy}{dx}\right)^2}\, dx$$

$$= \int_0^{\frac{9}{16}a} \sqrt{1 + \frac{a}{x}}\, dx$$

$$= \int_0^{\frac{9}{16}a} \sqrt{\frac{x+a}{x}}\, dx.$$

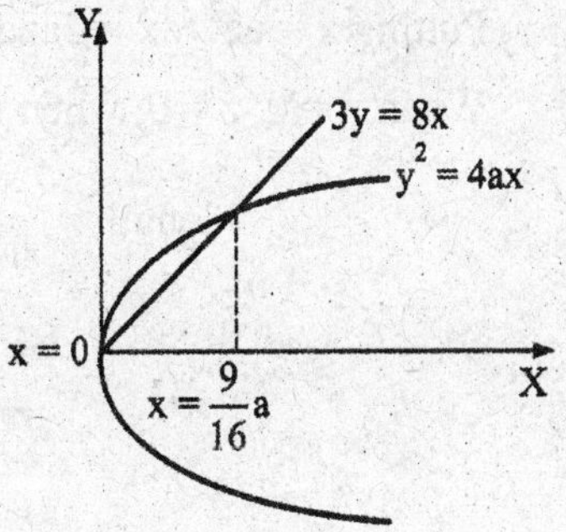

Putting $x = az^2$, $dx = 2az\,dz$; when $x = 0$, $z = 0$; when $x = \dfrac{9}{16}a$, $z = \dfrac{3}{4}$.

$$\therefore \quad \text{length} = \int_0^{\frac{3}{4}} \left[\frac{az^2 + a}{az^2}\right]^{\frac{1}{2}} 2az\, dz = 2a\int_0^{\frac{3}{4}} \sqrt{1+z^2}\, dz$$

$$= 2a\left[\frac{z\,.\,\sqrt{1+z^2}}{2} + \frac{1^2}{2}\log\frac{z+\sqrt{1+z^2}}{1}\right]_0^{3/4}$$

$$= 2a\left[\frac{3}{2\times 4}\sqrt{1+\frac{9}{16}} + \frac{1}{2}\log\left(\frac{3}{4} + \sqrt{1+\frac{9}{16}}\right)\right]$$

$$= \frac{3}{4}a \times \frac{5}{4} + a\log 2 = a\left(\frac{15}{16} + \log 2\right).$$

4. Find the length of an arc of the parabola $y^2 = 4ax$ measured from the vertex to one extremity of the latus rectum.

Here x varies from 0 to a.

$\therefore$ the required length of the arc OPL

$$= \int_0^a \sqrt{1 + (dy/dx)^2}\, dx.$$

The curve is $y^2 = 4ax$.

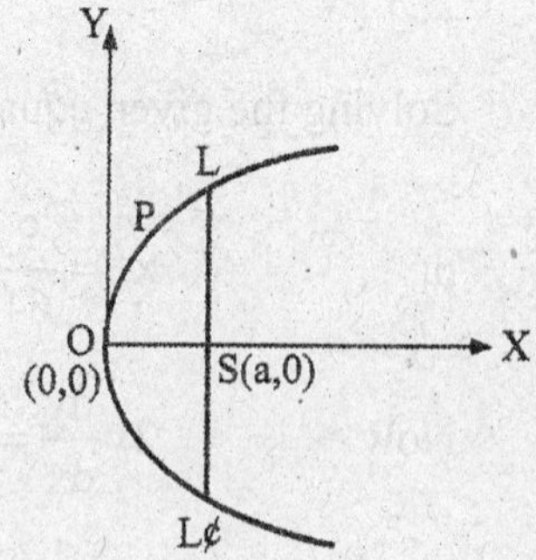

$$\therefore \qquad 2y\frac{dy}{dx} = 4a \quad \therefore \quad \frac{dy}{dx} = \frac{2a}{y}$$

$$\therefore \qquad 1 + \left(\frac{dy}{dx}\right)^2 = 1 + \frac{4a^2}{y^2} = 1 + \frac{4a^2}{4ax} = \frac{x+a}{x}.$$

$$\therefore \quad \text{length} = \int_0^a \sqrt{\frac{x+a}{x}}\,dx$$

Putting $x = az^2$, $dx = 2azdz$.

When $x = 0$, $z = 0$; when $x = a$, $z = 1$

$$\therefore \qquad \text{length} = \int_0^1 \sqrt{\frac{az^2 + a}{az^2}}\,2az\,dz = 2a\int_0^1 \sqrt{1+z^2}\,dz$$

$$= 2a\left[\frac{z\sqrt{1+z^2}}{2} + \frac{1^2}{2}\log\frac{z+\sqrt{1+z^2}}{1}\right]_0^1$$

$$= a\left[\sqrt{2} + \log\,(1+\sqrt{2})\right].$$

5. Find the length of the arc of the parabola $y^2 = 4ax$ cut off by its latus rectum.

   Write down the above solution and then add;

   the required length = 2 × length of arc OPL = $2a\left[\sqrt{2} + \log\ (1+\sqrt{2})\right]$.

6. Trace the curve $9y^2 = (x + 7)(x + 4)^2$ and obtain the length of its loop.

   Since even power of y occurs, so the curve is symmetrical about x-axis. The curve does not pass through the origin. Putting $y = 0$, $x = -7$, $x = -4$. So the curve cuts x-axis at $(-7, 0)$ and $(-4, 0)$.

   The curve will not extend to the left of $(-7, 0)$. For the values of x lying between -7 and -4, y has two equal and opposite values, hence a loop is formed, and for the loop, x varies from -7, to -4.

   The shape of the curve is as shown in the figure.

   Now the equation of the curve is

   $9y^2 = (x + 7)(x + 4)^2$

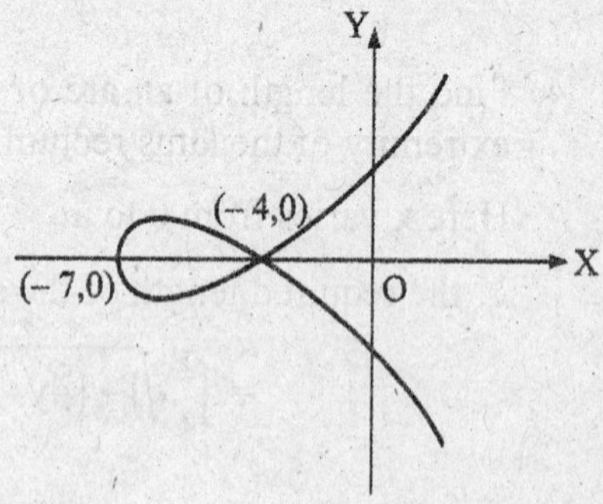

$$\therefore \qquad 18y\frac{dy}{dx} = (x+4)^2 + 2(x+4)\ (x+7)$$

$$= (x+4)\ \{x + 4 + 2x + 14\}$$

$= (x+4)\{3x+18\}$

$= 3(x+4)(x+6)$

$$\therefore \quad \left(\frac{dy}{dx}\right)^2 = \frac{9(x+4)^2(x+6)^2}{18\times 18y^2}$$

$$= 3\frac{(x+4)^2(x+6)^2}{36\times 9y^2} = \frac{(x+4)^2(x+6)^2}{12(x+7)(x+4)^2}$$

$$\left\{\because 9y^2 = (x+7)(x+4)^2\right\}$$

$$= \frac{(x+6)^2}{12(x+7)};$$

$$\therefore \quad 1+\left(\frac{dy}{dx}\right)^2 = 1+\frac{(x+6)^2}{12(x+7)} = \frac{(x+8)^2}{4(x+7)}.$$

$\therefore$ the length of the loop $= 2\times$ length of the upper half of the loop

$$= 2\int_{-7}^{-4}\sqrt{\frac{(x+8)^2}{4(x+7)}}dx = 2\int_{-7}^{-4}\frac{x+8}{2\sqrt{x+7}}dx = 2\int_{-7}^{-4}\frac{x+7+1}{2\sqrt{x+7}}dx$$

$$= 2\int_{-7}^{-4}\left[\sqrt{x+7}+\frac{1}{\sqrt{x+7}}\right]dx = \left[\frac{2}{3}(x+7)^{3/2}+2\sqrt{x+7}\right]_{-7}^{-4}$$

$$= \frac{2}{3}\cdot 3^{3/2}+2\cdot 3^{1/2} = 4\sqrt{3}$$

7. Find the length of the loop of the curve $9y^2 = (x-3)(x-6)^2$.

Here $\dfrac{dy}{dx} = \dfrac{3(x-6)(x-4)}{18y}$

When $y = 0$, $x = 3$, $x = 6$ (fig. as in previous example)

$$\therefore \quad 1+\left(\frac{dy}{dx}\right)^2 = 1+\frac{(x-4)^4}{4(x-3)} = \frac{(x-2)^2}{4(x-3)}$$

$\therefore$ the required length $= 2\int_3^6 \dfrac{x-2}{2\sqrt{x-3}}dx = \int_3^6 \dfrac{x-3+1}{\sqrt{x-3}}dx$

$$= \int_3^6\left[\sqrt{x-3}+\frac{1}{\sqrt{x-3}}\right]dx = 4\sqrt{3} \text{ (as above).}$$

8. Find the perimeter of the loop of the curve $9ay^2 = (x-2a)(x-5a)^2$

Even power of y occurs, so the curve is symmetrical about x-axis. Putting y = 0, x = 2a and x = 5a. When x < 2a, y is imaginary. So no part of the curve is in the left of (2a, 0).

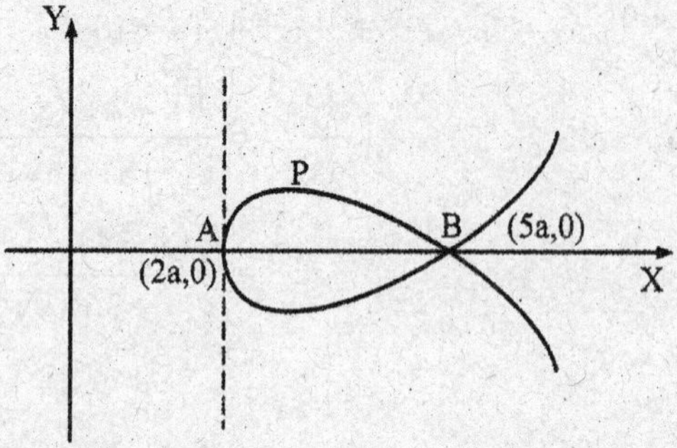

When 2a < x < 5a, y has two equal and opposite finite values. Thus a loop is formed within the interval.

When y > 5a, y has two equal and opposite values which increase in magnitude as $x \to \infty$. So, the shape of the curve is as shown in the figure.

The equation of the curve is $9ay^2 = (x-2a) \times (x-5a)^2$

$$\therefore \quad 18ay\frac{dy}{dx} = (x-2a)2(x-5a)+(x-5a)^2$$

$$= (x-5a)\{2(z-2a)+(x-5a)\} = (x-5a)\{2x-4a+x-5a\}$$

$$= (x-5a)(3x-9a) = 3(x-5a)(x-3a)$$

$$\therefore \quad \left(\frac{dy}{dx}\right)^2 = \frac{9(x-5a)^2(x-3a)^2}{18\times 18a^2y^2} = \frac{(x-5a)^2(x-3a)^2}{4a\cdot 9ay^2}$$

$$= \frac{(x-5a)^2(x-3a)^2}{4a(x-2a)(x-5a)^2} \qquad \{\because 9ay^2 = (x-2a)\times(x-5a)^2\}$$

$$= \frac{(x-3a)^2}{4a(x-2a)}$$

$$\therefore \quad 1+\left(\frac{dy}{dx}\right)^2 = 1+\frac{(x-3a)^2}{4a(x-2a)} = \frac{4ax-8a^2+x^2-6ax+9a^2}{4a(x-2a)}$$

$$= \frac{x^2-2ax+a^2}{4a(x-2a)} = \frac{(x-a)^2}{4a(x-2a)}$$

Now length of the loop

$= 2 \times$ length of the upper half APB of the loop

$$= 2\int_{2a}^{5a} \sqrt{\frac{(x-a)^2}{4a(x-2a)}}dx = \frac{1}{\sqrt{a}}\int_{2a}^{5a} \frac{x-a}{\sqrt{x-2a}}dx$$

$$= \frac{1}{\sqrt{a}}\int_{2a}^{5a} \frac{x-2a+a}{\sqrt{x-2a}}dx = \frac{1}{\sqrt{a}}\int_{2a}^{5a}\left[\sqrt{x-2a} + \frac{a}{\sqrt{x-2a}}\right]dx$$

$$= \frac{1}{\sqrt{a}}\left[\frac{2}{3}(x-2a)^{3/2} + 2a\sqrt{x-2a}\right]_{2a}^{5a}$$

$$= \frac{1}{\sqrt{a}}\left[\frac{2}{3}(3a)^{\frac{3}{2}} + 2a(3a)^{\frac{1}{2}}\right] = 4\sqrt{3}a.$$

9. Find the whole length of the loop of the curve $9y^2 = (x-2)(x-5)^2$.

Put a = 1 in the previous example.

The required length = $4\sqrt{3}$

10. Trace the curve $3ay^2 = x(x-a)^2$ and show that the length of the loop is $\frac{4a}{\sqrt{3}}$.

Also, if s be the length of an arc of $3ay^2 = x(x-a)^2$ measured from the origin to the point (x, y), show that $3s^2 = 4x^2 + 3y^2$.

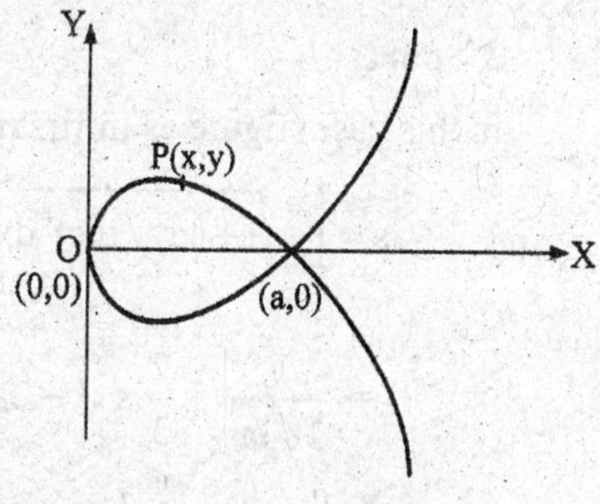

Even power of y occurs so the curve is symmetrical about x-axis.

The curve passes through the origin and the equation of tangent at origin is x = 0, i.e., y-axis (by equating the lowest-degree term to zero).

Putting y = 0, we get x = 0, x = a. So, the curve cuts x-axis at (0, 0) and (a, 0).

Thus a loop is formed between (0, 0) and (0, 0).

For –ve values of x, y^2 is –ve i.e., y becomes imaginary. Hence no part of the curve is on the left side of O.

As x increase from x = a, the value of y increase in +ve and –ve directions of y. The curve has no asymptote and is as shown in the figure.

If s be the length of the loop,

s = 2 × length of the upper half of the loop

$$= 2\int_0^a \sqrt{1+\left(\frac{dy}{dx}\right)^2}\,dx \qquad \ldots\ldots(1)$$

Now $3ay^2 = x\,(x-a)^2$.

$$\therefore \quad 6ay\frac{dy}{dx} = (x-a)^2 + x\times 2(x-a) = (x-a)^2 + 2x(x-a)$$

$$= (x-a)|x-a+2x| = (x-a)(3x-a)$$

$$\therefore \quad \left(\frac{dy}{dx}\right)^2 = \frac{(x-a)^2(3x-a)^2}{12a.3ay^2} = \frac{(x-a)^2(3x-a)^2}{12a.x(x-a)^2} = \frac{(3x-a^2)}{12ax}$$

$$\therefore \text{ from (1), } s = 2\int_0^a \sqrt{1+\left[\frac{(3x-a)^2}{12ax}\right]}\,dx = 2\int_0^a \frac{3x+a}{\sqrt{12ax}}\,dx$$

$$= \frac{2}{2\sqrt{3a}}\int_0^a \frac{3x+a}{\sqrt{x}}\,dx = \frac{1}{\sqrt{3a}}\int_0^a (3\sqrt{x}+\frac{a}{\sqrt{x}})\,dx$$

$$= \frac{1}{\sqrt{3a}}\left[3.\frac{2}{3}x^{3/2} + a2\sqrt{x}\right]_0^a = \frac{1}{\sqrt{3a}}\left[2a^{3/2} + 2a^{3/2}\right] = \frac{4a}{\sqrt{3}}$$

2nd part:

In this case (figure as in first part), the required length

$$s = \int_0^x \sqrt{1+(dy/dx)^2}\,dx = \frac{1}{2\sqrt{3a}}\int_0^x \left(3\sqrt{x}+\frac{a}{\sqrt{x}}\right)dx \text{ (as above)}$$

$$= \frac{1}{2\sqrt{3a}}\left[3.\frac{2}{3}.x^{3/2} + 2a\sqrt{x}\right]_0^x = \frac{1}{\sqrt{3a}}(x+a)\sqrt{x}$$

$$\therefore \quad 3as^2 = x\,(x+a)^2 \qquad \ldots\ldots(2)$$

$$\text{Again } a\left(4x^2+3y^2\right) = 4x^2a + x(x-a)^2 \left\{\because 3ay^2 = x(x-a)^2\right\}$$

$$= x\left(4ax + x^2 - 2ax + a^2\right) = x(x+a)^2 = 3as^2$$

[from (2)]

i.e., $3s^2 = 4x^2 + 3y^2$.

11. Show that the whole length of the curve $y^2(a^2 - y^2) = 8a^2x^2$ is $\pi a\sqrt{2}$.

The curve passes through the origin.

When x = 0, we get y = 0 and y = ±a .

The curve passes also through (0, a) and (0, a).

The curve is symmetrical about both axes as only even powers of x and y occur. When y > a or y < – a, x is imaginary.

Thus the curve consists of two loops as shown in the figure.

Total length of the curve

$$= 4 \times \text{length of the half loop OPA}$$

$$= 4\int_0^a \sqrt{1+\left(\frac{dx}{dy}\right)^2}\, dy.$$

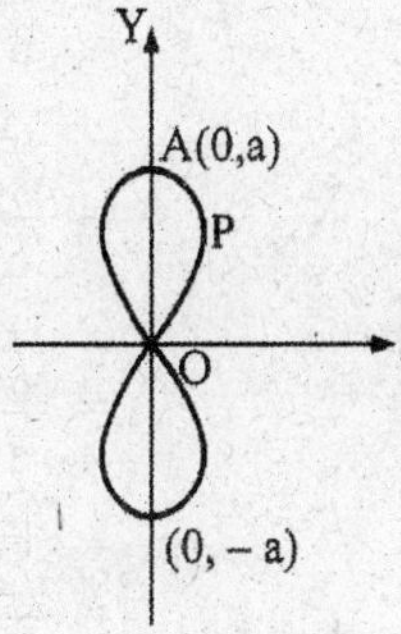

The given curve is $y^2\left(a^2 - y^2\right) = 8a^2x^2$

Or $\quad 2\sqrt{2}ax = y\sqrt{a^2 - y^2}$

$$\therefore \quad 2\sqrt{2}a\frac{dx}{dy} = \sqrt{a^2 - y^2} - \frac{y^2}{\sqrt{a^2 - y^2}}$$

$$= \frac{a^2 - 2y^2}{\sqrt{a^2 - y^2}}$$

$$\therefore \quad \left(\frac{dx}{dy}\right)^2 = \frac{1}{8a^2}\frac{\left(a^2 - 2y^2\right)^2}{a^2 - y^2}$$

$$\therefore \quad 1+\left(\frac{dx}{dy}\right)^2 = \frac{8a^2\left(a^2 - y^2\right) + \left(a^2 - 2y^2\right)^2}{8a^2\left(a^2 - y^2\right)}$$

$$= \frac{8a^4 - 8a^2y^2 + a^4 - 4a^2y^2 + 4y^4}{8a^2\left(a^2 - y^2\right)}$$

$$= \frac{9a^4 - 12a^2y^2 + 4y^4}{8a^2\left(a^2 - y^2\right)} = \frac{\left(3a^2 - 2y^2\right)^2}{8a^2\left(a^2 - y^2\right)}$$

$$\therefore\ s = 4\int_0^a \sqrt{1+\left(\frac{dx}{dy}\right)^2}\,dy = 4\int_0^a \frac{3a^2-2y^2}{2\sqrt{2a}\,\sqrt{\left(a^2-y^2\right)}}\,dy$$

$$= \frac{\sqrt{2}}{a}\int_0^a \frac{3a^2-2y^2}{\sqrt{a^2-y^2}}\,dy.$$

Putting $y = a\sin\theta$, $dy = a\cos\theta\, d\theta$.

When $y = 0$, $\theta = 0$; when $y = a$, $\theta = \pi/2$.

$$\therefore\quad s = \frac{\sqrt{2}}{a}\int_0^{\pi/2} \frac{3a^2 - 2a\sin^2\theta}{\sqrt{a^2 - a^2\sin^2\theta}}\, a\cos\theta\, d\theta$$

$$= \frac{\sqrt{2}}{a}\int_0^{\pi/2}\left(3a^2 - 2a^2\sin^2\theta\right)d\theta$$

$$= \frac{\sqrt{2}}{a}\left[3a^2\int_0^{\pi/2} d\theta - 2a^2\int_0^{\pi/2}\sin^2\theta\, d\theta\right]$$

$$= \frac{\sqrt{2}}{a}.3a^2.\frac{\pi}{2} - \frac{\sqrt{2}}{a}.2a^2.\int_0^{\pi/2}\frac{1-\cos 2\theta}{2}\,d\theta$$

$$= \frac{3}{\sqrt{2}}a\pi - \sqrt{2}a\left[\theta - \frac{\sin 2\theta}{2}\right]_0^{\pi/2}$$

$$= \frac{3}{\sqrt{2}}a\pi - \sqrt{2}a\left\{\frac{\pi}{2} - 0\right\}$$

$$= \frac{3}{\sqrt{2}}a\pi - \frac{a\pi}{\sqrt{2}} = \frac{2a\pi}{\sqrt{2}} = \sqrt{2}a\pi.$$

12. Find the length of the arc cut off by $y = 2x$ from the curve $y^3 = 8x^2$.

Solving the given equations,

$$y^3 = 8x^2 = 2.(2x) = 2.y^2;\quad \therefore\ y^2(y-2) = 0;\ y = 0, 2$$

$$\therefore\ \text{the required length } \int_0^2 \sqrt{1+\left(\frac{dx}{dy}\right)^2}\,dy.$$

Here $y^3 = 8x^2$; $\therefore\ 3y^2\dfrac{dy}{dx} = 16x$

$$\therefore \quad \left(\frac{dx}{dy}\right)^2 = \frac{9y^4}{16\times16x^2} = \frac{9y.y^3}{16\times16x^2} = \frac{9y.8x^2}{16\times16x^2} = \frac{9y}{32}$$

$$\therefore \text{ length} = \int_0^2 \sqrt{1+\frac{9y}{32}}\,dy$$

$$= \left[\frac{2}{3}\left(1+\frac{9y}{32}\right)^{3/2} \times \frac{32}{9}\right]_0^2 = \frac{64}{27}\left\{\left[1+\frac{9\times2}{32}\right]^{3/2} - 1\right\}$$

$$= \frac{64}{27}\left\{\frac{50}{32}\times\sqrt{\frac{50}{32}} - 1\right\} = \frac{64}{27}\times\left\{\frac{125}{64} - 1\right\}$$

$$= \frac{64}{27}\times\frac{61}{64} = \frac{61}{27}$$

13. Find the length of the arc of a semicubical parabola $ay^2 = x^3$ from the origin to the point (a, a).

The curve is $ay^2 = x^3$; $\quad \therefore \; 2ay\frac{dy}{dx} = 3x^2$

$$\therefore \quad \left(\frac{dy}{dx}\right)^2 = \frac{9x^4}{4a^2\,y^2} = \frac{9x^4}{4x^3a} = \frac{9}{a}\frac{x}{a}$$

$\therefore$ the required length

$$= \int_0^a \sqrt{1+(dy/dx)^2}\,dx$$

$$= \int_0^a \sqrt{1+\frac{9x}{4a}}\,dx = \left[\frac{\left(1+\frac{9x}{4a}\right)^{3/2}}{\frac{3}{2}\times\frac{a}{4a}}\right]_0^a$$

$$= \frac{8a}{27}\left[\left(1+\frac{9x}{4a}\right)^{3/2}\right]_0^a = \frac{8a}{27}\left[\frac{13^{3/2}}{2^3} - 1\right]$$

$$= \frac{8a}{27}\left[\frac{13\sqrt{13}}{8} - 1\right] = \frac{2}{27}a\left\{13\sqrt{13} - 8\right\}.$$

14. Prove that the length of the arc of the curve $27ay^2 = 4(x - 2a)^3$ from (2a, 0) to $\left(8a, 4\sqrt{2}a\right)$ is $2a\left(3\sqrt{3}-1\right)$.

The curve is $27ay^2 = 4(x - 2a)^3$

$$\therefore \quad 54ay\frac{dy}{dx} = 12(x-2a)^2; \quad \therefore \quad \frac{dy}{dx} = \frac{12(x-2a)^2}{54ay} = \frac{4(x-2a)^2}{18ay}$$

$$\text{or} \quad \left(\frac{dy}{dx}\right)^2 = \frac{4^2(x-2a)^2}{18^2a^2y^2} = \frac{4^2(x-2a)^2}{18\times 18a.\frac{4}{27}(x-2a)^3}$$

$$\left\{\because 27ay^2 = 4(x-2a)^3\right\}$$

$$= \frac{4\times 27}{18\times 18a}(x-2a) = \frac{x-2a}{3a}$$

$$\therefore \quad 1+\left(\frac{dy}{dx}\right)^2 = 1+\frac{x-2a}{3a} = \frac{x+a}{3a}$$

$$\therefore \quad \text{the required length} = \int_{2a}^{8a}\sqrt{1+\left(\frac{dy}{dx}\right)^2}\,dx$$

$$= \int_0^a\sqrt{\frac{x+a}{3a}}\,dx = \frac{1}{\sqrt{3a}}\left[\frac{2}{3}(x+a)^{3/2}\right]_{2a}^{8a}$$

$$= \frac{2}{3\sqrt{3a}}\left[(9a)^{3/2} - (3a)^{3/2}\right]$$

$$= \frac{2}{3\sqrt{3a}}\left\{27a^{3/2} - 3\sqrt{3}a^{3/2}\right\}$$

$$= 2a\left\{3\sqrt{3}-1\right\}.$$

15. Find the length of the arc of the curve $\sqrt{x} + \sqrt{y} = \sqrt{a}$ intercepted between the axes of coordinates.

Putting x = 0 in the equation of the curve, we get y = a, i.e., the curve cuts the y-axis at (0, a) and putting y = 0, we get x = a, i.e., the curve cuts the x-axis at (a, 0).

So, for the required length x varies from 0 to a.

$\therefore$ the required length $= \int_0^a \sqrt{1+(dy/dx)^2}\,dx.$

Here $\sqrt{x}+\sqrt{y}=\sqrt{a}$

$\therefore$ differentiating, we get $\dfrac{1}{2\sqrt{x}}+\dfrac{1}{2\sqrt{y}}\dfrac{dy}{dx}=0$

$$\therefore \quad \frac{dy}{dx}=-\sqrt{\frac{y}{x}}=-\frac{\sqrt{a}-\sqrt{x}}{\sqrt{x}}$$

$$\therefore \quad 1+\left(\frac{dy}{dx}\right)^2=1+\left(\frac{\sqrt{a}-\sqrt{x}}{\sqrt{x}}\right)^2=1+\frac{a+x-2\sqrt{ax}}{x}$$

$$=1+\frac{a}{x}+1-2\sqrt{\frac{a}{x}}=\frac{a}{2x}\left\{\frac{2x}{a}+2+\frac{2x}{a}-4\sqrt{\frac{x}{a}}\right\}$$

$$=\frac{a}{2x}\left\{1+\left(1+\frac{4x}{a}-4\sqrt{\frac{x}{a}}\right)\right\}=\frac{a}{2x}\left\{1+\left(1-2\sqrt{\frac{x}{a}}\right)^2\right\}$$

$$\therefore \text{ length}=\sqrt{\frac{a}{2}}\int_0^a\sqrt{1+\left(1-2\sqrt{\frac{x}{a}}\right)^2}\,\frac{dx}{\sqrt{x}}.$$

Putting $1-2\sqrt{\dfrac{x}{a}}=t,\ \dfrac{dx}{\sqrt{x}}=-\sqrt{a}\ dt$

When $x=0$, $t=1$; when $x=a$, $t=-1$.

$$\therefore \text{ length}=\sqrt{\frac{a}{2}}\int_1^{-1}\sqrt{1+t^2}\,-\sqrt{a}\ dt=-\frac{a}{\sqrt{2}}\int_1^{-1}\sqrt{1+t^2}\,dt$$

$$=-\frac{a}{\sqrt{2}}\left[\frac{1}{2}t\sqrt{1+t^2}+\frac{1}{2}\log\left\{t+\sqrt{1+t^2}\right\}\right]_1^{-1}$$

$$=-\frac{a}{\sqrt{2}}\left\{-\frac{1}{2}\sqrt{2}+\frac{1}{2}\log\left(\sqrt{2}-1\right)-\frac{1}{2}\sqrt{2}-\frac{1}{2}\log\left(1+\sqrt{2}\right)\right\}$$

$$=\frac{a}{2\sqrt{2}}\left\{2\sqrt{2}+\log\frac{\sqrt{2}+1}{\sqrt{2}-1}\right\}$$

$$= \frac{a}{2\sqrt{2}}\left\{2\sqrt{2} + \log\frac{(\sqrt{2}+1)\ (\sqrt{2}+1)}{(\sqrt{2}-1)\ (\sqrt{2}+1)}\right\}$$

$$= \frac{a}{2\sqrt{2}}\left\{2\sqrt{2} + \log(\sqrt{2}+1)^2\right\} = \frac{a}{2\sqrt{2}}\left\{2\sqrt{2} + 2\log\ (\sqrt{2}+1)\right\}$$

$$= \frac{a}{\sqrt{2}}\left\{\sqrt{2} + \log\ (\sqrt{2}+1)\right\}.$$

16. In the curve $x^{2/3} + y^{2/3} = a^{2/3}$ show that $\rho^2 + 4s^2 = 6as$, s being measured from the point for which $x = 0$.

The astroid is shown in the figure.

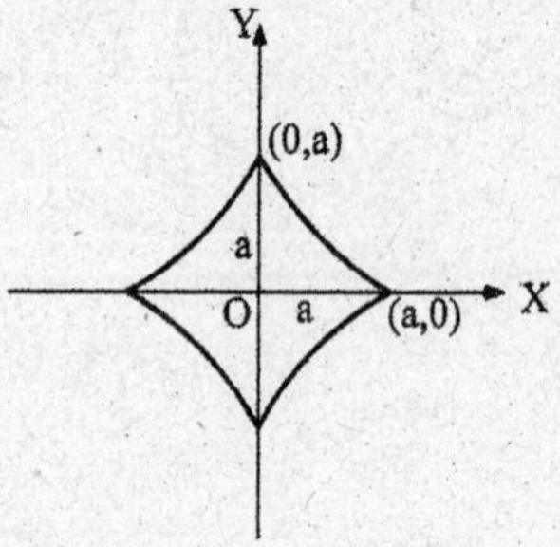

The given curve is $x^{2/3} + y^{2/3} = a^{2/3}$

Differentiating w.r.t. x,

$$\frac{2}{3}x^{-1/3} + \frac{2}{3}y^{-1/3}\frac{dy}{dx} = 0$$

$$\therefore \quad \frac{dy}{dx} = -\left(\frac{y}{x}\right)^{1/3} \qquad .....(1)$$

Again differentiating w.r.t. x,

$$\frac{d^2y}{dx^2} = -\frac{\frac{1}{3}y^{-2/3}\frac{dy}{dx}.x^{1/3} - \frac{1}{3}x^{-2/3}y^{1/3}}{x^{2/3}}$$

$$= -\frac{\frac{1}{3}y^{-2/3} \times \left\{-\left(\frac{y}{x}\right)^{1/3}\right\}x^{1/3} - \frac{1}{3}x^{-2/3}y^{1/3}}{x^{2/3}} \qquad \text{[using (1)]}$$

$$= \frac{\frac{1}{3}y^{-2/3} + \frac{1}{3}x^{-2/3}y^{1/3}}{x^{2/3}} = \frac{x^{2/3} + y^{2/3}}{3x^{4/3}.y^{1/3}}$$

multiplying the numerator and denominator by $x^{2/3} . y^{1/3}$.

$$\therefore \quad \frac{d^2y}{dx^2} = \frac{a^{2/3}}{3x^{4/3}y^{1/3}} \qquad \left(\because x^{2/3} + y^{2/3} = a^{2/3}\right) \qquad .....(2)$$

$$\therefore \qquad s = \int_0^x \sqrt{1+\left(\frac{dy}{dx}\right)^2}\,dx$$

$$= \int_0^x \sqrt{1+(y/x)^{2/3}}\,dx = \int_0^a \sqrt{\frac{x^{2/3}+y^{2/3}}{x^{2/3}}}\,dx$$

$$= \int_0^x \sqrt{\frac{a^{2/3}}{x^{2/3}}}\,dx = \int_0^x a^{1/3}x^{-1/3}\,dx$$

$$= a^{1/3}\left[\frac{x^{2/3}}{2/3}\right]_0^x = \frac{3}{2}a^{1/3}x^{2/3} \qquad \text{.....(3)}$$

Again, the entire length of the astroid

$$= 4 \times \text{length of the curve in the first quadrant}$$

$$= 4\int_0^a \sqrt{1+(dy/dx)^2}\,dx$$

(as x varies from 0 to a in the first quadrant)

$$= 4\int_0^a a^{1/3}x^{-1/3}\,dx = 4a^{1/3}\left[\frac{x^{2/3}}{2/3}\right]_0^a$$

$$= 4a^{1/3} \times \frac{3}{2}a^{2/3} = 6a \qquad \text{.....(4)}$$

which is the entire length of the curve.

$$\text{Now,} \qquad \rho = \frac{(1+y_1^2)^{3/2}}{y_2} = \frac{\left[1+\left(\frac{y}{x}\right)^{2/3}\right]^{3/2}}{y_2}$$

$$= \frac{(x^{2/3}+y^{2/3})^{3/2}}{xy_2} = \frac{(a^{2/3})^{3/2}}{xy_2} = \frac{a}{x+\dfrac{a^{2/3}}{3x^{4/3}y^{1/3}}} \qquad \text{[by (2)]} \qquad \text{.....(5)}$$

$$= 3a^{1/3}x^{1/3}y^{1/3}$$

$\therefore$ from (3) and (5),

$$\rho^2 + 4s^2 = 9a^{2/3}\,x^{2/3}\,y^{2/3} + 4 \times \frac{9}{4}.a^{2/3}\,x^{4/3}$$

$$= 9a^{2/3}\,x^{2/3}\left[y^{2/3}+x^{2/3}\right] = 9a^{2/3}x^{2/3}.a^{2/3} = 9a^{4/3}x^{2/3}$$

$$= 6a.\frac{3}{2}a^{1/3}.x^{2/3} = 6as \qquad \text{[by(3)]}.$$

17. Find the length of one quadrant of the curve

$$\left(\frac{x}{a}\right)^{2/3}+\left(\frac{y}{b}\right)^{2/3}=1$$

Parametric equations of the curve are

$$x = a\cos^3 t$$

$$y = b\sin^3 t.$$

$$\therefore \frac{dx}{dt} = -3a\cos^2 t\sin t\ dt, \quad \frac{dy}{dt} = 3b\sin^2 t\cos t\ dt.$$

Length of one quadrant of the curve

$$=\int_0^{\pi/2}\sqrt{\left(\frac{dx}{dt}\right)^2+\left(\frac{dy}{dt}\right)^2}$$

$$=\int_0^{\pi/2} 3\sin t\cos t\sqrt{a^2\cos^2 t+b^2\sin^2 t}\ dt$$

Putting $a^2\cos^2 t+b^2\sin^2 t=u^2$, $(-2a^2\cos t\ \sin t+2b^2\sin t\cos t)dt=2u\ du$

$$\therefore \quad \sin t\cos t\ dt=\frac{u\ du}{b^2-a^2}$$

When $t = 0$, $u = a$, when $t = \pi/2$, $u = b$.

$$\therefore \quad \text{length}=3\int_a^b u\frac{u\ du}{b^2-a^2}=\frac{3}{b^2-a^2}\int_a^b u^2\ du$$

$$=\frac{3}{b^2-a^2}\left[\frac{u^3}{3}\right]_a^b=\frac{b^3-a^3}{b^2-a^2}$$

$$=\frac{a^2+ab+b^2}{a+b}$$

Note: Perimeter of the four-cusped hypocycloid $= 4\dfrac{a^2+ab+b^2}{a+b}$

18. Show that the length of an arc of the curve $x\sin\theta+y\cos\theta=f'(\theta)$, $x\cos\theta-y\sin\theta=f''(\theta)$ is given by $s=f(\theta)+f''(\theta)+c$, where c is constant of integration.

Here $x\sin\theta+y\cos\theta=f'(\theta)$ .....(1)

$x\cos\theta-y\sin\theta=f''(\theta)$ .....(2)

(1)× sin θ + (2) × cos θ gives

$$x(\sin^2\theta+\cos^2\theta)=\sin\theta\, f'(\theta)+\cos\theta\, f''(\theta)$$

$$x=\sin\theta\, f'(\theta)+\cos\theta\, f''(\theta)$$

Similarly, $y=\cos t\, f'(\theta)-\sin\theta\, f''(\theta)$.

$$\therefore \qquad \frac{dx}{d\theta}=\cos\theta\, f'(\theta)+\sin\theta\, f''(\theta)-\sin\theta\, f''(\theta)+\cos\theta f''(\theta)$$

$$=\cos\theta\left[f'(\theta)+f'''(\theta)\right]$$

and $$\frac{dy}{d\theta}=-\sin\theta\left[f'(\theta)+f'''(\theta)\right]$$

∴ require length

$$=\int\sqrt{\left(\frac{dx}{d\theta}\right)^2+\left(\frac{dy}{d\theta}\right)^2}\,d\theta=\int\left[f'(\theta)+f'''(\theta)\right]\sqrt{\cos^2\theta+\sin^2\theta}\,d\theta$$

$$=\int\left[f'(\theta)+f'''(\theta)\right]d\theta=f(\theta)+f''(\theta)+C$$

19. If A denotes the area between the curve $y=c\cosh\left(\frac{x}{c}\right)$, the two ordinates $x = x_1$ and $x = x_2$ and the x-axis, and s stands for the length of intervening arc then prove that A = cs.

Here $$A=\int_{x_1}^{x_2} y\,dx=\int_{x_1}^{x_2} c\cosh\left(\frac{x}{c}\right)dx$$

$$=c^2\left[\sinh\frac{x}{c}\right]_{x_1}^{x_2}=c^2\left[\sinh\frac{x_2}{c}-\sinh\frac{x_1}{c}\right] \qquad .....(1)$$

and $$s=\int_{x_1}^{x_2}\sqrt{1+\left(\frac{dy}{dx}\right)^2}\,dx=\int_{x_1}^{x_2}\sqrt{1+\sinh^2\frac{x}{c}}\,dx$$

$$=\int_{x_1}^{x_2}\cosh\frac{x}{c}\,dx=c\left[\sinh\frac{x}{c}\right]_{x_1}^{x_2}$$

$$=c\left[\sinh\frac{x_2}{c}-\sinh\frac{x_1}{c}\right] \qquad .....(2)$$

From (1) and (2), A = cs.

20. Find the length of the arc of the curve $y = \log\dfrac{e^x - 1}{e^x + 1}$ from $x = 1$ to $x = 2$.

The curve is $y = \log(e^x - 1) - \log(e^x + 1)$

$$\therefore \quad \frac{dy}{dx} = \frac{e^x}{e^x - 1} - \frac{e^x}{e^x + 1} = \frac{e^x(e^x + 1) - e^x(e^x - 1)}{e^{2x} - 1}$$

$$= \frac{2e^x}{e^{2x} - 1}$$

$$\therefore \text{ required length} = \int_1^2 \sqrt{1 + \left(\frac{dy}{dx}\right)^2}\, dx$$

$$= \int_1^2 \sqrt{1 + \left(\frac{2e^x}{e^{2x} - 1}\right)^2}\, dx$$

$$= \int_1^2 \sqrt{\frac{(e^{2x} - 1)^2 + 4e^{2x}}{(e^{2x} - 1)^2}}\, dx$$

$$= \int_1^2 \sqrt{\frac{(e^{2x} + 1)^2}{(e^{2x} - 1)^2}}\, dx$$

$$= \int_1^2 \frac{e^{2x} + 1}{e^{2x} - 1}\, dx = \int_1^2 \frac{e^x + e^{-x}}{e^x - e^{-x}}\, dx$$

$$= \left[\log\ (e^x - e^{-x})\right]_1^2 = \log\left(e + \frac{1}{e}\right)$$

21. Show that the length of the arc of the curve $x^2 = a^2 (1 - e^{y/a})$ measured from the origin to the point (x, y) is $a \log\dfrac{a + x}{a - x} - x$.

From the equation of the curve,

$$e^{y/a} = 1 - \frac{x^2}{a^2} \text{ or } \frac{y}{a} = \log\left(1 - \frac{x^2}{a^2}\right)$$

or $$y = a \log\frac{a^2 - x^2}{a^2} = a\log(a^2 - x^2) - a \log a^2$$

$$\therefore \quad \frac{dy}{dx} = a\frac{-2x}{a^2 - x^2} = \frac{-2ax}{a^2 - x^2}$$

$$\therefore \text{ the require length } = \int_0^x \sqrt{1 + \left(\frac{dy}{dx}\right)^2}\, dx$$

$$= \int_0^x \sqrt{1 + \left(\frac{-2ax}{a^2 - x^2}\right)^2}\, dx$$

$$= \int_0^x \frac{a^2 + x^2}{a^2 - x^2}\, dx$$

$$= \int_0^x \left(\frac{2a^2}{a^2 - x^2} - 1\right) dx$$

$$= 2a^2 \int_0^x \frac{dx}{a^2 - x^2} - 2a^2 \int_0^x dx$$

$$= \left[2a^2 \times \frac{1}{2a}\log\frac{a+x}{a-x} - 2a^2 x\right]_0^x$$

$$= a\log\frac{a+x}{a-x} - x$$

22. Find the length of the curve $r = a\theta$ from $\theta = 0$ to $\theta = \pi/2$.

The curve is $r = a\theta$.

$$\therefore \quad \frac{dr}{d\theta} = a$$

$\therefore$ the required length

$$= \int_0^{\pi/2} \sqrt{r^2 + \left(\frac{dr}{d\theta}\right)^2}\, d\theta = \int_0^{\pi/2} \sqrt{a^2\,\theta^2 + a^2}\, d\theta$$

$$= a\int_0^{\pi/2} \sqrt{1 + \theta^2}\, d\theta$$

$$= a\left[\frac{1}{2}\theta\sqrt{1+\theta^2} + \frac{1}{2}\log\left\{\theta + \sqrt{1+\theta^2}\right\}\right]_0^{\pi/2}$$

$$= \frac{a}{2}\left[\frac{\pi}{2}\sqrt{1 + \frac{\pi^2}{4}} + \log\left\{\frac{\pi}{2} + \sqrt{1 + \frac{\pi^2}{4}}\right\}\right]$$

23. Find the circumference of the circle $r = 2a\cos\theta$.

The required circumference = 2 × upper half of the circumference

$$= 2\int_0^{\pi/2} \sqrt{r^2 + \left(\frac{dr}{d\theta}\right)^2}\, d\theta$$

$$= 2\int_0^{\pi/2} \sqrt{4a^2\cos^2\theta + (-2a\sin\theta)^2}\, d\theta$$

$$= 2.2a\int_0^{\pi/2} d\theta = 4a\left[\theta\right]_0^{\pi/2}$$

$$= 4a\frac{\pi}{2} = 2\pi a$$

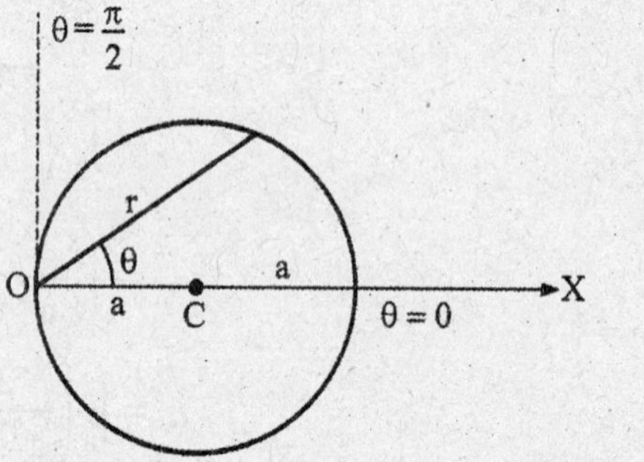

24. Find the entire length of the cardioid $r = a(1 + \cos\theta)$.

The curve is symmetrical about the initial line since if we put $-\theta$ for θ, the equation of the curve remains unchanged.

Starting with $\theta = 0$ to $\theta = \pi$, the corresponding values of r are obtained and the curve is roughly traced as shown.

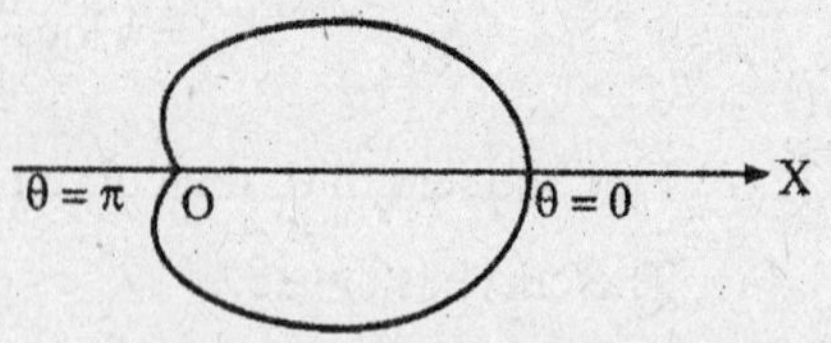

The required length

$= 2 \times$ length of the upper half

$$= 2\int_0^{\pi} \sqrt{r^2 + \left(\frac{dr}{d\theta}\right)^2}\, d\theta$$

Here $r = a(1 + \cos\theta)$

$$\therefore \quad \frac{dr}{d\theta} = -a\sin\theta\, d\theta$$

$$\therefore \quad r^2 + \left(\frac{dr}{d\theta}\right)^2 = a^2(1+\cos\theta)^2 + a^2\sin^2\theta = 2a^2(1+\cos\theta)$$

$$= 2a^2.2\cos^2\frac{\theta}{2}$$

$$\therefore \quad \text{length} = 2\int_0^{\pi} \sqrt{4a^2\cos^2\frac{\theta}{2}}\, d\theta$$

$$= 2.2a\int_0^{\pi} \cos\frac{\theta}{2}\, d\theta = 4a.2\left[\sin\frac{\theta}{2}\right]_0^{\pi} = 8a.$$

25. Prove that the total length of the cardioid $r = a(1 - \cos\theta)$ is 8a. Also show that the arc above the initial line is bisected by the line $\theta = \frac{2}{3}\pi$.

The curve is symmetrical about the initial line as its equation remains unchanged if θ be replaced by $-\theta$.

Corresponding to different values of θ the values of r are given as below.

θ	0	$\frac{\pi}{3}$	$\frac{\pi}{2}$	$\frac{2\pi}{3}$	π	$\frac{3\pi}{2}$	2π
r	0	$\frac{a}{2}$	a	$\frac{3}{2}a$	2a	a	0

With these values of r and θ the curve is traced as shown in the figure.

$\therefore$ the length of the whole curve = 2 × length of the upper half.

$$= 2\int_0^{\pi}\sqrt{r^2 + \left(\frac{dr}{d\theta}\right)^2}\,d\theta$$

(for the upper half θ varies from 0 to π).

$$= 2\int_0^{\pi}\sqrt{a^2(1-\cos\theta)^2 + a^2\sin^2\theta}\,d\theta.$$

Since here $r = a(1 - \cos\theta)$,

$dr = a\sin\theta\, d\theta$,

$$\text{length} = 2a\int_0^{\pi}\sqrt{2-2\cos\theta}\,d\theta$$

$$= 2a\int_0^{\pi} 2\sin\frac{\theta}{2}\,d\theta = 4a\left[-2\cos\frac{\theta}{2}\right]_0^{\pi} = 8a[-0+1] = 8a.$$

Again length of the arc from $\theta = 0$ to $\theta = \frac{2\pi}{3}$

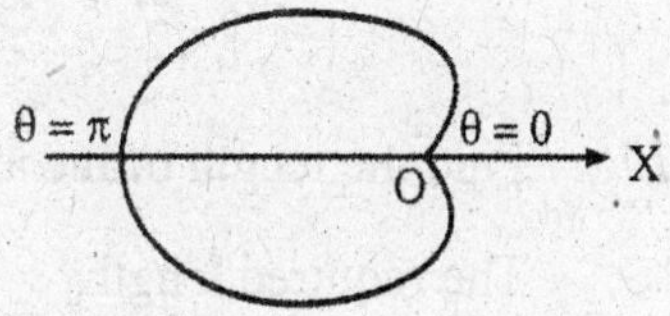

$$= \int_0^{2\pi/3}\sqrt{r^2 + \left(\frac{dr}{d\theta}\right)^2}\,d\theta$$

$$= \int_0^{2\pi/3} 2a\sin\frac{\theta}{2}\,d\theta = 2a\left[-2\cos\frac{\theta}{2}\right]_0^{2\pi/3} = -4a\left[\frac{1}{2}-1\right] = 2a.$$

26. Find the length of the arc of the curve $\frac{2a}{r} = 1 + \cos\theta$ from $\theta = 0$ to $\theta = \frac{\pi}{2}$.

Here S is the pole and SX is the initial line and the curve (parabola) is as shown in the figure.

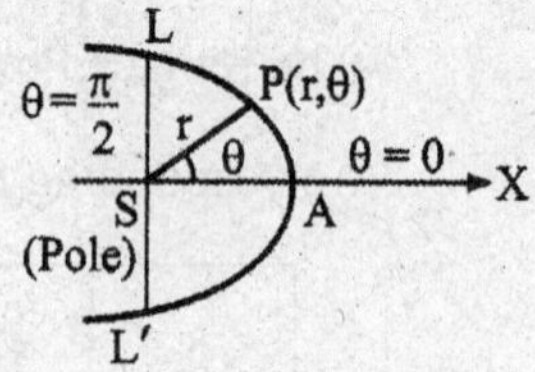

Now $r = \frac{2a}{1+\cos\theta} = a\sec^2\frac{\theta}{2}$

$$\therefore \quad \frac{dr}{d\theta} = 2a\sec\frac{\theta}{2}\sec\frac{\theta}{2}\tan\frac{\theta}{2}.\frac{1}{2}$$

$$= a\sec^2\frac{\theta}{2}\tan\frac{\theta}{2}$$

$\therefore$ length of the arc from $\theta = 0$ to $\theta = \pi/2$

$$s = \int_0^{\pi/2}\sqrt{r^2 + \left(\frac{dr}{d\theta}\right)^2}\,d\theta$$

$$= \int_0^{\pi/2}\sqrt{a^2\sec^4\frac{\theta}{2} + a^2\sec^4\frac{\theta}{2}\tan^2\frac{\theta}{2}}\,d\theta$$

$$= a\int_0^{\pi/2}\sec^2\frac{\theta}{2}\sqrt{1+\tan^2\frac{\theta}{2}}\,d\theta$$

Putting $\tan\frac{\theta}{2} = z$, $\sec^2\frac{\theta}{2}.\frac{1}{2}d\theta = dz$

When $\theta = 0$, $z = 0$; when $\theta = \pi/2$, $z = 1$.

$$\therefore \quad s = a\int_0^1 2\,dz\sqrt{1+z^2} = 2a\int_0^1\sqrt{1+z^2}\,dz$$

$$= 2a\left[\frac{z\sqrt{1+z^2}}{2} + \frac{1}{2}\log(z+\sqrt{1+z^2})\right]_0^1$$

$$= 2a\left[\frac{1}{\sqrt{2}} + \frac{1}{2}\log\left(1+\sqrt{2}\right)\right] = a\left[\sqrt{2} + \log\left(1+\sqrt{2}\right)\right].$$

27. Find the length of the arc of the parabola $\frac{2a}{r} = 1 + \cos\theta$ cut off by the latus rectum.

The required length

$$= 2 \times \text{length of the arc from } \theta = 0 \text{ to } \theta = \pi/2$$

$$= 2\int_0^{\pi/2}\sqrt{r^2 + \left(\frac{dr}{d\theta}\right)^2}\,d\theta = 2a\left\{\sqrt{2}\log\left(1+\sqrt{2}\right)\right\} \text{ (as above).}$$

(Give figure and calculation of the previous example.)

28. Find the length of the cardioid $r = a(1 - \cos\theta)$ lying outside the circle $r = a\cos\theta$.

Solving the equation of the curve, $a(1 - \cos\theta) = a\cos\theta$;

$\therefore \cos\theta = \frac{1}{2}; \therefore \theta = \frac{\pi}{3}.$

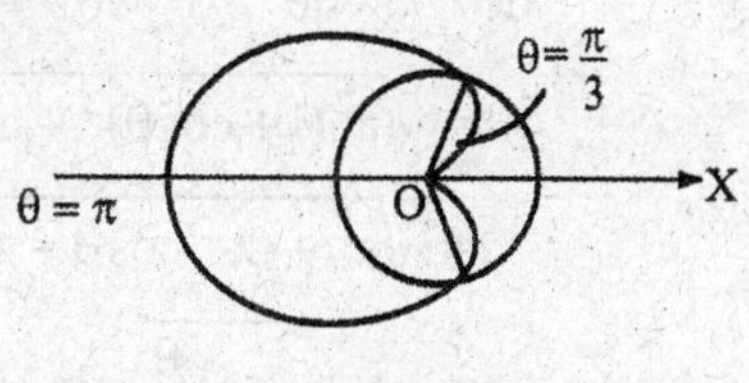

$\therefore$ the required length

$$= 2\int_{\pi/3}^{\pi}\sqrt{r^2 + \left(\frac{dr}{d\theta}\right)^2}\, d\theta$$

$$= 2a\int_{\pi/3}^{\pi} 2\sin\frac{\theta}{2}\, d\theta$$

$$= 4a\left[-2\cos\frac{\theta}{2}\right]_{\pi/3}^{\pi} = 8a\left(-\cos\frac{\pi}{2} + \cos\frac{\pi}{6}\right)$$

$$= 8a\left(0 + \frac{\sqrt{3}}{2}\right) = 4\sqrt{3}a.$$

29. Show that the arc of the spiral $r = ae^{\theta\cot\alpha}$ between any two radius vectors varies as the difference of these radii.

The curve is $r = ae^{\theta\cot\alpha}$.

$\therefore \log r = \log a + \theta\cot\alpha; \quad \therefore \frac{1}{r}\cdot\frac{dr}{d\theta} = \cot\alpha; \quad \therefore r\frac{d\theta}{dr} = \tan\alpha.$

If r_1 and r_2 be two radius vectors of the curve, the length of the arc between them,

$$s = \int_{r_1}^{r_2}\sqrt{1 + r^2\left(\frac{d\theta}{dr}\right)^2}\, dr = \int_{r_1}^{r_2}\sqrt{1 + \tan^2\alpha}\, dr$$

$$= \sec\alpha\int_{r_1}^{r_2} dr = (r_2 - r_1)\sec\alpha$$

$\therefore s \propto r_2 - r_1$, since $\sec\alpha$ is constant.

30. Determine the length of an arc of the cycloid $x = a(\theta + \sin\theta)$, $y = a(1 - \cos\theta)$ measured from the vertex.

The parameter θ increases from $-\pi$ to π in moving from A' to A of one arc, it being 0 at vertex O. The curve is symmetrical about OY, so

Are A'OA = 2 arc OA.

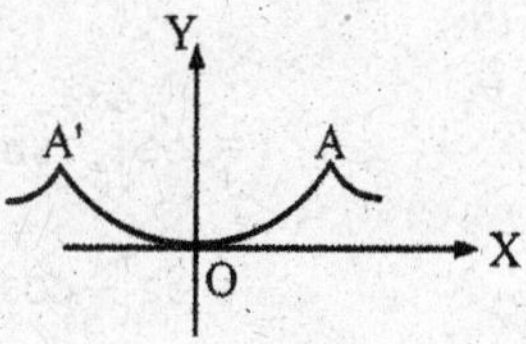

Here $x = a(\theta + \sin\theta)$, $y = a(1 - \cos\theta)$

$$\therefore \frac{dx}{d\theta} = a(1 + \cos\theta)$$

$$\frac{dy}{d\theta} = a(0 + \sin\theta)$$

$$\therefore \quad \frac{ds}{d\theta} = \sqrt{\left(\frac{dx}{d\theta}\right)^2 + \left(\frac{dy}{d\theta}\right)^2}$$

$$= \sqrt{a^2(1+\cos\theta)^2 + a^2\sin^2\theta}$$

$$= \sqrt{a^2 + 2a^2\cos\theta + a^2\cos^2\theta + a^2\sin^2\theta} = \sqrt{2a^2(1+\cos\theta)}$$

$$= \sqrt{4a^2\cos^2\frac{\theta}{2}} = 2a\cos\frac{\theta}{2}.$$

$$\therefore \text{ required length} = 2\int_0^{\pi} 2a\cos\frac{\theta}{2}\, d\theta = 8a\left[\sin\frac{\theta}{2}\right]_0^{\pi} = 8a.$$

31. Find the entire length of the astroid $x^{2/3} + y^{2/3} = a^{2/3}$.

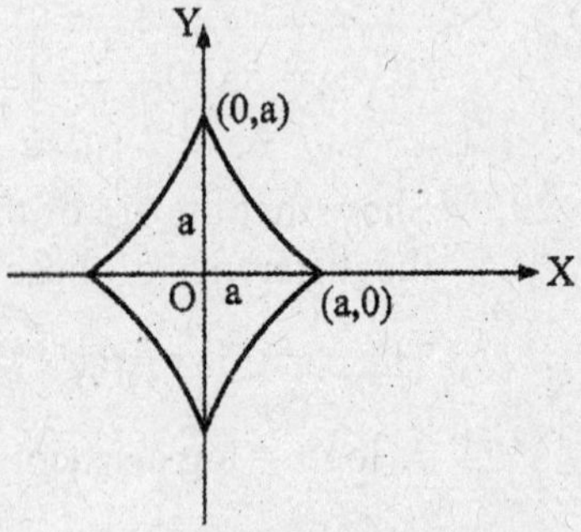

The curve is symmetrical about both axes and does not pass through the origin and cuts x-axis at (a, 0), (–a, 0) and y-axis at (0, a), (0, –a).

The parametric equation of the curve are $x = a\cos^3\theta$, $y = a\sin^3\theta$.

In the first quadrant, θ varies from 0 to $\pi/2$ and the curve is as shown in the figure.

We have $\dfrac{ds}{d\theta} = \sqrt{\left(\dfrac{dx}{d\theta}\right)^2 + \left(\dfrac{dy}{d\theta}\right)^2}$

$\therefore$ entire length of the astroid = 4 × length in 1st quadrant

$$= 4\int_0^{\pi/2} \sqrt{\left(\frac{dx}{d\theta}\right)^2 + \left(\frac{dy}{d\theta}\right)^2}\, d\theta$$

$$= 4\int_0^{\pi/2} \sqrt{(-3a\cos^2\theta\sin\theta)^2 + (3a\sin^2\theta\cos\theta)^2}\, d\theta$$

$$= 4\int_0^{\pi/2} 3a\sin\theta\cos\theta\sqrt{\cos^2\theta + \sin^2\theta}\, d\theta$$

$$= 6a\int_0^{\pi/2} \sin 2\theta\, d\theta = 6a\left[\frac{-\cos 2\theta}{2}\right]_0^{\pi/2}$$

$$= 3a[-\cos\pi + \cos 0] = 6a.$$

32. Find the intrinsic equation of the astroid $x^{2/3} + y^{2/3} = a^{2/3}$ where s is measured from the cusp or the point (a, 0) on x-axis.

If s be the required length,

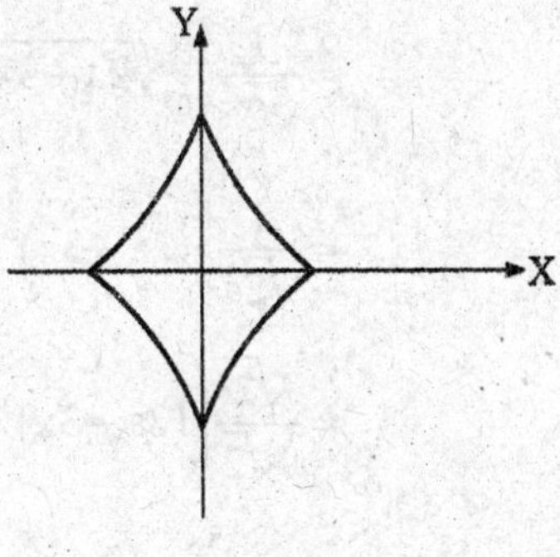

$$s = \int_0^t \sqrt{\left(\frac{dx}{dt}\right)^2 + \left(\frac{dy}{dt}\right)^2}\, dt \quad .....(1)$$

The parametric equations of the curve are

$$x = a\cos^3 t, \quad y = a\sin^3 t$$

$$\therefore \frac{dx}{dt} = -3a\cos^2 t \sin t, \quad \frac{dy}{dt} = 3a\sin^2 t \cos t$$

and $\tan\psi = \frac{dy}{dx} = \frac{dy/dt}{dx/dt} = -\tan t$;

$$\therefore \psi = -t \quad .....(2)$$

As s is measured from cusp where t = 0,

$$\therefore \text{from (1), } s = 3a\int_0^t \sqrt{\sin^2 t\cos^2 t\left(\sin^2 t + \cos^2 t\right)}d$$

$$= 3a\int_0^t \sin t\cos t\, dt = \frac{3}{2}a\int_0^t \sin 2t\, dt$$

$$= \frac{3}{2}a\left[-\frac{\cos 2t}{2}\right]_0^t = \frac{3a}{4}(1-\cos 2t)$$

$$= \frac{3}{2}a\sin^2 t \quad .....(3)$$

Eliminating t from (2) and (3), the intrinsic equation is $s = \frac{3}{2}a\sin^2\psi$.

33. Show that the intrinsic equation of the semicubical parabola

$$3ay^2 = 2x^3 \text{ is } 9s = 4a\left(\sec^3\psi - 1\right).$$

From the equation of the curve, $y = \sqrt{\frac{2}{3a}}x^{3/2}$

$$\therefore \frac{dy}{dx} = \sqrt{\frac{2}{3a}}.\frac{3}{2}x^{1/2} = \sqrt{\frac{3}{2a}}x^{1/2} = \tan\psi \quad .....(1)$$

Tracing its cusp (where x = 0) as fixed point, distance of any point from the cusp,

$$s=\int_0^x\sqrt{1+\left(\frac{dy}{dx}\right)^2}\,dx=\int_0^x\sqrt{1+\frac{3}{2a}x}\;dx$$

$$=\frac{1}{\sqrt{2a}}\int_0^x\sqrt{2a+3x}\;dx$$

$$=\frac{1}{\sqrt{2a}}\left[(2a+3x)^{3/2}.\frac{2}{3}.\frac{1}{3}\right]_0^x$$

$$=\frac{\sqrt{2}}{9\sqrt{2}}\left[(2a+3x)^{3/2}-(2a)^{3/2}\right] \qquad .....(2)$$

From (1), $x=\frac{2a}{3}\tan^2\psi$; substituting in (2),

$$s=\frac{\sqrt{2}}{9\sqrt{a}}\left[\left(2a+3.\frac{2}{3}a\tan^2\psi\right)^{3/2}-(2a)^{3/2}\right]$$

$$=\frac{\sqrt{2}.2^{3/2}a^{3/2}}{9\sqrt{a}}\left[(1+\tan^2\psi)^{3/2}-1\right]$$

$$=\frac{2^2a}{9}(\sec^3\psi-1),\text{ i.e., }\;9s=4a(\sec^3\psi-1)$$

34. Obtain the intrinsic equation of the parabola $y^2 = 4ax$, origin being taken as fixed point.

Let P(x, y) be any point on the curve. The tangent at origin is the y-axis. Let arc OP = s. Let the tangent at P make angle ψ with y-axis.

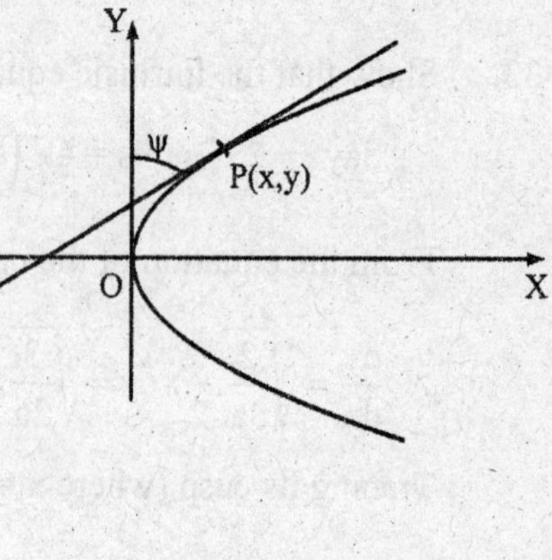

The parabola is $y^2 = 4ax$

$\therefore\; 2y\frac{dy}{dx}=4a$

$\therefore\; \frac{dy}{dx}=\frac{2a}{y}$

So, here $\tan\psi=\frac{dx}{dy}=\frac{y}{2a}$ .....(1)

Also, $s=\int_0^y\sqrt{1+\left(\frac{dx}{dy}\right)^2}\,dy$

$$=\frac{1}{2a}\left[\frac{y\sqrt{4a^2+y^2}}{2}+\frac{4a^2}{2}\log\frac{y+\sqrt{4a^2+y^2}}{2a}\right]$$

Eliminating y from (1) and (2), we get

$$s=\frac{1}{2a}\left[\frac{2a\tan\psi\sqrt{4a^2+4a^2\tan^2\psi}}{2}+2a^2\log\frac{2a\tan\psi+4a^2+4a^2\tan^2\psi}{2a}\right]$$

$$=a\left[\tan\psi\sec\psi+\log(\tan\psi+\sec\psi)\right],$$

Which is the required intrinsic equation.

35. Obtain the intrinsic equation of the catenary $y = a\cosh x/a$ taking the vertex (0, a) as the fixed point.

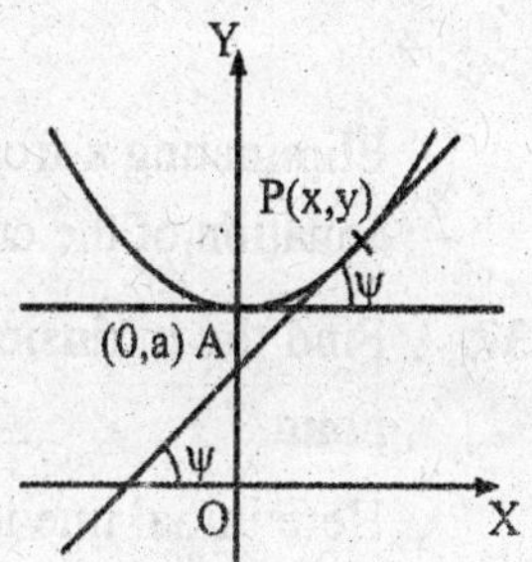

The tangent at A(0, a) is parallel to x-axis.

Let AP = s and tangent at P(x, y) makes an angle ψ with OX. Then

$$s=\int_0^x\sqrt{1+(dy/dx)^2}\,dx$$

$$=\int_0^x\sqrt{1+\sinh^2(x/a)}\,dx$$

$$=\int_0^x\cosh\frac{x}{a}dx=\left[a\sinh\frac{x}{a}\right]_0^x=a\sinh\frac{x}{a} \qquad .....(1)$$

Also, $\tan\psi=\frac{dy}{dx}=\sinh\frac{x}{a}$ .....(2)

Eliminating x from (1) and (2), we get

$s = a\tan\psi$ as the intrinsic equation of the curve.

36. Find the intrinsic equation of the curve $y=a\log\sec\frac{x}{a}$, the arc being measured from the origin.

Here $\frac{dy}{dx}=a.\frac{1}{\sec(x/a)}\sec\frac{x}{a}\tan\frac{x}{a}.\frac{1}{a}$,

i.e., $\tan\psi=\tan\frac{x}{a}$; $\therefore\ \psi=\frac{x}{a}$ .....(1)

Also $$s=\int_0^x \sqrt{1+\left(\frac{dy}{dx}\right)^2}\,dx$$

$$=\int_0^x \sqrt{1+\tan^2\frac{x}{a}}\,dx=\int_0^x \sec\frac{x}{a}\,dx$$

$$=a\left[\log\left\{\sec\frac{x}{a}+\tan\frac{x}{a}\right\}\right]_0^x$$

$$=a\left[\log\left(\sec\frac{x}{a}+\tan\frac{x}{a}\right)\right] \qquad .....(2)$$

Eliminating x from (1) and (2), $s = a\log(\sec\psi+\tan\psi)$ which is intrinsic equation of the curve

37. Find the intrinsic equation of the cardioid $r=a(1-\cos\theta)$, taking pole as the fixed point.

Here initial line is the tangent at pole. Let arc $OP = s$.

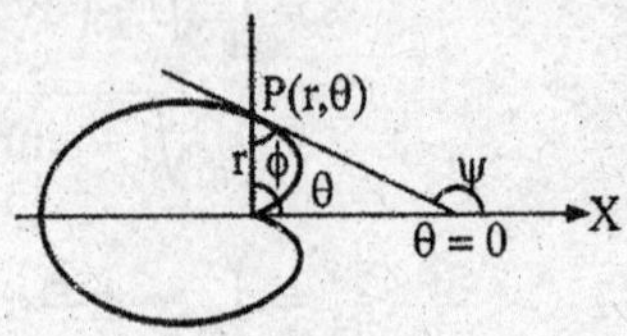

Here, we have $\frac{dr}{d\theta}=a\sin\theta$

$$\therefore \quad \tan\varphi=\frac{r\,d\theta}{dr}=\frac{a(1-\cos\theta)}{a\sin\theta}$$

$$=\frac{1-\cos\theta}{\sin\theta}$$

$$=\frac{2\sin^2\frac{\theta}{2}}{2\sin\frac{\theta}{2}\cos\frac{\theta}{2}}=\tan\frac{\theta}{2}; \quad \therefore \varphi=\frac{\theta}{2}.$$

Again, $$\psi=\theta+\varphi=\theta+\frac{\theta}{2}=\frac{3\theta}{2} \qquad ..... (1)$$

Also, $$s=\int_0^\theta \sqrt{r^2+\left(\frac{dr}{d\theta}\right)^2}\,d\theta=\int_0^\theta \sqrt{a^2(1-\cos\theta)^2+a^2\sin^2\theta}\,d\theta$$

$$=a\int_0^\theta \sqrt{1-2\cos\theta+\cos^2\theta+\sin^2\theta}\,d\theta=2a\int_0^\theta \sin\frac{\theta}{2}\,d\theta$$

$$= -4a\left[\cos\frac{\theta}{2}\right]_0^{\theta} = 4a\left[1-\cos\frac{\theta}{2}\right] \qquad \text{.....(2)}$$

Eliminating θ from (1) and (2),

$$s = 4a\left[1-\cos\frac{\psi}{3}\right],$$

which is the required intrinsic equation.

38. Find the intrinsic equation of the curve $r = a(1-\cos\theta)$ and hence or otherwise prove that $s^2+9\rho^2=16a^2$ where ρ is the radius of curvature of the curve at any point and s is the length of the arc up to this point measured from the point where $\theta=\pi$.

Here $\dfrac{dr}{d\theta} = a\sin\theta.$

If s be measured from $\theta=\pi$ then

$$s = \int_{\pi}^{\theta}\sqrt{r^2+\left(\frac{dr}{d\theta}\right)^2}\,d\theta$$

$$= \int_{\pi}^{\theta}\sqrt{a^2(1-\cos\theta)^2+a^2\sin^2\theta}\;d\theta$$

$$= \int_{\pi}^{\theta}\sqrt{a^2\left(1-2\cos\theta+\cos^2\theta\right)+a^2\sin^2\theta}\;d\theta$$

$$= \int_{\pi}^{\theta}\sqrt{2a^2(1-\cos\theta)}\;d\theta = 2a\int_{\pi}^{\theta}\sin\frac{\theta}{2}\,d\theta$$

$$= 2a\cdot 2\left[-\cos\frac{\theta}{2}\right]_{\pi}^{\theta} = 4a\left(-\cos\frac{\theta}{2}\right) \qquad \text{.....(1)}$$

Also, $\tan\varphi = r\dfrac{d\theta}{dr} = a(1-\cos\theta)\cdot\dfrac{1}{a\sin\theta}$

$$= \frac{2\sin^2\frac{\theta}{2}}{2\sin\frac{\theta}{2}\cos\frac{\theta}{2}} = \tan\frac{\theta}{2};$$

$$\therefore \qquad \varphi = \frac{\theta}{2} \qquad \text{.....(2)}$$

Also $\psi = \theta + \varphi$ .....(3)

Eliminating φ from (2) and (3),

$$\psi = \theta + \frac{\theta}{2} = \frac{3}{2}\theta; \;\therefore\; \theta = \frac{2}{3}\psi$$

$\therefore$ from (1), $$s = 4a \cdot \left(-\cos\frac{1}{3}\psi\right) \quad .....(4)$$

or $$s + 4a\cos\frac{1}{3}\psi = 0,$$

which is the intrinsic equation of the curve.

Differentiation of (4) gives

$$\frac{ds}{d\psi} = \frac{4a}{3}\sin\frac{1}{3}\psi, \text{ i.e., } \rho = \frac{4a}{3}\sin\frac{1}{3}\psi$$

or $$3\rho = 4a\sin\frac{1}{3}\psi \quad .....(5)$$

Squaring and adding (4) and (5),

$$s^2 + 9\rho^2 = 16a^2.$$

39. Obtain the intrinsic equation of the cycloid $x = a(\theta + \sin\theta), y = a(1-\cos\theta)$, the fixed point being the origin.

Here x-axis is the tangent at origin, the fixed point. Let $P(\theta)$ be a variable point on the cycloid.

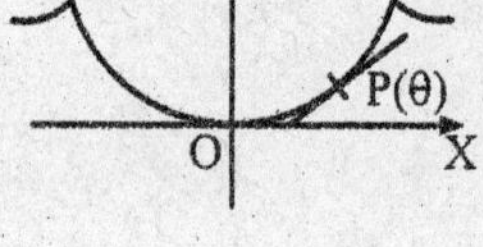

We have, $\frac{dx}{d\theta} = a(1+\cos\theta)$

and $$\frac{dy}{d\theta} = a\sin\theta.$$

$\therefore$ $$\tan\psi = \frac{dy}{dx} = \frac{\sin\theta}{1+\cos\theta}$$

$$= \frac{2\sin\frac{\theta}{2}\cos\frac{\theta}{2}}{2\cos^2\frac{\theta}{2}} = \tan\frac{\theta}{2};$$

$\therefore$ $$\psi = \frac{\theta}{2} \quad .....(1)$$

Also, $$s=\int_0^{\theta}\sqrt{\left(\frac{dx}{d\theta}\right)^2+\left(\frac{dy}{d\theta}\right)^2}\,d\theta=\int_0^{\theta}\sqrt{a^2(1+\cos\theta)^2+a^2\sin^2\theta}$$

$$=a\int_0^{\theta}\sqrt{1+\cos^2\theta+2\cos\theta+\sin^2\theta}=2a\int_0^{\theta}\sqrt{1+\cos\theta}\,d\theta$$

$$=2a\int_0^{\theta}\cos\frac{\theta}{2}\,d\theta=4a\left[\sin\frac{\theta}{2}\right]_0^{\theta}=4a\sin\frac{\theta}{2} \qquad .....(2)$$

From (1) and (2), eliminating θ, we get

$$s=4a\sin\psi \qquad .....(3)$$

as the required intrinsic equation of the cycloid.

Further, $$\rho=\frac{ds}{d\psi}=4a\cos\psi \qquad .....(4)$$

$$\therefore \quad \rho^2+s^2=16a^2\cos^2\psi+16a^2\sin^2\psi=16a^2$$

[from (3) and (4)].

40. Find the intrinsic equation of the evaluate of the circle, viz.,

$$\theta=\frac{\sqrt{r^2-a^2}}{a}-\cos^{-1}\frac{a}{r}$$

Taking (a, 0) as fixed point.

We have here

$$s=\int_a^r\sqrt{1+r^2\left(\frac{d\theta}{dr}\right)^2}\,dr \qquad .....(1)$$

Here $$\theta=\frac{\sqrt{r^2-a^2}}{a}-\cos^{-1}\frac{a}{r} \qquad .....(2)$$

$$\therefore \quad \frac{d\theta}{dr}=\frac{1}{2}\frac{2r}{a\sqrt{r^2-a^2}}+\frac{1}{\sqrt{1-(a/r)^2}}(-a/r^2)$$

$$=\frac{r}{a\sqrt{r^2-a^2}}-\frac{1}{\sqrt{r^2-a^2}}\cdot\frac{a}{r}=\frac{1}{\sqrt{r^2-a^2}}\cdot\frac{r^2-a^2}{r\cdot a}=\frac{\sqrt{r^2-a^2}}{ar}$$

$$\therefore \quad \tan\varphi=r\frac{d\theta}{dr}=r\cdot\frac{\sqrt{r^2-a^2}}{ar}=\frac{\sqrt{r^2-a^2}}{a} \qquad .....(3)$$

$$\therefore \quad \cos\varphi=\frac{a}{r}; \qquad \therefore \varphi=\cos^{-1}\frac{a}{r} \qquad .....(4)$$

So, with the help of (3) and (4), (2) can be written as

$$\theta = \tan\varphi - \varphi$$

or $\quad \tan\varphi = \theta + \varphi = \psi \qquad (\because \theta + \varphi = \psi)$

$$\therefore \text{ from (1), } s = \int_a^r \sqrt{1 + r^2 \cdot \frac{r^2 - a^2}{a^2 r^2}}\, dr$$

$$= \int_a^r \sqrt{1 + \frac{r^2 - a^2}{a^2}}\, dr = \int_a^r \frac{r}{a} dr$$

$$= \frac{1}{2a}(r^2 - a^2) = \frac{1}{2a}(a^2 \tan^2\varphi) \qquad \{\text{from (3)}\}$$

$$= \frac{a}{2}\psi^2 \qquad (\because \tan\varphi = \psi),$$

Which is the required intrinsic equation.

41. Show that the arc of the curve $p^2 (a^4 + r^4) = a^4 r^2$ between limits $r = b$, $r = c$ is equal in length to the arc of the hyperbola $xy = a^2$ between the limits $x = b$, $x = c$.

We have $s = \int \frac{r\, dr}{\sqrt{r^2 - p^2}}$.

Hence, the length of the curve between $r = b$ and $r = c$

$$= \int_b^c \frac{r\, dr}{\sqrt{r^2 - \dfrac{a^4 r^2}{a^4 + r^4}}} \qquad [\because p^2 (a^4 + r^4) = a^4 r^2]$$

$$= \int_b^c \frac{\sqrt{a^4 + r^4}}{r^2}\, dr \qquad \text{.....(1)}$$

Again for the hyperbola $xy = a^2$, $y = \frac{a^2}{x^4}$

$$\therefore \quad \frac{dy}{dx} = -\frac{a^2}{x^2}; \qquad \therefore \quad 1 + \left(\frac{dy}{dx}\right)^2 = 1 + \frac{a^4}{x^4} = \frac{a^4 + x^4}{x^4}$$

$$\therefore \quad \sqrt{1 + \left(\frac{dy}{dx}\right)^2} = \frac{\sqrt{a^4 + x^4}}{x^2}$$

$\therefore$ its length between $x = b$, $x = c$

$$= \int_b^c \frac{\sqrt{a^4 + x^4}}{x^2} dx \qquad \text{.....(2)}$$

From (1) and (2), the two lengths are equal.

42. Find the intrinsic equation of $p = r \sin \alpha$.

For pedal curves, $s = \int_0^r \frac{r\,dr}{\sqrt{r^2 - p^2}} = \int_0^r \frac{r\,dr}{\sqrt{r^2 - r^2 \sin^2 \alpha}}$

$$= \int_0^r \sin\alpha\, dr = r \sec\alpha \qquad .....(1)$$

Again $\frac{dp}{dr} = \sin\alpha$

$\therefore \quad \frac{ds}{d\psi} = r\frac{dr}{dp} = \frac{r}{\sin\alpha} = \frac{s}{\sec\alpha} \cdot \frac{1}{\sin\alpha}$ [using (1)]

or $\quad \frac{1}{s}\frac{ds}{d\psi} = \cot\alpha;$ or $\int \frac{ds}{s} = \int \cot\alpha\, d\psi$

or $\quad \log s = \psi \cot\alpha + \log c$, where $\log c$ is constant.

or $\quad \log\frac{s}{c} = \psi\cot\alpha$ or $s = ce^{\psi \cot \alpha}$,

which is the intrinsic equation of the curve.

Exercise 9.1

1. Find the length of the arc of the curve $y = \log \sec x$ from $x = 0$ to $x = \frac{\pi}{3}$.

[*Hint*: $\frac{dy}{dx} = \tan x$; required length $= \int_0^{\pi/3} \sec x\, dx$]

2. Find the arc length of the curve $y = \frac{1}{2}x^2 - \frac{1}{4}\log x$ from $x = 1$ to $x = 2$.

[*Hint*: $\frac{dy}{dx} = \frac{4x^2 - 1}{4x}$; required length $= \int_1^2 \frac{4x^2+1}{4x}dx = \int_1^2 \left(x + \frac{1}{4x}\right)dx$]

3. Find the perimeter of the loop of the curve $3ay^2 = x^2(a-x)$.

[*Hint*: see example 10. $\frac{dy}{dx} = \frac{1}{\sqrt{3a}}\left[x \cdot \frac{-1}{2\sqrt{a-x}} + \sqrt{a-x}\right]$

$$= \frac{2a - 3x}{2\sqrt{3a}\sqrt{a-x}}$$

$$\text{Length} = \int_0^a \sqrt{1+\left(\frac{dy}{dx}\right)^2}$$

$$= \frac{1}{\sqrt{3a}} \int_0^a \frac{4a-3x}{\sqrt{a-x}} dx = \frac{1}{\sqrt{3a}} \int_0^a \left[\frac{3(a-x)}{\sqrt{a-x}} + \frac{a}{\sqrt{a-x}}\right] dx$$

$$= \frac{1}{\sqrt{3a}} \left[-3\frac{2}{3}(a-x)^{3/2} - a\cdot 2(a-x)^{1/2}\right]_0^a \cdot]$$

4. Find the whole length of the curve given by parametric equations $x = a\cos^3 t$, $y = a\sin^3 t$.

[*Hint*: This is parametric form of example 16 or do like this:

$$\frac{dx}{dt} = -3a\cos^2 t \sin t, \frac{dy}{dt} = 3a\sin^3 \cos t\,;$$

$$\text{length} = 4\int_0^{\pi/2} \sqrt{\left(\frac{dx}{dt}\right)^2 + \left(\frac{dy}{dt}\right)^2}\, dx$$

$$= 4\int_0^{\pi/2} 3a\sin t\cos t\sqrt{\sin^2 t+\cos^2 t}\; dt$$

$$= 12a\int_0^{\pi/2} \sin t\cos t\; dt = 6a\int_0^{\pi/2} \sin^2 dt]$$

5. Find the length of the loop of the curve $x = t^2, y = t - \frac{1}{3}t^2$.

[*Hint*: Length $= 2\int_0^{\sqrt{3}} \sqrt{\left(\frac{dx}{dt}\right)^2 + \left(\frac{dy}{dt}\right)^2}\, dt = 2\int_0^{\sqrt{3}} \sqrt{(2t)^2 + (1-t^2)dt}$]

6. Find the length of the arc of the curve $x = e^t \sin t$, $y = e^t \cos t$ from $t = 0$ to $t = \pi/2$.

[*Hint*: Length $= \int_0^{\pi/2} \sqrt{\left(\frac{dx}{dt}\right)^2 + \left(\frac{dy}{dt}\right)^2}\, dt$

$$= \int_0^{\pi/2} e^t \sqrt{(\sin t+\cos t)^2 + (\cos t - \sin t)^2}\; dt$$

$$= \int_0^{\pi/2} \sqrt{2}e^t dt]$$

7. Find the perimeter of the curve $r = a(1 + \cos\theta)$ and show that arc of the upper half is bisected by $\theta = \frac{1}{3}\pi$.

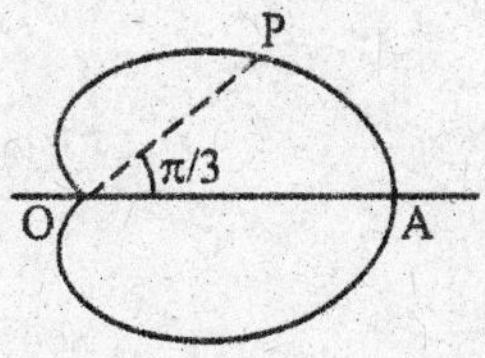

[*Hint*: See Example 24.

Here length of upper half $APO = 2a\int_0^{\pi} \cos\frac{\theta}{2}d\theta = 4a$. Length up to $\theta = \pi/3 = 2a\int_0^{\pi/3} \cos\frac{\theta}{2}d\theta = 2a$ = half the length of upper half].

8. Prove that the line $4r\cos\theta = 3a$ divides the cardiod $r = a(1 + \cos\theta)$ into two parts such that lengths of the arc on either side of the line are equal.

[*Hint*: For the point of intersection of the line and the cardiode we have (on eliminating r),

$$a(1+\cos\theta) = \frac{3a}{4\cos\theta}$$

or $$4\cos\theta + 4\cos^2\theta = 3 \text{ or } 4\cos^2\theta + 4\cos\theta - 3 = 0$$

or $$4\cos^2\theta + 6\cos\theta - 2\cos\theta - 3 = 0$$

or $$2\cos\theta\,(2\cos\theta + 3) - 1\,(2\cos\theta + 3) = 0$$

or $$(2\cos\theta - 1)(2\cos\theta + 3) = 0, \quad \therefore \cos\theta = \frac{1}{2} \text{ or } -\frac{3}{2}$$

i.e., $\theta = \pi/3$, the other value being impossible. So the line and the curve cut where $\theta = \pi/3$.

Now it becomes the above problem.]

9. Find the length of the arc of the parabola $\frac{l}{r} = 1 + \cos\theta$ cut off by its latus rectum.

[*Hint*: See example 27.]

10. Show that the intrinsic equation of $ay^2 = x^3$ taking its cusp as fixed point is $27s = 8a(\sec^3\psi - 1)$.

[*Hint*: Proceed just as example 33.]

Answers

1. $\log(2+\sqrt{3})$ 2. $\frac{3}{2}+\frac{1}{4}\log 2$ 3. $\frac{4a}{\sqrt{3}}$ 4. 6a

5. $4\sqrt{3}$ 6. $\sqrt{2}\left(e^{\pi/2}-1\right)$ 7. 8a 9. $l\left[\sqrt{2}+\log\left(\sqrt{2}+1\right)\right]$

CHAPTER 10

Volumes and Surfaces of Solids of Revolution

When a plane curve is revolved about a straight line to be called the axis of revolution in the plane of the curve, the body so generated is known as a solid of revolution.

The surface generated by the area of the curve is called the surface of revolution. The volume enclosed by the surface of revolution is known as the volume of the revolution.

For example- When a semicircle revolves about its diameter, a sphere is generated. The area of the semicircle generates the volume of the sphere and the arc covering the circumference of the semicircle generates the curved surface of the sphere i.e., surface of revolution.

Volume of a Solid of Revolution

Theorem: Show that the volume obtained by revolving about x-axis the arc of the curve y = f(x), intercepted between the points whose abscissae are a, b is

$$\int_a^b \pi y^2 dx = \pi\int_a^b [f(x)]^2 dx$$

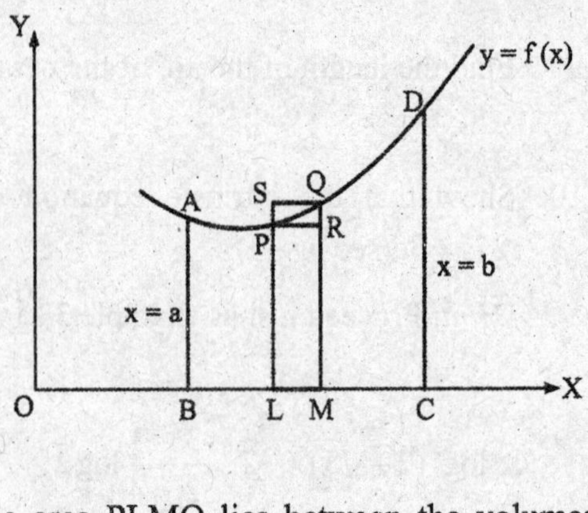

***Proof*:**

Let P(x, y) and Q(x + δx, y + δy) be any two neighbouring points on the curve y = f(x).

A solid is formed when the area ABCD is revolved about x-axis, where for A, x = a and for D, x = b.

Now, volume δV obtained by revolving the area PLMQ lies between the volumes obtained by revolving the two rectangles PLMR and SLMQ about x-axis, i.e., $\pi y^2 \delta x < \delta V < \pi(y+\delta y)^2 \delta y$

or $\pi y^2 < \dfrac{\delta V}{\delta x} < \pi(y+\delta y)^2$

$\therefore$ $\pi y^2 < \underset{\delta x \to 0}{\text{Lt}} \dfrac{\delta V}{\delta x} < \underset{\delta y \to 0}{\text{Lt}} \pi(y+\delta y)^2$

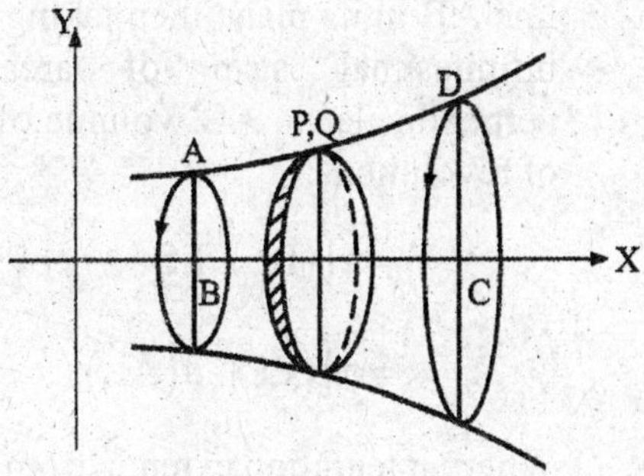

i.e., $\pi y^2 < \dfrac{dV}{dx} < \pi y^2$

$\therefore$ $\dfrac{dV}{dx} = \pi y^2;$ or $dV = \pi y^2\, dx;$

$\therefore$ $V = \int \pi y^2 dx$

Since here y = f(x) and limits of x are a and b

$\therefore$ required volume

$$V = \int_a^b \pi y^2 dx = \pi \int_a^b [f(x)]^2 dx$$

Note:

(i) If the ordinates of the points on the curve be decreasing, we can similarly find the volume.

(ii) If the curve x = f(y) revolves round y-axis between ordinates y = a, y = b then volume $= \int_a^b \pi x^2 dy$

(iii) If the curve be expressed by x = f(t), y = φ(t) then

$$V = \int_a^b \pi y^2\, dx = \pi \int_{t_1}^{t_2} [\varphi(t)]^2 f'(t) dt$$

where t_1, t_2 are values of t corresponding to x = a, x = b respectively.

(iv) If the equation of the curve be in polar form, the axis of revolution being initial line, and the portion of the volume be bounded by parallel planes normal to the initial line, we may still use the Cartesian formula, substituting x = r cos θ, y = r sin θ, taking the initital line as x-axis. So,

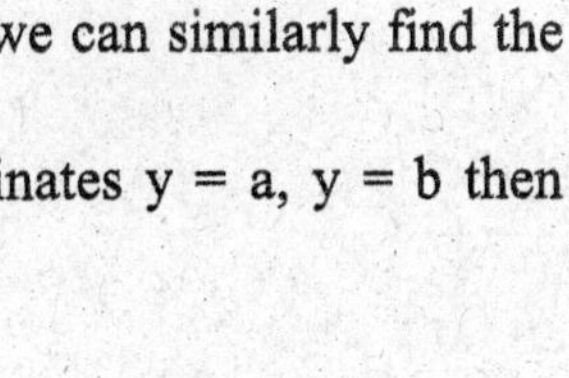

$$V = \pi \int_{x_1}^{x_2} y^2\, dx = \pi \int_{\theta_1}^{\theta_2} r^2 \sin^2\theta\, d(r\cos\theta)$$

A convenient formula for the volume of the solid generated by the revolution of the area bounded by the curve r = f(θ), $\theta = \theta_1$, $\theta = \theta_2$ about the initial line is

$$V = \frac{2}{3}\pi \int_{\theta_1}^{\theta_2} r^3 \sin\theta\, d\theta$$

(v) Volume when the axis of revolution is any line in the plane.

If the curve CD be revolved about any line AB in its plane then taking a general infinitesimal strip of area PQSL, perpendicular to AB, volume of the solid of revolution.

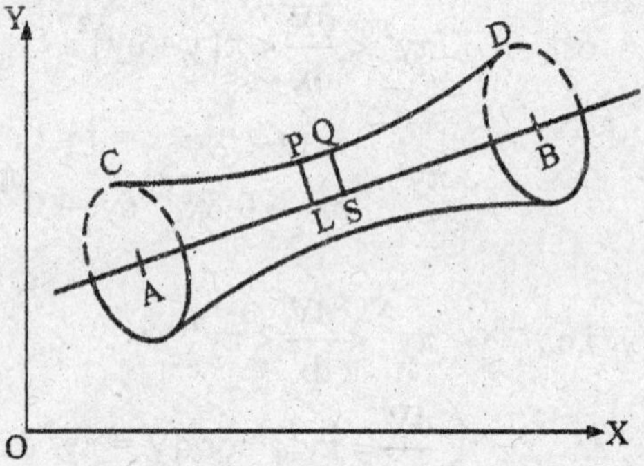

$$V = \pi \int (PL)^2 \, LS$$

$$= \pi \int (PL)^2 \, d(AL),$$

Where integration extends from A to B.

Surface of Revolution

When an arc of a plane curve is revolved about an axis in its plane, a surface of revolution is generated.

The area of the surface of the solid obtained by revolving about x-axis, the arc of the curve y = f(x) intercepted between the points whose abscissae are a, b is

$$S = \int_a^b 2\pi y \, ds \quad \text{or} \quad S = \int_a^b 2\pi y \frac{ds}{dx} dx,$$

i.e.,

$$2\pi \int_a^b f(x) \sqrt{1 + \left(\frac{dy}{dx}\right)^2} \, dx$$

If the axis of revolution be y-axis and we consider the portion of the surface bounded by $y = y_1$ and $y = y_2$ then

$$S = \int 2\pi x \, ds \quad \text{or} \quad S = 2\pi \int_{y_1}^{y_2} x \sqrt{1 + \left(\frac{dx}{dy}\right)^2} \, dy$$

In case of polar curve, we substitute

$$x = r\cos\theta,\ y = r\sin\theta,\ ds = \sqrt{r^2 + \left(\frac{dr}{d\theta}\right)^2} \, d\theta \ \text{ in } S = 2\pi \int y \, ds$$

In case of the curve x = f(t), y = φ(t) (parametric form)

$$S = \int 2\pi y \, ds = 2\pi \int_{t_1}^{t_2} y \sqrt{\left(\frac{dx}{dt}\right)^2 + \left(\frac{dy}{dt}\right)^2} \, dt,$$

where

$$ds = \sqrt{\left(\frac{dx}{dt}\right)^2 + \left(\frac{dy}{dt}\right)^2} \, dt$$

Remember

1. Volume = $\int \pi y^2 dx$ or $\int \pi x^2 dy$, when cartesian curve revolves about x-axis or y-axis. Put $x = f(t)$, $y = \varphi(t)$ in the above formulae if the curve is in parametric form. Put $x = r \cos\theta$, $y = r \sin\theta$ in (1), if the curve is in polar form revolving about initial line. $V = \frac{2}{3}\pi\int r^2 \sin\theta \, d\theta$ is also a convenient formula.

2. Volume when axis of revolution is any line

$$= \pi\int (PL)^2 \, d(AL)$$

3. Surface area = $\int 2\pi y \, ds$ or $\int 2\pi x \, ds$ where symbols have their usual meanings and range of integration to be adjusted according to the problem.

Solved Examples

1. Show that the volume of a right circular cone of height h and base of radius a is $\frac{1}{3}\pi a^2 h$.

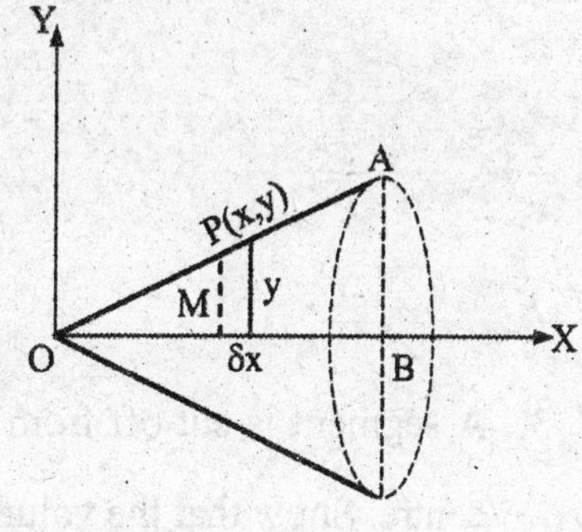

Let OB = h, the height of the cone; AB = a, radius of the circular base.

The cone is the solid generated by the revolution of the line OA about the line OB.

Consider an elementary strip of breadth δx parallel to the axis of y. Volume generated by revolving it about OB = $\pi y^2 \, \delta x$.

$\therefore$ the required volume $= \pi\int_0^h y^2 dx$

From similar triangles OPM and OAB, $\frac{PM}{AB} = \frac{OM}{OB}$

Or $\quad PM = \frac{OM}{OB}.AB = \frac{x}{h}a$; i.e., $y = \frac{ax}{h}$

$$\therefore \quad \text{volume} = \pi\int_0^h \frac{a^2}{h^2}x^2 dx = \pi\frac{a^2}{h^2}\left[\frac{x^3}{3}\right]_0^h$$

$$= \pi\frac{a^2}{3h^2}h^3 = \frac{1}{3}\pi a^2 h$$

2. Find the surface of a hemisphere of radius a.

A hemisphere is generated by the revolution of a quadrant of a circle $x^2 + y^2 = a^2$ about the x-axis.

$$\because \quad x^2 + y^2 = a^2$$

$$\therefore \quad y = \sqrt{a^2 - x^2}$$

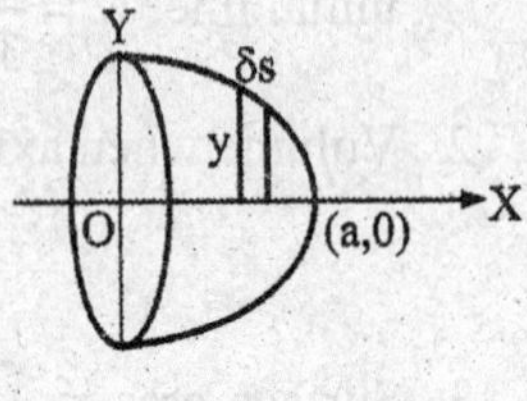

$$\therefore \quad \frac{dy}{dx} = -\frac{x}{\sqrt{a^2 - x^2}}$$

$$\therefore \quad \frac{ds}{dx} = \sqrt{1 + \left(\frac{dy}{dx}\right)^2}$$

$$= \sqrt{1 + \frac{x}{a^2 - x^2}} = \frac{a}{\sqrt{a^2 - x^2}}$$

If δs be an elementary arc at a distance y from the x-axis, parallel to the y-axis, surface of the hemisphere

$$= 2\pi \int y \, ds$$

$$= 2\pi \int_0^a \sqrt{a^2 - x^2} \cdot \frac{a}{\sqrt{a^2 - x^2}} dx = 2\pi a \int_0^a dx = 2\pi a^2$$

$$[= \frac{1}{2} \text{ (surface of a sphere)}].$$

3. A segment is cut off from a sphere of radius a by a plane at a distance $\frac{a}{2}$ from the centre. Show that the volume of the segment is 5/32 of the volume of the sphere.

Hence OC = $\frac{a}{2}$. The required volume is the volume generated by the revolution of the arc AB of the circle $x^2 + y^2 = a^2$ about the x-axis.

Volume of the segment

$$= \int_{\pi/2}^{a} \pi y^2 dx$$

$$= \pi \int_{\pi/2}^{a} \left(a^2 - x^2\right) dx = \pi \left[a^2 x - \frac{x^2}{3}\right]_{\pi/2}^{a}$$

$$= \pi\left[\frac{2a^3}{3} - \left(\frac{a^3}{3} - \frac{a^3}{3\times 8}\right)\right]$$

$$= \pi\left[\frac{2a^3}{3} - \frac{8a^3+3a^3}{24}\right] = \pi\left[\left(\frac{2a^3}{3} - \frac{11a^3}{24}\right)\right]$$

$$= \frac{5}{24}\pi a^3$$

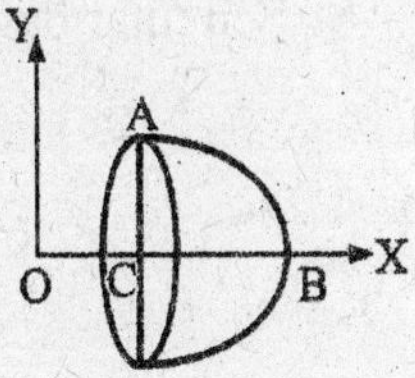

Also volume of the sphere $= \frac{4}{3}\pi a^3$

$\therefore$ volume of the segment : volume of the sphere

$$= \frac{5}{24}\pi a^3 : \frac{4}{3}\pi a^3 = \frac{5}{32}$$

$\therefore$ volume of the segment $= \frac{5}{32}$ volume of the sphere

4. The circle $x^2 + y^2 = a^2$ revolves round the x-axis. Find the surface and the volume of the solid generated (i.e., volume and surface area of a sphere of radius a)

Clearly, the solid generated is a sphere whose centre is the centre of the circle and whose radius is a.

Now, $$V = \pi\int_{-a}^{a} y^2 dx = \pi\int_{-a}^{a}\left(a^2 - x^2\right)dx$$

$$= \pi\left[a^2x - \frac{x^3}{3}\right]_{-a}^{a} = \pi\left[a^3 + a^3 - \frac{2a^3}{3}\right] = \frac{4}{3}\pi a^2$$

Surface area $= 2\pi\int y\, ds$

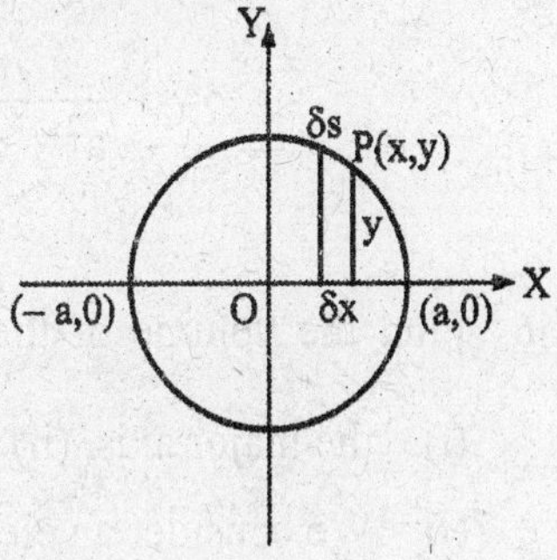

$$= 2\pi\int_{-a}^{a}\sqrt{a^2 - x^2}\sqrt{1+\left(\frac{dy}{dx}\right)^2}\,dx$$

$$= 2\pi\int_{-a}^{a}\sqrt{a^2 - x^2}\sqrt{1+\frac{x^2}{y^2}}\,dx$$

$$= 2\pi\int_{-a}^{a}\sqrt{a^2 - x^2}\,\frac{a}{\sqrt{a^2 - x^2}}\,dx$$

$$(\because x^2 + y^2 = a^2)$$

$$= 2\pi[x]_{-a}^{a} = 4\pi a^2$$

5. The intercept $\frac{x}{a}+\frac{y}{b}=1$ between the axes of coordinates revolves about y-axis. Find the area of the surface generated.

When the intercept of the given line, say AB, between the axes of coordinates revolves about the y-axis, a right circular cone of height OB = b and radius of the circular base a, is formed.

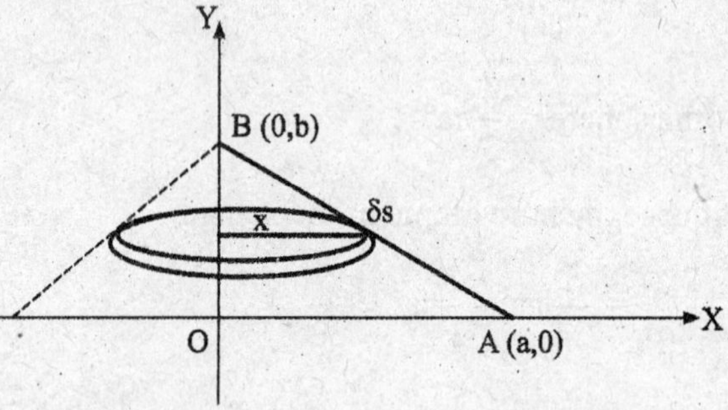

The surface area

$$S=\int 2\pi x\, ds=2\pi\int_0^b x\sqrt{1+\left(\frac{dx}{dy}\right)^2}\,dy$$

Here $\frac{x}{a}+\frac{y}{b}=1,\quad \therefore\ \frac{dx}{dy}=-\frac{a}{b},\quad 1+\left(\frac{dx}{dy}\right)^2=1+\frac{a^2}{b^2}=\frac{a^2+b^2}{b^2},$

$$\therefore\quad S=2\pi\int_0^b a\left(1-\frac{y}{b}\right)\sqrt{\frac{a^2+b^2}{b^2}}\,dy=2\pi a\sqrt{\frac{a^2+b^2}{b^2}}\left[y-\frac{y^2}{2b}\right]_0^b$$

$$=\frac{2\pi a}{b}\sqrt{a^2+b^2}\left(b-\frac{b^2}{2b}\right)=\frac{2\pi a}{b}\sqrt{a^2+b^2}\cdot\frac{b}{2}-\pi a\sqrt{a^2+b^2}.$$

6. Find the volume of the solid obtained by revolving the ellipse $\frac{x^2}{a^2}+\frac{y^2}{b^2}=1$ about

(i) the major axis, (ii) the minor axis.

(i) We consider an elementary strip parallel to the axis of y.

Volume generated when it is revolved about x-axis $=\pi y^2\delta x$

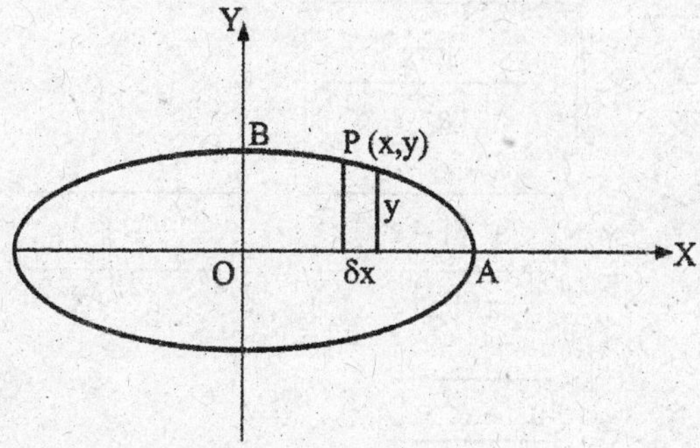

$\therefore$ required volume = 2 × volume obtained by the rotation of the quadrant OAB

$$= 2\int_0^a \pi y^2 dx$$

$$= 2\pi\int_0^a \frac{b^2}{a^2}\left(a^2 - x^2\right)dx = 2\pi\frac{b^2}{a^2}\left[a^2 x - \frac{x^3}{3}\right]_0^a = \frac{4}{3}\pi ab^2$$

(ii) In this case, the volume generated by the revolution about y-axis of the elementary strip parallel to x-axis = $\pi x^2\, \delta y$.

$\therefore$ the required volume

$$= 2\times\int_0^b \pi x^2 dy$$

$$= 2\pi\frac{a^2}{b^2}\int_0^b \left(b^2 - y^2\right)dy$$

$$= 2\pi\frac{a^2}{b^2}\left[b^2 y - \frac{y^3}{3}\right]_0^b = \frac{4}{3}\pi a^2 b.$$

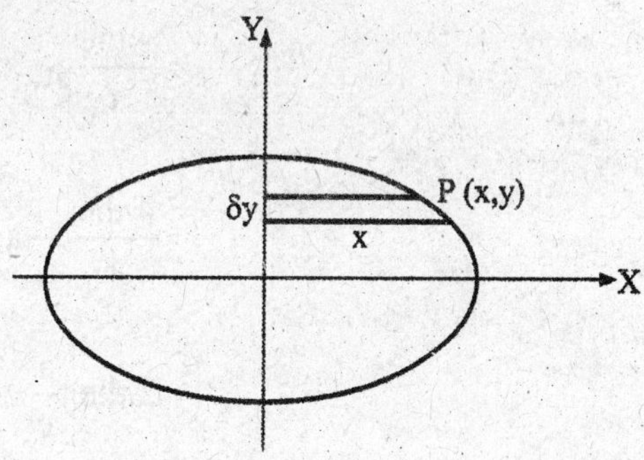

7. Find the area of the surface formed by the revolution of $x^2 + 4y^2 = 16$ about its major axis.

If the ellipse were

$$\frac{x^2}{a^2} + \frac{y^2}{b^2} = 1,$$

where $b^2 = a^2(1 - e^2)$ then

$$y = \frac{b}{a}\sqrt{a^2 - x^2},\quad \frac{dy}{dx} = -\frac{bx}{a\sqrt{a^2 - x^2}}$$

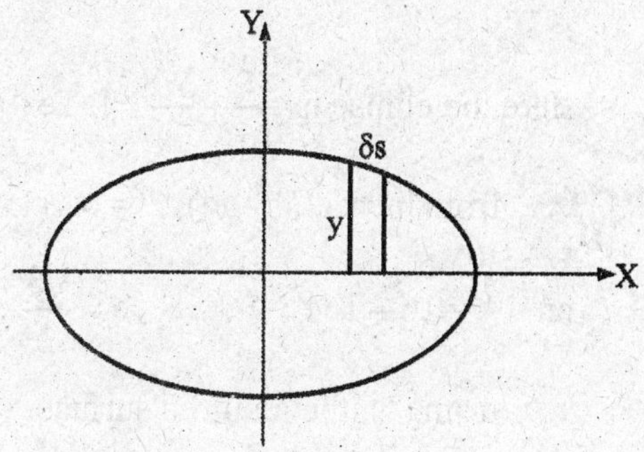

$$\frac{ds}{dx} = \sqrt{1+\left(\frac{dy}{dx}\right)^2} = \sqrt{\left[1+\frac{b^2x^2}{a^2\left(a^2-x^2\right)}\right]}$$

$$= \sqrt{1+\frac{a^2\left(1-e^2\right)x^2}{a^2\left(a^2-x^2\right)}} = \sqrt{1+\frac{\left(1-e^2\right)x^2}{a^2-x^2}}$$

$$= \sqrt{\frac{a^2-e^2x^2}{a^2-x^2}}$$

the required surface $= 2\int 2\pi y\, ds$

$$= 4\pi\int_0^x \frac{b}{a}\sqrt{a^2-x^2}\sqrt{\frac{a^2-e^2x^2}{a^2-x^2}}dx$$

$$= \frac{4\pi b}{a}\int_0^a \sqrt{a^2-e^2x^2}\,dx = \frac{4\pi be}{a}\int_0^a \sqrt{\frac{a^2}{e^2}-x^2}\;dx$$

$$= \frac{4\pi be}{a}\left[\frac{1}{2}x\sqrt{\frac{a^2}{e^2}-x^2}+\frac{1}{2}\frac{a^2}{e^2}\sin^{-1}\frac{ex}{a}\right]_0^a$$

$$= \frac{4\pi be}{a}\left[\frac{1}{2}a\sqrt{\frac{a^2}{e^2}-a^2}+\frac{1}{2}\frac{a}{e^2}\sin^{-1}e\right]$$

$$= 2\pi be\left[\frac{a}{e}\sqrt{1-e^2}+\frac{a}{e^2}\sin^{-1}e\right]$$

$$= 2\pi ab\left[\sqrt{1-e^2}+\frac{1}{e^2}\sin^{-1}e\right] \qquad .....(1)$$

Here the ellipse is $\frac{x^2}{16}+\frac{y^2}{4}=1$, i.e., $a=4, b=2$

$\therefore$ from $b^2 = a^2(1-e^2)$, $2^2 = 4^2(1-e^2)$

or $4 = 16 - 16e^2$; $\therefore$ $e = \frac{\sqrt{3}}{2}$

$\therefore$ from (1), the required surface

$$= 2\pi.4.2\left\{\sqrt{1-\frac{3}{4}}+\frac{2}{\sqrt{3}}\sin^{-1}\frac{\sqrt{3}}{2}\right\} = 16\pi\left[\frac{1}{2}+\frac{2}{\sqrt{3}}.\frac{\pi}{3}\right]$$

$$= 8\pi + \frac{32\pi^2}{3\sqrt{3}} = 8\pi + \frac{32\pi^2\sqrt{3}}{9}.$$

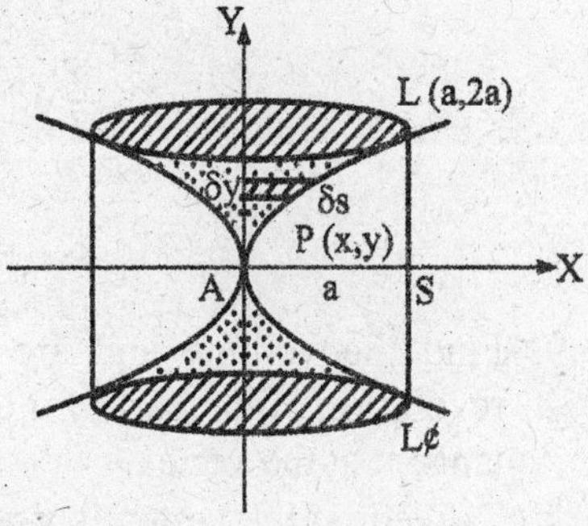

8. Find the volume and the area of the curved surface of the solid by the revolution of the part of the parabola $y^2 = 4ax$ bounded by the latus rectum about the tangent at the vertex.

Here the axis of revolution is the y-axis. The latus rectum is the line $x = a$,

$\therefore\ y^2 = 4a\,.\,a;\qquad \therefore\ y = \pm 2a$

We consider a strip parallel to x-axis. Volume generated by revolving it about y-axis $= \pi x^2\,\delta y$.

$\therefore$ required volume

$$= \pi\int x^2 dy$$

$$= \pi\int_{-2a}^{2a}\frac{y^4}{16a^2}dy$$

$$= \frac{\pi}{16a^2}\left[\frac{y}{5}\right]_{-2a}^{2a} = \frac{\pi}{80a^2}.64a^5 = \frac{4}{5}\pi a^3$$

Again, surface area

$$= 2\pi\int x\,ds = 2\pi\int_{-2a}^{2a} x\sqrt{1+\left(\frac{dx}{dy}\right)^2}\,dy = 2\pi\int_{-2a}^{2a}\frac{y^2}{4a}\sqrt{1+\frac{y^2}{4a^2}}\,dy$$

Putting $y = 2a\tan\theta$, $dy = 2a\sec^2\theta\,d\theta$

When $y = -2a$, $\theta = -\pi/4$; when $y = 2a$, $\theta = \pi/4$

$\therefore$ surface area

$$= 2\pi\int_{-\pi/4}^{\pi/4}\frac{4a^2}{4a}\tan^2\theta\sqrt{1+\frac{4a^2\tan^2\theta}{4a^2}}\,2a\sec^2\theta\,d\theta$$

$$= 4\pi a^2\int_{-\pi/4}^{\pi/4}\tan^2\theta\sec^3\theta\,d\theta = 4\pi a^2\int_{-\pi/4}^{\pi/4}\left(\sec^3\theta - 1\right)\sec^2\theta\,d\theta$$

$$= 4\pi a^2\int_{-\pi/4}^{\pi/4}(\sec^5\theta - \sec^3\theta)\,d\theta,$$

$$\text{using } \int \sec^n dx = \frac{\sec^{n-2} x \tan x}{n-1} + \frac{n-2}{n-1}\int \sec^{n-2} x\, dx$$

$$= 4\pi a^2 \left[\frac{1}{4}\tan\theta\sec^3\theta - \frac{1}{8}\tan\theta\sec\theta - \frac{1}{8}\log\tan\left(\frac{\pi}{4}+\frac{\theta}{2}\right)\right]_{-\pi/4}^{\pi/4}$$

$$= 4\pi a^2\left[\frac{3}{4}\sqrt{2} - \frac{1}{4}\log\cot\frac{\pi}{8}\right]$$

$$= \pi a^2\left[3\sqrt{2} - \log\left(\sqrt{2}+1\right)\right]$$

9. Find the volume and area of the curved surface of the solid generated by the revolution of the part of the parabola $y^2 = 4ax$ bounded by the latus rectum about the tangent at the vertex.

Tangent at the vertex is y-axis. So, LOL' revolves about y-axis. The latus rectum is $x = a$.

$\therefore \quad y^2 = 4a^2, \qquad \therefore \quad y = \pm 2a$

$\therefore$ the required volume

$$= \pi\int x^2 dy = \pi\int_{-2a}^{2a} \frac{y}{16a^2} dy \qquad (\because y^2 = 4ax)$$

$$= \frac{\pi}{80a^2}\left[y^5\right]_{-2a}^{2a} = \frac{\pi.64a^5}{80a^2} = \frac{4}{5}\pi a^3$$

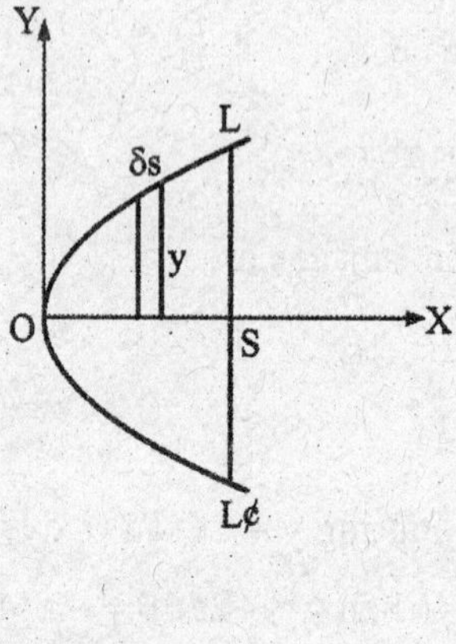

Surface area = $2\pi\int x\, ds$

$$= 2\pi\int_{-2a}^{2a} x\sqrt{1+\left(\frac{dx}{dy}\right)^2}\, dy$$

$$= 2\pi\int_{-2a}^{2a} \frac{y^2}{4a}\sqrt{1+\left(\frac{y}{2a}\right)^2}\, dy$$

Putting $y = 2a\tan\theta$, $dy = 2a\sec^2\theta\, d\theta$

When $y = -2a$, $\theta = -\pi/4$; when $y = 2a$, $\theta = \pi/4$

$$\therefore \quad \text{surface area} = 2\pi\int_{-\pi/4}^{\pi/4} \frac{4a^2\tan^2\theta}{4a}\sqrt{1+\frac{4a^2\tan^2\theta}{4a^2}}.2a\sec^2\theta\, d\theta$$

$$= 4\pi a^2\int_{-\pi/4}^{\pi/4} \tan^2\theta\sec^3\theta\, d\theta$$

$$= 4\pi a^2 \int_{-\pi/4}^{\pi/4} \left(\sec^5\theta - \sec^3\theta \right) d\theta$$

$$= 4\pi a^2 \left\{ \int_{-\pi/4}^{\pi/4} \sec^5\theta d\theta - \int_{-\pi/4}^{\pi/4} \sec^3\theta \, d\theta \right\}$$

$$= 4\pi a^2 \left\{ \left[\frac{\sec^3\theta \tan\theta}{4} \right]_{-\pi/4}^{\pi/4} + \frac{3}{4} \int_{-\pi/4}^{\pi/4} \sec^3\theta \, d\theta - \int_{-\pi/4}^{\pi/4} \sec^3\theta \, d\theta \right\},$$

Using reduction formula

$$\int \sec^n x dx = \frac{\sec^{n-2} x \tan x}{n-1} + \frac{n-2}{n-1} \int \sec^{n-2} x \, dx$$

$$\text{surface area} = 4\pi a^2 \left\{ \left[\frac{\sec^3\theta \tan\theta}{4} \right]_{-\pi/4}^{\pi/4} - \frac{1}{4} \int_{-\pi/4}^{\pi/4} \sec^3\theta \, d\theta \right\}$$

$$= 4\pi a^2 \left\{ \left[\frac{1}{4} \sec^3\theta \tan\theta - \frac{1}{8} \sec\theta \tan\theta \right]_{-\pi/4}^{\pi/4} - \frac{1}{8} \int_{-\pi/4}^{\pi/4} \sec\theta \, d\theta \right\}$$

by the same reduction formula

$$= 4\pi a^2 \left[\frac{1}{4} \sec^3\theta \tan\theta - \frac{1}{8} \sec\theta \tan\theta - \frac{1}{8} \log\tan\left(\frac{\pi}{4} + \frac{\theta}{3} \right) \right]_{-\pi/4}^{\pi/4}$$

$$= 4\pi a^2 \left[\frac{3}{4}\sqrt{2} - \frac{1}{4} \log\cot\frac{\pi}{8} \right]$$

$$= \pi a^2 \left[3\sqrt{2} - \log\left(\sqrt{2} + 1 \right) \right].$$

10. Find the volume formed by the revolution of the loop of the curve $y^2(a + x) = x^2 (a - x)$ about the x-axis.

The curve is symmetrical about x-axis as the equation contains even power of y. The curve passes through the origin and cuts x-axis at (a, 0) since by putting $y = 0$, we get $x = 0$, $x = a$. So, there is a loop between (0, 0) and (a, 0). Equation of the tangent at origin is $y = \pm x$, since by equating the lowest-degree term to zero, $ay^2 - ax^2 = 0$, i.e., $y = \pm x$.

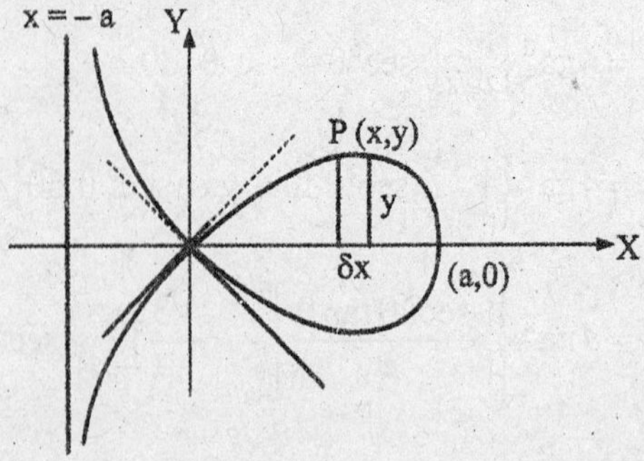

For x > a, y becomes imaginary, so no part of the curve is in the right of (a, 0). Equating to zero the coefficient of highest power of y, x + a = 0, so asymptote parallel to y-axis is x = –a and the curve is as shown in the figure.

∴ the required volume

$$= \int_0^a \pi y^2 dx = \pi\int_0^a \frac{x^2(a-x)}{a+x}dx = \pi\int_0^a \frac{ax^2 - x^2}{a+x}dx$$

$$= \pi\int_0^a \left[-x^2 + 2ax - 2a^2 + \frac{2a^3}{a+x}\right]dx \quad \text{(by actual division)}$$

$$= \pi\left[\frac{1}{3}x^3 + ax^2 - 2a^2x + 2a^3\log(a+x)\right]_0^a = \pi a^3\left[2\log 2 - \frac{4}{3}\right]$$

11. Find the volume formed by the revolution of the loop of the curve $y^2(a-x) = x^2(a+x)$.

The curve is symmetrical about x-axis since even power of y occurs. It passes through (0, 0) and (–a, 0). So, a loop is formed between (0, 0) and (–a, 0). Equating the coefficient of highest power of y, the asymptote is x – a = 0, i.e., x = a. Then, the curve is as shown in the figure

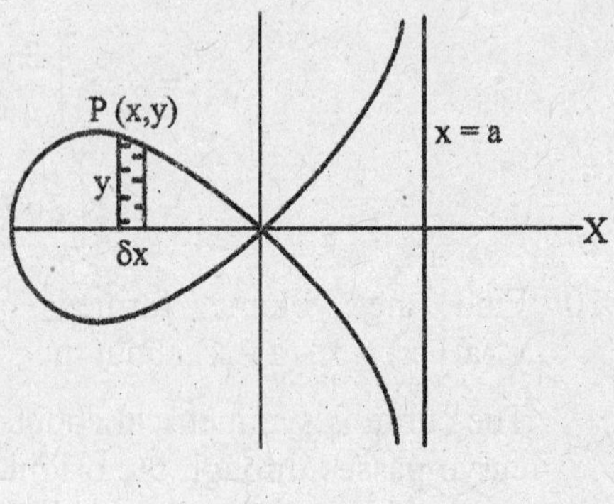

$$\text{The required volume} = \int_{-a}^{0} \pi y^2 dx$$

$$= \pi\int_{-a}^{0} x^2 \frac{a+x}{a-x}dx$$

$$= \pi\int_{-a}^{0}\left(-x^2 - 2xa - 2a^2 + \frac{2a^3}{a-x}\right)dx \quad \text{(by actual division)}$$

$$= \pi\left[-\frac{x^3}{3} - x^2a - 2a^2x - 2a^3\log(a-x)\right]_{-a}^{0}$$

$$= \pi a^3\left(2\log 2 - \frac{4}{3}\right)$$

12. Find the volume of the solid generated by the revolution of the upper half of the loop of the curve $y^2 = x^2(2-x)$ about x-axis.

The curve is symmetrical about x-axis as the equation of the curve remains unchanged by replacing y by – y. When y = 0, x = 0, x = 2. If x > 2, y becomes imaginary. Thus for the loop x = 0 and x = 2.

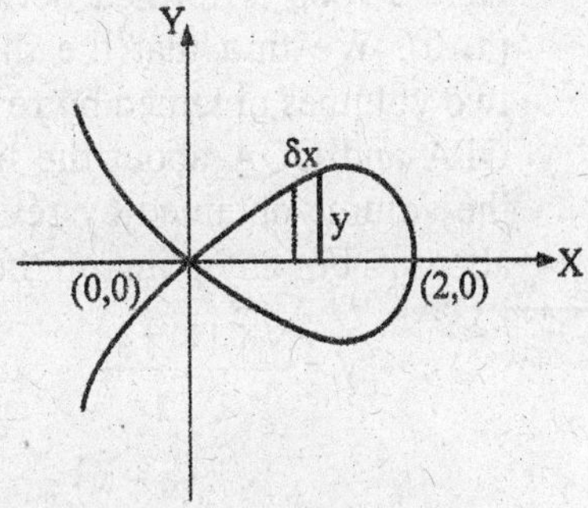

$$\text{the volume} = \int_0^2 \pi y^2 dx$$

$$= \pi\int_0^2 x^2(2-x)dx$$

$$= \pi\int_0^2 (2x^2 - x^3)dx$$

$$= \pi\left[\frac{2x^3}{3} - \frac{x^4}{4}\right]_0^2$$

$$= \pi\left[\frac{16}{3} - \frac{16}{4}\right] = \frac{4}{3}\pi.$$

13. The loop of the curve $2ay^2 = x(x-a)^2$ revolves about x-axis. Find the volume of the solid so generated.

The curve is symmetrical about x-axis as only even powers of y occur. The curve passes through the origin. Putting y = 0, x = 0, x = a. When x is negative y becomes imaginary and so no part of the curve is in the left of O. Thus a loop is formed between (0, 0), (a, 0).

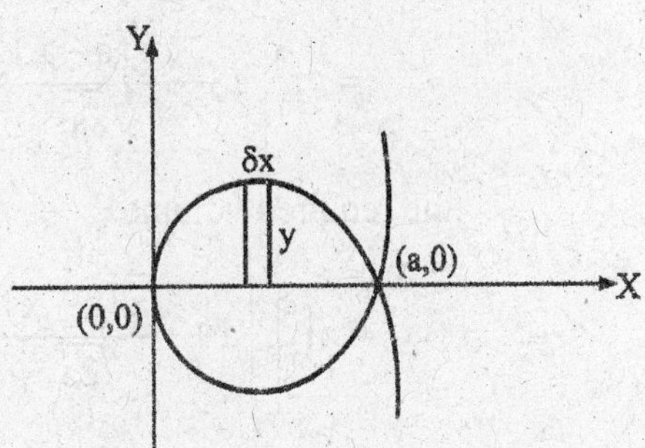

$$\text{The required volume} = \int_0^a \pi\, y^2 dx$$

$$= \pi\int_0^a \frac{x(x-a)^2}{2a}dx$$

$$= \pi\int_0^a \frac{x(x^2 - 2ax + a^2)}{2a}dx$$

$$=\frac{\pi}{2a}\left[\frac{x^4}{4}-\frac{2ax^3}{3}+\frac{a^2x^2}{2}\right]_0^a$$

$$=\frac{\pi}{2a}\left[\frac{a^4}{4}-\frac{2a^4}{3}+\frac{a^4}{2}\right]$$

$$=\frac{\pi}{24a}\left[3a^4-8a^4+6a^4\right]=\frac{\pi}{24}a^3.$$

14. The loop of the curve $2ay^2 = x(x-a)^2$ revolves about the straight line $y = a$; find the volume of the solid generated.

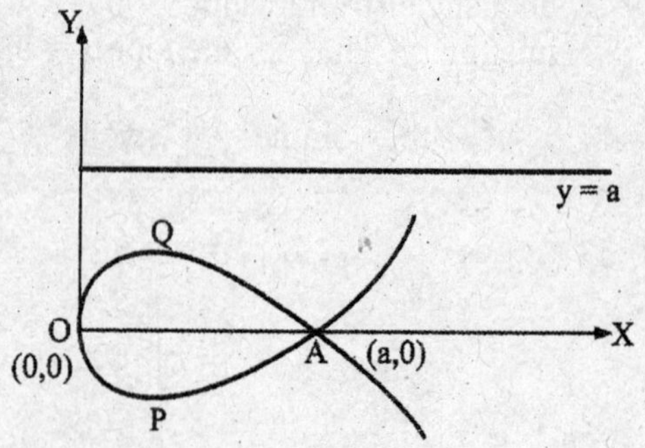

Here a loop is formed between (0, 0) and (a, 0). We find that the difference of the two volumes obtained by revolving the arcs OPA and OQA about the line $y = a$ gives the volume obtained by revolving the loop about it. For any point on the arc OPA.

$$y=\frac{\sqrt{x}.(x-a)}{\sqrt{2a}}$$

$$=-\frac{\sqrt{x}(a-x)}{\sqrt{2a}}$$

So, the volume obtained by revolving the arc OPA

$$=\pi\int_0^a\left[a+\frac{\sqrt{x}(a-x)}{\sqrt{2a}}\right]dx \qquad .....(I)$$

For any point on the arc OQA, $y=-\frac{\sqrt{x}(x-a)}{\sqrt{2a}}=\frac{\sqrt{x}(a-x)}{\sqrt{2a}}$

So, the volume obtained by revolving the arc OQA

$$=\pi\int_0^a\left[a-\frac{\sqrt{x}(a-x)}{\sqrt{2a}}\right]dx$$

$\therefore$ the required volume

$$=\pi\int_0^a\left[\left\{a+\frac{\sqrt{x}(a-x)}{\sqrt{2a}}\right\}^2-\left\{a-\frac{\sqrt{x}(a-x)}{\sqrt{2a}}\right\}^2\right]dx$$

$$= \frac{4a\pi}{\sqrt{2a}} \int_0^a \left[\sqrt{x}(a-x)dx = \frac{4a\pi}{\sqrt{2a}} \left[\frac{ax^{3/2}}{3/2} - \frac{x^{5/2}}{5/2} \right]_0^a \right]$$

$$= \frac{4a\pi}{\sqrt{2a}} \left[\frac{2}{3} a.a^{3/2} - \frac{2}{3} a^{5/2} \right] = \frac{4a\pi}{\sqrt{2}} \cdot \frac{4}{15} a^{\frac{5}{2}-\frac{1}{2}}$$

$$= \frac{8\sqrt{2}\pi a^3}{15}.$$

15. Find the area of the surface of revolution formed by revolving the curve r = 2a cos θ about the initial line.

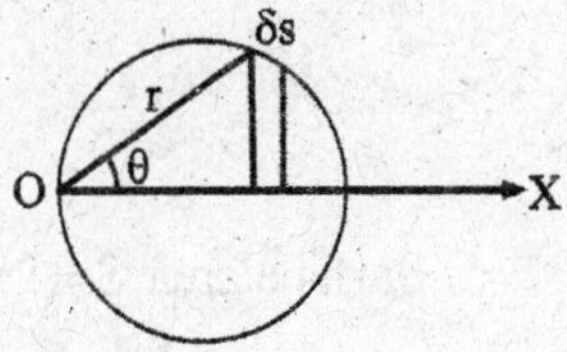

The given equation represents a circle of radius a with an end of the diameter as pole.

The required surface $S = 2\pi \int y \, ds$

For the upper part, θ varies from 0 to $\pi/2$

Here $r = 2a\cos\theta$

$$\therefore \quad \frac{dr}{d\theta} = -2a\sin\theta$$

$$\therefore \quad \frac{ds}{d\theta} = \sqrt{r^2 + \left(\frac{dr}{d\theta}\right)^2} = 2a$$

$$\therefore \quad S = 2\pi \int y \, ds = 2\pi \int_0^{\pi/2} r\sin\theta . 2a \, d\theta$$

$$= 2\pi \int_0^{\pi/2} 2a\cos\theta . \sin\theta \; 2a \, d\theta$$

$$= 4\pi a^2 \int_0^{\pi/2} \sin 2\theta \, d\theta = 4\pi a^2 \left[\frac{-\cos 2\theta}{2} \right]_0^{\pi/2}$$

$$= 4\pi a^2 \left[\frac{1}{2} - \left(-\frac{1}{2} \right) \right] = 4\pi a^2 .$$

16. The cardioid r = a(1 – cos θ) revolves about the initial line. Find the surface area and volume of the figure formed.

The cardioid is shown in the figure. Here limits of θ are from 0 to π.

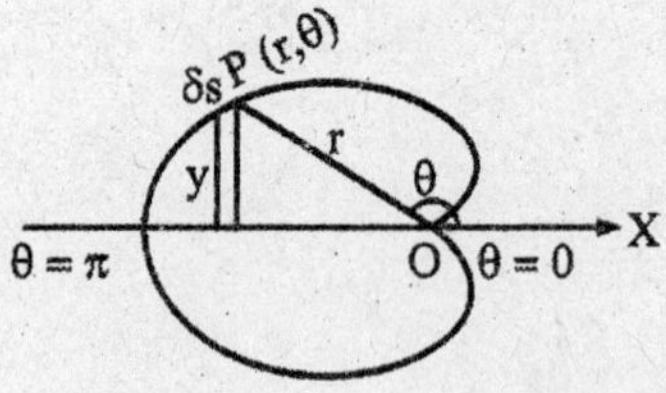

The volume $V = \frac{2}{3}\int_0^{\pi} r^3 \sin\theta \, d\theta$

$$= \frac{2}{3}\pi a^3 \int_0^{\pi} (1-\cos\theta)^2 \sin\theta \, d\theta$$

Putting $1 - \cos\theta = t$, $\sin\theta \, d\theta = dt$

When $\theta = 0, t = 0$; when $\theta = \pi, t = 2$

$$\therefore \quad V = \frac{2}{3}\pi a^3 \int_0^2 t^3 dt = \frac{2}{3}\pi a^3 \left[\frac{t^4}{4}\right]_0^2 = \frac{2}{3}\pi a^3.4 = \frac{8}{3}\pi a^3$$

Surface area $S = 2\pi\int y \, ds = 2\pi\int_0^{\pi} r\sin\theta\sqrt{r^2 + \left(\frac{dr}{d\theta}\right)^2} \, d\theta$

$$\because \quad r = a(1-\cos\theta); \quad \frac{dr}{d\theta} = a\sin\theta$$

$$\therefore \quad S = 2\pi a^2 \int_0^{\pi} \sin\theta(1-\cos\theta)\sqrt{(1-\cos\theta)^2 + \sin^2\theta} \, d\theta$$

Putting $1 - \cos\theta = t$, $\sin\theta \, d\theta = dt$

When $\theta = 0, t = 0$; when $\theta = \pi, t = 2$

$$\therefore \quad S = 2\pi a^2 \int_0^2 t\sqrt{t^2 + (2t - t^2)} dt$$

$$= 2\pi a^2 \int_0^2 t\sqrt{2t} dt$$

$$= 2\pi a^2.2^{\frac{1}{2}} \int_0^2 t^{\frac{3}{2}} dt = 2\pi a^2.2^{\frac{1}{2}} \left[\frac{t^{\frac{5}{2}}}{\frac{5}{2}}\right]_0^2$$

$$= \frac{4\sqrt{2}\pi a^2}{5}\left(2^{\frac{5}{2}}\right) = \frac{4}{5}\pi a^2.2^3 = \frac{32}{5}\pi a^2$$

17. Find the volume and the surface area of the solid generated by revolving the cardioid $r = a(1 + \cos\theta)$ about the initial line.

The cardioids is shown in the figure

The required volume $V = \pi\int_0^{2a} y^2 dx$ where $OA = 2a$

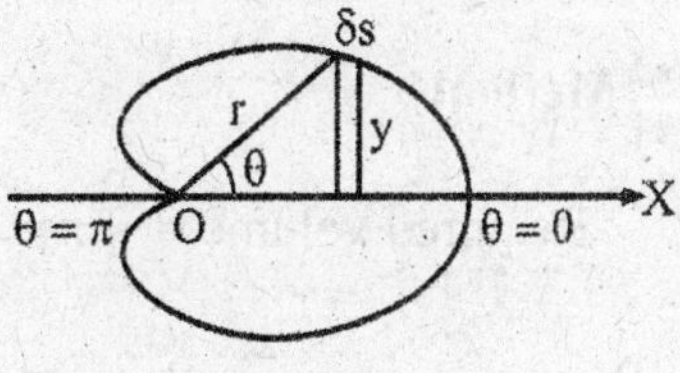

Changing to polar coordinates, we have

$x = r\cos\theta$

$y = r\sin\theta$

$\therefore \quad x = a(1+\cos\theta)\cos\theta$

$\quad\quad y = a(1+\cos\theta)\sin\theta$

$\therefore \quad dx = [-a\sin\theta(1+\cos\theta) - a\sin\theta \,.\, \cos\theta]\, d\theta$

$\quad\quad = -a(\sin\theta + 2\sin\theta\cos\theta)\, d\theta$

Also, when $x = 0,\ \theta = \pi$; when $x = 2a,\ \theta = 0$

$$\therefore \quad V = -\pi\int_\pi^0 a^2\sin^2\theta(1+\cos\theta)^2 .a\sin\theta(1+2\cos\theta)d\theta$$

$$= \pi a^3\int_0^\pi \sin^3\theta(1+\cos\theta)^2(1+2\cos\theta)d\theta$$

$$= \pi a^3\int_0^\pi\left(2\sin\frac{\theta}{2}\cos\frac{\theta}{2}\right)^3\left(2\cos^2\frac{\theta}{2}\right)^2\left\{1+2\left(2\cos^2\frac{\theta}{2}-1\right)^2\right\}d\theta$$

$$= \pi a^3\int_0^\pi 8\sin^3\frac{\theta}{2}\cos^3\frac{\theta}{2}.4\cos^4\frac{\theta}{2}\left(4\cos^2\frac{\theta}{2}-1\right)d\theta$$

$$= 128\pi a^3\int_0^\pi \sin^3\frac{\theta}{2}\cos^9\frac{\theta}{2}d\theta - 32\pi a^3\int_0^\pi \sin^3\frac{\theta}{2}\cos^7\frac{\theta}{2}\,d\theta$$

$$= 256\pi a^3\int_0^{\pi/2}\sin^3\alpha\cos^9\alpha\, d\alpha - 64\pi a^3\int_0^{\pi/2}\sin^3\alpha\cos^7\alpha\, d\alpha$$

where $\alpha = \frac{\theta}{2}$ so that $2d\alpha = d\theta$, when $\theta = 0,\ \alpha = 0$; when $\theta = \pi;\ \alpha = \frac{\pi}{2}$

$$\therefore \quad V = 256\pi a^3\frac{\Gamma(2)\Gamma(5)}{2\Gamma(7)} - 64\pi a^3\frac{\Gamma(2)\Gamma(4)}{2\Gamma(6)}$$

$$\because \quad \int_0^{\pi/2}\sin^m\theta\cos^n\theta d\theta = \frac{\Gamma\left(\frac{m+1}{2}\right)\Gamma\left(\frac{n+1}{2}\right)}{2\Gamma\left(\frac{m+n+2}{2}\right)}$$

$$\text{or} \quad V = 256\,\pi a^3\,\frac{1.4.3.1}{2.6.5.4.3.2.1} - 64\pi a^3\frac{1.3.2.1}{2.5.4.3.2.1}$$

$$(\because \Gamma(n) = (n-1)!)$$

$$= \frac{64\pi a^3}{15} - \frac{8\pi a^3}{5} = \frac{8}{3}\pi a^3$$

2nd Method

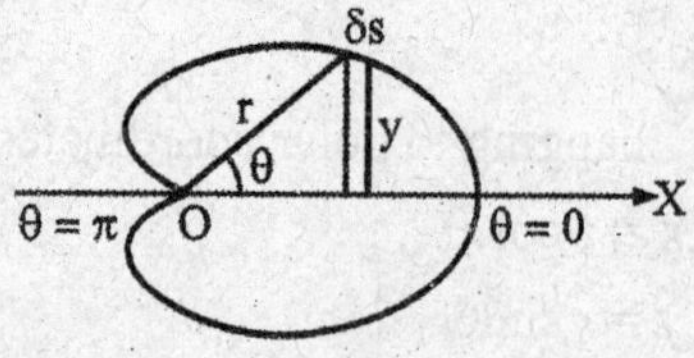

Required volume $= \dfrac{2}{3}\pi\int_0^{\pi} r^3 \sin\theta \, d\theta$

$$= \frac{2}{3}\pi\int_0^{\pi} a^3 (1+\cos\theta)^3 \sin\theta \, d\theta = -\frac{2}{3}\pi a^3 \int_1^{-1} (1+z)^3 \, dz,$$

where $z = \cos\theta$

$$= \frac{2}{3}\pi a^3 \left[\frac{(1+z)^4}{4}\right]_{-1}^{+1} = \frac{1}{6}\pi a^3 . 2^4 = \frac{8}{3}\pi a^3$$

The required surface area

$$S = 2\pi^2 \int y \, ds = 2\pi\int_0^{x} y \frac{ds}{d\theta} d\theta$$

$$= 4\pi a^2 \int_0^{\pi} \sin\theta (1+\cos\theta) \cos\frac{\theta}{2} d\theta, \qquad \text{....(1)}$$

since $y = r \sin\theta = a \sin\theta(1 + \cos\theta)$

and $$\frac{ds}{d\theta} = \sqrt{r^2 + \left(\frac{dr}{d\theta}\right)^2}$$

But $r = a(1 + \cos\theta)$; $\dfrac{dr}{d\theta} = -a\sin\theta$.

$$= \sqrt{a^2 (1+\cos\theta)^2 + a^2 \sin^2\theta}$$

$$= a\sqrt{1 + 2\cos\theta + \cos^2\theta + \sin^2\theta}$$

$$= 2a\cos\frac{\theta}{2}$$

$\therefore$ from (1), $$S = 4\pi a^2 \int_0^{\pi} 2\sin\frac{\theta}{2} \cos\frac{\theta}{2} . 2\cos^2\frac{\theta}{2} . \cos\frac{\theta}{2} d\theta$$

$$= 32\pi a^2 \int_0^{\pi/2} \sin\alpha \cos^4 d\alpha$$

where $\alpha = \dfrac{\theta}{2}$ so that $2d\alpha = d\theta$; when $\theta = 0$, $\alpha = 0$; when $\theta = \pi$, $\alpha = \pi/2$

$$\therefore \qquad S = -32\ \pi a^2 \int_0^{\pi/2} \cos^4 \alpha\, d(\cos \alpha)$$

$$= -32\ \pi\ a^2 \left[\frac{\cos^5 \alpha}{5}\right]_0^{\pi/2} = \frac{32}{5}\pi a^2$$

18. Find the surface of the solid obtained by revolving the curve $r^2 = a^2 \cos 2\theta$ about the initial line.

The curve is shown in the figure

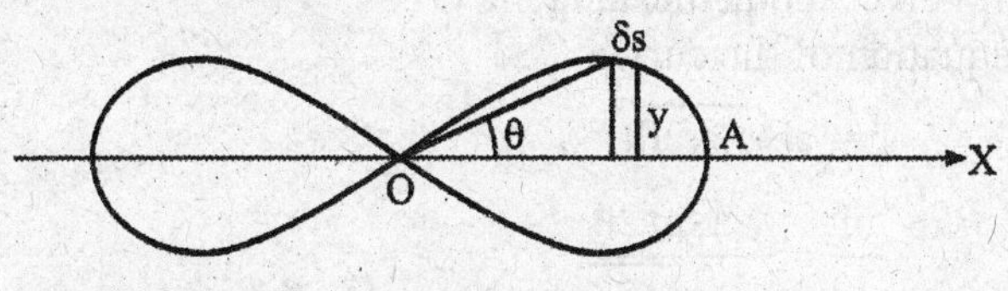

Now $y = r \sin \theta$ (see figure)

$$= a\sqrt{\cos 2\theta} \sin \theta,$$

$$\text{since } r = a\ a\sqrt{\cos 2\theta}\ ,$$

$$\therefore \qquad \frac{dr}{d\theta} = -\frac{a \sin 2\theta}{\sqrt{\cos 2\theta}}$$

$$\therefore \qquad \frac{ds}{d\theta} = \sqrt{r^2 + \left(\frac{dr}{d\theta}\right)^2}$$

$$= \sqrt{a^2 \cos 2\theta + \frac{a^2 \sin^2 2\theta}{\cos 2\theta}} = \frac{a}{\sqrt{\cos 2\theta}}$$

For arc AO, θ varies from 0 to $\pi/4$

So, the required surface

$$= 2\int 2\pi y\ ds = 4\pi \int_0^{\pi/4} \left\{ a\sqrt{\cos 2\theta}.\sin \theta.\frac{a}{\sqrt{\cos 2\theta}} \right\} d\theta$$

$$= 4\pi a^2 \int_0^{\pi/4} \sin \theta d\theta = 4\pi a^2 \left[-\cos \theta\right]_0^{\pi/4}$$

$$= 4\pi a^2 \left(-\frac{1}{\sqrt{2}} + 1\right) = 4\pi a^2 \left(1 - \frac{1}{\sqrt{2}}\right)$$

$$= 4\pi a^2 \left(\frac{\sqrt{2}-1}{\sqrt{2}}\right) = 2\sqrt{2}\ \pi a^2 \left(\sqrt{2}-1\right) = 2\pi a^2 \left(2-\sqrt{2}\right).$$

19. The lemniscate $r^2 = a^2 \cos 2\theta$ revolves about a tangent at the pole. Show that the surface of the solid generated is $4\pi a^2$.

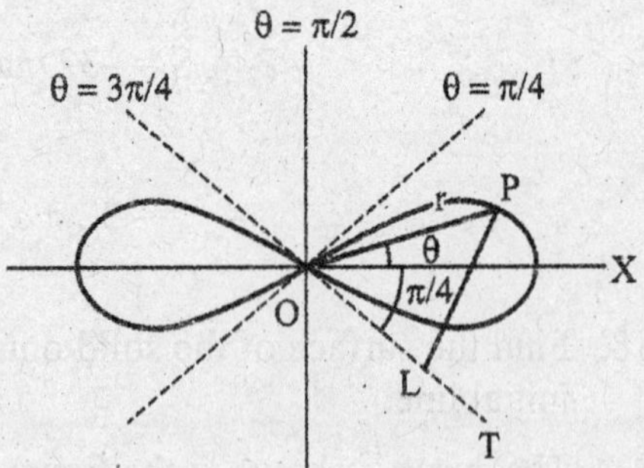

Since $r = 0$ when $\theta = \dfrac{\pi}{4}$ and $\dfrac{3\pi}{4}$, so, the tangents at the pole are given by

$$\theta = \frac{\pi}{4}, \theta = \frac{3\pi}{4}$$

Let the loops revolve about the tangent OT. Now form the equation of the curve

$$r = a\sqrt{\cos 2\theta}$$

$$\therefore \quad \frac{dr}{d\theta} = -\frac{a \sin 2\theta}{\sqrt{\cos 2\theta}}$$

$$\therefore \quad \frac{ds}{d\theta} = \sqrt{r^2 + \left(\frac{dr}{d\theta}\right)^2} = \sqrt{a^2 \cos 2\theta + \frac{a^2 \sin^2 2\theta}{\cos 2\theta}}$$

$$= \frac{a}{\sqrt{\cos 2\theta}}; \qquad \therefore \quad ds = \frac{a}{\sqrt{\cos 2\theta}} d\theta$$

$\therefore$ required surface $= 2\int 2\pi \,.\, PL \,.\, ds$,

where PL $\perp$ OT. From right triangle OPL.

$$PL = r\sin(\theta + \pi/4) = a\sqrt{\cos 2\theta}\,\sin(\theta + \pi/4)$$

$$\therefore \quad \text{surface} = 4\pi \int_{-\pi/4}^{\pi/4} a\sqrt{\cos 2\theta}\,\sin(\theta + \pi/4).\frac{a}{\sqrt{\cos 2\theta}} d\theta$$

$$= 4\pi a^2 \int_{-\pi/4}^{\pi/4} \sin(\theta + \pi/4)\, d\theta$$

$$= 4\pi a^2 \left[-\cos\left(\theta + \frac{\pi}{4}\right)\right]_{-\pi/4}^{\pi/4}$$

$$= 4\pi a^2$$

20. Show that if the area lying within the cardioids $r = 2a(1 + \cos\theta)$ and outside the parabola $r(1 + \cos\theta) = 2a$ revolves about the initial line, the volume generated is $18\pi a^3$.

For the point of intersection of the two curves,

$$2a(1 + \cos\theta) = \frac{2a}{1 + \cos\theta} \qquad \text{(eliminating r)}$$

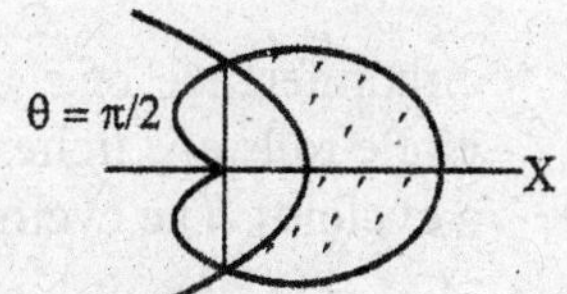

or $(1 + \cos\theta)^2 = 1$

or $1 + 2\cos\theta + \cos^2\theta = 1$

or $\cos\theta\,(2 + \cos\theta) = 0$

$\therefore$ $\cos\theta = 0$ since $\cos\theta \neq -2$; $\therefore\ \theta = \pi/2$

$\therefore$ the required volume

$$= \frac{2}{3}\int_0^{\pi/2} \pi(2a)^3 (1+\cos\theta)^3 \sin\theta\, d\theta - \frac{2}{3}\int_0^{\pi/2} \frac{\pi 8a^3}{(1+\cos\theta)^3} \sin\theta\, d\theta$$

(using volume $= \frac{2}{3}\pi \int_{\theta_1}^{\theta_2} r^2 \sin\theta\, d\theta$ for cardioid and the parabola)

$$= \frac{2}{3}\pi\int_0^{\pi/2}\left[8a^3(1+\cos\theta)^3 - \frac{8a^3}{(1+\cos\theta)^3}\right]\sin\theta\, d\theta$$

$$= \frac{16a^3\pi}{3}\int_0^{\pi/2}\left[(1+\cos\theta)^3 - \frac{1}{(1+\cos\theta)^3}\right]\sin\theta\, d\theta$$

Putting $\cos\theta = z$, $-\sin\theta\, d\theta = dz$; when $\theta = 0,\ z = 1$; when $\theta = \pi/2,\ z = 0$.

$\therefore$ volume $$= \frac{16a^3\pi}{3}\int_0^1\left[(1+z)^3 - (1+z)^{-3}\right] dz$$

$$= \frac{16a^3\pi}{3}\left[\frac{(1+z)^4}{4} - \frac{(1+z)^{-2}}{-2}\right]_0^1$$

$$= \frac{16a^3\pi}{3}\left[\left(4+\frac{1}{8}\right) - \left(\frac{1}{4}+\frac{1}{2}\right)\right]$$

$$= \frac{16a^3\pi}{3} \times \left(\frac{33}{8} - \frac{3}{4}\right) = \frac{16a^3\pi}{3} \times \frac{27}{8}$$

$$= 18\,\pi\, a^3.$$

21. Find the volume and surface of the solid generated by revolving (one arc of) the cycloid $x = a(\theta + \sin\theta)$, $y = a(1 + \cos\theta)$ about its base.

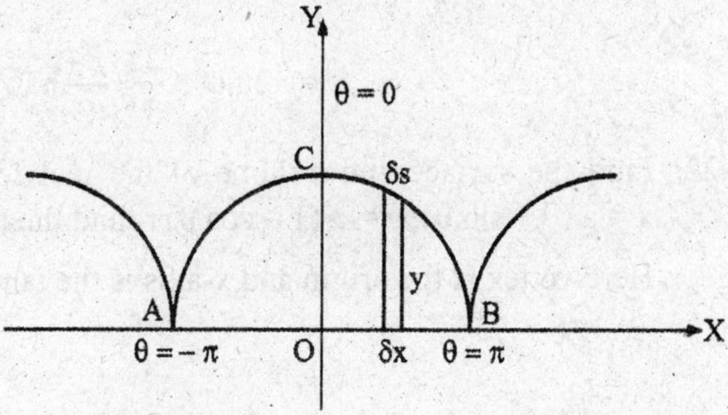

[The cycloid is the curve traced out by a point on the

circumference of a circle which rolls without sliding along a straight line. As the circle rolls on, arches are generated over and over again and any single arch is called a cycloid]. The cycloid is shown in the figure.

The volume of the solid obtained by revolving the arc ACB is double the volume of the solid obtained by revolving the arc BC.

$\therefore$ the required volume $V = 2\pi\int_0^{\pi/2} y^2 dx$

$$= 2\pi\int_0^{\pi} a^2(1+\cos\theta)^2 a(1+\cos\theta)d\theta = 2\pi a^3\int_0^{\pi}(1+\cos\theta)^3 d\theta$$

$$= 2\pi a^3\int_0^{\pi} 8\cos^6\frac{\theta}{2}d\theta = 32\pi a^3\int_0^{\pi/2}\cos^6\varphi\, d\varphi$$

where $\frac{\theta}{2} = \varphi$ so that $2d\varphi = d\theta$ when $\theta = 0, \varphi = 0$; when $\theta = \pi, \varphi = \pi/2$

$$\therefore \quad V = 32\pi a^3.\frac{5}{6}.\frac{3}{4}.\frac{1}{2}.\frac{\pi}{2} = 5\pi^2 a^3$$

Again, $$\frac{ds}{d\theta} = \sqrt{\left(\frac{dx}{d\theta}\right)^2 + \left(\frac{dy}{d\theta}\right)^2} = \sqrt{a(1+\cos\theta)^2 + a^2(-\sin\theta)^2}$$

$$= a\sqrt{4\cos^4\frac{\theta}{2} + 4\sin^2\frac{\theta}{2}.\cos^2\frac{\theta}{2}} = 2a\cos\frac{\theta}{2}$$

$\therefore$ surface $$S = 2\int 2\pi y\, ds = 4\pi\int_0^{\pi} a(1+\cos\theta)2a\cos\frac{\theta}{2}d\theta$$

$$= 4\pi\int_0^{\pi} 2a\cos^2\frac{\theta}{2}.2a\cos\frac{\theta}{2}d\theta$$

$$= 16\pi a^2\int_0^{\pi}\cos^3\frac{\theta}{2}d\theta = 32\pi a^2\int_0^{\pi/2}\cos^3\alpha\, d\alpha,$$

where $\frac{\theta}{2} = \alpha$ so that $d\theta = 2d\alpha$; when $\theta = 0, \alpha = 0$; when $\theta = \pi, \alpha = \frac{\pi}{2}$

$$\therefore \quad S = 32\,\pi a^2.\frac{2}{3} = \frac{64}{3}\pi a^2$$

22. Find the surface and volume of the reel formed by the revolution of the cycloid $x = a(\theta + \sin\theta)$, $y = a(1 - \cos\theta)$ round the tangent at the vertex.

Here vertex is the origin and x-axis is the tangent at vertex

The required volume = $\int 2\pi y^2 dx$

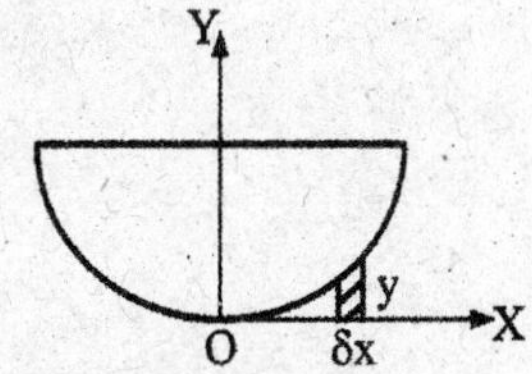

$$= 2\pi\int_0^{\pi} a^2(1-\cos\theta)^2 .a(1+\cos\theta)d\theta$$

$$= 2\pi a^3\int_0^{\pi}(1-\cos\theta)(1-\cos^2\theta)d\theta$$

$$= 2\pi a^3\int_0^{\pi}(1-\cos\theta)\{1-(1-\sin^2\theta)\}d\theta$$

$$= 2\pi a^3\int_0^{\pi}(1-\cos\theta)\sin^2\theta\, d\theta$$

$$= 2\pi a^3\int_0^{\pi}(\sin^2\theta - \sin^2\theta\cos\theta)\, d\theta$$

$$= 2\pi a^3\left\{\int_0^{\pi}\frac{1-\cos 2\theta}{2}d\theta - \int_0^{\pi}\sin^2\theta\, d(\sin\theta)\right\}$$

$$= 2\pi a^3\left[\frac{1}{2}\left(\theta - \frac{\sin^2\theta}{2}\right) - \frac{\sin^3\theta}{3}\right]_0^{\pi}$$

$$= 2\pi a^3\left[\frac{1}{2}(\pi - 0) - 0\right] = \pi^2 a^3$$

Again required surface = $2\int 2\pi y\, ds$

$$= 4\pi\int_0^{\pi} a(1-\cos\theta)2a\cos\frac{\theta}{2}d\theta$$

$$\left[\text{since}\frac{ds}{d\theta} = \sqrt{\left(\frac{dx}{d\theta}\right)^2 + \left(\frac{dy}{d\theta}\right)^2}\right]$$

[$\because$ $\dfrac{ds}{d\theta} = a\sqrt{(1+\cos\theta)^2 + \sin^2\theta}$

$$= a\sqrt{1+2\cos\theta+\cos^2\theta+\sin^2\theta}$$

$$= a\sqrt{2+2\cos\theta} = 2a\cos\frac{\theta}{2}\,]$$

i.e., surface = $8\pi a^2\int_0^{\pi} 2\sin^2\frac{\theta}{2}\cos\frac{\theta}{2}d\theta$

$$= 32\pi a^2 \int_0^{\pi} \sin^2 \frac{\theta}{2} . d. \left(\sin \frac{\theta}{2} \right)$$

$$= 32\pi a^2 \left[\frac{\sin^3 \frac{\theta}{2}}{3} \right]_0^{\pi} = \frac{32}{3} \pi a^2$$

23. Evaluate the surface area of the solid generated by revolving the cycloid $x = a(\theta - \sin\theta)$, $y = a(1 - \cos\theta)$ about the line $y = 0$.

The cycloid is shown in the figure

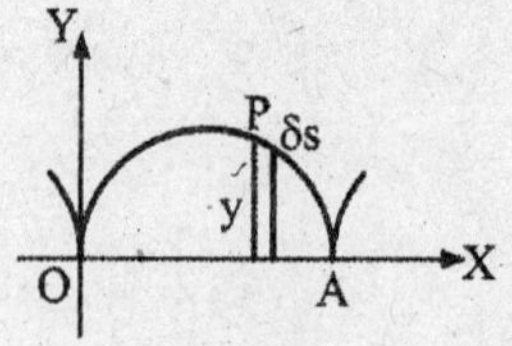

The required surface $s = 2\pi \int y \, ds$

$$= 2\pi \int_0^{2\pi} y . \frac{ds}{d\theta} d\theta,$$

since for the part OPA, θ varies from 0 to 2π

Also here
$$\frac{ds}{d\theta} = \sqrt{\left(\frac{dx}{d\theta}\right)^2 + \left(\frac{dy}{d\theta}\right)^2}$$

$$= \sqrt{a^2 (1 - \cos\theta)^2 + a^2 \sin^2\theta}$$

$$= a \sqrt{1 - 2\cos\theta + \cos^2\theta + \sin^2\theta}$$

$$= 2 a \sin \frac{\theta}{2}$$

$$\therefore \quad S = 2\pi \int_0^{2\pi} a(1 - \cos\theta) . 2a \sin\frac{\theta}{2} d\theta = 8\pi a^2 \int_0^{2\pi} \sin^3 \frac{\theta}{2} d\theta$$

$$= 16a^2 \int_0^{\pi} \sin^3 \alpha \, d\alpha, \quad \text{where } \frac{\theta}{2} = \alpha \text{ so that } 2d\alpha = d\theta$$

$$= 32\pi a^2 \int_0^{\pi/2} \sin^3 \alpha d\alpha = 32\pi a^2 . \frac{2}{3} = \frac{64}{3} \pi a^2 .$$

24. Find the volume of the spindle-shaped solid generated by revolving the astroid $x^{2/3} + y^{2/3} = a^{2/3}$ or $x = a \cos^3 t$, $y = a \sin^3 t$ about x-axis.

The parametric equation of the curve are $x = a \cos^3 t$, $y = a \sin^3 t$

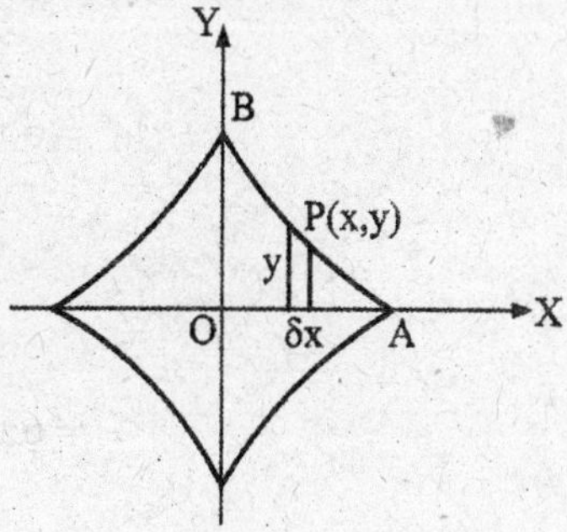

Considering an elementary strip of width δx parallel to y-axis.

The required volume = 2 × volume generated by the revolution of OABO for which t varies from 0 to 2π.

$$= 2\int_0^{\pi/2} \pi y^2 dx$$

$$= 2\pi\int_0^{\pi/2} a^2 \sin^6 t \, d\left(a\cos^3 t\right)$$

$$= 2\pi a^3 \int_0^{\pi/2} \sin^6 t \, 3\cos^2 t(-\sin t)dt$$

$$= 6\pi a^3 \int_0^{\pi/2} \sin^7 t \, \cos^2 t \, dt$$

(neglecting –ve sign)

$$= 6\pi a^3 \int_0^{\pi/2} \sin^7 \left(1-\sin^2 t\right) dt$$

$$= 6\pi a^3 \left\{\int_0^{\pi/2} \sin^7 t \, dt - \int_0^{\pi/2} \sin^9 t \, dt\right\}$$

$$\left(\text{if } J_n = \int_0^{\pi/2} \sin^n t \, dt \text{ then } J_n = \frac{n-1}{n} J_{n-2}\right.$$

$$\therefore \quad J_7 = \frac{6}{7}J_5, \ J_5 = \frac{4}{5}J_3, \ J_3 = \frac{2}{3}J_1, \ J_1 = \int_0^{\pi/2} \sin x \, dx = 1$$

$$\therefore \quad J_7 = \frac{6}{7}\times\frac{4}{5}\times\frac{2}{3}\times 1 = \frac{16}{35}; \text{ similarly } J_9 = 128/315)$$

$$= 6\pi a^3 . \left\{\frac{16}{35} - \frac{128}{315}\right\} = 6\pi a^3 \times \left(\frac{144-128}{315}\right)$$

$$= 6\pi a^3 . \frac{16}{315} = \frac{32}{105}\pi a^3$$

2nd Method (using gamma function)

$$\text{Volume} = 6\pi a^3 \int_0^{\pi/2} \sin^7 t \cos^2 t dt$$

$$= 6\pi a^3 \frac{\Gamma(4)\Gamma\left(\frac{3}{2}\right)}{2\Gamma\left(\frac{11}{2}\right)}$$

$$= 6\pi a^3 \frac{3.2.1.\frac{1}{2}\Gamma\left(\frac{1}{2}\right)}{2.\frac{9}{2}.\frac{7}{2}.\frac{5}{2}.\frac{3}{2}.\frac{1}{2}\Gamma\left(\frac{1}{2}\right)}$$

$$= \frac{32\pi a^3}{105}$$

25. Find the surface of the solid generated by the revolution of the astroid $x^{2/3} + y^{2/3} = a^{2/3}$ about the x-axis.

The curve (astroid) is shown in the figure

The parametric equations of the curve are $x = a\cos^3 t$, $y = a\sin^3 t$

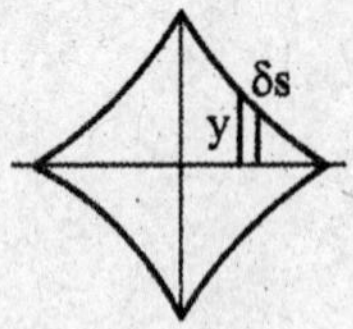

$$\frac{dx}{dt} = -3a\cos^2 t \sin t$$

$$\frac{dy}{dt} = 3a\sin^2 t \cos t$$

$$\therefore \quad ds = \sqrt{\left(\frac{dx}{dt}\right)^2 + \left(\frac{dy}{dt}\right)^2}\, dt$$

$$= 3a\sqrt{\cos^4 t \ \sin^2 t + \sin^4 t \ \cos^2 t}\ dt$$

$$= 3a\sin t\cos t\sqrt{\cos^2 t + \sin^2 t}\ dt = 3a\sin t \cos t\ dt$$

for the arc in the first quadrant, limits of t are from 0 to $\pi/2$

$\therefore$ the required surface $= 2\int 2\pi y\, ds$

$$= 4\pi\int_0^{\pi/2} a\sin^3 t.3a\sin t\cos t\ dt$$

$$= 12a^2\pi\int_0^{\pi/2} \sin^4 t\cos t\ dt$$

$$= 12a^2\pi\int_0^{\pi/2} \sin^4 t\ d(\sin t)$$

$$=12a^2\pi\left[\frac{\sin^5 t}{5}\right]_0^{\pi/2}=12a^2\pi\left(\frac{1}{5}-0\right)$$

$$=\frac{12}{5}\pi a^2.$$

26. Find the volume and the area of the surface generated by the revolution of the tractrix $x = a\cos t + \frac{1}{2}a\log\tan^2\frac{1}{2}t,\ y = s\sin t$ about its asymptote.

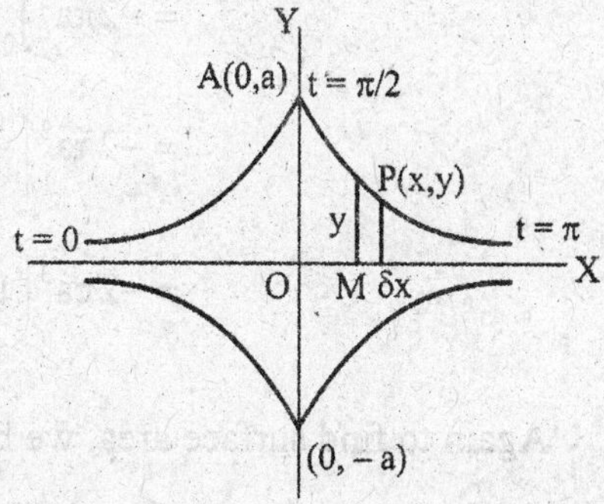

The tractrix is shown in the figure

Here OA = a, the portion of the tangent intercepted between the curve and the x-axis.
The x-axis is the asymptote. For the half of the upper part, t varies from 0 to $\pi/2$

Differentiating the equation of the curve, we have

$$\frac{dx}{dt}=-a\sin t+\frac{1}{2}a\frac{1}{\tan^2\frac{1}{2}t}\times 2\tan\frac{1}{2}t.\sec^2\frac{1}{2}t.\frac{1}{2}$$

$$=-a\sin t+a\frac{\sec^2\frac{1}{2}t}{\tan\frac{1}{2}t}.\frac{1}{2}$$

$$=-a\sin t+a.\frac{1}{2}\frac{1}{\cos^2\frac{1}{2}t}\times\frac{\cos\frac{1}{2}t}{\sin\frac{1}{2}t}$$

$$=-a\sin t+\frac{a}{2\sin\frac{1}{2}t\cos\frac{1}{2}t}=-a\sin t+\frac{a}{\sin t}$$

$$=\frac{a\left(1-\sin^2 t\right)}{\sin t}=\frac{a\cos^2 t}{\sin t}$$

or $$dx=\frac{a\cos^2 t}{\sin t}=dt$$

∴ required volume

$$= 2\int \pi y^2 dx = 2\pi\int_0^{\pi/2} a^2 \sin^2 t.\frac{a\cos^2 t}{\sin t} dt$$

$$= 2\pi a^3 \int_0^{\pi/2} \cos^2 t \sin t \, dt$$

$$= -2\pi a^3 \int_0^{\pi/2} \cos^2 t \, d(\cos\theta)$$

$$= -2\pi a^3 \left[\frac{\cos^3 t}{3}\right]_0^{\pi/2}$$

$$= -2\pi a^3 \left(0 - \frac{1}{3}\right) = \frac{2}{3}\pi a^3$$

Again to find surface area, we have $\frac{dy}{dt} = a\cos t$

$$\therefore \quad \frac{ds}{dt} = \sqrt{\left(\frac{dx}{dt}\right)^2 + \left(\frac{dy}{dt}\right)^2} = \sqrt{\frac{a^2\cos^4 t}{\sin^2 t} + a^2\cos^2 t}$$

$$= \frac{a\cos t}{\sin t}$$

∴ required surface

$$= 2\int 2\pi y \, ds$$

$$= 4\pi\int_0^{\pi/2} a\sin t.\frac{a\cos t}{\sin t} dt$$

$$= 4\pi a^2 \int_0^{\pi/2} \cos t \, dt$$

$$= 4\pi a^2 [\sin t]_0^{\pi/2} = 4\pi a^2$$

= surface of a sphere of radius a

27. Find the volume of the solid generated by the revolution of

$$y^2 = \frac{x^3}{a - x} \text{ i.e., } y^2(a - x) = x^3,$$

about its asymptote.

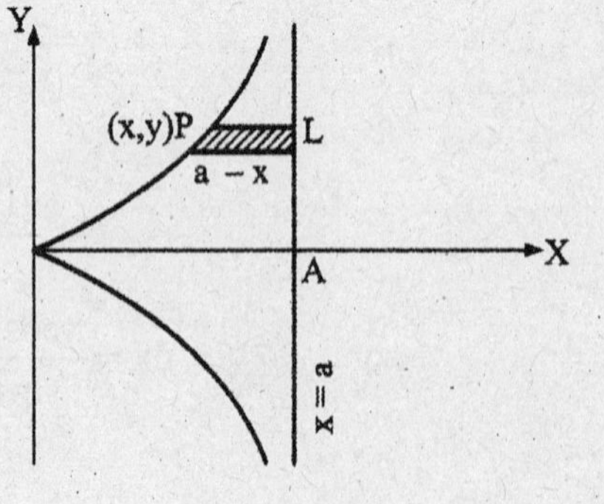

The curve is symmetrical about x-axis passes through the origin and equations of tangent at origin are y = 0, y = 0, i.e., x-axis. x = a is the asymptote, by equating to zero the coefficient of y^2, and the curve is as shown in the figure.

If P(x, y) be any point on the curve and PL perpendicular from P to the asymptote x = a then PL = a − x.

∴ the required volume

$$V = 2\int \pi(PL)^2\, d(AL)$$

$$= 2\pi\int_0^{\infty}(a-x)^2\, dy \qquad .....(I)$$

We have $$y^2 = \frac{x^2}{a-x}$$

∴ $$2y\frac{dy}{dx} = \frac{3x^2(a-x)+x^3}{(a-x)^2}$$

or $$2\frac{x^{\frac{3}{2}}}{\sqrt{a-x}}\frac{dy}{dx} = \frac{3x^2(a-x)+x^3}{(a-x)^2}, \qquad \left(\because y = \frac{x^{\frac{3}{2}}}{\sqrt{a-x}}\right)$$

or $$2\frac{dy}{dx} = \frac{3x^2(a-x)+x^3}{(a-x)^2}\times\frac{\sqrt{a-x}}{x^{3/2}}$$

$$= \frac{x^2\left[3(a-x)+x\right]\sqrt{a-x}}{x^{\frac{3}{2}}\times(a-x)^2} = \frac{\sqrt{x}(3a-3x+x)\sqrt{a-x}}{(a-x)^2}$$

$$= \frac{\sqrt{x}(3a-2x)\sqrt{a-x}}{(a-x)^2}$$

∴ $$dy = \frac{1}{2}\frac{(3a-2x)\sqrt{x}\sqrt{a-x}}{(a-x)^2}dx$$

∴ from (I), $$V = 2\pi\int_0^a \frac{(3a-2x)\sqrt{x}\sqrt{a-x}}{2(a-x)^2}\times(a-x)^2\, dx$$

[when x = 0, y = 0; when x = a; y = ∞]

$$= \pi\int_0^a (3a-2x)\sqrt{x}\sqrt{a-x}\, dx$$

Putting $x = a\sin^2\theta$, $dx = 2a\sin\theta\cos\theta\, d\theta$;

when $x = 0$, $\theta = 0$; when $x = a$, $\theta = \pi/2$

$$= \pi\int_0^{\pi/2}\left(3a - 2a\sin^2\theta\right)\sqrt{a\sin^2\theta}\sqrt{\left(a - a\sin^2\theta\right)}.2a\sin\theta\cos\theta\, d\theta$$

$$= \pi\int_0^{\pi/2}\left(3 - 2\sin^2\theta\right)\sqrt{a}\sin\theta\sqrt{a\cos^2\theta}.2a\sin\theta\cos\theta\, d\theta$$

$$= 2\pi a^3\int_0^{\pi/2}\left(3 - 2\sin^2\theta\right)\sin^2\theta\cos^2\theta\, d\theta$$

$$= 2\pi a^3\left\{3\int_0^{\pi/2}\sin^2\theta\cos^2\theta\, d\theta - 2\int_0^{\pi/2}\sin^4\theta\cos^2\theta\, d\theta\right\}$$

$$= 2\pi a^3\left\{\frac{3.1.1}{4.2}.\frac{\pi}{2} - 2\frac{3.1}{6.4.2}.\frac{\pi}{2}\right\} = 2\pi a^3 \times \frac{\pi}{8} = \frac{\pi^2 a^3}{4}$$

28. Find the volume of the solid obtained by the revolution of the cissoid $y^2(2a - x) = x^3$ about its asymptote.

The curve is symmetrical about x-axis, passes through the origin and equations of tangent at origin are $y = 0$, $y = 0$, i.e., x-axis. $x = 2a$ is the asymptote, by equating to zero the coefficient of y^2 and the curve is as shown in the figure.

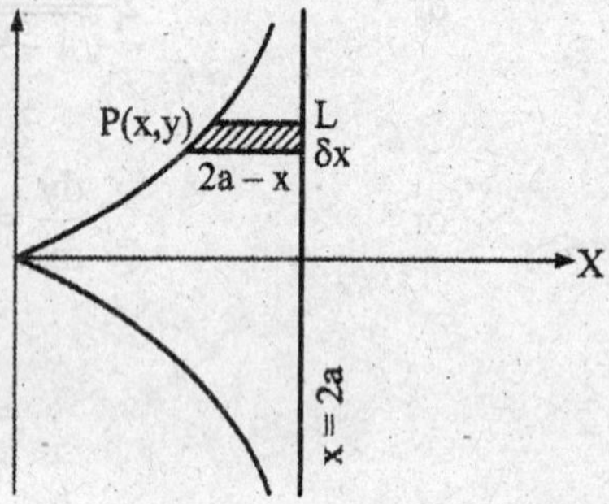

If P(x, y) be any point on the curve and PL perpendicular from P to the asymptote $x = 2a$ then $PL = 2a - x$.

∴ the required volume V

$$= 2\pi\int_0^{\infty}(PL)^2\, d(AL)$$

$$= 2\pi\int_0^{\infty}(2a - x)^2\, dy \qquad .....(1)$$

To change from y to x, we have

$$y^2 = \frac{x^3}{2a - x}; \quad 2y\frac{dy}{dx} = \frac{3x^2(2a - x) - x^3(-1)}{(2a - x)^2}$$

or

$$2\frac{x^{3/2}}{\sqrt{2a - x}}\frac{dy}{dx} = \frac{3x^2(2a - x) + x^3}{(2a - x)^2}$$

$$\therefore \quad 2dy = \frac{x^2\left[3(2a-x)+x\right]}{(2a-x)^2} \times \frac{\sqrt{2a-x}}{x^{3/2}} dx$$

$$= \frac{x^{1/2}(6a-3x+x)\sqrt{2a-x}}{(2a-x)^2} dx; \; dy = \frac{(3a-x)\sqrt{x}\sqrt{2a-x}}{(2a-x)^2} dx$$

From (1),

$$V = 2\pi\int_0^{2a} (2a-x)^2 \frac{(3a-x)\sqrt{x}\sqrt{2a-x}}{(2a-x)^2}$$

$\because$ when $y = 0, x = 0$; when $y = \infty, x = 2a$

$$= 2\pi\int_0^{2a} (3a-x)\sqrt{x}\sqrt{2a-x}\, dx$$

Putting $x = 2a\sin^2\theta$, $dx = 4a\sin\theta\cos\theta\, d\theta$;

when $x = 0, \theta = 0$; when $x = 2a$, $\theta = \pi/2$

$$\therefore \quad V = 16\pi a^3 \int_0^{\pi/2} (3 - 2\sin^2\theta)\sin^2\theta\cos^2\theta\, d\theta$$

$$= 16\pi a^3 \left\{3\int_0^{\pi/2} \sin^2\theta\cos^2\theta\, d\theta - 2\int_0^{\pi/2} \sin^4\theta\cos^2\theta d\theta\right\}$$

$$= 16\pi a^3 \left\{\frac{3.1.1}{4.2}.\frac{\pi}{2} - 2.\frac{3.1}{6.4.2}.\frac{\pi}{2}\right\} = 16\pi a^3 \times \frac{\pi}{8} = 2\pi^2 a^3$$

29. Find the volume of the solid formed by the revolution of the curve $(a-x)y^2 = a^2x$ about its asymptote.

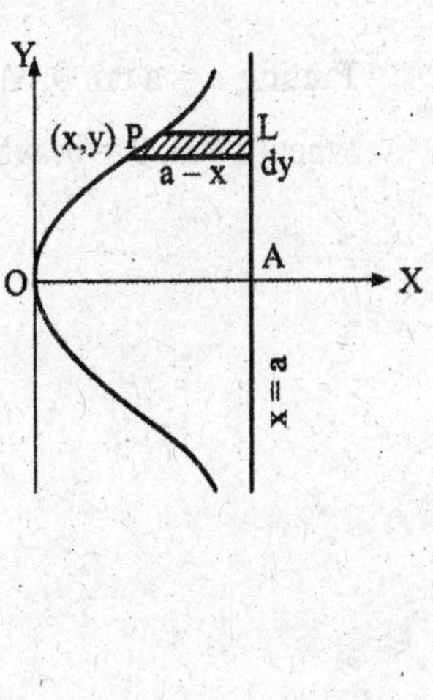

The curve is symmetrical about x-axis, passes through the origin; $x = 0$, i.e., y-axis is the tangent at origin, $x = a$ is the asymptote parallel to y-axis for $x > a$, y becomes imaginary, so no part of the curve is beyond $x = a$ and the curve is as shown in the figure.

If P(x, y) be any point on the curve, PL perpendicular from P to the asymptote then the length of the strip $PL = a - x$.

The required volume V

$$= 2\pi\int (PL)^2\, d(AL)$$

$$= 2\pi\int_0^{\infty} (a-x)^2\, dy$$

$$= 2\pi\int_0^{\infty}\left(a - \frac{ay^2}{y^2+a^2}\right)^2 dy \qquad [\because (a-x)\,y^2 = a^2 x]$$

$$= 2\pi a^6\int_0^{\infty}\frac{dy}{\left(y^2+a^2\right)^2}$$

Putting $y = a\tan\theta$ so that $dy = a\sec^2\theta\, d\theta$

$$V = 2\pi a^6\int_0^{\pi/2}\frac{a\sec^2\theta\, d\theta}{a^4\sec^4\theta} = 2\pi a^3\int_0^{\pi/2}\cos^2\theta\, d\theta$$

$$= 2\pi a^3.\frac{1}{2}.\frac{1}{2}\pi = \frac{1}{2}\pi^2 a^3$$

30. Find the volume of the solid formed by the revolution of the curve $y = \dfrac{a^3}{a^2+x^2}$ about its asymptote.

The curve is symmetrical about y-axis and does not pass through origin. Solving the equation of the curve $x^2 = \dfrac{a^2}{y}(a-y)$, no part of the curve is beyond y = a, y = 0, i.e., x-axis is the asymptote to the curve, and the curve is as shown in the figure.

The required volume

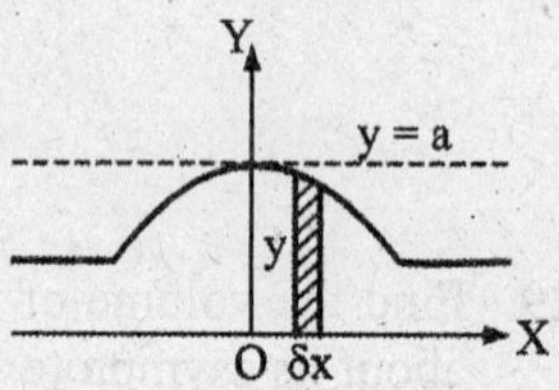

$$V = 2\int_0^{\infty}\pi y^2 dx$$

$$= 2\pi\int_0^{\infty}\frac{a^6}{\left(a^2+x^2\right)^2}dx$$

Putting $x = a\tan\theta$, $dx = a\sec^2\theta\, d\theta$

when $x = 0$, $\theta = 0$; when $x = \infty$, $\theta = \pi/2$

$$\therefore \qquad V = 2\pi\int_0^{\pi/2}\frac{a^2}{\left(a^2+a^2\tan^2\theta\right)^2}.a\sec^2 d\theta = 2\pi a^3\int_0^{\pi/2}\cos^2\theta\, d\theta$$

$$= 2\pi a^3\int_0^{\pi/2}\frac{1+\cos 2\theta}{2}d\theta$$

$$= \pi a^3\left[\theta + \frac{\sin 2\theta}{2}\right]_0^{\pi/2} = \pi a^3.\frac{\pi}{2} = \frac{1}{2}\pi a^2 a^3$$

31. The area cut off from the parabola $y^2 = 4ax$ by the chord joining the vertex to an end of the latus rectum, is rotated through four right angles about the chord. Find the volume of the solid generated.

The equation of the parabola is $y^2 = 4a$. LSL′ is the latus rectum. Here the area OPLO revolves about OL. PM is perpendicular to OL from P(x, y). Then volume required

$$= \int \pi(PM)^2 \, d(OM) \qquad .....(1)$$

To find PM, we see that the equation of OL is $y - 0 = \frac{2a-0}{a-0}(x-0)$, i.e., $y - 2x = 0$

$$\therefore \quad PM = \frac{y-2x}{\sqrt{1+2^2}} = \frac{y-2x}{\sqrt{5}} = \frac{2\sqrt{ax}-2x}{\sqrt{5}} \qquad (\because y^2 = 4ax)$$

$$= \frac{2}{\sqrt{5}}\left(\sqrt{ax}-x\right) \text{ i.e., } PM^2 = \frac{4}{5}\left(\sqrt{ax}-x\right)^2 \qquad .....(2)$$

Also $OM^2 = OP^2 - PM^2$

$$= \left\{(x-0)^2 + (y-0)^2\right\} - \frac{(y-2x)^2}{5}$$

$$= \frac{1}{5}\left(5x^2 + 5y^2 - y^2 + 4xy - 4x^2\right)$$

$$= \frac{1}{5}\left(x^2 + 4y^2 + 4xy\right) = \frac{1}{5}(x+2y)^2$$

$$= \frac{1}{5}\left(x + 2.2\sqrt{ax}\right)^2 = \frac{1}{5}\left(x + 4\sqrt{ax}\right)^2$$

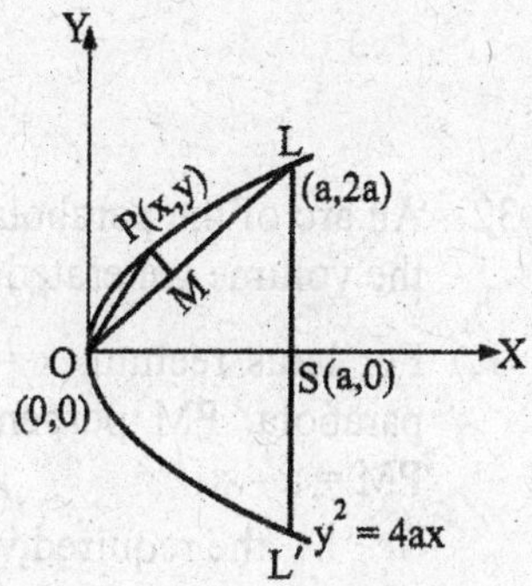

$$\therefore \quad OM = \frac{1}{\sqrt{5}}\left(x + 4\sqrt{ax}\right)$$

$$\therefore \quad d(OM) = \frac{1}{\sqrt{5}}\left(1 + 4\sqrt{a}.\frac{1}{2\sqrt{x}}\right)dx$$

$$= \frac{1}{\sqrt{5}}\left(1 + 2\sqrt{\frac{a}{x}}\right)dx \qquad .....(3)$$

$\therefore$ the required volume [from (1), (2) and (3)]

$$= \pi\int_0^a \frac{4}{5}\left(\sqrt{ax} - x\right)^2 . \frac{1}{\sqrt{5}}\left(1 + 2\sqrt{\frac{a}{x}}\right)dx$$

$$= \frac{4\pi}{5\sqrt{5}}\int_0^a \left(ax + x^2 - 2x\sqrt{ax}\right)\left(1 + \frac{2\sqrt{a}}{\sqrt{x}}\right)dx$$

$$= \frac{4\pi}{5\sqrt{5}}\int_0^a \left(ax + x^2 - 2\sqrt{a}x^{3/2}x^{1/2} + 2a^{3/2}x^{1/2} + 2\sqrt{a}x^{3/2} - 4ax \right)dx$$

$$= \frac{4\pi}{5\sqrt{5}}\left[\frac{-3ax^2}{2} + \frac{x^3}{3} + 2a^{3/2}.\frac{2}{3}x^{3/2}\right]_0^a$$

$$= \frac{4\pi}{5\sqrt{5}}\left[\frac{-3a^3}{2} + \frac{a^3}{3} + \frac{4a^3}{3}\right]$$

$$= \frac{4\pi a^3}{5\sqrt{5}}\left[-\frac{3}{2} + \frac{1}{3} + \frac{4}{3}\right]$$

$$= \frac{4\pi a^3}{5\sqrt{5}} \times \frac{-9+2+8}{6} = \frac{4\pi a^3}{5\sqrt{5}} \times \frac{1}{6}$$

$$= \frac{2}{15\sqrt{5}}\pi a^3 = \frac{2\sqrt{5}}{75}\pi a^3$$

32. An arc of the parabola is bounded at both ends by the latus rectum of length 4a. Find the volume generated when the arc is rotated about the latus rectum.

For latus rectum, x = a, $\therefore$ y = + 2a. If P(x, y) be any point on the arc of the parabola, PM perpendicular from P on the latus rectum then length of the strip PM = a – x.

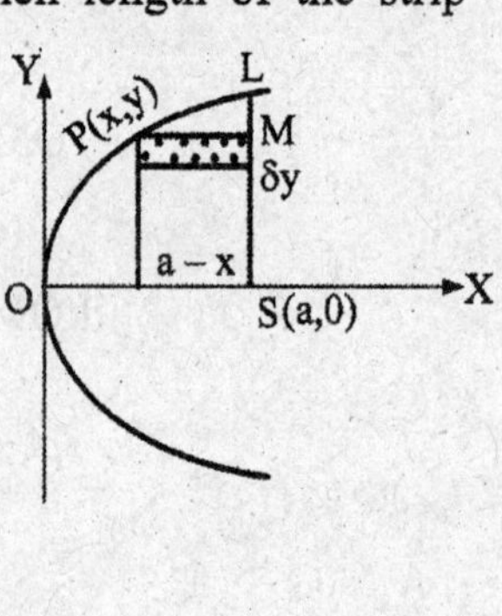

$\therefore$ the required volume

$$= \int \pi(PM)^2\, d(SM)$$

$$= \pi\int_{-2a}^{2a} (a - x)^2\, dy$$

But the parabola is $y^2 = 4ax$;

$\therefore \quad x = y^2/4a$

$$\therefore \text{ volume} = \pi\int_{-2a}^{2a}\left(a - y^2/4a\right)^2 dy = \pi\int_{-2a}^{2a}\frac{\left(4a^2 - y^2\right)^2}{16a^2}dy$$

$$= \frac{\pi}{16a^2}\int_{-2a}^{2a}\left(16a^4 - 8a^2y^2 + y^4\right)dy = \frac{\pi}{16a^2}\left[16a^4y - \frac{8a^2y^3}{3} + \frac{y^5}{5}\right]_{-2a}^{2a}$$

$$= \frac{\pi}{16a^2}\left[64a^5 - \frac{128}{3}a^5 + \frac{64a^5}{5}\right] = 4\pi a^3\left[1 - \frac{2}{3} + \frac{1}{5}\right] = \frac{32\pi a^3}{15}$$

33. The ellipse $b^2x^2 + a^2y^2 = a^2b^2$ is divided into two parts by the line $x = \frac{1}{2}a$ and the smaller part is rotated through four right angles about this line. Prove that volume generated is

$$\pi a^2 b\left(\frac{3\sqrt{3}}{4} - \frac{1}{3}\pi\right)$$

Here the smaller part, i.e., the area APDNA revolves round the line $x = \frac{1}{2}a$.

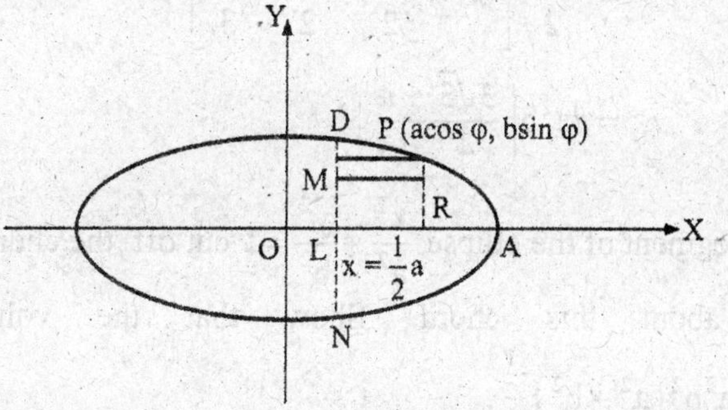

The parametric equations of the ellipse are $x = a\cos\varphi$, $y = b\sin\varphi$. Let P($a\cos\varphi$, $b\sin\varphi$) be a point on the arc AD. PM is perpendicular to DN.

The required volume

$$= 2\int \pi(PM)^2\, d(LM)$$

But, $PM = OR - OL = a\cos\varphi - \frac{1}{2}a, LM = b\sin\varphi$

To know the range of the parameter φ, we have $x = \frac{1}{2}a = a\cos\varphi$ or $\cos\varphi = \frac{1}{2}$;

$\therefore \varphi = \pi/3$

$$\therefore \quad \text{volume} = 2\pi\int_0^{\pi/3}\left(a\cos\varphi - \frac{1}{2}a\right)^2 d(b\sin\varphi)$$

$$= 2\pi a^2 b \int_0^{\pi/3} \left(\cos^2 \varphi - \cos\varphi + \frac{1}{4} \right) \cos\varphi d\varphi$$

$$= 2\pi a^2 b \int_0^{\pi/3} \left(\cos^3 \varphi - \cos^2 \varphi + \frac{1}{4} \cos\varphi \right) d\varphi$$

$$= 2\pi a^2 b \int_0^{\pi/3} \left(\frac{1}{4}(\cos 3\varphi + 3\cos\varphi - \frac{1}{2}(1 + \cos 2\varphi) + \frac{1}{4}\cos\varphi \right) d\varphi$$

$$= \frac{\pi a^2 b}{2} \int_0^{\pi/3} (\cos 3\varphi + 4\cos\varphi - 2\cos 2\varphi - 2) d\varphi$$

$$= \frac{\pi a^2 b}{2} \left[\frac{1}{3} \sin 3\varphi + 4\sin\varphi - \sin 2\varphi - 2 \right] d\varphi$$

$$= \frac{\pi a^2 b}{2} \left[\frac{1}{3} \sin 3\varphi + 4\sin\varphi - \sin 2\varphi - 2\varphi \right]_0^{\pi/3}$$

$$= \frac{\pi a^2 b}{2} \left[0 + 4.\frac{\sqrt{3}}{2} - \frac{\sqrt{3}}{2} - \frac{2\pi}{3} \right]$$

$$= \pi a^2 b \left(\frac{3\sqrt{3}}{4} - \frac{\pi}{3} \right)$$

34. The smaller segment of the ellipse $\frac{x^2}{a^2} + \frac{y^2}{b^2} = 1$ cut off the chord $\frac{x}{a} + \frac{y}{b} = 1$ revolves completely about this chord. Show that the volume generated is $\frac{1}{6}\pi(10 - 3\pi)a^2 b^2 (a^2 + b^2)^{-\frac{1}{2}}$.

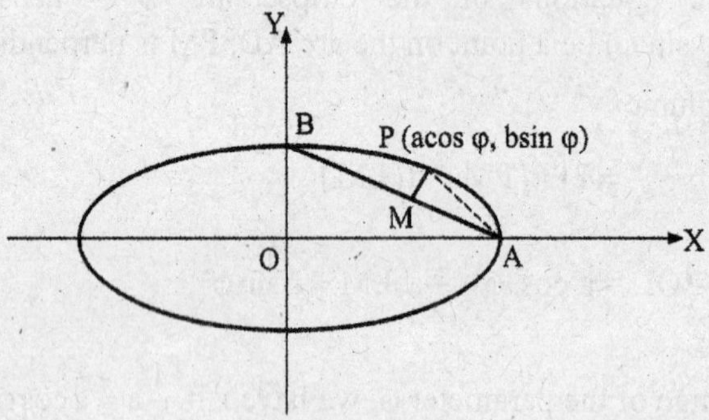

Let the given chord be APB. Let P (a cos φ, b sin φ) be a point on this chord. Let PM ⊥ AB. Then

$$PM = \frac{\frac{a\cos\varphi}{a} + \frac{b\sin\varphi}{b} - 1}{\sqrt{\frac{1}{a^2} + \frac{1}{b^2}}}$$

$$= \frac{ab(\cos\varphi + \sin\varphi - 1)}{\sqrt{a^2 + b^2}}$$

Also, $AM^2 = AP^2 - PM^2 = (a\cos\varphi - a)^2 + (b\sin\varphi)^2 - \frac{a^2b^2(\cos\varphi + \sin\varphi - 1)^2}{a^2 + b^2}$

$$= a^2(\cos\varphi - 1)^2 + b^2\sin^2\varphi$$

$$-\frac{a^2b^2}{a^2 + b^2}\left\{(\cos\varphi - 1)^2 + \sin^2\varphi + 2\sin\varphi(\cos\varphi - 1)\right\}$$

$$= \frac{a^4(\cos\varphi - 1)^2 + b^4\sin^2\varphi - 2a^2b^2\sin\varphi(\cos\varphi - 1)}{a^2 + b^2}$$

$$= \frac{\left\{a^2(\cos\varphi - 1) - b^2\sin\varphi\right\}^2}{a^2 + b^2}$$

$$\therefore \quad AM = \frac{a^2(1 - \cos\varphi) + b^2\sin\varphi}{\sqrt{a^2 + b^2}}$$

$$\therefore \quad d(AM) = \frac{a^2\sin\varphi + b^2\cos\varphi}{\sqrt{a^2 + b^2}}d\varphi$$

$\therefore$ the required volume $= \pi\int_0^{AB}(PM)^2\,d(AM)$

$$= \pi\int_0^{\pi/2}\frac{a^2b^2(\cos\varphi + \sin\varphi - 1)^2}{a^2 + b^2}\cdot\frac{a^2\sin\varphi + b^2\cos\varphi}{\sqrt{a^2 + b^2}}d\varphi$$

$$= \frac{\pi a^2b^2}{(a^2 + b^2)^{3/2}}\int_0^{\pi/2}(a^2\sin^2\varphi + b^2\cos\varphi)$$

$$\times(2 + 2\sin\varphi\cos\varphi - 2\sin\varphi - 2\cos\varphi)\,d\varphi$$

$$= \frac{2\pi a^2 b^2}{(a^2+b^2)^{3/2}} \int_0^{\pi/2} \left[a^2 \sin\varphi + b^2 \cos\varphi\right] + a^2 \sin^2\varphi \cos\varphi$$

$$+ b^2 \cos^2\varphi . \sin\varphi - (a^2 \sin^2\varphi + b^2 \cos^2\varphi)$$

$$-(a^2+b^2)\sin\varphi\cos\varphi]\, d\varphi$$

$$= \frac{2\pi a^2 b^2}{\left(a^2+b^2\right)^{3/2}} \left[\left(a^2+b^2\right) + \frac{a^2}{3} + \frac{b^2}{3} - \left(\frac{a^2\pi}{4} + \frac{b^2\pi}{4}\right) - \frac{a^2+b^2}{2}\right]$$

$$= \frac{\pi a^2 b^2}{\sqrt{a^2+b^2}} \left(\frac{5}{3} - \frac{\pi}{2}\right) = \frac{1}{6}\pi(10-3\pi)a^2b^2\left(a^2+b^2\right)^{-\frac{1}{2}}$$

35. A circular arc resolves about its chord. Find the volume and area of the surface generated. The arc subtends an angle 2α at the centre.

Let the arc be $x = a\cos\theta$, $y = a\sin\theta$,

i.e., $x^2 + y^2 = a^2$

The arc resolves about the chord BC

Let P($a\cos\theta$, $a\sin\theta$) be a point on the arc

Draw PM $\perp$ to OX and PL $\perp$ to AC

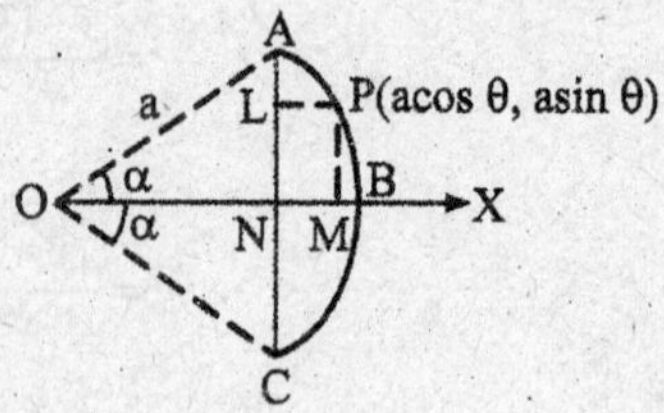

Required volume $= 2\int \pi(PL)^2\, d(LN)$

$$= 2\pi\int_0^{\alpha} (a\cos\theta - a\cos\alpha)^2\, d(a\sin\theta)$$

$$(\because\ PL = NM = OM - ON = a\cos\theta - a\cos\alpha)$$

$$= 2\pi a^3 \int_0^{\alpha} \left(\cos^2\theta - 2\cos\theta\cos\alpha + \cos^2\alpha\right)\cos\theta\, d\theta$$

$$= 2\pi a^3 \int_0^{\alpha} \left\{\left(1-\sin^2\theta\right)\cos\theta - \left(1+\cos 2\theta\right)\cos\alpha + \cos^2\alpha\cos\theta\right\} d\theta$$

$$= 2\pi a^3 \left[\sin\theta - \frac{\sin^3\theta}{3} - \theta\cos\alpha - \frac{1}{2}\sin 2\theta\cos x + \cos^2\alpha\sin\theta\right]_0^{\alpha}$$

$$= 2\pi a^3 \left[\left(\sin\alpha - \frac{1}{3}\sin^3\alpha - \alpha\cos\alpha - \frac{1}{2}\sin 2\alpha\cos\alpha + \cos^2\alpha . \sin\alpha\right) - 0\right]$$

$$= 2\pi a^3 \left[\sin\alpha - \frac{1}{3}\sin^3\alpha - \alpha\cos\alpha - \frac{1}{2}(2\sin\alpha\cos\alpha\cos\alpha + \cos^2\alpha.\sin\alpha \right]$$

$$= 2\pi a^3 \left[\sin\alpha - \frac{1}{3}\sin^3\alpha - \alpha\cos\alpha \right]$$

Required surface $= 2\int 2\pi(PL)ds$

But, $\frac{ds}{d\theta} = \sqrt{\left(\frac{dx}{d\theta}\right)^2 + \left(\frac{dy}{d\theta}\right)^2} = a$

$\therefore$ surface $= 4\pi\int_0^\alpha (a\cos\theta - a\cos\alpha)a\ d\theta$

$$= 4\pi a^2 [\sin\theta - \theta\cos\alpha]_0^\alpha$$

$$= 4\pi a^2 [\sin\alpha - \alpha\cos\alpha]$$

36. A quadrant of a circle of radius a revolves round its chord. Find the volume and the surface area of the solid spindle generated.

Let the arc APB revolves about AB. The parametric equations of the circle are $x = a\cos\theta$, $y = a\sin\theta$.

Let $P(a\cos\theta, a\sin\theta)$ be a point on the arc. Draw PM $\perp'$ to AB.

Equation of AB is $x + y = a$, i.e., $x + y - a = 0$.

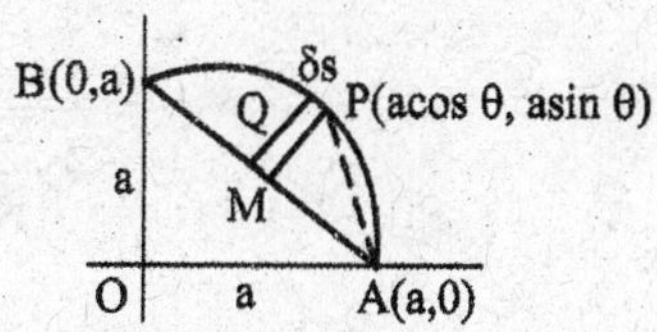

$\therefore$ length of $\perp$ PM $= \frac{a\cos\theta + a\sin\theta - a}{\sqrt{2}}$

$$= \frac{a}{\sqrt{2}}(\cos\theta + \sin\theta - 1)$$

Also, $AM = \sqrt{AP^2 - PM^2}$

$$= \sqrt{\left[(a\cos\theta - a)^2 + (a\sin\theta - 0)^2 - \frac{a^2}{2}(\cos\theta + \sin\theta - 1)^2\right]}$$

$$= a\sqrt{\begin{matrix}[\cos^2\theta + 1 - 2\cos\theta + \sin^2\theta \\ -\frac{1}{2}(\cos^2\theta + \sin^2\theta + 1 - 2\cos\theta - 2\sin\theta + 2\sin\theta\cos\theta)]\end{matrix}}$$

$$=\frac{a}{\sqrt{2}}\sqrt{\left[\cos^2\theta+\sin^2\theta+1-2\cos\theta-2\sin\theta-2\sin\theta\cos\theta\right]}$$

$$=\frac{a}{\sqrt{2}}\sqrt{(\sin\theta-\cos\theta+1)^2}=\frac{a}{\sqrt{2}}(\sin\theta-\cos\theta+1)$$

Elementary arc PQ = ds = $\sqrt{dx^2+dy^2}=a\ d\theta$. For the quadrant of a circle, limits of θ are from 0 to $\frac{\pi}{2}$.

$$\therefore \text{ required volume } =\int_0^{\pi/2}\pi PM^2 d(AM)$$

$$=\pi\int_0^{\pi/2}\frac{a^2}{2}(\cos\theta+\sin\theta-1)^2\, d\left[\frac{a}{\sqrt{2}}(\sin\theta-\cos\theta+1)\right]d\theta$$

$$=\frac{\pi a^2}{2}\int_0^{\pi/2}\left(\cos^2\theta+\sin^2\theta+1-2\cos\theta-2\sin\theta+2\sin\theta\cos\theta\right)$$

$$\frac{a}{\sqrt{2}}(\cos\theta+\sin\theta)d\theta$$

$$=\frac{\pi a^3}{2\sqrt{2}}\int_0^{\pi/2}\begin{bmatrix}\cos^3\theta+\sin^2\theta\cos\theta+\cos\theta-2\cos^2\theta-2\sin\theta\cos\theta\\ +2\sin\theta\cos^2\theta+\cos^2\theta\sin\theta+\sin^3\theta-2\sin\theta\cos\theta\\ -2\sin^2\theta+2\sin^2\theta\cos\theta\end{bmatrix}d\theta$$

$$=\frac{\pi a^3}{2\sqrt{2}}\int_0^{\pi/2}\begin{bmatrix}\cos^3\theta+\sin^3\theta+3\sin^2\theta\cos\theta+3\sin\theta\cos^2\theta\\ -2\left(\sin^2\theta+\cos^2\theta\right)+\sin\theta+\cos\theta-4\sin\theta\cos\theta\end{bmatrix}d\theta$$

$$=\frac{\pi a^3}{2\sqrt{2}}\int_0^{\pi/2}\begin{bmatrix}\int_0^{\pi/2}\cos^3\theta\ d\theta+\int_0^{\pi/2}\sin^3\theta\ d\theta+3\int_0^{\pi/2}\left(1-\cos^2\theta\right)\cos\theta\ d\theta\\ +3\int_0^{\pi/2}\left(1-\cos^2\theta\right)\cos\theta\ d\theta+3\int_0^{\pi/2}\sin\theta\left(1-\sin^2\theta\right)d\theta\\ -2\int_0^{\pi/2}d\theta+\int_0^{\pi/2}\sin\theta d\theta+\int_0^{\pi/2}\cos\theta\ d\theta-4\int_0^{\pi/2}\sin\theta d(\sin\theta)\end{bmatrix}$$

$$=\frac{\pi a^3}{2\sqrt{2}}\left\{\begin{array}{l}\frac{2}{3}+\frac{2}{3}+3[\sin\theta]_0^{\pi/2}-3\times\frac{2}{3}+3[-\cos\theta]_0^{\pi/2}\\ -3\times\frac{2}{3}-2\times\frac{\pi}{2}+[-\cos\theta]_0^{\pi/2}+[\sin\theta]_0^{\pi/2}-4\left[\frac{\sin^2\theta}{2}\right]_0^{\pi/2}\end{array}\right\}$$

$$=\frac{\pi a^3}{2\sqrt{2}}\left\{\begin{array}{l}\frac{4}{3}+3(1-0)-2+3(-0+1)-2-\pi+(-0+1)\\ +(1-0)-4\left(\frac{1}{2}-0\right)\end{array}\right\}$$

$$=\frac{\pi a^3}{2\sqrt{2}}\left\{\frac{4}{3}+3-2+3-2-\pi+1+1-2\right\}$$

$$=\frac{\pi a^3}{2\sqrt{2}}\left(\frac{4}{3}+2-\pi\right)=\frac{\pi a^3}{2\sqrt{2}}\left(\frac{4+6-3\pi}{3}\right)$$

$$=\frac{\pi a^3}{6\sqrt{2}}(10-3\pi)$$

The required surface area

$$= 2\pi\int PM\ ds$$

$$=\frac{2\pi a}{\sqrt{2}}\int_0^{\pi/2}(\cos\theta+\sin\theta-1)a\ d\theta$$

$$=\sqrt{2}\pi a^2\int_0^{\pi/2}(\cos\theta+\sin\theta-1)\ d\theta$$

$$=\sqrt{2}\pi a^2\left[\sin\theta-\cos\theta-\theta\right]_0^{\pi/2}$$

$$=\sqrt{2}\pi a^2\left[(1-0)-(0-1)-\frac{\pi}{2}\right]$$

$$=\sqrt{2}\pi a^2\left(2-\frac{\pi}{2}\right)$$

$$=2\sqrt{2}\pi a^2\left(1-\frac{\pi}{4}\right)$$

Exercise 10.1

1. Find the volume and surface area of the parabola generated by the revolution about the x-axis of the parabola $y^2 = 4ax$ from $x = 0$ to $x = h$.

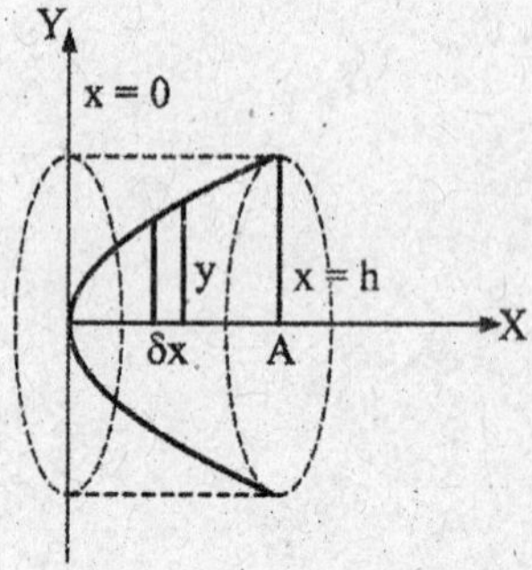

[*Hint*:

$$\text{Volume} = \int_0^h \pi y^2\,dx = \int_0^h \pi.\ 4ax\,dx = 4a\pi\left(\frac{1}{2}x^2\right)_0^b = 2a\pi h^2$$

$$\text{Surface} = 2\pi\int_0^h y\sqrt{1+\left(\frac{dy}{dx}\right)^2}\,dx$$

$$= 4\pi\sqrt{a}\int_0^h \sqrt{a+x}\,dx$$

$$= \frac{8}{3}\pi\sqrt{a}\left\{(a+h)^{3/2} - a^{3/2}\right\}]$$

2. Find the volume and surface area of the solid generated by the revolution of an arc of the catenary $y = c\cosh\left(\frac{x}{c}\right)$ about x-axis.

[*Hint*: Considering an arc from the vertex to a point (x, y),

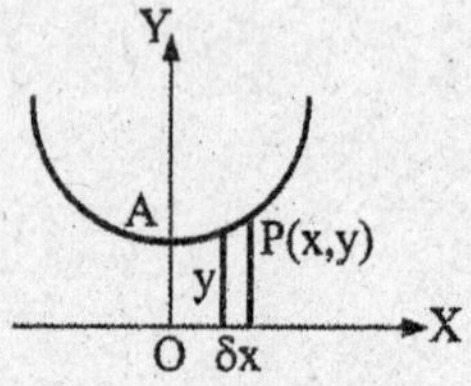

$$\text{Volume} = \int_0^x \pi y^2 dx$$

$$= \pi\int_0^x c^2\cosh^2\frac{x}{c}\,dx$$

$$= \frac{\pi c^2}{2}\int_0^x \left(1+\cosh\frac{2x}{c}\right)dx$$

$$= \frac{1}{2}\pi c^2\left[x + \frac{1}{2}c\sinh\frac{2x}{c}\right]$$

$$\text{Surface} = 2\pi\int y\,ds = 2\pi\int_0^x c\cosh\left(\frac{x}{c}\right)\sqrt{1+\left(\frac{dy}{dx}\right)^2}\,dx$$

$$= 2\pi c\int_0^x \cosh\left(\frac{x}{c}\right)\cosh\left(\frac{x}{c}\right)dx$$

$$\because \quad \frac{dy}{dx} = \sinh\left(\frac{x}{c}\right); \qquad \therefore \quad \sqrt{1+\left(\frac{dy}{dx}\right)^2} = \sqrt{1+\sinh^2\frac{x}{c}} = \cosh\frac{x}{c}$$

$$\therefore \quad \text{surface} = \pi c\int_0^x \left(1+\cosh\frac{2x}{c}\right)dx$$

$$= \pi c\left[x+\frac{1}{2}c\sinh\frac{2x}{c}\right]$$

$$= \pi c\left[x+c\sinh\frac{x}{c}\cosh\frac{x}{c}\right].]$$

3. The curve $y^2(a + x) = x^2 (3a - x)$ revolves about the axis of x. Find the volume generated by the loop.

[*Hint*: Loop is formed between (0, 0) and (3a, 0).

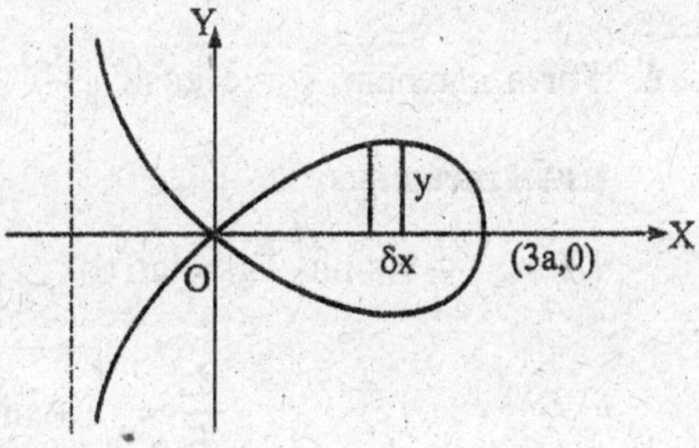

$$\text{Volume} = \int_0^{3a} \pi y^2 dx$$

$$= \pi\int_0^{3a} \frac{x^2(3a-x)}{a+x}dx$$

Put $a + x = t$]

4. Find the volume of the solid generated by the curve $x = 2a \sin^2 t$, $y = 2a \dfrac{\sin^3 t}{\cos t}$ about its asymptote.

[*Hint*: Figure as in example 28

$$\text{Volume} = 2\pi\int_0^{\infty} (2a-x)^2 dy$$

$$= 2\pi\int_0^{\pi/2} (2a-x)^2 \frac{dy}{dx}dt$$

$$= 2\pi\int_0^{\pi/2} 4a^2\cos^2 t\ 2a\frac{3\sin^2 t.\cos^2 t+\sin^4 t}{\cos^2 t}dt$$

$$=16\pi a^3\int_0^{\pi/2}\cos^2 t\sin^2 t\left(1+2\cos^2 t\right)dt$$

$$=16\pi a^3\left[\frac{\Gamma\left(\frac{2}{3}\right)\Gamma\left(\frac{2}{3}\right)}{2\Gamma(3)}+2\frac{\Gamma\left(\frac{5}{3}\right)\Gamma\left(\frac{3}{2}\right)}{2\Gamma(4)}\right]=2\pi^2 a^3]$$

5. Find the surface of the solid formed by the revolution about x-axis of the loop of the curve $x=t^2, y=t-\frac{1}{3}t^3$.

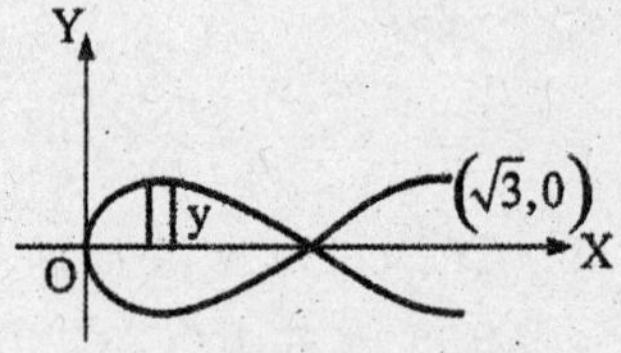

[*Hint*: $ds=\sqrt{dx^2+dy^2}dt=\left(1+t^2\right)dt$

$$\text{Surface}=2\pi\int y\,ds$$

$$=2\pi\int_0^{\sqrt{3}}\left(t-\frac{1}{3}t^3\right)\left(1+t^2\right)dt]$$

6. For a catenary $y=c\cosh\left(\frac{x}{c}\right)$, prove that $s=c\tan\psi$, where symbols have their usual meanings.

[*Hint*: $\frac{dy}{dx}=\sinh\left(\frac{x}{c}\right)=\tan\psi$

$$\frac{ds}{dx}=\sqrt{1+\sinh^2\left(\frac{x}{c}\right)}=\cosh\frac{x}{c}, ds=\cosh\frac{x}{c}dx \qquad .....(1)$$

$$\therefore \qquad S=\int_0^x\cosh\frac{x}{c}dx=c\sinh\frac{x}{c}; \quad \text{now use (1)}]$$

7. Find the area of the solid formed by revolution of the part of the curve $ay^2=x^3$ about y-axis from $x=0$ to $x=4a$.

[*Hint*: $y=\frac{1}{\sqrt{a}}x^{3/2}\quad \frac{ds}{dx}=\sqrt{1+\left(\frac{dy}{dx}\right)^2}=\sqrt{\frac{4a+9x}{4a}}$

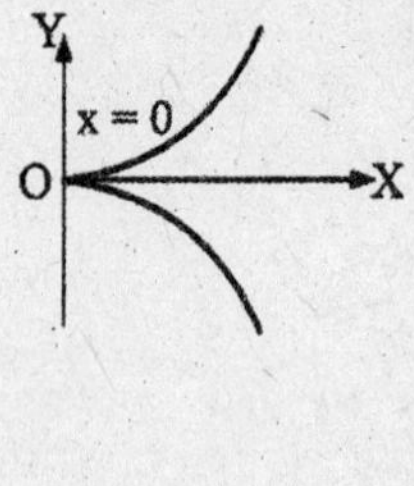

$$\text{Surface}=2\pi\int x\,dx$$

$$=\frac{\pi}{\sqrt{a}}\int_0^{4a}\sqrt{4a+9x}\,dx$$

Put $4a+9x=z$

$$\text{Surface}=\frac{2\pi}{81\sqrt{a}}\int_{\sqrt{4a}}^{\sqrt{40a}}\left(z^4-4az^2\right)dz]$$

8. Show that the volume of the solid formed by the revolution of the limacon $r = a + b\cos\theta$ $(a > b)$ about the initial line is $\frac{4}{3}\pi a\left(a^2 + b^2\right)$.

[*Hint*: Volume $= \frac{2\pi}{3}\int_0^{\pi} r^3 \sin\theta \, d\theta$

$$= \frac{2}{3}\int_0^{\pi}(a + b\cos\theta)^3 \sin\theta \, d\theta$$

Put $\cos\theta = z$

$$\text{Volume} = \frac{2}{3}\pi\int_{-1}^{1}(a + bz)^3 \, dz]$$

Answers

3. $\pi a^3(8\log 2 - 3)$

5. 3p

7. $\frac{128}{1215}\pi a^2\left\{125\sqrt{10} + 1\right\}$

CHAPTER 11

Centroid (Centre of Gravity) and Moment of Inertia

Centre of Gravity (CG)

We are familiar with the term 'centre of gravity'. Here we wish to use integral calculus to know the CG.

If δm be the mass of an elementary portion of the continuous body and (x, y) its coordinates referred to two rectangular axes then the coordinates $(\bar{x}, \bar{y})$ of the centre of gravity of the body are given by

$$\bar{x} = \frac{\int x\, dm}{\int dm}, \quad \bar{y} = \frac{\int y\, dm}{\int dm}$$

where the integration extends over the entire mass.

Formulae for CG in useful forms

(i) CG of a plane area

$$\bar{x} = \frac{\int x\, dA}{\int dA}, \quad \bar{y} = \frac{\int y\, dA}{\int dA}$$

as δm = mass of elementary area $\delta A = \rho \delta A$ where ρ is mass per unit area taken constant.

Case (i) The CG of a plane area bounded by the curve $y = f(x)$, the axis of x and ordinates $x = a$, $x = b$, is given by

$$\bar{x} = \frac{\int_a^b x\,(y\,dx)}{\int_a^b y\,dx},\ \bar{y} = \frac{\int_a^b \frac{y}{2}(y\,dx)}{\int_a^b y\,dx}$$

Case (ii) The CG of a plane area bounded by the curve $x = f(y)$, the axis of y and abscissae at $y = a,\ y = b$, is given by

$$\bar{x} = \frac{\int_a^b \frac{1}{2}x.(x\,dy)}{\int y\,dx},\ \bar{y} = \frac{\int_a^b y(x\,dy)}{\int_a^b x\,dy}$$

Case (iii) The CG of the sectorial area bounded by the curve $r = f(\theta)$ and radii vectors $\theta = \alpha,\ \theta = \beta$, is given by

$$\bar{x} = \frac{2}{3}\frac{\int_\alpha^\beta r^3 \cos\theta\,d\theta}{\int_\alpha^\beta r^2\,d\theta},\ \bar{y} = \frac{2}{3}\frac{\int_\alpha^\beta r^3 \sin\theta\,d\theta}{\int_\alpha^\beta r^2\,d\theta}$$

Case (iv) The CG of the area enclosed between the x-axis and two curves is given by

$$\bar{x} = \frac{\int \left(\frac{x_1 + x_2}{2}\right)(x_1 - x_2)dy}{\int (x_1 - x_2)dy},\ \bar{y} = \frac{\int y(x_1 - x_2)dy}{\int (x_1 - x_2)dy}$$

Case (v) The CG of the area enclosed between two curves is given by

$$\bar{x} = \frac{\int x(y_1 - y_2)dx}{\int (y_1 - y_2)dy},\ \bar{y} = \frac{\int \left(\frac{y_1+y_2}{2}\right)(y_1 - y_2)dx}{\int (y_1 - y_2)dx}$$

The limits of integration are given by the points of intersection.

(ii) CG of an arc

$$\bar{x} = \frac{\int x\,ds}{\int ds},\ \bar{y} = \frac{\int y\,ds}{\int ds} \text{ (replacing dm by } \rho ds\text{).}$$

(iii) CG of solid of revolution

About x-axis; $\bar{x} = \frac{\int xy^2\,dx}{\int y^2 dx},\ \bar{y} = 0$ about y-axis; $\bar{y} = \frac{\int yx^2\,dy}{\int x^2 dy},\ \bar{x} = 0$.

CG of solid of revolution always lies on the axis of revolution.

(iv) CG of surface of revolution

About x-axis: $\bar{x} = \dfrac{\int xy\, ds}{\int y\, dx}$, $\bar{y} = 0$; about y-axis; $\bar{y} = \dfrac{\int yx\, ds}{\int x\, dx}$, $\bar{x} = 0$;

Limits of integration being adjusted properly in each case.

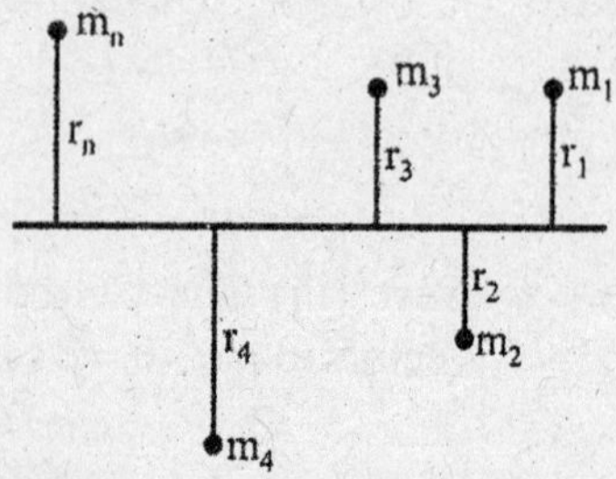

Moment of Inertia

The moment of inertia of a particle about a line is defined as the product of its mass and the square of its distance from the line.

If a system of particles of masses $m_1, m_2, \ldots,$ be situated as distances $r_1, r_2, \ldots,$ respectively from some fixed straight line then

$$\sum mr^2, \text{ i.e., } \quad m_1r_1^2 + m_2r_2^2 + \ldots$$

is the moment of inertia of the system about the given line.

If M be the total mass of the system, the moment of inertia (MI) of the system about any line is expressed in the form Mk^2, where k is the called the radius of gyration of the system about the given line.

Moment of inertia of a continuous distribution of mass may be written in the form $\int r^2 dm$, where the integration extends over the whole body.

The concept of MI can be extended to volumes, areas and arcs, and we have

$$\text{MI of a volume} = \int_v r^2 dv$$

$$\text{MI of an area} = \int_s r^2 dA$$

$$\text{MI of an arc} = \int_c r^2 ds$$

Theorem of Parallel and Perpendicular Axes

Moment of inertia of a body about a given line in its plane is equal to the about a parallel lines through its CG plus the product of the mass and square of the distance between the axes (lines).

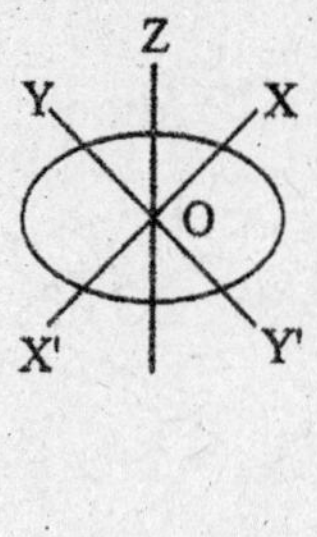

Thus $\quad I = I_g + Mr^2$

where $\quad I$ = MI about the given line,

I_g = MI about the line through CG,

M = total mass,

r = distance between two lines

If I_x and I_y be the moments of inertia of a lamina about two perpendicular axes OX and OY passing through O and lying in the plane then moment of inertia I_z of the lamina about a normal to the plane through O is given by $I_z = I_x + I_y$.

Solved Examples

1. Find the centroid of a uniform lamina in the form of quadrant of the ellipse $\frac{x^2}{a^2}+\frac{y^2}{b^2}=1$.

Let OAB be the quadrant whose CG is required. If $(\bar{x}, \bar{y})$ be the coordinates of CG, to find $\bar{x}$, we divide the lamina into vertical strips. The area of the strip $PL = y\,\delta x$, where (x, y) are the coordinates of P and δx is the width of the strip.

$\therefore \quad \delta m = y\rho\,\delta x,$

where ρ is the density of the lamina.

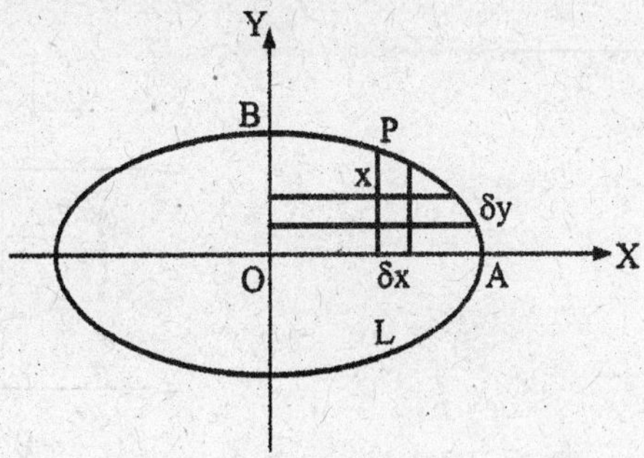

So,

$$\bar{x} = \frac{\int x\,dm}{\int dm} = \frac{\int_0^a xy\rho\,dx}{\int_0^a y\rho\,dx}$$

$$= \frac{\int_0^a x.\frac{b}{a}\sqrt{a^2-x^2}\,dx}{\int_0^a \frac{b}{a}\sqrt{a^2-x^2}\,dx}$$

Putting $x = a\sin\theta$. $dx = a\cos\theta\,d\theta$ and $\sqrt{a^2-x^2} = a\cos\theta$.

When $x = 0, \theta = 0$; when $x = a, \theta = \pi/2$

$$\therefore \quad \bar{x} = \frac{a^3\int_0^{\pi/2}\sin\theta\cos^2\theta\,d\theta}{a^2\int_0^{\pi/2}\cos^2\theta\,d\theta} = \frac{-a.\int_0^{\pi/2}\cos^2\theta(\cos\theta)}{\int_0^{\pi/2}\frac{1+\cos 2\theta}{2}d\theta}$$

$$= \frac{-2a\left[\frac{\cos^3\theta}{3}\right]_0^{\pi/2}}{\left[\theta+\frac{\sin 2\theta}{2}\right]_0^{\pi/2}} = \frac{4a}{3\pi}$$

To find $\bar{y}$, we divide the lamina into horizontal strips and as above,

$$\bar{y} = \frac{\int y\,dm}{\int dm} = \frac{\int_0^b y(x\rho\,dy)}{\int_0^b x\rho\,dy} = \frac{\int_0^b y.\frac{a}{b}\sqrt{b^2-y^2}\,dy}{\int_0^b \frac{a}{b}\sqrt{b^2-y^2}\,dy}$$

Putting $y = b\sin\theta$, $dy = b\cos\theta\,d\theta$ and $\sqrt{b^2-y^2} = b\cos\theta$.

When $y = 0, \theta = 0$; when $y = b, \theta = \pi/2$.

$$\therefore \quad \bar{y} = \frac{\int_0^{\pi/2} b\sin\theta\,.\,b\cos\theta\,b\cos\theta\,d\theta}{\int_0^{\pi/2} b\cos\theta.\,b\cos\theta\,d\theta}$$

$$= \frac{\int_0^{\pi/2} \sin\theta\cos^2\theta\,d\theta}{\int_0^{\pi/2}\cos^2\theta\,d\theta} = \frac{-b\int_0^{\pi/2}\cos^2\theta\,d(\cos\theta)}{\int_0^{\pi/2}\frac{1+\cos 2\theta}{2}\,d\theta}$$

$$= \frac{-2b\left[\frac{\cos^3\theta}{3}\right]_0^{\pi/2}}{[\theta]_0^{\pi/2} + \left\{\frac{\sin 2\theta}{2}\right\}_0^{\pi/2}}$$

$$= \frac{-2b(-1)}{3\left(\frac{\pi}{2}+0\right)} = \frac{4b}{3\pi}$$

2nd Method:

Proceed like above up to $\bar{x} = \frac{4a}{3\pi}$. As $\frac{x}{a}$ and $\frac{y}{b}$ are symmetrically placed in equation of the curve $\left(\frac{x}{a}\right)^2 + \left(\frac{y}{b}\right)^2 = 1$,

hence $\bar{y} = \frac{4b}{3\pi}$

2. Find the CG of the area enclosed by $y = mx$ and $y^2 = 4ax$.

Here the points of intersection of the line and the curve are

$O(0, 0)$ and $A\left(\frac{4a}{m^2}, \frac{4a}{m}\right)$.

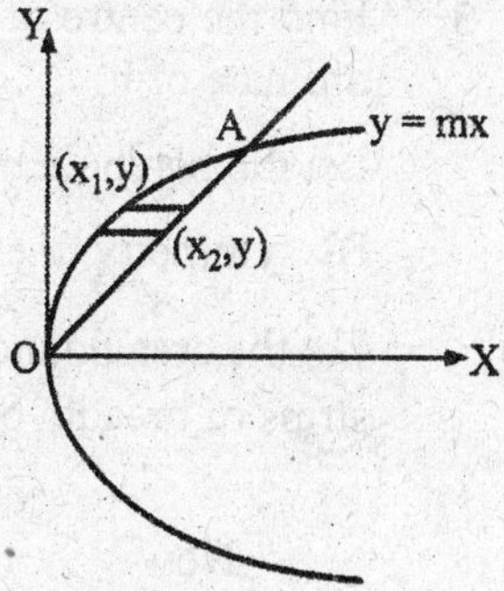

Area of the elementary strip

$$= (x_1 - x_2)\delta y$$

$$\therefore \quad \overline{x} = \frac{\int_0^{4a/m} \left(\frac{x_1 + x_2}{2}\right)(x_1 - x_2)dy}{\int_0^{4a/m} (x_1 - x_2)dy}$$

[see formula (iv) of I]

$$= \frac{1}{2}\frac{\int_0^{4a/m} (x_2^2 - x_1^2)dy}{\int_0^{4a/m} (x_2 - x_1)dy}$$

$$= \frac{1}{2}\frac{\int_0^{4a/m}\left[\left(\frac{y}{m}\right)^2 - \left(\frac{y}{4a}\right)^2\right]dy}{\int_0^{4a/m}\left[\left(\frac{y}{m} - \frac{y^2}{4a}\right)\right]dy} \quad \left[\text{since here } x_1 = \frac{y^2}{4a};\ x_2 = \frac{y}{m}\right]$$

$$= \frac{1}{2}\frac{\left[\frac{y^3}{3m^2} - \frac{y^5}{80a^2}\right]_0^{4a/m}}{\left[\frac{y^2}{2m} - \frac{y^3}{12a}\right]_0^{4a/m}} = \frac{1}{2}\frac{\left[\frac{64a^3}{3m^5} - \frac{64a^3}{5m^5}\right]}{\left[\frac{8a^2}{m^3} - \frac{16a^2}{3m^3}\right]}$$

$$= \frac{64a^3}{15m^5} \times \frac{3m^3}{8a^2} = \frac{8a}{5m^2}$$

$$\overline{y} = \frac{\int_0^{4a/m} y(x_1 - x_2)dy}{\int_0^{4a/m} (x_1 - x_2)dy} = \frac{\int_0^{4a/m} y(x_2 - x_1)dy}{\int_0^{4a/m} (x_2 - x_1)dy}$$

$$= \frac{\int_0^{4a/m} y(y/m - y^2/4a)dy}{\int_0^{4a/m} (y/m - y^2/4a)dy}$$

$$= \frac{\left[\frac{y^2}{3m} - \frac{y^4}{16a}\right]_0^{4a/m}}{\left[\frac{y^2}{2m} - \frac{y^3}{12a}\right]_0^{4a/m}} = \frac{\frac{64a^3}{3m^4} - \frac{16a^3}{m^4}}{\frac{16a^2}{2m^3} - \frac{64a^3}{12am^3}}$$

$$= \frac{16a^3}{3m^4} \times \frac{3m^3}{8a^2} = \frac{2a}{m}$$

3. Find the centre of gravity of a uniform lamina bounded by a parabola and a double ordinate of it.

Let the parabola be $y^2 = 4ax$ and the double ordinate be $x = k$.

By symmetry, the CG will be on x-axis hence $\overline{y} = 0$

Let the area bounded by the parabola and the double ordinate, $x = k$ be divided into strips of breadth δx parallel to y-axis.

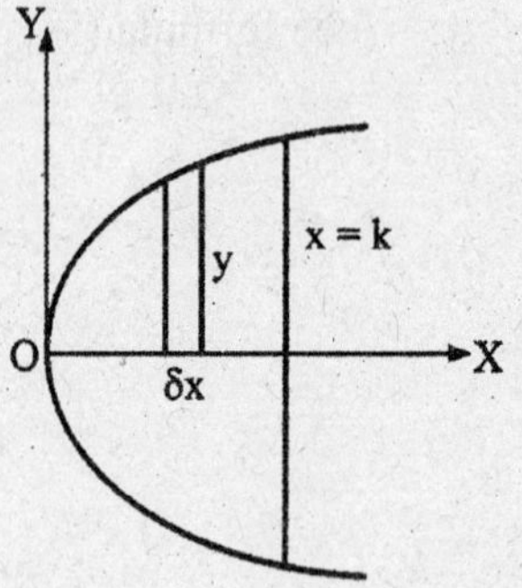

Now $$\overline{x} = \frac{\int_0^k x(2ydx)}{\int_0^k 2ydx}$$

$$= \frac{\int_0^k x\sqrt{4ax}dx}{\int_0^k \sqrt{4ax}dx} = \frac{\int_0^k x^{3/2}dx}{\int_0^k x^{1/2}dx}$$

$$= \frac{\left[\frac{x^{5/2}}{5/2}\right]_0^k}{\left[\frac{x^{3/2}}{3/2}\right]_0^k} = \frac{\frac{2}{5}k^{5/2}}{\frac{2}{3}k^{3/2}} = \frac{3}{5}k$$

$\therefore$ CG is $\left(\frac{3}{5}k, 0\right)$

4. Find the centroid of the area included between the parabola $y^2 = 4ax$ and its rectum.

Proceed like above example and put $k = a$

We get $$\overline{x} = \frac{3}{5}a, \overline{y} = 0$$

5. Find the CG of the area included between the curve $y^2(2a - x) = x^3$ and it asymptote.

The curve passes through the origin and is symmetrical about x-axis as even power of y occurs. So, $\overline{y} = 0$, where $(\overline{x}, \overline{y})$ are the coordinates of CG.

Equating to zero the coefficient of the highest power of y, the asymptote parallel to y-axis is $x = 2a$

$$\overline{x} = \frac{\int_0^{2a} x.2y\ dx}{\int_0^{2a} 2y\ dx}$$

Here area of the elementary strip is $2y\ \delta x$ and x varies from 0 to 2a

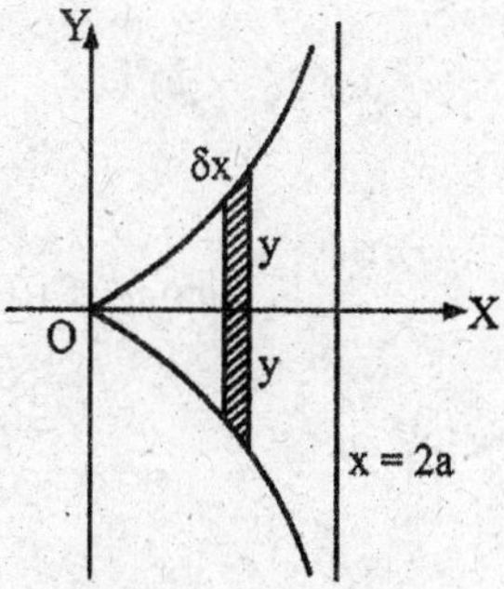

$$\therefore \quad \bar{x} = \frac{\int_0^{2a} xy\ dx}{\int_0^{2a} y\ dx} = \frac{\int_0^{2a} \frac{xx^{3/2}}{\sqrt{2a-x}} dx}{\int_0^{2a} \frac{x^{3/2}}{\sqrt{2a-x}} dx}$$

$$\left(\because y^2 = \frac{x^3}{2a-x}\right)$$

Putting $x = 2a \sin^2 \theta$, $dx = 4a\sin \theta \cos \theta\ d\theta$.
When $x = 0, \theta = 0$; when $x = 2a$, $\theta = \pi/2$

$$\therefore \quad \bar{x} = \frac{\int_0^{\pi/2} \frac{(2a)^{5/2} \sin^5 \theta\ 4a \sin\theta \cos\theta\ d\theta}{\sqrt{2a} \cos\theta}}{\int_0^{\pi/2} \frac{(2a)^{3/2} \sin^3 \theta 4a \sin\theta \cos\theta\ d\theta}{\sqrt{2a} \cos\theta}}$$

$$= \frac{16a^3 \int_0^{\pi/2} \sin^6 \theta\ d\theta}{8a^2 \int_0^{\pi/2} \sin^4 \theta\ d\theta} = \frac{2a.\frac{5}{6}.\frac{3}{4}.\frac{1}{2}.\frac{\pi}{2}}{\frac{3}{4}.\frac{1}{2}.\frac{1}{2}\pi} = \frac{5}{3}a$$

6. Find the centroid of the area bounded by the cardioid $r = a(1 + \cos \theta)$

Here the elementary area δs = area of the narrow sector $OPQ = \frac{1}{2} r^2\ \delta\theta$

and its CG is at G_1, where $OG_1 = \frac{2}{3}$ OP in limit $= \frac{2}{3}$ r.

Where $OG_1 = \frac{2}{3}$ OP in limit $= \frac{2}{3} r$

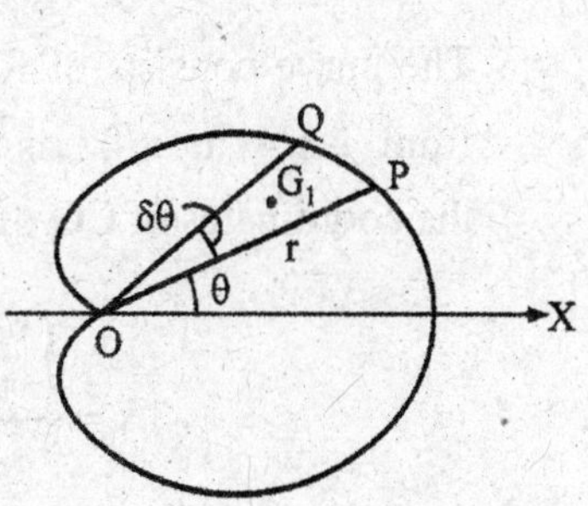

Coordinates of G_1 are $\left(\frac{2}{3} r \cos\theta, \frac{2}{3} r \sin\theta\right)$

So, if $(\bar{x}, \bar{y})$ be the required CG,

$$\bar{x} = \frac{\int_{-\pi}^{\pi} \left(\frac{2}{3} r \cos\theta\right) . \frac{1}{2} r^2 d\theta}{\int_{-\pi}^{\pi} \frac{1}{2} r^2 d\theta}$$

and $\bar{y} = 0$ (since CG lies on the initial line),

i.e., $$\bar{x} = \frac{2}{3}\frac{\int_0^{\pi} a^3(1+\cos\theta)^3 \cos\theta d\theta}{\int_0^{\pi} a^2(1+\cos\theta)^2 d\theta} \qquad [\because r = a(1+\cos\theta)]$$

(since if f(θ) is an even function of θ then $\int_{-a}^{a} f(\theta)d\theta = 2\int_0^{a} f(\theta)d\theta$)

$$= \frac{2}{3}\frac{a^3\int_0^{\pi}(1+3\cos\theta+3\cos^2\theta+\cos^3\theta)\cos\theta d\theta}{a^2\int_0^{\pi}(1+2\cos\theta+\cos^2\theta)d\theta}$$

$$= \frac{2}{3}a\frac{2\int_0^{\pi/2}(3\cos^2\theta+\cos^4\theta)d\theta}{\pi+2\int_0^{\pi/2}\cos^2\theta\, d\theta}$$

$\left(\because \int_0^{\pi}\cos^n\theta d\theta = 0 \text{ or } 2\int_0^{\pi/2}\cos^n\theta\, d\theta \text{ according as n is odd or even}\right)$

$$= \frac{4}{3}a\frac{\left[3.\frac{1}{2}\frac{\pi}{2}+\frac{3}{4}.\frac{1}{2}\frac{\pi}{2}\right]}{\pi+2.\frac{1}{2}\frac{\pi}{2}}$$

$\left(\text{as} \int_0^{\pi/2}\cos^n x dx = \frac{n-1}{n}.\frac{n-3}{n-2}...\frac{\pi}{2} \text{ when n is even}\right)$

$$= \frac{5\pi a}{4}\frac{3\pi a}{2} = \frac{5}{6}a \qquad \text{So CG is } \left(\frac{5}{6}a, 0\right)$$

7. Find the CG of a loop of the curve r = a cos 2θ.

The curve consists of four loops. Since r = 0 for $\theta = \pm\pi/4$ so for a loop θ varies from $-\pi/4$ to $\pi/4$, as this loop is symmetrical about the initial line. If $(\bar{x},\bar{y})$ be the coordinates of CG of this loop then $\bar{y} = 0$ and

$$\bar{x} = \frac{\int_{-\pi/4}^{\pi/4}\left(\frac{2}{3}r\cos\theta\right)\frac{1}{2}r^2 d\theta}{\int_{-\pi/4}^{\pi/4}\frac{1}{2}r^2 d\theta}$$

$$= \frac{2}{3}\frac{\int_{-\pi/4}^{\pi/4} r^3\cos\theta\, d\theta}{\int_{-\pi/4}^{\pi/4} r^2 d\theta}$$

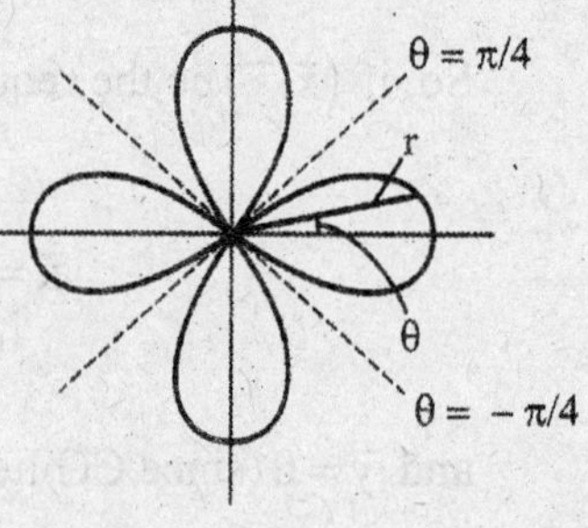

$$=\frac{2}{3}\frac{\int_{-\pi/4}^{\pi/4} a^3\cos^3 2\theta.\cos\theta\, d\theta}{\int_{-\pi/4}^{\pi/4} a^2\cos^2 2\theta\, d\theta}$$

The numerator of $\bar{x}=\frac{2}{3}\int_{-\pi/4}^{\pi/4} a^3\cos^3 2\theta\cos\theta\, d\theta$

$$=\frac{4}{3}a^3\int_0^{\pi/4}\cos^3 2\theta\cos\theta\, d\theta$$

Since $\int_{-a}^{a} f(\theta)d\theta=2\int_0^a f(\theta)d\theta$ when $f(\theta)$ is an even function of θ, i.e.,
when $f(\theta)=f(-\theta)$

$$=\frac{4}{3}a^3\int_0^{\pi/4}\left(1-2\sin^2\theta\right)^3\cos\theta\, d\theta$$

Put $\sqrt{2}\sin\theta=\sin\alpha$ so that $\sqrt{2}\cos\theta d\theta=\cos\alpha\, d\alpha$;
when $\theta=0,\alpha=0$; when $\theta=\pi/4,\ \alpha=\pi/2$

$\therefore$ numerator of $\bar{x}=\frac{4}{3}a^3.\frac{1}{\sqrt{2}}\int_0^{\pi/2}\left(1-\sin^2\alpha\right)^3\cos\alpha\, d\alpha$

$$=\frac{4}{3\sqrt{2}}a^3\int_0^{\pi/2}\cos^7\alpha\, d\alpha$$

$$=\frac{4}{3\sqrt{2}}a^3.\left[\frac{6}{7}.\frac{4}{5}.\frac{2}{3}\right]=\frac{32\sqrt{2}}{105}a^3$$

as $\int_0^{\pi/2}\cos^n x\, dx=\frac{n-1}{n}.\frac{n-3}{n-2}.\frac{n-5}{n-4}\ldots\frac{2}{3}.1$ when n is odd (see reduction formula)

The denominator of $\bar{x}=2\int_0^{\pi/2} a^2\cos^2 2\theta\, d\theta$

Putting $2\theta = z$ so that $2\, d\theta = dz$,

the denominator of $\bar{x}=a^2\int_0^{\pi/2}\cos^2 z\, dz=\frac{\pi a^2}{4}$

$$\therefore\quad \bar{x}=\frac{32\sqrt{2}}{105}a^3\times\frac{4}{\pi a^2}=\frac{128\sqrt{2}a}{105\pi}$$

$$\therefore\quad \text{CG is}\left(\frac{128\sqrt{2}a}{105\pi},0\right)$$

8. Find the CG of half of either loop of the lemniscate of Bernoulli $r^2 = a^2 \cos 2\theta$ bounded by the axis.

For half loop which is in the first quadrant, θ varies from $\theta = 0$ to $\theta = \pi/4$.

So if $(\bar{x}, \bar{y})$ be the coordinates of CG then

$$\bar{x} = \frac{\int_0^{\pi/4} \frac{2}{3} r \cos\theta . \frac{1}{2} r^2 d\theta}{\int_0^{\pi/4} \frac{1}{2} r^2 d\theta} \qquad \text{(formula)}$$

The denominator of

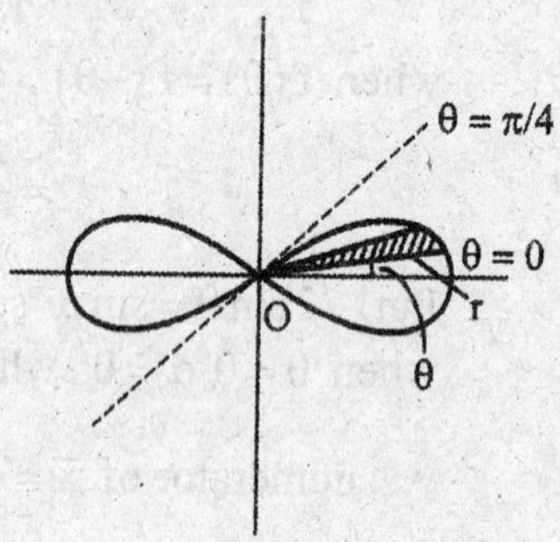

$$\bar{x} = \frac{1}{2}\int_0^{\pi/4} a^2 \cos 2\theta \, d\theta$$

$$= \frac{a^2}{4}[\sin 2\theta]_0^{\pi/4} = \frac{a^2}{4}$$

The numerator of $\bar{x} = \frac{1}{3}\int_0^{\pi/4} a^3 \cos^{3/2} 2\theta \cos\theta \, d\theta$

$$= \frac{a^3}{3}\int_0^{\pi/4} (1 - 2\sin\theta)^{3/2} \cos\theta \, d\theta$$

Put $\sqrt{2}\sin\theta = \sin\alpha$ so that $\sqrt{2}\cos\theta \, d\theta = \cos\alpha \, d\alpha$

When $\theta = 0, \alpha = 0$; when $\theta = \pi/4, \alpha = \pi/2$

$\therefore$ numerator of $\bar{x} = \frac{a^3}{3\sqrt{2}}\int_0^{\pi/2} \cos^4 \alpha \, d\alpha$

$$= \frac{a^3}{3\sqrt{2}} . \frac{3}{4} . \frac{1}{2} . \frac{\pi}{2} = \frac{\pi a^3}{16\sqrt{2}}$$

$$\therefore \qquad \bar{x} = \frac{\pi a^3}{16\sqrt{2}} \times \frac{4}{a^2} = \frac{\pi a}{4\sqrt{2}}$$

Again $\qquad \bar{y} = \frac{\int_0^{\pi/4} \frac{2}{3} r \sin\theta . \frac{1}{2} r^2 d\theta}{\int_0^{\pi/4} \frac{1}{2} r^2 d\theta} = \frac{\frac{a^3}{3}\int_0^{\pi/4} \cos^{3/2} 2\theta \sin\theta \, d\theta}{\frac{a^2}{4}}$

$$=\frac{\frac{a^3}{3}\int_0^{\pi/4}\left(2\cos^2\theta-1\right)^{3/2}\sin\theta\,d\theta}{\frac{a^4}{4}}$$

Put $\sqrt{2}\cos\theta=\sec\alpha$ so that $-\sqrt{2}\sin\theta\,d\theta=\sec\alpha\tan\alpha\,d\alpha$

When $\theta=0,\ \alpha=\pi/4;$ when $\theta=\pi/4,\ \alpha=0$

$$\therefore \qquad \bar{y}=-\frac{4}{3}a\int_{\pi/4}^{0}\frac{\tan^3\alpha\sec\alpha\tan\alpha\,d\alpha}{\sqrt{2}}$$

$$=\frac{2\sqrt{2}}{3}a\int_0^{\pi/4}\tan^4\alpha\sec\alpha\,d\alpha \qquad \text{.....(1)}$$

Integrating by parts,

$$I=\int_0^{\pi/4}\tan^{-4}\alpha\ \sec\alpha\,d\alpha=\left[\sec\alpha\tan^3\alpha\right]_0^{\pi/4}-3\int_0^{\pi/4}\sec^3\alpha\tan^2\alpha\,d\alpha$$

$$=\sqrt{2}-3\int_0^{\pi/4}\left(\tan^2\alpha+1\right)\sec\alpha\tan^2\alpha\,d\alpha$$

$$=\sqrt{2}-3\int_0^{\pi/4}\tan^4\alpha\sec\alpha\,d\alpha-3\int_0^{\pi/4}\sec\alpha\tan^2\alpha\,d\alpha$$

$$=\sqrt{2}-3I-3\int_0^{\pi/4}\tan^2\alpha\sec\alpha\,d\alpha$$

i.e., $\qquad 4I=\sqrt{2}-3I_1 \qquad$ .....(2)

where $\qquad I_1=\int_0^{\pi/4}\tan^2\alpha\sec\alpha\,d\alpha$

Integrating by parts,

$$I_1=\left[\sec\alpha\tan\alpha\right]_0^{\pi/4}-\int_0^{\pi/4}\sec^3\alpha\,d\alpha$$

$$=\sqrt{2}-I_1-\int_0^{\pi/4}\sec\alpha\,d\alpha$$

or $\qquad 2I_1=\sqrt{2}-\left[\log\tan\left(\frac{\alpha}{2}+\frac{\pi}{4}\right)\right]_0^{\pi/4}$

$$=\sqrt{2}-\log\tan\frac{3\pi}{8}$$

$\therefore$ from (2), $\qquad 4I=\sqrt{2}-\frac{3\sqrt{2}}{2}+\frac{3}{2}\log\tan\frac{3\pi}{8}$

$\therefore$ from (1), $\bar{y} = \frac{2\sqrt{2}}{3} a \times \frac{1}{4}\left\{\sqrt{2} - \frac{3\sqrt{2}}{2} + \frac{3}{2}\log\tan\frac{3\pi}{8}\right\}$

$$= \frac{a}{3\sqrt{2}}\left\{-\frac{1}{2} + \frac{3}{2}\log\tan\frac{3\pi}{8}\right\}$$

$$= a\left(-\frac{1}{6} + \frac{1}{2\sqrt{2}}\log\tan\frac{3\pi}{8}\right)$$

$$= \left\{\frac{1}{2\sqrt{2}}\log\left(1+\sqrt{2}\right) - \frac{1}{6}\right\}a$$

9. Find the centre of gravity of a circular arc subtending an angle 2α at the centre.

Let LMN be the circular arc subtending an angle 2α at the centre O.

Let OX be the axis of x and a line OY perpendicular to OX as the axis of y

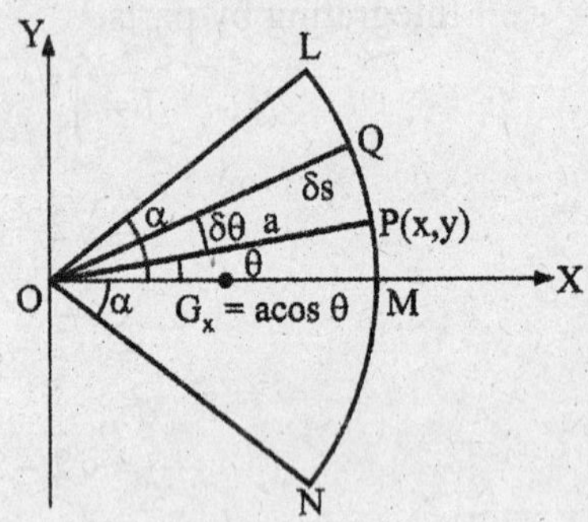

By symmetry, CG of the circular arc must lie at some point G on x-axis

$$\therefore \quad \bar{y} = 0$$

We consider an elementary arc PQ = δs subtending an angle δθ at O where ∠MOP = θ

Now, arc PQ = δs = a δθ; so ds=a, dθ, a being radius of the circle x = a cos θ

where (x, y) are the coordinates of P,

and θ extends from –α to α to cover the whole arc NML

$$\therefore \quad \bar{x} = \frac{\int x\ ds}{\int ds} = \frac{\int_{-\alpha}^{\alpha} a\cos\theta\ a\ d\theta}{\int_{-\alpha}^{\alpha} a\ d\theta} = \frac{a^2\left[\sin\theta\right]_{-\alpha}^{\alpha}}{a\left[\theta\right]_{-\alpha}^{\alpha}} = \frac{a.\sin\alpha}{\alpha}$$

So, CG lies on symmetrical radius OM at a distance

$$\frac{a\sin\alpha}{\alpha} \text{ from O}$$

10. Find the centroid of the semicircular arc in the first and fourth quadrants.
Put **$\alpha = \pi/2$** in example 9.

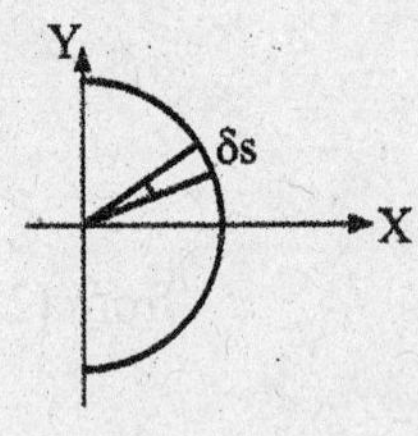

$$\bar{x} = \frac{a\sin\frac{\pi}{2}}{\frac{\pi}{2}} = \frac{2a}{\pi}$$

and the CG lies on the radius bisecting the arc

11. Find the CG of the sector of a circle

Let LON be the sector subtending an angle 2α at the centre O as the pole and OM its bisector.

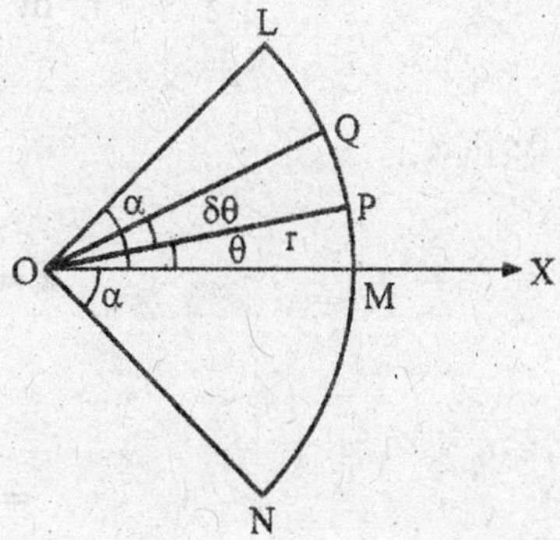

By symmetry, the CG lies on OM (x-axis)

$$\therefore \quad \bar{y} = 0$$

where $(\bar{x}, \bar{y})$ are the coordinates of CG

$$\bar{x} = \frac{\frac{2}{3}\int_{-\alpha}^{\alpha} r^3 \cos\theta \, d\theta}{\int_{-\alpha}^{\alpha} r^2 \, d\theta}$$

$$= \frac{2}{3} \frac{\int_{-\alpha}^{\alpha} a^2 \cos\theta \, d\theta}{\int_{-\alpha}^{\alpha} a^2 \, d\theta} \qquad \text{(putting } r = a\text{)}$$

since the equation of the circle is $r = a$ referred to O as pole

$$= \frac{2}{3} a \frac{[\sin\theta]_{-\alpha}^{\alpha}}{[\theta]_{-\alpha}^{\alpha}} = \frac{2}{3} a \frac{\sin\alpha}{\alpha}$$

12. Find the centre of gravity of a uniform semicircular disc.

Here $\angle LOM = 2\alpha = \pi \qquad \therefore \qquad \alpha = \pi/2$

Proceed like the above example

$$\bar{x} = \frac{2}{3} a \left(\frac{\sin \pi/2}{\pi/2} \right) = \frac{4a}{3\pi}, \quad \bar{y} = 0$$

13. Find the CG of the arc of the curve $\left(\frac{x}{a}\right)^{2/3} + \left(\frac{y}{a}\right)^{2/3} = 1$ lying in the first quadrant.

The curve (astroid) is shown in the figure

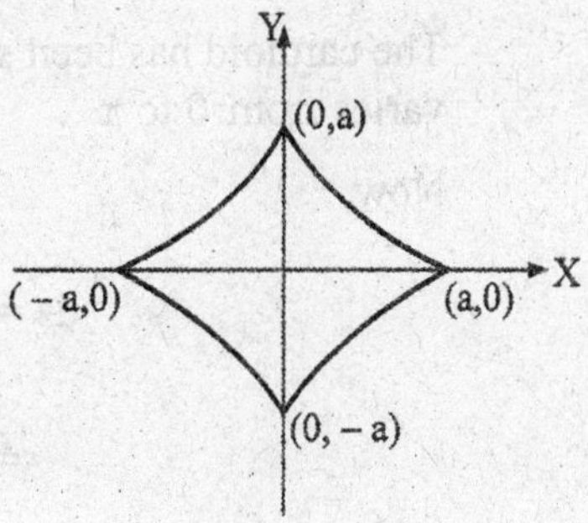

The parametric form of the equation of the curve is

$$\left. \begin{aligned} x &= a \sin^3 t \\ y &= a \cos^3 t \end{aligned} \right\}$$

$$\therefore \quad \frac{ds}{dt} = \sqrt{\left(\frac{dx}{dt}\right)^2 + \left(\frac{dy}{dx}\right)^2}$$

$$= \sqrt{\left(3a\sin^2 t\cos t\right)^2 + \left(3a\cos^2 t.\left(-\sin t\right)^2\right)} = 3a\sin t\cos t$$

$$\overline{x} = \frac{\int_0^{\pi/2} \rho x\, ds}{\int_0^{\pi/2} \rho\, ds} = \frac{\int_0^{\pi/2} x\frac{ds}{dt}dt}{\int_0^{\pi/2} \frac{ds}{dt}dt} = \frac{\int_0^{\pi/2} a\sin^3 t.3a\sin t\cos t\, dt}{\int_0^{\pi/2} 3a\sin t\cos t\, dt}$$

$$= \frac{a\int_0^{\pi/2} \sin^4 t\cos t\, dt}{\int_0^{\pi/2} \sin t\cos t\, dt} = \frac{a\int_0^{\pi/2} \sin^4 t\, d(\sin t)}{\int_0^{\pi/2} \sin t\, d(\sin t)}$$

$$= \frac{a\left[\frac{\sin^5 t}{5}\right]_0^{\pi/2}}{\left[\frac{\sin^2 t}{2}\right]_0^{\pi/2}}$$

$$= \frac{\frac{1}{5}a}{\frac{1}{2}} = \frac{2}{5}a$$

$$\therefore \quad \overline{y} = \frac{2a}{5}$$

Since x and y are symmetrically placed in the equation of the curve

Thus, CG is $\left(\frac{2a}{5}, \frac{2a}{5}\right)$

14. Find the position of CG of the arc of the cardioids $r = a(1 + \cos\theta)$ lying above the initial line.

The cardioid has been shown in the figure. For the arc lying above the initial line, θ varies from 0 to π

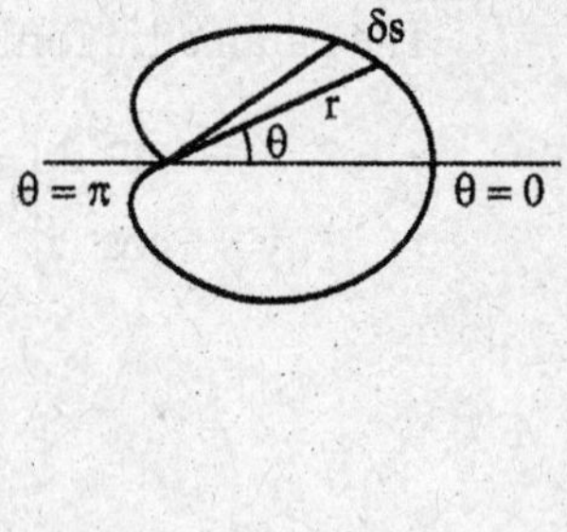

Now,

$$ds = \sqrt{r^2 + \left(\frac{dr}{d\theta}\right)^2}\, d\theta$$

$$= \sqrt{\left\{a^2\left(1+\cos\theta\right)^2 + \left(-a\sin\theta\right)^2\right\}}\, d\theta$$

$$= \sqrt{\left\{a^2\left(1+\cos^2\theta+2\cos\theta\right)+a^2\sin^2\theta\right\}}\,d\theta$$

$$= \sqrt{\left\{a^2+a^2\left(\cos^2\theta+\sin^2\theta\right)+2a^2\cos\theta\right\}}\,d\theta$$

$$= \sqrt{2a^2\left(1+\cos\theta\right)}\,d\theta = 2a\cos\frac{\theta}{2}\,d\theta$$

If $(\overline{x}, \overline{y})$ be the CG of the arc above the initial line then

$$\overline{x} = \frac{\int_0^x x\,ds}{\int_0^\pi ds} = \frac{\int_0^\pi r\cos\theta.2a\cos\frac{\theta}{2}\,d\theta}{\int_0^\pi 2a\cos\frac{\theta}{2}\,d\theta}$$

$$= \frac{\int_0^\pi a\left(1+\cos\theta\right).\cos\theta.2a\cos\frac{\theta}{2}\,d\theta}{4a\left[\sin\frac{\theta}{2}\right]_0^\pi}$$

$$= \frac{4a^2\int_0^\pi \cos^3\frac{\theta}{2}\left(2\cos^2\frac{\theta}{2}-1\right)d\theta}{4a}$$

$$= \frac{4a^2\int_0^{\pi/2}\cos^3\varphi\left(2\cos^2\varphi-1\right)2\,d\varphi}{4a}, \qquad \text{where } \frac{\theta}{2}=\varphi$$

$$= 2a\left[\int_0^{\pi/2} 2\cos^5\varphi\,d\varphi - \int_0^{\pi/2}\cos^3\varphi\,d\varphi\right]$$

$$= 2a\left[2.\frac{4}{3}.\frac{2}{3}\left(-\frac{2}{3}\right)\right] = \frac{4a}{5}$$

$$\left(\text{as } \int_0^{\pi/2}\cos^n x\,dx = \frac{n-1}{n}.\frac{n-3}{n-2}\ldots\frac{2}{3}.1, \qquad \text{where n is odd}\right)$$

Again $$\overline{y} = \frac{\int y\,ds}{\int ds} = \frac{\int_0^\pi r\sin\theta\left(2a\cos\frac{\theta}{2}\right)d\theta}{\int_0^\pi 2a\cos\frac{\theta}{2}\,d\theta}$$

$$=\frac{\int_0^{\pi} a(1+\cos\theta)\sin\theta .2a\cos\frac{\theta}{2}\,d\theta}{4a}$$

$$=\frac{\int_0^{\pi} a.2\cos^2\frac{\theta}{2}.2\sin\frac{\theta}{2}.\cos\frac{\theta}{2}.2a\cos\frac{\theta}{2}\,d\theta}{4a}$$

$$=\frac{8a^2\int_0^{\pi}\cos^4\frac{\theta}{2}\sin\frac{\theta}{2}\,d\theta}{4a}$$

$$=2a\times(-2)\left[\frac{\cos^5\frac{\theta}{2}}{5}\right]_0^{\pi}=-4a\times\left(-\frac{1}{5}\right)=\frac{4a}{5}$$

15. Show that the line joining the pole of the lemniscates $r^2 = a^2 \cos 2\theta$ to the centre of gravity of an arc PQ of the curve bisects the angle POQ.

Let us consider an element δs of arc PQ. Let R be the point (r, θ)

Differentiating the equation of the curve $2r\,dr = -a^2 \sin 2\theta \,.\, 2\,d\theta$;

i.e.,
$$\frac{dr}{d\theta}=-\frac{a^2}{r}\sin 2\theta$$

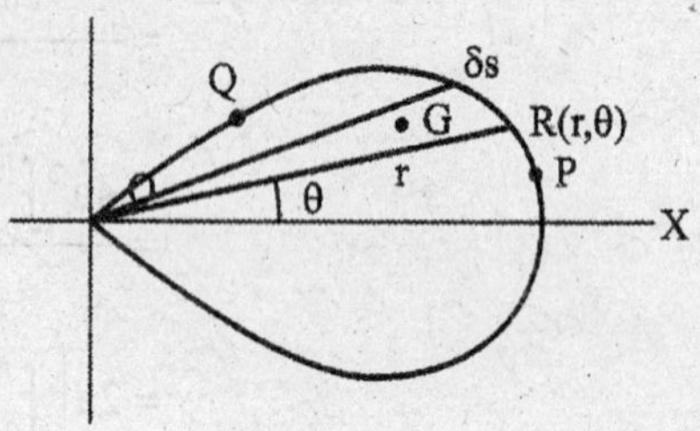

$$\therefore \quad \frac{ds}{d\theta}=\sqrt{r^2+\left(\frac{dr}{d\theta}\right)^2}$$

$$=\sqrt{\left\{r^2+\left(-\frac{a^2}{r}\sin 2\theta\right)^2\right\}}$$

$$=\frac{\sqrt{r^2+a^4\sin^2 2\theta}}{r}=\frac{\sqrt{a^4\cos^2 2\theta+a^4\sin^2 2\theta}}{r}$$

$$=\frac{a^2}{r}; \qquad \therefore \quad ds=\frac{a^2}{r}d\theta$$

If $\angle POX = \alpha$, $\angle QOX = \beta$ and $G(\bar{x}, \bar{y})$ be the CG of the arc PQ then

$$\overline{x} = \frac{\int x\,ds}{\int ds} = \frac{\int_{\alpha}^{\beta} r\cos\theta.\frac{a^2}{r}d\theta}{\int_{\alpha}^{\beta}\frac{a^2}{r}d\theta}$$

$$= \frac{\sin\beta - \sin\alpha}{\int_{\alpha}^{\beta}\frac{d\theta}{r}}$$

$$\overline{y} = \frac{\int y\,ds}{\int ds} = \frac{\int_{\alpha}^{\beta} r\sin\theta\frac{a^2}{r}d\theta}{\int_{\alpha}^{\beta}\frac{a^2}{r}d\theta} = \frac{\cos\alpha - \cos\beta}{\int_{\alpha}^{\beta}\frac{d\theta}{r}}$$

$$\therefore \quad \tan\angle GOX = \frac{\overline{y}}{\overline{x}} = \frac{\cos\alpha - \cos\beta}{\sin\beta - \sin\alpha} = \frac{2\sin\frac{\alpha+\beta}{2}\sin\frac{\beta-\alpha}{2}}{2\cos\frac{\beta+\alpha}{2}\sin\frac{\beta-\alpha}{2}}$$

$$\therefore \quad \angle GOX = \frac{1}{2}\{\angle POX + \angle QOX\}$$

i.e., OG bisects the angle POQ

16. Find the centroid of the arc of the cycloid $\mathbf{x = a(\theta + \sin\theta),\ y = a(1 - \cos\theta)}$ which lies in the positive quadrant.

For the arc OA in the positive quadrant, θ varies from 0 to π

$$\text{Now} \quad \frac{ds}{d\theta} = \sqrt{\left(\frac{dx}{d\theta}\right)^2 + \left(\frac{dy}{d\theta}\right)^2}$$

$$= \sqrt{\left\{a^2(1+\cos\theta)^2 + a^2\sin^2\theta\right\}} = 2a\cos\frac{\theta}{2}$$

If $(\overline{x}, \overline{y})$ be the CG of the arc OA then

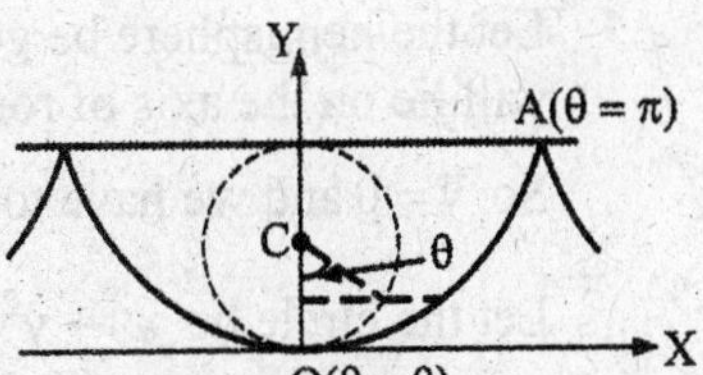

$$\overline{x} = \frac{\int \rho x\,ds}{\int \rho\,ds} = \frac{\int_0^a a(\theta + \sin\theta)2a\cos\frac{\theta}{2}d\theta}{\int_0^a 2a\cos\frac{\theta}{2}d\theta}$$

$$= \frac{2a^2 \int_0^{\pi} \left\{ \theta \cos\frac{\theta}{2} + 2\sin\frac{\theta}{2}\cos^2\frac{\theta}{2} \right\} d\theta}{4a\left[\sin\frac{\theta}{2}\right]_0^{\pi}}$$

$$= \frac{2a^2}{4a}\left[\theta\left(2\sin\frac{\theta}{2}\right) + 4\cos\frac{\theta}{2}\right]_0^{\pi} - 4\left[\frac{\cos^3\frac{\theta}{2}}{3}\right]_0^{\pi}$$

$$= \frac{a}{2}\left\{2\pi + 4(0-1) - \frac{4}{3}(0-1)\right\} = \frac{a}{2}\left(2\pi - 4 + \frac{4}{3}\right)$$

$$= \frac{a}{2}\left\{2\pi - \frac{8}{3}\right\} = \left(\pi - \frac{4}{3}\right)$$

Again $\quad \overline{y} = \dfrac{\int \rho y\, ds}{\int \rho\, ds} = \dfrac{\int_0^{\pi} a(1-\cos\theta)2a\cos\frac{\theta}{2}\,d\theta}{\int_0^{\pi} 2a\cos\frac{\theta}{2}\,d\theta}$

$$= \frac{2a^2\int_0^{\pi/2}(1-\cos 2\varphi)\cos\varphi . 2d\varphi}{4a} \quad \text{where } \frac{\theta}{2} = \varphi$$

$$= a\int_0^{\pi/2} 2\sin^2\varphi . \cos\varphi \, d\varphi = 2a\left[\frac{\sin^3\varphi}{3}\right]_0^{\pi/2}$$

$$= \frac{2}{3}a$$

17. Find the centre of gravity of a solid hemisphere of radius a.

Let the hemisphere be generated by the rotation of a semicircle about OX. The CG will lie on the axis of rotation OX (a line of symmetry)

So $\overline{y} = 0$ and we have to find $\overline{x}$ only

Let the circle be $x^2 + y^2 = a^2$

Let PQ be a narrow strip of width δx at a distance x from OY as shown in the figure

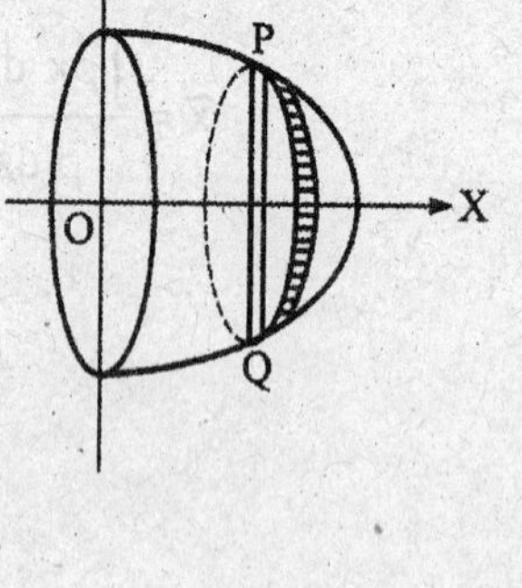

The rectangular strip PQ on rotating will generate a slab which can be considered as cylindrical as the width of the strip is very small

So, in limit, this volume = $\pi y^2 dx$

$\therefore$ volume of the hemisphere

$$= \int_0^a \pi y^2 dx$$

Moment of the cylindrical slab about $OY = \pi y^2 dx \times x$

$\therefore$ sum of moments of all such slabs about OY

$$= \int_0^a \pi y^2 x \, dx \qquad \text{.....(3)}$$

Also moment of the hemisphere about OY

$$= \bar{x} \times \int_0^a \pi y^2 dx \qquad \text{.....(4)}$$

$\therefore$ from (3) and (4), $= \bar{x} \int_0^a \pi y^2 dx = \int_0^a \pi y^2 x \, dx$

$$\therefore \quad \bar{x} = \frac{\int_0^a y^2 x \, dx}{\int_0^a y^2 dx} = \frac{\int_0^a (a^2 - x^2) x \, dx}{\int_0^a (a^2 - x^2) dx} = \frac{\left[\frac{a^2 x^2}{2} - \frac{x^4}{4}\right]_0^a}{\left[a^2 x - \frac{x^3}{3}\right]_0^a}$$

$$= \frac{\frac{a^4}{4}}{\frac{2}{3}a^3} = \frac{3}{8}a, \bar{y} = 0$$

Note: This method is very convenient sometimes

18. Find the CG of a solid right circular cone of height h.

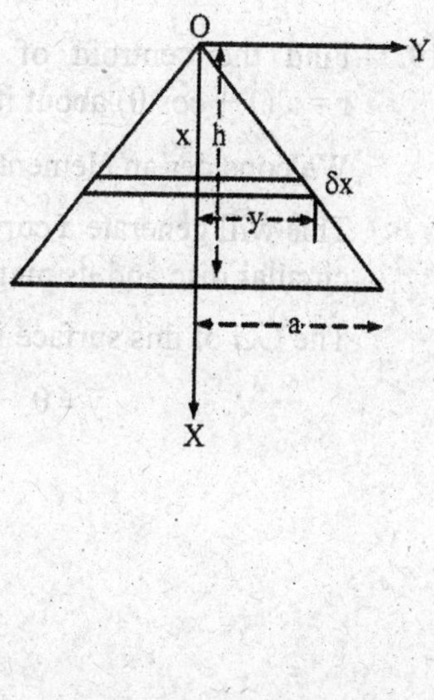

Let the cone be generated by revolving the triangle of the figure about x-axis. Since the cone is symmetric about x-axis (axis of the cone), its CG lies on it, so $\bar{y} = 0$ and we have only to find $\bar{x}$.

If we take a thin slice of thickness δ_a parallel to the base at a distance x from O as the element of volume

$$dV = \pi y^2 \, dx$$

$$V = \pi\int_0^h y^2 dx$$

$$= \pi\frac{a^2}{h^2}\int_0^h x^2 dx$$

$$\left(\because \frac{y}{a} = \frac{x}{h}, \therefore y = \frac{a}{h}, \text{ from similar triangles}\right)$$

$$= \frac{1}{3}\pi a^2 h$$

$\therefore$ moment of the cone about OY $= \bar{x} \times \frac{1}{3}\pi a^2 h$ .....(5)

Also moment of the slice about OY $= \pi y^2\, dx \times x$

$\therefore$ moment of all such slices about OY

$$= \int_0^h \pi y^2\, dx \times x$$

$$= \pi\frac{a^2}{h^2}\int_0^h x^3\, dx \qquad \left(\because y = \frac{a}{h}x\right)$$

$$= \frac{1}{4}\pi a^2 h^2 \qquad .....(6)$$

From (5) and (6), $\bar{x}\frac{1}{3}\pi a^2 h = \frac{1}{4}\pi a^2 h^2$

$$\therefore \qquad \bar{x} = \frac{\frac{1}{4}\pi a^2 h^2}{\frac{1}{3}\pi a^2 h} = \frac{3}{4}h$$

19. Find the centroid of the surface formed by the revolution of the cardioid $r = a(1 + \cos\theta)$ about its axis.

We consider an elementary arc δs of an curve at a distance y from the x-axis

This will generate a curved surface of an elementary circular disc and its magnitude will be $2\pi y\, \delta s$

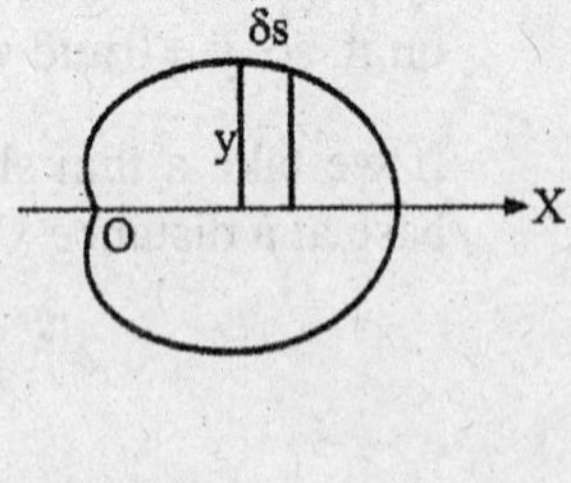

The CG of this surface is at a point (x, 0),

$$\therefore \qquad \bar{y} = 0$$

and $$\bar{x} = \frac{\int_0^{2\pi} x\rho(2\pi y\, ds)}{\int_0^{2\pi} 2\pi y\rho\, ds}, \quad \rho \text{ being the density}$$

$$= \frac{\int_0^{2\pi} xy\, ds}{\int_0^{2\pi} y\, ds} \quad .....(7)$$

The equation of the curve is $r = a(1 + \cos\theta)$ .....(8)

$$\therefore \quad \frac{dr}{d\theta} = -a \sin\theta$$

$$\therefore \quad \frac{ds}{d\theta} = \sqrt{r^2 + \left(\frac{dr}{d\theta}\right)^2} = \sqrt{a^2(1+\cos\theta)^2 + a^2\sin^2\theta} = 2a\cos\frac{\theta}{2}$$

$\therefore$ from (7)

$$\bar{x} = \frac{\int_0^{\pi} (r\cos\theta)(r\sin\theta)2a\cos\frac{\theta}{2}\, d\theta}{\int_0^{2\pi} r\sin\theta 2a\cos\frac{\theta}{2}\, d\theta} \quad (\because x = r\cos\theta,\ y = r\sin\theta)$$

$$= \frac{\int_0^{\pi} r^2 \cos\theta\sin\theta\cos\frac{\theta}{2}\, d\theta}{\int_0^{2\pi} r\sin\theta\cos\frac{\theta}{2}\, d\theta}$$

$$= \frac{\int_0^{\pi} a^2(1+\cos\theta)^2 \cos\theta\sin\theta\cos\frac{\theta}{2}\, d\theta}{\int_0^{\pi} a(1+\cos\theta)\sin\theta\cos\frac{\theta}{2}\, d\theta} \quad [\because r = a(1+\cos\theta)]$$

$$= \frac{a\int_0^{\pi}\left(2\cos^2\frac{\theta}{2}\right)^2\left(2\cos^2\frac{\theta}{2}-1\right)2\sin\frac{\theta}{2}\cos^2\frac{\theta}{2}\, d\theta}{\int_0^{\pi}(2\cos^2\frac{\theta}{2})2\sin\frac{\theta}{2}\cos^2\frac{\theta}{2}\, d\theta}$$

$$= \frac{8a\int_0^{\pi}\left\{2\cos^8\frac{\theta}{2}\sin\frac{\theta}{2} - \cos^6\frac{\theta}{2}\sin\frac{\theta}{2}\right\}d\theta}{4\int_0^{\pi}\cos^4\frac{\theta}{2}\sin\frac{\theta}{2}\theta\, d\theta}$$

$$= \frac{8a\left[-\frac{4}{9}\cos^9\frac{\theta}{2}+\frac{2}{7}\cos^7\frac{\theta}{2}\right]_0^{\pi}}{4\frac{(-2)}{5}\left[\cos^5\frac{\theta}{2}\right]_0^{\pi}}$$

$$= 8a\frac{\left(\frac{4}{9}-\frac{2}{7}\right)}{\frac{8}{5}} = \frac{\frac{80}{63}a}{\frac{8}{5}} = \frac{80}{63}a\times\frac{5}{8} = \frac{50}{63}a$$

Hence, CG is at the point $\left(\frac{50}{63}a,0\right)$

20. Find the position of the centre of gravity of the surface formed by the revolution of one loop of the lemniscate $r^2 = a^2 \cos 2\theta$ about the initial line.

Here $r = a\sqrt{\cos 2\theta}$; $\quad\therefore \quad \frac{dr}{d\theta} = \frac{-2a\sin 2\theta}{2\sqrt{\cos 2\theta}}$

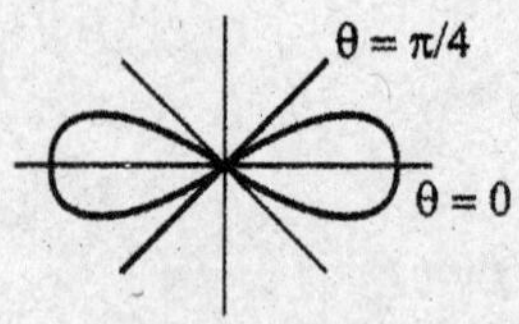

$$\therefore \quad \frac{ds}{d\theta} = \sqrt{r^2+\left(\frac{dr}{d\theta}\right)^2}$$

$$= \sqrt{a^2\cos 2\theta + a^2\frac{\sin^2 2\theta}{\cos 2\theta}}$$

$$= \frac{a}{\sqrt{\cos 2\theta}}$$

If $(\bar{x},\bar{y})$ be the coordinates of CG, $\bar{y}=0$ (since CG will lie on initial line about which it revolves)

$$\bar{x} = \frac{2\int_0^{\pi/4} r\cos\theta . r\sin\theta \frac{a}{\sqrt{\cos 2\theta}}d\theta}{2\int_0^{\pi/4} r\sin\theta . \frac{a}{\sqrt{\cos 2\theta}}d\theta}$$

Denominator $\quad = 2\int_0^{\pi/4} a\sqrt{\cos 2\theta}\sin\theta\frac{a}{\sqrt{\cos 2\theta}}d\theta$

$$= 2\int_0^{\pi/4} a^2 \sin\theta\, d\theta = -2a^2\left[\cos\theta\right]_0^{\pi/4}$$

$$= -2a^2\left(\frac{1}{\sqrt{2}}-1\right) = a^2\left(2-\sqrt{2}\right)$$

Numerator $\quad = 2\int_0^{\pi/4} a^2 \sin\theta\cos\theta\cos 2\theta \frac{a}{\sqrt{\cos 2\theta}} d\theta$

$$= a^3\int_0^{\pi/4} \sqrt{\cos 2\theta}\sin 2\theta \, d\theta$$

Put $\cos 2\theta = z^2$ so that $-2\sin 2\theta \, d\theta = 2z\,dz$

$\therefore$ numerator $\quad = -a^3\int_1^0 z^2 dz = \frac{a^3}{3}$

$\therefore \quad \bar{x} = \frac{a}{3\left(2-\sqrt{2}\right)} = \frac{a\left(2+\sqrt{2}\right)}{3(4-2)} = \frac{a}{6}\left(2+\sqrt{2}\right)$ and $\bar{y} = 0$

21. Find the centre of gravity of a thin right conical shell of uniform thickness and density.

Let the line OA ($y = mx$) revolve about OC (x-axis) to describe the curved surface of the thin conical shell

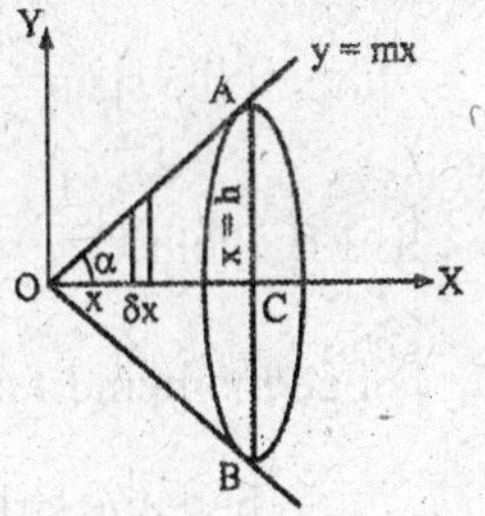

Here CG lies on the x-axis, line of symmetry

$\therefore \quad \bar{y} = 0$

and $\quad \bar{x} = \frac{\int x\rho 2\pi y \, ds}{\int \rho 2\pi y \, ds}$

$$= \frac{\int_0^h xy\frac{ds}{dx}dx}{\int_0^h y\frac{ds}{dx}dx} \quad .....(9)$$

Now $\quad \frac{ds}{dx} = \sqrt{1+\left(\frac{dy}{dx}\right)^2} = \sqrt{1+m^2} \quad \left(\because y = mx, \therefore \frac{dy}{dx} = m\right)$

$\therefore$ from (9),

$$\bar{x} = \frac{\int_0^h xmx\sqrt{1+m^2}\,dx}{\int_0^h mx\sqrt{1+m^2}\,dx}$$

$$= \frac{\int_0^h x^2 dx}{\int_0^h x\,dx} = \frac{\left[\frac{x^3}{3}\right]_0^h}{\left[\frac{x^2}{2}\right]_0^h} = \frac{2}{3}h$$

Moment of Inertia

22. Find the MI of a uniform rectangular lamina about an axis of symmetry through its centre.

Let ABCD be the rectangle

We shall find its MI about the axis OY

Let AD = 2a, AB = 2b

Consider a thin strip of width δy parallel to x-axis

Mass of the strip = $2a\delta y\rho$, ρ being the surface density

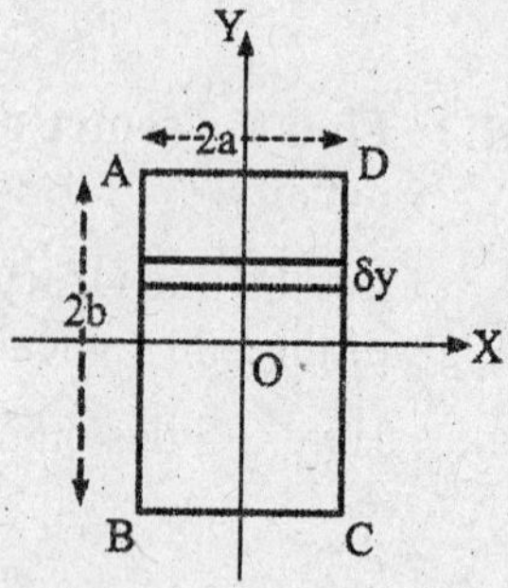

Its MI about OY $= 2a\delta y\rho\frac{a^2}{3}$

[since moment of a thin uniform straight rod

(here the strip) $= \frac{Ma^2}{3}$].

$\therefore$ MI of the rectangle about OY

$$= \frac{2a^3\rho}{3}\int_{-b}^{b} dy = \frac{2a^3\rho}{3}[y]_{-b}^{b} = \frac{2a^3\rho}{3}.2b$$

$$= \frac{4a^3b\rho}{3} = 4ab\rho.\frac{a^2}{3} = \frac{Ma^2}{3}$$

where M = mass of the rectangular lamina = $4ab\rho$

23. Find the moment of inertia of an isosceles triangle of base 2a and height h about its line of symmetry.

Let OAB be an isosceles triangle so arranged that its line of symmetry lies along OX

Given that AB = 2a and OC = h,

$\therefore$ AC = CB = a

We divide the triangle into large number of strips parallel to the base and consider one of the strips PQ at a distance x from O and of thickness δx.

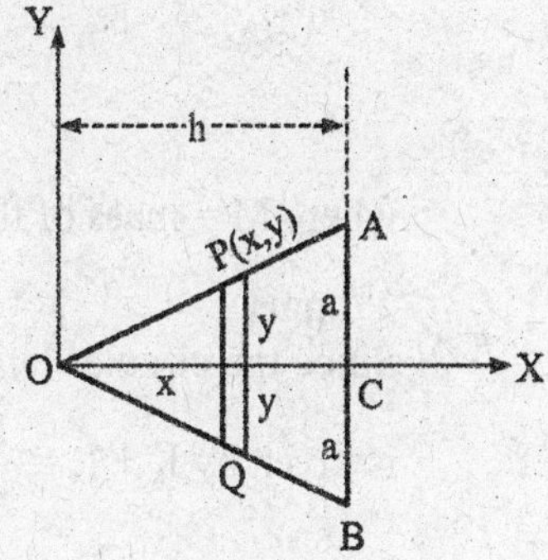

Now mass of the strip

= 2yρ δx, ρ being mass of unit area

$\therefore$ MI of the strip about OC = 2yr δx × $\dfrac{y^2}{3}$

[$\because$ MI of a rod of length 2a about a line perpendicular to the rod through its centre

= M(a²/3)

$$= \frac{2}{3}\frac{a^3}{h^2}x^3\rho\, dx$$

$$\left(\because \frac{PQ}{AB} = \frac{OD}{OC};\ \text{i.e.,}\ \frac{2y}{2a} = \frac{x}{h},\ \therefore\ y = x\frac{a}{h}\right)$$

$\therefore$ MI of the whole triangular lamina about OC

$$= \frac{2}{3}\frac{a^3}{h^3}\rho\int_0^h x^3 dx = \frac{2}{3}\frac{a^3}{h^3}\rho\left[\frac{x^4}{4}\right]_0^h$$

$$= \frac{2}{3}\frac{a^3}{h^3}\rho \times \frac{h^4}{4} = \frac{1}{6}a^3\rho h = \frac{1}{6}a^2.(\rho a h) = \frac{Ma^2}{6},$$

where M = mass of the triangle = ρha

24. Find the MI of a uniform circular lamina about an axis through its centre and perpendicular to the plane of the lamina. Also find the MI about a perpendicular diameter.

Let a be the radius of the circular lamina

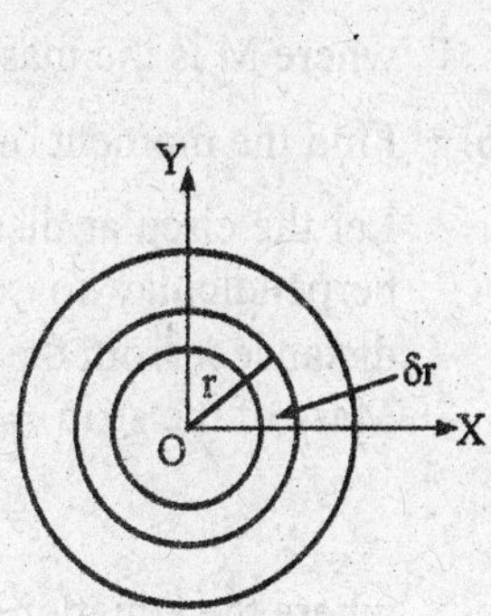

Let us consider a thin ring with O as centre and width δr at a distance r from the centre

The area of this ring is approximately $= 2\pi r\ \delta r\ \rho, \rho$ being area density

Its MI about OY $= (2\pi r\ \delta r\ \rho)r^2$

$\therefore$ moment of inertia about OY, the axis perpendicular to the plane of the lamina

$$= \int_0^a 2\pi\rho r^3 dr = 2\pi\rho \int_0^a r^3 dr$$

$$= 2\pi\rho \frac{a^4}{4} = \frac{\pi a^4 \rho}{2} = \frac{Ma^2}{2}$$

where M = mass of the lamina = $\pi a^2 \rho$

2nd part

Since the circular lamina is symmetrical about the diameter, so the required MI, I_x, is given by $I_x + I_x = \frac{Ma^2}{4}$

(Here $I_x = I_y \quad \therefore \quad I_x + I_x = I_z$) $\qquad \therefore \quad I_x = \frac{Ma^2}{4}$

25. Find the moment of inertia of a solid right circular cylinder about its axis.

Let the solid cylinder be divided into a large number of circular discs perpendicular to its axis. Let one of such discs be of radius a and thickness δx at a distance x from A.

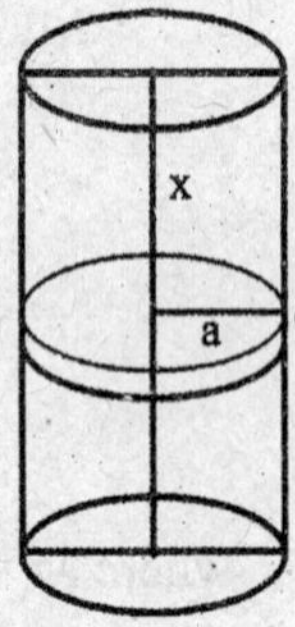

MI of this disc = $\left(\pi a^2 \rho\ \delta x\right)\frac{a^2}{2}$,

Since we know MI of a circular disc in such case = $M\frac{a^2}{2}$

$\therefore$ MI of the whole cylinder about the axis of the cylinder

$$= \int_0^h \pi a^2 \rho dx \frac{a^2}{2} = \frac{\pi\rho a^4}{2}\int_0^h dx = \frac{\pi\rho a^4}{2} \times h$$

$$= \left(\pi a^2 h\rho\right)\frac{a^2}{2} = M\frac{a^2}{2}$$

where M is the mass of the cylinder

26. Find the moment of inertia of a uniform circular disc about a tangent.

Let the circular disc be divided into large number of strips perpendicular to x-axis. Let one of the strips be at a distance x from the origin and of thickness δx.

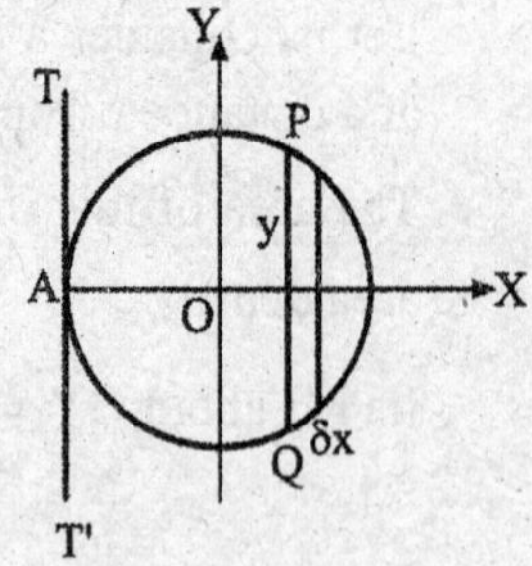

Mass of the strip = 2y δx.ρ

$$= 2\sqrt{a^2 - x^2}\,\delta x\rho$$

where the equation of the circle is $x^2 + y^2 = a^2$

MI of the strip PQ about OY $= \rho^2\sqrt{a^2 - x^2}\,\delta x.x^2$

$\therefore$ MI of the whole disc about OY

$$= I_y = 2\rho\int_{-a}^{a} x^2\sqrt{a^2 - x^2}\,dx$$

Putting x = a sin θ, dx = a cos θ dθ; when x = –a;

$$\theta = -\pi/2 \text{ and when } x = a,\ \theta = \pi/2$$

$$\therefore \quad I_y = 2\rho\int_{-\pi/2}^{\pi/2} a^2\sin^2\theta.a\cos\theta d\theta = 2\rho a^4\int_{-\pi/2}^{\pi/2}\sin^2\theta\cos^2\theta\,d\theta$$

$$= \frac{2\rho a^4}{4}\int_{-\pi/2}^{\pi/2}(\sin 2\theta)^2\,d\theta = \frac{\rho a^4}{2}\int_{-\pi/2}^{\pi/2}\frac{1-2\cos 4\theta}{2}\,d\theta$$

$$= \frac{\rho a^4}{4}\left[\theta - \frac{2\sin 4\theta}{4}\right]_{-\pi/2}^{\pi/2} = \frac{\rho a^4}{4}[\pi - 0] = \frac{\pi\rho a^4}{4}$$

$$= \frac{1}{4}\left(\pi a^2\rho\right)a^2 = \frac{1}{4}Ma^2$$

where M is the mass of the disc

Hence, MI about the tangent TAT′

$$= I_y + Ma^2 \quad \text{(by the theorem of parallel axis)}$$

$$= \frac{1}{4}Ma^2 + Ma^2 = \frac{5}{4}Ma^2$$

27. Find the moment of inertia of a thin (hollow) spherical shell of radius a and mass M about a diameter.

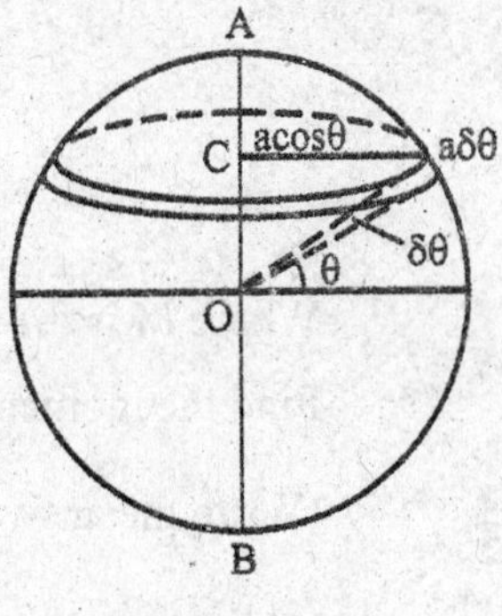

Consider an elementary ring of width $\delta x = a\delta\theta$ and radius a cos θ of the spherical shell where θ and δθ have been shown in the figure

Mass of this ring = $2\pi a\cos\theta..a\delta\theta\rho$,

ρ being density of the shell

Its MI about the diameter along

ACB = (2πa cos θ a δθ ρ) (a cos θ)^2

$\therefore$ MI of the whole spherical shell about the diameter ACB

$$= \int_{-\pi/2}^{\pi/2} 2\pi a\cos\theta\ a\rho\left(a\cos\theta\right)^2 d\theta = 2\pi a^4\rho\int_{-\pi/2}^{\pi/2}\cos^3\theta\,d\theta$$

$$=\frac{2\pi a^4\rho}{4}\int_{-\pi/2}^{\pi/2}(3\cos\theta+\cos 3\theta)\,d\theta$$

$$=\frac{\pi a^4\rho}{2}\left[3\sin\theta+\frac{\sin 3\theta}{3}\right]_{-\pi/2}^{\pi/2}$$

$$=\frac{\pi a^4\rho}{2}\left[6-\frac{2}{3}\right]=\frac{8\pi a^4\rho}{3}=\left(4\pi\rho a^2\right)\frac{2}{3}a^2=\frac{2}{3}Ma^2,$$

where M = mass of the shell

28. Find the moment of inertia of a solid sphere about a diameter.

Let a be the radius and ρ be the density of the sphere. Let the sphere be divided into thin spherical shells. We consider one such shell included between spheres of radii r and r + δr.

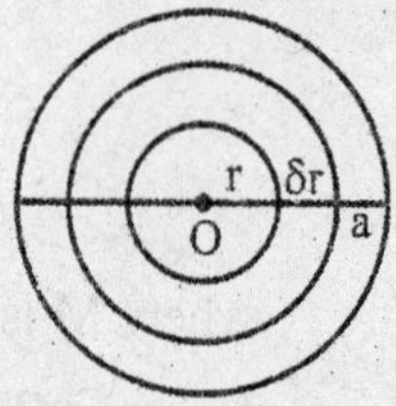

Its volume $=\frac{4}{3}\pi\left\{(r+\delta r)^3-r^3\right\}$

$=4\pi r^2\delta r$, neglecting small quantities

∴ mass of the shell $=4\pi r^2\delta\rho$

∴ MI of this shell about the diameter of the sphere

$$=\frac{2}{3}(4\pi r^2\delta r\rho)\times r^2 \qquad \left\{\text{since MI of a shell}=\frac{2}{3}Ma^2\right\}$$

∴ MI of the whole sphere about the diameter

$$=\int_0^a\frac{2}{3}.4\pi r^2\,dr\,\rho r^2=\frac{8}{3}\pi\rho\int_0^a r^4dr$$

$$=\frac{8}{5}\pi\rho\frac{a^5}{5}=\left(\frac{4}{3}\pi a^3\rho\right).\frac{2}{5}a^2=\frac{2}{3}Ma^2$$

Where M is the mass of the sphere

29. Find the moment of inertia of a uniform solid sphere about any tangent line.

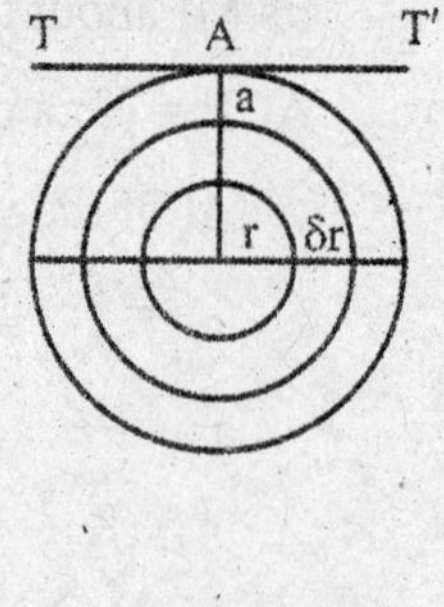

(Write the answer of the above example)

MI of the sphere about the diameter $=\frac{2}{5}Ma^2$,

where M is the mass of the sphere

∴ MI of the sphere about the tangent TAT′

$$= \frac{2}{5}Ma^2 + Ma^2$$

$$= \frac{7}{5}Ma^2$$

30. Prove that the moment of inertia about an axis through the centre perpendicular to the plane of a circular (annular) ring whose outer and inner radii are a and b is $\frac{1}{2}M(a^2 + b^2)$, where M denotes the mass of the ring.

Since a and b are the inner and outer radii of the ring, its mass $M = \pi(b^2 - a^2)\rho$, where ρ is the density of the material.

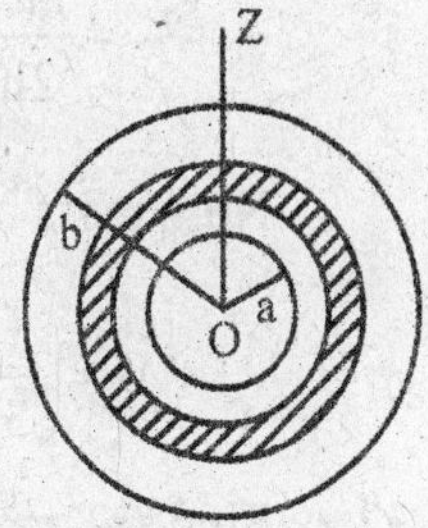

If δx be the width of an elementary circle of radius x with the same centre as that of the ring, the mass of this element $= 2\pi x\ \delta x\ \rho$

MI of this element about the line OZ which is perpendicular to the plane of the circular ring

$$= 2\pi x\ \delta x\ \rho . x^2 \qquad \text{(i.e., mass} \times \text{radius}^2\text{)}$$

∴ MI of the whole circular ring

$$= \int_a^b 2\pi x \rho . x^2 dx = 2\pi\rho\left[\frac{x^4}{4}\right]_a^b$$

$$= \frac{2\pi\rho}{4}\left(b^4 - a^4\right) = \frac{1}{2}\pi\rho\left(b^4 - a^4\right) = \frac{1}{2}\pi\rho\left(b^2 - a^2\right)\left(b^2 + a^2\right)$$

$$= \frac{1}{2}M\left(a^2 + b^2\right)$$

31. Find the moment of inertia of a solid right circular cone of height h and the semivertical angle α about its axis.

Let α be the semivertical angle, a be the radius of the base of the cone of height h. Divide the cone into plates parallel to the base. Let r be the radius and δy be the thickness of one of the plates at a distance y from C,

then its mass $= \pi r^2\ \delta y \rho$, ρ being density of the cone.

MI of this plate about the axis of the

$$\text{Cone} = \pi r^2 \rho \delta y \frac{r^2}{2}$$

$\therefore$ MI of the whole solid cone about its axis

$$= \int_0^h \frac{\pi\rho}{2} r^4 dy = \frac{\pi\rho}{2} \int_0^h \frac{a^4}{h^4} (h-y)^4 dy$$

$$\left[\because \frac{r}{h-y} = \frac{a}{h}, \ \therefore r = \frac{a}{h}(h-y) \right]$$

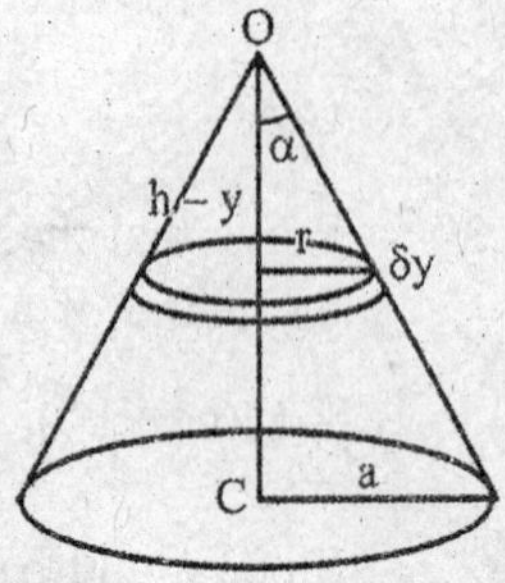

$$= \frac{\pi a^4 \rho}{2h^4} \int_0^h \left(h^4 - 4h^3 y + 6h^2 y^2 - 4hy^3 + y^4 \right) dy$$

$$= \frac{\pi a^4 \rho}{2h^4} \left[h^4 y - \frac{4h^3 y^2}{2} + \frac{6h^2 y^3}{3} - 4h \frac{y^4}{4} + \frac{y^5}{5} \right]_0^h$$

$$= \frac{\pi a^4 \rho}{2h^4} \left[h^5 - 2h^5 + 2h^5 - h^5 + \frac{h^5}{5} \right] = \frac{\pi \rho a^4}{10} h \qquad \text{.....(10)}$$

Now mass of the cone, $M = \frac{1}{3}\pi a^2 h\rho$ and $\tan\alpha = \frac{a}{h}$; $\because$ $a = h\tan\alpha$

$$\therefore \ MI = \left(\frac{1}{3}\pi a^2 h\rho \right) \cdot \frac{3}{10} a^2 = \frac{3}{10} Ma^2 = \frac{3}{10} Mh^2 \tan^2\alpha$$

32. Find the moment of inertia of a solid right circular cone about (i) an axis through its vertex parallel to its base and (ii) a line through its CG perpendicular to its axis.

Consider the cone as cut into plates of thickness δx. Let one of the plates be at a distance x from the vertex and of radius b. Then, its MI about its diameter

$$= \frac{1}{4}\pi\rho b^4 \delta x$$

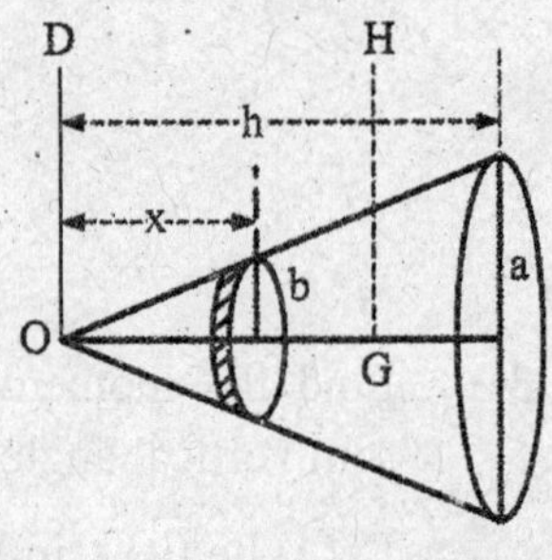

(since MI of a circular plate of radius r

about diameter $= \frac{1}{4}\pi\rho r^4$)

$\therefore$ by the theorem of parallel axes, MI of this plate

about $OD = \frac{1}{4}\pi\rho b^4 \delta x + \left(\rho\pi b^2 \delta x \right) x^2$

Taking limit of the sum of such expressions and writing it as an integral, the required moment of inertia

$$= \pi\rho \int_0^h \left(\frac{1}{4} b^4 + b^2 x^2 \right) dx$$

But $\frac{b}{a}=\frac{x}{h}$; $\quad\therefore\quad b=\frac{xa}{h}$

$$\therefore \qquad MI=\pi\rho\int_0^h\left(\frac{1}{4}\frac{x^4a^4}{h^4}+\frac{x^4a^2}{h^2}\right)dx$$

$$=\frac{\pi\rho a^2}{h^2}\left(\frac{a^2}{4h^2}+1\right)\int_0^h x^4dx=\frac{\pi\rho a^2}{h^2}\left(\frac{a^2}{4h^2+1}\right)\cdot\frac{h^5}{5}$$

$$=\frac{\pi\rho a^2h^3}{5}\left(\frac{a^2+4h^2}{4h^2}\right)\frac{\pi\rho a^2}{20}\left(a^2+4h^2\right)$$

Now $\frac{1}{3}\pi\rho a^2h$ = mass of the cone = M

$\therefore \qquad$ MI of the cone about OD $=\frac{3}{20}M\left(a^2+4h^2\right)$

2nd Part

If G be the CG of the cone and GH be the axis through G parallel to OD then since (by theorem of parallel axes)

MI of the cone about OD = MI of the cone about GH + M . OG^2,

$\therefore$ MI of the cone about GH = MI about OD – M . OG^2

$$=\frac{3}{20}M\left(a^2+4h^2\right)-M\left(\frac{3h}{4}\right)^2$$

(by 1st part and since $OG=\frac{3h}{4}$)

$$=\frac{3M}{80}\left(4a^2+16h^2-15h^2\right)$$

$$=\frac{3M}{80}\left(h^2+4a^2\right)$$

33. Find the moment of inertia of a uniform elliptic lamina about each of major and minor axes.

Let the ellipse be $\frac{x^2}{a^2}+\frac{y^2}{b^2}=1$

Let us divide the ellipse into thin strips parallel to the major axis. Let us consider an elementary strip of width δy at a distance y from the major axis

Its length $= 2x = \frac{2a}{b}\sqrt{b^2 - y^2}$

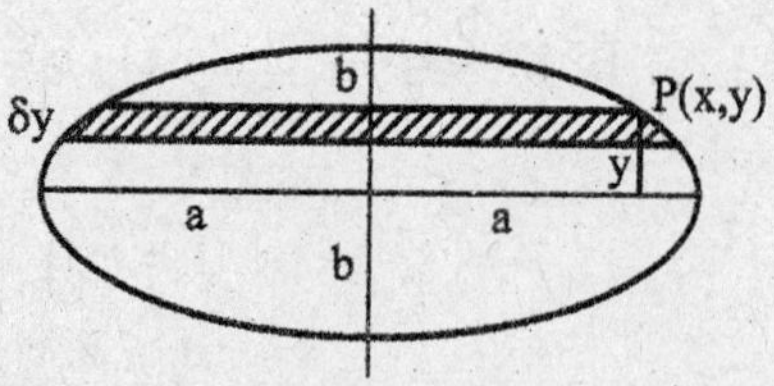

Its mass $= \frac{2a}{b}\sqrt{b^2 - x^2}\,\delta y,$

ρ being the surface density

∴ MI about major axis

$$= \int_{-b}^{b} y^2 . \frac{2a}{b}\sqrt{b^2 - y^2}\,dy\;\rho = \frac{4a\rho}{b}\int_0^b y^2\sqrt{b^2 - y^2}\,dy$$

Putting $y = b\sin\theta$, $dy = b\cos\theta\,d\theta$; when $y = 0$, $\theta = 0$; when $y = b$, $\theta = \pi/2$

$$\therefore \quad MI = 4ab^3\rho\int_0^{\pi/2} \sin^2\theta\cos^2\theta\,d\theta$$

$$= 4ab^3\rho\int_0^{\pi/2} \sin^2\theta\left(1 - \sin^2\theta\right)d\theta$$

$$= 4ab^3\rho\int_0^{\pi/2} (\sin^2\theta - \sin^4\theta)d\theta$$

$$= 4ab^3\rho\left\{\int_0^{\pi/2} \sin^2\theta d\theta - \int_0^{\pi/2} \sin^4\theta\,d\theta\right\}$$

$$= 4ab^3\rho\left[\frac{1}{2}.\frac{\pi}{2} - \frac{3}{4}.\frac{1}{2}\frac{\pi}{2}\right] = \frac{1}{4}ab^3\pi\rho$$

$$= \frac{1}{4}Mb^2$$

where M = mass of the ellipse = $\pi ab\rho$

Similarly, MI about minor axis $= \frac{1}{4}Ma^2$

34. Find the moment of inertia of a parabolic area (of latus rectum 4a) cut off by an ordinate at a distance h from the vertex about

(i) the tangent at the vertex,

(ii) its axis

The parabola is $y^2 = 4ax$

We divide the parabola into thin strips parallel to y-axis.

Let us take any elementary strip of width δx at a distance x from y-axis. Then mass of the strip $= 2y\delta x\rho, \rho$ being the surface density.

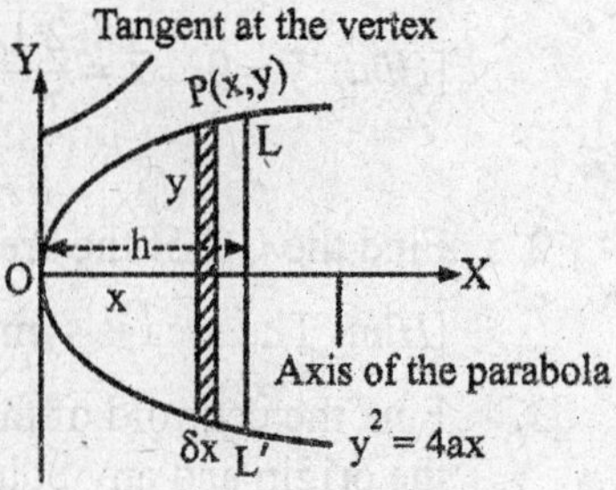

$\therefore$ mass of the area,

$$M = \int_0^h 2y\, dx\, \rho$$

$$= 2\rho\int_0^h \sqrt{4ax}\, dx$$

$$= 4\sqrt{a}\rho \int_0^h x^{\frac{1}{2}} dx = 4\sqrt{a}\ \rho\frac{2}{3}h^{3/2}$$

$$= \frac{8}{3}\rho\sqrt{ah^3} \qquad .....(11)$$

(i) MI about tangent at vertex (i.e., y-axis)

$$= \int_0^h 2y\, dx\, \rho x^2$$

$$= 2\rho\int_0^h \sqrt{4ax}.x^2 dx$$

$$= 4\sqrt{a}\rho\int_0^h x^{\frac{5}{2}} dx = \frac{8}{7}\sqrt{a}\ \rho\ h^{\frac{7}{2}}$$

$$= \frac{3}{7}Mh^2, \quad \text{where } M = \frac{8}{3}\rho\sqrt{ah^3} \quad \text{[from (11)]}$$

(ii) MI about the axis of the parabola

$$= \int_0^h 2y\, dx \times \rho\, .\, \frac{1}{3}y^2$$

$$= \frac{2}{3}\rho\int_0^h \sqrt{4ax}.\ 4ax.\ dx$$

$$= \frac{16}{3}\rho a^{\frac{2}{3}} \int_0^h x^{\frac{3}{2}} dx = \frac{16}{3}\rho a^{\frac{3}{2}} \times \frac{2}{5}h^{\frac{5}{2}}$$

$$= \frac{4}{5}Mah, \qquad \text{where} \quad M = \frac{8}{3}\rho\sqrt{ah^3}$$

Exercise 11.1

1. Find the position of the centroid of the area of the cardiod r = a(1 – cos θ)

[*Hint*: $\bar{y} = 0,\ \bar{x} = \frac{2}{3}\frac{\int_0^\pi a^3(1-\cos\theta)^3\cos\theta\, d\theta}{\int_0^\pi a^2(1-\cos\theta)^2 d\theta}$, proceed just as example 6]

2. Find the CG of the area of a loop of the lemiscate $r^2 = a^2 \cos 2\theta$.

[*Hint*: The area is symmetrical about the initial line. See example 8]

3. Find the centroid of the arc of the catenary y = ccosh x/c which is included between the origin and any point (x, y)

[*Hint*: $y = c\cosh\frac{x}{c}$, $\therefore\ \frac{dy}{dx} = \sin\ h\frac{x}{c}$

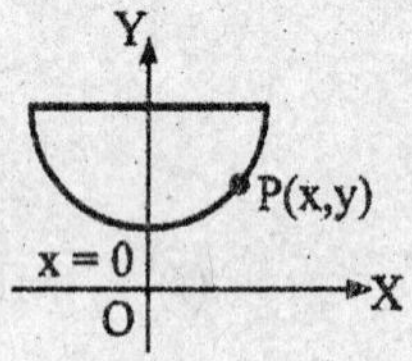

$$\frac{ds}{dx} = \sqrt{1+\left(\frac{dy}{dx}\right)^2}$$

$$= \sqrt{1+\sin\ h^2\frac{x}{c}} = \cosh\frac{x}{c}; s = c\sinh\frac{x}{c}$$

$$\bar{x} = \frac{\int x\, ds}{\int ds} = \frac{\int_0^x x\cosh\frac{x}{c}dx}{\int_0^x \cosh\frac{x}{c}dx}$$

$$= \frac{\left[x\left(c\sinh\frac{x}{c}\right)-\left(c^2\cosh\frac{x}{c}\right)\right]_0^x}{\left[c\sinh\frac{x}{c}\right]_0^x}$$

$$= \frac{x\sinh\frac{x}{c} - c\cosh\frac{x}{c} + c}{\sin\ h\frac{x}{c}}$$

$$= x - \frac{c^2\cosh\frac{x}{c} - c^2}{c\sinh\frac{x}{c}} = x - \frac{c\left(c\cosh\frac{x}{c} - c\right)}{s} = x - \frac{c(y-c)}{s}$$

$$\bar{y}=\frac{\int y\,ds}{\int ds}=\frac{\int_0^x \cosh^2\frac{x}{c}dx}{\int_0^x \cosh\frac{x}{c}dx}=\frac{\frac{c}{2}\int_0^x\left(1+\cosh\frac{2x}{c}\right)dx}{\left[c\sinh\frac{x}{c}\right]_0^x}$$

$$=\frac{1}{2}\frac{x+\left[\frac{c}{2}\sinh\frac{x}{c}\right]_0^x}{\sinh\frac{x}{c}}=\frac{1}{2}\frac{x+c\sinh\frac{x}{c}\cosh\frac{x}{c}}{\sinh\frac{x}{c}}$$

$$=\frac{1}{2}\frac{cx+c\sinh\frac{x}{c}.c\cosh\frac{x}{c}}{c\sinh\frac{x}{c}}=\frac{1}{2}\frac{cx+\delta y}{s}=\frac{1}{2}\left.y+\frac{cx}{2s}\right]$$

4. Find the centroid of the volume formed by the revolution of the cardioids $r = a(s + \cos\theta)$ about x-axis.

[*Hint*: Consider an elementary area $r\,\delta\theta\,\delta r$ at $P(r, \theta)$. A ring of volume $(2\pi\sin\theta)r\;\;\delta\theta\,\delta r$ is generated by the revolution of this elementary area about x-axis, where $r\sin\theta$ is the radius of the total circular arc of the ring.

The CG of the ring is at $(r\cos\theta, 0)$. If $(\bar{x}, \bar{y})$ be the coordinates of the required CG then $\bar{y}=0$ and

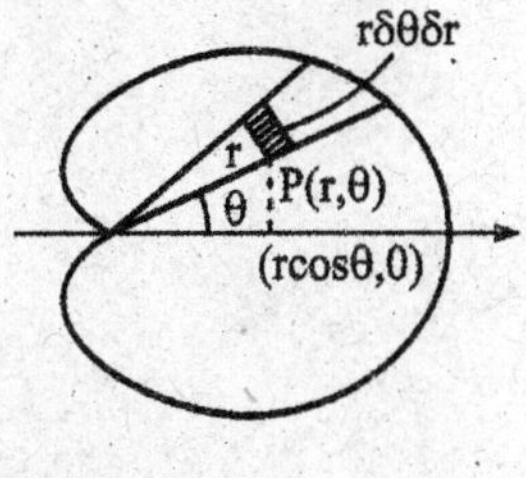

$$\bar{x}=\frac{\int_0^{\pi}\int_0^{r} r\cos\theta.2\pi\,r\sin\theta\,r\,d\theta\,dr}{\int_0^{\pi}\int_0^{r} 2\pi\,r\sin\theta\,r\,d\theta\,dr}$$

$$=\frac{\int_0^{\pi}\int_0^{a(1+\cos\theta)} r^3\sin\theta\cos\theta\,d\theta\,dr}{\int_0^{\pi}\frac{a^3}{3}(1+\cos\theta)^3\sin\theta\,d\theta}$$

$$=\frac{\int_0^{\pi}\left[\frac{r^4}{4}\right]_0^{a(1+\cos\theta)}\sin\theta\cos\theta d\theta}{\frac{a^3}{3}\int_0^2 z^3dz},\quad \text{where } 1+\cos\theta=z$$

$$=\frac{\frac{a^4}{4}\int_0^{\pi}(1+\cos\theta)^4\sin\theta\cos\theta d\theta}{\frac{a^3}{3}\left[\frac{z^4}{4}\right]_0^2}=\frac{\frac{a^4}{4}\int_2^0 -z^4(z-1)dz}{\frac{a^3}{3}\times 4}$$

$$=\frac{\frac{a^4}{4}\left[\frac{z^6}{6}-\frac{z^5}{5}\right]_0^2}{\frac{4}{3}a^3}=\frac{16}{15}a^4\times\frac{3}{4a^3}=\frac{4}{5}a$$

Answers

1. $\left(-\frac{5}{6}a,0\right)$

2. $\overline{y}=0,\overline{x}=\frac{\pi a\sqrt{2}}{8}$

CHAPTER 12

Beta and Gamma Functions

Definition

The first and second Eulerian integrals are termed respectively as beta function and gamma functions* which are defined as follows:

Beta function $B(l,m)=\int_0^1 x^{l-1}(1-x)^{m-1}\,dx$

Gamma function $\Gamma(n)=\int_0^\infty e^{-x}x^{n-1}dx$

where l, m, n are positive variables

I Prove B(l, m) = B (m, l) (Symmetrical property)

We have $$B(l,m)=\int_0^1 x^{l-1}(1-x)^{m-1}\,dx$$

$$=\int_0^1 (1-x)^{l-1}\{1-(1-x)\}^{n-1}dx$$

$$\left(\text{as} \int_0^a f(x)dx=\int_0^a f(a-x)dx\right)$$

$$=\int_0^1 (1-x)^{l-1}x^{m-1}dx$$

$$=\int_0^1 x^{m-1}(1-x)^{l-1}\,dx=B(m,l)$$

*Euler found a continuous function of the real variable x which reduces to x! when x is a positive integer and this led to the definition of gamma function. Gamma function can be defined in a number of ways but it is convenient to define it as a definite integral. Both gamma and beta functions have a wide application in physics, statistics and other applied sciences.

II Evaluation of beta function

We have $B(l,m)=\int_0^1 x^{l-1}(1-x)^{m-1}\,dx$

Let only m be a positive integer. Then integrating by parts, keeping $(1-x)^{m-1}$ as first function,

$$B(l,m)=\left[(1-x)^{m-1}\cdot\frac{x^l}{1}\right]+\int_0^1(x-1)(1-x)^{m-2}\frac{x^l}{l}\,dx$$

$$=\frac{m-1}{l}\int_0^1(1-x)^{m-2}x^l dx.$$

Again integrating by parts,

$$B(l,m)=\frac{(m-1)(m-2)}{1(l+1)}\int_0^1(1-x)^{m-3}x^{l+1}dx.$$

Applying this process again and again,

$$B(l,m)=\frac{(m-1)(m-2)\ldots 2.1}{1(l+1)\ldots(l+m-2)}\int_0^1 x^{l+m-2}dx$$

$$=\frac{(m-1)!}{l(l+1)\ldots(l+m-2)(l+m-1)}.$$

Let only l be positive integer. Then since $B(l, m) = B(m, l)$

$$B(l,m)=\frac{(l-1)!}{m(n+1)\ldots(m+l-1)}.$$

Let l and m both be positive integers. Then since

$$B(l,m)=\frac{(m-1)!}{l(l+1)\ldots(l+m-1)},$$

$$B(l,m)=\frac{(m-1)!(l-1)(l-2)\ldots 2\cdot 1}{1\cdot 2\ldots(l-1)l(l+1)\ldots(l+m-1)}.$$

Multiplying the numerator and denominator by $1.2\ldots.(l-1)$,

$$B(l,m)=\frac{(m-1)!(l-1)!}{(l+m-1)!}.$$

III Evaluation of gamma function

We have $$\Gamma(n)=\int_0^\infty x^{n-1}e^{-x}dx.$$

Integrating by parts, keeping x^{n-1} as first function,

$$\Gamma(n)=\left[-e^{-x}x^{n-1}\right]_0^\infty+(n-1)\int_0^\infty e^{-x}x^{n-2}dx$$

$$=(n-1)\int_0^\infty e^{-x}x^{n-2}dx$$

Since $$\underset{x\to 0}{\text{Lt}}\frac{x^{n-1}}{e^x}=\underset{x\to 0}{\text{Lt}}\frac{x^{n-1}}{1+x+\frac{x^2}{2}+\ldots+\frac{x^n}{n!}+\ldots}=0.$$

i.e., $\Gamma(n)=(n-1)\Gamma(n-1)$.

Applying this result again, $\Gamma(n)=(n-1)(n-2)\Gamma(n-2)$

Hence, if n is positive integer then repeating the process above,

$$\Gamma(n)=(n-1)(n-2)\cdot 3\cdot 2\cdot 1\,\Gamma(1)$$

i.e., $$\Gamma(1)=\int_0^\infty x^0e^{-x}dx=\left[-e^x\right]_0^\infty=1.$$

So $$\boxed{\Gamma(n)=(n-1)!}$$

Remember

$\Gamma(n)=(n-1)\Gamma(n-1)$,	for all n
$\Gamma(n)=(n-1)!$,	when n is positive integer
$\Gamma(1)=1,\Gamma(0)=\infty,\Gamma(-n)=\infty$,	where n is positive integer

IV Relation between beta and gamma functions

Prove that $$B(m,n)=\frac{\Gamma(m)\Gamma(n)}{\Gamma(m+n)}$$

We have $$\Gamma(m)=\int_0^\infty e^{-x}x^{m-1}dx$$

Putting zx for n where z is a constant with respect to x, we get

$$\Gamma(m)=\int_0^\infty e^{-zx}(zx)^{m-1}zdx$$

$$=\int_0^\infty e^{-zx}z^m x^{m-1}dx.$$

Multiplying both sides by $e^{-z}Z^{n-1}$, we get

$$\Gamma(m)e^{-z}z^{n-1}=\int_0^\infty e^{-z(1+x)}z^{m+n-1}.x^{m-1}\,dx$$

$$\therefore \qquad \Gamma(m)\int_0^\infty e^{-z}z^{n-1}\,dz=\int_0^\infty\left\{\int_0^\infty e^{-z(1+x)}z^{m+n-1}\,dz\right\}.x^{m-1}dx \qquad \text{.....(12.1)}$$

Putting z(1 + x) = u, $dz=\dfrac{du}{1+x}$

$$\therefore \qquad \int_0^\infty e^{-z(1+x)}z^{m+n-1}dz=\int_0^\infty e^{-u}\left(\frac{u}{1+x}\right)^{m+n-1}\frac{du}{1+x}$$

$$=\frac{1}{(1+x)^{m+n}}\int_0^\infty e^{-u}u^{m+n-1}\,du$$

$$=\frac{1}{(1+x)^{m+n}}\Gamma(m+n).$$

$\therefore$ from (12.1),

$$\Gamma(m)\int_0^\infty e^{-z}z^{n-1}dz=\int_0^\infty\frac{1}{(1+x)^{m+n}}\Gamma(m+n)x^{m-1}dx,$$

i.e.,

$$\Gamma(m)\Gamma(n)=\Gamma(m+n)\int_0^\infty\frac{x^{m-1}}{(1+x)^{m+n}}dx \qquad \text{.....(12.2)}$$

Further, we have $B(m,n)=\int_0^1 x^{m-1}(1-x)^{n-1}\,dx.$

Putting $\dfrac{1}{1+u}$ for x, we get

$$dx=-\frac{1}{(1+u)^2}du,\ 1-x=1-\frac{1}{1+u}=\frac{u}{1+u}$$

and $$B(m,n)=\int_{\infty}^{0}\frac{1}{(1+u)^{m-1}}\cdot\left(\frac{u}{1+u}\right)^{n-1}\left[-\frac{1}{(1+u)^2}\right]du$$

$$=\int_0^{\infty}\frac{u^{n-1}}{(1+u)^{m+n}}du$$

$$=\int_0^{\infty}\frac{x^{m-1}}{(1+x)^{m+n}}dx \qquad .....(12.3)$$

$$[\because \int_a^b f(x)dx=\int_a^b f(z)dz \text{ and } B(m,n)=B(n,m)]$$

$\therefore$ From (12.2) and (12.3), $\Gamma(m)\Gamma(n)=\Gamma(m+n).\ B(m,n)$

$\therefore$ $$B(m,n)=\frac{\Gamma(m)\Gamma(n)}{\Gamma(m+n)}.$$

V Prove that

(a) $\Gamma(n)\Gamma(1-n)=\dfrac{\pi}{\sin n\pi} \quad 0<n<1$

(b) $\Gamma\left(\dfrac{1}{2}\right)=\sqrt{\pi}.$

(a) We have $\Gamma(m)\Gamma(n)=\Gamma(m+n)\int_0^{\infty}\dfrac{x^{m-1}}{(1+x)^{m+n}}dx$ [from (12.2) of IV].

Putting m + n = 1,

$$\Gamma(n)\Gamma(1-n)=\int_0^{\infty}\frac{x^{n-1}}{1+x}dx=\frac{\pi}{\sin n\pi}$$

(b) Putting $n=\dfrac{1}{2}$ in (a),

$$\left\{\Gamma\left(\frac{1}{2}\right)\right\}^2=\pi;\ \therefore\ \Gamma\left(\frac{1}{2}\right)=\sqrt{\pi}$$

VI Find the value of

$$\Gamma\left(\frac{1}{n}\right)\Gamma\left(\frac{2}{n}\right)...\Gamma\left(\frac{n-1}{n}\right) \quad \text{n being a positive integer}$$

Let $$A=\Gamma\left(\frac{1}{n}\right)\Gamma\left(\frac{2}{n}\right)...\Gamma\left(\frac{n-1}{n}\right)$$

Writing it in reverse order,

$$A=\Gamma\left(1-\frac{1}{n}\right)\Gamma\left(1-\frac{2}{n}\right)...\Gamma\left(1-\frac{n-1}{n}\right).$$

Multiplying these, we get

$$A\times A=\left\{\Gamma\left(\frac{1}{n}\right)\Gamma\left(1-\frac{1}{n}\right)\right\}\left\{\Gamma\left(\frac{2}{n}\right)\Gamma\left(1-\frac{2}{n}\right)\right\}...\left\{\Gamma\left(\frac{n-1}{n}\right)\Gamma\left(1-\frac{n-1}{n}\right)\right\},$$

i.e., $$A^2=\frac{\pi}{\sin\frac{\pi}{n}}\cdot\frac{\pi}{\sin\frac{2\pi}{n}}\cdot....\frac{\pi}{\sin\left\{\frac{(n-1)\pi}{n}\right\}}. \qquad .....(12.4)$$

To find the right-hand side of (12.4), we have [by V(a)]

$$\frac{\sin n\theta}{\sin\theta}=2^{n-1}\sin\left(\theta+\frac{\pi}{n}\right)\sin\left(\theta+\frac{2\pi}{n}\right)...\sin\left(\theta+\frac{n-1}{n}\pi\right)$$

$$\therefore \quad \underset{\theta\to 0}{\text{Lt}}\frac{\sin n\theta}{\sin\theta}=2^{n-1}\sin\frac{\pi}{n}\sin\frac{2\pi}{n}...\sin\left(\frac{n-1}{n}\pi\right).$$

Put $$\underset{\theta\to 0}{\text{Lt}}\frac{\sin n\theta}{\sin\theta}=\underset{\theta\to 0}{\text{Lt}}\left\{n\frac{\sin n\theta}{n\theta}\cdot\frac{\theta}{\sin\theta}\right\}=n.$$

$$\therefore \quad n=2^{\pi-1}\sin\frac{\pi}{n}\sin\frac{2\pi}{n}...\sin\left(\frac{n-1}{n}\pi\right)$$

$$\therefore \quad \text{from (12.4)},\ A^2=\frac{\pi^{n-1}}{n/2^{n-1}}=\frac{(2\pi)^{n-1}}{n}$$

$$\therefore \quad A=\frac{(2\pi)^{\frac{n-1}{2}}}{n^{1/2}},$$

i.e., $$\Gamma\left(\frac{1}{n}\right)\Gamma\left(\frac{2}{n}\right)...\Gamma\left(\frac{n-1}{n}\right)=\frac{(2\pi)^{\frac{n-1}{2}}}{n^{1/2}}$$

VII. Prove that $\Gamma(n)\Gamma\left(n+\frac{1}{2}\right)=\frac{\sqrt{\pi}}{2^{2n-1}}\Gamma(2n).$ {Duplication formula)

We know that $B(m, n)=\int_0^1 x^{m-1}(1-x)^{n-1}\,dx$

Putting $x = \sin^2\theta$, $\quad dx = 2\sin\theta\cos\theta\, d\theta$

$$\int_0^1 x^{m-1}(1-x)^{n-1}\,dx = 2\int_0^{\pi/2}\sin^{2m-1}\theta\cos^{2n-1}\theta\, d\theta,$$

i.e., $\quad B(m,n) = 2\int_0^{\pi/2}\sin^{2m-1}\theta\cos^{2n-1}\theta\, d\theta$

But $\quad B(m,n) = \dfrac{\Gamma(m)\Gamma(n)}{\Gamma(m+n)}$

$$\therefore \quad \int_0^{\pi/2}\sin^{2m-1}\theta\cos^{2n-1}\theta\, d\theta = \frac{\Gamma(m)\Gamma(n)}{2\Gamma(m+n)} \qquad .....(12.5)$$

When $2m-1=0$, i.e., when $m=\dfrac{1}{2}$, $\quad$ from (12.5) we get

$$\int_0^{\pi/2}\cos^{2n-1}\theta\, d\theta = \frac{\sqrt{\pi}}{2}\cdot\frac{\Gamma(n)}{\Gamma\left(n+\frac{1}{2}\right)} \qquad .....(12.6)$$

Again when $m=n$, from (12.5) we get

$$\int_0^{\pi/2}\sin^{2n-1}\theta\cos^{2n-1}\theta = \frac{\Gamma(n)\Gamma(n)}{2\Gamma(2n)},$$

i.e., $\quad \displaystyle\int_0^{\pi/2}(\sin\theta\cos\theta)^{2n-1}\, d\theta = \frac{[\Gamma(n)]^2}{2\Gamma(2n)},$

i.e., $\quad \displaystyle\frac{1}{2^{2n-1}}\int_0^{\pi/2}(2\sin\theta\cos\theta)^{2n-1}d\theta = \frac{[\Gamma(n)]^2}{2\Gamma(2n)},$

i.e., $\quad \displaystyle\frac{1}{2^{2n-1}}\int_0^{\pi/2}(\sin 2\theta)^{2n-1}d\theta = \frac{[\Gamma(n)]^2}{2\Gamma(2n)}.$

Putting $2\theta=\alpha$, $2d\theta = d\alpha$, we get

$$\frac{1}{2^{2n}}\int_0^{\pi}\sin^{2n-1}\alpha\, d\alpha = \frac{[\Gamma(n)]^2}{2\Gamma(2n)}$$

or $$\frac{2}{2^{2n}}\int_0^{\pi/2}\sin^{2n-1}\alpha\,d\alpha=\frac{[\Gamma(n)]^2}{2\Gamma(2n)}$$

or $$2^{2n-2}\frac{[\Gamma(n)]^2}{\Gamma(2n)}=\int_0^{\pi/2}\sin^{2n-1}\alpha\,d\alpha=\int_0^{\pi/2}\cos^{2n-1}\alpha\,d\alpha$$

$$=\frac{\sqrt{\pi}}{2}\frac{\Gamma(n)}{\Gamma\left(n+\frac{1}{2}\right)}\quad[\text{from}(12.6)]$$

$\therefore$ $$\Gamma(n)\Gamma\left(n+\frac{1}{2}\right)=\frac{\sqrt{\pi}}{2^{2n-1}}\Gamma(2n).$$

Solved Examples

1. Show that (a) $\Gamma\left(\frac{7}{2}\right)=\frac{15}{8}\sqrt{\pi}$ (b) $\Gamma\left(\frac{1}{3}\right)\Gamma\left(\frac{2}{3}\right)=\frac{2}{\sqrt{3}}\pi$

(a) $$\Gamma\left(\frac{7}{2}\right)=\Gamma\left(\frac{5}{2}+1\right)=\frac{5}{2}\Gamma\left(\frac{5}{2}\right)\qquad[\text{since }\Gamma(n+1)=n\,\Gamma(n)]$$

$$=\frac{5}{2}\Gamma\left(\frac{3}{2}+1\right)=\frac{5}{2}\cdot\frac{3}{2}\Gamma\left(\frac{3}{2}\right)$$

$$=\frac{5}{2}\cdot\frac{3}{2}\Gamma\left(\frac{1}{2}+1\right)=\frac{5}{2}\cdot\frac{3}{2}\cdot\frac{1}{2}\Gamma\left(\frac{1}{2}\right)$$

$$=\frac{15}{8}\sqrt{\pi}\qquad\left(\text{since }\Gamma\left(\frac{1}{2}\right)=\sqrt{\pi}\right).$$

(b) $$\Gamma\left(\frac{1}{3}\right)\Gamma\left(\frac{2}{3}\right)=\Gamma\left(\frac{1}{3}\right)\Gamma\left(1-\frac{1}{3}\right)$$

$$=\frac{\pi}{\sin\frac{1}{3}\pi}\qquad[\text{by V(a)}]$$

$$=\frac{2}{\sqrt{3}}\pi.$$

2. Prove that $B(m,n)=B(m+1,n)+B(m,n+1)$ for $m>0, n>0$.

$$\text{RHS}=\frac{\Gamma(m+1)\Gamma(n)+\Gamma(m)\Gamma(n+1)}{\Gamma(m+n+1)}$$

$$= \frac{m\Gamma(m)\Gamma(n) + \Gamma(m)n\Gamma(n)}{(m+n)\Gamma(m+n)}$$

$$= \frac{(m+n)\Gamma(m)\Gamma(n)}{(m+n)\Gamma(m+n)} = \frac{\Gamma(m)\Gamma(n)}{\Gamma(m+n)} = B(m,n) = \text{LHS.}$$

3. Evaluate $\Gamma\left(\frac{1}{9}\right)\Gamma\left(\frac{2}{9}\right)\Gamma\left(\frac{3}{9}\right)\ldots\Gamma\left(\frac{8}{9}\right)$.

The given expression

$$= \Gamma\left(\frac{1}{9}\right)\Gamma\left(\frac{2}{9}\right)\Gamma\left(\frac{3}{9}\right)\cdots\Gamma\left(\frac{9-1}{8}\right).$$

We know that

$$\Gamma\left(\frac{1}{n}\right)\Gamma\left(\frac{2}{n}\right)\Gamma\left(\frac{3}{n}\right)\cdots\Gamma\left(\frac{n-1}{n}\right) = \frac{(2\pi)^{\frac{n-1}{2}}}{n^{\frac{1}{2}}}.$$

$\therefore$ the given expression

$$= \frac{(2\pi)^{\frac{9-1}{2}}}{9^{\frac{1}{2}}} = \frac{(2\pi)^4}{3} = \frac{16}{3}\pi^4.$$

4. Show that $2^n\Gamma\left(n+\frac{1}{2}\right) = 1\times3\times5\ldots(2n-1)\sqrt{\pi}$ where n is a positive integer.

$$\Gamma\left(n+\frac{1}{2}\right) = \left(n-\frac{1}{2}\right)\Gamma\left(n-\frac{1}{2}\right), \quad \text{since } \Gamma(n) = (n-1)\Gamma(n-1)$$

$$= \left(n-\frac{1}{2}\right)\left(n-\frac{3}{2}\right)\Gamma\left(n-\frac{3}{2}\right)$$

$$= \left(n-\frac{1}{2}\right)\left(n-\frac{3}{2}\right)\cdots\frac{3}{2}\cdot\frac{1}{2}\Gamma\left(\frac{1}{2}\right)$$

$$= \frac{2n-1}{2}\cdot\frac{2n-3}{2}\cdot\frac{2n-5}{2}\cdots\frac{3}{2}\cdot\frac{1}{2}\sqrt{\pi}$$

$$= \frac{1}{2n}(2n-1)(2n-3)\cdots3\cdot1\cdot\sqrt{\pi}$$

$$\therefore \quad 2^n\Gamma\left(n+\frac{1}{2}\right) = 1\cdot3\cdot5\cdots(2n\cdot1)\sqrt{\pi}.$$

5. Show that $\Gamma(p).\Gamma\left(\frac{1-p}{2}\right)=\dfrac{\sqrt{\pi}\,\Gamma\left(\frac{p}{2}\right)}{2^{1-p}\cos\left(\frac{1}{2}p\pi\right)}, 0<p<1.$

We know that $\Gamma(n)\Gamma(1-n)=\dfrac{\pi}{\sin n\pi}$ .....(12.7)

and $$\Gamma(n)\Gamma\left(n+\frac{1}{2}\right)=\frac{\sqrt{\pi}}{2^{2n+1}}\Gamma(2n) \quad .....(12.8)$$

Putting $n=\dfrac{p+1}{2}$ in (12.7),

$$\Gamma\left(\frac{p+1}{2}\right)\Gamma\left(\frac{1-p}{2}\right)=\frac{\pi}{\cos\left(\frac{1}{2}p\pi\right)} \quad .....(12.9)$$

Putting $2n=p$ in (12.8),

$$\Gamma\left(\frac{p}{2}\right)\Gamma\left(\frac{p+1}{2}\right)=\frac{\sqrt{\pi}}{2^{p-1}}\Gamma(p) \quad .....(12.10)$$

Dividing (12.9) by (12.10),

$$\frac{\Gamma\left(\frac{p+1}{2}\right)\Gamma\left(\frac{1-p}{2}\right)}{\Gamma\left(\frac{p}{2}\right)\Gamma\left(\frac{p+1}{2}\right)}=\frac{\pi}{\cos\left(\frac{1}{2}p\pi\right)}\times\frac{2^{p-1}}{\sqrt{\pi}\,\Gamma(p)}$$

or $$\frac{\Gamma\left(\frac{1-p}{2}\right)}{\Gamma\left(\frac{p}{2}\right)}=\sqrt{\pi}\cdot\frac{1}{2^{1-p}\cos\left(\frac{1}{2}p\pi\right)}\cdot\frac{1}{\Gamma(p)}'$$

i.e., $$\Gamma(p)\,\Gamma\left(\frac{1-p}{2}\right)=\frac{\sqrt{\pi}\,\Gamma\left(\frac{p}{2}\right)}{2^{1-p}\cos\left(\frac{1}{2}p\pi\right)}$$

6. **Show that** $\Gamma\left(n+\frac{1}{2}\right)=\frac{\Gamma(2n+1)\sqrt{\pi}}{2^{2n}\Gamma(n+1)}$

$$\Gamma\left(n+\frac{1}{2}\right)=\Gamma\left(\frac{2n+1}{2}\right)=\Gamma\left(\frac{2n-1}{2}+1\right)$$

$$=\frac{2n-1}{2}\Gamma\left(\frac{2n-1}{2}\right)=\frac{2n-1}{2}\Gamma\left(\frac{2n-3}{2}+1\right)$$

$$=\frac{2n-1}{2}\cdot\frac{2n-3}{2}\Gamma\left(\frac{2n-3}{2}\right)$$

$$=\frac{2n-1}{2}\cdot\frac{2n-3}{2}\cdot\frac{2n-5}{2}\cdot\ldots\cdot\frac{5}{2}\cdot\frac{3}{2}\cdot\frac{1}{2}\Gamma\left(\frac{1}{2}\right),$$

repeating the same result

$$=\frac{(2n-1)(2n-3)(2n-5)\ldots 5\cdot 3\cdot 1}{2^n}\sqrt{\pi} \qquad \ldots\ldots(12.11)$$

Multiplying the numerator and denominator of (12.11) by

$2n(2n-2)(2n-4)\ldots 4.2,$

$$\Gamma\left(n+\frac{1}{2}\right)=\frac{2n(2n-1)(2n-2)\ldots 5\cdot 4\cdot 3\cdot 2\cdot 1}{2^n\cdot 2\cdot n\cdot 2(n-1)\cdot 2(n-2)\ldots 2\cdot 2\cdot 2\cdot 1}\sqrt{\pi}$$

$$=\frac{\Gamma(2n+1)}{2^n\cdot 2^n\cdot n(n-1)(n-2)\ldots 2\cdot 1}\sqrt{\pi}$$

$$=\frac{\Gamma(2n+1)\sqrt{\pi}}{2^{2n}\,\Gamma(n+1)}$$

Note the above result can be written in the form

$$\Gamma\left(\frac{1}{2}\right)\Gamma(2n)=2^{2n-1}\Gamma(n)\,\Gamma\left(n+\frac{1}{2}\right).$$

7. Evaluate $\int_0^1\frac{dx}{\sqrt{1-x^n}}$

Putting $x^n=\sin^2\theta,\ x=\sin^{2/n}\theta$ $\qquad \therefore\ dx=\frac{2}{n}\sin^{\frac{2}{n}-1}\theta\cos\theta\,d\theta$

$$\therefore \quad \int_0^1 \frac{dx}{\sqrt{1-x^n}} = \int_0^{\pi/2} \frac{\frac{2}{n}\sin^{\frac{2}{n}-1}\theta\ \cos\theta\ d\theta}{\cos\theta}$$

$$= \frac{2}{n}\int_0^{\pi/2} \sin^{\frac{2}{n}-1}\theta\ d\theta = \frac{2}{n}\ \frac{\Gamma\left(\frac{2}{n}\right)\Gamma\left(\frac{1}{2}\right)}{2\Gamma\left(\frac{1}{n}+\frac{1}{2}\right)}$$

$$\because \quad \int_0^{\pi/2} \sin^m\theta\ \cos^n\theta\ d\theta = \frac{\Gamma\left(\frac{m+1}{2}\right)\Gamma\left(\frac{n+1}{2}\right)}{2\Gamma\left(\frac{m+n+2}{2}\right)};$$

here $m = \frac{2}{n} - 1,\ n = 0$ (see example 10)

$$= \frac{1}{n}\ \frac{\Gamma\left(\frac{2}{n}.\sqrt{\pi}\right)}{\Gamma\left(\frac{1}{n}+\frac{1}{2}\right)} \quad \left(\because \Gamma\left(\frac{1}{2}\right) = \sqrt{\pi}\right)$$

$$= \frac{\sqrt{\pi}}{n}\ \frac{\Gamma\left(\frac{2}{n}\right)}{\Gamma\left(\frac{1}{n}+\frac{1}{2}\right)}$$

8. Evaluate $\int_0^a \frac{dx}{\sqrt{(a^n - x^n)^{1/2}}}$

Put $x^n = a^n \sin^2\theta$, i.e., $x = a \sin^2/n\ \theta$,

so that $dx = \frac{2a}{n}\sin^{\left(\frac{2}{n}-1\right)}\theta\cos\theta\ d\theta.$

Also when $x = 0$, $\theta = 0$; when $x = a$, $\theta = \pi/2$.

$$\therefore \quad \int_0^a \frac{dx}{(a^n - x^n)^{1/n}} = \frac{2a}{n}\int_0^{\pi/2} \frac{\sin^{\frac{2}{n}-1}\theta\cos\theta\ d\theta}{a\cos^{\frac{2}{n}}\theta}$$

$$= \frac{2}{n}\int_0^{\pi/2} \sin^{\frac{2}{n}-1}\theta \cos^{1-\frac{2}{n}}\theta \, d\theta$$

$$= \frac{\frac{2}{n}\Gamma\left(\frac{1}{n}\right)\Gamma\left(1-\frac{1}{n}\right)}{2\Gamma(1)} = \frac{1}{n}\cdot\frac{\pi}{\sin \pi / n}$$

9. Evaluate $\int_0^1 \frac{x^{m-1}+x^{n-1}}{(1+x)^{m+n}}dx.$

Let the given integral = I, then

$$I = \int_0^1 \frac{x^{m-1}}{(1+x)^{m+n}}dx + \int_0^1 \frac{x^{n-1}}{(1+x)^{m+n}}dx.$$

Putting $x = \frac{1}{z}$ in the second integral, $dx = -\frac{1}{z^2}dz$. When n =0, z = ∞; when n =1, z = 1.

$$I = \int_0^1 \frac{x^{m-1}}{(1+x)^{m+n}}dx + \int_0^1 \frac{\left(\frac{1}{z}\right)^{n-1}}{\left(1+\frac{1}{z}\right)^{m+n}} \times \left(-\frac{1}{z^2}\right)dz$$

$$= \int_0^1 \frac{x^{m-1}}{(1+x)^{m+n}}dx + \int_1^\infty \frac{z^{m-1}}{(1+z)^{m+n}}dz$$

$$= \int_0^1 \frac{x^{m-1}}{(1+x)^{m+n}}dx + \int_1^\infty \frac{x^{m-1}}{(1+x)^{m+n}}dx$$

$$= \int_0^\infty \frac{x^{m-1}}{(1+x)^{m+n}}dx$$

$$= B(m, n) \quad \left\{\text{since } B(m,n) = \int_0^\infty \frac{x^{m-1}}{(1+x)^{m+n}} = \int_0^\infty \frac{x^{n-1}}{(1+x)^{m+n}}\right\}$$

$$= \frac{\Gamma(m)\Gamma(n)}{\Gamma(m+n)}$$

10. Show that $\int_0^{\pi/2} \sin^m\theta\cos^n\theta\, d\theta = \dfrac{\Gamma\left(\frac{m+1}{2}\right)\Gamma\left(\frac{n+1}{2}\right)}{2\Gamma\left(\frac{m+n+2}{2}\right)}$.

We know that

$$B(p, q) = \int_0^1 x^{p-1}(1-x)^{q-1}dx \qquad \text{.....(12.12)}$$

Putting $x = \sin^2\theta$, $dx = 2\sin\theta\cos\theta\, d\theta$; $1 - x = 1 - \sin^2\theta = \cos^2\theta$.

When $x = 0$, $q = 0$; when $x = 1$, $q = \pi/2$.

(12.12) becomes $B(p, q) = \int_0^{\pi/2} \sin^{2p-2}\theta\cos^{2q-2}\theta \cdot \sin\theta\cos\theta d\theta$

$$= 2\int_0^{\pi/2} \sin^{2p-1}\theta\cos^{2q-1}\theta\, d\theta$$

or $$\frac{\Gamma(p)\Gamma(q)}{2\Gamma(p+q)} = \int_0^{\pi/2} \sin^{2p-1}\theta\cos^{2q-1}\theta d\theta \quad [\text{by IV}]$$

Let $2p - 1 = m$, i.e., $p = \dfrac{m+1}{2}$ and $2q - 1 = n$, i.e., $q = \dfrac{n+1}{2}$

$$\int_0^{\pi/2} \sin^m\theta\cos^n\theta\, d\theta = \frac{\Gamma\left(\frac{m+1}{2}\right)\Gamma\left(\frac{n+1}{2}\right)}{2\Gamma\left(\frac{m+n+2}{2}\right)}$$

11. Show that $\int_0^{\pi/2} \tan^n\theta\, d\theta = \dfrac{\pi}{2}\sec\dfrac{n\pi}{2}$, indicating the restrictions on n.

$$\int_0^{\pi/2} \tan^n\theta\, d\theta = \int_0^{\pi/2} \sin^m\theta\cos^{-n}\theta\, d\theta$$

$$= \frac{\Gamma\left(\frac{n+1}{2}\right)\Gamma\left(\frac{1-n}{2}\right)}{2\Gamma(1)}$$

where $1 - n > 0$, i.e., $n < 1$

$$= \frac{1}{2}\Gamma\left(\frac{1+n}{2}\right)\Gamma\left(1 - \frac{1+n}{2}\right)$$

$$= \frac{1}{2}\frac{\pi}{\sin\frac{1+n}{2}\pi}, \qquad \text{provided } 0 < \frac{1+n}{2} < 1 \text{ , i.e., } 1 > n > -1$$

$$= \frac{1}{2}\pi \sec\frac{n\pi}{2}$$

12. Show that $\int_0^{\pi/2} \frac{dx}{\sqrt{\sin x}} \times \int_0^{\pi/2} \sqrt{\sin x}\, dx = \pi.$

The given integral

$$= \int_0^{\pi/2} \sin^{-\frac{1}{2}} x\, dx \times \int_0^{\pi/2} \sin^{-\frac{1}{2}} x\, dx$$

$$= \int_0^{\pi/2} \sin^{-\frac{1}{2}} x \cos^0 x\, dx \times \int_0^{\pi/2} \sin^{\frac{1}{2}} x \cos^0 x\, dx$$

$$= \frac{\Gamma\left(\frac{1}{4}\right)\Gamma\left(\frac{1}{2}\right)}{2\Gamma\left(\frac{3}{4}\right)} \times \frac{\Gamma\left(\frac{3}{4}\right)\Gamma\left(\frac{1}{2}\right)}{2\Gamma\left(\frac{5}{4}\right)}$$

$$= \frac{\Gamma\left(\frac{1}{4}\right)\sqrt{\pi}}{2\Gamma\left(\frac{3}{4}\right)} \times \frac{\Gamma\left(\frac{3}{4}\right)\sqrt{\pi}}{2\frac{1}{4}\Gamma\left(\frac{1}{4}\right)} = \pi.$$

13. Show that $\int_0^{¥} \cos\left(bz^{\frac{1}{x}}\right) dz = \frac{\Gamma(n+1)\cos\frac{1}{2}n\pi}{b^n}$

Let $z^{\frac{1}{n}} = t$ or $z = t^n,$ $\qquad \therefore dz = nt^{n-1}dt$

and $\qquad \int_0^{\infty} \cos\left(bz^{\frac{1}{n}}\right) dz = n\int_0^{\infty} t^{n-1} \cos(bt)\, dt$

$$= \text{real part of } n\int_0^{\infty} t^{n-1} e^{bti}\, dt$$

$$= \text{real part on } n\ \frac{\Gamma(n)}{(bt)^n}$$

$$= \text{real part of } \frac{n}{b^n}\Gamma(n)i^{-n}$$

$$= \text{real part of } \frac{n}{b^n}\Gamma(n)\left(\cos\frac{\pi}{2}+\sin\frac{\pi}{2}\right)^{-n}$$

$$= \text{real part of } \frac{\Gamma(n+1)}{b^n}\left(\cos\frac{n\pi}{2}-i\sin\frac{n\pi}{2}\right)$$

$$= \frac{\Gamma(n+1)}{b^n}\cos\frac{n\pi}{2}$$

Exercise 12.1

1. Show that $B(m, n) \cdot B(m + n, l) = B(n, l)\, B(n + l, m)$.

   [*Hint*: $\text{LHS} = \frac{\Gamma(m)\Gamma(n)}{\Gamma(m+n)} \cdot \frac{\Gamma(m+n)\Gamma(l)}{\Gamma(m+n+l)} = \frac{\Gamma(l)\Gamma(m)\Gamma(n)}{\Gamma(l+m+n)} = \text{RHS}$]

2. Evaluate $\Gamma(0.1)\Gamma(0.2)\Gamma(0.3)\ldots\Gamma(0.9)$

   [*Hint*: Expression = $\Gamma\left(\frac{1}{10}\right)\Gamma\left(\frac{2}{10}\right)\Gamma\left(\frac{3}{10}\right)\ldots\Gamma\left(\frac{9}{10}\right)$. Now use (VI)]

3. Show $\Gamma\left(\frac{3}{2}+x\right)=\Gamma\left(\frac{3}{2}-x\right)=\left(\frac{1}{4}-x^2\right)\pi\sec\pi x$, where $-1< 2x < 1$.

   [*Hint*: $\Gamma\left(\frac{3}{2}+x\right)=\left(\frac{1}{2}+x\right)\Gamma\left(\frac{1}{2}+x\right), \Gamma\left(\frac{3}{2}-x\right)=\left(\frac{1}{2}-x\right)\Gamma\left(\frac{1}{2}-x\right)$

$$\therefore\ \Gamma\left(\frac{3}{2}+x\right)\Gamma\left(\frac{3}{2}-x\right)=\left(\frac{1}{4}-x^2\right)\Gamma\left(\frac{1-2x}{2}\right)\Gamma\left(\frac{1+2x}{2}\right)$$

$$= \Gamma\left(\frac{1}{4}-x^2\right)\Gamma\left(\frac{1-2x}{2}\right)\Gamma\left(1-\frac{1-2x}{2}\right)$$

$$= \Gamma\left(\frac{1}{4}-x^2\right)\frac{\pi}{\sin(1-2x)\frac{\pi}{2}}$$

$$\left\{\because \Gamma(n)\Gamma(1-n)=\frac{\pi}{\sin n\pi}\right\}$$

$$= \left(\frac{1}{4}-x^2\right)\pi\sec\pi x \text{]}$$

4. Prove that $\int_0^1 \frac{dx}{\sqrt{1-x^4}} = \frac{\sqrt{2}}{8\sqrt{\pi}}\left\{\Gamma\left(\frac{1}{4}\right)\right\}^2$.

[*Hint*: See example 7. Put $x^4 = \sin^2\theta$.

$$I = \frac{1}{2}\frac{\Gamma\left(\frac{1}{4}\right)\Gamma\left(\frac{1}{2}\right)}{2\Gamma\left(\frac{3}{4}\right)} = \frac{1}{4}\frac{\left\{\Gamma\left(\frac{1}{4}\right)\right\}^2\sqrt{\pi}}{\Gamma\left(\frac{1}{4}\right)\Gamma\left(\frac{1}{4}+\frac{1}{2}\right)}$$

$$= \frac{\sqrt{\pi}}{4}\frac{\left\{\Gamma\left(\frac{1}{4}\right)\right\}^2}{\frac{\sqrt{\pi}}{2^{2\frac{1}{4}-1}}\Gamma\left(2\cdot\frac{1}{4}\right)}\left\{\text{since } \Gamma(n)\Gamma\left(n+\frac{1}{2}\right) = \frac{\sqrt{\pi}\Gamma(2n)}{2^{2n-1}}\right\}$$

$$= \frac{\sqrt{2}}{8\sqrt{\pi}}\left\{\Gamma\left(\frac{1}{4}\right)\right\}^2 \text{]}$$

Answer

2. $\frac{(2\pi)^{9/2}}{\sqrt{10}}$

CHAPTER 13

Multiple Integrals

Double and Triple Integrals

Let $f(x,y)$ be a continuous and single valued function of x and y within a region R bounded by a closed curve 'c' and upon the boundary c. Let the region R be subdivided in any manner into n subregions of areas $\delta R_1, \delta R_2, \ldots, \delta R_n$.

Let (x_i, y_i) be any point in the subregion of area δR_i.

Consider the sum

$$\sum_{i=1}^{n} f(x_i, y_i)\delta R_i$$

The limit of this sum as $n \to \infty$ $(i = 1, 2, \ldots)$ is defined as the double integral of $f(x, y)$ over the region R and is written as

$$\iint f(x,y)\,dA$$

$$\therefore \quad \iint_R f(x,y)\,dx\,dy = \underset{n\to\infty}{Lt} \sum_{i=1}^{n} f(x_i, y_i)\,\delta R_i$$

(a) Suppose the region R is described by the inequalities

$c \le y \le d$ and $g(y) \le x \le h(y)$

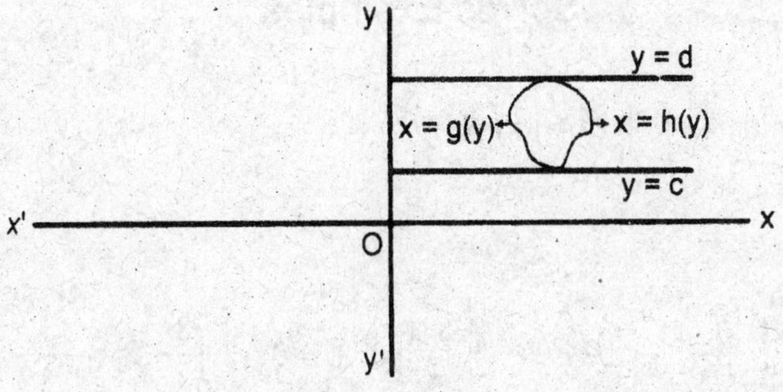

Fig. 13.1(a)

Then

$$\iint_R f(x,y)\,dy\,dx = \iint_R f(x,y)\,dx\,dy = \int_{y=c}^{f=d}\int_{x=g(y)}^{x=h(y)} f(x,y)\,dx\,dy$$

(b) If the region R is described by inequalities

$a \le x \le b$ and $g_1(x) \le y \le h_1(x)$

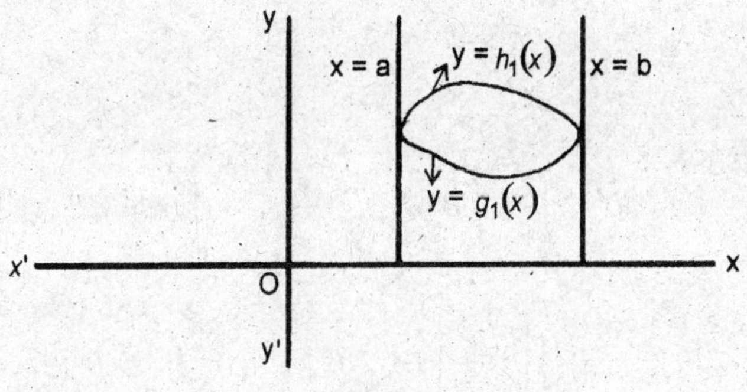

Fig. 13.1(b)

Then $$\iint_R f(x,y)\,dx\,dy = \iint_R f(x,y)\,dy\,dx = \int_{x=a}^{x=b}\int_{y=g_1(x)}^{y=h_1(x)} f(x,y)\,dx.dy$$

(c) If the region R is bounded by the lines $x = a,\ x = b,\ y = c,\ y = d$ (rectangle)

Then $$\iint_R f(x,y)\,dx\,dy = \int_a^b\int_c^d f(x,y)\,dy\,dx = \int_c^d\int_a^b f(x,y)\,dx\,dy$$

***Note*:** The order of integration is immaterial for constant limits

Solved Examples

***Example* 13.1:** Evaluate $\int_0^1 \int_1^2 (x^2 + y^2)\, dxdy$

Solution:

$$\int_0^1 \int_1^2 (x^2 + y^2) dx\, dy = \int_0^1 \left[\frac{x^3}{3} + y^2 x\right]_1^2 dy = \int_0^1 \frac{8}{3} + 2y^2 - \frac{1}{3} - y^2\, dy$$

$$= \int_0^1 \left[y^2 + \frac{7}{3}\right] dy = \left[\frac{y^3}{3} + \frac{7}{3} y\right]_0^1$$

$$= \frac{1}{3} + \frac{7}{3} = \frac{8}{3}$$

***Example* 13.2:** Evaluate $\int_1^4 \int_0^{\sqrt{4-x}} xydydx$

Solution:

$$\int_1^4 \int_0^{\sqrt{4-x}} xydydx = \int_{x=1}^4 \int_0^{y=\sqrt{4-x}} [xydy]\, dx = \int_{x=1}^4 \int_0^{y=\sqrt{4-x}} xydy\, .dx$$

$$= \int_{x=1}^4 \left[x \frac{y^2}{2}\right]_{y=0}^{y=\sqrt{4-x}} dx$$

$$= \int_{x=1}^4 \frac{x}{2}(4-x) dx = \left[\frac{4x^2}{4} - \frac{x^3}{6}\right]_0^4 = \frac{9}{2}$$

***Example* 13.3:** Evaluate $\int_0^1 \int_{\sqrt{y}}^{2-y} x^2 dx\, dy$

Solution:

$$\int_{y=0}^{y=1} \int_{x=\sqrt{y}}^{x=2-y} \left[x^2 dx\right] \text{dy} = \int_{y=0}^1 \left[\frac{x^3}{3}\right]_{x=\sqrt{y}}^{x=2-y} dy$$

$$= \int_{y=0}^{y=1} \frac{1}{3}(2-y)^3 - \frac{1}{3}\left(\sqrt{y}\right)^3 dy$$

$$= \frac{1}{3}\int_{y=0}^{y=1} 8 - y^3 - 12y + 6y^2 - y^{\frac{3}{2}}\, dy$$

$$= \frac{1}{3}\left[8y - \frac{y^4}{4} - \frac{12y^2}{2} + \frac{6y^3}{3} - \frac{y^{\frac{3}{2}+1}}{\frac{3}{2}+1}\right]_0^1$$

$$= \frac{1}{3}\left[8 - \frac{1}{4} - 6 + 2 - \frac{2}{5}\right] = \frac{67}{60}$$

***Example* 13.4:** Find the value of $\iint xy(x+y)dx\,dy$ taken over the region enclosed by the curves $y = x$ and $y = x^2$.

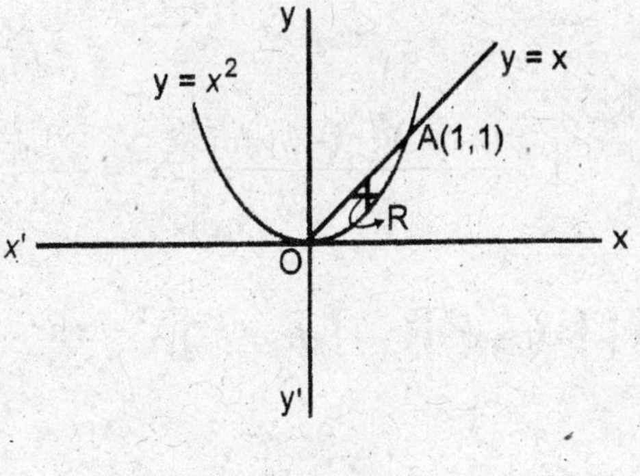

Fig 13.2

Solution:

R is the region bounded by the curves $y = x$ and $y = x^2$

$$I = \iint_R xy(x+y)dx\,dy = \int_{x=0}^{1}\left[\int_{y=x^2}^{y=x}\left(x^2y + xy^2\right)dy\right]dx$$

$$= \int_{x=0}^{1}\left[\frac{x^2y^2}{2} + \frac{xy^3}{3}\right]_{y=x^2}^{y=x} dx = \int_{x=0}^{x=1} \frac{x^4}{2} + \frac{x^4}{3} - \frac{x^6}{2} - \frac{x^7}{3} dx$$

$$= \left[\frac{x^5}{10} + \frac{x^5}{15} - \frac{x^7}{14} - \frac{x^8}{24}\right]_0^1 = \frac{1}{10} + \frac{1}{15} - \frac{1}{14} - \frac{1}{24} = \frac{3}{56}$$

***Example* 13.5:** Evaluate $\iint_A \frac{xy}{\sqrt{1-y^2}}\,dxdy$ where A is the area in the positive quadrant of the circle $x^2 + y^2 = 1$.

Solution:

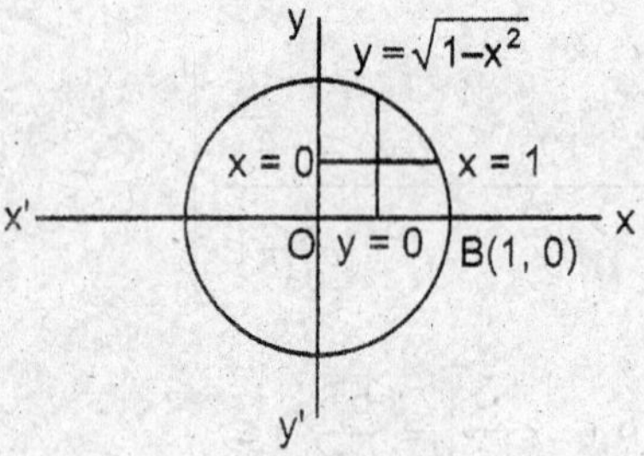

Fig 13.3

$$I = \iint_A \frac{xy}{\sqrt{1-y^2}}\,dxdy = \int_{x=0}^{x=1}\int_{y=0}^{y=\sqrt{1-x^2}} \frac{xydxdy}{\sqrt{1-y^2}}$$

$$= \int_{x=0}^{1} x\left[\int_{y=0}^{y=\sqrt{1-x^2}} \frac{(1-y^2)^{-\frac{1}{2}}(-2y)dy}{-2}\right]dx = -\frac{1}{2}\int_0^1 x\left[\frac{(1-y^2)^{-\frac{1}{2}}}{1/2}\right]_0^{\sqrt{1-x^2}} dx$$

$$= -\int_0^1 x\{1[-(1-x^2)]^{\frac{1}{2}} - 1\}\,dx = -\int_0^1 (x^2 - x)dx$$

$$= \left[\frac{-x^3}{3} + \frac{x^2}{2}\right]_0^1 = \frac{-1}{3} + \frac{1}{2} = \frac{1}{6}$$

***Example* 13.6:** Evaluate $\iint e^{ax+by}\,dx\,dy$ over the triangle $x = 0, y = 0, ax + by = 1$

Solution:

$$I = \iint e^{ax+by}\,dx\,dy = \int_{x=0}^{1/a}\left[\int_{y=0}^{y=\frac{1-ax}{b}} e^{ax+by}\,dy\right]dx$$

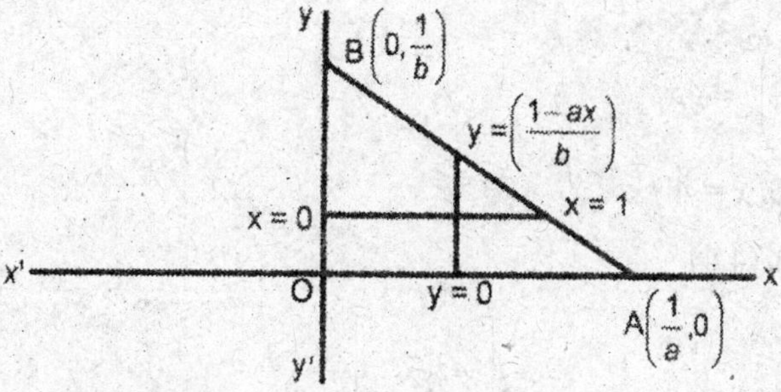

Fig. 13.4

$$= \int_{x=0}^{1/a} \left[\frac{e^{ax+by}}{b}\right]_0^{\frac{1-ax}{b}} dx = \frac{1}{b}\int_0^{\frac{1}{a}} \left[e^{ax+b\left(\frac{1-ax}{b}\right)} - e^{ax+0}\right] dx$$

$$= \frac{1}{b}\int_0^{\frac{1}{a}} \left(e^1 - e^{as}\right) dx = \frac{1}{b}\left[ex - \frac{e^{ax}}{a}\right]_0^{\frac{1}{a}}$$

$$= \frac{1}{b}\left[\frac{e}{a} - \frac{e}{a} - 0 + \frac{e^0}{a}\right] = \frac{1}{ab}$$

Example **13.7**: Evaluate $\iint_R y\,dx\,dy$ where R is the region bounded by the parabolas $y^2 = 4x$ and $x^2 = 4y$.

Solution:

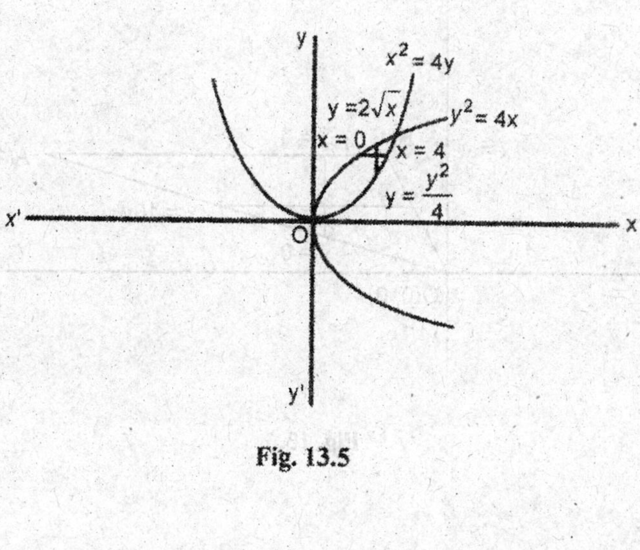

Fig. 13.5

Solving the curves $y^2 = 4x, x^2 = 4y$

$$\left(\frac{x^2}{4}\right)^2 = 4x$$

$$x = 0, x = 4$$

$$\therefore \quad (0,0), (4,4)$$

$$I = \iint_R y\,dx\,dy = \int_{x=0}^{x=4}\left[\int_{y=\frac{x^2}{4}}^{y=2\sqrt{x}} y\,dy\right]dx$$

$$I = \int_{x=0}^{x=4}\left[\frac{y^2}{2}\right]_{y=\frac{x^2}{4}}^{y=2\sqrt{x}} dx = \int_{x=0}^{x=4}\left(4x - \frac{x^4}{32}\right)dx$$

$$I = \left[x^2 - \frac{x^5}{160}\right]_0^4 = \frac{48}{5}$$

***Example* 13.8:** Evaluate $\iint_R \left(\sqrt{xy} - y^2\right)dydx$ where R is a triangle with vertices at (0, 0), (10, 1), (1, 1).

Solution:

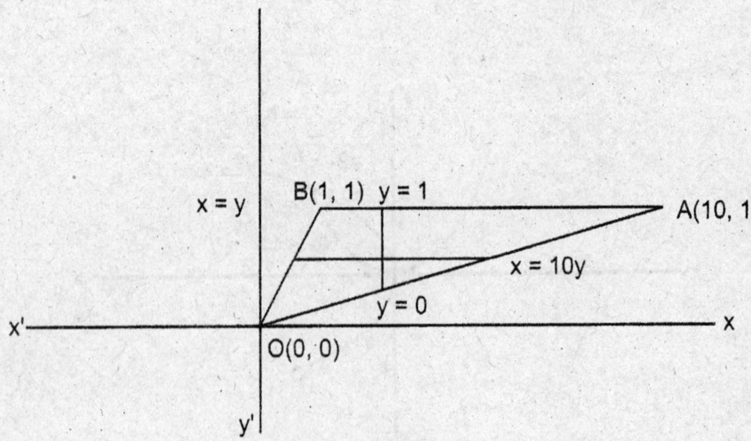

Fig. 13.6

The triangle OAB is within the limits $y \le x \le 10y$ and $0 \le y \le 1$

$$\iint_R \left(\sqrt{xy} - y^2 dydx\right) = \int_{y=0}^{1} \int_{x=y}^{x=10y} \left(\sqrt{xy} - y^2\right) dxdy$$

$$= \int_0^1 dy \left[\frac{2}{3}\frac{1}{y}\left(xy - y^2\right)^{\frac{3}{2}} \right]_y^{10y} = \int_0^1 \frac{2}{3}\frac{1}{y}\left(9y^2\right)^{3/2} dy$$

$$= 18\int_0^1 y^2 dy = 18\left(\frac{1}{3}\right) = 6$$

Exercise – 13.1

I. Evaluate the following double integrals:

1. $\int_0^2 \int_0^2 \left(x^2 + y^2\right) dxdy$ **[Ans:** $\frac{32}{3}$**]**

2. $\int_{-1}^{3} \int_{x^2}^{x+2} dydx$ **[Ans:** $\frac{8}{3}$**]**

3. $\int_1^2 \int_3^4 \frac{dxdy}{(x+y)^2}$ **[Ans:** $\log\frac{25}{24}$**]**

4. $\int_0^a \int_0^{\sqrt{a^2-x^2}} y^3 dydx$ **[Ans:** $\frac{25a^5}{15}$**]**

5. $\int_0^a \int_0^b \left(x^2 + y^2\right) dxdy$ **[Ans:** $\frac{ab}{3}\left(a^2 + b^2\right)$**]**

6. $\int_0^{2a} \int_0^{\sqrt{2x-x^2}} \left(x^2 + y^2\right) dydx$ **[Ans:** $\frac{3\pi a^4}{4}$**]**

7. $\int_0^1 \int_0^1 \frac{y^2 dxdy}{\sqrt{x^2 + y^2}}$ **[Ans:** $\frac{1}{3}\log\left(\sqrt{2} + 1\right)$**]**

8. $\int_0^a \int_{\frac{x^2}{a}}^{2a-x} xy\,dy\,dx$ **[Ans:** $\frac{3a^4}{8}$ **]**

9. Find the value of $\iint xy\,dxdy$ taken over the positive quadrant of the ellipse

$\frac{x^2}{a^2}+\frac{y^2}{b^2}=1$ **[Ans:** $\frac{a^2b^2}{8}$ **]**

10. $\int_0^a \int_0^{\sqrt{a^2-y^2}} \sqrt{a^2-x^2-y^2}\,dxdy$ **[Ans:** $\frac{\pi a^3}{6}$ **]**

11. $\int_{x=0}^{x=1} \int_{y=0}^{y=x} e^{\frac{y}{x}}\,dxdy$ **[Ans:** $\frac{e-1}{2}$ **]**

12. $\int_{x=0}^{1} \int_{y=0}^{\sqrt{1+x^2}} \frac{dxdy}{1+x^2+y^2}$ **[Ans:** $\frac{\pi}{4}\log\left(\sqrt{2}+1\right)$**]**

13. $\int_{\frac{x}{0}}^{\frac{\pi}{4}} \int_0^{\frac{\pi}{2}} sin(x+y)dxdy$ **[Ans: 1]**

14. Evaluate $\iint xy\,dxdy$ where R is the quadrant of the circle

$x^2 + y^2 = a^2$ when $x \geq 0, y \geq 0$. **[Ans:** $\frac{a^4}{8}$ **]**

15. Evaluate $\iint xy\,dxdy$ whose R is the region bounded by x-axis,

ordinate $x = 2a$ and the curve $x^2 = 4ay$. **[Ans:** $\frac{a^4}{3}$ **]**

16. Evaluate $\iint_R (x+y)dxdy$ whose R is the triangular region bounded by

$y = 2x$, $y = \frac{x}{2}$ and $y = 3 - x$.

17. Evaluate $\iint xy\ dxdy$ where R is the triangular region with vertices A(–6, 2), B (–1, 3) and C(9, –7). [Ans: $\frac{-1025}{27}$]

18. Evaluate $\iint xy\, dxdy$ over the region in the positive quadrant for which $x + y \leq 1$ [Ans: $\frac{1}{24}$]

13.1 Change of Order of Integration (Change of Variables)

The variables *x, y* in $\iint_R f(x,y)dxdy$ are changed to *u, v* with the help of the relations $x = f_1(u,v), y = f_2(u, v)$ then the double integral is transferred into

$$\iint_{R'} \{f_1(u,v) f_2(u,v)] |J| \} dudv$$

where $J = \begin{vmatrix} \frac{dx}{du} & \frac{dx}{dv} \\ \frac{dy}{du} & \frac{dy}{dv} \end{vmatrix}$ and 'R' is the region in the uv plane corresponding to the region R in the *xy* plane.

13.1.1 Changing from Cartesion to Polar co-ordinates

$$x = r\cos\theta,\ y = r\sin\theta$$

$$J = \begin{vmatrix} \frac{dx}{dr} & \frac{dx}{d\theta} \\ \frac{dy}{dr} & \frac{dy}{d\theta} \end{vmatrix} = \begin{vmatrix} \cos\theta & -r\sin\theta \\ \sin\theta & r\cos\theta \end{vmatrix} = r$$

$$\iint_R f(x,y)dxdy = \iint_{R'} f[r\cos\theta, r\sin\theta]\, |J|\, d\theta dr$$

$$= \iint_{R'} f[r\cos\theta, r\sin\theta] r\, d\theta dr$$

Note: In polar form *dxdy* is replaced by $rd\theta dr$

Solved Examples

***Example* 13.9:** Evaluate $\iint \sqrt{a^2 - x^2 - y^2}\, dxdy$ over the semi circle $x^2 + y^2 = ax$ in the positive quadrant.

Solution:

Substituting $x = r\cos\theta$,

$y = r\sin\theta$, in $x^2 + y^2 = ax$; $\quad r^2\cos^2\theta + r^2\sin^2\theta = ar\cos\theta \qquad r = a\cos\theta$

which is the equation of the given circle in polar co-ordinates.

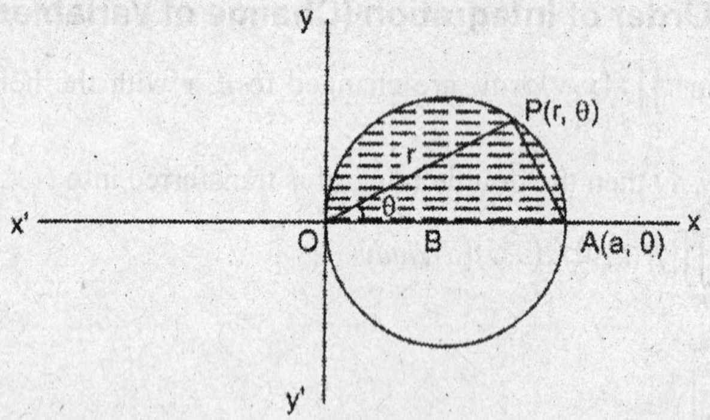

Fig. 13.7

From the figure it can be seen that

$$r = OP = a\cos\theta$$

For the upper half of the circle, θ varies from 0 to $\frac{\pi}{2}$ where as for any intermediatary value of θ, r varies from 0 to OP i.e., 0 to $a\cos\theta$.

$$\therefore \quad \iint_R \sqrt{a^2 - x^2 - y^2}\, dxdy = \int_{\theta=0}^{\frac{\pi}{2}} \int_{r=0}^{a\cos\theta} \sqrt{a^2 - r^2\cos^2\theta - r^2\sin^2} \, rdrd\theta$$

where R is the shaded region.

$$\int_0^{\frac{\pi}{2}} \int_0^{a\cos\theta} r\sqrt{a^2 - r^2}\, drd\theta = \int_0^{\frac{\pi}{2}} \left[-\frac{1}{3} \left(a^2 - r^2\right)^{\frac{3}{2}} \right]_0^{a\cos\theta} d\theta$$

$$= \frac{a^2}{3} \int_0^{\frac{\pi}{2}} \left(1 - \sin^2\theta\right) d\theta$$

$$= \frac{a^2}{3}\left\{[\theta]_0^{\frac{\pi}{2}} - \frac{2}{3}\right\}$$

$$= \frac{a^2}{3}\left(\frac{\pi}{2} - \frac{2}{3}\right)$$

***Example* 13.10:** Evaluate $\int_0^a \int_0^{\sqrt{a^2-y^2}} (x^2 + y^2)dxdy$ by changing into polar coordinates.

Solution:

From the given limit $x = \sqrt{a^2 - y^2}$ it is clear that one of the boundaries is the circle $x^2 + y^2 = a^2$, taking $x = r\cos\theta, y = r\sin\theta$. The given rigion is the first quadrant of the circle.

Here r varies from *'0'* to *'a'* and θ varies from *'0'* to $\frac{\pi}{2}$.

$$\int_0^a \int_a^{\sqrt{a^2-y^2}} \left(x^2 + y^2\right) dxdy = \int_{\theta=0}^{\frac{\pi}{2}} \int_{r=0}^{a} r^2 .rdrd\theta$$

$$= \int_0^{\frac{\pi}{2}} \left[\frac{r^4}{4}\right]_0^a d\theta = \frac{1}{4}\left[a^4\right]_0^{\frac{\pi}{2}} d\theta$$

$$= \frac{a^4}{4} \times \frac{\pi}{2} = \frac{\pi a^4}{8}$$

***Example* 13.11:** Transform the integral $\int_0^{\infty} \int_{r=0}^{\infty} e^{-\left(x^2+y^2\right)} dxdy$ to polar coordinates and evaluate it.

Solution:

The limits of x and y are both from *0* and '∞'. Therefore the region is in the first quadrant where r varies from *'0'* to '∞' and θ varies from *'0'* to $\frac{\pi}{2}$

Substituting $x = r\cos\theta$, $y = r\sin\theta$, and $dxdy = rdrd\theta$.

$$\int_0^\infty \int_0^\infty e^{-\left(x^2+y^2\right)} dxdy = \int_{\theta=0}^{\frac{\pi}{2}} \int_{r=0}^{\infty} e^{-r^2} rdrd\theta$$

$$= \int_0^{\frac{\pi}{2}} \left[\frac{e^{-r^2}}{-2}\right]_0^\infty d\theta = \frac{1}{2}\int_0^{\frac{\pi}{2}} d\theta = \frac{1}{2}[\theta]_0^{\frac{\pi}{2}} = \frac{\pi}{4}$$

13.1.2 Change of Order of Integration

The double integration can be integrated with respect to y first and then with respect to x or it can be integrated with respect to 'x' first and then with respect to 'y'. In the former case, the limits of integration are determined for the given region by drawing strips parallel to y-axis while in the second case by drawing strips parallel to x-axis.

Solved Examples

***Example* 13.12:** Evaluate by changing the order of integration $\int_0^a \int_{\frac{x^2}{a}}^{2a-x} xydxdy$

Solution:

The given limits are $x : 0 \to a, y : \frac{x^2}{a} \to 2a - x$

i.e., the region is bounded by $x = 0, x = a$ and $y = \frac{x^2}{a}, y = 2a - x$ i.e., $x + y = 2a$.

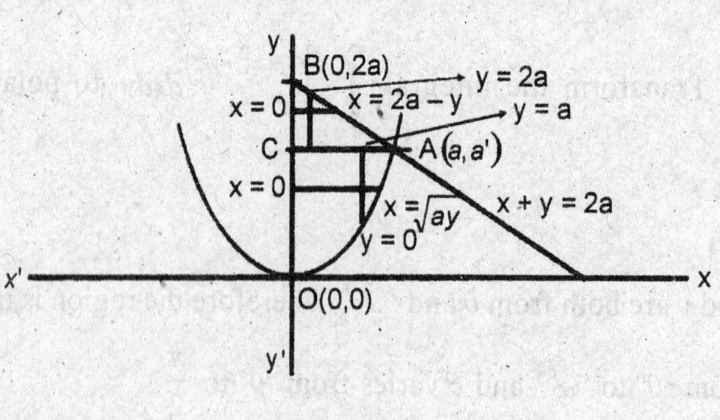

Fig. 13.8

$$\int_0^a \int_{x^2/a}^{2a-x} xy\,dx\,dy = \iint_{OACO} xy\,dx\,dy + \iint_{CABC} xy\,dx\,dy$$

$$= \int_{y=0}^{y=a} \int_{x=0}^{x=\sqrt{ay}} xy\,dx\,dy + \int_{y=a}^{y=2a} \int_{x=0}^{x=2a-y} xy\,dx\,dy$$

$$= \int_0^a dy \left[\frac{x^2}{2} y\right]_0^{\sqrt{ay}} + \int_a^{2a} dy \left[\frac{x^2}{2} y\right]_0^{2a-y}$$

$$= \frac{a}{2}\int_0^a y^2 dy + \frac{1}{2}\int_a^{2a} y(2a-y)^2 dy = \frac{3a^4}{8}$$

***Example* 13.13:** Change the order of integration and evaluate $\int_0^a \int_{\frac{x}{a}}^{\sqrt{\frac{x}{a}}} \left(x^2+y^2\right) dx\,dy$

Solution:

The given limits are $x = 0$, $x = a$ and $y = \dfrac{x}{a}$ and $y = \sqrt{\dfrac{x}{a}}$ i.e., $ay^2 = x$.

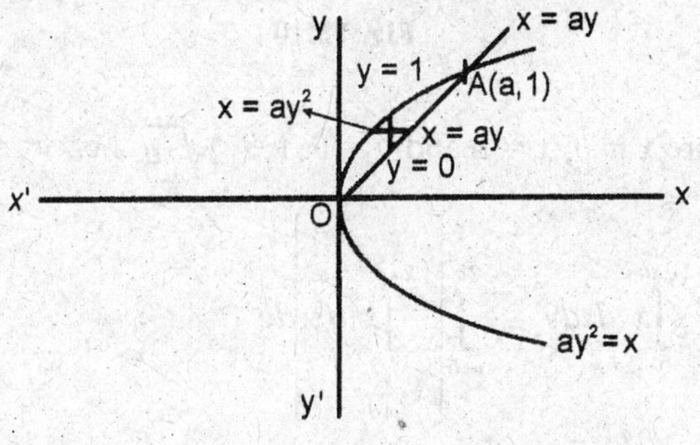

Fig. 13.9

$$\int_0^a \int_{\frac{x}{a}}^{\sqrt{\frac{x}{a}}} \left(x^2+y^2\right) dx\,dy = \int_{y=0}^{y=1} \left[\int_{x=ay^2}^{x=ay} \left(x^2+y^2\right) dx\right] dy$$

$$= \int_{y=0}^{1} dy \left[\frac{x^3}{3} + y^2 x\right]_{x=ay^2}^{x=ay}$$

$$= \int_0^1 \left(\frac{a^3}{3} y^3 \frac{-a^3}{3} y^6 + ay^3 - ay^4 \right) dy$$

$$= \frac{a^3}{28} + \frac{a}{20}$$

***Example* 13.14:** Evaluate $\int_0^a \int_0^{2\sqrt{xa}} x^2 dxdy$ by changing the order of integration.

Solution:

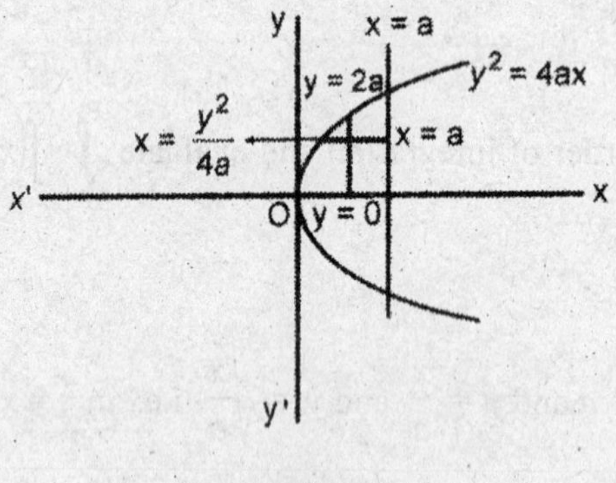

Fig. 13.10

The given limits are $x = 0$, $x = a$ and $y = 0$, $y = 2\sqrt{xa}$ i.e., $y^2 = 4ax$.

$$\int_0^a \int_0^{2\sqrt{xa}} x^2 dxdy = \int_{y=0}^{2a} \left[\int_{x=\frac{y^2}{4a}}^{x=a} x^2 dx \right] dy$$

$$= \int_0^{2a} \left[\frac{x^3}{3} \right]_{\frac{y^2}{4a}}^{a} dy$$

$$= \frac{1}{3} \int_0^{2a} \left[a^3 - \left(\frac{y^2}{4a} \right)^3 \right] dy = \frac{1}{3} \left[a^3 y - \frac{1}{64a^3} \times \frac{y^7}{7} \right]_0^{2a}$$

$$=\frac{1}{3}\left[2a^4-\frac{2a^4}{7}\right]=\frac{4a^4}{7}$$

***Example* 13.15:** Change the order of integration and evaluate $\int_0^4\int_y^4\frac{x}{x^2+y^2}dxdy$.

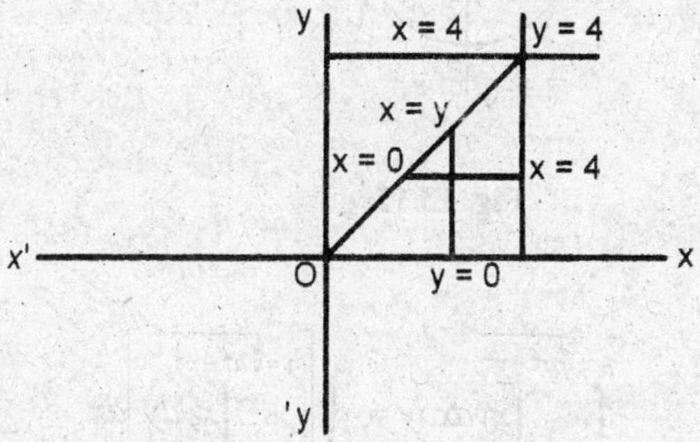

Fig. 13.11

Solution:

The given limits are $y=0,\ y=4$ and $x=y, x=4$.

$$\int_0^4\int_0^4\frac{xdxdy}{x^2+y^2}=\int_{x=0}^{x=4}\left[\int_{y=0}^{y=x}\frac{x}{x^2+y^2}dy\right]dx$$

$$=\int_0^4\left[tan^{-1}\left(\frac{y}{x}\right)\right]_0^x dx$$

$$=\int_0^4 tan^{-1}dx=\int_0^4\frac{\pi}{4}dx=\frac{\pi}{4}[x]_0^4=\pi$$

***Example* 13.16:** Evaluate by changing the order of integration $\int_0^a\int_0^{\sqrt{a^2-y^2}}xydxdy$

Solution:

The given limits are $y=0, y=$ a and $x=0, x=\sqrt{a^2-y^2}$ i.e., $x^2+y^2=a^2$

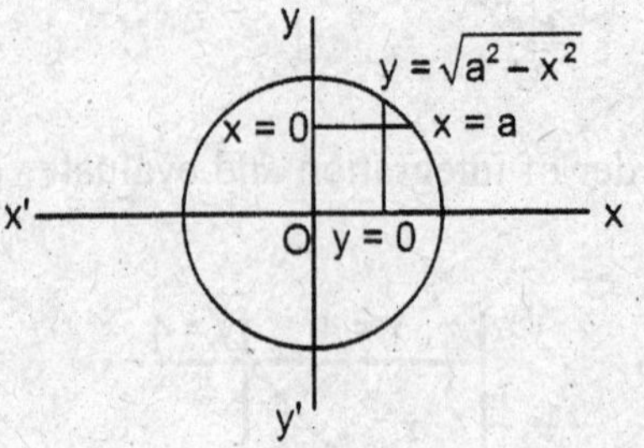

Fig. 13.12

$$\int_{y=0}^{a}\int_{0}^{\sqrt{a^2-y^2}} xy\,dx\,dy = \int_{x=0}^{a}\left[\int_{y=0}^{y=\sqrt{a^2-x^2}} xy\,dy\right]dx$$

$$= \int_{0}^{a}\left[x\frac{y^2}{2}\right]_{0}^{\sqrt{a^2-x^2}} dx = \frac{1}{2}\int_{0}^{a} x\left(\sqrt{a^2-x^2}\right)^2 dx$$

$$= \frac{1}{2}\left[a^2\frac{x^2}{2}-\frac{x^4}{4}\right]_{0}^{a}$$

$$= \frac{1}{2}\left[\frac{a^4}{2}-\frac{a^4}{4}\right] = \frac{a^4}{8}$$

13.2 Triple Integration

Let u = $f(x,y,z)$ be a single valued function of the independent variables x, y, z defined through the region V. Divde the region V into n subregions δV_1, δV_2 .... δV_n and $P(x_i, y_i, z_i)$ be any point inside or on the boundary of the subregion δV_i, then the sum under the limit $\underset{n\to\infty}{Lt}\sum_{i=1}^{n} f(x_i, y_i, z_i)dv_i$ is defined as triple integral.

$$\therefore \quad \underset{n\to\infty}{Lt}\sum_{i=1}^{n} f(x_i, y_i, z_i)dv_i = \iiint_V f(x,y,z)dv$$

Solved Examples

***Example* 13.17:** Evaluate $\int_{-1}^{1}\int_{0}^{z}\int_{x-z}^{x+z}(x+y+z)dxdydz$

Solution:

$$I = \int_{-1}^{1}dz\int_{0}^{z}dx\int_{x-z}^{x+z}(x+y+z)\,dy$$

$$= \int_{-1}^{1}dz\int_{0}^{z}dx\left[xy+\frac{y^2}{2}+zy\right]_{x-z}^{x+z}$$

$$= \int_{-1}^{1}dz\int_{0}^{z}dx\left[x(x+z)-x(x-z)+\left(\frac{x+z}{2}\right)^2-\left(\frac{x-z}{2}\right)^2+z(x+z)-z(x-z)\right]$$

$$= \int_{-1}^{1}dz\int_{0}^{z}[2z(x+z)+4xz]dx = \int_{-1}^{1}dz\left[2z\frac{x^2}{2}+2z^2x+\frac{4x^2}{2}z\right]_0^z$$

$$= \int_{-1}^{1}z^3+2z^3+2z^3dz = 5\int_{-1}^{1}z^3dz = 5\left[\frac{2^4}{4}\right]_{-1}^{1} = 0$$

***Example* 13.18:** Evaluate $\int_{0}^{2}\int_{0}^{x}\int_{0}^{x+y}e^{x+y+z}dxdydz$

Solution:

$$\int_0^2 e^x\left\{\int_0^x e^y\left(\int_0^{x+y}e^z dz\right)dy\right\}dx = \int_0^2 e^x\left[\int_0^x e^y.(e^z)_0^{x+y}dy\right]dx$$

$$= \int_0^2 e^x\left[\int_0^x e^y\left(e^{x+y}-1\right)dy\right]dx = \int_0^2 e^x\left[\int_0^x e^x.\frac{e^{2y}}{2}-e^y\right]_0^x dx$$

$$= \int_0^2 e^x\left[\left\{\frac{e^{3x}}{2}-e^x\right\}-\left\{e^x.\frac{1}{2}-1\right\}\right]dx = \int_0^2 \frac{e^{4x}}{2}-\frac{3}{2}e^{2x}+e^x dx$$

$$= \left[\frac{e^{4x}}{8}-\frac{3}{4}e^{2x}+e^x\right]_0^2 = \left[\frac{e^8}{8}-\frac{3}{4}e^2+e\right]-\left[\frac{1}{8}-\frac{3}{4}+1\right]$$

$$= \left[\frac{e^8}{8}-\frac{3}{4}e^2+e-\frac{3}{8}\right]$$

***Example* 13.19:** Evaluate $\int\limits_{y=0}^{1}\left\{\int\limits_{x=y^2}^{1}\left(\int\limits_{z=0}^{1-x} x\,dz\right)dx\right\}dy$

Solution:

$$I = \int\limits_{y=0}^{1}\int\limits_{x=y^2}^{1}[xz]_{z=0}^{1-x}dxdy$$

$$= \int\limits_{y=0}^{1}\left[\int\limits_{x=y^2}^{1}x(1-x)dx\right]dy = \int\limits_{y=0}^{1}\left[\frac{x^2}{2}-\frac{x^3}{3}\right]_{x=y^2}^{x=1}dy$$

$$= \int\limits_{y=0}^{1}\left(\frac{1}{6}-\frac{y^4}{2}+\frac{y^6}{3}\right)dy = \left[\frac{y}{6}-\frac{y^5}{10}+\frac{y^7}{21}\right]_0^1 = \frac{4}{35}$$

***Example* 13.20:** Evaluate $\int\limits_{z=0}^{4}\left[\int\limits_{x=0}^{2\sqrt{z}}\left(\int\limits_{y=0}^{\sqrt{4z-x^2}}dy\right)dx\right]dz$

Solution:

$$I = \int\limits_{z=0}^{4}\left\{\int\limits_{x=0}^{2\sqrt{z}}[y]_{y=0}^{\sqrt{4z-x^2}}dx\right\}dz = \int\limits_{z=0}^{4}\left\{\int\limits_{x=0}^{2\sqrt{z}}\sqrt{4z-x^2}\,dx\right\}dz$$

$$= \int\limits_{z=0}^{4}\left[\frac{x}{2}\sqrt{4z-x^2}+\frac{4z}{2}\sin^{-1}\left\{\frac{x}{\sqrt{4z}}\right\}\right]_{x=0}^{2\sqrt{z}}dz = \int_0^4 \pi z dz = \pi\left[\frac{z^2}{2}\right]_0^4 = 8\pi$$

***Example* 13.21:** Evaluate $\iiint_V xyz\,dz\,dy\,dx$ over the volume enclosed by three coordinate planes and the plane $x+y+z=1$.

Solution:

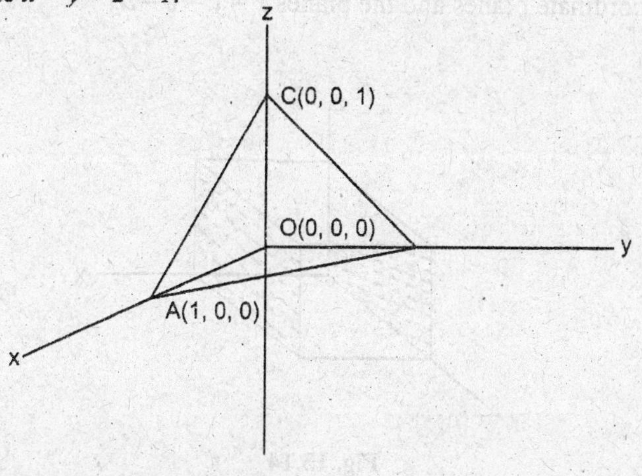

Fig. 13.13

The plane $x+y+z=1$ meets the coordinate axes in A(1,0,0), B(0,1,0) and C(0,0,1).

$$\iiint_V xyz\,dz\,dy\,dx = \int_{x=0}^{1}\int_{y=0}^{1-x}\left(\int_{z=0}^{1-x-y} xyz\,dz\right)dy\,dx$$

$$= \int_{x=0}^{1}\left(\int_{y=0}^{1-x} xy\left[\frac{z^2}{2}\right]_0^{1-x-y} dy\right)dx = \int_{x=0}^{1}\int_{y=0}^{1-x}\frac{xy}{2}\left[(1-x-y)^2-0\right]dy\,dx$$

$$= \frac{1}{2}\int_{x=0}^{1}\int_{y=0}^{1-x} xy\left(1+x^2+y^2-2x-2y+2xy\right)dy\,dx$$

$$= \frac{1}{2}\int_{x=0}^{1}\left[\frac{xy^2}{2}+\frac{1}{2}x^3y^2+\frac{1}{4}xy^4-x^2y^2-\frac{2}{3}xy^3+\frac{2}{3}x^2y^3\right]_0^{1-x} dx$$

$$= \frac{1}{24}\int_0^1\left(x+8x^2-30x^3+32x^4-11x^5\right)dx$$

$$= \frac{1}{24}\left[\frac{x^2}{2}+\frac{8x^3}{3}-30\frac{x^4}{4}+32\frac{x^5}{5}-11\frac{x^6}{6}\right]_0^1 = \frac{7}{21600}$$

***Example* 13.22:** Evaluate $\iiint_V (x^2 + y^2 + z^2)dzdydx$ where V is the volume of the cube bounded by the coordinate planes and the planes $x = y = z = a$.

Solution:

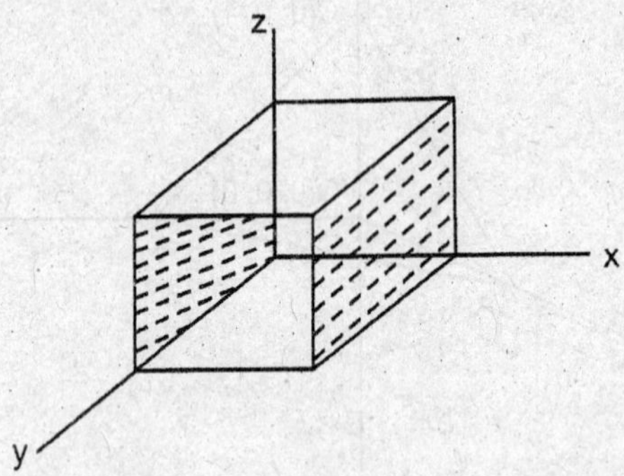

Fig. 13.14

$$\iiint_V (x^2 + y^2 + z^2)dzdydx = \int_{x=0}^{a}\int_{y=0}^{a}\int_{z=0}^{a}(x^2 + y^2 + z^2)dzdydx$$

$$= \int_{x=0}^{a}\int_{y=0}^{a}\left[x^2 z + y^2 z + \frac{z^3}{3}\right]_0^a dxdy$$

$$= \int_{x=0}^{a}\int_{y=0}^{a} x^2 a + y^2 a + \frac{a^3}{3} dydx$$

$$= \int_{x=0}^{a}\left[x^2 ay + \frac{y^3}{3}a + \frac{a^3}{3}y\right]_0^a dx$$

$$= \int_{x=0}^{a} x^2 a^2 + \frac{a^4}{3} + \frac{a^4}{4} dx$$

$$= \left[\frac{x^3}{3}a^2 + \frac{a^4}{3}x + \frac{a^4}{3}x\right]_0^a$$

$$= a^5$$

***Example* 13.23:** Evaluate by triple integration the volume of a hemisphere of radius a.

Solution:

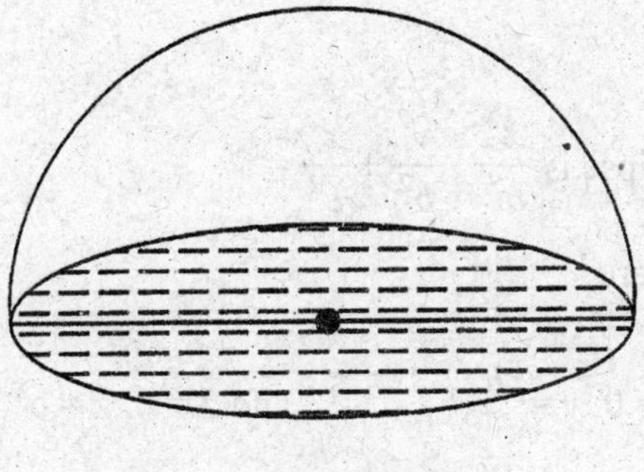

Fig. 13.15

The volume of the hemisphere

$$V = 4\int_{x=0}^{a}\int_{y=0}^{\sqrt{a^2-x^2}}\int_{z=0}^{\sqrt{a^2-x^2-y^2}} dz\,dy\,dx$$

$$= 4\int_{x=0}^{a}\int_{y=0}^{\sqrt{a^2-x^2}} [z]_0^{\sqrt{a^2-x^2-y^2}}\,dy\,dx$$

$$= 4\int_{x=0}^{a}\int_{y=0}^{\sqrt{a^2-x^2}} \sqrt{a^2-x^2-y^2}\,dy\,dx$$

$$= 4\int_{x=0}^{a}\left[\frac{y}{2}\sqrt{a^2-x^2-y^2}+\frac{1}{2}\left(a^2-x^2\right)\sin^{-1}\left(\frac{y}{\sqrt{a^2-x^2}}\right)\right]_{y=0}^{y=\sqrt{a^2-x^2}} dx$$

$$= 4\int_{x=0}^{a}\left[\frac{1}{2}\left(a^2-x^2\right)\frac{\pi}{2}\right]dx = \frac{2}{3}\pi a^3$$

***Example* 13.24:** Find the volume of the ellipsoid $\frac{x^2}{a^2}+\frac{y^2}{b^2}+\frac{x^2}{c^2}=1$ by triple integration.

Solution:

Given equation of the ellipse is $\frac{x^2}{a^2}+\frac{y^2}{b^2}+\frac{x^2}{c^2}=1$

The limits of integration are

$$x = 0, x = \text{a}, y = 0, y = \frac{b}{a}\sqrt{a^2-x^2}\text{, and } z = 0, z = c\sqrt{1-\frac{x^2}{a^2}-\frac{y^2}{b^2}}$$

Consider the first quadrant and the xy plane, x varies from $x = 0$ to $x = \text{a}$ and

y varies from $y = 0$ to $y = b\sqrt{1-\frac{x^2}{a^2}}$

$$\text{Volume} = 8\int_{x=0}^{a}\int_{y=0}^{b\sqrt{1-\frac{x^2}{a^2}}}\int_{z=0}^{c\sqrt{1-\frac{x^2}{a^2}-\frac{y^2}{b^2}}} dz\, dy\, dx = 8\int_{x=0}^{a}\int_{y=0}^{b\sqrt{1-\frac{x^2}{a^2}}} c\sqrt{1-\frac{x^2}{a^2}-\frac{y^2}{b^2}}\, dy\, dx$$

Substituting $\quad 1-\frac{x^2}{a^2}=\frac{t^2}{b^2}$

$$=8\int_{x=0}^{a}\int_{y=0}^{t}\frac{c}{b}\sqrt{t^2-y^2}\, dy\, dx = \frac{8c}{b}\int_{x=0}^{a}\frac{\pi}{4}b^2\left(1-\frac{x^2}{a^2}\right)dx$$

$$= 2\pi\, bc\int_{0}^{a}\left(1-\frac{x^2}{a^2}\right)dx = 2\pi\, bc\left[x-\frac{x^3}{3a^2}\right]_0^a$$

$$= 2\pi\, bc\left[a-\frac{a}{3}\right]=\frac{4\pi\, abc}{3}$$

***Example* 13.25:** Find the volume common to the cylinder $x^2 - y^2 = a^2$ and $x^2 + z^2 = \text{a}$

Solution:

From given limits of z from $-\sqrt{a^2-x^2}$ to $+\sqrt{a^2-x^2}$

y from $-\sqrt{a^2-x^2}$ to $+\sqrt{a^2-x^2}$ and

x from $-a$ to $+a$

$$v=\int_{x=-a}^{a}\int_{y=-\sqrt{a^2-x^2}}^{y=\sqrt{a^2-x^2}}\int_{z=-\sqrt{a^2-x^2}}^{z=\sqrt{a^2-x^2}} dx\,dy\,dz$$

$$=\int_{-a}^{a}\int_{y=-\sqrt{a^2-x^2}}^{y=\sqrt{a^2-x^2}} [z]_{-\sqrt{a^2-x^2}}^{\sqrt{a^2-x^2}}\,dy\,dx$$

$$=\int_{-a}^{a}\int_{-\sqrt{a^2-x^2}}^{\sqrt{a^2-x^2}} 2\sqrt{a^2-x^2}\,dy\,dx=2\int_{-a}^{a}\sqrt{a^2-x^2}\,[y]_{-\sqrt{a^2-x^2}}^{\sqrt{a^2-x^2}}\,dx$$

$$=4\int_{-a}^{a}\left(a^2-x^2\right)dx=4\left[a^2x-\frac{x^3}{3}\right]_{-a}^{a}$$

$$v=4\left(a^3-\frac{a^3}{3}+a^3-\frac{a^3}{3}\right)=\frac{16a^3}{3}$$

***Example* 13.26:** Find the volume of the wedge intercepted between the cylinder $x^2+y^2=2ax$ and the planes $z=x$, $z=2x$

Solution:

The limits of z from x to $2x$

y from $-\sqrt{2ax-x^2}$ to $+\sqrt{2ax-x^2}$

x from 0 to 2a

$$v=\int_{x=0}^{2a}\int_{-\sqrt{2ax-x^2}}^{\sqrt{2ax-x^2}}\int_{x}^{2x} dz\,dy\,dx$$

$$=2\int_{0}^{2a}\int_{0}^{\sqrt{2ax-x^2}}\int_{x}^{2x} dz\,dy\,dx=2\int_{0}^{2a}\int_{0}^{\sqrt{2ax-x^2}}[z]_{x}^{2x}\,dy\,dx$$

$$= 2\int_0^{2a}\int_0^{\sqrt{2ax-x^2}} x\,dy\,dx = 2\int_0^{2a} x[y]_0^{\sqrt{2ax-x^2}}\,dx = 2\int_0^{2a} x\sqrt{2ax-x^2}\,dx$$

Put $x = 2a\sin^2\theta$, $dx = 4a\sin\theta\cos\theta\,d\theta$; $x = 0 \Rightarrow \theta = 0$, $x = 2a \Rightarrow \quad \theta = \frac{\pi}{2}$

$$= 2\int_0^{\frac{\pi}{2}} 2a\sin^2\theta\,\sqrt{2a.2a\sin^2\theta - 4a^2\sin^4\theta}.4a\sin\theta\cos\theta\,d\theta$$

$$= 32a^3\int_0^{\pi/2} sin^4\theta\, cos^2\theta\, d\theta = \pi a^3$$

Exercise 13.2

1. Evaluate the following integrals by changing the order of integration:

1. $\int_0^3\int_1^{\sqrt{4-y}}(x+y)dxdy$ **[Ans:** $\frac{241}{60}$**]**

2. $\int_0^5\int_0^{x^2} x(x^2+y^2)dxdy$ **[Ans:** $\frac{29}{24}.5^6$**]**

3. $\int_0^{\frac{a}{\sqrt{2}}}\int_0^{\sqrt{a^2-x^2}} y^2dxdy$ **[Ans:** $\frac{a^4}{16}+\frac{\pi a^4}{32}$**]**

4. $\int_0^1\int_x^{2-x}\frac{x}{y}dxdy$ **[Ans:** $2\log 2 - 1$**]**

5. $\int_0^1\int_y^{\sqrt{y}} xy(x+y)dxdy$ **[Ans:** $\frac{3}{56}$**]**

6. $\int_{0}^{4a}\int_{x^2/4a}^{2\sqrt{ax}} dydx$ **[Ans:** $\frac{16a^2}{3}$ **]**

7. $\int_{1}^{2}\int_{3}^{4} x^2 + y^2 dxdy$ **[Ans: 10]**

8. $\int_{0}^{a}\int_{x}^{a} \frac{e^{-y}}{y} dy\, dx$ **[Ans: 1]**

9. Evaluate $\int_{-a}^{a}\int_{0}^{\sqrt{a^2-x^2}} \sqrt{x^2+y^2}\, dx\, dy$ by changing into polar coordinates. **[Ans:** $\frac{\pi a^3}{3}$ **]**

10. Evaluate $\int_{0}^{2}\int_{0}^{\sqrt{2x-x^2}} \left(\frac{x}{x^2+y^2}\right) dx\, dy$ by changing into polar coordinates. **[Ans:** $\frac{4}{3}$ **]**

II. Evaluate the following tripple integrals:

1. $\int_{0}^{a}\int_{0}^{x}\int_{0}^{y} xyzdxdydz$ **[Ans:** $\frac{a^6}{48}$ **]**

2. $\int_{0}^{1}\int_{0}^{1-x}\int_{0}^{1-x-y} (x^2+y^2+z^2) dxdydz$ **[Ans:** $\frac{1}{20}$ **]**

3. $\int_{0}^{1}\int_{1}^{1-y}\int_{0}^{1-x^2-y^2} dx\, dy\, dz$ **[Ans:** $\frac{1}{3}$ **]**

4. $\int_{0}^{1}\int_{0}^{2}\int_{1}^{2} x^2 yzdxdydz$ **[Ans: 1]**

5. $\int_{0}^{1}\int_{0}^{1}\int_{0}^{1} (x+y+z)dxdydz$ **[Ans:** $\frac{3}{2}$ **]**

6. $\int_0^1 \int_0^{\sqrt{1-x^2}} \int_0^{\sqrt{1-x^2-y^2}} xyz \, dxdydz$ **[Ans:** $\frac{1}{48}$ **]**

7. $\int_1^3 \int_{\frac{1}{x}}^1 \int_0^{\sqrt{xy}} xyz \, dxdydz$ **[Ans:** $\left(\frac{13}{9} - \frac{1}{6}\log 3\right)$ **]**

8. $\int_0^1 \int_0^{1-x} \int_0^{x+y} e^z \, dxdydz$ **[Ans:** $\frac{1}{2}$ **]**

9. $\int_0^4 \int_0^x \int_0^{\sqrt{x+y}} z \, dxdydz$ **[Ans: 16]**

10. $\int_0^a \int_{-\sqrt{a^2-x^2}}^{\sqrt{a^2-x^2}} \int_0^{mx} z^2 dxdydz$ **[Ans:** $\frac{4}{45}m^2a^2$ **]**

11. Evaluate $\iiint_v (x+y+z)dx\,dy\,dz$ where v is bounded by the plane $x + y + z = a$ $(a > 0)$ and the coordinate planes. **[Ans:** $\frac{a^4}{8}$ **]**

12. Evaluate $\iiint (x+y+z)dx\,dy\,dz$ taken over the volume bounded by the planes $x = 0$, $x = 1$, $y = 0$, $y = 1$ and $z = 0$, $z = 1$ **[Ans:** $\frac{3}{2}$ **]**

13. Evaluate $\iiint_v xy\,z\,dx\,dy\,dz$ where v is the volume bounded by the coordinate planes and $x + y + z = 1$. **[Ans:** $\frac{1}{720}$ **]**

CHAPTER 14

Vector Calculus

14.1 Vector Differentiation

14.1.1 Vector Point Function and Vector Field

Let P be any point in a region 'D' of space. Let **r** be the position vector of P. If there exists a vector function **F** corresponding to each P, then such a function **F** is called a vector point function and the region D is called a vector field.

Note: In what follows i, j, k are unit vectors along X, Y, Z axes respectively

For example, consider the vector function

$$\mathbf{F} = (x - y)\, i + xyj + yzk \quad .....(1)$$

Let P be a point whose position vector is

$$\boldsymbol{r} = 2i + j + 3k \quad \text{in the region D of space.}$$

At P, the value of **F** is obtained by putting $x = 2, y = 1, z = 3$ in $\vec{\mathbf{F}}$.

i.e., At P, $\mathbf{F} = \mathrm{i} + 2\mathrm{j} + 3\mathrm{k}$

Thus, to each point P of the region D, there corresponds a vector **F** given by the vector function (1).

Hence **F** is a vector point function (of scalar variables x, y, z) and the region D *is a vector field.*

Scalar point function and scalar field

If there exists a scalar 'f' given by a scalar function 'f' corresponding to each point P (with position vector $\boldsymbol{r}$) in a region D of space, 'f' is called a scalar point function and D is called a scalar field.

As an example, let P be a point whose position vector is $\boldsymbol{r} = 2i + j + 3k$.

Consider $f = xyz + xy + z$

Then the value of f at P is obtained by putting $x = 2, y = 1, z = 3$

i.e., At P, $f = 2.1.3 + 2.1 + 3 = 11$

Hence the scalar '11' is attached to the point P.

The function 'f' is a scalar point function (of scalar variables x, y, z), and D is a scalar field.

***Note*:** There can be vector and scalar function of one or more scalar variables.

14.1.2 Differentiation of a Vector

If $\boldsymbol{r}(u) = r_1(u)i + r_2(u)j + r_3(u)k$, (where r_1, r_2, r_3, are scalar functions of 'u') be a vector function of 'u', then,

$$\frac{dr}{du} = \underset{\delta u \to 0}{\text{Lt}} \frac{\delta r}{\delta u} = \underset{\delta u \to 0}{\text{Lt}} \frac{r(u+\delta u) - r(u)}{\delta u}$$

$$= \underset{\delta u \to 0}{\text{Lt}} \Sigma\left[\frac{r_1(u+\delta u) - r_1(u)}{\delta u} i\right]$$

$$= \Sigma \frac{dr_1}{du} i = \frac{dr_1}{du} i + \frac{dr_2}{du} j + \frac{dr_3}{du} k$$

EX

If $\boldsymbol{r}(u) = (3u^2 + 5u + 6)\, i + 3u^2 j - 4uk$, Find $\frac{dr}{du}$, when u = 1

$$\frac{dr}{du} = \left\{\frac{d}{du}\left(3u^2 + 5u + 6\right)\right\} i + \left\{\frac{d}{du}(3u^2)\right\} j + \left\{\frac{d}{du}(-4u)\right\} k$$

$$= (6u + 5)i + (bu)j - 4k = 11i + 6j - 4k, \ (\because u = 1)$$

***Note*:** We can apply the above rule of derivative to the case of partial derivatives also

Ex: If $\mathbf{A} = (x^2yz)i + (xy^2z)j - (3x^3y^2z^2)k$

find $\frac{\partial^2 A}{dxdy}$ at the point (1, –1, 2)

$$\frac{\partial A}{dy} = \left\{\frac{\partial}{\partial y}\left(x^2yz\right)\right\} i + \frac{\partial}{\partial y}\left\{xy^2z\right\} j - \frac{\partial}{\partial y}\left\{3x^3y^2z^2\right\} k$$

$$= (x^2z)i + (2xyz)j - (6x^3yz^2)\, k$$

$$\frac{\partial^2 A}{dxdy} = \left\{\frac{\partial}{\partial x}(x^2 z)\right\}i + \frac{\partial}{\partial x}\{2xyz\}j - \frac{\partial}{\partial x}\{(6x^3yz^2)\}k$$

$$= 2xz\,i + 2yz\,j - 18x^2yz^2\,k$$

At the point (1, –1, 2)

$$\frac{\partial^2 A}{dxdy} = 4i - 4j + 72k$$

14.1.3 Application to Space Curves

Let $r(t) = x(t)i + y(t)j + z(t)\,k$ represent the position vector of a point (x, y, z) on a space curve whose equations are given by $x = x(t)$, $y = y(t)$, $z = z(t)$, where 't' is time.

Then $$\frac{d\mathbf{r}}{dt} = \frac{dx}{dt}i + \frac{dy}{dt}j + \frac{dz}{dt}k$$

and $$\frac{d^2\mathbf{r}}{dt^2} = \frac{d^2x}{dt^2}i + \frac{d^2y}{dt^2}j + \frac{d^2z}{dt^2}k$$

(i) $\frac{d\mathbf{r}}{dt}$ represents the velocity vector v (or tangent vector) of the point (x, y, z)

(ii) $\frac{d^2\mathbf{r}}{dt^2}$ represents the acceleration vector **a** at the point (x, y, z)

EX If a particle moves along a curve $x = e^{-t}$, $y = 2\cos 2t$, $z = 2\sin 2t$, where 't' is time.

1) find velocity and acceleration at time t = 0, and

2) find also their magnitudes

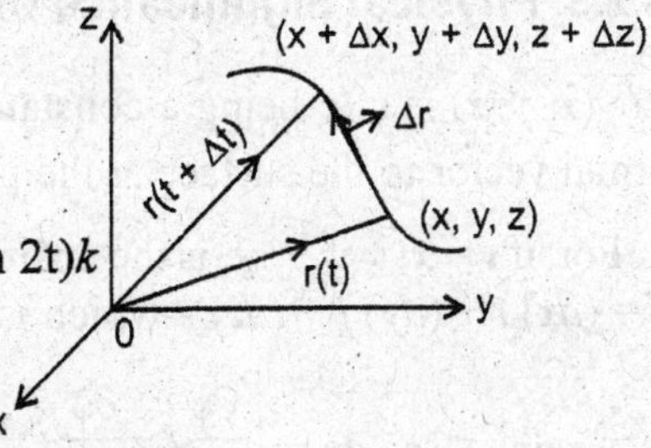

Sol : $r = xi + yj + zk$

$= (e^{-t})i + (2\cos 2t)\,j + (2\sin 2t)k$

$$v = \frac{dr}{dt} = \frac{d}{dt}(e^{-t})\,i + \frac{d}{dt}(2\cos 2t)j + \frac{d}{dt}(2\sin 2t)k$$

$= (-e^{-t})i - (4\sin 2t)j + (4\cos 2t)k$ ...(1)

$$a = \frac{d^2\mathbf{r}}{dt^2} = \frac{dv}{dt} = \frac{d}{dt}(-e^{-t})i - \frac{d}{dt}(4\sin 2t)j + \frac{d}{dt}(4\cos 2t)k$$

$= (e^{-t})\,i - (8\cos 2t)\,j - (8\sin 2t)\,k$ .... (2)

Putting t = 0 in (1), velocity as t = 0 is $v = -i + 4k$

Magnitude = $\sqrt{17}$

putting t = 0 in (2) acceleration at t = 0 is a = i – 8j

Magnitude = $\sqrt{65}$

14.2 Gradient of Scalar Function

14.2.1 The Vector Differential Operator 'DEL' or 'NABLA', denoted as '∇' is defined by

$$\nabla = i\frac{\partial}{\partial x} + j\frac{\partial}{\partial y} + k\frac{\partial}{\partial z} \qquad (i, j, k \text{ are unit vectors in } x, y, z \text{ directions})$$

This operator '∇' is used in defining the gradient, divergence and curl.

Properties of '∇' are similar to those of vectors. The operator is applied to both vector and scalar functions.

14.2.2 Gradient

If $\phi(x, y, z)$ is a scalar function, defined at each point (x, y, z) in a certain region of space and is differentiable, the gradient of ϕ (shortly written as grad ϕ) is defined as,

$$\text{grad } \phi = \nabla \phi = \left(i\frac{\partial}{\partial x} + j\frac{\partial}{\partial y} + k\frac{\partial}{\partial z}\right)\phi$$

$$= \left(\frac{\partial \phi}{\partial x}\right)i + \left(\frac{\partial \phi}{\partial y}\right)j + \left(\frac{\partial \phi}{\partial z}\right)k,$$

(which is a vector function)

$\therefore$ If ϕ defines a scalar field, 'grad ϕ' or $\nabla \phi$ defines a vector field.

14.2.3 Physical Significance of 'Grad ϕ'

If $\phi(x, y, z) = c$ (c being a constant) represents a surface, then 'grad ϕ' represents the normal vector to the surface at the point (x, y, z)

For, if $\mathbf{r} = xi + yj + zk$, is the position vector of the point (x, y, z) on the surface, we have, $d\mathbf{r} = (dx)\,i + (dy)\,j + (dz)\,k$ which is in the tangent plane to the surface of (x, y, z)

$$\text{Again, } \nabla \phi \,.\, d\mathbf{r} = \left[\frac{\partial \phi}{\partial x}i + \frac{\partial \phi}{\partial y}j + \frac{\partial \phi}{\partial z}k\right] . [\, dxi + dyj + dzk \,]$$

$$= \left(\frac{\partial \phi}{\partial x}\right)dx + \left(\frac{\partial \phi}{\partial y}\right)dy + \left(\frac{\partial \phi}{\partial z}\right)dz = d\phi = 0 \qquad (\because \phi = c)$$

$\therefore$ The vector '$\nabla \phi$' which is $\perp^r$ to the tangent plane is the normal vector to $\phi = c$ at (x, y, z)

14.2.4 Directional Derivative

If a be any vector, $\frac{\nabla\phi.\mathbf{a}}{|\mathbf{a}|}$ which represents the component of $\nabla\phi$ in the direction of **a** is known as the directional derivative of 'ϕ' in the direction of **a**,

(1) Physically the directional derivative is the rate of change of 'ϕ' in the direction of **a**.

(2) The directional derivative will be maximum in the direction of $\nabla\phi$ (i.e., $\mathbf{a} = \nabla\phi$) and the maximum value of the directional derivate $= \frac{\nabla\phi.\nabla\phi}{|\nabla\phi|} = |\nabla\phi|$.

14.2.5 Some basic properties of the Gradient

If ϕ and ψ are two scalar functions,

(1) grad $(\phi+\psi)$ = grad ϕ + grad ψ (or) $\nabla(\phi+\psi) = \nabla\phi + \nabla\psi$

Proof: grad $(\phi+\psi) = \nabla(\phi+\psi) = \left\{\frac{\partial}{\partial x}(\phi+\psi)\right\}i + \left\{\frac{\partial}{\partial y}(\phi+\psi)\right\}j + \left\{\frac{\partial}{\partial z}(\phi+\psi)\right\}k$

$$= \left(\frac{\partial\phi}{\partial x}i + \frac{\partial\phi}{\partial y}j + \frac{\partial\phi}{\partial z}k\right) + \left(\frac{\partial\psi}{\partial x}i + \frac{\partial\psi}{\partial y}j + \frac{\partial\psi}{\partial z}k\right)$$

$$= \nabla\phi + \nabla\psi$$

(2) grad $(\phi\psi) = \phi\,(grad\,\psi) + \psi(grad\phi)$ (or) $\nabla(\phi\psi) = \phi\,(\nabla\psi) + (\nabla\phi)\psi$

Proof: grad $(\phi\psi) = \nabla(\phi\psi)$

$$= \left\{\frac{\partial}{\partial x}(\phi\psi)\right\}i + \left\{\frac{\partial}{\partial y}(\phi\psi)\right\}j + \left\{\frac{\partial}{\partial z}(\phi\psi)\right\}k$$

$$= \left(\varphi\frac{\partial\psi}{\partial x} + \psi\frac{\partial\varphi}{\partial x}\right)i + \left(\varphi\frac{\partial\psi}{\partial y} + \psi\frac{\partial\varphi}{\partial y}\right)j + \left(\varphi\frac{\partial\psi}{\partial z} + \psi\frac{\partial\varphi}{\partial z}\right)k$$

$$= \phi\left(\frac{\partial\psi}{\partial x}i + \frac{\partial\psi}{\partial y}j + \frac{\partial\psi}{\partial z}k\right) + \psi\left(\frac{\partial\phi}{\partial x}i + \frac{\partial\phi}{\partial y}j + \frac{\partial\phi}{\partial z}k\right)$$

$$= \phi(\nabla\psi) + \psi(\nabla\phi) = \phi(grad\,\psi) + \psi(grad\,\phi)$$

(3) If $\psi \neq 0$, grad $\left(\frac{\phi}{\psi}\right) = \frac{\psi(grad\phi) - \phi(grad\psi)}{(\psi)^2}$ (Proof is left to the reader.)

Solved Examples

***Example* 14.1:** If $f = x^2yz$, find grad f at the point (1, –2, 1).

***Solution*:**

$$f = x^2yz\ ;\ \because\ \text{grad}\ \phi = \nabla\phi = \frac{\partial\phi}{\partial x}i + \frac{\partial\phi}{\partial y}j + \frac{\partial\phi}{\partial z}k,$$

$$\text{grad}\ f = \frac{\partial}{\partial x}(x^2yz)\ i + \frac{\partial}{\partial y}(x^2yz)j + \frac{\partial}{\partial z}(x^2yz)k$$

$$= (2xyz)i + (x^2z)j + (x^2y)k$$

$\therefore$ At the point (1, –2, 1), grad $\phi = -4i + j - 2k$

***Example* 14.2:** Find the unit normal to the surface $xy + yz + zx = 3$ at the point (1, 1, 1).

***Solution*:**

If $\phi = c$ is a surface, $\nabla\phi$ is the normal to it.

Here $\phi = xy + yz + zx$

$$\therefore \quad \text{normal to } \phi = \nabla\phi = \frac{\partial\phi}{\partial x}i + \frac{\partial\phi}{\partial y}j + \frac{\partial\phi}{\partial z}k.$$

$$= \left\{\frac{\partial}{\partial x}(xy + yz + zx)\right\}i + \left\{\frac{\partial}{\partial y}(xy + yz + zx)\right\}j + \left\{\frac{\partial}{\partial z}(xy + yz + zx)\right\}k$$

$$= (y + z)i + (x + z)j + (y + x)k$$

$\therefore$ normal at $(1,1,1) = 2i + 2j + 2k$

$$\therefore \quad \text{Unit normal} = \frac{2i + 2j + 2k}{\sqrt{2^2 + 2^2 + 2^2}} = \frac{i + j + k}{\sqrt{3}}$$

***Example* 14.3:** (a) Find the directional derivative of $f = 2e^{2x-y+z}$ at (1, 3, 1) in a direction towards the point (2, 1, 3).

***Solution*:**

$$f = 2e^{2x-y+z}\ ;\ \nabla f = 2e^{2x-y+z}(2i - j + k)$$

$$\nabla f\Big|_{(1,3,1)} = 2.e^{2-3+1}(2i - j + k) = 4i - 2j + 2k$$

Let A = (1, 3, 1) and B = (2, 1, 3),

AB = (2–1)i + (1–3)j + (3–1)k = i – 2j + 2k = **a** (say)

Directional derivative in the direction of a $= \dfrac{\nabla f.\mathbf{a}}{|\mathbf{a}|}$

$$= \frac{(4i-2j+2k).(i-2j+2k)}{\sqrt{1+4+4}} = \frac{(4+4+4)}{3} = 4$$

(b) In the 'Problem (a)' find the maximum value of the directional derivative

Maximum value of directional derivative $= |\nabla f|$

$$= |4i - 2j + 2k| = \sqrt{16+4+4} = 2\sqrt{6}$$

***Example* 14.4:** Find the acute angle between the surface $xy^2z = 2$ and $x^2 + y^2 + z^2 = 6$ at the point (2, 1, 1).

***Solution*:**

Let $f = xy^2z = 4$ be the surface (1)

Normal vector to (1) at (2, 1, 1) $= \nabla f\big|_{(2.1,1)} = |(y^2z)i + (2xyz)j + (xy^2)k|_{(2,\,1,\,1)}$

$= i + 4j + 2k = \mathbf{a}$ (say).

Let $g = (x^2 + y^2 + z^2) = 6$ be the surface (2)

Normal vector to (2) at (2,1,1) $= \nabla g\,|_{(2,1,1)} = (2xi + 2yj + 2zk)|_{(2,1,1)}$

$= 4i + 2j + 2k = \mathbf{b}$ (say)

∴ Angle between the surfaces

= Angle between the normals to them

= Angle between **a** and **b**

$$= \left|\cos^{-1}\left(\frac{\mathbf{a.b}}{ab}\right)\right| = \cos^{-1}\left|\frac{4+8+4}{\sqrt{1+16+4}\sqrt{16+4+4}}\right|$$

$$= \cos^{-1}\left(\frac{16}{\sqrt{21}\sqrt{24}}\right) = \cos^{-1}\left(\frac{16}{6\sqrt{14}}\right) = \cos^{-1}\left(\frac{8}{3\sqrt{14}}\right)$$

***Example* 14.5:** Find the constants p and q such that the surfaces $px^2 - qyz = (p + 2)x$ and $4x^2y + z^3 = 4$ are orthogonal at the point (1, –1, 2)

***Solution*:**

Let $f = px^2 - qyz - (p + 2)x = 0$ be surface (1), and

Let g = $4x^2y + z^3 = 4$ be surface (2)

Normal to (1) at (1, –1, 2) = $\nabla f\Big|_{(1,-1,2)} = \left(\dfrac{\partial f}{\partial x}i + \dfrac{\partial f}{\partial y}j + \dfrac{\partial f}{\partial z}k\right)\Big|_{(1,-1,2)}$

$$= [(2px - p - 2)i - (qz)j - (qy)k]|_{(1,-1,2)} = (p-2)i - (2q)j + (q)k = \mathbf{a} \text{ (say)}$$

Normal to (2) at (1, – 1, 2) = $\nabla g\,|_{(1,-1,2)}$

$$= [(8xy)i + (4x^2)j + (3z^2)k]\Big|_{(1,-1,2)} = -8i + 4j + 12k = \mathbf{b} \text{ (say)}$$

Since the surfaces (1) and (2) are orthogonal, **a.b** = 0

$\therefore -8(p-2) + 4(-2q) + 12(q) = 0$

$\Rightarrow -8p + 16 - 8q + 12q = 0$

$\Rightarrow -8p + 4q + 16 = 0$

$\Rightarrow 2p - q = 4$ ..... (1)

Since the point (1, –1, 2) lies on (1), we have,

$p + 2q - p - 2 = 0 \Rightarrow q = 1$

from (i) we get, $p = 5/2$ $\quad\therefore p = 5/2,\ q = 1$

***Example* 14.6:** If r = xi + yj + zk and r = |r| = $\sqrt{x^2+y^2+z^2}$ show that grad (r^3) = 3r r

***Solution*:**

Let $\phi = r^3 = (x^2 + y^2 + z^2)^{3/2}$

Then, $\dfrac{\partial \phi}{\partial x} = \dfrac{3}{2}(x^2 + y^2 + z^2)^{3/2-1}\ 2x = 3x\,r$

similarly $\dfrac{\partial \phi}{\partial y} = 3yr$; and $\dfrac{\partial \phi}{\partial z} = 3zr$

grad $\phi = \dfrac{\partial \phi}{\partial x}i + \dfrac{\partial \phi}{\partial y}j + \dfrac{\partial \phi}{\partial z}k$

$= (3xr)\,i + (3yr)\,j + (3zr)\,k$

$= 3r\,(xi + yj + zk) = 3r\,\mathbf{r}$

***Aliter*:** If $r^2 = x^2 + y^2 + z^2$, we have, $2r\dfrac{\partial r}{\partial x} = 2x \Rightarrow \dfrac{\partial r}{\partial x} = \dfrac{x}{r}$

Similarly $\dfrac{\partial r}{\partial y} = \dfrac{y}{r}$; $\dfrac{\partial r}{\partial z} = \dfrac{z}{r}$

$$\text{grad } \phi = \nabla r^3$$

$$= \frac{\partial}{\partial x}(r^3)i + \frac{\partial}{\partial y}(r^3)j + \frac{\partial}{\partial z}(r^3)k = 3r^2\frac{\partial r}{\partial x}i + 3r^2\frac{\partial r}{\partial y}j + 3r^2\frac{\partial r}{\partial z}k$$

$$= 3r^2\left[\frac{x}{r}i + \frac{y}{r}j + \frac{z}{r}k\right] = 3r\,\mathbf{r}$$

***Example* 14.7:** Evaluate grad r^n

***Solution*:**

Let $\phi = r^n = (x^2 + y^2 + z^2)^{n/2}$

$$\frac{\partial\varphi}{\partial x} = \frac{n}{2}\left(x^2 + y^2 + z^2\right)^{\frac{n}{2}-1}.2x = nx\left(x^2 + y^2 z^2\right)^{\frac{n}{2}-1}$$

similarly $\dfrac{\partial\phi}{\partial x} = ny\left(x^2 + y^2 + z^2\right)^{n/2-1}$ and $\dfrac{\partial\phi}{\partial z} = nz\left(x^2 + y^2 + z^2\right)^{n/2-1}$

$$\text{grad } r^n = \text{grad } \phi = \frac{\partial\phi}{\partial x}i + \frac{\partial\phi}{\partial y}j + \frac{\partial\phi}{\partial z}k$$

$$= n,\ \left(x^2 + y^2 + z^2\right)^{n/2-1}.\ (xi + yj + zk) = n\, r^{n-2}.\mathbf{r}$$

***Aliter*:** $$\text{grad } \phi = \left(\frac{\partial\phi}{\partial x}\right)i + \left(\frac{\partial\phi}{\partial y}\right)j + \left(\frac{\partial\phi}{\partial z}\right)k = \frac{\partial\phi}{\partial r}.\frac{\partial r}{\partial x}i + \frac{\partial\phi}{\partial r}.\frac{\partial r}{\partial y}j + \frac{\partial\phi}{\partial r}.\frac{\partial r}{\partial z}k$$

$$= nr^{n-1}\ .\ \frac{x}{r}i + nx^{n-1}\ .\ \frac{y}{r}j + nx^{n-1}\ .\ \frac{z}{r}k = nr^{n-2}.\mathbf{r}$$

***Example* 14.8:** If A is a constant vector prove that grad (r . A) = A

***Solution*:**

Let $A = A_1 i + A_2 j + A_3 k$ (A_1, A_2, A_3 being constant functions)

$r = xi + yj + zk$

$\mathbf{r}\ .\ \mathbf{A} = A_1 x + A_2 y + A_3 z = f$ (say)

$$\frac{\partial f}{\partial x} = A_1,\ \frac{\partial f}{\partial y} = A_2,\ \frac{\partial f}{\partial z} = A_3$$

$$\therefore \text{ grad } (\mathbf{r.A}) = \text{grad } f = \frac{\partial f}{\partial x}i + \frac{\partial f}{\partial y}j + \frac{\partial f}{\partial z}k = A_1 i + A_2 j + A_3 k = \mathbf{A}$$

***Example* 14.9:** Prove $\nabla\,(\phi(r)) = \dfrac{\phi^1(r).\mathbf{r}}{r}$

***Solution*:**

Let $f = \phi$ (r)

$$\frac{\partial f}{\partial x} = \phi^1(r)\frac{\partial r}{\partial x} = \phi^1(r).\frac{x}{r} \qquad \because \frac{\partial r}{\partial x} = \frac{x}{r}$$

$$\frac{\partial f}{\partial y} = \phi^1(r)\frac{\partial r}{\partial y} = \phi^1(r).\frac{y}{r} \qquad \frac{\partial r}{\partial y} = \frac{y}{r}$$

$$\frac{\partial f}{\partial z} = \phi^1(r)\frac{\partial r}{\partial z} = \phi^1(r).\frac{z}{r} \qquad \frac{\partial r}{\partial z} = \frac{z}{r} \text{ (see Aliter of ex. 14.6)}$$

$$\therefore\ \nabla\,(\phi\,(r)) = \nabla f = \frac{\partial f}{\partial x}i + \frac{\partial f}{\partial y}j + \frac{\partial f}{\partial z}k$$

$$= \frac{\phi^1(r)}{(r)}\,(xi + yj + zk) = \frac{\phi^1(r)\mathbf{r}}{(r)}$$

***Example* 14.10:** Find the equations for the tangent plane and normal line to the surface $z = x^2 + y^2$ at the point (2, 1, 5).

***Solution*:**

Let $\mathbf{r} = xi + yj + zk$ be the position vector of any point $P(x, y, z)$ on the surface.

Let $\mathbf{r}_1 = x_1 i + y_1 j + z_1 k$ be the position vector of fixed point $A(x_1, y_1, z_1)$ on the surface.

Then $\mathbf{AP} = (x - x_1)\,i + (y - y_1)\,j + (z - z_1)\,k = \mathbf{r} - \mathbf{r}_1$

Let $\mathbf{n}$ be the normal to the surface at A.

Then, since **AP** is perpendicular to **n**, we have,

$$(\mathbf{r} - \mathbf{r}_1)\,.\,\mathbf{n} = 0 \qquad \text{..... (1)}$$

which is the equation to the tangent plane at A.

Here, in the given problem

$\mathbf{r} - \mathbf{r}_1 = (x - 2)\,i + (y - 1)\,j + (z - 5)k$

and $\mathbf{n} = \nabla\,(x^2 + y^2 - z)$ at A (2, 1, 5)

$$= (2xi + 2yj - k)\big|_{(2,1,5)} = 4i + 2j - k$$

$\therefore$ The tangent plane at (2, 1, 5) is, (from (1)),

$$4(x-2)+2(y-1)-1(z-5)=0 \Rightarrow 4x+2y-z=5 \quad \text{.... (2)}$$

From (2), the direction ratios of the normal line at A are 4, 2, –1

∴ Equation to the normal line at A are,

$$\frac{x-x_1}{a}=\frac{y-y_1}{b}=\frac{z-z_1}{c} \text{ where}$$

$(x_1, y_1, z_1) = (2, 1, 5)$ and (a, b, c) = (4, 2, –1)

∴ The equations of normal line are $\frac{x-2}{4}=\frac{y-1}{2}=\frac{z-5}{-1}$

From the above example, we have to remember the following :

Let $\phi = c$ be any given surface and (x_1, y_1, z_1) be a point on it; then

(1) Equation to the tangent plane to $\phi = c$ at (x_1, y_1, z_1) is

$$(x-x_1)\frac{\partial\phi}{\partial x}+(y-y_1)\frac{\partial\phi}{\partial y}+(z-z_1)\frac{\partial\phi}{\partial z}=0$$

(2) Equations to Normal line at (x_1, y_1, z_1) are

$$\frac{x-x_1}{\partial\phi/\partial x}=\frac{y-y_1}{\partial\phi/\partial y}=\frac{z-z_1}{\partial\phi/\partial z}$$

Exercise 14.1

1. If $\phi = 2xz^3 - 3x^2yz$, find $\nabla\phi$ and $|\nabla\phi|$ at the point (2, 2, –1)

   [Ans : (i) $22i + 12j - 12k$ (ii) $2\sqrt{193}$ **]**

2. If $\mathbf{V} = 2x\,i - 3y^2\,j + z^3k$, and $\phi = 2xyz - 3z^2$, find $\mathbf{V}.\nabla\phi$ and $\mathbf{V}\times\nabla\phi$ at the point (1, 2, 3) **[Ans :** (1) – 426 (2) $6i + 352j + 156k$**]**

3. If $f = 2xyz$ and $g = x^2y + z$, fing $\nabla(f+g)$ and $\nabla(fg)$ at the point (1, –1, 0)

   [Ans : $-2i + j - k$; $2k$ **]**

4. Evaluate $\nabla\left(3r^2 - 4\sqrt{r} + \frac{6}{\sqrt[3]{r}}\right)$ **[Ans :** $(6 - 2r^{3/2} - 2r^{-\frac{7}{3}})\,\mathbf{r}$ **]**

5. If $\phi = r^2 e^{-r}$, show that grad $\phi = (2-r)\,e^{-r}\,\mathbf{r}$

6. Find a unit normal vector to the surface $z = x^2 + y^2$ at the point (1, –2, 5)

   [Ans : $\frac{1}{\sqrt{21}}(2i - 4j - k)$

7. Find the equations to the tangent plane and the normal line to the surface $xz^2 + x^2y = z - 1$ at the point (1, –3, 2)

[**Ans:** (i) $2x - y - 3z + 1 = 0$ (ii) $\frac{x-1}{-2} = \frac{y+3}{1} = \frac{z-2}{3}$

8. Find the equations to the Tangent plane and normal line to the surface $y = x^2 + z^2$ at the point (1, 5, –2)

[**Ans :** (i) $2x - y - 4z = 5$ (ii) $\frac{x-1}{2} = \frac{y-5}{-1} = \frac{z+2}{-4}$

9. Find the directional derivative of $U = 4xz^3 - 3x^2y^2z$ at (2, 1, –2) in the direction of $(3i - 2j + 6k)$. [**Ans:** $\frac{384}{7}$]

10. Find the directional derivative of $\phi = 4e^{2x-y+z}$ at the point (1, 1, –1) in a direction towards the point (2, 3, 1)

[**Ans :** 8/3]

11. Find the values of the constants a, b, c so that the directional derivative of $f = axy^2 + byz + cz^2x^3$ at (1, 2, –1) has a maximum magnitude 64 in a direction parallel to z-axis.

[*Hint*: f at (1, 2, –1) is $||^{lel}$ to z - axis.

∴ Equate coefficients of i and j to zero and $|\nabla f| = 64$. Thus get 3 equations in a, b, c and solve them]

[**Ans:** a =6, b = 24, c = – 8]

12. Find the acute angle between the surfaces $xy^2z = 3x + z^2$ and $3x^2 - y^2 + 2z = 1$ at the point (1, –2, 1) [**Ans :** $\cos^{-1}\frac{\sqrt{3}}{7\sqrt{2}}$]

13. Find grad ψ if $\mathbf{r} = xi + yj + zk$, $r = |\mathbf{r}|$ and

(i) $\psi = \text{Log } r$ (ii) $\psi = \frac{1}{r}$ (iii) $\psi = r$

[**Ans :** (i) $\mathbf{r}/r^2$ (ii) $-\mathbf{r}/r^3$ (iii) $\mathbf{r}/r$]

14. Find the directional derivative of $g = x^2y^2 + y^2z^2 + z^2x^2$ at the point (1, 1, –2) in the direction of the tangent to the curve $x = e^{-t}$, $y = 2 \sin t + 1$, $z = t - \cos t$ at $t = 0$.

[Hint : Tangent vector to the given curve is $\frac{dx}{dt}i + \frac{dy}{dt}j + \frac{dz}{dt}k$ [**Ans :** $\frac{2}{\sqrt{6}}$]

15. Find the acute angle between the normals to the surface $xy = z^2$ at the points

(1, 9, 3) and (3, 3, –3) [**Ans** : $\cos^{-1}\left(\frac{1}{\sqrt{177}}\right)$]

16. If $\mathbf{r} = xi + yj + zk$ and $\phi = x^3 + y^3 + z^3 - 3xyz$, show that $\mathbf{r}$. Grad $\phi = 3\phi$.

14.3 The Divergence of a Vector Function

14.3.1 Definition

If $\mathbf{A} = A_1 i + A_2 j + A_3 k$ is a vector function, defined and differentiable at each point (x, y, z) in a certain region of space [i.e., **A** defines a vector field], then the divergence of **A** (abbreviated as 'Div **A**') is defined as,

$$\text{Div } \mathbf{A} = \nabla . \mathbf{A}$$

$$= \left(i\frac{\partial}{\partial x} + j\frac{\partial}{\partial y} + k\frac{\partial}{\partial z}\right) . (A_1 i + A_2 j + A_3 k)$$

$$= \left(\frac{\partial A_1}{\partial x} i + \frac{\partial A_2}{\partial y} + \frac{\partial A_3}{\partial z}\right)$$

(since $i.i = j.j = k.k = 1$)

Note: (1) Div **A** is a scalar field

(2) $\nabla . \mathbf{A} \neq \mathbf{A} . \nabla$

14.3.2 Physical significance of the Divergence

If **A** represents the velocity of fluid in a fluid flow, Div **A** represents the rate of fluid flow through unit volume, (or) Div **A** gives the rate at which fluid is originating at a point per unit volume.

Similarly if **A** represents the Electric flux or heat flux, Div **A** represents the amount of electric flux or heat flux that diverges per unit volume in unit time.

14.3.3 Some properties of Divergence

If **A**, **B** are vector functions and 'f' is a scalar function, then, prove that

(1) Div (**A** + **B**) = Div **A** + Div **B** (i.e.,) $\nabla . (\mathbf{A} + \mathbf{B}) = \nabla . \mathbf{A} + \nabla . \mathbf{B}$

Proof: Let $\mathbf{A} = A_1 i + A_2 j + A_3 k$

$\mathbf{B} = B_1 i + B_2 j + B_3 k$

$\mathbf{A} + \mathbf{B} = (A_1 + B_1)\, i + (A_2 + B_2)\, j + (A_3 + B_3)\, k$

$$\text{Div}\,(\mathbf{A}+\mathbf{B}) = \frac{\partial}{\partial x}(A_1+B_1)+\frac{\partial}{\partial y}(A_2+B_2)+\frac{\partial}{\partial z}(A_3+B_3)$$

$$= \left(\frac{\partial A_1}{\partial x}+\frac{\partial A_2}{\partial y}+\frac{\partial A_3}{\partial z}\right)+\left(\frac{\partial B_1}{\partial x}+\frac{\partial B_2}{\partial y}+\frac{\partial B_3}{\partial z}\right)$$

$$= \text{Div}\,\mathbf{A} + \text{Div}\,\mathbf{B}.$$

(2) Prove that, Div $(f\mathbf{A})$ = (grad f) .$\mathbf{A}$ + f(Div $\mathbf{A}$) i.e., $\nabla.(f\mathbf{A}) = (\nabla f).\mathbf{A} + f(\nabla.\mathbf{A})$

Proof: Let $\mathbf{A} = A_1 i + A_2 j + A_3 k$ then $f\mathbf{A} = fA_1 i + fA_2 j + fA_3 k$

$$\nabla.(f\mathbf{A}) = \frac{\partial}{\partial x}(fA_1)+\frac{\partial}{\partial y}(fA_2)+\frac{\partial}{\partial z}(fA_3)$$

$$= f\frac{\partial A_1}{\partial x}+A_1\frac{\partial f}{\partial x}+f\frac{\partial A_2}{\partial y}+A_2\frac{\partial f}{\partial y}+f\frac{\partial A_3}{\partial z}+A_3\frac{\partial f}{\partial z}$$

$$= f\left[\frac{\partial A_1}{\partial x}+\frac{\partial A_2}{\partial y}+\frac{\partial A_3}{\partial z}\right]+\left[\left(\frac{\partial f}{\partial x}\right)A_1+\left(\frac{\partial f}{\partial y}\right)A_2+\left(\frac{\partial f}{\partial z}\right)A_3\right] \quad \text{..... (1)}$$

$$(\nabla f).\mathbf{A} = \left(\frac{\partial f}{\partial x}i+\frac{\partial f}{\partial y}j+\frac{\partial f}{\partial z}k\right).(A_1 i + A_2 j + A_3 k)$$

$$= \left(\frac{\partial f}{\partial x}\right)A_1+\left(\frac{\partial f}{\partial y}\right)A_2+\left(\frac{\partial f}{\partial z}\right)A_3 \quad \text{..... (2)}$$

$$f(\nabla.\mathbf{A}) = f\left(\frac{\partial A_1}{\partial x}+\frac{\partial A_2}{\partial y}+\frac{\partial A_3}{\partial z}\right) \quad \text{..... (3)}$$

(1), (2), (3) $\Rightarrow \nabla.(fA) = (\nabla f).A + f(\nabla.A)$

14.3.4 Solenoidal Vectors

A vector A is said to be solenoidal if Div A = 0

Solved Examples

***Example* 14.11**: If $\mathbf{A} = (x^2y)\,i + (xy^2z)\,j + (xyz)k$, find div $\mathbf{A}$ at the point (1, –1, 2).

Solution:

$$\mathbf{A} = (x^2y)\,i + (xy^2z)j + (xyz)k$$

$$\text{Div}\,\mathbf{A} = \frac{\partial A_1}{\partial x}+\frac{\partial A_2}{\partial y}+\frac{\partial A_3}{\partial z}$$

$$= \frac{\partial}{\partial x}(x^2 y) + \frac{\partial}{\partial y}(xy^2 z) + \frac{\partial}{\partial z}(xyz)$$

$$= 2xy + 2xyz + xy$$

$$= xy\,(2z + 3)$$

$\therefore$ At (1, –1, 2), Div **A** = (1) (–1) [4 + 3] = –7

***Example* 14.12:** If V = $2xyi + 3x^2yj - 3pyz\,k$ is solenoidal at (1, 1, 1), find 'p'.

Solution:

$$\text{Div V} = \frac{\partial}{\partial x}(2xy) + \frac{\partial}{\partial y}(3x^2 y) + \frac{\partial}{\partial z}(-3pyz)$$

$$= 2y + 3x^2 - 3py$$

At (1. 1, 1). Div **V** = 5 – 3p

Since V is solenoidal, Div **V** = 0 $\therefore p = \frac{5}{3}$

***Example* 14.13:** If r = $xi + yz + zk$, and r = |r|, show that Div $(r^3 r) = 6r^3$

Solution:

$r = \sqrt{x^2 + y^2 + z^2}$ $\quad\therefore r^3\mathbf{r} = (x^2 + y^2 + z^2)^{3/2}(xi + yj + zk)$

$$= A_1 i + A_2 j + A_3 k \text{ (say)}$$

then, $\quad A_1 = x(x^2 + y^2 + z^2)^{3/2}$

$$A_2 = y(x^2 + y^2 + z^2)^{3/2}$$

$$A_3 = z(x^2 + y^2 + z^2)^{3/2}$$

$$\frac{\partial A_1}{\partial x} = x\frac{3}{2}\left(x^2 + y^2 + z^2\right)^{\frac{3}{2}-1} 2x + \left(x^2 + y^2 + z^2\right)^{3/2} . 1$$

$$= 3x^2 r + r^3$$

similarly $\frac{\partial A_2}{\partial y} = 3y^2 r + r^3$; $\qquad \frac{\partial A_3}{\partial z} = 3z^2 r + r^3$

$$\text{Div A} = \frac{\partial A_1}{\partial x} + \frac{\partial A_2}{\partial y} + \frac{\partial A_3}{\partial z}$$

$$= 3r\,(x^2 + y^2 + z^2) + 3r^3 = 3r^3 + 3r^3 = 6r^3$$

Aliter: $\quad r^3\,\mathbf{r} = r^3\,xi + r^3\,yi + r^3 zk.$

$$\frac{\partial}{\partial x}(r^3 x) = r^3.1 + x\,.\,3r^2\,\frac{\partial r}{\partial x} = r^3 + x\,.\,3r^2\,.\,\frac{x}{r}$$

$$= r^3 + 3x^2 r \qquad \left(\because \frac{\partial r}{\partial x} = \frac{x}{r}\right)$$

Similarly $\frac{\partial}{\partial y}(r^3 y) = r^3 + 3y^2 r$ and $\frac{\partial}{\partial z}(r^3 z) = r^3 + 3z^2 r$

Div A = $6r^3$

***Example* 14.14** : Evaluate Div [r grad (r^{-3})] or $\nabla \cdot \left\{ r\nabla\left(\frac{1}{r^3}\right)\right\}$

Solution:

$$\text{grad}\,(r^{-3}) = \left\{\frac{\partial}{\partial x}(r^{-3})\, i\right\} + \left\{\frac{\partial}{\partial y}(r^{-3})\, j\right\} + \left\{\frac{\partial}{\partial z}(r^{-3})\, k\right\}$$

$$= -3r^{-4}\frac{\partial r}{\partial x} i - 3r^{-4}\frac{\partial r}{\partial y} j - 3r^{-4}\frac{\partial r}{\partial z} k$$

$$= -3r^{-4}\left(\frac{x}{r} i + \frac{y}{r} j + \frac{z}{r} k\right)$$

$$= -3r^{-5}(xi + yj + zk)$$

$$r\,\text{grad}\,(r^{-3}) = -3r^{-4}(xi + yj + zk)$$

$$= A_1 i + A_2 j + A_3 k \quad \text{(say)}$$

where $A_1 = -3r^{-4}x$, $A_2 = -3r^{-4}y$, $A_3 = -3r^{-4}z$

$$\frac{\partial A_1}{\partial x} = \frac{\partial}{\partial x}[-3r^{-4}x]$$

$$= -3r^{-4} \cdot 1 + x \cdot -3 \cdot -4\, r^{-5} \cdot \frac{\partial r}{\partial x} = -3r^{-4} + 12\,x\,r^{-5} \cdot \frac{x}{r}$$

$$= -3r^{-4} + 12\,x^2\,r^{-6}$$

Similarly, $\frac{\partial A_2}{\partial y} = -3r^{-4} + 12y^2 r^{-6}$

$$\frac{\partial A_3}{\partial z} = -3r^{-4} + 12z^2\,r^{-6}$$

$$\therefore \text{Div}\,(r\,\text{grad}\,r^{-3}) = \frac{\partial A_1}{\partial x} + \frac{\partial A_2}{\partial y} + \frac{\partial A_3}{\partial z}$$

$$= -9r^{-4} + 12\,r^{-6}(x^2 + y^2 + z^2) = -9r^{-4} + 12\,r^{-6} \cdot r^2 = 3r^{-4}$$

(Alternate method is left to the reader).

***Example* 14.15:** Show that $\nabla .(r^n \mathrm{r}) = (n + 3) r^n$. Hence show that r/r^3 is solenoidal.

***Solution*:**

$$r^n \mathbf{r} = r^n (xi + yj + zk)$$

$$\nabla .(r^n \mathbf{r}) = \frac{\partial}{\partial x}(xr^n) + \frac{\partial}{\partial y}(yr^n) + \frac{\partial}{\partial z}(zr^n)$$

$$= \left[r^n + xnr^{n-1}.\frac{\partial r}{\partial x}\right] + \left[r^n + y.nr^{n-1}\frac{\partial r}{\partial y}\right] + \left[r^n + z.n\, r^{n-1}\frac{\partial r}{\partial z}\right]$$

$$= 3r^n + n\, r^{n-1}\left(x.\frac{x}{r} + y.\frac{y}{r} + z.\frac{z}{r}\right)$$

$$= 3r^n + n\frac{r^{n-1}}{r}(x^2 + y^2 + z^2)$$

$$= 3r^n + n\, r^{n-2}.\, r^2 = (n + 3)\, r^n$$

If $n = -3$, $(r^{-3}\, r) = (-3 + 3)\, r^{-3} = 0$

$\therefore \frac{\mathbf{r}}{r^3}$ is solenoidal

***Example* 14.16:** Prove that Div $(C_1A + C_2B) = C_1$ Div A + C_2 Div B, where C_1, C_2 are constants.

***Solution*:**

Let $\mathbf{A} = A_1 i + A_2 j + A_3 k$

$\mathbf{B} = B_1 i + B_2 j + B_3 k$

$$C_1A + C_2B = (C_1A_1 + C_2B_1)\, i + (C_1A_2 + C_2B_2) j + (C_1A_3 + C_2B_3) k$$

$$\text{Div}\,(C_1\mathbf{A} + C_2\mathbf{B}) = \frac{\partial}{\partial x}(C_1A_1 + C_2B_1) + \frac{\partial}{\partial y}(C_1A_2 + C_2B_2) + \frac{\partial}{\partial z}(C_1A_3 + C_2B_3)$$

$$= C_1\left(\frac{\partial A_1}{\partial x} + \frac{\partial A_2}{\partial y} + \frac{\partial A_3}{\partial z}\right) + C_2\left(\frac{\partial B_1}{\partial x} + \frac{\partial B_2}{\partial y} + \frac{\partial B_3}{\partial z}\right)$$

$$= C_1 \text{ div } \mathbf{A} + C_2 \text{ div } \mathbf{B}$$

***Example* 14.17** If A = $2xi + 3yj + 5zk$ and $f = 2xyz$, find div (fA) at (1, 2, 3).

***Solution*:**

$$fA = 2xyz\,(2xi + 3yj + 5zk)$$
$$= (4x^2yz)i + (6xy^2z)j + (10xyz^2)k$$

$$\text{Div}\,(fA) = \frac{\partial}{\partial x}(4x^2yz) + \frac{\partial}{\partial y}(6xy^2z) + \frac{\partial}{\partial z}(10xyz^2)$$
$$= 40\,xyz$$

$\therefore$ At (1, 2, 3), div (fA) = 240

***Aliter*:** $\text{div}\,(fA) = \nabla f.\vec{A} = (\nabla f).\,\vec{A} + f(\nabla.\vec{A})$

$$\nabla f = 2yzi + 2xzj + 2xyk$$

$$\nabla f.A = 4xyz + 6xyz + 10xyz = 20xyz$$

$$f(\nabla .A) = 2xyz\,(2 + 3 + 5) = 20xyz$$

Hence $\nabla .(fA) = 40xyz$ and at (1,2,3) $\nabla .(fA) = 240$

***Example* 14.18:** If f, g are scalar fields show that $\nabla f \times \nabla g$ is solenoidal

***Solution*:**

$$\nabla f = \left(\frac{\partial f}{\partial x}\right)i + \left(\frac{\partial f}{\partial y}\right)j + \left(\frac{\partial f}{\partial z}\right)k$$

$$\nabla g = \left(\frac{\partial g}{\partial x}\right)i + \left(\frac{\partial g}{\partial y}\right)j + \left(\frac{\partial g}{\partial z}\right)k$$

$$\nabla f \times \nabla g = \begin{vmatrix} i & j & k \\ \partial f/\partial x & \partial f/\partial y & \partial f/\partial z \\ \partial g/\partial x & \partial g/\partial y & \partial g/\partial z \end{vmatrix}$$

$$= \Sigma\,(f_y g_z - f_z g_y)i \qquad \text{(Suffixes denote partial derivatives)}$$

$$\text{Div}\,(\nabla f \times \nabla g) = \Sigma\,\frac{\partial}{\partial x}\left(f_y g_z - f_z g_y\right)$$
$$= \Sigma\,(f_y\,g_{xz} + g_z\,f_{xy} - f_z\,g_{xy} - g_y\,f_{xz})$$
$$= 0$$

$\therefore\ \nabla f \times \nabla g$ is solenoidal

Exercise 14.2

1. If $\mathbf{V} = (x^2z)i - (2y^3z^2)j + (xy^2z)k$, find div A at the point (1, –1, 1)

[Ans: – 3]

2. If $\mathbf{r} = xi + yj + zk$, find div $\mathbf{r}$ [Ans: 3]

3. If $\mathbf{F} = (3xyz^2)i + (2xy^3)j - (x^2yz)k$, and $\phi = 3x^2 - yz$,

find (i) Div $\mathbf{F}$ (ii) Div ($\phi\mathbf{F}$) and (iii) Div (gradϕ); at the point (1, –1, 1)

[Ans: (i) 4 (ii) 1 (iii) 6]

4. If $\mathbf{V} = (3x^2y - z)i + (xz^3 + y^4)j - 2x^3z^2k$, find grad (Div $\mathbf{V}$) at the point (2, –1, 0)

[Ans: $-6i + 24j - 32k$]

5. Evaluate : (1) Div ($r^2\mathbf{r}$) (2) Div (r $\mathbf{r}$) (3) grad Div ($\mathbf{r}$/r) (4) div ($\mathbf{r}/r^3$)

[Ans. (1) $5r^2$ (2) 4r (3) $-2r/r^3$ (4) 0]

6. Show that $\mathbf{V} = 3y^4z^2i + 4x^3z^2j + 6x^2y^3k$ is solenoidal

7. Show that the vector $\mathbf{F} = (2x^2 + 8xy^2z)i + (3x^3y - 3xy)j - (4y^2z^2 + 2x^3z)k$ is not solenoidal, but $\mathbf{G} = xyz^2\,\mathbf{F}$ is solenoidal.

8. If $\mathbf{a}$ is a constant vector and $\mathbf{V} = \mathbf{a} \times \mathbf{r}$, where $\mathbf{r} = xi + yj + zk$, show that V is a solenoidal vector.

9. Determine the constant 'b' such that the vector, $\mathbf{V} = (2x + 3y)i + (by - 3z)j + (6x - 12z)k$ is solenoidal [Ans: 10]

10. If $\mathbf{r}_1$ and $\mathbf{r}_2$ are vectors joining fixed points A (x_1, y_1, z_1) and B(x_2, y_2, z_2) to a variable point P(x, y, z) prove that $\mathbf{r}_1 \times \mathbf{r}_2$ is solenoidal.

11. If $\mathbf{r} = xi + yj + zk$ and $r = |\mathbf{r}|$ show that, Div (grad r^n) = $n(n+1)r^{n-2}$.

12. If $g = r^{-2n}$, find div (grad g) and find 'n' such that 'g' is solenoidal.

[Ans: $\dfrac{2n(2n-1)}{r^{2n+2}}; n = \dfrac{1}{2}$]

14.4 Curl of a Vector Function

14.4.1 If A is a Differential Vector Function, then Curl A is defined as, Curl A = ∇ × A

If $\mathbf{A} = A_1i + A_2j + A_3k$, then Curl $\mathbf{A} = \begin{vmatrix} i & j & k \\ \partial/\partial x & \partial/\partial y & \partial/\partial z \\ A_1 & A_2 & A_3 \end{vmatrix}$

$$= \left(\frac{\partial A_3}{\partial y} - \frac{\partial A_2}{\partial z}\right)i + \left(\frac{\partial A_1}{\partial z} - \frac{\partial A_3}{\partial x}\right)j + \left(\frac{\partial A_2}{\partial x} - \frac{\partial A_1}{\partial y}\right)k$$

$$= \Sigma\left(\frac{\partial A_3}{\partial y} - \frac{\partial A_2}{\partial z}\right)i$$

Note: The curl of a vector is also a vector

14.4.2 Physical significance of Curl

Let $\mathbf{r} = xi + yj + zk$ be the position vector of a point $P(x, y, z)$ of a rigid body rotating about a fixed axis about the origin O with an angular velocity $\omega = \omega_1 i + \omega_2 j + \omega_3 k$. Then the velocity $\mathbf{V}$ of the particle P is given by,

$$\mathbf{V} = \omega \times \mathbf{r} = \begin{vmatrix} i & j & k \\ \omega_1 & \omega_2 & \omega_3 \\ x & y & z \end{vmatrix}$$

$$= (\omega_2 z - \omega_3 y)i + (\omega_3 x - \omega_1 z)j + (\omega_1 y - \omega_2 x)k$$

$$\text{Curl } \mathbf{V} = \begin{vmatrix} i & j & k \\ \partial/\partial x & \partial/\partial y & \partial/\partial z \\ \omega_2 z - \omega_3 y & \omega_3 x - \omega_1 z & \omega_1 y - \omega_2 x \end{vmatrix}$$

$$= i\left[\frac{\partial}{\partial y}(\omega_1 y - \omega_2 x) - \frac{\partial}{\partial z}(\omega_3 x - \omega_1 z)\right] + j\left[\frac{\partial}{\partial z}(\omega_2 z - \omega_3 y) - \frac{\partial}{\partial x}(\omega_1 y - \omega_2 x)\right]$$

$$+ k\left[\frac{\partial}{\partial x}(\omega_3 x - \omega_1 z) - \frac{\partial}{\partial y}(\omega_2 z - \omega_3 y)\right]$$

$$= i(\omega_1 + \omega_1) + j(\omega_2 + \omega_2) + k(\omega_3 + \omega_3) = 2\omega$$

Thus the curl of velocity vector is twice the angular velocity of rotation.

14.4.3 Irrotational Vector

A vector $\mathbf{V}$ whose curl is zero is said to be an irrotational vector.

14.4.4 Properties

(1) $\nabla \times (\mathbf{A} + \mathbf{B}) = \nabla \times \mathbf{A} + \nabla \times \mathbf{B}$ (or) curl $(\mathbf{A} + \mathbf{B})$ = curl $\mathbf{A}$ + curl $\mathbf{B}$

Proof: Let $\mathbf{A} = A_1 i + A_2 j + A_3 k$ and

$$\mathbf{B} = B_1 i + B_2 j + B_3 k \text{ so that}$$

$$\mathbf{A} + \mathbf{B} = (A_1 + B_1)i + (A_2 + B_2)j + (A_3 + B_3)k$$

$$\text{Curl } (\mathbf{A+B}) = \begin{vmatrix} i & j & k \\ \partial/\partial x & \partial/\partial y & \partial/\partial z \\ A_1 + B_1 & A_2 + B_2 & A_3 + B_3 \end{vmatrix}$$

$$= \begin{vmatrix} i & j & k \\ \partial/\partial x & \partial/\partial y & \partial/\partial z \\ A_1 & A_2 & A_3 \end{vmatrix} + \begin{vmatrix} i & j & k \\ \partial/\partial x & \partial/\partial y & \partial/\partial z \\ B_1 & B_2 & B_3 \end{vmatrix}$$

$$= \text{curl } \mathbf{A} + \text{curl } \mathbf{B}$$

(2) If ϕ is a scalar function and A is a vector function

$$\text{Curl } (\phi\mathbf{A}) = \phi(\text{curl } \mathbf{A}) + (\text{grad}\phi) \times \mathbf{A}$$

(or)

$$\nabla \times (\phi\mathbf{A}) = \phi(\nabla \times \mathbf{A}) + (\nabla \phi) \times \mathbf{A}$$

Proof: If $\mathbf{A} = A_1 i + A_2 j + A_3 k$, then $\phi\mathbf{A} = \phi A_1 i + \phi A_2 j + \phi A_3 k$

$$\nabla \times (\phi\mathbf{A}) = \begin{vmatrix} i & j & k \\ \partial/\partial x & \partial/\partial y & \partial/\partial z \\ \phi A_1 & \phi A_2 & \phi A_3 \end{vmatrix}$$

$$= i\left[\frac{\partial}{\partial y}(\phi A_3) - \frac{\partial}{\partial z}(\phi A_2)\right] + j\left[\frac{\partial}{\partial z}(\phi A_1) - \frac{\partial}{\partial x}(\phi A_3)\right] + k\left[\frac{\partial}{\partial x}(\phi A_2) - \frac{\partial}{\partial y}(\phi A_1)\right]$$

$$= \sum i\left[\phi\frac{\partial A_3}{\partial y} + A_3\frac{\partial \phi}{\partial y} - \phi\frac{\partial A_2}{\partial z} - A_2\frac{\partial \phi}{\partial z}\right]$$

$$= \phi\sum\left(\frac{\partial A_3}{\partial y} - \frac{\partial A_2}{\partial z}\right)i + \sum\left(A_3\frac{\partial \phi}{\partial y} - A_2\frac{\partial \phi}{\partial z}\right)i$$

$$= \phi\begin{vmatrix} i & j & k \\ \partial/\partial x & \partial/\partial y & \partial/\partial z \\ A_1 & A_2 & A_3 \end{vmatrix} + \begin{vmatrix} i & j & k \\ \partial\phi/\partial x & \partial\phi/\partial y & \partial\phi/\partial z \\ A_1 & A_2 & A_3 \end{vmatrix}$$

$$= \phi(\nabla \times \mathbf{A}) + (\nabla \phi) \times \mathbf{A}$$

14.4.5 Conservative Vector Field

A vector field **F**, which can be derived from a scalar field ϕ such that $\mathbf{F} = \nabla\phi$, is called a conservative vector field and ϕ is called the scalar potential of **F**.

Solved Examples

Example **14.19:** If A = $(xy)i + (yz)j + (zx)k$, find (a) curl A and (b) curl curl A at (1, 2, –3)

***Solution*:**

$$\mathbf{A} = xyi + yzj + zxk$$

(a) $$\text{curl } \mathbf{A} = \nabla \times \mathbf{A} = \begin{vmatrix} i & j & k \\ \partial/\partial x & \partial/\partial y & \partial/\partial z \\ xy & yz & zx \end{vmatrix}$$

$$= i\left[\frac{\partial}{\partial y}(zx) - \frac{\partial}{\partial z}(yz)\right] + j\left[\frac{\partial}{\partial z}(xy) - \frac{\partial}{\partial x}(zx)\right] + k\left[\frac{\partial}{\partial x}(yz) - \frac{\partial}{\partial y}(xy)\right]$$

$$= i(0 - y) + j(0 - z) + \text{k}\,(0 - x)$$

$$= -yi - zj - xk$$

$$\therefore \text{ curl } \mathbf{A} \text{ at } (1, 2, -3) = -2i + 3j - k$$

(b) $$\text{curl A} = \nabla \times (\nabla \times \mathbf{A})$$

$$= \nabla \times (-yi - zj - xk)$$

$$= \begin{vmatrix} i & j & k \\ \partial/\partial x & \partial/\partial y & \partial/\partial z \\ -y & -z & -x \end{vmatrix}$$

$$= i\left[\frac{\partial}{\partial y}(-x) - \frac{\partial}{\partial z}(-z)\right] + j\left[\frac{\partial}{\partial z}(-y) - \frac{\partial}{\partial x}(-x)\right] + k\left[\frac{\partial}{\partial x}(-z) - \frac{\partial}{\partial y}(-y)\right]$$

$$= i(0 - (-1) + j(0 - (-1)) + k(0 - (-1))$$

$$= \text{i} + \text{j} + \text{k}$$

$$\text{curl } \mathbf{A} \text{ at } (1, 2, -3) = i + j + k$$

Example 14.20: Show that V = $xi + y^2j + z^3k$ is irrotational

***Solution*:**

$$\text{Curl } \mathbf{V} = \nabla \times \mathbf{V}$$

$$= i\left[\frac{\partial}{\partial y}(z^3) - \frac{\partial}{\partial z}(y^2)\right] + j\left[\frac{\partial}{\partial z}(x) - \frac{\partial}{\partial x}(z^3)\right] + k\left[\frac{\partial}{\partial x}(y^2) - \frac{\partial}{\partial y}(x)\right]$$

$= 0$ $\qquad\therefore$ **V** is irrotational

***Example* 14.21:** If F = (4x + 3y + az)*i* + (*bx* – *y* + *z*)*j* + (2x + *cy* + *z*)*k* is irrotational, find the constants a, b, c

Solution:

$$\text{Curl } \mathbf{F} = \begin{vmatrix} i & j & k \\ \partial/\partial x & \partial/\partial y & \partial/\partial z \\ (4x+3y+az) & (bx-y+z) & (2x+cy+z) \end{vmatrix}$$

$$= i\left[\frac{\partial}{\partial y}(2x+cy+z)-\frac{\partial}{\partial z}(bx-y+z)\right]+j\left[\frac{\partial}{\partial z}(4x+3y+az)-\frac{\partial}{\partial x}(2x+cy+z)\right]$$

$$+k\left[\frac{\partial}{\partial x}(bx-y+z)-\frac{\partial}{\partial y}(4x+3y+az)\right]$$

$$= (c-1)i + (a-2)j + (b-3)k$$

Since **F** is irrotational, curl **F** = 0

$\therefore \quad c-1=0,\ a-2=0,\ b-3=0$

i.e., $a=2,\ b=3,\ c=1$

***Example* 14.22:** If r = *xi* + *yj* + *zk*, and r = |r|, find curl (r^n r)

Solution:

$\mathbf{r} = xi + yj + zk,\ r = |\mathbf{r}| = \sqrt{x^2+y^2+z^2}$

$r^n\,\mathbf{r} = (x^2+y^2+z^2)^{n/2}(xi+yj+zk)$

$$\text{curl}\,(r^n\,\mathbf{r}) = \begin{vmatrix} i & j & k \\ \partial/\partial x & \partial/\partial y & \partial/\partial z \\ x(x^2+y^2+z^2)^{n/2} & y(x^2+y^2+z^2)^{n/2} & z(x^2+y^2+z^2)^{n/2} \end{vmatrix}$$

$$= i\left[z.\frac{n}{2}\left(x^2+y^2+z^2\right)^{\frac{n}{2}-1}2y - y.\frac{n}{2}\left(x^2+y^2+z^2\right)^{\frac{n}{2}-1}.2z\right]$$

$$+j\left[x.\frac{n}{2}\left(x^2+y^2+z^2\right)^{\frac{n}{2}-1}.2z - z.\frac{n}{2}(x^2+y^2+z^2)^{\frac{n}{2}-1}.2x\right]$$

$$+k\left[y.\frac{n}{2}\left(x^2+y^2+z^2\right)^{\frac{n}{2}-1}.2x - x.\frac{n}{2}(x^2+y^2+z^2)^{\frac{n}{2}-1}.2y\right]$$

$$= i(0)+j(0)+k(0)$$

$$= 0$$

Aliter: $r^n \mathbf{r} = r^n (xi + yj + zk)$

$$\text{curl}(r^n r) = \begin{vmatrix} i & j & k \\ \partial/\partial x & \partial/\partial y & \partial/\partial z \\ xr^n & yr^n & zr^n \end{vmatrix}$$

$$= \Sigma i\left[\frac{\partial}{\partial y}(zr^n) - \frac{\partial}{\partial z}(yr^n)\right] = \Sigma i\left[z.nr^{n-1}\frac{\partial r}{\partial y} - y.nr^{n-1}\frac{\partial r}{\partial z}\right]$$

$$= \Sigma i\left[nzr^{n-1}\frac{y}{r} - y.n.r^{n-1}\frac{z}{r}\right]$$

$$= i(0) + j(0) + k(0) = 0$$

***Example* 14.23:** Prove that, if $F = (x + y + 1)i + j - (x + y)k$, F. curl F = 0

Solution:

$$\text{Curl } \mathbf{F} = \begin{vmatrix} i & j & k \\ \partial/\partial x & \partial/\partial y & \partial/\partial z \\ x+y+1 & 1 & -x-y \end{vmatrix}$$

$$= i[-1-0] + j[0+1] + k[0-1]$$

$$= -i + j - k$$

$\therefore$ $\mathbf{F}.\text{ curl }\mathbf{F} = -1(x + y + 1) + 1.1 + 1(x + y) = 0$

***Example* 14.24:** $P(x, y, z)$ is a variable point and $Q(x_1, y_1, z_1)$, $R(x_2, y_2, z_2)$ are fixed points.

If U = QP and V = RP; Prove that curl (U × V) is equal to 2(U – V).

***Solution*:**

$$P(x, y, z),\ Q = (x_1, y_1, z_1),\ R = (x_2, y_2, z_2)$$

$$\therefore \quad \mathbf{QP} = (x - x_1)i + (y - y_1)j + (z - z_1)k = \mathbf{U}$$

$$\mathbf{RP} = (x - x_2)i + (y - y_2)j + (z - z_2)k = \mathbf{V}$$

$$\mathbf{U} \times \mathbf{V} = \begin{vmatrix} i & j & k \\ (x-x_1) & (y-y_1) & (z-z_1) \\ (x-x_2) & (y-y_2) & (z-z_2) \end{vmatrix}$$

$$= \Sigma\{(y - y_1)(z - z_2) - (z - z_1)(y - y_2)\}\ i$$

$$= \Sigma\ \{y(z_1 - z_2) - z(y_1 - y_2) + (y_1 z_2 - y_2 z_1)\} i$$

$$\text{Curl}\,(\mathbf{U}\times\mathbf{V}) = \sum\left[\frac{\partial}{\partial y}\{x(y_1-y_2)-y(x_1-x_2)+(x_1y_2-x_2y_1)\}\right.$$

$$\left.-\frac{\partial}{\partial z}\{z(x_1-x_2)-x(z_1-z_2)+(z_1x_2-z_2x_1)\}\right]i$$

$$= \Sigma(-(x_1-x_2)-(x_1-x_2)i$$

$$= \Sigma -2(x_1-x_2)i = 2\Sigma(x_2-x_1)i$$

$$= 2(\mathbf{U}-\mathbf{V})$$

***Example* 14.25:** If F is a conservative vector field show that curl F = 0

***Solution*:**

F is a conservative vector field.

∴ There exists a scalar field 'ϕ' such that **F** = $\nabla\phi = \frac{\partial\phi}{\partial x}i+\frac{\partial\phi}{\partial y}j+\frac{\partial\phi}{\partial z}k$

$$\therefore\ \text{Curl}\,\mathbf{F} = \begin{vmatrix} i & j & k \\ \partial/\partial x & \partial/\partial y & \partial/\partial z \\ \partial\phi/\partial x & \partial\phi/\partial y & \partial\phi/\partial z \end{vmatrix}$$

$$= \Sigma i\left(\frac{\partial^2\phi}{\partial y\partial z}-\frac{\partial^2\phi}{\partial z\partial y}\right) = 0$$

***Example* 14.26:** Show that F = $(6xy + z^3)i + (3x^2 - z)\,j + (3xz^2 - y)k$ is irrotational. Find ϕ such that F = $\nabla\phi$

***Solution*:**

$$\text{Curl}\,\mathbf{F} = \begin{vmatrix} i & j & k \\ \partial/\partial x & \partial/\partial y & \partial/\partial z \\ 6xy+z^3 & 3x^2-z & 3xz^2-y \end{vmatrix}$$

$$= i\,(-1+1)+j\,(3z^2-3z^2)+k(6x-6x)$$

$$= 0$$

∴ **F** is irrotational

Let **F** = $\nabla\phi = \frac{\partial\phi}{\partial x}i+\frac{\partial\phi}{\partial y}j+\frac{\partial\phi}{\partial z}k$

$$\therefore \quad \frac{\partial\phi}{\partial x} = 6xy + z^3 \qquad \text{..... (1)}$$

$$\frac{\partial\phi}{\partial y} = 3x^2 - z \qquad \text{..... (2)}$$

$$\frac{\partial\phi}{\partial z} = 3xz^2 - y \qquad \text{..... (3)}$$

Integrating equation (1) with respect to x we get

$$\phi = 3x^2y + xz^3 + f(y, z) \qquad \text{..... (4)}$$

Differentiating (4) partially w.r.t 'y'

$$\frac{\partial\phi}{\partial y} = 3x^2 + \frac{\partial f(y,z)}{\partial y} \qquad \text{..... (5)}$$

From (2) and (5) we have

$$\frac{\partial f(y,z)}{\partial y} = -z \qquad \text{..... (6)}$$

Integrating (6) w.r.t 'y' we get

$$f(y, z) = -yz + h(z)$$

Hence $\phi = 3x^2 + zx^3 - yz + h(z)$ ..... (7)

differentiating (7) w.r.t 'z' we get

$$\frac{\partial\phi}{\partial z} = 3xz^2 - y + h^1(z) \qquad \text{..... (8)}$$

Comparing (3) and (8) we have

$h^1(z) = 0$, $\therefore$ $h(z) =$ constant, 'c' say

Hence $\phi = 3x^2y + xz^3 - yz + c$ (from 7)

Aliter:

$$\frac{\partial\phi}{\partial x} = 6xy + z^3 \qquad \text{.... (9)}$$

$$\frac{\partial\phi}{\partial y} = 3x^2 - z \qquad \text{.... (10)}$$

$$\frac{\partial\phi}{\partial z} = 3xz^2 - y \qquad \text{..... (11)}$$

Integrating the above equations respectively w.r.t x, y, and z, we get

$$\phi = 3x^2y + xz^3 + f(y, z) \qquad \text{.... (12)}$$

$$\phi = 3x^2y - yz + g(z, x) \qquad \text{..... (13)}$$

$$\phi = xz^3 - yz + h(x, y) \qquad \text{..... (14)}$$

ϕ should satisfy all the above three equations simultaneously

(i.e.) (12), (13) & (14)

$\therefore \phi = 3x^2y + xz^3 - yz + c$, where c is a numerical constant.

Note: [Here $f = -yz$, $g = xz^3$, $h = 3x^2y$ will satisfy (12), (13) & (14)

***Example* 14.27**: If f(r) is differentiable, show that f (r) $\bar{r}$ is irrotational

Solution:

$$f(r)\,\mathbf{r} = f(r)\,(xi + yj + zk)$$

$$\text{Curl}\,(f(r)\,\mathbf{r}) = \begin{vmatrix} i & j & k \\ \partial/\partial x & \partial/\partial y & \partial/\partial z \\ x f(r) & y f(r) & z f(r) \end{vmatrix}$$

$$= \sum i\left(\frac{\partial}{\partial y}\{z f(r)\} - \frac{\partial}{\partial z}\{y f(r)\}\right)$$

$$= \sum i\left(z.f'(r)\frac{y}{r} - y.f'(r)\frac{\partial r}{\partial z}\right)$$

$$= \sum i\left(z.f'(r)\frac{y}{r} - y.f'(r)\frac{z}{r}\right) = 0$$

$\therefore f(r)\mathbf{r}$ is irrotational

***Example* 14.28**: If U and V are irrotational, prove that U × V is solenoidal

Solution:

Let $\mathbf{U} = U_1 i + U_2 j + U_3 k$

$\mathbf{V} = V_1 i + V_2 j + V_3 k$

U is irrotational $\quad \therefore$ curl U = 0

$$\Rightarrow \sum i\left[\frac{\partial U_3}{\partial y} - \frac{\partial U_2}{\partial z}\right] = 0 \qquad \text{..... (1)}$$

similarly, $\Sigma i\left[\frac{\partial V}{\partial y}3-\frac{\partial V}{\partial z}2\right]=0$ ..... (2)

$$U \times V = \begin{vmatrix} i & j & k \\ U_1 & U_2 & U_3 \\ V_1 & V_2 & V_3 \end{vmatrix} = \Sigma i(U_2V_3 - U_3 V_2)$$

$$\text{Div } (U \times V) = \Sigma\frac{\partial}{\partial x}(U_2 V_3 - U_3V_2)$$

$$= \sum\left(U_2\frac{\partial V_3}{\partial x}+V_3\frac{\partial U_2}{\partial x}-U_3\frac{\partial V_2}{\partial x}-V_2\frac{\partial U_3}{\partial x}\right)$$ ..... (3)

From (1), we have, $\frac{\partial U_3}{\partial y}=\frac{\partial U_2}{\partial z};\frac{\partial U_1}{\partial z}=\frac{\partial U_3}{\partial x};\frac{\partial U_2}{\partial x}=\frac{\partial U_1}{\partial y}$ ..... (4)

and from (2), we have, $\frac{\partial V_3}{\partial v}=\frac{\partial V_2}{\partial z};\frac{\partial V_1}{\partial z}=\frac{\partial V_3}{\partial x};\frac{\partial V_2}{\partial x}=\frac{\partial V_1}{\partial y}$ ..... (5)

Substituting the equations of (4) & (5) in (3), we observe that all the terms of (3) will get cancelled. Hence Div (U × V) = 0

$\Rightarrow$ U × V is solenoidal

Exercise 14.3

1. If $\mathbf{V} = (2xz^2)i - (yz)j + (3xz^3)k$, and $f = x^2yz$, find the following at the point (1, 1, 1) (a) curl V (b) curl (fV) (c) curl (curl V)

 [**Ans:** (a) $i + j$ (b) $5i - 3j - 4k$ (c) $5i + 3k$]

2. If 'g' is a scalar function, show that curl (g grad g) = 0

3. Find the value of the constant 'p' for which the vector $\mathbf{V} = (pxy - z^3) + (p-2)\,x^2j + (1-p)\,xz^2k$ is irrotational.

 [**Ans:** 4]

4. Find the constants a, b, c so that the curl of

 the vector $\mathbf{A} = (x + 2y + az)i + (bx - 3y - z)j + (4x + cy + 2z)k$ is identically equal to zero

 [**Ans:** 4, 2, –1]

5. If $\mathbf{r} = xi + yj + zk$, $r = |\mathbf{r}|$, show that curl $\left(\frac{r}{r^2}\right) = 0$ and find a scalar function 'f'

such that $\frac{r}{r^2} = -\nabla f, f(a) = 0$ where $a > 0$.

[**Ans:** $f = \log\left(\frac{a}{r}\right)$]

6. If $\mathbf{r} = xi + yj + zk$, and $\boldsymbol{p}, \boldsymbol{q}$ are constant vectors,
show that (1) curl $[(\mathbf{r} \times \boldsymbol{p}) \times \boldsymbol{q}] = (\mathbf{p} \times \mathbf{q})$ and (2) curl $[(\mathbf{p.q})\mathbf{r})] = 0$.

14.5 Laplacian Operator : ∇^2

14.5.1 Definition

$$\nabla^2 = \nabla.\nabla = \left(i\frac{\partial}{\partial x} + j\frac{\partial}{\partial y} + k\frac{\partial}{\partial z}\right).\left(i\frac{\partial}{\partial x} + j\frac{\partial}{\partial y} + k\frac{\partial}{\partial z}\right)$$

$$= \left(\frac{\partial^2}{\partial x^2} + \frac{\partial^2}{\partial y^2} + \frac{\partial^2}{\partial z^2}\right) \text{ is called the Laplacian Operator}$$

'∇^2' can be applied to both scalar and vector functions as shown below.

$$\nabla^2\phi = \frac{\partial^2\phi}{\partial x^2} + \frac{\partial^2\phi}{\partial y^2} + \frac{\partial^2\phi}{\partial z^2}$$

where 'ϕ' is scalar function
f $\mathbf{A} = A_1 i + A_2 j + A_3 k$, is a vector function, then

$$\nabla^2 \mathbf{A} = \left(\frac{\partial^2}{\partial x^2} + \frac{\partial^2}{\partial y^2} + \frac{\partial^2}{\partial z^2}\right)(A_1 i + A_2 j + A_3 k)$$

$$= \left(\nabla^2 A_1\right)i + \left(\nabla^2 A_2\right)j + \left(\nabla^2 A_3\right)k$$

14.5.2 Vector Identities

we shall give below some vector identities with proofs.

1. If 'ϕ' is a scalar function curl grad $\phi = 0$, (or) $\nabla \times \nabla\phi = 0$

Proof: Curl grad $\phi = \nabla \times \left(\frac{\partial\phi}{\partial x}i + \frac{\partial\phi}{\partial y}j + \frac{\partial\phi}{\partial z}k\right)$

$$= \begin{vmatrix} i & j & k \\ \partial/\partial x & \partial/\partial y & \partial/\partial z \\ \partial\phi/\partial x & \partial\phi/\partial y & \partial\phi/\partial z \end{vmatrix}$$

$$= \sum i\left[\frac{\partial}{\partial y}\left(\frac{\partial\phi}{\partial z}\right) - \frac{\partial}{\partial z}\left(\frac{\partial\phi}{\partial y}\right)\right]$$

$$= \sum i\left(\frac{\partial^2\phi}{\partial y\partial z} - \frac{\partial^2\phi}{\partial z\partial y}\right)$$

= 0, assuming that 'ϕ' possesses continuous second order partial derivatives.

2. If **V** is a vector function, Div (Curl **A**) = 0 (or) $\nabla.(\nabla \times A) = 0$

Proof: Let $\mathbf{A} = A_1 i + A_2 j + A_3 k$

$$\text{Curl } \vec{A} = \sum i\left(\frac{\partial A_3}{\partial y} - \frac{\partial A_2}{\partial z}\right)$$

$$\text{Div (curl } \mathbf{A}) = \sum \frac{\partial}{\partial x}\left\{\frac{\partial A_3}{\partial y} - \frac{\partial A_2}{\partial z}\right\}$$

$$= \frac{\partial^2 A_3}{\partial x\partial y} - \frac{\partial^2 A_2}{\partial x\partial z} + \frac{\partial^2 A_1}{\partial y\partial z} - \frac{\partial^2 A_3}{\partial y\partial x} + \frac{\partial^2 A_2}{\partial z\partial x} - \frac{\partial^2 A_1}{\partial z\partial y}$$

$$= 0$$

3. If **A** is a vector function, curl (curl **A**) = grad (Div **A**) $-$ $\nabla^2 A$ (or) $\nabla \times (\nabla \times A) = \nabla(\nabla.A) - \nabla^2 A$

Proof: $\text{curl } \vec{A} = \sum i\left(\frac{\partial A_3}{\partial y} - \frac{\partial A_2}{\partial z}\right)$

$$\text{Curl (curl } \vec{A}) = \begin{vmatrix} i & j & k \\ \partial/\partial x & \partial/\partial y & \partial/\partial z \\ \left(\frac{\partial A_3}{\partial y} - \frac{\partial A_2}{\partial z}\right) & \left(\frac{\partial A_1}{\partial z} - \frac{\partial A_3}{\partial x}\right) & \left(\frac{\partial A_2}{\partial x} - \frac{\partial A_1}{\partial y}\right) \end{vmatrix}$$

$$= \sum\left[\frac{\partial}{\partial y}\left(\frac{\partial A_2}{\partial x} - \frac{\partial A_1}{\partial y}\right) - \frac{\partial}{\partial z}\left(\frac{\partial A_1}{\partial z} - \frac{\partial A_3}{\partial x}\right)\right]i$$

$$= \left[\left(-\frac{\partial^2 A_1}{\partial x^2}-\frac{\partial^2 A_1}{\partial y^2}-\frac{\partial^2 A_1}{\partial z^2}\right)i+\left(-\frac{\partial^2 A_2}{\partial x^2}-\frac{\partial^2 A_2}{\partial y^2}-\frac{\partial^2 A_2}{\partial z^2}\right)j\right.$$

$$\left.+\left(-\frac{\partial^2 A_3}{\partial x^2}-\frac{\partial^2 A_3}{\partial y^2}-\frac{\partial^2 A_3}{\partial z^2}\right)k\right]+\left[\left(\frac{\partial^2 A_1}{\partial x^2}+\frac{\partial^2 A_2}{\partial y\partial x}+\frac{\partial^2 A_3}{\partial z\partial x}\right)i\right.$$

$$\left.+\left(\frac{\partial^2 A_1}{\partial x\partial y}+\frac{\partial^2 A_2}{\partial y^2}+\frac{\partial^2 A_3}{\partial z\partial y}\right)j+\left(\frac{\partial^2 A_1}{\partial x\partial z}+\frac{\partial^2 A_2}{\partial y\partial z}+\frac{\partial^2 A_3}{\partial z^2}\right)k\right]$$

(adding and subtracting $\frac{\partial^2 A_1}{\partial x^2}i$, $\frac{\partial^2 A_2}{\partial y^2}j$ and $\frac{\partial^2 A_3}{\partial z^2}k$).

$$= -\left(\frac{\partial^2}{\partial x^2}+\frac{\partial^2}{\partial y^2}+\frac{\partial^2}{\partial z^2}\right)(A_1 i + A_2 j + A_3 k)$$

$$+\ i\frac{\partial}{\partial x}\left(\frac{\partial A_1}{\partial x}+\frac{\partial A_2}{\partial y}+\frac{\partial A_3}{\partial z}\right)+j\frac{\partial}{\partial y}\left(\frac{\partial A_1}{\partial x}+\frac{\partial A_2}{\partial y}+\frac{\partial A_3}{\partial z}\right)+k\frac{\partial}{\partial z}\left(\frac{\partial A_1}{\partial x}+\frac{\partial A_2}{\partial y}+\frac{\partial A_3}{\partial z}\right)$$

$$= -\ \nabla^2 A + \nabla\left(\frac{\partial A_1}{\partial x}+\frac{\partial A_2}{\partial y}+\frac{\partial A_3}{\partial z}\right)$$

$$= -\nabla^2 A + \nabla\left(\nabla.\vec{A}\right)$$

4. $\nabla.(\boldsymbol{A}\times\boldsymbol{B}) = \boldsymbol{B}.(\nabla\times\boldsymbol{A}) - \boldsymbol{A}.(\nabla\times\boldsymbol{B})$, (**A**, **B** are vector functions) (or)

Div (**A**×**B**)=**B**. curl **A** – **A**. curl **B**.

Proof: Let $\mathbf{A} = A_1 i + A_2 j + A_3 k$ and $\mathbf{B} = B_1 i + B_2 j + B_3 k$

$$\mathbf{A}\times\mathbf{B} = \begin{vmatrix} i & j & k \\ A_1 & A_2 & A_3 \\ B_1 & B_2 & B_3 \end{vmatrix}$$

$$= \sum i(A_2 B_3 - A_3 B_2)$$

$$\text{Div}\ (\mathbf{A}\times\mathbf{B}) = \sum \frac{\partial}{\partial x}(A_2 B_3 - A_3 B_2)$$

$$= \sum \left(A_2\frac{\partial B_3}{\partial x} + B_3\frac{\partial A_2}{\partial x} - A_3\frac{\partial B_2}{\partial x} - B_2\frac{\partial A_3}{\partial x}\right) \quad .....(1)$$

$$\nabla \times A = \begin{vmatrix} i & j & k \\ \partial/\partial x & \partial/\partial y & \partial/\partial z \\ A_1 & A_2 & A_3 \end{vmatrix} = \sum \left(\frac{\partial A_3}{\partial y} - \frac{\partial A_2}{\partial z} \right) i$$

$$B.(\nabla \times A) = \sum B_1 \left(\frac{\partial A_3}{\partial y} - \frac{\partial A_2}{\partial z} \right) \qquad \text{..... (2)}$$

Similarly,

$$A.(\nabla \times B) = \sum A_1 \left(\frac{\partial B_3}{\partial y} - \frac{\partial B_2}{\partial z} \right) \qquad \text{..... (3)}$$

Expanding the summations of (1), (2) & (3), we observe that,

$$\text{Div}(\mathbf{A} \times \mathbf{B}) = B.(\nabla \times \mathbf{A}) - \mathbf{A}.(\nabla \times \mathbf{B})$$

14.5.3 Operation of ∇ on Product of Two Functions

Suppose ϕ and ψ, are two scalar or vector point functions. When ∇ is operated on the product of ϕ and ψ, the following rule is useful.

$\nabla(\phi\psi) = \nabla(\phi_0\psi) + \nabla(\phi\psi_0)$ wherein the suffix '0' indicates that the function is not to be varied, that is, ∇ is not to be operated on that function. After the completion of operation of ∇ the suffixes are dropped.

While proving the identities the following are useful.

(a) $\mathbf{a.b} = \mathbf{b.a}$

(b) $\mathbf{a} \times \mathbf{b} = -\mathbf{b} \times \mathbf{a}$

(c) $\mathbf{a} \times \mathbf{a} = 0$

(d) $\mathbf{a}.(\mathbf{b} \times \mathbf{c}) = (\mathbf{a} \times \mathbf{b}).\mathbf{c}$

(e) $\mathbf{a} \times (\mathbf{b} \times \mathbf{c}) = (\mathbf{a.c})\mathbf{b} - (\mathbf{a.b})\mathbf{c}$

(f) $(\mathbf{a.c})\mathbf{b} = \mathbf{a} \times (\mathbf{b} \times \mathbf{c}) + (\mathbf{a.b})\mathbf{c}$

(g) $\mathbf{a.a} = \mathbf{a}^2$

(h) $\times$ operation is always between vectors only

vector identities using ∇ operation:

(i) $\nabla.(\mathrm{FG}) = \nabla.(\mathrm{FG_0}) + \nabla.(\mathrm{F_0G})$

$$= \nabla \mathrm{F.G_0} + \mathrm{F_0}\, \nabla.\mathbf{G}$$

$\therefore$ div (F G) = G. grad F + F div G

(ii) $\nabla \times (\mathrm{FG}) = \nabla \times (\mathrm{FG_0}) + \nabla \times (\mathrm{F_0G})$

$\nabla \times (\mathrm{FG_0}) = \nabla \mathrm{F} \times \mathrm{G_0}$ due to (h)

$\nabla \times (F_0\mathbf{G}) = (\nabla \times \mathbf{G})\, F_0 = F_0(\nabla \times \mathbf{G})$

$\therefore\ \nabla \times (F\mathbf{G}) = \nabla F \times \mathbf{G} + F(\nabla \times \mathbf{G})$

(i.e.,) curl $(F\mathbf{G}) = (\text{grad } F) \times \mathbf{G} + F \text{ curl } \mathbf{G}$

(iii) $\nabla \cdot (\mathbf{F} \times \mathbf{G}) = \nabla \cdot (\mathbf{F} \times \mathbf{G}_0) + \nabla \cdot (\mathbf{F}_0 \times \mathbf{G})$

$\nabla \cdot (\mathbf{F} \times \mathbf{G}_0) = (\nabla \times \mathbf{F}) \cdot \mathbf{G}_0$ due to (d)

$= \mathbf{G}_0 \cdot (\nabla \times \mathbf{F})$ due to (a)

$\nabla \cdot (\mathbf{F}_0 \times \mathbf{G}) = -\nabla \cdot (\mathbf{G} \times \mathbf{F}_0)$ due to (b)

$= -\nabla \times (\mathbf{G} \cdot \mathbf{F}_0)$ due to (d)

$= -(\nabla \times \mathbf{G}) \cdot \mathbf{F}_0$

$= -\mathbf{F}_0 (\nabla \times \mathbf{G})$ due to (a)

$\therefore\ \nabla \cdot (\mathbf{F} \times \mathbf{G}) = \mathbf{G} \cdot (\nabla \times \mathbf{F}) - \mathbf{F} \cdot (\nabla \times \mathbf{G})$

Thus div $(\mathbf{F} \times \mathbf{G}) = \text{G. curl } \mathbf{F} - \mathbf{F}. \text{ curl } \mathbf{G}$

(iv) $\nabla \times (\mathbf{F} \times \mathbf{G}) = \nabla \times (\mathbf{F} \times \mathbf{G}_0) + \nabla \times (\mathbf{F}_0 \times \mathbf{G})$

$\nabla \times (\mathbf{F} \times \mathbf{G}_0) = (\nabla \cdot \mathbf{G}_0)\, \mathbf{F} - (\nabla \cdot \mathbf{F})\, \mathbf{G}_0$ due to (e)

$= (\mathbf{G}_0 \cdot \nabla)\, \mathbf{F} - \mathbf{G}_0 (\nabla \cdot \mathbf{F})$ due to (a)

$\nabla \times (\mathbf{F}_0 \times \mathbf{G}) = (\nabla \cdot \mathbf{G})\, \mathbf{F}_0 - (\nabla \cdot \mathbf{F}_0)\, \mathbf{G}$ due to (e)

$= \mathbf{F}_0 (\nabla \cdot \mathbf{G}) - (\mathbf{F}_0 \cdot \nabla)\, \mathbf{G}$ due to (a)

$\therefore\ \nabla \times (\mathbf{F} \times \mathbf{G}) = (\mathbf{G} \cdot \nabla)\, \mathbf{F} - \mathbf{G}(\nabla \cdot \mathbf{F}) + \mathbf{F}(\nabla \cdot \mathbf{G}) - (\mathbf{F} \cdot \nabla)\, \mathbf{G}$

Thus curl $(\mathbf{F} \times \mathbf{G}) = (\mathbf{G} \cdot \nabla)\, \mathbf{F} - (\mathbf{F} \cdot \nabla)\, \mathbf{G} + \mathbf{F}(\nabla \cdot \mathbf{G}) - \mathbf{G}(\nabla \cdot \mathbf{F})$

(v) $\nabla (\mathbf{F} \cdot \mathbf{G}) = \nabla (\mathbf{F}_0 \cdot \mathbf{G}) + \nabla (\mathbf{F} \cdot \mathbf{G}_0)$

$\nabla (\mathbf{F}_0 \cdot \mathbf{G}) = \mathbf{F}_0 \times (\nabla \times \mathbf{G}) + (\mathbf{F}_0 \cdot \nabla)\, \mathbf{G}$ due to (f)

$\nabla (\mathbf{F} \cdot \mathbf{G}_0) = \nabla (\mathbf{G}_0 \cdot \mathbf{F}) = \mathbf{G}_0 \times (\nabla \times \mathbf{F}) + (\mathbf{G}_0 \cdot \nabla)\, \mathbf{F}$ due to (f)

$\therefore\ \nabla (\mathbf{F} \cdot \mathbf{G}) = \mathbf{F} \times (\nabla \times \mathbf{G}) + \mathbf{G} \times (\nabla \times \mathbf{F}) + (\mathbf{F} \cdot \nabla)\, \mathbf{G} + (\mathbf{G} \cdot \nabla)\, \mathbf{F}$

Thus grad $(\mathbf{F} \cdot \mathbf{G}) = \mathbf{F} \times \text{curl } \mathbf{G} + \mathbf{G} \times \text{curl } \mathbf{F} + (\mathbf{F} \cdot \nabla)\, \mathbf{G} + (\mathbf{G} \cdot \nabla)\, \mathbf{F}$

(vi) curl (grad $\mathbf{F}$) $= \nabla \times (\nabla \mathbf{F}) = (\nabla \times \nabla)\, \mathbf{F} = 0$ due to (c)

(vii) div (curl $\mathbf{F}$) $= \nabla \cdot (\nabla \times \mathbf{F}) = \nabla \times (\nabla \cdot \mathbf{F})$ due to (d)

$= (\nabla \times \nabla) \cdot \mathbf{F} = 0$ due to (c)

(viii) curl (curl $\mathbf{F}$) $= \nabla \times (\nabla \times \mathbf{F})$

$= (\nabla \cdot \mathbf{F})\nabla - (\nabla \cdot \nabla)\, \mathbf{F}$ due to (e)

$= \nabla (\nabla \cdot \mathbf{F}) - \nabla^2 \mathbf{F}$

(since ∇ must not appear at the end)

$= \text{grad (div } \mathbf{F}) - \nabla^2 \mathbf{F}$

Solved examples

Example **14.29**: If $f = x^2y^3z^2$, find $\nabla^2 f$ at (1, 2, 1)

Solution:

$$\nabla^2 f = \frac{\partial^2 f}{\partial x^2} + \frac{\partial^2 f}{\partial y^2} + \frac{\partial^2 f}{\partial z^2}$$

$$= 2y^3z^2 + 6x^2yz^2 + 2x^2y^3$$

$\therefore$ at (1, 2, 1), $\nabla^2 f = 16 + 12 + 16 = 44.$

Example **14.30:** Show that, if $r = xi + yj + zk$, $r = |r|$, then $\nabla^2 r^n = n(n+1)\, r^{n-2}$

Solution:

$$r^n = (x^2 + y^2 + z^2)^{n/2}, \qquad (\because r = \sqrt{x^2 + y^2 + z^2})$$

$$\frac{\partial}{\partial x}(r^n) = \frac{n}{2}(x^2 + y^2 + z^2)^{\frac{n}{2}-1}.2x = nx(x^2 + y^2 + z^2)^{\frac{n-2}{2}}$$

$$\frac{\partial^2}{\partial x^2}(r^n) = n\left[x.\frac{n-2}{2}(x^2 + y^2 + z^2)^{\frac{n-2}{2}-1}.2x + (x^2 + y^2 + z^2)^{\frac{n-2}{2}}\right]$$

$$= n\left[(n-2)x^2.(x^2 + y^2 + z^2)^{\frac{n-4}{2}} + (x^2 + y^2 + z^2)^{\frac{n-2}{2}}\right]$$

$$= n(n-2)\ x^2.r^{n-4} + nr^{n-2} \qquad \text{...(1)}$$

Similarly, $$\frac{\partial^2}{\partial y^2}(r^n) = n(n-2)y^2r^{n-4} + nr^{n-2} \qquad \text{...(2)}$$

and $$\frac{\partial^2}{\partial z^2}(r^n) = n(n-2)z^2r^{n-4} + nr^{n-2} \qquad \text{...(3)}$$

$$\nabla^2(r^n) = \left(\frac{\partial^2}{\partial x^2} + \frac{\partial^2}{\partial y^2} + \frac{\partial^2}{\partial z^2}\right)r^n$$

$$= 3nr^{n-2} + n(n-2)\, r^{n-4}\,(x^2 + y^2 + z^2) \qquad \text{adding (1), (2) \& (3)}$$

$$= 3nr^{n-2} + n(n-2)\, r^{n-2} \qquad (\because x^2 + y^2 + z^2 = r^2)$$

$$= n(n+1)\, r^{n-2}$$

Aliter: $$\frac{\partial}{\partial x}(r^n) = nr^{n-1}.\frac{\partial r}{\partial x} = nr^{n-1}.\frac{x}{r}$$

$$= nr^{n-2}\,.\,x \qquad \left(\because \frac{\partial r}{\partial x} = \frac{x}{r}\right)$$

$$\frac{\partial^2}{\partial z^2}\left(r^n\right)=n\left(r^{n-2}.1+x.(n-2)r^{n-3}.\frac{\partial r}{\partial x}\right)$$

$$=n\left(r^{n-2}+(n-2)r^{n-3}.\frac{x^2}{r}\right)$$

$$= nr^{n-2} + n(n-2)\, r^{n-4}.\, x^2 \qquad \text{...(1)}$$

Similarly, $\frac{\partial^2}{\partial y^2}\left(r^n\right)=nr^{n-2}+n(n-2)r^{n-4}.y^2$ ...(2)

and $\frac{\partial^2}{\partial z^2}\left(r^n\right)=nr^{n-2}+n(n-2)r^{n-4}.z^2$ ...(3)

Adding (1), (2) & (3) we get,

$$\left(\frac{\partial^2}{\partial x^2}+\frac{\partial^2}{\partial y^2}+\frac{\partial^2}{\partial z^2}\right)r^n=3nr^{n-2}+n(n-2)r^{n-4}\left(x^2+y^2+z^2\right)$$

$$\Rightarrow \quad \nabla^2 r^n=3nr^{n-2}+n(n-2)r^{n-4}.r^2$$

$$= r^{n-2}[3n + n^2 - 2n]$$

$$= n(n + 1)\, r^{n-2}$$

Note: If n = –1, we have $\nabla^2\left(\frac{1}{r}\right)= 0$, which means that $\left(\frac{1}{r}\right)$ satisfies the Laplace's equation

$$\frac{\partial^2\phi}{\partial x^2}+\frac{\partial^2\phi}{\partial y^2}+\frac{\partial^2\phi}{\partial z^2}=0 \text{ or } \nabla^2\phi=0$$

***Example* 14.31**: If $\nabla . U = 0$, $\nabla . V = 0$, $\nabla \times U = -\frac{\partial V}{\partial t}$, $\nabla \times V = \frac{\partial U}{\partial t}$, show that U and V satisfy the wave equation $\nabla^2 U = \frac{\partial^2 U}{\partial t^2}$

Solution:

$$\nabla\times(\nabla\times \mathbf{U})=\nabla\times\left(-\frac{\partial V}{\partial t}\right)=\frac{-\partial}{\partial t}(\nabla\times V)$$

$$=\frac{-\partial}{\partial t}\left(\frac{\partial U}{\partial t}\right)=\frac{-\partial^2 U}{\partial t^2} \qquad ...(1)$$

But $\nabla\times(\nabla\times \mathbf{U})=\nabla(\nabla.\mathbf{U})-\nabla^2\mathbf{U}$

$$=-\nabla^2\mathbf{U}\,(\because\ \nabla\cdot\vec{U}=0) \qquad ...(2)$$

from (1) & (2), $\frac{\partial^2\vec{U}}{\partial t^2}=\nabla^2\vec{U}$, which shows that U satisfies the wave equation.

Similarly,

$$\nabla\times(\nabla\times V)=\nabla\times\frac{\partial U}{\partial t}=\frac{\partial}{\partial t}(\nabla\times U)=\frac{-\partial^2 V}{\partial t^2} \qquad ...(3)$$

and $\nabla\times(\nabla\times \mathbf{V})=\nabla(\nabla.\mathbf{V})-\nabla^2\mathbf{V}=-\nabla^2\mathbf{V}(\because\ \nabla.\mathbf{V}=0)$ ...(4)

(3) and (4) $\Rightarrow$ V satisfies the wave equation.

***Example* 14.32:** If $\nabla.V=0$, show that $\nabla\times[\nabla\times\{\nabla\times(\nabla\times V)\}]=\nabla^4 V$

(or) curl [curl{curl (curl V)}] = ∇^4V

Solution:

We know that

$$\nabla\times(\nabla\times \mathbf{V})=\nabla(\nabla.\mathbf{V})-\nabla^2\mathbf{V}$$

$$=-\nabla^2\mathbf{V}\,(\because\ \nabla.\mathbf{V}=0)$$

$$=-\mathbf{U}\text{ (say)}$$

Then the given expression

$$=-\nabla\times(\nabla\times\mathbf{U})=\nabla^2\mathbf{U}-\nabla(\nabla.\mathbf{U})$$

$$=\nabla^2(\nabla^2\mathbf{V})-\nabla(\nabla.\mathbf{U}) \qquad (\because\ \nabla^2\mathbf{V}=\mathbf{U})$$

$$=\nabla^4\mathbf{V}-\nabla(\nabla.\mathbf{U})$$

$$\nabla\cdot\vec{U}=\nabla.(\nabla^2\mathbf{V})$$

$$=\left\{\sum i\frac{\partial}{\partial x}\right\}.\{\nabla^2\mathbf{v}\}$$

$$=\sum\left\{i\frac{\partial}{\partial x}\right\}.\left(\nabla^2(V_1 i+V_2 j+V_3 k)\right)$$

$$= \nabla^2\left(\frac{\partial V_1}{\partial x}+\frac{\partial V_2}{\partial y}+\frac{\partial V_3}{\partial z}\right)$$

$$= \nabla^2(Div V)=\nabla^2(0)=0 \qquad (\because \text{Div } \mathbf{V} = \nabla . \mathbf{V} = 0)$$

Exercise 14.4

(1) Show that $\nabla^2(\log r)=\frac{1}{r^2}$

(2) Prove that $\nabla^2(fg)=f(\nabla^2 g)+2(\nabla f).(\nabla g)+g(\nabla^2 f)$

(3) Show that $\nabla^2\left(\nabla.\frac{\boldsymbol{r}}{r^2}\right)=\frac{2}{r^4}$

(4) Show that $\nabla^2\phi(r)=\frac{d^2\phi}{dr^2}+\frac{2}{r}\frac{df}{dr}$

show also that, if $\nabla^2\phi=0$, then $\phi = C_1 + \frac{C_2}{r}$ where C_1, C_2 are constants

(5) If $\boldsymbol{r} = xi + yj + zk$, prove that, curl $(k \times \text{grad } \frac{1}{r}) + \text{grad } (k . \text{grad } \frac{1}{r}) = 0$

14.6 Vector Integration

14.6.1 Ordinary Integration of Vectors

(1) If $\mathbf{A}(u) = A_1(u)i + A_2(u)j + A_3(u)k$ be a vector function of a scalar variable 'u' ($A_1(u)$, $A_2(u)$, $A_3(u)$ assumed to be continuous in any given interval), the indefinite integral of $\mathbf{A}(u)$ is given by.

$$\int A(u)du = i\int A_1(u)du + j\int A_2(u)du + k\int A_3(u)du$$

(2) If there exists a vector $\mathbf{B}(u)$ such that $\mathbf{A}(u) = \frac{d}{du}(\mathbf{B}(u))$, we can write

$$\int A(u)du = \int \frac{d}{du}(\boldsymbol{B}(u))du = \boldsymbol{B}(u)+c$$

where c is a constant of integration independent of u.

(3) The definite integral of $\bar{\mathbf{A}}(u)$ between the limits 'a' and 'b' in the above, is written as,

$$\int_a^b A(u)\,du = \int_a^b \frac{d}{du}(\boldsymbol{B}(u))\,du = [\boldsymbol{B}(u)+c]_a^b = \boldsymbol{B}(b)-\boldsymbol{B}(a)$$

14.6.2 Line Integrals

Let **A**(x,y,z) be a continuous vector function defined in the entire region of space. Let c be any curve in the region. Divide c into n intervals by taking points

$$A = B_0, B_1, B_2 \dots B_n (= B).$$

Let P_i be any point in the interval $B_{i-1}\,B_i$

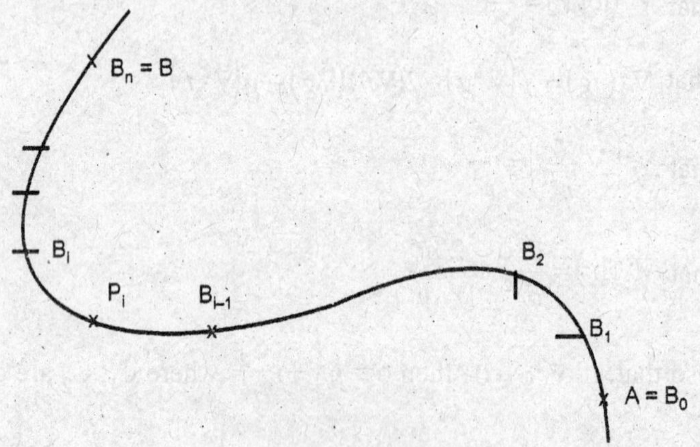

Let $r_0, r_1 \dots\dots r_n$ be the position vectors of points $B_0, B_1, B_2 \dots\dots B_n$ respectively. Let us consider the sum,

$$\sum A(P_i).\delta r_i$$

The limit of this sum as $n \to \infty$ and $|\delta r_i| \to 0$ is defined as the line integral of A along the curve c and is denoted symbolically by

$$\int_c A.d\mathbf{r} \quad \text{or} \quad \int_c A.\frac{dr}{dt}dt\,; \quad \text{which is a scalar.}$$

If c is a closed curve, the integral is written as $\oint A.dr$

Cartesian form of line integral

If $\quad \mathbf{A} = A_1 i + A_2 j + A_3 k$

$dr = (dx)i + (dy)j + (dz)k,$

$$\int_c A.dr = \int_c A_1 dx + A_2 dy + A_3 dz$$

Note: $\int_c \phi dr$, and $\int \mathbf{A} \times dr$ are also examples of line integrals.

14.6.3 Physical Applications

(1) Work done by a force

(1) If **A** represents a force and **dr** is an element of the path of the particle along a curve c, then the line integral

$$\int_P^Q A.dr \qquad \text{(P, Q are 2 points on c)}$$

represents the work done by force **A** in moving the particle from P to Q.

(2) Flow or circulation:

(2) If **A** is the electric field strength, the line integral given above i.e., $\int_c A.dr$, is called the flow of **A** along c. If c is a closed curve it is often referred to as circulation of A around c.

In general, the line integral $\int_P^Q A.dr$, will depend on the path from P to Q

14.6.4 Theorem

Prove that the necessary and sufficient condition for the integral $\int_c A.dr$, to be independent of the path c joining any two points is that **A** is a conservative vector field, (or) there exists a scalar field ϕ such that $\mathbf{A} = \nabla\phi$ (or) curl $\mathbf{A} = 0$, [i.e., the work done by the force A in moving the particle from one point to another is independent of the path if $\mathbf{A} = \nabla\phi$]

Proof: Let $P = (x_1, y_1, z_1)$, $Q = (x_2, y_2, z_2)$ be any two given points on the curve c. Let $\mathbf{A} = \nabla\phi$ where ϕ is single–valued and has continuous derivatives.

(1) Work done $= \int_P^Q A.dr$

$$= \int_P^Q \nabla\phi.dr = \int_P^Q \left(i\frac{\partial\phi}{\partial x} + j\frac{\partial\phi}{\partial y} + k\frac{\partial\phi}{\partial z}\right)(dxi + dyj + dzk)$$

$$= \int_P^Q \frac{\partial\phi}{\partial x}dx + \frac{\partial\phi}{\partial y}dy + \frac{\partial\phi}{\partial z}dz = \int_P^Q d\phi = \phi(Q) - \phi(P)$$

$$= \phi(x_2, y_2, z_2) - \phi(x_1, y_1, z_1)$$

i.e., the integral depends upon the two points only but not on the path joining them. (This is true only if ϕ is single–valued at all points P and Q).

(2) The integral is independent of the path. Then,

$$\phi(x,y,z) = \int_{(x_1,y_1,z_1)}^{(x,y,z)} A.dr = \int_{(x_1,y_1,z_1)}^{(x,y,z)} A.\frac{dr}{ds}ds$$

Differentiation with respect to 's' gives

$$\frac{d\phi}{ds} = A.\frac{dr}{ds} \qquad \text{...... (1)}$$

But
$$\frac{d\phi}{ds} = \nabla\phi.\frac{dr}{ds} \qquad \text{...... (2)}$$

(1) – (2) $\Rightarrow (A - \nabla\phi).\frac{dr}{ds} = 0$, which is true irrespective of $\frac{dr}{ds}$

$\therefore \quad A = \nabla\phi$

Solved Examples

***Example* 14.33:** $F(t) = (3t^2-t)i + (2-6t)j - 4tk$, find (a) $\int F(t)dt$ (b) $\int_2^4 F(t)dt$.

***Solution*:**

(a) $\int F(t)dt = i\int(3t^2 - t)dt + j\int(2-6t)dt - k\int 4tdt$

$$= \left(t^3 - \frac{t^2}{2}\right)i + \left(2t - 3t^2\right)j - 2t^2k + c$$

(b) $$\int_2^4 F(t)dt = \left[\left(t^3 - \frac{t^2}{2}\right)i + \left(2t - 3t^2\right)j - 2t^2k + c\right]_2^4 = 50i - 32j - 24k$$

***Example* 14.34** Evaluate: $\int_0^{\pi/2}[(3\sin\theta)i + (2\cos\theta)j]d\theta$

***Solution*:**

Given integral

$$= i\int_0^{\pi/2}(3\sin\theta)d\theta + j\int_0^{\pi/2}(2\cos\theta)d\theta \; [-3\cos\theta]_0^{\pi/2} i + +[2\sin\theta]_0^{\pi/2} j = 3i + 2j$$

***Example* 14.35:** The acceleration a of a particle at any time 't' $\geq$ 0 is given by,

$$\mathbf{a} = e^{-t}i - 6(t+1)j + (3\sin t)k$$

Find the velocity **V** and displacement **r** at any time 't' given that **V** = 0 when t = 0 and **r** = 0 when t = 0.

***Solution*:**

$$\mathbf{a} = e^{-t}i - 6(t+1)j + (3\sin t)k = \frac{d^2r}{dt^2}$$

$$\therefore \quad \mathbf{V} = \frac{dr}{dt} = \int a\,dt = -e^{-t}\,i - 6\left(\frac{t^2}{2}+t\right)j - (3\cos t)k + C_1 \qquad .....(1)$$

(C_1 is constant of integration)

But $\quad \mathbf{V} = 0$ when t = 0

$$\therefore \quad -i - 3k + C_1 = 0 \Rightarrow C_1 = i + 3k$$

Substituting in (1), we get

velocity $\quad \mathbf{V} = (1 - e^{-t})i - (3t^2 + 6t)j + (3 - 3\cos t)k$

Integrating,

$$\mathbf{r} = (t + e^{-t})i - (t^3 + 3t^2)j + (3t - 3\sin t)k + C_2 \qquad ..... (2)$$

(C_2 being constant of integration)

But $\quad \vec{r} = 0$ when t = 0

$\therefore$ (2) $\Rightarrow \quad +i + C_2 = 0 \qquad \therefore \; C_2 = -i$

From (2), $\quad \mathbf{r} = (t - 1 + e^{-t})i - (t^3 + 3t^2)j + (3t - 3\sin t)k$

***Example* 14.36:** Show that $\int F \times \frac{d^2F}{dt^2} = F \times \frac{dF}{dt} + c$

***Solution*:**

We know that

$$\frac{d}{dt}\left(F \times \frac{dF}{dt}\right) = F \times \frac{d}{dt}\left(\frac{dF}{dt}\right) + \frac{dF}{dt} \times \frac{dF}{dt}$$

$$= \mathbf{F} \times \frac{d^2\mathbf{F}}{dt^2} \qquad \left(\because \frac{d\mathbf{F}}{dt} \times \frac{d\mathbf{F}}{dt} = 0\right)$$

Hence the result.

***Example* 14.37:** If A = $2ti + 3t^2j - (4t + 1)k$, and B = $ti + 2j + t^2k$, find

(i) $\int_0^2 (\boldsymbol{A}.\boldsymbol{B})dt$ (ii) $\int_0^2 (\boldsymbol{A} \times \boldsymbol{B})dt$.

***Solution*:**

(i) **A.B** = $(2t)t + (3t^2)(2) - t^2(4t + 1) = 2t^2 + 6t^2 - 4t^3 - t^2 = 7t^2 - 4t^3$

$$\int_0^2 (\text{A.B})\,dt = \left[\frac{7t^3}{3} - t^4\right]_0^2 = \frac{8}{3}$$

(ii) $\mathbf{A} \times \mathbf{B} = \begin{vmatrix} i & j & k \\ 2t & 3t^2 & -(4t+1) \\ t & 2 & t^2 \end{vmatrix}$

$$= (3t^4 + 8t + 2)i + (-4t^2 - t - 2t^3)j + (4t - 3t^3)k$$

$$\int_0^2 (\boldsymbol{A} \times \boldsymbol{B})dt = \left[\left(\frac{3t^5}{5} + 4t^2 + 2t\right)i + \left(\frac{-4t^3}{3} - \frac{t^2}{2} - \frac{2t^4}{4}\right)j + \left(2t^2 - \frac{3t^4}{4}\right)k\right]_0^2$$

$$= \frac{196}{5}i - \frac{62}{3}j - 4k$$

***Example* 14.38:** If F(2) = $i + 2j - 2k$ and F(3) = $6i - 2j + 3k$,

evaluate $\int_2^3 \boldsymbol{F}.\frac{d\boldsymbol{F}}{du}du$

Solution:

From vector differentiation,

We get, $\frac{d}{du}(\boldsymbol{F}.\boldsymbol{F}) = \boldsymbol{F}.\frac{d\boldsymbol{F}}{du} + \frac{d\boldsymbol{F}}{du}.\boldsymbol{F} = 2\left(\boldsymbol{F}.\frac{d\boldsymbol{F}}{du}\right)$, so that

$$\mathbf{F}.\frac{d\boldsymbol{F}}{du} = \frac{1}{2}\frac{d}{du}(\boldsymbol{F}.\boldsymbol{F}) = \frac{1}{2}\frac{d}{du}|\boldsymbol{F}|^2$$

$$\therefore \quad \int_2^3 F.\frac{dF}{du}du = \frac{1}{2}\int_2^3\left\{\frac{d}{du}|F|^2\right\}du = \left[\frac{1}{2}|F|^2\right]_2^3$$

But $\quad \mathbf{F}(2) = i + 2j - 2k \Rightarrow |F|^2 = 1 + 4 + 4 = 9$, when u = 2

and $\quad \mathbf{F}(3) = 6i + 2j - 3k \Rightarrow |F|^2 = 36 + 4 + 9 = 49$, when u = 3

$$\therefore \quad \int_2^3 F.\frac{dF}{du}du = \frac{1}{2}[49 - 9] = 20$$

***Example* 14.39:** If F = $(x^2 - 2y)i - 6yzj + 8xz^2k$, evaluate $\int_c F.dr$ from the point (0,0,0) to the point (1,1,1) along the following paths

(1) $x = t$, $y = t^2$, $z = t^3$.

(2) the straight line from (0,0,0) to (1,0,0), then to (1,1,0) and then to (1,1,1) and

(3) the staight line joining (0,0,0) to (1,1,1).

***Solution*:**

(1) $x = t$; $\quad y = t^2$; $\quad z = t^3$;

$dx = dt$; $\quad dy = 2t\,dt$; $\quad dz = 3t^2\,dt$;

when t = 0, the point is (0,0,0), when t = 1, the point is (1,1,1)

$$\int_c \mathbf{F.dr} = \int_c (x^2 - 2y)dx - 6yz\,dy + 8xz^2\,dz$$

$$= \int_{t=0}^{t=1}(t^2 - 2t^2)dt - 6.t^2.t^3(2t)dt + 8t(t^3)^2(3t^2)dt = \int_0^1 -t^2dt - 12t^6dt + 24t^9dt$$

$$= \int_0^1 (24t^9 - 12t^6 - t^2)dt = \frac{24t^{10}}{10} = \left[\frac{24t^{10}}{10} - \frac{12t^2}{7} - \frac{t^3}{3}\right]_1^0$$

$$= \frac{12}{5} - \frac{12}{7} - \frac{1}{3} = \frac{252 - 180 - 35}{105} = \frac{37}{105}$$

***Aliter*:** Along C, $\boldsymbol{F} = (t^2 - 2t^2)i - 6.\ t^2\ t^3j + 8.t.t^6\ k = -t^2i - 6t^5j + 8t^7k$

$\mathbf{dr} = (dx)i + (dy)j + (dz)k = dt\ i + (2t\ dt)j + (3t^2\ dt)k$

$$\int_c F.dr = \int_{t=0}^{1} \left(-t^2 - 12t^6 + 24t^9\right) dt \qquad \text{(Taking dot product of } \mathbf{F} \text{ and } \mathbf{dr})$$

$$= \frac{37}{105}$$

(2) Let O = (0,0,0), P = (1,0,0), Q = (1,1,0), R = (1,1,1).

Then along OP, $y = 0$, $z = 0$, $dy = 0$, $dz = 0$ and x varies from 0 to 1.

$$\therefore \quad \int_{OP} F.dr = \int_{x=0}^{1} \left(x^2 - 2.0\right)dx - 6(0)(0)(0) + 8x(0)^2(0) = \int_0^1 x^2 dx = \frac{1}{3} \qquad \text{.... (1)}$$

Along PQ, $x = 1$, $z = 0$, $dx = 0$, $dz = 0$ and y varies from 0 to 1.

$$\therefore \quad \int_{PQ} F.dr = \int_{y=0}^{1} \left(1^2 - 2y\right)0 - 6y(0)dy + 8.1.(0)^2(0) = \int 0 = 0 \qquad \text{.... (2)}$$

Along QR, $x = 1$, $y = 1$, $dx = 0$, $dy = 0$ and z varies from 0 to 1.

$$\therefore \quad \int_{QR} F.dr = \int_{z=0}^{1} \left(1^2 - 2.1\right)(0) - 6.1.z(0) + 8.1.z^2.dz = \int_0^1 8z^2 dz = \frac{8}{3} \qquad \text{.... (3)}$$

Adding (1), (2), (3), $\int F.dr = \frac{1}{3} + 0 + \frac{8}{3} = 3$

(3) The equation of the straight line joining (0,0,0) to (1,1,1) is $\frac{x}{1} = \frac{y}{1} = \frac{z}{1} = t$ (say), so that $x = t$, $y = t$, $z = t$, $dx = dy = dz = dt$ 't' takes values from 0 to 1.

$$\therefore \quad \int_c F.dr = \int_0^1 \left[\left(t^2 2t\right) - 6.t.t + 8t.t^2\right] dt = \int_0^1 \left(8t^3 - 5t^2 - 2t\right) dt$$

$$= \left[2t^4 - \frac{5t^3}{3} - t^2\right]_0^1 = 2 - \frac{5}{3} - 1 = \frac{-2}{3}$$

***Example* 14.40:** Find the total work done by a force $F = 2xyi - 4zj + 5xk$ along the curve $x = t^2$,

$y = 2t + 1$, $z = t^3$, from the points $t = 1$ to $t = 2$.

Solution:

$x = t^2$, $dx = 2t\,dt$; $y = 2t + 1$, $dy = 2dt$; $z = t^3$, $dz = 3t^2\,dt$

Total work done = $\int_c F.dr$

$$= \int_c (2xyi - 4zj + 5xk).\{(dx)i + (dy)j + (dz)k\} = \int_c 2xy\,dx - 4z\,dy + 5x\,dz$$

$$= \int_{t=1}^{2} [2.t^2(2t+1)]2t\,dt - (4.t^3)2\,dt + (5t^2)3t^2\,dt = \int_1^2 (8t^4 + 4t^3 - 8t^3 + 15t^4)dt$$

$$= \int_1^2 (23t^4 - 4t^3)dt = \left[23\frac{t^5}{5} - t^4\right]_1^2 = \frac{23}{5}(32-1) - (16-1) = \frac{713}{5} - 15 = \frac{638}{5}$$

***Example* 14.41:** If c is the curve $y = 3x^2$ in the xy – plane and $F = (x + 2y)i - xyj$,

evaluate $\int F.dr$, from the point (0,0) to (1,3).

Solution:

Since c is a curve in xy – plane, we take $r = xi + yj$, so that

$$\mathbf{F.\,dr} = \{(x + 2y)i - xyj\} \,.\, \{(dx)i + (dy)j\} = (x + 2y)\,dx - xy\,dy.$$

***1st Method*:** By taking the curve in parametric coordinates as,

$x = t$, $y = 3t^2$, $dx = dt$, $dy = 6t\,dt$, so that t varies from 0 to 1 to get the points (0,0) & (1,3): We have

$$\int_{(0,0)}^{(1,3)} F.dr = \int_{t=0}^{1} (t + 6t^2)\,dt - t(3t^2)(6tdt) = \int_0^1 (t + 6t^2 - 18t^4)\,dt$$

$$= \left.\frac{t^2}{2} + \frac{6t^3}{3} - \frac{18t^5}{5}\right|_0^1 = \frac{1}{2} + 2 - \frac{18}{5} = \frac{5}{2} - \frac{18}{5} = -\frac{11}{10}$$

***2nd Method*:** $y = 3x^2$, $dy = 6x\,dx$ and x varies from 0 to 1.

$$\therefore \quad \int_c F.dr = \int_0^1 (x + 6x^2)dx - x.3x^2.6x.dx$$

$$= \int_0^1 (x + 6x^2 - 18x^4)dx = -\frac{11}{10}$$

***Example* 14.42:** Find the work done by the force F in moving a particle once around the circle 'C' in xy – plane, if the centre of the circle is origin and radius is '2' and

$$F = (x + y + z)i + (2x + y)j + (2x - y + z)k.$$

Solution:

xy – plane is $z = 0$,

$\therefore$ $\mathbf{F} = (x + y)i + (2x + y)j + (2x - y)k$ and $\mathbf{r} = xi + yj \Rightarrow \mathbf{dr} = (dx)i + (dy)j$

$\mathbf{F}.\,\mathbf{dr} = (x + y)dx + (2x + y)dy.$

The equation of the circle is $x = 2\cos\theta$, $y = 2\sin\theta$

$\therefore$ $dx = -2\sin\theta\, d\theta$, $dy = 2\cos\theta\, d\theta$

θ varies from 0 to 2π

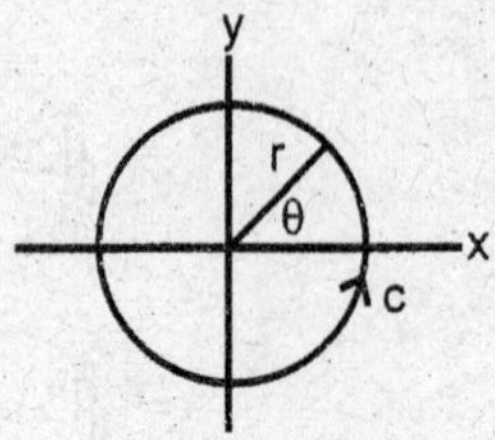

Work done = $\int F.dr$

$$= \int_0^{2\pi} (2\cos\theta + 2\sin\theta)(-2\sin\theta)d\theta + (2.2\cos\theta + 2\sin\theta)(2\cos\theta)d\theta$$

$$= \int_0^{2\pi} \left(-2\sin 2\theta - 4\sin^2\theta + 8\cos^2\theta + 2\sin 2\theta\right)d\theta = \int_0^{2\pi} \left(8\cos^2\theta - 4\sin^2\theta\right)d\theta$$

$$= \int_0^{2\pi} \left[4(1+\cos 2\theta) - 2(1-\cos 2\theta)\right]d\theta = \int_0^{2\pi} (2+\cos 2\theta)\,d\theta = \left[2\theta + 2\sin 2\theta\right]_0^{2\pi}$$

$$= 4\pi$$

***Example* 14.43:** Show that the necessary and sufficient condition for a vector field **V** to be conservative is curl **V** = 0

Solution:

(a) Necessary condition: If **V** is conservative, $\exists$ a 'ϕ' $\ni$ $\mathbf{V} = \nabla\phi$.

curl $\bar{V}$ = curl $(\nabla\phi) = 0$

(b) Sufficient condition: Let $\mathbf{V} = V_1 i + V_2 j + V_3 k$

$$\text{Curl } \mathbf{V} = \nabla \times \mathbf{V} = 0 \Rightarrow \begin{vmatrix} i & j & k \\ \partial/\partial x & \partial/\partial y & \partial/\partial z \\ V_1 & V_2 & V_3 \end{vmatrix} = 0$$

$$\Rightarrow \sum\left(\frac{\partial V_3}{\partial y} - \frac{\partial V_2}{\partial z}\right)i = 0$$

$$\Rightarrow \quad \frac{\partial V_3}{\partial y}=\frac{\partial V_2}{\partial z}; \quad \frac{\partial V_1}{\partial z}=\frac{\partial V_3}{\partial x}; \quad \frac{\partial V_2}{\partial x}=\frac{\partial V_1}{\partial y} \qquad \text{.... (1)}$$

The work done by the force field **V** in moving a particle from

(x_1,y_1,z_1) to (x,y,z) is $\int_c V.dr = \int_c V_1(x,y,z)dx + V_2(x,y,z)dy + V_3(x,y,z)dz$

where **V** is a path joining (x_1,y_1,z_1) to (x,y,z)

Let us choose a particular path consisting straight line segments from (x_1,y_1,z_1) to (x,y_1,z_1) to (x,y,z_1) to (x,y,z) and denote the work done along this path by a scalar function $\phi(x,y,z)$;

$$\therefore \quad \phi(x,y,z) = \int_{x_1}^{x} V_1(x,y_1,z_1)dx + \int_{y_1}^{y} V_2(x,y,z_1)dy + \int_{z_1}^{z} V_3(x,y,z)dz \qquad \text{..... (2)}$$

From (2), it can be seen that,

$$\frac{\partial\phi}{\partial z} = V_3(x,y,z) \qquad \text{.... (3)}$$

$$\frac{\partial\phi}{\partial y} = V_2(x,y,z_1) + \int_{z_1}^{z}\frac{\partial V_3}{\partial y}(x,y,z)dz = V_2(x,y,z_1) + \int_{z_1}^{z}\frac{\partial V_2}{\partial z}(x,y,z)dz \text{ [from(1)]}$$

$$= V_2(x,y,z_1) + V_2(x,y,z)\Big|_{z_1}^{z} = V_2(x,y,z_1) - V_2(x,y,z) - V_2(x,y,z_1)$$

$$= V_2(x,y,z) \qquad \text{.... (4)}$$

$$\frac{\partial\phi}{\partial x} = V_1(x,y_1,z_1) + \int_{y_1}^{y}\frac{\partial V_2}{\partial x}(x,y,z_1)dy + \int_{z_1}^{z}\frac{\partial V_3}{\partial x}(x,y,z_1)dz$$

$$= V_1(x,y_1,z_1) + \int_{y_1}^{y}\frac{\partial V_1}{\partial y}(x,y,z_1)dy + \int_{z_1}^{z}\frac{\partial V_1}{\partial z}(x,y,z) \qquad \text{from (1)}$$

$$= V_1(x,y_1,z_1) + \left[V_1(x,y,z_1)\right]_{y_1}^{y} + \left[V_1(x,y,z)\right]_{z_1}^{z}$$

$$= V_1(x,y_1,z_1) + V_1(x,y,z_1) - V_1(x,y_1,z_1) + V_1(x,y,z) - V_1(x,y_1,z_1)$$

$$= V_1(x,y,z) \qquad \text{.... (5)}$$

$$(3), (4), (5) \Rightarrow \frac{\partial \phi}{\partial x} i + \frac{\partial \phi}{\partial y} j + \frac{\partial \phi}{\partial z} k = V_1(x, y, z)i + V_2(x, y, z)j + V_3(x, y, z)k = V$$

$$\Rightarrow \mathbf{V} = \nabla \phi$$

Hence proved

***Example* 14.44**

(a) Show that $\mathbf{F} = y^2 i + (2xy + z^2)j + 2yzk$ is a conservative force field.

(b) Find its scalar potential.

(c) Find the work done in moving an object in this field from (1,2,1) to (3,1,4)

***Solution*:**

(a) $$\nabla \times \mathbf{F} = \begin{vmatrix} i & j & k \\ \partial/\partial x & \partial/\partial y & \partial/\partial z \\ y^2 & 2xy + z^2 & 2yz \end{vmatrix}$$

$$= i(2z - 2z) + j(0 - 0) + k(2y - 2y) = 0$$

$\therefore$ **F** is a conservative force field.

(b) Let 'ϕ' be the scalar potential of **F**.

***1st Method*:**

$$\mathbf{F} = \nabla \phi = \frac{\partial \phi}{\partial x} i + \frac{\partial \phi}{\partial y} j + \frac{\partial \phi}{\partial z} k$$

$$\therefore \frac{\partial \phi}{\partial x} = y^2 \quad .... (1) \qquad \frac{\partial \phi}{\partial y} = 2xy + z^2 \quad .... (2) \qquad \frac{\partial \phi}{\partial z} = 2yz \quad .... (3)$$

Integrating (1) w.r.t x, (2) w.r.t. y, & (3) w.r.t. z, respectively, we get,

$$\phi = xy^2 + f(y,z), \quad \phi = xy^2 + yz^2 + g(z,x), \quad \text{and} \quad \phi = yz^2 + h(x,y).$$

These equations will be consistent if f, g, h are taken as

$$f(y,z) = yz^2, \quad g(z,x) = 0, \quad h(x,y) = xy^2.$$

Hence $\phi = xy^2 + yz^2 +$ constant

2nd Method:

Since F is conservative, $\int_c F.dr$ is independent of path joining (x_1, y_1, z_1) to (x, y, z)

$$\phi = \int_{x_1}^{x}(y_1^2)dx + \int_{y_1}^{y}(2xy + z_1^2)dy + \int_{z_1}^{z} 2yzdz = xy_1^2\Big|_{x_1}^{x} + (xy^2 + z_1^2 y)\Big|_{y_1}^{y} + yz^2\Big|_{z_1}^{z}$$

$$= xy_1^2 - x_1y_1^2 + xy^2, + z_1^2y - x^2y_1 - z_1^2y_1 + yz^2 - yz_1^2$$

$$= xy^2 + yz^2 - x_1y_1^2 - z_1^2y_1 = xy^2 + yz^2 + \text{constant.}$$

3rd Method:

Since $\mathbf{F}.\,\mathbf{dr} = \nabla\phi.\boldsymbol{dr} = \frac{\partial\phi}{\partial x}dx + \frac{\partial\phi}{\partial y}dy + \frac{\partial\phi}{\partial z}dz = d\phi$,

$$d\phi = (y^2)dx + (2xy + z^2)dy + (2yz)dz = (y^2dx + 2xydy) + (z^2dy + 2yz)dz$$

$$= d(xy^2) + d(yz^2) = d(xy^2 + yz^2)$$

$$\Rightarrow \phi = xy^2 + yz^2 + \text{constant}$$

(c)work done = $\int_c \boldsymbol{F.dr} = \int_{P_1=(1,2,1)}^{P_2=(3,1,4)} d\phi$

$$= \phi(P_2) - \phi(P_1) = xy^2 + xz^2 \Big|_{(1,2,1)}^{(3,1,4)}$$

$$= (3.1^2 + 3.4^2) - (1.2^2 + 1.1^2) = 51 - 5 = 46.$$

***Example* 14.45:** Evaluate $\int_c (x^3dy + y^2dx)$ where c is the boundary of the triangle whose vertices are (0,0), (1,0), (0,1).

Solution:

Let $\quad I = \int_c (x^3dy + y^2dx)$

$$I = I_1 + I_2 + I_3, \text{ where } I_1 = \int_{OA}, I_2 = \int_{AB}, I_3 = \int_{BO}$$

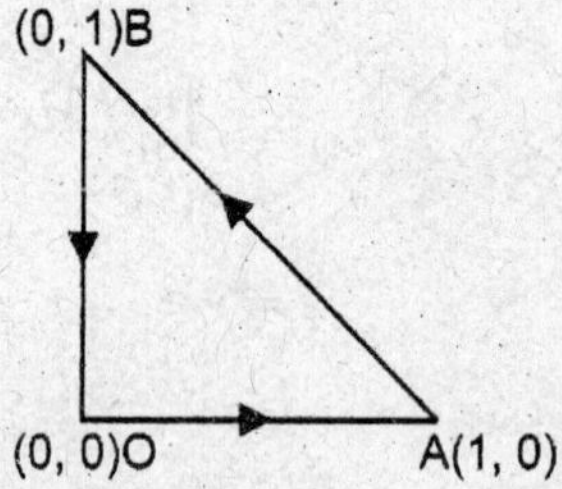

(i) Along OA, $y = 0, \Rightarrow dy = 0$

$$\therefore I_1 = \int x^3.0 + 0^2.dx = 0$$

(ii) Along AB, $x + y = 1$, $y = 1 - x \Rightarrow dy = -dx$, x varies from 1 to 0.

$$\therefore I_2 = \int_1^0 x^3(-dx) + (1-x)^2 dx$$

$$= \int_0^1 (x^2 - x^3 + 1 - 2x)dx = -\frac{x^3}{3} - \frac{x^4}{4}\Bigg|_1^0 = 0 - \left(\frac{1}{3} - \frac{1}{4}\right) = -\frac{1}{12}$$

(iii) Along BO, $x = 0 \Rightarrow dx = 0 \quad \therefore \; I_3 = \int 0.dy + y^2.0 = 0$

$$\therefore \; I = I_1 + I_2 + I_3 = -\frac{1}{12}$$

***Example* 14.46:** If f = xy^2 z^2 evaluate $\int_c f d\mathbf{r}$ where the curve 'c' is given by $x = t, y = t^2$, $z = t^3$ from $t = 0$ to 1.

***Solution*:**

$$f = xy^2z^2 = t(t^2)^2.(t^3)^2 = t^{11}$$

$$dr = dx\, i + dy\, j + dz\, k$$

$$dr = (dt)i + (2tdt)j + (3t^2dt)k = (i + 2tj + 3t^2k)dt$$

$$\int_c fdr = \int_{t=0}^1 t^{11}(i + 2tj + 3t^2k)dt = i\int_0^1 t^{11} + dt + j\int_0^1 2t^{12}dt + k\int_0^1 3t^{13}dt$$

$$= i\frac{t^{12}}{12}\Big|_0^1 + j\frac{2t^{13}}{13}\Big|_0^1 + k\frac{3t^{14}}{14}\Big|_0^1 = \frac{1}{12}i + \frac{2}{13}j + \frac{3}{14}k$$

$$3zi - 2xj + yk$$

***Example* 14.47:** If $A = 3zi - 2xj + yk$, and c is the curve given by *evaluate* $\int_c A \times dr$ *from* $t = 0$ *to* $t = \frac{\pi}{2}$.

***Solution*:**

$$A \times dr = \begin{vmatrix} i & j & k \\ 3z & -2x & y \\ dx & dy & dz \end{vmatrix}$$

$$= (-2x\,dz - y\,dy)i + (y\,dx - 3z\,dz)j + (3z\,dy + 2x\,dx)k$$

$$x = \cos t, y = \sin t, z = 2\cos t$$

$$\Rightarrow dx = (-\sin t)\,dt, dy = (\cos t)\,dt, dz = (-2\sin t)\,dt$$

$\therefore$ (1) $\Rightarrow$ $\mathbf{A} \times d\mathbf{r} = i[(-2\cos t)(-2\sin t) - \sin t \,.\, \cos t]dt + j\,[(\sin t)(-\sin t)$

$- 3(2\cos t)(-2\sin t)]dt + k[3(2\cos t)(\cos t) + (2\cos t)(-\sin t)]dt.$

$= i(3\sin t\cos t)\,dt + j[(12\sin t\cos t) - \sin^2 t)]\,dt + k(6\cos 2t - 2\sin t\cos t)\,dt.$

$$\therefore \int_c A \times dr = i\int_0^{\pi/2} \frac{3}{2}\sin 2t\,dt + j\int_0^{\pi/2}\left[6\sin 2t - \frac{1}{2}(1 - \cos 2t)\right]dt$$

$$+ k\int_0^{\pi/2}[3(1 + \cos 2t) - \sin 2t]dt$$

$$= i\frac{3}{2}\left(\frac{-\cos 2t}{2}\right)_0^{\pi/2} + j\left[-3\cos 2t - \frac{t}{2} + \frac{\sin 2t}{4}\right]_0^{\pi/2} + k\left[3t + \frac{3}{2}\sin 2t + \frac{\cos 2t}{2}\right]_0^{\pi/2}$$

$$= i\left(\frac{-3}{4}\right)(-1-1) + j\left[-3(-1-1) - \frac{\pi}{4} + 0\right] + k\left[\frac{3\pi}{2} + 0 + \frac{1}{2}(-1-1)\right]$$

$$= \frac{3}{2}i + \left(6 - \frac{\pi}{4}\right)j + \left(\frac{3\pi}{2} - 1\right)k$$

Exercise 14.5

1. If $\mathbf{U}(t) = (2t^2 - 1)i + 3tj + (2 - t)k$. Find (a) $\int U(t)dt$, (b) $\int_2^4 U(t)dt$

[**Ans:** (a) $\left(\frac{2t^3}{3} - t\right)i + \frac{3t^2}{2}j + \left(2t - \frac{t^2}{2}\right)k$, (b) $\frac{106}{3}i + 18j - 2k$]

2. Evaluate: $\int_0^{\pi/2} (6\sin u)i - (3\cos u)j + uk$

[**Ans:** $6i - 3j + \frac{\pi^2}{8}k$]

3. If $\mathbf{A}(s) = si - s^2j + (s + 1)k$, $\mathbf{B}(s) = 2si + 6sk$, find (a) $\int_0^2 \mathbf{A.B}ds$, (b) $\int_0^2 \mathbf{A} \times \mathbf{B}ds$.

[**Ans:** (a) $\frac{100}{3}$, (b) $-48i - 12j + 16k$]

4. If $\mathbf{A} = ti - j + 2tk$, $\mathbf{B} = t^2i - tj + 2k$, $\mathbf{C} = i - 2j + 2k$, evaluate

(a) $\int_0^1 \{A.(B \times C)\}dt$, (b) $\int_0^1 \{A \times (B \times C)\}dt$

[**Ans:** (a) $-\frac{1}{3}$, (b) $-\frac{5}{2}i + \frac{37}{6}j + \frac{8}{3}k$]

5. The acceleration of a particle a at any time $t \geq 0$ is given by $a = e^t i + (2\cos 2t)j + (2\sin 2t)k$. If the velocity **V** and displacement **r** are both zero at $t = 0$, find **V** and **r** at any time t.

[**Ans:** $\mathbf{V} = (e^t - 1)i + (\sin 2t)j + (1 - \cos 2t)k$,

$r = (e^t - t - 1)i + \frac{1}{2}(1 - \cos 2t)j + \left(t - \frac{1}{2}\sin 2t\right)k$

6. Evaluate $\int_2^3 A.\frac{dA}{du}\,du$, given that, $\vec{A}(2) = 4i - 2j + 3k$ and $\vec{A}(3) = 2i + j + 2k$

[**Ans:** -10]

Exercise 14.6

1. If $\phi = xyz$, evaluate $\int_c \phi dr$, where c is the curve $x = t^3, y = t^2, z = t$, from t = 0 to 1

[Ans: $\frac{1}{3}i+\frac{1}{4}j+\frac{1}{7}k$]

2. If $\mathbf{F} = xi - yzj + z^2k$, and c is the curve given by $x = t, y = t^3, z = t^2$, evaluate (i) $\int_c F \times dr$ and (ii) $\int_c F.dr$, from t = 0 to t = 1.

[Ans: (i) $-\frac{5}{7}i-\frac{7}{15}j+\frac{11}{12}k$, (ii) $\frac{23}{24}$]

3. If $\mathbf{A} = (2x + 3)i + xyj + (zx - y)k$, evaluate $\int_c F.dr$ along c where c is

(a) the curve $x = t^3, y = 2t^2, z = t$ from t = 0 to 1.
(b) The straight lines from (0,0,0) to (1,0,0), then to (1,0,1) and then to (1,2,1)
(c) The straight line joining (0,0,0) and (1,2,1).

[Ans: (a) $\frac{491}{105}$, (b) $\frac{13}{2}$, (c) $\frac{14}{3}$]

4. If $\mathbf{A} = (3xy - 2y^2)i + (x - y)j$, evaluate $\int_c A.dr$ along the curve c in xy – plane given by $y = x^3$ from the point (0,0) to (2,8).

[Ans: $\frac{1308}{35}$]

5. If $\mathbf{F} = (2x - y)i + (x - 2y)j$, evaluate $\int_c A.dr$ where c is the closed curve shown in the figure below.

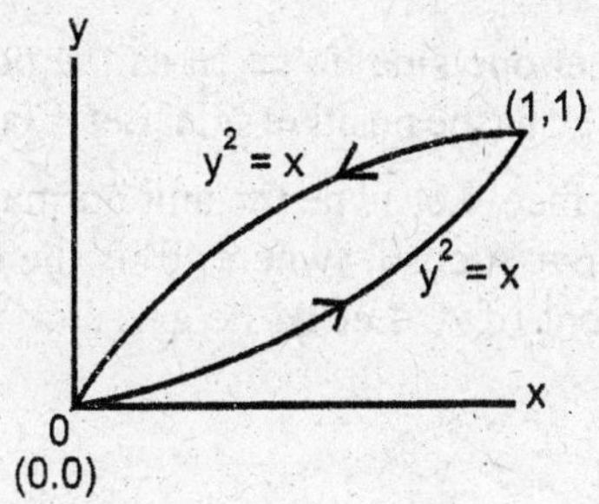

[Ans: $(\frac{1}{3})$]

6. If $\mathbf{A} = (3x + 2y)i + (x + y)j$, and c is the boundary of the tringle whose vertices are (0,0), (1,0), (0,1), evaluate $\int_c A.dr$.

[Ans: $\frac{3}{2}$]

7. If $\mathbf{A} = (2x + y)i + (3x - 2y)j$, compute the circulation of **A** about the circle **C**: $x^2 + y^2 = 4$, traversed in the positive direction.

[Ans: 8π]

8. Find the work done in moving a particle in the force field. $\mathbf{F} = 2x^2 i + (2yz - x)j + yk$, along (a) the straight line from (0,0,0) to (3,1,2) (b) the space curve $x = 3t^2$, $y = t$, $z = 3t^2 - t$ from t = 0 to t = 1

[Ans: (a) $\frac{113}{6}$, (b) $\frac{58}{3}$]

9. (a) Prove that $\mathbf{V} = (2x \sin y - 3)i + (x^2 \cos y + z^2)j + 2(yz + 1)k$ is a conservative force field. (b) Find the scalar potential of $\vec{V}$. (c) Find the work done in moving an object in this field from (1,0,–1) to $(2, \frac{\pi}{2}, 1)$.

[Ans: (b) $x^2 \sin y + yz^2 - 3x + 2z$ + constant, (c) $\frac{\pi}{2} + 5$]

10. If $\mathbf{A} = (9x^2y - 2xz^3)i + 3x^3j - 3x^2z^2k$, (a) prove that $\int_c A.dr$ is independent of the curve 'C' joining two given points. (b) show that there is a differentiable function ϕ such that $\mathbf{A} = \nabla\phi$ and find it.

[Ans: $\phi = 3x^3y - x^2z^3 + c$]

14.7 Surface Integrals

14.7.1 Definition

Let S be a two–sided surface. Let one side be taken as the positive side. If S is a closed surface, the outer side is considered as the positive side. Let **A** be a vector function. Consider an element of area 'ds' in the surface. Let $\vec{n}$ be the unit normal vector to ds in the positive direction. It can be seen that $\mathbf{A.n} = A\cos\theta$. (where '$\theta$' is the angle between **A** and **n** and $A = |\mathbf{A}|$) is the normal component of A. Let **ds** be a vector whose magnitude is ds and whose direction is that of **n**.

$\therefore$ **ds** = **n** ds.

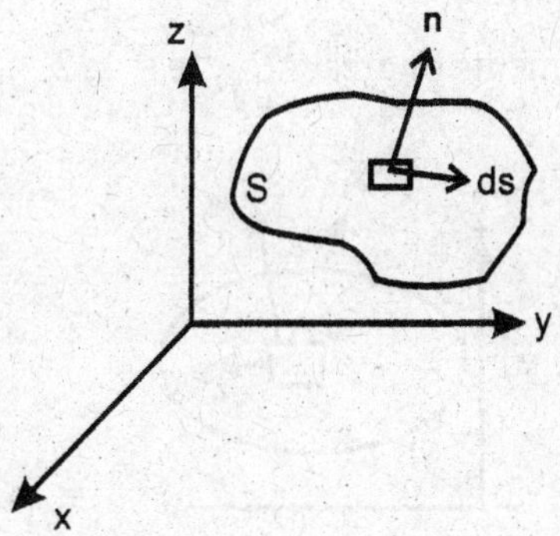

Then the integral,

$$\iint_s A.ds = \iint_s A.n.ds \qquad ..... (1)$$

is an example of a surface integral which is also called as flux of **A** over s.
If 'ϕ' is a scalar function,

$$\iint_s \phi ds \qquad ..... (2)$$

$$\iint_s \phi n ds \qquad ..... (3)$$

$$\iint_s A\times ds = \iint_s A\times n\, ds \qquad ..... (4)$$

are some other examples of surface integrals.

***Note*:** (1) The surface integrals can also be defined in terms of limits of sums. (see 7.7.2)

(2) The notation $\oiint_s$ is also used to denote a surface integral over the closed surface s.

(3) Sometimes the notation $\oint_s$ may also be used for surface integrals.

(4) Surface integrals can be conveniently evaluated by expressing them as double integrals over the projected area of s on one of the coordinate planes (see 14.7.3)

14.7.2 Definition of Surface Integral as the Limit of a Sum

The area S is divided into 'L' elements of area ΔS_m, m = 1, 2, ... L,

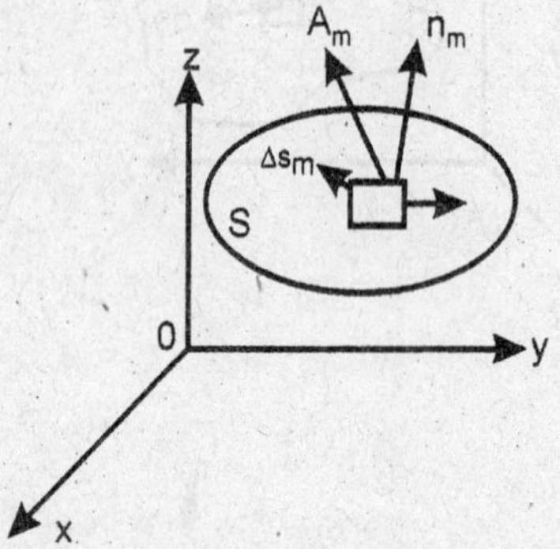

let $P_m = (x_m, y_m, z_m)$ be any point in ΔS_m. Let $A(x_m, y_m, z_m) = A_m$

Let $\boldsymbol{n}_m$ be the positive unit normal to ΔS_m at P_m. Then $(A_m.\boldsymbol{n}_m)$ is the normal component of A_m at P_m. Consider the sum,

$$\sum_{m=1}^{L} A_m.\boldsymbol{n}_m \Delta S_m \qquad \text{..... (1)}$$

The limit of the sum (1) as L $\to \infty$ such that the largest dimension of each $\Delta S_m \to 0$ (if the limit exists) is known as the surface integral of the normal component of A over S and denoted by $\iint_S A.n.ds$

14.7.3 Evaluation of a Surface Integral

To evaluate surface integrals, it is convenient to express them as double integrals taken over the projected area of the surface S on one of the coordinate planes (*xy,yz*, or *zx* planes).

If R is the projection of S on the *xy* – plane, it can be shown that

$$\iint_S A.\boldsymbol{n}ds = \iint_R A.\boldsymbol{n}\frac{dxdy}{|\vec{n}.k|}$$

From (14.7.2), the surface integral is the limit of the sum

$$\sum_{m=1}^{L} A_m.\boldsymbol{n}_m \Delta S_m \qquad \text{..... (1)}$$

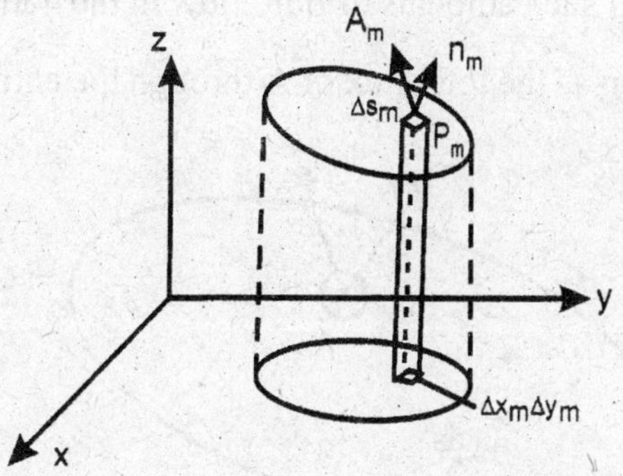

The projection of ΔS_m on the xy – plane is $|\boldsymbol{n}_m \Delta S_m .k|$ (or) $|\boldsymbol{n}_m .k|\Delta S_m$ which is equal to $\Delta x_m \Delta y_m$

$$\therefore \Delta S_m = \frac{\Delta x_m \Delta y_m}{|\vec{n}_m .k|}$$

∴ The sum (1) becomes

$$\sum_{m=1}^{L} \boldsymbol{A}_m .\boldsymbol{n}_m \frac{\Delta x_m \Delta y_m}{|\vec{n}_m .k|}$$

By the fundamental theorem of integral calculus the limit of this sum as $L \to \infty$ in such a manner that the largest Δx_m and Δy_m approach zero is

$$\iint_R A.\boldsymbol{n} \frac{dxdy}{|\vec{n}.k|}$$ which is the required result.

Note: Similarly if R is the projection of S on yz and zx planes respectively, it can be seen as

$$\iint_S A.\boldsymbol{n}ds = \iint_R A.\boldsymbol{n} \frac{dydz}{|\vec{n}.i|}$$

and

$$\iint_S A.\boldsymbol{n}ds = \iint_R A.\boldsymbol{n} \frac{dzdx}{|\vec{n}.j|}$$

14.7.4 Physical Interpretation of Surface Integrals

Let **A** denote the velocity of a moving fluid. Let S be a fixed surface in the fluid. Let ds be an element of surface. Then **A.n** ds = **A.dS** represents the amount of fluid that passes normally through dS in unit time at any point. If the direction of $\vec{n}$ is outward or positive, the amount of fluid flow is positive. Similarly if dS' is another element for which $\vec{n}$ is in the negative direction, the fluid flow is negative at that point.

The algebric sum of all such amounts of fluid flow in the normal directions is equal to $\int_S A.nds$ and it is known as the total flux of **A** through the entire surface S.

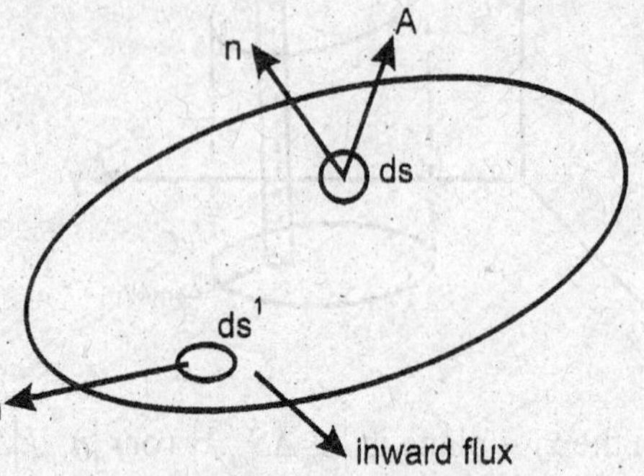

A can be a vector denoting physical quantities such as electric force, magnetic force, flux of heat or gravitational force etc. In all these cases, $\int_S A.nds$ denotes total flux of **A** through S.

Sloved Examples

Example **14.48:** Evaluate $\iint_s A.nds$ where $A = (x + y^2)i - 2xj + 2yzk$ and S is the surface of the plane $2x + y + 2z = 6$ in the first octant.

Solution:

$$\mathbf{A} = (x + y^2)i - 2xj + 2yzk$$

Let $\phi = 2x + y + 2z - 6$

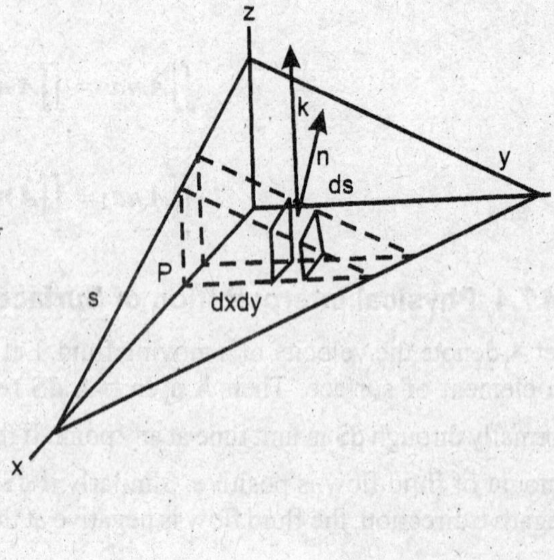

$$\nabla\phi = \frac{\partial\phi i}{\partial x} + \frac{\partial\phi j}{\partial y} + \frac{\partial\phi k}{\partial z} = 2i + j + 2k$$

$$\text{Unit normal n to S} = \frac{\nabla\phi}{|\nabla\phi|} = \frac{2i + j + 2k}{\sqrt{2^2 + 1^2 + 2^2}}$$

$$= \frac{2i + j + 2k}{3}$$

$$\mathbf{A.n} = \frac{1}{3}\left[2\left(x+y^2\right)-2x+4yz\right]$$

$$= \frac{1}{3}\left[2x+2y^2-2x+2y(6-2x-y)\right]$$

[Substituting $2z = 6 - 2x - y$ $\quad (\because 2x + y + 2z = 6)$]

$$= \frac{1}{3}(12y-4xy)$$

If R be the projection of S on the xy – plane.

$$|n.k| = \frac{2}{3} \Rightarrow ds = \frac{dxdy}{|n.k|} = \frac{3}{2}dxdy$$

$$\therefore \quad \iint_R A.nds = \iint_R (A.n)\frac{dxdy}{|n.k|}$$

$$= \iint_R \frac{1}{3}(12y-4xy)\frac{3}{2}dxdy$$

$$= \iint_R (6y-2xy)dxdy \qquad \text{..... (1)}$$

To evaluate this double integral over R,

(i) Keep x fixed and integrate w.r.t. y from $y = 0$ to $(6 - 2x)$, (ii) and then integrate w.r.t. x from $x = 0$ to $x = 3$.

∴ Given integral

$$\int_{x=0}^{3}\int_{y=0}^{6-2x}(6y-2xy)dydx$$

$$= \int_{x=0}^{3}\left[3y^2-xy^2\right]_{y=0}^{6-2x}dx = \int_{x=0}^{3}(3-x)(6-2x)^2dx = \int_{x=0}^{3}\left(108-108x+36x^2-4x^3\right)dx$$

$$= \left[108x-54x^2+12x^3-x^4\right]_0^3 = 324-486+324-81 = 81$$

***Example* 14.49:** If F = $4xzi - y^2j + yzk$, evaluate $\iint_s F.nds$ where S is the surface of the cube bounded by $x = 0, x = 1, y = 0, y = 1, z = 0$ and $z = 1$.

***Solution*:** The surface S can be divided into 6 faces (see figure).

(i) S_1: Face EPFA

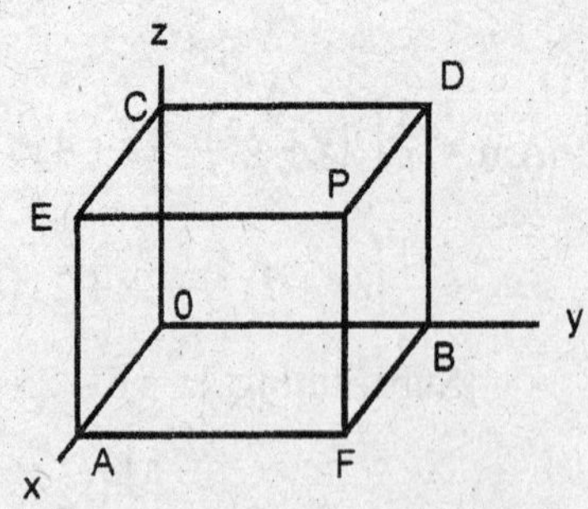

(ii) S_2: Face OBDC

(iii) S_3: Face PFBD

(iv) S_4: Face OCEA

(v) S_5: Face PDCE

(vi) S_6: Face OBFA

$$\iint_S \boldsymbol{F}.nds = \iint_{S_1} \boldsymbol{F}.nds + \iint_{S_2} \boldsymbol{F}.nds + \iint_{S_3} \boldsymbol{F}.nds + \iint_{S_4} \boldsymbol{F}.nds + \iint_{S_5} \boldsymbol{F}.nds + \iint_{S_6} \boldsymbol{F}.nds$$

On S_1: $\mathbf{n} = i,\ x = 1$

$$\iint_{S_1} \boldsymbol{F}.nds = \int_0^1 \int_0^1 \left(4zi - y^2 j + yzk\right).idydz = \int_0^1 \int_0^1 4zdydz = \int_0^1 2z^2\Big|_0^1 dy = \int_0^1 2dy = 2$$

On S_2: $\mathbf{n} = -i,\ x = 0$

$$\iint_{S_2} \boldsymbol{F}.n.ds = \int_0^1 \int_0^1 \left(-y^2 j + yzk\right)(-i)dydz = \int_0^1 \int_0^1 (0)dydz = 0$$

On S_3: $\mathbf{n} = j,\ y = 1$

$$\iint_{S_3} \boldsymbol{F}.n.ds = \int_0^1 \int_0^1 (4xzi - j + zk).jdxdz = \int_0^1 \int_0^1 (-1)dxdz = \int_0^1 -z\Big|_0^1 dx = \int_0^1 -1dx = -1$$

On S_4: $\mathbf{n} = -j,\ y = 0$

$$\therefore \iint_{S_4} \boldsymbol{F}.n.ds = \int_0^1 \int_0^1 (4xzi).(-j)dxdz = \int_0^1 \int_0^1 (0)dydz = 0$$

On S_5: $\mathbf{n} = k,\ z = 1$

$$\therefore \iint_{S_5} \boldsymbol{F}.n.ds = \int_0^1 \int_0^1 \left(4xi - y^2 j + yk\right).kdxdy = \int_0^1 \int_0^1 ydxdy = \int_0^1 \frac{y^2}{2}\Big|_0^1 dx = \int_0^1 \frac{1}{2}dx = \frac{1}{2}$$

On S_6: $\mathbf{n} = -\mathrm{k},\ \mathrm{z} = 0$

$$\iint_{S_6} \boldsymbol{F}.n.ds = \int_0^1 \int_0^1 \left(-y^2 j\right).(-k)dxdy = 0$$

$$\therefore \iint_S \boldsymbol{F}.n.ds = 2 + 0 - 1 + 0 + \frac{1}{2} + 0 = \frac{3}{2}$$

***Example* 14.50:** If A = $z^2i + x^2j - y^2zk$, and S is the surface of the cylinder $x^2 + y^2 = 16$ included in the first octant between $z = 0$ and $z = -5$, evaluate $\iint_S A.n.ds$

***Solution*:**

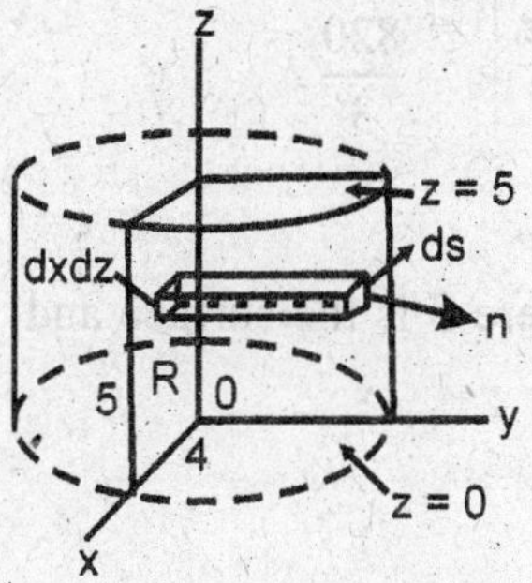

Project s on *xz* plane and let the projection be R. (See figure)

$$\iint_S A.n.ds = \iint_R A.n \frac{dxdy}{|n.j|} \qquad ..... (1)$$

The normal to $x^2 + y^2 = 16$ is

$$\nabla(x^2 + y^2) = \frac{\partial}{\partial x}(x^2 + y^2)i + \frac{\partial}{\partial y}(x^2 + y^2)j$$

$$= 2xi + 2yj$$

Unit normal $\vec{n} = \dfrac{2xi + 2yj}{\sqrt{(2x)^2 + (2y)^2}} = \dfrac{2(xi + yj)}{2\sqrt{x^2 + y^2}} = \dfrac{xi + yj}{4} \qquad (\because x^2 + y^2 = 16)$

$$\mathbf{A.n} = \frac{z^2x + x^2y}{4}$$

$$\mathbf{n.j} = \frac{y}{4}$$

$\therefore$ From (1) $\displaystyle\iint_S A.nds = \iint_R \frac{z^2x + x^2y}{4} \frac{dxdz}{y/4}$

$$= \iint_R \frac{(z^2x + x^2y)}{y} dxdz = \int_{z=0}^{5}\left[\int_{x=0}^{4}\left(\frac{xz^2}{\sqrt{16 - x^2}} + x^2\right)dx\right]dz$$

$$= \int_{z=0}^{5}\left(4z^2 \frac{64}{3}\right)dz \quad + \quad \left[\because \int_0^4 \frac{x}{\sqrt{16-x^2}}dx = 4, \quad \int_0^4 x^2 dx = \frac{64}{3}\right]$$

$$= \left[\frac{4z^3}{3} + \frac{64z}{3}\right]_0^5 = \frac{820}{3}$$

Example 14.51: Evaluate $\iint_S \phi n ds$ where S is the surface and $f = \dfrac{xyz}{8}$

Solution:

We have $\iint_S \phi n ds = \iint_R \phi \vec{n} \dfrac{dxdz}{|\vec{n}.j|}$

using $\mathbf{n} = \dfrac{xi + yj}{4}$, $n.j = \dfrac{y}{4}$, the integral on R.H.S. becomes,

$$\iint_R \frac{1}{8} xyz \frac{(xi + yj)}{4} . \frac{dxdz}{y/4}$$

$$\frac{1}{8}\int_{z=0}^{5}\int_{x=0}^{4} xz(xi + yj)dxdz = \frac{1}{8}\int_{z=0}^{5}\left[\int_0^4 \left(x^2 zi + xz\sqrt{16-x^2}\right) jdx\right]dz$$

$$\frac{1}{8}\int_0^5 \left(\frac{64z}{3}i + \frac{64z}{3}j\right)dz \qquad \left[\because \int_0^4 x\sqrt{16-x^2}\,dx = \frac{64}{3} \int_0^4 x^2 dx = \frac{64}{3}\right]$$

$$\frac{1}{8}.\frac{64}{3}\left(\frac{z^2}{2}i + \frac{z^2}{2}j\right)\Big|_0^5 = \frac{8}{3}.\frac{25}{2}(i + j) = \frac{100}{3}(i + j)$$

***Example* 14.52:** If A = $yi + (x - 2xz)j - xyk$, evaluate $\iint_S (curl A).nds$ where S is the surface of the sphere $x^2 + y^2 + z^2 = 4$ above the xy plane

Solution:

$$\text{Curl } \vec{A} = \begin{vmatrix} i & j & k \\ \partial/\partial x & \partial/\partial y & \partial/\partial z \\ y & x - 2xz & -xy \end{vmatrix}$$

$$= i\{\frac{\partial}{\partial y}(-xy) - \frac{\partial}{\partial z}(x-2xz)\} + j\{\frac{\partial}{\partial z}(y) - \frac{\partial}{\partial x}(-xy)\} + k\{\frac{\partial}{\partial x}(x-2xz) - \frac{\partial}{\partial y}(y)\}$$

$$= xi + yj - 2zk$$

The normal to the surface is $\nabla(x^2 + y^2 + z^2) = 2xi + 2yj + 2zk$

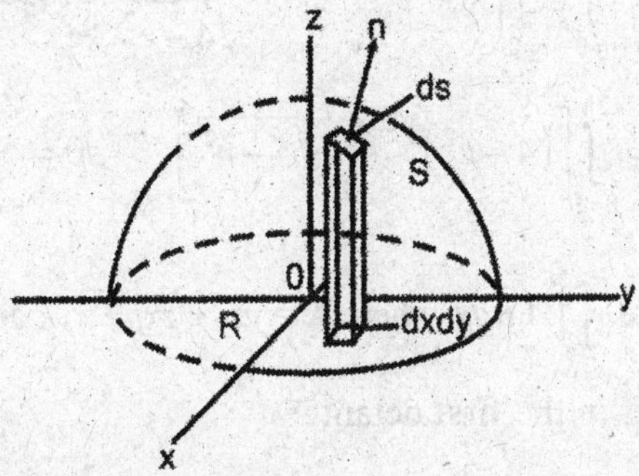

Unit normal $\quad \mathbf{n} = \dfrac{2xi + 2yj + 2zk}{\sqrt{(2x)^2 + (2y)^2 + (2z)^2}}$

$$= \frac{xi + yj + zk}{2} \qquad (\because x^2 + y^2 + z^2 = 4)$$

(Curl **A**). $\quad \mathbf{n} = \dfrac{x^2 + y^2 - 2z^2}{2}$

The projection of S on xy plane is the circle $x^2 + y^2 = 4$, $z = 0$ (see figure)

$$\therefore \iint_S (curl\ \mathbf{A.n})ds = \iint_R (curl\ \mathbf{A.n})\frac{dxdy}{|\mathbf{n.k}|}$$

$$= \iint_R \frac{(x^2 + y^2 - 2z^2)}{2}\frac{dxdy}{z/2}$$

$$= \int_{x=-2}^{2}\int_{y=-\sqrt{4-x^2}}^{4-x^2} \frac{x^2 + y^2 - 2(4 - x^2 - y^2)}{\sqrt{4 - x^2 - y^2}}dxdy \quad (\because z^2 = 4 - x^2 - y^2)$$

$$= \int_{x=-2}^{2}\int_{y=-\sqrt{4-x^2}}^{4x^2} \frac{3(x^2 + y^2) - 8}{\sqrt{4 - x^2 y^2}}dxdy$$

Changing into polar coordinates by taking $x = r\cos\theta$, $y = r\sin\theta$, $dx\,dy = r\,dr\,d\theta$, the integral becomes

$$\int_{\theta=0}^{2\pi}\int_{r=0}^{2}\left(\frac{3\left(r^2-4\right)+4}{\sqrt{4-r^2}}\right)r\,drd\theta$$

$$=\int_{\theta=0}^{2\pi}[\int_{r=0}^{2}\left(-3r\sqrt{4-r^2}+\frac{4r}{\sqrt{4-r^2}}\right)dr]d\theta$$

$$=\int_{\theta=0}^{2\pi}\left[\left(4-r^2\right)^{3/2}-4\sqrt{4-r^2}\right]_{r=0}^{2}d\theta=\int_{\theta=0}^{2}(8-8)d\theta=0$$

***Example* 14.53:** Evaluate $\iint_S A.nds$ where $A = yzi + zxj + xyk$ and S is the part of the sphere $x^2 + y^2 + z^2 = 9$ which lies in the first octant.

***Solution*:**

$$\text{Unit normal } \mathbf{n} \text{ to } S=\frac{\nabla\left(x^2+y^2+z^2\right)}{\left|\nabla\left(x^2+y^2+z^2\right)\right|}$$

$$=\frac{2xi+2yj+2zk}{\sqrt{4x^2+4y^2+4z^2}}=\frac{xi+yj+zk}{3} \qquad (\because x^2+y^2+z^2=9)$$

$$\mathbf{A.n} = 3xyz; \qquad \mathbf{n}.k=\frac{z}{3}$$

If R is the projection of S on xy – plane, we have,

$$\iint_S A.nds=\iint_R A.n\frac{dxdy}{|n.k|}$$

$$=\iint_R 3xyz\frac{dxdy}{z/3}=9\iint_R xy\,dxdy$$

The region R is bounded by x–axis, y–axis and the circle $x^2 + y^2 = 9$; $z = 0$ changing to polar coordinates, the last integral becomes

$$9\int_{\theta=0}^{\pi/2}\int_{r=0}^{3}(r\cos\theta)(r\sin\theta)rdrd\theta=9\int_0^{\pi/2}[\int_0^3 r^3\cos\theta\sin\theta dr]d\theta$$

$$=9\int_0^{\pi/2}\left[\left(\frac{r^4}{4}\right)\right]_0^3\cos\theta\sin\theta\,d\theta=9.\frac{81}{4}\int_0^{\pi/2}\cos\theta\sin\theta d\theta$$

$$=\frac{729}{4}.\frac{1}{2}=\frac{729}{8} \qquad \left(\because\int_0^{\pi/2}\cos\theta\sin\theta d\theta=\frac{1}{2}\right)$$

Exercise 14.3

1. If $\mathbf{F} = 18zi - 12xj + 3yk$, evaluate $\iint_S F.nds$ where S is that part of the plane $2x + 3y + 6z = 12$ which is located in the first octant

   (*Hint*: Take projection of S on the xy plane and $ds = \frac{dxdy}{|n.k|}$)

   [Ans: 24]

2. Evaluate $\iint_S V.nds$, where $\mathbf{V} = yi + 2xj - zk$ and s is the surface of the plane $2x + y = 6$ in the first octant cut off by the plane $z = 4$

   [Ans: 108]

3. If $\mathbf{r} = xi + yj + zk$ find the value of the integral $\iint_S r.nds$ over

   a) the surface S of the unit cube bounded by the coordinate planes and the planes $x = y = z = 1$ and
   b) the surface of the sphere of radius 'a' with centre at the origin

   [Ans: a) 3, b) $4\pi a^3$]

4. Evaluate $\iint_S F.nds$ over the entire surface of the region above the xy plane bounded by the cone $z^2 = x^2 + y^2$ and the plane $z = 4$, if $\vec{F} = 4xzi + xy^2zj + 3zk$

   [Ans: 320π]

5. If S is the surface of the parabolic cylinder $y^2 = 8x$ in the first octant bounded by the planes $y = 4$ and $z = 6$, evaluate $\iint_S F.nds$, where $\mathbf{F} = 2yi - zj + x^2k$

   [*Hint*: $\iint_S F.nds = \iint_R F.n\frac{dydz}{|n.i|}$, where R is the projection of S on the yz plane]

   [Ans: 132]

6. Evaluate $\iint_S V.nds$ over the entire surface S of region bounded by the cylinder $x^2 + z^2 = 9$, $x = 0$, $y = 0$, $z = 0$ and $y = 8$ if $\mathbf{V} = 6zi + (2x + y)j - xk$

   [Ans: 18π]

7. Evaluate $\iint_S A.nds$ where $\mathbf{A} = zi + xj - 3y^2zk$ and S is the surface of the cylinder $x^2 + y^2 = 16$ included in the first octant between $z = 0$ and $z = 5$

   [Ans: 90]

8. If $\mathbf{V} = yzi + zxj + xyk$ and S is that part of the sphere $x^2 + y^2 + z^2 = 1$ which lies in the first octant, find the value of $\iint_S V.nds$.

   [Ans: $\frac{3}{8}$]

9. Evaluate $\iint_S \{(x^3 - yz^2)i - (2x^2y)j + 2k\}.ds$ over a cube with edges of length 'r' parallel to the coordinate axes

[**Ans:** $\frac{1}{2}$]

10. If $\mathbf{V} = xi + yj + zk$, and S is the triangle with vertices at (1,0,0), (0,1,0) and (0,0,1), find the value of $\iint_S V.ds$. [**Ans:** $\frac{1}{2}$]

11. If $A = xi - yj + (z^2 - 1)k$, find the value of $\iint_S A.nds$, where S is the closed surface bounded by the planes z = 0, z = 1 and the cylinder $x^2 + y^2 = 1$ [**Ans:** π]

14.8 Volume Integrals

Consider a closed surface in space enclosing a volume V. Then, integrals of the form $\iiint_V A dv$ and $\iiint_V \phi dv$, [**A** is a vector function, f is a scalar function] are examples of volume integrals.

14.8.1 Expression of Volume Integral as the Limit of a Sum

Let **A** be a continuous vector function. Let S be a surface enclosing the region D. Divide this region D into a finite number of subregions D_i ..., = 1, 2, ...n.

Let Δv_i be the volume of the subregion D_i enclosing any point whose position vector is $\mathbf{r}_i$. Consider the sum

$$V = \sum_{i=1}^{n} A(r_i)\Delta v_i$$

The limit of this sum as n $\to \infty$ such that $\Delta v_i \to 0$, is called the volume integral of A over D and denoted by $\iiint_D A dv$

If $\mathbf{A} = A_1(x, y, z)i + A_2(x, y, z)j + A_3(x, y, z)k$,

so that dv = dx dy dz

$$\iiint_D A dv = i\iiint_D A_1(x,y,z)dx\,dy\,dz + j\iiint_D A_2(x,y,z)dx\,dy\,dz +$$

$$k\iiint_D A_3(x,y,z)dx\,dy\,dz$$

Solved Examples

***Example* 14.54:** If F = $(2x^2 - 3z)i - 2xyj - 4xk$, evaluate $\iiint_V \nabla.F dv$ where V is the closed region bounded by the planes $x = 0, y = 0, z = 0$ and $2x + 2y + z = 4$.

***Solution*:**

$$\nabla.F = \frac{\partial}{\partial x}(2x^2 - 3z) - \frac{\partial}{\partial y}(2xy) - \frac{\partial}{\partial z}(4x) = 4x - 2x = 2x$$

$$\therefore \iiint_V \nabla.F dv = \int_{x=0}^{2}\int_{y=0}^{2-x}\int_{z=0}^{4-2x-2y} 2xdzdydx = \int_{x=0}^{2}\int_{y=0}^{2-x}\int_{z=0}^{4-2x-2y} 2xdz\,]dydx$$

$$= \int_{x=0}^{2}\int_{y=0}^{2-x} 2xz\Big|_{z=0}^{4-2x-2y} dydx = \int_{x=0}^{2}\int_{y=0}^{2-x} 2x(4-2x-2y)dydx = \int_{x=0}^{2}\int_{y=0}^{2-x}(8x - 4x^2 - 4xy)dy]dx$$

$$= \int_{x=0}^{2} 8xy - 4x^2y - 2xy^2\Big|_{y=0}^{2-x} dx \quad = \int_{x=0}^{2}[8x(2-x) - 4x^2(2-x) - 2x(2-x)^2]dx$$

$$= \int_{x=0}^{2}(8x - 8x^2 + 2x^3)dx = (4x^2 - \frac{8x^3}{3} + \frac{2x^4}{4})\Big|_0^2 \quad = 16 - \frac{64}{3} + 8 = \frac{8}{3}$$

***Example* 14.55:** Evaluate $\iiint_V (\nabla.A)dv$ over the region bounded by $x^2 + y^2 = 4$, $z = 0$ and $z = 3$, where A = $4xi - 2y^2j + z^2k$.

***Solution*:**

$$\nabla.A = \frac{\partial}{\partial x}(4x) - \frac{\partial}{\partial y}(2y^2) + \frac{\partial}{\partial z}(z^2) = 4 - 4y + 2z$$

$$\therefore \iiint_V (\nabla.A)dv = \iiint_V (4 - 4y + 2z)dv = \int_{x=-2}^{x=2}\int_{y=-\sqrt{4-x^2}}^{y=\sqrt{4-x^2}} [\int_{z=0}^{3}(4 - 4y + 2z)dz]dydx$$

$$= \int_{x=-2}^{2}\int_{y=-\sqrt{4-x^2}}^{y=\sqrt{4-x^2}} (4z - 4yz + z^2)\Big|_0^3 \, dydx = \int_{x=-2}^{2}[\int_{y=-\sqrt{4-x^2}}^{y=\sqrt{4-x^2}} (21 - 12y)dy]dx$$

$$= \int_{x=-2}^{2} (21y - 6y^2)\Big|_{y=-\sqrt{4-x^2}}^{y=\sqrt{4-x^2}} dx = \int_{x=-2}^{2} 42\sqrt{4-x^2}\,dx = 84\int_{0}^{2}\sqrt{4-x^2}\,dx$$

$$= 84[\frac{x}{2}\sqrt{4-x^2} + 2\sin^{-1}(\frac{x}{2})]_0^2 = 84[0 + 2(\frac{\pi}{2}) - 0] = 84\pi$$

***Example* 14.56:** Evaluate $\iiint_V \phi dv$ taken over the rectangular parallelopiped $0 \le x < a$, $0 \le y < b$, $0 \le z < c$ and $\phi = 2(x + y + z)$

Solution:

$$\iiint_V \phi dv = \iiint_V 2(x+y+z)dv = \int_{x=0}^{a}\int_{y=0}^{b}\left[\int_{z=0}^{c} 2(x+y+z)\right]dydx$$

$$= \int_{x=0}^{a}\int_{y=0}^{b} 2xz + 2yz + z^2\Big|_{z=0}^{c} dydx = \int_{x=0}^{a}[\int_{y=0}^{b}(2cx + 2cy + c^2)dy]dx$$

$$= \int_{x=0}^{a} 2cxy + cy^2 + c^2y\Big|_0^b dx = \int_{x=0}^{a}(2bcx + cb^2 + c^2b)dx = bcx^2 + (b^2c + bc^2)x\Big|_0^a$$

$$= a^2bc + a(b^2c + bc^2) = abc(a + b + c)$$

***Example* 14.57:** If $\phi = 4y + 2xz$, evaluate $\iiint_V \phi dv$ over the region in the first octant bounded by $x^2 + y^2 = 9$, $z = 0$, $z = 2$.

Solution:

$$\iiint_V \phi dv = \iiint_V (4y + 2xz)dv$$

$$= \int_{x=0}^{x=3}\int_{y=0}^{y=\sqrt{9-x^2}}[\int_{z=0}^{z=2}(2xz + 4y)dz]dydx = \int_0^3\int_0^{\sqrt{9-x^2}}(4yz + xz^2)\Big|_0^2 dydx$$

$$= \int_0^3 \int_0^{\sqrt{9-x^2}} [8y + 4x)dy]dx = \int_0^3 (4y^2 + 4xy)\Big|_0^{\sqrt{9-x^2}} dx$$

$$= \int_0^3 [4(9-x^2) + 4x\sqrt{9-x^2}]dx = 108$$

***Example* 14.58:** Evaluate $\iiint_V F dv$ where $\vec{F} = xzi - 2xj + 2y^2k$ and V is the region bounded by the surfaces $x = 0$, $y = 0$, $y = 6$, $z = x^2$, and $z = 4$

***Solution*:**

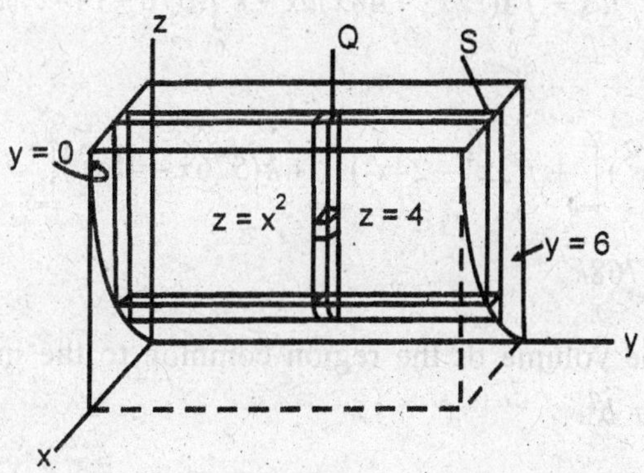

The region V is covered by (see the figure) (a) keeping x and y fixed and integrating from $z = x^2$ to $z = 4$ (base top of column PQ) (b) then by keeping x fixed and integrating from $y = 0$ to $y = 6$ (R to S in the slab) and (c) finally integrating from $x = 0$ to $x = 2$ (where $z = x^2$ meets $z = 4$).

∴ The required integral is

$$\int_{x=0}^{2} \int_{y=0}^{6} \int_{z=x^2}^{4} (xzi - 2xj + 2y^2k)dz\,]dydx$$

$$= i \int_{x=0}^{2} \int_{y=0}^{6} \int_{z=x^2}^{4} xzdz\, dydx + j \int_{x=0}^{2} \int_{y=0}^{6} \int_{z=x^2}^{4} -2xdz\, dydx + k \int_{x=0}^{2} \int_{y=0}^{6} \int_{z=x^2}^{4} 2y^2dz]\, dydx$$

$$= i\int_{x=0}^{2}\int_{y=0}^{6}\frac{xz^2}{2}\Big|_{z=x^2}^{4} dy\,dx + j\int_{x=0}^{2}\int_{y=0}^{6} -2xz\Big|_{z=x^2}^{4} dydx + k\int_{x=0}^{2}\int_{y=0}^{6} 2y^2z\Big|_{z=x^2}^{4} dydx$$

$$= i\int_{x=0}^{2}[\int_{y=0}^{6}(8x-\frac{x^5}{2})dy]dx + j\int_{x=0}^{2}[\int_{y=0}^{6}(2x^3-8x)dy]dx + k\int_{x=0}^{2}[\int_{y=0}^{6}(8y^2-2x^2y^2)dy]dx$$

$$= i\int_{x=0}^{2} 8xy-\frac{x^5y}{2}\Big|_{y=0}^{6} dx + j\int 2x^3y-8xy\Big|_{y=0}^{6} dx + k\int_{x=0}^{2}\frac{8y^3}{3}-\frac{2x^2y^3}{3}\Big|_{y=0}^{6} dx$$

$$= i\int_0^2(48x-3x^5)dx + j\int_0^2(12x^3-48x)dx + k\int_0^2(576-144x^2)dx$$

$$= i(24x^2-\frac{1}{2}x^6)\Big|_{x=0}^{2} + j(3x^4-24x^2)\Big|_{x=0}^{2} + k(576x-48x^3)\Big|_{x=0}^{2}$$

$$= 64i - 48j + 768k$$

***Example* 14.59:** Find the volume of the region common to the intersecting cylinders $x^2 + y^2 = b^2$ and $x^2 + z^2 = b^2$.

***Solution*:**

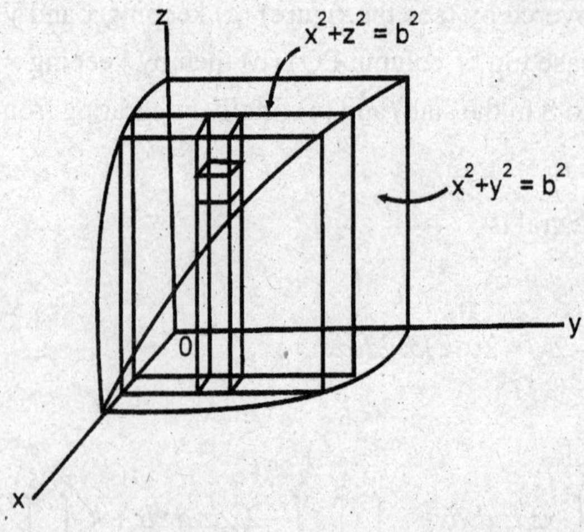

Required volume is equal to 8 times the volume of the region shown in the figure (as the axes cut the volume into 8 equal parts one in each octant)

$$\text{Volume} = 8\int_{x=0}^{b}\int_{y=0}^{\sqrt{b^2-x^2}}\left(\int_{z=0}^{\sqrt{b^2-x^2}} dz\right)dydx = 8\int_{x=0}^{b}\left(\int_{y=0}^{\sqrt{b^2-x^2}}\sqrt{b^2-x^2}\,dy\right)dx$$

$$= 8\int_{x=0}^{b}\left(b^2-x^2\right)dx \quad = 8\left[\left(b^2x-\frac{x^3}{3}\right)_0^3\right] = 16\frac{b^2}{3}.$$

Exercise 14.8

1. Evaluate $\iiint_v (2x+y)\,dv$ where **V** is the closed region bounded by the cylinder $z=4-x^2$ and the planes $x=0, y=2$ and $z=0$. [**Ans:** $\frac{80}{3}$]

2. If $\mathbf{A}=\left(2x^2-3z\right)i-2xyj-4xk$, and V is the closed region bounded by the planes x=0,y=0,z=0and $2x+2y+z=4$,findthevalueof $\iiint_v (\text{curl}\mathbf{A})\,dv$

   [**Ans:** $\frac{8}{3}(j-k)$]

3. Evaluate $\iiint_v f\,dv$ where $f=45x^2y$ and v is the closed region bounded by the planes $4x+2y+z=8, x=0, y=0$ and $z=0$ [**Ans:** 128]

4. If $\mathbf{A}=2xzi-xj+y^2k$ and v is the region bounded by the surfaces $x=0, y=0, y=6,$ $z=x^2$ and z = 4, find the value of $\iiint_v \mathbf{A}dv$ **Ans:** $128i-24j+384k$]

5. Evaluate $\iiint_v (\text{Div}\,\mathbf{A})dv$ taken over the rectangular parallelepiped, $0\le x\le 1,\ 0\le y\le 2,$ $0\le z\le 3$, where $\mathbf{A}=\left(x^2-yz\right)i+\left(y^2-zx\right)j+\left(z^2-xy\right)k$ [**Ans:**36]

6. Evaluate $\iiint_v (\text{Div}\,\mathbf{F})dv$ for the volume of a cube with edges of length unity parallel to the coordinate axes where $\text{F}=\left(x^3-yz^2\right)i-\left(2x^2y\right)j+2k$

[**Ans:** 1/3]

14.9 Green's Theorem in the Plane

14.9.1 Green's Theorem

Let

(i) R be a closed region of the xy plane bounded by a simple closed curve C

(ii) P(x,y) and Q(x,y) be continuous functions of x and y having continous derivatives in R. Then $\int_c Pdx + Qdy = \iint_R \left(\frac{\partial Q}{\partial x} - \frac{\partial P}{\partial y}\right)dxdy$

where C is in the positive direction.

***Proof*:**

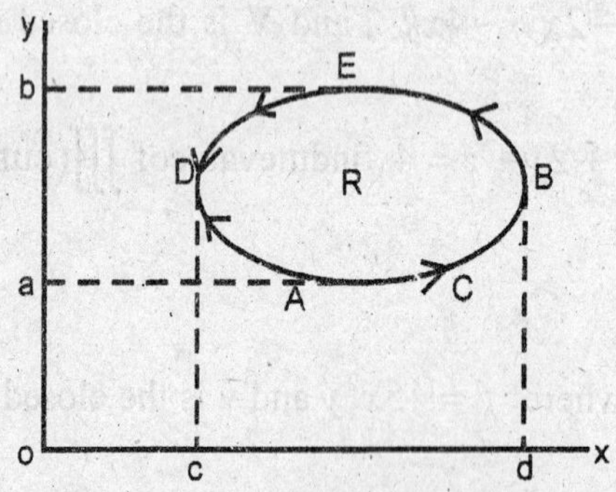

Let the equations of the curves DAB and DEB (see the figure) be respectively $y = f_1(x)$ and $y = f_2(x)$. Let R be the region bounded by the curve C. Then,

$$\iint_R \frac{\partial P}{\partial y} dxdy = \int_{x=c}^{d} \left[\int_{y=f_1(x)}^{f_2(x)} \frac{\partial P}{\partial y} dy \right] dx$$

$$= \int_{x=c}^{d} P(x,y)\Big|_{y=f_1(x)}^{y=f_2(x)} dx = \int_{x=c}^{d} [P(x,f_2) - P(x,f_1)]\,dx = -\int_c^d P(x,f_1)\,dx - \int_d^c P(x,f_2)\,dx$$

$$= -\oint_c P(x,y)dx$$

i.e., $$\oint_c P(x,y)dx = -\iint_R \frac{\partial P}{\partial y} dxdy \qquad \text{..... (1)}$$

Similarly if the equations of the curves ADE be and ABE be taken respectively as $x = g_1(y)$ and $x = g_2(y)$, we have,

$$\iint_R \frac{\partial Q}{\partial x} dxdy = \int_{y=a}^{b} \left[\int_{x=g_1(y)}^{g_2(y)} \frac{\partial Q}{\partial x} dx \right] dy$$

$$= \int_a^b [Q(g_2, y) - Q(g_1, y)] dy = \int_b^a Q(g_1, y) dy + \int_a^b Q(g_2, y) dy$$

$$= \oint_c Q(x, y) dy$$

Thus $\oint_c Q(x, y) dy = \iint_R \frac{\partial Q}{\partial x} dxdy$ ..... (2)

Adding (1) and (2),

$$\oint_c P dx + Q dy = \iint_R \left(\frac{\partial Q}{\partial x} - \frac{\partial P}{\partial y} \right) dxdy$$

14.9.2 Vector Notation of Green's Theorem

Green's Theorem in the plane can be put in vector notation in the following way.

Let $\mathbf{F} = P(x, y)i + Q(x, y)j$

and $\mathbf{r} = xi + yj$, so that

$d\mathbf{r} = (dx)i + (dy)j$

$\therefore$ $\mathbf{F}.d\mathbf{r} = P\,dx + Q\,dy$

Again, $$\nabla \times \mathbf{F} = \begin{vmatrix} i & j & k \\ \partial/\partial x & \partial/\partial y & \partial/\partial z \\ P & Q & 0 \end{vmatrix}$$

$$= \left(\frac{\partial Q}{\partial x} - \frac{\partial P}{\partial y} \right) k, \text{ so that}$$

$$(\nabla \times \vec{F}) k = \frac{\partial Q}{\partial x} - \frac{\partial P}{\partial y}$$

Taking dR = dxdy, Green's theorem in the plane can be stated in the vector form as,

$$\oint_c \mathbf{F}.d\mathbf{r} = \iint_R (\nabla \times \mathbf{F}) k dR$$ ..... (A)

14.9.3 Physical interpretation of Green's Theorem

1. Let **F** denote the force field acting on a particle.

   Then $\oint_C F.dr$ represents the work done in moving the particle around a closed curve C.

   ∴ From (A) it follows that the work done is determined by curl $\vec{F} = \nabla \times \vec{F}$.

2. In particular, if $\nabla \times \mathbf{F} = 0$ i.e., if **F** is conservative (or $\mathbf{F} = \nabla f$)

   Then $\oint_C F.dr = 0$. i.e., the work done is independent of the path.

3. Conversely, if the integral is independent of the path, i.e., if

   $$\oint_C F.dr = 0, \text{ then } \nabla \times \mathbf{F} = 0$$

   In the plane, $\nabla \times \mathbf{F} = 0$ is equivalent to saying that $\frac{\partial P}{\partial y} = \frac{\partial Q}{\partial x}$ where $\mathbf{F} = Pi + Qj$.

14.9.4 Application of Green's theorem to the evaluation of Area of a simple Closed Curve

The area bounded by a simple closed curve C = $\frac{1}{2}\oint_C xdy - ydx$

Proof: By Green's theorem, we have,

$$\oint_C Pdx + Qdy = \iint_R \left(\frac{\partial Q}{\partial x} - \frac{\partial P}{\partial y}\right) dxdy$$

if we put $P = -y$, and $Q = x$,

i.e., $\frac{\partial P}{\partial y} = -1, \quad \frac{\partial Q}{\partial x} = 1, \quad \frac{\partial Q}{\partial x} - \frac{\partial P}{\partial y} = 2,$

∴ We get,

$$\oint_C xdy - ydx = \iint_R 2dxdy = 2A,$$ where A is the required area.

i.e., $A = \frac{1}{2}\oint_C xdy - ydx$.

Solved Examples

***Example* 14.60:** Find the area of the ellipse $\frac{x^2}{a^2}+\frac{y^2}{b^2}=1$.

***Solution*:**

Parametric equations of the ellipse are,

$$x = a\cos\theta, \qquad y = b\sin\theta$$

$$dx = -a\sin\theta\, d\theta, \quad dy = b\cos\theta\, d\theta.$$

$\therefore$ By Green's theorem,

$$\text{Area of the ellipse} = \frac{1}{2}\oint_c xdy - ydx$$

$$=\frac{1}{2}\int_0^{2\pi}(a\cos\theta\, b\cos\theta + b\sin\theta\, a\sin\theta)d\theta$$

$$=\frac{1}{2}\int_0^{2\pi} ab\,d\theta = \pi ab$$

***Example* 14.61:** Evaluate $\oint_c (y-\sin x)dx + (\cos x)dy$, a) directly and b) using Green's theorem, where c is the boundary of the triangle in *xy*–plane whose vertices are (0,0), $\left(\frac{\pi}{2},0\right)$ and $\left(\frac{\pi}{2},1\right)$ traversed in the positive direction.

***Solution*:**

Let $\quad I = \oint_c (y-\sin x)dx + (\cos x)dy$

(a) along OA: $\quad y = 0, \quad dy = 0$

$$\therefore \int_{OA}(y-\sin x)dx + (\cos x)dy = \int_0^{\pi/2} -\sin x dx = \cos x\Big|_0^{\pi/2} = -1$$

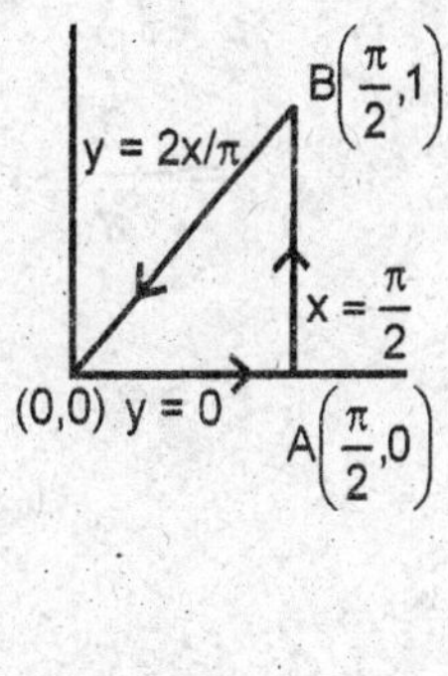

along AB: $x = \frac{\pi}{2}, \quad dx = 0$

$$\therefore \quad \int_{AB} (y - \sin x)dx + (\cos x)dy = \int 0dy = 0$$

along BO: Equation of OB is $y = \dfrac{2x}{\pi}$, $dy = \dfrac{2}{\pi}dx$,

$$\therefore \quad \int_{BO} (y - \sin x)dx + \cos x dy = \int_{\pi/2}^{0} \left[\left(\frac{2x}{\pi} - \sin x \right) + \frac{2}{\pi}\cos x \right] dx$$

$$= \frac{x^2}{\pi} + \cos x + \frac{2}{\pi}\sin x \Big|_{\pi/2}^{0} = (0 + 1 + 0) - \left(\frac{\pi}{4} + 0 + \frac{2}{\pi} \right) = 1 - \frac{\pi}{4} - \frac{2}{\pi}.$$

$\therefore$ I = Sum of the integrals along OA, AB and BO

$$= -1 + 0 + 1 - \frac{\pi}{4} - \frac{2}{\pi} = -\frac{\pi}{4} - \frac{2}{\pi}$$

(b) By Green's theorem,

$$\oint_c Pdx + Qdy = \iint_R \left(\frac{\partial Q}{dx} - \frac{\partial P}{dy} \right) dx\, dy, \quad \text{Here } P = y - \sin x, \; Q = \cos x$$

$$\frac{\partial Q}{\partial x} = -\sin x, \qquad \frac{\partial P}{\partial y} = 1$$

$$\therefore \; I = \iint_R (-\sin x - 1)\, dx\, dy = \int_{x=0}^{\pi/2} \left[\int_{y=0}^{2x/\pi} (-\sin x - 1)\, dy \right] dx = \int_{x=0}^{\pi/2} -y \sin x - y \int_0^{2x/\pi} dx$$

$$= \int_0^{\pi/2} \left(-\frac{2x}{\pi}\sin x - \frac{2x}{\pi} \right) dx = -\frac{2}{\pi} \int_{x=0}^{\pi/2} x \sin x\, dx - \frac{2}{\pi} \int_0^{\pi/2} x dx$$

$$= -\frac{2}{\pi} \left[-x \cos x + \sin x + \frac{x^2}{2} \right]_0^{\pi/2} - \frac{2}{\pi} \left[0 + 1 + \frac{\pi^2}{8} \right] = -\frac{2}{\pi} - \frac{\pi}{4}$$

$$\oint_c (3x + 4y)dx + (2x - 3y)dy$$

***Example* 14.62:** Evaluate $\oint_C (3x+4y)\,dx+(2x-3y)\,dy$ where C is the circle in *xy* plane with centre at origin and radius 2 units.

Solution:

By Green's theorem,

$$\oint P dx + Q dy = \iint_R \left(\frac{\partial Q}{\partial x} - \frac{\partial P}{\partial y}\right) dxdy$$

Here $\quad P = 3x+4y,\ Q = 2x-3y\ ;\quad \frac{\partial P}{\partial y} = 4,\ \frac{\partial Q}{\partial x} = 2$

$\therefore$ The given integral $= \iint_R (2-4)\,dxdy$

$= -2\iint_R dxdy = -2\text{A},$ where A is the area of the circle C.

$= -2 \text{ x } \pi \text{ x } 2^2 = -8\pi$

***Example* 14.63:** Verify Green's theorem in the plane for the integral $\oint_c (3x^2-8y^2)dx+(4y-6xy)dy$ where C is the boundary of the region given by

(1) $y = \sqrt{x},\ y = x^2$

(2) $x = 0, y = 0,\ \text{x}+\text{y} = 1$

Solution:

1. The given region is shown in the figure below.

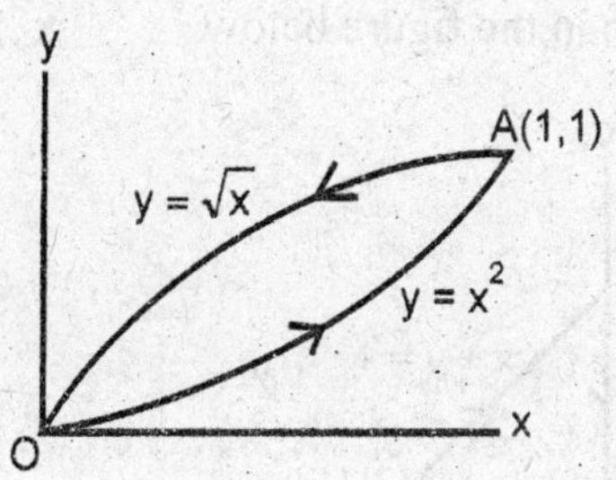

let $I = \oint (3x^2 - 8y^2)dx + (4y - 6xy)dy$

The points of intersection of $y = x^2$ and $y = \sqrt{x}$ are (0, 0) and (1, 1).

We have to integrate I along

(1) $y = x^2$ from 0 to A.

(2) along $y = \sqrt{x}$ from A to O, and add the two values.

Along $y = x^2$, $dy = 2xdx$

$$\therefore I = \int_{AO} (3x^2 - 8x^4)dx + (4x^2 - 6x^3)2xdx = \int_{x=0}^{x=1} (3x^2 - 20x^4 + 8x^3)dx = x^3 - 4x^5 + 2x^4 \Big|_0^1 = -1$$

Along $y = \sqrt{x}$; $x = y^2$, $dx = 2ydy$

$$\therefore I = \int_{AO} (3y^4 - 8y^2)dy + (4y - 6x^3)dy = \int_{y=1}^{y=0} (6y^5 - 22y^3 + 4y)(2y)dy = y^6 - \frac{11}{2}y^4 + 2y^2 \Big|_1^0 = \frac{5}{2}$$

$$\therefore I = -1 + \frac{5}{2} = \frac{3}{2}$$

By Green's Theorem,

$$\oint_c P\,dx + Q\,dy = \iint_R \left(\frac{\partial Q}{\partial x} \cdot \frac{\partial P}{\partial y}\right) dxdy$$

Here, $P = 3x^2 - 8y^2$; $Q = 4y - 6xy$

$$\frac{\partial Q}{\partial x} - \frac{\partial P}{\partial y} = -6y - (-16y) = 10y$$

$$\therefore I = \int\int_R 10ydxdy = \int_{x=0}^{1}\left[\int_{y=x^2}^{\sqrt{x}} 10ydy\right]dx = \int_{x=0}^{1} 5y^2 \Big|_{x^2}^{\sqrt{x}} dx = \int_{x=0}^{1} (5x - 5x^4)dx = \frac{3}{2}$$

Green's theorem is verified.

2. The given region is shown in the figure below.

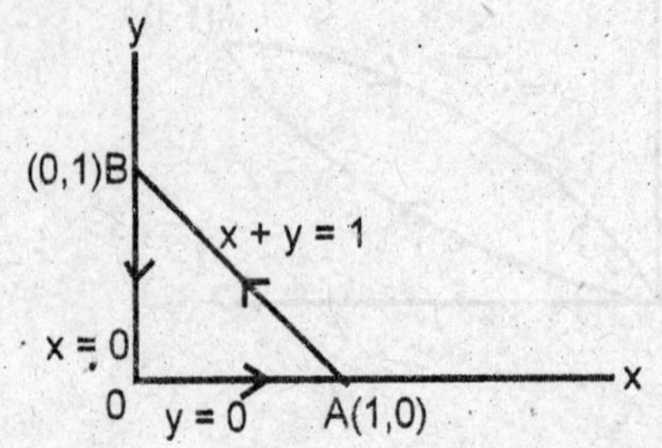

Along OA, $y = 0 \Rightarrow dy = 0$

$\therefore$ Given integral $= \int_{x=0}^{1} 3x^2 dx = x^3 \Big|_0^1 = 1$

Along AB, $y = 1 - x \Rightarrow dy = -dx$

$\therefore$ Given integral

$$= \int_{x=1}^{0} \{3x^2 - 8(1-x)^2\} dx + \{4(1-x) - 6x(1-x)(-dx)\} = \int_{x=1}^{0} (-11x^2 + 26x - 12)\, dx$$

$$= -\frac{11}{3}x^3 + 13x^2 - 12x \Big|_1^0 = \frac{11}{3} - 13 + 12 = \frac{8}{3}$$

Along BO, $x = 0 \Rightarrow dx = 0$

Given integral $= \int_{y=1}^{y=0} 4y\,dy = 2y^2 \Big|_1^0 = -2$

$\therefore$ The given integral $= 1 + \frac{8}{3} - 2 = \frac{5}{3}$

By Green's theorem,

$$\oint_C P dx + Q dy = \iint_R \left(\frac{\partial Q}{\partial x} . \frac{\partial P}{\partial y} \right) dx dy$$

Here, $P = 3x^2 - 8y^2$, $Q = 4y - 6xy$

$$\frac{\partial Q}{\partial x} . \frac{\partial P}{\partial y} = -6y - (-16y) = 10y$$

$\therefore$ The given integral $= \iint_R 10\, y dx dy = \int_{x=0}^{1} \left[\int_{y=0}^{y=1-x} 10\, y dy \right] dx = \int_{x=0}^{1} 5y^2 \Big|_0^{1-x} dx = \int_0^1 5(1-x)^2 dx$

$$= \frac{5(1-x)^3}{3x-1} \Big|_0^1 = -\frac{5}{3}(0-1) = \frac{5}{3}$$

Hence the theorem is verified.

***Example* 14.64:** Evaluate $\int_{(0,0)}^{(2,1)} (10x^4 - 2xy^3)\ dx\ -\ 3x^2y^2\ dy$ along the path $x^4 - 6xy^3 - 4y^2 = 0$

Solution:

$$P = 10x^4 - 2xy^3,\ Q = 3x^2y^2$$

$$\frac{\partial P}{\partial y} = -6xy^2 = \frac{\partial Q}{\partial x}$$

The integral is independent of the path. Hence we can use any path. For example, if we use the path from points (0, 0) to (2, 0) and then from (2, 0) to (2, 1); we can evaluate the integral.

(i) From (0, 0) to (2, 0) ; $y = 0,\quad dy = 0$

$$\therefore \text{The integral} = \int_0^2 10x^4\ dx - 2x^5\Big|_0^2 = 64$$

(ii) From (2, 0) to (2, 1) ; $x = 2,\quad dx = 0$

$$\therefore \text{The integral} = \int_0^2 10x^4\ dx = 2x^5 \int_0^2 = 64$$

$$\int_0^2 12y^2\ dy = -4y^3\Big|_0^2 = -4$$

$\therefore$ The value of the integral = 64 – 4 = 60

Aliter : Since $\frac{\partial P}{\partial y} = \frac{\partial Q}{\partial x}$, we know that $\{10x^4 - 2xy^3)dx - 3x^2y^2\ dy\}$ is an exact differential of $(2x^5 - x^2y^3)$.

$$\text{The given integral} = \int_{(0,0)}^{(2,1)} d\left(2x^5 - x^2y^3\right) = 2x^5 - x^2y^3 \int_{(0,0)}^{(2,1)} = 2.2^5 - 2^2.1^3 = 60$$

***Example* 14.65:** Verify Green's theorem for $\int_c (e^{-x}\sin y)\,dx + (e^{-x}\cos y)\,dy$ where C is the boundary of the rectangle whose vertices are (0, 0) (p, 0) (p, $\frac{\pi}{2}$) and (0, $\frac{\pi}{2}$) traversed in the +ve direction.

***Solution*:**

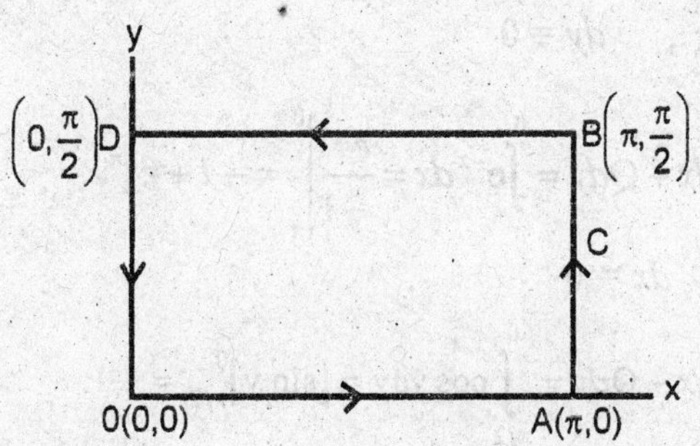

By Green's theorm,

$$\oint_c P dx + Q dy = \iint_R \left(\frac{\partial Q}{\partial x} - \frac{\partial P}{\partial y}\right) dxdy$$

Here, $\quad P = e^{-x}\sin y, \quad Q = e^{-x}\cos y$

$$\frac{\partial Q}{\partial x} - \frac{\partial P}{\partial y} = -e^{-x}\cos y - e^{-x}\cos y = -2e^{-x}\cos y$$

$$\therefore \quad \oint_c P dx + Q dy = \iint_R -2e^{-x}\cos y\, dxdy$$

$$= -2\int_{x=0}^{\pi}\left[\int_{y=0}^{\pi/2} e^{-x}\cos dy\right]dx = -2\int_{x=0}^{\pi}\left[e^{-x}\sin y\right]_0^{\pi/2} dx$$

$$= -2\int_c^{\pi} e^{-x}dx = \left[2e^{-x}\right]_0^{\pi} = 2(e^{-\pi}-1) \qquad .....(i)$$

Again, $\quad \oint_c P\,dx + Q\,dy = \int_{OA} + \int_{AB} + \int_{BD} + \int_{DO}$

Along OA ; $y = 0, \quad dy = 0$

$\therefore \quad \int_{OA} Pdx + Qdy = 0$

Along AB : $x = \pi, \quad dx = 0$

$$\int_{AB} Pdx + Qdy = \int_0^{\pi/2} e^{-\pi} \cos y dy = \left[e^{-\pi} \sin y\right]_0^{\pi/2} = e^{-\pi}$$

Along BD ; $y = \frac{\pi}{2}, \quad dy = 0$

$$\int_{BD} P\,dx + Qdy = \int_\pi^0 e^{-x} dx = \left.\frac{e^{-x}}{-1}\right|_\pi^0 = -1 + e^{-\pi}$$

Along DO ; $x = 0, \quad dx = 0$

$$\therefore \quad \int_{DO} P\,dx + Qdy = \int_{\pi/2}^0 \cos y dy = [\sin y]_{\pi/2}^0 = -1$$

$$\therefore \quad \int_c P\,dx + Q\,dy = 0 \; + e^{-\pi} - 1 + e^{-\pi} - 1 = 2(e^{-\pi} - 1) \qquad \text{.....(ii)}$$

$\therefore$ From (i) and (ii) it is proved that

$$\int_c P\,dx + Qdy = \iint_R \left(\frac{\partial Q}{\partial x} - \frac{\partial P}{\partial y}\right) dxdy$$

Hence the theorem is verified.

***Example* 14.66:** Apply Green's theorem to obtain the area bounded by the curve

$x^{2/3} + y^{2/3} = a^{2/3}, \; a > 0$.

***Solution*:**

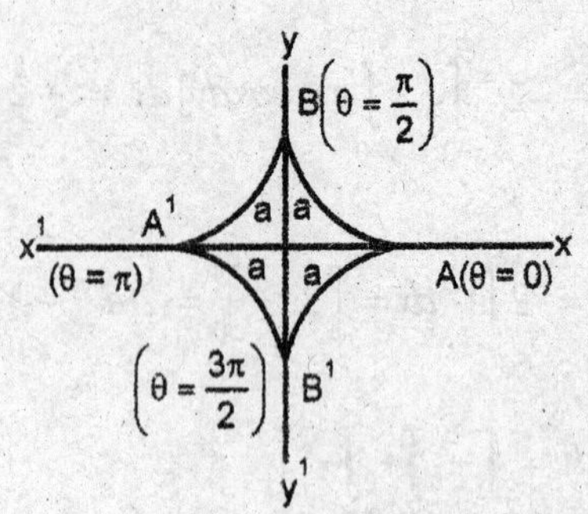

The parametric equations of the curve are $x = a\cos^3\theta$, $y = \sin^3\theta$. A rough sketch of the curve is given below :

$$x = a\cos^3\theta, \qquad y = a\sin^3\theta$$

$$dx = -3\, a\cos^2\theta\, \sin\theta\, d\theta,\ dy = 3a\, \sin^2\theta\cos\theta\, d\theta$$

By Green's theorem, the area bounded by a simple closed curve C is given by

$$\frac{1}{2}\int_C (xdy - ydx)$$

∴ The area bounded by the given curve C

$$= \frac{1}{2}\int_{0=0}^{2\pi}\{a\cos^3\theta\, .\, 3a\sin^2\theta\cos\theta + a\sin^3\theta . 3a\cos^2\theta\sin\theta\}d\theta$$

$$= \frac{3a^2}{2}\int_0^{2\pi}\cos^2\theta\sin^2\theta\,(\cos^2\theta + \sin^2\theta)d\theta$$

$$= \frac{3a^2}{2}\int_0^{2\pi} cos^2\theta\, sin^2\theta\, d\theta = \frac{3a^2}{8}\int_0^{2\pi} sin^2 2\theta\, d\theta = \frac{3a^2}{16}\int_0^{2x}(1 - cos4\theta)\, d\theta = \frac{3\pi a^2}{8}$$

***Example* 14.67:** Verify Green's theorem for $\int_C (2xy - x^2)dx + (x + y^2)dy$, where C is the closed curve in *xy*-plane bounded by the curves $y = x^2$ and $y^2 = x$.

***Solution*:**

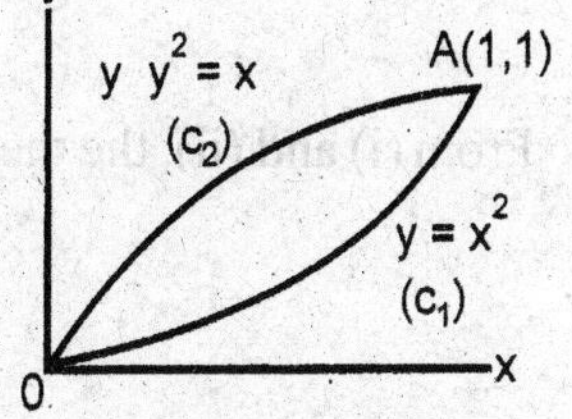

By Green's theorem,

$$\int_C Pdx + Qdy = \iint_R \left(\frac{\partial Q}{\partial x} - \frac{\partial P}{\partial y}\right) dxdy$$

Here, $\quad P = 2xy - x^2,\ Q = x + y^2,\ \dfrac{\partial Q}{\partial x} - \dfrac{\partial P}{\partial y} = 1 - 2x$

$$\iint_R \left(\frac{\partial Q}{\partial x} - \frac{\partial P}{\partial y}\right) dxdy = \int_{x=0}^{1}\left[\int_{y=x^2}^{y=\sqrt{x}}(1-2x)dy\right] dx = \int_0^1 [y - 2xy]_{y=x^2}^{y=\sqrt{x}}\, dx$$

$$= \int_0^1 (\sqrt{x} - 2x\sqrt{x} - x^2 + 2x^3)dx = \frac{2}{3}x^{\frac{3}{2}} - 2.\left[\frac{2}{5}x^{\frac{5}{2}} - \frac{x^3}{3} + 2.\frac{x^4}{4}\right]\Bigg]_0^1$$

$$= \frac{2}{3} - \frac{4}{5} - \frac{1}{3} + \frac{1}{2} = \frac{1}{30} \qquad \text{..... (i)}$$

Again $\int_C Pdx + Qdy = \int_{C_1}(Pdx + Qdy) + \int_{C_2}(Pdx + Qdy)$

Along C_1, $y = x^2, dy = 2xdx$, "x" varies from 0 to 1

$$\therefore \int_{C_1} Pdx + Qdy = \int_{x=0}^{1}(2x^3 - x^2)dx + (x + x^4)2xdx = \int_0^1 (2x^3 + x^2 + 2x^5)dx$$

$$= \frac{x^4}{2} + \frac{x^3}{3} + \frac{x^6}{3}\Big|_0^1 = \frac{7}{6}$$

Along C_2, $x = y^2,\ dx = 2ydy,$ 'y' varies from 1 to 0

$$\therefore \int_{C_2} Pdx + Qdy = \int_{y=1}^{0}(2y^3 - y^4)2ydy + (y^2 + y^2)dy = \int_1^0 (4y^4 - 2y^5 + 2y^2)dy$$

$$= \frac{4}{5}y^5 - \frac{2}{6}y^6 + \frac{2}{3}y^3\Big|_1^0 = -\frac{4}{5} + \frac{1}{3} - \frac{2}{3} = -\frac{17}{15}$$

$$\therefore \int_C Pdx + Qdy = \frac{7}{6} - \frac{17}{15} = \frac{1}{30} \quad .....(ii)$$

From (i) and (ii), the theorem is verified.

Exercise 14.9

1. Evaluate $\oint (x^2 + y^2)dx + 3xy^2dy$, (a) directly (b) by Green's theorem, where c is the circle $x^2 + y^2 = 4$, traversed in the +ve direction.

(Ans : 12π)

2. Evaluate $\oint (x^2 + 2xy)dx + (x^2y + 3)dy$ around the boundary C of the region given by $y^2 = 8x$ and $x = 2$, (a) directly and (b) by Green's theorem.

(Ans : 128)

3. Verify Green's theorem for the integral $\oint(3x^2+2y)dx-(x+3\cos y)dy$ where C is the boundary of the parallelogram with vertices at (0, 0), (2, 0), (3, 1) and (1, 1)

(Ans : –6)

4. Evaluate $\int_{(0,0)}^{(\pi,2)}(6xy-y^2)dx+(3x^2-2xy)dy$ along the cycloid $x = \theta - \sin\theta$, $y = 1-\cos\theta$. Verify the result by using Green's theorem.

(Ans : $6\pi^2 - 4\pi$)

5. Using Green's theorem find the area bounded by one arch of the cycloid $x=a(\theta-\sin\theta),\ y=a(1-\cos\theta), a>0,$ and x – axis.

(Ans: $3\pi a^2$)

6. Evaluate $\oint_c\left(2x^2-y^2\right)dx+\left(x^2+y^2\right)dy$ by Green' theorem where C is the boundary of the surface in the xy plane enclosed by x axis and the semi–circle $y=\sqrt{1-x^2}$

(Ans: $4/3$)

7. Evaluate $\oint_c(\cos x\sin y-xy)\,dx+\sin x\cos y\ dy$, using Green's theorem where c is the circle $x^2+y^2=1$

(Ans: 0)

8. Verify Green's Theorem in the plane for $\oint_c\left(x^2-xy^3\right)dx+\left(y^2-2xy\right)dy$ where C is the square with vertices at (0,0) , (2,0) and (0 ,2)

(Ans: 8)

9. Evaluate $\int_{(0,0)}^{(2,1)}\left(12x^3-2xy^3\right)dx-3x^2y^2dy$ along the path $x^3-y^3+y-4xy=0$

Ans: $3x^4-x^2y^3\Big|_{(0,0)}^{(2,1)}=44$)

14.10 Gauss Divergence Theorem

Let (1) V be the volume bounded by a closed surface S

(2) **A** be a vector function of position with continuous derivatives.

Then, $\iiint\limits_{v} \nabla.\mathbf{A}.dv = \iint\limits_{s} \mathbf{A}.\mathbf{n}ds = \oint\limits_{s} \mathbf{A}.ds$

where **n** is positive unit normal (outside drawn normal) to S.

Proof:

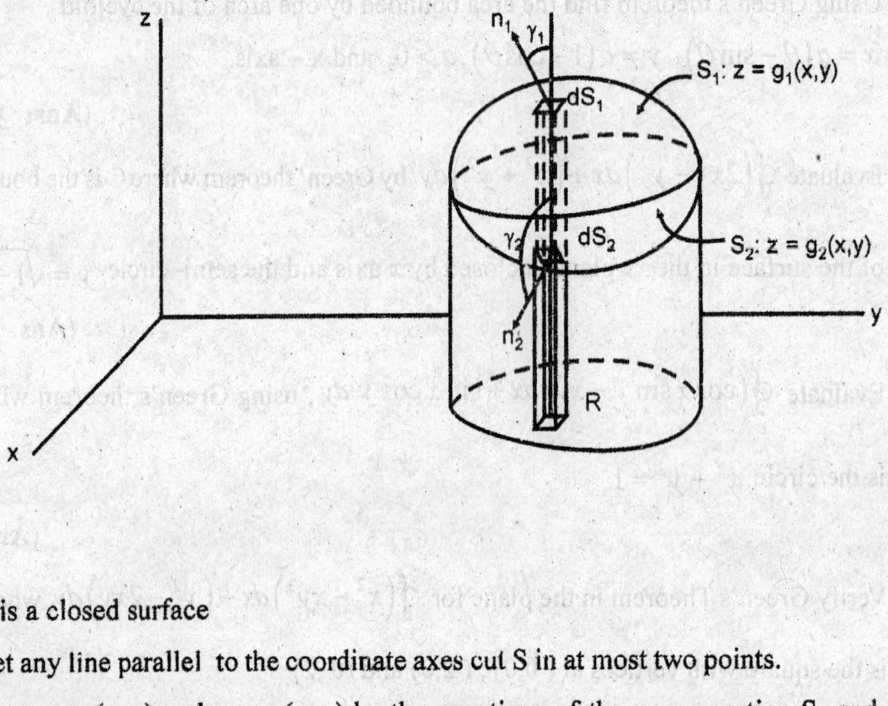

S is a closed surface

Let any line parallel to the coordinate axes cut S in at most two points.

Let $z = g_1(x, y)$ and $z = g_2(x, y)$ be the equations of the upper portion S_1 and lower portion S_2 respectively.

Let R be the projection of S on the *xy*-plane.

If $\mathbf{A} = A_1 i + A_2 j + A_3 k$

$$\nabla.\mathbf{A} = \frac{\partial A_1}{\partial x} + \frac{\partial A_2}{\partial y} + \frac{\partial A_3}{\partial z}$$

Then,

$$\iiint_v \frac{\partial A_3}{\partial z}\, dv = \iiint_V \frac{\partial A_3}{\partial z}\; dzdydx$$

$$= = \iint_R \left[\int_{z=g_2(x,y)}^{g_1(x,y)} S\frac{\partial A_3}{\partial z} dz \right] dydx = \iint_R \left[A_3(x,y,z) \right]_{z=g_2}^{z=g_1} dydx$$

$$= \iint_R [A_3(x,y,g_1) - A_3(x,y,g_2)]dydx$$

For the upper portion S_1, $dy\,dx = (\cos\gamma_1)dS_1 = k.\, \mathbf{n}_1\, dS_1$

since the normal $\mathbf{n}_1$ to S_1 makes an acute angle γ_1 with k.

For the lower portion S_2, $dydx = (-\cos\gamma_2)\, dS_2 = -k.\mathbf{n}_2\, dS_2$,

since the normal $\mathbf{n}_2$ to S_2 makes an obtuse angle γ_2 with k.

$\therefore \iint_R A_3(x,y,g_1)dydx = \iint_{s_1} A_3 k.\boldsymbol{n}_1\, dS_1$ and

$\iint_R A_3(x,y,g_2)dydx = \iint_{s_2} A_3 k.\boldsymbol{n}_2\, dS_2$,

So that, $\iint_R [A_3(x,y,g_1) - A_3\;(x,\; y,\; g_2)\;]dydx = \iint_{S_1} A_3 k.\boldsymbol{n}_1 dS_1 + \iint_{s_2} A_3 k.\boldsymbol{n}_2 dS_2$

$$= \iint_S A_3 k.\boldsymbol{n} dS \qquad \text{..... (1)}$$

Similarly, by projecting S on the other coordinate planes, we get,

$$\iiint_v \frac{\partial A_1}{\partial x}\, dv = \iint_s A_1 i.\boldsymbol{n}\, ds \qquad \text{..... (2)}$$

and $\iiint_v \frac{\partial A_2}{\partial y}\, dv = \iint_s A_2 j.\boldsymbol{n}\, ds$ ..... (3)

(1) + (2) + (3) $\Rightarrow$

$$\iiint_v \left(\frac{\partial A_1}{\partial x} + \frac{\partial A_2}{\partial y} + \frac{\partial A_3}{\partial z}\right) dv = \iint_s (A_1 i + A_2 j + A_3 k).\boldsymbol{n} dS$$

i.e. $\iiint_v \nabla.\mathbf{A}\, dv = \iint_s \boldsymbol{A}.\boldsymbol{n}\, dS$ which proves the theorem.

Ex (a) Express Gauss' Theorem in words and (b) obtain its Cartesian form.

(a) Gauss' Theorem states that "The surface integral of the normal component of vector **A** taken over a closed surface is equal to the integral of the divergence of **A** taken over the volume enclosed by the surface.

(b) Let $A = A_1 i + A_2 j + A_3 k$

Then $\text{div } A = \nabla.\vec{A} = \frac{\partial A_1}{\partial x} + \frac{\partial A_2}{\partial x} + \frac{\partial A_3}{\partial z}$

Let the unit normal $\boldsymbol{n}$ to S make angles α, β, γ with the +ve coordinate axes so that, $\cos\alpha = n.\ i, \cos\beta = n.\ j, \cos\gamma = n.k$ and $\cos\alpha$, $\cos\beta$, $\cos\gamma$ are the direction cosines of $\boldsymbol{n}$

Then if $\boldsymbol{n} = n_1 i + n_2 j + n_3 k$.

$$\boldsymbol{n} = (\cos\alpha)i + (\cos\beta) + (\cos\gamma) \qquad [\boldsymbol{n}.i = n_1 = \cos\alpha \text{ etc.}]$$

$$\therefore\ \mathbf{A}.\boldsymbol{n} = A_1 \cos\alpha + A_2 \cos\beta + A_3 \cos\gamma$$

Hence the divergence theorem can be written in Cartesian form as,

$$\iiint_v \left(\frac{\partial A}{\partial x} + \frac{\partial A}{\partial y} + \frac{\partial A}{\partial z}\right) dxdydz = \iint_S (A_1 \cos\alpha + A_2 \cos\beta + A_3 \cos\gamma)\, ds$$

$$= \iint_S A_1 dydz + A_2 dzdx + A_3 dzdy$$

Solved Examples

***Example* 14.68:**

Evaluate using the divergence theorem $= \iint_S \mathbf{F}.n\, ds$ where $\mathbf{F} = 2xyi + yz^2 j + xzk$ and S is the surface of the parallelepiped bounded by $x = 0$, $y = 0$, $z = 0$, $x = 2$, $y = 1$ and $z = 3$.

Solution

By the divergence theorem: $\iint_S \mathbf{F}.n\, ds = \iiint_v \nabla.\mathbf{F} dv$

$$\text{Here, } \nabla.\mathbf{F} = \frac{\partial}{\partial x}(2xy) + \frac{\partial}{\partial y}(yz^2) + \frac{\partial}{\partial y}(xz)$$

$$= (2y + z^2 + x)$$

$$\therefore\ \iint_S F.n\, ds = \iiint_v (2y + z^2 + x) dv$$

$$= \int_{x=0}^{2}\int_{y=0}^{1}\int_{z=0}^{3}(2y+z^2+x)dxdydz = \int_{x=0}^{2}\int_{y=0}^{1}[\int_{z=0}^{3}(2y+z^2+x)dz]\,dxdy$$

$$= \int_0^2\int_0^1 \left.2yz+\frac{z^3}{3}+xz\right|_0^3 dxdy = \int_0^2\int_0^1(6y+9+3x)dx\,dy$$

$$= \int_{x=0}^{2}[\int_{y=0}^{1}(6y+9+3x)dy]dx = \int_0^2 \left.3y^2+9y+3xy\right|_0^1 dx = \int_0^2(12+3x)\,dx = 12x+\left.\frac{3x^2}{2}\right|_0^2$$

$$= 24+6=30$$

***Example* 14.69:** Verify the divergence theorem for $\vec{F} = (4xy)i - (y^2)j + (xz)k$, over the cube bounded by $x = 0$, $x = 1$, $y = 0$, $y = 1$, $z = 0$ and $z = 1$.

Solution:

By the divergence theorem

$$\iint_s \boldsymbol{F}.\boldsymbol{n}\,dS = \iiint_v \nabla.\boldsymbol{F}\,dv$$

Here $\vec{F} = (4xy)i - (y^2)j + (xz)k$

$$\nabla.\boldsymbol{F} = \frac{\partial}{\partial x}(4xy) - \frac{\partial}{\partial y}(y^2) + \frac{\partial}{\partial z}(xz) = 4y - 2y + x = x + 2y$$

$$\iiint_v \nabla.\mathbf{F}dv = \int_{x=0}^{1}\int_{y=0}^{1}\int_{z=0}^{1}(x+2y)dzdydx = \int_{x=0}^{1}\int_{y=0}^{1}[\int_{z=0}^{1}(x+2y)dz]dydx$$

$$= \int_{x=0}^{1}\int_{y=0}^{1}[xz+2yz]_0^1\,dydx$$

$$= \int_{x=0}^{1}[\int_{y=0}^{1}(x+2y)\,dy]\,dx = \int_{x=0}^{1}\left.xy+y^2\right|_0^1 dx = \int_{x=0}^{1}(x+1)dx = \frac{x^2}{2}+x\Big|_0^1 = \frac{3}{2}$$

To evaluate $\iint_S F.n dS$

The surface S contains 6 faces (see figure)

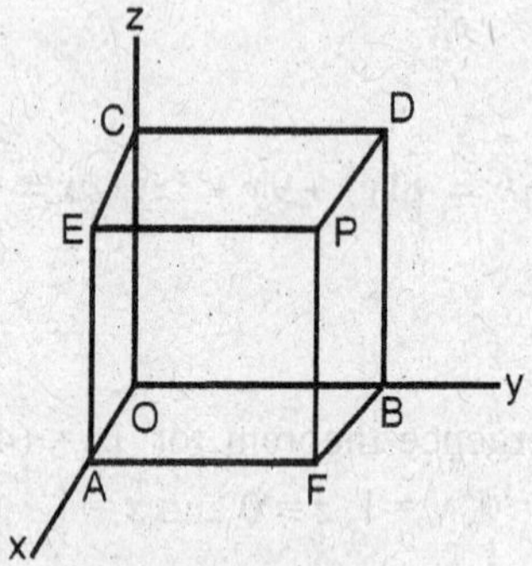

S_1 - Face EPFA, S_2 - Face OBDC

S_3 - Face PFBD, S_4 - Face OCEA

S_5 - Face PDCE, S_6 - Face OBFA

The surface integral $\iint_S F.n dS$ is equal to the sum of the surface integrals on the above 6 faces.

on S_1, $\mathbf{n} = i$, $\quad x = 1$, $\mathbf{F.n} = 4y$; $dS = dy\, dz$

$$\therefore \iint_S F.ndS = \int_{y=0}^{1}\int_{z=0}^{1} 4y dy dz = \int_{y=0}^{1}[\int_{z=0}^{1} 4y\, dz]\, dy = \int_{y=0}^{1} 4yz\Big|_0^1 dy = \int_0^1 4y dy = 2y^2\Big|_0^1 = 2$$

on S_2, $\mathbf{n} = -i$, $x = 0$, $\mathbf{F.n} = 0$

$$\therefore \iint_{S_2} \mathbf{F.n\ dS} = 0$$

on S_3, $\mathbf{n} = j$, $y = 1$, $\mathbf{F.n} = -1$, $dS = dx\, dz$.

$$\therefore \iint_{S_3} F.n\, dS = \int_{x=0}^{1}[\int_{z=0}^{1} -1\, dz]\, dx = \int_{x=0}^{1} -z\Big|_0^1 dx = \int_{x=0}^{1} -dx = -x\Big|_0^1 = -1$$

on S_4, $\mathbf{n} = -j$, $y = 0$, $\mathbf{F.n} = 0$

$$\therefore \iint_{S_4} F.n\, dS = 0$$

on S_5, $n = k$, $z = 1$, $\mathbf{F.n} = x$; $dS = dx\,dy$

$$\therefore \iint_{S_5} F.n dS = \int_{x=0}^{1}\int_{y=0}^{1} x dx dy = \int_{x=0}^{1}\int_{y=0}^{1} x\, dy]\, dx = \int_{x=0}^{1} xy\Big|_0^1 dx = \int_{x=0}^{1} x dx = \frac{x^2}{2}\Big|_0^1 = \frac{1}{2}$$

on S_6, $\mathbf{n} = k$, $z = 0$, $\mathbf{F.n} = 0$

$$\therefore \iint_{S_6} F.n dS = 0$$

Hence $\therefore \iint_S F.n dS = 2 - 1 + \frac{1}{2} = \frac{3}{2}$

The divergence theorem is verified

***Example* 14.70:** If $r = xi + yj + zk$, evaluate $\therefore \iint_S r.n\, dS$ where S is any closed surface.

***Solution*:**

By the divergence theorem

$$\therefore \iint_S r.n\, dS = \iiint_V \nabla.r\, dv$$,where V is the volume enclosing S.

$$\nabla.r = \frac{\partial}{\partial x}(x) + \frac{\partial}{\partial y}(y) + \frac{\partial}{\partial z}(z) = 3$$

$$\therefore \iint_s r.n\, ds = \iiint_v 3\, dv = 3v$$

***Example* 14.71:** If 'ψ' is a scalar function, prove that $\iiint_V \nabla\psi dv = \iint_S \psi\, n\, ds$

***Solution*:**

By divergence theorem

$$\iiint_v \nabla.A dv = \iint_s A.n\, ds \qquad \text{..... (1)}$$

Take $\mathbf{A} = \psi\mathbf{c}$, $\mathbf{C} = c_1 i + c_2 j + c_3 k$ is any constant vector.

$$\therefore (1) \Rightarrow \iiint_V \nabla.(\psi \mathbf{c})\, dv = \iint_s (\psi \mathbf{c}).\boldsymbol{n} ds \qquad ..... (2)$$

But,

$$\nabla.(\psi \mathbf{c}) = (\nabla \psi).\mathbf{c} + \psi(\nabla.\mathbf{c})$$

$$= (\nabla \psi).\mathbf{c} \qquad [\because \nabla.c = \frac{\partial}{\partial x}(c_1) + \frac{\partial}{\partial y}(c_2) + \frac{\partial}{\partial z}(c_3) = 0,$$

$$= \mathbf{c}.\nabla \psi \qquad c_1, c_2, c_3, \text{ being constant}]$$

and $(\psi \mathbf{c}).\mathbf{n} = \mathbf{c}.(\psi \mathbf{n})$ $[\because \mathbf{a}.\mathbf{b} = \mathbf{b}.\mathbf{a}]$

so that (2) becomes,

$$\iiint_v (c.\nabla\psi)\, dv = \iint_s \{c.(\psi \boldsymbol{n})\} ds$$

$$\Rightarrow c.\iiint_v \nabla\psi dv = c.\iint_s \psi \boldsymbol{n}\, ds$$

$$\Rightarrow \iiint_V \nabla\psi dv = \iint_s \psi \boldsymbol{n}\, ds \qquad (\because \vec{c} \text{ is arbitrary})$$

***Example* 14.72:** If s is a closed surface enclosing a volume V and if $F = (lx)i + (my)j + (nz)k$, l, m, n being constants, evaluate $\iint_s \boldsymbol{F}.\boldsymbol{n}\, ds$

***Solution*:**

By Gauss' theorem

$$\iint_s \boldsymbol{F}.\boldsymbol{n}\, ds = \iiint_v \nabla.\boldsymbol{F} dv$$

$$\nabla.\boldsymbol{F} = \frac{\partial}{\partial x}(lx) + \frac{\partial}{\partial y}(my) + \frac{\partial}{\partial z}(nz) = l + m + n$$

$$\therefore \iint_s \boldsymbol{F}.\boldsymbol{n} ds = \iiint_v (l + m + n)\, dv$$

$$= (l + m + n)\, V$$

***Example* 14.73:** If E = curl A, evaluate $\iint_s \boldsymbol{E.n}\, ds$ where s is any closed surface

Solution:

By Gauss' theorem,

$\iint_s \boldsymbol{E.n}\, ds = \iiint_V (\text{div } \mathbf{E})\ dv$ where V is the volume enclosed by s

But Div $\vec{E}$ = div (curl $\vec{A}$) = 0 [See 14.5.2 (2)]

$\therefore \iint_s \boldsymbol{E.n}\, ds = 0$

***Example* 14.74:** If s any closed surface and n is unit +ve normal to s, show that $\iint_s \boldsymbol{n}\, ds = 0$

***Solution*:**

Consider a constant vector

a = a_1i + a_2j + a_3k (i.e., a_1, a_2, a_3 are constants)

Then, $\boldsymbol{a}.[\iint_s \boldsymbol{n} ds] = \iint_s (\boldsymbol{a.n})\, ds$

$= \iiint_V (\nabla.\boldsymbol{a})\ dv$ (by divergence theorem)

$= 0$ $[\because \nabla.\boldsymbol{a} = \frac{\partial a_1}{\partial x} + \frac{\partial a_2}{\partial y} + \frac{\partial a_3}{\partial z} = 0]$

Thus $\boldsymbol{a}.[\iint_s \boldsymbol{n}\, ds] = 0$ is true where **a** is any arbitrary vector.

$\therefore \iint_s \boldsymbol{n} ds = 0$

***Example* 14.75:** Evaluate $\iint_S (\boldsymbol{F.n}) ds$, where s is the region bounded by

$y^2 = 8x$, $x = 2$, $z = 0, z = 3$, and $F = 2xi + 3yj + \frac{1}{3} z^3 k$, using Gauss' theorem

SOLUTION

By the divergence theorem,

$$\iint_S (\mathbf{F}.n)\, ds = \iiint_v (\nabla.\mathbf{F}) dv \ ,$$

$$\nabla.\mathbf{F} = \frac{\partial}{\partial x}(2x)+\frac{\partial}{\partial y}(3y)+\frac{\partial}{\partial y}\left(\frac{z^3}{3}\right)=5+z^2$$

From (1), $\iint_S (\mathbf{F}.n)\, ds = \iiint_v (5+z^2) dv$,

$$= \int_{x=0}^{2}\int_{y=2\sqrt{2x}}^{2\sqrt{2x}}\left[\int_{z=0}^{3}\left(5+z^2\right)dz\right]dydx$$

$$= \int_{x=0}^{2}\int_{y=2\sqrt{2x}}^{2\sqrt{2x}} 5z+\frac{z^3}{3}\Bigg|_0^3 dydx$$

$$= \int_{x=0}^{2}\left[\int_{y=2\sqrt{2x}}^{2\sqrt{2x}} 24dy\right]\int_{x=0}^{2} 24\, y\Big|_{-2\sqrt{2x}}^{2\sqrt{2x}} dx$$

$$= \int_{x=0}^{2} 96\sqrt{2}x dx = 96\sqrt{2}\frac{x^2}{2}\Bigg|_0^2 = 192\sqrt{2}$$

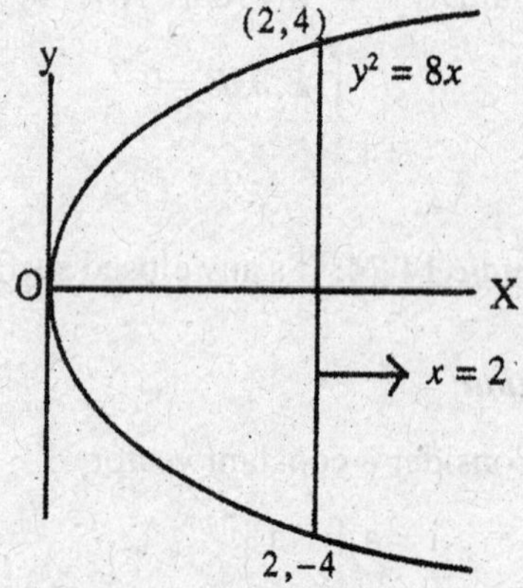

EXAMPLE 76

Evaluate $\iint_S \left(2xi - 3y^2 j + z^2 k\right).n\ ds$ over the surface bounded by $x^2+y^2=1$, $z=0$, $z=2$, using Gauss' theorem.

SOLUTION

By the divergence theorem,

$$\iint_S (\mathbf{F}.n)\, ds = \iiint_v (\nabla.\mathbf{F}) dv,$$

Here, $\mathbf{F} = 2xi - 3y^2 j + z^2 k$

$$\text{Div } \mathbf{F} = \frac{\partial}{\partial x}(2x)-\frac{\partial}{\partial y}(3y^2)+\frac{\partial}{\partial y}(z^2)$$

$$= 2-6y+2z$$

$$\therefore \iint (\mathbf{F}.n) ds = \int_{x=1}^{+1}\int_{y=\sqrt{1-x^2}}^{\sqrt{1-x^2}}\int_0^2 (2-6y+2z)\, dzdydx$$

$$= \int_{x=-1}^{+1} \int_{y=-\sqrt{1-x^2}}^{\sqrt{1-x^2}} \left[\int_{z=0}^{2} (2-6y+2z)dz \right] dydx = \int_{x=-1}^{+1} \int_{y=-\sqrt{1-x^2}}^{\sqrt{1-x^2}} 2z - 6yz + z^2 \Big|_{-1}^{1} dydx$$

$$= \int_{x=-1}^{+1} \left[\int_{y=-\sqrt{1-x^2}}^{\sqrt{1-x^2}} (8-12y)dy \right] dx = \int_{x=-1}^{1} 8y - 6y^2 \Big|_{-\sqrt{1-x^2}}^{\sqrt{1-x^2}} dx = \int_{-1}^{1} 16\sqrt{1-x^2}\, dx$$

$$= 16\left[\frac{x}{2}\sqrt{1-x^2} + \frac{1}{2}\sin^{-1} x \right]_{-1}^{1} = 8\pi$$

***Example* 14.77:** Evaluate using the divergence theorem $\iint_s (\boldsymbol{F.n})\boldsymbol{dS}$ where S is the surface of the sphere $x^2 + y^2 + z^2 = b^2$ in the first octant and F = $yi + zj + xk$

Solution:

By divergence theorem,

$$\therefore \iint_s \boldsymbol{F.n}\, ds = \iiint_V \nabla.\boldsymbol{F}\, dv \qquad \text{..... (1)}$$

$\mathbf{F} = yi + zj + xk$

$\nabla.\boldsymbol{F} = 0$

$$\therefore \iint_v (\nabla.\boldsymbol{F})\, dv = 0 \qquad \text{..... (2)}$$

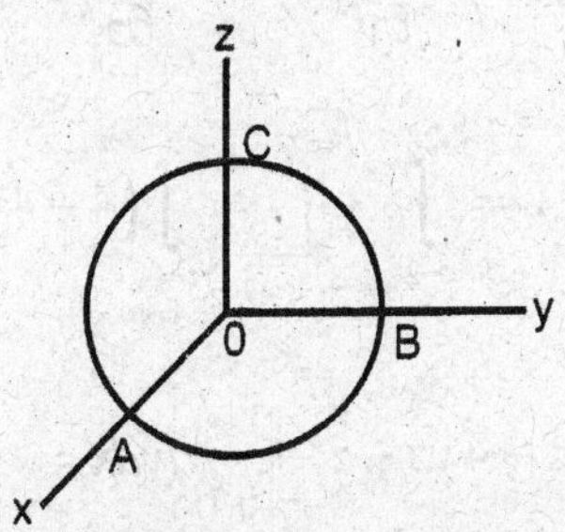

Let us evaluate the surface integrals over the faces OAB, OBC and OCA.

$$\iint_{OAB} \boldsymbol{F.\vec{n}}\, ds = -\int_{x=0}^{b} \int_{y=0}^{\sqrt{b^2-x^2}} x\, dx\, dy \qquad (\because\ \vec{n} = -k)$$

$$= -\int_0^b xy\Big|_0^{\sqrt{b-x^2}} dx = -\int_0^b x\sqrt{b^2-x^2}\, dx = -\pi\frac{b^3}{3}$$

Similarly $\iint\limits_{OAB} \mathbf{F.n}ds = \iint\limits_{OAB} \mathbf{F.n}ds + \iint\limits_{ABC} \mathbf{F.n}ds + \iint\limits_{OCA} \mathbf{F.n}ds + \iint\limits_{ABC} \mathbf{F.n}ds$

$$= -\pi b^3 + \iint\limits_{ABC} \mathbf{F.n}ds \quad ;$$

From (i), (ii), and (iii), we get

$$0 = -\pi b^3 + \iint\limits_{ABC} \mathbf{F.n}ds$$

$$\iint\limits_{ABC} \mathbf{F.n}ds = \pi b^3$$

Example 14.78:

Verify divergence theorem for $\mathbf{F} = 4xi - 2y^2 j + z^2 k$ taken over the region bounded by $x^2 + y^2 = 4$, $z = 0$ and $z = 3$.

SOLUTION

By the divergence theorem, we have

$$\iiint\limits_{v} Div\vec{F}dv = \iint\limits_{S} \vec{F}.\vec{n}\, ds \qquad .....(1)$$

$$(1)\ Div\vec{F} = \frac{\partial}{\partial x}(4x) - \frac{\partial}{\partial y}(2y^2) + \frac{\partial}{\partial z}(z^2) = 4 - 4y + 2z$$

$$\therefore \text{ L.H.S. of (1)} = \int_{x-2}^{2} \int_{y=-\sqrt{4-x^2}}^{\sqrt{4-x^2}} \left[\int_{z=0}^{3} (4 - 4y + 2z)\,dz \right] dydx$$

$$= \int_{x=-z}^{2} \int_{y=-\sqrt{4-x^2}}^{\sqrt{4-x^2}} \left[4z - 4yz + z^2\right]_{z=0}^{3} dydx = \int_{x=-2}^{2} \left[\int_{y=-\sqrt{4-x^2}}^{\sqrt{4-x^2}} (21 - 12y)\,dy \right] dx$$

$$= 42 \int_{x=-2}^{2} \sqrt{4 - x^2}\,dx = 84\pi$$

[Do the integration w.r.t.x yourself, taking $x = 2\sin\theta$]

(2) To evaluate surface integral $\iint\limits_{S} \vec{F}.\vec{n}\, ds$

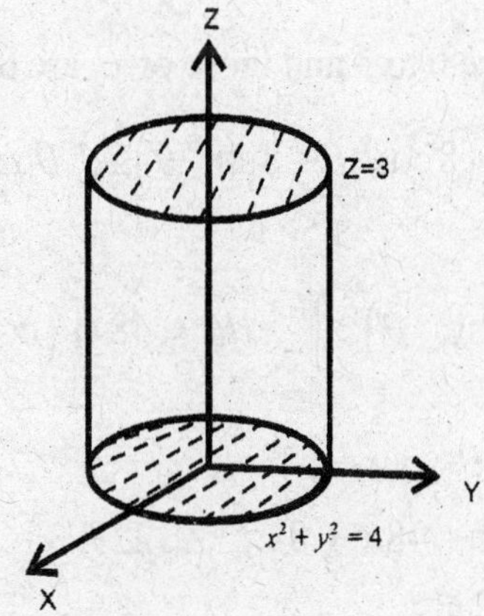

The given surface of the cylinder can be divided into 3parts, namely

(a) S_1 : the circular surface z = 0

(b) S_2 : the surface z = 3 (circular) and

(c) S_3 : the cylindrical portion of $S: x^2 + y^2 = 4,\ z = 0,\ z = 3$

we now find $\iint \vec{F}.\vec{n}\ ds$ over S_1 , S_2 , S_3 .If we add them, we get R.H.S of (1).

(a) on $S_1 : z = 0;\ \vec{n} = -k\ ;\ \vec{F}.\vec{n} = -\left(4xi - 2y^2 j\right).k = 0\ ;\ \therefore\ \iint_{S_1} \vec{F}.\vec{n}\ ds = 0.$

(b) on $S_2 : z = 3\ ;\ \vec{n} = k\ ;\ \vec{F}.\vec{n} = \left(4xi - 2y^2 j + 9k\right).k = 9;\ ds = \dfrac{dxdy}{\left|\vec{n}.k\right|} = dxdy$

$\therefore\ \iint_{S_2} \vec{\mathbf{F}}.\vec{\mathbf{n}}\ ds = \iint_{S_2} 9dxdy = 9A$, where A is the area of the circle

$x^2 + y^2 = 4,\ = 9\pi\left(2^2\right) = 36\pi$

(c) on S_3 : Let $\phi = x^2 + y^2 - 4 = 0;$

$$\vec{\mathbf{n}} = \frac{\nabla\phi}{\left|\nabla\phi\right|} = \frac{2\left(xi + yi\right)}{2\sqrt{x^2 + y^2}} = \frac{xi + yi}{2}\left(\text{since } x^2 + y^2 = 4\right)$$

$$\vec{\mathbf{F}}.\vec{\mathbf{n}} = \frac{4x^2 - 2y^3}{2} = 2x^2 - y^3;$$

To evaluate $\iint_{S_3} \vec{\mathbf{F}}.\vec{\mathbf{n}}\ ds$, take $x = 2\cos\theta, y = 2\sin\theta,$

and $ds = 2d\theta\, dz$; limits of z are 0 to 3 and those of θ are 0 to 2π .

Hence $\iint_{S_3} \vec{\mathbf{F}}.\vec{\mathbf{n}}\, ds = \int_{\theta=0}^{2\pi}\int_{z=0}^{3}\left(8\cos^2\theta - 8\sin^3\theta\right)2d\,\theta\, dz$

$$= 16\int_{\theta=0}^{2\pi}\left[\left(\cos^2\theta - \sin^3\theta\right)z\right]_{z=0}^{3} d\theta = 48\int_{\theta=0}^{2\pi}\left(\cos^2\theta - \sin^3\theta\right)d\theta = 48\pi$$

[Do the integration w.r.t. θ yourself]

$\therefore$ R.H.S of (1) $= 0 + 36\pi + 48\pi = 84\pi$; $\therefore$ L.H.S = R.H.S

Hence the theorem is verified.

Eercise 14.10

1. Verify Gauss's divergence theorem for $\mathbf{A} = (x^2 - yz)i + (y^2 - zx)j + (z^2 - xy)k$ taken over the rectangular parallalopiped $0 \le x \le 2, 0 \le y \le 3, 0 \le z \le 1$. **[Ans:** 36]

2. Use the divergence theorem to find $\iint_S \mathbf{F}.ndS$, where $f = (3x + 2z^2)i - (z^2 - 2y)j + (y^3 - 2z)k$ and S is the surface of the sphere with centre at (2,-1,3) and radius 2 units. **[Ans:** 32π]

3. Verify the divergence theorem for the vector function, $\mathbf{A} = (4xz)i - (y^2)j + (yz)k$, taken over the unit cube bounded by $x = 0, x = 1, y = 0, y = 1, z = 0$ and $z = 1$. **[Ans:** 3/2]

4. If $\mathbf{r} = xi + yj + zk$, and S is the surface of the rectangular parallelepiped bounded by planes $x = 0, y = 0, z = 0, x = a, y = b$ and $z = c$, find the value pf $\iint_S \mathbf{r.n}dS$ using Gauss's theorem Verify your answer by direct evaluation of the integral. **[Ans:** 3abc]

5. Use the divergence theorem to evaluate $\iint_S \mathbf{A.n}ds$ for $A = (2x)i - (2y)j + (3z)k$ where s is the sphere given by $(x-1)^2 + y^2 + z^2 = 1$ **[Ans:** 4π]

6. If $\mathbf{V} = (lx)i + (my)j + (nz)k$, l, m, n being constants show that

$\iint_S \mathbf{V.ds} = \frac{32\pi}{3}(l + m + n)$. where S is the surface the sphere

$(x-3)^2 + (y-2)^2 + (z-1)^2 = 4$.

14.11 Stoke's Theorem

Let (1) S be an open, two-sided surface bounded by a simple closed curve C.

(2) $\vec{A}$ be a vector function having continuous derivatives

Then, $\oint A.dr = \iint_S (\nabla \times A).n ds = \iint_S (\nabla \times A).ds$

where C travels in the +ve direction and $\vec{n}$ is the unit +ve (outward drawn) normal to S.

Proof :

Let S be the surface. Let the projections of S on the coordinate planes be regions bounded by simple closed curves.

Let 'R' the projection of S on xy plane be bounded by C^1. (see the figure above)

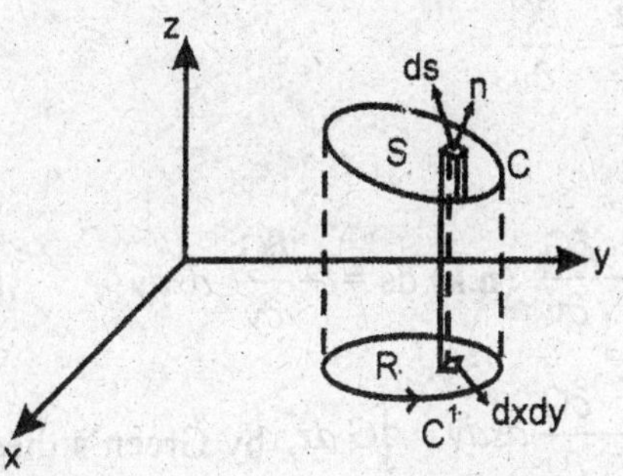

Let the equation of S be $z = \phi_1(x, y)$ where ϕ_1 is a single valued, continuous and differentiable function.

Let $\quad \mathbf{A} = A_1 i + A_2 j + A_3 k$

Then $\quad \nabla \times (A_1 i) = \begin{vmatrix} i & j & k \\ \partial/\partial x & \partial/\partial y & \partial/\partial z \\ A_1 & 0 & 0 \end{vmatrix} = \dfrac{\partial A_1}{\partial z} j - \dfrac{\partial A_1}{\partial y} k$

$$\{\nabla \times (A_1 i)\}.n \, ds = \left(\frac{\partial A_1}{\partial z} - \frac{\partial A_1}{\partial y}\right).nds = \left\{\frac{\partial A_1}{\partial z}(n.j) - \frac{\partial A_3}{\partial y}(n.k)\right\}ds \qquad ..... (1)$$

The position vector r of any point on S can be taken as

$$r = xi + yj + zk$$

$$= xi + yj + \phi_1(x, y)k$$

and $\dfrac{\partial \boldsymbol{r}}{\partial y} = j + \dfrac{\partial \phi_1}{\partial y} k$

But $\dfrac{\partial \boldsymbol{r}}{\partial y}$ being the vector tangent to S, it is perpendicular to $\vec{n}$.

$$\therefore \frac{\partial \boldsymbol{r}}{\partial y} . \mathbf{n} = 0 \Rightarrow \boldsymbol{n}.j = -\frac{\partial \phi_1}{\partial y} (\boldsymbol{n}.k)$$

$$\therefore (1) \Rightarrow \{ \nabla \times A_1 i) \} . \mathbf{n}\, ds = \left(-\frac{\partial A_1}{\partial z}\frac{\partial \phi_1}{\partial y}\boldsymbol{n}.k - \frac{\partial A_1}{\partial y}\boldsymbol{n}.k\right)ds$$

$$= -\left(\frac{\partial A_1}{\partial z}\frac{\partial z}{\partial y} + \frac{\partial A_1}{\partial y}\right)(\boldsymbol{n}.k)ds \qquad (\because z = \phi_1) \qquad \text{..... (ii)}$$

on S, $A_1(x\, y, z) = A_1(x, y, \phi_1(y)) = G(x, y)$ (say)

$$\frac{\partial A_1}{\partial y} + \frac{\partial A_1}{\partial z} . \frac{\partial z}{\partial y} = \frac{\partial G}{\partial y}$$

$\therefore$ (ii) becomes,

$$[\nabla \times (A_1 i)] .\mathbf{n}\, ds = -\frac{\partial G}{\partial y} (\mathbf{n}.\mathrm{k})\, ds = -\frac{\partial G}{\partial y}\, dxdy \qquad [\because \boldsymbol{n}.k.\ ds = dxdy]$$

$\iint\limits_{S} (\nabla \times A_1 i)\} .\boldsymbol{n}\, ds = \iint\limits_{R} -\dfrac{\partial G}{\partial y} dxdy = \oint\limits_{C} G\, dx$, by Green's theorem in the plane. Now at each point (x, y) of C, the value of G is the same as the value of A_1 at each point (x, y, z) of C, and since dx is same for both curves, we have

$$\oint\limits_{C} G dx = \oint\limits_{C} A_1\, dx$$

Hence,
$$\iint\limits_{S} \{\nabla \times (A_1 i)\} .n\, ds = \oint\limits_{C} A_1 dx \qquad \text{..... (3)}$$

Similarly by projecting S on yz and zx planes, it can be shown that,

$$\iint\limits_{S} \{\nabla \times (A_2 i)\} .\vec{n}\, ds = \oint\limits_{C} A_2 dy \qquad \text{..... (4)}$$

$$\iint\limits_{S} \{\nabla \times (A_3 k)\} .\vec{n}\, ds = \oint\limits_{C} A_3 dz \qquad \text{..... (5)}$$

Adding (iii), (iv) and (v), $\iint_S \nabla \times \boldsymbol{A}.\boldsymbol{n}\, ds = \oint_C \boldsymbol{A}.\, d\boldsymbol{r}$

(since $\mathbf{A}.\,d\mathbf{r} = A_1\, dx + A_2 dy + A_3 dz$)

Hence the theorem is proved.

Ex (a) Express Stoke's theorem in words and (b) obtain its cartesian form (c) deduce Green's theorem from stoke's theorem.

Sol: (a) The line integral of the tangential component of a vector $\vec{A}$ taken around a simple closed curve C is equal to the surface integral of the normal component of curl $\vec{A}$ taken over a surface S having C as its boundary.

(b) As in 14.10.2(b)

$\mathbf{A} = A_1 i + A_2 j + A_3 k$

$\vec{n} = (\cos\alpha) i + (\cos\beta) j + (\cos\gamma) k$

Then, $\nabla \times \boldsymbol{A} = \begin{vmatrix} i & j & k \\ \partial/\partial x & \partial/\partial y & \partial/\partial z \\ A_1 & A_2 & A_3 \end{vmatrix}$

$$= \left(\frac{\partial A_3}{\partial y} - \frac{\partial A_2}{\partial z}\right) i + \left(\frac{\partial A_1}{\partial z} - \frac{\partial A_3}{\partial x}\right) j + \left(\frac{\partial A_2}{\partial x} - \frac{\partial A_1}{\partial y}\right) k$$

$$(\nabla \times \boldsymbol{A}).\boldsymbol{n} = \left(\frac{\partial A_3}{\partial y} - \frac{\partial A_2}{\partial z}\right)\cos\alpha + \left(\frac{\partial A_1}{\partial z} - \frac{\partial A_3}{\partial x}\right)\cos\beta + \left(\frac{\partial A_2}{\partial x} - \frac{\partial A_1}{\partial y}\right)\cos\gamma$$

$\mathbf{A}.\,d\mathbf{r} = (A_1 i + A_2 j + A_3 k).\,(dx\, i + dy\, j + dz\, k) = A_1 dx + A_2 dy + A_3\, dz$

Hence the cartesian form of the Stoke's theorem can be stated as,

$$\iint_S \left[\left(\frac{\partial A_3}{\partial y} - \frac{\partial A_2}{\partial z}\right)\cos\alpha + \left(\frac{\partial A_1}{\partial z} - \frac{\partial A_3}{\partial x}\right)\cos\beta + \left(\frac{\partial A_2}{\partial x} - \frac{\partial A_1}{\partial y}\right)\cos\gamma\right] ds$$

$$= \oint_C A_1 dx + A_2 dy + A_3 dz$$

(c) Suppose that the surface 'S' lies in *xy*-plane i.e., Z = 0. Then the normal to the surface will be along Z-axis.

$\therefore\ \vec{n} = k;\quad$ since $Z = 0,\quad \vec{F} = F_1 i + F_2 \hat{j}$

$$\nabla \times \vec{F} = \begin{vmatrix} i & j & k \\ \partial/\partial x & \partial/\partial y & \partial/\partial z \\ F_1 & F_2 & 0 \end{vmatrix} = -\frac{\partial F_2}{\partial z} i + \frac{\partial F_1}{\partial z} j + \left(\frac{\partial F_2}{\partial x} - \frac{\partial F_1}{\partial y}\right) k$$

$$\left(\nabla\times\vec{F}\right).\vec{n} = \frac{\partial F_2}{\partial z} - \frac{\partial F_1}{\partial y}$$; In xy-plane, $ds = dx\,dy$

∴ Stoke's theorem states

$$\int_C \vec{F}.dr = \iint_S (\nabla\times\vec{F}).\vec{n}\,ds$$

i.e., $$\int_C \vec{F}.dr = \iint_S \left(\frac{\partial F_2}{\partial x} - \frac{\partial F_1}{\partial y}\right) dx\,dy,$$ which is Green's theorem

Solved Examples

***Example* 14.79:** Verify Stoke's Theorem for $\mathbf{A} = (x - 2y)i + yz^2j + y^2zk$, where S is the upper half of the sphere $x^2 + y^2 + z^2 = 1$ and C is its boundary.

***Solution*:**

The boundary of the projection of S in the xy-plane is a circle with centre at origin and unit radius. Its parametric equations are $x = \cos\theta$, $y = \sin\theta$, $z = 0$, $0 \leq \theta < 2\pi$

∴ $dx = (-\sin\theta)d\theta,\ dy = (\cos\theta)d\theta.$

∴ $$\oint_C A.dr = \oint_C (x-2y)dx + yz^2dy + y^2zdz$$

$$= \int_{\theta=0}^{2\pi} [\cos\theta - 2\sin\theta]\,(-\sin\theta)d\theta \qquad (\because z = 0)$$

$$= \int_0^{2\pi} [\frac{-\sin 2\theta}{2} + 1 - \cos 2\theta]d\theta = \frac{\cos 2\theta}{4} + \theta - \frac{\sin 2\theta}{2}\Bigg|_0^{2\pi} = 2\pi$$

$$\nabla\times\mathbf{A} = \begin{vmatrix} i & j & k \\ \partial/\partial x & \partial/\partial y & \partial/\partial z \\ (x-2y) & yz^2 & y^2z \end{vmatrix} = 2k$$

$$\iint_S (\nabla\times A).nds = \iint_S 2(n.k)ds = 2\iint_R dxdy$$

($n.k\ ds = dxdy$ and R is the projection of S on the xy plane)

$$\iint_S (\nabla\times A).nds = 2\int_{x=-1}^{+1}\int_{y=-\sqrt{1-x^2}}^{y=\sqrt{1-x^2}} dy\,dx = 4\int_{-1}^{+1}\sqrt{1-x^2}\,dx = 8\int_0^1\sqrt{1-x^2}dx = 2\pi$$

Hence Stoke's theorem is verified.

Example 14.80

EXAMPLE 80

Prove that a necessary and sufficient condition that $\oint_c F.dr = 0$ for every closed curve C is that $\nabla \times \mathbf{F} = 0$.

Proof: (a) The condition is necessary : Let $\nabla \times F = 0$;

Then by Stoke's theorem, $\int_C \mathbf{F}.dr = \iint_S (\nabla \times \mathbf{F}).\mathbf{n}\, ds = 0$

(b) The condition is sufficient:

Suppose that $\oint_c F.dr = 0$ around every closed path c.

Assume that $\nabla \times \mathbf{F} \neq 0$ at some point P. then, assuming that $\nabla \times F$ is continuous, there exists a region with P as its interior point where $\nabla \times F = 0$.Let S be surface contained in this region and let the normal n to S at each point has the same direction as $\nabla \times F$.

Then $\nabla \times F = a\, \boldsymbol{n}$, (a being a +ve constant);Let C be the boundary of S.

Then by Stokes theorem, $\int_c F.dr = \iint_s (\nabla \times F).n\, ds = a \iint_s n.nds > 0$

Which is a contradiction to the hypothesis that $\int F.dr = 0$; $\therefore \nabla \times F = 0$

Note: It follows that $\nabla \times F = 0$ is also a necessary and sufficient condition for the line integral $\int_{n}^{P_2} F.dr$ to be independent of path joining the points P_1 and P_2.(see 8.6.4)

Example 14.81

If $\vec{r} = xi + yj + zk$, $r = xi + yj + zk$, show that $\int_c r.dr = 0$

$$\text{But Curl } \vec{r} = \begin{vmatrix} i & j & k \\ \frac{\partial}{\partial x} & \frac{\partial}{\partial y} & \frac{\partial}{\partial z} \\ x & y & z \end{vmatrix} = 0 \qquad .....(1)$$

$$\therefore (1) \Rightarrow \int_C \mathbf{r.dr} = 0$$

***Example* 14.82:** If '*f*' and '*g*' are scalar functions, show that

$$\int_C f(grad\ g).\ \mathbf{dr} = -\int_C g(grad\ f).\mathbf{dr}$$

***Solution*:**

By stoke's theorem,

$$\int_C \{\text{grad } (fg)\}.\ \mathbf{dr} = \iint_S [\text{curl}\{\text{grad}(fg)\}].\mathbf{n}\ ds = 0, \text{ since curl grad } (fg) = 0$$

$$\therefore \int_C \{grad\ (fg)\}.\mathbf{dr} = 0 \qquad ..... (1)$$

But grad$(fg) = f(grad\ g) + g(grad\ f)$ ..... (2)

Hence the result [from (1) and (2)]

***Example* 14.83:** If A is any vector function, prove by stoke's theorem that div curl A = 0.

***Solution*:**

Let V be any volume enclosed by a closed surface S. Then by Gauss' divergence theorem. We get,

$$\iiint_v \nabla.(curl\ \mathbf{A})dv = \iint_s (curl\ \mathbf{A}).\mathbf{n}ds \qquad ..... (1)$$

Divide the surface S into two portions S_1 and S_2 by a closed curve C.

Then $\iint_S (curl\ \mathbf{A}).\mathbf{n}\ ds$

$$= \iint_{S_1} (curl\ \mathbf{A}).\mathbf{n}\ ds_1 + \iint_{S_2} (curl\ \mathbf{A}).\mathbf{n}\ ds_2 = \int_C \mathbf{A.dr} - \int_C \mathbf{A.dr}$$

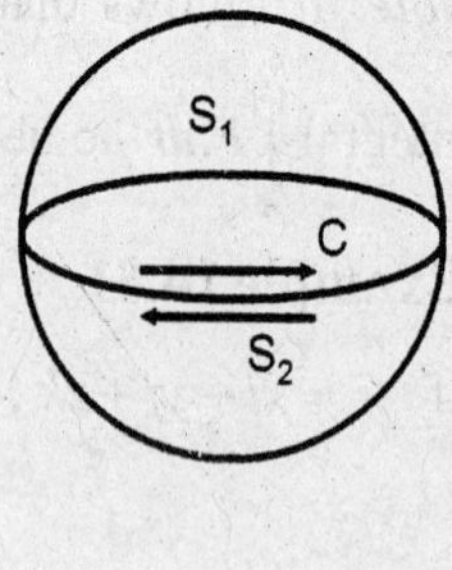

= 0, by Stoke's theorem, since the +ve directions along the boundaries of S_1 and S_2 are opposite.

$$\therefore (1) \Rightarrow \iiint \nabla.(curl\ A)dv = 0$$

Since this is true for all volume elements V, we have,

$$\nabla \,.\, \text{curl } \mathbf{A} = 0 \Rightarrow \text{div}(\text{curl } \vec{A}) = 0$$

***Example* 14.84:** Use Stoke's theorem and prove that curl grad f = 0, where 'f' is a scalar function.

***Solution*:**

If S is a surface enclosed by a simple closed curve C, we have, by Stoke's theorem,

$$\iint_s \{curl(grad\ f)\}.\boldsymbol{n}\ ds = \int_C (grad\ f).\boldsymbol{dr} \qquad .... (1)$$

Now, grad f. **dr**

$$= (\frac{\partial f}{\partial x} i + \frac{\partial f}{\partial y} j + \frac{\partial f}{\partial z} k).(dx\, i + dy\, j + dz\, k)$$

$$= \frac{\partial f}{\partial x} dx + \frac{\partial f}{\partial y} dy + \frac{\partial f}{\partial z} dz = df$$

$$\therefore \int_C (grad\ f).\boldsymbol{dr} = \int_A^A df$$

= f(A) – f(A) = 0, where A is any point on C

$$\therefore (1) \Rightarrow \iint_s \{curl(grad\ f)\}.\boldsymbol{n}ds = 0$$

Since this equation is true for all surface elements S,

we have, curl (grad f) = 0

***Example* 14.85:** Verify stoke's theorem for **A** = $y^2 j - 2xyj$ taken round the rectangle bounded by $x = \pm b$, $y = 0$, $y = a$

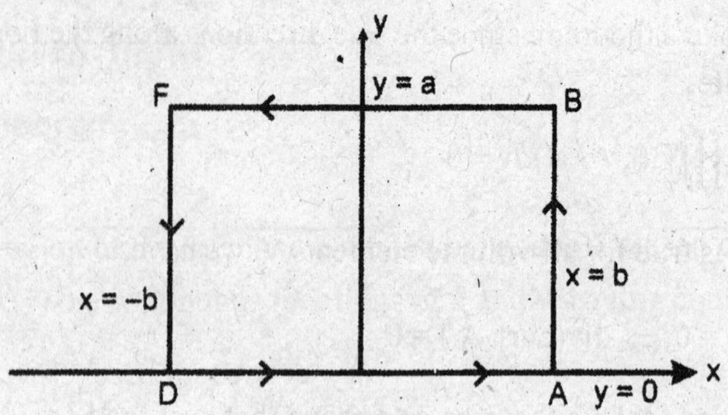

***Solution*:**

$$\text{curl } \vec{A} = \begin{vmatrix} i & j & k \\ \partial/\partial x & \partial/\partial y & \partial/\partial z \\ y^2 & -2xy & 0 \end{vmatrix} = -4yk$$

For the given surface S, **n** = k

∴ (curl **A**). **n** = – 4*y*

Hence $\iint_s (curl A).n\, ds$

$$= \iint_s -4y\, dx\, dy = \int_{y=0}^{a} [\int_{x=-b}^{b} -4y dx] \text{dy} = \int_0^a -4xy\Big|_{-b}^{b} dx = \int_0^a -8by\, dy$$

$$= -4by^2\Big|_0^a = -4a^2b \qquad \text{..... (1)}$$

$$\int_C A.dr = \int_{DA} + \int_{AB} + \int_{BF} + \int_{FD}$$

$$\int A.dr = y^2 dx - 2xydy$$

Along DA, y = 0, dy = 0, $\Rightarrow \int_{DA} A.dr = 0$ $\qquad$ (∵ A.dr = 0)

Along AB, $x = \text{b}, dx = 0$

$$\therefore \int_{AB} A.dr = \int_{y=0}^{a} -2by\,dy = -by^2\Big|_0^a = -a^2b$$

Along BF, $y = a,\ dy = 0$

$$\therefore \int_{BF} A.dr = \int_{b}^{-b} a^2\ dx = -2a^2b$$

Along FD, $x = -b,\ dx = 0$

$$\therefore \int_{FD} A.dr = \int_{a}^{0} 2by\,dy = -by^2\Big|_a^0 = -a^2b$$

$$\therefore \int_C A.dr = 0 - a^2b - 2a^2b - a^2b = -4a^2b \qquad \text{..... (2)}$$

From (1) and (2), $\int_C A.dr = \iint_S curl(A).nds$

Hence the theorem is verified

***Example* 14.86:** Use stoke's theorem to evaluate the integral $\int_C A.dr$ where $A = 2y^2i + 3x^2j - (2x+z)k$, and C is the boundary of the triangle whose vertices are (0, 0, 0), (2, 0, 0), (2, 2, 0).

***Solution*:**

$$\text{Curl } \mathbf{A} = \begin{vmatrix} i & j & k \\ \partial/\partial x & \partial/\partial y & \partial/\partial z \\ 2y^2 & 3x^2 & -2x-z \end{vmatrix}$$

$$= 2j + (6x - 4y)k$$

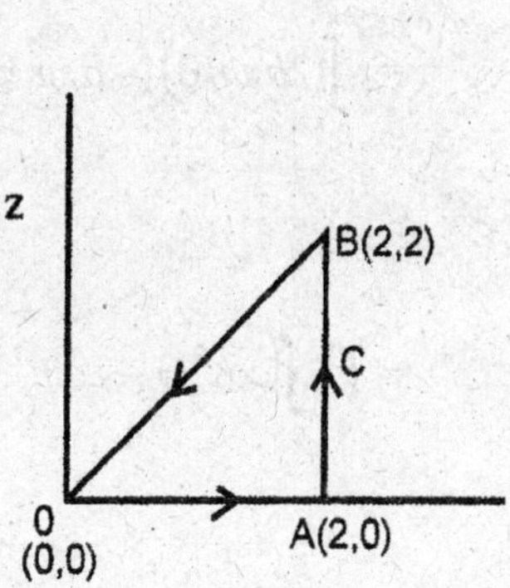

Since the z–coordinate of each vertex of the triangle is z xy–plane.

$\therefore n = k.$

$\therefore$ (curl **A**). **n** $= 6x - 4y$

consider the triangle in xy-plane.

Equation of the straight line OB is $y = x$

By Stoke's theorem,

$$\int_C A.dr = \iint_S (Curl A).nds$$

$$= \int_{x=0}^{2}\int_{y=0}^{y=x}(6x-4y)dxdy = \int_{x=0}^{2}\left[\int_{y=0}^{x}(6x-4y)dy\right]dx$$

$$= \int_{x=0}^{2} 6xy - 2y^2 \Big|_0^x dx = \int_0^2 (6x^2 - 2x^2)dx = 4\frac{x^3}{3}\Big|_0^2 = \frac{32}{3}.$$

***Example* 14.87:** Use Stoke's theorem to evaluate $\iint (curl A).nds$, where $A = 2yi + (x - 2zx)j + xyk$, and S is the surface of the sphere $x^2 + y^2 + z^2 = b^2$ above the xy-plane.

Solution:

The boundary C of the surface S is the circle $x^2 + y^2 + z^2 = b^2$, z = 0.

The parametric equations of C are $x = b\cos\theta$, $y = b\sin\theta$, $z = 0$, $0 \le \theta < 2\pi$

By Stoke's theorem, we have,

$$\iint_S (Curl A).nds = \int_C A.dr$$

$$= \int_C 2ydx + (x - 2zx)dy + xydz = \int_C 2ydx + xdy \qquad (\because z = 0, dz = 0 \text{ on } C)$$

$$= \int_0^{2\pi} (2b\sin\theta)(-b\sin\theta)d\theta + b\cos\theta.b\cos\theta.d\theta \qquad [\because x = b\cos\theta \Rightarrow dx = -b\sin\theta\, d\theta$$

$$\text{and } y = b\sin\theta \Rightarrow dy = b\cos\theta\, d\theta]$$

$$= b^2\int_0^{2\pi}(\cos^2\theta - 2\sin^2\theta)d\theta = b^2\int_0^{2\pi}\left[\frac{1+\cos 2\theta}{2} - (1-\cos 2\theta)\right]d\theta$$

$$= \frac{b^2}{2}\int_0^{2\pi}(-1+3\cos 2\theta) = +\frac{b^2}{2}\left[-\theta+\frac{3\sin 2\theta}{2}\right]_0^{2\pi} = -\frac{b^2}{2}.2\pi = -\pi b^2$$

***Example* 14.88:** Apply Stoke's theorem to evaluate $\int A.dr$ .where $A = (x-y)i + (2y+z)j + (y-z)k$ and C is the boundary of the triangle whose vertices are $\left(\frac{1}{6},0,0\right)$ $\left(0,\frac{1}{3},0\right)$ and $\left(0,0,\frac{1}{2}\right)$.

***Solution*:**

Let $A = \left(\frac{1}{6},0,0\right)$ $B = \left(0,\frac{1}{3},0\right)$ $C = \left(0,0,\frac{1}{2}\right)$.

The equation of the plane ABC is (by intercept form),

$$\frac{x}{1/6}+\frac{y}{1/3}+\frac{z}{1/2} = 1 \Rightarrow 6x+3y+2z = 1 \qquad \text{..... (1)}$$

The direction ratios of the normal to (1) are 6, 3, 2

$\therefore$ Direction cosines are $\frac{6}{7},\frac{3}{7},\frac{2}{7}$ $\qquad \left(\because \sqrt{6^2+3^2+2^2}=7\right)$

If $\vec{n}$ is the unit normal to the plane, $\vec{n} = \frac{6}{7}i+\frac{3}{7}j+\frac{2}{7}k$

$$\mathbf{A} = (x-y)i + (2y+z)j + (y-z)k$$

$$\nabla \times \boldsymbol{A} = \begin{vmatrix} i & j & k \\ \partial/\partial x & \partial/\partial y & \partial/\partial z \\ x-y & 2y+z & y-z \end{vmatrix} = k$$

$$\therefore \ (\nabla \times \boldsymbol{A}).\boldsymbol{n} = \frac{2}{7}$$

∴ By Stoke's theorem,

$$\int_C \mathbf{A}.dr = \iint_S (\nabla \mathbf{F}).\mathbf{n}\, ds = \frac{2}{7}\iint_S ds = \frac{2}{7}.(\text{Area of triangle ABC})$$

To find the area of triangle ABC:

$$AB = \sqrt{\left(\frac{1}{6}\right)^2 + \left(\frac{1}{3}\right)^2} = \frac{\sqrt{5}}{6},$$

$$AC = \sqrt{\left(\frac{1}{6}\right)^2 + \left(\frac{1}{2}\right)^2} = \frac{\sqrt{10}}{6}$$

Direction ratios of AB are $\frac{-1}{6}, \frac{1}{3}, 0$.

Direction Cosines of AB are $\frac{-1}{\sqrt{5}}, \frac{2}{\sqrt{5}}, 0$

Direction ratios of AC are $\frac{-1}{6}, 0, \frac{1}{2}$

Direction cosines of AC are $\frac{-1}{\sqrt{10}}, 0, \frac{3}{\sqrt{10}}$

$$\cos C\hat{A}B = \left(\frac{-1}{\sqrt{5}}\right)\left(\frac{-1}{\sqrt{10}}\right) + 0 + 0 = \frac{1}{\sqrt{50}}$$

$$\sin C\hat{A}B = \frac{7}{\sqrt{50}}$$

$$\therefore \text{ Area of triangle } ABC = \frac{1}{2} AB.AC \sin C\hat{A}B = \frac{1}{2}.\frac{\sqrt{5}}{6}.\frac{\sqrt{10}}{6} = \frac{7}{72}$$

$$\int_c \mathbf{A}.\overline{dr} = \frac{2}{7} \times \frac{7}{72} = \frac{1}{36} \text{ [from(2)]}$$

***Example* 14.89**

Evaluate $\iint_s (curl\ A).n\, ds$ taken over the portion s of the surface $x^2 + y^2 + z^2 - 2fx + fz = 0$ above the xy *plane* $z = 0$, if $A = \Sigma(x^2 + y^2 - z^2)i$ and verify Stoke's theorem.

Solution:

Let 'S' denote the portion of the surface, $x^2 + y^2 + z^2 - 2fx + fz = 0$ above the *xy-plane* $z = 0$.

The surface S meets the xy-plane in the circle 'C', whose equations are $x^2 + y^2 - 2fx = 0, z = 0$.

$\Rightarrow (x-f)^2 + y^2 = f^2, z = 0$

$\therefore$ The parametric equations of 'C' can be taken as

$x = f + f\cos\theta, \qquad y = f\sin\theta, \qquad z = 0 \qquad (0 \le \theta < 2\pi)$

Let S_1 denote the plane region bounded by C. If S^1 is the surface consisting of S and S_1, S^1 is a closed surface.

$\therefore$ From divergence theorem, we have

$$\iint_{s^1} (curl\ A.n)\, ds = 0$$

i.e., $\iint_S (Curl\ A).n\, ds + \iint_{S_1} (Curl\ A).n\, ds = 0 \qquad [\because S^1 \text{ consists of S and } S_1]$

i.e., $\iint_S (Curl\ A).n\, ds - \iint_{S_1} (curl\ A).k ds = 0 \qquad [\because n = -k \text{ on } S_1]$

$\therefore \iint_S (Curl\ A).n\, ds = \iint_{S_1} (curl\ A).k ds \qquad \ldots\ldots (1)$

$$\text{Now, curl } \mathbf{A} = \begin{vmatrix} i & j & k \\ \partial/\partial x & \partial/\partial y & \partial/\partial z \\ x^2+y^2-z^2 & y^2+z^2-x^2 & z^2+x^2-y^2 \end{vmatrix}$$

$$= i(-2y-2z) + j(-2z-2x) + k(-2x-2y)$$

$\therefore$ (curl **A**). $k = -2(x+y)$

$\therefore$ From (1), $\iint_S (curl\ A).n ds = -2\iint_{s^1} (x+y) ds$

Polar equation of S_1 is $r = 2f\cos\theta$.

$\therefore$ changing to polar coordinates,

$$\iint_{S_1} (curl\ A).n ds = -2\int_{\theta=0}^{\pi}\int_{r=0}^{2f\cos\theta} (r\cos\theta + r\sin\theta)\, r d\theta dr$$

$$= -2\int_{\theta=0}^{\pi}\left[\int_{r=0}^{2f\cos\theta} (\cos\theta+\sin\theta) r^2 .dr\right] d\theta = -2\int_{\theta=0}^{\pi} (\cos\theta+\sin\theta) \frac{r^3}{3}\Big|_0^{2f\cos\theta} d\theta$$

$$= -2\times\frac{8f^3}{3}\int_0^{\pi}(\cos\theta+\sin\theta)\cos^3\theta d\theta = -\frac{16f^3}{3}\int_0^{\pi}\cos^4\theta d\theta \quad [\because \int_0^{\pi}\cos^3\theta\sin\theta d\theta = 0]$$

$$= \frac{16f^3}{3}\times 2\int_0^{\pi/2}\cos^4\theta d\theta = -\frac{32f^3}{3}.\frac{3}{4}\times\frac{1}{2}\times\frac{\pi}{2} = -2\pi f^3 \qquad \text{..... (2)}$$

Again, $\int_C A.dr$

$$= \int_C (x^2+y^2-z^2)dx + (y^2+z^2-x^2)dy + (z^2+x^2-y^2)dz$$

$$= \int_C (x^2+y^2)dx + (y^2-x^2)dy \qquad [\because \text{ on C } \quad z=0, \quad dz=0]$$

$$= \int_0^{2\pi}\left[(f+f\cos\theta)^2 + f^2\sin^2\theta\right](-f\sin\theta)d\theta$$

$$+ \int_0^{2\pi}\left[f^2\sin^2\theta - (f+f\cos\theta)^2\right](f\cos\theta)d\theta$$

$$f^3\left[\int_0^{2\pi}\left[\{(1+\cos\theta)^2+\sin^2\theta\}(-\sin\theta) + \{\sin^2\theta - (1+\cos\theta^2)\}\cos\theta\right]d\theta\right.$$

$$= f^3\left[\int_0^{2\pi}-2\sin\theta d\theta + \int_0^{2\pi}-2\sin 2\theta d\theta + \int_0^{2\pi}-2\cos\theta d\theta + \int_0^{2\pi}-\cos^3\theta d\theta\right.$$

$$\left. - \int_0^{2\pi}(1+\cos 2\theta)d\theta + \int_0^{2\pi}\sin^2\theta\cos\theta d\theta\right] \qquad \text{(on simplification)}$$

$$= f^3(-2\pi) = -2\pi f^3 \qquad \text{(all other integrals vanish)} \qquad \text{.....(3)}$$

from (2) and (3), we have

$$\int_S\int(Curl A).nds = \int_C A.dr$$

which verifies Stoke's theorem.

Exercise 14.11

1. If $F = (xe^x)i + (3y^2)j - (z)dz$, and C is the curve $x^2 + y^2 = 9, z = 2$, evaluate $\oint \vec{F}.d\vec{r}$ using Stoke's theorem.

[Ans: 0]

2. Apply Stoke's theorem to obtain the value of the integral $\oint_C V.dr$,where $V = (3y^2)i + (2x^2)j - (x + 2z)k$ and C is the boundary of the triangle whose vertices are (0,0,0),(1,0,0) and (1,1,0)

[Ans: 1]

3. Verify Stoke's theorem for $F = (2x - y)i - (yz^2)j - (y^2z)k$ if S is the upper half surface of the sphere $x^2 + y^2 + z^2 = 1$ and C is the boundary .

[Ans: π]

4. If $F = (y - z + 2)i + (yz + 4)j - (xz)k$ and S represents the surface of the cube $x = 0, y = 0, z = 0, x = 2, y = 2, z = 2$ above the *xy plane* , verify that $\iint_s (Curl\ F).ds = \oint_C F.dr$,C being the boundary of S traversed in the +ve direction.

[Ans: Each integral =–4]

5. Find the value of the integral $\oint_c (yz)dx + (zx)dy + (xy)dz$, using Stoke's theorem where C is the Curve $x^2 + y^2 = 4, z = y^2$

[Ans: 0]

6. Verify Stoke's theorem for the function $V = (3x^2)i + (2xy)j$, integrated along the square x =0, y=0, x= 1,y=1 in the $xy - plane$. [Ans: 0]

7. Evaluate by Stoke's theorem the integral $\oint_C A.dr$ where $A = (2 \sin z)i - (3 \cos x)j +$ $(\sin y)\ k$; where C is the boundary of the rectangle $0 \le x \le \pi, 0 \le y \le 1, z = 0$

[Ans: 6]

Exercise 14.12 (Miscellaneous problems)

1. If $f(x,y,z)=x^l y^m z^n -1$, find the directional derivative of f at (1,1,1) in the direction of (i+2j+2k)

[**Ans:** $\frac{1}{3}(l+2m+2n)$]

2. Find the acute angle between the surfaces

$x^2+y^2+z^2=6$ and $3xyz+y^2z-xy+3=0$ at (1,-1,2)

[**Ans:** $\cos^{-1}\left(\frac{\sqrt{5}}{3}\right)$]

3. If $\boldsymbol{r} = xi + yj + zk$, and $\boldsymbol{p}$, $\boldsymbol{q}$ are constant vectors show that $Div\{(\boldsymbol{r}\times\boldsymbol{p})\times\boldsymbol{q}\}=-2(\boldsymbol{p}\cdot\boldsymbol{q})$

4. If $\boldsymbol{F}=(x^2y)i-(y^2z)j+(z^2x)k$, find curl curl **F** at (1,-2,3)

5. If $f=xyz(x+y+z)$, prove that $curl\ grad\ f=0$

6. A fluid motion is given by $V=(z^3)i-(y^2)j+(3xz^2)k$ .Show that it is irrotational. Find its velocity potential ϕ such that $V=\nabla\phi$

[Ans: $\phi = xz^3 - \frac{y^3}{3} + c$]

7. Evaluate $\int_{(0,0)}^{(2,1)}(4x^3-12x^2y^2)dx-(8x^3y)dy$ along the path $x^3-3xy^2=2y^3$

[Ans: 16]

8. If $V = (xy)i - (yz)j + (zx^2)k$ and S is the surface of the cube bounded by $x=0, x=2, y=0, y=2, z=0, z=2$, evaluate $\int_s V.n\,ds$

[Ans:32/3]

9. If $f=4x+yz$, evaluate $\iiint_V f dv$ over the region in the first octant bounded by

$x^2+y^2=1, z=0, z=3$ [Ans:11/2]

10. Verify Green's theorem in the plane for $\int_C (x^3 - y^2)dx + (x^2 - 2xy)dy$,where C is the boundary of the square bounded by $0 \le x \le 1, 0 \le y \le 1$ [Ans:1]

11. Using Gauss divergence theorem, prove that $\iint_s F.nds = \pi r^2 l^2$, where $F = (y^2z)i + (xz)j + (z^2)k$, and S is the surface bounded by $x^2 + y^2 = r^2, z = 0,$ and $z = l$

12. Show that the Stoke's theorem , when restricted to the $xy - plane$, is Green's theorem in the plane .
(Hint : In Stoke's theorem , take $\mathbf{A} = Pi + Qj$; $\boldsymbol{n} = k$; and $ds = dxdy$)

Summary

1. Grad $\phi = \nabla\phi = i\frac{\partial\phi}{dx} + j\frac{\partial\phi}{\partial y} + k\frac{\partial\phi}{\partial z}$ $\left[\nabla = i\frac{\partial}{\partial x} + j\frac{\partial}{\partial y} + k\frac{\partial}{\partial z}\right]$ represents the normal vector to the surface $\phi = c$.

2. Directional derivative of ϕ in the direction of $\vec{a} = \frac{\nabla\phi.\vec{a}}{|\vec{a}|}$.

3. Divergence of $\vec{F} = Div\vec{F} = \nabla.F = \frac{\partial F_1}{\partial x} + \frac{\partial F_2}{\partial y} + \frac{\partial F_3}{\partial z}$, (where $\vec{F} = F_1 i + F_2 j + F_3 k$) is a scalar.

4. $\vec{F}$ is solenoidal if $Div\,\vec{F} = 0$.

5. Curl $\vec{F} = \nabla \times \vec{F} = \begin{vmatrix} i & j & k \\ \frac{\partial}{\partial x} & \frac{\partial}{\partial y} & \frac{\partial}{\partial z} \\ F_1 & F_2 & F_3 \end{vmatrix}$ is a vector where $\vec{F} = F_1 i + F_2 j + F_3 k$.

6. $\vec{F}$ is irrotational if curl $\vec{F} = \vec{O}$.

Part III

CHAPTER 15

Fourier Series

15.0 Introduction

"Fourier series", which was named after the French Mathematician "Jean-Baptiste Joseph Fourier" (1768-1830) has many interesting applications in Engineering problems as well as Mathematical physics. Fourier series has been an important tool in solving problems in many fields like current and voltage in an alternating circuit, conduction of heat in solids, electrodynamics etc. In this chapter, the basic concepts relating to Fourier series and expansion of several functions as Fourier series are discussed.

15.1 Periodic Functions

A function $f : R \to R$ is said to be periodic if there exists a positive number T such that $f(x + T) = f(x)$ for all $x \in R$. T is called the period of $f(x)$.

The smallest positive period T of a function 'f is sometimes called the *primitive period of f.*

Examples

1. Trigonometric functions sin x, cos x, cosec x are periodic functions with primitive period 2π.
2. sin $2x$, cos $2x$ are periodic functions with primitive period π.
3. tan x has a primitive period π.
4. sin nx is periodic with primitive period $\frac{2\pi}{n}$.

5. $\tan nx$ is periodic with primitive period $\frac{\pi}{n}$.
6. If n is any integer, then $f(x + nT) = f(x)$ shows that any integral multiple of T is also a period.
7. If f_1, and f_2 are periodic functions having the same period T, then, $f(x) = c_1 f_1(x) + c_2 f_2(x)$ [c_1, c_2 being constants] also has the period T.
8. If T is the period of 'f' then $f(c_1x + c_2)$ also has the period T. (c_1, c_2 are constants)

15.1.1 Even and odd functions

Even functions :

A function $f(x)$ is said to be even if $f(-x) = f(x)$.

Examples

(i) $f(x) = \cos x$ is even.

(ii) $f(x) = x^2$ is even.

Odd functions :

A function $f(x)$ is said to be odd if $f(-x) = -f(x)$.

Examples

(i) $f(x) = \sin x$ is odd.

(ii) $f(x) = x^3$ is odd.

Note:

1. There may be some functions which are neither even nor odd.

Example

$$f(x) = 4\sin x + 3\tan x - e^x$$

2. (a) The product of two even functions is again even.
   (b) The product of two odd functions is even.
   (c) The product of an even and an odd functions is odd.

Solved Examples

***Example* 15.1:** If f and g are periodic functions with same period T, show that $(c_1f + c_2 g)$ is also a periodic function of period T, where c_1 and c_2 are real numbers.

***Solution*:**

Since f and g are periodic functions with period T, we have,

$$f(x + T) = f(x) \quad .....(1)$$

and $$g(x + T) = g(x) \text{ for all } x. \quad .....(2)$$

$$(c_1f + c_2g)(x + T) = c_1f(x + T) + c_2g(x + T)$$
$$= c_1f(x) + c_2g(x) \qquad \text{from (1) \& (2)}$$
$$= (c_1f + c_2g)(x)$$

$\therefore$ $(c_1f + c_2g)$ is periodic with period T.

***Example* 15.2:** Find the primitive periods of the functions

(i) $\sin 3x$ (ii) $\cos 2\pi x$

(iii) $\sin\left(\dfrac{2\pi x}{k}\right)$ (iv) $\tan 2x$

***Solution*:**

(i) Let T be the period of $\sin 3x$.

Then $\sin 3x = \sin\{3(x + T)\}$ .....(1)

But $\sin 3x = \sin(3x + 2\pi)$

$$= \sin\left\{3\left(x + \frac{2\pi}{3}\right)\right\} \qquad .....(2)$$

From (1) and (2), we get, $T = \dfrac{2\pi}{3}$.

$\therefore$ period of $\sin 3x = \dfrac{2\pi}{3}$.

(ii) Let T be the period of $\cos 2\pi x$

$\therefore$ $\cos 2\pi x = \cos\{2\pi(x + T\}$ .....(3)

But $\cos 2\pi x = \cos(2\pi x + 2\pi)$

$= \cos\{2\pi(x + 1\}$ .....(4)

from (3) and (4),

$T = 1$

$\therefore$ Period of $\cos 2\pi x = 1$.

(iii) Let T be the period of $\sin\left(\dfrac{2\pi x}{K}\right)$

$$\therefore \quad \sin\frac{2\pi x}{K} = \sin\left\{\frac{2\pi(x + T)}{K}\right\} \qquad .....(5)$$

But $\sin\dfrac{2\pi x}{K} = \sin\left(\dfrac{2\pi x}{K} + 2\pi\right)$

$$= \sin\left\{\frac{2\pi}{K}(x+K)\right\} \qquad .....(6)$$

(5) and (6) $\Rightarrow T = K$

$$\therefore \text{ Period of } \sin\left(\frac{2\pi x}{K}\right) = K$$

(iv) Let T be the period of tan $2x$.

$$\therefore \quad \tan 2x = \tan\{2(x + T)\} \qquad .....(7)$$

But
$$\tan 2x = \tan\left\{2\left(x+\frac{\pi}{2}\right)\right\} \qquad .....(8)$$

from (7) and (8), $T = \dfrac{\pi}{2}$

$$\therefore \text{ Period of } \tan 2x = \frac{\pi}{2}$$

***Example* 15.3:** Express the function $f(x)$ as the sum of an even function and an odd function.

***Solution*:**

$f(x)$ can be written as,

$$f(x) = \frac{1}{2}[f(x) + f(x)]$$

$$= \frac{1}{2}[f(x) + f(-x)] + \frac{1}{2}[f(x) - f(-x)] \qquad .....(1)$$

Let $\quad \dfrac{1}{2}[f(x) + f(-x)] = g(x)$

and $\quad \dfrac{1}{2}[f(x) - f(-x)] = h(x)$,

so that, $f(x) = g(x) + h(x)$ ......(2)

(from (1))

Now, $\quad g(-x) = \dfrac{1}{2}[f(-x) + f(x)] = g(x)$

i.e., $g(x)$ is even.

and, $h(-x) = \frac{1}{2}[f(-x) - f(x)] = -h(x)$

i.e., $h(x)$ is odd.

$\therefore$ (2) gives the required representation.

***Example* 15.4:** Find whether the following functions are even, or odd or neither of them.

(i) $x \sin x + \cos x + x^2 \cosh x$

(ii) $\text{Log}\left[\frac{1-x}{1+x}\right]$

(iii) $x \cosh x + x^3 \sinh x$

Solution:

(i) Let $f(x) = x \sin x + \cos x + x^2 \cosh x$

$$f(-x) = (-x) \sin(-x) + \cos(-x) + (-x)^2 \cosh(-x)$$
$$= x \sin x + \cos x + x^2 \cosh x$$
$$= f(x) \qquad [\because \sin(-x) = -\sin x$$
$$\cos(-x) = \cos x$$
$$\cosh(-x) = \cosh x]$$

$\therefore f(x)$ is even

(ii) Let $f(x) = \log\left[\frac{1-x}{1+x}\right]$

$$f(x) = \log(1-x) - \log(1+x)$$
$$f(-x) = \log(1+x) - \log(1+x)$$
$$= -[\log(1-x) - \log(1+x)]$$
$$= -f(x)$$

$\therefore f(x)$ is odd.

(iii) $f(x) = x \cosh x + x^3 \sinh x$

$$f(-x) = -x \cosh(-x) + (-x)^3 \sinh(-x)$$
$$= -x \cosh x + x^3 \sinh x$$

$\therefore f(x)$ is neither even nor odd.

Exercise – 15(a)

I. Find the primitive periods of the following functions.

1. $\tan 3x$ 2. $\sec 4x$ 3. $\cos \frac{\pi}{3} x$

[Ans : 1. $\frac{\pi}{3}$ 2. $\frac{\pi}{2}$ 3. 6]

II. Find whether the following functions are even, or odd, or neither of them.

(i) $4x^2 \cos x + x^3 \sinh x - x^4$

[Ans : even**]**

(ii) $x \sin x + x^3 \cos x$

[Ans : neither even nor odd**]**

(iii) $x^3 \cosh x - x$

[Ans : odd**]**

15.2 Fourier Series

15.2.1 Introduction

Representation of periodic functions in terms of simple periodic functions is of great importance in engineering problems. We now discuss to express various functions of period 2π in terms of simple functions as 1, $\sin x$, $\cos x$, $\sin 2x$, $\cos 2x$, ...

15.2.2 Trigonometric series

A series of the form, $a_0 + (a_1 \cos x + a_2 \cos 2x + ... + a_n \cos nx + ...) + (b_1 \sin x + b_2 \sin 2x + ... + b_2 \sin nx + ...) = a_0 + \sum_{n=1}^{\infty} a_n \cos nx + \sum_{n=1}^{\infty} b_n \sin nx$, where a_0, a_n, b_n are real constants is called a *trigonometric series* and a_0, a_n, b_n are coefficients of the series.

Since each term of the above series is a function of period 2π, it can be observed that if the series is convergent then its sum is also a function of period 2π.

15.2.3 Theorem

Let (1) $f(x)$ be a periodic function of period 2π.

(2) The function $f(x)$ be expressed as a trigonometric series

$$f(x) = \frac{a_0}{2} + \sum_{n=1}^{\infty} a_n \cos nx + \sum_{n=1}^{\infty} b_n \sin nx \qquad .....(1)$$

Then,

$$a_0 = \frac{1}{\pi} \int_{-\pi}^{\pi} f(x) dx \qquad .....(A)$$

$$a_m = \frac{1}{\pi} \int_{-\pi}^{\pi} f(x) \cos mx \, dx \qquad .....(B)$$

$$b_m = \frac{1}{\pi} \int_{-\pi}^{\pi} f(x) \sin mx \, dx \qquad .....(C)$$

$$m = 1, 2, 3, ...$$

Proof : To prove the theorem, the following results of integration are used.

$$\int_{-\pi}^{\pi} \sin nx\, dx = \int_{-\pi}^{\pi} \cos nx\, dx = 0 \quad (\text{if } n \neq 0) \qquad .....(2)$$

$$\int_{-\pi}^{\pi} \cos mx \cos nx\, dx = 0 \quad \text{if } m \neq n \qquad .....(3)$$

$$\int_{-\pi}^{\pi} \cos^2 nx\, dx = \pi \qquad .....(4)$$

$$\int_{-\pi}^{\pi} \sin mx \cos nx\, dx = 0 \quad \text{for all } m, n \qquad .....(5)$$

$$\int_{-\pi}^{\pi} \sin mx \sin nx\, dx = 0 \quad \text{if } m \neq n \qquad .....(6)$$

$$\int_{-\pi}^{\pi} \sin^2 nx\, dx = \pi, \text{ if } n \neq 0 \qquad .....(7)$$

Assuming that the series (1) can be integrated term by term, and integrating it from $-\pi$ to π, we get,

$$\int_{-\pi}^{\pi} f(x)\, dx = \int_{-\pi}^{\pi} \frac{a_0}{2}\, dx + \sum_{n=1}^{\infty} a_n \int_{-\pi}^{\pi} \cos nx\, dx + \sum_{n=1}^{\infty} b_n \int_{-\pi}^{\pi} \sin nx\, dx$$

$$= a_0 \pi \qquad \text{(using (2))}$$

$$\therefore \quad a_0 = \frac{1}{\pi} \int_{-\pi}^{\pi} f(x)\, dx \qquad \text{proved (A)}$$

Multiplying (1) both sides by $\cos mx$ and integrating term by term from $-\pi$ to π, we get,

$$\int_{-\pi}^{\pi} f(x) \cos mx\, dx = \frac{a_0}{2} \int_{-\pi}^{\pi} \cos mx\, dx + \sum_{n=1}^{\infty} a_n \int_{-\pi}^{\pi} \cos nx \cos mx\, dx + \sum_{n=1}^{\infty} b_n \int_{-\pi}^{\pi} \sin nx \cos mx dx$$

$$= a_m \int_{-\pi}^{\pi} \cos^2 mx\, dx \qquad \text{[using (2), (3) and (5)]}$$

$$= a_m \cdot \pi \qquad \text{(from (4))}$$

$$\therefore \quad a_m = \frac{1}{\pi}\int_{-\pi}^{\pi} f(x)\cos mx\,dx \qquad \text{proved (B)}$$

Again, multiplying (1) both sides by sin mx and integrating from $-\pi$ to π, we get,

$$\int_{-\pi}^{\pi} f(x)\sin mx\,dx = \frac{a_0}{2}\int_{-\pi}^{\pi}\sin mx\,dx + \sum_{n=1}^{\infty} a_n \int_{-\pi}^{\pi}\cos nx \sin mx\,dx$$

$$+ \sum_{n=1}^{\infty} b_n \int_{-\pi}^{\pi}\sin mx \sin mx\,dx$$

$$= b_m \int_{-\pi}^{\pi}\sin^2 mx\,dx \qquad \text{[using (2), (5) and (6)]}$$

$$= b_m \pi \qquad \text{(from (7))}$$

$$\therefore \quad b_m = \frac{1}{\pi}\int_{-\pi}^{\pi} f(x)\sin mx\,dx \qquad \text{\_\_\_\_\_\_ proved (C)}$$

Note 1 :

Euler's Formulae :

The formulae for the coefficients a_0, a_n, b_n given by [in theorem 15.2.3]

$$a_0 = \frac{1}{\pi}\int_{-\pi}^{\pi} f(x)dx$$

$$a_n = \frac{1}{\pi}\int_{-\pi}^{\pi} f(x)\cos nx\,dx$$

$$b_n = \frac{1}{\pi}\int_{-\pi}^{\pi} f(x)\sin nx\,dx$$

are known as Euler's formulae.

15.2.4 Fourier Series

The series $f(x) = \frac{a_0}{2} + \sum_{n=1}^{\infty} a_n \cos nx + \sum_{n=1}^{\infty} b_n \sin nx$, where a_0, a_n, b_n are given by Euler's formulae [15.2.3 Note (1)] is called the Fourier series of $f(x)$ and a_0, a_n, b_n are called the Fourier coefficients or Fourier constants.

Note 2 :

Let $f(x)$ be a periodic function with period 2π. Then we can obtain the Fourier series in any interval of length 2π, say $(0, 2\pi)$, $[0, 2\pi]$, $(-\pi, \pi)$, $[-\pi, \pi]$ etc.

If the interval is taken as $(c, c + 2\pi)$, (c being a constant), then the Euler's formulae for the Fourier constants are given by,

$$a_0 = \frac{1}{\pi} \int_c^{c+2\pi} f(x)\,dx$$

$$a_n = \frac{1}{\pi} \int_c^{c+2\pi} f(x) \cos nx\,dx$$

and
$$b_n = \frac{1}{\pi} \int_c^{c+2\pi} f(x) \sin nx\,dx$$

15.2.5 Convergence of Fourier Series

The theorem 15.2.3 states that if $f(x)$ can be expanded as a Fourier series, the Fourier coefficients are given by the Euler's formulae. However, for any given function $f(x)$, it is not necessary that its Fourier series need converge to $f(x)$.

But most of the functions occurring in Engineering mathematics are represented by their Fourier series and the following theorem (known as Dirichlet's theorem) gives certain conditions under which a function $f(x)$ can be expanded as a Fourier series.

15.2.6 Dirichlet's Theorem

A function $f(x)$ defined in the interval $a_1 \leq x \leq a_2$ can be represented as a Fourier series if $f(x)$ satisfies the following conditions in the interval.

(i) $f(x)$ and its integrals are finite and single - valued.

(ii) $f(x)$ has a finite number of discontinuities

and (iii) $f(x)$ has a finite number of maxima and minima.

Then the Fourier series converges to $f(x)$ at all points where $f(x)$ is continuous. Also the series converges to the average of the left limit and the right limit of $f(x)$ at each point of discontinuity of $f(x)$.

Note :

The conditions given above are known as Dirichlet's conditions.

Examples

1. "$\sin^{-1} x$" cannot be expanded as Fourier series since it is not single-valued.
2. "$\tan x$" cannot be expanded as Fourier series in $(0, 2\pi)$ since $\tan x$ is infinite at $x = \frac{\pi}{2}$ and $x = \frac{3\pi}{2}$.

15.2.7 Summary

Representation of a function $f(x)$ as a Fourier series in an interval of length 2π.

1. *Working Rule* :

Let $f(x)$ be a given periodic function with period 2π. Let the interval be $(-\pi, \pi)$. Then,

$$f(x) = \frac{a_0}{2} + \sum_{n=1}^{\infty} a_n \cos nx + \sum_{n=1}^{\infty} b_n \sin nx,$$

where,
$$a_0 = \frac{1}{\pi} \int_{-\pi}^{\pi} f(x)\,dx$$

$$a_n = \frac{1}{\pi} \int_{-\pi}^{\pi} f(x) \cos nx\,dx$$

$$b_n = \frac{1}{\pi} \int_{-\pi}^{\pi} f(x) \sin nx\,dx,$$

gives the Fourier expansion of $f(x)$.

2. Some results of integration such as the following are to be remembered.

Result 1 : Generalised rule of integration by parts.

$$\int uv\,dx = uv_1 - u'v_2 + u''v_3 - u'''v_4 + \ldots..$$

[suffixes represent successive integrals and dashes represent successive derivatives]

Result 2 :

$$\int e^{ax} \sin bx\, dx = \frac{e^{ax}}{a^2+b^2} [a \sin bx - b \cos bx]$$

Result 3 :

$$\int e^{ax} \cos bx\, dx = \frac{e^{ax}}{a^2+b^2} [a \cos bx + b \sin bx]$$

Solved Examples

***Example* 15.5:** Obtain a Fourier series to represent $f(x) = x^2$ in $(0, 2\pi)$ and hence deduce that

$$\frac{1}{1^2} - \frac{1}{2^2} + \frac{1}{3^2} - \frac{1}{4^2} + \ldots\ldots = \frac{\pi^2}{12}$$

***Solution*:**

Let $f(x) = x^2$

Let $$x^2 = \frac{a_0}{2} + \sum_{n=1}^{\infty} a_n \cos nx + \sum_{n=1}^{\infty} b_n \sin nx \qquad \ldots..(1)$$

be the Fourier series of $f(x)$ in $(0, 2\pi)$).

Then $$a_0 = \frac{1}{\pi}\int_0^{2\pi} x^2 dx = \frac{1}{\pi}\left[\frac{x^3}{3}\right]_0^{2\pi} = \frac{8\pi^2}{3}$$

$$\therefore \quad \frac{a_0}{2} = \frac{4\pi^2}{3}$$

$$a_n = \frac{1}{\pi}\int_0^{2\pi} x^2 \cos nx\, dx$$

$$= \frac{1}{\pi}\left[x^2\left(\frac{\sin nx}{n}\right) - 2x\left(\frac{-\cos nx}{n^2}\right) + 2\left(\frac{-\sin nx}{n^3}\right)\right]_0^{2\pi}$$

(using generalised rule of integration by parts)

$$= \frac{1}{\pi}\left[\frac{4\pi}{n^2}\right] = \frac{4}{n^2} \qquad \left\{\begin{array}{l} \because \sin 2\pi n = \sin 0 = 0 \\ \cos 2\pi n = (-1)^{2n} = 1 \end{array}\right\}$$

$$b_n = \frac{1}{\pi}\int_0^{2\pi} x^2 \sin nx\, dx$$

$$= \frac{1}{\pi}\left[x^2\left(\frac{-\cos nx}{n}\right) - 2x\left(\frac{-\sin nx}{n^2}\right) + 2\left(\frac{\cos nx}{n^3}\right)\right]_0^{2\pi}$$

$$= \frac{1}{\pi}\left[\left(\frac{-4\pi^2}{n} + \frac{2}{n^3}\right) - \frac{2}{n^3}\right]$$

$$= -\frac{4\pi}{n}$$

Substituting the values of a_0, a_n, b_n in (1), we get,

$$x^2 = \frac{4\pi^2}{3} + \sum_{n=1}^{\infty}\frac{4}{n^2}\cos nx - \sum_{n=1}^{\infty}\frac{4\pi}{n}\sin nx$$

$$= \frac{4\pi^2}{3} + 4\left\{\frac{\cos x}{1^2} + \frac{\cos 2x}{2^2} + \frac{\cos 3x}{3^2} + ...\right\}$$

$$- 4\pi\left\{\frac{\sin x}{1} + \frac{\sin 2x}{2} + \frac{\sin 3x}{3} + ...\right\} \qquad .....(2)$$

is the Fourier expansion of $f(x) = x^2$ in $(0, 2\pi)$

Take $x = \pi$ in (2), so that,

$$\pi^2 = \frac{4\pi^2}{3} + 4\left\{\frac{-1}{1^2} + \frac{1}{2^2} - \frac{1}{3^2} + \frac{1}{4^2} + ...\right\}$$

$$\Rightarrow \quad \frac{-\pi^2}{3} = -4\left\{\frac{1}{1^2} - \frac{1}{2^2} + \frac{1}{3^2} - \frac{1}{4^2} + ...\right\}$$

$$\Rightarrow \quad \frac{\pi^2}{12} = \frac{1}{1^2} - \frac{1}{2^2} + \frac{1}{3^2} - \frac{1}{4^2} + ...$$

***Example* 15.6:** Expand $f(x) = x + x^2$ in $-\pi < x < \pi$ in a Fourier series. Hence show that

$$1 + \frac{1}{2^2} + \frac{1}{3^2} + \frac{1}{4^2} + \ldots\ldots = \frac{\pi^2}{6}$$

***Solution*:**

Let $$(x + x^2) = \frac{a_0}{2} + \sum_{n=1}^{\infty} a_n \cos nx + \sum_{n=1}^{\infty} b_n \sin nx \qquad \ldots\ldots (1)$$

be the Fourier series of $f(x)$.

Then, $$a_0 = \frac{1}{\pi}\int_{-\pi}^{\pi}(x + x^2)\,dx$$

$$= \frac{1}{\pi}\left[\frac{x^2}{2} + \frac{x^3}{3}\right]_{-\pi}^{\pi}$$

$$= \frac{1}{\pi}\left[\left(\frac{\pi^2}{2} + \frac{\pi^3}{3}\right) - \left(\frac{\pi^2}{2} - \frac{\pi^3}{3}\right)\right]$$

$$= \frac{1}{\pi}\left[\frac{2\pi^3}{3}\right]$$

$$= \frac{2\pi^2}{3}$$

$$a_n = \frac{1}{\pi}\int_{-\pi}^{\pi}(x + x^2)\cos nx\,dx$$

$$= \frac{1}{\pi}\left[(x + x^2)\left(\frac{\sin nx}{n}\right) - (1 + 2x)\left(\frac{-\cos nx}{n^2}\right) + 2\left(\frac{-\sin nx}{n^3}\right)\right]_{-\pi}^{\pi}$$

$$= \frac{1}{\pi}\left[(1 + 2\pi)\frac{\cos n\pi}{n^2} - (1 - 2\pi)\frac{\cos(-n\pi)}{n^2}\right]$$

$$= \frac{1}{\pi}\left[\frac{4\pi}{n^2}(-1)^n\right]$$

$$= \frac{4}{n^2}(-1)^n \qquad \{\because \sin n\pi = \sin(-n\pi) = 0, \text{ and}$$

$$\cos n\pi = \cos(-n\pi) = (-1)^n\}$$

$$b_n = \frac{1}{\pi}\int_{-\pi}^{\pi}(x+x^2)\sin nx\, dx$$

$$= \frac{1}{\pi}\left[(x+x^2)\left(\frac{-\cos nx}{n}\right) - (1+2x)\left(\frac{-\sin nx}{n^2}\right) + 2\left(\frac{\cos nx}{n^3}\right)\right]_{-\pi}^{\pi}$$

$$= \frac{1}{\pi}\left[\frac{(\pi+\pi^2)(-1)^{n+1}}{n} - \frac{(-\pi+\pi^2)(-1)^{n+1}}{n} + \frac{2}{n^3}\left((-1)^n - (-1)^n\right)\right]$$

$$= \frac{2}{n}(-1)^{n+1}$$

Substituting in (1), the required Fourier series is,

$$x + x^2 = \frac{\pi^2}{3} + 4\sum_{n=1}^{\infty}\frac{(-1)^n}{n^2}\cos nx + 2\sum_{n=1}^{\infty}\frac{(-1)^{n+1}}{n}\sin nx$$

$$= \frac{\pi^2}{3} - 4\left(\frac{\cos x}{1^2} - \frac{\cos 2x}{2^2} + \frac{\cos 3x}{3^2} - +.....\right)$$

$$+ 2\left(\frac{\sin x}{1} - \frac{\sin 2x}{2} + \frac{\sin 3x}{3} - +.....\right) \qquad .....(2)$$

(2) is true for all values of x in $-\pi < x < \pi$. But it does not represent $f(x) = x + x^2$ at $x = \pm\pi$ which are the points of discontinuity of $f(x)$.

$\therefore$ At $x = \pi$, the sum of the series

$$= \frac{1}{2}[f(\pi+0) + f(\pi-0)]$$

$$= \frac{1}{2}[\pi^2 - \pi + \pi^2 + \pi]$$

$$= \pi^2$$

$$\therefore \quad \pi^2 = \frac{\pi^2}{3} - 4\left(\frac{\cos\pi}{1^2} - \frac{\cos 2\pi}{2^2} + \frac{\cos 3\pi}{3^2} - +.....\right) + 2(0)$$

{substituting $x = \pi$ in R.H.S of (2)}

(or) $\pi^2 - \frac{\pi^2}{3} = 4\left(\frac{1}{1^2}+\frac{1}{2^2}+\frac{1}{3^2}+.....\right)$

(or) $\frac{1}{1^2}+\frac{1}{2^2}+\frac{1}{3^2}+..... = \frac{\pi^2}{6}$

***Example* 15.7:** Obtain the Fourier expansion of $f(x) = \cos ax$ in $(0, 2\pi)$ and hence deduce that $\pi \operatorname{cosec} a\pi = \frac{1}{a} + \sum(-1)^n \left\{\frac{1}{a+n}+\frac{1}{a-n}\right\}$

***Solution*:**

Let $f(x) = \cos ax$

$$= \frac{a_0}{2} + \sum_{n=1}^{\infty} a_n \cos nx + \sum_{n=1}^{\infty} b_n \sin nx \qquad .....(1)$$

be the Fourier expansion of $f(x)$.

Then, $a_0 = \frac{1}{\pi}\int_0^{2\pi} \cos ax\, dx$

$$= \frac{1}{\pi}\left[\frac{\sin ax}{a}\right]_0^{2\pi}$$

$$= \frac{\sin 2a\pi}{a\pi}$$

$$a_n = \frac{1}{\pi}\int_0^{2\pi} \cos ax \cos nx\, dx$$

$$= \frac{1}{2\pi}\int_0^{2\pi}\{\cos(a+n)x + \cos(a-n)x\}dx$$

[$\because$ $2\cos A \cos B = \cos(A + B) + \cos(A - B)$]

$$= \frac{1}{2\pi}\left[\frac{\sin(a+n)x}{a+n}+\frac{\sin(a-n)x}{a-n}\right]_0^{2\pi}$$

$$= \frac{1}{2\pi}\left[\frac{1}{a+n}.\sin(a+n)2\pi + \frac{1}{a-n}\sin(a-n)2\pi\right]$$

$$= \frac{1}{2\pi}\left[\frac{\sin 2a\pi}{a+n} + \frac{\sin 2a\pi}{a-n}\right], \quad \{\because \sin(a \pm m)2\pi = \sin 2a\pi \cos 2x\pi$$

$$\pm \cos 2a\pi \sin 2x\pi$$

$$= \sin 2a\pi\}$$

$$= \frac{a\sin 2a\pi}{\pi(a^2 - n^2)}$$

$$b_n = \frac{1}{\pi}\int_0^{2\pi} \cos ax \sin nx\, dx$$

$$= \frac{1}{2\pi}\int_0^{2\pi} \{\sin(n+a)x + \sin(n-a)x\} dx$$

$$(\because 2\sin A \cos B = \sin(A+B) + \sin(A-B))$$

$$= \frac{1}{2\pi}\left[-\frac{1}{n+a}\cos(n+a)x - \frac{1}{n-a}\cos(n-a)x\right]_0^{2\pi}$$

$$= \frac{1}{2\pi}\left[-\frac{1}{n+a}\cos(n+a)2\pi - \frac{1}{n-a}\cos(n-a)2\pi + \frac{1}{n+a} + \frac{1}{n-a}\right]$$

$$= \frac{1}{2\pi}\left[\frac{-1}{n+a}\cos 2a\pi - \frac{1}{n-a}\cos 2a\pi + \frac{1}{n+a} + \frac{1}{n-a}\right]$$

$$= \frac{(1-\cos 2a\pi)}{2\pi}\left[\frac{1}{a+n} - \frac{1}{a-n}\right]$$

$$= \frac{-n}{\pi}\frac{(1-\cos 2a\pi)}{a^2 - n^2}$$

Substituting the values of a_0, a_n, b_n, (2) we get

$$\cos ax = \frac{\sin 2a\pi}{2a\pi} + \frac{a\sin 2a\pi}{\pi}\sum_{n=1}^{\infty}\frac{\cos nx}{a^2 - n^2}$$

$$-\frac{1-\cos 2a\pi}{\pi}\sum_{n=1}^{\infty}\frac{n\sin nx}{a^2-n^2} \qquad .....(2)$$

is the required Fourier series.

Now, taking $x = \pi$, (2) gives,

$$\cos a\pi = \frac{2\sin a\pi\cos a\pi}{2a\pi} + \frac{2a\sin a\pi\cos a\pi}{\pi}\sum_{n=1}^{\infty}\frac{\cos n\pi}{a^2-n^2}$$

$$(\because \sin n\pi = 0)$$

Multiplying both sides by π cosec $a\pi$ sec $a\pi$, we get,

$$\pi \operatorname{cosec} a\pi = \frac{1}{a} + \sum_{n=1}^{\infty}(-1)^n\frac{2a}{a^2-n^2}$$

$$= \frac{1}{a} + \sum(-1)^n\left\{\frac{1}{a+n}+\frac{1}{a-n}\right\}$$

which is the required result.

***Example* 15.8:** Find the Fourier series for $f(x)$ where $f(x) = \frac{(\pi-x)^2}{4}$ in $(0, 2\pi)$ and hence deduce that

(i) $\frac{1}{1^2}+\frac{1}{2^2}+\frac{1}{3^2}+ ..... = \frac{\pi^2}{6}$, and

(ii) $\frac{1}{1^2}+\frac{1}{3^2}+\frac{1}{5^2}+ ..... = \frac{\pi^2}{8}$

***Solution*:**

Let $$\frac{(\pi-x)^2}{4} = \frac{a_0}{2} + \sum_{n=1}^{\infty}a_n\cos nx + \sum_{n=1}^{\infty}b_n\sin nx$$

be the Fourier series representing the given function in $(0, 2\pi)$.

Then $$a_0 = \frac{1}{\pi}\int_0^{2\pi}\frac{(\pi-x)^2}{4}dx$$

$$= \frac{1}{4\pi}\left[\frac{(\pi - x)^3}{-3}\right]_0^{2\pi}$$

$$= \frac{1}{4\pi}\left[\frac{\pi^3}{3} + \frac{\pi^3}{3}\right]$$

$$= \frac{\pi^2}{6}$$

$$a_n = \frac{1}{\pi}\int_0^{2\pi}\frac{(\pi - x)^2}{4}\cos nx\, dx$$

$$= \frac{1}{4\pi}\left[(\pi - x)^2\left(\frac{\sin nx}{n}\right) - 2(\pi - x)(-1)\left(\frac{-\cos nx}{n^2}\right)\right.$$

$$\left. + 2\left(\frac{-\sin nx}{n^3}\right)\right]_0^{2\pi}$$

$$= \frac{1}{4\pi}\left[\frac{-2(\pi - 2\pi)}{n^2} + \frac{2\pi}{n^2}\right] = \frac{1}{n^2}$$

$$b_n = \frac{1}{\pi}\int_0^{2\pi}\frac{(\pi - x)^2}{4}\sin nx dx$$

$$= \frac{1}{4\pi}\left[\frac{(\pi - x)^2(-\cos nx)}{n} - 2(\pi - x)(-1)\frac{(-\sin nx)}{n^2} + 2\frac{\cos nx}{n^3}\right]_0^{2\pi}$$

$$= \frac{1}{4\pi}\left[-\frac{1}{n}(\pi - 2\pi)^2 + (\pi - 0)^2\frac{1}{n} + \frac{2}{n^3}(1-1)\right]$$

$$= 0$$

$\therefore$ (1) $\Rightarrow$

$$\frac{(\pi - x)^2}{4} = \frac{\pi^2}{12} + \sum_{n=1}^{\infty}\frac{\cos nx}{n^2} \qquad .....(2)$$

Put $x = 0$ in (2) to get,

$$\frac{\pi^2}{4} = \frac{\pi^2}{12} + \frac{1}{1^2} + \frac{1}{2^2} + \frac{1}{3^2} + ...$$

(or) $$\frac{1}{1^2} + \frac{1}{2^2} + \frac{1}{3^2} + ... = \frac{\pi^2}{6} \quad ..... \text{(proved)} \qquad .....(3)$$

Put $x = \pi$ in (2) to get,

$$0 = \frac{\pi^2}{12} - \left(\frac{1}{1^2} - \frac{1}{2^2} + \frac{1}{3^2} - + ...\right)$$

$$\Rightarrow \quad \frac{1}{1^2} - \frac{1}{2^2} + \frac{1}{3^2} - + ... = \frac{\pi^2}{12} \qquad .....(4)$$

Adding (3) and (4), we get

$$2\left[\frac{1}{1^2} + \frac{1}{3^2} + \frac{1}{5^2} + ...\right] = \frac{\pi^2}{4}$$

(i.e.,) $$\frac{1}{1^2} + \frac{1}{3^2} + \frac{1}{5^2} + ... = \frac{\pi^2}{8} \text{ (proved)}$$

***Example* 15.9:** Obtain the Fourier series expansion of $f(x) = e^{-x}$ in $(0, 2\pi)$.

***Solution*:**

Let $$e^{-x} = \frac{a_0}{2} + \sum_{n=1}^{\infty} a_n \cos nx + \sum_{n=1}^{\infty} b_n \sin nx \qquad .....(1)$$

be the required series.

Then, $$a_0 = \frac{1}{\pi}\int_0^{2\pi} e^{-x} dx = \frac{1}{\pi}\left[\frac{e^{-x}}{-1}\right]_0^{2\pi}$$

$$= \frac{1}{\pi}\left[-e^{-2\pi} + 1\right] = \frac{1 - e^{-2\pi}}{\pi}$$

$$a_n = \frac{1}{\pi}\int_0^{2\pi} e^{-x} \cos nx \, dx$$

$$= \frac{1}{\pi}\left[\frac{e^{-x}}{n^2+1}(-\cos nx + n\sin nx)\right]_0^{2\pi}$$

$$\left\{\int e^{ax}\cos bx\,dx = e^{ax}\left(\frac{a\cos bx + b\sin bx}{a^2+b^2}\right)\right.$$

$$\left.\text{Here } a=-1, b=n\right\}$$

$$= \frac{1}{\pi(n^2+1)}\left[-e^{-2\pi}+1\right] = \frac{1-e^{-2\pi}}{\pi(n^2+1)}$$

$$b_n = \frac{1}{\pi}\int_0^{2\pi} e^{-x}\sin nx\,dx$$

$$= \frac{1}{\pi}\left[\frac{e^{-x}}{n^2+1}(-\sin nx - n\cos nx)\right]_0^{2\pi}$$

$$\left\{\because \int e^{ax}\sin bx\,dx = \frac{e^{ax}(a\sin bx - b\cos bx)}{a^2+b^2};\quad \begin{matrix} a=-1 \\ b=n \end{matrix}\right\}$$

$$= \frac{1}{\pi(n^2+1)}\left[e^{-2\pi}(0-n[\cos 2n\pi - 1])\right] = 0$$

$$\therefore \quad e^{-x} = \frac{1-e^{-2\pi}}{2\pi} + \frac{1-e^{-2\pi}}{\pi}\sum_{n=1}^{\infty}\frac{\cos nx}{n^2+1} \qquad \text{(From (1))}$$

***Example* 15.10:** Expand the function, $f(x) = x\sin x$ as a Fourier series in $(0, 2\pi)$.

***Solution*:**

Let the required series be

$$x\sin x = \frac{a_0}{2} + \sum_{n=1}^{\infty} a_n\cos nx + \sum_{n=1}^{\infty} b_n \sin nx \qquad .....(1)$$

Then, $$a_0 = \frac{1}{\pi}\int_0^{2\pi} x\sin x\,dx$$

$$= \frac{1}{\pi}\left[x(-\cos x) - 1(-\sin x)\right]_0^{2\pi}$$

$$= \frac{1}{\pi}\left[2\pi(-1)\right] = -2$$

$$a_n = \frac{1}{\pi}\int_0^{2\pi} x\sin x \cos nx\, dx$$

$$= \frac{1}{2\pi}\int_0^{2\pi} x\{\sin(1+n)x + \sin(1-n)x\}dx$$

$$= \frac{1}{2\pi}\int_0^{2\pi} x\sin(1+n)x\, dx + \frac{1}{2\pi}\int_0^{2\pi} x\sin(1-n)x\, dx$$

$$= \frac{1}{2\pi}\left[-x\frac{\cos(1+n)x}{1+n} + \frac{\sin(1+n)x}{(1+n)^2}\right]_0^{2\pi}$$

$$+ \frac{1}{2\pi}\left[-x\frac{\cos(1-n)x}{1-n} + \frac{\sin(1-n)x}{(1-n)^2}\right]_0^{2\pi}$$

$$= \frac{1}{2\pi}\left[-\frac{2\pi\cos(1+n)2\pi}{1+n} - \frac{2\pi\cos(1-n)2\pi}{1-n}\right]$$

$$= \left[-\frac{1}{1+n} - \frac{1}{1-n}\right]$$

$$= \left(\frac{1}{n-1} - \frac{1}{n+1}\right)$$

$$= \frac{2}{(n^2-1)} \qquad [n \neq 1], \text{ i.e., } n = 2, 3, \ldots\ldots$$

When $n = 1$,

$$a_1 = \frac{1}{\pi}\int_0^{2\pi} x\sin x\cos x\, dx$$

$$= \frac{1}{2\pi}\int_0^{2\pi} x\sin 2x\,dx$$

$$= \frac{1}{2\pi}\left[x\frac{(-\cos 2x)}{2} + \frac{\sin 2x}{4}\right]_0^{2\pi}$$

$$= \frac{-1}{4\pi}[2\pi.1 - 0]$$

$$= \frac{-1}{2}$$

$$b_n = \frac{1}{\pi}\int_0^{2\pi} x\sin x\sin nx\,dx$$

$$= \frac{1}{2\pi}\int_0^{2\pi} x\{\cos(n-1)x - \cos(n+1)x\}dx$$

$$= \frac{1}{2\pi}\int_0^{2\pi} x\cos(n-1)x\,dx - \frac{1}{2\pi}\int_0^{2\pi} x\cos(n+1)x\,dx$$

$$= \frac{1}{2\pi}\left[x\frac{\sin(n-1)x}{n-1} + \frac{\cos(n-1)x}{(n-1)^2}\right]_0^{2\pi}$$

$$- \frac{1}{2\pi}\left[x\frac{\sin(n+1)x}{n+1} + \frac{\cos(n+1)x}{(n+1)^2}\right]_0^{2\pi}$$

$$= \frac{1}{-2\pi}\left[\frac{\cos(n-1)2\pi}{(n-1)^2} - \frac{1}{(n-1)^2} - \frac{\cos(n+1)2\pi}{(n+1)^2} + \frac{1}{(n+1)^2}\right]$$

$= 0,$ provided $n \neq 1$;

$\therefore \quad b_2 = b_3 = b_4 = \ldots\ldots = 0$

$$b_1 = \frac{1}{\pi}\int_0^{2\pi} x\sin x\sin x\,dx$$

$$= \frac{1}{2\pi}\int_0^{2\pi} x(1-\cos 2x)dx$$

$$= \frac{1}{2\pi}\left[x\left(x - \frac{\sin 2x}{2}\right) - 1\left(\frac{x^2}{2} + \frac{\cos 2x}{4}\right)\right]_0^{2\pi}$$

$$= \frac{1}{2\pi}\left[4\pi^2 - 1\left\{2\pi^2 + \frac{1}{4} - \frac{1}{4}\right\}\right] = \pi$$

$\therefore$ (1) gives,

$$x \sin x = -1 - \frac{1}{2}\cos x + \pi \sin x + \sum_{n=2}^{\infty} + \frac{2}{n^2 - 1}\cos nx$$

$$= -1 + \pi \sin x - \frac{1}{2}\cos x + \frac{2}{2^2 - 1}\cos 2x +$$

$$+ \frac{2}{3^2 - 1}\cos 3x + \frac{2}{4^2 - 1}\cos 4x + \ldots..$$

***Example* 15.11:** Find the Fourier series for $f(x)$, where

$$f(x) = \begin{cases} 1, & 0 < x \leq \pi \\ 2 - \dfrac{x}{\pi}, & \pi \leq x \leq 2\pi \end{cases}$$

***Solution*:**

[Note that here the interval is split into 2 subintervals. Hence the integrals also are to be found in the respective subintervals.]

$$f(x) = \frac{a_0}{2} + \sum_{n=1}^{\infty} a_n \cos nx + \sum_{n=1}^{\infty} b_n \sin nx \qquad \ldots..(1)$$

$$\therefore \quad a_0 = \frac{1}{\pi}\int_0^{2\pi} f(x)dx$$

$$= \frac{1}{\pi}\left[\int_0^{\pi} 1dx + \int_{\pi}^{2\pi}\left(2 - \frac{x}{\pi}\right)dx\right]$$

$$= \frac{1}{\pi}\left[x\Big|_0^{\pi} + \left(2x - \frac{x^2}{2\pi}\right)_{\pi}^{2\pi} \right]$$

$$= \frac{1}{\pi}\left[\pi + (4\pi - 2\pi) - \left(2\pi - \frac{\pi}{2}\right) \right]$$

$$= \frac{3}{2}$$

$$a_n = \frac{1}{\pi}\int_0^{2\pi} f(x)\cos nx\, dx$$

$$= \frac{1}{\pi}\left[\int_0^{\pi} \cos nx\, dx + \int_{\pi}^{2\pi} \left(2 - \frac{x}{\pi}\right)\cos nx\, dx \right]$$

$$= \frac{1}{\pi}\left[\frac{-\sin nx}{n}\Big|_0^{\pi} + \left\{ \left(2 - \frac{x}{\pi}\right)\frac{\sin nx}{n} + \frac{1}{\pi}\left(\frac{-\cos nx}{n^2}\right) \right\}_{\pi}^{2\pi} \right]$$

$$= \frac{1}{\pi}\left[\frac{-1}{\pi n^2}(\cos 2n\pi - \cos n\pi) \right]$$

$$= -\frac{1}{\pi^2 n^2}\left\{1 - (-1)^n\right\}$$

$$b_n = \frac{1}{\pi}\int_0^{2\pi} f(x)\sin nx\, dx$$

$$= \frac{1}{\pi}\left[\int_0^{\pi} \sin nx\, dx + \int_{\pi}^{2\pi} \left(2 - \frac{x}{\pi}\right)\sin nx\, dx \right]$$

$$= \frac{1}{\pi}\left[\frac{-\cos nx}{n}\Big|_0^{\pi} + \left\{ \left(\frac{2-x}{\pi}\right)\left(\frac{-\cos nx}{n}\right) + \frac{1}{\pi}\left(\frac{-\sin nx}{n^2}\right) \right\}_{\pi}^{2\pi} \right]$$

$$= \frac{1}{\pi}\left[\frac{(-1)^{n+1} + 1}{n} + 0 + \frac{(-1)^n}{n} \right]$$

$$= \frac{1}{\pi}\left[\frac{(-1)^n}{n}(-1+1)+\frac{1}{n}\right]$$

$$= \frac{1}{n\pi}$$

$\therefore \quad (1) \Rightarrow$

$$f(x) = \frac{3}{4} - \frac{1}{\pi^2}\sum_{n=1}^{\infty}\left\{\frac{1-(-1)^n}{n^2}\right\}\cos nx + \frac{1}{\pi}\sum_{n=1}^{\infty}\frac{\sin nx}{n}$$

$$= \frac{3}{4} - \frac{2}{\pi^2}\left[\frac{\cos x}{1^2} + \frac{\cos 3x}{3^2} + \frac{\cos 5x}{5^2} + ...\right]$$

$$+ \frac{1}{\pi}\left[\frac{\sin x}{1} + \frac{\sin 2x}{2} + \frac{\sin 3x}{3} + ...\right]$$

***Example* 15.12:** Express the function,

$$f(x) = -\pi, \qquad -\pi < x < 0$$
$$= x, \qquad 0 < x < \pi,$$

as a Fourier series. Hence deduce that

$$\frac{1}{1^2} + \frac{1}{3^2} + \frac{1}{5^2} + ... = \frac{\pi^2}{8}$$

***Solution*:**

Let $\qquad f(x) = \frac{a_0}{2} + \sum_{n=1}^{\infty} a_n \cos nx + \sum_{n=1}^{\infty} b_n \sin nx \qquad .....(1)$

Then, $\qquad a_0 = \frac{1}{\pi}\int_{-\pi}^{\pi} f(x)dx$

$$= \frac{1}{\pi}\int_{-\pi}^{0} -\pi dx + \frac{1}{\pi}\int_{0}^{\pi} x dx$$

$$= \frac{1}{\pi}[-\pi x]_{-\pi}^{0} + \frac{1}{\pi}\left[\frac{x^2}{2}\right]_0^{\pi}$$

$$= \frac{1}{\pi}\left(0-\pi^2\right) + \frac{1}{\pi}\left(\frac{\pi^2}{2}\right)$$

$$= \frac{-\pi}{2}$$

$$a_n = \frac{1}{\pi}\int_{-\pi}^{\pi} f(x)\cos nx\,dx$$

$$= \frac{1}{\pi}\left[\int_{-\pi}^{0} -\pi\cos nx\,dx + \int_{0}^{\pi} x\cos nx\,dx\right]$$

$$= \frac{1}{\pi}\left[\frac{-\pi\sin nx}{n}\Bigg|_{-\pi}^{0} + \left\{x\frac{\sin nx}{n} + 1\left(\frac{\cos nx}{n^2}\right)\right\}_0^{\pi}\right]$$

$$= \frac{1}{\pi}\left[\frac{(-1)^n - 1}{n^2}\right]$$

$$b_n = \frac{1}{\pi}\int_{-\pi}^{\pi} f(x)\sin nx\,dx$$

$$= \frac{1}{\pi}\left[\int_{-\pi}^{0} -\pi\sin nx\,dx + \int_{0}^{\pi} x\sin nx\,dx\right]$$

$$= \frac{1}{\pi}\left[\frac{\pi\cos nx}{n}\Bigg|_{-\pi}^{0} + \left\{\frac{x(-\cos nx)}{n} - (1)\left(\frac{-\sin nx}{n^2}\right)\right\}_0^{\pi}\right]$$

$$= \frac{1}{\pi}\left[\frac{\pi}{n}\left(1-(-1)^n\right)\right] + \frac{1}{\pi}\left[\pi\frac{(-1)^{n+1}}{n}\right]$$

$$= \frac{1}{\pi}\left[\frac{\pi}{n}+\frac{(-1)^n}{n}(-\pi-\pi)\right]$$

$$= \frac{1}{n}\left[1-(-1)^n 2\right]$$

$\therefore$ (1) $\Rightarrow$ $$f(x) = -\frac{\pi}{4}+\frac{1}{\pi}\sum_{n=1}^{\infty}\left\{\frac{(-1)^n-1}{n^2}\right\}\cos nx+\sum_{n=1}^{\infty}\left\{\frac{1-2(-1)^n}{n}\right\}\sin nx$$

$$= \frac{-\pi}{4}-\frac{2}{\pi}\left(\frac{\cos x}{1^2}+\frac{\cos 3x}{3^2}+\frac{\cos 5x}{5^2}+.....\right)$$

$$+\left(\frac{3\sin x}{1}-\frac{\sin 2x}{2}+\frac{3\sin 3x}{3}-.....\right) \qquad .....(2)$$

is the Fourier expansion of $f(x)$

$f(x)$ is not defined when $x = 0$.

$\therefore$ $$f(0) = \frac{1}{2}\left[\underset{x\to 0+}{Lt} f(x)+\underset{x\to 0-}{Lt} f(x)\right]$$

$$= \frac{1}{2}(0-\pi) = -\frac{\pi}{2}$$

$\therefore$ Substituting $x = 0$

(2) $\Rightarrow$

$$\frac{-\pi}{2} = \frac{-\pi}{4}-\frac{2}{\pi}\left(\frac{1}{1^2}+\frac{1}{3^2}+\frac{1}{5^2}+.....\right)$$

$$\frac{-\pi}{4} = \frac{-2}{\pi}\left(\frac{1}{1^2}+\frac{1}{3^2}+\frac{1}{5^2}+.....\right)$$

(or) $$\frac{1}{1^2}+\frac{1}{3^2}+\frac{1}{5^2}+..... = \frac{\pi^2}{8}$$

Example **15.13:** Represent the function

$$f(x) = -x^2, \qquad -\pi \le x \le 0$$
$$= x^2, \qquad 0 \le x \le \pi$$

as a Fourier series.

***Solution*:**

Let $$f(x) = \frac{a_0}{2} + \sum_{n=1}^{\infty} a_n \cos nx + \sum_{n=1}^{\infty} b_n \sin nx \qquad .....(1)$$

Then $$a_0 = \frac{1}{\pi}\left[\int_{-\pi}^{0} -x^2 dx + \int_{0}^{\pi} x^2 dx\right]$$

$$= \frac{1}{\pi}\left[\left.\frac{-x^3}{3}\right|_{-\pi}^{0} + \left.\frac{x^3}{3}\right|_{0}^{\pi}\right] = 0$$

$$a_n = \frac{1}{\pi}\left[\int_{-\pi}^{0} -x^2 \cos nx\, dx + \int_{0}^{\pi} x^2 \cos nx\, dx\right]$$

$$= \frac{1}{\pi}\left[-x^2 \frac{\sin nx}{n} - (-2x)\left(\frac{-\cos nx}{n^2}\right) + (-2)\left(\frac{-\sin nx}{n^3}\right)\right]_{-\pi}^{0}$$

$$+ \frac{1}{\pi}\left[x^2 \frac{\sin nx}{n} - 2x\left(\frac{-\cos nx}{n^2}\right) + 2\left(\frac{-\sin nx}{n^3}\right)\right]_{0}^{\pi}$$

$$= \frac{2}{\pi}\frac{(-1)^n(-\pi)}{n^2} + \frac{1.2}{\pi}\frac{(-1)^n(\pi)}{n^2} = 0$$

$$b_n = \frac{1}{\pi}\left[\int_{-\pi}^{0} -x^2 \sin nx\, dx + \int_{0}^{\pi} x^2 \sin nx\, dx\right]$$

$$= \frac{1}{\pi}\left[-x^2\left(\frac{-\cos nx}{n}\right) + 2x\left(\frac{-\sin nx}{n^2}\right) - 2\left(\frac{\cos nx}{n^3}\right)\right]_{-\pi}^{0}$$

$$+ \frac{1}{\pi}\left[x^2\left(\frac{-\cos nx}{n}\right) - 2x\left(\frac{-\sin nx}{n^2}\right) + 2\left(\frac{\cos nx}{x^3}\right)\right]_{0}^{\pi}$$

$$= \frac{1}{\pi}\left[\frac{-\pi^2(-1)^n}{n} - 2\left(\frac{1}{n^3} - \frac{(-1)^n}{n^3}\right) - \frac{\pi^2}{n}(-1)^n + 2\left(\frac{(-1)^n}{n^3} - \frac{1}{n^3}\right)\right]$$

$$= \frac{-2\pi}{n}(-1)^n - \frac{4}{\pi n^3} + \frac{4}{\pi n^3}(-1)^n$$

$\therefore$ (1) gives that,

$$f(x) = \sum_{n=1}^{\infty}\left\{-\frac{2\pi}{n}(-1)^n - \frac{4}{\pi n^3} + \frac{4(-1)^n}{\pi n^3}\right\} \sin nx$$

$$= 2\left(\pi - \frac{4}{\pi}\right)\sin x - \pi \sin 2x + \frac{2}{3}\left(\pi - \frac{4}{9\pi}\right)\sin 3x + ...$$

***Example* 15.14:** An alternating current after passing through a rectifier has the form,
$i = I_0 \sin\theta,\ 0 \le \theta \le \pi = 0,\ \pi \le \theta \le 2\pi,$
where I_0 is the maximum current and the period is 2π. Express it as a Fourier series. Deduce that

$$\frac{1}{2} = \frac{1}{1.3} + \frac{1}{3.5} + \frac{1}{5.7} + ...$$

Solution:

Here i is a function of θ.

Let $$i = \frac{a_0}{2} + \sum_{n=1}^{\infty} a_n \cos n\theta + \sum_{n=1}^{\infty} b_n \sin n\theta \qquad .....(1)$$

be the Fourier series that represents i.

Now, $$a_0 = \frac{1}{\pi}\int_0^{\pi} I_0 \sin\theta \, d\theta$$

$$= \frac{I_0}{\pi}\left(-\cos\theta\right)_0^{\pi}$$

$$= \frac{I_0}{\pi}(1+1)$$

$$= \frac{2I_0}{\pi}$$

$$a_n = \frac{1}{\pi}\int_0^{\pi} I_0 \sin\theta \cos n\theta \, d\theta$$

$$= \frac{I_0}{2\pi}\left[\int_0^{\pi}\sin(1+n)\theta\,d\theta + \int_0^{\pi}\sin(1-n)\theta\,d\theta\right]$$

$$= \frac{I_0}{2\pi}\left[\frac{-\cos(1+n)\theta}{1+n}\Bigg|_0^{\pi} - \frac{\cos(1-n)\theta}{1-n}\Bigg|_0^{\pi}\right]$$

$$= \frac{I_0}{2\pi}\left[\frac{-1}{1+n}\{\cos(1+n)\pi - 1\} - \left\{\frac{\cos(1-n)\pi - 1}{1-n}\right\}\right]$$

$$= \frac{I_0}{2\pi}\left[\frac{(-1)^n + 1}{1+n} + \frac{(-1)^n + 1}{1-n}\right]$$

$$\left\{\begin{aligned}\because \cos(1+n)\pi &= \cos\pi\cos n\pi = (-1)^{n+1}\\ &= \cos(1-n)\pi\end{aligned}\right\}$$

$$= \frac{I_0}{2\pi}\{(-1)^n + 1\}\left(\frac{2}{1-n^2}\right)$$

$$= -\frac{I_0}{\pi}\{(-1)^n + 1\}\frac{1}{(n+1)(n-1)} \quad ; (n \neq 1)$$

$$\Rightarrow \qquad a_2 = \frac{-I_0}{\pi}.\frac{2}{1.3} \ ; \ a_3 = 0 \ ;$$

$$a_4 = \frac{-I_0}{\pi}.\frac{2}{3.5}, \text{ etc., ; and } a_1 = \frac{1}{\pi}\int_0^{\pi}\sin\theta\cos\theta\,d\theta = 0$$

Again, $b_n = \dfrac{1}{\pi}\displaystyle\int_0^{\pi} I_0 \sin\theta \sin n\theta\, d\theta$

$$= \frac{I_0}{2\pi}\int_0^{\pi}\{\cos(n-1)\theta - \cos(n+1)\theta\}d\theta$$

$$= \frac{I_0}{2\pi}\left[\frac{\sin(n-1)\theta}{n-1} - \frac{\sin(n+1)\theta}{n+1}\right]_0^{\pi}$$

$= 0,$ for all $n \neq 1$.

When $n = 1$,

$$b_1 = \frac{1}{\pi}\int_0^{\pi} I_0 \sin^2\theta\, d\theta$$

$$= \frac{I_0}{2\pi}\int_0^{\pi}(1-\cos 2\theta)d\theta$$

$$= \frac{I_0}{2\pi}\left[\theta - \frac{\sin 2\theta}{2}\right]_0^{\pi}$$

$$= \frac{I_0}{2}$$

$\therefore$ (1) gives that,

$$i = \frac{I_0}{\pi} - \frac{2I_0}{\pi}\left\{\frac{\cos 2\theta}{1.3} + \frac{\cos 4\theta}{3.5} + ...\right\} + \frac{I_0}{2}\sin\theta \qquad .....(2)$$

which is the required series.

Substituting $\theta = \pi$ and i = 0 in (2),

We get,

$$0 = \frac{I_0}{\pi} - \frac{2I_0}{\pi}\left\{\frac{1}{1.3} + \frac{1}{3.5} + \frac{1}{5.7} + ...\right\}$$

Dividing by $\frac{I_0}{\pi}$ both sides, we get,

$$2\left\{\frac{1}{1.3} + \frac{1}{3.5} + \frac{1}{5.7} + ...\right\} = 1$$

$$\Rightarrow \qquad \frac{1}{1.3} + \frac{1}{3.5} + \frac{1}{5.7} + ... = \frac{1}{2}$$

Exercise – 15.2

1. Obtain the Fourier series expansions for the following functions in the intervals given against them.

(a) $f(x) = x$ ..... in $(-\pi, \pi)$

Hence show that $1 - \frac{1}{3} + \frac{1}{5} - + ... = \frac{\pi}{4}$

$$\left[\textbf{Ans : } f(x) = 2\sum_{n=2}^{\infty}\frac{(-1)^{n+1}}{n}\sin nx; \text{put } x = \frac{\pi}{2}\right]$$

(b) $f(x) = e^x$ in $(0, 2\pi)$

$$\left[\textbf{Ans : } f(x) = \frac{e^{2\pi}-1}{\pi}\left[\frac{1}{2}+\sum_{n=1}^{\infty}\frac{\cos nx}{n^2+1}-\sum_{n=1}^{\infty}\frac{n\sin nx}{n^2+1}\right]\right]$$

(c) $f(x) = e^x$ ..... $(-\pi, \pi)$

$$\left[\textbf{Ans : } f(x) = \frac{\sinh\pi}{\pi}\left[1+2\sum_{n=1}^{\infty}\frac{(-1)^n\cos nx}{n^2+1}-\frac{n(-1)^n\sin nx}{n^2+1}\right]\right]$$

(d) $f(x) = \cosh x$ ..... in $(-\pi < x < \pi)$

$$\left[\textbf{Ans : } f(x) = \frac{2\sinh\pi}{\pi}\left[\frac{1}{2}+\sum_{n=1}^{\infty}\frac{(-1)^n-1}{n^2+1}\cos nx\right]\right]$$

(e) $f(x) = \cos px$ ..... in $(-\pi, \pi)$
('p' not an integer)

$$\left[\textbf{Ans : } \frac{2p\sin p\pi}{\pi}\left[\frac{1}{2p^2}+\frac{\cos x}{1^2-p^2}-\frac{\cos 2x}{2^2-p^2}+.....\right]\right]$$

(f) $f(x) = x\sin x$ ..... in $(-\pi, \pi)$

$$\left[\textbf{Ans : } 1-\frac{\cos x}{2}-2\left\{\frac{\cos 2x}{1.3}-\frac{\cos 3x}{2.4}+\frac{\cos 4x}{3.5}+.....\right\}\right]$$

(g) $f(x) = \sqrt{1-\cos x}$ ..... in $(-\pi, \pi)$.

Hence show that $\frac{1}{1.3}+\frac{1}{3.5}+\frac{1}{5.7}+..... = \frac{1}{2}$

$$\left[\text{Ans : } \frac{2\sqrt{2}}{\pi}-\frac{4\sqrt{2}}{\pi}\sum_{n=1}^{\infty}\frac{\cos nx}{4n^2-1}\right]$$

(h) $f(x) = x\cos x$ ..... in $(0, 2\pi)$

$$\left[\text{Ans : } \pi\cos x-\frac{1}{2}\sin x-2\sum_{n=2}^{\infty}\frac{1}{n^2-1}\sin nx\right]$$

(i) $f(x) = (x - x^2)$ ..... in $(0, 2\pi)$

$$\left[\text{Ans : } \frac{-\pi^2}{3}+4\left(\frac{\cos x}{1^2}-\frac{\cos 2x}{2^2}+\frac{\cos 3x}{3^2}-+.....\right)+2\left(\frac{\sin x}{1}-\frac{\sin 2x}{2}+\frac{\sin 3x}{3}-+.....\right)\right]$$

2. Show that for $-\pi < x < \pi$,

$$\frac{x(\pi^2-x^2)}{12}=\frac{\sin x}{1^3}-\frac{\sin 2x}{2^3}+\frac{\sin 3x}{3^3}-+.....$$

3. If $f(x) = x^3$ is defined in $\pi < x < \pi$ and is a periodic function with period 2π, prove that

$$x^3 = 2\left[\left(\frac{\pi^2}{1}-\frac{6}{1^3}\right)\sin x-\left(\frac{\pi^2}{2}-\frac{6}{2^3}\right)\sin 2x+\left(\frac{\pi^2}{3}-\frac{6}{3^3}\right)\sin 3x-+.....\right]$$

4. Obtain the Fourier series for the functions given below.

(a) $f(x) = \begin{cases} 0, & \text{in} -\pi<x<0 \\ 1, & \text{in} \ 0<x<\pi \end{cases}$

$$\left[\text{Ans : } \frac{1}{2}+\frac{2}{\pi}\left[\frac{\sin x}{1}+\frac{\sin 3x}{3}+\frac{\sin 5x}{5}+.....\right]\right]$$

(b) $f(x) = \begin{cases} x & \text{in} -\pi\le x\le 0 \\ 0 & \text{in} \ 0<x<\pi \end{cases}$

$$\left[\text{Ans : } \frac{-\pi}{4}+\frac{2}{\pi}\sum_{n=1}^{\infty}\frac{\cos(2n-1)x}{(2n-1)^2}+\sum_{n=1}^{\infty}\frac{(-1)^{n+1}\sin nx}{n}\right]$$

(c) $f(x) = \begin{cases} 0 & \\ \sin x & \text{in } 0 \le x \le \pi \end{cases}$

Hence show that $\frac{1}{1.3}+\frac{1}{3.5}+\frac{1}{5.7}+..... = \frac{1}{2}$ and $\frac{1}{1.3}-\frac{1}{3.5}+\frac{1}{5.7}\;..... = \frac{\pi-2}{4}$

[Ans : $\frac{1}{\pi}-\frac{2}{\pi}\sum_{n=1}^{\infty}\frac{1}{4n^2-1}\cos 2nx+\frac{1}{2}\sin x$; put $x=0$ and $x=\frac{\pi}{2}$]

(d) $f(x) = \begin{cases} x & \text{in } 0 \le x \le \pi \\ (2\pi - x) & \text{in } \pi \le x \le 2\pi \end{cases}$

[Ans : $\frac{\pi}{2}-\frac{4}{\pi}\left\{\frac{\cos x}{1^2}+\frac{\cos 3x}{3^2}+\frac{\cos 5x}{5^2}+.....\right\}$]

(e) $f(x) = \begin{cases} (\pi + x) & \text{in } -\pi < x < 0 \\ (\pi - x) & \text{in } 0 < x < \pi \end{cases}$

[Ans : $\frac{\pi}{2}+\frac{4}{\pi}\left\{\frac{\cos x}{1^2}+\frac{\cos 3x}{3^2}+\frac{\cos 5x}{5^2}+.....\right\}$]

5. Obtain the Fourier series of sinh ax in $(-\pi < x < \pi)$

[Ans : $\frac{2}{\pi}\sinh a\pi\sum_{n=1}^{\infty}\frac{(-1)^{n+1}.n\sin nx}{n^2+a^2}$]

6. Expand $f(x) = (\pi - x)^2$ as a Fourier series in $0 < x < 2\pi$ and hence find the sum of the series $\frac{1}{1^2}+\frac{1}{2^2}+\frac{1}{3^2}+.....$

[Ans : $\frac{\pi^2}{3}+4\sum_{n=1}^{\infty}\frac{\cos nx}{n^2};\frac{\pi^2}{6}$]

7. Find the Fourier series to represent the function,

$f(x) = \begin{cases} -c, & -\pi<x<0 \\ c, & 0<x<\pi \end{cases}$,

'c' being a constant. Also deduce that $1-\frac{1}{3}+\frac{1}{5}-\frac{1}{7}+..... = \frac{\pi}{4}$

[Ans : $f(x) = \frac{4c}{\pi}\left[\sin x+\frac{\sin 3x}{3}+\frac{\sin 5x}{5}+.....\right]$, put $x=\frac{\pi}{2}$]

8. A sinusoidal voltage $E \sin t$ is passed through a half wave rectifier which clips the negative portion of the wave. The resulting periodic function ϕ is defined by

$$\phi(t) = \begin{cases} E\sin t, & 0 \le t \le \pi \\ 0, & \pi \le t \le 2\pi \end{cases}$$

Show that,

$$\phi(t) = \frac{E}{\pi} + \frac{1}{2}E\sin t - \frac{2E}{\pi}\sum \frac{\cos 2nt}{4n^2 - 1}$$

15.3 Change of Interval [Functions with Arbitrary Periods]

So far, the Fourier expansions of functions having a period 2π have been studied. The study of functions having arbitrary periods (need not necessarily be 2π) will be taken up now.

15.3.1 Euler's Formulae for Functions having a Period '$2l$' where l is any Positive Number

Let $f(x)$ be defined in $(-l, l)$ let $y = \frac{\pi x}{l}$, so that $x = \frac{ly}{\pi}$.

Further, when $x = -l$, we have $y = -\pi$ and when $x = l$, $y = \pi$; so that the function,

$F(y) = f\left(\frac{ly}{\pi}\right)$ is defined in the interval $(-\pi, \pi)$.

$\therefore$ The Fourier expansion of $F(y)$ is given by,

$$F(y) = \frac{a_0}{2} + \sum_{n=1}^{\infty}(a_n \cos ny + b_n \sin ny) \qquad .....(1)$$

where $$a_0 = \frac{1}{\pi}\int_{-\pi}^{\pi} F(y)dy$$

$$a_n = \frac{1}{\pi}\int_{-\pi}^{\pi} F(y)\cos ny\,dy \qquad .....(2)$$

$$b_n = \frac{1}{\pi}\int_{-\pi}^{\pi} F(y)\sin ny\,dy$$

$\because$ $$F(y) = f\left(\frac{ly}{\pi}\right)$$

(1) and (2) can be written as,

$$f\left(\frac{ly}{\pi}\right) = \frac{a_0}{2} + \sum_{n=1}^{\infty}(a_n \cos ny + b_n \sin ny) \qquad .....(3)$$

$$a_0 = \frac{1}{\pi}\int_{-\pi}^{\pi} f\left(\frac{ly}{\pi}\right)dy$$

and $$a_n = \frac{1}{\pi}\int_{-\pi}^{\pi} f\left(\frac{ly}{\pi}\right)\cos ny\,dy \quad .....(4)$$

$$b_n = \frac{1}{\pi}\int_{-\pi}^{\pi} f\left(\frac{ly}{\pi}\right)\sin ny\,dy$$

Substituting $y = \frac{\pi x}{l}$, $dy = \frac{\pi}{l}dx$ (and changing the limits accordingly) the above equations (3) and (4) can be written as

$$f(x) = \frac{a_0}{2} + \sum_{n=1}^{\infty} a_n \cos\left(\frac{n\pi x}{l}\right) + \sum_{n=1}^{\infty} b_n \sin\left(\frac{n\pi x}{l}\right) \quad .....(5)$$

where $$a_0 = \frac{1}{l}\int_{-l}^{l} f(x)dx$$

$$a_n = \frac{1}{l}\int_{-l}^{l} f(x)\cos\left(\frac{n\pi x}{l}\right)dx$$

$$b_n = \frac{1}{l}\int_{-l}^{l} f(x)\sin\left(\frac{n\pi x}{l}\right)dx$$

(4) and (5) give the Fourier series expansion and the Euler's formulae for a function $f(x)$ defined in $(-l, l)$ with a period $2l$.

Note : The above formulae are valid for any interval of length '$2l$', namely $(c, c + 2l)$.

Solved Examples

***Example* 15.15:** If a function $f(x)$ is defined as

$$f(x) = \begin{cases} -1 \text{ in } -2 < x < 0 \\ 1 \text{ in } 0 < x < 2 \end{cases}$$, find the Fourier series of $f(x)$.

***Solution*:**

Here the interval is $(-2, 2)$ i.e., $l = 2$

Let $f(x) = \frac{a_0}{2} + \sum_{n=1}^{\infty} a_n \cos\left(\frac{n\pi x}{2}\right) + \sum_{n=1}^{\infty} b_n \sin\left(\frac{n\pi x}{2}\right)$ .....(1)

where $a_0 = \frac{1}{2}\int_{-2}^{2} f(x)dx$

$$= \frac{1}{2}\left[\int_{-2}^{0} -1dx + \int_{0}^{2} 1dx\right]$$

$$= \frac{1}{2}\left[-x\Big|_{-2}^{0} + x\Big|_{0}^{2}\right]$$

$$= \frac{1}{2}\left[0-(2)+(2-0)\right]$$

$$= 0$$

$$a_n = \frac{1}{2}\int_{-2}^{2} f(x)\cos\left(\frac{n\pi x}{2}\right)dx$$

$$= \frac{1}{2}\left[\int_{-2}^{0} -\cos\left(\frac{n\pi x}{2}\right)dx + \int_{0}^{2}\cos\left(\frac{n\pi x}{2}\right)dx\right]$$

$$= \frac{1}{2}\left[\frac{-\sin\frac{n\pi x}{2}}{\frac{n\pi}{2}}\Bigg|_{-2}^{0} + \frac{\sin\frac{n\pi x}{2}}{\frac{n\pi}{2}}\Bigg|_{0}^{2}\right] = 0 \qquad [\because \sin n\pi = \sin(-n\pi) = 0]$$

$$b_n = \frac{1}{2}\int_{-2}^{2} f(x)\sin\left(\frac{n\pi x}{2}\right)dx$$

$$= \frac{1}{2}\left[\int_{-2}^{0} -\sin\left(\frac{n\pi x}{2}\right)dx + \int_{0}^{2}\sin\left(\frac{n\pi x}{2}\right)dx\right]$$

$$= \frac{1}{2}\left[\frac{\cos\left(\frac{n\pi x}{2}\right)}{\frac{n\pi}{2}}\Bigg|_{-2}^{0} - \frac{\cos\left(\frac{n\pi x}{2}\right)}{\frac{n\pi}{2}}\Bigg|_{0}^{2}\right]$$

$$= \frac{1}{2}\left[\frac{2}{n\pi}\left(1-(-1)^n\right)-\frac{2}{n\pi}\left((-1)^n-1\right)\right]$$

$$= \frac{1}{2}.\frac{2}{n\pi}\left[2-2(-1)^n\right]$$

$$= \frac{2}{n\pi}\left(1-(-1)^n\right)$$

$\therefore$ (1) gives that

$$f(x) = \frac{2}{\pi}\sum_{n=1}^{\infty}\frac{\left(1-(-1)^n\right)}{n}\sin\frac{n\pi x}{2}$$

$$= \frac{4}{\pi}\left[\frac{\sin\left(\frac{x\pi}{2}\right)}{1}+\frac{\sin\left(\frac{3x\pi}{2}\right)}{3}+\frac{\sin\left(\frac{5x\pi}{2}\right)}{5}+\ldots..\right]$$

(or) $$f(x) = \frac{4}{\pi}\sum_{n=0}^{\infty}\frac{1}{2n+1}\sin\left\{\frac{(2n+1)\pi x}{2}\right\}$$

***Example* 15.16:** Obtain the Fourier series of the function, $f(x) = \pi x$ in $-K<x<K$, K being a positive number.

***Solution*:**

Here $l = K$

Let $$f(x) = \frac{a_0}{2}+\sum_{n=1}^{\infty}a_n\cos\frac{n\pi x}{K}+\sum_{n=1}^{\infty}b_n\sin\frac{n\pi x}{K} \qquad .....(1)$$

is the Fourier series required.

$$a_0 = \frac{1}{l}\int_{-l}^{l}f(x)dx$$

$$= \frac{1}{K}\int_{-K}^{K}\pi x\,dx$$

$$= \left.\frac{\pi x^2}{2K}\right|_{-K}^{K} = 0$$

$$a_n = \frac{1}{K}\int_{-K}^{K} f(x)\cos\left(\frac{n\pi x}{K}\right)dx$$

$$= \frac{1}{K}\int_{-K}^{K} \pi x\cos\left(\frac{n\pi x}{K}\right)dx$$

$$= \frac{1}{K}\left[\pi x\frac{\sin\left(\frac{n\pi x}{K}\right)}{\frac{n\pi}{K}} - \pi\left(\frac{-\cos\left(\frac{n\pi x}{K}\right)}{\left(\frac{n\pi}{K}\right)^2}\right)\right]_{-K}^{K}$$

$$= \frac{\pi}{K}.\frac{K^2}{n^2\pi^2}[\cos n\pi - \cos(-n\pi)] = 0$$

$$b_n = \frac{1}{K}\int_{-K}^{K} \pi x\sin\left(\frac{n\pi x}{K}\right)dx$$

$$= \frac{1}{K}\left[\pi x.\frac{-\cos\left(\frac{n\pi x}{K}\right)}{\left(\frac{n\pi}{K}\right)} + \frac{\pi\sin\left(\frac{n\pi x}{K}\right)}{\left(\frac{n\pi}{K}\right)^2}\right]_{-K}^{K}$$

$$= \frac{-1}{K}\left[\frac{\pi K}{n\pi}\{K\cos n\pi - (-K\cos(-n\pi))\}\right]$$

$$= \frac{-2K}{n}\cos n\pi = \frac{2K}{n}(-1)^{n+1}$$

$\therefore$ (1) $\Rightarrow$

$$f(x) = \sum_{n=1}^{\infty}\frac{2K}{n}(-1)^{n+1}\sin\left(\frac{n\pi x}{K}\right)$$

$$= 2K\left[\sin\left(\frac{\pi x}{K}\right) - \frac{1}{2}\sin\left(\frac{2\pi x}{K}\right) + \frac{1}{3}\sin\left(\frac{3\pi x}{K}\right) - + ...\right]$$

***Example* 15.17:** Obtain the Fourier series for $f(x) = (x-1)^2$ in $0 < x < 1$.

***Solution*:**

Here the interval is (0, 1).

i.e., $2l = 1 \quad \Rightarrow \quad l = \dfrac{1}{2},$

Let $$f(x) = \frac{a_0}{2} + \sum_{n=1}^{\infty} a_n\left(\cos\frac{n\pi x}{l}\right) + \sum_{n=1}^{\infty} b_n \sin\frac{n\pi x}{l} \qquad \text{.....(1)}$$

$$a_0 = \frac{1}{l}\int_0^{2l} f(x)dx$$

$$= 2\int_0^1 (x-1)^2\,dx = 2\left.\frac{(x-1)^3}{3}\right|_0^1$$

$$= 2\left(0+\frac{1}{3}\right) = \frac{2}{3}$$

$$a_n = \frac{1}{l}\int_0^{2l} f(x)\cos\left(\frac{n\pi x}{l}\right)dx$$

$$= 2\int_0^1 (x-1)^2.\cos(2n\pi x)dx$$

$$= 2\left[(x-1)^2\frac{\sin(2n\pi x)}{2n\pi} - 2(x-1)\left\{\frac{-\cos(2n\pi x)}{(2n\pi)^2}\right\}\right.$$

$$\left. + 2\left\{\frac{-\sin(2n\pi x)}{(2n\pi)^3}\right\}\right]_0^1$$

$$= 2\left[\frac{2}{4n^2\pi^2}(0+1)\right]$$

$$= \frac{1}{n^2\pi^2}$$

$$b_n = \frac{1}{l}\int_0^{2l} f(x)\sin\left(\frac{n\pi x}{l}\right)dx$$

$$= 2\int_0^1 (x-1)^2 \sin(2n\pi x)dx$$

$$= 2\left[(x-1)^2\left\{\frac{-\cos(2n\pi x)}{2n\pi}\right\} - 2(x-1)\left\{\frac{-\sin(2n\pi x)}{(2n\pi)^2}\right\}\right.$$

$$\left. + 2\left\{\frac{\cos(2n\pi x)}{(2n\pi)^3}\right\}\right]_0^1$$

$$= 2\left[\frac{-1}{2n\pi}(0-1) + \frac{2}{8n^3\pi^3}(1-1)\right]$$

$$= \frac{1}{n\pi}$$

$\therefore$ (1) $\Rightarrow$

$$f(x) = \frac{1}{3} + \sum_{n=1}^{\infty}\frac{\cos(2n\pi x)}{n^2\pi^2} + \sum_{n=1}^{\infty}\frac{\sin(2n\pi x)}{n\pi}$$

***Example* 15.18:** Find a Fourier series expansion for $f(x) = x - x^2$, $-1 < x < 1$.

***Solution*:**

Here $l = 1$

Let $$f(x) = \frac{a_0}{2} + \sum_{n=1}^{\infty} a_n \cos n\pi x + \sum_{n=1}^{\infty} b_n \sin n\pi x \qquad .....(1)$$

$$a_0 = \int_{-1}^{1}\left(x - x^2\right)dx$$

$$= \left(\frac{x^2}{2} - \frac{x^3}{3}\right)\Bigg|_{-1}^{1}$$

$$= \frac{-2}{3}$$

$$a_n = \int_{-1}^{1}\left(x - x^2\right)\cos(n\pi x)dx$$

$$= \left[\left(x - x^2\right)\frac{\sin n\pi x}{n\pi} - (1-2x)\left(\frac{-\cos n\pi x}{n^2\pi^2}\right) - 2\left(\frac{-\sin n\pi x}{n^3\pi^3}\right)\right]_{-1}^{1}$$

$$= \frac{(1-2)(-1)^n}{n^2\pi^2} - \frac{(1+2)(-1)^n}{n^2\pi^2}$$

$$= \frac{-4(-1)^n}{n^2\pi^2}$$

$$b_n = \int_{-1}^{1}\left(x - x^2\right)\sin(n\pi x)dx$$

$$= \left[\left(x - x^2\right)\frac{(-\cos n\pi x)}{n\pi} - (1-2x)\left(\frac{-\sin n\pi x}{n^2\pi^2}\right) - 2\left(\frac{\cos n\pi x}{n^3\pi^3}\right)\right]_{-1}^{1}$$

$$= \left[\frac{-1}{n\pi}\left\{0 + 2(-1)^n\right\} - \frac{2}{n^3\pi^3}\left\{(-1)^n - (-1)^n\right\}\right]$$

$$= \frac{-2}{n\pi}(-1)^n$$

$\therefore$ (1) $\Rightarrow$

$$f(x) = -\frac{1}{3} - \frac{4}{\pi^2}\sum_{n=1}^{\infty}\frac{(-1)^n}{n^2}\cos(n\pi x) - \frac{2}{\pi}\sum_{n=1}^{\infty}\frac{(-1)^n}{n}\sin(n\pi x)$$

***Example* 15.19:** If $f(x)$ is defined as given below, find its Fourier series.

$$f(x) = \begin{cases} 0, & -2 \le x \le -1 \\ 1+x, & -1 \le x \le 0 \\ 1-x, & 0 \le x \le 1 \\ 0, & 1 \le x \le 2 \end{cases}$$

***Solution*:**

Here the interval is $(-2, 2)$

$\therefore \qquad l = 2$

Let $$f(x) = \frac{a_0}{2} + \sum_{n=1}^{\infty} a_n \cos\left(\frac{n\pi x}{2}\right) + \sum_{n=1}^{\infty} b_n \sin\left(\frac{n\pi x}{2}\right) \quad .....(1)$$

be the Fourier series.

Now, $$a_0 = \frac{1}{2}\int_{-2}^{2} f(x)dx$$

$$= \frac{1}{2}\left[\int_{-1}^{0}(1+x)\,dx + \int_{0}^{1}(1-x)\,dx\right], \qquad \text{(since } f(x) = 0 \text{ elsewhere)}$$

$$= \frac{1}{2}\left[\left(x+\frac{x^2}{2}\right)\Bigg|_{-1}^{0} + \left(x-\frac{x^2}{2}\right)\Bigg|_{0}^{1}\right]$$

$$= \frac{1}{2}\left[\left(0+\frac{1}{2}\right)+\left(\frac{1}{2}-0\right)\right]$$

$$= \frac{1}{2}$$

$$a_n = \frac{1}{2}\left[\int_{-1}^{0}(1+x)\cos\left(\frac{n\pi x}{2}\right)dx + \int_{0}^{1}(1-x)\cos\left(\frac{n\pi x}{2}\right)dx\right]$$

$$= \frac{1}{2}\left[\left\{(1+x)\frac{\sin\left(\frac{n\pi x}{2}\right)}{\frac{n\pi}{2}} - 1.\frac{-\cos\left(\frac{n\pi x}{2}\right)}{\left(\frac{n\pi}{2}\right)^2}\right\}_{-1}^{0}\right.$$

$$+\left\{(1-x)\frac{\sin\left(\frac{n\pi x}{2}\right)}{\frac{n\pi}{2}}+1.\frac{-\cos\left(\frac{n\pi x}{2}\right)}{\left(\frac{n\pi}{2}\right)^2}\right\}_0^1\Bigg]$$

$$=\frac{1}{2}\left[\frac{4}{n^2\pi^2}\left\{1-\cos\left(\frac{n\pi}{2}\right)\right\}-\frac{4}{n^2\pi^2}\left\{\cos\left(\frac{n\pi}{2}\right)-1\right\}\right]$$

$$=\frac{1}{2}\left[\frac{8}{n^2\pi^2}-\frac{8}{n^2\pi^2}\cos\left(\frac{n\pi}{2}\right)\right]$$

$$=\frac{4}{n^2\pi^2}\left(1-\cos\left(\frac{n\pi}{2}\right)\right)$$

$$b_n=\frac{1}{2}\left[\int_{-1}^{0}(1+x)\sin\left(\frac{n\pi x}{2}\right)dx+\int_{0}^{1}(1-x)\sin\left(\frac{n\pi x}{2}\right)dx\right]$$

$$=\frac{1}{2}\Bigg[\left\{(1+x).\frac{-\cos\left(\frac{n\pi x}{2}\right)}{\frac{n\pi}{2}}+\frac{\sin\left(\frac{n\pi x}{2}\right)}{\left(\frac{n\pi}{2}\right)^2}\right\}_{-1}^0$$

$$+\left\{(1-x).\frac{-\cos\left(\frac{n\pi x}{2}\right)}{\left(\frac{n\pi}{2}\right)}-\frac{\sin\left(\frac{n\pi x}{2}\right)}{\left(\frac{n\pi}{2}\right)^2}\right\}_0^1\Bigg]$$

$$=\frac{1}{2}\left[\frac{-2}{n\pi}\{1-0\}+\left\{\frac{-2}{n\pi}(0-1)\right\}\right]=0$$

$\therefore$ (1) gives that,

$$f(x)=\frac{1}{4}+\frac{4}{\pi^2}\sum_{n=1}^{\infty}\frac{1}{n^2}\left\{1-\cos\left(\frac{n\pi}{2}\right)\right\}\cos\left(\frac{n\pi x}{2}\right)$$

***Example* 15.20:** If $f(x)=\begin{cases}0,-5<x<0\\3,0<x<5\end{cases}$, find the Fourier series of $f(x)$.

Solution:

Here the interval is $(-5, 5)$

$\therefore \quad l = 5$

$$\therefore \text{ Let } f(x)=\frac{a_0}{2}+\sum_{n=1}^{\infty}a_n\cos\left(\frac{n\pi x}{5}\right)+\sum_{n=1}^{\infty}b_n\sin\left(\frac{n\pi x}{5}\right) \quad .....(1)$$

be the Fourier series of $f(x)$.

$$a_0=\frac{1}{l}\int_{-l}^{l}f(x)dx=\frac{1}{5}\int_0^5 3dx=3$$

$$a_n=\frac{1}{5}\int_0^5 3\cos\left(\frac{n\pi x}{5}\right)dx=\frac{3}{5}.\left.\frac{\sin\left(\frac{n\pi x}{5}\right)}{\frac{n\pi}{5}}\right|_0^5=0$$

$$b_n=\frac{1}{5}\int_0^5 3\sin\left(\frac{n\pi x}{5}\right)dx=\frac{3}{5}.\left.\frac{-\cos\left(\frac{n\pi x}{5}\right)}{\frac{n\pi}{5}}\right|_0^5$$

$$=\frac{-3}{n\pi}\left[(-1)^n-1\right]$$

$$=\frac{6}{n\pi} \text{ if } n \text{ is odd and } \quad \text{'0' if } n \text{ is even.}$$

$$\therefore \quad (1)\Rightarrow f(x)=\frac{3}{2}+\frac{6}{\pi}\left[\frac{\sin\left(\frac{\pi x}{5}\right)}{1}+\frac{\sin\left(\frac{3\pi x}{5}\right)}{3}+\frac{\sin(\pi x)}{5}+.....\right]$$

Exercise - 15.3

1. Find the Fourier series of $f(x) = e^{-x}$ in $(-l, l)$.

$$\textbf{Ans : } \sinh l\left[\frac{1}{a}+\sum_{n=1}^{\infty}\frac{(-1)^n}{n^2\pi^2+l^2}\left\{2l\cos\left(\frac{n\pi x}{l}\right)+2n\pi\sin\left(\frac{n\pi x}{l}\right)\right\}\right]$$

2. If $f(x) = f(x)=\begin{cases}2-x, 0<x<4\\ x-6, 4<x<8\end{cases}$, find the Fourier expansion of $f(x)$.

$$\textbf{Ans : } \frac{16}{\pi^2}\sum_{n=1}^{\infty}\frac{1}{n^2}\cos\left(\frac{n\pi x}{4}\right) \text{ where } n \text{ is odd.}$$

3. Expand $f(x) = x + x^2$ as a Fourier series in $-1 < x < 1$.

$$\textbf{Ans : } \frac{1}{3}+\frac{2}{\pi}\left[\sum_{n=1}^{\infty}(-1)^n\left\{\frac{2}{\pi n^2}\cos(n\pi x)-\frac{1}{n}\sin(n\pi x)\right\}\right]$$

4. If $f(x) = \begin{cases}a-x, \; 0<x<a\\ 0, \; a<x<2a\end{cases}$, obtain the Fourier series of $f(x)$.

$$\textbf{Ans : } \frac{l}{4}+\frac{2l}{\pi^2}\left\{\cos\left(\frac{\pi x}{l}\right)+\frac{\cos\left(\frac{3\pi x}{l}\right)}{3^2}+.....\right\}+\frac{l}{\pi}\left\{\sin\left(\frac{\pi x}{l}\right)+\frac{\sin\left(\frac{2\pi x}{l}\right)}{2}+\frac{\sin\left(\frac{3\pi x}{l}\right)}{3}+.....\right\}$$

5. Find the Fourier series of $f(x) = 2x - x^2$ in $(0, 3)$.

$$\textbf{Ans : } \frac{-9}{\pi^2}\sum_{n=1}^{\infty}\frac{\cos\left(\frac{2n\pi x}{3}\right)}{n^2}+\frac{3}{\pi}\sum_{n=1}^{\infty}\frac{\sin\left(\frac{2n\pi x}{3}\right)}{3}$$

6. If $f(x) = \begin{cases}\pi x, & 0\le x\le 1\\ \pi(2-x), & 1\le x\le 2\end{cases}$, find the Fourier expansion of $f(x)$

$$\textbf{Ans : } \frac{\pi}{2}-\frac{4}{\pi}\left\{\frac{\cos(\pi x)}{1^2}+\frac{\cos(3\pi x)}{3^2}+\frac{\cos(5\pi x)}{5^2}+.....\right\}$$

7. Expand $f(x) = 9 - x^2$ as a Fourier series in (–3, 3).

$$\textbf{Ans} : 6+\frac{36}{\pi^2}\left\{\frac{1}{1^2}\cos\left(\frac{\pi x}{3}\right)-\frac{1}{2^2}\cos\left(\frac{2\pi x}{3}\right)+\frac{1}{3^2}\cos\left(\frac{3\pi x}{3}\right)+.....\right\}$$

8. If $f(x) = \begin{cases} 0, -2 < x < -1 \\ c, -1 < x < 1 \\ 0, 1 < x < 2 \end{cases}$, where 'c' is a constant, find the Fourier series of $f(x)$.

$$\textbf{Ans} : \frac{c}{2}+\frac{2c}{\pi}\left\{\cos\left(\frac{\pi x}{2}\right)-\frac{1}{3}\cos\left(\frac{3\pi x}{2}\right)+\frac{1}{5}\cos\left(\frac{5\pi x}{2}\right)+.....\right\}$$

9. A sinusoidal voltage "E sin (wt)" is passed through a half-wave rectifier which clips the negative portion of the wave. Expand the resulting periodic function $U(t)$ defined below as a Fourier series.

$$U(t) = 0, \quad -\frac{T}{2} < t < 0$$

$$U(t) = E \sin (wt), \; 0 < t < \frac{T}{2},$$

where $$T = \frac{2\pi}{w}$$

[*Hint*: Here $l = \frac{T}{2} = \frac{\pi}{w}$, $a_0 = \frac{2E}{\pi}$; $a_n = \frac{-2E}{\pi(n^2-1)}$ $(n \neq 1)$ and n is even ;

$a_n = 0$ when $n = 1$ and also n is odd ; $b_n = 0$ $(n \neq 1)$ and $b_1 = \frac{E}{2}$]

$$\textbf{Ans} : U(t) = \frac{E}{\pi} + \frac{E}{2}\sin wt - \frac{2E}{\pi}\left(\frac{\cos 2wt}{1.3}+\frac{\cos 4wt}{3.5}+\frac{\cos 6wt}{5.7}+.....\right)$$

15.4 Fourier Series of Even and Odd Functions

Let $f(x)$ be defined in an interval $(-l, l)$. Then the Fourier series of $f(x)$ is given by

$$f(x) = \frac{a_0}{2} + \sum_{n=1}^{\infty} a_n \cos\left(\frac{n\pi x}{l}\right) + \sum_{n=1}^{\infty} b_n \sin\left(\frac{n\pi x}{l}\right) \quad .....(1)$$

where $$a_0 = \frac{1}{l}\int_{-l}^{l} f(x)dx$$

$$a_n = \frac{1}{l}\int_{-l}^{l} f(x)\cos\left(\frac{n\pi x}{l}\right)dx \quad .....(2)$$

$$b_n = \frac{1}{l}\int_{-l}^{l} f(x)\sin\left(\frac{n\pi x}{l}\right)dx$$

15.4.1 Even Functions

Then, $f(x)\cos\left(\frac{n\pi x}{l}\right)$ is even and $f(x)\sin\left(\frac{n\pi x}{l}\right)$ is odd.

$\therefore$ The Euler formulae given in (2) above will become,

$$a_0 = \frac{1}{l}\int_{-l}^{l} f(x)dx = \frac{2}{l}\int_{0}^{l} f(x)dx \quad .....(3)$$

$$a_n = \frac{1}{l}\int_{-l}^{l} f(x)\cos\left(\frac{n\pi x}{l}\right)dx$$

$$= \frac{2}{l}\int_{0}^{l} f(x)\cos\left(\frac{n\pi x}{l}\right)dx \quad .....(4)$$

and $$b_n = \frac{1}{l}\int_{-l}^{l} f(x)\sin\left(\frac{n\pi x}{l}\right)dx = 0 \quad .....(5)$$

$\left(\because \int_{-a}^{a} \varnothing(x)dx = 0 \text{ if } \varnothing(x) \text{ is odd}\right)$

$\therefore$ when $f(x)$ is even, its Fourier series will be

$$f(x) = \frac{a_0}{2} + \sum_{n=1}^{\infty} a_n \cos\left(\frac{n\pi x}{l}\right)$$

Where a_0 and a_n are given by (3) and (4), namely,

$$a_0 = \frac{2}{l}\int_0^l f(x)dx \text{, and}$$

$$a_n = \frac{2}{l}\int_0^l f(x)\cos\left(\frac{n\pi x}{l}\right)dx$$

Note:

1. If $l = \pi$, the above formulae holds good in $(-\pi, \pi)$
2. If $f(x)$ is even, the Fourier series contains only cosine terms ($\because b_n = 0$).

15.4.2 Odd Functions

$$f(x) \text{ is odd} \Rightarrow f(x)\cos\left(\frac{n\pi x}{l}\right) \text{ is odd.}$$

$$\therefore \quad a_0 = 0 \text{ and } a_n = 0$$

$$f(x)\sin\left(\frac{n\pi x}{l}\right) \text{ is even.}$$

$$\therefore \quad b_n = \frac{1}{l}\int_{-l}^{l} f(x)\sin\left(\frac{n\pi x}{l}\right)dx$$

$$= \frac{2}{l}\int_0^l f(x)\sin\left(\frac{n\pi x}{l}\right)$$

$\therefore$ when $f(x)$ is odd, its Fourier series is given by

$$f(x) = \sum_{n=1}^{\infty} b_n \sin\left(\frac{n\pi x}{l}\right)$$

where $\quad b_n = \frac{2}{l}\int_0^l f(x)\sin\left(\frac{n\pi x}{l}\right)dx$

Note:

1. If $f(x)$ is odd, its Fourier series contains only sine terms ($\because a_0 = a_n = 0$).
2. The above two articles (15.4.1 and 15.4.2) will help in simplifying the procedure of calculation of a_0, a_n, and b_n.

In fact, in some of the problems in the previous articles namely 15.2 and 15.3, these concepts of even and odd functions can be used.

Solved Examples

***Example* 15.21:** Obtain the Fourier series of $f(x) = x$ in $(-\pi, \pi)$.

***Solution*:**

$f(x) = x$ is an odd function. So $a_0 = 0$, $a_n = 0$ and the Fourier series of $f(x)$ is given by

$$f(x) = \sum_{n=1}^{\infty} b_n \sin(nx) \qquad .....(1)$$

where $$b_n = \frac{2}{\pi}\int_0^{\pi} f(x)\sin nx\,dx$$

$$= \frac{2}{\pi}\int_0^{\pi} x\sin nx\,dx$$

$$= \frac{2}{\pi}\left[x\left(\frac{-\cos nx}{n}\right) + \frac{\sin nx}{n^2}\right]_0^{\pi}$$

$$= \frac{2}{\pi}\left[\frac{-1}{n}(\pi\cos n\pi - 0)\right] = \frac{2}{n}(-1)^{n+1}$$

$\therefore$ The Fourier series of $f(x)$ is $\displaystyle\sum_{n=1}^{\infty}\frac{2}{n}(-1)^{n+1}\sin nx$ (from (1))

***Example* 15.22:** Obtain the Fourier series of $f(x) = x^2$ in $(-l, l)$.

***Solution*:**

$f(x) = x^2$ is an even function.

$\therefore$ $b_n = 0$

The Fourier series of $f(x)$ is

$$f(x) = \frac{a_0}{2} + \sum_{n=1}^{\infty} a_n \cos\left(\frac{n\pi x}{l}\right) \quad .....(1)$$

$$a_0 = \frac{2}{l}\int_0^l f(x)dx = \frac{2}{l}\int_0^l x^2 dx$$

$$= \frac{2}{l}\left[\frac{x^3}{3}\right]_0^l = \frac{2l^2}{3}$$

$$a_n = \frac{2}{l}\int_0^l f(x)\cos\left(\frac{n\pi x}{l}\right)dx$$

$$= \frac{2}{l}\int_0^l x^2 \cos\left(\frac{n\pi x}{l}\right)dx$$

$$= \frac{2}{l}\left[x^2 \frac{\sin\left(\frac{n\pi x}{l}\right)}{n\frac{\pi}{l}} - (2x).\frac{-\cos\left(\frac{n\pi x}{l}\right)}{\left(\frac{n\pi}{l}\right)^2} + 2.\frac{-\sin\left(\frac{n\pi x}{l}\right)}{\left(\frac{n\pi}{l}\right)^3}\right]_0^l$$

$$= \frac{2}{l}.\frac{2l^2}{n^2\pi^2}.(l\cos n\pi - 0)$$

$$= \frac{4l^2}{n^2\pi^2}(-1)^n$$

$$\therefore \quad (1) \Rightarrow \quad f(x) = \frac{1}{3}l^2 + \frac{4l^2}{\pi^2}\sum_{n=1}^{\infty}\frac{(-1)^n}{n^2}\cos\left(\frac{n\pi x}{l}\right),$$

which is the required Fourier series.

Example **15.23:** Find the Fourier series of the function $f(x)$ defined by

$$f(x) = x + \frac{\pi}{2}, \qquad -\pi < x < 0$$

$$= \frac{\pi}{2} - x, \qquad 0 < x < \pi$$

Hence show that

$$\frac{1}{1^2}+\frac{1}{3^2}+\frac{1}{5^2}+.....=\frac{\pi^2}{8}$$

Solution:

$$f(-x) = -x + \frac{\pi}{2} \text{ in } (-\pi, 0)$$

$$= \frac{\pi}{2} - x$$

$$= f(x) \text{ in } (0, \pi)$$

$\therefore$ $f(x)$ is an even function.

Hence $$f(x) = \frac{a_0}{2}+\sum_{n=1}^{\infty} a_n \cos nx \qquad (\because b_n = 0) \qquad .....(1)$$

$$a_0 = \frac{2}{\pi}\int_0^{\pi} f(x)dx$$

$$= \frac{2}{\pi}\int_0^{\pi}\left(\frac{\pi}{2}-x\right)dx$$

$$= \frac{2}{\pi}\left[\frac{\pi x}{2}-\frac{x^2}{2}\right]_0^{\pi}$$

$$= \frac{2}{\pi}\left(\frac{\pi^2}{2}-\frac{\pi^2}{2}\right) = 0$$

$$a_n = \frac{2}{\pi}\int_0^{\pi}\left(\frac{\pi}{2}-x\right)\cos nx\, dx$$

$$= \frac{2}{\pi}\left[\left(\frac{\pi}{2}-x\right)\frac{\sin nx}{n}+\frac{-\cos nx}{n^2}\right]_0^{\pi}$$

$$= \frac{-2}{\pi}\left[\frac{\cos n\pi - 1}{n^2}\right] = \frac{-2}{\pi n^2}\left((-1)^n - 1\right)$$

$$= \frac{2}{\pi n^2}\left[1-(-1)^n\right]$$

$\therefore$ (1) $\Rightarrow$ $$f(x) = \sum_{n=1}^{\infty}\frac{2}{\pi n^2}\left\{1-(-1)^n\right\}\cos nx$$

$$= \frac{4}{\pi}\left[\frac{\cos x}{1^2}+\frac{\cos 3x}{3^2}+\frac{\cos 5x}{5^2}+.....\right] \qquad .....(2)$$

Since $f(x)$ is discontinous at $x = 0$, we have,

$$f(x) = \frac{1}{2}\left[\lim_{x\to 0+} f(x) + \lim_{x\to 0-} f(x)\right]$$

$$= \frac{1}{2}\left[\frac{\pi}{2}+\frac{\pi}{2}\right] = \frac{\pi}{2}$$

$\therefore$ Putting $x = 0$ in (2), we have,

$$\frac{\pi}{2} = \frac{4}{\pi}\left[\frac{1}{1^2}+\frac{1}{3^2}+\frac{1}{5^2}+.....\right]$$

$$\Rightarrow \quad \frac{1}{1^2}+\frac{1}{3^2}+\frac{1}{5^2}+..... = \frac{\pi^2}{8}$$

***Example* 15.24:** If $f(x)$ is defined by, $f(x) = \begin{cases} -c \text{ when } -\pi < x < 0 \\ c \text{ when } 0 < x < \pi \end{cases}$,

and $f(x + 2\pi) = f(x)$, obtain the Fourier series of $f(x)$. Deduce that $1-\frac{1}{3}+\frac{1}{5}-\frac{1}{7}+..... = \frac{\pi}{4}$

***Solution*:**

Clearly $f(x)$ is odd.

$\therefore$ $a_0 = a_n = 0$ and

$$f(x) = \sum_{n=1}^{\infty} b_n \sin nx$$

$$b_n = \frac{2}{\pi}\int_0^{\pi} f(x)\sin nx dx$$

$$= \frac{2}{\pi}\int_0^{\pi} c\sin nx\, dx = \frac{2}{\pi}\left[\frac{-c\cos nx}{n}\right]_0^{\pi}$$

$$= -\frac{2c}{\pi n}[\cos n\pi - 1]$$

$$= \frac{2c}{\pi n}\left[1-(-1)^n\right]$$

$$\therefore \quad b_n = \frac{4c}{n\pi} \quad \text{if } n \text{ is odd and}$$

$$= 0 \quad \text{if } n \text{ is even}$$

$$\therefore (1) \Rightarrow \quad f(x) = \frac{4c}{\pi}\left[\frac{\sin x}{1} + \frac{\sin 3x}{3} + \frac{\sin 5x}{5} + ...\right] \qquad .....(2)$$

Putting $x = \frac{\pi}{2}$,

$$(2) \Rightarrow \quad c = \frac{4c}{\pi}\left[1 - \frac{1}{3} + \frac{1}{5} - \frac{1}{7} + ...\right]$$

$$\Rightarrow \quad 1 - \frac{1}{3} + \frac{1}{5} - \frac{1}{7} + ... = \frac{\pi}{4}$$

***Example* 15.25:** Obtain the Fourier series representing the function $f(x) = |x|$ in $-\pi < x < \pi$ and deduce that $\frac{1}{1^2} + \frac{1}{3^2} + \frac{1}{5^2} + ... = \frac{\pi^2}{8}$

***Solution*:**

Since $f(x)$ is an even function, $b_n = 0$, and $f(x) = \frac{a_0}{2} + \sum_{n=1}^{\infty} a_n \cos nx$ .....(1)

where $a_0 = \frac{2}{\pi}\int_0^{\pi} f(x) dx$ and

$$a_n = \frac{2}{\pi}\int_0^{\pi} f(x)\cos nx\, dx$$

$$\therefore \qquad a_0 = \frac{2}{\pi}\int_0^{\pi} x dx = \frac{2}{\pi}\left[\frac{x^2}{2}\right]_0^{\pi} = \pi$$

$$a_n = \frac{2}{\pi}\int_0^{\pi} x\cos nx\, dx$$

$$= \frac{2}{\pi}\left[x\frac{\sin nx}{n} - 1\frac{-\cos nx}{n^2}\right]_0^{\pi}$$

$$= \frac{2}{\pi}\left[\frac{\cos nx}{n^2}\right]_0^{\pi} \qquad (\because \sin n\pi = \sin 0 = 0)$$

$$= \frac{2}{\pi}\left[\frac{(-1)^n - 1}{n^2}\right]$$

$$= \frac{-4}{\pi n^2} \qquad \text{if } n \text{ is odd and}$$

$$= 0 \qquad \text{if } n \text{ is even.}$$

$$\therefore \ (1) \Rightarrow \quad f(x) = \frac{\pi}{2} - \frac{4}{\pi}\left[\frac{\cos x}{1^2} + \frac{\cos 3x}{3^2} + \frac{\cos 5x}{5^2} + \ldots\right] \qquad \ldots(2)$$

Which is the Fourier series for $f(x)$ Put $x = 0$ to get the result

$$\frac{1}{1^2} + \frac{1}{3^2} + \frac{1}{5^2} + \ldots = \frac{\pi^2}{8}.$$

***Example* 15.26:** Expand $f(x) = x \cos x$ as a Fourier series in $-\pi < x < \pi$.

***Solution*:**

$f(x) = x \cos x$ is *an* odd function.

$$\therefore \qquad a_0 = 0, \text{ and } a_n = 0$$

Let

$$f(x) = \sum_{n=1}^{\infty} b_n \sin nx \qquad \ldots(1)$$

where $$b_n = \frac{2}{\pi}\int_0^{\pi} f(x)\sin nx\,dx$$

$$b_n = \frac{2}{\pi}\int_0^{\pi} x\cos x\sin nx\,dx$$

$$= \frac{1}{\pi}\int_0^{\pi} x\{\sin(n+1)x + \sin(n-1)x\}dx$$

$$(\because \sin(A + B) + \sin(A - B) = 2\sin A\cos B)$$

$$= \frac{1}{\pi}\left[\int_0^{\pi} x\sin\{(n+1)x\}dx + \int_0^{\pi} x\sin\{(n-1)x\}dx\right]$$

$$= \frac{1}{\pi}\left[x.\frac{-\cos\{(n+1)x\}}{n+1} + \frac{\sin\{(n+1)x\}}{(n+1)^2}\right]_0^{\pi}$$

$$+ \frac{1}{\pi}\left[x.\frac{-\cos\{(n-1)x\}}{n-1} + \frac{\sin\{(n-1)x\}}{(n-1)^2}\right]_0^{\pi}$$

$$= -\frac{1}{\pi}\left[x.\frac{\cos\{(n+1)x\}}{n+1} + \frac{x\cos\{(n-1)x\}}{n-1}\right]_0^{\pi}$$

(The other terms vanish because

$\sin(n + 1)\,\pi = \sin(n - 1)\,\pi = \sin 0 = 0$)

$$= -\frac{1}{\pi}\left[\pi\frac{\cos\{(n+1)\pi\}}{n+1} + \frac{\pi\cos\{(n-1)\pi\}}{n-1}\right]$$

$$= -\left[\frac{\cos n\pi\cos\pi}{n+1} + \frac{\cos n\pi\cos\pi}{n-1}\right]$$

$$= -\left[(-1)^{n+1}\left\{\frac{1}{n+1} + \frac{1}{n-1}\right\}\right]$$

$$= \frac{(-1)^n}{n^2-1}, \qquad (n \neq 1).$$

when $n = 1$,

$$b_1 = \frac{2}{\pi}\int_0^{\pi} x\cos x\sin x\,dx = \frac{1}{\pi}\int_0^{\pi} x\sin 2x\,dx$$

$$= \frac{1}{\pi}\left[x.\frac{-\cos 2x}{2} + \frac{\sin 2x}{4}\right]_0^{\pi}$$

$$= -\frac{1}{\pi}\left[\frac{\pi}{2} - 0\right] = \frac{-1}{2}$$

$$\therefore (1) \Rightarrow \quad f(x) = -\frac{1}{2}\sin x + \sum_{n=2}^{\infty}\frac{(-1)^n}{n^2-1}\sin nx,$$

which is the required Fourier series of $f(x)$.

Exercise - 15.4

1. Find the Fourier series of $f(x) = \begin{cases} 1+\dfrac{2x}{\pi}, -\pi \le x \le 0 \\ 1-\dfrac{2x}{\pi}, 0 \le x \le \pi \end{cases}$

Ans : $\frac{8}{\pi^2}\left[\frac{\cos x}{1^2} + \frac{\cos 3x}{3^2} + \frac{\cos 5x}{5^2} + ...\right]$

2. If $f(x) = x \sin x$ in $(-\pi, \pi)$, find the Fourier series of $f(x)$.

Ans : $1 - \frac{\cos x}{2} - 2\left[\frac{\cos 2x}{1.3} - \frac{\cos 3x}{2.4} + \frac{\cos 4x}{3.5} - + ...\right]$

3. Show that, $\frac{2\sin p\pi}{\pi}\left[\frac{\sin x}{1^2-p^2} - \frac{2\sin 2x}{2^2-p^2} + \frac{3\sin 3x}{3^2-p^2} - + ...\right]$ is the Fourier expansion of $\sin px$ in $(-\pi, \pi)$.

4. Expand $f(x) = |\sin x|$ as a Fourier series in $(-\pi, \pi)$.

Ans : $\frac{2}{\pi} - \frac{4}{\pi}\left(\frac{\cos 2x}{1.3} + \frac{\cos 4x}{3.5} + ...\right)$

5. Obtain the Fourier series expansion of $f(x) = x^2$ in $(-\pi, \pi)$.

$$\textbf{Ans : } \frac{\pi^2}{3} + \sum_{n=1}^{\infty} 4\frac{(-1)^n}{n^2}\cos nx$$

6. Expand $f(x) = 4 - x^2$ in $(-2, 2)$ as Fourier series.

$$\textbf{Ans : } \frac{8}{3} + \frac{16}{\pi^2}\left\{\cos\frac{\pi x}{2} - \frac{1}{2^2}\cos\frac{2\pi x}{2} + \frac{1}{3^2}\cos\frac{3\pi x}{2} - + \ldots\right\}$$

7. Expand $f(x) = \sin 3x$ as a Fourier series in $(-l, l)$.

$$\textbf{Ans : } 2\pi \sin 3l\left[\frac{1}{\pi^2 - 9l^2}\sin\frac{\pi x}{l} - \frac{2}{2^2\pi^2 - 9l^2}\sin\frac{2\pi x}{l} + \ldots\right]$$

8. Find a Fourier series for $f(t) = 1 - t^2$ where $-1 \le t \le 1$.

$$\textbf{Ans : } \frac{2}{3} + \frac{4}{\pi^2}\sum_{n=1}^{\infty}\frac{(-1)^{n+1}}{n^2}\cos n\pi t$$

9. Expand $\cosh x$ as a Fourier series in $(-\pi, \pi)$.

$$\textbf{Ans : } \frac{2\sinh\pi}{\pi}\left\{\frac{1}{2} + \sum_{n=1}^{\infty}\left\{\frac{(-1)^n - 1}{n^2 + 1}\right\}\cos nx\right\}$$

15.5 Half Range Series

In many engineering and physical applications, it may sometimes be required to have the Fourier series expansion of a function in the interval $[0, l]$ where 'l' is half the periods. Such a series is known as half range Fourier series.

15.5.1 Half Range Sine Series

Let $f(x)$ be a function defined in the interval $[0, l]$. Let us define a new function $g(x)$ as follows.

$$g(x) = \begin{cases} f(x), 0 \le x < l \\ -f(-x), -l \le x \le 0 \end{cases} \quad \ldots\ldots(1)$$

Then $g(x)$ is clearly an odd function in $[-l, l]$ and hence the Fourier series of $g(x)$ contains only sine terms.

Also, in $[0, l]$, $g(x) = f(x)$.

$\therefore$ The sine series of $g(x)$ gives the half range sine series of $f(x)$ in $[0, l]$.

But the Fourier series of $g(x)$ is given by $g(x) = \sum_{n=1}^{\infty} b_n \sin\left(\frac{n\pi x}{l}\right)$, where

$$b_n = \frac{2}{l}\int_0^l g(x).\sin\left(\frac{n\pi x}{l}\right)dx \quad \text{[from 15.4.2]}$$

Hence the half range sine series of $f(x)$ in $[0, l]$ is given by

$$\left.\begin{aligned} f(x) &= \sum b_n \sin\left(\frac{n\pi x}{l}\right) \\ \text{where} \quad b_n &= \frac{2}{l}\int_0^l f(x)\sin\left(\frac{n\pi x}{l}\right)dx \end{aligned}\right\} \quad .....(2)$$

Note 1:

Geometrically, the definetion of $g(x)$ of (1) is obtained as follows.

Define $f(x)$ in $[0, l]$ and take its reflection in x - axis (see figure below).

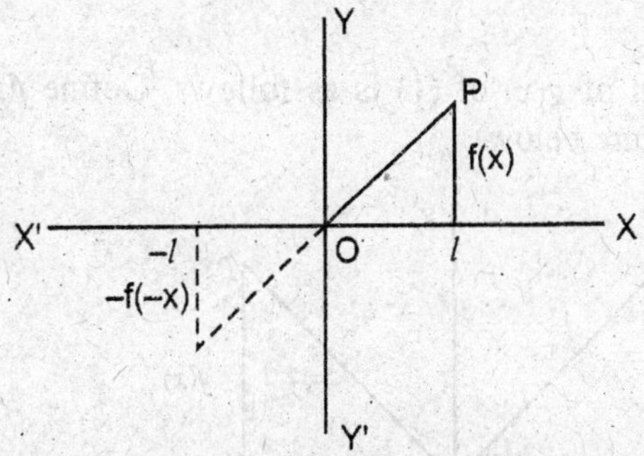

Note 2:

If we put $l = \pi$ in (2), we get the corresponding half range sine series in $[0, \pi]$.

15.5.2 Half Range Cosine Series

If $f(x)$ is defined in $[0, l]$, we define a function $g(x)$ as follows

$$g(x) = \begin{cases} f(x), & 0 \le x \le l \\ f(-x), & -l \le x \le 0 \end{cases} \quad .....(1)$$

so that $g(x)$ is an even function in $[-l, l]$.

Hence the Fourier series of $g(x)$ will contain only cosine terms, and

$$g(x) = \frac{a_0}{2} + \sum_{n=1}^{\infty} a_n \cos\left(\frac{n\pi x}{l}\right)$$

where $$a_0 = \frac{2}{l}\int_0^l g(x)dx$$

$$a_n = \frac{2}{l}\int_0^l g(x)\cos\left(\frac{n\pi x}{l}\right)dx$$

Since $g(x) = f(x)$ in $[0, l]$, the cosine series of $g(x)$ will be the half range cosine series of $f(x)$ in $[0, l]$.

$\therefore$ The half range cosine series of $f(x)$ in $[0, l]$ is given by,

$$f(x) = \frac{a_0}{2} + \sum a_n \cos\left(\frac{n\pi x}{l}\right) \quad .....(2)$$

where $a_0 = \frac{2}{l}\int_0^l f(x)dx$, and

$$a_n = \frac{2}{l}\int_0^l f(x)\cos\left(\frac{n\pi x}{l}\right)dx$$

***Note* 1 :**

Geometrically the definition of $g(x)$ of (1) is as follows. Define $f(x)$ in $[0, l]$ and take its reflection in y - axis (see figure below).

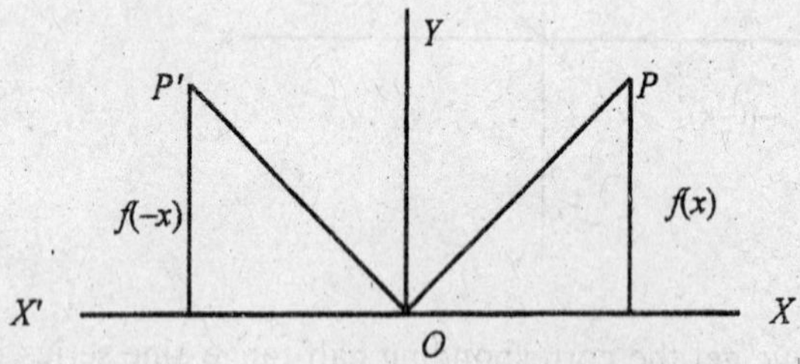

***Note* 2 :**

If we put $l = \pi$ in (2), we get the corresponding half range cosine series of $f(x)$ in $[0, \pi]$

Solved Examples

***Example* 15.27:** Find a half range sine series for $f(x) = c$ in $0 < x <$ p, 'c' being a constant.

***Solution*:**

The Fourier sine series for $f(x) = c$ in $(0, \pi)$ is given by,

$$f(x) = \sum_{n=1}^{\infty} b_n \sin(nx) \quad .....(1)$$

where $b_n = \frac{2}{\pi}\int_0^{\pi} f(x)\sin(nx)dx$

$$= \frac{2}{\pi}\int_0^{\pi} c\sin(nx)dx$$

$$= \frac{2c}{\pi}\left[\frac{-\cos(nx)}{n}\right]_0^{\pi}$$

$$= -\frac{2c}{\pi}\left[\frac{\cos n\pi - \cos 0}{n}\right]$$

$$= \frac{2c}{\pi n}\left[1-(-1)^n\right]$$

$\therefore \quad b_n = 0 \qquad$ if n is even

$= \dfrac{4c}{n\pi} \qquad$ if n is odd

$\therefore \quad (1) \Rightarrow \quad f(x) = \dfrac{4c}{\pi}\displaystyle\sum_{n=1}^{\infty}\frac{\sin(nx)}{n} \qquad [n \text{ odd}]$

$$= \frac{4c}{\pi}\left[\frac{\sin x}{1}+\frac{\sin 3x}{3}+\frac{\sin 5x}{5}+.....\right]$$

is the required sine series of $f(x)$.

***Example* 15.28:** Expand $f(x) = x$ as (1) a half range sine series in $[0, \pi]$ and (2) cosine series, in $(0, \pi)$. Hence find the sum of the series (1) $\frac{1}{1^2}+\frac{1}{3^2}+\frac{1}{5^2}+.....$, and (2) $1-\frac{1}{3}+\frac{1}{5}-\frac{1}{7}+.....$

***Solution*:**

1. ***Sine series :*** Half range sine series of

$$f(x) = \sum_{n=1}^{\infty} b_n \sin(nx) \qquad .....(1)$$

where $\quad b_n = \dfrac{2}{\pi}\displaystyle\int_0^{\pi} f(x)\sin(nx)dx$

$$= \frac{2}{\pi}\int_0^{\pi} x\sin(nx)dx$$

$$= \frac{2}{\pi}\left[x.\frac{-\cos(nx)}{n}+\frac{\sin(nx)}{n^2}\right]_0^{\pi}$$

$$= -2\frac{(-1)^n}{n} = \frac{2(-1)^{n+1}}{n}$$

$\therefore$ From (1), the half range sine series of $f(x)$ is,

$$x = 2\sum_{n=1}^{\infty}\frac{(-1)^{n+1}}{n}\sin(nx)dx$$

$$= 2\left[\frac{\sin x}{1} - \frac{\sin 2x}{2} + \frac{\sin 3x}{3} - +.....\right]$$

Put $x = \frac{\pi}{2}$ in the above result to get,

$$\frac{\pi}{2} = 2\left[1 - \frac{1}{3} + \frac{1}{5}.....\right]$$

(or) $$1 - \frac{1}{3} + \frac{1}{5} - +..... = \frac{\pi}{4}$$

2. ***Cosine series :*** The cosine series of $f(x)$ is give by

$$f(x) = \frac{a_0}{2} + \sum_{n=1}^{\infty} a_n \cos nx$$

$$a_0 = \frac{2}{\pi}\int_0^{\pi} f(x)dx$$

$$= \frac{2}{\pi}\int_0^{\pi} xdx = \frac{2}{\pi}\left[\frac{x^2}{2}\right]_0^{\pi} = \pi$$

and, $$b_n = \frac{2}{\pi}\int_0^{\pi} f(x)\cos(nx)dx$$

$$= \frac{2}{\pi}\int_0^{\pi} x\cos(nx)dx$$

$$= \frac{2}{\pi}\left[x\frac{\sin(nx)}{n} - \frac{1.(-\cos(nx))}{n^2}\right]_0^{\pi}$$

$$= \frac{2}{\pi n^2}\left[(-1)^n - 1\right]$$

$\Rightarrow \quad b_n = 0$ when n is even and

$= \dfrac{-4}{\pi n^2}$ when n is odd.

$\therefore$ (2) $\Rightarrow x = \dfrac{\pi}{2} - \dfrac{4}{\pi}\left[\dfrac{\cos x}{1^2} + \dfrac{\cos 3x}{3^2} + \dfrac{\cos 5x}{5^2} + ...\right]$ is the cosine series of $f(x)$.

Putting $x = 0$ in the above result, we get

$$0 = \frac{\pi}{2} - \frac{4}{\pi}\left[\frac{1}{1^2} + \frac{1}{3^2} + \frac{1}{5^2} + ...\right]$$

$$\Rightarrow \quad \frac{1}{1^2} + \frac{1}{3^2} + \frac{1}{5^2} + ... = \frac{\pi^2}{8}$$

***Example* 15.29:** Find the (half range) sine series of $f(x) = \begin{cases} 2x, & 0 < x < 1 \\ 4 - 2x, & 1 < x < 2 \end{cases}$

***Solution*:**

The sine series of $f(x)$ in $(0, l)$ is given by

$$f(x) = \sum b_n \sin\left(\frac{n\pi x}{l}\right)$$

where $\quad b_n = \dfrac{2}{l}\displaystyle\int_0^l f(x)\sin\left(\frac{n\pi x}{l}\right)dx$

Here $\quad l = 2$;

$$\therefore \quad b_n = \frac{2}{2}\int_0^2 f(x)\sin\left(\frac{n\pi x}{2}\right)dx$$

$$= \int_0^1 2x\sin\left(\frac{n\pi x}{2}\right)dx + \int_1^2 (4-2x)\sin\left(\frac{n\pi x}{2}\right)dx$$

$$= \left[2x\left\{\frac{-\cos\left(\frac{n\pi x}{2}\right)}{\frac{n\pi}{2}}\right\} + 2.\frac{\sin\left[\frac{n\pi x}{2}\right]}{\left(\frac{n\pi}{2}\right)^2} \right]_0^1$$

$$+ \left[(4-2x)\left\{\frac{-\cos\left(\frac{n\pi x}{2}\right)}{\frac{n\pi}{2}}\right\} + 2.\frac{\sin\left\{\frac{n\pi x}{2}\right\}}{\left(\frac{n\pi}{2}\right)^2} \right]_1^2$$

$$= \left[-\frac{4}{n\pi}\cos\left(\frac{n\pi}{2}\right) + \frac{4}{n\pi} + \frac{8}{n^2\pi^2}\sin\left(\frac{n\pi}{2}\right) + \right.$$

$$\left. \frac{4}{n\pi}\cos\left(\frac{n\pi}{2}\right) - \frac{8}{n^2\pi^2}\sin\left(\frac{n\pi}{2}\right) \right]$$

$$= \frac{4}{n\pi}$$

$\therefore$ The sine series is given by $\sum_{n=1}^{\infty} \frac{4}{n\pi}\sin\left(\frac{n\pi x}{2}\right)$

***Example* 15.30:** Find a Fourier sine series for $f(x) = cx + d$ in $0 < x < l$ for, c, d being constants.

***Solution*:**

The sine series of $f(x)$ is given by

$$f(x) = \sum b_n \sin\frac{n\pi x}{l} \qquad .....(1)$$

where $$b_n = \frac{2}{l}\int_0^l f(x)\sin\left(\frac{n\pi x}{l}\right)dx$$

$$b_n = \frac{2}{l}\int_0^l (cx+d)\sin\left(\frac{n\pi x}{l}\right)dx$$

$$= \frac{2}{l}\left[-(cx+d)\frac{\cos\left(\frac{n\pi x}{l}\right)}{\left(\frac{n\pi}{l}\right)} + c\frac{\sin\left(\frac{n\pi x}{l}\right)}{\left(\frac{n\pi}{l}\right)^2}\right]_0^l$$

$$= \frac{2}{l}\left[\frac{-l(cl+d)}{n\pi}(-1)^n + \frac{dl}{n\pi}\right]$$

$$= \frac{2}{\pi}\left[\frac{d-(cl+d)(-1)^n}{n}\right]$$

$\therefore$ From (1), the sine series of $f(x)$ is given by

$$cx + d = \frac{2}{\pi}\sum_{n=1}^{\infty}\left[\frac{d-(cl+d)(-1)^n}{n}\sin\left(\frac{n\pi x}{l}\right)\right]$$

***Example* 15.31:** Obtain the half range cosine series for $f(x) = x - x^2$, in $0 \le x \le 1$.

Solution:

The half range cosine series for $f(x)$ in $0 \le x \le 1$ is given by.

$$f(x) = \frac{a_0}{2} + \sum_{n=1}^{\infty} a_n \cos(n\pi x)$$

where $a_0 = \frac{2}{1}\int_0^1 f(x)dx$

and $a_l = \frac{2}{1}\int_0^1 f(x)\cos(n\pi x)dx$

Now, $a_0 = 2\int_0^1 (x - x^2)dx$

$$= 2\left[\frac{x^2}{2} - \frac{x^3}{3}\right]_0^1 = \frac{1}{3}$$

$$a_n = 2\int_0^1 (x - x^2)\cos(n\pi x)dx$$

$$= 2\left[(x - x^2)\frac{\sin(n\pi x)}{n\pi} + (1-2x)\frac{\cos(n\pi x)}{(n\pi)^2}\right]_0^1$$

$$= 2\left[(-1)\frac{\cos n\pi}{n^2\pi^2} - \frac{1}{n^2\pi^2}\right]$$

$$= 2\left[\frac{(-1)^{n+1} - 1}{n^2\pi^2}\right]$$

$\therefore$ The cosine series of $f(x)$ is [from (1)] given by,

$$f(x) = \frac{1}{6} + \frac{2}{\pi^2}\sum_{n=1}^{\infty}\left\{\frac{(-1)^{n+1} - 1}{n^2}\right\}\cos n\pi x$$

$$= \frac{1}{6} - \frac{4}{\pi^2}\left\{\frac{\cos 2\pi x}{2^2} + \frac{\cos(4\pi x)}{4^2} + ...\right\}$$

***Example* 15.32:** If $f(x)$is defined as

$$f(x) = \begin{cases} \frac{\pi}{3}, 0 \le x < \frac{\pi}{3} \\ 0, \frac{\pi}{3} \le x \le \frac{2\pi}{3} \\ -\frac{\pi}{3}, \frac{2\pi}{3} < x < \pi \end{cases},$$

find the half range cosine series of $f(x)$.

***Solution*:**

The half range cosine series for $f(x)$ is given by

$$f(x) = \frac{a_0}{2} + \sum_{n=1}^{\infty} a_n \cos(nx) \qquad .....(1)$$

where $\quad a_0 = \frac{2}{\pi}\int_0^{\pi} f(x)dx$

and $\quad a_n = \frac{2}{\pi}\int_0^{\pi} f(x)\cos(nx)dx$

$$\therefore \qquad a_0 = \frac{2}{\pi}\left[\int_0^{\frac{\pi}{3}} \frac{\pi}{3}dx + \int_{\frac{2\pi}{3}}^{\pi} -\frac{\pi}{3}dx\right]$$

$$= \frac{2}{\pi}\left[\frac{\pi x}{3}\bigg|_0^{\frac{\pi}{3}} - \frac{\pi x}{3}\bigg|_{\frac{2\pi}{3}}^{\pi}\right]$$

$$= \frac{2}{\pi}\left[\frac{\pi^2}{9} - \left(\frac{\pi^2}{3} - \frac{2\pi^2}{9}\right)\right] = 0$$

$$a_0 = \frac{2}{\pi}\left[\int_0^{\frac{\pi}{3}} \frac{\pi}{3}\cos(nx)dx - \int_{\frac{2\pi}{3}}^{\pi} \frac{\pi}{3}\cos(nx)dx\right]$$

$$= \frac{2}{\pi}\left[\frac{\pi}{3}.\frac{\sin(nx)}{n}\bigg|_0^{\frac{\pi}{3}} - \frac{\pi}{3}\frac{\sin(nx)}{n}\bigg|_{\frac{2\pi}{3}}^{\pi}\right]$$

$$= \frac{2}{3n}\left[\sin\left(\frac{n\pi}{3}\right) + \sin\left(\frac{2n\pi}{3}\right)\right]$$

Put $\quad n = 1, 2, 3, \ldots..,$ we get

$$a_1 = \frac{2}{3}\left[\sin\left(\frac{\pi}{3}\right) + \sin\left(\frac{2\pi}{3}\right)\right] = \frac{2}{\sqrt{3}}$$

$$a_2 = \frac{2}{2.3}\left[\sin\left(\frac{2\pi}{3}\right) + \sin\left(\frac{4\pi}{3}\right)\right] = 0$$

$$a_3 = \frac{2}{3.3}[\sin\pi + \sin 2\pi] = 0$$

$$a_4 = \frac{2}{3.4}\left[\sin\left(\frac{4\pi}{3}\right) + \sin\left(\frac{8\pi}{3}\right)\right] = 0$$

$$a_5 = \frac{2}{3.5}\left[\sin\left(\frac{5\pi}{3}\right) + \sin\left(\frac{10\pi}{3}\right)\right] = \frac{-2}{5\sqrt{3}}$$

$$a_6 = 0$$

$$a_7 = \frac{2}{3.7}\left[\sin\left(\frac{7\pi}{3}\right) + \sin\left(\frac{14\pi}{3}\right)\right] = \frac{2}{7\sqrt{3}} \text{ etc..}$$

$\therefore$ $(1) \Rightarrow f(x) = \frac{2}{\sqrt{3}}\left[\cos x - \frac{\cos 5x}{5} + \frac{\cos 7x}{7} + ...\right]$ is required cosine series.

***Example* 15.33:** Obtain the half range cosine series for $f(x) = \sin x$, in $0 < x < \pi$

***Solution*:**

The half range cosine series of $f(x)$ is given by

$$f(x) = \frac{a_0}{2} + \sum_{n=1}^{\infty} a_n \cos(nx) \qquad .....(1)$$

where $a_0 = \frac{2}{\pi}\int_0^{\pi} f(x)dx$, and

$$a_n = \frac{2}{\pi}\int_0^{\pi} f(x)\cos(nx)dx$$

Here, $a_0 = \frac{2}{\pi}\int_0^{\pi} \sin x\, dx$

$$= \frac{2}{\pi}[-\cos x]_0^{\pi} = \frac{4}{\pi}$$

$$a_n = \frac{2}{\pi}\int_0^{\pi} \sin x \cos(nx) dx$$

$$= \frac{1}{\pi}\int_0^{\pi} [\sin\{(1+n)x\} + \sin\{(1-n)x\}] dx$$

$$= \frac{1}{\pi}\left[-\frac{\cos\{(1+n)x\}}{(1+n)} - \frac{\cos\{(1-n)x\}}{1-n}\right]_0^{\pi}$$

$$= -\frac{1}{\pi}\left[\frac{\cos\{(1+n)\pi\}-1}{1+n} + \frac{\cos\{(1-n)\pi\}-1}{1-n}\right]$$

$$= -\frac{1}{\pi}\{(-1)^{n+1} - 1\}\left\{\frac{1}{1+n} + \frac{1}{1-n}\right\}$$

$[\because \cos\{(1+n)\pi\} = \cos\{(1-n)\pi\} = (-1)^{n+1}]$

$$= \frac{2}{\pi}\{1-(-1)^{n+1}\}\frac{1}{1-n^2}, (n \neq 1)$$

when $n = 1$,

$$a_1 = \frac{2}{\pi}\int_0^{\pi} \sin x \cos x \, dx$$

$$= \frac{1}{\pi}\int_0^{\pi} \sin 2x \, dx$$

$$= \frac{-1}{\pi}\left[\frac{\cos 2x}{2}\right]_0^{\pi}$$

$$= 0$$

$$\therefore \ (1) \Rightarrow \quad f(x) = \frac{2}{\pi} - \frac{2}{\pi}\left[\sum_{n=1}^{\infty} \frac{\{1-(-1)^{n+1}\}}{n^2-1} \cos nx\right]$$

$$= \frac{2}{\pi}\left[1 - \frac{2\cos 2x}{1.3} - \frac{2\cos 4x}{3.5}.....\right]$$

$$= \frac{4}{\pi}\left[\frac{1}{2} - \frac{\cos 2x}{1.3} - \frac{\cos 4x}{3.5} - \frac{\cos 6x}{5.7}.....\right]$$

Exercise – 15.5

1. Obtain the (half range) sine series for the following functions :

(a) $f(x) = \begin{cases} x, & 0 < x \le \frac{\pi}{2} \\ \pi - x, & \frac{\pi}{2} \le x < \pi \end{cases}$

Ans : $f(x) = \frac{4}{\pi}\left\{\sin x - \frac{1}{9}\sin 3x + \frac{1}{25}\sin 5x.....\right\}$

(b) $f(x) = e^x,\ 0 < x < 1$

Ans : $f(x) = \sum_{n=1}^{\infty} \frac{2n\pi}{n^2\pi^2 + 1}\{1 - (-1)^n e\}\sin(n\pi x)$

(c) $f(x) = x^2$ in $(0, \pi)$

Ans : $f(x) = \frac{2}{\pi}\left[\left(\pi^2 - 4\right)\sin x - \frac{1}{2}\pi^2 \sin 2x + \frac{1}{3}\left[\pi^2 - \frac{4}{3^2}\right]\sin 3x + ...\right]$

(d) $f(x) = \begin{cases} \frac{1}{4} - x, & 0 < x < \frac{1}{2} \\ x - \frac{3}{4}, & \frac{1}{2} < x < 1 \end{cases}$

Ans : $f(x) = \sum_{n=1}^{\infty}\left[\frac{\{1 - (-1)^n\}}{2n\pi} - \frac{4}{n^2\pi^2}\sin\left(\frac{n\pi}{2}\right)\right]\sin(n\pi x)$

$$= \frac{1}{\pi^2}\left[(\pi - 4)\sin \pi x + \frac{(3\pi + 4)}{3^2}\sin(3\pi x) + \frac{(5\pi - 4)}{5^2}\sin(5\pi x).....\right]$$

(e) $f(x)\begin{cases} = \frac{2x}{l}, & 0 \le x \le \frac{l}{2} \\ = \frac{2}{l}(l - x), & \frac{l}{2} \le x \le l \end{cases}$

Ans : $f(x) = \frac{8}{\pi^2}\sum_{n=1}^{\infty}\frac{1}{n^2}\sin\left(\frac{n\pi}{2}\right)\sin\left(\frac{n\pi x}{l}\right)$

2. Obtain the (half range) cosine series for the following functions :

(a) $f(x) = \begin{cases} x, & 0 < x \le \dfrac{\pi}{2} \\ \pi - x, & \dfrac{\pi}{2} \le x < \pi \end{cases}$

$$\textbf{Ans : } f(x) = \frac{\pi}{4} - \frac{2}{\pi}\left(\cos 2x + \frac{\cos 6x}{9} + \frac{\cos 10x}{25} + ...\right)$$

(b) $f(x) = \begin{cases} 1, & 0 < x < \dfrac{3}{2} \\ 1 - x, & \dfrac{3}{2} < x < 3 \end{cases}$

$$\textbf{Ans : } f(x) = -\frac{1}{8} + \frac{3}{\pi^2}\left[(\pi + 2)\cos\left(\frac{\pi x}{3}\right) - \cos\left(\frac{2\pi x}{3}\right) + \frac{1}{9}(2 - 3\pi)\cos\left(\frac{3\pi x}{3}\right) + ...\right]$$

(c) $f(x) = l - x$ in $0 < x < l$

$$\textbf{Ans : } f(x) = \frac{l}{2} + \frac{4l}{\pi^2}\left[\frac{\cos\left(\frac{\pi x}{l}\right)}{1^2} + \frac{\cos\left(\frac{3\pi x}{l}\right)}{3^2} + ...\right]$$

(d) $f(x) = e^x$, in $0 < x < \pi$

$$\textbf{Ans : } f(x) = \frac{(e^\pi - 1)}{\pi} + \frac{2}{\pi}\sum_{n=1}^{\infty}\frac{\left[e^\pi(-1)^n - 1\right]}{n^2 + 1}\cos nx$$

(e) $f(x) = (x - 1)^2$; in $0 < x < 1$

$$\textbf{Ans : } f(x) = \frac{1}{3} + \frac{4}{\pi^2}\sum_{n=1}^{\infty}\frac{\cos n\pi x}{n^2}$$

Part IV

CHAPTER 16

Differential Equations of First Order and their Applications

16.1 Introduction

Differential equations play an important role in many applications in the field of science and engineering, such as (i) problems relating to motion of particles (ii) problems involving bending of beams (iii) stability of electric system, etc. For example, Newton's law of cooling states that the rate of change of temperature of a body varies as the excess temperature of the body to that of its surroundings. If $\theta(t)$ is the temperature of the body at time 't' and θ_0 is the temperature of the room in which the body is kept, then $d\theta/dt$ gives the rate of change of temperature with time.

$$\frac{d\theta}{dt} = K(\theta - \theta_0) \text{ ; K is constant}$$

Similarly Newton's second law of motion for a particle of mass m moving in a straight line can be written as

$$m\frac{d^2x}{dt^2} = F$$

where m is the mass, x is the distance of the particle at time 't' measured from a fixed origin and F the external impressed force.

A differential equation is an equation involving an unknown function and its derivatives. If there is only one independent variable and one dependent variable the equation is called *an ordinary differential equation.*

If there are more than one independent variable the equation is called a *partial differential equation* as this involves partial derivatives.

For example :

$$4\frac{d^3y}{dx^2}+3x\frac{dy}{dx}-y=e^x \qquad \text{....(a)}$$

$$\left(\frac{d^3y}{dx^3}\right)^4+\left(\frac{d^2y}{dx^2}\right)+\left(\frac{dy}{dx}\right)^{12}+y^6=x^8 \qquad \text{....(b)}$$

$$\frac{d^2y}{dx^2}+4y=0 \qquad \text{....(c)}$$

$$\left[1+\left(\frac{dy}{dx}\right)^2\right]^{3/2}=4\frac{d^2y}{dx^2} \qquad \text{....(d)}$$

$$\frac{\partial^2 u}{\partial x^2}=a^2\frac{\partial^2 u}{\partial t^2} \qquad \text{....(e)}$$

$$\frac{\partial u}{\partial t}=e^2\frac{\partial^2 u}{\partial x^2} \qquad \text{....(f)}$$

$$\frac{\partial^2 u}{\partial x^2}+\frac{\partial^2 u}{\partial y^2}=\frac{1}{c^2}\left(\frac{\partial^2 u}{\partial t^2}\right) \qquad \text{....(g)}$$

The first four equations (a), (b), (c) and (d) are ordinary differential equations and the remaining three are partial differential equations.

Order of a differential equation : The order of a differential equation is the order of the highest ordered derivative appearing in the equation.

Degree of a differential equation : The degree of a differential equation is the power to which the highest ordered derivative appears in the equation after clearing the radicals if any.

In the above example:

(a) is a differential equation of order 3 and degree 1.

(b) is of third order and fourth degree differential equation.

(c) is a second order, first degree differential equation.

(d) is a second order, second degree differential equation.

[**Note:** Topics 16.2 to 16.3 are given for the reference of students as a revision]

16.2 Formation of an Ordinary D.E

The differential equations are formed by eliminating all the arbitrary constants that are involved in the functional relationship between the dependent and independent variables.

$y = cx^2 + c^2$ where c is an arbitrary constant. .... (1)

To eliminate 'c': (only one constant)

From (1) $$\frac{dy}{dx} = c.2x + 0$$

$$c = \frac{1}{2x}\frac{dy}{dx}$$

Substitution of c in (1) gives

$$y = \frac{1}{2x}\frac{dy}{dx}x^2 + \frac{1}{4x^2}\left(\frac{dy}{dx}\right)^2$$

$$\left(\frac{dy}{dx}\right)^2 + 2x^3\frac{dy}{dx} - 4x^2y = 0$$

is the required D.E and $y = cx^2 + c^2$ is called the solution of the *D.E.*

Note: Depending on the number of constants in the given equation, differentiate it as many number of times successively. Then the elimination of the arbitrary constants from the resulting equations and the given equation gives the required differential equation whose order is equal to the number of constants.

Solved Examples

***Example* 16.1:** Eliminate the arbitrary constants a, b from $xy + x^2 = ae^x + be^{-x}$ and form the differential equation.

Solution:

The given equation is $xy + x^2 = ae^x + be^{-x}$ .....(1)

The number of arbitrary constants is two. Differentiating (1) w.r., to 'x' two times successively.

$$x\frac{dy}{dx} + y + 2x = ae^x - be^{-x} \quad .....(2)$$

$$x\frac{d^2y}{dx^2} + \frac{dy}{dx} + \frac{dy}{dx} + 2.1 = ae^x + be^{-x} \quad .... (3)$$

From (1), (2) and (3) elimination of a, b gives the D.E.

From (1) and (3) we get

$$x\frac{d^2y}{dx^2} + \frac{2dy}{dx} + 2 = xy + x^2 \text{ is the required D.E.}$$

***Example* 16.2:** Form the differential equation by eliminating the constants a and b from $ax^2 + by^2 = 1$

Solution:

Differentiating

$$ax^2 + by^2 = 1 \text{ w.r.t '}x\text{'} \quad .... (1)$$

$$2ax + 2by\frac{dy}{dx} = 0 \quad .... (2)$$

Again differentiating w.r.t., 'x'

$$2a + 2by\frac{d^2y}{dx^2} + 2b\frac{dy}{dx}.\frac{dy}{dx} = 0 \quad .... (3)$$

Elimination of a, b from (1), (2) and (3) gives

$$\begin{vmatrix} x^2 & y^2 & -1 \\ x & yy_1 & 0 \\ 1 & (yy_2 + y_1^2) & 0 \end{vmatrix} = 0$$

Expanding the determinant we get

$$x\frac{d^2y}{dx^2} + x\left(\frac{dy}{dx}\right)^2 - y\left(\frac{dy}{dx}\right) = 0$$

***Example* 16.3:** Form the differential equation by eliminating the constants from $y = a \sec x + b \tan x$

Solution:

Given equation is

$$y = a\sec x + b\tan x \quad .... (1)$$

Differentiating w.r. to 'x'

$$\frac{dy}{dx} = \text{a}\sec x \tan x + b\sec^2 x \qquad \text{.... (2)}$$

$$\frac{dy}{dx} = \sec x[\text{a} \tan x + b \sec x] \qquad \text{.... (3)}$$

Further differentiation gives

$$\frac{d^2y}{dx^2} = a \sec x \tan^2 x + \text{a} \sec^3 x + b(2\sec^2 x \tan x)$$

i.e., $$\frac{d^2y}{dx^2} = \text{a}\sec x \tan^2 x + \text{b}\sec^2 x\tan x + \text{b}\sec^2 x\tan x + \text{a}\sec^3 x$$

i.e., $$\frac{d^2y}{dx^2} = \sec x \tan x(a\tan x + b\sec x) + \sec^2 x(b\tan x + a\sec x) \qquad \text{.....(4)}$$

Substituting

$$\text{a}\sec x + \text{b}\tan x = y \qquad \text{from (1)}$$

and $$\text{a}\tan x + \text{b}\sec x = \frac{\left(\frac{dy}{dx}\right)}{\sec x} \qquad \text{from (2)}$$

in (3) we get $$\left(\frac{d^2y}{dx^2}\right) = \sec x \tan x \frac{\left(\frac{dy}{dx}\right)}{\sec x} + \sec^2 x(y)$$

i.e., $$\frac{d^2y}{dx^2} - \tan x \frac{dy}{dx} - y\sec^2 x = 0$$

***Example* 16.4:** Form the differential equation of all circles passing through the origin and having their centres on the x - axis.

Solution:

Take any tangent to the circle as y-axis, the centre lies on x-axis. Let 'a' be the radius of the circle.

Then centre is (a, 0)

$\therefore$ Equation of the circle is $(x - \text{a})^2 + y^2 = \text{a}^2$ .... (1)

Differentiating (1) w.r. to 'x'

$$2(x-a)+2y\frac{dy}{dx}=0$$

$$x-a=-y\frac{dy}{dx}$$

$$a=x+y\frac{dy}{dx} \qquad \text{.... (2)}$$

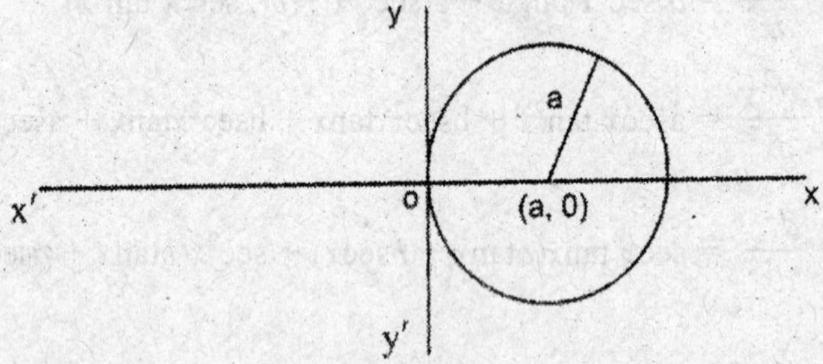

Fig. 16.1

From (1) and (2)

$$\left[x-\left(x+y\frac{dy}{dx}\right)\right]^2+y^2=\left[x+y\frac{dy}{dx}\right]^2$$

$$y^2\left(\frac{dy}{dx}\right)^2+y^2=x^2+y^2\left(\frac{dy}{dx}\right)^2+2xy\left(\frac{dy}{dx}\right)$$

$2xy\dfrac{dy}{dx}+x^2-y^2=0$ is the required D.E

***Example* 16.5 :** Form the differential equation of all central conics whose axes coincide with the axes of coordinates.

Solution:

The equation of all central conics whose axes coincide with the axes is

$$ax^2+by^2=1 \qquad \text{.... (1)}$$

Differentiating (1) w.r.t., x

$$2ax+2by\frac{dy}{dx}=0 \qquad \text{.... (2)}$$

Differentiating (2) w.r.t., 'x' again

$$a+by\frac{d^2y}{dx^2}+b\left(\frac{dy}{dx}\right)^2=0 \qquad .... (3)$$

Eliminating a, b from (1), (2) and (3)

$$\begin{vmatrix} x^2 & y^2 & 1 \\ x & y\frac{dy}{dx} & 0 \\ 1 & y\frac{d^2y}{dx^2}+\left(\frac{dy}{dx}\right)^2 & 0 \end{vmatrix}=0$$

$$\Rightarrow \qquad xy\frac{d^2y}{dx^2}+x\left(\frac{dy}{dx}\right)^2-y\left(\frac{dy}{dx}\right)=0 \text{ is the required D.E.}$$

Exercise 16.1

1. **Eliminate the arbitrary constants from the following and find the corresponding differential equation :**

(i) $y = mx + c$ (m, c arbitrary constants)

[**Ans :** $\frac{d^2y}{dx^2}=0$]

(ii) $y = a\,e^{2x} + b\,e^{-2x}$ (a, b arbitrary - constants)

[**Ans :** $\frac{d^2y}{dx^2}-4y-0$]

(iii) $y = ax + bx^2$ (a, b arb. constants)

[**Ans :** $x^2\frac{d^2y}{dx^2}-2x\frac{dy}{dx}+2y=0$]

(iv) $(x-h)^2+(y-k)^2=a^2$, (h, k a, arb. constants)

[**Ans :** $\left[1+\left(\frac{dy}{dx}\right)^2\right]^3=a^2\left(\frac{d^2y}{dx^2}\right)$]

(v) $y = (a + bx)e^{-x}$, (a, b arb. constants)

[*Ans :* $\frac{d^2y}{dx^2}+\frac{2dy}{dx}+y=0$]

(vi) $y = a\sin x + b\cos x$ (a, b arb. constants) [**Ans :** $\frac{d^2y}{dx^2}+y=0$]

2. Find the differential equation of all circles with centre on the line $y = x$ and having radius '1'.

$$\text{[Ans : } \left[1+\frac{dy}{dx}\right]^2 = (x-y)^2\left[1+\left(\frac{dy}{dx}\right)^2\right]\text{]}$$

3. Form the differential equation of all the circles with centre on the line $y = -x$ and passing through the origin.

$$\text{[Ans. : } (x^2+y^2)\left(\frac{dy}{dx}-1\right) = 2(y-x)\left(x+y\frac{dy}{dx}\right)\text{]}$$

4. Find the differential equation of all the parabolas with vertex at the origin and foci on the x - axis.

$$\text{[Ans : } y^2 - 2xy\frac{dy}{dx} = 0\text{]}$$

5. Find the differential equation of all parabolas with the origin as focus and axis along x-axis.

$$\text{[Ans : } 2x\frac{dy}{dx} + y\left(\frac{dy}{dx}\right)^2 - y = 0\text{]}$$

16.3 The Differential Equations of the First Order and of the First Degree

16.3.1 Separation of Variables

Sometimes the differential equation

$$\frac{dy}{dx} = q(x, y)$$

can be written as

$$f(x)\,dx + g(y)dy = 0 \qquad \text{.... (1)}$$

if the variables can be separated.

Integration of (1) gives the solution of the equation.

i.e., $$\int f(x)dx + \int g(y)\,dy = c$$

where c is an arbitrary constant

***Example* 16.6:** Solve $e^x \tan y\, dx + (1-e^x)\sec^2 y\, dy = 0$

Solution:

The given equation

$$e^x \tan y\, dx + (1-e^x)\sec^2 y\, dy = 0$$

can be rearranged as

$$\frac{e^x}{1-e^x}dx + \frac{\sec^2 y}{\tan y}\, dy = 0$$

Integrating $$\int\frac{e^x}{1-e^x}dx + \int\frac{\sec^2 y}{\tan y}\, dy = c$$

$$-\log(e^x-1) + \log \tan y = c$$

i.e., $$\frac{\tan y}{e^x-1} = c \text{ or } \tan y = c(e^x-1)$$

***Example* 16.7:** Solve $\frac{dy}{dx} = 1 + x^2 + y^2 + x^2y^2$

Solution:

The given differential equation can be written as

$$\frac{dy}{dx} = (1+x^2)(1+y^2)$$

$$\frac{dy}{1+y^2} = (1+x^2)dx \Rightarrow \tan^{-1} y = x + \frac{x^3}{3} + C$$

***Example* 16.8:** Solve $\frac{dy}{dx} - 2xy = x$, where $y(0) = 1$

Solution:

$$\frac{dy}{dx} = x(1+2y)$$

$$\frac{dy}{1+2y} = xdx$$

On integration

$$\frac{1}{2}\log(1+2y) = \frac{x^2}{2} + C \qquad \text{.... (1)}$$

Given $y = 1$ when $x = 0$

Substituting in (1)

$$\frac{1}{2}\log(3) = 0 + C$$

$$\therefore \qquad C = \frac{1}{2}\log 3 = \log\left(\sqrt{3}\right)$$

Hence the solution is

$$\frac{1}{2}\log(1 + 2y) = \frac{x^2}{2} + \log\sqrt{3} \text{ (i.e.,) } \log\left(\frac{1+2y}{3}\right) = x^2$$

***Example* 16.9 :** Solve $\dfrac{dy}{dx} = (4x + y + 1)^2$

***Solution*:**

$$\frac{dy}{dx} = (4x + y + 1)^2 \qquad \text{.... (1)}$$

Substituting $4x + y + 1 = t$ in (1)

$$4 + \frac{dy}{dx} = \frac{dt}{dx}$$

i.e.,

$$\frac{dy}{dx} = \frac{dt}{dx} - 4$$

$$\frac{dt}{dx} - 4 = t^2$$

Integrating

$$\int\frac{dt}{t^2+4} = \int dx \Rightarrow \frac{dt}{t^2+2^2} = x + c$$

i.e.,

$$\frac{1}{2}\tan^{-1}\frac{t}{2} = x + c$$

$$\tan^{-1}\left(\frac{4x+y+1}{2}\right) = 2(x + c)$$

The solution can also be written as

$$4x + y + 1 = 2\tan(2x + c)$$

where c is an arbitrary constant.

Exercise 16.2

Solve the Following Differential Equations

1. $\frac{dy}{dx} = e^{x+y}$ [**Ans :** $e^x + e^{-y} = c$]

2. $(2 - x)dy - (3 + y)\,dx = 0$ [**Ans** $(3 + y)(2 - x) = c$]

3. $\frac{dy}{dx} = e^{x-y} + x^2e^{-y}$ [**Ans :** $e^y = e^x + \frac{x^3}{3} + c$]

4. $y - x\frac{dy}{dx} = a\left(y^2 + \frac{dy}{dx}\right)$ [**Ans :** $yc = (a + x)(1 - ay)$]

5. $3e^x \tan y\,dx + (1-e^x)\sec^2 y\,dy = 0$ [**Ans :** $\tan y = c(1- e^x)^3$]

6. $(x-y)^2\frac{dy}{dx} = a^2$ [**Ans :** $(x - y) + \log\left(\frac{x-y-a}{x-y+a}\right) = x + c$]

7. $\frac{dy}{dx} = (3x + 4y + 1)^2$ [**Ans :** $2(3x + 4y + 1) = \sqrt{3}\tan^{-1} 2\sqrt{3}x + c$]

8. $\frac{dy}{dx} = (2x + y + 1)^2$ [**Ans :** $\frac{1}{\sqrt{2}}\tan^{-1}\left(\frac{2x+y+1}{\sqrt{2}}\right) = x + c$]

9. $\frac{dy}{dx} = \tan(x + y)$ [**Ans :** $\log[\sin(x + y) + \cos(x + y)] = x - y + c$]

10. $\frac{dy}{dx} = \frac{2}{(x+2y-3)}$ [**Ans :** $(x + 2y - 3) - 4\log(x + 2y + 1) = x + c$]

16.3.2 Homogeneous Equations

The differential equation of the form

$$\frac{dy}{dx} = \frac{f(x,y)}{g(x,y)}$$

where f, g are homogeneous functions of same degree in x, y is called a homogenous differential equation.

Such a differential equation can be written as

$$\frac{dy}{dx}=\frac{x^n\psi\left(\frac{y}{x}\right)}{x^n\phi\left(y/x\right)}=\frac{\psi\left(\frac{y}{x}\right)}{\phi\left(y/x\right)} \qquad \text{.... (1)}$$

Substituting $y = vx$

$$\frac{dy}{dx}=v+x\frac{dv}{dx}$$

The D.E (1) becomes

$$v+x\frac{dv}{dx}=\frac{\psi(v)}{\phi(v)}$$

$$\frac{dx}{x}+\frac{\phi(v)dv}{v\phi(v)-\psi(v)}=0$$

Integration yields the solution

$$\int\frac{dx}{x}+\int\frac{\phi(v)dv}{v\phi(v)-\psi(v)}=c \text{ where } v=\frac{y}{x}$$

***Example* 16.10:** Solve $(x^2 + y^2)\dfrac{dy}{dx} = xy$

Solution:

$$\frac{dy}{dx}=\frac{xy}{x^2+y^2}$$

Substituting $y = vx$, $\quad \dfrac{dy}{dx}=v+x\dfrac{dv}{dx}$

$$\therefore \quad v+x\frac{dv}{dx}=\frac{xvx}{x^2+v^2x^2}=\frac{v}{1+v^2}$$

$$x\frac{dv}{dx}=\frac{v}{1+v^2}-v=\frac{-v^3}{1+v^2}$$

$$\frac{dx}{x}+\frac{1+v^2}{v^3}\,dv = 0$$

$$\int\frac{dx}{x}+\int\frac{1}{v^3}dv+\int\frac{1}{v}dv=c$$

$$\log x+\frac{v^{-2}}{-2}+\log v = c$$

$$\log (xv)-\frac{1}{2v^2} = c$$

$$\log\left(x\left(\frac{y}{x}\right)\right)\frac{-x^2}{2y^2} = c \Rightarrow 2y^2\log y - x^2 = 2cy^2$$

i.e., $$2y^2\log y = 2cy^2 + x^2$$

***Example* 16.11:** Solve $x\frac{dy}{dx}+\frac{y^2}{x}=y$

Solution:

$$x^2\frac{dy}{dx}+y^2 = xy$$

i.e., $$\frac{dy}{dx}=\frac{xy-y^2}{x^2}$$

Substituting $y = vx$,

$$\frac{dy}{dx}=v+x\frac{dv}{dx}$$

$\therefore$ $$v+x\frac{dv}{dx}=\frac{xvx-v^2x^2}{x^2}$$

$$v+x\frac{dv}{dx}=v-v^2$$

$$x\frac{dv}{dx} = -v^2$$

$$\frac{dx}{x} + \frac{dv}{v^2} = 0$$

Integration yields

$$\log x + \frac{v^{-2+1}}{-2+1} = c \Rightarrow \log x - \frac{1}{v} = c$$

$$\log x - \frac{x}{y} = c$$

$$y\log x - x = cy \qquad \text{or} \qquad y\log x = x + cy$$

***Example* 16.12:** Solve $\left[x + y\sin\left(y/x\right)\right]dx = x\sin\left(y/x\right)dy$

***Solution*:**

The given differential equation can be written as

$$\frac{dy}{dx} = \frac{x + y\sin\left(y/x\right)}{x\sin\left(y/x\right)}$$

Substituting $y = vx \Rightarrow \frac{dy}{dx} = v + x \cdot \frac{dy}{dx}$

$$\therefore \qquad v + x\frac{dv}{dx} = \frac{x + vx\sin v}{x\sin\sin(v)}$$

$$v + x\frac{dv}{dx} = \frac{1 + v\sin}{\sin v}$$

$$x\frac{dv}{dx} = \frac{1 + v\sin v}{\sin v} - v = \frac{1}{\sin v}$$

$$\sin v\, dv = \frac{dx}{x}$$

Integrating

$$\Rightarrow \quad -\cos v = \log x + c$$

$$\Rightarrow \quad \log x + \cos\left(\frac{y}{x}\right) = c$$

EXERCISE 16.3

Solve the Following Equations

1. $\dfrac{dy}{dx} = \dfrac{y^2}{xy - x^2}$ **[Ans :** $y = ce^{y/x}$ **]**

2. $(2 - 2xy)dx = (x^2 - 2xy)dy$ **[Ans :** $xy(y - x) = c$ **]**

3. $2xy + (y^2 - x^2)\dfrac{dy}{dx} = 0$ **[Ans :** $x^2 + y^2 = cy$ **]**

4. $(x + y)^2\, dx = 2x^2 dy$ **[Ans :** $\log x = 2\tan^{-1}\left(\frac{y}{x}\right) + c$ **]**

5. $xdy - ydx = \sqrt{x^2 + y^2}\, dx$ **[Ans :** $y + \sqrt{x^2 + y^2} = cx^2$ **]**

6. $x\sin\left(\dfrac{y}{x}\right) dy = \left(y\sin\dfrac{y}{x} - x\right) dx$ **[Ans :** $\cos\dfrac{y}{x} = \log cx$ **]**

7. $x\cos\left(\frac{y}{x}\right)(ydx + xdy) = y\sin\left(\frac{y}{x}\right)(xdy - ydx)$ **[Ans :** $\sec\left(\frac{y}{x}\right) = yxc$**]**

8. $(x^2y - x^3)dy - (y^3 + xy^2)\, dx = 0$ **[Ans :** $y\sqrt{x^2 + y^2} = cx.e^{\tan^{-1}\left(\frac{y}{x}\right)}$ **]**

9. $(x^2 + y^2)\, dx = 2xydy$ **[Ans :** $(x^2 - y^2) = xc$ **]**

10. $x\dfrac{dy}{dx} = y[\log y - \log x + 1]$ **[Ans :** $y = xe^{cx}$ **]**

16.3.3 Non-Homogeneous Differential Equations

The D.E of the form

$$\frac{dy}{dx}=\frac{ax+by+c}{Ax+By+C} \quad .... (1)$$

where *a, b, c, A, B, C* are constants, is called a *non-homogeneous differential equation.*

Case (i)

If $\frac{a}{A}\neq\frac{b}{B}$

Substituting $x=X+h,\ Y=y+k$

where *h, k* are (constants) to be chosen so as to satisfy.

$ah+bk+c=0,\ Ah+Bk+C=0$

Solving these equations, values of *h, k* are obtained.

The given D.E then reduces to a homogeneous D.E.

$$\frac{dY}{dX}=\frac{aX+bY}{AX+BY}$$

which is then solved taking Y = VX and then

substitute X = $x-h$ and Y = $y-k$ in the solution.

Case (ii)

If $\frac{a}{A}=\frac{b}{B}$

Then the differential equation will be of the form

$$\frac{dy}{dx}=\frac{(ax+by)+c}{m(ax+by)+C} \quad .... (2)$$

since Ax + By will be constant *m* times $ax+by$.

Now substitute

$$ax+by=t,$$

Differentiation gives

$$a+b\frac{dy}{dx}=\frac{dt}{dx}$$

i.e., $$\frac{dy}{dx} = \frac{\frac{dt}{dx} - a}{b}$$

D.E (2) then reduces to

$$\frac{\frac{dt}{dx} - a}{b} = \frac{t + c}{mt + C}$$

Then the solution is obtained by using the method of separation of variables.

***Example* 16.13:** Solve $\frac{dy}{dx} = \frac{x + 2y - 3}{2x + y - 3}$

Solution:

$$\frac{dy}{dx} = \frac{x + 2y - 3}{2x + y - 3} \qquad \text{.... (1)}$$

Substituting

$x = X + h,\ y = Y + k$

where h, k are chosen to satisfy $h + 2k - 3 = 0$ and $2h + k - 3 = 0$

solving we get

$$h = 1,\ k = 1 \qquad \text{.... (2)}$$

i.e., we take $x = X + 1,\ y = Y + 1$

The D.E (1) reduces to

$$\frac{dY}{dX} = \frac{X - 2Y}{2X + Y}$$

Substituting $Y = VX,\ \frac{dY}{dX} = V + X\frac{dV}{dX}$

$$\therefore \qquad V + X\frac{dV}{dX} = \frac{X + 2VX}{2X + VX}$$

$$V + X\frac{dV}{dX} = \frac{1 + 2V}{2 + V} \Rightarrow X\frac{dV}{dX} = \frac{1 - V^2}{2 + V}$$

$$\frac{2+v}{1-v}dv = \frac{dX}{X}$$

$$\frac{dX}{X} + \left(1 + \frac{3}{v-1}\right) dv = 0$$

$\log X + v + 3\log(v-1) = \log c$

$v + \log(v-1)^3 + \log X = \log c$

$v - \log(v-1)^3 X = \log c$

$$\frac{Y}{X} + \log\left(\frac{Y-x}{X}\right)^3 X = \log c$$

i.e., $\log(Y - X)^3 + XY = X\log c$

$\log(y - 1 - x + 1)^3 + (x - 1)(y - 1) = (x - 1)\log c$

$\log(y - x)^3 = (x - 1)(\log c - y + 1)$

***Example* 16.14:** Solve $(2x + 3y + 1)\,dx + (2y - 3x + 5)dy = 0$

Solution:

$$\frac{dy}{dx} = -\frac{2x+3y+1}{2y-3x+5}$$

Substituting $x = X + h$, $y = Y = k$, $\dfrac{dy}{dx} = \dfrac{dY}{dX}$

Choosing h, k so that $2h + 3k + 1 = 0$, $3h - 3k - 5 = 0$

we get $h = 1$, $k = -1$

The given differential equation reduces to

$$\frac{dV}{dX} = \frac{2X+3Y}{3X-2Y}$$

Substituting

$$Y = VX$$

$$\frac{dV}{dX} = V + X\frac{dV}{dX}$$

$$\therefore \quad V + X\frac{dV}{dX} = \frac{2X + 3VX}{3X - 2VX}$$

$$V + X\frac{dV}{dX} = \frac{2 + 3V}{3 - 2V}$$

$$X\frac{dV}{dX} = \frac{2(1 + V^2)}{3 - 2V}$$

$$\left(\frac{3 - 2V}{1 + V^2}\right)dV = \frac{2dX}{X}$$

Integrating

$$3\int\frac{dv}{1+v^2} - \int\frac{2v}{1+v^2}dv = 2\int\frac{dX}{X} + c$$

$$3\tan^{-1}(V) - \log(1 + v^2) = 2\log X + c$$

$$3\tan^{-1}\left(\frac{y}{x}\right) - \log\left(\frac{X^2 + Y^2}{X^2}\right) = 2\log X + c$$

But $X = x - h = x - 1$

$Y = y - k = y + 1$

$$3\tan^{-1}\left(\frac{y+1}{x-1}\right) - \log\left[\frac{(x-1)^2 + (y+1)^2}{(x-1)^2}\right] = 2\log(x - 1) + c$$

***Example* 16.15:** Solve $\dfrac{dy}{dx} = \dfrac{x - y + 1}{2x - 2y}$

Solution:

$$\frac{dy}{dx} = \frac{(x-y)+1}{2(x-y)}$$

Substituting $x - y = V$

$$\frac{dy}{dx} = 1 - \frac{dV}{dx}$$

$$1 - \frac{dV}{dx} = \frac{V+1}{2V}$$

$$\frac{dV}{dx} = 1 - \frac{V+1}{2V} = \frac{V-1}{2V}$$

$$\frac{2V}{V-1}\, dV = dx$$

Integrating

$$2\int \frac{V-1+1}{V-1}\, dV = x + c$$

$$2[V + \log(V-1)] = x + c$$

$$2[x - y + \log(x - y - 1)] = x + c$$

$$2 \log (x - y - 1) = 2y - x + c$$

or $$(x - y - 1)2 = e^{2y-x} C$$

***Example* 16.16:** Solve $\frac{dy}{dx} = -\frac{4x+6y+1}{2x+3y-5}$

Solution:

$$\frac{dy}{dx} = -\frac{2(2x+3y)+1}{(2x+3y)-5}$$

Substituting

$$2x + 3y = V$$

$$2+3\frac{dy}{dx} = \frac{dV}{dx}$$

$$\frac{dy}{dx} = \frac{1}{3}\left[\frac{dV}{dx} - 2\right]$$

The differential equation reduces to

$$\frac{1}{3}\left(\frac{dV}{dx} - 2\right) = -\frac{-2V+1}{V-5}$$

$$\frac{dV}{dx} = 2 + \frac{6V+3}{5-V} = \frac{10-2V+6V+3}{5-v} = \frac{13+4V}{5-V}$$

$$\frac{5-V}{4V+13}\, dV = dx$$

Integrating $\int\left(1-\frac{33}{4V+13}\right) dv = -4\int dx + c$

$$v - \frac{33}{4}\log(4V + 13) = -4x + c$$

$$(2x + 3y) - \frac{33}{4}\log[4(2x + 3y) + 13] = -4xm + C$$

i.e., $4(6x + 3y - c) = 33\log[4(2x + 3y) + 13]$

Exercise 16.4

Solve the Following Equations

1. $\frac{dy}{dx} = \frac{y-x+5}{y+x+3}$

[**Ans :** $\tan^{-1}\left(\frac{y+4}{x-1}\right)+\frac{1}{2}\log\left[\left(\frac{y+4}{x-1}\right)^2+1\right] = \log(x-1) + \log c$]

2. $\frac{dy}{dx} = \frac{x+2y-3}{2x+3y-5}$

[**Ans :** $(2+\sqrt{3})\log\left[\frac{y-1}{x-1}-\frac{1}{\sqrt{3}}\right] - (2-\sqrt{3})\log\left[\frac{y-1}{x-1}+\frac{1}{\sqrt{3}}\right]+\frac{2}{\sqrt{3}}\log x = \log c$]

3. $\frac{dy}{dx} = \frac{2y-x-4}{y-3x+3}$

[**Ans :** $(x-2)^2 - 5(x-2)(y-3) + (y-3)^2 = c\left\{\frac{2(y-3)-[5+\sqrt{21}(x-2)]}{2(y-3)-(5-\sqrt{21}(x-2))}\right\}$]

4. $(2x + 5y + 1)dx - (5x + 2y - 1)dy = 0$

[**Ans :** $(x + y)^7 = c\left(x-y-\frac{2}{3}\right)^3$]

5. $\frac{dy}{dx} = \frac{y-x+1}{y+x+5}$

[**Ans :** $\tan^{-1}\left(\frac{y-2}{x-3}\right) + \frac{1}{2}\log[(y-2)^2 + (x-3)^2] = \log c$]

6. $\frac{dy}{dx} = \frac{3y+2x+4}{4x+6y+5}$

[**Ans :** $\frac{2}{7}(2x+3y) - \frac{9}{49}(14x+21y+22) = x + c$]

7. $(4x - 6y - 1)\,dx + (3y - 2x - 2)dy = 0$

[**Ans :** $x - y + \frac{3}{4}\log(8x - 12y - 5) = c$]

8. $\frac{dy}{dx} = \frac{x-y+3}{2x-2y+5}$

[**Ans :** $(x - y) + \log[2 + x - y] = \frac{x}{2} + c$]

16.4 Linear Differential Equations

A differential equation of the form $\frac{dy}{dx} + py = q$ where p, q are functions of 'x', alone or constants is said to be a linear differential equation of first order. Multiplying both sides of the equation by $e^{\int pdx}$ [called the *integrating factor (I.F)*] we get,

$$e^{\int pdx}\frac{dy}{dx} + e^{\int pdx}py = q.e^{\int pdx} \qquad \text{.... (i)}$$

The left hand side is the differential coefficient of $y.\,e^{\int pdx}$

$\therefore$ (i) can be written as

$$d\left(y.e^{\int pdx}\right) = q\,e^{\int pdx}$$

Integrating

$$ye^{\int p dx} = \int q e^{\int p dx} dx + c$$

which gives the desired solution.

Note:

In some cases a differential equation can be reduced to the linear form by taking 'y' as independent variable and x as the dependent variable.

The D.E is written as

$$\frac{dx}{dy} + p_1 x = q_1$$

p_1, q_1 are functions of y or constants

Now the I.F = $e^{\int p_1 dy}$

∴ Solution is

$$x.e^{\int p_1 dy} = \int q_1.e^{\int p_1 dy} + c$$

Solved Examples

***Example* 16.17:** Solve $(1 + x^2)\dfrac{dy}{dx} + 2yx - 6x^2 = 0$

Solution:

Rearranging the given differential equation to the form

$$\frac{dy}{dx} + py = q$$

We have

$$\frac{dy}{dx} + \frac{2x}{1+x^2} y = \frac{6x^2}{1+x^2}$$

Here $p = \dfrac{2x}{1+x^2}, q = \dfrac{6x^2}{1+x^2}$

$$\text{I.F} = e^{\int p dx} = e^{\int \frac{2x}{1+x^2} dx}$$

$$\text{I.F} = e^{\log(1+x^2)} = (1 + x^2)$$

Solution is given by

$$y(\text{I.F}) = \int p(I.F)dx + c$$

$$\therefore \quad y(1 + x^2) = \int \frac{6x^2}{1+x^2} \times (1 + x^2)dx + c$$

$$y(1 - x^2) = 6\frac{x^3}{3} + c$$

i.e., $y(1 - x^2) = 2x^3 + c$

***Example* 16.18:** Solve $x \log x \dfrac{dy}{dx} + y = 2\log x$

Solution:

$$\frac{dy}{dx} + \frac{y}{x\log x} = \frac{2}{x}$$

Here $\quad p = \dfrac{1}{x\log x}$ and $q = \dfrac{2}{x}$

$$\text{I.F} = e^{\int p dx} = e^{\int \frac{1}{x\log x}dx} = e^{\log(\log x) = \log x}$$

Solution is

$$y \log x = \int \frac{2}{x} \log x dx + c$$

$$y \log x = 2\frac{(\log x)^2}{2} + c$$

***Example* 16.19:** Solve $x \cos x \dfrac{dy}{dx} + y(x\sin x + \cos x) = 1$

Solution:

$$\frac{dy}{dx} + y.\frac{x\sin + \cos x}{x\cos x} = \frac{1}{x\cos x}$$

$$p = \frac{x\sin x + \cos x}{x\cos x}, q = \frac{1}{x\cos x}$$

$$\text{I.F} = e^{\int \frac{x\sin x + \cos x}{x\cos x}dx}$$

$$= e^{\int\left(\tan x + \frac{1}{x}\right)dx}$$

$$= e^{(\log(x\sec x))}$$

$$\text{I.F} = e^{\log(x \sec x)}$$

$$= x\sec x$$

Solution is

$$y(x\sec x) = \int \frac{1}{x\cos x} \times x \sec x \, dx + c$$

$$xy \sec x = \int \sec^2 x dx \in c$$

$$xy \sec x = \tan x + c$$

***Example* 16.20 :** Solve $\frac{dy}{dx} + 2y\tan x = \sin x$ given that $y = 0$ where $x = \frac{\pi}{3}$

***Solution*:**

$$\frac{dy}{dx} + 2y\tan x = \sin x$$

$$p = 2\tan x,\ q = \sin x$$

$$\text{IF} = e^{\int 2\tan x dx}$$

$$dx = e^{2\log\sec x} = \sec^2 x$$

$$y^{\text{IF}} = \int q \times IF.dx + c$$

$$y\sec^2 x = \int \sec x.\sec^2 x dx + c$$

$$y\sec^2 x = \int \sec x \tan x dx + c$$

$$y\sec^2 x = \sec x + c \quad \text{.... (1)}$$

Given that $y = 0$ when $x = \pi/3$

$$\therefore \quad 0 = \sec\frac{\pi}{3} + c \Rightarrow c = -2$$

Substituting $c = -2$ in (1)

$$y\sec^2 x = \sec x - 2$$

is the required solution

***Example* 16.21:** Solve $(x + 2y^2)\dfrac{dy}{dx} = y$

Solution:

$$x + 2y^2 = y\frac{dy}{dx}$$

$$\therefore \quad \frac{dx}{dy} - \frac{x}{y} = 2y^2$$

is in the form of

$$\frac{dx}{dy} + p_1 y = q_1$$

$$P_1 = \frac{-1}{y}, \quad q_1 = 2y^2$$

$$\text{IF} = e^{\int p_1\, dy}$$

$$= e^{\int -\frac{1}{y} dy}$$

$$= \frac{1}{y}$$

$$x(\text{IF}) = \int q_1 (IF) dy + c$$

$$x\frac{1}{y} = \int 2y^2 \times \frac{1}{y} dy + c$$

$$\frac{x}{y} = y^2 + c$$

***Example* 16.22:** Solve $(x + y + 1)\dfrac{dy}{dx} = 1$

Solution:

$$\frac{dx}{dy} = x + y + 1$$

$$\frac{dx}{dy} - x = y + 1$$

$$P_1 = -1,\ q_1 = y + 1$$

$$\text{IF} = e^{\int -1 dy} = e^{-y}$$

Solution is given by

$$x(\text{IF}) = \int q_1 (IF) dy + c$$

$$xe^{-y} = \int (y+1) e^{-y}\, dy + c$$

$$xe^{-y} = -(y + 1)e^{-y} - e^{-y} + c$$

i.e., $e^{-y}[x + y + 2] = c$ or $x + y + 2 = ce^{y}$

Exercise 16.5

Solve the Following Differential Equations

1. $(1 + y^2)dx = (\tan^{-1}y - x)dx$ **[Ans :** $xe^{\tan^{-1} y} = \tan^{-1}y e^{\tan^{-1}y} - e^{\tan^{-1} y} + c$]

2. $\cos^2 x \dfrac{dy}{dx} + y = \tan x$ **[Ans :** $y = ce^{-\tan x} - \tan x - 1$]

3. $x\dfrac{dy}{dx} + 2y - x^2 \log = 0$ **[Ans :** $y = \dfrac{c}{x^2} + \dfrac{1}{4} x^2 \log x - \dfrac{x^2}{16}$]

4. $\dfrac{dy}{dx} + y\cot x = 4x \operatorname{cosec} x$, if $y = 0$, when $x = \pi/2$

   [Ans : $y\sin x = 2x^2 - \dfrac{\pi^2}{2}$]

5. $ye^{y}dx = (y + 2xe^{y})dy$ **[Ans :** $xy^{-2} = c - e^{-y}$]

6. $(x + 3y^3) \dfrac{dy}{dx} = y$ **[Ans :** $x = \dfrac{3}{2}y^3 + cy$]

7. $(xy - 1) \dfrac{dy}{dx + y^3} + y^3 = 0$ **[Ans :** $x = ce^{\frac{1}{y}} + \dfrac{1}{y} + 1$]

8. $\dfrac{dy}{dx} = x^3 - 2xy$ if $y = 2$ when $x = 1$ **[Ans :** $2y - x^2 + 1 = 4\, e^{1 - x^2}$]

9. $(1+x^3)\dfrac{dy}{dx} + 6x^2y = e^x$ **[Ans :** $y(1+x^3)^2 = e^x(x^3 - 3x^2 + 6x - 5) + c$**]**

10. $\dfrac{dy}{dx} = -\dfrac{x + y\cos x}{1+\sin x}$ **[Ans :** $y(1+\sin x) = c - \dfrac{x^2}{2}$ **]**

16.4.1 Non-linear Differential Equation of First Order

Bernoulli's Equation

The differential equation of the form

$$\frac{dy}{dx} + py = qy^n \qquad \text{.... (1)}$$

where p, q are functions of x alone is said to be a Bernoulli's differential equation.

Dividing (1) throughout by y^n

$$y^{-n}\frac{dy}{dx} + py^{-n+1} = q \qquad \text{.... (2)}$$

Substituting y^{-n+1}(coefficient of p) = v

$$(1-n)\,\frac{1}{y^n}\frac{dy}{dx} = \frac{dv}{dx}$$

(2) reduces to

$$\frac{1}{(1-n)}\frac{dv}{dx} + pv = q$$

$$\frac{dv}{dx} + (1-n)pv = (1-n)q$$

which is linear in v.

The above D.E can be solved by using the method given in *example* 16.17.

Solved Examples

***Example* 16.23:** Solve $\dfrac{dy}{dx} - y\tan x = y^2\sec x$

Solution:

$$\frac{dy}{dx} - y\tan x = y^2\sec x \qquad \text{.... (1)}$$

$$y^{-2}\frac{dy}{dx} - \frac{1}{y}\tan x = \sec x$$

Substituting

$$-\frac{1}{y} = v,$$

$$+\frac{1}{y^2}\frac{dy}{dx} = \frac{dv}{dx}$$

(1) reduces to

$$\frac{dv}{dx} + v\tan x = \sec x$$

is linear in v.

Here $p = \tan x,\ q = \sec x$

$$\text{IF} = e^{\int p dx}$$

$$= e^{\int \tan x dx}$$

$$= e^{\log\sec x}$$

$$= \sec x$$

Solution of the D.E (1) is therefore

$$v(\text{IF}) = \int q(IF)dx + c$$

$$v \,.\, \sec x = \int \sec x.\sec x dx + c$$

$$= \int \sec^2 x dx + c$$

Substituting

$$v = -\frac{1}{y}$$

$$-\frac{1}{y}\sec x = \tan x + c$$

i.e., $y(\tan x + c) + \sec x = 0$

***Example* 16.24:** Solve $(3xy + y^2)\, dx - 3x^2 dy = 0$

Solution:

$$3x^2 \frac{dy}{dx} - 3xy = y^2$$

is in Bernoulli's form

$$\frac{1}{y^2}\frac{dy}{dx} - \frac{1}{xy} = \frac{1}{3x^2}$$

Substituting

$$\frac{-1}{y} = v$$

D.E reduces to

$$\frac{dv}{dx} - \frac{1}{xy} v = \frac{1}{3x^2} \qquad \text{.... (1)}$$

which is linear in v

Here $\quad p = \frac{1}{x}, \qquad q = \frac{1}{3x^2}$

$$\text{IF} = e^{\int p dx}$$

$$= e^{\int \frac{1}{x} ds}$$

$$= x$$

Solution of the D.E (1) is

$$v(\text{IF}) = \int q(\text{IF}) dx + c$$

$$-\frac{1}{y} x = \int \frac{1}{3x^2} \, . \, x dx + c$$

$$-\frac{x}{y} = \frac{1}{3} \log x + c$$

i.e., $\quad y\left(\frac{1}{3}\log x + c\right) + x = 0$

***Example* 16.25:** Solve $\frac{dy}{dx} + (2x\tan^{-1}y - x^3)(1 + y^2) = 0$

Solution:

Let $\tan^{-1}y = v$

$$\frac{1}{1+y^2}\frac{dy}{dx} = \frac{dv}{dx}$$

The given differential equation becomes

$$\frac{dv}{dx} + 2xv = x^3 \qquad \text{.... (1)}$$

which is linear in v

Here $p = 2x, \qquad q = x^3$

$$\text{IF} = e^{\int 2x\,dx} = e^{x^2}$$

Solution of the D.E (1) is

$$v(\text{IF}) = \int q(\text{IF})dx + c$$

Writing $x^2 = \text{t}$

$$xdx = \frac{1}{2}dt$$

$$v.e^{x^2} = \int t.e^t \frac{dt}{2} + c$$

$$\tan^{-1}y.e^{x^2} = \frac{1}{2}[\text{t.e}^\text{t} - \text{e}^\text{t}] + \text{c}$$

$$e^{x^2}\tan^{-1}y = \frac{1}{2}\left[x^2e^{x^2} - e^{x^2}\right] + c$$

***Example* 16.26:** Solve $\tan y \frac{dy}{dx} + \tan x = \cos y \cos^3 x$

Solution:

$$\tan y\frac{dy}{dx} + \tan x = \cos y \cos^3 x$$

Dividing by cosy throughout

$$\sec y \tan y \frac{dy}{dx} + \sec y \tan x = \cos^3 x$$

Substituting $\sec y = v$, we get

$$\sec y \tan y \frac{dy}{dx} = \frac{dv}{dx}$$

$$\frac{dv}{dx} + v \,.\, \tan x = \cos^3 x \quad \text{.... (1)}$$

is linear in v

Here $p = \tan x, \qquad q = \cos^3 x$

$$\text{IF} = e^{\int p dx} = e^{\int \tan x dy} = \sec x$$

Solution of the D.E (1) is

$$v(\text{IF}) = \int q(\text{IF})dx + c$$

$$v.\ \sec x = \int \cos^3 x \times \sec x dx + c$$

$$\sec y \sec x = \int \cos^2 x dx + c$$

$$\sec x \sec y = \int \frac{1+\cos 2x}{2} + c$$

$$\sec x \sec y = \frac{x}{2} + \frac{\sin 2x}{4} + c$$

***Example* 16.27**: Solve $\frac{dy}{dx} = (\sin x - \sin y)\ \frac{\cos x}{\cos y}$

Solution:

$$\cos y \frac{dy}{dx} + \sin y \cos x = \sin x \cos x$$

Substituting $\sin y = v$

$$\cos y \frac{dy}{dx} = \frac{dv}{dx}$$

The given equation reduces to

$$\frac{dv}{dx} + v \cos x = \sin x \cos x \quad \text{.... (1)}$$

is linear in v

Here $\quad p = \cos x, \quad q = \sin x \cos x$

$$\text{IF} = e^{\int p dx} = e^{\int \cos x dx} = e^{\sin x}$$

Solution of the D.E (1) is

$$v.e^{\sin x} = \int \sin x \cos x . e^{\sin x} dx + c$$

write $\quad \sin x = t \Rightarrow \cos x \, dx = dt$ in the RHS

$$v \,.\, e^{\sin x} = \int t.e^t dt + c$$

$$\sin y \, e^{\sin x} = [te^t - e^t] + c$$

$$\sin y \, e^{\sin x} = e^{\sin x}[\sin x - 1] + c$$

Exercise 16.6

Solve the Following Differential Equations

1. $(y\log x - 1)y dx = x dy)$ $\quad$ [Ans : $\frac{1}{y} = 1 + \log x + cx$]

2. $\frac{dy}{dx}\cos x + y\sin x = \sqrt{y \sec x}$ $\quad$ [Ans : $2y \; y^{1/2} \sqrt{\sec x} = \tan x + 2c$]

3. $\frac{dy}{dx} + x\sin 2y = x^3\cos^2 y$ $\quad$ [Ans : $\frac{1}{y^2} = - \sin^2 x - \sin x - \frac{1}{2} + ce^{2\sin x}$]

4. $\frac{dy}{dx} - \frac{\tan y}{1+x} = (1+x)\, e^x \sec y$ $\quad$ [Ans : $\sin y = (1 + x)\,(e^x + c)$]

5. $x\frac{dy}{dx} + y\log y = xye^x$ $\quad$ [Ans : $x\log y = (x - 1)e^x + c$]

6. $3\frac{dy}{dx} + \frac{2}{x+1}y = \frac{x^3}{y^2}$ $\quad$ [Ans : $(x + 1)^2 y^3 = \frac{x^6}{6} + \frac{2x^5}{5} + \frac{x^4}{4} + c$]

7. $\frac{dy}{dx} + y\tan x = y^3\sec x$ $\quad$ [Ans : $\cos^2 x = y^2(c + 2\sin x)$]

8. $\frac{dy}{dx} = \frac{1}{x^2y^3 + xy}$ $\quad$ [Ans : $-\frac{1}{x}e^{\frac{y^2}{2}} = 2\left(\frac{y^2}{2} - 1\right)e^{\frac{y^2}{2}} + c$]

16.5 Exact Differential Equations

Let us consider the differential equation Mdx + Ndy = 0 where M, N are functions of x, y.

If this equation is to be exact, then it must have been derived by directly differentiating some function of x, y.

Hence Mdx + Ndy = du, say .... (1)

But from differential calculus

$$du = \frac{\partial u}{\partial x}dx + \frac{\partial u}{\partial y}dy \qquad \text{.... (2)}$$

From (1) and (2) we get

$$M = \frac{\partial u}{\partial x}, \quad N = \frac{\partial u}{\partial y}$$

Now
$$\frac{\partial M}{\partial y} = \frac{\partial^2 u}{\partial y \partial x} \quad \text{and} \quad \frac{\partial N}{\partial x} = \frac{\partial^2 u}{\partial x \partial y}$$

$\Rightarrow$ $\frac{\partial M}{\partial y} = \frac{\partial N}{\partial x}$ is the condition for exactness.

$\therefore$ The differential equation Mdx + Ndy = 0

is exact if $\frac{\partial M}{\partial y} = \frac{\partial N}{\partial x}$

Then the solution is expressed in the form

$$\int M dx \quad + \quad \int N dy = c$$

(treating y as constant integrate w.r.t x) + (integrate w.r.t y those terms that are independent of x)

Note : If N has no term independent of x then the solution is

$$\int M dx = c$$

Solved Examples

***Example* 16.28:** Solve (x + 2y – 3) dy – (2x – y + 1)dx = 0

Solution:

$(x + 2y - 3)dy - (2x - y + 1)\,dx = 0$

$M = -(2x - y + 1)$

N = (x + 2y – 3)

$$\frac{\partial M}{\partial y} = 1$$

$$\frac{\partial N}{\partial x} = 1$$

∴ The given differential equation is exact

The solution is

$$\therefore \quad -\int(2x - y + 1)dx + \int(x + 2y - 3)\,dy = c$$

$$\frac{-2x^2}{2} + yx - x + \frac{2y^2}{2} - 3y = c$$

$$\Rightarrow \quad y^2 - x^2 + xy - x - 3y = c$$

***Example* 16.29:** Solve $(x^2 + y^2)dx + 2xy\,dy = 0$

Solution:

$$(x^2 + y^2)\,dx + 2xy\,dy = 0$$

$$M = x^2 + y^2 \qquad N = 2xy$$

$$\frac{\partial M}{\partial y} = 2y \qquad \frac{\partial N}{\partial x} = 2y$$

∴ The given differential equation is exact

∴ Solution is $\int(x^2 + y^2)dx + \int 2xy dy = c$

$$\frac{x^3}{3} + y^2 x = c$$

***Example* 16.30:** Solve $(1 + e^{x/y})dx + e^{x/y}\left(1 - \frac{x}{y}\right)dy = 0$

Solution:

$$(1 + e^{x/y})\,dx + e^{x/y}\left(1 - x/y\right)dy = 0$$

$$M = 1 + e^{x/y}, \quad N = e^{x/y}\left(1 - \frac{x}{y}\right)$$

and $$\frac{\partial N}{\partial x} = e^{x/y}\left(-\frac{1}{y}\right) + \left(1 - \frac{x}{y}\right)e^{x/y} \times \frac{1}{y}$$

$$\frac{\partial M}{\partial y} = \frac{-x}{y^2}\ e^{x/y}$$

$$\frac{\partial M}{\partial y} = \frac{\partial N}{\partial x}$$

$\therefore$ The given differential equation is exact

$\therefore$ Solution is

$$\int\left(1 + e^{x/y}\right)dx + \int e^{x/y}\left(1 - \frac{x}{y}\right)dy = c$$

$$x + e^{x/y}\ (y) = c$$

Exercise 16.7

Solve the Following Differential Equations

1. $(e^y + 1)\cos x\, dx + e^y \sin x dy = 0$ **[Ans :** $(e^y + 1)\sin x = c$ **]**
2. $(y\cos x + \sin y + y)dx + (\sin x + x\cos y + x)dy = 0$ **[Ans :** $y\sin x + (\sin y + y)x = c$ **]**
3. $(x^2 - ay)dx = (ax - y^2)dy$ **[Ans :** $x^3 + y^3 - 3axy = c$ **]**
4. $(ax + hy + g)dx + (hx + by + f)dy = 0$

   [Ans : $\frac{ax^2}{2} + (hy + g)x + \frac{by^2}{2} + fy = c$ **]**
5. $(x^2 + y^2 - a^2)xdx + (x^2 - y^2 - b^2)ydy = 0$

   [Ans : $x^4 + 2x^2y^2 - 2a^2x^2 - y^4 - 2b^2y^2 = 4c$ **]**

16.5.1 Integrating Factors

If the differential equation Mdx + Ndy = 0 is not exact, it can be made exact by multiplying it with some function of x, y. Such a function is called an *integrating factor.*

***Rules for finding the integrating factors* :**

1. *Integrating factors found by inspection*

Ex. Solve $x\,dy - ydx = 0$

Solution:

$$xdy - ydx = 0$$

Dividing by x^2

$$\frac{xdy - ydx}{x^2} = 0$$

$$d\left(\frac{y}{x}\right) = 0$$

On integration

$$\frac{y}{x} = c$$

16.5.2 First Method to find an Integrating Factor

If the differential equation $Mdx + Ndy$ is not exact, but is homogenous and $Mx + Ny \neq 0$, then the integrating factor is $\frac{1}{Mx + Ny}$. Multiply the differential equation by IF. The DE becomes exact.

Ex. Solve $(x^2y - 2xy^2)dx - (x^3 - 3x^2y)dy = 0$

Solution: $(x^2y - 2xy^2)dx - (x^3 - 3x^2y)dy = 0$ .... (1)

The differential equation (1) is homogeneous

$M = x^2y - 2xy^2$ $\qquad$ $N = -(x^3 - 3x^2y)$

$\frac{\partial M}{\partial y} = x^2 - 4xy$ $\qquad$ $\frac{\partial N}{\partial x} = -3x^2 + 6xy$

$\therefore$ The DE is not exact and

$$Mx + Ny = x^2y^2 \neq 0$$

$$\therefore \qquad IF = \frac{1}{Mx + Ny} = \frac{1}{x^2y^2}$$

Multiplying the DE by the integrating factor $\frac{1}{x^2y^2}$

$$\left(\frac{x^2y - 2xy^2}{x^2y^2}\right)dx - \left(\frac{x^2 - 3x^3y}{x^2y^2}\right)dy = 0 \qquad \text{.... (2)}$$

write $$M_1 = \frac{1}{y} - \frac{2}{x} \quad N_1 = \frac{-x}{y^2} + \frac{3}{y}$$

then $$\frac{\partial M_1}{\partial y} = \frac{\partial N_1}{\partial x} = -\frac{1}{y^2}$$

$\therefore$ DE (2) is exact

$\therefore$ Solution is $$\int\left(\frac{1}{y} - \frac{2}{x}\right)\partial x - \int\left(\frac{x}{y^2} - \frac{3}{y}\right)\partial y = c$$

$$\frac{x}{y} - 2\log x + 3\log y + c$$

i.e., $$\frac{x}{y} = \log x^2 y^3 c$$

or $$x^2 y^3\, c = e^{\frac{x}{y}}$$

16.5.3 Second Method to Find the Integrating Factor

If the differential equation Mdx + Ndy = 0 is not exact and is of the form

$$f(xy)ydx + g(xy)xdy = 0$$

and $$\text{M}x - \text{N}y \neq 0$$

then $$\frac{1}{\text{M}x - \text{N}y} \text{ is an integrating factor}$$

***Example* 16.31:** Solve $(x^2y^2 + xy + 1)ydx + (x^2y^2 - xy + 1)xdy = 0$

Solution:

$$(x^2y^2 + xy + 1)ydx + (x^2y^2 - xy + 1)xdy = 0 \qquad \text{.... (1)}$$

$$\text{M} = x^2y^3 + xy^2 + y, \quad \text{N} = x^2y^2 - x^2y + x$$

$$\text{M}x - \text{N}y = 2x^2y^2 \neq 0$$

Hence the $$\text{IF} = \frac{1}{\text{M}x - \text{N}y} = \frac{1}{2x^2y^2}$$

Multiplying (1) by IF

$$\frac{(x^2y^2+xy+1)}{2x^2y^2}+\frac{(x^2y^2-xy+1)}{2x^2y^2}=0$$

$$\frac{1}{2}\left\{y+\frac{1}{x}+\frac{2}{x^2y}\right\}dx+\frac{1}{2}\left\{x-\frac{1}{y}+\frac{1}{xy^2}\right\}\text{dy}=0 \quad \text{.... (2)}$$

$$\frac{\partial M_1}{\partial y}=\frac{1}{2}-\frac{1}{x^2y^2}$$

$$\frac{\partial N_1}{\partial x}=\frac{1}{2}-\frac{1}{x^2y^2}$$

∴ DE (2) is exact

Solution of the DE (1) is

$$\frac{1}{2}\int\left(y+\frac{1}{x}+\frac{1}{x^2y}\right)dx+\frac{1}{2}\int\left(x-\frac{1}{y}+\frac{1}{xy^2}\right)dy=c$$

$$\frac{1}{2}\left[xy+\log x-\frac{1}{xy}\right]-\frac{1}{2}\log y=c$$

***Example* 16.32:** Solve $(xy\sin xy+\cos xy)ydx+(xy\sin xy-\cos xy)xdy=0$

Solution:

$(xy\sin xy+\cos xy)dx+(xy\sin xy-\cos xy)xdy=0$ .... (1)

$M=(xy\sin xy+\cos xy)y,\quad N=(xy\sin xy-\cos xy)x$

$Mx-Ny=2xy\cos xy\neq 0$

$$\therefore \qquad \text{IF}=\frac{1}{Mx-NY}=\frac{1}{2xy\cos xy}$$

Multiplying the DE by $\dfrac{1}{2xy\cos xy}$

The DE reduces to

$$\frac{1}{2}\left(y\tan xy+\frac{1}{x}\right)dx+\frac{1}{2}\left(x\tan xy-\frac{1}{y}\right)dy=c \quad \text{.... (2)}$$

which is exact (verify)

$\therefore$ Solution is

$$\frac{1}{2}\int\left(y\tan xy+\frac{1}{x}\right)dx+\frac{1}{2}\int\left(x\tan xy-\frac{1}{y}\right)dy=c$$

$$\frac{1}{2}\log\sec xy+\frac{1}{2}\log x-\frac{1}{2}\log y=c$$

i.e., $\log x\sec xy = 2c + \log y.$ Taking $2c$ as $\log A$

$\Rightarrow$ $x\sec xy = Ay$

Exercise 16.8

Solve the Following Differential Equations

1. $(x^3y^3 + x^2y^2 + xy + 1)ydx + (x^3y^3 - x^2y^2 - xy + 1)xdy = 0$

[**Ans** : $xy - \dfrac{1}{xy} - 2\log y = c$]

2. $(x^2y^2 + xy + 1)ydx + (x^2y^2 - xy + 1)xdy = 0$

[**Ans** : $xy + \log x - \log y - \dfrac{1}{xy} = c$]

3. $(xy^2 + 2x^2y^3)dx + (x^2y - x^3y^2)ydx = 0$

[**Ans** : $\log x^2 - \log y - \dfrac{1}{xy} = c$]

4. $(x^4y^4 + x^2y^2 + xy)y\,dx + (x^4y^4 - x^2y^2 + xy)xdy = 0$

[**Ans** : $\dfrac{1}{2}x^2y^2 - \dfrac{1}{xy} - \log x - \log y = c$]

16.5.4 Third Method to find the Integrating Factor

If $Mdx + Ndy = 0$ is not exact and $\dfrac{\dfrac{\partial M}{\partial y}-\dfrac{\partial N}{\partial x}}{N}$ is a function of x alone say $f(x)$,

then the IF $= e^{\int f(x)dx}$

***Example* 16.33:** Solve $(x^2 + y^3 + 6x)dx + y^2xdy = 0$

Solution:

$$(x^2 + y^3 + 6x)dx + y^2xdy = 0 \qquad .....(1)$$

$M = x^2 + y^3 + 6x \qquad N = y^2x$

$$\frac{\partial M}{\partial y} = 3y^2, \quad \frac{\partial N}{\partial x} = y^2$$

$$\frac{\dfrac{\partial M}{\partial y} - \dfrac{\partial N}{\partial x}}{N} = \frac{3y^2 - y^2}{y^2x} = \frac{2}{x}$$

is a function of x alone

$$\therefore \qquad IF = e^{\int \frac{2}{x}dx} = x^2$$

Multiplying the given DE by x^2

$$x^2(x^2 + y^3 + 6x)dx + y^2x^3dy = 0 \qquad .....(2)$$

$$\frac{\partial M_1}{\partial y} = 3x^2y^2$$

$$\frac{\partial N_1}{\partial x} = 3x^2y^2$$

$\therefore$ DE (2) is exact

$\therefore$ Solution is

$$\int (x^4 + x^2y^3 + 6x^3)dx + \int y^2x^2dy = c$$

$$\frac{x^5}{5} + \frac{x^3}{3}y^3 + \frac{6x^4}{4} = c$$

***Example* 16.34:** Solve $(x^2 + y^2)\,dx - 2xydy = 0$

Solution:

$$(x^2 + y^2)dx - 2xydy = 0 \qquad .....(1)$$

$M = x^2 + y^2,\ N = -2xy$

$$\frac{\partial M}{\partial y} = 2y, \quad \frac{\partial N}{\partial x} = -2y$$

$$\frac{\dfrac{\partial M}{\partial x} - \dfrac{\partial N}{\partial x}}{N} = \frac{2y + 2y}{-2xy} = \frac{-2}{x} \text{ is a function of } x \text{ alone}$$

$$\therefore \qquad IF = e^{\int -\frac{2}{x}dx} = \frac{1}{x^2}$$

The given DE reduces to

$$\left(1+\frac{y^2}{x^2}\right)dx - \frac{2y}{x}dy = 0 \qquad .....(2)$$

$$\frac{\partial M_1}{\partial y} = \frac{2y}{x^2}, \qquad \frac{\partial N_1}{\partial x} = \frac{2y}{x^2}$$

(2) is exact

Solution of (1) is $\quad x - \dfrac{y^2}{x} = c$

16.5.5 Fourth Method to Find the Integrating Factor

If the differential equation Mdx + Ndy = 0 is not exact and $\left(\dfrac{\frac{\partial N}{\partial x} - \frac{\partial M}{\partial y}}{M}\right)$ is a function of y alone, say $f(y)$

Then $\quad \text{IF} = e^{\int f(y)dy} \quad \int\left(1+\frac{y^2}{x^2}\right)dx = c$

***Example* 16.35:** Solve $(xy^3 + y)\,dx + 2(x^2y^2 + x + y^4)dy = 0$

***Solution*:**

$$(xy^3 + y)dx + 2(x^2y^2 + x + y^4)dy = 0$$

$$M = xy^3 + y \qquad N = 2(x^2y^2 + x + y^4)$$

$$\frac{\partial M}{\partial y} = 3xy^2 + 1; \qquad \frac{\partial N}{\partial x} = 4xy^2 + 2$$

$$\left(\frac{\frac{\partial N}{\partial x} - \frac{\partial M}{\partial y}}{M}\right) = \frac{(4xy^2+2)-(3xy^2+1)}{xy^3+y}$$

$$= \frac{(xy^2+1)}{y(xy^2+1)} = \frac{1}{y}$$

is a function of y alone

$$F = e^{\int \frac{1}{y} dy} = y$$

Multiplying the differential equation by I.F $= y$

$$(xy^4 + y^2)\, dx + 2(x^2y^3 + xy + y^5)dy = 0 \qquad .....(1)$$

$$\frac{\partial M_1}{\partial y} = 4xy^3 + 2y; \qquad \frac{\partial N_1}{\partial x} = 4xy^3 + 2y$$

$\therefore$ The differential equation (1) is exact

Solution is $\int (xy^4 + y^2)dx + 2\int (x^2y^3 + xy + y^5)\, dy = 0$

$$\frac{x^2y^4}{2} + y^2x + \frac{2.y^6}{6} = c$$

$$3x^2y^4 + 6y^2x + 2y^6 = c$$

***Example* 16.36:** Solve $(3x^2y^4 + 2xy)dx + (2x^3y^3 - x^2)dy = 0$

***Solution*:**

$$(3x^2y^4 + 2xy)dx + (2x^3y^3 - x^2)dy = 0$$

$$\frac{\partial M}{\partial y} = 12x^2y^3 + 2x; \qquad \frac{\partial N}{\partial x} = 6x^2y^3 - 2x$$

$$\frac{\frac{\partial N}{\partial x} - \frac{\partial M}{\partial y}}{} = \frac{(6x^2y^3 - 2x) - (12x^2y^3 + 2x)}{3x^2y^4 + 2x} = \frac{6x^2y^3 - 4x}{y(3x^2y^3 + 2x)} = -\frac{2}{y}$$

is a function of y alone

$$\therefore \qquad IF = e^{-\int \frac{2}{y} dy} = \frac{1}{y^2}$$

Multiplying the DE by $\frac{1}{y^2}$, the DE reduces to

$$\left(3x^2y^2 + \frac{2x}{y}\right)dx + \left(2x^3y - \frac{x^2}{y^2}\right)dy = 0$$

$$\frac{\partial M_1}{\partial y} = 6x^2y - \frac{2x}{y^2} \qquad \frac{\partial N_1}{\partial x} = 6xy - \frac{2x}{y^2}$$

$\therefore$ DE (1) is exact

Solution of the DE is

$$\left(3x^2y^2 + \frac{2x}{y}\right)dx + \left(2x^3y - \frac{x^2}{y^2}\right)$$

$$3\frac{x^3}{3}y^2 + \frac{2}{y}\frac{x^2}{2} = c$$

$$xy^3 + x^2 = cy$$

Exercise 16.9

Solve the Following Differential Equations

1. $(x + y^3 + 1)dx + y^2dy = 0$

   [Ans : $\dfrac{x^2e^{3x}}{3} - \dfrac{2x}{9}e^{3x} + \dfrac{2}{9}\dfrac{e^{3x}}{3} + (y^3+1)\dfrac{e^{3x}}{3} = c$ **]**

2. $(x^2 + y^2 + 2x)dx + 2ydy = 0$ **[Ans :** $(x^2 + y^2)e^x = c$ **]**

3. $2xydy - (x^2 + y^2 + 1)dx = 0$ **[Ans :** $y^2 - x^2 + 1 = cx$ **]**

4. $\left(y + \dfrac{y^3}{3} + \dfrac{x^2}{2}\right)dx + \dfrac{1}{4}(x + xy^2)dy = 0$ **[Ans :** $3x^4y + x^4y^3 + x^6 = c$ **]**

5. $(y^4 + 2y)dx + (xy^3 + 2y^4 - 4x)dy = 0$ **[Ans :** $xy + \dfrac{2x}{y^2} + y^2 = c$ **]**

6. $(xy^2 - x^2)dx + (3x^2y^2 + x^2y - 2x^3)dy = 0$ **[Ans :** $\dfrac{x^2}{2}y^2e^{6y} - e^{6y} \cdot \dfrac{x^3}{3} = c$ **]**

16.5.6 Fifth Method for finding Integrating Factor

If the differential equation

$$Mdx + Ndy = 0 \qquad \text{.... (1)}$$

is not exact and is of the form

$$x^ay^b(mydx + nxdy) + x^r y^s(pydx + qxdy) = 0$$

when a, b, r, s, m, n, p, q are constants

Then the $\quad$ IF $= x^hy^k$

where h, k are constants such that the equation (1) after multiplying with IF becomes exact.

***Example* 16.37:** Solve $xy^3(ydx + 2xdy) + 3ydx + 5xdy) = 0$

***Solution*:**

$$xy^3(ydx + 2xdy) + (3ydx + 5xdy) = 0 \qquad .....(1)$$

The $\quad \text{IF} = x^h y^k$

Multiplying (1) by IF $= x^h y^k$ it must become exact

$$(x^{h+1} y^{k+4} + 3x^h y^{k+1})dx + (2x^{h+2} y^{k+3} + 5x^{h+1} y^k)dy = 0 \qquad .....(2)$$

is exact if

$$\frac{\partial M}{\partial y} = (4 + k)x^{h+1}y^{k+3} + 3x^h(k + 1)y^k$$

$$\frac{\partial N}{\partial x} = 2(h + 2)\, x^{h+1}y^{k+3} + 5(h + 1)x^h y^k$$

For DE (2) to be exact

$$\frac{\partial M}{\partial y} = \frac{\partial N}{\partial x}$$

$\therefore \quad (4 + k)x^{h+1}y^{k+3} + 3x^h(k + 1)y^k = 2(h + 2)x^{h+1}y^{k+3} + 5(h+1)x^h y^k$

Comparing the coefficients on both sides

$$2h - k = 0 \text{ and } 5h - 3k = -2$$

Solving $\quad h = 2, k = 4$

Substituting $h = 2,\ k = 4$ in (2) DE (1) reduces to

$$(x^3y^8 + 3x^2y^5)\, dx + (2x^4y^7 + 5x^3y^4)dy = 0 \text{ which is exact}$$

$\therefore$ Solution is $\dfrac{x^4}{4}y^8 + \dfrac{3x^3}{3}y^5 = c \Rightarrow x^4y^8 + 4y^5x^3 = c$

***Example* 16.38:** Solve $(3x + 2y^2)ydx + 2x(2x + 3y^2)dy = 0$

***Solution*:**

$$(3x + 2y^2)ydx + 2x(2x + 3y^2)dy = 0 \qquad .... (1)$$

$$\text{IF} = x^h y^k$$

Multiplying (1) by IF

$$(3x^{h+1} y^{k+1} + 2x^h y^{k+3})dx + (4x^{h+2}y^k + 6x^{h+1} y^{k+2})dy = 0 \qquad .... (2)$$

$$\frac{\partial M}{\partial y} = 3(k+1)\, x^{h+1}\, y^k + 2(k+3)x^h\, y^{k+2}$$

$$\frac{\partial N}{\partial x} = 4(h+2)\, x^{h+1}\, y^k + 6(h+1)x^h\, y^{k+2}$$

$$\frac{\partial M}{\partial y} = \frac{\partial N}{\partial x} \Rightarrow 4h + 8 = 3k + 3$$

$$6h + 6 = 2k + 6$$

$$\begin{aligned} 3h - k &= 0 \\ 4h - 3k &= -5 \end{aligned} \quad \Rightarrow \quad h = 1,\ k = 3$$

$\therefore$ (2) $\Rightarrow (3x^2 y^4 + 2x\, y^6)dx + (4x^3 y^3 + 6x^2 y^5)dy = 0$, which is exact

$\therefore$ solution is $x^3 y^4 + x^2 y^6 = c$

Exercise 16.10

I Solve the following differential equations

1. $x(3ydx + 2xdy) + 8y^4(ydx + 3xdy) = 0$ **[Ans :** $x^2y^3(x + 4y^4) = c$**]**

2. $(3x^2y - y^4)dx + (6x^3 + xy^3)dy = 0$ **[Ans :** $\dfrac{3x^{-4/3+1}y^{-2/3}}{-\frac{4}{3}+1} - \dfrac{x^{\frac{-10}{3}+1}}{-\frac{10}{3}+1}y^{7/3} = c$ **]**

3. $(y^2 + 2x^2y)dx + (2x^3 - xy)dy = 0$ **[Ans :** $4x^{1/2}y^{1/2} - \frac{2}{3}x^{-3/2}y^{3/2} = cx$ **]**

4. $(2x^2y - 3y^4)dx + (3x^3 + 2xy^3)dy = 0$ **[Ans :** $5x^{-36/13}y^{24/13} - 12x^{-10/13}y^{-15/13} = c$ **]**

16.6 Equations of First Order but of Higher Degree

A differential equation in which we have only first differential coefficient, i.e., $\frac{dy}{dx}$ (not higher) will be of first order as already defined, but since it is of higher degree it may involve $\frac{dy}{dx}$ raised to any power. It will be convenient for us to denote $\frac{dy}{dx}$ by p. Thus the general form of the above type of equation is

$$P^n + A_3 p^{n-1} + A_2 p^{n-2} + \ldots + A_{n-1}p + A_n = 0 \qquad \ldots..(1)$$

where $p = \frac{dy}{dx}$ and $A_1, A_2, \ldots A_n$ stand for functions of x and y

Below we shall give various methods of solving the above type of equations in different cases.

16.6.1 Equations which can be solved for p

Rule: In this case resolves the given equation (1) into linear factors of the type

$$(p - \beta_1)(p - \beta_2)(p - \beta_3) \ldots..(p - \beta_n) = 0$$

Equate each factor equal to zero and replace p by $\frac{dy}{dx}$ and solve the differential equation of first order and first degree whose solution will be of the type

$$f_1(x, y, c_1) = 0, \quad f_2(x, y, c_2) = 0$$

$$f_n(x, y, c_n) = 0$$

We have already stated that constants occurring in the differential equation can be made to take any value whatsoever we like. Hence there is no loss of generality if we replace all the constants $c_1, c_2, \ldots c_n$ by another constant c. Hence the solution can be taken to be

$$f_1(x, y, c) = 0, f_2(x, y, c) = 0$$

$$f_n(x, y, c) = 0$$

Combining the above equations, the solution of differential equation (1) can be written as

$$f_1(x, y, c), f_2(x, y, c) \ldots. f_n(x, y, c) = 0$$

***Example* 16.39:** Solve $p^2 + 2py \cot x = y^2$

$$p^2 + 2py \cot x - y^2 = 0$$

$$p = \frac{-2y\cot x \pm \sqrt{(4y^2 \cot^2 x + 4y^2)}}{2}$$

$$= -y \cot x \pm y \text{ cosec } x$$

$$\therefore \quad \frac{dy}{dx} = -y\left(\frac{\cos x + 1}{\sin x}\right)$$

and $$\frac{dy}{dx} = -y\left(\frac{\cos x - 1}{\sin x}\right)$$

We shall now solve each of the above differential equations by separating the variables

$$\therefore \quad \frac{dy}{-y} = \frac{(\cos x + 1)}{\sin x}dx = \frac{(1-\cos^2 x)}{\sin x(1-\cos x)}dx = \frac{\sin x}{1-\cos x}dx$$

Integrating, we get

$$-\log y + c = \log(1 - \cos x) \text{ or } \log k = \log y + \log(1 - \cos x)$$

or $\log k = \log y(1 - \cos x)$ or $y(1 - \cos x) = k$

Now taking the other differential equation, we have

$$\frac{dy}{y} = \frac{1-\cos x}{\sin x}dx = \frac{1-\cos^2 x}{\sin x(1+\cos x)}dx = \frac{\sin x}{1+\cos x}dx$$

Integrating, we get

$$\log y = -\log(1 + \cos x) + c \quad \text{or} \quad \log y + \log(1 + \cos x) = \log k$$

or $\log y(1 + \cos x) = \log k$ or $y(1 + \cos x) = k$

Hence the solutions are $y(1 \pm \cos x) = k$

***Example* 16.40:** $x^2\left(\frac{dy}{dx}\right)^2 + xy\frac{dy}{dx} - 6y^2 = 0$

$$p^2x^2 + pxy - 2y^2 = 0 \text{ or } (px + 3y)(px - 2y) = 0$$

$$\therefore \quad \frac{dy}{dx} = -\frac{3y}{x} \text{ or } \frac{2y}{x}$$

or $$\frac{dy}{y}+3\frac{dx}{x}=0 \text{ and } \frac{dy}{y}-\frac{2}{x}dx=0$$

Integrating

$$\log y + 3\log x = c \text{ and } \log y - 2\log x = c$$

or $$\log yx^3 = \log k \text{ and } \log \frac{y}{x^2} = \log k$$

$\therefore$ $$yx^2 - k = 0 \text{ and } y - kx^2 = 0$$

Hence the combined solution is $(yx^3 - k)(y - kx^2) = 0$

***Example* 16.41:** $x^2\left(\frac{dy}{dx}\right)^2 + 3xy\frac{dy}{dx} + 2y^2 = 0$

Here $p = -\frac{y}{x}$ or $-\frac{2y}{x}$

Ans. $(xy - k)(x^2y - k) = 0$

Exercise 16.11

Solve the following differential equations

1. $p^2 - 5p + 6 = 0$

Ans. $(y - 2x - c)(y - 3x - c) = 0$

2. $4y^2p^2 + 2pxy(3x + 1) + 3x^2 = 0$

or $(2py + x)(2py + 3x^2) = 0$ etc

Ans. $(x^2 + 2y^2 - c)(x^3 + y^2 - c) = 0$

3. $p^2(x + 2y) + 3p^2(x + y) + (y + 2x)p = 0$

$p(p + 1)\{p(x + 2y) + (y + 2x) = 0$

Solutions are $y = c$, $y = -x + c$ and $(x + 2y)\,dy + (y + 2x)\,dx = 0$ which is exact or may be solved as a homogeneous equation and whose solution is $xy + x^2 + y^2 = c$

(b) $xyp^2 + p(3x^2 - 2y^2) - 6xy = 0$

or $(py + 3x)(y^2 + 3x^2 + c) = 0$

Ans. $(y - cx^2)(y^2 + 3x^2 + c) = 0$

4 (a) $(p + y + x)(xp + y + x)(p + 2x) = 0$

(i) $\frac{dy}{dx} + y = -x$. It is linear and I.F = $e^{\int 1.dx} =$

Multiplying both sides by e^{∞} and integrating, we get

$$ye^{x} = \int -xe^{x}.dx = -e^{\infty}(x-1) - c \text{ or } y + x - 1 + ce^{-\infty} = 0$$

(ii) $\frac{dy}{dx} + \frac{y}{x} = -1.$

It is also linear I.F. = $e^{\int 1/x\, dx} = e^{\log x} = x.$

Multiplying both sides by x and integrating, we get

$$xy = \int -xdx = -\frac{x^2}{2} + k \text{ or } 2xy + x^2 + c = 0$$

(iii) $\frac{dy}{dx} + 2x = 0$ or $dy + 2x = 0$

Integrating, $y + x^2 + c = 0$

Hence the solution is

$$(y + x - 1 + ce^{-\infty})(2xy + x^2 + c)(y + x^2 + c) = 0$$

(b) $p^2 + x^3y - x^3p - yp = 0$

or $(p - x^3)(p - y) = 0$

Ans. $(y - ce^{\infty})(4y - x^4 + c) = 0$

(c) $p^2 + px - xy - y^2 = 0$

$(p^2 - y^2) + x(p - y = 0)$ or $(p - y)(p + y + x) = 0$

Both these have been solved in (a) and (b) parts

Ans. $(y + x - 1 + ce^{\infty})(y - ce^{\infty})$

(d) $p(p - y) = x(x + y)$

$p^2 - py - x(x + y) = 0$ or $(p + x)(p - y - x) = 0$

$(2y + x^2 + c)(y + x + 1 - ce^{\infty}) = 0$

(e) $(x^2 + x)p^2 + (x^3 + x - 2xy - y)p + y^2 - xy = 0$

The above equation can be factorized as

$[xp + x - y][(x + 1)p - y] = 0$

$$\therefore \qquad x\frac{dy}{dx}+x-y=0 \text{ and } (x+1)\frac{dy}{dx}-y=0$$

or
$$xdy - ydx + xdx = 0 \text{ and } \frac{dx}{x+1}-\frac{dy}{y}=0$$

or
$$\frac{xdy-ydx}{x^2}+\frac{1}{x}dx=0 \text{ and } \frac{dx}{x+1}-\frac{dy}{y}=0$$

Integrating,

$$\frac{y}{x} + \log x + \log c = 0 \text{ and } \log (x+1) - \log y = \log c$$

$$y + x \log cx = 0 \text{ and } x + 1 - cy = 0$$

Hence general solution is

$$(y + x \log cx)(x + 1 - cy) = 0$$

***Example* 16.42 (a):** $xp^2 + (y - x) p - y = 0$

Ans. $(y - x + c)(xy + c) = 0$

(b) $x^2p^2 + 2xp - 3x^2 = 0$

$(p + 3x)(p - x) = 0$

Ans. $(2y - x^2 + c)(2y + 3x^2 + c) = 0$

***Example* 16.43:** $yp^2 + (x - y) p - x = 0$

Here $p = 1$ or $-\dfrac{x}{y}$

Ans. $(y - x - c)(x^2 + y^2 - c) = 0$

***Example* 16.43:** $p^3 - (x^2 + xy + y^2) p^2 + (x^3y + x^2y^2 + xy^3) p - x^3y^3 = 0$

or $(p - xy)(p - x^2)(p - y^2) = 0$

Ans. $(x^3 - 3y + c)\left(e^{x^2/2} + cy\right)(xy + cy + 1) = 0$

(b) $p^3 + 2xp^2 - y^2p^2 - 2xy^2p = 0$

or $p(p + 2x)(p - y^2) = 0$

Ans. $(y - c)(y + x^2 - c)(xy + cy + 1) = 0$

(c) $p^2 - 2p \cosh x + 1 = 0$

or $p^2 - p\left(e^{\infty+}e^{-\infty}\right)+1=0$ or $(p - e^{\infty})\ (p - e^{-\infty}) = 0$

or $y = e^{\infty} + c$ and $y = -e^{-\infty} + c$

or $(y - e^{\infty} - c)(y + e^{-\infty} - c) = 0$

***Example* 16.45:** $x^2\left(\frac{dy}{dx}\right)^2 - 2xy\frac{dy}{dx} + 2y^2 - x^2 = 0$

$$p^2 x^2 - 2xyp + 2y^2 - x^2 = 0$$

$$\therefore \quad p = \frac{dy}{dx} = \frac{2xy \pm \sqrt{(4x^2y^2 - 8x^2y^2 + 4x^4)}}{2x^2} = \frac{y \pm \sqrt{(x^2 - y^2)}}{x}$$

Above is homogeneous and therefore putting y = vx, we get

$$v + x = \frac{dy}{dx} = \frac{v \pm \sqrt{(1-v^2)}}{1} \quad \text{or} \quad x\frac{dy}{dx} = \pm\sqrt{(1-v^2)}$$

$$\text{or} \qquad \frac{dv}{\sqrt{(1-v^2)}} = \pm\frac{dx}{x}$$

Integrating, we get

$$\sin^{-1} v = + \log x \pm \log c = \pm \log cx \text{ or } \sin^{-1} y/x = \pm \log cx$$

***Example* 16.46:** $3p^2y^2 - 2xyp + 4y^2 - x^2 = 0$

Multiplying throughout by 3, we get

$$9p^2y^2 - 6xyp + 12y^2 - 3x^2 = 0 \qquad .....(1)$$

$$\text{or} \qquad (3yp - x)^2 - 4(x^2 - 3y^2) = 0$$

$$\text{Put} \qquad x^2 - 3y^2 = y^2$$

$$\therefore \qquad 2x - 6y\frac{dy}{dx} = 2v\frac{dv}{dx} \quad \text{or} \quad (x - 3py) = v\frac{dv}{dx}$$

Putting in (1), we get

$$v^2\left(\frac{dv}{dx}\right)^2 - 4y^2 = 0; \quad \therefore \quad \frac{dv}{dx} = \pm 2$$

$$\text{or} \quad v = c \pm 2x \text{ or } v^2 = (c \pm 2x)^2 \text{ or } x^2 - 3y^2 = (c + 2x)^2$$

***Example* 16.47:** $2p^3 - (2x + 4 \sin x - \cos x)\, p^2 - (x \cos x - 4x \sin x + \sin 2x)\, p + x \sin 2x = 0$

Clearly if we put p = x, the equation is clearly satisfied and hence p – x is a factor

$$\therefore \qquad (p - x)\,(2p^2 - (4 \sin x - \cos x)\, p - \sin 2x) = 0$$

$$\text{or} \qquad (p - x)\,[2p\,(p - 2 \sin x) + \cos x\,(p - 2 \sin x) = 0$$

or $\quad (p - x)(p - 2\sin x)(2p + \cos x) = 0$

Ans. $\quad (2y - x^2 - c)(y + \sin x - c)(y + 2\cos x - c) = 0$

***Example* 16.48:** $xy^2(p^2 + 2) = 2py^2 + x^2$

Arrange as a quadratic in p, and we get

$$p = \frac{x}{y} \text{ or } \frac{2y^2 - x^2}{xy}; \quad \therefore \quad \frac{dy}{dx} = \frac{x}{y} \text{ and } \frac{dy}{dx} = \frac{2y^2 - x^2}{xy}$$

One of them is solved by separating the variables and the other is homogeneous

Ans. $\quad (x^2 - y^2 + c)(x^2 - y^2 + cx^4) = 0$

***Example* 16.49:** $p^3 - (y + 2x - e^{\infty - v})p^2 + (2xy - 2xe^{\infty - v} - ye^{\infty - v})p + 2xye^{\infty - y} = 0$

Rewriting the equation, we get

$$p^2(p - y) - 2px(p - y) + pe^{\infty - v}(p - y) - 2xpe^{\infty - v}(p - y) = 0$$

or $$(p - y)\{p(p - 2x) + e^{\infty - y}(p - 2x)\} = 0$$

or $$(p - y)(p - 2x)(p + e^{\infty - y}) = 0$$

$$p - y = 0 \text{ gives } \frac{dy}{dy} = dx \text{ or } \log y = x + \log c; \therefore y = ce^{\infty}$$

$$p - 2x = 0 \text{ gives } dy = 2x\,dx \text{ or } y = x^2 + c$$

$$p + e^{\infty - y} = 0 \text{ give } e^y\,dy + e^{\infty}\,dx = 0 \text{ or } e^y + e^{\infty} = c$$

Hence the solution is

$$(y - ce^{\infty})(y - x^2 - c)(e^y + e^x - c) = 0$$

***Example* 16.50:** $(a^2 - x^2)p^3 + bx(a^2 - x^2)p^2 - p - bx = 0$

or $$(p - bx)\left\{p\sqrt{(a^2 - x^2)} - 1\right\}\left\{p\sqrt{(a^2 - x^2)} + 1\right\} = 0.$$

Ans. $$\left(y - \frac{1}{2}bx^2 - c\right)\left\{\frac{x^2}{a^2} - \sin^2(y - c)\right\} = 0.$$

***Example* 16.51:** Find the complete primitive of

$$4p^2x(x - a)(x - b) = \{3x^2 - 2x(a + b) + ab)^2$$

$$\frac{dy}{dx} = p = \mp\frac{3x^2 - 2x(a + b) + ab}{2\sqrt{x^3 - x^2(a + b) + abx]}}$$

Integrating

$$y = \pm\sqrt{\left[x^3 - x^2(a+b) + abx\right]} + c$$

$$\because \qquad \int \frac{f'(x)}{2\sqrt{f(x)}}\,dx = \sqrt{\{f(x)\}}$$

or $\quad (y-c)^2 = x(x-a)(x-b)$

(b) Find the complete primitive of

$$p^2 x(x-1)(x-2) = (3x^2 - 6x + 2)^2$$

Ans. $\quad (y-c)^2 = x(x-1)(x-2)$

(c) Find the complete primitive of

$$4p^2 x - (3x-a)^2 = 0$$

Ans. $\quad (y-c)^2 = x(x-a)^2$

16.6.2 Equations Solvable for y

Since the equation is solvable for y, we will have a relation of the type

$$y = f(x, p) \qquad \text{.....(1)}$$

Differentiate above w.r.t x and denote $\frac{dy}{dx}$ by p we will have another relation

$$p = F\left(x, p, \frac{dp}{dx}\right) \qquad \text{.....(2)}$$

This equation when solved will give us a relation of the form

$$\Psi(x, p, c) = 0 \qquad \text{.....(3)}$$

Eliminating p between (1) and (3), we will get a relation between x, y and c which will be the required solution. However, it is not possible for us to eliminate p between (1) and (3), then express the values of x and y in terms of p and c and these together will be the solution of the given differential equation in terms of parameters p.

16.6.3 Equation Solvable for x

In this case the equation will be of the type

$$x = f(y, p) \qquad \text{.....(1)}$$

Differentiating w.r.t y and denote $\frac{dx}{dy}$ by $\frac{1}{p}$ and you will have another differential equation which when solved will give a relation between y, p and c. Then eliminate y between this relation and the given relation to get the required solution.

Examples based on solvable for y

***Example* 16.52 (a):** $y + px = x^4 p^2$

or $\quad y = -px + x^4 p^2 \qquad \text{.....(1)}$

Differentiating w.r.t, x and denoting $\frac{dy}{dx}$ by p, we get

$$p = -p - x\frac{dp}{dx} + x^4 2p\frac{dp}{dx} + p^2 4x^3$$

or $\quad 2p + x\frac{dp}{dx} = 2px^3\left(2p + x\frac{dp}{dx}\right)$

or $\quad \left(2p + x\frac{dp}{dx}\right)(1 - 2px^3) = 0$

$\therefore \quad 2p + x\frac{dp}{dx} = 0$

or $\quad \frac{2}{x}dx + \frac{dp}{p} = 0$

or $\quad \log x^2 + \log p = \log c$

or $\quad px^2 = c \quad \therefore \quad p = \frac{c}{x^2} \qquad \text{.....(2)}$

Eliminating p between (1) and (2), we get

$$y = -x, \frac{c}{x^2} + x^4\frac{c^2}{x^4}$$

or $\quad xy = -c + c^2x$

(b) $\quad x^3p^2 + x^2py + a^3 = 0$

Proceeding as above, we get $p = c/x^2$

Eliminating p, we get $c^2 + cxy + a^3x = 0$

***Example* 16.53:** $y = 2px + p^4x^2$

Differentiating w.r.t x and denoting $\frac{dy}{dx}$ by p

$$p = 2p + 2x\frac{dp}{dx} + 2p^4x + 4p^3x^2\frac{dp}{dx}$$

or $$\left(p+2x\frac{dp}{dx}\right)(1+2p^2x)=0$$

$$\therefore \quad p+2x\frac{dp}{dx}=0 \text{ or } 2\frac{dp}{p}+\frac{dx}{x}=0$$

or $\log p^2+\log x=\log c$ or $p^2x=c$; $p^2=\dfrac{c}{x}$

Putting in the given equation, thereby eliminating p, we get

$$y-c^2=2px \text{ or } (y-c^2)^2=4x^2\frac{c}{x}$$

or $$(y-c^2)^2=4cx$$

***Example* 16.54:** $y=2px-p^2$

Proceeding as above, we get

$$p=2p+2x\frac{dp}{dx}-2p\frac{dp}{dx} \quad \text{or} \quad p\frac{dx}{dp}+2x-2p=0$$

or $$\frac{dx}{dp}+\frac{2}{p}x=2 \text{ (linear)}$$

$$\text{I.F}=e^2\log p=p^2$$

Multiplying both sides by p^2 and integrating, we get

$$x.p^2=\int 2p^2dp+c=\frac{2}{3}p^3+c$$

or $$x=cp^{-2}+\frac{2}{3}p \qquad \text{....(1)}$$

Putting for x in the equation, we get

$$y=2cp^{-1}+\frac{1}{3}p^2 \qquad \text{.....(2)}$$

The p-eliminant of (1) and (2) gives the solution of given equation or we may say (1) and (2) together give the solution

***Example* 16.55:** $x-yp=ap^2$

or $$y=\frac{x}{p}-ap$$

$$\therefore \quad p=\frac{1}{p}-\frac{x}{p^2}\frac{dp}{dx}-a\frac{dp}{dx}$$

or $$\frac{dp}{dx}(ap^2 + x) = p(1 - p^2)$$

or $$\frac{dx}{dp} - x.\frac{1}{p\left(1-p^2\right)} = \frac{ap}{1-p^2} \quad \text{[Linear]}$$

$$\int p\,dp = \int -\frac{1}{p\left(1-p^2\right)}dp = -\int\left\{\frac{1}{p} + \frac{p}{\left(1-p^2\right)}\right\}dp$$

$$= -\left[\log p - \frac{1}{2}\log\left(1-p^2\right)\right] = \log\frac{\sqrt{\left(1-p^2\right)}}{p}$$

$\therefore$ $$\text{I.F.} = e^{\int p\,dp} = \frac{\sqrt{\left(1-p^2\right)}}{p}$$

Multiplying both sides by I.F and integrating

$$x\frac{\sqrt{\left(1-p^2\right)}}{p} = c + \int\frac{ap}{1-p^2}\frac{\sqrt{\left(1-p^2\right)}}{p}dp = c + a\sin^{-1}p$$

$\therefore$ $$x = \frac{p}{\sqrt{\left(1-p^2\right)}}\left[c + a\sin^{-1}p\right] \quad .....(2)$$

Putting for x in (1), we get

$$y = \frac{1}{\sqrt{\left(1-p^2\right)}}\left[c + a\sin^{-1}p\right] - ap \quad .....(3)$$

Equating (2) and (3) together give the solution

Particular Case $p^2 - py + x = 0$

Put $a = -1$ in above

(b) $x - yp = ap^2$

$y = -\frac{x}{p} + ap$ Differentiating and proceeding as above in (a) we get

$$\frac{dx}{dp} - \frac{x}{p\left(p^2+1\right)} = \frac{ap}{p^2+1} \quad \text{(linear)}$$

Ans. $$x\frac{\sqrt{\left(p^2+1\right)}}{p} = \left(a\sinh^{-1}p + c\right)$$

***Example* 16.56:** $y = 3x + \log p$

Proceeding as usual, $p(p-3) = \frac{dp}{dx}$

or $$dx = \frac{1}{3}\left[\frac{1}{p-3} - \frac{1}{p}\right]dp$$

Integrating, $$x = \frac{1}{3}\log\frac{p-3}{p} - \frac{1}{3}\log c$$

$$\therefore \quad e^{3\infty} = \frac{p-3}{pc} \text{ or } p = \frac{3}{1-ce^{3\infty}}$$

Putting in the given equation, we get

$$y = 3x + \log\frac{3}{1-ce^{3\infty}}$$

***Example* 16.57:** $y - 2px = f(xp^2)$ .....(1)

Differentiating w.r.t x, we get

$$p = 2p + 2x\frac{dp}{dx} + f'(xp^2).\left\{p^2 + x.2p\frac{dp}{dx}\right\}$$

or $$0 = \left(p + 2x\frac{dp}{dx}\right) + pf'(xp^2).\left\{p + 2x\frac{dp}{dx}\right\}$$

or $$0 = \{1 + pf'(xp^2)\}\left\{p + 2x\frac{dp}{dx}\right\} = 0$$

$$\therefore \quad p + 2x\frac{dp}{dx} = 0 \text{ or } \frac{dx}{x} + \frac{2}{p}dp = 0$$

or $$\log x + \log p^2 = \log k \text{ or } p^2 x = k; \;\; \therefore \;\; p = \frac{k}{\sqrt{x}}$$

Putting the value of p in (1), we get

$$y = 2x\frac{k}{\sqrt{x}} + f\left(x.\frac{k^2}{x}\right) \text{ or } y = 2k\sqrt{x + f(k^2)}$$

Particular case $y = 2xp\ xp^2$

***Example* 16.58:** $p^3 + mp^2 = a(y + mx)$. Solvable for y

$$ay = p^3 + mp^2 - amx$$

Differentiating w.r.t x

$$ap = (3p^2 + 2mp)\frac{dp}{dx} - am$$

or $$a(p + m) = p(3p + 2m)\frac{dp}{dx}$$

or $$a\,dx = \frac{p(3p+2m)}{(p+m)}dp - \left(3p - m + \frac{m^2}{p+m}\right)dp$$

Integrating, we get

$$ax = c + 3\frac{p^2}{2} - mp + m^2 \log(p + m) \qquad .....(2)$$

Putting for x in the given equation, we get

$$ay = -m\left[c + \frac{3}{2}p^2 - mp + m^2\log(p+m)\right] + mp^2 + p^2 \qquad .....(3)$$

Equations (2) and (3) together give the required solution

***Example* 16.59:** $4p^3 + 3px = y$

Ans. $$y = -\frac{8}{7}p^3 + cp^{-1/2}, x = -\frac{12}{7}p^2 + \frac{c}{3}p^{-3/2}$$

***Example* 16.60:** $y = \frac{1}{\sqrt{(1+p^2)}} + b$

Ans. $$(x + c)^2 + (y - b)^2 = 1$$

***Example* 16.61:** $y = p \tan p + \log \cos p$, solvable for y

Ans. $x = \tan p + c$, $y = p \tan p + \log \cos p$

***Example* 16.62:** $y = p^2 x + p^4$

Differentiating w.r.t x, we get

$$p = p^2 + (2px + 4p^2)\frac{dp}{dx} \text{ or } (p - p^2)\frac{dx}{dp} - 2px = 4p^2$$

or $$\frac{dx}{dp} - \frac{2}{1-p}x = \frac{4p^2}{1-p} \qquad \text{Linear} \qquad \text{I.F.} = (1-p)^2$$

Multiplying by I.F and integrating, we get

$$x(1-p)^2 = \int 4p^2(1-p)dp + c = \frac{4}{3}p^3 - p^4 + c$$

$$\therefore \qquad x = \frac{1}{(1-p)^2}\left[\frac{4}{3}p^3 - p^4 + c\right] \qquad \text{.....(1)}$$

Putting for x in the given equation, we get the value of y

***Example* 16.63:** $xp^2 - 2yp + x = 0$

$$2yp = x(p^2 + 1) \text{ or } 2y = x\left(p + \frac{1}{p}\right)$$

Differentiating w.r.t. x, we get

$$2p = \left(p + \frac{1}{p}\right).1 + x\left(1 - \frac{1}{p^2}\right)\frac{dp}{dx}$$

$$\text{or} \qquad \left(p - \frac{1}{p}\right) = x\left(1 - \frac{1}{p^2}\right)\frac{dp}{dx} \text{ or } p = x\frac{dp}{dx}$$

$$\therefore \qquad \frac{dp}{p} = \frac{dx}{x} \text{ or } \log p = \log x + \log c \text{ or } p = cx$$

Putting for p in the given equation, we get

$$2y \,.\, cx = x\,(c^2 x^2 + 1) \text{ or } 2cy = c^2 x^2 + 1$$

(b) $xp^2 - 2yp + 4x = 0$

Example 16.64: $3xy = 2p\,(x^2 - p)$

$$y = \frac{2}{3}py - \frac{2p^2}{3x} \text{ etc}$$

Examples based on Equations solvable for x

***Example* 16.65:** $y^2 \log y = xyp + p^2$

$$\text{or} \qquad x = \frac{y \log y}{p} - \frac{p}{y} \qquad \text{.....(1)}$$

Differentiating w.r.t. y and denote $\frac{dx}{dy}$ by $\frac{1}{p}$

$$\frac{1}{p} = \frac{p(1+\log y) - y\log y.\frac{dp}{dy}}{p^2} - \frac{y.\frac{dp}{dy} - p}{y^2}$$

or $$\frac{1}{p}=\frac{1}{p}+\frac{1}{p}\log y-\frac{y}{p^2}\log y\frac{dp}{dy}-\frac{1}{y}.\frac{dp}{dy}+\frac{p}{y^2}$$

or $$\frac{1}{y}\frac{dp}{dy}\left(1+\frac{y^2}{p^2}\log y\right)=\frac{p}{y^2}\left(1+\frac{y^2}{p^2}\log y\right)$$

or $$\frac{dp}{dy}=\frac{p}{y}$$

On cancelling the common factor on either side

$$\therefore \quad \frac{dp}{p}=\frac{dy}{y}; \qquad \therefore \qquad \log p-\log y=\log k$$

or $$\frac{p}{y}=k \text{ or } p=ky$$

Substituting the value of p in the given equation, we get

$$\log y=kx+x^2$$

***Example* 16.66:** $x^2+p^2x=yp$

or $$y=\frac{x^2}{p}+px$$

Differentiating w.r.t x, we get

$$p=\frac{p2x-x^2\frac{dp}{dx}}{p^2}+p+x\frac{dp}{dx}$$

First cancel p and then x from both sides

$$0=2p-x\frac{dp}{dx}+p^2\frac{dp}{dx}$$

or $$2p\frac{dx}{dp}-x=-p^2$$

or $$\frac{dx}{dp}-\frac{x}{2p}=-\frac{p}{2} \qquad .....(1)$$

Above is a linear differential equation

$$\int -\frac{1}{2p}dp = e^{-\frac{1}{2}\log p} = e^{\log p^{-1/2}} = p^{-1/2}$$

and I.F. = e

Multiplying both sides by I.F. and integrating, we get

$$xp^{-1/2} = \int \frac{-p}{2}\frac{1}{\sqrt{p}}dp = -\frac{1}{2}.\frac{2}{3}p^{3/2} + c$$

$$\therefore \quad x = \left[-\frac{1}{2}p^2 + c\sqrt{(p)}\right] \quad .....(2)$$

Hence it is not possible for us to eliminate p and as such we find the values of x and y in terms of p. Putting for x in (1), we get

$$y = \frac{\left(c\sqrt{p} - \frac{1}{3}p^2\right)^2}{p} + p\left(c\sqrt{p} - \frac{1}{3}p^2\right) \quad .....(1)$$

Equations (2) and (3) which represent the values of x and y in terms of parameter p, together give the required solution.

***Example* 16.67:** $y = 2px + y^2 p^3$

$$2x = \frac{y}{p} - y^2p^2 \qquad \text{Differentiate w.r.t y}$$

$$\therefore \quad \frac{2}{p} = \frac{1}{p} - \frac{y}{p^2}\frac{dp}{dy} - 2yp^2 - 2py^2\frac{dp}{dy}$$

$$\left(\frac{1}{p} + 2yp^2\right) = -\frac{y}{p}\frac{dp}{dy}\frac{1}{p} + 2yp^2 = 0$$

$$\therefore \quad 1 = -\frac{y}{p}\frac{dp}{dy} \text{ or } \frac{dy}{y} + \frac{dp}{p} = 0 \quad \text{or} \quad \log py = \log c$$

$$\therefore \quad py = c \quad \text{or} \quad p = \frac{c}{y}$$

Putting for p in the given equation, we get

$$y = 2\frac{c}{y}.x + y^2\frac{c^3}{y^3}$$

or $\qquad y^3 = 2cx + c^3$

***Example* 16.68:** $yp^2 - 2xp + y = 0$

$$2x = yp + \frac{y}{p}$$

$\therefore$ As above $py = c$ etc, $y^2 = 2cx - c^2$

***Example* 16.69:** $y = 2px + p^2 y$

$$y^2 = 2cy + c^2$$

***Example* 16.70:** $p^3 - 4xyp + 8y^2 = 0$

It is solvable for x and you will get $\frac{dp}{dy} = \frac{p}{2y}$ which gives

$$p^2 = cy \text{ or } p = \sqrt{c}\sqrt{y}$$

Eliminating p between the above two, we get

$$c^{3/2}y^{3/2} - 4c^{1/2}xy + 8y^{3/2} = 0$$

or $$\frac{1}{2}\sqrt{c}\left(\frac{c}{4} - x\right) = -y^{1/2}$$

Squaring $\frac{c}{4}\left(\frac{c}{4} - x\right)^2 = y$ or $y = k\,(k - x)^2$

***Example* 16.71 (a) :** $ayp^2 + (2x - b)\,p - y = 0$

or $$2x = \frac{y}{p} + b - apy$$

Differentiating w.r.t. y etc., we will get

$$\frac{y}{p}\frac{dp}{dy} + 1 = 0 \text{ or } \frac{dp}{p} + \frac{dy}{y} = 0 \text{ or } \log p + \log y = \log k$$

or $$py = k$$

Eliminating p, we get $ak^2 + (2x - b)\,k - y^2 = 0$

(b) $y - 2px + ayp^2 = 0$

$$2x = \frac{y}{p} - apy \text{ etc}$$

***Example* 16.72:** $p = \tan\left(x - \frac{p}{1+p^2}\right)$, or $x = \tan^{-1}p + \frac{p}{1+p^2}$

Differentiating w.r.t y

$$\frac{1}{p} = \left[\frac{1}{1+p^2} + \frac{1}{1+p^2} - \frac{2p^2}{\left(1+p^2\right)}\right]\frac{dp}{dy} \text{ or } dy = \frac{2p}{\left(1+p^2\right)^2}dp$$

Ans. $y = c - \frac{1}{1+p^2}, x = \tan^{-1}p + \frac{p}{1+p^2}$

***Example* 16.73:** $x + \frac{p}{\sqrt{\left(1+p^2\right)}} - a$

Differentiate w.r.t y

Ans. $(y + c)^2 + (x - a)^2 = 1$

***Example* 16.74:** $x = y + a \log p$

Ans. $y = c - a \log (p - 1), x = c + a \log \frac{p}{p-1}$

16.6.4 Clairaut's Equation

The equation of the type $y = px + F(p)$ is known as Clairaut's equation.

$$y = px + f(p)$$

Differentiating w.r.t x, we get

$$p = p + x\frac{dp}{dx} + f'(p)\frac{dp}{dx} = 0$$

or $\frac{dp}{dx}\{x + f'(p)\} = 0$ $\therefore$ $\frac{dp}{dx} = 0 \quad p = c$

Eliminating p, we get $y = cx + f(c)$ as the required solution. If however we eliminate p between the given equation and the relation $x + f'(p) = 0$, we get another solution which is called singular solution and will be discussed later on.

Rule In order to solve Clairaut's equation, replace p by c

Exercise 16.12

1. Solve $y = px + a/p$ and obtain the singular solution

Differentiating w.r.t x., we get

$$p = p + x\frac{dp}{dx} - \frac{a}{p^2}.\frac{dp}{dx}$$

$$\therefore \quad \frac{dp}{dx}\left(x - \frac{a}{p^2}\right) = 0$$

Taking $dp/dx = 0$, we get $p = c$ and putting in the given equation we get $y = cx + (a/c)$ as the solution,

Taking $x - (a/p^2) = 0$ or $p = \sqrt{(a/x)}$

Eliminating p between this and the given equation we shall get the singular solution as $y = x\sqrt{(a/x)} + a\sqrt{(x/a)}$

or $\qquad y = 2\sqrt{(ax)}$ or $y^2 = 4ax$

2. Solve $y = px + p - p^2$ and obtain the singular solution as wel

Ans. $y = cx + c - c^2$, $4y = (x+1)^2$

3 (a) $y = x\dfrac{dy}{dx} + \left(\dfrac{dy}{dx}\right)^2$ **(b)** $y = px - 1/p^2$

Ans. (a) $y = cx + c^2$ (b) $y = cx - 1/c^2$

(c) $xp^2 - yp + 2 = 0$ or $y = px + \dfrac{2}{p}$

Ans. $y = cx + 2/c$

(d) $p^2x = py = 1$ or $y = px + \dfrac{1}{p}$

Ans. $y = cx + \dfrac{1}{c}$

4. $(y - px)(p - 1) = p$ or $y = px + \dfrac{p}{p-1}$

Ans. $y = cx + \dfrac{c}{c-1}$

5. (a) $p = \log(px - y)$ or $y = px - e^p$

Ans. $y = cx - e^c$

(b) $xp^3 - (y + 3)p^2 + 4 = 0$

$$Y = px + \frac{4}{p^2} - 3 \qquad \text{Clairaut's form}$$

Ans. $y = cx + 4/c^2 - 3$

6. (a) $(y - px)^2/(1 + p^2) = a^2$

Ans. $(y - cx)^2/(1 + c^2) = a^2$

(b) $p^2(x^2 - a^2) - 2pxy + y^2 - b^2 = 0$

or $(y - px)^2 = a^2p^2 + b^2$ or $y = px \pm \sqrt{a^2p^2 + b^2}$

Ans. $(y - cx)^2 = a^2c^2 + b^2$

7. (a) $\sin px \cos y = \cos px \sin y + p$

$$\sin(px - y) = p$$

or $y = px - \sin^{-1} p$

Ans. $y = cx - \sin^{-1} c$

(b) $p^2x(x - 2) + p(2y - 2xy - x + 2) + y^2 + y = 0$

The given equation can be written as

$$(y^2 - 2pxy + p^2x^2) + 2p(y\ px) + (y - px + 2p) = 0$$

$$(y - px)^2 + (y - px)(2p + 1) + 2p = 0$$

$$(y - px + 1)(y - px + 2p) = 0$$

or $y = px - 1$ and $y = px - 2p$

Each of them is of Clairaut's form and hence the solution is obtained by putting $p = c$ and is

$$(y - cx + 2c)(y - cx + 1) = 0$$

8. $(x - a)p^2 + (x - y)p - y = 0$

The equation can be written as

$$y(p+1) = xp\,(p+1) - ap^2$$

or $y = px - \dfrac{ap^2}{p+1}$ which is clairaut's form

$\therefore$ $y = cx - \dfrac{ac^2}{c+1}$ is the solution

9. $$y^2 + x^2\left(\frac{dy}{dx}\right)^2 - 2xy\frac{dy}{dx} = 4\left(\frac{dx}{dy}\right)^2$$

or $(y - px)^2 = \dfrac{4}{p^2}$ where $p = \dfrac{dy}{dx}$

$\therefore$ $y = px \pm \dfrac{2}{p}$ [Clairaut's form]

$\therefore$ $y = cx \pm \dfrac{2}{c}$ or $(y - cx)^2 = \dfrac{4}{c^2}$

16.6.5 Lagrange's Equations

If the equation is of the form $y = xf_1(p) + f_2(p)$, then it is not of the form of Clairauts equation. Differentiating w.r.t x,

We get

$$p = f_1(p) + x\,f_1'(p)\frac{dp}{dx} + f_2'(p)\frac{dp}{dx}$$

$$\{p - f_1(p)\}\frac{dx}{dp} - xf_1'(p) = f_2'(p)$$

or $$\frac{dx}{dp} - \frac{f_1'(p)}{p - f_1(p)}x = \frac{f_2'(p)}{p - f_1(p)}$$

Above is a linear differential equation with p as independent variable and is solved as discussed.

***Example* 16.75:** $y = (1 + p)\,x + p^2$ .....(1)

It is not Clairaut's form. Differentiating w.r.t x, we get

$$p = (1+p) + x\frac{dp}{dx} + 2p\frac{dp}{dx}$$

or $$\frac{dx}{dp}+x=-2p \quad \text{which is linear}$$

$$I.F = e^{\int 1dp} = e^{v}$$

Multiplying both sides by I.F and integrating, we get

$$xe^{v} = -2\int pe^{v}dp = -2(p-1)e^{v}+c$$

or $$x = 2(1-p)+ce^{-v} \qquad .....(2)$$

Putting for x in (1), we get the value of y as

$$y = (1+p)\{2(1-p)+ce^{-v}\}+p^2 = 2-2p^2+c(1+p)e^{-p}+p^2$$

or $$y = 2p^2+c(1+p)e^{-v} \qquad .....(3)$$

Equations (2) and (3) give the solution to given differential equation.

Note: We shall show that sometimes by change of variables in a suitable manner the given equation reduces to clairaut's form.

16.6.6 Equations reducible to Clairaut's Form

***Example* 16.76:** $x^2(y-px) = p^2y$

We shall reduce the above equation to Clairaut's form by change of variables

Put $x^2 = u$ and $y^2 = v$

$\therefore$ $2x\,dx = du$ and $2y\,dy = dv$

or $$\frac{y\,dy}{x\,dx}=\frac{dv}{du} \text{ or } \frac{y}{x}p=\frac{dv}{du}$$

Putting for p in the given equation, we get

$$x^2\left(y-x.\frac{x\,dv}{y\,du}\right)=\frac{x^2}{y^2}\left(\frac{dv}{du}\right)^2 y$$

or $$\left(y^2-x^2\frac{dv}{du}\right)=\left(\frac{dv}{du}\right)^2 \qquad [\text{Put } x^2 = u \text{ and } y^2 = v]$$

$\therefore$ $$v = u\frac{dv}{du}+\left(\frac{dv}{du}\right)$$

If we take $\frac{dv}{du} = P$, then

$$v = Pu + P^2$$

which is Clairaut's form, and its solution is

$$v = cu + c^2 \text{ or } y^2 = cx^2 + c^2$$

Another form

$$y^2 - pxy = \frac{p^2y^2}{x^2}$$

Cancel y and it becomes $x^2(y - px) = p^2y$

Example **16.77:** Use the transformation $u = x^2$ and $v = y^2$ to solve $(px - y)(py + x) = h^2p$

Just as in Ex. 16.74. we have on putting $x^2 = u$ and $y^2 = y$

$$\frac{y}{x}p = \frac{dv}{du} \text{ or } p = \frac{x}{y}\frac{dv}{du}$$

Putting for p in the given equation, we get

$$p^2 xy + p(x^2 - y^2 - h^2) - xy = 0$$

$$xy.\frac{x^2}{y^2}\left(\frac{dv}{du}\right)^2 + \frac{x}{y}\left(\frac{dv}{du}\right)\left(x^2 - y^2 - h^2\right) - xy = 0$$

or $$x^2\left(\frac{dv}{du}\right)^2 + \left(\frac{dv}{du}\right)\left(x^2 - y^2 - h^2\right) - y^2 = 0 \qquad \left[\text{put } \frac{dv}{du} = P\right]$$

or $$uP^2 + P(u - v - h^2) - v = 0$$

or $$uP(P + 1) - v(P + 1) - Ph^2 = 0$$

or $$v = uP - \frac{P}{P+1}h^2$$

which is of the form $y = px + f(p)$ i.e., Clairaut's form

Hence the solution is $v = uc - \frac{c}{c+1}h^2$

or $$y^2 = cx^2 - \frac{c}{c+1}h^2$$

16.7 Applications to Geometry, Law of Natural Growth and Newton's Law of Cooling

When some action is applied on a quantity the changes and the action affects all the parts equally. The rate of change depends on the original quantity.

For example, the total population 'p' of a country increases with respect to time 't', say, then its rate of change with time is $\frac{dp}{dt}$. Under ideal conditions the rate of change of the population will be proportional to the total population at any given time and is called the law of natural growth.

$\therefore$ The growth of the population satisfies the differential equation

$$\frac{dp}{dt} = kp$$

where $k > 0$ is a constant

Solving the differential equation we have

$$p = ce^{kt}$$

The rate at which a quantity x is decreasing is given by $\frac{dx}{dt}$ and this is negative. This rate of decrease or decay is found to be proportional to x itself.

$$\therefore \quad \frac{dx}{dt} = -kx \qquad \text{(where } k > 0 \text{ is a constant)}$$

is the law of decomposition

Solving the DE $\qquad x = ce^{-kt}$

Newton's law of cooling:

Newton's law of cooling states that the rate of decrease of the temperature of a body is proportional to the difference of the temperature of the body and that of its surrounding medium.

Let θ be the temperature of the body at time 't' and θ_0 the temperature of its surrounding medium.

$\therefore$ Difference between the temperatures $= \theta - \theta_0$ $(\theta > \theta_0)$ and the rate of decrease of the temperature of the body is $-\frac{d\theta}{dt}$, since the body is cooling proportionally.

Using Newton's law of cooling $-\frac{d\theta}{dt} \propto (\theta - \theta_0)$

or $\dfrac{-d\theta}{dt} = k(\theta - \theta_0)$ where $k > 0$ is a constant of proportionality

$$\frac{d\theta}{dt} = -k\,(\theta - \theta_0)$$

$$\int \frac{d\theta}{\theta - \theta_0} = -k \int dt + c$$

$$\log(\theta - \theta_0) = -kt + \text{c} \quad \text{or} \quad \theta = \theta_0 + Ce^{-kt}$$

Solved Examples

***Example* 16.78:** The mass of crystalline deposit increases at a rate which is proportional to its mass at that time. The deposit has started around a crystal seed of 5 grams. Find an expression of its mass at time 't'. If in 30 minutes the mass of the deposit increases by '1' gram, what will be the mass of the deposit after 10 hours.

Solution:

Let m be the mass of crystalline at time 't' then by law of growth

$$\frac{dm}{dt} = km; \qquad \frac{dm}{km} = dt$$

on integration

$$\log m = kt + c \qquad \text{.... (1)}$$

initially when t = 0, m = 5

(1) ⇒ $\log 5 = 0 + c$

∴ $c = +\log 5$

Substituting c value in (1)

$$\log m = kt + \log 5$$

i.e., $$kt = \log\left(\frac{m}{5}\right) \qquad \text{.... (2)}$$

When t = 30 minutes mass deposit increases by '1' gram

∴ m = 5 + 1 = 6 grams

Substituting $t = \frac{1}{2}$ (hours), m = 6 in (2)

$$k\left(\frac{1}{2}\right) = \log\left(\frac{6}{5}\right)$$

$$k = \log\left(\frac{6}{5}\right)^2$$

Substituting in (2)

$$\log\left(\frac{6}{5}\right)^2 t = \log\frac{m}{5} \quad \text{.... (3)}$$

Now to find the mass after 10 hours (i.e t = 10)

from (3) we get

$$\log\left(\frac{6}{5}\right)^2 \times 10 = \log\frac{m}{5} \Rightarrow \log\left(\frac{m}{5}\right) = \log\left(\frac{6}{5}\right)^{20}$$

$$m = 5\left(\frac{6}{5}\right)^{20} \text{ grams}$$

***Example* 16.79:** The rate at which a certain substance decomposes in a certain solution at any instant is proportional to the amount of it present in the solution at that instant. Initially, there are 27 grams and three hours later, it is found that 8 grams are left. How much substance will be left after one more hour.

Solution:

If m grams is the amount of the substance left in the solution at time 't', then the rate at which it decomposes is $\frac{dm}{dt}$, which is proportional to m.

∴ By law of decay

$$\frac{dm}{dt} = -km \quad (k > 0)$$

$$\int\frac{dm}{m} = -k\int dt + c$$

$$\log m = -kt + c \quad \text{.... (1)}$$

Initially when $t = 0$, $m = 27$

From (1) we have

$$\log 27 = -k.0 + c$$

$\Rightarrow$ $c = \log 27$

Substitution of 'c' in (1) gives

$$\log m = -kt + \log 27$$

$$\therefore \quad \log\left(\frac{m}{27}\right) = -kt \qquad \text{.... (2)}$$

It is given that m = 8 when t = 3

$\therefore$ From (2)

$$\log\left(\frac{8}{27}\right) = -k.3$$

$$\Rightarrow \quad -k = \log\left(\frac{8}{27}\right)^{1/3}$$

$$-k = \log\frac{2}{3}$$

Then (2) becomes

$$\log\frac{m}{27} = \log\left(\frac{2}{3}\right)t \qquad \text{... (3)}$$

when $t = 4$

$$\log\left(\frac{m}{27}\right) = 4\log\frac{2}{3}$$

$$\log\frac{m}{27} = \log\left(\frac{2}{3}\right)^4$$

$$m = 27 \times \left(\frac{2}{3}\right)^4 \text{ grams}$$

$$m = \frac{16}{3} \text{ grams.}$$

***Example* 16.80:** The number x of bacteria in a culture grow at a rate proportional to x. The value of x was initially 50 and increased to 150 in one hour; what will be the value of x after $1\frac{1}{2}$ hour.

Solution:

$$\frac{dx}{dt} = kx$$

$$\frac{dx}{x} = kdt$$

$$\log x = kt + c \qquad \text{.... (1)}$$

c is the constant of integration

when $t = 0, x = 50$

$$\therefore \quad \log 50 = k.0 + c$$

or $$c = \log 50$$

(1) $\Rightarrow$ $\log x = kt + \log 50$

$$\therefore \quad \log\frac{x}{50} = kt$$

$$x = 150, \qquad \text{when } t = 1 \qquad \text{.....(2)}$$

$$\therefore \quad \log\frac{150}{50} = \text{k}.1$$

or $$k = \log 3$$

(2) then gives

$$\log\left(\frac{x}{50}\right) = t\log 3$$

we want to find x when $t = \frac{3}{2}$

$$\log\left(\frac{x}{50}\right) = \frac{3}{2}\log 3$$

$$\frac{x}{50} = (3)^{\frac{3}{2}}$$

$$x = 50\,(3)^{\frac{3}{2}} \text{ grams}$$

***Example* 16.81:** The rate of cooling of a body is proportional to the difference between the temperature of the body and the surrounding air. If the air temperature is 20°C and the body cools for 20 minutes from 140°C to 80°C, find when the temperature will be 35°C.

Solution:

If θ is the temperature of the body at time 't' then from Newton's law of cooling

$$\frac{-d\theta}{dt}\,\alpha\,(\theta - 20) \Rightarrow \frac{d\theta}{dt} = -k(\theta - 20)$$

$$\therefore \quad \int \frac{d\theta}{\theta - 20} = -k\int tdt + c$$

$$\log(\theta - 20) = -kt + c \qquad \text{.... (1)}$$

Initially when

$$t = 0, \theta = 140$$

$$\log(\theta - 20) = 0 + c,$$

$$\therefore \quad c = \log(120),$$

(1) reduces to

$$\log(14\theta - 20) = -kt + \log 120$$

or $\quad +kt = \log(120) - \log(\theta - 20) \qquad \text{.... (2)}$

It is given that $\theta = 80$ when $t = 20$ minutes

$$k\,.\,20 = \log 120 - \log(80 - 20)$$

$$k = \frac{1}{20}\log\left(\frac{120}{60}\right)$$

$$k = \frac{1}{20}\log 2$$

Substituting in (2)

$$\left(\frac{1}{20}\log 2\right)t = \log 120 - \log(\theta - 20)$$

It is required to find t when

$\theta = 35°$c

$$t = \frac{\log 120 - \log(35 - 20)}{\frac{1}{20}\log 2} = \frac{20\log\left(\frac{120}{15}\right)}{\log 2}$$

$$t = 20\frac{\log(8)}{\log 2} = 20\frac{\log 2^3}{\log 2} = 60\frac{\log 2}{\log 2} = 60 \text{ minutes}$$

Exercise 16.13

1. In a certain reaction, the rate of conversion of a substance at time 't' is proportional to the quantity of the substance still untransformed at that instant. At the end of one hour 60 grams, while at the end of four hours 21 grams remain. How many grams of the first substance was there initially?

   [Ans : 85 grams approximately]

2. The rate of growth of a bacteria is proportional to the number present. If initially there were 100 bacteria and the amount doubles in '1' hour, how many bacteria will be there after $2\frac{1}{2}$ hours. [Ans : 564]

3. Under certain conditions cane sugar in water is converted into dextrose at a rate which is proportional to the amount unconverted at any time. If 75 grams was there at time t = 0.0 and 8 grams are converted during the first 30 minutes find the amount converted in $1\frac{1}{2}$ hour. [Ans : 21.5 gm]

4. The rate of cooling of a body is proportional to the difference between the temperature of the body and the surrounding air. If the surrounding air is kept at 30°C and the body cools from 80°C to 60°C in 20 minutes, find the temperature of the body after 40 minutes. [Ans : 48°C]

5. If the air is maintained at 30°C and the temperature of the body cools from 80°C to 60°C in 20 minutes. Find the temperature of the body after 40 minutes.

   [Ans : 48°C]

6. The rate at which a heated body cools in air is proportional to the difference between the temperature of the body and that of the surrounding air. A body originally at 80° C cools down to 60°C in 20 minutes the temperature of the air being 40°C what will be the temperature of the body after 40 minutes from the original temperature.

   [Ans : 50°C]

7. The rate at which bacteria multiply is proportional to the instantaneous number present. If the original number doubles in 2 hours in how many hours will it triple.

   [Ans : $\frac{2\log 3}{\log 2}$]

8. Water at temperature 100°C cools in 10 minutes to 80°C in a room of temperature 25°C. Find the temperature of water after 20 minutes. [Ans : 65.5°C]

9. A cup of coffee at temperature 100°C is placed in a room whose temperature is 15°C and it cools to 60°C in 5 minutes, find its temperature after a further interval of 5 minutes. [Ans : 38.8°C]

16.8 Applications to Electrical Circuits

***Example* 16.82:** The current i in a circuit containing an inductance L, resistance R and voltage E sin ω t is given by

$$L\frac{di}{dt}+Ri=E\sin\omega t$$

If i = 0 at t = 0, show that at any time t

$$i=\frac{E}{\sqrt{R^2+L^2\omega^2}}\left[\sin(\omega t-\phi)+\sin\phi\, e^{-\frac{Rt}{L}}\right] \quad \text{where } \tan\phi=\frac{L\omega}{R}$$

***Solution*:** The given differential equation can be written as

$$\frac{di}{dt}+\frac{R}{L}i=\frac{E}{L}\sin\omega t \qquad \text{which is linear in i.} \qquad .....(1)$$

Here $\quad \text{I.F.}=e^{\int\frac{R}{L}dt}=e^{\frac{Rt}{L}}$, hence the solution of (1) as

$$ie^{\frac{Rt}{L}}=\int\frac{E}{L}e^{\frac{Rt}{L}}\sin wt\,dt=\frac{E}{L}\frac{e^{\frac{Rt}{L}}}{\omega^2+\frac{R^2}{L^2}}\left[\frac{R}{L}\sin\omega t-\omega\cos\omega t\right]+C$$

or $\quad i=Ce^{-\frac{Rt}{L}}+\frac{E}{\omega^2L^2+R^2}[R\sin\omega t-\omega L\cos\omega t]$

where C is the constant of integration. .....(2)

If we write here $R=k\cos\phi$ and $\omega L=k\sin\phi$

so that $k^2=R^2+\omega^2L^2$ and $\tan\phi=\frac{\omega L}{R}$

then $\quad i=Ce^{-\frac{Rt}{L}}+\frac{E}{\omega^2L^2+R^2}\sin(\omega t-\phi)$

To determine C we apply the initial conditions that i = 0 at t = 0 and get

$$C=\frac{E\sin\phi}{\sqrt{\omega^2L^2+R^2}}$$

$$i=\frac{E\sin\phi e^{\frac{-Rt}{L}}}{\sqrt{\omega^2L^2+R^2}}+\frac{E}{\sqrt{L^2\omega^2+R^2}}\sin(\omega t-\phi)$$

$$=\frac{E}{\sqrt{\omega^2L^2+R^2}}[\sin\phi e^{-\frac{Rt}{L}}+\sin\ (\omega t-\phi)]$$

***Example* 16.83:** A condenser of capacitance C is charged through a resistance R by a steady voltage V. The charge Q on the plate is given by

$$R\frac{dQ}{dt}+\frac{Q}{C}=V.$$

If initially the plate is chargeless find the charge and the current at time 't'.

***Solution*:** The given equation can be written as $\frac{dQ}{dt}+\frac{Q}{RC}=\frac{V}{R}$

which is linear in Q. Here I.F. $= e^{\int\frac{dt}{RC}} = e^{\frac{t}{RC}}$

$\therefore$ General solution is $Qe^{\frac{t}{RC}} = \int\frac{V}{R}e^{\frac{t}{RC}}dt = \frac{V}{R}RCe^{\frac{t}{RC}} + C'$

or $\quad Q = VC + C'e^{\frac{-t}{RC}}$ where C' is the arbitrary constant of integration.

Initially $\quad Q = 0$ at $t = 0$ hence $0 = VC + C' \quad \therefore\ C' = -VC$

Therefore $\quad Q = VC(1 - e^{\frac{-t}{RC}})$ which gives the charge at any time t.

Further, current $= i = \frac{dQ}{dt} = \frac{VC}{RC}e^{\frac{-t}{RC}}$ or $i = \frac{V}{R}e^{\frac{-t}{RC}}$

16.9 Applications to Heat Conduction (or Heat Flow) Problems

Following are the basic rules to be followed while dealing the heat flow problems:

(i) Heat flows from higher temperature to lower temperature

(ii) The amount of heat in a body is proportional to its mass and temperate

(iii) The amount of heat Q flowing per second through a slab of surface area A and thickness δx whose faces are at temperatures t and $t + \delta t$, is given by

$$Q = -kA\frac{dt}{dx}$$

where k is the co-efficient of thermal conductivity of the material of the body.

***Example* 16.84:** A pipe 20 cm in diameter contains steam at 160°C and is protected with a covering 5 cm thick for which k, the coefficient of thermal conductivity = 10025. If the temperature of the outer surface of the covering is 50 °C, find the temperature at half way through the covering under steady conditions.

Solution: We have $Q = -kA\frac{dt}{dx} = -k(2\pi\, x)\frac{dt}{dx}$,

since A is the surface area of the pipe of length 1 cm.

or $$dt = \frac{-Q}{2k\pi}\frac{dx}{x} \qquad \text{....(1)}$$

we are given that t = 100°C at x = 10 cm and t = 50°C at x = 15 cm.

Integrating (1) throughout, gives $\int_{160}^{50} dt = \frac{-Q}{2\pi k}\int_{10}^{15}\frac{dx}{x}$

or $$-110^\circ = \frac{-Q}{2\pi k}\log\frac{15}{10} \quad \text{or} \quad \frac{Q}{2\pi k} = \frac{110^\circ}{\log 1.5} \qquad \text{....(2)}$$

Let t = T when x = 12.5 (that is, half way through the covering), we have

$$\int_{160^0}^{T} dt = \frac{-Q}{2\pi k}\int_{10}^{125}\frac{dx}{x} \Rightarrow T - 160^\circ = \frac{-Q}{2\pi k}[\log x]_{10}^{125}$$

Using (2), we get $T = 160^\circ - 110^\circ\frac{\log 1.25}{\log 1.5} = 99.5\ ^\circ C$

***Example* 16.85:** Heat flows through the wall of a hollow sphere of inner and outer radii r_1 and r_2 respectively, in a steady state. The temperature u at a distance r ($r_1 < r < r_2$) from the centre of the sphere, is given by

$r\frac{d^2u}{dr^2} + 2\frac{du}{dr} = 0$. If u_1 and u_2 are the temperatures of the inner and the outer surfaces of the sphere, find u in terms of r.

Solution: Let us write $\frac{du}{dr} = z$ then the given equation becomes

$$r\frac{dz}{dr} + 2z = 0 \text{ or } \frac{dz}{z} + \frac{2dr}{r} = 0 \qquad \text{....(1)}$$

which on integration, gives log z + 2 log r = log C

or $z\,r^2 = C$ where C is an arbitrary constant.

or $r^2\frac{du}{dr} = C$ or $du = C\frac{dr}{r^2}$

which on integration throughout, gives

$$u = \frac{-C}{r} + C' \qquad \text{....(2)}$$

where C' is another constant of integration.

Since $u = u_1$ at $r = r_1$ and $u = u_2$ at $r = r_2$, we get

$$u_1 = \frac{-C}{r_1} + C' \quad \text{and} \quad u_2 = \frac{-C}{r_2} + C'$$

$$\Rightarrow \quad C = \frac{r_1 r_2 (u_1 - u_2)}{r_1 - r_2} \quad \text{and} \quad C' = \frac{u_1 r_1 - u_2 r_2}{r_1 - r_2}$$

Putting these values of C and C' in (2), we get

$$u = \frac{1}{r_1 - r_2}\left[(u_1 r_1 - u_2 r_2) - \frac{r_1 r_2 (u_1 - u_2)}{r}\right]$$

as the required relation.

16.10 Applications to Problems of Newton's Law of Cooling

***Example* 16.86:** If the temperature of a body drops from 100 °C to 60 °C in one minute when the surrounding temperature is 20 °C, find the time when the temperature will come down to 30 °C.

Solution: The Newton's law of cooling states that:

"The temperature of a body changes at a rate proportional to the difference in temperatures of the surrounding medium and that of the body itself."

Thus, if θ is the temperature of the body at a time t and θ_0 is the temperature of the surrounding medium, then

$$\frac{d\theta}{dt} \alpha -(\theta - \theta_0) \text{ or } \frac{d\theta}{dt} = -k(\theta - \theta_0) \qquad \text{....(1)}$$

where k is the constant of proportionality.

$$\therefore \quad \int \frac{d\theta}{\theta - \theta_0} = -\int k\, dt \text{. Here } \theta_0 = 20\ °C$$

$$\therefore \log(\theta - 20°) = -kt + C \qquad \text{....(2)}$$

where C is the constant of integration.

Initially at $t = 0$, $\theta = 100°$ hence $C = \log 80°$

therefore (2) becomes $\log \frac{\theta - 20}{80} = -kt$ ....(3)

Next, we are given that $\theta = 60^\circ$ when t = 1 min, therefore from (3), we have

$$\log\frac{60-20}{80} = -k \quad \text{or } k = \log 2$$

Substituting this value of k in (3), we get $\log\frac{\theta-20}{80} = -t\log 2$ ....(4)

We are to find t when q becomes 30°

Therefore after 3 minutes the body temperature will come down to 30 °C.

16.11 Applications to Problems on Chemical Reactions

***Example* 16.87:** Uranium disingrates at a rate proportional to the amount present at that instant. If m_1 and m_2 grams of uranium are present at times t_1 and t_2 respectively, find the half-life of uranium.

***Solution*:** The differential equation of disintegration of uranium is $\frac{dm}{dt} = -km$ ...(1)

where m is the mass of uranium at any time t and k is the constant of proportionality can be written as $\frac{dm}{m} = -kdt$ which on integration, gives

$$\int_{m_1}^{m_2} \frac{dm}{m} = -\int_{t_1}^{t_2} k\,dt \quad \text{or} \quad \log m_2 - \log m_1 = -k(t_2 - t_1)$$

or
$$k = \frac{\log\left(\frac{m_1}{m_2}\right)}{t_2 - t_1} \quad \text{....(2)}$$

Now, we want to compute time T in which the uranium will disintegrate to half its original amount M, say. As such

$$\int_{M_1}^{M/2} \frac{dm}{m} = -k\int_0^T dt \quad \text{or} \quad \log\left(\frac{M/2}{M}\right) = -kT \quad \text{or} \quad kT = \log 2$$

$$\Rightarrow \qquad T = \frac{\log 2}{k} = \frac{(t_2 - t_1)\log 2}{\log\left(\frac{m_1}{m_2}\right)}$$

which is the required time for the uranium to disintegrate to half its original amount.

Exercise 16.14

1. A particle is projected vertically upwards in the gravitational field under a resistance equal to k times the square of its velocity per unit mass. If V_0 is the initial velocity find the maximum height attained.

2. The differential equation for the motion of a particle moving in a central orbit is $\frac{d^2x}{dt^2} = -\left(x + \frac{a^4}{x^3}\right)$.

   If it starts from test at a distance 'a', find the time it will take to arrive at the origin.

3. A car starts from rest and its acceleration at a time t is equal to $k\left(1 - \frac{t}{T}\right)$ where k is a constant and T is the time taken to attain the highest speed which is maintained constant after that time. Find the highest speed and the distance travelled till this speed is attained.

4. Show that the curve for which the optional of the tangent intercepted between the co-ordinate axes is bisected at the point of contact, is a rectangular hyperbola.

5. Find the equation of the curve whose subtangent is twice the abscissa of the point of contact and passes through the point (1, 2).

6. Find the orthogonal trajectories of a system of co.focal and co-axial parabolas.

7. In an electric circuit containing an inductance L, a resistance R and a battery E, the current i builds up at a rate given by $L\frac{di}{dt} + Ri = E$. Find the current i in terms of time t. Also, calculate the time taken before the current takes half its peak value.

8. A voltage Ea^{-at} is applied initially to an electric circuit containing inductance L and resistance R. Obtain the current developed at any time t.

9. The temperature q of a body satisfies the equation $\frac{d\theta}{dt} = k(\theta - \theta_0)$ where θ_0 is the temperature of the surrounding medium. A thermometer which reads 75° F indoors is taken outdoors. After 5 minutes it reads 65 °F and after another 5 minutes, it reads 60 °F.

   What is the outside temperature?

10. The rate of which certain mass of radium decomposes is at any instant proportional to the mass then present and the mass decreases to half its original values in 1500 years. If the initial mass be 100 mg, in how many years will it be 80 mg?

CHAPTER 17

Higher Order Linear Differential Equations and their Applications

17.1 Linear Differential Equation

Let (i) the differential operator "$\frac{d}{dx}$" be donoted by 'D'

(ii) $p_1, p_2, p_3, ..., p_n$ be either functions of x or constants

(iii) R be a function of x. Then the general form of a linear differential equation (L.D.E) of order n is given by

$$D^n y + p_1 D^{n-1} y + p_2 D^{n-2} y + ... p_n y = R \quad .... (1)$$

or simply $(D^n + p_1 D^{n-1} + ..., + p_n)y = 0$, If $p_1, p_2, ..., p_n$ are constants then (1) is called a L D E with constant coefficients.

Denoting the differential operator $(D^n + p_1 D^{n-1} + ... ,+ p_n)$ by f (D),

(1) can be written as

$$f(D)y = R \quad .... (2)$$

Ex. Consider the D.E, $\frac{dy}{dx} + Py = 0$

Separating the variables $\frac{dy}{y} = -P(x)\,dx$

Integration yields $\log y = \int P(x)dx \Rightarrow y = Ce^{-\int P(x)dx} + \log c$ ..... (3)

Solution is $y = Ce^{-\int Pdx}$

Suppose p is a constant = –m, say, then solution is $y = Ce^{mx}$

(i.e.,) the solution of $(D - m)y = 0$ is $y = Ce^{mx}$ .... (4)

17.1.1 Complementary Function (C.F)

If R = 0 then equation (2) becomes $f(D)y = 0$

The solution of this equation is called the C.F. of (2)

We shall take, here afterwards, $f(m) = 0$ as the auxiliary equation, (A.E).

To find C.F.

Consider a second order L.D.E

$$(D^2 + a_1D + a_2)y = 0 \qquad ..... (6)$$

A.E. is $m^2 + a_1m + a_2 = 0$.

Let m_1, m_2 be the roots of this equation.

Now four cases arise.

(i) m_1, m_2 are real and distinct

(ii) m_1, m_2 are real and equal

(iii) m_1, m_2 are complex and distinct

(iv) m_1, m_2 are complex and equal

Case I: (6) can be written as

$$[D^2 - (m_1 + m_2)D + m_1m_2]\,y = 0 \text{ or } (D - m_2)(D - m_1)\,y = 0 \qquad .... (7)$$

Call $(D - m_1)\,y = Y$

(7) now becomes $(D - m_2)\,Y = 0$

From (4) it follows that $Y = Ce^{m_2x}$

From (7) $(D - m_1)y = Ce^{m_2x}$

$\text{I.F.} = e^{-m_1x}$

Solution is $y\, e^{-m_1x} = C \int e^{(m_2-m_1)x}\, dx + C_1$ .... (8)

$$y.e^{-m_1x} = \frac{C}{m_2 - m_1} e^{(m_2-m_1)x}\, dx + C_1, \text{ Call } \frac{C}{m_2 - m_1} \text{ as } C_2.$$

(i.e.,) $y = C_2\, e^{m_2x} + C_1\, e^{m_1x}$ where C_1 and C_2 are arbitrary constants

Similarly if $m_1, m_2, \ldots, m_n$ are real and dinstinct roots of $f(m) = 0$ then

$$y = C_1\, e^{m_1x} + C_2\, e^{m_2x} + \ldots + C_n\, e^{m_nx} \quad \ldots (I)$$

(I) is the solution of $f(D)y = 0$ where $C_1, C_2, \ldots, C_n$ are arbitrary constants.

***Case* II:** $m_1 = m_2$,

From (8) it follows that the solution of (6) is given by

$$y.e^{-m_1x} = C \int e^{0.x} dx + C_1 = Cx + C_1$$

$$\therefore \quad y = (Cx + C_1)\, e^{m_1x}$$

Similarly if m_1 is repeated say 'r' times then solution of $(D - m_1)^r y = 0$ is

$$y = (C_1 + C_2x + C_3x^2 + \ldots + C_rx^{r-1})\, e^{m_1x} \quad \ldots (II)$$

(II) is the solution corresponding to a root m_1 repeated r times;

i.e., II is the C.F

***Case* III:** m_1, m_2 are complex, say $(\alpha \pm i\beta)$

Then solution of (6) is given by

$$y = A\, e^{(\alpha + i\beta)x} + B\, e^{(\alpha - i\beta)x} \quad \text{where A and B are arbitrary constants}$$

(i.e.,) $y = Ae^{\alpha x}.\, e^{i\beta x} + Be^{\alpha x}.e^{-i\beta x} = e^{\alpha x}\,[A(\cos\beta x + i\sin\beta x) + B\,(\cos\beta x - i\sin\beta x)]$

$$= e^{\alpha x}\,[C_1 \cos\beta x + C_2 \sin\beta x] \quad \ldots (III)$$

where $A + B = C_1$ and $i(A - B) = C_2$

(III) gives the solution corresponding to two complex conjugate roots $(\alpha \pm i\beta)$, i.e., III is C.F.

***Case* IV:** If $(\alpha \pm i\beta)$ are repeated say 's' times then the corresponding solution (follows from case (II) and (III) is given by

$y = e^x[(C_1 + C_2x + C_3x^2 + \ldots + C_sx^{s-1})]\cos\beta x + (d_1 + d_2x + d_3x^2 + \ldots + d_sx^{s-1}) + \sin\beta x]$....

(IV) is the C.F

Solved Examples

***Example* 17.1:** Solve $\frac{d^2y}{dx^2}+5\frac{dy}{dx}+6y=0$

***Solution*:**

A.E is $m^2 + 5m + 6 \Rightarrow (m + 2)(m + 3) = 0$

$\therefore$ $m = -2, m = -3.$

Solution is $y = Ae^{-2x} + B\,e^{-3x}$

***Example* 17.2:** Solve $\frac{d^2y}{dx^2}-11\frac{dy}{dx}+30y=0$

***Solution*:**

A.E is $m^2 - 11m + 30 \Rightarrow (m - 5)(m - 6) = 0$

Solution is $y = Ae^{5x} + B\,e^{6x}$

***Example* 17.3:** Solve $2\,\frac{d^2y}{dx^2}+5\frac{dy}{dx}-12y=0$

***Solution*:**

A.E is $2m^2 + 5m - 12 \Rightarrow (2m - 3)(m + 4) = 0$

$m = 3/2, m = -4$

$\therefore$ Solution is $y = C_1 e^{\frac{3}{2}x} + c_2\,e^{-4x}$

***Example* 17.4:** Solve $\frac{d^3y}{dx^3}+4\frac{d^2y}{dx^2}-6y=0$

***Solution*:**

$m^3 + 4m^2 + m - 6 = 0$

$(m - 1)(m + 2)(m + 3) = 0$

$\therefore$ Solution is $y = c_1e^{x} + c_2e^{-2x} + C_3e^{-3x}$

Exercise 17.1

Solve the following D.E:

1. $\frac{d^2y}{dx^2}+2\frac{dy}{dx}-3y=0$ **Ans :** $y = Ae^x + Be^{-3x}$

2. $\frac{d^2y}{dx^2}-3y=0$ **Ans :** $y = Ae^x + Be^{-x}$

3. $\frac{d^3y}{dx^3}+2\frac{d^2y}{dx^2}-5\frac{dy}{dx}-6y=0$ **Ans :** $y = Ae^{-x} + Be^{2x} + Ce^{-3x}$

4. $9\frac{d^2y}{dx^2}+18\frac{dy}{dx}-16y=0$ **Ans :** $y = Ae^{\frac{2}{3}x} Be^{-\frac{8}{3}x}$

5. $\frac{d^2y}{dx^2}+3\frac{dy}{dx}+2y=0$ **Ans :** $y = Ae^{-x} + Be^{-2x}$

6. Solve $\frac{d^2y}{dx^2}+10\frac{dy}{dx}+25y=0$ **Ans :** $y = e^{-5x}(Ax + B)$

7. $\frac{d^3y}{dx^3}-\frac{d^2y}{dx^2}+8\frac{dy}{dx}-4y=0$ **Ans :** $y = (Ax + B)e^{2x} + Ce^x$

8. Solve $\frac{d^3y}{dx^3}-\frac{d^2y}{dx^2}+8\frac{dy}{dx}-4y=0$ **Ans :** $y = (Ax + B)e^{2x} + Ce^x$

9. Solve $\frac{d^3y}{dx^3}-3\frac{d^2y}{dx^2}+3\frac{dy}{dx}-y=0$ **Ans :** $y = (Ax^2 + Bx + C)e^x$

10. Solve $\frac{d^2y}{dx^2}+7y=0$ **Ans :** $y = A\cos\sqrt{7}x + B\sin\sqrt{7}x$

11. Solve $2\frac{d^2y}{dx^2}+4\frac{dy}{dx}+3y=0$ **Ans :** $y = e^{-x}\left(A\cos\frac{x}{\sqrt{2}} + B\sin\frac{x}{\sqrt{2}}\right)$

12. Solve $\frac{d^4y}{dx^4}+y=0$

Ans : $y = e^{x/\sqrt{2}}\left(C_1\cos\frac{x}{\sqrt{2}}+C_2\sin\frac{x}{\sqrt{2}}\right) + e^{x/\sqrt{2}}\left(C_3\cos\frac{x}{\sqrt{2}}+C_4\sin\frac{x}{\sqrt{2}}\right)$

13. Solve $\frac{d^4y}{dx^4}-64y=0$

Ans : $y = C_1e^{2x} + C_2e^{-2x} + e^{-x}(C_3\cos\sqrt{3}x + C_4\sin\sqrt{3}x) +$
$e^x(C_3\cos\sqrt{3}x + C_6\sin\sqrt{3}x)$

Solved Examples

***Example* 17.5:** Solve $\frac{d^2y}{dx^2}+6\frac{dy}{dx}+9y=0$

Solution:

A.E is $m^2 + 6m + 9 = 0$

i.e. $(m + 3)^2 = 0$, $m = -3, -3$

$\therefore$ Solution is $y = e^{-3x}(A + Bx)$

***Example* 17.6:** Solve $\frac{d^3y}{dx^3}+\frac{d^2y}{dx^2}=0$

Solution:

A.E is $m^3 + m^2 = 0 \Rightarrow m^2(m + 1) = 0$

$\therefore$ $m = 0, 0, -1$

Solution is $y = (A + Bx)e^{0.x} + C e^{-x} = (A + Bx) + Ce^{-x}$

***Example* 17.7:** Solve $\frac{d^4y}{dx^4}+18\frac{d^2y}{dx^2}+81y=0$

Soltion:

A.E. is $m^4 - 18m^2 + 81 = 0$

$\therefore$ $(m^2 - 9)(m^2 - 9) = 0$

$\therefore$ $m = 3, 3, -3, -3$

Solution is $y = (A + Bx)e^{-3x} + (C + xD)e^{3x} = (C_1x + C_2)e^{-3x} + (C_3x + C_4)e^{3x}$

***Example* 17.8:** Solve $\frac{d^2y}{dx^2}+4\frac{dy}{dx}+9y=0$

Solution:

A.E. is $m^2 + 4m + 9 = 0$

$m = -2 \pm i\sqrt{5}$

$\alpha = -2$ and $\beta = \sqrt{5}$

$\therefore$ Solution is $y = e^{-2x}(A\cos x\sqrt{5} + B\sin\sqrt{5}x)$

***Example* 17.9:** Solve $\frac{d^3y}{dx^3}+y=0$

while the solution of $f(D)\,y=0$

is called the complementary function (C.F.) ..... (2)

Suppose y_1 is C.F and y_2 is P.I for (1)

Then $f(D)\,y_1=0$ from (2)

and $f(D)\,y_2=R$ from (1)

Hence $f(D)\,[y_1+y_2]=f(D)\,y_1+f(D)\,y_2$

$$=0+R=R$$

Which shows that y_1+y_2 is also a solution of (1). y_1+y_2 (i.e.) C.F. + P.I is called the most general solution of (1).

***EX*:** Calculation of $\frac{1}{D-m_1}R$

$\frac{1}{D-m_1}R=y$, say, Operating both sides with $(D-m_1)$, we get $(D-m_1)\,y=R$

(i.e.,) $\frac{dy}{dx}-m_1y=R$

$$\text{I.F. is } e^{\int -m_1dx}=e^{m_1x}$$

∴ Solution is given by

$$y.e^{m_1x}=\int Re^{-m_1x}dx+c$$

(i.e.,) $y=Ce^{m_1x}+e^{m_1x}\int Re^{m_1x}dx=\text{C.F.}+\text{P.I}$

$$\text{P.I.}=e^{m_1x}\int Re^{m_1x}dx$$

If $m_1=0,\frac{1}{D}R=\int R\,dx$

Thus the operator $\frac{1}{D}$ stands for integration

Solution:

A.E. is $m^3 + 1 = 0$

$\therefore (m + 1)(m^2 - m + 1) = 0$

$$\therefore \qquad m = -1, m = \frac{1+\sqrt{3}i}{2}$$

Solution is $y = C_1 e^{-x} + e\left(C_2 \cos\frac{\sqrt{3}}{2} + C_3 \sin\frac{\sqrt{3}}{2}x\right)$

***Example* 17.10:** Solve $\frac{d^4y}{dx^4} + 2\frac{d^3y}{dx^3} + 3\frac{d^2y}{dx^2} + 2\frac{dy}{dx} + y = 0$

A.E. is $m^4 + 2m^3 + 3m^2 + 2m + 1 = 0$

$(m^2 + m + 1)^2 = 0$

$$\therefore \qquad m = \frac{-1+\sqrt{3}i}{2}, \frac{-1-\sqrt{3}i}{2}$$

Solution is $y = e^{\frac{1}{2}x}\left[(C_1 + C_2x)\cos\frac{\sqrt{3}}{2}x + (C_3 + C_4x)\sin\frac{\sqrt{3}}{2}x\right]$

17.2 Particular Integral

Consider the D.E. $f(D)y = R$ .....(1)

Operating both sides with $\frac{1}{f(D)}$ (called inverse operator)

We have $$\frac{1}{f(D)}[f(D)]y = \frac{1}{f(D)R}$$

or $$y = \frac{1}{f(D)}R$$

This solution is called the particular integral of (1)

Solved Examples

***Example* 17.11:** Find the value of $\frac{1}{D-5}x$

Solution:

$$\frac{1}{D-5}x = e^{5x}\int e^{-5x}.xdx = e^{5x}\left[-x\frac{e^{-5x}}{5}+\int\frac{e^{-5x}}{5}dx\right] = e^{5x}\left[-\frac{xe^{-5x}}{5}-\frac{xe^{-5x}}{25}\right]$$

$$= -\frac{x}{5}-\frac{1}{25}$$

***Example* 17.12:** Find $\frac{1}{D^2-5D+6}e^x$

Solution:

$$\frac{1}{D^2-5D+6}e^x = \frac{1}{(D-2)(D-3)}.(e^x)$$

$$= \frac{1}{(D-2)}\cdot\frac{1}{(D-3)}(e^x) = \frac{1}{(D-2)}e^{3x}\int e^{-3x}.e^x dx = -\frac{1}{(D-2)}\cdot e^{3x}\frac{e^{-2x}}{2}$$

$$= -\frac{1}{2}\frac{1}{(D-2)}(e^x) = -\frac{1}{2}e^{2x}\int e^{-2x}.e^x dx = \frac{1}{2}e^{2x}\int e^{-x}dx = \frac{1}{2}e^{2x}.e^{-x} = \frac{e^x}{2}$$

***Example* 17.13:** Solve $(D^2 + a^2)\,y = \tan ax$

Solution:

A.E is $m^2 + a^2 = 0 \Rightarrow m = \pm ai$

C.F. is $y = C_1 \cos ax + C_2 \sin ax$

$$P.I = \frac{1}{(D+ai)(D-ai)}\tan ax = \frac{1}{2ai}\left[\frac{1}{D-ai}-\frac{1}{D+ai}\right]\tan ax$$

$$= \frac{1}{2ai}\left[e^{iax}\int e^{-aix}\tan ax dx - e^{-iax}\int e^{aix}\tan ax dx\right]$$

$$\int e^{-aix}\tan ax dx = \int(\cos ax - i\sin ax)\frac{\sin ax}{\cos ax}dx = \int \sin ax - i\frac{(1-\cos^2 ax)}{\cos ax}dx$$

$$= -\frac{\cos ax}{a} - \frac{i}{a}\log(\sec ax + \tan ax) + \frac{i\sin ax}{a}$$

Similarly $\int e^{aix}\tan ax dx = \frac{\cos ax}{a} + \frac{i}{a}\log(\sec ax + \tan ax) - \frac{i\sin ax}{a}$

$$I_1 = \left(\frac{\cos ax + i\sin ax}{a}\right)[-\cos ax + i\sin ax - i\log(\sec ax + \tan ax]$$

$$= \frac{1}{a}[-1 - i\cos ax\log(\sec ax + \tan ax) + \sin ax\log(\sec ax + \tan ax)]$$

$$\therefore \text{P.I.} = \frac{1}{2a^2 i}[-2i\cos ax\log(\sec ax + \tan ax)]$$

$$= -\frac{\cos ax}{a}\log(\sec ax + \tan ax)$$

∴ Most general solution is

$$y = C_1\cos ax + C_2\sin ax - \frac{\cos ax}{a^2}\log(\sec ax + \tan ax)$$

***Example* 17.14:** Solve $(D^2 - 5D + 6)y = e^x$

***Solution*:**

A.E. is $m^2 - 5m + 6 = 0$

C.F. is $y = C_1e^{2x} + C_2e^{3x}$

P.I. is $y = \frac{1}{D^2 - 5D + 6}(e^x) = \frac{e^x}{2}$

∴ Complete solution (C.S.) is y = C.F. + P.I.

(i.e.) $y = C_1e^{2x} + C_2e^{3x} + \frac{e^x}{2}$

Exercise 17.2

Find the values of :

1. $\frac{1}{D}\cos x$ 2. $\frac{1}{D^2}(e^{-x})$ 3. $\frac{1}{D}(x^2)$

4. $\frac{1}{D-2}\sin x$ 5. $\frac{1}{D^3-D^2}x$

Ans: (1) $+\sin x$ (2) e^{-x} (3) $\frac{x^3}{3}$ (4) $-\frac{(2\sin x+\cos x)}{5}$

(5) $-\left(\frac{x^3}{6}+\frac{x^2}{2}\right)$

Solve the following :

6. $\frac{d^2y}{dx^2}-4\frac{dy}{dx}+3y=e^{2x}$ **Ans :** $y = C_1e^{3x} + C_2e^{x} + \frac{e^{2x}}{2}$

7. $\frac{d^2y}{dx^2}+y=5x+3$ **Ans :** $y = C_1\cos x + C_2\sin x + (5x+3)$

8. $\frac{d^2y}{dx^2}-3\frac{dy}{dx}+2y=e^{-2x}$ **Ans :** $y = C_1e^{x} + C_2e^{2x} + \frac{e^{-2x}}{2}$

9. $\frac{d^2y}{dx^2}-\frac{dy}{dx}-2y=x$ **Ans :** $y = \left(C_1e^{-x}+C_2e^{2x}-\left(\frac{x}{2}-\frac{1}{4}\right)\right)$

10. $\frac{d^2y}{dx^2}-4\frac{dy}{dx}+4y=x^2+1$ **Ans :** $y = (C_1x + C_2)\, e^{2x} + (4x^2 + 8x + 10)$

11. $(D^2 + a^2)y = \sec ax$ **Ans :** $y = C_1\cos ax + C_2\sin ax + C_2\sin ax + \frac{x\sin ax}{a}+\frac{\cos ax}{a^2}\log\cos ax$

12. $(D^2 + 9)\, y = \tan 3x$ **Ans :** $y = C_1\cos 3x + C_2\sin 3x - \frac{\cos 3x}{9}\log(\sec 3x + \tan 3x)$

17.2.1 Methods of Finding P.I.

We shall now consider the methods for finding P.I. where R is of some special form

Case (i) When R is of the form e^{ax}

(a) if $f(a) \neq 0$

Since $De^{ax} = ae^{ax}$; $D^2e^{ax} = a^2 e^{ax}$ .... $D^n(e^{ax}) = a^n.e^{ax}$

$$f(D)\, e^{ax} = (D^n + K_1D^{n-1} + K_2D^{n-2} + ...K_n)e^{ax}$$

$$= (a^n + K_1a^{n-1} + ... + K_n)e^{an} = f(a)\, e^{ax}$$

Operating with $\frac{1}{f(D)}$ on both sides

$$\frac{1}{f(D)}\left[F(D)e^{ax}\right] = \frac{1}{f(D)}\,[\,f(a)e^{ax}]$$

(i.e.,) $e^{ax} = \frac{1}{f(D)}f(a)\,e^{ax} = f(a).\left[\frac{1}{f(D)}e^{ax}\right]$

(or) $\frac{1}{f(D)}e^{ax} = \frac{1}{f(a)}e^{ax}$

Note: If K is a constant then $\frac{1}{f(D)}(K) = \frac{1}{f(D)}K.e^{ax} = \frac{1}{f(m)}.K$

for example $\frac{1}{D^2-2D+1}e^{3x} = \frac{1}{9-6+1}\,e^{3x} = \frac{1}{4}\,e^{3x}$

(b) If $f(a) = 0$ then it is possible to write $f(D)$ as $\phi(D)\,(D-a)^r$ where $\phi(a) \neq 0$
Suppose r = 1 then

$$P.I. = \frac{1}{f(D)}e^{ax} = \frac{1}{D-a}\frac{1}{\phi(D)}\,e^{ax}$$

$$= \frac{1}{D-a}\,\frac{1}{\phi(a)}\,e^{ax} = \frac{1}{\phi(a)}\,\frac{1}{D-a}\,e^{ax}$$

$$= \frac{1}{\phi(a)}e^{ax}\int e^{-ax}.e^{ax}dx = \frac{1}{\phi(a)}e^{ax}dx = \frac{1}{\phi(a)}e^{ax}.x \quad \text{.... (1)}$$

if r = 2 then $\frac{1}{f(D)}\,e^{ax} = \frac{1}{(D-a)^2}\,\frac{1}{\phi(D)}\,e^{ax} = \frac{1}{\phi(a)}\,\frac{1}{(D-a)^2}\,e^{ax}$

$$= \frac{1}{\phi(a)}\,\frac{1}{D-a}\,[xe^{ax}]$$

$$P.I. = \frac{1}{f(D)}\,e^{ax} = \frac{1}{\phi(a)}\,e^{ax}.\,\frac{x^2}{\angle 2}$$

In general, if (D–a) is repeated 'r' times then

$$P.I. = \frac{1}{f(D)}\,e^{ax} = \frac{1}{\phi(a)}\,e^{ax}.\,\frac{x^r}{r!}$$

Solved Examples

Example **17.15:** Solve $\frac{d^2y}{dx^2} + 4\frac{dy}{dx} + 5y = 13e^{3x}$

***Solution*:**

$(D^2 + 4D + 5) = 13e^{3x}$

$\therefore$ A.E. is $m^2 + 4m + 5 = 0$, giving $m = -2 \pm i$

$\therefore$ C.F. is $y = e^{-2x}(A\cos x + B\sin x)$

$$\text{P.I.} = \frac{1}{D^2+4D+5}(13e^{3x})$$

replace 'D' by '3'

$$= \frac{13e^{3x}}{9+12+5} = \frac{13e^{3x}}{26} = \frac{e^{3x}}{2}$$

$\therefore$ Complete solution is y = C.F. + P.I. $= e^{-2x}(A\cos x + B\sin x) + \frac{e^{3x}}{2}$

Example **17.16:** Solve $\frac{d^2y}{dx^2} + 4\frac{dy}{dx} + 5y = 2e^{-2x}$, given that $x = 0$, $y = 1$ and $\frac{dy}{dx} = -2$

***Solution*:**

Given equation is $f(D)y = 2e^{-2x}$ were $f(D) = D^2 + 4D + 5$

$\therefore$ A.E. is $m^2 + 4m + 5 = 0$, giving $m = -2 \pm i$

$\therefore$ C.F. is $y = e^{-2x}(A\cos x + B\sin x)$

$$\text{P.I.} = \frac{1}{D^2+4D+5}2e^{-2x} = \frac{2e^{-2x}}{4-8+5} = 2e^{-2x}$$

$\therefore$ Complete solution is y = C.F. + P.I. $= e^{-2x}(A\cos x + B\sin x) + 2e^{-2x}$ .... (4)

Now $y(0) = 1$, (given)

Substituting in (4)

$y(0) = 1(A) + 2;\ A = -1.$

$y^1(x) = e^{-2x}(-A\sin x + B\cos x) - 2e^{-2x}(A\cos x + B\sin x) - 4e^{-2x}$

$y^1(0) = B - 2A - 4,\ y^1(0) = -2$ (given)

$\therefore \quad -2 = B - 2A - 4, \Rightarrow B = 0$

$\therefore \quad y(x) = e^{-2x}(-\cos x) + 2e^{-2x}$

$= 2e^{-2x} - e^{-2x}\cos x$

***Example* 17.17:** Solve $\frac{d^3y}{dx^3} + 2\frac{d^2y}{dx^2} + \frac{dy}{dx} = e^{2x}$

Soution:

A.E. is $m^3 + 2m^2 + m = 0$; $m(m+1)^2 = 0$

$\therefore$ m = 0, –1, –1.

C.F. = $C_1e^{0x} + (C_2 + C_3x)e^{-x}$

$$\text{P.I.} = \frac{1}{D^3 + 2D^2 + D}e^{2x} = \frac{e^{2x}}{18}$$

$\therefore$ Complete solution is $y = C_1 + (C_2 + C_3x)e^{-x} + \frac{e^{2x}}{18}$

***Example* 17.18:** Solve $\frac{d^2y}{dx^2} - 3\frac{dy}{dx} + 2y = e^x$

Solution:

$f(D) = D^2 - 3D + 2$

A.E. is $m^2 - 3m + 2 = 0 \Rightarrow (m-2)(m-1) = 0$; m = 1, 2

C.F. is $C_1e^x + C_2e^{2x}$ here f(1) = 0

$$\therefore \frac{1}{(D-1)}\frac{1}{(D-2)}e^x = \frac{1}{(D-1)}\frac{1}{(1-2)}e^x = \frac{1}{(D-1)}e^x = -e^x\frac{x}{1!}$$

$\therefore$ Most general solution is

$\therefore y = C_1e^x + C_2e^{2x} - xe^n$

Case(ii) R is of the form sin ax or cos ax

$D^2(\sin ax) = (-a^2)^2 \sin ax$

$(D^2)^2(\sin ax) = (-a^2)^2 \sin ax$

$(D^2)^3(\sin ax) = (-a^2)^3 \sin ax$

....

....

....

$(D^2)^n (\sin ax) = (-a^2)^n \sin ax$

Hence $f(D^2) \sin ax = f(-a^2) \sin ax$, provided $f(-a^2) \neq 0$

$$\therefore \quad \frac{1}{f(D^2)} f(D^2) \sin ax = \frac{1}{f(D^2)} f(-a^2) \sin ax$$

$$\text{or} \quad \sin ax = \frac{1}{f(D^2)} f(-a^2) \sin ax$$

$$\frac{1}{f(D^2)} \sin ax = \frac{\sin ax}{f(-a^2)}$$

Similarly it can be shown that $\frac{1}{f(-D^2)} \cos ax \quad ax = \frac{1}{f(-a^2)} \cos ax$

$$\text{Thus} \quad \frac{1}{f(D)} [\sin ax \text{ or } \cos ax] = \frac{1}{\phi_1(D^2) + D\phi_2(D^2)} [\sin ax \text{ or } \cos ax]$$

$$= \frac{1}{\phi_1(-a^2) + \phi_2(-a^2)D} [\sin ax \text{ or } \cos ax]$$

If $f(D)$ vanishes when D^2 is replaced by $-a^2$ then we write

$$\frac{1}{f(D)} [\sin ax] = \frac{1}{f(D)} [\text{Imaginary part of } e^{iax}]$$

$$= \text{I.P of } \frac{1}{f(D)} e^{iax} \quad = \frac{-x}{2a} \cos ax = \frac{-x}{2a} \cos ax$$

$$\text{Similarly} \quad \frac{1}{f(D)} [\cos ax] = \text{R.P. of } \frac{1}{f(D)} e^{iax} = \frac{x}{2a} \sin ax = \frac{x}{2a} \sin ax$$

Solved Examples

***Example* 17.19:** Solve $\frac{d^2y}{dx^2} - 3\frac{dy}{dx} + 2y = \sin 3x$

***Solution*:**

A.E. is $m^2 - 3m + 2 = 0$; $m = 1, 2$

$\therefore$ C.F. $= C_1e^x + C_2e^{2x}$

$$\text{P.I. } \frac{1}{D^2-3D+2}(\sin 3x) = \frac{1}{-9-3D+2}(\sin 3x) = \frac{-1}{(3D+7)}(\sin 3x) = \frac{(3D-7)}{9D^2-49}(\sin 3x)$$

$$= \frac{(3D-7)}{-81-49}(\sin 3x) = -\frac{9\cos 3x - 7\sin 3x}{D^2-3D+2}$$

Hence complete solution is

$$y = C_1e^x + C_2e^{2x} - \frac{1}{130}(9\cos 3x + 7\sin 3x)$$

***Example* 17.20:** Solve $(D^3 - D)\,y = \sin x$

***Solution*:**

A.E. is $m^3 - m = 0$

$m = 0, m = \pm 1$

C.F. is $C_1e^{0x} + C_2e^x + C_3e^{-x} = C_1 + C_2e^x + C_3e^{-x}$

$$\text{P.I.} = \frac{1.\sin x}{D^3 - D} = \frac{1}{D^2-1} \cdot \frac{1}{D}(\sin x)$$

$$= -\frac{1}{D^2-1}\,.\cos x = \frac{-\cos x}{-1-1} = -\frac{x}{2}\sin x$$

The complete solution is

$$y = C_1 + C_2e^x + C_3e^{-x} - \frac{x}{2}\sin x$$

***Example* 17.21:** Solve $(D^2 - D + 1)\,y = \cos 2x$

***Solution*:**

A.E. is $m^2 - m + 1 = 0 \Rightarrow m = \dfrac{1 \pm \sqrt{3}i}{2}$

$\therefore$ C.F. $= e^{\frac{1}{2}x}\left(A\cos\dfrac{\sqrt{3}}{2}x + B\sin\dfrac{\sqrt{3}}{2}x\right)$

$$\text{P.I.} = \frac{1}{D^2-D+1}\cos 2x = \frac{1}{-4-D+1}\cos 2x = \frac{1}{-3-D}\cos 2x = -\frac{(3-D)}{9-D^2}\cos 2x$$

$$= -\frac{(3-D)\cos 2x}{13} = -\frac{1}{13}(2\sin 2x + 3\cos 2x)$$

$$\therefore \text{C.F.} + \text{P.I.} = e^{\frac{x}{2}}\left(A\cos\frac{\sqrt{3}}{2}x + B\sin\frac{\sqrt{3}}{2}x\right) - \frac{1}{13}(2\sin 2x + 3\cos 2x)$$

***Example* 17.22:** Solve $D^2(D^2+9) = \sin 2x + 5$

***Solution*:**

A.E. is $m^2(m^2+9) = 0 \Rightarrow m = 0, 0, +3i$

C.F. = $(C_1 + C_2x) + C_3\cos 3x + C_4\sin 3x$

$$\text{P.I.} = \frac{1}{D^2(D^2+9)}\sin 2x + \frac{5.e^{0.x}}{(D^2+9)D^2}$$

$$= \frac{\sin 2x}{-4(-4+9)} + \frac{5}{9}.\frac{1}{D^2}(1) = -\frac{\sin 2x}{20} + \frac{5x^2}{18}$$

Complete solution is $y = (C_1 + C_2x) + C_3\cos_3 x + C_4\sin 3x - \frac{\sin 2x}{20} + \frac{5x^2}{18}$

***Example* 17.23:** Solve $(D^4 - 2D^3 + 2D^2 - 2D + 1)\,y = e^x + \sin^2(x/2)$

***Solution*:**

A.E. is $f(D) = D^4 - 2D^3 + 2D^2 - 2D + 1$

$m^4 - 2m^3 + 2m^2 - 2m + 1 = 0$

$= (m-1)^2(m^2+1) = 0$

$\therefore m = 1, 1, m = i, -i$

C.F. is $(C_1+C_2x)e^x + (C_3\cos x + C_4\sin x)$

$$\text{P.I.} = \frac{1}{f(D)}e^x + \frac{1}{f(D)}\sin^2(x/2)$$

$$\frac{1}{f(D)}e^x = \frac{1}{(D-1)^2(D^2+1)}e^x = \frac{x^2}{2!}\,\frac{e^x}{2}$$

$$\frac{1}{f(D)}\sin^2(x/2) = \frac{1}{(D-1)^2(D^2+1)}\sin^2(x/2)$$

$$= \frac{1}{(D-1)^2(D^2+1)}\left(\frac{1-\cos x}{2}\right)$$

$$= \frac{1}{2}\left[\frac{1}{(D-1)^2(D^2+1)}(1-\cos x)\right] = \frac{1}{2}\left[-1\frac{1}{(D^2-2D+1)(D^2+1)}\cos x\right]$$

$$= \frac{1}{2} - \frac{1}{-4D}.\frac{1}{D^2+1}(\cos x)$$

$$= \frac{1}{2} + \frac{1}{4}\cdot\frac{1}{D^2+1}\cdot\frac{D}{(D^2)}(\cos x)$$

$$= \frac{1}{2}+\frac{1}{4}.\frac{1}{D^2+1}.\frac{D}{(-1)}(-\sin x)$$

$$= \frac{1}{2} + \frac{1}{4}\cdot\frac{1}{D^2+1}\cdot \sin x$$

$$\frac{1}{D^2+1}\sin x = \text{I.P of } \frac{1}{D^2+1}e^{ix} = \text{I.P of } \frac{1}{(D+i)(D-i)}e^{ix}$$

$$= \text{I.P of } \frac{1}{2i}xe^{ix} = \frac{x}{2}\text{ I.P. of } -i[\cos x + i\sin x] = -\frac{x}{2}\cos x$$

$\therefore$ P.I. $= \frac{1}{2} - \frac{1}{8}x\cos x$

$\therefore$ Hence the complete solution is

$$y = (C_1 + C_2x)\,e^x + C_3\cos x + C_4\sin x + \frac{1}{2}.\frac{x^2}{2!}e^x + \frac{1}{2} - \frac{x}{8}\cos x$$

Exercise 17.3

Solve the following

1. $(D^2+4)y = e^x + \sin 2x$ **Ans :** $C_1\cos 2x + C_2\sin 2x - \frac{x}{2}\cos 2x$

2. $(D^2+D+1)y = \sin 2x$

Ans : $e^{\frac{x}{2}}\left(C_1\cos\frac{\sqrt{3}}{2}x + C_2\sin\frac{\sqrt{3}}{2}x - \frac{1}{13}(2\cos 2x + 3\sin 2x)\right)$

3. $(D^2+1)\,y = \sin\frac{3}{2}x.\sin\frac{x}{2}$

Ans: $(C_1\cos x + C_2\sin x) + e^{\frac{\sqrt{3}x}{2}}(C_3\cos\frac{x}{2} + C_4\sin\frac{x}{2}) + e^{\frac{\sqrt{3}x}{2}}(C_5\cos\frac{x}{2} + C_6\sin\frac{x}{2})$

$$+\frac{x}{12}\sin x + \frac{1}{126}\cos 2x$$

4. $(D^3+4D)^2\,y = \text{Sin}2x$ **Ans:** $y = C_1 + C_2\cos 2x + C_3\sin 2x - \frac{x\sin x}{8} - \frac{\cos 2x}{16}$

Case (iii) When 'R' is of the form x^m

$$\frac{1}{f(D)}\,x^m = [f(D)]^{-1}.x^m$$

Now expand $[f(D)]^{-1}$ in ascending powers of D and retain as far as D^m and then operate on x^m

Solved Examples

***Example* 17.24:** Solve $(D^2+5D+6)\,y = x$

Solution:

A.E. is $m^2+5m+6=0 \Rightarrow m=-2,-3$

C.F. is $C_1e^{-2x}+C_2e^{-3x}$

$$\text{P.I.} = \frac{1}{D^2+5D+6}\,x = \frac{1}{6}\left(1+\frac{D^2+5D}{6}\right)^{-1} x$$

$$= \frac{1}{6}\,[1-\frac{5D}{6}]\,x = \frac{1}{6}\,[x-\frac{5}{6}] = \frac{x}{6} - \frac{5}{36}$$

Solution is therefore $y = C_1e^{-2x}+C_2e^{-3x}+\frac{x}{6}-\frac{5}{36}$

***Example* 17.25:** Solve $\frac{d^3y}{dx^3} - 3\frac{dy}{dx} - 2y = x^2$

Solution:

A.E is $m^3-3m-2=0 \Rightarrow (m-2)(m+1)^2=0$

C.F. is $C_1e^{2x} + (C_2 + C_3x)e^{-x}$

$$\text{P.I.} = \frac{x^2}{D^3 - 3D - 2} = -\frac{1}{2}\left(1 - \left(\frac{D^3 - 3D}{2}\right)\right)^{-1} x^2$$

$$= -\frac{1}{2}\left[1 + \frac{D^3 - 3D}{2} + \left(\frac{D^2 - 3D}{2}\right)^2 + \ldots\ldots\right] x^2$$

$$= \frac{-1}{2}\left[1 - \frac{3D}{2} + \frac{9}{4}D^2\right] x^2 = -\frac{1}{2}\left[x^2 - \frac{6x}{2} + \frac{9}{2}\right] = \frac{1}{4}\left[2x^2 - 6x + 9\right]$$

Complete Solution is $y = C_1e^{2x} + (C_2 + C_3x)e^{-x} - \frac{1}{4}(2x^2 - 6x + 9)$

Exercise 17.4

1. Solve $(D^3 - 3D^2 + 2)y = x$

Ans : $(C_1+C_2x)e^{x} + C_3e^{-2x} + \frac{1}{2}\left(x + \frac{3}{2}\right)$

2. Solve $(D^2 + 2D + 3)y = x + x^2$

Ans : $y = e^{-x}\left(C_1 \cos\sqrt{2}x + C_2 \sin\sqrt{x}\right) + \frac{14 - 6x}{27}$

3. Solve $(D^2 + 2D + 1)\,y = \cos 2x + x^2$

Ans : $y = (C_1x + C_2e^{-x} - \frac{1}{13}(3\cos 2x - 2\sin 2x) + (x^2 - 2x)$

4. $(D^2 - 4D + 4)\,y = x^2 + \sin 2x + e^{3x}$

Ans : $y = (C_1 + C_2x)e^{2x} + \frac{2x^2 + x + 3}{8}\ \frac{\cos 2x}{8} + e^{3x}$

Case (iv) To find $\frac{1}{f(D)}$ $(e^{ax}V)$ where V is a function of x

$$D(e^{ax}V_1) = e^{ax}D(V_1) + ae^{ax}V_1 = e^{ax}(D+a)V_1$$

Similarly $\quad D^2(e^{ax}V_1) = e^{ax}(D+a)^2V_1$

$$D^3(e^{ax}V_1) = e^{ax}(D+a)^3V_1$$

....

$$D_n(e^{ax}V_1) = e^{ax}(D+a)^nV_1$$

$\therefore \qquad f(D)\,(e^{ax}V_1) = e^{ax}f(D+a)V_1$

Calling $\qquad f(D+a)V_1 = V$, we get $V_1 = \dfrac{1}{f(D+a)}V$

$\therefore \qquad f(D)\,[e^{ax}\dfrac{1}{f(D+a)}V] = e^{ax}V$ or $e^{ax}\dfrac{1}{f(D+a)}V = \dfrac{1}{f(D)}e^{ax}V$

$\therefore \qquad \dfrac{1}{f(D)}e^{ax}V = e^{ax}\dfrac{1}{f(D+a)}V$

Note :

1. While finding $\dfrac{1}{f(D)}e^{ax}$ when $f(a) = 0$ we can employ the above method for

$$\frac{1}{f(D)}e^{ax} = \frac{1}{f(D)}[e^{ax}\,.\,e^{0.x}] = e^{ax}\,.\,\frac{1}{f(D+a)}[e^{0.x}]$$

2. Further $\dfrac{1}{f(D^2)}$ [cos ax or sin ax] when $f(-a^2) = 0$ we can write

$$\frac{1}{f(D^2)}[\cos ax \text{ or } \sin ax]$$

$$= \frac{1}{f(D^2)}[\text{R.P. or I.P of } e^{iax}] = \text{R.P. or I.P. of } e^{iax}\,.\,\frac{1}{f(D+ia)^2}.1$$

Solved Examples

***Example* 17.26:** Solve $\dfrac{d^2y}{dx^2} - 2\dfrac{dy}{dx} + 4y = e^x\cos x$

Solution:

A.E. is $m^2 - 2m + 4 = 0$

$m = 1 \pm \sqrt{3}i$

C.F. is $e^x(C_1\cos\sqrt{3}x + C_2\text{Sin}\sqrt{3}x)$

$$\text{P.I.} = \frac{1}{D^2-2D+4}e^x\cos x = e^x\frac{1}{(D+1)^2-2(D+1)+4}\cos x$$

$$= e^x\frac{1}{D^2+2D+1-2D-2+4}(\cos x)$$

$$= e^x \frac{1}{D^2+3}(\cos x) = e^x.\frac{\cos x}{-1^2+3} = \frac{e^x \cos x}{2}$$

$\therefore$ Complete solution is y = C.F. + P.I.

$$y = e^x(C_1 \cos\sqrt{3}x + C_2 \sin\sqrt{3}x) + \frac{e^x \cos x}{2}$$

***Example* 17.27:** Solve $(D^2 - 2)y = e^x\cos x$

***Solution*:**

A.E. is $m^2 - 2 = 0$ $m = \pm\sqrt{2}$

$\therefore$ C.F. is $C_1e^{\sqrt{2}x} + C_2e^{-\sqrt{2}x}$

$$\text{P.I.} = \frac{1}{(D^2-2)} e^x \cos x = e^x . \frac{1}{(D+1)^2 - 2} \cos x = e^x.\frac{1}{D^2+2D-1} . \cos x$$

$$= e^x . \frac{1}{2D-2} . \cos x$$

$$= e^x . \frac{(2D+2)}{4D^2-4} \cos x = \frac{e^x(-2\sin x + 2\cos x)}{-4-4} = \frac{e^x}{4}(\sin x - \cos x)$$

$$\therefore \quad y = C_1e^{\sqrt{2}x} + C_2e^{-\sqrt{2}x} + \frac{e^x}{4}(\sin x - \cos x)$$

***Example* 17.28:** Solve $\frac{d^3y}{dx^3} - 7\frac{dy}{dx} - 6y = e^{2x} + xe^{2x}$

***Solution*:**

A.E. is $m^3 - 7m - 6 = 0$ $\therefore$ $m = -1, 3, -2$

$\therefore$ C.F. $= C_1e^{-x} + C_2e^{3x} + C_3e^{-2x}$

$$\text{P.I.} = \frac{(1+x)}{D^3-7D-6}e^{2x} = e^{2x}\frac{1}{(D+2)^3 - 7(D+2) - 6}(1+x)$$

$$= e^{2x}\frac{1}{D^3 - 6D^2 + 5D - 12} = \frac{-e^{2x}}{12}\left[1 - \left(\frac{D^3 - 6D^2 + 5D}{12}\right)\right]^{-1}(1+x)$$

$$= \frac{-e^{2x}}{12}(1 + \frac{5D}{12})(1+x) = \frac{-e^{2x}}{12}[1 + x + \frac{5}{12}] = \frac{-e^{2x}}{12}[x + \frac{17}{12}]$$

$$\therefore \quad y = C_1e^{-x} + C_2e^{3x} + C_3e^{-2x} - \frac{e^{2x}}{12}[x + \frac{17}{12}]$$

***Example* 17.29:** Solve $\frac{d^3y}{dx^3} - 3\frac{d^2y}{dx^2} + 3\frac{dy}{dx} - y = e^x(1+x)$

***Solution*:**

A.E. is $m^3 - 3m^2 + 3m - 1 = 0$ or $(m-1)^3 = 0$ or $m = 1, 1, 1$.

C.F. – $(C_1 x^2 + C_2 x + C_3)e^x$

$$\text{P.I.} = \frac{1}{(D-1)^3} \cdot (x+1)\, e^x = \frac{e^x.(x+1)}{(D+1-1)^3}$$

$$= e^x . \frac{1}{D^3}(x+1) = e^x . \frac{1}{D^2}\left(\frac{x^2}{2} + x\right)$$

$$= e^x \frac{1}{D}\left(\frac{x^3}{6} + \frac{x^2}{2}\right) = e^x\left(\frac{x^4}{24} + \frac{x^3}{6}\right)$$

Complete solution is $y = (C_1 x^2 + C_2 x + C_3)\, e^x + e^x\left(\frac{x^4}{24} + \frac{x^3}{6}\right)$

***Example* 17.30:** Solve $\frac{d^2y}{dx^2} - 4\frac{dy}{dx} + y = e^{2x}\sin 2x$

***Solution*:**

A.E. is $m^2 - 4m + 1 = 0$, $m = 2 \pm \sqrt{3}$

$$\text{C.F.} = C_1 e^{(2+\sqrt{3})x} + C_2 e^{(2-\sqrt{3})x}$$

$$\text{P.I.} = \frac{1}{D^2 - 4D + 1}\, e^{2x}\sin 2x = \frac{e^{2x}}{(D+2)^2 - 4(D+2) + 1}\sin 2x$$

$$= \frac{e^{2x}}{D^2 - 3}\sin 2x = \frac{e^{2x}\sin 2x}{-4-3} = -\frac{1}{7}\, e^x \sin 2x$$

Complete solution is

$$y = \text{C.F.} + \text{P.I.} = C_1 e^{(2+\sqrt{3})x} + C_2 e^{(2-\sqrt{3})x} - \frac{1}{7} e^x \sin 2x$$

Exercise 17.5

1. Solve $\frac{d^2y}{dx^2} - 2\frac{dy}{dx} + y = x^2 e^{3x}$

Ans : $y = (C_1 + C_2 x)\, e^x + \frac{e^{3x}}{8}(2x^2 - 4x + 3)$

2. Solve $(D^3 - 7D - 6)\, y + e^{2x} + x^2e^{3x}$

$$\textbf{Ans : } y = C_1e^{-x} + C_2e^{-2x} + C_3e^{3x} - e^{2x}\frac{1}{12}\left(x^2 + \frac{5}{6}x + \frac{169}{72}\right)$$

3. Solve $(D^2 - 2D + 4)y = e^x\cos^2 x$

$$\textbf{Ans : } y = e^x(C_1\cos 3\sqrt{x} + C_2\sin 3\sqrt{x}) + \frac{1}{8}e^x - \frac{1}{2}\sin 2x$$

4. Solve $\dfrac{d^2y}{dx^2} + y = e^x\sin x$

$$\textbf{Ans : } y = C_1\cos x + C_2\sin x - \frac{1}{5}e^x(2\cos x - \sin x)$$

5. Solve $\dfrac{d^2y}{dx^2} + y = e^{2x}\sin x$

$$\textbf{Ans : } y = C_1e^{-x} + e^{x/2}\left(C_2\cos\frac{\sqrt{3}}{2}x + C_3\sin\frac{\sqrt{3}}{2}x\right) - \frac{1}{130}e^{2x}.(11\cos x - 3\sin x)$$

Case (v) To find $\dfrac{1}{f(D)}[xv]$ where v is a function of x

$D(xv_1) = xDv_1 + v_1$

$D^2(xv_1) = D(xDv_1 + v_1) = xD^2v_1 + 2Dv_1 = xD^2v_1 + (D^2)'\, v_1$

Similarly $D^3(xv_1) = xD^3v_1 + (D^3)'\, v_1$

in general $D^n((xv_1) = xD^nv_1 + (D^n)'\, v_1$

(here 1 indicates differentiation w.r.t 'D')

Thus, $f(D)(xv_1) = x_1f(D)v_1 + f^1(D)v_1$ .....(1)

now let $f(D)v_1 = v$

$$v_1 = \frac{1}{f(D)}.v$$

From (1) $f(D)\left[x.\dfrac{1}{f(D)}v\right] = xv + f^1(D)\left[\dfrac{1}{f(D)}v\right]$

Operating both sides with $\dfrac{1}{f(D)}$

$$x\frac{1}{f(D)}v = \frac{1}{f(D)}(xv) + \frac{1}{f(D)}.f^1(D)\left[\frac{1}{f(D)}.v\right]$$

$$\frac{1}{f(D)}(xv) = x\,.\frac{1}{f(D)}v - \frac{1}{f(D)}\,.\,[f^1(D).\frac{1}{f(D)}.v]$$

$$\frac{1}{f(D)}(xv) = [x - \frac{1}{f(D)}f^1(D)]\;[\frac{1}{f(D)}v]$$

Solved Examples

***Example* 17.31:** Solve $(D^2 + 9)\,y = x\sin x$

Solution:

A.E. is $m^2 + 9 = 0 \Rightarrow m = \pm\, 3i$ $\quad\therefore$ CF $= c_1\cos 3x + c_2 \sin 3x$

$$\text{P.I.} = \frac{1}{D^2+9}(x\sin x) = \{x - \frac{1}{D^2+9}.2D\}.\frac{1}{D^2+9}.\sin x$$

[using the result $\frac{x.V}{(D)} = \frac{x.L.v}{f(D)} = f'(D)\left[\frac{1}{[f(D)]^2}\right]V$]

$$= \{x - \frac{1}{D^2+9}2D\}\,.\,\frac{1}{8}\sin x = \frac{x}{8}\sin x - \frac{1}{D^2+9}\,\frac{2}{8}\cos x$$

$$= \frac{x}{8}\sin x - \frac{2}{8}\,.\,\frac{1}{8}\,.\cos x = \frac{x}{8}\sin x - \frac{1}{32}\cos x$$

Complete solution is $y = C_1\cos 3x + C_2\sin 3x + \frac{x}{8}\sin x - \frac{1}{32}\cos x$

***Example* 17.32:** Find the solution of $\frac{d^2y}{dx^2} + 9y = x\cos 2x$

Solution:

A.E. is $m^2 + 9 = 0 \Rightarrow m = \pm\, i\sqrt{3}$

C.F.is $y = C_1\cos\sqrt{3}\,x + C_2\sin\sqrt{3}\,x$

and $\quad$ P.I. $= \frac{1}{D^2+9}(x\cos 2x)$

$$= \text{Real part of } \frac{1}{D^2+9}\,x(\cos 2x + i\sin 2x)$$

$$= \text{R.P. of } \frac{1}{D^2+9}\,xe^{2ix} = \text{R.P. of } e^{2ix}\,\frac{1}{(D+2i)^2+9}\,x = \text{R.P. of } e^{2ix}\,.\frac{1}{D^2+4iD+5}\,x$$

$$= \text{R.P of } \frac{e^{2ix}}{5}\left(1+\frac{D^2+4iD}{5}\right)^{-1} x = \text{R.P. } \frac{e^{2ix}}{5}\left(1-\frac{D^2+4iD}{5}\right) x$$

$$= \text{R.P. of } \frac{e^{2ix}}{5}\ [1-\frac{4iD}{5}-\frac{D^2}{5}+---]x$$

$$= \text{R.P. of } \frac{e^{2ix}}{5}\ [x-\frac{4i}{5}]$$

$$= \frac{1}{5}\ [x\cos 2x + \frac{4}{5}\ \sin 2x]$$

Most general solution is

$$y = C_1\cos 3x + C_2\sin 3x + \frac{1}{5}\ [x\cos 2x + \frac{4}{5}\ \sin 2x]$$

***Example* 17.33:** Solve $(D^2 - 4D + 4)y = 8e^{2x}\, x^2\sin 2x$

***Solution*:**

A.E is $m^2 - 4m + 4 = 0 \Rightarrow m = 2, 2$

C.F. is $(C_1x + C_2)e^{2x}$

$$\text{P.I.} = \frac{8}{(D-2)^2}\ e^{2x}\,.\, x^2\sin 2x = 8e^{2x}\ \frac{1}{(D+2-2)^2}\ x^2\sin 2x = 8e^{2x}\frac{1}{D^2}x^2\sin 2x$$

$$= 8e^{2x}\ [\text{I.P of } \frac{1}{D^2}x^2\ e^{2ix}] = 8e^{2x}[\text{I.P. of } e^{2ix}.\ \frac{1}{(D-2i)}x^2]$$

$$= 8e^{2x}[\text{I.P of } e^{2ix}\ \frac{1}{-4+4iD+D^2}x^2\ ]$$

$$= 8e^{2x}\ [\text{ I.P of } e^{2ix}\ (\frac{-1}{4})\ [1-\frac{4iD+D^2}{4}]^{-1}\ x^2$$

$$= -2e^{2x}[\text{I.P. of } e^{2ix}\ 1-\frac{4iD+D^2}{4}\ ]^{-1}\ x^2$$

$$= -2e^{2x}\ [\text{I.P. of } e^{2ix}\ [1+\frac{4iD+D^2}{4}+(\frac{4iD+D^2}{4})^2]x^2$$

$$= -2e^{2x}\ [\text{I.P of } e^{2ix}\ [x^2+\frac{8ix+2}{4}+\frac{16i^2D^2}{4}(x^2)]$$

$$= -\frac{e^{2x}}{2} \text{ [I.P. of } (\cos 2x + i\sin 2x)(4x^2 + 8ix - 30)$$

$$= \frac{-e^{2x}}{2} \text{ .I.P. of } [4x^2\cos 2x + 8ix\cos 2x - 30\cos 2x + i4x^2\sin 2x - 8x\sin 2x - 30i\sin 2x]$$

$$= -e^{2x}[4x\cos 2x + 2x^2\sin 2x - 15\sin 2x]$$

General solution is y = C.F. + P.I.

$$y = (C_1x + C_2)e^{2x} - e^{2x}[2x^2\sin 2x + 4x\cos 2x - 15\sin 2x]$$

Exercise 17.6

1. Solve $(D^2 - 1)\, y = x^2\cos 3x$

 Ans: $y = C_1e^x + C_2e^{-x} - \frac{1}{10}\left(x^2\cos 3x - \frac{26}{50}\cos 3x - \frac{6}{5}x\sin 3x\right)$

2. Solve $(D^2 + 4)y = x\sin^2 x$

 Ans: $y = C_1\cos 2x + C_2\sin 2x + \frac{x}{4}\left[\frac{1}{2} + \sin 2x\right]$

3. Solve $(D^4 + 2D^2 + 1)y = x^2\cos x$

 Ans: $y = (C_1x + C_2)\cos x + (C_3x + C_4)\sin x - \frac{1}{48}(x^4 - 9x^2)\cos x + \frac{1}{12}x^3\sin x$

4. Solve $(D^2 - 2D + 1)y = xe^x\sin x$

 Ans: $C_1\cos 3x + C_2\sin 3x + \frac{1}{5}(x\cos 2x + \frac{4}{5}\sin 2x)]$

5. Solve $(D^2 - 6D + 13)y = 8e^{3x}\sin 2x$

 Ans: $y = e^{3x}[C_1\cos 2x + C_2\sin 2x] - 2e^{3x}\, x\cos 2x$

6. Solve $(D^4 + D^2 + 1)y = e^{\frac{-x}{2}}\cos\left(\frac{x\sqrt{3}}{2}\right)$

 Ans: $y = e^{\frac{-x}{2}}\left[C_1\cos\left(\frac{1}{2}x\sqrt{3}\right) + C_2\sin\left(\frac{1}{2}x\sqrt{3}\right)\right] + e^{\frac{-x}{2}}\left[C_3\cos\left(\frac{1}{2}\sqrt{3}\right)\right.$

 $\left. + C_4\sin\left(\frac{1}{2}x\sqrt{3}\right)\right] + \frac{e^{\frac{-x}{2}}}{4}\left[x\cos\left(\frac{1}{2}x\sqrt{3}\right) + \frac{1}{\sqrt{3}}x\sin\left(\frac{1}{2}x\sqrt{3}\right)\right]$

7. Solve $\frac{d^2y}{dx^2} + y = e^x\sin x$

 Ans: $C_1\cos x + C_2\sin x - \frac{1}{5}e^x(2\cos x - \sin x)$

17.3 Cauchy's Homogenious Linear Equation

Most general form of the homogeneous equation is

$$(x^nD^n + K_1x^{n-1} D^{n-1} + ... + K_{n-1} xD)y + = X \quad .... (1)$$

Where $k_1, k_2, ... k_n$ are constants and 'X' is either a constant or a function of x

To solve such equations it is convenient to transform them into linear equations with constant coefficients with the substitution $x = e^z$ i.e., $z = \log x$.

$$\frac{dy}{dx} = \frac{dy}{dz} \cdot \frac{dz}{dx} = \frac{1}{x}\frac{dy}{dz} \quad .... (2)$$

$$\frac{d^2y}{dx^2} = \frac{d}{dx}\left(\frac{dy}{dx}\right) = \frac{1}{x}\frac{d}{dz}\left(\frac{1}{x}\frac{dy}{dz}\right) = \frac{1}{x}\left(\frac{-1}{x^2}\frac{dx}{dz} \cdot \frac{dy}{dz} + \frac{1}{x}\frac{d^2y}{dz^2}\right)$$

$$= \frac{1}{x}\left[-\frac{1}{x}\frac{dy}{dz} + \frac{1}{x}\frac{d^2y}{dz^2}\right] \qquad (\because \frac{dx}{dz} = x)$$

$$= \frac{1}{x^2}\left(\frac{d^2y}{dz^2} - \frac{dy}{dz}\right) \quad .....(3)$$

Similarly $$\frac{d^3y}{dx^3} = \frac{1}{x^3}\left(\frac{d^3y}{dz^3} - 3\frac{d^2y}{dz^2} + 2\frac{dy}{dz}\right) \quad ..... (4)$$

By taking 'D' to stand for $\frac{d}{dz}$,

$$(2), (3), (4) \Rightarrow x\frac{dy}{dx} = \frac{dy}{dz} = Dy$$

$$x^2\frac{d^2y}{dx^2} = D(D-1)y$$

$$x^3\frac{d^3y}{dx^3} = D(D-1)(D-2)y \quad ..... (5)$$

$$x^n\frac{d^ny}{dx^n} = D(D-1)(D-2) .... (D-n+1)y$$

By substituting (5) in the given equation (1) it transforms into a linear differential equation with constant coefficients.

Solved Examples

***Example* 17.34:** Solve $x^2\frac{d^2y}{dx^2} - x\frac{dy}{dx} + 4y = 0$

Solution:

Writing $x = e^z$ (i) reduces to

$$[D(D-1) - D + 4]y = 0, \text{ where } D = \frac{d}{dz}$$

$$(D^2 - 2D + 4)y = 0$$

A.E. is $m^2 - 2m + 4 = 0 \Rightarrow m = 1 \pm \sqrt{3}\,i$

$\therefore$ C.F. is given by $y = \left(C_1 \cos\sqrt{3}z + C_2 \sin\sqrt{3}z\right)e^z$

substituting $z = \log x$,

$$y = C_1\cos\left(\sqrt{3}\log x\right) + C_2\sin\left(\sqrt{3}\log x\right) e^{\log x}$$

$$y = [C_1\cos\left(\sqrt{3}\log x\right) + C_2\sin\left(\sqrt{3}\log x\right)]x$$

***Example* 17.35**: Solve $x^2\dfrac{d^2y}{dx^2} - 3x\dfrac{dy}{dx} - 4y = 2x^2$ .....(i)

Solution:

Writing $x = e^z$ (i) reduces to

$$[D(D-1) + 3D - 4]y = 2e^{2z},$$

$$(D^2 - D + 3D - 4)y = 2e^{2z}$$

A.E. is $m^2 + 2m - 4 = 0 \Rightarrow m = -1 \pm \sqrt{5}$

$\therefore$ C.F. is $y = \left(C_1 e^{-(1+\sqrt{5})z} + C_2 e^{-(1-\sqrt{5})z}\right)$

$$\text{P.I.} = \frac{2}{D^2 + 2D - 4}(e^{2z}) = \frac{2}{4+4-4}e^{2z} = \frac{2e^{2z}}{4} = \frac{e^{2z}}{2}$$

Most general solution is

$$y = \left(C_1 e^{-(1+\sqrt{5})z} + C_2 e^{-(1-\sqrt{5})z} + \frac{e^{2z}}{2}\right)$$

$$= \left(C_1 e^{-(1+\sqrt{5})\log x} + C_2 e^{-(1-\sqrt{5})\log x} + \frac{e^{2\log x}}{2}\right)$$

$$= C_1\frac{1}{x^{(1+\sqrt{5})}} + \frac{C_2}{x^{(1-\sqrt{5})}} + \frac{x^2}{2}$$

***Example* 17.36**: Solve $x^2\dfrac{d^2y}{dx^2} - 2y = x^2 + \dfrac{1}{x}$

Solution:

Taking $x = e^z$ given equation reduces to

$[D(D-1)-2]y = e^{2x} + e^{-z}$

A.E. is $m^2-m-2=0 \Rightarrow m=2,-1$

$\therefore$ C.F. is $y = C_1e^{2x} + C_2e^{-z}$

$$\text{P.I.} = \frac{1}{(D-2)(D+1)}e^{2z} + \frac{1}{(D-2)(D+1)}e^{-z}$$

$$= \frac{1}{(D-2)(2+1)}e^{2z} + \frac{1}{(-1-2)(D+1)}e^{-z}$$

$$= \frac{1}{3(D-2)}e^{2z} - \frac{1}{3(D+1)}e^{-z}$$

$$= e^{2z}\frac{1}{3(D+2-2)}(1) - e^{-3}\frac{1}{3(D-1+1)}(1)$$

$$= \frac{1}{3}e^{2z}\cdot\frac{1}{D}(1) - \frac{1}{3}e^{-z}\frac{1}{D}(1)$$

$$= \frac{1}{3}ze^{2x} - \frac{1}{3}ze^{-z}$$

Complete solution C.F + P.I.

i.e. $$y = C_1e^{2z} + C_2e^{-z} + \frac{1}{3}z(e^{2x} - e^{-z})$$

$$y = C_1e^{2\log x} + C_2e^{-\log x} + \frac{1}{3}(e^{2\log x} - e^{-\log x})\log x$$

$$y = C_1x^2 + C_2\frac{1}{x} + \frac{1}{3}(x^2 - \frac{1}{x})\log x$$

***Example* 17.37:** Solve $x^3\dfrac{d^3y}{dx^3} + 2x^2\dfrac{d^2y}{dx^2} + 2y = \left(10x + \dfrac{10}{x}\right)$

***Solution*:**

Taking $x = e^z$ the equation reduces to

$$[D(D-1)(D-2) + 2D(D-1) + 2]y = 10e^z + 10e^{-z}$$

$$(D^3 - D^2 + 2)y = 10(e^z + e^{-z})$$

A.E. is $m^3 - m^2 + 2 = 0$

$\Rightarrow (m+1)(m^2-2m+2) = 0$

(i.e.,) $m = -1, m = 1 \pm i$

$\therefore$ C.F. is $y = C_1 e^{-z} + e^z(C_2\cos z + C_2 \sin z)$

$$\text{P.I.} = \frac{10}{D^3 - D^2 + 2} e^z + \frac{10}{D^3 - D^2 + 2} e^{-z}$$

$$= \frac{10e^z}{1-1+2} + \frac{10e^{-z}}{(D+1)(D^2 - 2D + 2)}$$

$$= 5e^{-z} + \frac{10}{5(D+1)} e^{-z} = 5e^{-z} + 2e^{-z} . \frac{1}{(D-1)+1} 1$$

$$= 5e^{-z} + 2e^{-z} \frac{1}{D}(1) = 5e^{-z} + 2e^{-z} . z$$

Complete solution = C.F. + P.I.

$$= C_1 e^{-z} + e^z(C_2\cos z + C_3 \sin z) + 5e^{-z} + 2e^{-z} . z \frac{\log x}{x}$$

$$= \frac{C_1}{x} + x(C_2\cos(\log x) + C_3\sin(\log x)) + 5x + 2$$

***Example* 17.38:** Solve $x^2 \dfrac{d^2y}{dx^2} + x\dfrac{dy}{dx} + y = \log x \sin(\log x)$

***Solution*:**

Writing $z = \log x$ the equation reduces to

$$[D(D-1) + D + 1]y = z\sin z$$

$$(D^2 + 1)y = z\sin z$$

A.E. is $m^2 + 1 = 0 \Rightarrow m = \pm i$

$\therefore$ C.F. is $y = C_1\cos z + C_2 \sin z$

$$\text{P.I.} = \frac{1}{D^2 + 1} z\sin z$$

$$= \text{I.P. of } \frac{1}{D^2+1} ze^{iz} = \text{I.P. of } e^{iz} \frac{1}{(D+i)^2 + 1} z$$

$$= \text{I.P. of } e^{iz} \frac{1}{D^2 - 1 + 2Di + 1} z$$

$$= \text{I.P. of } e^{iz} \frac{1}{D^2 + 2Di} z$$

$$= \text{I.P of } e^{iz}\ \frac{1}{2Di}\left(1+\frac{D}{2i}\right)^{-1} z$$

$$= \text{I.P. of } e^{iz}\ \frac{1}{2Di}\left(1-\frac{D}{2i}\right) z$$

$$= \text{I.P. of } e^{iz}\ \frac{1}{2Di}\left(z-\frac{1}{2i}\right)$$

$$= \text{I.P of } \frac{-e^{iz}.i}{2}\left(\frac{z^2}{2}-\frac{1}{2i}z\right)$$

$$= \text{Imaginary part of } \left[\frac{-i}{2}(\cos z + i\sin z)\left(\frac{z^2}{2}+\frac{iz}{2}\right)\right]$$

$$= -\frac{z^2}{2}\cos z - \frac{1}{4} z\sin z = -\frac{z}{4}(\sin z + 2z\cos z)$$

Hence complete solution is

$$y = C_1\cos z + C_2\sin z - \frac{z}{4}(\sin z + 2z\cos z)$$

$$= C_1\cos(\log x) + C_2\sin(\log x) - \frac{\log x}{4}\ [\sin(\log x) + 2(\log x)\cos(\log x)]$$

***Example* 17.39:** Solve$(x^2D^2 - xD + 4)y = \cos(\log x) + x\sin(\log x)$

***Solution*:**

Taking $x = e^z$ or $z = \log x$ the equation reduces to

$\therefore \quad [D(D-1) - D + 4]y = \cos z + e^z \sin z.$

A.E. is $m^2 - 2m + 4 = 0 \Rightarrow m = 1 \pm \sqrt{3}i$

$\therefore \quad \text{C.F.} = e^z (C_1\cos\sqrt{3}z + C_2\sin\sqrt{3}z)$

$$\text{P.I.} = \frac{1}{D^2-2D+4}\cos z + \frac{1}{D^2-2D+4} e^z \sin z$$

$$= \frac{1}{3-2D}.\cos z + \frac{e^z}{(D+1)^2-2(D+1)+4}\sin z$$

$$\frac{(3+2D)}{9+4}\cos z + \frac{e^z\sin z}{3} = \frac{3\cos z - 2\sin z}{13} + \frac{e^z\sin z}{3}$$

The complete solution is y = C.F. + P.I.

$$y = e^{z}(C_1\cos\sqrt{3}z + C_2\sin\sqrt{3}z) + \frac{1}{13}(3\cos z - 2\sin z) + \frac{e^{z}\sin z}{3}$$

$$= x[C_1\cos\sqrt{3}(\log x) + C_2\sin\sqrt{(\log x)}] + \frac{1}{13}[3\cos(\log x) - 2\sin(\log x)] + \frac{x\sin(\log x)}{3}$$

17.4 Legendre's Linear Equation

An equation of the form

$$(ax+b)^n\frac{d^n y}{dx^n} + K_1(ax+b)^{n-1}\frac{d^{n-1}y}{dx^{n-1}} + \ldots\ldots + k_n y = X$$

Where K's are constants and X is either a constant or a function of x is called Legendre's linear equation. To solve such an equation we transform it into linear equation with constant coeficients by the substitution.

$$ax + b = e^z \text{ or } z = \log(ax+b)$$

Now $\frac{dy}{dx} = \frac{dy}{dz}\cdot\frac{dz}{dx} = \frac{a}{ax+b}\cdot\frac{dy}{dz}$

$\therefore\quad (ax+b)^2\frac{dy}{dx} = a\frac{dy}{dz} = aDy \qquad$ where $D = \frac{d}{dz}$

$$\frac{d^2y}{dx^2} = \frac{d}{dx}\left(\frac{a}{ax+b}\frac{dy}{dz}\right) = \frac{a^2}{(ax+b)^2}\cdot\frac{dy}{dz} + \frac{a^2}{(ax+b)^2}\cdot\frac{d^2y}{dz^2}$$

$$(ax+b)^2\frac{d^2y}{dx^2} = a^2\left(\frac{d^2y}{dz^2} - \frac{dy}{dz}\right) = a^2(D(D-1)y$$

Similarly $(ax+b)^3\frac{d^3y}{dx^3} = a^3 D(D-1)(D-2)y$

----

by substituting these values, the given equation reduces to linear equation with constant coefficients.

Solved Examples

***Example* 17.40:** Solve $(1+x)^2\frac{d^2y}{dx^2} + (1+x)\frac{dy}{dx} + y = \sin 2[\log(1+x)]$ .... (1)

***Solution*:**

Write $1 + x = e^z \Rightarrow \log(1+x) = z$

$(1+x)\frac{dy}{dx} = Dy,\ (1+x)^2\frac{d^2y}{dx^2} = D(D-1)y \qquad$ where $D = \frac{d}{dz}$

Substituting these values (1) reduces to

$$(D^2 - D + D + 1)y = \sin 2z \; ; \; \text{(i.e)} \; (D^2 + 1)y = \sin 2z$$

A.E. is $m^2 + 1 = 0 \Rightarrow m = \pm i$

$$\text{C.F.} = C_1 \cos z + C_2 \sin z$$

$$\text{P.I.} = \frac{1}{D^2+1} \sin 2z = \frac{\sin 2z}{-4+1} = \frac{\sin 2z}{-3}$$

Complete solution is = C.F + P.I = $C_1 \cos z + C_2 \sin z - \dfrac{\sin 2z}{3}$

$$= C_1 \cos[\log(1+x)] + C_2 \sin[\log(1+x)] - \frac{\sin 2(\log(1+x))}{3}$$

Example **17.41:** Solve $(2x+3)^2 \dfrac{d^2y}{dx^2} + 6(2x+3)\dfrac{dy}{dx} + 6y = \log(2x+3)$

***Solution*:**

Writing $2x+3 = e^z$ or $z = \log(2x+3)$

the given equation reduces to $[2^2.D(D-1) + 6.2D + 6]y = z$

i.e. $\quad (4D^2 + 8D + 6)y = z \qquad$ where $D = \dfrac{d}{dz}$

A.E. is $4m^2 + 8m + 6 = 0 \Rightarrow 2m^2 + 4m + 3 = 0$

$$\Rightarrow m = \frac{-2 \pm \sqrt{2}i}{2} = -1 + \frac{1}{\sqrt{2}} i$$

C.F. is $y = e^{-z} [A\cos \frac{1}{\sqrt{2}} z + B\sin \frac{1}{\sqrt{2}} z]$

$$\text{P.I.} = \frac{1}{4D^2+8D+6} z = \frac{1}{6} \frac{1}{1+\frac{4D}{3}+\frac{2D^2}{3}} .z = \frac{1}{6} [1+\frac{4D}{3}+\frac{2D^2}{3}]^{-1} z$$

$$= \frac{1}{6} [1-\frac{4D}{3}] z = \frac{1}{6} [z-\frac{4}{3}] = \frac{z}{6} - \frac{2}{9}$$

Complete solution is y = C.F. + P.I.

$$= e^{-z} [A\cos \frac{1}{\sqrt{2}} z + B \sin \frac{1}{\sqrt{2}} z] + \frac{z}{6} - \frac{2}{9}$$

$$= \frac{1}{2x+3} [A\cos(\frac{1}{\sqrt{2}} \log(2x+3)) + B\sin(\frac{1}{\sqrt{2}} \log(2x+3))] + \frac{1}{6} \log(2x+3) - \frac{2}{9}$$

Exercise 17.7

Solve the following differential equations

1. $(x^2D^2 - 3xD + 4)\, y = 2x^2$

Ans: $(C_1 + C_2\log x)x^2 + x^2 (\log x)^2$

2. $(x^2D^2 - 3xD + 5)y = \sin(\log(x))$

Ans: $x^2(C_1\cos\log x + C_2\sin\log x) + (\cos\log x + \sin\log x)$

3. $(x^2D^2 - xD + 2)y = x\log x$

Ans: $x(C_1\cos\log x + C_2\sin\log x) + x\log x$

4. $(x^2D^2 - 3xD + 5)y = x^2 \sin\log x$

Ans: $x^2(C_1\cos\log x + C_2\sin\log x) - \frac{1}{2} x^2 \log x\cos\log x$

5. $(x^2D^2 - 3xD + 1)y = \dfrac{(\sin\log x)+1}{x}$

Ans: $x^2[C_1\cosh(\sqrt{3}\log x) + C_2\sinh(\sqrt{3}\log x) + \frac{1}{6x} + \frac{1}{61x}[5\sin(\log x) - 6\cos(\log x)]$

6. $(x^3D^3 + x^2D^2)y = x^2 + x + 1$

Ans: $C^1+(C^2+C^3\log x) + x + \frac{(\log x)^2}{2} . x + \frac{x^2}{2}$

7. $(x^4D^4 + 2x^3D^3 + x^2D^2 - xD+1)y = x + \log x$

Ans: $y = x[C_1+C_2\log x + C_3(\log x)^2 + C_4(\log x)^3] + \frac{(\log x)^4}{4!} . x^4 + \log x + 4$

8. $(x^2D^2 + xD + 1)y = (\log x)^2 + \sin(\log x)$

Ans: $C_1\cos\log x + C_2\sin\log x + (\log x)^2 - x[2\cos\log x - \sin\log x]$

9. $(2x-1)^2 \dfrac{d^2y}{dx^2} + (2x-1)\dfrac{dy}{dx} + 2y = 0$

Ans. $y = C_1(2x-1) + \dfrac{C_2}{\sqrt{2x+1}}$

10. $(5+2x)^2 \dfrac{d^2y}{dx^2} - 6(5+2x)\dfrac{dy}{dx} + 8y = 6x$

Ans: $y = C_1(5+2x)^{2+\sqrt{2}} + C_2(5+2x)^{2+\sqrt{2}} - \frac{3x}{2} - \frac{45}{8}$

11. $(2x+1)^2 \dfrac{d^2y}{dx^2} - 6(2x+1)\dfrac{dy}{dx} + 16y = 8(1+2x)^2$

Ans: $y = C_1 + C_2 \log(2x+1) + [\log(2x+1)^2](2x+1)^2$

12. $(2x+3)\dfrac{d^2y}{dx^2} - (2x+3)\dfrac{dy}{dx} - 12y = 6x$

Ans: $C_1(2x+3)^2 + C_2(2x+3)^{-\frac{1}{2}} - 2x + \dfrac{3}{2}$

17.5 Linear Differential Equations of Second Order– Method of Variation of Parameters

The general form is $\dfrac{d^2y}{dx^2} + P\dfrac{dy}{dx} + Qy = R$ ..... (1)

where P, Q, R are functions of x.

Let the C.F. of the equation

$$\frac{d^2y}{dx^2} + P\frac{dy}{dx} + Qy = R$$

be $y = Au + Bv$ ..... (2)

where A and B are constants and u, v are two independent solutions of

$$\frac{d^2y}{dx^2} + P\frac{dy}{dx} + Qy = 0 \quad ..... (3)$$

such that $u_2 + Pu_1 + Qu = 0$ and $v_2 + Pv_1 + Qv = 0$

where $u_2 = \dfrac{d^2u}{dx^2}$, $u_1 = \dfrac{du}{dx}$ etc.

In order to obtain the solution of the equation (1) the arbitrary constants A and B are treated as arbitrary functions of x and are chosen in such a way that

$y = A(x)u + B(x)v$ satisfies (1), ($\therefore$ it is P.I of (1)) .... (4)

$\therefore \dfrac{dy}{dx} = Au_1 + uA_1 + Bv_1 + vB_1$ (suffixes indicating the order of the derivative)

Now we chose A and B such that $A_1u + B_1v = 0$ .... (5)

So, $\dfrac{dy}{dx} = Au_1 + Bv_1$ .... (6)

Again differentiating

$$\frac{d^2y}{dx^2} = (Au_2 + u_1A_1) + (Bv_2 + v_1B_1)$$

substituting the values of y, dy/dx, d^2y/dx^2 in (1)

$$[Au_2 + Bv_2 + A_1u_1 + B_1v_1] + P[Au_1 + Bv_1] + Q(Au + Bv] = R$$

(or) $\quad A(u_2 + Pu_1 + Qu) + B(v_2 + Pv_1 + Qv) + A_1u_1 + B_1v_1 = R$

$\therefore \quad A_1u_1 + B_1v_1 = R$ (since u and v are solutions of (3) .... (7)

Solving (5) and (7) we get

$$\frac{dA}{dx} = A_1 = \frac{-vR}{uv_1 - u_1v} \text{ and } \frac{dB}{dx} = B_1 = \frac{uR}{uv_1 - u_1v}$$

integrating we obtain

$$A(x) = \int \frac{-vR\,dx}{(uv_1 - u_1v)} + C_1;\ B(x) = \int \frac{uR}{uv_1 - u_1v} + C_2 \quad .... (8)$$

where C_1 and C_2 are arbitrary constants

Substituting (8) in (4) we get the complete general solution of the equation (1)

Working Rule: First find two independent solutions u and v of (3). Then C.F. is given by $y = Au + Bv$ where A and B are arbitrary constants. Treating A and B as functions of x, we have the solution of (1) as $y = A_1(x)\,u + B_1(x)\,v$ where A(x) and B(x) are given by (8).

Solved Examples

***Example* 17.42:** Solve by the method of variation of parameters $\frac{d^2y}{dx^2} - y = \frac{2}{1+e^x}$

***Solution*:**

A.E. is $m^2 - 1 = 0 \Rightarrow m = \pm 1$

C.F. is $y = Ae^x + Be^{-x}$

Assuming A and B as functions of x such that the given equation is satisfied by $y = Ae^x + Be^{-x}$ we have

$$\frac{dy}{dx} = Ae^x - Be^{-x} + e^x\frac{dA}{dx} + e^{-x}\frac{dB}{dx} = Ae^x - Be^{-x} \quad .... (1)$$

Choosing A and B so that $e^x\frac{dA}{dx} + e^{-x}\frac{dB}{dx} = 0$ .... (2)

also $\dfrac{d^2y}{dx^2} = Ae^x + Be^{-x} + e^x\dfrac{dA}{dx} + e^{-x}\dfrac{dB}{dx} = 0$ .... (3)

Substituting (1) to (3) in the given equation, we get

$$e^x\frac{dA}{dx} + e^{-x}\frac{dB}{dx} = \frac{2}{1+e^x} \qquad .... (4)$$

Solving (2) and (4) we get $\dfrac{dA}{dx} = \dfrac{e^{-x}}{e^x+1}$ .... (5)

and $\dfrac{dB}{dx} = -\dfrac{e^x}{e^x+1}$ .... (6)

$$A = \int\frac{e^{-x}}{e^x+1}\,dx = \int 1+\frac{e^{-x}}{e^x+1} - 1\,dx = \int\frac{e^x+1+e^{-x}}{e^x+1} - 1\,dx = \int\frac{e^x}{e^x+1} + \frac{e^{-x}(e^x+1)}{e^x+1} - 1\,dx$$

$= \log(e^x+1) - e^{-x} - x$

Integrating (5) and (6) $A = -e^{-x} + \log(1+e^x) - x + C_1$, $B = -\log(1+e^x) + C_2$

Hence the solution is

$$y = [-e^{-x} + \log(e^x+1) - x + C_1]\,e^x + [C_2 - \log(1+e^x)]e^{-x}$$

(or) $y = C_1e^x + C_2e^{-x} - 1 + e^x[\log(e^x+1) - x] - e^{-x}\log(e^x+1)]$

Remarks : If (i) $1 + P + Q = 0$ then $y = e^x$

(ii) $1 - P + Q = 0$ then $y = e^{-x}$

and (iii) $P + Qx = 0$ then $y = x$ are solutions of $D^2y + PDy + Qy = 0$

***Example* 17.43:** Solve by the method of variation of parameters

$$\frac{d^3y}{dx^3} + (1-\cot x)\frac{dy}{dx} - \cot x = \sin^2 x$$

***Solution*:**

$$\frac{d^3y}{dx^3} + (1-\cot x)\frac{dy}{dx} - \cot x = \sin^2 x \qquad .... (1)$$

$$\frac{d^3y}{dx^3} + (1-\cot x)\frac{dy}{dx} - \cot x = 0 \qquad .... (2)$$

C.F.P $= 1 - \cot x$ $\qquad$ $Q = -\cot x$ comparing with original eqn.

$\therefore\ 1 - P + Q = 1 - 1 + \cot x - \cot x = 0$

showing that $v = e^{-x}$ is a solution of (2)

To find another independent solution of (2)

let $u = e^{-x} \,.\, w$

By this substitution, the equation (2) reduces to

$$\frac{d^2w}{dx^2}+\frac{dw}{dx}(1-\cot x-\frac{2}{e^{-x}}e^{-x})=0$$

or $$\frac{d^2w}{dx^2}=(1+\cot x)\frac{dw}{dx} \Rightarrow \frac{\frac{d}{dx}\left(\frac{dw}{dx}\right)}{\frac{dw}{dx}}=1+\cot x$$

$$\log\frac{dw}{dx}=x+\log\sin x \qquad \text{or} \qquad \frac{dw}{dx}=e^x\sin x$$

$$\therefore\ w=\int e^x\sin x\,dx=-\frac{e^x}{2}(\cos x-\sin x)$$

$$u=e^{-x}\,[-\frac{e^x}{2}(\cos x-\sin x)]=-\frac{1}{2}(\cos x-\sin x)$$

The second independent solution can be taken as $\cos x-\sin x$

Thus solution of (2) is

$$y=A(\cos x-\sin x)+Be^{-x}$$

For finding the solution of (1) we treat A and B as functions of x

i.e., $y=A(x)(\cos x-\sin x)+B(x).e^{-x}$ is the solution of (1) where $A(x)$ and $B(x)$ are obtained by solving

$$(\cos x-\sin x)A^1+e^{-x}.B^1=0 \qquad \text{.... (4)}$$

and $$-(\sin x+\cos x)A^1-e^{-x}B^1=\sin^2 x \qquad \text{.... (5)}$$

solving (4) and (5) $$\frac{dA}{dx}=-\frac{1}{2}\sin x$$

and $$\frac{dB}{dx}=e^x\frac{\sin 2x}{4}-e^x\frac{(1-\cos 2x)}{4}$$

integrating we get

$$A=-\frac{1}{2}\int\sin x\,dx+C_1=\frac{\cos x}{2}+C_1 \qquad \text{.... (6)}$$

and $$B=\frac{-e^x}{20}(2\cos 2x-\sin 2x)-\frac{e^x}{4}+\frac{-e^x}{20}(\cos 2x+2\sin 2x)+C_2 \qquad \text{.... (7)}$$

Hence the complete solution is

$$y=C_1(\cos x-\sin x)+C_2e^{-x}-\frac{1}{10}(\sin 2x-2\cos 2x)$$

***Example* 17.44:** Solve by the method of variation of parameters

$$(1-x)\frac{d^2y}{dx^2}+x\frac{dy}{dx}-y=(1-x)^2$$

Solution:

The given equation can be written as

$$\frac{d^2y}{dx^2}+\frac{x}{1-x}\frac{dy}{dx}-\frac{1}{1-x}y=1-x \qquad \text{.... (1)}$$

$$P=\frac{x}{1-x};\ Q=\frac{1}{1-x} \text{ and } X=1-x$$

Clearly $P + Qx = 0$

Hence $y = x$ is a solution of $\frac{d^2y}{dx^2}+\frac{x}{1-x}\frac{dy}{dx}-\frac{1}{1-x}y=0$ .... (2)

To obtain the second independent solution of (2)
take $y = vx$

Then (2) reduces to $\frac{d^2v}{dx^2}+\frac{dv}{dx}\left(\frac{x}{1-x}+\frac{2}{x}.1\right)=0$

$$\frac{d}{dx}\left[\frac{dv}{dx}\right]+\frac{dv}{dx}\left[\frac{x}{1-x}+\frac{2}{x}\right]=0$$

$$\frac{\frac{d}{dx}\frac{dv}{dx}}{\frac{dv}{dx}}=-\frac{x}{1-x}-\frac{2}{x}=+\left(\frac{1-x-1}{1-x}\right)-\frac{2}{x}=+1-\frac{1}{1-x}-\frac{2}{x}$$

$$\log\left(\frac{dv}{dx}\right)=x+\log(x-1)-\log x^2$$

$$\frac{dv}{dx}=\frac{x-1}{x^2}.e^x=e^x\left[\frac{1}{x}+\frac{(-1)}{x^2}\right]$$

$$\therefore v=e^x.\frac{1}{x}$$

The second independent solution is $xv = e^x$

solution of equation (2) is $y = Ae^x + Bx$ .... (3)

To find the solution of (1) treat A and B as functions of x such that

$$e^x\frac{dA}{dx}+x\frac{dB}{dx}=0 \qquad \text{.... (4)}$$

and, $e^x \dfrac{dA}{dx} + x\dfrac{dB}{dx} = 1 - x$ .... (5)

Solving (4) and (5) we get $\dfrac{dA}{dx} = -xe^{-x}$ and $\dfrac{dB}{dx} = 1$

integrating, $A = -\int xe^{-x}dx + C_1 = e^{-x}(1 + x) + C_1$; $B = x + C_2$

Hence the complete solution is $y = A\,e^x + B.x$

$$= [C_1 + e^{-x}(1 + x)]e^x + (x + C_2)\,x = C_1e^x + C_2x + x^2 + (1 + x)$$

Exercise 17.8

Solve the following by the method of variation of parameters

1. $\dfrac{d^2y}{dx^2} + a^2y = \sec ax$

Ans: $y = C_1\cos ax + C_2\sin ax + \dfrac{x}{a}\sin ax + \dfrac{1}{a}\cos ax \log(\cos ax)$

2. $\dfrac{d^2y}{dx^2} + y = \tan x$

Ans: $y = -[\log(\sec x + \tan x)]\cos x + C_1\cos x + C_2\sin x$

3. $\dfrac{d^2y}{dx^2} + 4y = \operatorname{cosec} 2x$

Ans: $y = C_1\cos 2x + C_2\sin 2x + x\cos 2x + \sin 2x \log(\sin 2x)$

4. $x^2\dfrac{d^2y}{dx^2} - 2x(1+x)\dfrac{dy}{dx} + 2(1 + x)y = x^3$

Ans: $y = C_1\,xe^{2x} + C_2x - \dfrac{x^2}{4} - \dfrac{x}{4}$

17.6 Applications

17.6.1 Bending of Beams

A bar or a rod of a rectangular or circular cross section, with its length very much greater than its thickness is called a beam.

The end conditions of beam are of four types.

1. Both ends are fixed (built in)
2. One end fixed and the other freely supported hinged or pin jointed.

3. One end fixed and the other free (cantilever)
4. Both ends freely supported or pin jointed.

If the beam is fixed only at one end and loaded at the other end it is called cantilever beam.

A beam of rectangular cross section fixed at one end and loaded at the other end so as to be bent a little, with its upper surface becoming slightly convex and the lower one concave (within the elastic limits). All the longitudinal filaments in the upper half of the beam thus get extended or lengthened under tension, all those in the lower half get compressed or shortened under pressure. In between these two there is a region of the beam where the fibres are neither stretched nor compressed. This particular region is called the neutral surface of the beam and the curve of any particular fibre in this surface is called the elastic curve or deflection curve of the beam. The line in which any plane section of the beam cuts the neutral surface is called the neutral axis of the section.

A beam or a bar of a homogeneous, isotropic material, subjected to an axial push or compressive stress is called a strut.

The amount, by which a beam deflects depends upon the cross-section and bending of the beam.

If we consider a cross section of a beam AB at a distance 'x' from one end A, M is the bending moment about AB, of all external forces acting on either side of the two portions of the beam separated by the cross section.

The Bernoulli-Euler law is $M = \dfrac{EI}{\rho}$

where E = the modulus of elasticity of the beam.

I = moment of interia of the cross section of the beam about neutral axis.

ρ = Radius of curvature of the elastic curve at $p(x, y)$

$$\rho = \frac{\left(1+\left(\frac{dy}{dx}\right)^2\right)^{\frac{3}{2}}}{\frac{d^2y}{dx^2}}$$

If the deflection of the beam is small, its slope $\dfrac{dy}{dx}$ will be small and hence we can neglect it.

$$\therefore \quad \rho = \frac{1}{\frac{d^2y}{dx^2}}$$

$\therefore \quad M = EI.\dfrac{d^2y}{dx^2}$ is the differential equation of the elastic curve.

17.6.2 Boundary Conditions

(i) If the beam is freely supported at an end A, the deflection y at A is zero.

i.e., if $x = x_a$ at A, $y(x_a) = 0$

and there is no bending moment.

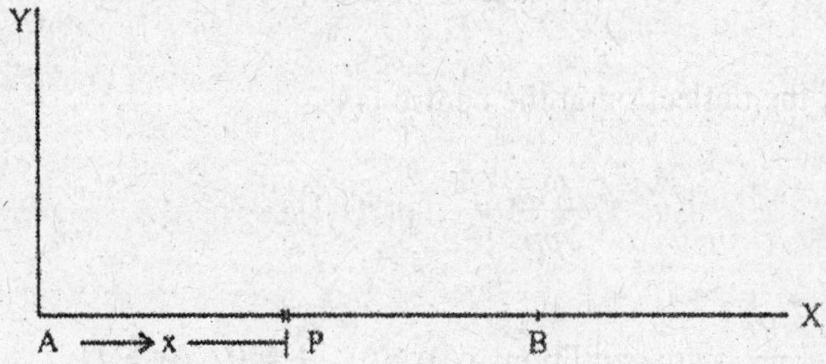

i.e., $\dfrac{d^2y}{dx^2} = 0$ at A.

(ii) If the beam is fixed at one end horizontally (clamped) at A, then

at $x = x_a$, $y(x_a) = \left.\dfrac{dy}{dx}\right|_{x_a} = 0$

(iii) If the beam end A is perfectly free, there is no bending moment or shear force at A.

$$\therefore \quad \frac{d^2y}{dx^2} = \frac{d^3y}{dx^3} = 0$$

The shear force along the cross-section of the beam is EI $\dfrac{d^3y}{dx^3}$

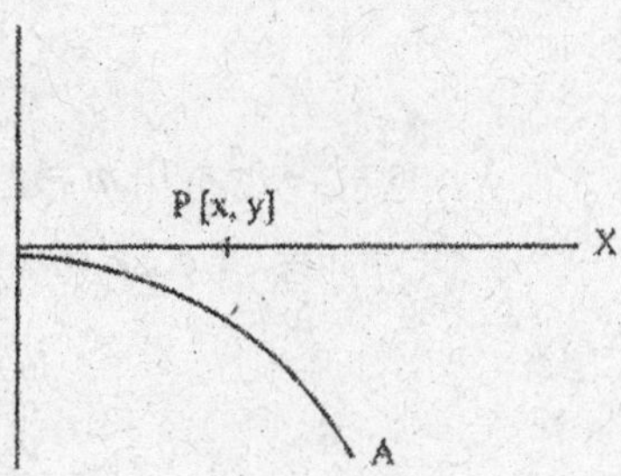

and load per unit length is

EI $\dfrac{d^4y}{dx^4}$ (i.e., the intensity of loading).

Solved Examples

***Example* 17.45:** A horizontal tie rod of length $2l$ with concentrated load W at the centre and ends freeely hinged satisfies the differential equation.

$$\text{EI}\,\frac{d^2y}{dx^2}=py-\frac{\omega}{2}x$$

Prove that the deflection at the centre is

$$\frac{\omega}{2pn}(nl-\tan h\,n\,l)$$

where $n^2=\dfrac{\text{P}}{\text{EI}}$, with conditions $y(0)=0$, $\dfrac{dy}{dx}=0$ at $x=l$.

***Solution*:**

Given $\text{EI}\,\dfrac{d^2y}{dx^2}-py=-\dfrac{\omega}{2}x$ and $n^2=\dfrac{\text{p}}{\text{EI}}$

Subject to $y=0$ when $x=0$, $\dfrac{dy}{dx}=0$ when $x=l$.

$$\frac{d^2y}{dx^2}-\frac{\text{p}}{\text{EI}}y=-\frac{\omega}{2\text{EI}}x$$

Substituting $\text{EI}=\dfrac{p}{n^2}$, $\dfrac{d^2y}{dx^2}-n^2y=-\dfrac{\omega n^2}{2p}x$

$$\left(D^2-n^2\right)y=-\frac{\omega\,n^2}{2p}x$$

A.E is $m^2-n^2=0$; $m=\pm n$; $\therefore$ C.F is

$$y=\text{C}_1\cosh(n\,x)+\text{C}_2\sinh(n\,x)$$

$$\text{P.I. is}\quad y=\frac{-\dfrac{\omega\,n^2}{2p}x}{D^2-n^2}=-\frac{\omega\,n^2}{2p}\left(D^2-n^2\right)^{-1}x$$

$$=-\frac{\omega\,n^2}{2p}\left(-n^2\right)^{-1}\left[1-\frac{D^2}{n^2}\right]^{-1}x=\frac{\omega}{2p}\left[1+\frac{D^2}{n^2}+....\right]x=\frac{\omega}{2p}x$$

The solution of (1) is, y = C.F + P.I

i.e., $y = C_1 \cosh(n x) + C_2 \sinh(n x) + \frac{\omega x}{2p}$

The condition $x = 0, y = 0$ gives $C_1 = 0$

$$\therefore \quad y = C_2 \sin h(n x) + \frac{\omega x}{2p} \qquad \text{..... (1)}$$

Differentiating *w.r.* to x

$$\frac{dy}{dx} = C_2 n \cos h(n x) + \frac{\omega}{2p}$$

Applying the condition $\frac{dy}{dx} = 0$, $x = l$ in the above equation, we get,

$$C_2 n \cos h(n l) + \frac{\omega}{2p} = 0, \; C_2 = \frac{-\omega}{2pn \cos h(nl)}$$

Substituting the value of C_2 in the solution (1)

$$y = \frac{-\omega \sin h(nx)}{2pn \cos h(nl)} + \frac{\omega x}{2p}$$

when $x = l$, the deflection at the centre is

$$y = \frac{\omega l}{2p} - \frac{\omega \tan h(n l)}{2pn}$$

$$\therefore \quad y = \frac{\omega}{2pn}[n l - \tan h(n l)]$$

***Example* 17.46:** A light horizontal strut of AB length 'l' is freely pinned at A and B. It is under the action of equal and opposite compressive forces p at its ends and it carries a load ω at its centre, and so it satisfies the differential equation.

$$EI \frac{d^2y}{dx^2} + py + \frac{\omega x}{2} = 0 \qquad \left(0 \le x \le l/2\right)$$

also when x = 0, y = 0 and x = $l/2$, $\frac{dy}{dx} = 0$

Prove that $$y = \frac{\omega}{2p}\left[\frac{\sin nx}{n\cos\frac{nl}{2}} - x\right], \qquad \left(0 \le x \le l/2\right)$$

Solution:

Given $$\text{EI}\,\frac{d^2y}{dx^2} + py = -\frac{\omega x}{2}, \qquad \left(0 \le x \le l/2\right)$$

and $$y(0) = 0,\; y'\left(\frac{l}{2}\right) = 0$$

$$\frac{d^2y}{dx^2} + \frac{p}{\text{EI}}y = \frac{-\omega x}{2\text{EI}}$$

$$\frac{d^2y}{dx^2} + n^2y = -\frac{-\omega n^2}{2p}x \qquad \left(\because n^2 = \frac{p}{\text{EI}}\right)$$

$$\left(D^2 + n^2\right)y = -\frac{\omega n^2}{2p}x \qquad \text{....(1)}$$

A.E is $m^2 + n^2 = 0$, $m = + ni$,

C.F is $y = C_1\cos(n\,x) + C_2\sin(n\,x)$ .....(2)

$$\text{P.I is}\quad y = \frac{-\frac{\omega n^2 x}{2p}}{D^2 + n^2} = \left(D^2 + n^2\right)^{-1}\left(-\frac{\omega n^2 x}{2p}\right)$$

$$= -\frac{\omega n^2}{2p}\left(n^2\right)^{-1}\left[1 + \frac{D^2}{n^2}\right]^{-1} x$$

$$= -\frac{\omega}{2p}\left[1 - \frac{D^2}{n^2} + \ldots\ldots\right]x = -\frac{\omega x}{2p}$$

From (2) and (3) the solution of (1) is

$$y = c_1\cos nx + c_2\sin nx - \frac{\omega}{2p}x \qquad \text{.....(4)}$$

$$y(0) = 0 \Rightarrow c_1 = 0 \Rightarrow y = c_2\sin nx - \frac{\omega}{2p}x \qquad \text{.....(5)}$$

Differentiating w.r to 'x'

$$\frac{dy}{dx} = c^2 n \cos nx \quad \frac{-\omega}{2p}$$

$$\frac{dy}{dx} = 0 \text{ at } x = \frac{l}{2} \Rightarrow 0 = c_2\, n \cos n\,\frac{l}{2} - \frac{\omega}{2p}$$

$$\therefore \quad c_2 = \frac{\omega}{2pn\cos\frac{nl}{2}};$$

$$\therefore (5) \Rightarrow \quad y = \frac{\omega}{2pn\cos\frac{nl}{2}} \sin n\,x - \frac{\omega}{2p}\,x$$

$$\text{(or)} \quad y = \frac{\omega}{2p}\left[\frac{\sin nx}{2n\cos\frac{nl}{2}} - x\right] \text{ is the solution}$$

***Example* 17.47:** A horizontal, simply supported, uniform beam of length l meters bends under its own weight, which is ω kg/metres. Find the equation of its elatic curve and the maximum deflection.

***Solution*:**

Let the beam OA of length l meters be supported at the ends O and A. The total weight of the beam is w l kg and hence each end O and A supports a load of $\frac{\omega l}{2}$ .kg, let P(x, y) be any point on elastic curve.

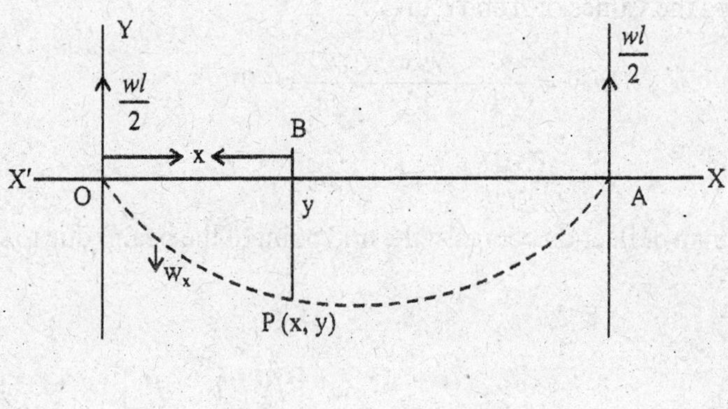

Consider a section OP of the beam. The forces acting on this section are the thrust $\frac{\omega l}{2}$ kg upward at the end 'O' and a weight wx kg at a distance $\frac{x}{2}$ metres from the end 'O'.

Now the moments M about the point p(x, y) is

$$M = wx\frac{x}{2} - \frac{wl}{2}x$$

The equation of the elastic curve is

$$M = EI\frac{d^2y}{dx^2}$$

$$\therefore \qquad EI\frac{d^2y}{dx^2} = \frac{wx^2}{2} - \frac{wl}{2}x \qquad .....(1)$$

The boundary conditions in this case are $y = 0$ when $x = 0$ and $x = l$

We know that at the ends O and A of the beam there is no deflection

Integrating (1) w.r.t 'x'

$$EI\frac{dy}{dx} = \frac{wx^3}{6} - \frac{wlx^2}{4} + a$$

Again integrating w.r.t 'x'

$$EIy = \frac{wx^4}{24} - \frac{wlx^3}{12} + ax + b \qquad .....(2)$$

where a, b are constants of integration

Now applying the conditions in (2)

$y = 0$ when $x = 0$, $\Rightarrow$ b = 0 also $y = 0$ when $x = 1$

$$0 = \frac{wl^4}{24} - \frac{wl^4}{12} + al + 0, \qquad \therefore \qquad a = \frac{wl^3}{24}$$

Substituting the values of a and b in (2)

$$EI\ y = \frac{wx^4}{24} - \frac{wlx^3}{12} + \frac{wl^3}{24}x + 0$$

$$\therefore \qquad y = \frac{w}{24EI}\left[x^4 - 2x^3l + xl^3\right] \qquad .....(3)$$

The maximum deflection occurs at the mid point of the beam (due to symmetry of the curve), i.e. when $x = \frac{l}{2}$;

$$\therefore (3) \Rightarrow \text{Maximum deflection of } y = \frac{w}{24EI}\left[\frac{l^4}{16} - \frac{l^4}{4} + \frac{l^4}{2}\right] = \frac{5}{384}\frac{wl^4}{EI}$$

***Example* 17.48:** A horizontal beam of length $2l$, with a uniform load w per unit length is fixed horzontally at both the ends. If there is an additional vertical load W at the middle of the beam, find the equation of the elastic curve and the maximum deflection.

***Solution*:**

Let the beam OA be of length $2l$ fixed at both ends O and A, P(x, y) be any point on the elastic curve. Since the beam is fixed horizontally there must be a couple of a constant moment exerted by the wall at each end.

Also each end supports half the load on the beam i.e., load at each end = $\frac{W}{2}+wl$.

Now the moments of the forces acting on OP about the point P are given by,

$$EI\frac{d^2y}{dx^2} = R - \left(\frac{W}{2}+wl\right)x + wx.\frac{x}{2}$$

$$EI\frac{d^2y}{dx^2} = R - \left(\frac{W}{2}+wl\right)x + \frac{wx^2}{2} \qquad .....(1)$$

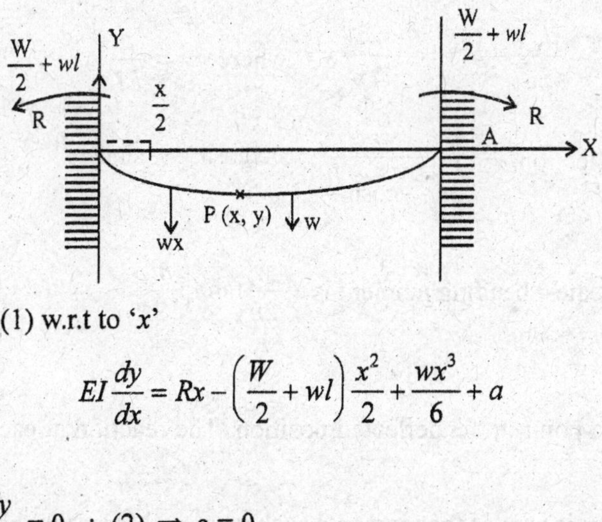

Integrating (1) w.r.t to 'x'

$$EI\frac{dy}{dx} = Rx - \left(\frac{W}{2}+wl\right)\frac{x^2}{2} + \frac{wx^3}{6} + a \qquad .....(2)$$

At $x = 0$, $\frac{dy}{dx} = 0$, ∴ (2) ⇒ a = 0

Also at $x = l$, $\frac{dy}{dx} = 0$

$$\therefore \qquad O = Rl - \left(\frac{W}{2}+wl\right)\frac{l^2}{2} + \frac{wl^3}{6}$$

$$\therefore \qquad R = \frac{W}{4}l + \frac{wl^2}{3}$$

$$\therefore (2) \Rightarrow \quad EI\frac{dy}{dx} = \frac{W}{4}(lx - x^2) + \frac{w}{6}(2l^2x - 3lx^2 + x^3) \qquad .....(3)$$

Integrating w.r. to 'x'

$$EIy = \frac{W}{4}\left(\frac{lx^2}{2} - \frac{x^3}{3}\right) + \frac{w}{6}\left(l^2x^2 - lx^3 + \frac{x^4}{4}\right) \qquad .....(4)$$

is the equation of the elastic curve.

The maximum deflection occurs at $x = l$ (by symmetry)

$$\therefore (4) \Rightarrow \quad EIy = \frac{W}{4}\left(\frac{l^3}{2} - \frac{l^3}{3}\right) + \frac{w}{6}\left[l^4 - l^4 + \frac{l^4}{4}\right]$$

(or) $\quad y = \dfrac{l^3}{24EI}\left[W + wl\right]$ is the maximum deflection.

***Example* 17.49:** A light uniform tie-rod of length l and flexural regidity EI is subjected to a horizontal tension p at its ends and carries a concentrated load w at its mid-point. Show that

(i) the deflection at a distance x from one end is given by the equation

$$\frac{d^2y}{dx^2} - n^2y = -\frac{wn^2}{2p}x, \quad \text{where } n^2 = \frac{p}{EI}$$

(ii) The deflection $y = \dfrac{w}{2np - \cosh\dfrac{nl}{2}}\left(nx\cosh\dfrac{nl}{2} - \sinh nx\right)$

(iii) The greatest bending moment is $-\dfrac{w}{2n}\tanh\left(\dfrac{nl}{2}\right)$

***Solution*:**

Let R(x, y) be a point in the deflected position. The reaction at each support is $\dfrac{w}{2}$

$EI\dfrac{d^2y}{dx^2}$. Algebraic sum of the moments of the external forces in the portion

$$OR = Py - \frac{w}{2}x \qquad .....(1)$$

i.e., $\quad \dfrac{d^2y}{dx^2} = n^2y - \dfrac{w}{2}\dfrac{n^2}{p}x, \qquad$ since $\dfrac{P}{EI} = n^2$

$$\frac{d^2y}{dx^2} - n^2y = -\frac{wn^2}{2p}x \qquad .....(2)$$

$$\therefore \quad (D^2 - n^2)y = \frac{-wn^2}{2p}x$$

A.E is $m^2 - n^2 = 0$; $m = \pm n$

$\therefore$ C.F is $\quad y = c_1 \cos h\, nx + c_2 \sin h\, nx$

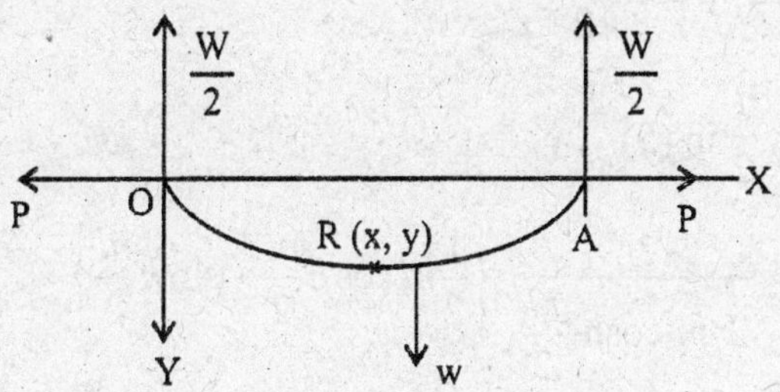

The P.I. is $$y = \frac{-\frac{wn^2}{2p}x}{(D^2 - n^2)} = \frac{-wn^2 x}{2p} \times \frac{-1}{n^2} = \frac{wx}{2p}$$

$\therefore$ The complete solution of (2) is, $y =$ C.F + P.I

i.e. $$y = c_1 \cosh nx + c_2 \sinh nx + \frac{wx}{2p} \quad .....(3)$$

This equation holds good over the left half of the deflection curve.

The boundary conditions are y = 0 when x = 0 and $\frac{dy}{dx} = 0$ when $x = \frac{l}{2}$

Substituting the boundary conditions $y = 0$ when $x = 0$ in (3), $c_1 = 0$

$$\therefore (3) \Rightarrow \quad y = c_2 \sinh nx + \frac{wx}{2p} \quad .....(4)$$

Differentiating (4) w.r. to 'x'

$$\frac{dy}{dx} = c_2 n \cosh nx + \frac{w}{2p}$$

Now applying the condition $\frac{dy}{dx} = 0$, when $x = \frac{l}{2}$

$$0 = c_2 n \cos nl + \frac{w}{2p}, \quad c_2 = \frac{-w}{2pn\cosh\frac{nl}{2}}$$

$\therefore$ The solution (4) reduces to $y = \dfrac{-w\sinh nx}{2pn\cosh\dfrac{nl}{2}} + \dfrac{wx}{2p}$

$$\therefore \qquad y = \frac{w}{2pn\cosh\dfrac{nl}{2}}\left[nx\cosh\frac{nl}{2} - \sinh nx\right] \qquad .....(5)$$

Substituting $\qquad x = \dfrac{l}{2}$ in (5)

Maximum value of $\quad y = \dfrac{w}{2pn\cosh\dfrac{nl}{2}}\left\{\dfrac{1}{2}nl\cos h\dfrac{nl}{2} - \sin h\dfrac{nl}{2}\right\} \qquad .....(6)$

Now substituting $\qquad x = \dfrac{l}{2}$ in (1)

The greatest bending moment

$$EI\,\frac{d^2y}{dx^2} = py - \frac{w}{2}\cdot\frac{l}{2}$$

substituting the value of y from 6

$$EI\,\frac{d^2y}{dx^2} = p\left(\frac{w}{2Pn\cosh\dfrac{nl}{2}}\right)\left\{\frac{nl}{2}\cos h\frac{nl}{2} - \sin h\frac{nl}{2}\right\} - \frac{wl}{4}$$

$$EI\,\frac{d^2y}{dx^2} = \frac{w}{2n\cosh\dfrac{nl}{2}}\left\{\frac{nl}{2}\cos h\frac{nl}{2} - \sin h\frac{nl}{2}\right\} - \frac{wl}{4}$$

$$= -\frac{w}{2n}\tan h\,\frac{nl}{2} \text{ is the greatest bending moment}$$

Exercise 17.9

1. Solve the equation $EI\,\dfrac{d^2y}{dx^2} + py = -\dfrac{wl^2}{8}\sin\dfrac{\pi x}{l}$ for a strut of length l, freely hinged at each end. Prove that bending moment at the centre is $\dfrac{-wl^2}{8}\,\dfrac{Q}{Q-P}$ where $Q = EI\,\dfrac{\pi^2}{l^2}$.

2. The deflection of a strut of length 'l' with one end (x = 0) built-in and the other supported and subjected to end thrust P, satisfies the equation $\frac{d^2y}{dx^2}+a^2y=\frac{a^2R}{P}(l-x)$.

   Prove that the deflection curve is

$$y=\frac{R}{P}\left(\frac{\sin ax}{a}-l\cos ax+l-x\right) \qquad \text{where } al=\tan al.$$

3. A horizontal strut of weight w per unit length is freely hinged at its ends. It is under the action of equal and opposite forces P at its ends. The differential equation satisfied by strut is given by $\frac{d^2y}{dx^2}+n^2y=\frac{w}{2El}(x^2-lx)$ where $n^2=\frac{P}{EI}$. Show that if

$$y=0 \text{ when } x=0 \text{ and } x=l,\ y=w\,\frac{\cos n\left(x-\frac{l}{2}\right)}{pn^2\cos n\frac{l}{2}}+\frac{W}{2P}\left(x^2-lx-\frac{2}{n^2}\right)$$

4. A beam of length $2l$ meters has its ends fixed horizontally carrying a load of w kg per meter. Find the equation of the elastic curve, and maximum deflection.

$$\left[Ans:\ y=\frac{wx^2}{24\,EI}\left(4l^2+x^4-4lx^3\right)\text{and maximum deflection}=\frac{wl^4}{24EI}.\right]$$

5. A uniform horizontal strut of length l freely supported at both and ends, carries a uniformly distributed load w per unit length. If the thrust at each end is P. Prove that the maximum deflection is $\frac{w}{p}a^2\left(\sec\frac{al}{2}-1\right)-\frac{wl^2}{8P}$ where $\frac{P}{EI}=a^2$ prove also that the maximum bending moment is of the magnitude $\frac{w}{a^2}\left(\sec\frac{al}{2}-1\right)$

17.6.3 Electrical Circuits

Consider the linear differential equation which governs the flow of electricity in simple series circuits. A simple electric circuit consists of electric energy such as a battery or a generator, a resistance, such as an electric bulb or an electric appliance, which uses this energy.

A series circuit consists of four elements whose action is described below.

(i) A source of electromotive Force (E.M.F) 'E': This may be a battery or generator. E may be a constant or function of time.

(ii) The resistance 'R': By Ohm's law resistance opposes the flow current I by producing a drop in electromotive force of magnitude IR.

(iii) Inductance 'L': This opposes any change in current by producing a drop in E.M.F of magnitude $L\frac{dI}{dt}$.

(iv) The capacitance C: This stores the charge Q. The change accumulated by the capacitor resists the flow of additional charge and the drop in E.M.F arising in this way is $\frac{1}{c_1}Q$ where $I=\frac{dQ}{dt}$.

Relation between current I and voltage drop across various elements of an electric circuit

Let us consider an electric circuit containing a resistance R, an inductance L and a capacitance C. If I is the current in the circuit at any time t, then

(i) The potential drop E_R acrosss the resistance is given by $E_R = IR$.

(ii) The potential drop E_L across the inductance is given by $E_L = L\frac{dI}{dt}$

(iii) The potential drop E_c across the capacitance is given by $E_C = \frac{Q}{C} = \frac{1}{C}\int I\,dt$

where $Q = \frac{dI}{dt}$ is the charge in circuit.

The kirchoff's law states that "In an electric circuit the sum of voltage drop (potential drops) is equal to the supplied voltage (E.M.F)".

$$\therefore \quad IR + L\frac{dI}{dt} + \frac{Q}{c} = E$$

$$\therefore \quad L\frac{dI}{dt} + IR + \frac{Q}{c} = E$$

$$\text{or} \quad L\frac{d^2Q}{dt^2} + R\frac{dQ}{dt} + \frac{Q}{C} = E$$

which is a second order linear ordinary differential equation.

Notations and Symbols

	Physical quantity	Symbol	Unit	Diagrammatic Representation
1	E.M.F or voltage	E or V	volt	
2.	Current	I or i	Ampere	 
3.	Charge	Q or q	Coulomb	–
4.	Resistance	R	ohm	
5.	Inductance	L	Henry	
6.	Capacitance	C	Farad	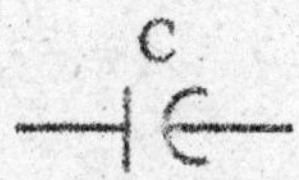
7.	Key or switch			

Solved Examples

***Example* 17.50:** Find the current I in the RLC circuit assuming zero initial current and charge I. If R = 80 Ohms, L = 20 Henrys, C = 0.01 Farads and E = 100 volts.

Solution:

The differential equation for the circuit is $L\frac{d^2Q}{dt^2}+R\frac{dQ}{dt}+\frac{Q}{C}=E$ ..... (1)

$$20\frac{d^2Q}{dt^2}+80\frac{dQ}{dt}+\frac{1}{0.01}=100 \Rightarrow \therefore \frac{d^2Q}{dt^2}+4\frac{dQ}{dt}+5Q=5 \quad ..... (2)$$

or $(D^2 + 4D + 5)Q = 5$

The A.E. is $m^2 + 4m + 5 = 0$; $\therefore$ $m = -2 \pm i$

C.F. is, $Q = e^{-2t}\,[c_1 \cos t + c_2 \sin t]$ ..... (3)

$$P.I = \frac{5}{D^2+4D+5} = \frac{5e^{ot}}{D^2+4D+5} = \frac{5e^{ot}}{0^2+4.0+5} = 1 \qquad \text{...... (4)}$$

$\therefore$ The solution of the differential equation (2) is Q = CF + PI

$$Q = e^{-2t}\,[c_1 \cos t + c_2 \sin t] \qquad \text{(from 3 and 4)} \qquad \text{..... (5)}$$

$$\text{Now } \frac{dQ}{dt} = e^{-2t}\,[-c_1 \sin t + c_2 \cos t] - 2e^{-2t}\,[c_1 \cos t + c_2 \sin t] \qquad \text{..... (6)}$$

Given initial conditions are when t = 0, Q = 0, $\frac{dQ}{dt} = 0$

(5) $\Rightarrow$ $c_1 + 1 = 0$ $\Rightarrow$ $c_1 = -1$

(6) $\Rightarrow$ $0 = c_2 - 2c_1$ $\therefore$ $c_2 = -2$

(5) $\Rightarrow$ $Q = 1 + e^{-2t}\,[-\cos t - 2\sin t] = 1 - e^{-2t}\,[\cos t + 2\sin t]$

and substituting $c_1 = -1$, $c_2 = -2$ in (6)

$$I = \frac{dQ}{dt} = 5e^{-2t} \sin t$$

Example **17.51:** Find the current I in the RLC circuit assuming zero initial current and charge and R = 160 ohms, L = 20 Henrys, C = 2.10^{-3} Farads and E = 48.1 sin(10 t) volts.

Solution:

Given R = 160 ohms, L = 20 Henrys, C = 2.10^{-3} Farads, E = 481 sin (10 t) volts and Q = 0, $\frac{dQ}{dt} = 0$ when t = 0

The differential equation for the charge Q is $L\frac{d^2Q}{dt^2} + R\frac{dQ}{dt} + \frac{Q}{c} = E$

$$\therefore \quad 20\frac{d^2Q}{dt^2} + 160\frac{dQ}{dt} + \frac{10^3}{2}Q = 481 \sin(10t) \qquad \text{..... (1)}$$

A.E is, $20m^2 + 160m + 500 = 0$

$\Rightarrow$ $m^2 + 8m + 25 = 0$, $m = -4 \pm 3i$

$\therefore$ $C.F = e^{-4t}(c_1 \cos 3t + c_2 \sin 3t)$ ..... (2)

$$P.I = \frac{481 \sin(10t)}{20D^2 + 160D + 520} = \text{I.P. of } \frac{481\, e^{10ti}}{20D^2 + 160D + 500}$$

$$= 481\left[\frac{Im\,a - part\ of\ e^{10ti}}{20(10i)^2 + 160(10i) + 500}\right]$$

$$= 481\left[\frac{Im\,a - part\ of\left(\cos 10\,ti + i\sin 10\,ti\right)}{-1500 + 1600i}\right]$$

$$\text{P.I} = \frac{-15\sin 10t - 16\cos 10t}{100} \qquad \text{..... (3)}$$

$\therefore$ (2), (3) $\Rightarrow$ The solution of the differential equation is

$$Q = e^{-4t}\left[c_1\cos 3t + c_2\sin 3t\right] - \left(\frac{15\sin 10t + 16\cos 10t}{100}\right) \qquad \text{..... (4)}$$

and $I = \dfrac{dQ}{dt} = -4e^{-4t}(c_1\cos 3t + c_2\sin 3t) + 3e^{-4t}(-c_1\sin 3t + c_2\cos 3t)$

$$-\left(\frac{15\cos 10t + 16\sin 10t}{10}\right) \qquad \text{..... (5)}$$

$\therefore$ (4) $\Rightarrow$ $0 = c_1 - \dfrac{-16}{100}$; $c_1 = \dfrac{4}{25}$

$t = 0$, $\dfrac{dQ}{dt}$ give (5) $\Rightarrow$ $-4c_1 + 3c_2 - \dfrac{15}{10} = 0$

$$3c_2 = \frac{16}{25} + \frac{3}{2};\quad c_2 = \frac{107}{150}$$

$$\therefore \quad I = e^{-4t}\left[\frac{3}{2}\cos 3t - \frac{10}{3}\sin 3t\right] - \left(\frac{15\cos 10t + 16\sin 10t}{10}\right)$$

***Example* 17.52:** Show that the frequency of free vibrations in a closed electrical circuit with inductance L and capacitance C in series is given by $\dfrac{30}{\pi\sqrt{LC}}$ per minute.

***Solution*:**

Let Q be the charge on the capacitor after time 't' seconds. The voltage drops across the inductance and capacitor are $\dfrac{d^2Q}{dt^2}$ *and* $\dfrac{Q}{C}$ respectively

The differential equation for the circuit is $L\,\dfrac{d^2Q}{dt^2}+\dfrac{Q}{C}=0 \quad \dfrac{d^2Q}{dt^2}=-\dfrac{Q}{LC}$

$= -w^2Q \qquad$ where $\dfrac{1}{LC}=w^2 \qquad$ .... (1)

The above equation represents an oscillatory circuit with frequency η given by

$$\eta=\frac{w}{2\pi}\text{ per second }=\frac{1}{2\pi\sqrt{LC}}\times 60\text{ per minute }=\eta=\frac{30}{\pi\sqrt{LC}}\text{ per minute}$$

***Example* 17.53:** For an RLC circuit, the charge Q on a plate of the condensor is given by

$L\dfrac{d^2Q}{dt^2}+Q\,\dfrac{dQ}{dt}+\dfrac{Q}{C}=E\sin\omega t$ where $I=\dfrac{dQ}{dt}$.

The circuit is tuned to resonance so that $w^2=\dfrac{1}{LC}$. If $CR^2 < 4L$

and initially Q = 0, I = 0, show that $Q=\dfrac{E}{R\omega}\left[e^{\frac{-Rt}{2L}}\left(\cos pt+\dfrac{R}{2LP}\sin Pt\right)-\cos wt\right]$

and $\qquad I=\dfrac{E}{R}\left[\sin\omega t-\dfrac{1}{P\sqrt{LC}}e^{\frac{-Rt}{2L}}\sin pt\right]$ where $P^2=\dfrac{1}{LC}-\dfrac{R^2}{4L^2}$

***Solution*:**

Given the differential equation is $L\,\dfrac{d^2Q}{dt^2}+R\dfrac{dQ}{dt}+\dfrac{Q}{c}=E\sin\omega t$

$$\frac{d^2Q}{dt^2}+\frac{R}{L}\frac{dQ}{dt}+\frac{Q}{CL}=\frac{E}{L}\sin\omega t$$

Denoting $\dfrac{R}{L}=2\lambda$ and $\dfrac{1}{CL}=\mu^2$, we get $\dfrac{d^2Q}{dt^2}+2\lambda\,\dfrac{dQ}{dt}+Q\mu^2=\dfrac{E}{L}\sin\omega t$

$\therefore \qquad (D^2+2\lambda\,D+\mu^2) \qquad Q=\dfrac{E}{L}\sin wt \qquad$ ..... (1)

A.E is $m^2+2\,\lambda\,m+\mu^2=0$

$$m=-2\lambda\pm\frac{\sqrt{4\lambda^2-4\mu^2}}{2}=-\lambda\pm\sqrt{\lambda^2-\mu^2}$$

Now, $\lambda^2 - \mu^2 = \frac{R^2}{4L^2} - \frac{1}{LC} = \frac{CR^2 - LC}{4L^2C} \quad < 0, \qquad (\because CR^2 < 4L)$

$\therefore \quad m = -\lambda + i\sqrt{\mu^2 - \lambda^2} = -\lambda \pm ip$

where $\quad p^2 = \frac{1}{LC} - \frac{R^2}{4L^2} = \mu^2 - \lambda^2$

$\therefore$ The C.F $= e^{-\lambda t}(c_1 \cos pt + c_2 \sin pt)$

$$= e^{\frac{-Rt}{2L}}[c_1 \cos pt + c_2 \sin pt] \qquad \left[\therefore \lambda = \frac{R}{2L}\right] \qquad \text{..... (2)}$$

$$\text{P.I} = \frac{1}{D^2 + 2\lambda D + \mu^2} \cdot \frac{E}{L} \sin \omega t = \frac{E}{L} \frac{1}{(-\omega^2 + 2\lambda D + \mu^2)} \sin \omega t$$

$$= \frac{E}{L} \frac{1}{\left(\mu^2 - w^2\right) + 2\lambda D} \sin \omega t = \frac{E}{L} \frac{1}{2\lambda D} \sin wt \quad \left(\because \omega^2 = \frac{1}{LC} = \mu^2 \text{ and } 2\lambda = \frac{R}{L}\right)$$

$$= \frac{-E}{L\omega} \cdot \frac{\cos \omega t}{\frac{R}{L}} = \frac{-E}{Rw} \cdot \cos wt \qquad \text{..... (3)}$$

From (2), (3) the solution of the differential equation is

Q = CF + PI

$$Q = e^{\frac{-Rt}{2L}}[c_1 \cos pt + c_2 \sin pt] - \frac{E}{R\omega} \cos \omega t \qquad \text{..... (4)}$$

Initially when t = 0, a = 0, the solution (4) reduces to

$$0 = c_1 - \frac{E}{R\omega}; \qquad \therefore \qquad c_1 = \frac{E}{R\omega}$$

Differentiating (4) w.r.to 't'

$$\frac{dQ}{dt} = I = e^{\frac{-Rt}{2L}}[-p c_1 \sin pt + p c_2 \cos pt]$$

$$-\frac{R}{2L} e^{\frac{-RT}{2L}}\left(c_1 \cos pt + c_2 \sin pt\right) + \frac{E}{R} \sin \omega t$$

using the initial conditions when t = 0, $\frac{dQ}{dt} = 0 = I$ we get,

$$Pc_2 - \frac{R}{2L} c_1 = 0$$

$$c_2 = \frac{R}{2PL}c_1 = \frac{R}{2PL}\frac{E}{R\omega} = \frac{E}{2PL\omega}$$

$$\therefore \quad (4) \quad \Rightarrow \quad \frac{E}{R\omega}\left[e^{\frac{RT}{2L}}\left(\cos pt + \frac{R}{2PL}\sin pt\right) - \cos wt\right] \qquad \text{.....(5)}$$

Differentiating (5) w.r. to 't'

$$\frac{dQ}{dt} = \frac{E}{R\omega}\left[e^{\frac{-RL}{2L}}\left(-p\sin pt + \frac{R}{2L}\cos pt\right) - \frac{R}{2L}e^{\frac{Rt}{2L}}\left(\cos pt + \frac{R}{2PL}\sin pt\right) + \omega\sin wt\right]$$

$$I = \frac{E}{RW}\left[e^{\frac{-Rt}{2L}}\left\{\left(\frac{R}{2L}\cos pt - \frac{R}{2L}\cos pt\right) - \left(p\sin pt + \frac{R^2}{4PL^2}\sin pt\right)\right\} + \omega\sin wt\right]$$

$$= \frac{E}{RW}\left[\omega\sin wt - e^{\frac{-Rt}{2L}}\left(\frac{4P^2L^2 + R^2}{4PL^2}\right)\sin pt\right]$$

$$= \frac{E}{RW}\left[\omega\sin wt - e^{\frac{-Rt}{2L}}\left(\frac{1}{LCP}\right)\sin wt\right] \qquad \left\{\because \frac{4P^2L^2 + R}{4L^2} = \frac{1}{Lc}\right\}$$

$$I = \frac{E}{R}\left[\sin wt - e^{\frac{-Rt}{2L}}\frac{\sqrt{LC}}{LCP}\sin pt\right] \qquad \left[w^2 = \frac{1}{LC}\right]$$

$$I = \frac{E}{R}\left[\sin wt - \frac{1}{P\sqrt{Lc}}E^{\frac{Rt}{2l}}\sin pt\right]$$

is the current in the circuit at any time 't'.

Exercise 17.10

1. Find the current in RLC circuit assuming zero initial conditions and

   (i) R = 6 ohms, L = 1 henry, C = 0.04 Farad and E = 24 cos 5t volts.

   (ii) R = 40 ohms, L = 10 Henrys, C = 0.02 Farad and E = 800 cos 5t volts.

Ans: (i) $I = e^{-3t}(3\sin 4t - 4\cos 4t) + 4\cos 5t$

(ii) $I = 10(\cos 5t - 2\sin 5t_2 - e^{-2t}\cos t)$

2. If L = 0.2 Henry and C = 0.8×10^{-6} Farads, determine the resistance R so that the circuit is critically damped. **Ans:** (R = 10^3 ohms)

3. An E.M.F of E sin pt is applied at t = 0 to a circuit containing a capacitance C and inductance L. If $P^2 = \frac{1}{LC}$ and initially the current and change are each zero, show that the current at time t is $\frac{Et}{2L}$ sin pt.

4. An unchanged condenser of capacity C is charged by applying an EMF-

   $E \sin\left(\frac{t}{\sqrt{LC}}\right)$ through leads of self inductance L and negligible resistance.

   Prove that at time 't' the charge on one of the plates is

   $$\frac{EC}{2}\left[\sin\left(\frac{t}{\sqrt{LC}}\right)-\frac{t}{\sqrt{LC}}\cos\left(\frac{t}{\sqrt{LC}}\right)\right]$$

5. An electric circuit consists of an inductance of 0.1 henry, a resistance of 20 ohms and a condenser of capacitance 25 micro farads. Find the charge Q and the current i at any time t, given that at t = 0, Q = 0.05 coulomb, $i = \frac{dQ}{dt}$, when t = 0.

   Ans: $Q = e^{-100t}$ (0.05 cos 624.5t + 0.008 sin 624 .5t) ; $I = 0.32\, e^{-100t}$ sin 624.5t)

6. An E.M.F – E sin pt is applied at t = 0 to an electric circuit containing a condenser C and inductance L in series. The current I satisfies the equation,

   $$L\frac{dI}{dt}+\frac{1}{c}\int I dt = E \sin pt$$

   If $p^2 = \frac{1}{LC}$ and initially the current I and the charge Q are zero, show that the current in the circuit at time t is given by $\frac{Et}{2L}$ sin pt.

17.6.4 Simple Harmonic Motion

The kind of motion, in which a particle moves in a straight line in such a way that its acceleration is always directed towards a fixed point on the line (called the centre of force) and varies as the distance of the particle from the fixed point is called simple harmonic motion.

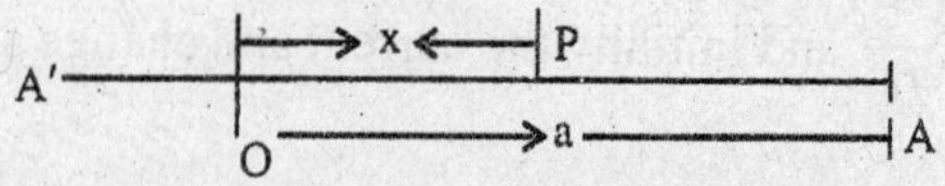

Let a particle move along a straight line AA^1 such that its acceleration is always directed towards a fixed point 'O' and varies as its distance from this fixed point. The point 'O' is called centre of attraction. If the particle starts from rest at the point A and be at a point P at any time 't' such that $OP = x$, then the equation of motion is

$$\frac{d^2x}{dt^2} = -\mu^2 x \qquad \text{.....(1)}$$

where μ^2 is the constant of proportionality, where the negative sign has been taken because the force acting on P is towards O. The differential equation (1) gives the acceleration of the particle of any position.

The equation (1) can be written as $(D^2 + \mu^2)\, x = 0$;

$$\text{A.E. is } m^2 + \mu^2 = 0 \Rightarrow m = \pm \mu i$$

The solution of the equation (1) is $x = c_1 \cos \mu t + c_2 \sin \mu t$ .....(2)

Let $OA = a$, so that $t = 0, x = a$; $\therefore$ (2) gives $c_1 = a$

Differentiating (2) w.r. to 't', $v = \dfrac{dx}{dt} = -c_1\, \mu \sin \mu t + c_2\, \mu \cos \mu t$ .....(3)

where v is the velocity of the particle at the point P.

Initially, when $t = \theta$, $v = 0$; $\therefore$ (3) $\Rightarrow c_2 = a$

Hence (2) $\Rightarrow$ $x = a \cos \mu t$ .....(4)

which is the equation of motion and gives displacement of the particle from the fixed point O, at any time t

Differentiating (4) w.r. to 't', $v = \dfrac{dx}{dt} = -a\mu \sin \mu t$ .....(5)

$$= -\mu a \sqrt{1 - \cos^2 \mu t} = -\mu a \sqrt{1 - \frac{x^2}{a^2}}$$

$$\therefore \quad v = \frac{dx}{dt} = -\mu \sqrt{a^2 - x^2} \qquad \text{.....(6)}$$

This equation (6) gives velocity of the particle at any time 't' when its displacement from the fixed point 'O' is x (velocity v decreases as x increases).

From equations (4) and (5), it is evident that the motion is oscillatory and periodic of period $\frac{2\pi}{\mu}$; usually denoted by T is called the time period of the oscillation.

The particle oscillates between A^1 and A which are 'a' units from the origin on either side with a period T.

The motion is called periodic motion or simple harmonic motion (simply S.H.M)

The particle starts from the point A, with an acceleration directed towards O, the velocity at the point A being zero. As the particle moves towards 'O' the acceleration decreases and vanishes at O. The velocity becomes maximum at the point 'O', as it is evident from equation (6). Due to this velocity the particle moves towards the point A^1 and the acceleration increases, becoming maximum at the point A^1. At the point A^1 ($OA^1 = a$), the velocity vanishes and the particle moves towards the point O and the motion continues in this manner.

The point 'O' is called centre of motion or the mean position. The maximum distance 'a' which the particle covers on both sides of the mean position is called the amplitude of the motion.

The number of oscillations made by the particle in unit time is called the frequency of the motion and is given by $\frac{1}{T} = \frac{\mu}{2\pi}$. It may be observed that the time period of motion T is independent of the amplitude 'a'

Some Important Results about S.H.M

(i) If the centre of force is at origin the equation of S.H.M is $\frac{d^2x}{dt^2} = -\mu^2 x$

(ii) If the centre of force is not at origin but is at the point $x = \pm b$, then the equation of motion is $\frac{d^2x}{dt^2} = -\mu^2 (x \mp b)$

(iii) The velocity v at a distance x from the centre and the distance x from the centre at time t are respectively given by

$$v^2 = \mu^2 (a^2 - x^2) \text{ and } x = a \cos \mu t$$

where a is the amplitude and the time t has been measured from the extreme position in the positive direction.

(iv) Maximum acceleration = $\mu^2 a$ (at extreme points)

(v) Maximum velocity = μa (at the centre)

(vi) Periodic time $T = \dfrac{2\pi}{\mu}$.

Solved Examples

***Example* 17.54:** In a S.H.M the amplitude of motion is 5 meters and the period is 4 seconds. Find the time required by the particle in passing between the points which are at distance of 4 meters and 2 meters from the centre of the force and are on same side of it. Also find the velocities at these points.

***Solution*:**

The equation of S.H.M. is $\dfrac{d^2x}{dt^2} = -\mu^2 x$ .....(1)

The time period = $\dfrac{2\pi}{\mu} = 4$ (given) $\Rightarrow \mu = \dfrac{\pi}{2}$

We have the solution of (1) is $x = a \cos \mu t$

$$a = 5,\ \mu = \frac{\pi}{2} \Rightarrow x = 5 \cos \frac{\pi}{2} t \quad .....(2)$$

Let the times when the particle is at distances of 4 metres and 2 metres from the centre of motion respectively be t_1 and t_2 (sec).

$$\therefore \quad 4 = 5 \cos\left(\frac{\pi}{2} t_1\right) \Rightarrow t_1 = \frac{2}{\pi} \cos^{-1}\left(\frac{4}{5}\right)$$

and

$$2 = 5 \cos\left(\frac{\pi}{2} t_2\right) \Rightarrow t_2 = \frac{2}{\pi} \cos^{-1}\left(\frac{2}{5}\right)$$

Time required in passing through these points

$$t_2 - t_1 = \frac{2}{\pi}\left(\cos^{-1}\frac{2}{5} - \cos^{-1}\frac{4}{5}\right) = 0.33 \text{ sec}$$

Differentiating (2) w.r. to 't'

$$\frac{dx}{dt} = -5\,\frac{\pi}{2}\,\sin\frac{\pi}{2}\,t = \frac{-5\pi}{2}\sqrt{1-\frac{x^2}{25}} \qquad \left(\because x = 5\cos\frac{\pi}{2}\,t\right)$$

$$\therefore \quad \frac{dx}{dt} = \frac{-\pi}{2}\sqrt{25-x^2}$$

$\therefore\ v = \dfrac{-\pi}{2}\sqrt{5^2-4^2} = +4.71$ m/sec (numerically) is the velocity when $x = 4$ metres.

Similarly when $x = 2$ metres, $V = \dfrac{\pi}{2}\sqrt{21}$ m/sec

***Example* 17.55:** A particle is moving with S.H.M and while making an excursion from one position of rest to the other, its distances from the middle point of its path at three consecutive seconds are observed to be x_1, x_2, x_3; prove that the time of a complete oscillation is

$$\frac{2\pi}{\cos^{-1}\left(\dfrac{x_1+x_3}{2x_2}\right)}$$

***Solution*:**

The equation of S.H.M is $x = a\cos\mu t$ .....(1)

If the distances of the moving point from the mean position are x_1, x_2, x_3 at the end of $t, t+1, t+2$ seconds respectively, we get using (1),

$$x_1 = a\cos\mu t \qquad .....(2)$$

$$x_2 = a\cos\mu(t+1) \qquad .....(3)$$

$$x_3 = a\cos\mu(t+2) \qquad .....(4)$$

(2) and (4) $\Rightarrow$ $\quad x_1 + x_3 = a[\cos\mu t + \cos\mu(t+2)]$

$$= 2a\cos\left(\frac{\mu t+\mu t+2\mu}{2}\right)\cos\left(\frac{\mu t-\mu t-2\mu}{2}\right)$$

$$= 2a\cos\{\mu(t+1)\}\cos\mu, \qquad [\because \cos(-\theta) = \cos\theta]$$

$$\Rightarrow \quad x_1 + x_3 = 2\, x_2 \cos \mu, \qquad \text{from (3)}$$

$$\therefore \quad \mu = \cos^{-1}\left(\frac{x_1 + x_3}{2x_2}\right)$$

$\therefore$ Hence period of a complete oscillation $T = \dfrac{2\pi}{\mu} = \dfrac{2\pi}{\cos^{-1}\left(\dfrac{x_1 + x_3}{2x_2}\right)}$

***Example* 17.56:** At what distance from the centre the velocity in a S.H.M will be half of the maximum.

***Solution*:**

Let centre of motion be the origin and 'a' be the amplitude

In a S.H.M the velocity v of the particle at a distance $|x|$ from the centre is

$$v = -\mu^2 \sqrt{(a^2 - x^2)}\,; \quad \therefore \quad v^2 = \mu^2\,(a^2 - x^2) \qquad \text{.... (1)}$$

From (1), *v* is maximum when $x = 0$

$\therefore$ The maximum velocity $v = a\mu$

Let x_1 be the distance from the centre of the point when the velocity is $\frac{1}{2} a\mu$

Substituting $x = x_1$ and $v = \frac{1}{2} a\mu$ in (1),

$$\frac{1}{4}\, a^2\mu^2 = \mu^2\,(a^2 - x_1^2)$$

$$x_1^2 = \frac{3a^2}{4} \qquad \therefore \qquad x_1 = \pm\, \frac{a\sqrt{3}}{2}$$

$\therefore$ When the velocity is half of the maximum, there are two points each at a distance $\dfrac{a\sqrt{3}}{2}$ from the centre and on either side of it.

***Example* 17.57:** A particle is performing a S.H.M of period T about a centre O and it passes through a point P where OP = b with velocity v in the direction OP; prove that the time which elapses before it returns to P is $\dfrac{T}{\pi}\tan^{-1}\left(\dfrac{vT}{2\pi b}\right)$.

Solution:

Let the equation of the S.H.M with centre 'O' as origin be $\frac{d^2x}{dt^2} = -\mu^2 x$ and the time

period $T = \frac{2\pi}{\mu}$

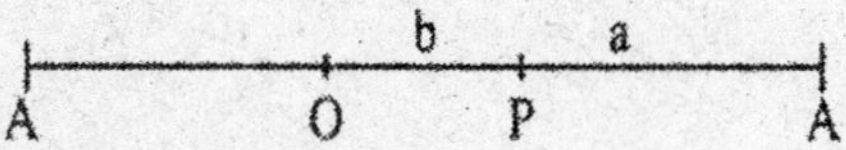

$$v^2 = \mu^2 (a^2 - x^2) \quad .....(1)$$

When the particle passes through P its velocity is v in the direction OP. Also OP = b

Substituting $x = b$ in (1)

$$v^2 = \mu^2 (a^2 - b^2) \quad .....(2)$$

Let A be an extremely of the motion. From P the particle comes to instantaneous rest at A and then returns back to P. In S.H.M the time from P to A equal to the time from A to P.

$\therefore$ Motion from A to P is $\frac{dx}{dt} = -\mu\sqrt{a^2 - x^2}$

$$\therefore \qquad dt = -\frac{1}{\mu}\frac{dx}{\sqrt{a^2 - x^2}} \quad .....(3)$$

Let t_1 be the time from A to P. Then at A, t = a and at P, $t = t_1$ and x = b

Integrating (3)

$$\int_0^{t_1} dt = \frac{1}{\mu}\int_a^b \frac{-dx}{\sqrt{a^2 - x^2}}$$

$$t_1 = \frac{1}{\mu}\left[\cos^{-1}\left(\frac{x}{a}\right)\right]_a^b = \frac{1}{\mu}\cos^{-1}\left(\frac{b}{a}\right)$$

$$\therefore \text{ The required time} = 2t_1 = \frac{2}{\mu}\cos^{-1}\left(\frac{b}{a}\right) = \frac{2}{\mu}\tan^{-1}\left(\frac{\sqrt{a^2 - b^2}}{b}\right)$$

$$= \frac{2}{\mu}\tan^{-1}\left(\frac{v}{b\mu}\right); \qquad \text{using (2)}$$

$$= \frac{2}{\frac{2\pi}{T}}\tan^{-1}\left(\frac{v}{b\left(\frac{2\pi}{T}\right)}\right) \qquad \left(\because T = \frac{2\pi}{\mu}\right)$$

$$= \frac{T}{\pi}\tan^{-1}\left(\frac{vT}{2\pi b}\right)$$

***Example* 17.58:** The speed v of the point p which moves in a line is given by the relation $v^2 = a + 2bx - cx^2$ where x is the distance of the point P from a fixed point on the path and a, b, c are constants. Show that the motion is simple harmonic if c is positive, determine the period and the amplitude of the motion.

Solution:

Given $v^2 = a + 2bx - cx^2$ .....(1)

Differentiating w.r. to '*x*', we get

$$\therefore \quad v\frac{dv}{dx} = c\left(\frac{b}{c} - x\right)$$

$$\therefore \quad \frac{d^2x}{dt^2} = v\frac{dv}{dx} = -c\left(x - \frac{b}{c}\right) \quad .....(2)$$

Given c is positive

$\therefore$ Equation (2) represents a S.H.M. with centre of force at the point $x = \frac{b}{c}$

$\therefore$ Given equation (1) represents S.H.M of period $T = \frac{2\pi}{\mu} = \frac{2\pi}{\sqrt{c}}$

substituting v = 0 in (1), we get

$$a + 2bx - cx^2 = 0 \qquad \therefore \; cx^2 - 2bx - a = 0$$

$$x = \frac{+b \pm \sqrt{b^2 - ac}}{c}$$

$\therefore$ The distances of the two positions of instantaneous rest *A* and A^1 from the fixed point *O* are given by

$$OA = \frac{b + \sqrt{b^2 + ac}}{c} \quad \text{and} \quad OA^1 = \frac{b - \sqrt{b^2 + ac}}{c}$$

The distance of any of these two positions from the centre $x = \frac{b}{c}$ is the amplitude of the motion.

$$\therefore \text{The amplitude} = OA = \frac{b + \sqrt{b^2 + ac}}{c} - \frac{b}{c}.$$

$$\text{The amplitude} = \frac{\sqrt{b^2 + ac}}{c}$$

Exercise 17.11

1. At the ends of three successive seconds, the distances of a moving point with S.H.M. from its mean position measured in the same direction are 1, 5, 5, the intervals of time being measured from the mean positions. Show that the period of a complete oscillation is $\dfrac{2\pi}{\cos^{-1}\left(\frac{3}{5}\right)}$

2. A point executes S.H.M. such that in two of its positions the velocities are u and v and the corresponding accelerations are α, β. Show that the distance between the points is $\dfrac{v^2-u^2}{\alpha+\beta}$

3. A particle is moving in a straight line and when the distance of the particle from the equilibrium position has the values x_1 and x_2 the corresponding velocities are u_1 and u_2 respectively. Show that the period of oscillation is $2\pi\left\{\dfrac{(x_2^2-x_1^2)}{(u_1^2-u_2^2)}\right\}^{\frac{1}{2}}$

4. A point moves with a S.H.M. When the point is 0.75 metres from mid path, its velocity is 11 meters per second and when 2 metres from the centre of the path, its velocity is 3 m/sec. Find the time period of motion and also the greatest acceleration of the particle.

   (**Ans:** Time period = 1.1 m/sec ; acceleration = 67.3 m/sec^2)

5. The time peroid of a particle executing S.H.M. is 16 sec; 2 seconds after it passes through the centre of oscillation, its velocity is 2 m/sec. Find its amplitude.

   (**Ans:** 7.2 m)

6. A particle moves with S.H.M. of period 12 sec travels 8 cm from the position of rest in 2 sec. Find the amplitude, the maximum velocity and the velocity at the end of 2 sec.

   (**Ans:** 16 cm, 4.619 cm/sec, $\dfrac{4\sqrt{3}\pi}{3}$ cm/sec)

7. A particle moving in a straight line with S.H.M. with amplitude 1.6 metres and the time period 4 sec. Find the time required by the particle in passing between points which are at distances 1.2 and 0.6 metres from the centre of force and are on the same side on it.

CHAPTER 18

Partial Differential Equations

18.1 Partial Differential Equations

18.1.1 Introduction

Let x, y be two independent variables and z be a dependent variable given by the relation $z = f(x, y)$.

A Partial Differential Equation is a differential equation involving partial derivatives,

$$\frac{\partial z}{\partial x} = p; \quad \frac{\partial^2 z}{\partial x^2} = r; \quad \frac{\partial^2 z}{\partial x \partial y} = \frac{\partial^2 z}{\partial y \partial x} = s;$$

$$\frac{\partial z}{\partial y} = q; \quad \frac{\partial^2 z}{\partial y^2} = t; \text{ etc...}$$

Partial derivatives occur only when there are two or more variables. In general if $z = f(x_1, x_2, x_3, \dots x_n)$, $x_1, x_2, \dots x_n$ are independent variables, and z is the dependent variable then we use the following notations;

$$P_1 = \frac{\partial z}{\partial x_1}, P_2 = \frac{\partial z}{\partial x_2}, P_3 = \frac{\partial z}{\partial x_3}, \dots P_n = \frac{\partial z}{\partial x_n}$$

The order of a Partial Differential equation is the order of the highest partial differential coefficient occurring in it, and the degree of a partial differential equation is the degree of the derivative of highest order occurring in it. The partial differential equation is called linear if it is of first degree in the dependent variable and its partial derivatives. An equation which is not linear is called a non-linear partial differential equation.

18.1.2 Formation of a Partial Differential Equation by the elimination of Arbitrary Constants

Consider the equation $f(x, y, z, a, b) = 0$ .....(1)

where a, b are arbitrary constants.

Differentiating (1) partially w.r.t. to 'x' and 'y' respectively, we get

$$\frac{\partial f}{\partial x} + \frac{\partial f}{\partial z}\frac{\partial z}{\partial x} = 0 \quad \text{i.e.,} \quad \frac{\partial f}{\partial x} + p\frac{\partial f}{\partial z} = 0 \qquad .....(2)$$

and

$$\frac{\partial f}{\partial y} + \frac{\partial f}{\partial z}\frac{\partial z}{\partial y} = 0 \quad \text{i.e.,} \quad \frac{\partial f}{\partial y} + q\frac{\partial f}{\partial z} = 0 \qquad .....(3)$$

Elimination of the two constants 'a' and 'b' from (1), (2), (3) gives an equation of the form $\phi(x, y, z, p, q) = 0$ which is a first order partial differential equation. If however the number of constants be greater than the number of independent variables, then the result of eliminating the constants will give rise to a partial differential equation of higher order than the first.

Solved Examples

***Example* 18.1:** Form a partial differential equation by eliminating the arbitrary constants 'a' 'b' from $z = ae^{-b^2t} \cos bx$.

Solution:

$$z = ae^{-b^2t}\cos bx \qquad .....(1)$$

Differentiate (1) partially w.r.t. to 'x' and 't'

$$\frac{\partial z}{\partial x} = ae^{-b^2t}(-\sin bx)b \qquad .....(2)$$

$$\frac{\partial z}{\partial t} = (a\cos bx)e^{-b^2t}. -b^2 \qquad .....(3)$$

Differentiate (2) partially w.r.t. to 'x'

$$\frac{\partial^2 z}{\partial x^2} = -ab^2e^{-b^2t}\cos bx \qquad .....(4)$$

Elimination of a, b from (3) and (4) gives

$$\frac{\partial z}{\partial t} = \frac{\partial^2 z}{\partial x^2}$$

***Example* 18.2:** Eliminate the constants a, b from $z = (x^2 + a)(y^2 + b)$, and form the partial differential equation.

Solution:

$$z = (x^2 + a)(y^2 + b) \qquad .....(1)$$

Differentiate (1) partially w.r.t. to 'x' and 'y'

$$\frac{\partial z}{\partial x} = (y^2 + b)(2x) \qquad .....(2)$$

and $$\frac{\partial z}{\partial y} = (x^2 + a)(2y) \qquad .....(3)$$

From (2) & (3)

$$\left(\frac{\partial z}{\partial x}\right)\left(\frac{\partial z}{\partial y}\right) = 4xy(x^2 + a)(y^2 + b) \qquad .....(4)$$

Substituting for z from (1) in (4) we get $pq = 4xyz$.

***Example* 18.3:** Find a partial differential equation by eliminating constants a, b, c from

$$\frac{x^2}{a^2} + \frac{y^2}{b^2} + \frac{z^2}{c^2} = 1 \qquad .....(1)$$

Solution:

Differentiate (1) partially. w.r.t. to 'x' and 'y'

$$\frac{2x}{a^2} + \frac{2z}{c^2}\frac{\partial z}{\partial x} = 0 \qquad .....(2)$$

and $$\frac{2y}{b^2} + \frac{2z}{c^2}\frac{\partial z}{\partial y} = 0 \qquad .....(3)$$

Differentiating (2) partially w.r.t. to 'x'

$$\frac{1}{a^2} + \frac{1}{c^2}\left[\left(\frac{\partial z}{\partial x}\right)^2 + z\frac{\partial^2 z}{\partial x^2}\right] = 0 \Rightarrow \frac{c^2}{a^2} + \left(\frac{\partial z}{\partial x}\right)^2 + z\frac{\partial^2 z}{\partial x^2} = 0 \qquad .....(4)$$

Substituting $\frac{c^2}{a^2} = -\frac{z}{x}\frac{\partial z}{\partial x}$ from (2) in the equation (4)

$$\left(-\frac{z}{x}\frac{\partial z}{\partial x}\right) + \left(\frac{\partial z}{\partial x}\right)^2 + z\frac{\partial^2 z}{\partial x^2} = 0 \text{ (i.e.,) } zx\frac{\partial^2 z}{\partial x^2} + x\left(\frac{\partial z}{\partial x}\right)^2 - z\frac{\partial z}{\partial x} = 0$$

***Example* 18.4:** Find the partial differential equation by eliminating a, b from

$$a z + b = a^2 x + y.$$

Solution:

$$az + b = a^2 x + y \qquad .....(1)$$

Differentiating (1) partially w.r.t. to 'x' and 'y'

$$a\frac{\partial z}{\partial x} = a^2 \quad \text{(i.e.,)} \quad \frac{\partial z}{\partial x} = a \qquad .....(2)$$

and

$$a\frac{\partial z}{\partial y} = 1 \qquad .....(3)$$

From (2) and (3) $\frac{\partial z}{\partial x}\frac{\partial z}{\partial y} = 1$ (i.e.,) $\quad pq = 1$

***Example* 18.5:** Find the partial differential equation by eliminating h, k from

$$(x - h)^2 + (y - k)^2 + z^2 = r^2.$$

Solution:

$$(x - h)^2 + (y - k)^2 + z^2 = r^2 \qquad .....(1)$$

Differentiating (1) partially w.r.t. to 'x' and 'y'

$$2(x - h) + 2z\frac{\partial z}{\partial x} = 0 \quad \text{(i.e.,)} \quad x - h = -zp \qquad .....(2)$$

and

$$2(y - k) + 2z\frac{\partial z}{\partial y} = 0 \quad \text{(i.e.,)} \quad y - k = -zq \qquad .....(3)$$

Substituting for $x - h$, $y - k$ from (2), (3) in (1) we get

$$z^2(p^2 + q^2 + 1) = r^2$$

Exercise – 18.1

Form the partial differential equations by eliminating the arbitrary constants indicated in brackets from the following equations :

1. $z = (x - a)^2 + (y - b)^2$ $\quad (a, b)$

   Ans : $p^2 + q^2 = 4z$

2. $z = ax^2 + by^2$ $\quad (a, b)$

   Ans : $2z = px + qy$

3. $z = ax + by + ab$ $\quad (a, b)$

   Ans : $z = px + pq$

4. $z = ae^{pt} \sin px$ $\quad (p, a)$

   Ans : $\frac{\partial^2 z}{\partial x^2} + \frac{\partial^2 z}{\partial t^2} = 0$

5. $z = ax + a^2y^2 + b$ (a, b)

Ans : $q = 2yp^2$

6. $z = ax + (1 - a)y + b$ (a, b)

Ans : $p + q = 1$

7. $z = a(x + y) + b(x - y) + abt + c$ (a, b, c)

Ans : $p^2 - q^2 = 4\dfrac{\partial z}{\partial t}$

8. $z = xy + y\sqrt{x^2 - a^2 - b^2}$ (a, b)

Ans : $px + qy - pq = 0$

18.2. Formation of the Partial Differential Equations by the Elimination of Arbitrary Functions

Derive a partial differential equation by the elimination of the arbitrary function ϕ from $\phi(u, v) = 0$ where u, v are functions of x, y and z.

$$\phi(u, v) = 0 \qquad .....(1)$$

Differentiating (1) partially w. r. t. to 'x' and 'y'

$$\frac{\partial \phi}{\partial u}\left(\frac{\partial u}{\partial x} + \frac{\partial u}{\partial z}.\frac{\partial z}{\partial x}\right) + \frac{\partial \phi}{\partial v}\left(\frac{\partial v}{\partial x} + \frac{\partial v}{\partial z}.\frac{\partial z}{\partial x}\right) = 0$$

(i.e.,)
$$\frac{\partial \phi}{\partial u}\left(\frac{\partial u}{\partial x} + p.\frac{\partial u}{\partial z}\right) + \frac{\partial \phi}{\partial v}\left(\frac{\partial v}{\partial x} + p\frac{\partial v}{\partial z}\right) = 0 \qquad .....(2)$$

and
$$\frac{\partial \phi}{\partial u}\left(\frac{\partial u}{\partial y} + q\frac{\partial u}{\partial z}\right) + \frac{\partial \phi}{\partial v}\left(\frac{\partial v}{\partial y} + q\frac{\partial v}{\partial z}\right) = 0 \qquad .....(3)$$

Eliminating $\dfrac{\partial \phi}{\partial u}$ and $\dfrac{\partial \phi}{\partial v}$ from (2) and (3)

$$\left(\frac{\partial u}{\partial x} + \frac{\partial u}{\partial z}p\right)\left(\frac{\partial v}{\partial y} + \frac{\partial v}{\partial z}q\right) = \left(\frac{\partial u}{\partial y} + \frac{\partial u}{\partial z}q\right)\left(\frac{\partial v}{\partial x} + \frac{\partial v}{\partial z}p\right)$$

(i.e.,)
$$\left(\frac{\partial u}{\partial y}\frac{\partial v}{\partial z} - \frac{\partial u}{\partial z}\frac{\partial v}{\partial y}\right)p + \left(\frac{\partial u}{\partial z}\frac{\partial v}{\partial x} - \frac{\partial u}{\partial x}\frac{\partial v}{\partial z}\right)q = \frac{\partial u}{\partial x}\frac{\partial v}{\partial y} - \frac{\partial u}{\partial y}\frac{\partial v}{\partial x}$$

is the partial differential equation after the elimination of ϕ from $\phi(u, v) = 0$.

Written in a simpler form

$$\frac{\partial(u,v)}{\partial(y,z)}p+\frac{\partial(u,v)}{\partial(z,x)}q=\frac{\partial(u,v)}{\partial(x,y)}$$

Above equation is generally written as $pP + qQ = R$ where

$$P=\frac{\partial u}{\partial y}\frac{\partial v}{\partial z}-\frac{\partial u}{\partial z}\frac{\partial v}{\partial y}; \qquad Q=\frac{\partial u}{\partial z}\frac{\partial v}{\partial x}-\frac{\partial u}{\partial x}\frac{\partial v}{\partial z}; \qquad R=\frac{\partial u}{\partial x}\frac{\partial v}{\partial y}-\frac{\partial u}{\partial y}\frac{\partial v}{\partial x}$$

Solved Examples

***Example* 18.6:** Form the partial differential equation by eliminating the arbitrary function from $z = x^2 f(x - y)$.

***Solution*:**

$$z = x^2 f(x-y) \qquad .....(1)$$

Differentiating (1) partially w.r.t. x and y

$$\frac{\partial z}{\partial x} = x^2 f'(x-y) + f(x-y) \,.\, 2x \qquad .....(2)$$

$$\frac{\partial z}{\partial y} = x^2 f'(x-y)(-1) \qquad .....(3)$$

Substituting $f(x-y) = \frac{z}{x^2}$ from (1) and

$x^2 f'(x-y) = -\frac{\partial z}{\partial y}$ from (3) in (2) gives

$$\frac{\partial z}{\partial x} = -\frac{\partial z}{\partial y} + 2x \,.\, \frac{z}{x^2}$$

(i.e.,) $$x\frac{\partial z}{\partial x} + x\frac{\partial z}{\partial y} = 2z$$

***Example* 18.7:** Eliminate the arbitrary function f and hence obtain the partial differential equation from $z = x + y + f(x, y)$.

***Solution*:**

$$z = x + y + f(x, y) \qquad .....(1)$$

Differentiate (1) partially w.r.t. 'x' and 'y'

$$p = 1 + y f'(x y)$$
$$q = 1 + x f'(x y)$$

Eliminating $f'(x, y)$ from (2) & (3)

$$\frac{p-1}{q-1} = \frac{y}{x} \Rightarrow px - qy = x + y$$

***Example* 18.8:** Form the partial differential equation by eliminating the arbitrary function 'f' from

$$z = e^{ax+by} f(ax - by) \qquad .....(1)$$

***Solution*:**

Differentiating (1) Partially w.r.t. 'x' and 'y'

$$p = e^{ax+by} f'(ax - by)\, a + ae^{ax+by} f(ax - by) \qquad .....(2)$$

$$q = e^{ax+by} f'(ax - by)(-b) + be^{ax+by} f(ax - by) \qquad .....(3)$$

From (2) & (1) we get $p - az = ae^{ax+by} f'(ax - by)$

and from (3) & (1) we get $q - bz = -be^{ax+by} f'(ax - by)$

$$\frac{p-az}{q-bz} = \frac{a}{-b} \qquad \text{(i.e.,)} \quad bp + aq = 2abz$$

***Example* 18.9:** Form the partial differential equation by eliminating the arbitrary function 'f' from

$$f(lx + my + nz, x^2 + y^2 + z^2) = 0 \qquad .....(1)$$

***Solution*:**

Let $u = lx + my + nz$, $v = x^2 + y^2 + z^2$ then (1) reduces to $f(u, v) = 0$ .....(2)

elimination of 'f' from (2) gives the partial differential equation

$$pP + qQ = R \qquad .....(3)$$

where $p = \dfrac{\partial z}{\partial x}, q = \dfrac{\partial z}{\partial y}$ and

$$P = \frac{\partial u}{\partial y}\frac{\partial v}{\partial z} - \frac{\partial u}{\partial z}\frac{\partial v}{\partial y} = m\,(2\,z) - n\,(2\,y)$$

$$Q = \frac{\partial u}{\partial z}\frac{\partial v}{\partial x} - \frac{\partial u}{\partial x}\frac{\partial v}{\partial z} = n\,(2\,x) - l\,(2\,z)$$

$$R = \frac{\partial u}{\partial x}\frac{\partial v}{\partial y} - \frac{\partial u}{\partial y}\frac{\partial v}{\partial x} = l\,(2\,z) - m\,(2\,x)$$

Substituting P, Q, R values in (3) we get

$$(mz - ny)p + (nx - lz)q = ly - mx$$

***Example* 18.10:** Form the partial differential equation by eliminating the arbitrary function 'f' from $f(x^2 + y^2 - z^2, x + y + z) = 0$.

Solution:

$$f(x^2 + y^2 - z^2, x + y + z) = 0 \qquad .....(1)$$

Let $u = x^2 + y^2 - z^2$, $v = x + y + z$; then (1) is in the form

$$f(u, v) = 0 \qquad .....(2)$$

Eliminating 'f' from (2) the partial differential equation is

$$pP + qQ = R \qquad .....(3)$$

where $p = \dfrac{\partial z}{\partial x}$, $q = \dfrac{\partial z}{\partial y}$ and

$$P = \frac{\partial u}{\partial y}\frac{\partial v}{\partial z} - \frac{\partial u}{\partial z}\frac{\partial v}{\partial x} = 2y \,.\, 1 + 2z \,.\, 1$$

$$Q = \frac{\partial u}{\partial z}\frac{\partial v}{\partial x} - \frac{\partial u}{\partial x}\frac{\partial v}{\partial z} = -2z \,.\, 1 - 2x \,.\, 1$$

$$R = \frac{\partial u}{\partial x}\frac{\partial v}{\partial y} - \frac{\partial u}{\partial y}\frac{\partial v}{\partial x} = 2x \,.\, 1 - 2y \,.\, 1$$

Substituting the values of P, Q, R in (3) we get

$$(y + z)p - (x + z)\, q = x - y$$

***Example* 18.11:** Eliminate arbitrary functions f, g from $z = f(x + iy) + g(x - iy)$ and hence obtain the partial differential equation.

Solution:

$$z = f(x + iy) + g(x - iy) \qquad .....(1)$$

Differentiating (1) partially w.r.t. 'x' and 'y' we get

$$\frac{\partial z}{\partial x} = f'(x + iy) + g'(x - iy)$$

and $$\frac{\partial z}{\partial y} = f'(x + iy)(i) + g'(x - iy)(-i)$$

$$\frac{\partial^2 z}{\partial x^2} = f''(x+iy) + g''(x-iy) \qquad .....(2)$$

$$\frac{\partial^2 z}{\partial y^2} = f''(x+iy)i^2 + g''(x-iy)(i^2) \qquad .....(3)$$

From (2) and (3) we have $\dfrac{\partial^2 z}{\partial x^2} + \dfrac{\partial^2 z}{\partial y^2} = 0$

Exercise - 18.2

Eliminate the arbitrary functions and hence find the partial differential equation.

1. $z = y^2 + 2f\left(\dfrac{1}{x} + \log y\right)$

   Ans : $px^2 + qy = 2y^2$

2. $z = xy + f\left(x^2 + y^2\right)$

   Ans : $py - qx = y^2 - x^2$

3. $z = f(x - at) + g(x + at)$

   Ans : $\dfrac{\partial^2 z}{\partial t^2} = a^2 \dfrac{\partial^2 z}{\partial x^2}$

4. $z = f\left(\dfrac{y}{x}\right)$

   Ans : $px + qy = 0$

5. $z = f(x + ay)$

   Ans : $ap - q = 0$

6. $f(x^2 + y^2 + z^2, z^2 - 2xy) = 0$

   Ans : $z(p - q) = y - x$

7. $x + y + z = f(x^2 + y^2 + z^2)$

   Ans : $(y - z)\,p + (z - x)\,q = x - y$

8. $z = f(x^2 + y^2 + z^2)$

   Ans : $py - qx = 0$

9. $z = f(x - y)$

Ans : $p + q = 0$

10. $f(x^2 + y^2 + z^2, xyz) = 0$

Ans : $x(y^2 - z^2)p + y(z^2 - x^2)q = z(x^2 - y^2)$

11. $z = e^{iy}f(x + ay)$

Ans : $q = ap + iz$

18.3 First Order linear Partial Differential Equations

18.3.1 Lagrange's Equation

The partial differential equation of the type $Pp + Qq = R$ .....(1)

is called Lagarange's equation where P, Q, R are functions of x, y, z

and $$p = \frac{\partial z}{\partial x}, \quad q = \frac{\partial z}{\partial y}$$

In (18.2). we have seen that the differential equation of the type (1) is obtained by eliminating the arbitrary function ϕ from $\phi(u, v) = 0$ .....(2)

where $$P = \frac{\partial u}{\partial y}\frac{\partial v}{\partial z} - \frac{\partial u}{\partial z}\frac{\partial v}{\partial y}, \quad Q = \frac{\partial u}{\partial z}\frac{\partial v}{\partial x} - \frac{\partial u}{\partial x}\frac{\partial v}{\partial z}, \quad R = \frac{\partial u}{\partial x}\frac{\partial v}{\partial y} - \frac{\partial u}{\partial y}\frac{\partial v}{\partial x}$$

The general integral of (1) is given by (2) and we have to find 'u' and 'v'.

Suppose that $u = c_1$ and $v = c_2$ are two solutions of the differential equation (1)

Differentiating partially w.r.t. x, y, z gives

$$\frac{\partial u}{\partial x}dx + \frac{\partial u}{\partial y}dy + \frac{\partial u}{\partial z}dz = 0 \quad .....(3)$$

$$\frac{\partial v}{\partial x}dx + \frac{\partial v}{\partial y}dy + \frac{\partial v}{\partial z}dz = 0 \quad .....(4)$$

Solving (3) and (4) we get

$$\frac{dx}{\frac{\partial u}{\partial y}\frac{\partial v}{\partial z} - \frac{\partial u}{\partial z}\frac{\partial v}{\partial y}} = \frac{dy}{\frac{\partial u}{\partial z}\frac{\partial v}{\partial x} - \frac{\partial u}{\partial x}\frac{\partial v}{\partial z}} = \frac{dz}{\frac{\partial u}{\partial x}\frac{\partial v}{\partial y} - \frac{\partial u}{\partial y}\frac{\partial v}{\partial x}}$$

(i.e.,) $$\frac{dx}{P} = \frac{dx}{Q} = \frac{dz}{R} \quad .....(5)$$

which are called Lagrange's auxiliary equations. Solution of (5) gives two independent solutions $u = c_1$, $v = c_2$.

The general solution of (1) is $\phi(u, v) = 0$ or $u = \phi(v)$ or $v = \phi(u)$

Solved Examples

***Example* 18.12:** Solve $\frac{y^2 z}{x} p + xzq = y^2$

***Solution*:**

Given $\frac{y^2 z}{x} p + xzq = y^2$ .....(1)

Lagrange's auxiliary equations are

$$\frac{dx}{\frac{y^2 z}{x}} = \frac{dy}{xz} = \frac{dz}{y^2} \quad .....(2)$$

From (2), taking first two fractions and integrating

$$\frac{dx}{\frac{y^2 z}{x}} = \frac{dy}{xz}$$

$$x^2 dx - y^2 dy = 0$$

$$x^3 - y^3 = c_1 \quad .....(3)$$

From (2), taking first and last fractions

$$\frac{dx}{\frac{y^2 z}{x}} = \frac{dy}{y^2}$$

$$\therefore \quad xdx - zdz = 0$$

$$x^2 - z^2 = c_2 \quad .....(4)$$

From (3) and (4)

General solution of (1) is

$$f(x^3 - y^3, x^2 - z^2) = 0 \text{ or } x^3 - y^3 = f(x^2 - y^2)$$

Example 18.13: Solve $p \tan x + q \tan y = \tan z$.

***Solution*:**

The Langrange's auxiliary equations are

$$\frac{dx}{\tan x} = \frac{dy}{\tan y} = \frac{dz}{\tan z} \quad .....(1)$$

Integrating $\dfrac{dx}{\tan x} = \dfrac{dy}{\tan y}$

$$\int \cot x\,dx - \int \cot y\,dy = \log c_1$$

$$\log \sin x - \log \sin y = \log c_1$$

$$\Rightarrow \quad \frac{\sin x}{\sin y} = c_1 \qquad .....(2)$$

Integrating $\dfrac{dy}{\tan y} = \dfrac{dz}{\tan z}$ we get

$$\frac{\sin y}{\sin z} = c_2 \qquad .....(3)$$

The general solution, from (2) and (3), is

$$\phi\left(\frac{\sin x}{\sin y}, \frac{\sin y}{\sin z}\right) = 0$$

***Example* 18.14:** Solve $3pxy + qz^2 + yz = 0$

***Solution*:**

Given $3pxy + qz^2 + yz = 0$ .....(1)

The Lagrange's auxiliary equations are

$$\frac{dx}{3xy} = \frac{dy}{z^2} = \frac{dz}{-yz} \qquad .....(2)$$

From $\dfrac{dx}{3xy} = \dfrac{dz}{-yz}$ of (2) we have

$$\frac{dx}{x} = -3\frac{dz}{z}$$

$$\log x + 3 \log z = \log c_1 \text{ (i.e.,) } xz^3 = c_1 \qquad .....(3)$$

Again from $\dfrac{dy}{z^2} = \dfrac{dz}{-yz}$ we have

$$ydy + zdz = 0 \quad \text{(i.e.,)} \; y^2 + z^2 = c_2 \qquad .....(4)$$

From (3) and (4) the general solution is

$$f \qquad (xz^3, y^2 + z^2) = 0$$

***Example* 18.15:** Solve $p + 3q = 5z + \tan(y - 3x)$

***Solution*:**

The Lagrange's auxiliary equations are

$$\frac{dx}{1} = \frac{dy}{3} = \frac{dz}{5z + \tan(y - 3x)} \qquad .....(1)$$

Integrating we get $y - 3x = c_1$ .....(2)

Substituting $y - 3x = c_1$ we get from (1)

$$\frac{dx}{1} = \frac{dz}{5z + \tan c_1}$$

(i.e) $$x - \frac{1}{5}\log(5z + \tan c_1) = c_2$$

$$5x - \log(5z + \tan c_1) = c_2 \quad \text{(i.e.,)} \; 5x - \log[5z + \tan(y - 3x)] = c_2 \qquad .....(3)$$

From (2) and (3), the general solution is

$$\phi\{(y - 3x), 5x - \log[5z + \tan(y - 3x)]\} = 0$$

***Example* 18.16:** Solve $p - q = x + y + z$

***Solution*:**

The Lagrange's auxiliary equations are

$$\frac{dx}{1} = \frac{dy}{-1} = \frac{dz}{x + y + z} \qquad .....(1)$$

From $\dfrac{dx}{1} = \dfrac{dy}{-1}$ we have $x + y = c_1$ .....(2)

Substituting $x + y = c_1$ we get from (1)

$$\frac{dy}{-1} = \frac{dz}{z + c_1}$$

$$y + \log(z + c_1) = c_2 \quad \Rightarrow y + \log(z + x + y) = c_2 \qquad .....(3)$$

From (2) and (3) the general solution is

$$\phi[(x + y), y + \log(x + y + z)] = 0$$

***Example* 18.17:** Solve $z(p-q)=z^2+(x+y)^2$

Solution:

Lagrange's auxiliary equations are

$$\frac{dx}{z}=\frac{dy}{-z}=\frac{dz}{z^2+(x+y)^2} \quad \text{.....(1)}$$

Solving $\frac{dx}{z}=\frac{dy}{-z}$ we get $x+y=c_1$ .....(2)

Substituting $x+y=c_1$ in (1)

$$\frac{dx}{z}=\frac{dz}{z^2+(x+y)^2}$$

we get $dx=\frac{zdz}{z^2+c_1^2}$. Solving we get

$$x-\frac{1}{2}\log\left(z^2+c_1^2\right)=c_2 \quad \text{(i.e.,)}\ 2x-\log[z^2+(x+y)^2]=c_2 \quad \text{.....(3)}$$

From (2) and (3), the general solution is

$$\phi\left\{x+y, 2x-\log\left[z^2+(x+y)^2\right]\right\}=0$$

Exercise – 18.3

Solve :

1. $xp+yq=z$

Ans : $f(x/z, y/z)=0$

2. $p+q=1$

Ans : $f(x-y, y-z)=0$

3. $zp+x=0$

Ans : $f(x^2+z^2, y)=0$

4. $p\sqrt{x}+q\sqrt{y}=\sqrt{z}$

Ans : $f\left(\sqrt{x}-\sqrt{y}, \sqrt{y}-\sqrt{z}\right)=0$

5. $yzp+zxq=xy$

Ans : $f(x^2-y^2, y^2-z^2)=0$

6. $p+q=\sin x$

Ans : $f(y-x, z+\cos x)=0$

7. $x^2p+y^2q=z^2$

Ans : $f((y-x)/xy, (z-y)/zy)=0$

8. $p\sin x + q\cos y = \tan z$

Ans : $\phi\left(\dfrac{\operatorname{cosec} x - \cot x}{\sec y + \tan y}, \dfrac{\sec y + \tan y}{\sin z}\right) = 0$

9. $p - 2q = 3x^2 \sin(y + 2x)$

Ans : $f(2x + y, x^2\sin(2x + y)) = 0$

10. $p - q = \log(x + y)$

Ans : $f(x + y, x - (z/\log(x + y))) = 0$

11. $\dfrac{\partial z}{\partial x} + \dfrac{\partial z}{\partial y} = x + y + z$

Ans : $f[x + y, e^{-y}(x + y + z + 2)] = 0$

12. $py + qx = xyz^2(x^2 - y^2)$

Ans : $f([x^2 - y^2, (x^2 - y^2)/x^2 + (2/z)] = 0$

13. $xzp + yzq = xy$

Ans : $f(xy - z^2, x/y) = 0$

18.3.2 Multipliers Method to solve the Lagrange's Partial Differential Equations

$pP + qQ = R$ is Lagrange's partial differential equation and its auxiliary equations are

$$\frac{dx}{P} = \frac{dy}{Q} = \frac{dz}{R} \qquad .....(1)$$

In the above equation (1) if any two of the fractions are not integrable by any method choose two sets of suitable multipliers α, β, γ and α', β', γ' which, after using algebraic methods on (1), gives

$$\frac{\alpha dx + \beta dy + \gamma dz}{\alpha P + \beta Q + \gamma R} = \frac{\alpha' dx + \beta' dy + \gamma' dz}{\alpha' P + \beta' Q + \gamma' R} \qquad .....(2)$$

The multipliers are so choosen such that

$$\alpha P + \beta Q + \gamma R = 0 \text{ and } \alpha' P + \beta' + \gamma' = 0$$

After integration, (2) gives two independent solutions $u = c_1$ and $v = c_2$, so that the complete solution is $F(u, v) = 0$.

Solved Examples

***Example* 18.18:** Solve $x^2(y - z)p + y^2(z - x)q = z^2(x - y)$

Solution:

Lagrange's auxiliary equations are

$$\frac{dx}{x^2(y-z)} = \frac{dy}{y^2(z-x)} = \frac{dz}{z^2(x-y)} \qquad .....(1)$$

Choosing $\frac{1}{x^2}, \frac{1}{y^2}, \frac{1}{z^2}$ as multipliers

$$\text{each fraction of (1)} = \frac{\frac{1}{x^2}dx + \frac{1}{y^2}dy + \frac{1}{z^2}dz}{(y-z)+(z-x)+(x-y)}$$

$$\Rightarrow \qquad \frac{1}{x^2}dx + \frac{1}{y^2}dy + \frac{1}{z^2}dz = 0$$

(i.e.) $\frac{1}{x} + \frac{1}{y} + \frac{1}{z} = c_1$ .....(2)

Choosing $\frac{1}{x}, \frac{1}{y}, \frac{1}{z}$ as multipliers each fraction of (1) $= \dfrac{\frac{1}{x}dx + \frac{1}{y}dy + \frac{1}{z}dz}{x(y-z)+y(z-x)+z(x-y)}$

$$\Rightarrow \qquad \frac{1}{x}dx + \frac{1}{y}dy + \frac{1}{z}dz = 0$$

Solving we get $\log x + \log y + \log z = \log c_2$

(i.e.,) $xyz = c_2$ .....(3)

From (2) and (3), the complete solution is

$$\phi\left(\frac{1}{x} + \frac{1}{y} + \frac{1}{z}, xyz\right) = 0$$

***Example* 18.19:** Solve $\frac{(b-c)}{a}yzp + \frac{(c-a)}{b}zxp = \frac{(a-b)}{c}xy$

***Solution*:**

Lagrange's auxiliary equations are

$$\frac{dx}{\left(\frac{(b-c)}{a}\right)yz} = \frac{dy}{\left(\frac{c-a}{b}\right)zx} = \frac{dz}{\left(\frac{a-b}{c}\right)xy} \qquad .....(1)$$

Choosing ax, by, cz as multipliers, each fraction of (1)

$$= \frac{ax\,dx + bydy + czdz}{(b-c)xyz + (c-a)yzx + (a-b)xyz}$$

$$\Rightarrow \qquad axdx + bydy + czdz = 0$$

$$\therefore \qquad ax^2 + by^2 + cz^2 = c_1$$

Choosing a^2x, b^2y, c^2z as multipliers, each fraction of (1)

$$= \frac{a^2\ xdx + b^2ydy + c^2zdz}{a(b-c)xyz + b(c-a)xyz + c(a-b)xyz}$$

$$\Rightarrow \quad a^2\,xdx + b^2\,ydy + c^2\,zdz = 0$$

$$a^2x^2 + b^2y^2 + c^2z^2 = c_2 \qquad .....(3)$$

From (2) and (3), the complete solution is

$$\phi\left(ax^2 + by^2 + cz^2, a^2x^2 + b^2y^2 + c^2z^2\right) = 0$$

***Example* 18.20:** Solve $x(y^2 - z^2)p + y(z^2 - x^2)\,q = z(x^2 - y^2)$

***Solution*:**

Lagrange's auxiliary equations are

$$\frac{dx}{x(y^2 - z^2)} = \frac{dy}{y(z^2 - x^2)} = \frac{dz}{z(x^2 - y^2)} \qquad .....(1)$$

Choosing $x,\ y,\ z$ as multipliers each fraction of (1)

$$= \frac{xdx + ydy + zdz}{x^2(y^2 - z^2) + y^2(z^2 - x^2) + z^2(x^2 - y^2)}$$

$$\Rightarrow \quad xdx + ydy + zdz = 0$$

$$\therefore \quad x^2 + y^2 + z^2 = c_1 \qquad .....(2)$$

Choosing $\frac{1}{x}, \frac{1}{y}, \frac{1}{z}$ as multipliers, each fraction of (1)

$$= \frac{\frac{1}{x}dx + \frac{1}{y}dy + \frac{1}{z}dz}{(y^2 - z^2) + (z^2 - x^2) + (x^2 - y^2)}$$

$$\Rightarrow \quad \frac{1}{x}dx + \frac{1}{y}dy + \frac{1}{z}dz = 0$$

$$\therefore \quad xyz = c_2 \qquad .....(3)$$

From (2) and (3) the complete solution is

$$\phi\left(x^2 + y^2 + z^2, xyz\right) = 0$$

***Example* 18.21**: Solve $(x + 2z)\,p + (4zx - y)q = 2x^2 + y$

***Solution*:**

Lagrange's auxiliary equations are

$$\frac{dx}{x+2z} = \frac{dy}{4zx-y} = \frac{dz}{2x^2+y} \qquad .....(1)$$

Chossing $y, x, -2x$ as multipliers, each fraction of (1)

$$= \frac{ydx + xdy - 2zdz}{y(x+2z) + x(4zx-y) - 2z(2x^2+y)}$$

$\Rightarrow \quad ydx + xdy - 2zdz = 0$

$d(xy) - 2zdz = 0$

$\therefore \quad xy - z^2 = c_1 \qquad .....(2)$

Choosing $2x, -1, -1$ as multipliers

$$\text{each fraction of (1)} = \frac{2xdx - dy - dz}{2x(x+2z) - (4zx-y) - (2x^2+y)}$$

$\Rightarrow \quad x^2 - y - z = c_2 \qquad .....(3)$

From (2) and (3) the complete solution is $f(xy - z^2, x^2 - y - z) = 0$

***Example* 18.22**: Find the integral surface of $x(y^2 + z)\,p - y(x^2 + z)q = z(x^2 - y^2)$, which contains the straight line $x + y = 0$, $z = 1$.

***Solution*:**

Lagrange's auxiliary equations are

$$\frac{dx}{x(y^2+z)} = \frac{dy}{-y(x^2+z)} = \frac{dz}{z(x^2-y^2)} \qquad .....(1)$$

Using $x, y, -1$ as multipliers each fraction of (1)

$$= \frac{xdx + ydy - dz}{x^2(y^2+z) - y^2(x^2+z) - z(x^2-y^2)}$$

$\Rightarrow \quad xdx + ydy - dz = 0$

$\therefore \quad x^2 + y^2 - 2z = c_1 \qquad .....(2)$

Using $\frac{1}{x}, \frac{1}{y}, \frac{1}{z}$ as multipliers

$$\text{each fraction of (1)} = \frac{\frac{1}{x}dx + \frac{1}{y}dy + \frac{1}{z}dz}{(y^2+z)-(x^2+z)+(x^2-y^2)}$$

$$\Rightarrow \quad \frac{1}{x}dx + \frac{1}{y}dy + \frac{1}{z}dz = 0$$

$$xyz = c_2 \qquad .....(3)$$

substitute $z = 1$ in (2) and (3) to get $x^2 + y^2 - 2 = c_1$ and $xy = c_2$

Again, $c_1 + 2c_2 = (x+y)^2 - 2 = -2$ since $x + y = 0$

$\therefore$ $c_1 + 2c_2 + 2 = 0 \Rightarrow$ The required surface is

$x^2 + y^2 - 2z + 2xyz + 2 = 0.$

***Example* 18.23:** Solve $x(z^2 - y^2)p + y(x^2 - z^2)q = z(y^2 - x^2)$

***Solution*:**

The Lagrange's equations are

$$\frac{dx}{x(z^2-y^2)} = \frac{dy}{y(x^2-z^2)} = \frac{dz}{z(y^2-x^2)}$$

Using x, y, z as multipliers, each fraction of (1)

$$= \frac{xdx + ydy + zdz}{x^2(z^2-y^2)+y^2(x^2-z^2)+z^2(y^2-x^2)}$$

$$\Rightarrow \quad xdx + ydy + zdz = 0$$

$$x^2 + y^2 + z^2 = c_1 \qquad .....(2)$$

Using $\frac{1}{x}, \frac{1}{y}, \frac{1}{z}$ as mulitpliers, each fraction of (1)

$$= \frac{\frac{1}{x}dx + \frac{1}{y}dy + \frac{1}{z}dz}{(z^2-y^2)+(x^2-z^2)+(y^2-x^2)}$$

$$\Rightarrow \quad \frac{1}{x}dx + \frac{1}{y}dy + \frac{1}{z}dz = 0$$

$$\therefore \quad xyz = c_2 \qquad .....(3)$$

From (2) and (3) the complete solution is

$$\phi(x^2 + y^2 + z^2, xyz) = 0$$

***Example* 18.24:** Solve $(y^2 + z^2 - x^2)p - 2xyq = -2zx$

Solution:

Lagrange's auxiliary equations are

$$\frac{dz}{y^2+z^2-x^2} = \frac{dy}{-2xy} = \frac{dz}{-2zx} \qquad .....(1)$$

$$\frac{dy}{-2xy} = \frac{dz}{-2zx} \text{ gives } \log y = \log z + \log c_1$$

(i.e.,) $$\frac{y}{z} = c_1 \qquad .....(2)$$

Using x, y, z as multipliers each fraction of (1)

$$= \frac{xdx + ydy + zdz}{x(y^2+z^2-x^2) + y(-2xy) + z(-2zx)}$$

$$\therefore \quad \frac{2(xdx + ydy + zdz)}{2x(x^2+y^2+z^2)} = \frac{dy}{2xy} \text{ gives}$$

$$x^2 + y^2 + z^2 = c_2 y$$

$$\Rightarrow \quad \frac{x^2+y^2+z^2}{y} = c_2 \qquad .....(3)$$

from (2) and (3), the complete solution is

$$\phi\left(\frac{y}{z}, \frac{x^2+y^2+z^2}{y}\right) = 0$$

***Example* 18.25:** Solve $(2x^2 + y^2 + z^2 - 2yz - zx - xy)p + (x^2 + 2y^2 + z^2 - yz - 2zx - xy)$ $q = (x^2 + y^2 + 2z^2 - yz - 2xy)$

Solution:

The Lagrange's auxiliary equations are

$$\frac{dx}{2x^2+y^2+z^2-2yz-zx-xy} = \frac{dy}{x^2+2y^2+z^2-yz-2zx-xy}$$

$$= \frac{dz}{x^2+y^2+2z^2-yz-zx-2xy} \qquad .....(1)$$

From (1) $$\frac{dx-dy}{(x-y)(x+y+z)}=\frac{dy-dz}{(y-z)(x+y+z)}$$

$\therefore$ $$\frac{dx-dy}{x-y}=\frac{dy-dz}{y-z}$$

$\Rightarrow$ $$\frac{x-y}{y-z}=c_1 \qquad .....(2)$$

Similarly $$\frac{dy-dz}{(y-z)(x+y+z)}=\frac{dz-dx}{(z-x)(x+y+z)} \Rightarrow \frac{dy-dz}{y-z}=\frac{dz-dx}{z-x}$$

$\Rightarrow$ $$\frac{y-z}{z-x}=c_2 \qquad .....(3)$$

From (2) and (3), the complete solution is

$$\phi\left(\frac{x-y}{y-z},\frac{y-z}{z-x}\right)=0$$

***Example* 18.26:** Solve $yp - xq = -\ xe^{x^2+y^2}$

***Solution*:**

Lagrange's auxiliary equations are

$$\frac{dx}{y}=\frac{dy}{-x}=\frac{dz}{-xe^{x^2+y^2}} \qquad .....(1)$$

From the first two fractions we get

$$x\,dx + y\,dy = 0 \Rightarrow x^2 + y^2 = c_1 \qquad .....(2)$$

Also from (1) we have

$$\frac{dy}{x}=\frac{dz}{xe^{x^2+y^2}} \qquad .....(3)$$

Substituting $x^2 + y^2 = c_1$ from (2), equation (3) gives

$$dy = \frac{dz}{e^{c_1}} \Rightarrow e^{c_1}dy = dz$$

$$e^{c_1}y - z = c_2 \text{ (i.e.,) } ye^{x^2+y^2} - z = c_2 \qquad .....(4)$$

From (2) and (3), the complete solution is

$$\phi\left(x^2+y^2, ye^{x^2+y^2} - z\right) = 0$$

Exercise - 18.4

Solve :

(1) $x(y-z)p + y(z-x)q = z(x-y)$

Ans : $\phi(x+y+z,\ xyz) = 0$

(2) $(mz-ny)p + (nx-lz)q = ly - mx$

Ans : $\phi\left(lx+my+nz,\ x^2+y^2+z^2\right) = 0$

(3) $(x^2-yz)p + (y^2-zx)q = z^2 - xy$

Ans : $\varphi\left\{\frac{(z-x)}{y-z}, \frac{x-y}{y-z}\right\} = 0$

(4) $(y+z)p + (z+x)q = x + y$

Ans : $\phi\{(x-y)(y-z),\ (x-y)\sqrt{x+y+z}\} = 0$

(5) $xzp + yzq = xy$

Ans : $\phi\{(x/y),\ xy - z^2) = 0$

(6) $\left(\frac{y-z}{yz}\right)p + \left(\frac{z-x}{zx}\right)q = \left(\frac{x-y}{xy}\right)$

Ans : $\phi(x+y+z,\ xyz) = 0$

(7) $(y^2+z^2)p - xyq = -zx$

Ans : $\phi(y/z,\ x^2+y^2+z^2) = 0$

(8) $(y-zx)p + (x+yz)q = x^2 + y^2$

Ans : $\phi(x^2-y^2+z^2,\ xy-z) = 0$

18.4 Non-linear Partial Differential Equations of Order One

18.4.1 Definitions

(a) ***Complete Integral :***

The equation $\phi(x, y, z, a, b) = 0$ .....(1)

which contains as many constants as the number of independent variables is called the complete integral of the partial differential equation

$$F(x, y, z, p, q) = 0 \qquad .....(2)$$

(b) ***Particular Integral :***

A particular integral of (2) is obtained by giving particular values to 'a' and 'b' in the complete integral (1).

(c) ***Singular Integral :***

The equation of the envelope of the surface represented by the complete integral of a given partial differential equation (2) is called its singular integral.

If $\phi(x, y, z, a, b) = 0$ is the complete integral then the envelope of the above surface is obtained by eliminating 'a' and 'b' from $\phi(x, y, z, a, b) = 0$, $\frac{\partial \phi}{\partial a} = 0$ and $\frac{\partial \phi}{\partial b} = 0$.

The eliminant which will be the envelope of $\phi(x, y, z, a, b) = 0$ is called the singular integral.

(d) ***General Integral :***

Assume that in the complete integral $\phi(x, y, z, a, b)$ one of the constants is a function of the other (say $b = f(a)$) then the complete integral becomes

$$f(x, y, z, a, f(a)) = 0 \qquad .....(3)$$

The general integral is obtained by eliminating 'a' between (3) and $\frac{\partial \phi}{\partial a} = 0$

18.4.2 Standard Form I : (Equations involving *p* and *q* only)

$$\phi(p, q) = 0 \qquad .....(1)$$

The complete integral of (1) is given by $z = ax + by + c$ .....(2)

where 'a' and 'b' are connected by $\phi(a, b) = 0$ .....(3)

From the relation $\phi(a, b) = 0$, we can write b interms of 'a', say, $b = \psi(a)$ and putting in (2) the complete integral which contains the arbitrary constants 'a' and 'c' is obtained

i.e., $$z = ax + \phi(a)y + c \qquad .....(4)$$

is the complete integral (number of arbitrary constants are equal to number of independent variables).

Differentiating (4) partially w.r.t 'a' and 'c' we get

$$0 = x + \psi'(a)y + 0 \qquad .....(5)$$

and $$0 = 1 \text{ (which is absurd)} \qquad .....(6)$$

The equation (6) shows that such equations will have no singular integral as the singular integral is obtained by eliminating 'a' and 'c' from (4), (5), (6).

Put $c = \aleph(a)$ in the complete integral, then we have

$$Z = ax + \psi(a)y + \aleph(a) \quad .....(7)$$

Differentiate (7) partially w.r.t. to 'a' giving

$$0 = x + \phi'(a)y + \aleph'(a) \quad .....(8)$$

The general integral is obtained by eliminating 'a' from (7) and (8)

Solved Examples

***Example* 18.27:** Solve $p^2 + q^2 = m^2$

Solution:

$$p^2 + q^2 = m^2 \quad .....(1)$$

the complete integral is $z = ax + by + c$ where 'a', 'b' are connected by the relation $a^2 + b^2 = m^2$

Put $b = \sqrt{m^2 - a^2}$ in $z = ax + by + c$

$$z = ax + \sqrt{m^2 - a^2}\ y + c \quad .....(2)$$

is the complete integral

Partially differentiate (2) w.r.t. 'a' and 'c'

$$0 = x - \left(\frac{ay}{\sqrt{m^2 - a^2}}\right) + 0 \quad .....(3)$$

$$0 = 1 \quad .....(4)$$

Elimination of 'a' and 'c' from (2), (3) and (4) gives the singular integral.

Equation (4) shows that the equations will have no singular integral. Put $c = \psi(a)$ in (2) then we have

$$z = ax + \sqrt{m^2 - a^2} + \psi(a) \quad .....(5)$$

Differentiating (5) partially w.r.t. 'a'

$$0 = x - \frac{ay}{\sqrt{m^2 - a^2}} + \psi(a) \quad .....(6)$$

General integral is obtained by eliminating 'a' from (5) and (6)

***Example* 18.28:** Solve $\sqrt{p} + \sqrt{q} = 1$

Solution:

$$\sqrt{p} + \sqrt{q} = 1 \quad .....(1)$$

the complete integral is $z = ax + by + c$

where 'a' and 'b' are connected by the relation $\sqrt{a} + \sqrt{b} = 1$

substituting $b = (1 - \sqrt{a})^2$ in $z = ax + by + c$ we get the complete Integral as

$$z = ax + (1 - \sqrt{a})^2 y + c$$

***Example* 18.29:** Solve $x^2p^2 + y^2q^2 = z^2$

Solution:

$$x^2p^2 + y^2 q^2 = z^2 \qquad .....(1)$$

We now reduce the given equation to the form $f(p, q) = 0$

(1) can be written as

$$\left(x\frac{\partial z}{\partial x}\right)^2 + \left(y\frac{\partial z}{\partial y}\right)^2 = z^2$$

$$\left(\frac{x}{z}\frac{\partial z}{\partial x}\right)^2 + \left(\frac{y}{z}\frac{\partial z}{\partial y}\right)^2 = 1 \qquad .....(2)$$

Substituting $\log x = X$, $\log y = Y$, $\log z = Z$

$$\Rightarrow \qquad \frac{\partial X}{\partial x} = \frac{1}{x}, \frac{\partial Y}{\partial y} = \frac{1}{y}, \frac{\partial Z}{\partial z} = \frac{1}{z}$$

$$\frac{\partial Z}{\partial X} = \frac{x}{z}\frac{\partial z}{\partial x}$$

Similarly $\quad \dfrac{\partial Z}{\partial Y} = \dfrac{y}{z} \cdot \dfrac{\partial z}{\partial y}$

Equation (2) can be written as

$$\left(\frac{\partial Z}{\partial X}\right)^2 + \left(\frac{\partial Z}{\partial Y}\right)^2 = 1 \qquad .....(3)$$

which is in the form $f(p, q) = 0$ as $P^2 + Q^2 = 1$ where $P = \dfrac{\partial Z}{\partial X}$, $Q = \dfrac{\partial Z}{\partial Y}$

The complete integral is $Z = aX + bY + c \qquad .....(4)$

where a, b are connected by the relation $a^2 + b^2 = 1$. Substituting $b = \pm\sqrt{1-a^2}$ in (4) we get the complete integral $Z = aX \pm \sqrt{1-a^2}\, Y + c$

$\therefore$ From (3), the complete integal of the given $p.d.e$ is

$$\log z = a\log x \pm \sqrt{1-a^2}\, \log y + c$$

***Example* 18.30:** Find complete integral of $pq = x^m y^n z^l$

***Solution*:**

$$\left(\frac{\partial z}{\partial x}\right)\left(\frac{\partial z}{\partial y}\right) = x^m y^n z^l \qquad .....(1)$$

This can be written as

$$\left(\frac{\partial z}{x^m z^{\frac{l}{2}} \partial x}\right)\left(\frac{\partial z}{y^n z^{\frac{l}{2}} \partial y}\right) = 1 \qquad .....(2)$$

Introducing $\frac{\partial X}{\partial x} = x^m,\ \frac{\partial Y}{\partial y} = y^n,\ \frac{\partial Z}{\partial z} = \frac{1}{z^{l/2}}$

(i.e.,) $\frac{x^{m+1}}{m+1} = X,\ \frac{y^{n+1}}{n+1} = Y,\ \left(\frac{Z^{\frac{-l}{2}+1}}{\frac{-l}{2}+1}\right) = Z$

The given equation reduces to

$$\left(\frac{\partial Z}{\partial X}\right)\left(\frac{\partial Z}{\partial Y}\right) = 1$$

The complete integral is

$$Z = aX + bY + c \qquad .....(4)$$

where a, b are connected by $ab = 1$.

substituting $b = \frac{1}{a}$ in (4), the complete integral is

$$Z = aX + \frac{1}{a} Y + c$$

Using (3) in (4) we get

$$\frac{z^{-1/2}+1}{-\frac{l}{2}+1}=a\frac{x^{m+1}}{m+1}+\frac{1}{a}\frac{y^{n+1}}{n+1}+c$$

(i.e)
$$\frac{2\left(z^{\frac{-1}{2}}+1\right)}{2-l}=a\frac{x^{m+1}}{m+1}+\frac{1}{a}\frac{y^{n+1}}{n+1}+c$$

Ecercise 18.5

Find complete integrals of the following equations:

1. $q=e^{-p/\alpha}$

   Ans : $z=ax+e^{-p/a}y+c$

2. $q=3p^2$

   Ans : $z=ax+3a^2y+c$

3. $q=2p^2+1$

   Ans : $z=ax+\frac{1}{\sqrt{2}}\left(\sqrt{a-1}\right)y+c$

4. $p^2+p=q^2$

   Ans : $z=ax+\left(\sqrt{a^2+a}\right)y+c$

5. $p+q=pq$

   Ans : $z=ax+\frac{a}{a-1}y+c$

6. $p^2q^3=1$

   Ans : $z=ax+\left(\frac{1}{a^2}\right)^{1/3}y+c$

18.4.3 Standard Form II

$$f(x, y, p, q)=0$$

Arrange the given equation in the form $f_1(x, p)=f_2(y, q)$.

Let $f_1(x, p)=f_2(y, q)=a$

where 'a' is an arbitrary constant.

From $\quad f_1(x, p) = a$, and $f_2(y, q) = a$

we get $\quad p = \phi_1(x, a)$ and $q = \phi_2(y, a)$ .....(1)

Now
$$dz = \frac{\partial z}{\partial x}dx + \frac{\partial z}{\partial y}dy$$
$$dz = \phi_1(x, a)x + \phi_2(y, a)dy$$

integrating

$$z = \int \phi_1(x,a)dx + \int \phi_2(y,a)dy + c$$

is the complete integral. The general integral is obtained as usual, whereas there will be no singular integral as in the case of standard form I.

Solved Examples

***Example* 18.31:** Solve $pq = xy$

***Solution*:**

$$pq = xy$$

$$\frac{p}{x} = \frac{y}{q} = a, \text{ where 'a' is an arbitrary constant.}$$

$$\frac{p}{x} = a \Rightarrow p = ax, \; \frac{y}{q} = a \Rightarrow q = \frac{y}{a}$$

Substituting for p, q in $dz = pdx + qdy$

$$dz = ax \, . \, dx + \frac{y}{a}dy$$

Integrating we get

$$z = a\frac{x^2}{2} + \frac{1}{a}\frac{y^2}{2} + c \quad .....(1)$$

which is the complete integral.

Differentiating (1)partially w.r.t. 'a' and 'c'

$$0 = \frac{x^2}{2} - \frac{1}{a^2}\frac{y^2}{2} + 0 \quad .....(2)$$

and $\quad 0 = 0 + 0 + 1$ .....(3)

Result (3) is absurd.

Hence there is no singular integral.

To find General integral, put $c = f(a)$ in (1)

$$z = a\frac{x^2}{2} + \frac{1}{a}\frac{y^2}{2} + f(a) \qquad \text{.....(4)}$$

Differentiating (4) partially w.r.t. 'a'

$$0 = \frac{x^2}{2} - \frac{1}{a^2}\frac{y^2}{2} + f'(a) \qquad \text{.....(5)}$$

The general integral is obtained by eliminating 'a' from (4) and (5)

***Example* 18.32:** Solve $p^2 + q^2 = x^2 + y^2$

Solution:

$$p^2 + q^2 = x^2 + y^2 \qquad \text{.....(1)}$$

Rewriting the given equation we have

$p^2 - x^2 = y^2 - q^2 = a^2$, where '$a$' is an arbitrary constant.

$$p = \sqrt{a^2 + x^2} \text{ and } q = \sqrt{y^2 - a^2}$$

substituting for p, q in $dz = pdx + qdy$

$$dz = \sqrt{a^2 + x^2}\, dx + \sqrt{y^2 - a^2}\, dy$$

Integrating

$$z = \frac{x}{2}\sqrt{a^2 + x^2}\, dx + \frac{a^2}{2}\sinh^{-1}(x/a) + \frac{y}{2}\sqrt{y^2 + a^2} - \frac{a^2}{2}\cosh^{-1}(x/a) + c$$

is the complete integral.

***Example* 18.33:** Find the integral of $z^2 (p^2 + q^2) = x^2 + y^2$

Solution:

$$z^2 (p^2 + q^2) = x^2 + y^2$$

(i.e.,) $$z^2\left[\left(\frac{\partial z}{\partial x}\right)^2 + \left(\frac{\partial z}{\partial y}\right)^2\right] = x^2 + y^2$$

$$\Rightarrow \quad \left(\frac{z\partial z}{\partial x}\right)^2 + \left(\frac{z\partial z}{\partial y}\right)^2 = x^2 + y^2 \qquad \text{.....(1)}$$

substituting $\frac{z\partial z}{\partial x} = \frac{\partial Z}{\partial x}$ and $\frac{z\partial z}{\partial y} = \frac{\partial Z}{\partial y}$ we have

$$\left(\frac{\partial Z}{\partial x}\right)^2 + \left(\frac{\partial Z}{\partial y}\right)^2 = x^2 + y^2$$

$$p^2 + q^2 = x^2 + y^2$$

is of the form $f(p, q, x, y) = 0$

$$p^2 + x^2 = y^2 - q^2 = a^2$$

where 'a' is an arbitrary constant

$$\therefore \quad p = \sqrt{a^2 + x^2} \text{ and } q = \sqrt{y^2 - a^2}$$

substituting in $dZ = pdx + qdy$

$$\therefore \quad dZ = \sqrt{a^2 + x^2}\, dx + \sqrt{y^2 - a^2}\, dy$$

Integrating

$$z^2 = x\sqrt{a^2 + x^2} + a^2 \sinh^{-1}(x/a) + y^2\sqrt{y^2 - a^2} - a^2 \cosh^{-1}(x/a) + c$$

is the complete integral.

Exercise - 18.6

Solve :

1. $x^2p^2 = yq^2$

   Ans : $z = a\log x + 2a\sqrt{y + c}$

2. $p - 3x^2 = q^2 - y$

   Ans : $z = x^3 + ax \pm 2/3(y - a)^{3/2} + c$

3. $pq = ay$

   Ans : $2az = a^2x^2 + y^2 + c$

4. $p^{1/3} - q^{1/3} = 3x - 2y$

   Ans : $z = 3x^3 - 3ax^2 + a^2x + 2y^4 - 4ay^3 + 3a^2y^2 - a^3y + c$

5. $yp = 2yx + \log q$

   Ans : $az = ax^2 + a^2 + e^{ay} + ac$

6. $p^2 = q + x$

   Ans : $3z = 3ay \pm 2(x + a)^{3/2} + 3c$

7. $q = xyp^2$

   Ans : $(2z - ay^2 - 2b)^2 = 16ax.$

8. $p + q = \sin x + \sin y$

   Ans : $z = a(x - y) - (\cos x + \cos y) + c$

18.4.4 Standard Form III

Consider the Partial differential equations of the form $f(z, p, q) = 0$
i.e., those equations which donot contain x and y

$$f(z, p, q) = 0 \qquad .....(1)$$

substitute $\quad q = ap \qquad .....(2)$

solving (1) and (2) p is obtained in terms of z, say $p = \phi(z)$ $\qquad .....(3)$

from (2) and (3) we obtain $q = a\phi(z)$ $\qquad .....(4)$

substituting for p, q in $dz = pdx + qdy$

$$dz = \phi(z)\, dx + a\phi(z)dy \quad \text{(i.e.,)} \quad \frac{dz}{\phi(z)} = dx + ady$$

Integrating

$$x + ay = F(z) + c \quad \text{where} \qquad F(z) = \int \frac{dz}{\phi(z)}$$

is the complete integral of (1). Singular and general integrals are then obtained by usual methods.

Solved Examples

***Example* 18.34:** Solve $p^3 + q^3 = 3pqz$

***Solution*:**

$$p^3 + q^3 - 3pqz = 0 \qquad .....(1)$$

take $\quad q = ap \qquad .....(2)$

Then (1) reduces to

$$p^3 + a^3p^3 - 3p^2az = 0$$

$$\Rightarrow \quad p^2\{p(1 + a^3) - 3az\} = 0$$

$$\Rightarrow \quad p = \frac{3az}{1+a^3}, \text{ from (2) } q = \frac{3a^2z}{1+a^3} \qquad .....(3)$$

Substituting for p, q in $dz = pdx + qdy$

$$dz = \frac{3az}{1+a^3}dx + \frac{3a^2z}{1+a^3}dy \quad \Rightarrow \quad \frac{1+a^3}{3a}\frac{dz}{z} = dx + ady$$

Integrating we get $\dfrac{1+a^3}{3a}\log z = (x + ay) + c$ which is the complete integral.

***Example* 18.35:** Solve $p^2 = z^2(1 - pq)$

***Solution*:**

$$p^2 = z^2(1 - pq) \qquad .....(1)$$

take $\quad q = ap \qquad .....(2)$

from (1), (2)

$$p^2 = z^2(1 - ap^2)$$

$$p^2 + az^2p^2 = z^2$$

$$\therefore \quad p = \frac{z}{\sqrt{1+az^2}} \text{ and } q = \frac{az}{\sqrt{1+az^2}}$$

substituting for p, q in the identity $dz = pdx + qdy$

$$dz = \frac{z}{\sqrt{1+az^2}}dx + \frac{az}{\sqrt{1+az^2}}dy$$

(i.e.,)
$$\frac{\sqrt{1+az^2}}{z}dz = dx + ady \qquad .....(3)$$

substituting $1 + az^2 = t^2 \quad \Rightarrow 2azdz = 2tdt$

$$\Rightarrow \quad \frac{dz}{z} = \frac{tdt}{az^2} = \frac{tdt}{t^2-1}$$

equation (3) reduces to $t.\dfrac{tdt}{t^2-1} = dx + ady \qquad .....(4)$

integrating (4) we have $\displaystyle\int\frac{t^2dt}{t^2-1} = x + ay + c$

$$\int\left(1+\frac{1}{t^2-1}\right)dt = x + ay + c$$

$$t + \frac{1}{2}\log\frac{t-1}{t+1} = x + ay + c$$

(i.e.,)
$$\sqrt{1+az^2} + \frac{1}{2}\log\frac{\sqrt{1+az^2}-1}{\sqrt{1+az^2}+1} = (x + ay + c)$$

The above equation is the Complete Integral of equation (1)

***Example* 18.36:** Solve $p(1 + q^2) = q(z - a)$

***Solution*:**

$$p(1 + q^2) = q(z - a) \quad .....(1)$$

from (1) substituting $q = ap$ we get

$$p(1 + a^2p^2) = ap(z - a) \quad \Rightarrow \quad 1 + a^2p^2 = a(z - a)$$

$$a^2p^2 = a(z - a) - 1 \qquad \text{(i.e.,)} \quad ap = \sqrt{a(z-a)-1}$$

$$p = \frac{\sqrt{a(z-a)-1}}{a} \quad .....(2)$$

$$\therefore \quad q = \sqrt{a(z-a)-1}$$

substituting for p, q in $dz = pdx + qdy$

$$dz = \frac{\sqrt{a(z-a)-1}}{a} dx + \sqrt{a(z-a)-1} \, . \, dy$$

$$\frac{a}{\sqrt{a(z-a)-1}} dz = dx + ady \quad .....(3)$$

Integrating (3) we get

$$2\sqrt{a(z-a)-1} = x + ay + c$$

Squaring gives

$$4[a(z - a) - 1] = (x + ay + c)^2$$

which is the complete integral.

Exercise - 18.7

Solve :

1. $p^3 + q^3 = 27z$ **Ans :** $z^2(1 + a^3) = 8(x + ay + c)^3$
2. $9(p^2z + q^2) = 4$ **Ans :** $(z + a^2)3 = (x + ay + c)^2$
3. $z^2(p^2 + q^2 + 1) = c^2$ **Ans :** $(a^2 + 1)(c^2 - z^2) = (x + ay + b)^2$
4. $p(1 + q) = qz$ **Ans :** $\text{Log}(az - 1) = (x + ay + c)$
5. $z = pq$ **Ans :** $(x + ay + c)^2 = 4az$
6. $z^2p^2 + q^2 = p^2q$ **Ans :** $z = a \tan(x + ay + c)$
7. $z = p^2 - q^2$ **Ans :** $4(1 - a^2) z = (x + ay + c)$

18.4.5 Standard Form –IV (CLAIRAUT'S TYPE)

Consider a partial differential equation of the form

$$z = px + qy + f(p, q) \qquad .....(1)$$

The complete integral of the equation of the type (1) is

$$z = ax + by + f(a, b) \qquad .....(2)$$

Writing $b = \phi(a)$ in (2) gives

$$z = ax + \phi(a)y + f\{a, \phi(a)\} \qquad .....(3)$$

and then differentiating w.r.t. 'a'

$$0 = x + \phi'(a)y + \frac{\partial f}{\partial a} \qquad .....(4)$$

The general integral is obtained by eliminating 'a' between (3) and (4)

To find the singular integral, differentiate (2) partially w.r.t. 'a' and 'b'

$$0 = x + \frac{\partial f}{\partial a} \qquad .....(5)$$

$$0 = y + \frac{\partial f}{\partial b} \qquad .....(6)$$

Elimination of 'a' and 'b' between (2), (5), (6) gives the singular integral.

Solved Examples

***Example* 18.37:** Solve $z = px + qy + p^2 + q^2$

***Solution*:**

$$z = px + qy + p^2 + q^2 \qquad .....(1)$$

is of the form $z = px + qy + f(p, q)$

$\therefore$ The complete integral of the equation (1) is $z = ax + by + a^2 + b^2$.

$$.....(2)$$

Differentiating (2) partially w.r.t. 'a' and 'b'

$$0 = x + 2a \qquad .....(3)$$

and $$0 = y + 2b \qquad .....(4)$$

eliminating 'a' from (3), (4) and substituting for a and b in (2)

$$z = \left(-\frac{x}{2}\right)x + \left(-\frac{y}{2}\right)y + \frac{x^2}{4} + \frac{y^2}{4}$$

$x^2 + y^2 + 4z = 0$ is the singular integral

substituting $b = \phi(a)$ in the complete integral (2) we have

$$z = ax + \phi(a)y + a^2 + [\phi(a)]^2 \quad .....(5)$$

differentiating (5) partially w.r.t 'a' we get

$$0 = x + \phi'(a)y + 2a + 2\phi(a)\,\phi'(a) \quad .....(6)$$

To obtain the general integral eliminate 'a' from (5) and (6).

***Example* 18.38:** Solve $z = px + qy - 2\sqrt{pq}$

***Solution*:**

$$z = px + qy - 2\sqrt{pq} \quad .....(1)$$

The complete integral of (1) is

$$z = ax + by - 2\sqrt{ab} \quad .....(2)$$

differentiating (2) partially w.r.t. 'a' and 'b'

$$0 = x - \sqrt{\frac{b}{a}} \quad .....(3)$$

$$0 = y - \sqrt{\frac{a}{b}} \quad .....(4)$$

eliminating 'a', 'b' from (3), (4) we get

$$x = \sqrt{\frac{b}{a}},\ y = \sqrt{\frac{a}{b}}$$

$\Rightarrow xy = 1$ is the singular integral

To find the general integral put $b = \phi(a)$ in (2)

$$z = ax + \phi(a)\,y - 2\sqrt{a}\ .\ \phi(a) \quad .....(5)$$

differentiating (5) partially w.r.t. 'a' we have

$$0 = x + \phi'(a)y - \frac{\left[a\varphi'(a)+\varphi(a)\right]}{\sqrt{a\varphi(a)}} \quad .....(6)$$

The general integral is obtained by eliminating 'a' from (5) and (6)

***Example* 18.39:** Solve $z = px + qy + c\sqrt{1+p^2+q^2}$

***Solution*:**

$$z = px + qy + c\sqrt{1+p^2+q^2} \qquad .....(1)$$

the complete integral of (1) is

$$z = ax + by + c\sqrt{1+a^2+b^2} \qquad .....(2)$$

to obtain singular integral, differentiate (2) partially w.r.t. to '*a*' and '*b*'

$$0 = x + c\frac{a}{\sqrt{1+a^2+b^2}} \qquad .....(3)$$

$$0 = y + c\frac{b}{\sqrt{1+a^2+b^2}} \qquad .....(4)$$

from (3) and (4) we have

$$x^2 + y^2 = \frac{c^2(a^2+b^2)}{(1+a^2+b^2)} \text{ and } c^2 - x^2 - y^2 = c^2 - \frac{c^2(a^2+b^2)}{(1+a^2+b^2)}$$

$$c^2 - x^2 - y^2 = \frac{c^2}{(1+a^2+b^2)} \qquad .....(5)$$

From (3) $\quad a = -\frac{x}{c}\sqrt{1+a^2+b^2} \quad \Rightarrow \quad a = \frac{-x}{\sqrt{c^2-x^2-y^2}}$ by (5)

From (4) $\quad b = -\frac{y}{c}\sqrt{1+a^2+b^2} \quad \Rightarrow \quad b = \frac{-y}{\sqrt{c^2-x^2-y^2}}$ by (5)

substituting the values of a, b in (2)

$$z = -\frac{x^2}{\sqrt{c^2-x^2-y^2}} - \frac{y^2}{\sqrt{c^2-x^2-y^2}} + c\,.\,\frac{c}{\sqrt{c^2-x^2-y^2}}$$

(i.e.,) $\quad z = \sqrt{c^2-x^2-y^2} \quad \Rightarrow \quad x^2 + y^2 + z^2 = c^2$

is the singular integral.

To find the general integral, put $b = \phi(a)$ in (2) giving

$$z = ax + \phi(a)y + c\sqrt{1+a^2+[\phi(a)]^2} \qquad .....(6)$$

differentiating (6) partially w.r.t. 'a'

$$0 = x + \phi^{|}(a)y + c\frac{|a+\phi(a)\phi^{|}(a)|}{\sqrt{1+a^2+[\phi(a)]^2}} \qquad .....(7)$$

eliminating 'a' from (6) and (7) we get the general integral.

Exercise - 18.7

Solve :

1. $z = px + qy + pq$

   Ans : C.I. is $z = ax + by + ab$
   S.I. is $z + xy = 0$

2. $z = px + qy + \sqrt{\alpha p^2 + \beta q^2 + \gamma}$

   Ans : C.I. is $Z = ax + by + \sqrt{\alpha a^2 + \beta b^2 + \gamma}$

   S.I. is $\frac{x^2}{\alpha} + \frac{y^2}{\beta} + \frac{z^2}{\gamma} = 1$

3. $z = px + qy - p^2q$

   Ans : C.I. is $z = ax + by - a^2b$
   S.I. is $z^2 = x^2y$

4. $(p - q)(z - px - qy) = 1$

   Ans : C.I. is $z = ax + by + \frac{1}{a-b}$

5. $z = px + qy + 4p^2q^2$

   Ans : C.I. is $z = ax + by + 4a^2 b^2$

6. $z = px + qy + \sqrt{\frac{pq}{p+q}}$

   Ans : C.I. is $z = ax + by + \sqrt{\frac{ab}{a+b}}$

Summary

1. Formation of Partial differential equation by the elimination of arbitrary constants and arbitrary functions.

(a) ***Elimination of arbitrary constants***

Let $f(x, y, z, a, b) = 0$ ..... (1)

be the equation where a, b are arbitrary constants.

Differentiating partially w.r.to x and y

$$\frac{\partial f}{\partial x} + \frac{\partial f}{\partial z}.\frac{\partial z}{\partial x} = 0 \quad \Rightarrow \quad \frac{\partial f}{\partial x} + p\frac{\partial f}{\partial z} = 0 \quad \text{..... (2)}$$

$$\frac{\partial f}{\partial y} + \frac{\partial f}{\partial z}\frac{\partial z}{\partial y} = 0 \quad \Rightarrow \quad \frac{\partial f}{\partial y} + q\frac{\partial f}{\partial z} = 0 \quad \text{..... (3)}$$

Elimination of two constants a, b from (1), (2), (3) gives an equation of the form $\phi\ (x, y, z, p, q) = 0$ which is the first order P.D.E.

If the numbers of constants are more than the number of independent variables, then the result of eliminating the constants will give rise to a P.D.E of higher order than the first.

(b) ***Elimination of arbitrary functions*:**

Let $\phi\ (u, v) = 0$ ..... (1)

be the equation where u, v are functions of x, y, z and ϕ be the arbitrary function.

Differentiating (1) partially with respect to x and y

$$\frac{\partial \phi}{\partial u}\left(\frac{\partial u}{\partial x} + \frac{\partial u}{\partial z}\frac{\partial z}{\partial x}\right) + \frac{\partial \phi}{\partial v}\left(\frac{\partial v}{\partial x} + \frac{\partial v}{\partial z}\frac{\partial z}{\partial x}\right) = 0 \quad \text{..... (2)}$$

and

$$\frac{\partial \phi}{\partial u}\left(\frac{\partial u}{\partial y} + \frac{\partial u}{\partial z}\frac{\partial z}{\partial y}\right) + \frac{\partial \phi}{\partial v}\left(\frac{\partial v}{\partial y} + \frac{\partial v}{\partial z}.\frac{\partial z}{\partial y}\right) = 0 \quad \text{..... (3)}$$

Eliminating $\frac{\partial \phi}{\partial u}, \frac{\partial \phi}{\partial v}$ from (2), (3)

$$\left(\frac{\partial u}{\partial y}\frac{\partial v}{\partial z} - \frac{\partial u}{\partial z}\frac{\partial v}{\partial y}\right).\frac{\partial z}{\partial x} + \left(\frac{\partial u}{\partial z}\frac{\partial v}{\partial x} - \frac{\partial u}{\partial x}\frac{\partial v}{\partial z}\right)\frac{\partial z}{\partial y} = \frac{\partial u}{\partial x}\frac{\partial v}{\partial y} - \frac{\partial u}{\partial y}\frac{\partial v}{\partial x}$$

$$\frac{\partial z}{\partial x} = p, \quad \frac{\partial z}{\partial y} = q$$

$\therefore$ Pp + Qq = R is the required P.D.E

where $$P = \frac{\partial u}{\partial y}\frac{\partial v}{\partial z} - \frac{\partial u}{\partial z}\frac{\partial v}{\partial y}$$

$$Q = \frac{\partial u}{\partial z}\frac{\partial v}{\partial x} - \frac{\partial u}{\partial x}\frac{\partial v}{\partial z}$$

$$R = \frac{\partial u}{\partial x}\frac{\partial v}{\partial y} - \frac{\partial u}{\partial y}\frac{\partial v}{\partial x}$$

2. *Lagrange's Linear P.D.E*

The P.D.E $$Pp + Qq = R \qquad \text{..... (1)}$$

is called Lagrange's first order partial differential equation

where $$P = \frac{\partial u}{\partial y}\frac{\partial v}{\partial z} - \frac{\partial u}{\partial z}\frac{\partial v}{\partial y}$$

$$Q = \frac{\partial u}{\partial z}\frac{\partial v}{\partial x} - \frac{\partial u}{\partial x}\frac{\partial v}{\partial z}$$

$$R = \frac{\partial u}{\partial x}\frac{\partial v}{\partial y} - \frac{\partial u}{\partial y}\frac{\partial v}{\partial x}$$

To solve (1), first write Lagrange's auxiliary equation (subsidary equation)

$$\frac{\partial x}{P} = \frac{\partial y}{Q} = \frac{\partial z}{R} \qquad \text{..... (2)}$$

Auxiliary equation gives two independent solutions u = c_1 and v = c_2 where u, v are functions of *x, y, z*

From these two solutions, the general solution is $\phi\ (u, v) = 0$

3. *Non linear partial differential equations of order one*

(i) *Complete integral*

If $F(x, y, z, p, q) = 0$ ..... (1)

is the non linear partial differential equation of first order then the equation

$$\phi\ (x, y, z, a, b) = 0 \qquad \text{..... (2)}$$

which contains as many constants as the number of independent variables is called the complete integral.

(ii) *Particular integral*: A particular integral of (1) is obtained by giving particular values to a and b in the complete integral.

(iii) *Singular integral*: Differentiating the complete integral ϕ (x, y, z, a, b)= 0 ..... (2)

Partially w. r. to 'a' and 'b' and then equate to zero

$$\frac{\partial \phi}{\partial a} = 0 \qquad \text{..... (3)}$$

$$\frac{\partial \phi}{\partial b} = 0 \qquad \text{.... (.4)}$$

Elimination of 'a' and 'b' from (2), (3), (4), gives an equation of the form f (x, y, z) = 0 is called singular integral.

There are four standard forms of non-linear first order partial differential equations.

(i) ***Standard form* I**: The equation of the form f (p, q) = 0 (i.e., the equation in terms of p and q only) is called standard type I.

The solution of equation is

$$z = ax + by + c \qquad \text{...(1)}$$

$$\frac{\partial z}{\partial x} = a \, , \quad \frac{\partial z}{\partial y} = b$$

Now replacing $p = \dfrac{\partial z}{\partial x} = a \text{ and } q = \dfrac{\partial z}{\partial y} = b$ in the given P.D.E, we get

$$f(a, b) = 0 \Rightarrow b = \phi(a)$$

Substituting $b = \phi(a)$ in (1)

$z = ax + \phi\,(a)y + c$ is called complete integral

(ii) **Standard Form II**: The equation of the form $f(x, y, p, q) = 0$ ..... (1)
is called standard type II

Arrange (1) in the form $f_1(x, p) = f_2(y, q) = a$ (constant)

From these two equations we get $p = \phi_1(x, a)$ and $q = \phi_2(y, a)$

Substituting in $dz = \dfrac{\partial z}{\partial x}dx + \dfrac{\partial z}{\partial y}dy = p\,dx + q\,dy$

Integrating $\int dz = \int p dx + \int q dy + c$

$$\int dz = \int \phi_1(x, a)\,dx + \int \phi_2(y, a)\,dy$$

$z = \int \phi_1(x, a)\,dx + \int \phi_2(y, a)dy + c$ is the complete integral

(iii) ***Standard form* III**: The equation of the form $f(z, p, q) = 0$ ..... (1)

Substituting q = ap ..... (2)

In (1) we get

$$p = \phi(z) \qquad \text{.....(3)}$$

from (2) ,(3) $\quad q = a\,\phi(z) \qquad$ .....(4)

From (3), (4) substituting values of p, q in

$$dz = p.\,dx + q\,dy$$

$$dz = \phi(z)\,dx + a\,\phi(z)\,dy$$

$$\therefore \quad \frac{dz}{\phi(z)} = dx + a\,dy$$

Integrating $F(z) = x + ay + c$ is the complete integral

(iv) ***Standard form* IV**: (Clairaut's equation)

The P.D.E of the form $z = px + qy + f(p, q)$ ...(1)

is called Clairaut's equation

The complete integral of (1) is

$$z = ax + by + f(a, b) \qquad \text{.....(2)}$$

To find the singular integral differentiate (2) partially with respect to 'a' and 'b'

$$0 = x + \frac{\partial f}{\partial a} \qquad \text{.....(3)}$$

$$0 = y + \frac{\partial f}{\partial b} \qquad \text{.....(4)}$$

The eliminant of a, b from (3) (4) gives the singular integral

Solved University Questions (JNTU)

1. Form the partial differential equation by eliminating the arbitrary constants a and b from $z = ax + by + a^2 + b^2$

Solution:

Given $z = ax + by + a^2 + b^2$ .....(1)

Differentiating (1) partially with respect to x and y

$$\frac{\partial z}{\partial x} = a = p$$

$$\frac{\partial z}{\partial y} = b = q$$

Substituting $a = p = \frac{\partial z}{\partial x},\ b = q = \frac{\partial z}{\partial y}$ in (1)

$z = px + qy + p^2 + q^2$ is the required Partial Differential equation

2. Form the P.D.E by eliminating the arbitrary constants a and b from

$$z = ax + by + \left(\frac{a}{b}\right) - b$$

Solution:

Given P.D.E is $z = ax + by + \frac{a}{b} - b$ ..... (1)

Differentiating partially with respect to x and y

$$\frac{\partial z}{\partial x} = p = a$$

$$\frac{\partial z}{\partial y} = q = b$$

Substituting these values in (1)

$$z = px + qy + \frac{p}{q} - q$$

3. Form P.D.E by eliminating a, b from $2z = \sqrt{x+a} + \sqrt{y-a} + b$

Solution:

Given $2z = \sqrt{x+a} + \sqrt{y-a} + b$ ....(1)

Differentiating (1) partially with respect to x and y

$$2\frac{\partial z}{\partial x} = \frac{1}{2\sqrt{x+a}} \quad \Rightarrow \quad \sqrt{x+a} = \frac{1}{4p}$$

and

$$2\frac{\partial z}{\partial y} = \frac{1}{2\sqrt{y-a}} \quad \Rightarrow \quad \sqrt{y-a} = \frac{1}{4q}$$

$$\therefore \quad x + a = \frac{1}{16p^2} \quad \Rightarrow \quad a = \frac{1}{16p^2} - x \qquad \text{..... (2)}$$

and

$$y - a = \frac{1}{16q^2} \quad \Rightarrow \quad a = y - \frac{1}{16q^2} \qquad \text{..... (3)}$$

from (2), (3)

$$\frac{1}{16p^2} - x = y - \frac{1}{16q^2}$$

$$16(x+y) = \frac{1}{p^2} + \frac{1}{q^2}$$

4. Form the P.D.E by eliminating constants a and b from $z = a \log\left[\frac{b(y-1)}{1-x}\right]$

Solution:

$$z = a \log\left[\frac{b(y-1)}{1-x}\right]$$

$$z = a\,[\log b\,(y-1) - \log(1-x)]$$

$$z = a \log b + a \log(y-1) - a \log(1-x) \qquad \text{..... (1)}$$

Differentiating (1) partially with respect to x and y

$$\frac{\partial z}{\partial x} = p = 0 + 0 + \frac{a}{1-x}$$

$$p = \frac{a}{1-x} \qquad \text{..... (2)}$$

$$\frac{\partial z}{\partial y} = q = 0 + \frac{a}{y-1} \qquad \text{..... (3)}$$

From (2) and (3); $a = p(1-x)$ and $a = q(y-1)$

$$\therefore \quad p(1-x) = q(y-1)$$

$$px + qy = p + q$$

5. Form the P.D.E by eliminating the arbitrary functions f and g from $z = f(x) + e^y g(x)$

Solution:

Given $\quad z = f(x) + e^y g(x) \qquad$ ..... (1)

Differentiate (1) partially with respect to $\underline{x}$ and y

$$\frac{\partial z}{\partial x} = p = f'(x) + e^y g'(x) \quad \text{..... (2)}$$

$$\frac{\partial z}{\partial y} = q = e^y g(x) \quad \text{..... (3)}$$

Differentiate (3) partially with respect to 'y'

$$\frac{\partial^2 z}{\partial y^2} = e^y g(x) \quad \text{..... (4)}$$

From (3) and (4)

$$\frac{\partial^2 z}{\partial y^2} = \frac{\partial z}{\partial y}$$

$$t - q = 0$$

6. Form a partial differential equation by eliminating the arbitrary function f from $z = f(x^2 - y^2)$

Solution:

Given $z = f(x^2 - y^2)$ ..... (1)

Differentiate (1) partially with respect to x and y

$$\frac{\partial z}{\partial x} = p = f'(x^2 - y^2).2x \quad \text{..... (2)}$$

$$\frac{\partial z}{\partial y} = q = f'(x^2 - y^2)(-2y) \quad \text{..... (3)}$$

From (2) and (3)

$$\frac{p}{q} = \frac{f'(x^2 - y^2).2x}{f'(x^2 - y^2)(-2y)}$$

$$py + qx = 0$$

7. Form the P.D.E by eliminating function f from $xyz = f(x^2 + y^2 + z^2)$

Solution:

Given $xyz = f(x^2 + y^2 + z^2)$ ..... (1)

Differentiate (1) partially with respect to x and y

$$yz + xy\frac{\partial z}{\partial x} = f'(x^2 + y^2 + z^2)\left[2x + 2z\frac{\partial z}{\partial x}\right] \quad \text{..... (2)}$$

$$xz + xy\,\frac{\partial z}{\partial y} = f'\left(x^2+y^2+z^2\right)\left[2y + 2z\frac{\partial z}{\partial y}\right] \qquad \text{..... (3)}$$

From (2) and (3)

$$\frac{yz + xy\,p}{xz + xy\,q} = \frac{f'\left(x^2+y^2+z^2\right)2(x+zp)}{f'\left(x^2+y^2+z^2\right)2(y+zq)}$$

$$\frac{yz + xy\,p}{xz + xy\,q} = \frac{x+zp}{y+zq}$$

$(x + zp)\,(xz + xy\,q) = (y + zq)\,(yz + xy\,p)$ is the required P.D.E

8. Form the P.D.E by eliminating the arbitrary function f from $z = xy + f(x^2 + y^2)$

Solution:

$$z = xy + f(x^2+y^2) \qquad \text{..... (1)}$$

Differentiate (1) partially with respect to x and y

$$\frac{\partial z}{\partial x} = p = y + f'\left(x^2+y^2\right)2x$$

$$\therefore \quad p - y = 2x\; f'\left(x^2+y^2\right) \qquad \text{..... (2)}$$

$$\frac{\partial z}{\partial y} = q = x + f'\left(x^2+y^2\right)2y$$

$$\therefore \quad q - x = 2y\,f'\left(x^2+y^2\right) \qquad \text{..... (3)}$$

From (2) and (3)

$$\frac{p-y}{q-x} = \frac{2xf'\left(x^2+y^2\right)}{2y\,f'\left(x^2+y^2\right)}$$

$$y(p-y) = x(q-x)$$

$$x^2 - y^2 = qx - py$$

9. Form the Partial differential equation by eliminating f from $z = y^2 + 2f\left(\frac{1}{x} + \log y\right)$

Solution:

Given $\quad z = y^2 + 2f\left(\frac{1}{x} + \log y\right) \qquad \text{..... (1)}$

Differentiate (1) partially with respect to x and y

$$\frac{\partial z}{\partial x} = p = 0 + 2\,f'\left(\frac{1}{x} + \log y\right)\left(-\frac{1}{x^2}\right) \qquad \text{..... (2)}$$

$$\frac{\partial z}{\partial y} = q = 2y + 2f'\left(\frac{1}{x} + \log y\right) \cdot \left(\frac{1}{y}\right) \quad \text{..... (3)}$$

From (2) $\quad -px^2 = 2f'\left(\frac{1}{x} + \log y\right)$

Substituting the value of $2f'\left(\frac{1}{x} + \log y\right) = -px^2$ in (3)

$\therefore \quad q = 2y - px^2\left(\frac{1}{y}\right) \quad \Rightarrow \quad px^2 + qy - 2y^2 = 0$

10. Form the P.D.E from F $(xy + z^2, x + y + z) = 0$

Solution:

Given F $(xy + z^2, x + y + z) = 0$ ..... (1)

Consider $u = xy + z^2$ and $v = x + y + z$

$\therefore$ The given (1) becomes

F $(u, v) = 0$ ..... (2)

Differentiate (2) partially with respect to

$$\frac{\partial F}{\partial u}\frac{\partial u}{\partial x} + \frac{\partial F}{\partial u}\frac{\partial u}{\partial z} \cdot \frac{\partial z}{\partial x} + \frac{\partial F}{\partial v}\frac{\partial v}{\partial x} + \frac{\partial F}{\partial v}\frac{\partial v}{\partial z}\frac{\partial z}{\partial x} = 0$$

$$\frac{\partial F}{\partial u}(y) + \frac{\partial F}{\partial u}(2z)p + \frac{\partial F}{\partial v}(1) + \frac{\partial F}{\partial v}(1)p = 0$$

$$(y + 2zp)\frac{\partial F}{\partial u} + (1 + p)\frac{\partial F}{\partial v} = 0 \quad \text{.... (3)}$$

Differentiate (2) partially with respect to y

$$\frac{\partial F}{\partial u}\frac{\partial u}{\partial y} + \frac{\partial F}{\partial u}\frac{\partial u}{\partial z} \cdot \frac{\partial z}{\partial y} + \frac{\partial F}{\partial v} \cdot \frac{\partial v}{\partial y} + \frac{\partial F}{\partial v}\frac{\partial v}{\partial z}\frac{\partial z}{\partial y} = 0$$

$$\frac{\partial F}{\partial u}(x) + \frac{\partial F}{\partial u}(2z)q + \frac{\partial F}{\partial v}(1) + \frac{\partial F}{\partial v}(1)q = 0.$$

$$(x + 2zq)\frac{\partial F}{\partial u} + (1 + q)\frac{\partial F}{\partial v} = 0 \quad \text{..... (4)}$$

Using determinant method of eliminating $\frac{\partial F}{\partial u}$ and $\frac{\partial F}{\partial v}$ between (3) and (4)

$$\begin{vmatrix} y+2pz & 1+p \\ x+2qz & 1+q \end{vmatrix} = 0$$

$$(y + 2pz)(1 + q) - (x + 2q z)(1 + p) = 0$$

$$y + 2pz + qy + 2pqz - x - 2qz - px - 2pqz = 0$$

$p(x - 2z) + q(2z - y) = y - x$ is the required equation

11. Form P.D.E from $\phi\left(\frac{y}{x}, x^2 + y^2 + z^2\right) = 0$ by eliminating arbitrary function ϕ.

Solution:

Given $\phi\left(\frac{y}{x}, x^2 + y^2 + z^2\right) = 0$

Let $u = \frac{y}{x}$, $v = x^2 + y^2 + z^2$

$\therefore$ $\phi(u, v) = 0$ ..... (1)

Differentiate (1) partially with respect to 'x'

$$\frac{\partial \phi}{\partial u}\frac{\partial u}{\partial x} + \frac{\partial \phi}{\partial u}\frac{\partial u}{\partial z}.\frac{\partial z}{\partial x} + \frac{\partial \phi}{\partial v}\frac{\partial v}{\partial x} + \frac{\partial \phi}{\partial v}\frac{\partial v}{\partial z}\frac{\partial z}{\partial x} = 0$$

$$\frac{\partial \phi}{\partial u}\left(-\frac{y}{x^2}\right) + \frac{\partial \phi}{\partial u}(0) + \frac{\partial \phi}{\partial v}(2x) + \frac{\partial \phi}{\partial v}(2z)p = 0$$

$$\frac{\partial \phi}{\partial u}\left(-\frac{y}{x^2}\right) + (2x + 2zp)\frac{\partial \phi}{\partial v} = 0 \qquad \text{..... (2)}$$

Similarly differentiate (1) partially with respect to 'y'

$$\frac{\partial \phi}{\partial u}\frac{\partial u}{\partial y} + \frac{\partial \phi}{\partial u}\frac{\partial u}{\partial z}\frac{\partial z}{\partial y} + \frac{\partial \phi}{\partial v}\frac{\partial v}{\partial y} + \frac{\partial \phi}{\partial v}\frac{\partial v}{\partial z}\frac{\partial z}{\partial y} = 0$$

$$\frac{\partial \phi}{\partial u}\left(\frac{1}{x}\right) + \frac{\partial \phi}{\partial u}(0)q + \frac{\partial \phi}{\partial v}(2y) + \frac{\partial \phi}{\partial v}(2z)q = 0$$

$$\frac{\partial \phi}{\partial u}\left(\frac{1}{x}\right) + (2y + 2zq)\frac{\partial \phi}{\partial v} = 0 \qquad \text{..... (3)}$$

Eliminating $\frac{\partial\phi}{\partial u}, \frac{\partial\phi}{\partial v}$ using determinant method

$$\begin{vmatrix} \frac{-y}{x^2} & 2x+2zp \\ \frac{1}{x} & 2y+2zq \end{vmatrix} = 0$$

$$(2y+2zq)\left(\frac{-y}{x^2}\right) - \frac{1}{x}(2x+2zp) = 0$$

$$y^2 + zqy + x^2 + zpx = 0$$

$$z(px+qy) + x^2 + y^2 = 0$$

12. Form P.D.E by eliminating the arbitrary functions f_1 and f_2 from $f_1(y+2x) + f_2(y-3x)$

***Solution*:**

$$z = f_1(y+2x) + f_2(y-3x) \qquad \ldots..(1)$$

Differentiating (1) partially w.r. to x and y

$$\frac{\partial z}{\partial x} = f_1^1(y+2x)(2) + f_2^1(y-3x)(-3) \qquad \ldots..(2)$$

$$\frac{\partial z}{\partial y} = f_1^1(y+2x) + f_2^1(y-3x) \qquad \ldots..(3)$$

Differentiating (2) partially w. r.to 'x'

$$\frac{\partial^2 z}{\partial x^2} = f_1''(y+2x)4 + f_2''(y-3x)(9) \qquad \ldots..(4)$$

Differentiating (3) partially w. r. to 'y'

$$\frac{\partial^2 z}{dy^2} = f_1''(y+2x) + f_2''(y-3x) \qquad \ldots..(5)$$

Differetiating (2) partially w. r. to y

$$\frac{\partial^2 z}{dx\partial y} = f_1''(y+2x)(2) + f_2''(y-3x)(-3) \qquad \ldots..(6)$$

Multiplying (6) by (3) and adding (4)

[i.e., (6) × 3 + (4) ⇒]

$$\frac{\partial^2 z}{\partial x^2} + 3\frac{\partial^2 z}{\partial x \partial y} = 10 f_1''(y+2x) \quad \text{..... (7)}$$

Similarly (6) × (– 2) + (4) ⇒

$$\frac{\partial^2 z}{\partial x^2} - 2\frac{\partial^2 z}{\partial x \partial y} = 15 f_2''(y-3x) \quad \text{..... (8)}$$

From (7), (8) substituting the values of $f_1''(y+2x)$ and $f_2''(y-3x)$ in (5)

$$\frac{\partial^2 z}{dy^2} = \frac{1}{10}\left(\frac{\partial^2 z}{\partial x^2} + 3\frac{\partial^2 z}{\partial x \partial y}\right) + \frac{1}{15}\left(\frac{\partial^2 z}{\partial x^2} - 2\frac{\partial^2 z}{\partial x \partial y}\right)$$

$$\frac{\partial^2 z}{dy^2} = \frac{1}{6}\frac{\partial^2 z}{\partial x^2} + \frac{1}{6}\frac{\partial^2 z}{\partial x \partial y}$$

$$\therefore \quad \frac{\partial^2 z}{dx^2} + \frac{\partial^2 z}{\partial x \partial y} - 6\frac{\partial^2 z}{\partial y^2} = 0 \quad \text{or} \quad r + s - 6t = 0$$

is the required partial differential equation

13. Form the P.D.E by eliminating arbitrary functions from $z = x f_1(x + t) + f_2(x + t)$

***Solution*:**

Given $z = x f_1(x + t) + f_2(x + t)$ ..... (1)

Differentiating (1) partially w. r. to x and 't'

$$\frac{\partial z}{\partial x} = x f_1' + f_1 + f_2' \quad \text{..... (2)}$$

and

$$\frac{\partial z}{\partial t} = x f_1' + f_2' \quad \text{..... (3)}$$

Differentiating (2) and (3) partially w. r. to x and 't' respectively

$$\frac{\partial^2 z}{\partial x^2} = x f_1'' + f_1' + f_1' + f_2'' \quad \text{..... (4)}$$

$$\frac{\partial^2 z}{\partial t^2} = x f_1'' + f_2'' \qquad \text{..... (5)}$$

Differentiating (2) partially w. r. to 't'

$$\frac{\partial^2 z}{\partial x \partial t} = x f_1'' + f_1' + f_2'' \qquad \text{..... (6)}$$

Adding (4) (5)

$$\frac{\partial^2 z}{\partial x^2} + \frac{\partial^2 z}{\partial t^2} = 2x f_1'' + 2f_1' + 2f_2''$$

$$= 2\left[x f_1'' + f_1' + f_2'\right]$$

from (6)

$$\frac{\partial^2 z}{\partial x^2} + \frac{\partial^2 z}{\partial t^2} = 2\left[\frac{\partial^2 z}{\partial x \partial t}\right]$$ is the required partial differential equation

14. Form partial differential equation xy + yz + zx = $f\left(\frac{z}{x+y}\right)$ by eliminating 'f'

***Solution*:**

$$xy + yz + zx = f\left(\frac{z}{x+y}\right) \qquad \text{..... (1)}$$

Differentiating (1) partially w. r. to x

$$y + yp + xp + z = f'\left(\frac{z}{x+y}\right)\left[\frac{(x+y)p - z}{(x+y)^2}\right]$$

$$\therefore \quad \left[\frac{(x+y)^2}{(x+y)p - z}\right]\left[p(x+y) + (y+z)\right] = f' \qquad \text{..... (2)}$$

Differentiating (1) partially w. r. to 'y'

$$x + yq + z + xq = f'\left(\frac{z}{x+y}\right)\left[\frac{(x+y)q - z}{(x+y)^2}\right]$$

$$\left[\frac{(x+y)^2}{(x+y)q-z}\right]\left[q(x+y)+(x+z)\right]=f' \qquad \text{..... (3)}$$

From (2) and (3)

$$\left[\frac{(x+y)^2}{(x+y)p-z}\right]\left[p(x+y)+(y+z)\right]=\left[\frac{(x+y)^2}{(x+y)q-z}\right]\left[q(x+y)+(x+z)\right]$$

$$\therefore \qquad [(x+y)q-z][p(x+y)+(y+z)]=[(x+y)p-z][q(x+y)+(x+z)]$$

on simplification

$$p(x+2z)-q(y+2z)=\frac{z(y-x)}{x+y}$$

15. Solve $x(y-z)p+y(z-x)q=z(x-y)$

Solution:

Lagrange's subsidiary equation is

$$\frac{dx}{x(y-z)}=\frac{dy}{y(z-x)}=\frac{dz}{z(x-y)} \qquad \text{..... (1)}$$

From (1) each fraction equal to $\dfrac{dx+dy+dz}{x(y-z)+y(z-x)+z(x-y)}$

$$\Rightarrow \qquad dx+dy+dz=0$$

Integrating $x+y+z=c_1$ ..... (2)

(1) can be written as

$$\frac{\frac{dx}{x}}{y-z}=\frac{\frac{dy}{y}}{z-x}=\frac{\frac{dz}{z}}{x-y}=\frac{\frac{dx}{x}+\frac{dy}{y}+\frac{dz}{z}}{y-z+z-x+x-y}$$

$$\Rightarrow \qquad \frac{dx}{x}+\frac{dy}{y}+\frac{dz}{z}=0$$

Integrating $\log x+\log y+\log z=\log c_2$

$$xyz=c_2 \qquad \text{.... (3)}$$

From (2) and (3)

The general solution is $\phi(x+y+z, xyz)=0$

16. Solve $xp - yq = y^2 - x^2$

Solution:

Lagrange's Auxiliary equation is

$$\frac{dx}{x} = \frac{dy}{-y} = \frac{dz}{y^2 - x^2} \qquad \text{..... (1)}$$

From first two fractions of (1)

$$\frac{dx}{x} = \frac{dy}{-y}$$

$\Rightarrow \quad \log x = -\log y + \log c_1 \qquad$ (integrating)

$\Rightarrow \quad yx = c_1 \qquad$ ..... (2)

using $x, y, 1$ as multipliers

each fraction of (1) $= \dfrac{xdx + ydy + dz}{x^2 - y^2 + y^2 - x^2} = \dfrac{xdx + ydy + dz}{0}$

$\therefore \quad x\,dx + y\,dy + dz = 0$

Integrating

$$\frac{x^2}{2} + \frac{y^2}{2} + z = c$$

$$x^2 + y^2 + 2z = c_2 \qquad \text{..... (3)}$$

from (2) (3) the general solution is

$$\phi\,(xy, x^2 + y^2 + 2z) = 0$$

17. Solve $z^2\,(p^2\,x^2 + q^2) = 1$

Solution:

Given $\quad z^2\,(p^2\,x^2 + q^2) = 1 \qquad$ ..... (1)

$$z^2\left[\left(x\frac{\partial z}{\partial x}\right)^2 + \left(\frac{\partial z}{\partial y}\right)^2\right] = 1$$

Take $X = \log x \Rightarrow dX = \dfrac{1}{x}dx \quad$ in (1)

$$z^2\left[\left(\frac{\partial z}{\partial X}\right)^2 + \left(\frac{\partial z}{\partial y}\right)^2\right] = 1$$

$$z^2\,(p^2 + q^2) = 1 \qquad \text{..... (2)}$$

take $\quad q = ap \qquad$ ..... (3)

$$z^2 (p^2 + a^2 p^2) = 1$$

$$z^2 p^2(1 + a^2) = 1$$

$$p = \frac{1}{z\sqrt{1+a^2}}$$

$$q = ap = \frac{a}{z\sqrt{1+a^2}}$$

we know $dz = pdX + q\,dy$ $\left(\because p = \frac{dz}{dx}\right)$

$$dz = \frac{1}{z\sqrt{1+a^2}} dX + \frac{ady}{z\sqrt{1+a^2}} \qquad (\because q = ap)$$

$$z\sqrt{1+a^2}\, dz = dX + ady \qquad \text{integrating}$$

$$\sqrt{1+a^2}\,\frac{z^2}{2} = X + ay + c$$

$$\frac{z^2}{2}\sqrt{1+a^2} = \log x + ay + c \qquad (\because X = \log x)$$

18. Solve $y^3 \frac{\partial z}{\partial x} + x^2 \frac{\partial z}{\partial y} = 0$

Solution:

Given partial differential equation is $y^3 p + x^2 q = 0$

$$\therefore \qquad \frac{P}{x^2} = \frac{-q}{y^3} = a \qquad \text{(constant)}$$

$$p = ax^2 \,;\; q = -ay^3$$

Substituting p, q in $dz = p\,dx + q\,dy$

$$dz = ax^2\,dx + (-ay^3)dy \qquad \text{integrating}$$

$$z = \frac{ax^3}{3} - a\frac{y^4}{4} + c \qquad \text{is the complete solution}$$

19. Solve $pqz = p^2(qx + p^2) + q^2(py + q^2)$

Solution:

Given $$pqz = p^2q\left(x + \frac{p^2}{q}\right) + pq^2\left(y + \frac{q^2}{p}\right)$$

$$z = p\left(x + \frac{p^2}{q}\right) + q\left(y + \frac{q^2}{P}\right)$$

$$z = (px + qy) + \left(\frac{p^3}{q} + \frac{q^3}{p}\right) \qquad \text{..... (1)}$$

is in the form of $z = px + qy + f(p, q)$

$\therefore$ The complete solution of (1) is by replacing $p = a$, $q = b$ in (1)

$$z = ax + by + \left(\frac{a^3}{b} + \frac{b^3}{a}\right)$$

20. Solve $(x + pz)^2 + (y + qz)^2 = 1$

Solution:

Given $\quad (x + pz)^2 + (y + qz)^2 = 1$

$$z^2 = Z \Rightarrow \quad 2z\,dz = dZ$$

$$z dz = \frac{dZ}{2}$$

$$\therefore \quad \left(x + \frac{1}{2}P\right)^2 + \left(y + \frac{Q}{2}\right)^2 = 1$$

where $\quad P = \frac{dZ}{dx}; \quad Q = \frac{dZ}{dy}$

$$\therefore \quad \left(x + \frac{P}{2}\right)^2 = 1 - \left(y + \frac{Q}{2}\right)^2$$

is in the form of $f_1(p, x) = f_2(q, y)$

$$\left(x + \frac{P}{2}\right)^2 = 1 - \left(y + \frac{Q}{2}\right)^2 = a^2 \qquad \text{..... (1)}$$

$$\left(x + \frac{P}{2}\right)^2 = a^2$$

$$\therefore \quad \frac{P}{2} = a - x \qquad \text{..... (2)}$$

$$P = 2(a - x)$$

From (1) $$1-\left(y+\frac{Q}{2}\right)^2=a^2$$

$$\left(y+\frac{Q}{2}\right)^2=1-a^2$$

$\therefore$ $$y+\frac{Q}{2}=\sqrt{1-a^2}$$

$$Q=2\sqrt{1-a^2}-2y \qquad \text{..... (3)}$$

From (2), (3) substituting the values of P, Q in

$$dZ=P\,dx+Q\,dy$$

$dZ=2(a-x)\,dx+2\left(\sqrt{1-a^2}-y\right)dy$; integrating we get

$$Z=2\left(ax-\frac{x^2}{2}\right)+2\left(y\sqrt{1-a^2}-\frac{y^2}{2}\right)+c$$

substituting $Z=z^2$

$\therefore$ $$z^2=(2ax-x^2)+\left(2y\sqrt{1-a^2}-y^2\right)+c$$

is the complete solution

21. Solve Partial Differential equation $p^2z^2\sin^2x+q^2z^2\cos^2y=1$

Solution:

Given Partial Differential equation can be written as

$$(pz)^2\sin^2x+(qz)^2\cos^2y=1$$

$$\left(z\frac{\partial z}{dx}\right)^2\sin^2x+\left(z\frac{\partial z}{\partial y}\right)^2\cos^2y=1$$

$$z^2=Z\Rightarrow \qquad 2\,z\,dz=dZ$$

$\therefore$ $$z\,dz=\frac{\partial Z}{z}$$

$\therefore$ $$\left(\frac{1}{2}\frac{\partial Z}{\partial x}\right)^2\sin^2x+\left(\frac{1}{2}\frac{\partial Z}{\partial y}\right)^2\cos^2y=1$$

$$\frac{1}{4}\,P^2 \sin^2 x + \frac{1}{4}\,Q^2 \cos^2 y = 1$$

$$(P^2 \sin^2 x) = (4 - Q^2 \cos^2 y)$$

is in the form of $f_1(p, x) = f_2(q, y)$

$$\therefore \qquad P^2 \sin^2 x = (4 - Q^2 \cos^2 y) = a^2$$

$$P^2 \sin^2 x = a^2$$

$$P = \frac{a}{\sin x} \qquad \text{..... (1)}$$

$$4 - Q^2 \cos^2 y = a^2$$

$$Q \cos y = \sqrt{4 - a^2}$$

$$Q = \sqrt{\frac{4 - a^2}{\cos y}} \qquad \text{..... (2)}$$

Substituting values of P, Q in

$dZ = P\,dx + Q\,dy$, we get

$$dZ = a \operatorname{cosec} x\,dx + \sqrt{4 - a^2}\,\sec y\,dy$$

integrating

$$Z = a\int \operatorname{cosec} x dx + \sqrt{4 - a^2}\int \sec y\,dy + c$$

substituting $Z = z^2$

$$z^2 = a \log(\operatorname{cosec} x - \cot x) + \sqrt{4 - a}\,\log(\sec y + \tan y) + c$$

is the complete solution

22. Solve $\frac{p}{x^2} + \frac{q}{y^2} = z$

Solution:

$$\frac{p}{x^2} + \frac{q}{y^2} = z$$

$$\frac{1}{x^2}\left(\frac{\partial z}{\partial x}\right)+\frac{1}{y^2}\left(\frac{\partial z}{\partial y}\right)=z \quad \text{..... (1)}$$

$$x^3 = X \quad ; \quad y^3 = Y$$

$$3x^2\,dx = dX \quad ; \quad 3y^2\,dy = dY$$

$$x^2\,dx = \frac{dX}{3} \quad ; \quad y^2\,dy = \frac{dY}{3}$$

Substituting in (1) we get

$$3\frac{\partial z}{\partial X}+3\frac{dz}{dY}=z$$

$$3P + 3Q = z \quad \text{..... (2)}$$

is in the form of $f(z, p, q) = 0$

take Q = aP in (2) $\therefore$

$$3P + 3aP = z \qquad \therefore \qquad P = \frac{z}{3(1-a)}$$

$$\therefore \qquad Q = \frac{a\,.\,z}{3(1-a)} \qquad (\because\ Q = aP)$$

Substituting the values of P, Q in dz = P dX + Q dY

$$dz = \frac{z}{3(1-a)}dX + \frac{az}{3(1-a)}dY$$

$$3(1-a)\frac{dz}{z} = dX + a\,dY,$$

integrating we get

$$3(1-a)\log z = X + aY + c \qquad (\because\ X = x^3,\ Y = y^3)$$

$$3(1-a)\log z = x^3 + ay^3 + c$$

23. Solve $(mz - ny)\,p + (nx - lz)\,q = ly - mx$

Solution:

The given equation is in the form Pp + Qq = R, where P = mz – ny, Q = nx – lz, R = ly – mx

The auxiliary equations are given by, $\frac{dx}{P} = \frac{dy}{Q} = \frac{dz}{R}$

$$\text{i.e.,} \quad \frac{dx}{mz - ny} = \frac{dy}{nx - lz} = \frac{dz}{ly - mx} \qquad .....(1)$$

If we choose x, y, z as multipliers, we get,

Each fraction of (1) = $\frac{xdx + ydy + zdz}{0}$ $\Rightarrow$ x dx + y dy + z dx = 0, which on integration

gives $\frac{x^2}{2} + \frac{y^2}{2} + \frac{z^2}{2} = a$ (or) $x^2 + y^2 + z^2 = a_1$

Again, if we choose l, m, n as multipliers,

Each fraction of (1) = $\frac{l\,dx + m\,dy + n\,dz}{0}$

$\Rightarrow l\,dx + m\,dy + n\,dz = 0$, which on integration gives $lx + my + nz = a_2$

$\therefore$ The general solution of the given equation if

$f(x^2 + y^2 + z^2, lx + my + nz) = 0$

24. Solve $p^2 + q^2 = x^2 + y^2$

Solution:

We can write the given equation as

$p^2 - x^2 = y^2 - q^2 = c$ (say), ('c' being constant)

$p^2 - x^2 = c \quad \Rightarrow \quad p = \sqrt{x^2 + c}$ and $q = \sqrt{y^2 - c}$

since $dz = p\,dx + q\,dy$, we get,

$$dz = \sqrt{x^2 + c}\,dx + \sqrt{y^2 - c}\,dy$$

Integration of this equation gives,

$$z = \int \sqrt{x^2 + c}\,dx + \int \sqrt{y^2 - c}\,dy$$

$$\text{i.e.,} \quad z = \frac{x}{2}\sqrt{x^2 + c} + \frac{c}{2}\sinh^{-1}\left(\frac{x}{\sqrt{c}}\right) + \frac{y}{2}\sqrt{y^2 - c} - \frac{c}{2}\cosh^{-1}\left(\frac{y}{\sqrt{c}}\right) + k$$

$$\therefore \quad z = \frac{1}{2}\left[x\sqrt{x^2 + c} + y\sqrt{y^2 - c}\right] + \frac{c}{2}\left[\sinh^{-1}\left(\frac{x}{\sqrt{c}}\right) - \cosh^{-1}\left(\frac{y}{\sqrt{c}}\right)\right] + k$$

is the solution of the given equation.

Objective Type Questions

I. Chose the Correct Answer

1. The partial differential equation after the elimination of the constants from $\log(az-1) = x + ay + b$ is

[*Ans:* (c)]

(a) $q(p+1) = zp$ (b) $p(q-1) = zq$

(c) $p(q+1) = zq$ (d) None

2. The complete integral of the equation $z = px + qy + p^2q^2$ is

[*Ans:* (b)]

(a) $z = ax + by + c$ (b) $z = ax + by + a^2 + b^2$

(c) $z = ax + by - a^2b^2$ (d) None

3. The solution of the partial differential equation $p\sqrt{x} + q\sqrt{y} = \sqrt{z}$ is

[*Ans:* (a)]

(a) $\phi(x-y, y-z) = 0$ (b) $\phi\left(\sqrt{x}-\sqrt{y}, \sqrt{y}-\sqrt{z}\right) = 0$

(c) $\phi\left(\frac{1}{\sqrt{x}}-\frac{1}{\sqrt{y}}, \frac{1}{\sqrt{y}}-\frac{1}{\sqrt{z}}\right) = 0$ (d) None

4. The complete integral of $3p^2 - 2q^2 = 4pq$ is

[*Ans:* (a)]

(a) $z = ax \pm a\left[1+\frac{\sqrt{10}}{2}\right]y + c$ (b) $z = ax \pm \left[\frac{\sqrt{10}}{2}\right]y + c$

(c) $z = ax \pm ay + c$ (d) None

5. The solution of the differential equation $(1+y)p + (1+x)q = z$ is

[*Ans:* (c)]

(a) $\phi\left[x(y-z), \frac{x+y+z}{2}\right] = 0$ (b) $\phi\left[y(x-z), \frac{x+y+z}{2}\right] = 0$

(c) $\phi\left[z(y-x), \frac{x+y+z}{2}\right] = 0$ (d) None

6. The solution of $(x^2 - yz)p + (y^2 - zx)q = z^2 xy$ is

[*Ans:* (d)]

(a) $\phi\left(\frac{1}{x}+\frac{1}{y}+\frac{1}{2}, xy+yz+zx\right)=0$ (b) $\phi\left(xy+yz+zx, xyz\right)=0$

(c) $\phi\left(xy+yz+zx, x+y+z\right)=0$ (d) None

7. If the number of arbitrary constants to be eliminated is equal to the number of independent variables then the P.D.E of .... order [*Ans*: (a)]

(a) First (b) Second (c) Third (d) None

8. Eliminating the arbitrary constants a, b from $2z=\sqrt{x+a}+\sqrt{y+b}$ the P.D.E is .....

[*Ans:* (b)]

(a) p + q = 2 pqz (b) p + q − 8 pqz = 0 (c) p + q = pqz (d) None

9. Eliminating the arbitrary constants a, b from $2z=\frac{x^2}{a^2}+\frac{y^2}{b^2}$, the P.D.E is ... [*Ans*: (a)]

(a) $2z = xp + yq$ (b) $z = xp + yq$ (c) $yp + xq = x$ (d) None

10. Eliminating the arbitrary function '*f*' from $z=x^n f\left(\frac{y}{x}\right)$ the P.D.E is .... [*Ans*: (b)]

(a) $z = px + qy$ (b) $xp + yq = nz$ (c) $xp + qy = \frac{z}{n}$ (d) None

11. Eliminating the arbitrary function *f* from $z = x + y + f(xy)$, the P.D.E is .... [*Ans*: (c)]

(a) $xq - yp = x - y$ (b) $xp - yq = x + y$

(c) $xp - yq = x - y$ (d) None

12. The partial differential equation Pp + qQ = R is called equation [*Ans*: (c)]

(a) Auxiliary (b) Non linear (c) Linear (d) None

13. If $\frac{y}{z}=c_1$ is one solution of the partial differential equation $px + qy = z$ and second solution is .... [Ans: (a)]

(a) $\frac{x}{z}=c_2$ (b) $xy = c_2$ (c) $xz = c_2$ (d) None

14. Pp + qQ = R is the Lagrange's partial differential equation then $\frac{dx}{P} = \frac{dy}{Q} = \frac{dz}{R}$ is called...... equation [*Ans*: (a)]

(a) Auxiliary (b) Non linear (c) linear (d) None

15. The solution of the partial differential equation $p - x^2 - q - y^2 = 0$ is ...... [*Ans*: (b)]

(a) $z = 3ax + ay + x^3 - y^3 + c$ (b) $z = 3(ax + ay) + x^3 - y^3 + c$

(c) $z = x^3 - y^3 + c$ (d) None

16. The general solution of the equation $pe^y - qe^x = 0$ is .... [*Ans*: (a)]

(a) $z = a(e^x + e^y) + c$ (b) $e^x + e^y = az + c$

(c) $z = ae^x + ce^y$ (d) None

II. Fill in the blanks

1. The partial differential equation after elimination of the arbitrary constants *a*, *b* from $z = ax^3 + by^3$ is ______.

[*Ans*: $px + qy = 3z$]

2. The complete integral of the equation $p + q - pq = 0$ is ______.

[*Ans*: $z = ax + \frac{ay}{a-1} + b$]

3. Solution of the partial differential equation $z(x - y) = x^2 p - y^2 q$ is ______.

[*Ans*: $F\left[\frac{1}{x} + \frac{1}{y}, \frac{z}{x+y}\right] = 0$]

4. One subsidiary solution of the partial differential equation $(x^2 - y^2)\, p + (y^2 - zx)q = x^2 - xy$ is ________.

[*Ans*: $\frac{x-y}{y-2} = c_1$ or $xy + yz + zx = c_2$]

5. If $xy - z^2 = c_1$ is the one of the subsidiary solution of the partial differential equation. $(x + 2z)\,p + (4xz - y)\,q = x^2 + y$, then corresponding Lagrange's multipliers are ______.

[*Ans:* multipliers are $y, x, -2z$]

6. The order of a partial differential equation is the order of ____________ derivative that appears in the P.D.E

[*Ans*: the highest]

7. If the dependent variable and its derivatives appears to the first degree only in the P.D.E, then the P.D.E is called ____________

[*Ans*: Linear]

8. The order of the partial differential equation does not depend on the number of ____________

[*Ans*: Arbitrary constants]

9. The P.D.E by eliminating the arbitrary constants a, b, c from $ax + by + cz = 0$ is ____________

[**Ans:** $z = px + qy$]

10. The P.D.E by eliminating arbitrary constants a, b from $z = ax^2 + by^2 + 2ab\,xy$ is ____________

$$\left[\textit{Ans}: x\frac{\partial^2 z}{\partial x^2} = \frac{\partial^2 z}{\partial y^2}\right]$$

11. The P.D.E by eliminating arbitrary function 'f' from $z = f(x + 2y) + g(x + y)$ is ____________

[*Ans*: $2z_x = z_y$]

12. The P.D.E by eliminating arbitrary functions f, g from $z = f(x - y) + g(x + y)$ is ____________

$$\left[\textit{Ans}: \frac{\partial^2 z}{\partial x^2} = \frac{\partial^2 z}{\partial y^2}\right]$$

13. The P.D.E by eliminating arbitrary function f from $z = x f(y - x)$ is _______________

[*Ans*: $z = px + qx$]

14. The partial differential equation $pP + qQ = R$ where P, Q, R are functions of x, y, z and is called _______________ equation

[*Ans*: *Lagrange's*]

15. The solution of the P.D.E $pP + qQ = R$ is the form of _______________

[*Ans*: $f(u, v) = 0$ when u, v are functions of x, y, z]

16. The general solution of the P.D.E $p + yq = x$ is _______________

$$\left[\textbf{Ans}: z^2 - x^2 = f\left(\frac{x+z}{y}\right)\right]$$

17. The general solution of the P.D.E $x\frac{\partial z}{\partial x} + y\frac{\partial z}{\partial y} = nz$ is _______________

$$\left[\textbf{Ans}: z = x^{n} f\left(\frac{y}{x}\right)\right]$$

18. The solution of the P.D.E $yzp + zxq = xy$ is _______________

[*Ans*: $f(x^2 - y^2, y^2 - z^2) = 0$]

19. The solution of the P.D.E $P - 2q = 3x^2 \sin(y + 2x)$ is _______________

[*Ans*: $z = x^3 \sin(y + 2x) + f(y - 2x)$]

20. One solution of the P.D.E $(mz - ny) P + (nx - lz) q = ly - mx$ is $x^2 + y^2 + z^2 = c_1$ and another solution is _______________

[*Ans*: $lx + my + nz = c_2$]

21. Clairaut's form of partial differential equation is _______________

[*Ans*: $z = px + qy + f(p, q)$]

22. The general solution of the P.D.E $p^2 + q^2 = m^2$ is ______________

$$\left[Ans: z = ax + \sqrt{m^2 - a^2}\,y + c\right]$$

23. The general solution of the P.D.E $pq = xy$ is ______________

$$\left[Ans: 2z = ax^2 + \frac{y^2}{a} + c\right]$$

24. The general solution of the equation $z = pq$ is ______________

[*Ans:* $4az = (x + ay + c)^2$]

25. The general solution $\cot x + q \cot y = \cot z$ is ______________

$$\left[Ans: F\left(\frac{\cos y}{\cos x}, \frac{\cos x}{\cos y}\right) = 0\right]$$

III. True or False

1. The partial differential equation after elimination of the arbitrary constants a, b from $z = a\,(x + \log y) - \frac{x^2}{2} - b$ is $p + qy = x$

[*Ans:* False]

2. The complete integral of $p^3 + q^3 = 8z$ is $3z^2\,(1 + a^3) = 4(x + ay + b)^3$

[*Ans:* True]

3. Solution of the partial differential equation $(y^2 + z^2)\,p - xyq = -xz$ is $\phi\left[\frac{x}{y}, x^2 + y^2 + z^2\right] = 0$

[*Ans:* False]

4. The singular integral of $p^3 + q^3 = 8z$ is $z = 0$

[*Ans:* True]

5. The complete integral of $(1 - x)\,p + (2 - y)\,q = 3 - z$ is $z = ax + by - a - 2b + 3$

[*Ans:* True]

6. The complete integral of $x^2 p^2 = yq^2$ is $z = ax + 2a\sqrt{y} + b$

[*Ans:* False]

7. The solution of the partial differential equation $z(xp - yq) = y^2 - x^2$ is

$$f(xy, x^2 + y^2 + z^2) = 0$$

[*Ans:* True]

8. Lagrange's equation $Pp + qQ = R$ is non-linear partial differential equation

[*Ans*: False]

9. The solution of a linear partial differential equation of nth order contains (n – 1) arbitrary functions.

[*Ans*: False]

10. Non linear equations are those in which the dependent variable, its derivatives and their powers are linear

[*Ans*: False]

11. Eliminating a, b from $a(x^2 + y^2) + bz^2 = 1$ the partial differential equation is $y\frac{\partial z}{\partial x} = x\frac{\partial z}{\partial y}$

[*Ans*: True]

12. Eliminating arbitrary constants a and α from $z = a e^{-\alpha^2 t}$ cos (αt), the partial differential equation is $\frac{\partial^2 z}{\partial t^2} = \frac{\partial z}{\partial x}$

[*Ans*: False]

13. Eliminating the arbitrary function f from $z = f(\sin x + \cos y)$ the partial differential equation is p $\sin y + q \cos x = 0$

[*Ans*: True]

14. The general solution of the partial differential equation $p^3 - q^3 = 0$ is is $z = ax + by + c$

[*Ans*: True]

15. The general solution of the partial differential equation $z = px + qy + f(pq)$ is $z = ax + by + c$

[*Ans*: False]

16. The general solution of the partial differential equation $pq - xy = 0$ is $z = ax^2 + \frac{y^2}{a} + c$

[*Ans*: False]

17. The equation $\phi(x, y, z, a, b) = 0$ which contains as many constants as the number of independent variables is called complete integral of the partial differential equation

[*Ans*: True]

18. The P.D.E $z = px + qy + f(p, q)$ is Clairaut's equation

[*Ans*: True]

19. The complete integral of the partial differential equation $z = px + qy + p^2 + q^2$ is $z = ax + by + c$

[*Ans*: False]

Chapter 19

Solution of Partial Differential Equations

19.1 Partial Differential Equations – Applications

19.1.1 Some of the important partial differential equations which occur in the study of Engineering and Physical problems are

1. ***Wave Equation :***

$$\frac{\partial^2 y}{\partial t^2} = c^2 \frac{\partial^2 y}{\partial x^2}$$

which occurs in the problems involving vibrations of strings, rods, etc.,

2. ***One-dimensional heat flow equation :***

$$\frac{\partial u}{\partial t} = c^2 \frac{\partial^2 u}{\partial x^2}$$

(Diffusion equation in one dimension)

3. ***Two dimensional heat flow equation (Laplace's equation in two dimensions) :***

That which occurs in problems of two dimensional heat flow in steady state, i.e., in which temperature does not change with time. (In the conduction of heat in a plate in steady state).

$$\frac{\partial^2 u}{\partial x^2} + \frac{\partial^2 u}{\partial y^2} = 0$$

4. *Telegraph equations :*

$$\frac{\partial^2 v}{\partial x^2} = RC\frac{\partial v}{\partial t} \text{ and } -\frac{\partial v}{\partial x} = RI$$

$$\frac{\partial^2 I}{\partial x^2} = RC\frac{\partial I}{\partial t} \text{ and } -\frac{\partial I}{\partial t} = c\frac{\partial v}{\partial t}$$

Problems which satisfy certain initial and boundary conditions, are called boundary value problems. A method suitable for such problems is the method of separation of variables also known as product method. This is a very convenient method in solving the above partial differential equations.

It involves a solution which breaks up into a product of functions each of which involves only one of the independent variables. In this way the solution of the partial differential equation is converted into the solution of ordinary differential equations.

The method is illustrated in the solved examples.

Solved Examples

***Example* 19.1:** Solve the equation $\frac{\partial u}{\partial x} = 4\frac{\partial u}{\partial y}, u(0, y) = 8e^{-3y}$ by the method of separation of variables.

Solution: The given differential equation is $\frac{\partial u}{\partial x} = 4\frac{\partial u}{\partial y}$ .....(1)

Assume that the complete solution of the partial differential equation can be taken as

$$u = X(x)\, Y(y) = XY \quad .....(2)$$

$$\frac{\partial u}{\partial x} = YX' \text{ and } \frac{\partial u}{\partial y} = XY'$$

(1) becomes

$$YX' = 4\, XY'$$

$$\Rightarrow \quad \frac{X'}{X} = 4\frac{Y'}{Y} = K, \text{ say} \quad .....(3)$$

$$\frac{X'}{X} = K, \frac{4Y'}{Y} = K$$

$$\Rightarrow \quad \frac{dX}{dx} - KY = 0 \text{ and } \frac{4dY}{dy} - YK = 0$$

$$(D - K)X = 0 \text{ and } (4D' - K)Y = 0$$

where $D = \frac{d}{dx}, \quad D' = \frac{d}{dy}$

Solving we get $X = c_1 e^{kx}$, and $Y = c_2 e^{k/4y}$

Substituting these in (2), we get

$$u(x, y) = (c_1 e^{kx})(c_2 e^{k/4y})$$

writing $c_1 c_2 = c$

we have $u(x, y) = c\, e^{k\left(x+\frac{y}{4}\right)}$ .....(4)

is the solution of the differential equation (1)

The given initial conditions are $u(0, y) = 8e^{-3y}$

i.e., $u = 8e^{-3y}$ when $x = 0$

Substituting the above conditions in (4) we find the constants c and k

$$8e^{-3y} = c\, e^{k\left(0+\frac{y}{4}\right)} \qquad c = 8,\ k = -12$$

Substituting $c = 8, k = -12$, (4) reduces to

$u(x, y) = 8e^{-12x-3y}$ which is the required solution.

Note: The answer can be verified by substituting it in the given equation along with the condition

***Example* 19.2:** Solve the equation $\frac{\partial^2 u}{\partial x \partial t} = e^{-t} \cos x$ given that $u = 0$, when $t = 0$ and $\frac{\partial u}{\partial t} = 0$ when $x = 0$ by the method of separation of variables.

Solution: $\frac{\partial^2 u}{\partial x \partial t} = e^{-t} \cos x$ .....(1)

Let $u = X(x)\, T(t)$ .....(2)

be the solution of the differential equation such that (2) satisfies (1)

Differentiating (2) partially w.r.t to x and t

$$\frac{\partial^2 u}{\partial x \partial t} = X'T'$$

Substituting

$$\frac{\partial^2 u}{\partial x \partial t} = X'T' \text{ in } (1)$$

$$X'T' = e^{-t} \cos x \Rightarrow \frac{X'}{\cos x} = \frac{e^{-t}}{T'}$$

$$\frac{X'}{\cos x} = \frac{e^{-t}}{T'} = K$$

where K is constant

$$\Rightarrow \quad X' - K \cos x = 0 \text{ and } T'K - e^{-t} = 0$$

$$\Rightarrow \quad \left(\frac{dx}{dx} - K \cos x\right) = 0 \text{ and } \left(\frac{dT}{dt}K - e^{-t}\right) = 0$$

$$dX = K\cos x dx, \qquad dT = \frac{e^{-t}}{K} dt$$

Solving the above two differential equations

$$\Rightarrow \quad X = K \sin x + C_1,\ T = \frac{-e^{-t}}{K} + C_2$$

Substituting for X, T in (2)

Solution (2) becomes

$$u(x, t) = (K \sin x + C_1)\left(\frac{-e^{-t}}{K} + C_2\right) \quad .....(3)$$

where C_1, C_2 and K are constants

$$0 = C_1\left(\frac{-e^{-t}}{K} + C_2\right) \Rightarrow C_1 = 0$$

Solution (3) also can be written as

$$u(x, t) = -e^{-t} \sin x - \frac{C_1 e^{-t}}{K} + C_2 K \sin x + C_1 C_2$$

$$u(x, t) = -e^{-t} \sin x - Ae^{-t} + B \sin x + C \quad .....(4)$$

Eq. (4) is the solution where A, B, C are constants. Imposing the given condition $u = 0$ when $t = 0$ on solution (4), we get

$$0 = -\sin x - A + B\sin x + C \quad .....(5)$$

differentiating (4) partially w.r.t to 't' we get

$$\frac{\partial u}{\partial t} = e^{-t}\sin x + Ae^{-t}$$

It is given that $\frac{\partial u}{\partial t} = 0$, when $x = 0$

$\therefore$ $0 = Ae^{-t} \Rightarrow A = 0$ .....(6)

Substituting $A = 0$ in (5)

$-\sin x + B\sin x + C = 0$

$(B-1)\sin x + C = 0$

Clearly $C = 0 \quad \therefore \quad B = 1$ .....(7)

Substituting $A = 0, B = 1, C = 0$ in (5)

The solution (5) reduces to

$$u(x, t) = \sin x - e^{-t}\sin x$$

19.2 One-dimensional Wave Equation

The partial differential equation $\frac{\partial^2 y}{\partial t^2} = C^2\frac{\partial^2 y}{\partial x^2}$ .....(1)

represents the wave equation (vibration of a string) where x is the length of the string, t is the time and y is the displacement of string. To solve (1) we use the method of separation of variables.

Let $y = X(x)T(t)$ .....(2)

be the solution of the equation (1) such that (2) satisfies (1).

Substituting (2) in the equation (1) we get

$$XT'' = C^2TX''$$

$$\frac{X''}{X} = \frac{T''}{C^2T} = K, \text{ say} \quad .....(3)$$

(Since x, t are independent variables, each expression must be a constant).

From (3) we have two ordinary differential equations.

$$X'' - KX = 0 \quad .....(4)$$

And $\qquad T'' - KC^2T = 0 \qquad$ .....(5)

***Case* (i) :**

If $K < 0$ then let $K = -n^2$, then (4) and (5) become

$$X'' + n^2X = 0 \text{ and } T'' + n^2C^2T = 0$$

(i.e.,) $\qquad \dfrac{d^2X}{dx^2} + n^2 x = 0$ and $(D^2 + n^2 C^2)T = 0$, where $D = \dfrac{d}{dx}$, $D' = \dfrac{d}{dt}$

We get $\qquad X = C_1 \cos nx + C_2 \sin nx$

and $\qquad T = C_3 \cos nCt + C_4 \sin nCt$

Substituting for X, T in (2) we have

$$y = (C_1 \cos nx + C_2 \text{ Sin } nx)\,(C_3 \cos nct + C_4 \sin nCt) \qquad .....(6)$$

***Case* (ii) :**

If $K > 0$ let $K = n^2$, then (4) and (5) become

$$X'' - n^2X = 0 \text{ and } T'' - n^2C^2T = 0$$

(i.e) $\qquad (D^2 - n^2)X = 0$ and $(D'^2 - n^2C^2)T = 0$

$$D = \pm\, n \text{ and } D' = \pm\, nc$$

$\therefore \qquad X = c_5\, e^{nx} + c_6\, e^{-nx}$

and $\qquad T = c_7\, e^{nct} + c_8\, e^{-nct}$

Substituting for X, T in the solution (2) we get

$$y = (C_5 e^{nx} + C_6 e^{-nx})\,(C_7 e^{nct} + C_8 e^{-nct}) \qquad .....(7)$$

***Case* (iii) :**

If $K = 0$, then (4), (5) become

$$X'' = 0 \text{ and } T'' = 0$$

Integrating the above equations we get

$$X = C_9 X + C_{10} \text{ and } T = C_{11} t + C_{12}$$

Substituting for X, T in the solution (2)

$$y = (C_9 x + C_{10})(C_{11} t + C_{12}) \qquad .....(8)$$

Equations (6), (7), (8) are the three possible solutions of the wave equation.

Since it is a problem of vibrating string, every position of which is repeated as time passes, the solution should be a periodic function of time 't'. The solution (6) is in terms of sine and cosine functions which are periodic by nature, hence only (6) can be taken as the solution of wave equation.

Solved Examples

***Example* 19.3:** A tightly stretched string with fixed end points $x = 0$ and $x = l$ is initially in a position given by $y(x, 0) = y_0 \sin^3\left(\frac{\pi x}{l}\right)$. If it is released from rest from this position, find the displacement 'y' at any distance x from one end and at any time 't'.

Solution:

The partial differential equation

$$\frac{\partial^2 y}{\partial t^2} = C^2 \frac{\partial^2 y}{\partial x^2} \qquad .....(1)$$

is similar to (1) of (19.1.1). 'x' is the length of the string, t is the time, y is the displacement of the string .

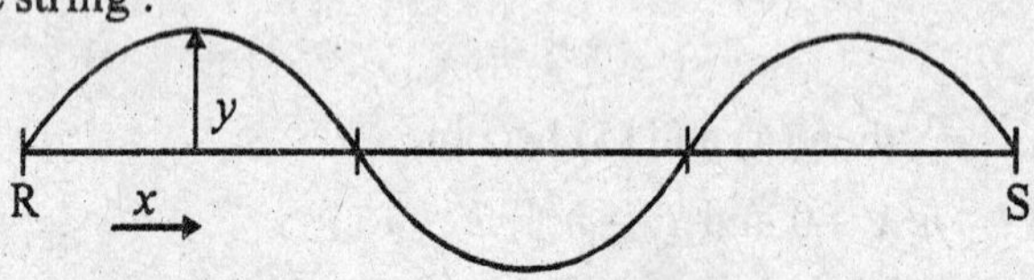

The given conditions are :

$y(0, t) = 0$, i.e., at 'R' $x = 0, t = t, y = 0$

$y(l, t) = 0$, i.e., at 'S' $x = l, t = t, y = 0$

Further $\quad \frac{\partial y}{\partial t} = 0$, at $t = 0$, and $y(x, 0) = y_0 \sin^3 \frac{\pi x}{l}$

The solution of the vibration of string (using the variables separables method) is

$$y(x, t) = (C_1 \cos nx + + C_2 \sin nx)(C_3 \cos nct + C_4 \sin nct) \qquad .....(2)$$

where C_1, C_2, C_3, C_4 and n are constants.

Now applying the first condition $y(0, t) = 0$, the solution (2) becomes

$$0 = (C_1 \cos 0 + C_2 \sin 0)(C_3 \cos nct + C_4 \sin nct)$$

$$\Rightarrow \quad C_1 = 0$$

The solution (2) reduces to

$$y(x, t) = C_2 \sin nx(C_3 \cos nct + C_4 \sin nct)$$

$$y(x, t) = \sin nx(A \cos nct + B \sin nct) \text{ [Taking } C_2C_3 = A, C_2C_4 = B\text{]} \qquad .....(3)$$

Applying the second condition $y(l, t) = 0$

$$0 = (\sin nl)(A \cos nct + B \sin nct) \quad \Rightarrow \quad \sin nl = 0$$

$$\therefore \qquad nl = p\pi \qquad (p = 0, 1, 2 \ldots)$$

Substituting $n = \dfrac{p\pi}{l}$ in (3)

$$y(x, t) = \sin\left(\frac{p\pi x}{l}\right)\left[A\cos\frac{p\pi ct}{l} + B\sin\frac{p\pi ct}{l}\right]$$

$$\therefore \qquad y(x, t) = \sin\left(\frac{p\pi x}{l}\right)\left[\sum_0^\infty A_p \cos\frac{p\pi ct}{l} + \sum_0^\infty B_p \sin\frac{p\pi ct}{l}\right] \qquad \ldots\ldots(4)$$

is the solution.

Differentiating (4) partially w.r.to t

$$\frac{\partial y}{\partial t} = \sin\left(\frac{p\pi x}{l}\right)\left[\sum_0^\infty A_p\left(-\sin\frac{p\pi ct}{l}\right).\frac{p\pi c}{l} + \sum_0^\infty B_p\left(\cos\frac{p\pi ct}{l}\right)\frac{p\pi c}{l}\right]$$

Applying $\dfrac{\partial y}{\partial t} = 0$ at $t = 0$ we get

$$0 = \sin\left(\frac{p\pi x}{l}\right)\left[\sum_0^\infty - A_p \sin 0 \frac{p\pi c}{l} + \sum_0^\infty B_p \cos 0.\frac{p\pi c}{l}\right] \Rightarrow B_p = 0$$

Substituting $B_p = 0$ (4) reduces to

$$y(x, t) = \sum A_p \sin\frac{p\pi x}{l}\cos\frac{p\pi ct}{l} \qquad \ldots\ldots(5)$$

Using the last condition

$$y(x, 0) = y_0 \sin^3\frac{\pi x}{l}$$

$$y_0 \sin^3\frac{\pi x}{l} = \sum A_p \sin\frac{p\pi x}{l}\cos 0$$

$$\sum A_p \sin\frac{p\pi x}{l} = y_0\left[\frac{3}{4}\sin\frac{\pi x}{l} - \frac{1}{4}\sin\frac{3\pi x}{l}\right]$$

$$A_0 + A_1 \sin\frac{\pi x}{l} + A_2 \sin\frac{2\pi x}{l} + A_3 \sin\frac{3\pi x}{l} + ... = \frac{3y_0}{4}\sin\frac{\pi x}{l} - \frac{y_0}{4}\sin\frac{3\pi x}{l}\ \pi x$$

Equating the coefficient of $\sin \pi x$ and $\sin 3\pi x$

$$A_0 = 0,\ A_1 = \frac{3y_0}{4},\ A_2 = 0,\ A_3 = \frac{-y_0}{4},\ A_4 = A_5 = ... = 0$$

Solution (5) can be written as

$$y(x, t) = A_0 . 0 + A_1 \sin\frac{\pi x}{l}\cos\frac{\pi ct}{l} + A_2 \sin 2\frac{\pi x}{l}\cos 2\frac{\pi ct}{l} +$$

$$+ A \sin 3\frac{\pi x}{l}\cos 3\frac{\pi ct}{l} + ...$$

Substituting the values of A_0, A_1 .....

$$y(x, t) = 0 + \frac{3}{4}y_0 \sin\frac{\pi x}{l}\cos\frac{\pi ct}{l} + 0 + \frac{-y_0}{4}\sin 3\frac{\pi x}{l}\cos 3\frac{\pi ct}{l}$$

$$y(x, t) = \frac{3y_0}{4}\sin\frac{\pi x}{l}\cos\frac{\pi ct}{l} - \frac{y_0}{4}\sin\frac{3\pi x}{l}\cos\frac{3\pi ct}{l}$$

is the required solution.

***Example* 19.4:** A tightly stretched string with fixed ends at $x = 0$ and $x = l$ is initially in equilibrium position. It is set vibrating by giving to each of its points an initial velocity

$$\left(\frac{\partial y}{\partial t}\right) = y_0 \sin^3\left(\frac{\pi x}{l}\right), \qquad t = 0$$

Find the displacement $y(x, t)$.

Solution: The vibration of a string is given by

$$\frac{\partial^2 y}{\partial t^2} = C^2\frac{\partial^2 y}{\partial x^2} \qquad .....(1)$$

The solution of the vibration of a string is

$$Y = (C_1 \cos nx + C_2 \sin nx)\ (C_3 \cos nct + C_4 \sin nct) \qquad .....(2)$$

One end R of the String is fixed throughout the motion i.e., $x = 0, y = 0$ for all 't'

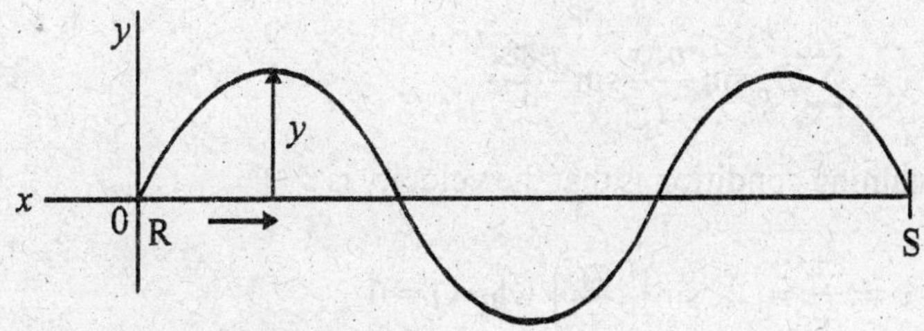

$\therefore$ the solution (2) reduces to

$$0 = (C_1 + 0)(C_3 \cos nct + C_4 \sin nct)$$

$\therefore$ $C_1 = 0$, substituting in (2)

$$\therefore \quad y = C_2 \sin nx\,(C_3 \cos nct + C_4 \sin nct)$$

$$y = \sin nx[A\cos nct + b \sin nct] \quad .....(3)$$

where $A = C_2C_3,\; B = C_2C_4$

The other end S is fixed throughout the motion i.e., when $x = l$, $y = 0$ for all t

$\therefore$ the solution (3) reduces to

$$0 = \sin nl\,(A\cos nct + B \sin nct)$$

$$\Rightarrow \quad nl = p\pi \quad \therefore \quad n = \frac{p\pi}{l}; \qquad (p = 0, 1, 2, ………)$$

substituting in (3)

$$y = \sin\frac{p\pi x}{l}\left[A_p \cos\frac{p\pi ct}{l} + B\sin\frac{p\pi ct}{l}\right]$$

$$\therefore \quad y = \sin\frac{p\pi x}{l}\left[\sum A_p \cos\frac{p\pi ct}{l}\sum B_p \sin\frac{p\pi ct}{l}\right] \quad .....(4)$$

Initially the string is stretched along the x-axis and is at rest in its equilibrium position.

i.e., $y = 0$ for all 'x' when $t = 0$,

$\therefore$ from(4) we get

$$0 = \sin\frac{p\pi x}{l}\left[\sum A_p + 0\right]$$

$$\Rightarrow \quad \sum A_p = 0$$

substituting in the solution (4) we get

$$y = \sum B_p \sin\frac{p\pi x}{l}\sin\frac{p\pi ct}{l} \qquad .....(5)$$

Now, the remaining condition is that the velocity is

$$v = \frac{\partial y}{\partial t} = y_0 \sin^3\left(\frac{\pi x}{l}\right) \text{ when } t = 0$$

Differentiating (5) partially w.r.t. to 't'

$$\frac{\partial y}{\partial t} = \sum B_p \sin\frac{p\pi x}{l}.\cos\frac{p\pi ct}{l}.\frac{p\pi c}{l}$$

$$y_0 \sin^3\left(\frac{\pi x}{l}\right) = \sum B_p \sin\frac{p\pi x}{l}.\frac{p\pi c}{l}$$

(using the given condition)

writing $\sin^3\theta = \frac{3}{4}\sin\theta - \frac{1}{4}\sin 3\theta$ we have

$$y_0\left[\frac{3}{4}\sin\frac{\pi x}{l} - \frac{1}{4}\sin\frac{3\pi x}{l}\right] = \sum B_p \frac{p\pi c}{l}.\sin\frac{p\pi x}{l}$$

$$= 0 + B_1\left(\frac{\pi c}{l}\right)sin\frac{\pi x}{l} + B_2\left(\frac{2\pi c}{l}\right)\sin\left(\frac{2\pi x}{l}\right)$$

$$+ B_3\left(\frac{3\pi c}{l}\right)\sin\left(\frac{3\pi x}{l}\right) + B_4\left(\frac{4\pi c}{l}\right)\sin\frac{4\pi x}{l} + ...$$

Equating the coefficients from both sides we get

$$B_1 = \frac{3ly_0}{4\pi c},\ B_2 = 0,\ B_3 = -\frac{ly_0}{12\pi c},\ B_4 = B_5 = ... = 0$$

Substituting these values in the solution (5)

$$y = \frac{3ly_0}{4\pi c}\sin\frac{\pi x}{l}\sin\frac{\pi ct}{l} - \frac{ly_0}{12\pi c}\sin\frac{3\pi x}{l}\sin\frac{3\pi ct}{l}$$

***Example* 19.5:** A tightly stretched string with fixed end points $x = 0$ and $x = l$ is initially at rest in its equilibrium position. If it is set vibrating by giving to each of its points a velocity $\frac{\partial y}{\partial t} = 3(lx - x^2)$ at $t = 0$, find $y(x, t)$.

Solution: The partial differential equation

$$\frac{\partial^2 y}{\partial t^2} = c^2 \frac{\partial^2 y}{\partial x^2} \qquad .....(1)$$

represents the equation of the vibration of a string

The solution of the equation (1) is

$$Y = (C_1 \cos nx + C_2 \sin nx)(C_3 \cos nct + C_4 \sin nct) \qquad .....(2)$$

The initial and boundary conditions are

$$y(0, t) = 0,\ y(l, t) = 0,\ y(x, 0) = 0, \text{ and } \left.\frac{\partial y}{\partial t}\right|_{t=0} = 3(lx - x^2) \qquad .....(3)$$

Substituting $x = 0, y = 0$ for all 't' in the solution (2)

$$0 = (C_1 + 0)(C_3 \cos nct + C_4 \sin nct) \Rightarrow C_1 = 0, \text{(2) now becomes}$$

$$y = (A \cos nct + B \sin nct) \sin nx \qquad .....(4)$$

where $\quad A = C_1C_3,\ B = C_1C_4$

Applying the condition $y(l, t) = 0$ i.e., $x = l, y = 0$ for all 't' on the solution (4)

$$0 = (A \cos nct + B \sin nct) \sin nl$$

$$\therefore \quad \sin nl = 0 \Rightarrow nl = p\pi \quad (p = 0, 1, 2, ...)$$

substituting $n = \frac{p\pi}{l}$ in (4)

$$y = \sin \frac{p\pi x}{l}\left[\sum A_p \cos \frac{p\pi ct}{l} + \sum B_p \sin \frac{p\pi ct}{l}\right] \qquad .....(5)$$

Imposing the condition $y(x, 0) = 0$ on (5)

$$0 = \sin \frac{p\pi x}{l}\left[\sum A_p.1 + \sum B_p.0\right] \Rightarrow \sum A_p = 0$$

then (5) reduces to

$$y = \sum B_p \sin\frac{p\pi x}{l}\sin\frac{p\pi ct}{l} \qquad .....(6)$$

Differentiating (6) partially w.r to 't'

$$\frac{\partial y}{\partial t} = \sum B_p \sin\frac{p\pi x}{l}\left(\cos\frac{p\pi ct}{l}\right).\frac{p\pi c}{l}$$

Now using the last condition $\frac{\partial y}{\partial t} = 3(lx - x^2)$ at $t = 0$ we get

$$3(lx - x^2) = \sum B_p\left(\frac{p\pi c}{l}\right)\sin\frac{p\pi x}{l} \qquad .....(7)$$

But the expansion of $f(x) = 3(lx - x^2)$ into half range fourier sine series in $(0, l)$ is

$$f(x) = 3(lx - x^2) = \sum B_p \sin\frac{p\pi x}{l} \qquad .....(8)$$

where $\quad B_p = \frac{2}{l}\int_0^l 3\left(lx - x^2\right)\sin\frac{p\pi x}{l}\,dx$

(i.e.,) $\quad B_p = \frac{2}{l}\left[\Sigma 3\left(lx - x^2\right)\left(\frac{-l}{p\pi}\cos\frac{p\pi x}{l}\right) - 3(l - 2x)\left(\frac{-l^2}{p^2\pi^2}\sin\frac{p\pi x}{l}\right) - 3(0 - 2)\left(\frac{l^3}{p^3\pi^3}\cos\frac{p\pi x}{l}\right) + 0\right]_0^l$

$$B_p = \frac{2}{l}\left[0 - 3.0 - \frac{6l^3}{p^3\pi^3}(\cos p\pi - \cos 0)\right]$$

$$B_p = \frac{12l^2}{p^3\pi^3}\,[1 - \cos p\pi]$$

$$B_p = \frac{12l^2}{p^3\pi^3}\left[1-(-1)^p\right]$$

$\therefore$ The halfrange fourier sine series of $f(x)$ in $(0, l)$ is given by

$$f(x) = 3(lx - x^2) = \sum \frac{12l^2}{p^3\pi^3}\left[1-(-1)^p\right]\sin\frac{p\pi x}{l}$$

Substituting this in (7) we have

$$\sum \frac{12l^2}{p^3\pi^3}\left[1-(-1)^p\right]\sin\frac{p\pi x}{l} = \sum B_p \frac{p\pi c}{l}\sin\frac{p\pi x}{l}$$

equating the coefficient of $\sin\left(\frac{p\pi x}{l}\right)$

$$\sum B_p = \sum \frac{12l^3}{cp^4\pi^4}\left[1-(-1)^p\right]$$

$$B_1 = \frac{24l^3}{c\pi^4},\ B_2 = 0,\ B_3 = \frac{24l^3}{c3^4\pi^4},$$

$$B_4 = 0,\ B_5 = \frac{24l^3}{c5^4\pi^4},\ B_6 = 0,\ \text{etc.....}$$

substituting these values in the solution (6)

$$y = \frac{24l^3}{c\pi^4}\sin\frac{\pi x}{l}\sin\frac{\pi ct}{l} + \frac{24l^3}{c3^4\pi^4}\sin\frac{3\pi x}{l}\sin\frac{3\pi ct}{l} +$$

$$\frac{24l^3}{c5^4\pi^4}\sin\frac{5\pi x}{l}.\sin\frac{5\pi ct}{l} + \ldots..$$

$$\therefore \quad y = \frac{24l^3}{c\pi^4}\left\{\sin\frac{\pi x}{l}\sin\frac{\pi ct}{l} + \frac{1}{3^4}\sin\frac{3\pi x}{l}\sin\frac{3\pi ct}{l} + \frac{1}{5^4}\sin\frac{5\pi x}{l}\sin\frac{5\pi ct}{l} + \ldots..\right\}$$

***Example* 19.6:** A tightly stretched string with fixed end points $x = 0$, $x = l$ in the shape defined by $y = \lambda x\,(l - x)$ where λ is a constant, released from this position of rest. Find $y(x, t)$.

Solution: The equation of the vibration of a string is

$$\frac{\partial^2 y}{\partial t^2} = c^2\frac{\partial^2 y}{\partial x^2} \qquad \ldots..(1)$$

The solution of above equation using the variables separables method is

$$Y = (C_1 \cos nx + C_2 \sin nx)(C_3 \cos n\,ct + C_4 \sin nct) \qquad .....(2)$$

and the given initial and boundary conditions are

$$y(0, t) = 0,\ y(l, t) = 0,\ \left.\frac{\partial y}{\partial t}\right|_{t=0} = 0,\ y(x, 0) = \lambda x(l - x) \qquad .....(3)$$

Using the first three conditions of (3) in (2), the solution becomes

$$y = \sum A_p \sin\frac{p\pi x}{l}\cos\frac{p\pi ct}{l} \qquad .....(4)$$

(Ref: Example 19.3)

Now applying the condition $y(x, 0) = \lambda x(l - x)$

i.e., substituting $y = \lambda x(l - x)$, $x = x$, $t = 0$ in (4) we get

$$\lambda x(l - x) = \sum A_p \sin\left(\frac{p\pi x}{l}\right).\cos 0 \qquad .....(5)$$

But Fourier half-range sine series of $f(x) = \lambda x(l - x)$ in the range $(0, l)$ is

$$\lambda x(l - x) = \sum \frac{4\lambda l^2}{p^3\pi^3}\left[1 - (-1)^p \sin\frac{p\pi x}{l}\right] \qquad .....(6)$$

From (5) and (6) we get

$$\sum \frac{4\lambda l^2}{p^3\pi^3}\left[1 - (-1)^p\right]\sin\frac{p\pi x}{l} = \sum A_p \sin\frac{p\pi x}{l}$$

$$\Rightarrow \quad A_p = \frac{4\lambda l^2}{p^3\pi^3}\left[1 - (-1)^p\right]$$

$$A_1 = \frac{8\lambda l^2}{\pi^3}, \quad A_2 = 0$$

$$A_3 = \frac{8\lambda l^2}{3^3\pi^3}, \quad A_4 = 0$$

$$A_5 = \frac{8\lambda l^2}{5^3\pi^3}, \quad A_6 = 0, \text{ etc.,.......}$$

Substituting these values in the solution (4)

$$y = \frac{8\lambda l^2}{\pi^3} \sin\frac{\pi x}{l} \cos\frac{\pi ct}{l} + \frac{8\lambda l^2}{3^3\pi^3} \frac{\sin 3\pi x}{l} \frac{\cos 3\pi ct}{l} + \ldots..$$

$$y = \frac{8\lambda l^2}{\pi^3}\left[\sin\frac{\pi x}{l}\cos\frac{\pi ct}{l} + \frac{1}{3^3}\frac{\sin 3\pi x}{l}\cos\frac{3\pi ct}{l} + \frac{1}{5^3}\sin\frac{5\pi x}{l}\cos\frac{5\pi ct}{l} + \ldots..\right.$$

Exercise 19.1

1. A string is stretched between two fixed points distance 'l' apart and is released from rest from the initial position given by

$$y = \frac{2k}{l}x, \quad 0 \le x \le \frac{l}{2}$$

$$y = \frac{2k}{l}(l - x), \quad \frac{l}{2} \le x \le l$$

Show that the deflection $y(x, t)$ of the string at any time 't' is given by

$$y(x, t) = \frac{8k}{\pi^2}\left[\sin\frac{\pi x}{l}\cos\frac{\pi ct}{l} - \frac{1}{3^2}\frac{\sin 3\pi x}{l}\frac{\cos 3\pi ct}{l} + \frac{1}{5^2}\frac{\sin 5\pi x}{l}\frac{\cos 5\pi ct}{l}\ldots..\right]$$

2. A string is stretched and fastened to two points distance 'l' apart. Motion is started by displacing the string in the form $y = 4\sin^3\left(\frac{\pi x}{l}\right)$ from which it is released at time $t = 0$. Find the displacement $y(x, t)$.

Ans : $y(x, t) = 3\sin\frac{\pi x}{l}\cos\frac{\pi ct}{l} - \sin\frac{3\pi x}{l}\cos\frac{3\pi ct}{l}$

3. A string of length 'a' is drawn aside at its mid point through a distance 'p', so that its initial position is given by

$$y(x, 0) = \frac{2px}{a}, \qquad 0 \le x \le \frac{a}{2}$$

$$= 2p - \frac{2px}{a}, \qquad \frac{a}{2} \le x = a$$

It is initially at rest (so that $\frac{\partial y}{\partial t} = 0, t = 0$) and is suddenly released from this position.

If its end points are kept fixed through ($y(0, t) = y(a, t) = 0$) find $y(x, t)$

Ans : $y(x, t) = \sum_{n=1}^{\infty} \frac{8p(-1)^{n+1}}{(2n-1)^2 \pi^2} \sin \frac{(2n-1)\pi x}{a} . \cos \frac{(2n-1)\pi ct}{a}$

19.3 One Dimensional Heat Equation

The partial differential equation

$$\frac{\partial u}{\partial t} = c^2 \frac{\partial^2 u}{\partial x^2} \qquad .....(1)$$

is the one dimensional heat equation.

where u is the temperature in the rod at any point 'x' and at any time 't' and $c^2 = \frac{K}{s\rho}$ is the coefficient of conductivity where s is the specific heat and ρ is the density of the rod.

We shall now solve the above equation (1) by the method of separation of variables.

Let $\qquad u = X(x)\, T(t) = XT \qquad .....(2)$

be the solution of (1).

Then $\frac{\partial u}{\partial t} = XT'$ and $\frac{\partial^2 u}{\partial x^2} = TX''$ where $T' = \frac{dT}{dt}, X'' = \frac{d^2X}{dx^2}$

Substituting these in (1), we get,

$$XT' = C^2 TX''$$

$$\frac{X''}{X} = \frac{1}{C^2}\frac{T'}{T} = K \text{ (say)} \qquad .....(3)$$

where K is a constant which can be negative, positive or zero

(i) When K is negative, say, $K = -n^2$ then from (3)

$$\frac{X''}{X} = \frac{1}{C^2}\frac{T'}{T} = -n^2$$

$\therefore \quad X'' + n^2x = 0 \quad$ and $\quad T + n^2c^2T = 0$

(i.e.,) $\quad \dfrac{d^2x}{dx^2} + n^2x = 0 \quad$ and $\quad \dfrac{dT}{dt} + n^2c^2 = 0$

(i.e.,) $\quad (D^2 + n^2)\,x = 0 \quad$ and $\quad (D_1 + n^2c^2)\,T = 0 \quad$ where $D_1 = \dfrac{d}{dt}$

The solution is

$$X = c_1 \cos nx + c_2 \sin nx, \text{ and } T = c_3e^{-n^2c^2t}$$

Substituting these values in the solution (2) the complete solution is

$$u = (c_1 \cos nx + c_2 \sin nx)\, c_3e^{-n^2c^2t} \qquad .....(4)$$

(ii) When K is positive say, $K = n^2$, from (3) we get

$$\frac{X''}{X} = n^2 \text{ and } \frac{T'}{c^2T} = n^2$$

$$\frac{d^2X}{dx^2} - n^2x = 0 \text{ and } \frac{dT}{dt} - n^2c^2T = 0$$

$$(D^2 - n^2)\,X = 0,\ X = c_4e^{nx} + c_5\,e^{-nx} \quad \text{and} \quad T = c_6\,e^{n^2c^2t}$$

Substituting these in $u = XT$ the solution (2) becomes

$$u = (c_4e^{nx} + c_5e^{-nx})c_6\,e^{n^2c^2t} \qquad .....(5)$$

(iii) when $K = 0$, then from (3)

$$\frac{X''}{X} = \frac{1}{c^2}\frac{T'}{T} = 0$$

$$X'' = 0 \text{ and } T' = 0$$

(i.e.,) $\quad \dfrac{d^2X}{dx^2} = 0$ and $\dfrac{dT}{dt} = 0$

$$\frac{dX}{dx} = c_7 \text{ and } T = c_8$$

$\Rightarrow \quad X = c_7x + c_9$

Substituting these in $u = XT$.

The solution (2) becomes

$$u = (c_7x + c_9)c_8 \qquad .....(6)$$

Of these solutions (4), (5), (6) we have to choose that solution which is consistent with the physical nature of the problem.

If u is finite when $t \to \infty$ which is the case in steady state, we find that (5) is not a possible solution in steady state. Under these conditions the solution is given by (4)

i.e., $u = (A \text{ con } nx + B \sin nx)\, e^{-n^2c^2t}$ where $A = c_1 c_3$, $B = c_2 c_3$

Solved Examples

***Example* 19.7:** Find the solution of the one dimensional heat equation $\dfrac{\partial u}{\partial t} = c^2 \dfrac{\partial^2 u}{\partial x^2}$ under the boundary conditions $u(0, t) = 0$, $u(l, t) = 0$ and $u(x, 0) = x$, $(0 < x < l)$, l being the length of the rod.

***Solution*:** One dimensional heat equation is

$$\frac{\partial u}{\partial t} = c^2 \frac{\partial^2 u}{\partial x^2} \quad .....(1)$$

The solution of (1) using variable separables method is

$$u = (c_1 \cos nx + c_2 \sin nx)\, e^{-n^2c^2t} \quad .....(2)$$

and the given initial conditions are at the end 'A' $u(0, t) = 0$, and at end B of the rod $u(l, t) = 0$ and $u(x, 0) = x$; $(0 < x, l)$.

Using the initial condition $u(0, t) = 0$ in the solution (2).

(i.e.,) substituting the $u = 0, x = 0, t = t$ in solution (2)

$$0 = (c_1 1 + 0)\, e^{-n^2c^2t} \quad \Rightarrow \quad c_1 = 0$$

Substituting $c_1 = 0$ in the soultion in (2)

$$u = c_2 \sin nx\, e^{-n^2c^2t} \quad .....(3)$$

Using the initial condition $u(l, t) = 0$ in (3)

i.e., $u = 0, x = l, t = t$

$$0 = c_2 \sin nl \; e^{-n^2c^2t} \quad \Rightarrow \quad \sin nl = 0$$

$$\therefore \quad nl = p\pi \quad (p = 0, 1, 2.....)$$

$$n = \frac{p\pi}{l}$$

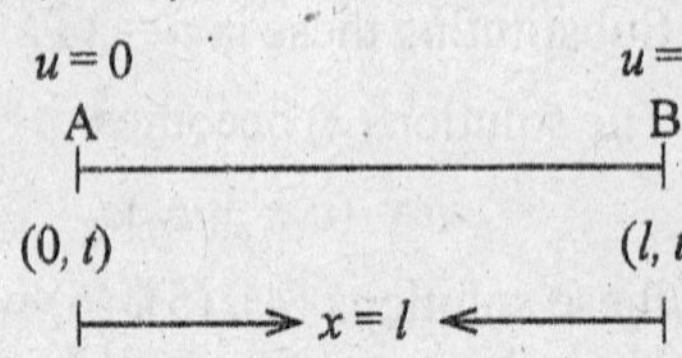

The solution (3) becomes

$$u = c_2' \sin\frac{p\pi x}{l} e^{-\frac{p^2\pi^2c^2t}{l^2}} \quad \text{where } p = 0, 1, 2.....$$

Hence the solution is

$$u = \sum B_p \sin\frac{p\pi x}{l} e^{-\frac{p^2\pi^2c^2t}{l^2}} \qquad .....(4)$$

Using the condition $u(x, 0) = x$ in the solution (4)

i.e., substituting $u = x, x = x, t = 0$ in the solution (4)

$$x = \sum B_p \sin\left(\frac{p\pi x}{l}\right).e^0 \qquad .....(5)$$

But, the Fourier half-range sine series for the function

$f(x) = x$ in $(0, l)$ is

$$x = \sum A_p \sin\frac{p\pi x}{l} \qquad .....(6)$$

where $$\sum A_p = \frac{2}{l}\int_0^l x \sin\frac{p\pi x}{l} dx$$

$$= \frac{2}{l}\left[(x)\left(-\cos\frac{p\pi x}{l}\right).\frac{l}{p\pi} - (1)\left(-\sin\frac{p\pi x}{l}\right).\frac{l^2}{p^2\pi^2}\right]_0^l$$

$$= \frac{2}{l}\left[-\frac{l^2}{p\pi}\cos p\pi - 0 - 0\right]$$

$$\sum A_p = \frac{2l}{p\pi}(-1)^{p+1}$$

Then (6) becomes

$$x = \sum A_p \sin\frac{p\pi x}{l}$$

(i.e.,) $x = \sum \frac{2l}{p\pi}(-1)^{p+1} \sin\frac{p\pi x}{l}$ is the half - range sine series of the function $f(x) = x$.

Substituting this in L.H.S of (5)

$$\sum \frac{2l}{p\pi}(-1)^{p+1} \sin \frac{p\pi x}{l} = \sum B_p \sin \frac{p\pi x}{l}$$

$$\Rightarrow \quad \sum B_p = \frac{2l}{p\pi}(-1)^{p+1}$$

Substituting this value in the solution (4)

$$u = \sum_{p=1}^{\infty} \frac{2l}{p\pi}(-1)^{p+1} \sin \frac{p\pi x}{l} e^{-\frac{p^2\pi^2c^2t}{l^2}}$$

Therefore the solution is

$$u = \frac{2l}{\pi}\left\{\frac{1}{1}.\sin\frac{\pi x}{l}.e^{-\frac{\pi^2c^2t}{l^2}} - \frac{1}{2}\sin\frac{2\pi x}{l}e^{-\frac{4\pi^2c^2t}{l^2}} + \frac{1}{3}\sin\frac{3\pi x}{l}e^{-\frac{9\pi^2c^2t}{l^2}} - \ldots\ldots\right\}$$

***Example* 19.8:** Solve the equation $\frac{\partial u}{\partial t} = c^2 \frac{\partial^2 u}{\partial x^2}$ for the conduction of heat along a rod, without radiation, subject to the following conditions :

(a) u is not infinite for $t \to +\infty$

(b) $\frac{\partial u}{\partial x} = 0$ for $x = 0$ and $x = l$ (i.e., no heat can pass through the ends of the rod).

(c) $u = lx - x^2$ for $t = 0$ between $x = 0$ and $x = l$.

Prove that

$$u = \frac{l^2}{6} - \frac{l^2}{\pi^2}\sum_{p=1}^{\infty}\frac{1}{p^2}\cos\frac{2\pi px}{l}.e^{\frac{-4\pi^2p^2c^2t}{l^2}}$$

***Solution*:** One dimensional heat equation is

$$\frac{\partial u}{\partial t} = c^2 \frac{\partial^2 u}{\partial x^2} \qquad \ldots\ldots(1)$$

The solution using the method of separation of variables satisfying the given condition (a) is

$$u = (c_1 \cos nx + c_2 \sin nx)\, e^{-c^2 n^2 t} \qquad \text{.....(2)}$$

Now applying the condition (b)

$$\left.\frac{\partial u}{\partial x}\right|_{x=0} = \left.(-c_1 n \sin nx + c_2 n \cos nx)\, e^{-c^2 n^2 t}\right|_{x=0} = 0$$

$$\Rightarrow \qquad n \,.\, c_2 \,.\, e^{-n^2 c^2 t} = 0$$

$$\Rightarrow \qquad c_2 = 0$$

Substituting $c_2 = 0$ in (2) it reduces to

$$u = c_1 \cos nx \; e^{-c^2 n^2 t} \qquad \text{.....(3)}$$

and

$$\left.\frac{\partial u}{\partial x}\right|_{x=l} = \left. - c_1\, n \sin nx \,.\, e^{-c^2 n^2 t}\right|_{x=l} = 0$$

$$\Rightarrow \qquad -c_1 n \sin nl \; e^{-c^2 n^2 t} = 0$$

$$\Rightarrow \qquad \sin nl = 0$$

$$nl = p\pi \qquad (p = 0, 1, 2.....)$$

$$\therefore \qquad n = \frac{p\pi}{l}$$

Substituting $n = \frac{p\pi}{l}$ in the solution (3)

$$u = c_1 \cos \frac{p\pi x}{l}\; e^{\frac{-c^2 p^2 \pi^2}{l^2} t} \qquad \text{.....(4)}$$

Imposing the condition (*c*) $u = lx - x^2$ for $t = 0$ on (4)

$$lx - x^2 = \sum A_p \cos \frac{p\pi x}{l} e^0 \qquad \text{.....(5)}$$

Using the Fourier half-range cosine series in the range (0, *l*) the expansion of $lx - x^2$ is.

$$lx - x^2 = \frac{l^2}{6} - \frac{l^2}{\pi^2}\left(\cos\frac{2\pi x}{l} + \frac{1}{2^2}\cos\frac{4\pi x}{l} + \frac{1}{3^2}\cos\frac{6\pi x}{l} + ...\right)$$

substituting this value in L.H.S of (5)

$$\frac{l^2}{6}-\frac{l^2}{\pi^2}\left(\cos\frac{2\pi x}{l}+\frac{1}{2^2}\cos\frac{4\pi x}{l}+\frac{1}{3^2}\cos\frac{6\pi x}{l}+.....\right)=\sum A_p\cos\frac{p\pi x}{l}$$

comparing the coefficients of like terms on both sides we get

$$A_0=\frac{l^2}{6}\qquad A_1=0, A_3=0\ A_5=0\ .....$$

$$A_2=\frac{-l^2}{\pi^2};$$

$$A_4=-\frac{1}{2^2}\frac{l^2}{\pi^2};$$

$$A_6=-\frac{1}{3^2}\frac{l^2}{\pi^2}\ .....$$

Substituting these in the solution (4)

$$u=\frac{l^2}{6}-\frac{l^2}{\pi^2}\left\{\cos\frac{2\pi x}{l}e^{\frac{4c^2\pi^2 t}{l}}+\frac{1}{2^2}\cos\frac{4\pi x}{l}e^{\frac{16c^2\pi^2 t}{l}}+.....\right\}$$

$$\therefore\qquad u=\frac{l^2}{6}-\frac{l^2}{\pi^2}\sum_{p=1}^{\infty}\frac{1}{p^2}\cos\frac{2\pi px}{l}e^{\frac{-4c^2p^2\pi^2 t}{l}}$$

***Example* 19.9:** A homogenous rod of conduction material of length 'l' has ends kept at zero temperature and the temperature at the centre is 'T' and falls uniformly to zero at the two ends. Find the temperature at any time.

Solution:

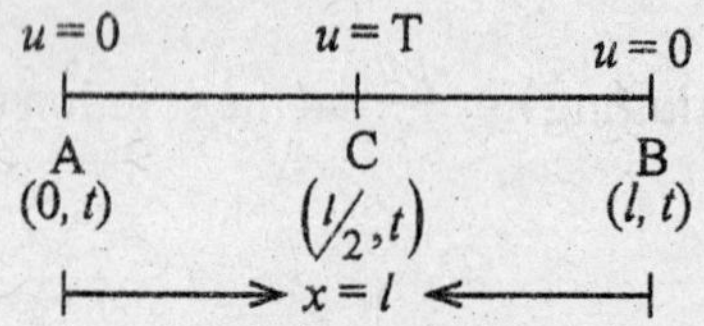

One dimensional heat equation is

$$\frac{\partial u}{\partial t}=c^2\frac{\partial^2 u}{\partial x^2}\qquad .....(1)$$

The solution of the equation (1) using the variable separable method is

$$u=\left(c_1\cos nx+c^2\sin nx\right)e^{-n^2c^2t}\qquad .....(2)$$

The given initial conditions are at the end 'A' of the rod $u(0, t) = 0$ and .....(i)

at the end of B of the rod $u(l, t) = 0$ .....(ii)

When $t = 0$, it is given that its temperature falls uniformly to zero at the ends (the heat flow was independent of time i.e., steady state condition).

$\therefore$ The temperature depends only upon 'x' and not on 't', hence (1) reduces to $\dfrac{\partial^2 u}{\partial x^2} = 0$

The general solution is then $u = ax + b$ where a, b are constants.

For the portion AC of the rod we have at A

'A' using $u = 0, x = 0$ in $u = ax + b$

$0 = a \cdot 0 + b$

$\Rightarrow$ $b = 0$, then $u = ax + b$ reduces to

$$u = ax \qquad \text{.....(iii)}$$

At 'C' using $u = T, x = \dfrac{l}{2}$ in (iii)

$$T = a \cdot \frac{l}{2} \quad \Rightarrow a = \frac{2T}{l}$$

Substituting $a = \dfrac{2T}{l}$, $b = 0$ in the solution $u = ax + b$ we get

$$u = \frac{2T}{l}x, \qquad (0 \le x \le l/2) \qquad \text{.....(3)}$$

For the portion CB of the rod

At 'C' using $u = T, x = l/2$ in $u = ax + b$ we get

$$T = a\frac{l}{2} + b \qquad \text{.....(4)}$$

At 'B' using $u = 0, x = l$ in $u = ax + b$

$$0 = al + b \qquad \text{.....(5)}$$

From (4) and (5), we get

$$a = \frac{2T}{l}, b = 2T$$

Substituting for a, b in $u = ax + b$ we get

$$u = \frac{2T}{l}x + 2T$$

(i.e.,) $u = \dfrac{2T}{l}(l - x), \quad \dfrac{l}{2} \le x \le l$

Thus $u = \dfrac{2T}{l}x, \quad 0 \le x \le l/2$

$$= \frac{2T}{l}(l - x), \qquad l/2 \le x \le l \qquad \text{.....(iv)}$$

Using the condition (i) in the solution (2)

$$0 = (c_1 + 0)\, e^{-n^2c^2t}$$

$\Rightarrow$ $c_1 = 0$, the solution (2) reduces to

$$u = c_2 \sin nx\; e^{-n^2c^2t} \qquad \text{.....(6)}$$

Using the condition (ii) in the solution (6)

$$0 = c_2 \sin nl \,.\, e^{-n^2c^2t}$$

$$\Rightarrow \qquad \sin nl = 0$$

$$\therefore \qquad nl = p\pi$$

$$n = \frac{p\pi}{l}, \qquad (p = 0, 1, 2.....)$$

The solution (6) reduces to

$$u = c_2 \sin \frac{p\pi x}{l} \,.\, e^{-\frac{p^2\pi^2c^2t}{l^2}}$$

The general solution is therefore

$$u = \sum_{p=1}^{\infty} B_p \sin \frac{p\pi x}{l} e^{-\frac{p^2\pi^2c^2t}{l^2}} \qquad \text{.....(7)}$$

using the condition (iii)

$$u = \frac{2T}{l}x, \qquad 0 \le x \le \tfrac{l}{2}$$

$$= \frac{2T}{l}(l - x), \qquad \tfrac{l}{2} \le x \le 0)$$

at $\qquad t = 0$

$$u = \sum B_p \sin \frac{p\pi x}{l} .e^0 \qquad \text{.....(8)}$$

where $\qquad u = \frac{2T}{l}x\;; \qquad (0 \le x \le \frac{l}{2})$

$$= \frac{2T}{l}(l - x), \qquad \left(\tfrac{l}{2} \le x \le l\right)$$

Now the Fourier half range sine series of $f(x)$ is

$$f(x) = \sum A_p \sin \frac{p\pi x}{l} \qquad .....(9)$$

where $A_p = \frac{2}{l}\int_0^l f(x) \sin \frac{p\pi x}{l}\, lx$

$$A_p = \frac{l}{2}\int_0^{l/2} f(x) \sin \frac{p\pi x}{l}\, dx + \int_{l/2}^{l} f(x) \sin \frac{p\pi x}{l}\, dx$$

(i.e.,) $A_p = \frac{2}{l}\left\{\int_0^{l/2} \frac{2T}{l} x \sin \frac{p\pi x}{l}\, dx + \int_{l/2}^{l} \frac{2T}{l}(l-x) \sin \frac{p\pi x}{l}\, dx\right\}$

$$A_p = \frac{4T}{l^2}\left[(x)\left(-\cos\frac{p\pi x}{l}\right)\cdot\frac{l}{p\pi} - (1)\left(-\sin\frac{p\pi x}{l}\right)\frac{l^2}{p^2\pi^2}\right]_0^{l/2}$$

$$+ \frac{4T}{l^2}\left[(l-x)\left(-\cos\frac{p\pi x}{l}\right)\cdot\frac{l}{p\pi} - (0-1)\left(-\sin\frac{p\pi x}{l}\right)\left(\frac{l^2}{p^2\pi^2}\right)\right]_{l/2}^{l}$$

$\therefore$ $A_p = \frac{8T}{\pi^2 p^2} \sin\left(\frac{p\pi}{2}\right)$

Substituting this in (9) we get $f(x) = \sum_{p=1}^{\infty} \frac{8T}{p^2\pi^2} \sin\frac{p\pi}{2} \sin\frac{p\pi x}{l}$ is the Fourier

half-range sine series of $f(x) = u(x) = \frac{2T}{l}x, \left(0 \le x \le \frac{l}{2}\right)$

$$= \frac{2T}{l}(l-x), \left(\frac{l}{2} \le x \le l\right)$$

Substituting this in L.H.S of (8)

$$\sum \frac{8T}{p^2\pi^2} \sin\frac{p\pi}{2} \sin\frac{p\pi x}{l} = \sum B_p \sin\frac{p\pi x}{l}.e^0$$

$$\Rightarrow \quad \sum_{p=1}^{\infty} B_p = \sum_{p=1}^{\infty} \frac{8T}{p^2\pi^2} \sin\frac{p\pi}{2}$$

Substituting this in the solution (7)

$$u = \sum_{p=1}^{\infty} \frac{8T}{p^2\pi^2} \sin\frac{p\pi}{2} \sin\frac{p\pi x}{l} e^{-\frac{p^2\pi^2c^2t}{l^2}}$$

$\therefore$ The solution is

$$u = \frac{8T}{\pi^2} \sum_{p=1}^{\infty} \frac{1}{p^2} \sin\frac{p\pi}{2} \sin\frac{p\pi x}{l} e^{-\frac{p^2\pi^2c^2t}{l^2}}$$

***Example* 19.10:** A rod of length 'l' has its ends A and B kept at 0 °C and 100 °C respectively until steady state conditions prevail. If the temperature at B is reduced suddenly to 0 °C and kept so while that of 'A' is maintained. Find the temperature $u(x, t)$ at a distance x from A at time t.

***Solution*:** One dimensional heat equation is

$$\frac{\partial u}{\partial t} = c^2 \frac{\partial^2 u}{\partial x^2} \qquad .....(1)$$

The solution of (1) using variables separable method is

$$u = (c_1 \cos nx + c_2 \sin nx)\, e^{-n^2c^2t} \qquad .....(2)$$

The given initial conditions are at A

$$u(0, t) = 0$$

and at B $\quad u(l, t) = 0$

Before the temperature at B is reduced to 0 °C, the temperature at any point of the rod did not change with time (steady state conditions prevailed). When it depends on 'x' only, the equation of heat conduction (1) reduces to

$$\frac{\partial^2 u}{\partial x^2} = 0$$

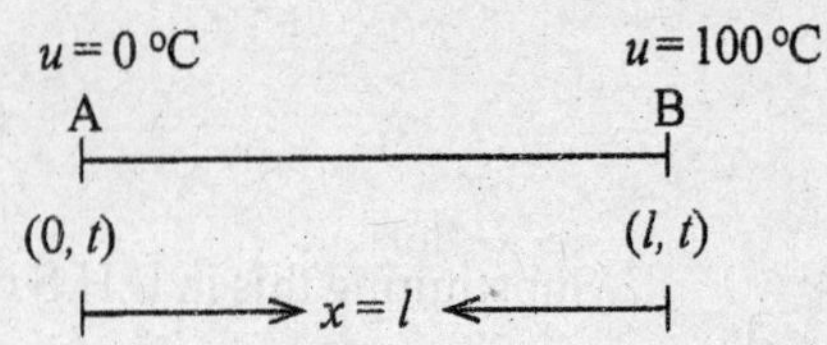

The general solution of the above equation is

$$u = ax + b$$

When $\quad x = 0,\ u = 0$ (the end A)

$\therefore \quad 0 = a \cdot 0 + b \quad \Rightarrow \quad b = 0$

$\therefore \quad u = ax$

$u = 100\ ^oC, x = l$ (the end B)

$$100\ ^oC = a\,.\,l \quad \Rightarrow \quad a = \frac{100}{l}$$

Using the values $a = \frac{100}{l}$ and $b = 0$ in

$$u = ax + b \text{ we get } u = \frac{100}{l}x$$

$\therefore$ The initial condition is $u(x, 0) = \frac{100x}{l}$

$$\therefore \quad u(0, t) = 0$$
$$u(l, t) = 0$$
$$u(x, 0) = \frac{100x}{l}$$

are the three initial conditions.

Imposing the condition $u(0, t) = 0$ on (2)

$$0 = (c_1 + 0)\, e^{-n^2c^2t}$$

$$\Rightarrow \quad c_1 = 0$$

Substituting this in (2), we get

$$u = (c_2 \sin nx)\, e^{-n^2c^2t} \quad \text{.....(3)}$$

Imposing the condition $u(l, t) = 0$ on (3)

$$0 = c_2 \sin nl\ .\ e^{-n^2c^2t}$$

$$\Rightarrow \quad \sin nl = 0$$

$$\therefore \quad nl = p\pi, \quad (p = 0, 1, 2.....)$$

$$n = \frac{p\pi}{l}$$

The solution (3) reduces to

$$u = \sum_{p=1}^{\infty} A_p \sin\frac{p\pi x}{l}.e^{-\frac{p^2\pi^2c^2t}{l^2}} \quad \text{.....(4)}$$

Now imposing the condition $u(x, 0) = \frac{100x}{l}$ on (4)

$$\frac{100x}{l} = \sum A_p \sin\frac{p\pi x}{l}.e^0 \qquad .....(5)$$

But the half-range Fourier sine series of $f(x)$ in the range $(0, l)$

$$f(x) = \frac{100x}{l} = \sum B_p \sin\frac{p\pi x}{l} \qquad .....(6)$$

where $$B_p = \frac{2}{l}\int_0^l f(x)\sin\frac{p\pi x}{l}dx$$

$$= \frac{2}{l}\int_0^l \left(\frac{100x}{l}\right)\sin\frac{p\pi x}{l}dx$$

$$B_p = \frac{2}{l}\left[\left(\frac{100x}{l}\right)\left(-\cos\frac{p\pi x}{l}\right)\left(\frac{l}{p\pi}\right) - \left(\frac{100}{l}\right)\left(-\sin\frac{p\pi x}{l}\right)\left(\frac{l^2}{p^2\pi^2}\right)\right]_0^l$$

$$= \frac{2}{l}\left[-\frac{100l}{p\pi}\cos p\pi\right] = \frac{-200}{p\pi}\cos p\pi$$

(using this in (6) we get $\frac{100x}{l} = \sum -\frac{200}{p\pi}\cos p\pi \sin\frac{p\pi x}{l}$

is the Fourier sine series of

$$f(x) = \frac{100x}{l}$$

Substituting in L.H.S of (5) it reduces to

$$\sum -\frac{200}{p\pi}\cos p\pi \sin\frac{p\pi x}{l} = \sum A_p \sin\frac{p\pi x}{l}$$

$$\Rightarrow \qquad \sum A_p = \sum -\frac{200}{p\pi}\cos p\pi$$

(i.e.,) $$\sum A_p = \frac{200}{\pi}\sum\frac{(-1)^{p+1}}{p}$$

Substituting in the solution (4) reduces to

$$u = \frac{200}{\pi}\sum\frac{(-1)^{p+1}}{p}\sin\frac{p\pi x}{l}e^{-\frac{p^2\pi^2c^2t}{l^2}}$$

19.4 Heat Equation with Non-homogenous Boundary Conditions

If the ends of the rod are maintained at temperatures different from zero, then the boundary conditions are non-homogenous. In such cases we break the problem into two parts,

$$u(x, t) = u_s(x) + u_t(x, t) \qquad .....(1)$$

where $u_s(x)$ is called the steady state solution of the heat equation satisfying the prescribed boundary conditions, i.e., if

$$u(0, t) = A$$

$$u(l, t) = B$$

where l is the length of the rod are the given boundary conditions then $u_s(x)$ is the solution of

$$\frac{\partial^2 u_s}{\partial x^2} = 0,\ u_s(0) = A,\ u_s(l) = B \qquad .....(2)$$

$u_t(x, t)$ is called the transient solution which satisfies the heat equation with zero boundary conditions and the initial conditions

$$u_t(x, 0) = u(x, 0) - u_s(x)$$

i.e., $u_t(x, t)$ is the solution of

$$\frac{\partial u_t}{\partial t} = c^2\frac{\partial^2 u_t}{\partial x^2}$$

$$u_t(0, t) = 0$$

$$u_t(l, t) = 0$$

$$u_t(x, 0) = u(x, 0) - u_s(x)$$

Then $u(x, t) = u_s(x) + u_t(x, t)$ satisfies

$$\frac{\partial u}{\partial t} = c^2 \frac{\partial^2 u}{\partial x^2}, \; u(0, t) = A, \; u(l, t) = B$$

and $u(x, 0)$ as prescribed.

Solved Examples

***Example* 19.11:** The ends A and B of a rod of length 50 cm have the temperatures 30 °C and 80 °C respectively till steady state prevails. Then the temperature of the ends are changed to 40 °C and 80 °C respectively and are maintained. Find the temperature distribution in the rod at any time 't'.

***Solution*:** In the steady state $\frac{\partial u}{\partial t} = 0$

$\therefore$ The heat equation $\frac{\partial u}{\partial t} = c^2 \frac{\partial^2 u}{\partial x^2}$ reduces to $\frac{\partial^2 u}{\partial x^2} = 0$.

The solution of the above equation is

$$u(x) = ax + b \qquad \text{(independent of time 't')} \qquad .....(1)$$

Using the initial conditions

$u(0) = 30^o$ i.e., $u = 30$ °C when $x = 0$

$u(50) = 80^o$C i.e., $u = 80$ °C when $x = 50$

Substituting $u = 30, x = 0$

$30 = a \cdot 0 + b \Rightarrow b = 30$

and substituting $u = 80^o, x = 50$ and $b = 30$ in (1)

$80 = a \cdot 50 + 30$

$a = 1$

$\therefore$ (1) gives

$$u(x, 0) = x + 30 \qquad .....(2)$$

Thus we have a problem with non-homogenous boundary conditions

$$u(x, t) = u_s(x) + u_t(x, t) \qquad .....(3)$$

where $$\frac{\partial^2 u_s}{\partial x^2} = 0,$$

$\Rightarrow$ $u_s(x) = ax + b$ under the conditions $u_s(0) = 40$, $u_s(50) = 80$

Using $u = 40, x = 0$ in $u_s(x) = ax + b$

$$40 = a \,.\, 0 + b \quad \Rightarrow \quad b = 40$$

$$\therefore \quad 80 = a \,.\, 50 + 40 \quad \Rightarrow \quad a = \frac{4}{5}$$

Thus $$u_s(x) = \frac{4}{5}x + 40 \qquad .....(4)$$

From (3), when $t = 0$ we get

$$u_t(x, 0) = u(x, 0) - u_s(x)$$

Substituting from (2) and (4)

$$u_t(x, 0) = (x + 30) - \left(\frac{4}{5}x + 40\right)$$

$$u_t(x, 0) = \frac{x}{5} - 10 \qquad .....(5)$$

Now the transient solution $u_t(x, t)$ which decreases with the increase of time is given by using variables separable method.

$$u_t(x, t) = (c_1 \cos nx + c_2 \sin nx)\, e^{-c^2 n^2 t} \qquad .....(6)$$

With the boundary conditions

$$u_t(0, t) = 0$$

$$u_t(50, t) = 0$$

and $$u_t(x, 0) = \frac{x}{5} - 10$$

Using the condition $u_t(0, t)$ in (6)

$$0 = (c_1 \cos n \,.\, 0 + c_2 \sin n \,.\, 0)\, e^{-c^2 n^2 t}$$

$\Rightarrow$ $c_1 = 0$, then solution (6) reduces to

$$u_t(x, t) = c_2 \sin nx\; e^{-n^2 c^2 t} \qquad .....(7)$$

Imposing the condition $u_t(50, t) = 0$ on (7)

$$0 = c_2 \sin (n \cdot 50) e^{-n^2 c^2 t}$$

$$\Rightarrow \qquad 50n = p\pi \qquad (p = 0, 1, 2.....)$$

$\therefore$ substituting $n = \dfrac{p\pi}{50}$ in the solution (7)

$$u_t(x, t) = \sum A_p \sin \frac{p\pi x}{50} . e^{\frac{-p^2\pi^2c^2t}{50^2}} \qquad .....(8)$$

Using the last condition $u_t(x, 0) = \dfrac{x}{5} - 10$ in (8)

$$\frac{x}{5} - 10 = \sum A_p \sin \frac{p\pi x}{50} . e^{-0} \qquad .....(9)$$

Now the Half-range Fourier sine series of $f(x)$ in the range (0, 50)

$$f(x) = \frac{x}{5} - 10 = \sum B_p \sin \frac{p\pi x}{50} \qquad .....(10)$$

where $$B_p = \frac{2}{50} \int_0^{50} f(x) \sin \frac{p\pi x}{50} dx$$

$$B_p = \frac{2}{50} \int_0^{50} \left(\frac{x}{5} - 10\right) \sin \frac{p\pi x}{50} dx$$

$$= \frac{2}{50} \left\{ \left(\frac{x}{5} - 10\right)\left(-\cos \frac{p\pi x}{50}\right)\left(\frac{50}{p\pi}\right) - \left(\frac{1}{5} - 0\right)\left(-\sin \frac{p\pi x}{50}\right)\left(\frac{50}{p\pi}\right)^2 \right\}_0^{50}$$

$$= \frac{2}{50} \left\{ (10)\left(\frac{50}{p\pi}\right) \right\}$$

$$B_p = \frac{20}{p\pi}$$

The half-range Fourier sine series of $f(x)$ in the range (0, 50) is, from (10) is

$$f(x) = \frac{x}{5} - 10 = \sum_{p=1}^{\infty} \frac{20}{p\pi} \sin \frac{p\pi x}{50}$$

Substituting this in L.H.S of (9)

$$\sum \frac{20}{p\pi} \sin\frac{p\pi x}{50} = \sum A_p \sin\frac{p\pi x}{50}.e^0$$

$$\Rightarrow \quad \sum A_p = \sum \frac{20}{p\pi}$$

Using $\sum A_p = \sum \frac{20}{p\pi}$ in the solution (8)

$$u_t(x, t) = \sum \frac{20}{p\pi} \sin\frac{p\pi x}{50} e^{\frac{-p^2\pi^2c^2t}{50^2}} \qquad .....(11)$$

$\therefore$ The complete solution $u(x, t) = u_s(x) + u_t(x, t)$ from (4) and (11) is

$$u(x, t) = \left(\frac{4}{5}x + 40\right) + \sum \frac{20}{p\pi} \sin\frac{p\pi x}{50}.e^{\frac{-p^2\pi^2c^2t}{50^2}}$$

$$u(x, t) = \frac{4}{5}x + 40 + \frac{20}{\pi}\sum \frac{1}{p} \sin\frac{p\pi x}{50} e^{\frac{-p^2\pi^2c^2t}{50^2}}$$

***Example* 19.12:** A bar with insulated sides is initially at temperature 0 °C throughout. The end $x = 0$ is kept at 0 °C and the heat is suddenly applied so that $\frac{\partial u}{\partial x} = 10$ at $x = l$ for all time. Find the temperature function $u(x, t)$.

***Solution*:** The differential equation of one dimensional heat flow is

$$\frac{\partial u}{\partial t} = c^2 \frac{\partial^2 u}{\partial x^2} \qquad .....(1)$$

The possible solutions of this equation are

$$u = (c_1 \cos nx + c_2 \sin nx)\, e^{-n^2c^2t} \qquad .....(2)$$

$$u = c_3x + c_4 \qquad .....(3)$$

The given initial conditions are

$$u(x, 0) = 0$$

$$\frac{\partial u}{\partial x} = 10 \text{ at } x = l \text{ for all } t$$

Since the above conditions of the problem are such that any one of the above solutions does not satisfy them, we use the combination of the solutions to satisfy the given conditions, i.e., sum of (2) and (3)

Thus the solution of (1) is taken as

$$u = (c_3\, x + c_4) + (c_1 \cos nx + c_2 \sin nx)\, e^{-n^2c^2t} \qquad .....(4)$$

Now using the condition $u(x, 0) = 0$ on (4)

$$0 = (c_3 \,.\, 0 + c_4) + (c_1 + 0)\, e^{-n^2c^2t}$$

$\Rightarrow$ $c_1 = 0$ and $c_4 = 0$. Hence the solution (4) reduces to

$$u = c_3\, x + c_2 \sin nx \,.\, e^{-n^2c^2t} \qquad .....(5)$$

Using the condition $\dfrac{\partial u}{\partial x} = 10$ at $x = l$ for all 't'

$$\frac{\partial u}{\partial x} = c_3 + c_2 \,.\, n \cos nx \,.\, e^{-n^2c^2t}$$

(i.e.,) $\quad 10 = c_3 + c_2 \,.\, n \,.\, \cos nl \,.\, e^{-n^2c^2t}$

$\Rightarrow \quad c_3 = 10$ and $\cos nl = 0$

$\therefore \quad nl = (2p + 1)\,\dfrac{\pi}{2}, \qquad (p = 0, 1, 2.....)$

$\therefore \quad n = \dfrac{(2p+1)\pi}{2l}$

Substituting for c_3 and n in the solution (5)

$$u = 10x + \sum B_p \sin \frac{(2p+1)\pi}{2l} x.e^{\frac{-(2p+1)\pi^2c^2t}{4l^2}} \qquad .....(6)$$

Using the condition $u(x, 0) = 0$ on (6)

$$0 = 10 \,.\, x + \sum B_p \sin \frac{(2p+1)\pi}{2l} x.e^0$$

$$-10\, x = \sum B_p \sin \frac{(2p+1)\pi x}{2l} \qquad .....(7)$$

But the half-range Fourier sine series of $f(x)$ in $(0, l)$ is

$$f(x) = -10x = \sum A_p \sin\frac{(2p+1)}{2l}\pi x \qquad .....(8)$$

where $A_p = \frac{2}{l}\int_0^l f(x).\sin\frac{(2p+1)\pi x}{2l}dx$

$$= \frac{2}{l}\int_0^l (-10x)\sin\frac{(2p+1)\pi x}{2l}dx$$

$$= \left\{\frac{2}{l}(-10x)\left[-\cos\frac{(2p+1)\pi x}{2l}\times\frac{2l}{(2p+1)\pi}\right]\right.$$

$$\left. -(-10)\left[-\sin\frac{(2p+1)\pi x}{2l}\times\frac{4l^2}{(2p+1)^2\pi^2}\right]\right\}_0^l$$

$$= \frac{2}{l}\left[\frac{-40l^2}{(2p+1)^2\pi^2}\right]$$

$$= \frac{-80l}{(2p+1)^2\pi^2}$$ Hence (8) becomes

$$-10x = \sum\frac{-80l}{(2p+1)^2\pi^2}\sin\frac{(2p+1)\pi x}{2l}$$

Substituting this in L.H.S of (8)

$$\sum\frac{-80l}{(2p+1)^2\pi^2}\sin\frac{(2p+1)\pi x}{2l} = \sum B_p\sin\frac{(2p+1)\pi x}{2l}$$

$$\Rightarrow \quad B_p = \sum\frac{-80l}{(2p+1)^2\pi^2}$$

$\therefore$ The solution (6) reduces to

$$u = 10x + \sum \frac{-80l}{(2p+1)^2 \pi^2} \sin\left(\frac{2p+1}{2l}\right)\pi x.e^{-\frac{(2p+1)^2 \pi^2 c^2 t}{l^2}}$$

(i.e.,) $$u = 10x - \frac{80l}{\pi^2} \sum \frac{1}{(2p+1)^2} \sin \frac{(2p+1)}{2l} \pi x.e^{-\frac{(2p+1)^2 \pi^2 c^2 t}{l^2}}$$

Exercise 19.2

1. Solve the equation $\frac{\partial u}{\partial t} = c^2 \frac{\partial^2 u}{\partial x^2}$ subject to the following conditions

(i) $u(0, t) = 0$ (ii) $\frac{\partial u}{\partial t} = 0$, for $x = l$ (iii) $u(x, 0) = \frac{u_0 x}{l}$ for $0 \le x \le l$

$$\textbf{Ans : } u = \frac{8u_0}{x^2} \sum \frac{(-1)^n}{(2n+1)^2} \sin\left(\frac{2n+1}{2l}\right)\pi x.e^{-\frac{(2n+1)^2 \pi^2 c^2}{4l^2}}$$

2. A homogenous rod of conducting material of length 100 cm has its ends kept at zero degrees and the temperature initially is $u(x, 0) = x$, $0 \le x \le 50$

$= 100 - x$, $50 \le x \le 100$

find the temperature $u(x, t)$.

$$\textbf{Ans : } u = \frac{100}{\pi^2} \sum \frac{(-1)^n}{(2n+1)^2} \sin \frac{(2n+1)\pi x}{100}.e^{\frac{-(2n+1)^2 c^2 \pi^2}{(100)^2}}$$

3. A homogenous rod of conducting material of length 'l' has its ends kept at 0 °C and the initial temperature in the rod is $u(x, 0) = 0$, $0 \le x \le \frac{1}{2}$

$= l - x$, $\frac{1}{2} \le x \le l$

Find the temperature distribution $u(x, t)$ in the rod at any time 't'.

$$\textbf{Ans : } u(x, t) = \frac{4l}{\pi^2} \sum \frac{\sin \frac{n\pi}{2}}{n^2} \sin\left(\frac{n\pi x}{l}\right).e^{-\frac{n^2 \pi^2 c^2 t}{l^2}}$$

4. The ends A and B of a rod 30 cm long have their temperatures kept at 20 °C and at 80 °C until steady state conditions prevail. The temperature of the end B is suddenly

reduced to 60 °C and kept so while the end A is raised to 40 °C. Find the temperature distribution of the rod at time 't'

$$\textbf{Ans : } 40 + \frac{2x}{3} - \frac{40}{\pi}\sum \frac{1}{n}\sin\frac{n\pi x}{15}.e^{-\frac{n^2\pi c^2 t}{225}}$$

5. The ends A and B of a rod of 20 m length have temperatures at 30 °C and 80 °C until steady state prevails. The temperatures of the ends are then suddenly changed to 40 °C and 60 °C respectively. Find the temperature distribution of the rod.

$$\textbf{Ans : } 40 + x - \frac{20}{\pi}\sum\left(\frac{1+2\cos n\pi}{n}\right)\sin\left(\frac{n\pi x}{20}\right).e^{-\frac{n^2\pi^2 c^2 t}{400}}$$

6. Find the solution of $\frac{\partial u}{\partial t} = c^2 \frac{\partial^2 u}{\partial x^2}$ for the following conditions

(i) $u = 0$ when $x = 0, x = l$ for all values of 't'

(ii) $u = 2\sin\frac{p\pi x}{l}$ where p is integer, for all values of x, when $t = 0$

$$\textbf{Ans : } u = 2\sum \sin\frac{p\pi x}{l}.e^{-\frac{p^2c^2\pi^2 t}{l^2}}$$

19.5 Two Dimensional Heat Flow (Laplace Equation in Two Dimensions)

The partial differential equation

$$\frac{\partial^2 u}{\partial x^2} + \frac{\partial^2 u}{\partial y^2} = c^2\frac{\partial u}{\partial t} \qquad .....(1)$$

represents two dimensional heat flow.

In the steady state, i.e., when temperature u doesnot change with time, $\frac{\partial u}{\partial t} = 0$ and equation (1) reduces to the form

$$\frac{\partial^2 u}{\partial x^2} + \frac{\partial^2 u}{\partial y^2} = 0 \qquad .....(2)$$

which is known as Laplace's Equation in two dimensions.

We shall now solve the equation (2) by the method of separation of variables.

Let the trial solution be

$$u = X(x)\,Y(y) = XY \qquad .....(3)$$

Such that the solution (3) satisfies (2)

$$\therefore \quad \frac{\partial u}{\partial x} = X'Y \text{ and } \frac{\partial^2 u}{\partial x^2} = X''Y$$

$$\frac{\partial u}{\partial y} = XY' \text{ and } \frac{\partial^2 u}{\partial y^2} = XY''$$

Hence from (3)

$$X''Y + X.Y'' = 0$$

$$\therefore \quad \frac{X''}{X} = -\frac{Y''}{Y} = K \text{ (say)}$$

$$X'' - KX = 0 \text{ and } Y'' + KY = 0 \quad .....(4)$$

The three possible solutions of equations (4) are

$$u = (c_1 e^{nx} + c_2 e^{-nx})(c_3 \cos ny + c_4 \sin ny) \text{ ..... for } K = n^2 \quad .....(5)$$

$$u = (c_5 \cos nx + c_6 \sin nx)(c_7 e^{ny} + c_8 e^{ny}) \text{ ..... for } K = -n^2 \quad .....(6)$$

$$u = (c_9 x + c_{10})(c_{11} y + c_{12}) \text{ .... for } K = 0 \quad .....(7)$$

With the help of given boundary conditions we choose a suitable solution from the solutions (5), (6) and (7).

Solved Examples

***Example* 19.13:** A rectangular plate of width 'a' cm with insulated surface has its temperature $u(x, y)$ equal to zero on both the long sides and one of the short sides so that $u(0, y) = u(a, y) = u(x, \infty) = 0$. Also $u(x, 0) = kx$. Show that the steady state temperature within the plate is given by

$$u(x, t) = \frac{2ak}{\pi} \sum_{1}^{\infty} \frac{(-1)^{n+1}}{n} \sin\left(\frac{n\pi x}{a}\right).e^{-\frac{n\pi y}{a}}.$$

Solution:

The two dimensional heat Equation is

$$\frac{\partial^2 u}{\partial x^2} + \frac{\partial^2 u}{\partial y^2} = 0 \quad .....(1)$$

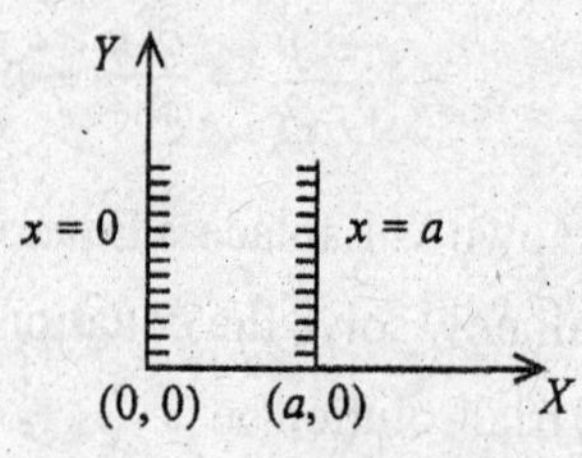

The given initial conditions are

$$\left.\begin{aligned} u(0, y) &= 0 \\ u(a, y) &= 0 \\ u(x, \infty) &= 0 \\ u(x, 0) &= kx \end{aligned}\right\} \qquad \text{.....(2)}$$

Now using the method of separation of variables

$$u = (c_1 \cos nx + c_2 \sin nx)(c_3 e^{ny} + c_4 e^{-ny}) \qquad \text{.....(3)}$$

under the conditions (2)

Now imposing the condition $u(0, y) = 0$ on (3)

$$0 = (c_1 + c_2 \,.\, 0)(c_3 e^{ny} + c_4 e^{-ny})$$

$$\Rightarrow \qquad c_1 = 0$$

Putting $c_1 = 0$ in (3) we get

$$u = c_2 \sin nx\,(c_3 e^{ny} + c_4 e^{-ny}) \qquad \text{.....(4)}$$

Applying the condition $u(a, y) = 0$

$$0 = c_2 \sin na\,(c_3 e^{ny} + c_4 e^{-ny})$$

$$\Rightarrow \qquad \sin na = 0$$

$$na = p\pi, \qquad (p = 0, 1, 2 \,.....)$$

$$n = \frac{p\pi}{a}, \qquad (p = 0, 1, 2 \,.....)$$

Substituting $n = \dfrac{p\pi}{a}$ in the solution (4)

$$u = \sin \frac{p\pi x}{a}\left[\sum A_p e^{\frac{p\pi y}{a}} + \sum B_p e^{-\frac{p\pi y}{a}}\right] \qquad \text{.....(5)}$$

Using the condition $u(x, \infty) = 0$

$$u = \sin \frac{p\pi x}{a}\left[\sum A_p e^{\frac{p\pi y}{a}} + \sum B_p e^{\frac{-p\pi y}{a}}\right]$$

$$\Rightarrow \qquad \sum A_p = 0$$

$\therefore$ solution (5) gives

$$u = \sum B_p \sin\frac{p\pi x}{a} e^{\frac{-p\pi y}{a}} \qquad .....(6)$$

Applying the last condition $u(x, 0) = kx$ on (6)

$$kx = \sum B_p \sin\frac{p\pi x}{a} . e^0 \qquad .....(7)$$

But half-range Fourier sine series of $f(x)$ in the range $(0, a)$ is

$$f(x) = kx = \sum A_p \sin\left(\frac{p\pi x}{a}\right) \qquad .....(8)$$

$$\text{where} A_p = \frac{2}{a}\int_0^a f(x)\sin\left(\frac{p\pi x}{a}\right)dx$$

(i.e.,) $$A_p = \frac{2}{a}\int_0^a (kx)\sin\left(\frac{p\pi x}{a}\right)dx$$

$$= \frac{2}{a}\left\{(kx)\left(-\cos\frac{p\pi x}{a}\right)\left(\frac{a}{p\pi}\right) - (k)\left(-\sin\frac{p\pi x}{a}\right)\left(\frac{a^2}{p^2\pi^2}\right)\right\}_0^a$$

$$= \frac{-2ak}{p\pi} \cos p\pi$$

(i.e.,) $$A_p = \frac{2ak(-1)^{p+1}}{p\pi}$$

substituting in (8)

$$kx = \sum \frac{2ak}{p\pi}(-1)^{p+1} \sin\frac{p\pi x}{a}$$

Substituting in L.H.S of (7)

$$\sum \frac{2ak}{p\pi}(-1)^{p+1} \sin\frac{p\pi x}{a} = \sum B_p \sin\frac{p\pi x}{a} . e^0$$

$$\Rightarrow \qquad B_p = \sum \frac{2ak(-1)^{p+1}}{p\pi}$$

Substituting in the solution (6)

$$u = \frac{2ak}{\pi}\sum \frac{(-1)^{p+1}}{p\pi}\sin\frac{p\pi x}{a}.e^{\frac{-p\pi y}{a}} \text{ is the desired solution.}$$

***Example* 19.14:** A rectangular metal plate with insulated surfaces is of width 'a' and too long compared to its breadth that it can be considered infinite in length without introducing an appreciable error. If the temperature along one short edge $y = 0$ is given by $u(x, 0) = u_0 \sin\left(\frac{\pi x}{a}\right)$ for $0 < x < a$ and other long edges $x = 0$, $x = a$ and the short edge are kept at zero degrees temperature, find $u(x, y)$ describing the steady state.

***Solution*:** In the steady state the heat distribution function $u(x, y)$ satisfies Laplace equation.

$$\frac{\partial^2 u}{\partial x^2} + \frac{\partial^2 u}{\partial y^2} = 0 \qquad \text{.....(1)}$$

Its solution using the method of separation of variables is

$$u = (c_1 \cos nx + c_2 \sin nx)\,(c_3\, e^{ny} + c_4 e^{-ny}) \qquad \text{.....(2)}$$

The given initial conditions are

$$\lim_{y\to\infty} (x, y) = 0$$

$$u(0, y) = 0$$

$$u(a, y) = 0$$

$$u(x, 0) = u_0 \sin\left(\frac{\pi x}{a}\right),\ 0 < x < a$$

Now using the first initial condition in (2)

$$0 = (c_1 \cos nx + c_2 \sin nx)\,(c_3\, e^{\infty} + c_4\, e^{-\infty})$$

$\Rightarrow$ $c_3 = 0$. (2) then gives

$$u = (c_1 \cos nx + c_2 \sin nx)(c_4\, e^{-ny}) \qquad \text{.....(3)}$$

Now using $u(0, y) = 0$ in the solution (3)

$$0 = (c_1 + c_2 \cdot 0)\,(c_4 e^{-ny})$$

$\Rightarrow$ $c_1 = 0$. (3) now reduces to

$$u = c_2\, c_4 \sin nx\, e^{-ny}$$

$\Rightarrow$ $$u = A \sin nx\, e^{-ny} \qquad .....(4)$$

where $c_1\, c_2 = A$

Imposing the condition

$u(a, y) = 0$ on the solution (4)

$$0 = A \sin na\, e^{-ny}$$

$\Rightarrow$ $$\sin na = 0$$

$$na = p\pi \qquad (p = 0, 1, 2 ......)$$

$$n = \frac{p\pi}{a} \qquad (p = 0, 1, 2 .....)$$

Substituting $n = \frac{p\pi}{a}$ in the solution (4)

$$u = \sum_{p=1}^{\infty} A_p \sin \frac{p\pi x}{a} e^{-\frac{p\pi y}{a}} \qquad .....(5)$$

Now applying the condition

$u = (x, 0) = u_0 \sin \frac{\pi x}{a}$ in (5)

$$u_0 \sin \frac{\pi x}{a} = \sum A_p \sin \frac{p\pi x}{a} . e^0$$

$$u_0 \sin \frac{\pi x}{a} = A_1 \sin \frac{\pi x}{a} + A_2 \sin \frac{2\pi x}{a} + A_3 \sin \frac{3\pi x}{a} + ...$$

$\Rightarrow$ $$A_1 = u_0, A_2 = A_3 = A_4 = ... = 0$$

Substituting these in the solution (5) we get

$$u = u_0 \sin \frac{\pi x}{a} . e^{\frac{-\pi y}{a}}$$

Example **19.15:** A rectangular plate with insulated surfaces is 8 cm wide and is so long compared to its width that it may be considered infinite in length. If the temperature along the short edge $y = 0$ is given by $u(x, 0) = 100 \sin\left(\frac{\pi x}{8}\right)$, $0 < x < 8$, while the two long edges $x = 0$ and $x = 8$, as well as the other short edge are kept at 0 °C, find the steady temperature $u(x, y)$.

Solution: In the steady state, the temperature u satisfies the Laplaces equation

$$\frac{\partial^2 u}{\partial x^2} + \frac{\partial^2 u}{\partial y^2} = 0 \qquad .....(1)$$

The boundary conditions are

$$u(0, y) = 0$$
$$u(8, y) = 0$$

$$\lim_{y \to \infty} u(x, y) = 0$$

$$u(x, 0) = 100 \sin \frac{\pi x}{8}$$

The solution of the equation using method of separation of variables

$$u = (c_1 \cos nx + c_2 \sin nx)(c_3 e^{ny} + c_4 e^{-ny}) \qquad .....(2)$$

under the boundary conditions.

Using the condition $u(0, y) = 0$ in (2)

$$0 = (c_1 + c_2 \,.\, 0)(c_3 e^{ny} + c_4 e^{-ny}) \Rightarrow c_1 = 0$$

Substituting this in the solution (2)

$$u = c_2 \sin nx\,(c_3 e^{ny} + c_4 e^{-ny})$$

$$\Rightarrow \quad u = \sin nx\,(A e^{ny} + B e^{-ny}) \qquad .....(3)$$

where $\quad c_2 c_3 = A$

$$c_2 c_4 = B$$

Using the condition $u(8, y) = 0$ in (3)

$$0 = \sin n\,8\,(Ae^{ny} + Be^{-ny})$$

$$\Rightarrow \quad \sin 8n = 0$$

$$\therefore \quad 8n = p\pi \qquad (p = 0, 1, 2,.....)$$

$$n = p\pi/8 \qquad (p = 0, 1, 2,.....)$$

The solution (3) reduces to

$$u = \sin \frac{p\pi x}{8}\left(\sum A_p . e^{\frac{p\pi}{8}y} + \sum B_p e^{\frac{-p\pi}{8}y}\right) \qquad .....(4)$$

Now using the condition

$$\lim_{y\to\infty} u(x, y) = 0 \text{ in (4)}$$

$$0 = \sin \frac{p\pi x}{8}\left[\sum A_p e^{\infty} + \sum B_p e^{-\infty}\right] \Rightarrow \sum A_p = 0$$

Substituting $\sum A_p = 0$ in (4) the solution (4) becomes.

$$u = \sum B_p \sin \frac{p\pi x}{8} . e^{-\frac{p\pi}{8}y} \qquad .....(5)$$

Applying the condition

$$u(x, 0) = 100\sin \frac{\pi x}{8} \text{ in (5)}$$

$$100 \sin \frac{\pi x}{8} = \sum B_p \frac{\sin p\pi x}{8} . e^0$$

$$\therefore \quad 100 \sin \frac{\pi x}{8} = B_1 \sin \frac{\pi x}{8} + B_4 \sin \frac{2\pi x}{8} + ...$$

$$B_1 = 100, B_2 = B_3 = B_4 = ... = 0$$

substituting in (5)

$$u = 100 \sin\left(\frac{\pi x}{8}\right) e^{\frac{-\pi}{8}y} \text{ is the desired solution.}$$

Exercise 19.3

1. An infinitely long plane uniform plate is bounded by two parallel edges and an end at right angles to them. The breadth is π; this end is maintained at a temperature u_0 at all points and other edges are at zero temperature. Find the temperature at any point of the plate is steady state.

   Ans : $u(x, y) = 4\frac{u_0}{\pi}\left[e^{-y}\sin x + \frac{1}{3}e^{-3y}\sin 3x + ...\right]$

2. A rectangular plate with insulated surfaces is 10 cm wide and so long compared to its width that it may be considered infinite in length without introducing any appreciable error. If the temperature along the short edge $y = 0$ is given by

   $$u(x, 0) = 20x; \quad 0 < x < 5$$
   $$= 20(10 - x); \ 5 < x < 10$$

   while the two long edges $x = 0$, $x = 10$ as well as the other short edge are kept at 0 °C. Find the steady state temperature at any point (x, y) of the plate.

$$\textbf{Ans : } u = \frac{800}{\pi}\left(e^{-\frac{\pi y}{100}}\sin\frac{\pi x}{10} - \frac{1}{3}e^{-\frac{3\pi y}{10}}\sin\frac{3\pi x}{10} + ...\right)$$

3. Find the steady state temperature distribution in a rectangular plate bounded by $x = 0, x = a, y = 0, y = b$ subject to the boundary conditions

$$u(0, y) = u(a, y) = u(x, b) = 0$$

$$u(x, 0) = x; \qquad 0 \le x \le a/2$$

$$= (a - x); \; a/2 \le x \le a$$

$$\textbf{Ans : } u = \frac{4a}{\pi^2}\sum \frac{1}{p^2}\frac{\sin\frac{p\pi}{2}}{\sinh\frac{p\pi b}{a}}\sinh\frac{p\pi(b-y)}{a}.\sin\frac{p\pi x}{a}$$

4. A square plate of side p lying along $0 \le x \le p, 0 \le y \le \pi$ has its edges $x = 0, x = \pi$, $y = p$ insulated. The side $y = 0$ is kept at temperature $u(x, 0) = x$. Find the steady state temperature distribution in the plate.

$$\textbf{Ans : } u(x, y) = \frac{\pi}{2} - \frac{4}{\pi}\sum_{1}^{\infty}\frac{\cosh(2n-1)(y-\pi)\cos(2n-1)x}{(2n-1)^2\cosh(2n-1)\pi}$$

5. Find the solution of the equation $\frac{\partial^2 u}{\partial x^2} + \frac{\partial^2 u}{\partial y^2} = 0$ for the following conditions.

(i) $u = 0$ for all values of y when $x = 0$ and 1

(ii) $u \to 0$ as $y \to \infty$

(iii) $y = 0, u = \sin^3 \pi x$ in $0 < x < 1$

$$\textbf{Ans : } u = \frac{3}{4}e^{-\pi y}.\sin\pi x - \frac{1}{4}e^{-3\pi y}.\sin 3\pi x$$

CHAPTER 20

Solution in Series

20.1 Ordinary and Singular Points of a Differential Equation

Consider the equation $P_0 \frac{d^2y}{dx^2} + P_1 \frac{dy}{dx} + P_2 y = 0.$

where P_0, P_1 and P_2 are functions of x.

x = a is an ordinary point of the above equation if P_0 does not vanish for x = a, whereas x = a will be a singular point if P_0 vanishes for x = a.

Here we shall consider the solutions of the differential equations exprered as a series.

20.2 Working Rule when x = 0 is a Singular Point

Assume $y = x^k \left(a_0 + a_1 x + a_2 x^2 + ...\right)$

Find the values of $\frac{dy}{dx}$ and $\frac{d^2y}{dx^2}$ and put in the given differential equation.

The value of k will be determined by equating to zero the coefficient of lowest power of x. This coefficient will be a quadratic in k and is called indical equation.

The values of various coefficients a_1, a_2 ... are determined in terms of a_0 by equating to zero the coefficients of various powers of x. We shall illustrate the above by taking suitable examples.

***Example* 20.1:** Solve in series the equation

$$x^2 \frac{d^2y}{dx^2} + x \frac{dy}{dx} + \left(x^2 - n^2\right) y = 0.$$

Let $\quad y = x^k\left[a_0 + a_1x + a_2x^2 + a_3x^3 + ...\right].$

$$\frac{dy}{dx} = x^{k-1}\left[a_0k + a_1(k+1)x + a_2(k+2)x^2 + ...\right],$$

$$\frac{d^2y}{dx^2} = x^{k-2}\left[a_0k(k-1) + a_1(k+1)kx + a_2(k+2)(k+1)x^2 + ...\right]$$

Putting the values of y, $\frac{dy}{dx}$ and $\frac{d^2y}{dx^2}$ in the given equation, we get after collecting the terms of $x^k, x^{k+1}, x^{k+2}...,$

$$a_0\left[k(k-1) + k - n^2\right]x^k + a_1\left[(k+1)^2 - n^2\right]x^{k+1}$$

$$+a_2\left[\left\{(k+2)^2 - n^2\right\} + a_0\right]x^{k+2} + ... = 0.$$

Equating to zero the coefficients of various powers of x,

$$k^2 - n^2 = 0 \quad \therefore k = +n, -n.$$

$a_1\left\{(k+1)^2 - n^2\right\} = 0$ gives $a_1 = 0$,

$a_2\left\{(k+2)^2 - n^2\right\} + a_0 = 0$ gives $a_2 = \frac{a_0}{n^2 - (k+2)^2}$

$$a_3 = \frac{a_1}{n^2 - (k+3)^2} = 0 \qquad \because \quad a_1 = 0.$$

Similarly $a_5 = a_7 = 0$.

$a_d = \frac{a_2}{n^2 - (k+4)^2} = \frac{1}{n^2 - (k+4)^2} \cdot \frac{1}{n^2 - (k+2)^2} a_0$ and so on

$$\therefore \quad y = a_0x^k\left[1 + \frac{1}{n^2 - (k+2)^2}x^2 + \frac{1}{n^2 - (k+4)^2 n^2 - (k+2)^2}x^4 + ...\right]$$

Now put k = n and – n and take a_0 = a and a' respectively in the above to get two solutions.

$$ax^n\left[1-\frac{1}{4(n-1)}x^2+\frac{1}{4.8.(n+2)(n+1)}.x^4...\right]=au \qquad \text{say}$$

and $$a'x^n\left[1-\frac{1}{4(-n+1)}x^2+\frac{1}{4.8(-n+2)(-n+1)}x^4+...\right]=a'v$$

$\therefore\ y=au+a'v$ is the complete solution.

***Example* 20.2:** Solve completely in series the equation

$$\left(2x+x^2\right)\frac{d^2y}{dx^2}-\frac{dy}{dx}-6xy=0.$$

As in last example assume

$$y=x^k\left(a_0+a_1x+a_2x^2+...\right),$$

$$\frac{dy}{dx}=x^{k-1}\left[a_0k+a_1(k+1)x+a_2(k+2)x^2+....\right]$$

$$\frac{d^2y}{dx^2}=x^{k-2}\ [a_0k(k-1)+a_1(k+1)kx+a_2(k+2)(k+1)x^2+...]$$

Putting in the given equation and collecting the various coefficients and then equating to zero the coefficients of x^{k-1}, x^k, x^{k+1} etc., we get

$$2a_0k(k-1)-a_0k-0$$

or $$a_0k(2k-3)=0 \ \therefore\ k=0,\ \frac{2}{3}.$$

$$a_1(k+1)k.2-a_1(k+1)=0$$

or $$a_1(k+1)(2k-1)=0. \qquad \because a_1=0.$$

$$a_0k(k-1)+2a_2(k+2)(k+1)-a_2(k+2)-6a_0=0$$

or $$a_2(k+2)\{2k+2-1\}+a_0\left(k^2-k-6\right)=0$$

$$a_2(k+2)(2k+1)+a_0(k-3)(k+2)=0$$

or $$a_2(2k+1)+a_0(k-3)=0.$$

Similarly $a_3(2k+3)+a_1(k-2)=0$,

$$a_4(2k+5)+a_2(k-1)=0 \text{ and so on.}$$

Since $a_1 = 0$, $\quad \therefore a_3 = a_5 = a_7 = \ldots 0.$

$$a_2 = \frac{k-3}{2k+1}, a_4 = -\frac{k-1}{2k+5}, a_2 = \frac{(k-1)(k-3)}{(2k+5)(2k+1)} \text{etc.}$$

$$\therefore \quad y = a_0 x^k \left[1 - \frac{k-3}{2k+1}x^2 + \frac{(k-1)(k-3)}{(2k+5)(2k+1)}x^4 \ldots\right]$$

Now put k = 0 and $\frac{3}{2}$ and put a_0 = a and a′ respectively and we get independent solutions as

$$a\left(1+3x^2+\frac{3}{5}x^4-\frac{1}{15}x^6\ldots\right)=au$$

and $$a'x^{3/2}\left[1+\frac{3}{8}x^2-\frac{1.3}{8.16}x^4+\frac{1.3.5}{8.16.24}x^6\ldots\right]=a'v.$$

$$\therefore \quad y = au + a'v.$$

***Note*:** When the values of k given by indical equation are equal, then the following procedure will be adopted.

Suppose k = a, a; then find the solution in term of k and call it v, then the two solutions will be obtained by putting k = a in y and $\frac{\partial y}{\partial k}$. It will be clear from the following example.

***Example* 20.3:** Solve completely $x\frac{d^2y}{dx^2}-\frac{dy}{dx}+x^2y=0.$

Let $\quad y = x^k\left[c_0 + c_1x + c_2x^2 + \ldots\right].$

Calculate $\frac{dy}{dx}$ and $\frac{d^2y}{dx^2}$ and put in the given equation and equate to zero the coefficients of $x^{k-1}, x^k, x^{k+1}\ldots$

Coefficient of x^{k-1} is $\quad k^2c_0 = 0, \quad \therefore k = 0, 0.$

Coefficient of x^k is $\quad (k+1)^2 c_1 = 0 \quad \therefore c_1 = 0.$

Coefficient of x^{k+1} is $\quad (k+2)^2 c_2 = 0, \quad \therefore c_2 = 0.$

Coefficient of x^{k+2} is $\quad (k+3)^2 c_a + c_0 = 0$

$$\therefore \quad c_3 = \frac{-c_0}{(k+3)^2}$$

Similarly $c_4 = -\dfrac{1}{(k+4)^2} \; c_1 = 0,\; c_5 = 0,\; c_1 = 0,\; c_2 = 0.$

$$c_6 = -\frac{1}{(k+6)^2} = \frac{1}{(k+3)^2(k+6)^2}, \quad c_7 = 0, \quad c_8 = 0.$$

$$\therefore \quad y = c_0 x^k\left[1 - \frac{1}{(k+3)^2}x^3 + \frac{1}{(k+3)^2(k+6)^2}x^6 - \frac{1}{(k+3)^2(k+6)^2(k+9)^2}x^9 + \ldots\right] \quad \ldots\ldots(1)$$

$$\frac{\partial y}{\partial k} = c_0 x^k \log x\;[\;]$$

$$+\, c_0 x^k\left[\frac{2}{(k+3)^2}x^2 - \left\{\frac{2}{(k+3)^3(k+6)^2} + -\frac{2}{(k+3)^2(k+6)^3}\right\}x^6\ldots\right] \quad \ldots\ldots(2)$$

or
$$\frac{\partial y}{\partial k} = y\log x + 2c_0 x^k\left[\frac{1}{(k+3)^3}x^3 - \left\{\frac{1}{(k+3)^3(k+6)^2} + \frac{1}{(k+3)^2(k+6)^3}x^6\ldots\right\}\right].$$

Now put $k = 0$ and $c_0 = c$ and c' in (1) and (2) and the two solutions are

$$y = c\left[1 - \frac{1}{3^2}x^3 + \frac{1}{3^4(2\,!)^2}x^6 - \frac{1}{3^6(3\,!)^2}x^9 + \ldots\right] = cu \text{ say}$$

$$\frac{\partial y}{\partial k} = c'u\log x + 2c'$$

$$\left[\frac{1}{3^3}x^3 - \frac{1}{3^6(3\,!)^2}\left(1+\frac{1}{2}\right)x^6 + \frac{1}{3^7(3\,!)}\left(1+\frac{1}{2}+\frac{1}{3}\right)x^9 + \ldots\right] = c'v \quad \text{say.}$$

Hence the complete solution is

$$y = cu + c'v.$$

***Example* 20.4:** Obtain a general solution in a series of powers of x of the equation

$$x\frac{d^2y}{dx^2}+\frac{dy}{dx}+xy=0$$

Let $y = x^k (c_0 + c_1x + c_2x^2 + c_3x^3 + \dots)$.

Calculate $\frac{dy}{dx}$ and $\frac{d^2y}{dx^2}$ and put in the above equation.

Now equate to zero the coefficients of $x^{k-1}, x^k, x^{k+1} \dots$

Coefficient of x^{k-1} is $c_0 [k (k + 1) + k] = 0$

$\therefore \quad c_0k^2 = 0 \qquad \therefore \quad k = 0, 0.$

Coefficient of x^k is $c_1 [k (k + 1) + k + 1] = 0$

or $\quad c_1 (k + 1)^2 = 0. \quad \therefore \quad c_1 = 0.$

Coefficient of x^{k+1} is $c_2 \{k (k - 1) + 2 (2k + 1) + k + 2)\} + c_0$

$$\therefore \quad c_2=\frac{-c_0}{(k+2)^2}$$

Coefficient of x^{k+2} is $c_3 (k + 3)^2 + c_1 = 0.$

$$\therefore \quad c_3=\frac{-c_1}{(k+2)^2}=0; \qquad \because c_1=0.$$

Coefficient of x^{k+3} is $c_4 (k + 4)^2 + c_2 = 0$

$$\therefore \quad c_4=\frac{-c_2}{(k+4)^2}=\frac{c_0}{(k+4)^2(k+2)^2\dots}.$$

$$\therefore \quad y=c_0x^k\left[1-\frac{1}{(k+2)^2}x^2+\frac{1}{(k+4)^2(k+2)^2}x^4+\dots\right]. \qquad \dots(3)$$

$$\frac{\partial y}{\partial k}=c_0x^k.\log x[\ ]$$

$$+\ c_0x^k\left[\frac{2}{(k+2)^3}x^2-\left\{\frac{2}{(k+2)^3(k+4)^2}+\frac{2}{(k+2)^2(k+4)^3}\right\}x^4+\dots\right. \qquad \dots(4)$$

Putting $k = 0$ and $c_0 = c$ and c' in (3) and (4), we get the two independent solutions as

$$c\left(1-\frac{1}{2^2}x^2-\frac{1}{2^24^2}x^4-\frac{1}{2^2.4^2.6^2}x^6+\dots\right)=cu \text{ say}$$

and $\quad c'u \log x + c'\left[\frac{1}{2^2}x^2 - \frac{1}{2^2:4^2}\left(1+\frac{1}{2}\right)x^4 + \frac{1}{2^2.4^2 6^2} \times \left(1+\frac{1}{2}+\frac{1}{3}\right)x^6 ...\right] = c'v$ say.

Hence the general solution is

$$y = cu + c'v.$$

***Example* 20.5:** Integrate in a series the equation

$$(x - x^2)\frac{d^2y}{dx^2} + (1-5x)\frac{dy}{dx} - 4y = 0.$$

Proceeding as before, we get on equating to zero the various coefficients,

$$c_0\{k(k-1)+k\} = 0, \therefore k^2 = 0, \text{ i.e., } k = 0, 0$$

$$c_1\{(k+1)k+k+1\} - c_0\{k(k-1)+5k-4\} = 0$$

or $\quad c_1(k+1)^2 - c_0(k+2)^2 = 0$

$$\therefore \quad c_1 = \left(\frac{k+2}{k+1}\right)^2 c_0.$$

$$c_2 = \left(\frac{k+3}{k+1}\right)^2 c_0, \; c_3 = \left(\frac{k+4}{k+1}\right)^2 c_0 ...$$

$$\therefore \quad y = c_0 x^n\left[1+\left(\frac{k+2}{k+1}\right)^2 x + \left(\frac{k+3}{k+1}\right)^2 x^2 + \left(\frac{k+4}{k+1}\right)^2 x^3 + ...\right] \quad .....(5)$$

$$\frac{\partial y}{\partial x} = c_0 x^n \log x[\;].$$

$$+c_0 x^n\left[2\frac{k+2}{k+1}.\frac{-1}{(k+1)^2}x + 2\frac{k+2}{k+1}.\frac{-2}{(k+1)^2}x^2 + 2\frac{k+4}{k+1}.\frac{-3}{(k+1)^2}x^3 + ...\right] \quad ....(6)$$

Putting $k = 0$ and $c_0 = c$ and c' in (5) and (6) above we get the independent solutions as

$$c\left(1^2 + 2^2x + 3^2x^2 + 4^2x^3 ...\right) = cu \text{ say}$$

and $\quad c'u \log x - 2c'\left(1.2x + 2.3x^2 + 3.4x^3 ....\right) = cv \quad$ say.

Hence the general solution is

$$y = cu + c'v.$$

***Example* 20.6:** Integrate in series Legendre's equation

$$\left(1-x^2\right)\frac{d^2y}{dx^2}-2x\frac{dy}{dx}+n(n+1)y=0$$

where n is a + ve integer.

Let $\quad y=x^k\left(c_0+c_1x+c_2x^2+...\right).$

Calculate $\frac{dy}{dx}$ and $\frac{d^2y}{dx^2}$ and put in the given equation.

$$\therefore \quad \left(1-x^2\right)x^{k-1}[k(k-1)c_0+(k+1)kc_1x+(k+2)(k+1)c_1x^2$$
$$+(k+3)(k+2)c_3x^3+...]$$
$$-\ 2x.x^{k-2}\left[kc_0+(k+1)c_1x+(k+2)c_2x^2+...\right]$$
$$+n(n+1)x^k(c_0+c_1x+c_2x^2+...]=0$$

Equating to zero the coefficient of $x^{k-2}, x^{k-1}, x^{k+1}$ ...

$$c_0k(k-1)=0;\ \therefore\ k=0 \text{ or } 1 \text{ from } x^{k-2}, \qquad .....(7)$$

$c_1k(k+1)=0$ when k = 0, it makes c_1 indeterminate.

But when $\quad k=1, c_1=0.$

$$c_1(k+2)(k+1)-c_0\ \{k(k-1)+2k-n(n+1)\}=0$$

or $\quad c_2(k+2)(k+1)-c_0\ \{k(k-1)+2k-n(n+1)\}=0 \qquad .....(8)$

and $\quad c_3(k+3)(k+2)-c_1\ c_1\{(k+1)(k+2)-n(n+1)\}=0 \qquad .....(9)$

Now put k = 0 and we get

$$c_2=-\frac{n(n+1)}{1.2}c_0$$

$$c_3=-\frac{n(n+1)-2}{3!}c_1 \quad \text{or} \quad c_3=-\frac{(n+2)(n-1)}{3!}c_1.$$

Similarly $c_4=\frac{n(n-2)(n+1)(n+3)}{4!}c_0$ and so on,

Hence the solution containing two arbitrary constants c_0 and c_1 is obtained on putting the values of coefficients.

$$\therefore \quad y = c_0 + c_1 x - \frac{n(n+1)}{2!} c_0 x^2 - \frac{(n+2)(n-1)}{3!} c_1 x^3$$

for $\quad k = 0$

$$+ \frac{n(n-2)(n+1)(n+3)}{4!} c_0 x^{4-} \ldots$$

$$= c_0 \left[1 - \frac{n(n+1)}{2!} x^2 + \frac{n(n-2)(n+1)(n+3)}{4!} x^{5-} \ldots\right]$$

$$+ c_1 \left[x - \frac{(n-2)(n+2)}{3!} x^3 - \frac{(n-1)(n-3)(n+2)(n+4)}{5!} x^{5-} \ldots\right]$$

Note: If we put n = 1 we will get the corresponding series for the differential equation

$$\left(1 - x^2\right) \frac{d^2y}{dx^2} - 2x \frac{dy}{dx} + y = 0.$$

***Example* 20.7:** Integrate in series the differential equation

$$\left(1 - x^2\right) \frac{d^2y}{dx^2} + 2x \frac{dy}{dx} + y = 0.$$

Proceeding as before taking $y = x^k \left(c_0 + c_1 x + c_2 x^2 \ldots\right)$, we get on equating to zero coefficients of

$x^{k-2}, x^{k-1}, x^k \ldots$

$c_0 k(k-1)k = 0. \therefore k = 0$ or 1. $\qquad \because c_0 \neq 0.$

$c_1 (k+1)k = 0$. It makes c_1 indeterminate when k = 0. However $c_1 = 0$ when k = 1,

$c_2 (k+2)(k+1) - c_0 \{k(k+1) - 2k - 1\} = 0.$

$c_3 (k+3)(k+2) - c_1 \{k(k+1) - 2(k+1) - 1\} = 0$ and so on

Putting k = 0, we get

$c_2 = -\frac{1}{2} c_0, \; c_3 = -\frac{1}{2} c_1, \; c_4 = -\frac{1}{3} c_0, \; c_5 = -\frac{c_1}{40}, \; c_6 = -\frac{c_0}{80}$ and so on.

Hence the solution containing two arbitrary constants c_0 and c_1 is obtained by putting the values of various coefficients in above.

$$\therefore \quad x = c_0\left[1-\frac{1}{2}x^2+\frac{1}{8}x^4+\frac{1}{80}x^6...\right]+c_1\left[x-\frac{1}{2}x^3+\frac{1}{40}x^5...\right],$$

***Example* 20.8:** Integration in series

(a) $$\frac{d^2y}{dx^2}+x^2y=0.$$

Ans. $$y=c_0\left[1-\frac{1}{3.4}x^4+\frac{1}{3.4.7.8}x^8-.\right]+c_1\left[x-\frac{1}{4.5}x^6+\frac{1}{4.5.8.9}x^9...\right]$$

(b) $$\left(2+x^2\right)\frac{d^2y}{dx^2}+x\frac{dy}{dx}+(1+x)y=0.$$

Ans. $$y=c_0\left[1-\frac{1}{4}x^3-\frac{1}{12}x^4+\frac{5}{96}x^5\right]+c_1\left[x-\frac{1}{8}x^2+\frac{1}{24}x^5...\right]$$

20.3 Solution of Differential Equation in a Series when x = 0 is an Ordinary Point. i.e., when P_0 does not Vanish for x = 0

Here assume $y=c_0+c_1x+c_2x^2+...$

Find $\frac{dy}{dy}$ and $\frac{d^2y}{dx^2}$ and put their values in the given equation.

Then equate to zero the coefficients of various powers of x and you will get a series in terms of two arbitrary constants.

***Example* 20.9:** Solve $\frac{d^2y}{dx^2}-x^3\frac{dy}{dx}-y=0$ in powers of x.

Here x = 0 is an ordinary point and as such we assume

$$y=c_0+c_1x+c_2x^2+c_3x^3+...+c_nx^n+...$$

$$\frac{dy}{dx}=c_1+2c_2x+3c_2x^2+...nc_nx^{n-1}+...$$

$$\frac{d^2y}{dx^2} = 2c_2 + 6c_3x + ...n(n-1)c_nx^{n-3} + ...$$

Putting these values in the given equation, we get

$$(2c_2 - c_0)(6c_3 - c_1)x + (12c_4 - c_1 - c_2)x^2$$
$$+(20c_6 - 2c_2 - c_3) + ... + (n+2)(n+1)c_{n+2}$$
$$-(n-1)c_{n-1} - c_n]\, x^n + ... = 0.$$

Equating to zero the coefficients of various powers of x we, get

$$2c_9 - c_0 = 0; \;\therefore\; c_2 = \frac{1}{2}c_0.$$

$$6c_3 - c_1 = 0; \;\therefore\; c_3 = \frac{1}{6}c_1$$

$$12c_4 - c_1 - c_2 = 0; \;\therefore\; c_4 = \frac{1}{24}c_0 + \frac{1}{12}c_1, \text{ and so on.}$$

$$(n+2)(n+1)c_{n+2} - (n-1)c_{n-1}\; c_n = 0$$

$$\therefore \quad c_{n+2} = \frac{(n-1)c_{n-1}}{(n+1)(n+2)} + \frac{c_n}{(n+1)(n+2)}$$

$$\therefore \quad y = c_0 + c_1x + x + \frac{1}{2}c_0x^2 + \frac{1}{6}c_1x^2 + \left(\frac{1}{24}c_0 + \frac{1}{12}c_1\right)x^4 + ...$$

$$y = c_0\left(1 + \frac{1}{2}x^2 + \frac{1}{24}x^4 + ...\right) + c_1\left(x + \frac{1}{6}x^3 + \frac{1}{12}x^4 + ...\right).$$

Above is the complete solution containing two constants c_0 and c_1

***Example* 20.10:** Solve $\frac{d^2y}{dx^2} - 2x^3\frac{dy}{dx} + 4xy = x^2 + 2x + 2$ in powers of x.

Here again x = 0 is an ordinary point and assume that

$$y = c_0 + c_1x + c_2x^2 + c_3x^3 + ...$$

Calculate $\frac{dy}{dx}$ and $\frac{d^2y}{dx^2}$ and put in the given equation.

$$y=(2c_2-2)+(6c_3-4c_0-2)x+(12c_4+2c_1-1)x^3$$
$$+20c_5x^3+...+[(n+2)(n+1)c_{n+2}$$
$$-2(n-1)c_{n-1}+4c_{n-1}]x^n,$$

Equate to zero the coefficients of various powers of x,

$$2c_2-2=0;\ \therefore\ c_2=1$$

$$6c_3+4c_0-2=0;\ \therefore\ c_3=\frac{1}{3}-\frac{2}{3}c_0, c_4=\frac{1}{1_2}=\frac{1}{6}c_1, c_5=0$$

and $$c_{n+2}=\frac{2(n-3)}{(n+1)(n+2)}c_{n-1}.$$

Hence the complete solution is

$$y=c_0\left(1-\frac{2}{3}x^3-\frac{3}{4_5}x^6...\right)+c_1\left(x-\frac{1}{6}x^4-\frac{1}{8_8}x^7...\right).$$
$$+x^2+\frac{7}{3}x^3+\frac{1}{1_2}x^4+\frac{1}{4_8}x^6+...$$

Part V

CHAPTER 21

Functions of a Complex Variable

21.1 Introduction

In the attempts made to solve algebraic equations such as $x^2 + 1 = 0$, the concept of imaginary numbers was introduced. Euler (1707 – 1783) was the first to denote the imaginary unit i, defined as $i^2 = -1$.

As many of the algebraic equations do not have solutions in the real number system ***R***, the necessity to introduce the set of complex numbers (of the form a + ib) arose. The theory of functions of a complex variable, which is one of the most useful branches of Mathematics, was finally placed on a sound foundation by the mathematicians Cauchy, Riemann, Weierstrass, Gauss and others.

In the light of this theory many mathematical concepts got clear clarifications. Further the theory found large number of applications in Physics, Applied Mathematics and Engineering problems.

Note: In sections 21.1.1 to 21.1.10, some introductory concepts on complex numbers are given as revision.

21.1.1 Complex Numbers

The number $(a+bi), a,b \in R$ is called a ***Complex Number***. It is denoted by 'z' and written as

$z = a + bi$; a, b are called real and imaginary parts of z respectively and denoted as a = Re (z) and b = Im(z)

z = 3 + 4i, 2–3i, 2i, 4 are all examples of complex numbers. $z = x + iy; x, y \in R$ is called the ***Complex variable.***

In a complex number $a + bi$, if $b = 0$ then it is purely real. Example : 4, 3, –1 etc.

If $a = 0$, it is purely imaginary. Example: 2i, –i ,3i etc.

The complex number $z = a + bi$ is also defined as an ordered pair of real numbers a, b and is denoted by z = (a, b).

Thus the complex numbers $2 + 3i$, $3 - 4i$, 1, $-4i$, respectively can be denoted by (2, 3), (3, –4), (1, 0), and (0, –4)

21.1.2 The Four Fundamental Operations of Complex Numbers

Let $z_1 = (x_1 + iy_1)$, $z_2 = (x_2 + iy_2)$ be two complex numbers. Then,

1. ***Addition :*** $z_1 + z_2 = (x_1 + x_2) + i(y_1 + y_2)$
2. ***Subtraction :*** $z_1 - z_2 = (x_1 - x_2) + (y_1 - y_2)$
3. ***Multiplication :*** $z_1 z_2 = (x_1 + iy_1)(x_2 + iy_2) = (x_1x_2 - y_1y_2) + i(x_1y_2 + x_2y_1)$
4. ***Division:***

Let $z_2 \neq 0,\ ie.,\ x_2 \neq 0$ and $y_2 \neq 0$

Then,
$$\frac{z_1}{z_2} = \frac{x_1 + iy_1}{x_2 + iy_2} = \frac{(x_1 + iy_1)(x_2 - iy_2)}{(x_2 + iy_2)(x_2 - iy_2)}$$
$$= \frac{(x_1x_2 + y_1y_2)}{x_2^2 + y_2^2} + i\frac{[x_2y_1 - x_1y_2]}{x_2^2 + y_2^2}$$

[$x - iy$ called the complex conjugate of ($x + iy$) is denoted by $\bar{z}$]

Note : $z_1 = z_2 \Leftrightarrow x_1 = x_2$ and $y_1 = y_2$

21.1.3 Graphical Representation of Complex Numbers

Fig. 21.1(a) denotes the xy plane $(R \times R)$ in which the two perpendicular lines $X'0X, Y'0Y$ are the X axis and Y axis respectively, O being the origin. In this plane we can locate any point by an ordered pair (a, b) of real numbers called rectangular coordinates of the point. Points A,B,C,D,E,F are indicated in this plane by coordinates (3.5, 0),(2, 3), (-1, 5), (-2, –1), (3, -4), and (0, –3), respectively.

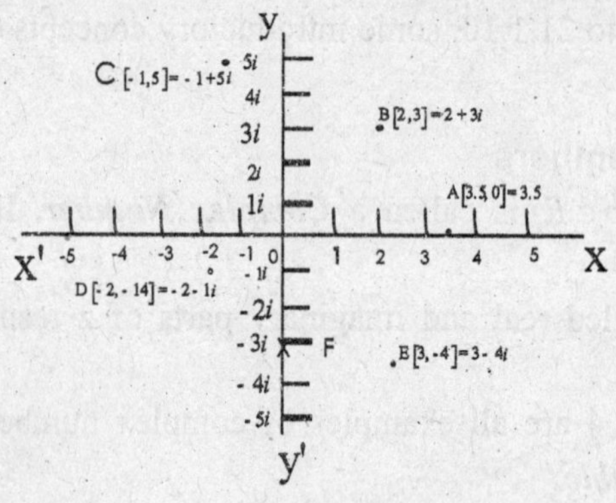

Fig. 21.1(a)

Corresponding to each of these points one and only one complex number can be associated as 3.5 + 0i, 2 + 3i, –1 + 5i, –2 –i, 3 – 4i and 0 – 3i respectively. Conversely, any complex number $z=(a+ib)$ can be represented by one and only one point (a,b) of this plane.

As a result of this, the complex number z is referred to as the point 'z' of this plane. It is for this reason, this plane which represents the complex numbers is called as ***Complex Plane,*** or z-plane. It is also known as ***Argand Plane*** or ***Gaussian Plane.***

The points on X axis are purely real and those on the Y-axis are purely imaginary. Hence they are respectively called real and imaginary axes. The set of complex numbers is denoted by 'C'.

Complex Conjugate

Let $P=a+ib=(a,b)$ be a point in the z-plane. [Fig. 21(b)]

Then its image in the real axis is $P'=(a-ib)=(a,-b)$

P' is called the Complex Conjugate of P.

If z $=(a+ib)$, then $(a-ib)$ is denoted by $\bar{z}$.

Note : Let $\quad z=a+ib,\ \bar{z}=a-ib$

Then 1. $z\bar{z}=a^2+b^2$

2. $\text{Re}(z)=a=\dfrac{z+\bar{z}}{2}$

3. $\text{Im}(z)=b=\dfrac{z-\bar{z}}{2i}$

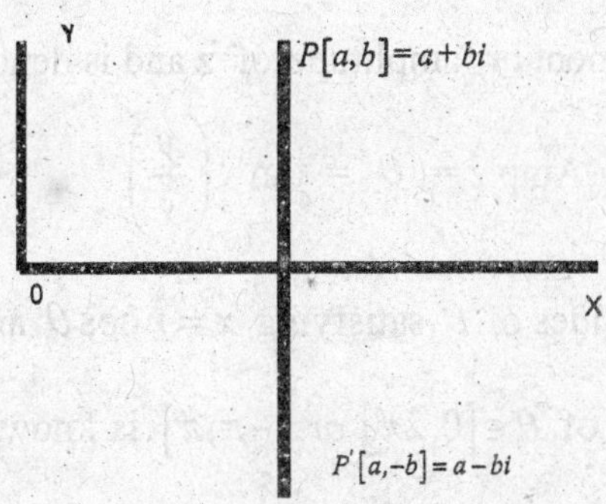

Fig. 21.1(b)

21.1.4 Polar Form or Mod-Amp form of Complex Numbers

In Fig. 21.2, the point P (x,y) represents the complex number z = $x+iy$ in the z-plane.

Then, we have, $x=r\cos\theta,\ y=r\sin\theta,$

$$r=\sqrt{x^2+y^2},\quad \theta=\tan^{-1}\left(\frac{y}{x}\right),\ (\text{'r' is + ve})$$

$$z=x+iy=r(\cos\theta+i\sin\theta)=rcis\theta\ \ (\text{short form})$$

is called the mod-amp form of 'z'

[r, θ are polar coordinates]

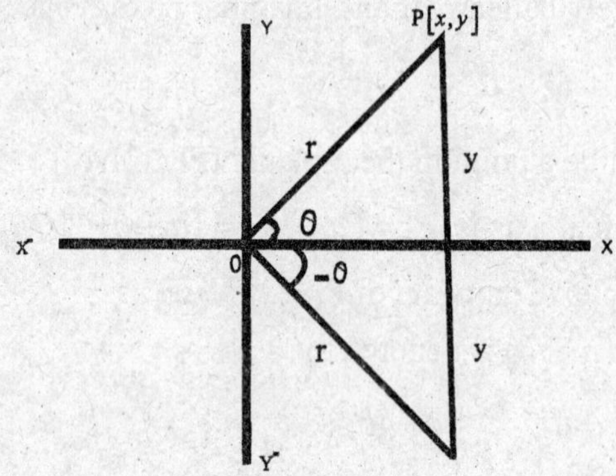

Fig. 21.2

1. 'r' is called the modulus or magnitude of z and is denoted by,

$$|z|=r=\sqrt{x^2+y^2} \quad .....(1)$$

2. 'θ' is called the argument or 'amplitude' of z and is denoted by,

$$\text{Arg } z=\text{ Amp } z=\ \theta\ =\tan^{-1}\left(\frac{y}{x}\right) \quad .....(2)$$

There can be many values of θ satisfying $x=r\cos\theta$ and $y=r\sin\theta$

Therefore, the values of $\theta\in[0,2\pi]$ or $[-\pi,\pi]$ is known as the principal value of Arg z.

Another Representation of Polar Form

We know that, $e^x=1+\frac{x}{1!}+\frac{x^2}{2!}+.......$ Let $x=i\theta$

Then $$e^{i\theta}=1+\frac{i\theta}{1!}+\frac{i^2\theta^2}{2!}+........=\left(1-\frac{\theta^2}{2!}+\frac{\theta^4}{4!}+...\right)+i\left(\frac{\theta}{1!}-\frac{\theta^3}{3!}+...\right)$$

Thus $$e^{i\theta}=(\cos\theta+i\sin\theta) \qquad \text{(from trigonometry)}$$

Therefore, the polar form of $z=x+iy$ can also be written as,

$$z=r\,\text{cis}\,\theta=r\,e^{i\theta}$$

3. Let P^1 (x, –y) represent the complex number $\bar{z}$ = $x-iy$, the conjugate of z= $x+iy$

Then $$\bar{z}=r(\cos\theta-i\sin\theta)=r\text{cis}(-\theta);|\bar{z}|=r;$$

and $$\text{Arg }\bar{z}=-\theta=-\text{Arg}(z)$$

4. Let $$z_1=x_1+iy_1=r_1(\cos\theta_1+i\sin\theta_2)$$

and, $$z_2=x_2+iy_2=r_2(\cos\theta_2+i\sin\theta_2),$$

so that $$|z_1|=r_1,\ \text{Arg }z_1=\theta_1,\ \text{and }|z_2|=r_2,\ \text{Arg }z_2=\theta_2$$

then $$z_1z_2=r_1e^{i\theta_1}.r_2e^{i\theta_2}=r_1r_2e^{i(\theta_1+\theta_2)}=r_1r_2cis(\theta_1+\theta_2)$$

$$\frac{z_1}{z_2}=\frac{r_1e^{i\theta_1}}{r_2e^{i\theta_2}}=\frac{r_1}{r_2}e^{i(\theta_1-\theta_2)}=\frac{r_1}{r_2}cis(\theta_1-\theta_2)$$

Therefore, we can write,

$$|z_1z_2|=r_1r_2=|z_1||z_2| \text{ and } \left|\frac{z_1}{z_2}\right|=\frac{r_1}{r_2}=\frac{|z_1|}{|z_2|}$$

$$\text{Arg }(z_1z_2)=\theta_1+\theta_2=\text{Arg }z_1+\text{Arg }z_2,$$

and $$\text{Arg}\left(\frac{z_1}{z_2}\right)=\theta_1-\theta_2=\text{Arg }z_1-\text{Arg }z_2$$

In general, $$\text{Arg }(z_1z_2.....z_n)=\text{Arg }z_1+\text{Arg }z_2+.....+\text{Arg }z_n$$

21.1.5 Some More Properties of Moduli and Arguments

$z_1, z_2,.....$ are complex numbers. Then,

1. $|z_1+z_2|\le|z_1|+|z_2|$

2. $\left|\sum_{j=1}^{n} z_j\right| \le \sum_{j=1}^{n} |z_j|$ $\quad (n = 1, 2, 3, ...)$

3. $|z_1 - z_2| \ge |z_1| - |z_2|$

4. $|z^n| = |z|^n, \quad n \in N$

5. Arg $z^n = n$ Arg $z, \quad n \in N$

21.1.6 De Moivre's Theorem

1. If n is an integer (+ve or –ve), then,

$$(\cos\theta + i\sin\theta)^n = \cos n\theta + i\sin n\theta \text{ or } (cis\theta)^n = cisn\theta$$

2. If n is any rational number, then $(\cos n\theta + i\sin n\theta)$ is one of the value of $(cis\theta)^n$

This theorem is used to solve complex equations and to find n^{th} roots of a complex number.

21.1.7 Vector Representation of Complex Numbers

Let $z = x + iy$ be a complex number and let it represented by the point P (x, y) in the z – plane, (see Fig. 21.3), then z can be considered as the vector OP with initial point at origin O and terminal point at P. $\overline{OP} = x + iy$ is also called as the position vector of P.

If AB = OP and AB || OP, then we write, $\overline{AB}$ and $\overline{OP} = x + iy$

(i.e., vectors $\overline{AB}$ and $\overline{OP}$ are equal vectors).

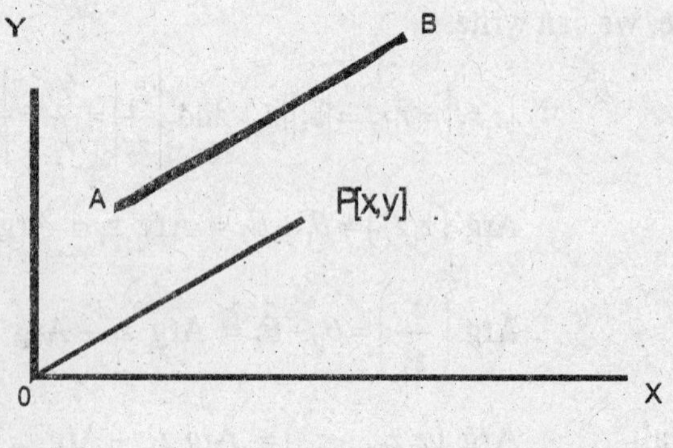

Fig. 21.3

Addition of Complex Numbers

Let z_1 and z_2 be represented by P, Q [Fig. 21.4], then $(z_1 + z_2)$ is represented by 'R' by parallelogram law of vectors.

Similarly, $(z_1 - z_2)$ is represented by vector $\overline{QP}$

If $\quad z_1 = a_1 + ib_1 \ , \ z_2 = a_2 + ib_2 \ ,$

Then $\quad |z_1 - z_2| = \sqrt{(a_1 - a_2)^2 + (b_1 - b_2)^2}$

which is the distance between P and Q . (or Length of vector $\overline{QP}$)

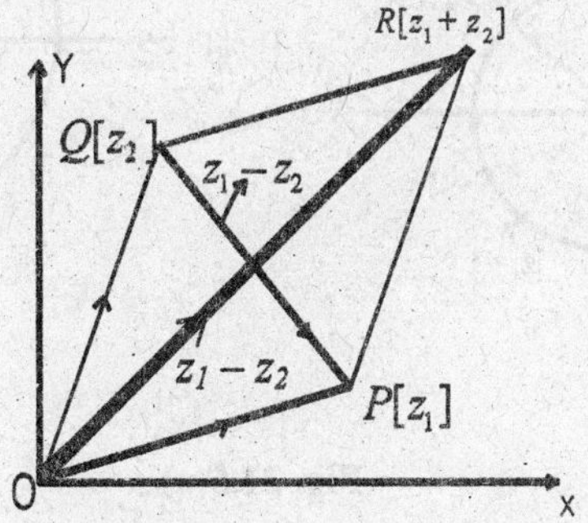

Fig. 21.4

21.1.8 Loci of 'z'

In real variables, we know that,

(i) $x = 0$ *is equation of x axis*

(ii) $y = 0$ *is equation of y axis*

(iii) $ax + by + c = 0$ *represents a straight line*

(iv) $y = mx$ *is a line through the origin*

(v) $x^2 + y^2 + 2gx + 2fy + c = 0$ *represents a circle,*

so on and so forth.

Thus, given an equation in x and y, we are able to recognize it as a curve (in the xy – plane) which represents the locus of x, y. In a similar way we can find the Loci of 'z' denoted by complex equations (equations involving 'z').

Let us examine some such situations in this section.

1. $|z| = 1$ is the unit circle , for , if $z = x + iy$, then

$$|z| = 1 \Rightarrow \sqrt{x^2 + y^2} = 1 \Rightarrow x^2 + y^2 = 1$$

which is a circle with centre at origin and radius equal to unity. $|z| > 1$ is the exterior of the unit circle and $|z| < 1$ is its interior [Fig. 21.5(a)]. $|z| = 1$ can be written in polar form as $z = e^{i\theta}$.

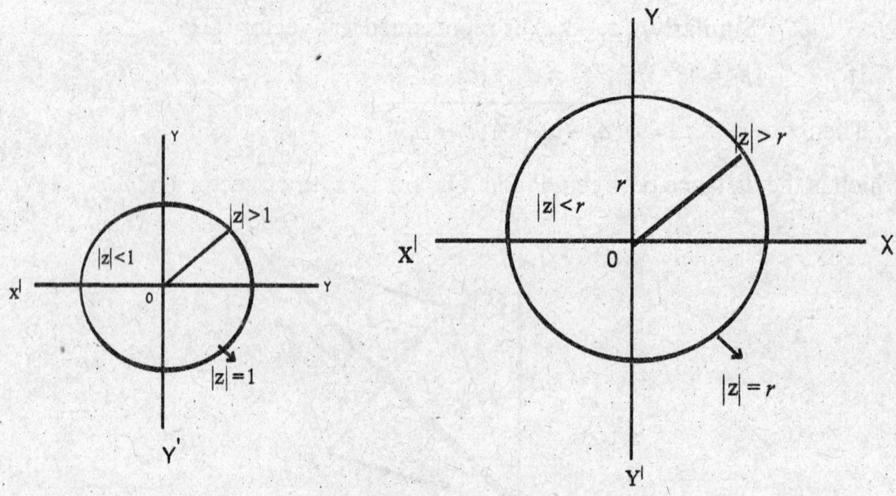

Fig. 21.5(a)

2. $|z| = r$ represents a circle with origin as centre and radius 'r' units.

$\Rightarrow$ $|z| = r$ is the circle $\quad x^2 + y^2 = a^2$

$|z| < r$, is the interior

and $|z| > r$ is the exterior of above circle. (Fig. 21.5(a))

3. Let $a = c + id$, (c, d $\in$ R)

then $|z - a| = r \Rightarrow |(x - c) + i(y - d)| = r$

$$\Rightarrow (x-c)^2 + (y-d)^2 = r^2$$

$\therefore$ $|z - a| = r$ represents a circle with centre at 'A' or (c, d) and radius 'r' (Fig. 21.5(b)).

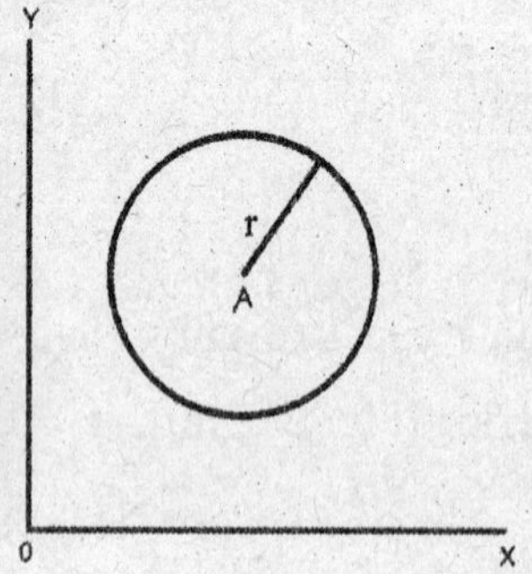

Fig. 21.5(b)

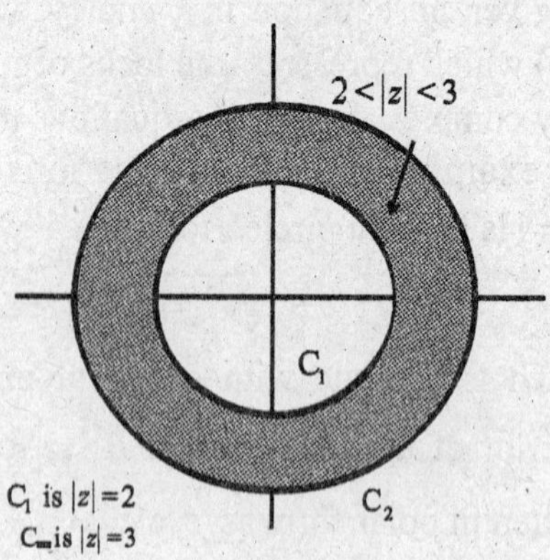

Fig. 21.5(c)

4. $2 < |z| < 3$, represents the annulus between circles $|z| = 2$ and $|z| = 3$ (Fig. 21.5(c))

5. Consider the inequality, $\frac{\pi}{4} < \text{Arg } z < \frac{\pi}{3}$

$$\text{Arg } z = \frac{\pi}{4}, \Rightarrow \tan^{-1}\frac{y}{x} = \frac{\pi}{4} \Rightarrow y = x$$

which is the ray $(\vec{OA})$ $\theta = \frac{\pi}{4}$ [Fig . 21.5(d)]

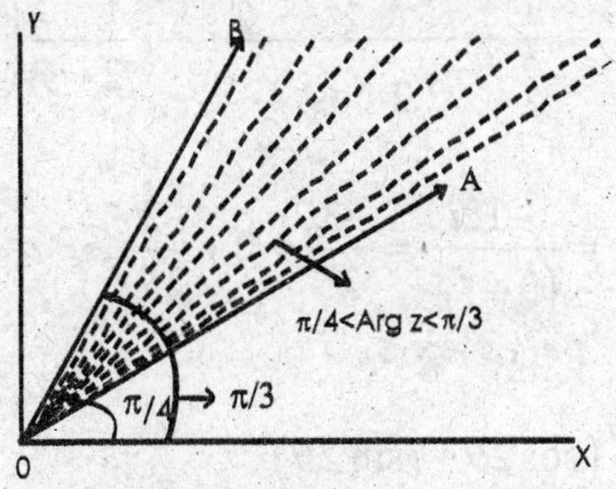

Fig. 21.5(d)

$$\text{Arg } z = \frac{\pi}{3} \text{ represents the ray } \theta = \frac{\pi}{3} \quad (\vec{OB})$$

Hence the given inequality represents the infinite region between the above two rays [shaded portion in Fig. 21.5(d)].

Solved Examples

***Example* 21.1:** Express the following in the form $(a + ib)$ [or rationalize the denominators]

1. $\frac{3-4i}{3+4i}$ 2. $\frac{1}{1+\cos\theta+i\sin\theta}$ 3. $\frac{1}{(3+i)^2} - \frac{1}{(3-i)^2}$

Solution:

1. $\frac{3-4i}{3+4i} = \frac{(3-4i)^2}{(3+4i)(3-4i)} = \frac{9-16-24i}{3^2+4^2} = \frac{-7}{25} + \left(\frac{-24}{25}\right)i$

2. $$\frac{1}{1+\cos\theta+i\sin\theta}=\frac{(1+\cos\theta-i\sin\theta)}{(1+\cos\theta+i\sin\theta)(1+\cos\theta-i\sin\theta)}$$

$$=\frac{(1+\cos\theta-i\sin\theta)}{(1+\cos\theta)^2+\sin^2\theta}=\frac{(1+\cos\theta)-i\sin\theta}{2.(1+\cos\theta)}$$

$$=\frac{1}{2}-\frac{\sin\theta}{2.(1+\cos\theta)}i$$

$$=\frac{1}{2}-\frac{2.\sin\frac{\theta}{2}\cos\frac{\theta}{2}}{2.2.\cos^2\frac{\theta}{2}}i=\frac{1}{2}-\frac{1}{2}\left(\tan\frac{\theta}{2}\right)i$$

3. $$\frac{(3-i)^2-(3+i)^2}{[(3+i)(3-i)]^2}=\frac{-12i}{(9+1)^2}=\frac{-3i}{25}=0+\frac{-3}{25}i$$

***Example* 21. 2:** Simplify

$$\frac{(\cos\theta+i\sin\theta)^{10}(\cos 2\theta+i\sin 2\theta)^4}{(\cos\theta+i\sin\theta)^3}$$

***Solution*:** Numerator, $N_r=\left(e^{(i\theta)}\right)^{10}\left(e^{i(2\theta)}\right)^4=e^{i(10\theta)}e^{i(8\theta)}=e^{i(18\theta)}$, $N_r=cis(18\theta)$

Denominator, $D_r=e^{(i3\theta)}$

$\therefore$ Given expression $=\dfrac{e^{i(18\theta)}}{e^{i(3\theta)}}=e^{i(15\theta)}=\cos 15\theta+i\sin 15\theta$

***Example* 21.3:** Solve $z^6-1=0$, (or) Find the 6th roots of unity (or) Find all the values of $(1)^{1/6}$.

***Solution*:** $z^6-1=0 \Rightarrow z=(1)^{1/6}$

$$1=\cos(0)+i\sin(0)=\cos(0+2m\pi)+i\sin(0+2m\pi)$$

$$=\cos(2m\pi)+i\sin(2m\pi),\ m=0,1,2,......$$

$$z=(1)^{1/6}=\left[\cos(2m\pi)+i\sin(2m\pi)\right]^{1/6}$$

$$z=\cos\left(\frac{2m\pi}{6}\right)+i\sin\left(\frac{2m\pi}{6}\right)$$ (by De Moivre's Theorem .)

Here $m = 0, 1, 2, 3, 4, 5$

Substituting these values for 'm' we get the 6th roots of unity, as,

$$z_1 = \cos(0) + i\sin(0) = 1 \qquad (m = 0)$$

$$z_2 = \cos\left(\frac{\pi}{3}\right) + i\sin\left(\frac{\pi}{3}\right) \qquad (m = 1)$$

$$z_3 = \cos\left(\frac{2\pi}{3}\right) + i\sin\left(\frac{2\pi}{3}\right) \qquad (m = 2)$$

$$z_4 = \cos\pi + i\sin\pi = -1 \qquad (m = 3)$$

$$z_5 = \cos\left(\frac{4\pi}{3}\right) + i\sin\left(\frac{4\pi}{3}\right) \qquad (m = 4)$$

$$= \cos\left(\frac{-2\pi}{3}\right) + i\sin\left(\frac{-2\pi}{3}\right) = \cos\left(\frac{2\pi}{3}\right) - i\sin\left(\frac{2\pi}{3}\right)$$

$$z_6 = \cos\left(\frac{5\pi}{3}\right) + i\sin\left(\frac{5\pi}{3}\right) = \cos\left(\frac{-\pi}{3}\right) + i\sin\left(\frac{-\pi}{3}\right)$$

Note:

1. In short the above 6 values can be given as,

$$z_m = cis\left(\frac{m\pi}{3}\right) \qquad (m = 0 \text{ to } 5)$$

2. Plotting these 6 points in the z – plane, (Fig. 21.6), we can note that the 6th roots of unity are the six vertices of a regular hexagon of side unity (centre at origin).

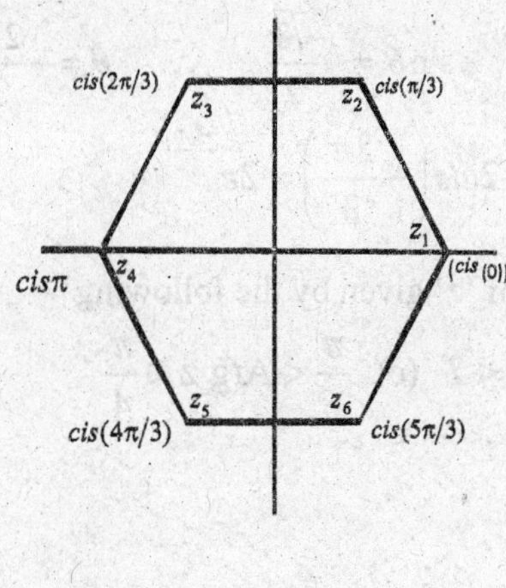

***Example* 21.4:** Express the following in Polar Form

(a) $-i$ (b) $1-i$ (c) $-1-\sqrt{3}i$

***Solution*:**

(a) Let $-i = x + iy = r(\cos\theta + i\sin\theta)$, $(r > 0)$

$\therefore$ then $r\cos\theta = 0$ and $r\sin\theta = -1$

$$r^2 = 0^2 + (-1)^2 = 1 \quad \Rightarrow \quad r = 1$$

$$\sin\theta = -1 \quad \text{and} \quad \cos\theta = 0 \quad \Rightarrow \quad \theta = \frac{-\pi}{2}$$

$$-i = \cos\left(\frac{-\pi}{2}\right) + i\sin\left(\frac{-\pi}{2}\right) = cis\left(\frac{-\pi}{2}\right) = e^{-i\pi/2}$$

(b) $1 - i = r(\cos\theta + i\sin\theta)$, then

$$r\cos\theta = 1 \quad \text{and} \quad r\sin\theta = -1$$

$$r^2 = 1^2 + (-1)^2 = 2 \Rightarrow r = \sqrt{2}$$

$$\sin\theta = -\frac{-1}{\sqrt{2}} \quad \text{and} \quad \cos\theta = \frac{1}{\sqrt{2}} \quad \Rightarrow \quad \theta = \frac{3\pi}{4}$$

Hence, $1 - i = \sqrt{2}cis\left(\frac{3\pi}{4}\right) = \sqrt{2}e^{\frac{3\pi}{4}i}$

(c) Let $-1 - \sqrt{3i} = rcis\theta$, then

$$r\cos\theta = -1 \quad \text{and} \quad r\sin\theta = -\sqrt{3}$$

$$r^2 = 1 + 3 = 4 \quad \Rightarrow \quad r = 2$$

$$\cos\theta = -\frac{1}{2} \quad \text{and} \quad \sin\theta = -\frac{\sqrt{3}}{2} \quad \therefore \quad \theta = -\frac{2\pi}{3}$$

$$\therefore \quad -1 - \sqrt{3}\,i = 2cis\left(-\frac{2\pi}{3}\right) = 2e^{\frac{-2\pi}{3}i}$$

***Example* 21.5:** Find the Locus of 'z' given by the following

(a) $\left|\frac{z-2}{z+2}\right| = 2$ (b) $1 < |z| < 2$ (c) $\frac{\pi}{6} < \text{Arg } z < \frac{\pi}{4}$

***Solution*:**

(a) Let $z = x + iy$, then

$$\left|\frac{z-2}{z+2}\right| = 2 \Rightarrow |z-2| = 2|z+2|$$

$$\Rightarrow \quad |x+iy-2| = 2|x+iy+2|$$

$$\Rightarrow \quad (x-2)^2+y^2 = 4\left[(x+2)^2+y^2\right] \Rightarrow 3x^2+3y^2+20x+12=0$$

$$\Rightarrow \quad x^2+y^2+\frac{20}{3}x+4=0$$

which represents a circle with centre at $(-\frac{10}{3},0)$ and radius equal to $\sqrt{\frac{100}{9}-4}=\frac{8}{3}$

$\therefore$ Locus of 'z' is a circle with centre at $z=-\frac{10}{3}$ and radius $\frac{8}{3}$

(b) $|z|=1 \Rightarrow x^2+y^2=1$, which is unit circle with centre at origin.

$|z|=2 \Rightarrow x^2+y^2=4$, which is a circle with centre at origin and radius '2'

$\therefore$ $1<|z|<2$ represents the annular region between these two circles. [Fig. 21.7]

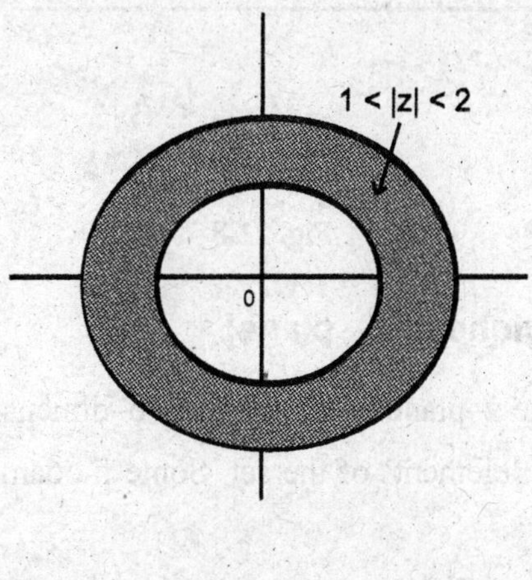

Fig. 21.7

(c) Let $z=r(\cos\theta+i\sin\theta)=x+iy$

$$\therefore \quad \text{Arg } z = \frac{\pi}{6} \Rightarrow \theta=\frac{\pi}{6} \text{ or } \tan^{-1}\left(\frac{y}{x}\right)=\frac{\pi}{6}$$

or $\quad y = \tan\left(\frac{\pi}{6}\right)x \quad \Rightarrow \quad y = \frac{1}{\sqrt{3}}x$

$\therefore$ Locus of z is the ray $\theta = \frac{\pi}{6}$ $(\overrightarrow{0A})$

Similarly Arg z = $\frac{\pi}{4}$ is the ray $\theta = \frac{\pi}{4}$ $(\overrightarrow{0B})$

$\Rightarrow$ Locus of z is the region between these two rays.
[shaded portion of Fig. 21.8]

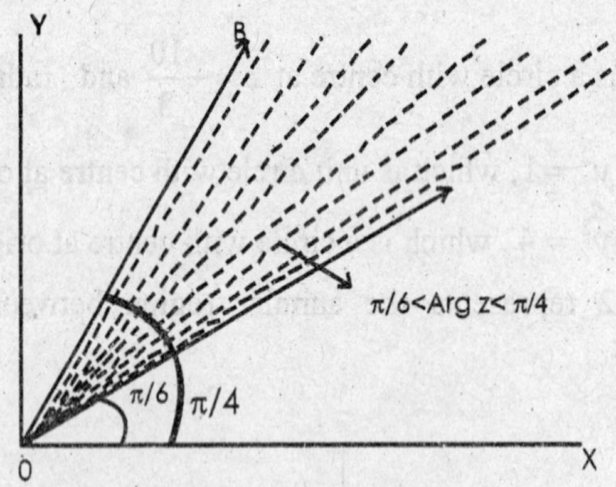

Fig. 21.8

21.1.9 Point Sets [Regions in z – plane]

A collection of points in the z–plane is known as two–dimensional point set. Each point of the set is a 'member' or 'element' of the set. Some fundamental definitions regarding such sets are noted below.

1. ***Neighborhood:*** *Let* $\delta > 0$ *and* $a \in C$, *then the inequality* $|\, z - a \,| < \delta$ *representing the set of all points in the circle with centre at 'a' and radius 'δ' s known as a 'δ - neighborhood' of the point 'A'. It is denoted by* $N_\delta(a)$ *'or'* $S(a, \delta)$. *Thus* $N_\delta(a) = S(a, \delta) = \{z : |z - a| < \delta\}$. *It is an 'open disc' of radius 'δ'* *(shaded portion of* Fig. 21.9(a)).

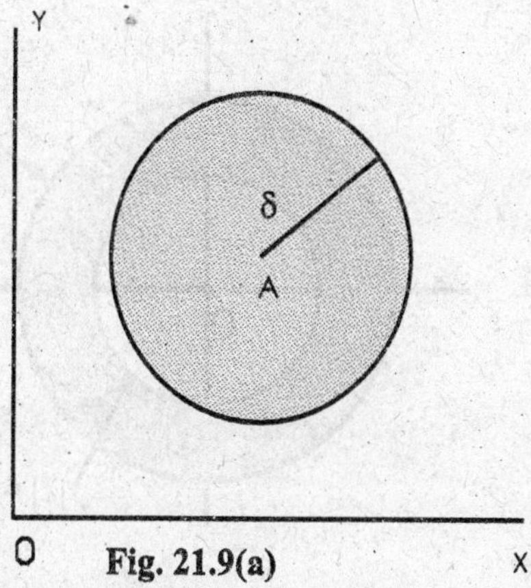

Fig. 21.9(a)

2. ***Deleted Neighborhood:*** *If in the above neighborhood the point 'a' is deleted, it is known as the deleted neighborhood of 'a': i.e.,* $0 < |z-a| < \delta$.

3. ***Limit Point:*** *A point 'a' is called a limit point of a point set 'S' if every* δ *- neighborhood of 'a' contains points of* S. *Since* δ *is any positive number, S has infinitely many points. Further, the point 'a' may belong to S or may not belong to it.*

4. ***Closed Set:*** *A set S is closed if it contains all its limit points.*

   $|z| \leq 1$ *is a closed set.*

5. ***Bounded Set:*** *If there exists a constant M such that* $|z| < M$, *for all* $z \in S$, *then S is bounded.*

   A set which is not bounded is said to be unbounded.

   A set which is bounded and closed is sometimes called compact.

6. ***Interior Point:*** *Let S be a point set. If there exists a* $N_\delta(a)$ *such that all its points belong to S, then 'a' is called an interior point of S.*

7. ***Boundary Point:*** *If every* $N_\delta(a)$ *contains points belonging to S and also points not belonging to S, then 'a' is called a boundary point of S.*

8. ***Exterior Point:*** *If a point is neither an interior point nor a boundary point of a set S, it is an exterior point of S.*

9. ***Open Set:*** *If a set consists of only of interior points, then it is an open set, for example,* $|z| < 1$ *is open.*

10. ***Connected Set:*** Let S be an open set. If we can join any two points of S by a path consisting of straight line segments all points of which belong to S, then S is called a connected set.

The shaded portion between $|z| = 3$ *and* $|z| = 2$ *is a connected set* (Fig. 21.9(b))

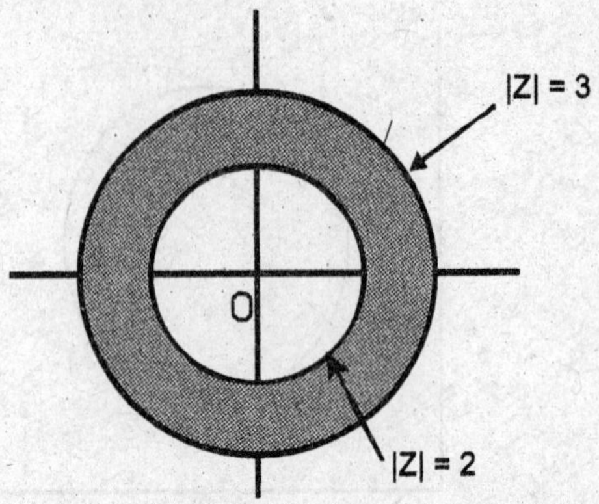

Fig. 21.9(b)

11. ***Open Region or Domain:*** *An open connected set is called an open region or domain.*

Eg: $|z| < 1$ *is a domain while* $|z| \le 1$ *is not.*

The region within $|z| = 1$ *is an open connected set.*

12. ***Closure of a Set:*** *Let S be a set. If we add to it all its limit points, the resulting set is called the closure of S and is a closed set.*

13. ***Closed Region:*** *The closure of an open region or domain is called a closed region.*

14. ***Regions :*** *If to an open region or domain , some, all or none of its limit points are added, we get a set called 'region' . If all the limit points are added, the region is closed.*

If none of the limit points are added, the region is open.

In general, whenever the word is used without qualifying, it means open region or domain.

15. ***Types of Regions:***

(a) ***Simply connected Region:***

Let R be a given region. [Fig. 21.9(d)] Let a simple closed curve c lie in R. If c can be made to shrink to a point with out leaving 'R', then R is said to be simply connected.

C_1 and C_2 within R can be shrunk to a point [Fig .21.9(c)]

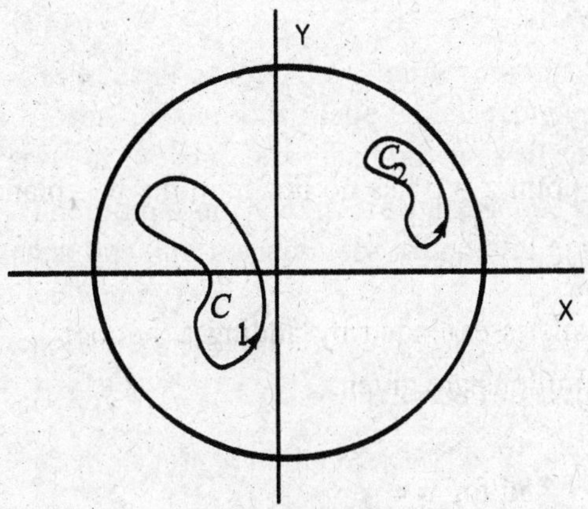

Fig. 21.9(c)

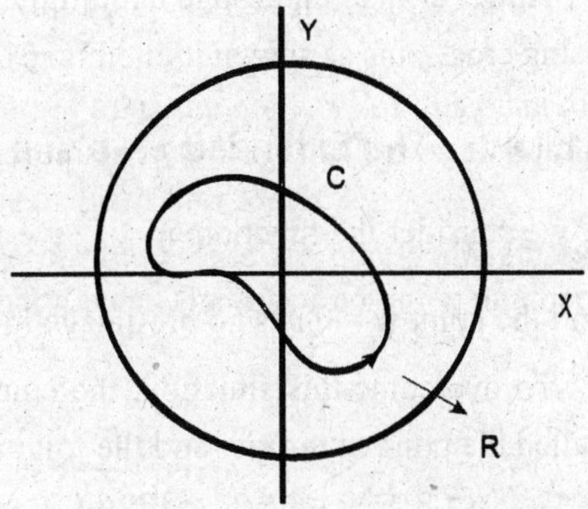

Fig. 21.9(d)

(b) ***Multiply-connected Region:***

A region which is not simply connected is said to be multiply-connected.

We can say intuitively that

1. *a simply–connected region is one which does not have any 'holes' in it, while*
2. *a multiply–connected region is one which has holes in it.*

Fig. 21.9(e) has one hole, and Fig. 21.9(f) has three holes.

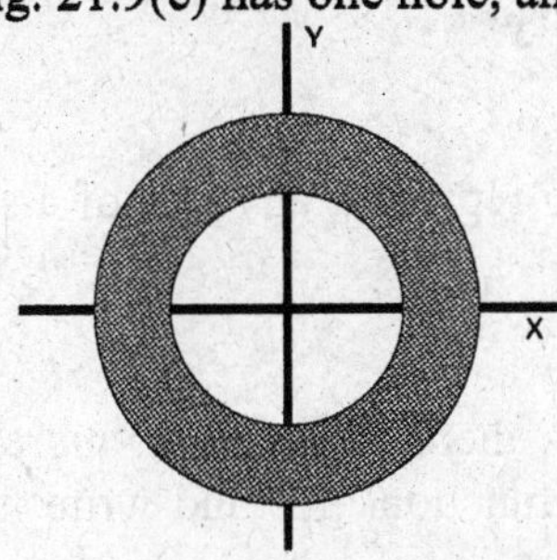

(e)

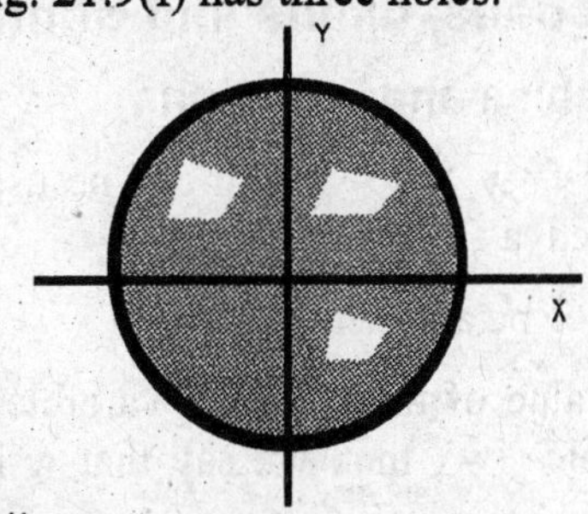

(f)

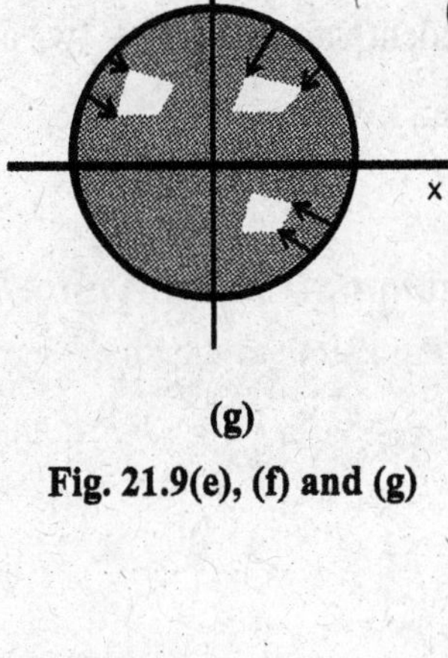

(g)

Fig. 21.9(e), (f) and (g)

However, we can change a multiply-connected region into a simply connected one by using cross-cuts as shown in the Fig. 21.9(g)

21.1.10 The Extended z – plane

If we consider the function $w=\dfrac{1}{z}$, we find the point z = 0 has no image in the w – plane and the point w = 0 has no pre-image in z –plane.

To overcome this situation, the complex plane is extended by adding a symbol '∞' called the point at infinity and the following definitions are given.

1. $a+\infty=\infty+a=\infty$ *for all* $a\in C\cup\{0\}$
2. $a\infty=\infty a=\infty$ *for all* $a\in C-\{0\}$ and for $a=\infty$
3. $\dfrac{a}{\infty}=0$ *for all* $a\in C-\{0\}$
4. $\dfrac{a}{0}=\infty$ *for all* $a\in C-\{0\}$

$C\cup\{\infty\}$ is called the ***Extended Complex Plane.***

In the extended complex plane every straight line can be assumed to pass through ∞.

21.2 Functions, Limits and Continuity

21.2.1 Variables and Functions

The symbol 'z' (= $x + iy$) which can be used to represent any one of a set of complex numbers is called a ***Complex Variable.***

Let w = $u + iv$ be another variable.

If to each value of the complex variable 'z', there exist one or more values of the complex variable 'w', then we say that w is a function of z and write it as w = f (z), w = g (z), etc.

The variable 'z' is an independent variable while 'w' is a dependent one.

Thus, if $f(z)=z^2+i$, then

(i) $f(i)=i^2+i=-1+i$,

(ii) $f(3)=3^2+i=9+i$, and

(iii) $f(a)=a^2+i$, etc.

21.2.2 Single and Multiple-valued Functions

If there exists only one value of w corresponding to each value of z, we say that, $w = f(z)$ is single-valued.

Eg : $w = f(z) = z^3$ is single valued.

If there exist more than one value of w corresponding to each value of z, we say that, $w = f(z)$ is multiple-valued.

Eg: $w = f(z) = z^{1/3}$ is a multiple-valued function since each

z gives rise to 3 values of 'w'

Note: Unless otherwise stated, here afterwards, 'function means single-valued function'.

21.2.3 Inverse Functions

If $w = f(z)$, then $z = g(w) = f^{-1}(w)$ is an inverse function corresponding to w.

21.2.4 Transformations

Let $w = f(z) = u + iv$, where

$z = x + iy,\ x, y \in R$; and $w = u + iv,\ \ u, v \in R$ be a single valued function of z then,

we can write, $f(x + iy) = u + iv$

By equating real and imaginary parts,

$$u = u(x, y),\ v = v(x, y) \qquad .....(1)$$

(1) enables us to transform points P (x , y) of z-plane into points $P'(u, v)$.

To avoid confusion the points (u , v) are plotted on another plane, namely, w-plane, [see Fig 21.10(a) and Fig 21.10(b)].

The equations (1) or equivalently, the equation $w = f(z)$ is called a transformation. The point P is said to be mapped to the point P′ by the transformation $w = f(z)$ and P′ is called the image of P and P is the pre – image P′.

Under a given transformation $w = f(z)$, in general, the points on a curve PAQ [Fig. 21.10(a)] in z – plane are mapped into points on the curve $P'A'Q'$ [Fig. 21.10(b)] of w- plane.

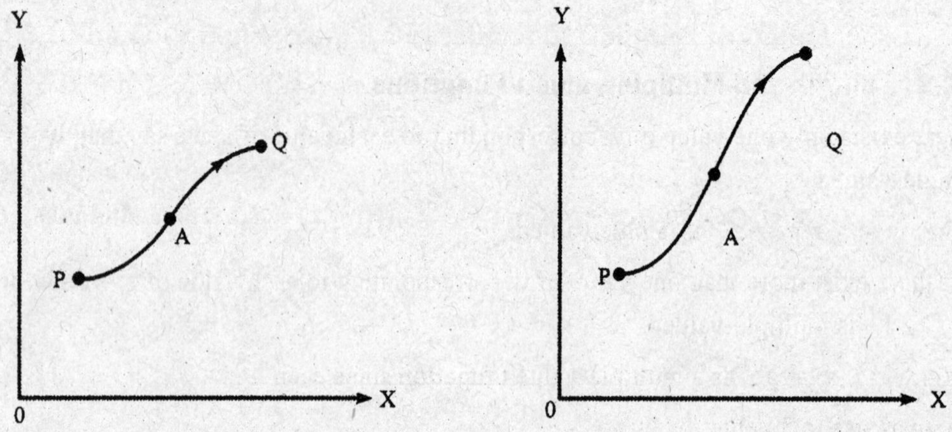

Fig. 21.10(a) Fig. 21.10(b)

21.2.5 Limits

1. Let f(z) be a single valued function defined in a δ – neighborhood $N_{\delta}(a)$ of $z = a$ with a possible exception of z =a itself (i.e ., in a deleted neighbourhood of 'a').
2. Let there exist a complex number 'l' such that $|f(z) - l| < \in$ wherever $0 < |z - a| < \delta$ for any positive number $\in$ (however small). Then, 'l' is said to be the limit of f(z) as z approaches 'a' and is written as $\underset{z \to a}{Lt}\left(f(z)\right) = l$

***Remark*:** Note here the difference between the real valued functions and complex valued functions.

In functions of a real variable, say, *f(x)*, *x* can approach a point *'a'* along two paths only and they are the straight lines from left and right of *'a'*. So we get the left hand limit and the right hand limit. If both these limits are equal, we say that the limit exists and the common value is the limit of the function. But the situation is quite different in case of complex functions. Here z can approach *'a'* in many ways. This can happen along number of paths (see figures 21.11(a) and 21.11(b)) straight or curved.

But along whatever path does $z \to a$, the limit must be unique.

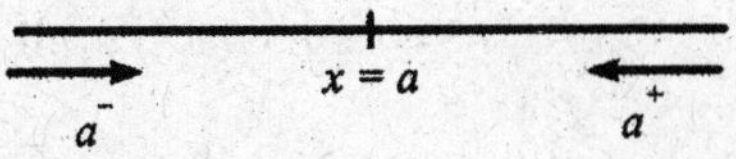

Fig. 21.11(a)

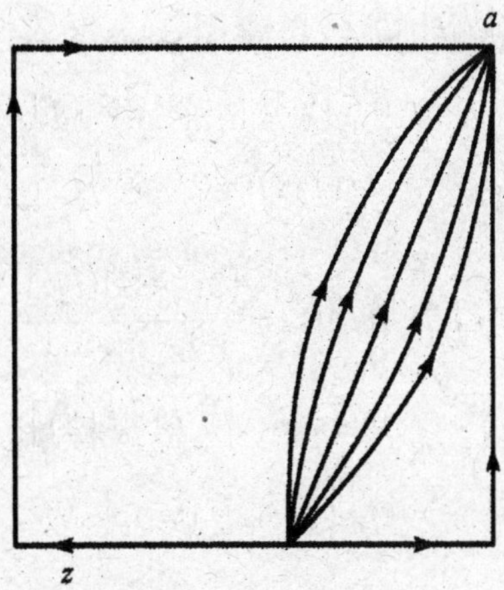

Fig. 21.11(b)

21.2.6 Theorems on Limits (Without Proof)

Some elementary theorems on limits are given below which can be verified from definition.

Let $\underset{z\to a}{Lt} f(z)=l$ or $\underset{z\to a}{Lt}\, g(z)=m$ then

1. $\underset{z\to a}{Lt}\left[f(z)\pm g(z)\right]=\underset{z\to a}{Lt}\left(f(z)\right)\pm\underset{z\to a}{Lt}\left(g(z)\right)=l\pm m$
2. $\underset{z\to a}{Lt}\left[f(z)g(z)\right]=\left[\underset{z\to a}{Lt}\left(f(z)\right)\right]\left[\underset{z\to a}{Lt}\left(g(z)\right)\right]=lm$
3. $\underset{z\to a}{Lt}\left[\dfrac{f(z)}{g(z)}\right]=\dfrac{l}{m}$ (provided $m\neq 0$)
4. If 'k' is 'a' complex constant and $\underset{z\to a}{Lt}\left(f(z)\right)=l$ then $\underset{z\to a}{Lt}\left[kf(z)\right]=kl$

Solved Examples

***Example* 6:** Show That $\underset{z\to 1}{Lt}\left[\dfrac{z^2-1}{z-1}\right]=2$, from Definition

***Solution*:** Let

$$f(z) = \frac{z^2-1}{z-1}$$

Here, *f (z) is not defined at z =1; When* $z \neq 1$, $f(z) = z+1$

$$\therefore \quad |f(z)-2| = |(z+1)-2| = |z-1| \text{ when } z \neq 1$$

For any given $\in > 0$, we choose $\delta = \in$

Then $\quad 0 < |z-1| < \delta \Rightarrow |f(z)-2| < \in$

$\therefore \quad \underset{z\to 1}{Lt}(f(z)) = 2$

***Example* 21.7:** If $f(z) = \dfrac{(3z^2+2)(i+1)}{2z^2+z+3}$,

find $\quad \underset{z\to i}{Lt}\left[f(z)\right]$

Solution: $\quad \underset{z\to i}{Lt}(3z^2+2) = 3i^2+2 = -3+2 = -1$

$$\underset{z\to i}{Lt}(i+1) = i+1$$

$$\underset{z\to i}{Lt}(2z^2+z+3) = (2i^2+i+3) = 1+i$$

$$\therefore \quad \underset{z\to i}{Lt}(f(z)) = \frac{-(i+1)}{(1+i)} = -1$$

***Example* 21.8:** Evaluate $\quad \underset{z\to e^{i\pi/3}}{Lt}\left[\dfrac{z^3-1}{z^4+z^2+1}\right]$

Solution: $\quad \underset{z\to e^{i\pi/3}}{Lt}(z^3-1) = \left(e^{i\pi/3}\right)^3 - 1 = -2 \qquad (\text{since, } e^{i\pi} = -1)$

$$\underset{z\to e^{i\pi/3}}{Lt}(z^4+z^2+1) = -\left(e^{i\pi/3}\right) + \left(e^{2i\pi/3}\right) + 1$$

$$= -\left(\cos\left(\frac{\pi}{3}\right) + i\sin\left(\frac{\pi}{3}\right)\right) + \left(\cos\left(\frac{2\pi}{3}\right) + i\sin\left(\frac{2\pi}{3}\right)\right) + 1$$

$$= -\frac{1}{2} - \frac{\sqrt{3}i}{2} + \frac{1}{2} - \frac{\sqrt{3}i}{2} + 1 = 1 - \sqrt{3}i$$

$$\therefore \text{ Required Limit} = -\frac{2}{1-\sqrt{3}i} = -\frac{2(1+\sqrt{3}i)}{(1+\sqrt{3}i)(1-\sqrt{3}i)}$$

$$= -\frac{2(1+\sqrt{3}i)}{4} = -\frac{1}{2} - \frac{\sqrt{3}i}{2}$$

***Example* 21.9:** Evaluate $\underset{z\to i}{Lt}\left[\frac{z^2+1}{z^6+1}\right]$

***Solution*:** Both the numerator and the denominator $\to 0$ as $z \to i$. Therefore, we factorize the denominator.

$$\underset{z\to i}{Lt}\left[\frac{z^2+1}{z^6+1}\right] = \underset{z\to i}{Lt}\left[\frac{(z^2+1)}{(z^2+1)(z^4-z^2+1)}\right]$$

$$= \underset{z\to i}{Lt}\left[\frac{1}{z^4-z^2+1}\right]$$

$$= \frac{1}{i^4-i^2+1} = \frac{1}{1-(-1)+1} = \frac{1}{3}$$

***Aliter*:**

$$\underset{z\to i}{Lt}\left[\frac{z^2+1}{z^6+1}\right] = \left(\frac{0}{0}\, form\right)$$

$$= \underset{z\to i}{Lt}\frac{\frac{d}{dz}(z^2+1)}{\frac{d}{dz}(z^6+1)} \text{[by L'Hospital Rule]}$$

$$= \underset{z\to i}{Lt}\left[\frac{2z}{6z^5}\right] = \underset{z\to i}{Lt}\left[\frac{1}{3z^4}\right] = \frac{1}{3i^4} = \frac{1}{3}.$$

21.2.7 Continuity

A function $f(z)$ is said to be continuous at a point $z = a$, if

$$\underset{z\to a}{Lt}\left(f(z)\right) = f(a)$$

or if, for any given $\in > 0$, there exists a $\delta > 0$ such that,

$|z-a| < \delta \Rightarrow |f(z)-f(a)| < \in$, then $f(z)$ is continuous at 'a'.

***Note*:** The definition of continuity of $f(z)$ at $z = a$ implies the following

1. *$f(z)$ is defined at $z = a$*
2. $\underset{z\to a}{Lt}\left(f(z)\right)$ *exists, and*
3. $\underset{z\to a}{Lt}\left(f(z)\right) = f(a)$

If a function $f(z)$ *is continuous at each point of a region, then* $f(z)$ *is said to be continuous in the region.*

21.2.8 Theorems on Continuity (without proof)

Some elementary theorems on continuous functions are given below.

1. Let $f(z) = a_0 z^n + a_1 z^{n-1} + \ldots\ldots + a_n, (a_0 \neq 0, a_1, a_2, \ldots$ are complex constants) be a polynomial function of z of degree 'n', then $f(z)$ is continuous in C.
2. A constant function $f(z) = k$, is continuous . [k is a complex constant].
3. If $f(z)$ and $g(z)$ are continuous in a region, so are $f(z) \pm g(z)$, $[f(z)\, g(z)]$ *and*

   $f(z)/g(z)$, $[g(z) \neq 0]$
4. Let $f(z) = u(x, y) + iv(x, y)$ be continuous in a region. Then $u(x, y)$ and $v(x, y)$ are also continuous in a region. Conversely, if u, v are continuous, so is $(u + iv)$.

Points of Discontinuity

If $f(z)$ is not continuous at z = a, then the point z = a is called a point of discontinuity of $f(z)$.

Solved Examples

***Example* 21.10:** Show that $f(z) = \bar{z}$ is continuous in C.

***Solution*:** Let $a \in C, |f(z) - f(a)| = |\bar{z} - \bar{a}| = |z - a|$

$$(\text{ since } |\bar{z} - \bar{a}| = |\overline{z - a}| = |z - a|)$$

For any given $\in > 0$, we choose $\delta = \in$ .Then we get,

$$|z - a| < \delta \Rightarrow |f(z) - f(a)| < \in$$

Therefore, $f(z)$ is continuous at a. Since a is arbitrary, f is continuous at every point $\in C$.

Therefore, f is continuous.

***Example* 21.11:** Show that $f(z) = |z|$ is continuous.

Solution: $|f(z) - f(a)| = ||z| - |a|| \leq |z - a|$

For any given $\in > 0$ we chose $\delta = \in$

Then, we get,

$$|z - a| < \delta \quad \Rightarrow \quad |f(z) - f(a)| < \in$$

Therefore, f is continuous at a. Since a is arbitrary, f is continuous at every point

Therefore, f is continuous.

***Example* 21.12:** Discuss the continuity of *f (x, y)* at the origin, where

$$f(x,y)=\frac{x^2-y^2}{x^2+y^2} \qquad (x,y)\neq(0,0)$$

$$=0, \qquad (x,y)=(0,0)$$

***Solution*:** In the Fig. 21.12 the point A(*x, y*) approaches the point O (0, 0) along two paths.

1. Along ABO and 2. Along ACO.

Let us find $\underset{(x,y)\to(0,0)}{Lt}\left[f(x,y)\right]$ along these two paths.

1. Along AB, y → 0 and along BO, $x \to 0$.

$$\underset{(x,y)\to(0,0)}{Lt}\left[f(x,y)\right]=\underset{x\to0}{Lt}\left[\underset{y\to0}{Lt}\left\{\frac{x^2-y^2}{x^2+y^2}\right\}\right]$$

$$=\underset{x\to0}{Lt}\left(\frac{x^2}{x^2}\right)=\underset{x\to0}{Lt}(1)=1$$

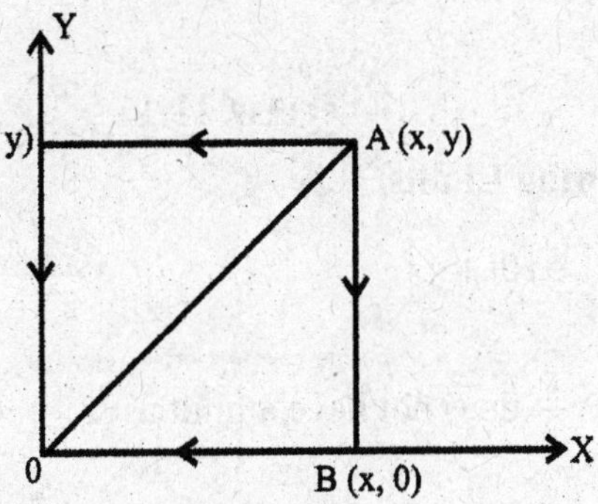

Fig. 21.12

2. Along AC , $x \to 0$ and along CO, y → 0

$$\underset{(x,y)\to(0,0)}{Lt}\left[f(x,y)\right]=\underset{y\to0}{Lt}\left[\underset{x\to0}{Lt}\left\{\frac{x^2-y^2}{x^2+y^2}\right\}\right]$$

$$\underset{y\to0}{Lt}\left[\frac{-y^2}{y^2}\right]=\lim_{y\to0}(-1)=-1$$

Since the two limits are unequal, the limit does not exist. Hence $f(x, y)$ is not continuous at the origin.

***Example* 21.13:** Find the points of discontinuity of the following functions:

(a) $f(z) = \dfrac{3z^2 + 4}{z^4 - 16}$ (b) $f(z) = \cot z$

***Solution*:**

(a) Both numerator and denominator are polynomials in z. Therefore they are continuous. But $f(z)$ is continuous only when the denominator is not equal to zero. Therefore the points of discontinuity of $f(z)$ are given by.

$$z^4 - 16 = 0, \qquad \text{i.e.,} \qquad (z^2 + 4)(z^2 - 4) = 0$$

i.e., $z = \pm 2i$ and $z = \pm 2$

Therefore, points of discontinuity of $f(z)$ are $z = \pm 2, \pm 2i$

(b) $f(z) = \dfrac{\cos z}{\sin z}$; $f(z)$ is not continuous if $\sin z = 0$

i.e. if $z = k\pi, \quad k = 0, \pm 1, \pm 2, \ldots\ldots$

$\therefore$ points of discontinuity are $z = k\pi, \quad k = 0, \pm 1, \pm 2, \ldots\ldots$

Exercise 21.1

I Evaluate the following Limits:

1. $\underset{z \to 2i}{Lt}\left[iz^4 + 3z^2 - 10i + 4\right]$ 2. $\underset{z \to e^{\pi i/4}}{Lt}\left[\dfrac{z^2}{z^4 + z + 1}\right]$

II Prove that $f(z) = \dfrac{\bar{z}}{z}$ does not have a limit as $z \to 0$.

***Hint*:** Let $z \to 0$ along the line $y = mx$, then $\underset{z \to 0}{Lim}(f(x)) = \dfrac{1 - im}{1 + im}$ which is not unique.

III Find whether the following limits exist or not.

1. $\underset{z \to 0}{Lt}\left[\dfrac{\text{Re}(z)}{|z|}\right]$ 2. $\underset{z \to 0}{Lt}\left[\dfrac{\text{Im}(z)}{|z|}\right]$

***Hint*:** Find the limits of these function as $z \to 0$ along x-axis and y-axis. The two limits will not be equal. $\therefore$ Limits do not exist.

IV Let $f(z) = \dfrac{z^2+4}{z-2i}$, when $z \neq 2i$

$= 3+4i$, when $z = 2i$

Discuss the continuity of $f(z)$ at (i) $z = i$, $(ii) z = 2i$

V If $f(x,y) = \dfrac{x^3 - y^3}{x^3 + y^3}$, when $(x,y) \neq (0,0) = 0$,

when $(x,y) = (0,0)$, discuss the continuity of f at the origin.

[***Hint :*** Proceed as in ex.3.7]

VI Find the points of discontinuity of the following

1. $f(z) = \dfrac{2z-3}{z^2+2z+2}$ 2. $f(z) = \dfrac{1}{z} - Sec\, z$ 3. $f(z) = \dfrac{\tanh z}{z^2+1}$

Answers

I 1. $4-18i$ 2. $\dfrac{1}{\sqrt{2}}(1+i)$ **IV.** (i) continuous (ii) not continuous.

V. not continuous

VI. 1. $z = -1 \pm i$ 2. $z = 0, z = n\dfrac{\pi}{2}$ (n being an odd integer)

3. $z = \pm i, z = n\pi i (n \in N)$

21.3 Differentiability – Analytic Functions

21.3.1 Differentiability-Derivative

Let (i) $f(z)$ be a complex function defined in a region D, and (ii) let $z \in D$.

Then f is differentiable at 'z', if, $\underset{h\to 0}{Lt}\left\{\dfrac{f(z+h)-f(z)}{h}\right\}$ exists and is finite.

This limit, denoted by $f^1(z)$ or $\dfrac{df}{dz}$, is known as the derivative of $f(z)$ at z.

$f(z)$ is differentiable in the region D if it is differentiable for all $z \in D$.

Thus, $$f^1(z) = \frac{df}{dz} = \underset{h\to 0}{Lt}\left\{\frac{f(z+h)-f(z)}{h}\right\} \quad .....(1)$$

Note:

(i) *In the definition (1) above, we may use sometimes 'Δz' in place of h.*

(ii) *Differentiability of $f(z)$ implies continuity, though the converse need not be true.*

21.3.2 Differentiability ⇒ Continuity

Prove that if $f(z)$ is differentiable, it is continuous.

***Proof*:** Let $f(z)$ be differentiable at $z \in D$ i.e., $f^1(z)$ exists and is finite.

Since, $\{f(z+h)-f(z)\} = \left\{\frac{f(z+h)-f(z)}{h}\right\}.h$, (where h $\neq 0$)

We have,

$$\underset{h\to 0}{Lt}\{f(z+h)-f(z)\} = \underset{h\to 0}{Lt}\left\{\frac{f(z+h)-f(z)}{h}\right\}.\underset{h\to 0}{Lt}(h)$$

$$= f^1(z).0 = 0 \text{ [by definition of } f^1(z)\text{]}$$

Thus $\underset{h\to 0}{Lt}\{f(z+h)-f(z)\} = 0$

$\Rightarrow$ $\underset{h\to 0}{Lt}\{f(z+h)\} = f(z)$, which shows that $f(z)$ is continuous.

21.3.3 Elementary Properties of Derivatives

The rules of differentiation of complex functions are similar to those of real functions. We give below some of them.

Let $f(z), g(z), h(z)$ be differentiable functions of z. Then the following rules of differentiation (similar to those of elementary calculus) hold good.

1. $\frac{d}{dz}[f(z)\pm g(z)] = \frac{df(z)}{dz} \pm \frac{dg(z)}{dz} = f^1(z) \pm g^1(z)$

2. If k is a complex constant, then $\frac{d}{dz}[kf(z)] = k\frac{df}{dz} = kf^1(z)$

3. $\dfrac{d}{dz}\left[f(z)g(z)\right]=f(z)\dfrac{d}{dz}\{g(z)\}+g(z)\dfrac{d}{dz}\{f(z)\}$

$$=f(z)g^{1}(z)+g(z)f^{1}(z)$$

4. $\dfrac{d}{dz}\left[\dfrac{f(z)}{g(z)}\right]=\dfrac{g(z)\dfrac{d}{dz}\{f(z)\}-f(z)\dfrac{d}{dz}\{g(z)\}}{\{g(z)\}^{2}},\qquad \{g(z)\neq 0\}$

$$=\frac{g(z)f^{1}(z)-f(z)g^{1}(z)}{\{g(z)\}^{2}}$$

5. Let $w=f(\eta)$, where $\eta=g(z)$, then,

$$\frac{dw}{dz}=\frac{dw}{d\eta}\frac{d\eta}{dz}=f^{1}(\eta)\frac{d\eta}{dz}=f^{1}(g(z)).g^{1}(z)$$

6. Let $w=f(\eta)$, where $\eta=g(\zeta)$, and $\zeta=\eta(z)$ then,

$$\frac{dw}{dz}=\frac{dw}{d\eta}\frac{d\eta}{d\zeta}\frac{d\zeta}{dz}$$ [(5) and (6) are called chain rules]

7. If $w=f(z)$, then $z=f^{-1}(w)$

$\dfrac{dw}{dz}\, and\, \dfrac{dz}{dw}$ are related by, $\dfrac{dw}{dz}=\dfrac{1}{\left(\dfrac{dz}{dw}\right)}$

8. Parametric Differentiation

If $w=f(t)$ and $z=g(t)$, then,

$$\frac{dw}{dz}=\frac{dw/dt}{dz/dt}=\frac{f^{1}(t)}{g^{1}(t)}$$

9. If $w=c$, where 'c' is a complex constant, then $\dfrac{dw}{dz}=0$

10. If n is a +ve integer, then $\dfrac{d}{dz}\left(z^{n}\right) = nz^{(n-1)}$

11. If n is a +ve integer and $n\neq 0$, then $\dfrac{d}{dz}\left(z^{-n}\right) = \dfrac{-n}{z^{n+1}}$

21.3.4 Analytic Functions

Definition:

1. *A function f (z) is said to be analytic at a point z = a if there exists a neighborhood* $|z-a| < d$ *at all points of which* $f'(z)$ *exists.*

2. *A function f (z) for which* $f'(z)$ *exists at all points in a region 'R' is said to be analytic in R. Sometimes the terms 'regular' and 'holomorphic' are used as synonyms for 'analytic'.*

   A function which is analytic for all $z \in C$ is called an **entire function.**

21.3.5 Cauchy–Riemann Equations [C–R Equations]

Theorem: Let $w = f(z) = u(x,y) + iv(x,y)$ be a complex function. The necessary and sufficient conditions for *f (z)* to be analytic in a region 'R' are,

1. ***Necessary:*** *u and v satisfy the equations,*
   $\partial u/\partial x = \partial v/\partial y$; $\partial u/\partial y = -\partial v/\partial x$, known as C–R equations; and

2. ***Sufficient:*** *The partial derivatives,* $\partial u/\partial x, \partial v/\partial y, \partial u/\partial y$ and $\partial v/\partial x$ are continuous in R.

Proof:

1. **Necessary:**

   Let $w = f(z) = u(x,y) + iv(x,y)$ be analytic in R. We shall prove that C- R equations are satisfied.

$$z = x + iy, \qquad \Delta z = \Delta x + i\Delta y$$

Since *f (z)* is analytic , $f^1(z)$ exists in R. Therefore,

$$\underset{\Delta z \to 0}{Lt}\left(\frac{f(z+\Delta z)-f(z)}{\Delta z}\right) = f^1(z)$$

$$\underset{\substack{\Delta x \to 0 \\ \Delta y \to 0}}{Lim}\left(\frac{\left(u(x+\Delta x, y+\Delta y)+iv(x+\Delta x, y+\Delta y)\right)-\left(u(x,y)+iv(x,y)\right)}{\Delta x + i\Delta y}\right) \qquad .....(1)$$

must exist in whatever manner $\Delta z \to 0$, i.e., Δx and $\Delta y \to 0$

Let us consider two paths along which $\Delta z \to 0$,

Case 1:

Let $\Delta y = 0$ and $\Delta x \to 0$ then, the limit in (1) will be

$$f^1(z) = \lim_{\Delta x \to 0}\left[\left(\frac{u(x+\Delta x, y) - u(x,y)}{\Delta x}\right) + i\left(\frac{v(x+\Delta x, y) - v(x,y)}{\Delta x}\right)\right]$$

$$= \frac{\partial u}{\partial x} + i\frac{\partial v}{\partial x} \quad \text{(provided the partial derivatives exist)} \qquad .....(2)$$

Case 2:

Let $\Delta x = 0$ and $\Delta y \to 0$, then the limit in (1) will be

$$f^1(z) = \lim_{\Delta y \to 0}\left[\left(\frac{u(x, y+\Delta y) - u(x,y)}{i\Delta y}\right) + i\left(\frac{v(x, y+\Delta y) - v(x,y)}{i\Delta y}\right)\right]$$

$$= \frac{1}{i}\frac{\partial u}{\partial y} + \frac{\partial v}{\partial y} = -i\frac{\partial u}{\partial y} + \frac{\partial v}{\partial y} \qquad .....(3)$$

Since $f\ (z)$ is analytic, the values of $f^{'}(z)$ obtained in the above two cases [i.e., (2) and (3)], must be equal.

$$\frac{\partial u}{\partial x} + i\frac{\partial v}{\partial x} = -i\frac{\partial u}{\partial y} + \frac{\partial v}{\partial y}$$

Equating real and imaginary parts we get,

$$\frac{\partial u}{\partial x} = \frac{\partial v}{\partial y} \quad \text{and} \quad \frac{\partial v}{\partial x} = -\frac{\partial u}{\partial y} \quad \text{which are C – R equations..}$$

Sufficiency: Let $f(z) = u(x,y) + iv(x,y)$ satisfy the conditions (1) and (2) given in the theorem. We shall prove that $f\ (z)$ is analytic, i.e., $f^{|}(z)$ exists.

Since, $\frac{\partial u}{\partial x}, \frac{\partial u}{\partial y}$ are assumed to be continuous, we have,

$$\Delta u = u(x+\Delta x, y+\Delta y) - u(x,y)$$

$$= \{u(x+\Delta x, y+\Delta y) - u(x, y+\Delta y)\} + \{u(x, y+\Delta y) - u(x,y)\}$$

$$\left(\frac{\partial u}{\partial x} + \epsilon_1\right)\Delta x + \left(\frac{\partial u}{\partial y} + \epsilon_2\right)\Delta y = \frac{\partial u}{\partial x}\Delta x + \frac{\partial u}{\partial y}\Delta y + \epsilon_1\Delta x + \epsilon_2\Delta y.$$

where $\epsilon_1, \epsilon_2 \to 0$ as $\Delta x, \Delta y \to 0$.

Similarly supposing that $\frac{\partial v}{\partial y}, \frac{\partial v}{\partial x}$ be continuous, we have,

$$\Delta v = \left(\frac{\partial v}{\partial x} + \eta_1\right)\Delta x + \left(\frac{\partial v}{\partial y} + \eta_2\right)\Delta y = \frac{\partial v}{\partial x}\Delta x + \frac{\partial v}{\partial y}\Delta y + \eta_1 \Delta x + \eta_2 \Delta y$$

where $\eta_1, \eta_2 \to 0$ as $\Delta x, \Delta y \to 0$

$$\Delta w = \Delta u + i\Delta v = \left(\frac{\partial u}{\partial x} + i\frac{\partial v}{\partial x}\right)\Delta x + \left(\frac{\partial u}{\partial y} + i\frac{\partial v}{\partial y}\right)\Delta y + \in \Delta x + \eta \Delta y \quad .....(4)$$

where $\in = \in_1 + i\eta_1 \to 0$ and $\eta = \in_2 + i\eta_2 \to 0$ as $\Delta x, \Delta y \to 0$

From C – R equations, (4) can be written as ,

$$\Delta w = \left(\frac{\partial u}{\partial x} + i\frac{\partial v}{\partial x}\right)\Delta x + \left(\frac{-\partial v}{\partial x} + i\frac{\partial u}{\partial x}\right)\Delta y + \in \Delta x + \eta \Delta y$$

$$= \left(\frac{\partial u}{\partial x} + i\frac{\partial v}{\partial x}\right)(\Delta x + i\Delta y) + \in \Delta x + \eta \Delta y$$

$$\therefore \quad \frac{\Delta w}{\Delta z} = \left(\frac{\partial u}{\partial x} + i\frac{\partial v}{\partial x}\right) + \frac{\in \Delta x + \eta \Delta y}{\Delta x + i\Delta y}$$

Taking limits as $\Delta z \to 0$, we get,

$$f^1(z) = \frac{dw}{dz} = \underset{\Delta z \to 0}{Lim}\left(\frac{\Delta w}{\Delta z}\right) = \frac{\partial u}{\partial x} + i\frac{\partial v}{\partial x}$$

$\therefore$ $f^{|}(z)$ exists and is unique. Therefore, $f^{|}(z)$ is analytic.

Hence the theorem

We follow the following notation for partial derivatives.

$$\frac{\partial u}{\partial x} = u_x; \quad \frac{\partial u}{\partial y} = u_y; \quad \frac{\partial^2 u}{\partial x^2} = u_{xx};$$

$$\frac{\partial^2 u}{\partial y^2} = u_{yy}; \quad \frac{\partial^2 u}{\partial x \partial y} = u_{xy}; \quad \frac{\partial^2 u}{\partial y \partial x} = u_{yx}; \text{ (similarly for v)}$$

Note:

1. *In the analytic function $f(z) = u(x, y) + iv(x,y)$, u , v are called conjugate functions.*

2. *The values of* $f^1(z)$ *obtained in equations (2) and (3) of theorem 21.3.5 can be remembered as follows.*

$$f^1(z) = \frac{\partial u}{\partial x} + i\frac{\partial v}{\partial x} = u_x + iv_x$$

$$= \frac{\partial v}{\partial y} + i\frac{\partial v}{\partial x} = v_y + iv_x = v_y - iu_y$$

$$= \frac{\partial u}{\partial x} - i\frac{\partial u}{\partial y} = u_x - iu_y \text{ (from } C-R \text{ equations)}$$

3. *C – R equations (or) C – R conditions (in Cartesian form) are* $u_x = v_y$ and $u_y = -v_x$

21.3.6 Cauchy–Riemann Equations in Polar Coordinates

The coordinates of a point (x, y) can be expressed in polar coordinates as

$$x = r\cos\theta, \qquad \text{and} \qquad y = r\sin\theta,$$

$$z = x + iy = r(\cos\theta + i\sin\theta) = re^{i\theta}$$

$$\left[\text{since } e^{i\theta} = \cos\theta + i\sin\theta\right]$$

Let $w = f(z) = u(r,\theta) + iv(r,\theta)$ be an analytic function of z.

Therefore, we have, $f\left(re^{i\theta}\right) = u(r,\theta) + iv(r,\theta)$ .....(1)

Since 'f' is analytic, differentiating (1) w.r.t. 'r', we get,

$$\left[f'\left(re^{i\theta}\right)\right]\left(e^{i\theta}\right) = \partial u/\partial r + i\,\partial v/\partial r \quad .....(2)$$

Similarly, differentiating (1) w.r.t. 'θ' we have,

$$\left[f'\left(re^{i\theta}\right)\right]\left(rie^{i\theta}\right) = \partial u/\partial\theta + i\,\partial v/\partial\theta \quad .....(3)$$

(2) and (3) will give,

$$ir\left\{\frac{\partial u}{\partial r} + i\frac{\partial v}{\partial r}\right\} = \frac{\partial u}{\partial\theta} + i\frac{\partial v}{\partial\theta} \quad \text{or} \quad iru_r - rv_r = u_\theta + iv_\theta$$

Equating real and imaginary parts, we get,

$$u_r = \frac{1}{r}v_\theta \qquad \text{and} \qquad u_\theta = -rv_r$$

or

$$\partial u/\partial r = (1/r)\;\partial v/\partial\theta \qquad \text{and} \qquad \partial u/\partial\theta = (-r)\;\partial v/\partial r$$

These equations are known as C–R equations in polar coordinates.

Note: Further, $r(u_r + iv_r) = r\left[(u_x x_r + u_y y_r) + i(v_x x_r + v_y y_r)\right]$

$$= r\left[(u_x \cos\theta + u_y \sin\theta) + i(v_x \cos\theta + v_y \sin\theta)\right]$$

$$\left[\text{since, } x = r\cos\theta, x_r = \cos\theta, y = r\sin\theta, y_r = \sin\theta\right]$$

$$= (r\cos\theta)(u_x + iv_x) + i(r\sin\theta)(v_y - iu_y)$$

$$= xf^1(z) + iyf^1(z) \quad \left[\text{since } f^1(z) = u_x + iv_x = v_y - iu_y\right]$$

$$= (x + iy) f^1(z) = zf^1(z)$$

Therefore, the formula for $f^{|}(z)$ in polar coordinates can be taken as,

$$f^1(z) = \frac{r}{z}(u_r + iv_r) \qquad \text{.....(4)}$$

21.3.7 Complex form of C–R Equations

Let $f(z) = u(x, y) + iv(x, y)$ be analytic in R.

Then the C–R equations can be put in the complex form, as $f_x = -if_y$

Proof: $f(z) = u(x, y) + iv(x, y)$, then, $f_x = u_x + iv_x$; $f_y = u_y + iv_y$

Now $f_x = -if_y \Leftrightarrow u_x + iv_x = -i(u_y + iv_y)$

$$\Leftrightarrow u_x + iv_x = -iu_y + v_y \Leftrightarrow u_x = v_y \text{ and } u_y = -v_x,$$

which are C – R equations. Thus the two are C – R equations equivalent to the equation $f_x = -if_y$

21.3.8 Harmonic Functions

1. $\nabla^2 = \frac{\partial^2}{\partial x^2} + \frac{\partial^2}{\partial y^2}$, is the operator called the Laplacian.

2. Let $f(x, y)$ possess continuous second order partial derivatives and satisfy the Laplace's Equation,

$$\nabla^2 f = \left(\frac{\partial^2}{\partial x^2} + \frac{\partial^2}{\partial y^2}\right) f = 0 \quad \text{(or)} \quad f_{xx} + f_{yy} = 0$$

Then $f(x, y)$ is said to be a ***Harmonic Function.***

21.3.9 Harmonic Conjugates

Prove that the Real and Imaginary Parts of an Analytic Function are Harmonic

Proof :

Let $f(z) = u(x,y) + iv(x,y)$ be an analytic function .....(1)

Then $u_x = v_y$

and $u_y = -v_x$ .....(2)

((2) and (3) are C–R equations.)

From (2) and (3) we get , $u_{xx} = v_{xy}$ [differentiating (2) partially w.r.t .x]

And $u_{yy} = -v_{yx}$ [differentiating (3) partially w.r.t .y]

so that , $u_{xx} + u_{yy} = v_{xy} - v_{yx} = 0,$ which shows that u is harmonic.

Similarly, $v_{xx} = -u_{xy}$ and $v_{yy} = u_{yx}$

So that $v_{xx} + v_{yy} = 0$ $\Rightarrow$ v is harmonic

Note:

1. *If f(z) = u + iv is analytic, then v is said to be a harmonic conjugate of u.*
2. *If u (or v) is harmonic, it can be the real (or imaginary) part of an analytic function.*

21.3.10 Singular Point (Singularity)

If $f(z)$ ceases to be analytic at a point z = a, then the point z = a is called the singular point or singularity of $f(z)$

Ex : For $f(z) = \frac{1}{z}, z = 0$ is a singular point. Since $f(z)$ is not defined at z = '0' it is not analytic at z = '0'

21.3.11 Maximum – Modulus Theorem

Let $f(z)$

(i) be analytic in a bounded region D and on its boundary D_1

(ii) be a non constant in D and also on D_1. Then $|f(z)|$ attains its maximum only on D_1 and not at any point of D. Further if $f(z) \neq 0$ then the minimum of $|f(z)|$ is also attained on D_1.

Proof: Let us prove the theorem by method of contradiction

Assume that $|f(z)|$ attains a maximum M at z_0 of D

Since f(z) is not a constant , so is $|f(z)|$.

Now construct a circle $C: |z - z_0| = r$ such that the interior of C is in D.

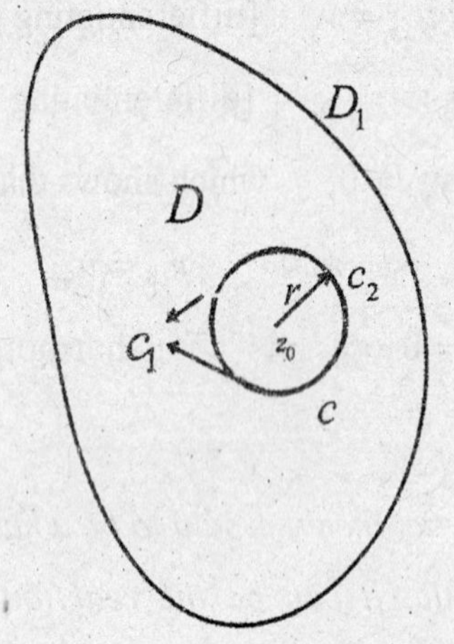

Let' P' be a point on C and C_1 be an arc containing 'P'.

Since $|f(z_0)|$ is maximum, we have for any z on C_1 .

$$|f(z)| \leq M - \in (\in > 0)$$

Calling the complimentary of C_1 of C as C_2 and taking the length of arc C_1 as l_1, the arc length of C_2 is $2\pi r - l_1$

Here we use the inequality called ***Ml* inequality** which states that if l is the length of C and $|f(z)| \leq M$ every where on C, then

$$\left|\int_c f(z)dz\right| \leq Ml \qquad .....(1)$$

Now $M = |f(z_0)| \le \frac{1}{2\pi}\left|\int_{c_1} \frac{f(z)}{z-z_0}dz\right| + \frac{1}{2\pi}\left|\int_{c_2} \frac{f(z)}{z-z_0}dz\right|$ .....(2)

Using (1) we have from (2) that

$$\le \frac{1}{2\pi}(M-\in)\frac{l_1}{r} + \frac{1}{2\pi}M.\frac{2\pi r - l_1}{r}$$

$\le M - \frac{\in l_1}{2\pi r} < M$ Which is a contradiction. Hence our assumption that $|f(z)|$ has a maximum at a point of *D* is false.

i.e., $|f(z)|$ Attains maximum only on D_1 (the boundary of *D*) .....(3)

if f(z) $\neq 0$ in *D*, then $\frac{1}{f(z)}$ is analytic in *D*.

Hence from (3) it follows that $\left|\frac{1}{f(z)}\right|$ attains a maximum on D_1

i.e., $|f(z)|$ attains a minimum on D_1.

Solved Examples

***Example* 21.14:** If $f(z) = z^3 + 2z$, find $f^1(z)$

Solution: $f^1(z) = \underset{h\to 0}{Lt}\left(\frac{f(z+h)-f(z)}{h}\right) = \underset{h\to 0}{Lt}\left(\frac{(z+h)^3 + 2(z+h) - z^3 - 2z}{h}\right)$

$$= \underset{h\to 0}{Lt}\left(\frac{3z^2h + 3zh^2 + h^3 + 2h}{h}\right)$$

$$= \underset{h\to 0}{Lt}\left(3z^2 + 3zh + h^2 + 2\right) = 3z^2 + 2$$

***Example* 21.15:** "Continuity does not Imply Differentiability" . Give an example to verify the truth of this statement.

***Solution*:** Consider the function, $f(z) = \bar{z}$; $f(z)$ is continuous in C [See Ex 21.10].

$$f^1(z) = \underset{h\to 0}{Lt}\left(\frac{f(z+h)-f(z)}{h}\right) = \underset{h\to 0}{Lt}\left(\frac{\overline{z+h}-\bar{z}}{h}\right) = \underset{h\to 0}{Lt}\left(\frac{\bar{h}}{h}\right)$$

Since this limit does not exist, $f^1(z)$ does not exist [see problem II of Exercise – 3(a)].

$\therefore$ $f(z)$ is not differentiable .

or $f(z) = \bar{z} = x - iy = u + iv$; $u = x, v = -y$

$u_x = 1$; $u_y = 0$; $v_x = 0$; $v_y = -1$

Since $u_x \neq v_y$, C – R Equations are not satisfied. Hence $f(z)$ is not differentiable.

***Example* 21.16:** Prove that an analytic function with a constant real part or constant imaginary part is constant.

***Solution*:** Let $f(z) = u + iv$ be an analytic function.

$\therefore$ By C –R equations, we have, $u_x = v_y$ and $u_y = -v_x$ ......(1)

***Case* 1:**

Let $u = c_1$, where c_1 is a constant. $\therefore$ $u_x = 0$, and $u_y = 0$

From (1), we get,

$v_y = 0$ and $v_x = 0$ $\Rightarrow$ v is a constant say c_2

$\therefore$ $f = c_1 + ic_2$ = constant

***Case* 2:**

Let $v = a_1$ where a_1 is a constant, $\therefore$ $v_x = 0$, $v_y = 0$,

Hence (1) $\Rightarrow u_x = 0$, $u_y = 0 \Rightarrow u$ is a constant.

Let $u = a_2$; then $f = a_2 + ia_1$ = constant.

***Example* 21.17:** Let $f(z)$ be an analytic function. Show that $f(z)$ is constant, in each of the following cases.

(i) $f^1(z) = 0$, **(ii)** $|f^1(z)|$ is constant and **(iii)** [Arg $f(z)$] is constant.

***Solution*:** Let $f(z) = u + iv$, be analytic then,

***Case* 1:**

$f^1(z) = 0 \Rightarrow$ $u_x + iv_x = v_y - iu_y = 0$ [using formulae for $f^1(z)$]

$\Rightarrow$ $u_x = u_y = v_x = v_y = 0$

$\Rightarrow$ u and v are constants. $\Rightarrow$ $f(z)$ is constant.

***Case* 2 :** $|f(z)| = \sqrt{u^2 + v^2}$ = constant (given)

$\therefore$ Let $u^2 + v^2 = c$ ......(1)

differentiating (1) partially w.r.t .x and w.r.t.y,we get,

$$uu_x + vv_x = 0 \quad ......(2)$$

and $$uu_y + vv_y = 0 \quad ......(3)$$

But we know that $u_x = v_y$ and $u_y = -v_x$ [C – R equations] ......(4)

Using (4) in (2) and (3), we get,

$$uu_x - vu_y = 0 \quad ......(5)$$

and $$uu_y + vu_x = 0 \quad ......(6)$$

$$(5)\times u + (6)\times v \Rightarrow (u^2 + v^2)u_x = 0 \quad \Rightarrow u \text{ is constant.}$$

Similarly, we can show that v is constant. Hence *f (z)* is constant.

***Case* 3** $[\text{Arg} f(z)] = \tan^{-1}(v/u) = k$ (say)

$\therefore$ $(v/u) = \tan k = c$ (say) (k and c are constants) $\therefore\ v = cu$

$\Rightarrow$ $v_x = cu_x$ and $v_y = cu_y$

$\Rightarrow$ $v_x/u_x = c = v_y/u_y \quad \Rightarrow \quad u_x v_y = u_y v_x$

$\Rightarrow$ $v_y^2 + v_x^2 = 0$ [using C–R Equations]

$\Rightarrow$ $v_x = 0, v_y = 0 \quad \Rightarrow v$ is constant.

Similarly, it can be shown that u is constant. Hence *f (z)* is constant.

***Example* 21.18**: If f (z) and $\overline{f(z)}$ are analytic, show that f (z) is constant.

Solution: Let *f (z)* = $u + iv$ then $\overline{f(z)} = u - iv = u + i(-v)$

Applying C – R equations for both *f (z)* and $\overline{f(z)}$, (since they are analytic), we get,

$$u_x = v_y,\ u_y = -v_x, \quad \text{and } u_x = -v_y,\ u_y = v_x$$

$\Rightarrow$ $2u_x = 0 = 2u_y \Rightarrow u_x = u_y = 0$ [adding the above equations]

Similarly, we get, $v_x = v_y = 0$

$$f^1(z) = u_x + iv_x = 0 \Rightarrow f(z) \text{ is constant.}$$

***Example* 21.19:** Examine whether the following functions are analytic or not.

(a) $f(z) = Re(z)$ **(b)** $f(z) = 1/(z-2)$

***Solution*:**

(a) $f(z) = x$ i.e., $u + iv = x$; $u = x$, $v = 0$

$\Rightarrow u_x = 1,\ u_y = 0,\quad v_x = 0,\quad v_y = 0 \Rightarrow u_x \neq v_y$

Therefore, C – R equations are not satisfied. Hence, $f(z)$ is not analytic.

(b) $f(z) = 1/(z-2)$ is not analytic at z = 2 [since $f(z)$ is not defined at z = 2]

$f^1(z) = \dfrac{-1}{(z-2)^2}$, exists if $z \neq 2$. Therefore, $f(z)$ is analytic at all $z \neq 2$

Example 21.20: Verify C–R equations for $f(z) = z^3$

***Solution*:**

$$f(z) = z^3 = (x+iy)^3 = (x^3 - 3x^2y) + i(3x^2y - y^3)$$

$$f(z) = u + iv \quad \Rightarrow \quad u = x^3 - 3xy^2,\quad v = 3x^2y - y^3$$

$$u_x = 3x^2 - 3y^2 = v_y;\quad u_y = -6xy = -v_x$$

Hence C –R equations are verified.

Example 21.21: If f(z) = u(x, y)+ iv (x, y) is analytic show that u (x, y) = a, v(x, y) = b , constitute an orthogonal system of curves. (a, b being constants)

Solution: Let $a = a_1$ and $b = b_1$ give any two particular members of the system. Let 'm_1' and 'm_2' denote the slopes of the tangents to these curves respectively at their points of intersection.

If $m_1m_2 = -1$, the tangents will be at right angles showing that the curves are orthogonal.

From calculus, we have if $f(x,y) = c$, then $dy/dx = -f_x/f_y$

Therefore, $u(x,y) = a_1 \Rightarrow m_1 = -u_x/u_y$

Similarly, $m_2 = -v_x/v_y$ $\therefore$ $m_1m_2 = (-u_x/u_y)(-v_x/v_y) = -1$

[since from C – R equations $u_x = v_y, u_y = -v_x$]

Hence u = a, v = b are orthogonal.

***Example* 21.22:** Find the constant 'a' for which the function $f(x, y) = \left(ax^2 + 4xy - 3y^2\right)$ is harmonic.

***Solution*:** $f(x, y) = \left(ax^2 + 4xy - 3y^2\right)$, $f_x = 2ax + 4y$, $f_{xx} = 2a$

$f_y = 4x - 6y$; $f_{yy} = -6$; f is harmonic.

$\therefore \quad f_{xx} + f_{yy} = 0 \Rightarrow 2a - 6 = 0 \Rightarrow a = 3$

***Example* 21.23:** Let $f = u + iv$ be an analytic function in a region R. Then v is a harmonic conjugate of u if u is a harmonic conjugate of $-v$.

Solution: Let v be a harmonic conjugate of u. Then, $f = u + iv$ is analytic. Therefore, $if = iu - v$ is also analytic. Hence u is a harmonic conjugate of $-v$.

Conversely, let u be a harmonic conjugate of $-v$. Hence $g = -v + iu$ is also analytic.

Therefore, $-ig = u + iv$ is analytic. Hence, v is the harmonic conjugate of u.

***Example* 21.24:** Prove that any two harmonic conjugates of a given harmonic function in a region differ by a constant.

***Solution*:** Let u be harmonic in a region R. Let v and v^1 be two harmonic conjugates of u.

Then, $(u + iv)$ and $(u + iv^1)$ are analytic in R.

Hence by C – R equations, we have,

$$u_x = v_y = v_y^1 \quad \text{and} \quad u_y = -v_x = -v_x^1$$

i.e., $$v_y = v^1{}_y \quad \text{and} \quad v_x = v^1{}_x$$

i.e., $$\frac{\partial}{\partial x}\left(v - v^1\right) = 0 \quad \text{and} \quad \frac{\partial}{\partial y}\left(v - v^1\right) = 0,$$

Therefore, $v - v^1 = c$ where 'c' is a constant.

***Example* 21.25:** If $f(z) = u(r, \theta) + iv(r, \theta)$ is analytic, show that,

$$u_{rr} + \frac{1}{r}u_r + \frac{1}{r^2}u_{\theta\theta} = 0$$

***Solution*:** Since f is analytic, we have, from C–R equations,

$$ru_r = v_\theta \quad \text{......(1)}$$

and $$rv_r = -u_\theta \quad \text{......(2)}$$

Differentiating (1) partially w.r.t. 'r', and (2) partially w.r.t. θ, we get,

$$v_{r\theta} = ru_{rr} + u_r \quad \text{......(3)}$$

and $\quad v_{\theta r} = -\frac{1}{r} u_{\theta\theta}$ ......(4)

But $\quad v_{r\theta} = v_{\theta r}$

$\therefore \quad$ (3) and (4) $\Rightarrow \quad r u_{rr} + u_r + \frac{1}{r} u_{\theta\theta} = 0$

$$\Rightarrow \quad u_{rr} + \frac{1}{r} u_r + \frac{1}{r^2} u_{\theta\theta} = 0$$

[This equation is known as Laplace equation in polar coordinates]

Example 21.26: Let *f(x, y)* and *g (x, y)* be two harmonic functions and let $u = f_y - g_x$ and $v = f_x + g_y$. Prove that $(u + iv)$ is an analytic.

Solution: $\quad f$ and g are harmonic.

$\therefore \quad f_{xx} + f_{yy} = 0$ ......(1)

and $\quad g_{xx} + g_{yy} = 0$ ......(2)

Again, $\quad u = f_y - g_x \quad$ and $\quad v = f_x + g_y$ (given)

$\Rightarrow \quad u_x = f_{xy} - g_{xx}; \quad u_y = f_{yy} - g_{yx}$

$v_x = f_{xx} + g_{xy} = -f_{yy} + g_{xy}$ from (1)

and $\quad v_y = f_{yx} + g_{yy} = f_{yx} - g_{xx}$ from (2)

Therefore, $u_x = v_y$ and $u_y = -v_x$ i.e., *u, v* satisfy C–R equations.

Further, since *f* and *g* are harmonic, u_x, u_y, v_x and v_y are continuous. Hence $(u + iv)$ is analytic.

***Example* 21.27:** If $(u + iv)$ is an analytic function, show that uv is harmonic.

Solution: Since $(u + iv)$ is analytic, we have,

$u_{xx} + u_{yy} = 0$ .....(1)

$v_{xx} + v_{yy} = 0$ [since u, v are harmonic] ....(2)

Also by C – R equations

$u_x = v_y$ .....(3)

and $\quad u_y = -v_x$ .....(4)

Let $\quad f = uv$ then, $f_x = uv_x + vu_x$; $f_{xx} = uv_{xx} + 2u_x v_x + vu_{xx}$

$$f_y = uv_y + vu_y\,;\; f_{yy} = uv_{yy} + 2u_y v_y + vu_{yy}$$

$$= uv_{yy} - 2u_x v_x + vu_{yy} \text{ from (3) and (4)}$$

$\therefore$ $$f_{xx} + f_{yy} = u(v_{xx} + v_{yy}) + v(u_{xx} + u_{yy}) = 0,\ \text{from (1) \& (2)}$$

$\Rightarrow$ $f = uv$ is harmonic.

***Example* 21.28:** Let $f(z) = (u + iv)$ be a analytic function. Let $g(x, y)$ be any function of x and y possessing partial derivatives of 1st and 2nd order; then prove that, $(g_x)^2 + (g_y)^2 = \{(g_u)^2 + (g_v)^2\}|f^1(z)|^2$

Solution: We have, $\quad g_x = g_u u_x + g_v v_x$ .....(1)

and $\quad g_y = g_u u_y + g_v v_y = -g_u v_x + g_v u_x$ [from C–R equations] .....(2)

Squaring and adding (1) and (2), we get,

$$g_x^2 + g_y^2 = g_u^2\left(u_x^2 + v_x^2\right) + g_v^2\left(u_x^2 + v_x^2\right) = \left(g_u^2 + g_v^2\right)\left(u_x^2 + v_x^2\right)$$

$$= \{(g_u)^2 + (g_v)^2\}|f^1(z)|^2 \qquad \left(\text{since } f^1(z) = u_x + iv_x\right)$$

Example 21.29: If $f(z) = u + iv$ is analytic, show that, $\left(\dfrac{\partial^2}{\partial x^2} + \dfrac{\partial^2}{\partial y^2}\right)|f(z)|^2 = 4|f^1(z)|^2$

Solution: From $f(z) = u + iv$ we have, $|f(z)|^2 = u^2 + v^2 = \phi(x, y)$ (say)

$\therefore$ $$\phi_x = 2(uu_x + vv_x),\ \phi_{xx} = 2\left[uu_{xx} + u_x^2 + vv_{xx} + v_x^2\right] \quad .....(1)$$

$$\phi_y = 2\left(uu_y + vv_y\right),\ \phi_{yy} = 2\left[uu_{yy} + u_y^2 + vv_{yy} + v_y^2\right] \quad .....(2)$$

Adding (1) and (2), we get

$$\phi_{xx} + \phi_{yy} = 2\left[u\left(u_{xx} + u_{yy}\right) + v\left(v_{xx} + v_{yy}\right) + u_x^2 + v_x^2 + u_y^2 + v_y^2\right] \quad .....(3)$$

But since $\quad f(z) = u + iv$ is analytic,

we have, $\left(u_{xx} + u_{yy}\right) = 0$ and $\left(v_{xx} + v_{yy}\right) = 0$ [u, v are harmonic]

Also, $u_x = v_y$ and $u_y = -v_x$ [C–R equations]

Using these equations, (3) becomes,

$$\phi_{xx} + \phi_{yy} = 4\left(u_x^2 + v_x^2\right)$$

i.e., $$\left(\frac{\partial^2}{\partial x^2} + \frac{\partial^2}{\partial y^2}\right)|f(z)|^2 = 4|f^1(z)|^2 \quad \left(\text{since } f^1(z) = u_x + iv_x\right)$$

***Example* 21.30:** Let $f(z) = r^{1/2}e^{i\theta/2}$ where $r > 0$ and $0 < \theta < 2\pi$. Show that f is differentiable and find $f^1(z)$.

***Solution*:** $$f(z)=\sqrt{r}\left\{\cos\left(\frac{\theta}{2}\right)+i\sin\left(\frac{\theta}{2}\right)\right\}=u+iv \text{ (say)}$$

then $$u=\sqrt{r}\cos\left(\frac{\theta}{2}\right); \qquad v=\sqrt{r}\sin\left(\frac{\theta}{2}\right)$$

$$u_r=\frac{1}{2\sqrt{r}}\cos\left(\frac{\theta}{2}\right); \qquad v_r=\frac{1}{2\sqrt{r}}\sin\left(\frac{\theta}{2}\right);$$

$$u_\theta=\frac{-\sqrt{r}}{2}\sin\left(\frac{\theta}{2}\right) \qquad v_\theta=\frac{\sqrt{r}}{2}\cos\left(\frac{\theta}{2}\right)$$

$\therefore$ we get, $u_r=\frac{1}{r}v_\theta$ and $v_r=-\frac{1}{r}u_\theta$ which shows that u and v satisfy C–R equations.

Further all the first order partial derivatives $(u_r, u_\theta, v_r \text{ and } v_\theta)$ are continuous. Hence $f^1(z)$ exists. i.e., $f(z)$ is differentiable.

From (4) of 3.3.6, $$f^1(z)=\frac{r}{z}(u_r+iv_r)$$

$$=\frac{r}{z}\frac{1}{2\sqrt{r}}\left[\cos\left(\frac{\theta}{2}\right)+i\sin\left(\frac{\theta}{2}\right)\right]$$

$$=\frac{\sqrt{r}}{2z}e^{i\theta/2}=\frac{\sqrt{re^{i\theta}}}{2z}=\frac{\sqrt{z}}{2z}=\frac{1}{2\sqrt{z}} \quad \left(\text{since } z=re^{i\theta}\right)$$

***Example* 21.31:** Show that the following functions satisfy C–R equations at the origin, yet they are not differentiable at that point.

(i) $$f(z)=\begin{cases}\frac{xy}{x^2+y^2}, & z\neq 0\\ 0, & z=0\end{cases}$$ (ii) $f(z)=\sqrt{xy}$

***Solution*:**

(i) Here $u=\frac{xy}{x+y^2}$, $v=0$ (taking $f(z) = u + iv$)

$$u_x(0,0) = \underset{h\to 0}{Lt}\left[\frac{u(h,0)-u(0,0)}{h}\right] = \underset{h\to 0}{Lt}\left[\frac{0-0}{h}\right] = 0$$

Similarly, $u_y(0,0) = 0$, $v_x(0,0) = 0$, and $v_y(0,0) = 0$

Hence C–R equations are satisfied at (0 , 0) .

Again, $f(z) = \dfrac{xy}{x^2+y^2}$

Let z → 0 along the path $y = mx$

Then $\underset{z\to 0}{Lt} f(z) = \underset{x\to 0}{Lt}\left(\dfrac{x.mx}{x^2+m^2x^2}\right) = \dfrac{m}{1+m^2}$ $(x \neq 0)$,

which is different for values of *m*. Hence *f (z)* has no limit as z → 0.

$\therefore$ *f (z)* is not continuous at z = 0 . $\therefore$ $f^1(z)$ does not exist at origin.

(ii) Left as an exercise. Proceed exactly as in (i)

***Example* 21.32:** Let *f (z)* = *u* + *iv* , where,

$$f(z) = \begin{cases} \dfrac{x^3(1+i)-y^3(1-i)}{x^2+y^2}, & z \neq 0 \\ 0, & z = 0 \end{cases}$$

then show that, (1) *f (z)*is continuous at all z . (2) *f (z)* satisfies C–R equations at z = 0, and

(3) $f^1(0)$ does not exist.

Solution: **1.** If *f (z)* = *u* + *iv* , then $u = \dfrac{x^3-y^3}{x^2+y^2}$, $v = \dfrac{x^3+y^3}{x^2+y^2}$

When $z \neq 0$, *u* and *v* are rational functions with non- vanishing denominators. Therefore, *f (z)* is continuous when $z \neq 0$.

Continuity at z = 0: Put $x = r\cos\theta$, $y = r\sin\theta$

Then $u = r\left(\cos^3\theta - \sin^3\theta\right)$; $v = r\left(\cos^3\theta + \sin^3\theta\right)$. Both *u* and *v* →0 as r → 0, whatever value 'θ' may take.

Again $f(0) = 0 \Rightarrow u = v = 0$ at z = 0.

Thus the actual and the limiting values of *u* and *v* are equal at the origin.

Therefore, *u, v* are continuous at origin and so is *f (z)*.

Hence $f(z)$ is continuous for all z.

(2) At the origin,

$$u_x = \underset{x\to 0}{Lt}\left[\frac{u(x,0)-u(0,0)}{x}\right] = \underset{x\to 0}{Lt}\left[\frac{x-0}{x}\right] = 1$$

Similarly, $u_y = \underset{y\to 0}{Lt}\left[\frac{0-y}{y}\right] = -1,\ v_x = 1, v_y = 1$

Hence, $u_x = v_y$, $u_y = -v_x$. Therefore, C–R equations are satisfied at origin.

(3) $$f^{1}(0) = \underset{z\to 0}{Lt}\left[\frac{f(z)-f(0)}{z}\right] = \underset{z\to 0}{Lt}\left[\frac{(x^3-y^3)+i(x^3+y^3)}{x^2+y^2}\right]\left(\frac{1}{x+iy}\right)$$

Let $z \to 0$ along the line $y = x$ then,

$$f^1(0) = \underset{x\to 0}{Lt}\left(\frac{2i}{2(1+i)}\right) = \frac{i(1-i)}{(1+i)(1-i)} = \frac{1+i}{2}$$

Again, let $z \to 0$ along x – axis ($y = 0$) then,

$$f^1(0) = \underset{x\to 0}{Lt}\left(\frac{x^3(1+i)}{x^3}\right) = 1+i$$

Since both values of $f^1(0)$ are unequal, $f^1(0)$ does not exist.

***Example* 21.33:** If 'm' is a positive integer, show that $f(z) = z^m$ is an entire function. Also find $f^1(z)$

***Solution*:** Taking $z = rcis\theta$, (i.e., converting to polar coordinates)

$f(z) = (rcis\theta)^m = r^m cis(m\theta)$, by Demoivre's theorem.

Then $u = r^m \cos m\theta$, $v = r^m \sin m\theta$, $u_r = mr^{m-1}\cos m\theta$;

$v_r = mr^{m-1}\sin m\theta$, $u_\theta = -mr^m \sin m\theta$, $v_\theta = mr^m \cos m\theta$

Therefore, $u_r = \frac{1}{r}v_\theta$ and $v_r = \frac{-1}{r}u_\theta$

Hence C–R equations are satisfied. Further u_r, u_θ, v_r and v_θ are continuous.

Hence, $f(z) = u + iv$ is analytic for all $z \in C$ i .e ., $f(z)$ is an entire function.

Again, $f^1(z)=\frac{r}{z}\left(u_r+iv_r\right)=\frac{r}{z}\left[mr^{m-1}\left(\cos m\theta+i\sin m\theta\right)\right]$

$$=\frac{r}{re^{i\theta}}mr^{m-1}cism\theta=mr^{m-1}cis\{(m-1)\theta\}=mz^{m-1}$$

***Example* 21.34:** If $f(z) = l\left(x^2-y^2\right)+imxy$, ($l$, m being constants) is analytic for all $z\in C$, show that $\frac{m+l}{m-l}=3$

Solution: Let $f(z)=l\left(x^2-y^2\right)+imxy=u+iv$

$\therefore$ $u=l\left(x^2-y^2\right)$; $v=mxy$.

Since $f(z)$ is analytic, we have, by C–R equations,

$$u_x=v_y \quad \text{and} \quad u_y=-v_x \qquad .....(1)$$

But $u_x=2lx$; $u_y=-2ly$ and $v_x=my$; $v_y=mx$

Clearly, C – R equations (1) are satisfied if $m = 2l$.

$$\therefore \quad \frac{m+l}{m-l}=\frac{2l+l}{2l-l}=3$$

***Example* 21.35:** Show that $u=x^2-y^2$ and $v=\frac{-y}{x^2+y^2}$ are harmonic, but (u + iv) is not analytic.

Solution: $u=x^2-y^2 \Rightarrow u_x=2x \quad u_y=-2y$

$u_{xx}=2$; $u_{yy}=-2$; $u_{xx}+u_{yy}=0 \Rightarrow u$ is harmonic

$$v=\frac{-y}{x^2+y^2}\Rightarrow v_x=\frac{\left(x^2+y^2\right)0-(-y)2x}{\left(x^2+y^2\right)^2}=\frac{2xy}{\left(x^2+y^2\right)^2}$$

$$v_{xx}=\frac{\left(x^2+y^2\right)^2 2y-2xy.2\left(x^2+y^2\right)2x}{\left(x^2+y^2\right)^4}=\frac{2x^2y+2y^3-8x^2y}{\left(x^2+y^2\right)^3}=\frac{2y^3-6x^2y}{\left(x^2+y^2\right)^3}$$

$$v_y=\frac{\left(x^2+y^2\right)(-1)+y.2y}{\left(x^2+y^2\right)^2}=\frac{-x^2+y^2}{\left(x^2+y^2\right)^2}$$

$$v_{yy} = \frac{(x^2+y^2)^2 2y - (-x^2+y^2).2(x^2+y^2).2y}{(x^2+y^2)^4}$$

$$= \frac{2x^2y + 2y^3 + 4x^2y - 4y^3}{(x^2+y^2)^3} = \frac{-2y^3 + 6x^2y}{(x^2+y^2)^3}$$

Therefore, $v_{xx} + v_{yy} = 0 \quad \Rightarrow \quad v$ is harmonic.

Thus u and v are both harmonic. But the C–R equations are not satisfied.

Therefore, $(u + iv)$ is not analytic.

Exercise 21.2

1. **Find the derivatives of the following (from first principles):**

   (a) $z^2 + 3z + 2$ (b) $\frac{1+z}{1-z}(z \neq 0)$

2. **Determine the Singular Points for the following functions:**

   (a) $\frac{z}{z+i}$ (b) $\frac{z-1}{z^2+5z+6}$ (c) $\frac{3z-2}{z^2+2z+5}$

3. **Verify whether the following functions are harmonic or not:**

   (a) $f(x,y) = e^x \cos y$ (b) $f(x,y) = 3x^2 + 2xy - 2y^2$

4. **Verify C–R Equations for the following :**

   (a) $f(z) = iz + 4$ (b) $f(z) = e^{-x}[\cos y - i \sin y]$

5. **Prove the following functions are nowhere analytic:**

   (a) $f(z) = xy + iy$ (b) $f(z) = lm(z)$ (c) $f(z) = |z|$

   [*Hint*: Show that C R equations are not satisfied.]

6. **Prove the following functions entire:**

   (a) $f(z) = 3x - 4y + i(4x + 3y)$ (b) $f(z) = iz + 3$

7. **Find the constants *l* and *m* so that the functions *f (z)* should be analytic, where** $f(z) = (x^2 + ly^2 - 2xy) + i(mx^2 - y^2 + 2xy)$

   [*Hint*: Apply C–R equations].

8. If $f(z) = (u + iv)$ is analytic in a region R and $v = u^2$, show that $f(z)$ is a constant

9. If $f(z) = \log z$, show that f is analytic for all $z \neq 0$

10. Show that $f(z) = \left(x^2 + y^2\right)$ is continuous everywhere but differentiable only at the origin.

11. If $f(z) = \begin{cases} \dfrac{x^2 y^5 (x + iy)}{x^4 + y^{10}}, & z \neq 0 \\ 0, & z = 0 \end{cases}$ show that C–R equations are satisfied at the origin but $f(z)$ is not analytic at that point.

12. Prove that the functions $u(x, y) = x^3 - 3xy^2 + xy$, and $v(x, y) = x^2 - y^2 + e^x \cos y$ are both harmonic, and ($u + iv$) is not analytic.

Answers

1. (a) $2z + 3$ (b) $\dfrac{2}{(1 - z)^2}$

2. (a) $z = -i$ (b) $z = -2$, $z = -3$ (c) $z = -1 \pm 2i$

3. (a) Harmonic (b) Not harmonic **4.** $l = -1; m = 1$

21.4 Methods of Construction of an Analytic Function f(z) = u + iv

Let $w = f(z) = u + iv$ be an analytic function. In this section we shall discuss the methods to find (1) v, given u(2) u, given v, (3) Milne – Thompson method to find $f(z)$ directly given u or v and (4) to find $f(z)$ given $(u + v)$ or $(u - v)$.

21.4.1 Method of finding *v* given *u*

Let $w = f(z) = u(x, y) + iv(x, y)$ be analytic. Let $u(x, y)$ be given. Then the following procedure is adopted to find v . Since u is given, u_x and u_y can be found. Since $f(z)$ is analytic, C –R equations hold good.

$\therefore$ $v_x = -u_y$ and $v_y = u_x$

$\therefore$ $$dv = v_x dx + v_y dy = -u_y dx + u_x dy \quad .....(1)$$

Let $M = -u_y$, $N = u_x$ then $\dfrac{\partial M}{\partial y} = -u_{yy}, \dfrac{\partial N}{\partial x} = u_{xx}$

Since u is harmonic, we have $u_{xx} + u_{yy} = 0$

$$\therefore \quad \frac{\partial M}{\partial y} = \frac{\partial N}{\partial x} \Rightarrow Mdx + Ndy = 0 \text{ is exact}$$

Hence from (1), $dv = 0 \Rightarrow Mdx + Ndy = 0$

which being an exact differential equation can be solved and hence v can be found.

***Aliter* 1:**

Alternately, $v_x = -u_y$ .....(1) $v_y = -u_x$ .....(2)

Integrating (1) w.r.t.x we get,

$$v = \int (-u_y)\, dx + \phi(y) \quad .....(3)$$

$$[\phi(y) \text{ is constant with respect to } x]$$

Differentiating (3) partially w.r.t.y and equating with (2) we get $\phi(y)$. Substituting $\phi(y)$ in (3), We get v.

***Aliter* 2:**

We can first find *f (z)* from which we get v. We know that,

$$f^1(z) = u_x + iv_x = u_x - iu_y \quad .....(1)$$

Integrating (1), *f (z)* is found from which v can be found.

21.4.2 To Find *u* Given *v*

Proceeding as in 21.4.1, we have,,

$$du = u_x dx + u_y dy = v_y dx - v_x dy \ .$$

which is again exact and can be solved.

Note: Aliters 1 and 2 given in 21.4.1can also be used here suitably.

21.4.3 Milne–Thompson Method

Using this method, given, u(or v) we can find directly *f (z)*.

Method :

1. Given u to find *f (z)*

   Let *f (z)* = $u + iv$ be analytic.

   Then $f^1(z) = u_x + iv_x = u_x - iu_y$ .....(1)

Let $u_x(x,y)=\phi_1(x,y)$ and $u_y(x,y)=\phi_2(x,y)$

$$z=x+iy,\bar{z}=x-iy,\Rightarrow x=\frac{z+\bar{z}}{2},y=\frac{z-\bar{z}}{2i}$$

$$\therefore\ (1)\Rightarrow f^1(z)=\phi_1\left\{\frac{z+\bar{z}}{2},\frac{z-\bar{z}}{2i}\right\}-i\phi_2\left\{\frac{z+\bar{z}}{2},\frac{z-\bar{z}}{2i}\right\}$$

Putting $\bar{z}=z$, we get, $f^1(z)=\phi_1(z,0)-i\phi_2(z,0)$

Integrating w.r.t. z, we get $f(z)=\int[\phi_1(z,0)-i\phi_2(z,0)]dz+c$ where c is constant

2. Given v, to find *f (z)*
   Here we take, $\psi_1(x,y)=v_y$; $\psi_2(x,y)=v_x$

   Then, we get on proceeding as in 1.

$$f^1(z)=v_y+iv_x=\psi_1(z,0)+i\psi_2(z,0)$$

$$\therefore\quad f(z)=\int[\psi_1(z,0)+i\psi_2(z,0)]dz+c$$

21.4.4 Given (u – v) or (u + v), to find f (z)

Let $f(z)=u+iv$ ......(1)

be analytic, then, $i[f(z)]=-v+iu$ ......(2)

$$(1)+(2)\ \Rightarrow(1+i)f(z)=(u-v)+i(u+v) \quad .....(3)$$

Let $F(z)=(1+i)f(z)$; $U=u-v$; $V=(u+v)$

Then (3) $\Rightarrow F(z)=U+iV$. is also analytic .....(4)

Applying Milne–Thompson method to (4) we can get F (z)

Hence $f(z)=F(z)/(1+i)$

21.4.5 Application of Analytic Functions

1. **Velocity potential and stream function**

   Problems in electrostatic field in xy – plane and two dimensional fluid flow involve the application of analytic functions. Here we have the complex potential 'W' given by,

$$W=\phi+i\psi \quad \text{(which is analytic)}$$

where, $\phi(x, y)$ is the velocity potential and $\psi(x, y)$ is the stream function.

Given one of these two functions, we can find the other, using the methods given in 21.4.1 to 21.4.4

2. Orthogonal trajectories

Let $f(z) = u + iv$ be an analytic function. Then we know that the family of curves given by equations $u(x, y)$ = constant, cut the family of curves $v(x, y)$ = constant orthogonally. (see ex. 21.21)

Thus we can obtain the orthogonal trajectories of a given family of curves (the study of which was done in "differential equations") using the theory of analytic functions.

$\therefore$ If $u = (x, y) = C_1$ represents a family of curves, then $v = (x, y) = C_2$ represents its orthogonal trajectory where 'v' is harmonic conjugate of 'u'.

Solved Examples

***Example* 21.36:** Show that $u = \log\sqrt{x^2 + y^2}$ can be the real part of an analytic function $f(z) = u + iv$. Find v and hence the function $f(z)$ in terms of z.

Solution: $u = \log\sqrt{x^2 + y^2} = ½ \log(x^2 + y^2)$

can be the real part of an analytic function $f(z) = u + iv$, if it is harmonic.

$$u_x = \frac{1}{2}\frac{1}{x^2 + y^2}2x = \frac{x}{x^2 + y^2}; \quad u_{xx} = \frac{(x^2 + y^2).1 - x.2x}{(x^2 + y^2)^2} = \frac{y^2 - x^2}{(x^2 + y^2)^2}$$

$$u_y = \frac{1}{2}\frac{1}{x^2 + y^2}2y = \frac{y}{x^2 + y^2};$$

$$u_{yy} = \frac{(x^2 + y^2).1 - y.(2y)}{(x^2 + y^2)^2} = \frac{x^2 - y^2}{(x^2 + y^2)^2} = -u_{xx}$$

$\therefore$ $u_{xx} + u_{yy} = 0 \Rightarrow u$ is harmonic. So it can be the real part of $f(z) = u + iv$.

$$dv = v_x dx + v_y dy = -u_y dx + u_x dy \qquad \text{[from C – R equations]}$$

$$= \frac{-y}{x^2 + y^2}dx + \frac{x}{x^2 + y^2}dy \qquad ......(1)$$

which is exact.

Therefore, integrating, we get,

$$v=\int\frac{-y}{x^2+y^2}dx+c=\int\frac{1}{1+\left(y^2/x^2\right)}\left[-\frac{y}{x^2}\right]dx+c$$

$$=\int\frac{1}{1+\left(y/x\right)^2}d\left[\frac{y}{x}\right]=\tan^{-1}(y/x)+c$$

$$f(z)=u+iv=(1/2)\log\, z^2+ic=\log z+ic \qquad \text{[c is constant]}$$

Note: *Integration of (1) is done by the method of solving an "exact differential equation: Mdx + Ndy = 0.*

i.e., $\int$ Mdx (keeping y constant) + (terms of N free from x)dy = c.

Aliter **(1) :** To find *v* {using Aliter (1) of 21.4.1}

$$v_x=-u_y = \frac{-y}{x^2+y^2} \qquad .....(2)$$

$$v_y = u_x = \frac{x}{x^2+y^2} \qquad .....(3)$$

Integrating (2) w.r.t. x, we get

$$v = \int\frac{-y}{x^2+y^2}dx = \tan^{-1}\left(\frac{y}{x}\right)+\phi(y) \qquad .....(4)$$

Differentiating (4) w.r.t. y partially, we get,

$$v_y=\frac{1}{1+y^2/x^2}\frac{1}{x}+\phi^1(y)=\frac{x}{x^2+y^2}+\phi^1(y) \qquad .....(5)$$

Comparing (3) and (5), we get, $\phi^1(y)=0\Rightarrow\phi(y)=c$ [c is constant]

Therefore (4) $\Rightarrow$ $v=\tan^{-1}y/x+c$

Aliter **(2) :** { Using Aliter (2) of 21.4.1} To find $f(z)$ first.

$$f^1(z)=u_x-iu_y \text{ (Using note (2) of 21.3.5) } = (x-iy)/x^2+y^2$$

Putting $x=z, y=0,$ we have $f^1(z)=\frac{1}{z}$, Integrating we get, $f(z)=logz+k$

Example **21.37:** If $u=x^3y-xy^3+2x^2-2y^2$ is the real part of an analytic function, $f(z)=u+iv$; find *v*. Find also $f(z)$ in terms of *z*.

Solution: $u = x^3y - xy^3 + 2x^2 - 2y^2, \quad u_x = 3x^2y - y^3 + 4x, \quad u_y = x^3 - 3xy^2 - 4y$

$$dv = v_x dx + v_y dy = -u_y dx + u_x dx \qquad \text{[from C-R equations]}$$

i.e., $$dv = -(x^3 - 3xy^2 - 4y)dx + (3x^2y - y^3 + 4x)dy \qquad .....(1)$$

which is exact. Therefore, integrating (1), we get,

$$v = \int\left(-x^3 + 3xy^2 + 4y\right)dx + \int -y^3 dy + c$$

$$= -\frac{x^4}{4} + \frac{3}{2}x^2y^2 + 4xy - \frac{y^4}{4} + c$$

$$\therefore \quad f(z) = u + iv$$

$$= (x^3y - xy^3 + 2x^2 - 2y^2) + i\left(-\frac{x^4}{4} + \frac{3x^2y^2}{2} + 4xy - \frac{y^4}{4}\right) + c$$

Substituting $x = z$, $y = 0$, we get $f(z) = 2z^2 - iz^4/4 + c$

Aliter: Milne-Thompson Method to find $f(z)$

If $$u_x = \phi_1(x, y), u_y = \phi_2(x, y),$$

$$f(z) = \int[\phi_1(z,0) - i\phi_2(z,0)]dz + c_1 = \int\left[4z - i\{z^3\}\right]dz + c_1$$

$$= 2z^2 - \frac{iz^4}{4} + c_1 \quad (c_1 \text{ being constant})$$

***Example* 21.38:** Show that $u = x^3 - 3xy^2 + 3x^2 - 3y^2 + 1$, is harmonic. Find its harmonic conjugate. Find also the corresponding analytic function *f(z)* in terms of *z*.

Solution: $u = x^3 - 3xy^2 + 3x^2 - 3y^2 + 1, \quad u_x = 3x^2 - 3y^2 + 6x;$

$$u_{xx} = 6x + 6; \quad u_y = -6xy - 6y; \quad u_{yy} = -6x - 6$$

$$u_{xx} + u_{yy} = 0 \Rightarrow \text{u is harmonic.}$$

Let *v* be the harmonic conjugate of *u*, then,

$$dv = v_x dx + v_y dy = -u_y dx + u_x dx \qquad \text{(by C-R equations)}$$

$$dv = (6xy + 6y)dx + (3x^2 - 3y^2 + 6x)dy, \qquad \text{which is exact.}$$

Therefore, integrating,

$$v=\int(6xy+6y)\,dx+\int -3y^2dy+c=3x^2y+6xy-y^3+c$$

$$f(z)=\left(x^3-3xy^2+3x^2-3y^2+1\right)+i(3x^2y+6xy-y^3+c)$$

$$=z^3+3z^2+1+ic=z^3+3z^2+k \quad (k=1+ic)$$

***Example* 21.39:** If $u=e^{2x}\left[x\cos\ 2y-y\sin\ 2y\right]$ is the real part of an analytic function *f(z), find f(z)*

***Solution*:** [since we need not find *v*, we use Milne-Thompson Method]

$$u_x=2e^{2x}\left[x\cos\ 2y-y\sin\ 2y\right]+e^{2x}\cos 2y$$

$$u_y=e^{2x}\left[-2x\sin\ 2y-\sin\ 2y-2y\cos 2y\right]$$

Let $\quad u_x=\phi_1(x,y)$ and $u_y=\phi_2(x,y)$

Then $\quad \phi_1(z,0)=2ze^{2z}+e^{2z}$ and $\phi_2(z,0)=0$

By Milne-Thompson Method,

$$f(z)=\int\left[\phi_1(z,0)-i\phi_2(z,0)\right]dz+c=\int(2ze^{2z}+e^{2z})dz+c$$

$$=2\left\{z.\frac{e^{2z}}{2}-\int\frac{e^{2z}}{2}1dz\right\}+\frac{e^{2z}}{2}+c \quad \text{[integrating 1}^{\text{st}}\text{ term by parts]}$$

$$=2\left\{\frac{ze^{2z}}{2}-\frac{e^{2z}}{4}\right\}+\frac{e^{2z}}{2}+c=ze^{2z}+c$$

***Example* 21.40:** Show that $v(x,y)=3x^2y-y^3$ can be the imaginary part of an analytic function *f(z)* = *u* + *i v* and find *u* .Hence find $f(z)$ in terms of z.

Solution: $\quad v=3x^2y-y^3$

$\therefore \quad v_x=6xy$; and $v_{xx}=6y$

$v_y=3x^2-3y^2,\ v_{yy}=-6y;\ v_{xx}+v_{yy}=0\Rightarrow v$ is harmonic

$\Rightarrow \quad v$ can be the imaginary part of an analytic function $f(z)=u+iv$

To find *u* :

$$du=u_xdx+u_ydx=v_ydx-v_xdy \qquad \text{(by C – R equations)}$$

$$\therefore \quad du=\left(3x^2-3y^2\right)dx-6xydy \qquad .....(1)$$

which is exact

$\therefore$ Integrating, $u = \int (3x^2 - 3y^2) dx + c = x^3 - 3xy^2 + c$

$$f(z) = u + iv = (x^3 - 3x^2y + c) + i(3x^2y - y^3)$$

$$= z^3 + c \qquad \text{[using } x = z,\ y = 0\text{]}$$

Aliter **(1) To find *u* :**

$$u_x = v_y = 3x^2 - 3y^2 \qquad \text{.....(2)}$$

$$u_y = -v_x = -6xy \qquad \text{.....(3)}$$

Integrating (2), we get,

$$u = x^3 - 3xy^2 + f(y) \qquad \text{.....(4)}$$

Differentiating (4) w.r.t.y partially, we get,

$$u_y = -6xy + f^1(y) \qquad \text{.....(5)}$$

Comparing (3) and (5),

$$f^1(y) = 0 \quad \Rightarrow \quad f(y) = c$$

$$\therefore \quad u = x^3 - 3x^2y + c \qquad \text{from (4)}$$

$$f(z) = u + iv = (x^3 - 3x^2y + c) + i(3x^2y - y^3)$$

$$= z^3 + c \qquad \text{(using } x = z, y = 0\text{)}$$

Example **21.41:** If $v = e^x \sin y$ is the imaginary part of an analytic function, $f(z) = u + iv$, find u and hence find $f(z)$ in terms of z.

Solution: $v = e^x \sin y;\quad v_x = e^x \sin y;\qquad v_y = e^x \cos y;$

$$du = u_x dx + u_y dy = v_y dx - v_x dy \qquad \text{[by C-R equations]}$$

$$= (e^x \cos y) dx - (e^x \sin y) dy \qquad \text{.....(1)}$$

which is exact

Integrating (1), we get,

$$u = \int (e^x \cos y) dx + c = e^x \cos y + c$$

Therefore $f(z) = u + iv = e^x \cos y + c + i(e^x \sin y)$

i.e., $f(z) = e^z + c$ [putting x = z, y = 0]

Note: Try by other methods also

***Example* 21.42:** Find the analytic function *f(z)* whose imaginary part is $\left(\dfrac{x-y}{x^2+y^2}\right)$

Solution: Let $f(z)=u+iv$ be analytic. Hence $v=\dfrac{x-y}{x^2+y^2}$

$$v_x = \frac{(x^2+y^2)(1)-(x-y)2x}{(x^2+y^2)^2}=\frac{y^2-x^2+2xy}{(x^2+y^2)^2}=\psi_2(x,y)(say)$$

and,
$$v_y = \frac{(x^2+y^2)(-1)-(x-y)2y}{(x^2+y^2)^2}=\frac{-x^2+y^2-2xy}{(x^2+y^2)^2}=\psi_1(x,y)$$

then,
$$\psi_1(z,0)=-\frac{1}{z^2} \text{ and } \psi_2(z,0)=-\frac{1}{z^2}$$

By Milne–Thompson Method,

$$f(z)=\int[\psi_1(z,0)+i\psi_2(z,0)]dz+c=\int-\frac{(1+i)}{z^2}dz+c = \frac{(1+i)}{z}+c$$

***Example* 21.43:** Find the analytic function *f(z)* whose imaginary part is *sin x sinh y* . Also find its real part.

***Solution*:** Let $f(z)=u+iv$ be analytic; Given $v=\sin x\sinh y$

$$v_x=\cos x\sinh y=\psi_2(x,y) \quad \text{(say)}$$

and
$$v_y=\sin x\cosh y=\psi_1(x,y) \quad \text{(say)}$$

By Milne–Thompson method, we get,

$$f(z)=\int\{\psi_1(z,0)+i\psi_2(z,0)\}dz+c=\int(\sin z)dz+c =-\cos z+c$$
$$=-\cos x\cos iy+\sin x\sin iy+c$$
$$=-\cos x\cosh y+i\sin x\sinh y+c$$

$$\therefore \quad u=-\cos x\cosh y+c$$

***Example* 21.44:** Find the analytic function f(z) = u + iv, given that $u-v=e^x(\cos y-\sin y)$

Solution: Let $f(z)=u+iv \Rightarrow (1+i)f(z)=(u-v)+i(u+v)$

i.e., F (z) = U + iV is analytic,

where F(z) = (1 + *i*) *f(z)*; U = $u-v$ and V = $u+v$

Now, $\quad U = e^x(\cos y - \sin y)$

$$U_x = e^x(\cos y - \sin y) = \phi_1(x, y) \qquad \text{(say)}$$

and $$U_y = e^x(-\sin y - \cos y) = \phi_2(x, y) \qquad \text{(say)}$$

then, $\quad \phi_1(z, 0) = e^z; \quad \phi_2(z, 0) = -e^z$

$\therefore$ By Milne–Thompson method, we get,

$$F(z) = \int \{\phi_1(z, 0) - i\phi_2(z, 0)\} + c$$

$$= (1+i)e^z + c \Rightarrow f(z) = e^z + k \quad \text{where } k = \frac{c}{1+i}$$

***Example* 21.45:** If $u - v = \dfrac{\cos x + \sin x - e^{-y}}{2\cos x - e^y - e^{-y}}$, Find an analytic function $f(z) = u + iv$, given that $f(\pi/2) = 0$

Solution: Let $f(z) = u + iv$ so that $(1+i)f(z) = (u - v) + i(u + v)$

Taking $\quad (1+i)f(z) = F(z), U = (u - v), V = (u + v)$,

we have, $\quad F(z) = U + iV$, which is analytic.

Given that $$U = \frac{\cos x + \sin x - e^{-y}}{2\cos x - e^y - e^{-y}} = (1/2)\left[1 + \frac{2\cos x + 2\sin x - 2e^{-y}}{2\cos x - e^y - e^{-y}} - 1\right]$$

$$= (1/2)\left[1 + \frac{2\sin x + e^y - e^{-y}}{2\cos x - e^y - e^{-y}}\right] = (1/2)\left[1 + \frac{2\sin x + 2\sinh y}{2\cos x - 2\cosh y}\right]$$

$$\left[\text{since } \frac{e^y - e^{-y}}{2} = \sinh y; \quad \frac{e^y + e^{-y}}{2} = \cosh y\right]$$

$$= (1/2)\left[1 + \frac{\sin x + \sinh y}{\cos x - \cosh y}\right]$$

Differentiating partially w.r.t.x and y separately, we get,

$$U_x = \frac{\cos x(\cos x - \cosh y) + \sin x(\sin x + \sinh y)}{2(\cos x - \cosh y)^2}$$

$$= \frac{1 - \cos x \cosh y + \sin x \sinh y}{2(\cos x - \cosh y)^2} = \phi_1(x, y) \qquad \text{(say)}$$

$$U_y = \frac{\cosh y(\cos x - \cosh y) + \sinh y(\sin x + \sinh y)}{2(\cos x - \cosh y)^2}$$

$$= \frac{1}{2}\left[\frac{\cos x \cosh y + \sin x \sinh y - 1}{(\cos x - \cosh y)^2}\right] = \phi_2(x, y) \quad \text{(say)}$$

Then $\phi_1(z,0) = \dfrac{1-\cos z}{2(\cos z - 1)^2} = \dfrac{-1}{2(\cos z - 1)}$

And $\phi_2(z,0) = \dfrac{(\cos z - 1)}{2(\cos z - 1)^2} = \dfrac{1}{2(\cos z - 1)}$

By Milne–Thompson Method,

$$\text{F}(z) = \int [\phi_1(z,0) - i\phi_2(z,0)]dz + c = \int (-1/2)\frac{(1+i)}{(\cos z - 1)}dz + c$$

$$= \frac{(1+i)}{2}\int \frac{1}{2\sin^2\left(\frac{z}{2}\right)}dz + c = \frac{(1+i)}{2}\int (1/2)\operatorname{cosec}^2\left(\frac{z}{2}\right)dz + c$$

$$= -\frac{(1+i)}{2}\cot\left(\frac{z}{2}\right) + c$$

$\therefore \quad f(z) = \dfrac{F(z)}{1+i} = (-1/2)\cot\dfrac{z}{2} + k \quad$ where $\quad k = \dfrac{c}{1+i}$

$$f(\pi/2) = 0 \Rightarrow 0 = (-1/2)\cot\frac{\pi}{4} + k \Rightarrow k = 1/2$$

$\therefore \quad f(z) = (1/2)\left(1 - \cot\dfrac{z}{2}\right)$

***Example* 21.46:** Find the analytic function $f(z) = u + iv$, given that, $u + v = \dfrac{\sin 2x}{\cosh 2y - \cos 2x}$

Solution: Let $f(z) = u + iv$ so that $(1+i)f(z) = (u - v) + i(u + v)$

Taking, $(1+i)f(z) = F(z), U = (u - v), V = (u + v)$,

we have, $F(z) = U + iV$

Given that $V = \dfrac{\sin 2x}{\cosh 2y - \cos 2x}$

$$V_x = \frac{(\cosh 2y - \cos 2x)2\cos 2x - 2\sin^2 2x}{(\cosh 2y - \cos 2x)^2}$$

$$= \frac{2(\cos 2x \cosh 2y - 1)}{(\cosh 2y - \cos 2x)^2} = \psi_2(x, y)$$

$$V_y = \frac{(\cos 2y - \cos 2x)(0) - 2\sin 2x \sinh 2y}{(\cosh 2y - \cos 2x)^2}$$

$$= \frac{-2\sin 2x \sinh 2y}{(\cosh 2y - \cos 2x)^2} = \psi_1(x, y)$$

$$\psi_1(z, 0) = 0, \quad \psi_2(z, 0) = \frac{2(\cos 2z - 1)}{(1 - \cos 2z)^2}$$

$$= \frac{-2}{1 - \cos 2z} = \frac{-2}{2\sin^2 z/2} = -\cos ec^2\left(\frac{z}{2}\right)$$

Therefore, by Milne-Thompson Method, we have,

$$F(z) = \int [\psi_1(z,0) + i\psi_2(z,0)]\,dz + c = \int -i\cos ec^2\left(\frac{z}{2}\right) dz + c$$

$$= 2i\cot\frac{z}{2} + c$$

$$f(z) = \frac{F(z)}{1+i} = \frac{2i}{1+i}\cot\frac{z}{2} + k \qquad \text{where } k = \frac{c}{1+i}$$

$$f(z) = (1+i)\cot\frac{z}{2} + k \qquad \left(\text{since } \frac{2i}{1+i} = 1+i\right)$$

***Example* 21.47:** If $u + v = (x - y)(x^2 + 4xy + y^2)$, find the analytic function $f(z) = u + iv$

***Solution*:** Let $f(z) = u + iv$, so that $(1+i)f(z) = (u - v) + i(u + v)$

Taking $(1+i)f(z) = F(z), U = (u - v), V = (u + v)$, we have, $F(z) = U + iV$

Given $V = (x - y)(x^2 + 4xy + y^2)$

$$V_x = (x-y)(2x+4y)+x^2+4xy+y^2 = \psi_2(x,y)$$

$$V_y = (x-y)(4x+2y)-(x^2+4xy+y^2) = \psi_1(x,y)$$

$$\psi_1(z,0) = 3z^2 \quad \text{and} \quad \psi_2(z,0) = 3z^2$$

By Milne-Thompson Method,

$$F(z) = \int [\psi_1(z,0)+i\psi_2(z,0)]\,dz + c$$

$$= (1+i)\int 3z^2 dz + c \;=\; (1+i)z^3 + c$$

$$f(z) = \frac{F(z)}{1+i} = z^3 + k \quad \text{where} \quad k = \frac{c}{1+i}$$

Example **21.48:** If $\varnothing = x/(x^2+y^2)$ is the velocity potential of two dimensional fluid flow, find the stream function ψ.

***Solution*:**

$$\varnothing = x/(x^2+y^2)\ ; \quad \varnothing_x = \frac{(x^2+y^2).1 - x.2x}{(x^2+y^2)^2} = \frac{y^2-x^2}{(x^2+y^2)^2}$$

Similarly $\quad \varnothing_y = \dfrac{-2xy}{(x^2+y^2)^2}$ {verify}

$$d\psi = \psi_x dx + \psi_y dy = -\varnothing_y dx + \varnothing_x dy \qquad \text{[from C-R equations]}$$

$$= \frac{2xy}{\left(x^2+y^2\right)^2}dx + \frac{y^2-x^2}{\left(x^2+y^2\right)^2}dy$$

Integrating, we get, $\psi = \dfrac{-y}{\left(x^2+y^2\right)} + c$, is the stream function.

Example **21.49:** Show that $u(r,\theta) = -r^3 \sin 3\theta$ is harmonic. Find its harmonic conjugate.

***Solution*:** The equation, $r^2 u_{rr} + r u_r + u_{\theta\theta} = 0$ .....(1)

is the Laplce equation in polar coordinates (see Ex: 21.24)

If 'u' satisfies (1) then u will be harmonic.

Now, $u_r = -3r^2 \sin 3\theta$; $u_{rr} = -6r \sin 3\theta$; $u_\theta = -3r^3 \cos 3\theta$ and $u_{\theta\theta} = 9r^3 \sin 3\theta$;

$$\therefore \; r^2 u_{rr} + r u_r + u_{\theta\theta} = r^2\left(-6r\sin 3\theta\right) + r\left(-3r^2 \sin 3\theta\right) + 9r^3 \sin 3\theta = 0$$

$\therefore$ 'u' is harmonic ; Let 'v' be its harmonic conjugate.

By C – R equations, $v_r = -\frac{1}{r} u_\theta = \frac{-1}{r}\left(-3r^3 \cos 3\theta\right) = 3r^2 \cos 3\theta$

Integrating partially w.r.t 'r', we get, $v = r^3 \cos 3\theta + C(\theta)$ .....(2)

Differentiating (2) w.r.t. 'θ' partially, $v_\theta = -3r^3 \sin 3\theta + C'(\theta)$ .....(3)

But by C – R equations, $v_\theta = r u_r = -3r^3 \sin 3\theta$ .....(4)

Comparing (3) and (4), $C^1(\theta) = 0 \Rightarrow C(\theta) =$ constant.

$\therefore \; v = r^3 \cos 3\theta + k$ from (2), where 'k' is constant.

***Example* 21.50:** Show that $v(r,\theta) = r^2 \cos 2\theta - r\cos\theta + 4$, is harmonic and find its harmonic conjugate.

Solution: $v_r = 2r\cos 2\theta - \cos\theta$, $v_\theta = -2r^2 \sin 2\theta + r\sin\theta$

$$v_{rr} = 2\cos 2\theta; \; v_{\theta\theta} = -4r^2 \cos 2\theta + r\cos\theta$$

$$r^2 v_{rr} + r v_r + v_{\theta\theta} = \left[2r^2 \cos 2\theta + 2r^2 \cos 2\theta - r\cos\theta - 4r^2 \cos 2\theta + r\cos\theta\right] = 0$$

$\therefore$ v satisfies Laplace equation $\Rightarrow$ v is harmonic.

Let u be its harmonic conjugate. By C–R equations, we have,

$$u_r = \frac{1}{r} v_\theta = -2r\sin 2\theta + \sin\theta \qquad \text{......(1)}$$

Integrating (1) partially w.r.t. r, we get,

$$u = -r^2 \sin 2\theta + r\sin\theta + c(\theta) \qquad \text{.....(2)}$$

Differentiating (2) w.r.t. 'θ', we get

$$u_\theta = -2r^2 \cos 2\theta + r\cos\theta + c^1(\theta) \qquad \text{......(3)}$$

But $\quad u_\theta = -rv_r = -2r^2\cos 2\theta + r\cos\theta \quad$ (By C–R equations) ......(4)

Comparing (3) and (4), $c^1(\theta) = 0 \quad \Rightarrow \; c(\theta) =$ constant.

$\therefore$ (2) $\Rightarrow u = -r^2\sin 2\theta + r\sin\theta + k \quad$ (k is constant)

***Example* 21.51:** Find the orthogonal trajectories of the family of curves given by $e^{-x}\cos y + xy = c$

Solution: Let $\quad u = e^{-x}\cos y + xy \quad$ ......(1)

$$u_x = -e^{-x}\cos y + y\,;\; u_y = -e^{-x}\sin y + x$$

$$u_{xx} = e^{-x}\cos y\,; \qquad u_{yy} = -e^{-x}\cos y$$

Let $v(x,y) = c_2$ be the orthogonal trajectory of (1)

Then 'v' is the harmonic conjugate of u; if $(u + iv)$ is analytic; i.e., if u is harmonic, which is true since $u_{xx} + u_{yy} = 0$; $dv = v_x dx + v_y dy = -u_y dx + u_x dx$ (by C–R equations)

i.e., $\quad dv = \left(e^{-x}\sin y - x\right)dx + \left(-e^{-x}\cos y + y\right)dy$, (which is exact)

$\therefore \quad v = \int\left(e^{-x}\sin y - x\right)dx(\text{ keeping 'y' constant }) + \int y\,dy = c_2$

$$-e^{-x}\sin y - \frac{x^2}{2} + \frac{y^2}{2} = c_2$$

$\therefore$ The required orthogonal trajectory is

$$\Rightarrow -e^{-x}\sin y - \frac{x^2}{2} + \frac{y^2}{2} = c_2 \qquad \text{(or)}$$

$$\Rightarrow 2e^{-x}\sin y + x^2 - y^2 = k, \qquad (k \text{ constant})$$

***Example* 21.52:** Find the orthogonal trajectory of the curves given by, $\left(r^2+1\right)\cos\theta = Cr$

Solution: Let $u = \frac{1}{r}\left(r^2+1\right)\cos\theta = C$ (or) $u = \left(r + \frac{1}{r}\right)\cos\theta = C$ be the given system of curves

$$u_r = \left(1 - \frac{1}{r^2}\right)\cos\theta;\ u_{rr} = \frac{2}{r^3}\cos\theta;\ u_\theta = -\left(r + \frac{1}{r}\right)\sin\theta\ ;$$

$$u_{\theta\theta} = -\left(r + \frac{1}{r}\right)\cos\theta$$

$$\therefore \qquad r^2 u_{rr} + r u_r + u_{\theta\theta} = \cos\theta\left[\frac{2}{r} + r - \frac{1}{r} - r - \frac{1}{r}\right] = 0$$

$\therefore$ u is harmonic. Let v be its harmonic conjugate.

By C – R conditions, $v_r = \frac{-1}{r} u_\theta = \left(1 + \frac{1}{r^2}\right)\sin\theta$ ....(1)

Integrating (1) partially w.r.t. 'r', we get , $v = \left(r - \frac{1}{r}\right)\sin\theta + c(\theta)$ ....(2)

Differentiating (2) w.r.t. 'θ' partially, we get , $v_\theta = \left(r - \frac{1}{r}\right)\cos\theta + c^1(\theta)$ ....(3)

But by C – R conditions, $v_\theta = r u_r = \left(r - \frac{1}{r}\right)\cos\theta$ ....(4)

(3) and (4) $\Rightarrow c^1(\theta) = 0 \Rightarrow c =$ constant.

$\therefore\ v = \left(r - \frac{1}{r}\right)\sin\theta = k$ (k constant) . The required orthogonal trajectory is

$$r - \frac{1}{r} = k\operatorname{cosec}\theta$$

Exercise 21.3

1. Show that, each of the following functions $u\ (x, y)$ can be the real part of an analytic function $f\ (z) = u + iv$; Find the conjugate function $v(x, y)$ and the function $f(z)$ (in terms of z) .

(a) $u(x,y) = x^2 - y^2 - y$ (b) $u(x,y) = \sin x \sinh y$

(c) $u(x,y) = 2x - x^3 + 3xy^2$ (d) $u(x,y) = 2x(1-y)$

(e) $u(x,y) = x^2 - y^2 - 5x + y + 2$ (f) $u(x,y) = 2xy + 3y$

(g) $u(x,y) = x^2 - y^2$ (h) $u(x,y) = \cos x \cosh y$

2. Show that in each of the following, *v (x, y)* harmonic. Find its harmonic conjugate *u (x, y)* and the corresponding analytic function, $f(z) = u + iv$. [in terms of z]
   (a) $v = 3x^2y - y^3$ (b) $v = x^4 - 6x^2y^2 + y^4$
   (c) $v = \sin x \sinh y$ (d) $v = 4xy - x^3 + 3xy^2$

3. Find the analytic function $f(z)$ whose real part is given by
   (a) $u = e^x(x\cos y - y\sin y)$
   (b) $u = 3x^2y + 2x^2 - 2y^2 - y^3$
   (c) $u = \left(\sin x \cosh y + 2\cos x \sinh y + x^2 - y^2 + 4xy\right)$
   (d) $u = \dfrac{2\cos x \cosh y}{\cos 2x + \cosh 2y}$

4. Find the analytic function $f(z) = u + iv$, whose imaginary part is given by
   (a) $v = e^{-x}(x\cos y - y\sin y)$
   (b) $v = ye^x \cos y + xe^x \sin y$
   (c) $v = e^{-x}\left[2xy\cos y + (y^2 - x^2)\sin y\right]$
   (d) $v = x^3 - 3xy^2 + 3x^2 - 3y^2 + 1$
   (e) $v = \log(x^2 + y^2) + x - 2y$

5. Find the analytic function, $f(z) = u + iv$, given that
   (a) $u - v = (x - y)(x^2 + 4xy + y^2)$
   (b) $u - v = \dfrac{e^y - \cos x + \sin x}{\cosh y - \cos x}$
   (c) $u + v = \dfrac{x}{x^2 + y^2}$, given $f(1) = 1$
   (d) $u + v = x^3 + 3x^2y - 3xy^2 - y^3$

6. If $\psi = (3xy^2 - x^3)$ represents the stream function of an electrostatic field in xy – plane, find the potential function ϕ.

7. Show that $v(r,\theta) = r^3 \cos 3\theta$ is harmonic. Find its harmonic conjugate and the corresponding analytic function.

8. Find the orthogonal trajectories of the following family of curves.

(a) $e^x \cos y = c$ (b) $e^x \sin y = c$

(c) $r^{-3} \cos 3\theta = c$ (d) $r^2 \cos 2\theta = c$

Answers

1. (a) $2xy + x\ ;\ z^2 + iz$ (b) $\cos x \cosh y; i \cos z$

(c) $y^3 - 3x^2 y + 2y; 2z - z^3$ (d) $x^2 - y^2 + 2y; 2z + iz^2$

(e) $2xy - x - 5y; z^2 - (5 + i) z$ (f) $y^2 - x^2 - 3x; -i(z^2 + 3z)$

(g) $2xy; z^2$ (h) $-\sin x \sinh y; \cos z$

2. (a) $-y^3 - 3xy^2; z^3$

(b) $4xy^3 - 4x^3 y; iz^4$

(c) $-\cos x \cosh y; -\cos z$

(d) $2(x^2 - y^2) - y^3 + 3x^2 y;\ 2z^2 - iz^3$

3. (a) ze^z (b) $2z^2 - iz^3$

(c) $(1 - 2i)(\sin z + z^2)$ (d) $\sec z$

4. (a) ize^{-z} (b) ze^z

(c) $z^2 e^{-z}$ (d) $i[z^3 + 3z^2]$

(e) $i(z + 2\log z) - 2z$

5. (a) $-iz^3$ (b) $\cot \dfrac{z}{2}$

(c) $\dfrac{1+i}{2z} + \dfrac{i-1}{2}$ (d) z^3

6. $-iz^3$

7. $u = -r^3 \sin 3\theta\ ;\ u + iv = ir^3 e^{3i\theta} + c = iz^3 + c$

8. (a) $e^x \sin y = k$ (b) $e^x \cos y = k$

(c) $\sin 3\theta = cr^3$ (d) $r^2 \sin 2\theta = c$

21.5 Elementary Functions

Elementary complex functions such as exponential, trigonometric, hyperbolic, logarithmic, and inverse trigonometric functions etc., play an important part in many fields of engineering. The basic definitions and properties of these functions are given in this section.

21.5.1 Polynomial Functions

$$w = a_0 z^n + a_1 z^{n-1} + a_2 z^{n-2} + \dots\dots + a_{n-1} z + a_n = f(z)$$

where $a_0 \neq 0, a_1, a_2 \dots\dots a_n$ are complex constants and n is positive integer is called a polynomial function of nth degree in z.

Note : $w = a_0 z + a_1$ is a linear function

21.5.2 Rational Algebraic Functions

$w = \dfrac{F(z)}{G(z)}$, where F (z), G(z) are polynomials is a rational function.

Note: The function $w = \dfrac{az+b}{cz+a}$, $(ad - bc \neq 0)$ is called a bilinear (or fractional linear) transformation.

21.5.3 Exponential Function, (e^z)

$$w = \exp z = e^z = e^{x+iy} = e^x . e^{iy} = e^x \{\cos y + i \sin y\} \qquad \dots..(1)$$

is known as the complex exponential function where $e = 2.71828\dots.$ is the natural base of logarithms.

This function is of utmost importance, as all other functions are defined from it.

Properties of e^z

1. If z_1, z_2, are two complex numbers; we can prove that, $e^{z_1} . e^{z_2} = e^{z_1+z_2}$; and $\dfrac{e^{z_1}}{e^{z_2}} = e^{z_1-z_2}$
2. $\left(e^z\right)^n = e^{nz}$; $e^0 = 1$
3. If $z = x, y = 0$, (1) $\Rightarrow e^z = e^x$

4. If $x = 0$, $z = iy \Rightarrow e^z = e^{iy}$; $e^{iy} = \cos y + i\sin y$, is Euler's formula;

$$\left|e^{iy}\right| = \sqrt{\cos^2 y + \sin^2 y} = 1$$

5. $e^z = e^x(\cos y + i\sin y)$.

Since, $\sin y = \sin(y + 2k\pi)$, $\cos y = \cos(y + 2k\pi)$, $k = 0, 1, 2, \ldots$

We have, $e^z = e^{z+2k\pi i}$, $\qquad k = 0, 1, 2, \ldots$

$\therefore$ e^z is a periodic function of period 2π.

6. $\dfrac{d}{dz}(e^z) = e^z \Rightarrow e^z$ is analytic for all z. i.e., e^z is an entire function.

21.5.4 Trigonometric Functions

The definitions of trigonometric (or circular functions) sinz, cosz, etc are given in terms of exponential functions as follows.

$$e^{iz} = \cos z + i\sin z\,;\ e^{-iz} = \cos z - i\sin z$$

$$\therefore \qquad \cos z = \frac{e^{iz} + e^{-iz}}{2} \text{ and } \sin z = \frac{e^{iz} - e^{-iz}}{2i}$$

Similarly, $\sec z = \dfrac{1}{\cos z} = \dfrac{2}{e^{iz} + e^{-iz}}$, $\operatorname{cosec} z = \dfrac{1}{\sin z} = \dfrac{2i}{e^{iz} - e^{-iz}}$

$$\tan z = \frac{\sin z}{\cos z} = \frac{e^{iz} - e^{-iz}}{i\left(e^{iz} + e^{-iz}\right)},\ \cot z = \frac{\cos z}{\sin z} = \frac{i\left(e^{iz} + e^{-iz}\right)}{e^{iz} - e^{-iz}}$$

Properties

Many properties of real circular functions hold in the case of complex trigonometric functions

For example,

1. $\sin^2 z + \cos^2 z = 1$; $\sec^2 z - \tan^2 z = 1$; $\operatorname{cosec}^2 z - \cot^2 z = 1$

   $\sin(-z) = -\sin z$; $\cos(-z) = \cos z$;

   $\sin(z_1 \pm z_2) = \sin z_1 \cos z_2 \pm \cos z_1 \sin z_2$

$$\cos(z_1 \pm z_2) = \cos z_1 \cos z_2 \mp \sin z_1 \sin z_2$$

$$\tan(z_1 \pm z_2) = \frac{\tan z_1 \pm \tan z_2}{1 \mp \tan z_1 \tan z_2}$$

2. sin z and cos z are periodic functions of period 2π
3. tan z and cot z have period π.
4. sin 2z = 2sin z cos z

$$\cos 2z = \cos^2 z - \sin^2 z \;=\; 2\cos^2 z - 1 \;=\; 1 - 2\sin^2 z$$

$$\tan 2z = \frac{2\tan z}{1-\tan^2 z} \;;\; \sin 3z = 3\sin z - 4\sin^3 z$$

$$\cos 3z = 4\cos^3 z - 3\cos z \;;\; \tan 3z = \frac{3\tan z - \tan^3 z}{1 - 3\tan^2 z}$$

5. $\frac{d}{dz}(\sin z) = \cos z ; \quad \frac{d}{dz}(\cos z) = -\sin z$

$$\frac{d}{dz}(\tan z) = \sec^2 z$$

$$\frac{d}{dz}(cosec\ z) = -cosec\ z \cot z$$

$$\frac{d}{dz}(\sec z) = \sec z \tan z \;;\; \frac{d}{dz}(\cot z) = -\cos ec^2 z....$$

6. sin z and cos z are analytic for all z.
7. tan z and sec z are analytic for all z except when cos z = 0
8. cot z and cosec z are analytic for all z except when sin z = 0
9. ***Zeroes :*** Zeroes of sin z are given by sin z = 0.

   Therefore, zeroes of sin z are given by, $z = k\pi, k = 0, \pm 1, \pm 2, ...$

   Similarly, zeroes of cos z are $z = (k + 1/2)\pi$, where $k = 0, \pm 1, \pm 2, ...$

21.5.5 Hyperbolic Functions

Definitions of hyperbolic functions of z are given below.

$$\sinh z = \frac{e^z - e^{-z}}{2} ; \; \cosh z = \frac{e^z + e^{-z}}{2} \;;\; \tanh z = \frac{\sinh z}{\cosh z} = \frac{e^z - e^{-z}}{e^z + e^{-z}}$$

$$\cos ec\,h\, z = \frac{1}{\sinh z} = \frac{2}{e^z - e^{-z}}\; ; \; \text{sech}\, z = \frac{1}{\cosh z} = \frac{2}{e^z + e^{-z}}$$

$$\text{cothh}\, z = \frac{1}{\tanh z} = \frac{\cosh z}{\sinh z} = \frac{e^z + e^{-z}}{e^z - e^{-z}}$$

***Properties*:**

1. Some identies [which can be easily verified using above definitions] to be remembered are given below.

$$\cosh^2 z - \sinh^2 z = 1;\; \text{sech}^2 z = 1 - \tanh^2 z\; ;\; \text{cosech}^2 z = \coth^2 z - 1\; ;$$

$$\sinh(-z) = -\sin z\; ;\; \cosh(-z) = \cosh z\; ;\; \tanh(-z) = -\tanh z\; ;$$

$$\sinh(z_1 \pm z_2) = \sinh z_1 \cosh z_2 \pm \cosh z_1 \sinh z_2$$

$$\cosh(z_1 \pm z_2) = \cosh z_1 \cosh z_2 \pm \sinh z_1 \sinh z_2$$

$$\tanh(z_1 \pm z_2) = \frac{\tanh z_1 \pm \tanh z_2}{1 \pm \tanh z_1 \tanh z_2}$$

2. Relations between hyperbolic and trigonometric functions and vice – versa are given below:

$$\sin iz = i \sinh z\; ; \quad \cos iz = \cosh z \quad ; \quad \tan iz = i \tanh z\; ;$$

$$\sinh iz = i \sin z\; ; \quad \cosh iz = \cos z \quad ; \quad \tanh iz = i \tan z$$

3. To express hyperbolic and trigonometric functions in the form $u(x,y) + iv(x,y)$:

(i) $\sinh z = \sinh(x + iy) = \sinh x \cosh(iy) + \cosh x \sinh(iy)$

$$= \sinh x \cos y + i \cosh x \sin y$$

(ii) $\cosh z = \cosh(x + iy) = \cosh x \cosh(iy) + \sinh x \sinh(iy)$

$$= \cosh x \cos y + i \sinh x \sin y$$

(iii) $\sin z = \sin(x + iy) = \sin x \cos(iy) + \cos x \sin(iy)$

$$= \sin x \cosh y + i \cos x \sinh y$$

(iv) $\cos z = \cos x \cosh y - i \sin x \sinh y$ and so on.

***Derivatives*:**

4. $\dfrac{d}{dz}(\sinh z) = \cosh z\; ;\; \dfrac{d}{dz}(\cosh z) = \sinh z$

$$\frac{d}{dz}(\tanh z) = \text{sech}^2 z\; ;\; \frac{d}{dz}(\cos \text{ech} z) = -\cos \text{ech} z \coth z$$

$$\frac{d}{dz}(\text{sechz}) = -\sec hz \tanh z \; ; \; \frac{d}{dz}(\coth hz) = -\cos\text{ech}^2 z...$$

5. Analyticity

(a) sinh z and cosh z are analytic everywhere. i.e., They are entire functions.

(b) sech z and tanh z are analytic at all z except at points where cosh z = 0.

(c) cosech z and coth z are analytic everywhere except at points where sinh z = 0.

6. Periodicity

(i) The functions sinh z, cosh z, cosech z and sech z are periodic and have a period '$2\pi i$'

For Example :

$$\begin{aligned}\sinh(z+2\pi i) &= \frac{1}{i}\sin\left[i(z+2\pi i)\right] \\ &= -i\sin\{i(z+2\pi i)\} = -i\sin(iz-2\pi) \\ &= -i(\sin iz\cos 2\pi - \cos iz\sin 2\pi) = -i\sin iz \\ &= -i.i\sinh z = \sinh z\end{aligned}$$

(ii) tanh z and coth z are periodic with period 'πi'

21.5.6 Logarthmic Function: (Inverse of Exponential Functions)

The logarithmic function (log z) is also defined from the exponential function.

1. Definition of log z:

Let z = $re^{i\theta}$ and w be two complex numbers and $z \neq 0$;

Let z = e^w, then we write

$$w = \log z = \log(re^{i\theta}) = \log r + i(\theta + 2k\pi) \quad ......(1)$$

$$k = 0, \pm 1, \pm 2......$$

and 'w' is called the natural logarithm of 'z'.

Equation (1) shows that log z is multiple- valued [in fact infinitely many valued] function. Out of these infinite values, taking $k = 0$ in (1), we get

$$\log z = \log r + i\theta, \quad 0 \leq \theta < 2\pi \quad ......(2)$$

is known as the principal – value of log z. (sometimes other intervals of length 2π, such as $-\pi < \theta \leq \pi$ are also used.)

Note (1) log z can be also defined for real bases other than *'e'* as,

$$z = a^w \left[a > 0\right] \Rightarrow w = \log_a z \;\left(\text{where } a \neq 0,1\right) \Rightarrow w = \frac{\log z}{\log a}$$

2. The principal value of log z given in (2) is sometimes denoted by Log z. Thus

$$\text{Log z} = \log r + i\theta, \quad r > 0, \quad 0 < \theta \leq 2\pi \text{, and}$$

Log z is single valued.

3. If we write $w = u + iv$, $z = x + iy$, $w = \text{Log}\, z$

then, $$u + iv = \log r + i\theta = \log\sqrt{x^2 + y^2} + i\tan^{-1}(y/x)$$

so that, $$u = \log\sqrt{x^2 + y^2}\;;$$

$v = \tan^{-1}(y/x)$ are respectively the real and imaginary parts of Log z.

Properties:

1. $\log z$ is multiple valued while Log z is single valued.
2. $\log z$ is not defined at z = 0.
3. w = Log z is analytic at all $z \neq 0$; $\frac{d}{dz}(\text{Log z}) = 1/z$
4. If $z_1, z_2 \neq 0$ are complex numbers, we have, $\log(z_1 z_2) = \log z_1 + \log z_2$

$$\log\frac{z_1}{z_2} = \log z_1 - \log z_2$$

$$\log 1 = 0 \quad \text{so that} \quad \log(1/z) = -\log z$$

5. $\log z$ has a branch point at z = 0.

21.5.7 Inverse Trigonometric Functions

1. If z = sin w, then w = $\sin^{-1} z$ is called inverse sine of z. Similarly, the other inverse trigonometric functions such as $\cos^{-1} z, \tan^{-1} z, \text{cosec}^{-1} z$ etc are defined.

2. All these functions are multiple valued and can be written in terms of logarithmic functions are given below. In all these formulae, an additive constant, $2k\pi, k = 0, \pm1, \pm2, \ldots$ is omitted in the logarithms.

(a) $\sin^{-1} z = \frac{1}{i}\log\left[iz + \sqrt{1 - z^2}\right]$ (b) $\cos^{-1} z = \frac{1}{i}\log\left[z + \sqrt{z^2 - 1}\right]$

(c) $\tan^{-1} z = \frac{1}{2i} \log\left[\frac{1+iz}{1-iz}\right]$ (d) $\text{cosec}^{-1} z = \frac{1}{i} \log\left[\frac{i+\sqrt{z^2-1}}{z}\right]$

(e) $\sec^{-1} z = \frac{1}{i} \log\left[\frac{1+\sqrt{1-z^2}}{z}\right]$ (f) $\cot^{-1} z = \frac{1}{2i} \log\left[\frac{z+i}{z-i}\right]$

3. Derivatives

$$\frac{d}{dz}\left(\sin^{-1} z\right) = \frac{1}{\sqrt{1-z^2}} \qquad \frac{d}{dz}\left(\cos^{-1} z\right) = \frac{-1}{\sqrt{1-z^2}}$$

$$\frac{d}{dz}\left(\tan^{-1} z\right) = \frac{1}{1+z^2} \qquad \frac{d}{dz}\left(\text{cosec}^{-1} z\right) = \frac{-1}{z\sqrt{z^2-1}}$$

$$\frac{d}{dz}\left(\sec^{-1} z\right) = \frac{1}{z\sqrt{z^2-1}} \qquad \frac{d}{dz}\left(\cot^{-1} z\right) = \frac{-1}{1+z^2}$$

21.5.8 Inverse Hyperbolic Functions

1. $z = \sinh w \Rightarrow w = \sinh^{-1} z$

   Similarly, other inverse hyperbolic functions can be defined.
2. Inverse hyperbolic functions are multiple valued.
3. They are expressed in terms of natural logarithms.

   [An additive constant $2k\pi$, $k = 0, \pm 1, \pm 2, ...$is omitted in them] as given below.

   (a) $\sinh^{-1} z = \log\left[z+\sqrt{z^2+1}\right]$ (b) $\cosh^{-1} z = \log\left[z+\sqrt{z^2-1}\right]$

   (c) $\tanh^{-1} z = ½ \log\left[\frac{1+z}{1-z}\right]$ (d) $\text{cosech}^{-1} z = \log\left[\frac{1+\sqrt{z^2+1}}{z}\right]$

   (e) $\text{sech}^{-1} z = \log\left[\frac{1+\sqrt{1-z^2}}{z}\right]$ (f) $\coth^{-1} z = ½ \log\left[\frac{z+1}{z-1}\right]$

4. **Derivatives**

$$\frac{d}{dz}\left[\sinh^{-1} z\right] = \frac{1}{\sqrt{1+z^2}} \qquad \frac{d}{dz}\left[\cosh^{-1} z\right] = \frac{1}{\sqrt{z^2-1}}$$

$$\frac{d}{dz}\left[\tanh^{-1} z\right] = \frac{1}{1-z^2} \qquad \frac{d}{dz}\left[\text{cosech}^{-1} z\right] = \frac{-1}{z\sqrt{z^2+1}}$$

$$\frac{d}{dz}\left[\text{sech}^{-1} z\right] = \frac{-1}{z\sqrt{1-z^2}} \qquad \frac{d}{dz}\left[\coth^{-1} z\right] = \frac{1}{1-z^2}$$

21.5.9 The Exponential Functions, z^a

Let a, z, be any two complex numbers. Then,

$$w = e^{a\log z} = z^a \qquad \text{......(1)}$$

is the exponent function of z. If $z = re^{i\theta}$, we know that,

$$\log z = \log r + i(\theta + 2k\pi) \qquad \text{......(2)}$$

$k = 0, \pm 1, \pm 2, \ldots..$ where 'θ' is principal value of arg z,

then (1) $\Rightarrow$ $z^a = e^{[a\{\log r + i(\theta + 2k\pi)\}]} = e^{[a\log r + ia\theta + 2ka\pi i]}$

$$= e^{a\log r} \times e^{ia\theta} \times e^{2ka\pi i} \qquad \text{......(3)}$$

(3) $\Rightarrow$ z^a is in general infinite valued, and different values are obtained by giving $k = 0, \pm 1, \pm 2, \ldots$ in (3)

Let us discuss the properties of z^a for different values of 'a'.

(a) Let $a = n$, be an integer. Then

(3) $\Rightarrow$ $z^n = e^{n\log r} \times e^{in\theta} \times e^{2kn\pi i}$

which is the single valued n^{th} power of z and is analytic.

(b) Let $a = 1/n, n$ being an integer then ,

(3) $\Rightarrow$ $z^{(1/n)} = e^{(1/n)\log r} \times e^{i(1/n)\theta} \times e^{2k(1/n)\pi i}$

which gives distinct values for different values of k and analytic.

(c) Let $a = (m/n), m, n \in N$ then,

(3) $\Rightarrow$ $z^{(m/n)} = e^{(m/n)\log r} \times e^{i(m/n)\theta} \times e^{2k(m/n)\pi i}$

which again is multiple – valued.

(d) In general, $\dfrac{d}{dz}(z^a) = az^{a-1}$

Solved Examples

***Example* 21.53:** Express the following functions in the form $u + iv$. [or, separate the following functions into real and imaginary parts.]

(a) $z^2 e^z$ (b) e^{z^2}

Solution:

(a) Since

$$z = x + iy,$$

$$z^2 = (x+iy)^2 = x^2 - y^2 + 2ixy$$

$$e^z = e^{(x+iy)} = e^x \times e^{iy} = e^x(\cos y + i \sin y)$$

$$z^2 e^z = (x^2 - y^2 + 2ixy)e^x(\cos y + i\sin y)$$

$$= e^x\left[\left\{\left(x^2-y^2\right)\cos y - 2xy\sin y\right\} + i\left\{\left(x^2-y^2\right)\sin y + 2xy\cos y\right\}\right]$$

$$= u(x, y) + iv(x, y)$$

where $u(x,y) = e^x\left[\left(x^2-y^2\right)\cos y - 2xy\sin y\right]$

and $v(x,y) = e^x\left[\left(x^2-y^2\right)\sin y + 2xy\cos y\right]$

(b) $$e^{z^2} = e^{(x+iy)^2} = e^{x^2-y^2+2ixy} = e^{x^2-y^2}e^{2ixy}$$

$$= e^{x^2-y^2}\left[\cos 2xy + i\sin 2xy\right] = u(x,y) + iv(x,y)$$

where $u(x,y) = e^{x^2-y^2}\cos 2xy$; $v(x,y) = e^{x^2-y^2}\sin 2xy$

CHAPTER 22

Complex Integration

22.1 Introduction of Line Integral

In this unit we define the integral of complex functions f(z) along a curve. The integration of a real function $f(x)$ on the real line is known as the Riemann integration. The integration of a complex function along a path is known as the Riemann Stieltges integration.

Let z(t) = x(t) + i y(t) , $a \le t \le b$, represent a simple and smooth curve 'C', and f(z) be a continuous function defined at each point of C. The interval $a \le t \le b$ is divided into partitions as

$$a = t_o < t_1 < t_2 < \ldots\ldots < t_n = b$$

Corresponding to these values of 't' the points on the curve are , say, $z_o, z_1 \ldots\ldots z_n$

i.e. $\quad Z_r = z(t_r) \qquad$ r = 0,1,...........n

Let n_r be any arbitrary point in the interval (Z_{r-1}, Z_r) and Let s_n denote the sum

$$\sum_{r=1}^{n} f(n_r)\delta Z_r \quad \text{where } \delta Z_r = Z_r - Z_{r-1}$$

Here $|\delta Z_r| = |Z_r - Z_{r-1}|$ is the length of the chord AB

As $n \to \infty$ the number of sub divisions increases such that the largest of $|\delta Z_r|$ approaches to zero, that is, max$|\delta Z_r| \to 0$

If the sum $S_n \to l$ say, as $n \to \infty$, then f(z) is said to be integrable along the curve 'C' and the limit of S_n denoted by

$$\underset{\max|\Delta Z_r|\to 0}{Lt} S_n = \int_C f(z)dz$$

is called the contour integral or the line integral of f(z) along the curve C. The above result also holds when 'C' is a piecewise smooth curve and f(z) is a piecewise continuous function defined at each point of C.

22.1.1 Properties of Contour Integrals

Suppose f(z) and $\phi(z)$ are two continuous functions defined on a piecewise smooth curve 'C' and a, b are arbitrary complex constants then

1. $\int_C f(z)dz = \int_{C_1} f(z)dz$ where C_1 denotes the curve traversed in the opposite direction of 'C'.

2. $\int_C [af(z)+b\phi(z)]dz = \int_C f(z)dz + b\int_C \phi(z)dz$

3. $\int_C f(z)dz = \int_{C_1} f(z)dz + \int_{C_2} f(z)dz + \dots\dots\dots + \int_{C_n} f(z)dz$

   where 'C' is the curve consisting of 'n' piecewise smooth arcs $C_1, C_2, \dots\dots C_n$ (the terminal point Z_r of C_r coincides with the initial point of C_{r+1} for r=1,2......n-1)

4. $\int_C f(z)dz = \int_C [u(x,y)+iv(x,y)](dx+idy)$

   $\int_C f(z)dz = \int_C u(x,y)dx - v(x,y)dy + i\int_C v(x,y)dx + u(x,y)dy$

   Note : The above complex line integral can be expressed in terms of two real line integrals which can be integrated by using the rules of integration for real functions

Solved Examples

***Example* 22.1:** Evaluate $\int_C (y - x - 3x^2 i)dz$ where 'C'

(i) is the line segment z = 0 to z = 1 + i

(ii) consist of two line segments, one from z = 0 to z = i and other from z = i to z = 1 + i

***Solution*:**

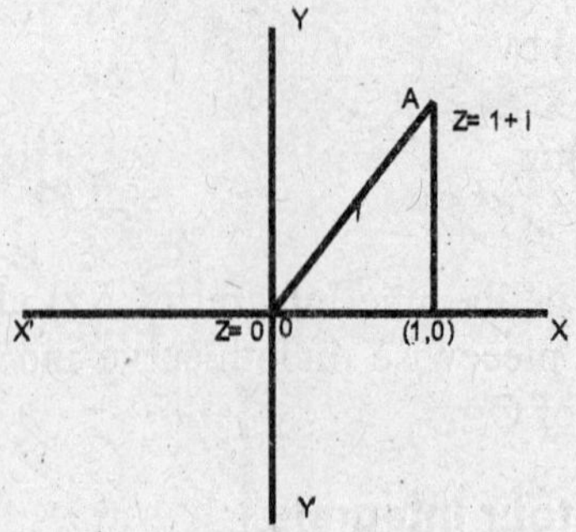

Fig. 22.1(a)

(i) C is the line joining the points z = 0 to z = 1 + i

i.e., O(0, 0), to A(1,1)

Equation of the line is y = x; dy = dx; and x:0 → 1

$$\int_C (y - x - 3x^2 i)(dx + idy) = \int_0^1 \left(x - x - 3x^2 i\right)(dx + idx)$$

$$= (1+i)\int_0^1 -3x^2 i dx = -3i(1+i)\frac{x^3}{3}\Bigg]_0^1 = -i(i+1) = 1 - i$$

(ii) C consists of the two lines joining the points O (0, 0) to A(0,1) and to B(1,1)

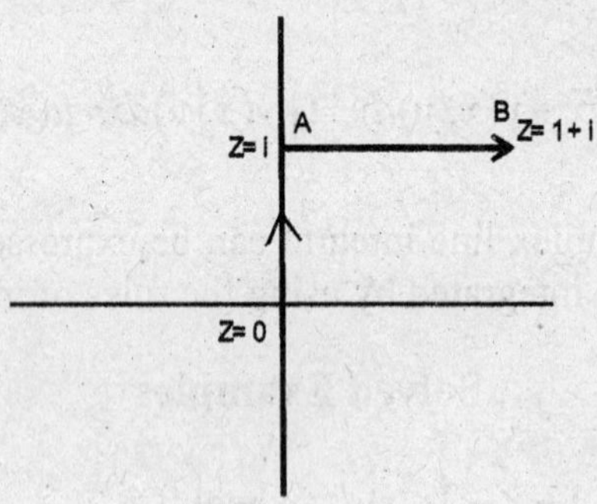

Fig. 22.1(b)

$$\int_C \left(y-x-3x^2 i\right)dz = \int_{OA} \left(y-x-3x^2 i\right)(dx+idy)+\int_{AB} \left(y-x-3x^2 i\right)(dx+idy)$$

$$= \int_{y=0}^{1} (y-0-0)(0+idy)+\int_{x=0}^{1} \left(1-x-3x^2 i\right)(dx+i.0)$$

(Since along OA, $x = 0$, $dx = 0$ and along AB, $y = 1$, $dy = 0$)

$$= \left[i\frac{y^2}{2}\right]_0^1 + \left[x-\frac{x^2}{2}-\frac{3x^3 i}{3}\right]_0^1 = (i/2)+1-(1/2)-i$$

$$= \frac{1-i}{2}$$

Example 22.2: Evaluate $\int_c \frac{z+2}{z}dz$ where C is

(i) The semi circle $z = 2e^{i\theta} (0 \le \theta \le \pi)$

and (ii) The circle $z = 2e^{i\theta} (0 \le \theta \le 2\pi)$

Solution: (i) C is the semi circle $z = 2e^{i\theta}$ $\therefore dz = 2ie^{i\theta} d\theta$, $\theta : 0 \to \pi$

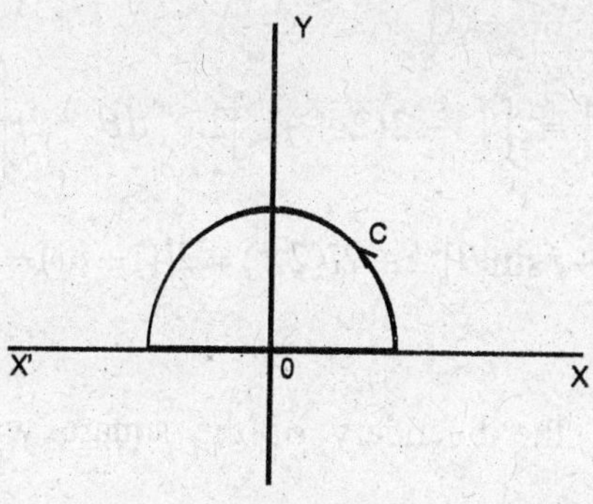

Fig. 22.2(a)

$$\therefore \quad \int_c \frac{z+2}{z}dz = \int_{\theta=0}^{\pi} \left(1+\frac{2}{2e^{i\theta}}\right) 2ie^{i\theta} d\theta$$

$$= \int_0^{\pi} 2ie^{i\theta} d\theta + \int_{\theta}^{\pi} 2id\theta$$

$$= 2i\left[\frac{e^{i\theta}}{i}\right]_0^{\pi} + 2i[\theta]_0^{\pi}$$

$$= 2[(\cos\pi - \cos 0] + 2i\pi$$

$$= -\ 4 + 2\pi i$$

(ii) C is the circle $z = 2e^{i\theta}, \theta : 0 \to 2\pi$, $dz = 2ie^{i\theta}d\theta$

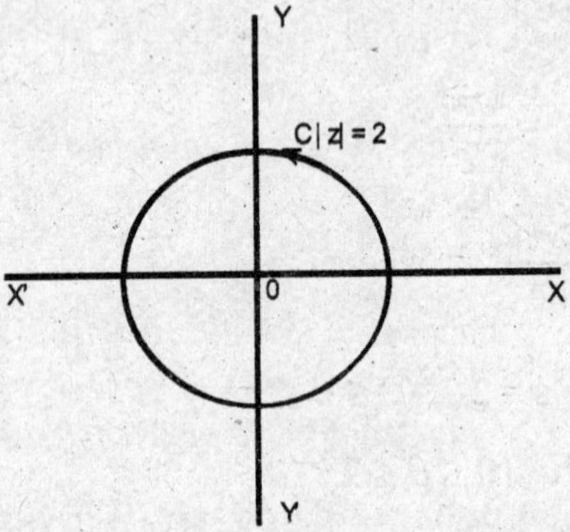

Fig. 22.2(b)

$$\int_C [1 + (2/z)dz] = \int_0^{2\pi} \left[1 + 2(2e^{i\theta})^{-1}\right] 2ie^{i\theta}d\theta = 2i\left[\frac{e^{i\theta}}{i}\right]_0^{2\pi} + 2i[\theta]_0^{2\pi}$$

$$= 2[\cos\theta + i\sin\theta]_0^{2\pi} + 2i(2\pi) = 2[(1+i0) - (1+i0)] + 4\pi i = 4\pi i$$

***Example* 22.3:** Show that

$\int_C (3z+1)dz = 0$ where C is the boundary of the square with vertices at the points $z = 0, z = 1,\ z = 1 + i,\ z = i$ and $z = i$

Solution: $\int_C (3z+1)dz = \int_{OABCO} (3z+1)dz = \int_{OABCO} [3(x+iy)+1](dx+idy)$

$$= \int_{OABCO} [(3x+1)dx - 3ydy] + i[3ydx + (3x+1)dy]$$

The curve OABCO is formed by the lines OA, AB, BC, CO

Along 'OA' : Equation of OA is y = 0; dy = 0; x : $0 \to 1$

Along 'AB' : Equation of AB is x = 1; dx = 0; y : $0 \to 1$

Along 'BC: Equation of BC is y = 1; dy = 0; x: 1→0

Along 'CO' : Equation of CO is x = 0; dx = 0; y: 1→0

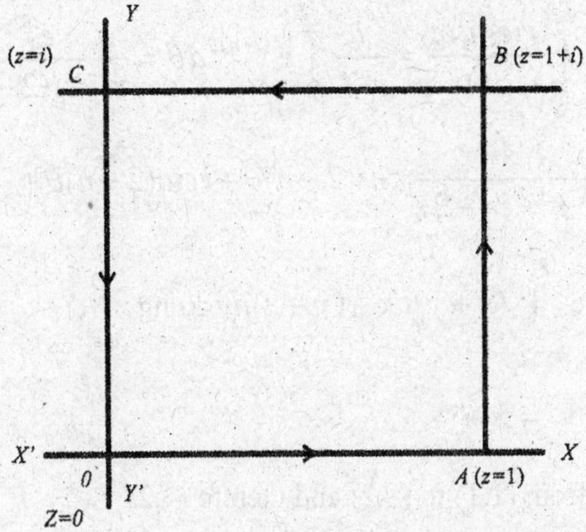

Fig. 22.3

$$\therefore \int_C (3z+1)dz = \int_{x=0}^{1} [(3x+1]dx + \int_{y=0}^{1} (4+3iy)idy + \int_{x=1}^{0} (3x+1+3i)dx +$$

$$+\int_{y=1}^{0} (i-3y)dy$$

$$= \left[\frac{3x^2}{2}+x\right]_0^1 + \left[4iy - \frac{3y^2}{2}\right]_0^1 + \left[\frac{3x^2}{2}+x+3ix\right]_1^0 + \left[iy - \frac{3y^2}{2}\right]_1^0$$

$$= (5/2)+(4i-3/2)+((-3/2)-1-3i)+(-i+3/2) = 0$$

***Example* 22.4**: Evaluate (i) $\int_C \frac{dz}{z-z_0}$ (ii) $\int_C \frac{dz}{(z-z_0)^{n-1}}$ where C is the circle $|z-z_0| = r$

Solution:

(i) Let where c is the circle of $|z-z_0| = r$

i.e., $z - z_0 = re^{i\theta}$, $dz = rie^{i\theta}d\theta$, and $\theta: 0 \to 2\pi$

$$\text{I} = \int_0^{2\pi} \frac{rie^{i\theta}d\theta}{re^{i\theta}} = i[\theta]_0^{2\pi} = 2\pi i$$

(ii) $$\int_C \frac{dz}{(z-z_0)^{n-1}} = \int_0^{2\pi} \frac{rie^{i\theta}d\theta}{(re^{i\theta})^{n-1}} = \frac{i}{r^{n-2}}\int_0^{2\pi} e^{i(2-n)\theta}d\theta = \frac{i}{r^{n-2}}\left[\frac{e^{i(2-n)\theta}}{i(2-n)}\right]_0^{2\pi}$$

$$= \frac{-1}{r^{n-2}(n-2)}[\cos(2-n)\theta + i\sin(2-n)\theta]_0^{2\pi} = 0$$

***Example* 22.5:** Evaluate $\int_{(1,1)}^{(4,2)} (x+y)dx + (y-x)dy$ along

(a) the parabola $y^2 = x$

(b) straight lines from (1,1) to (1,2) and then to (4,2)

(c) the curve $x = 2t^2 + t + 1, y = t^2 + 1$

***Solution*:** (a) given $\text{I} = \int_{(1,1)}^{(4,2)} (x+y)dx + (y-x)dy$; To integrate along the parabola

$y^2 = x$

Substituting $\text{x} = y^2$, $dx = 2ydy$

$$\text{I} = \int_1^2 (y^2+y)2ydy + (y-y^2)dy = \left[\frac{2y^4}{4} + \frac{2y^3}{3} + \frac{y^2}{2} - \frac{y^3}{3}\right]_1^2 = \frac{34}{3}.$$

(b) Integrating along straight lines from (1, 1) to (1,2) and then (4,2)

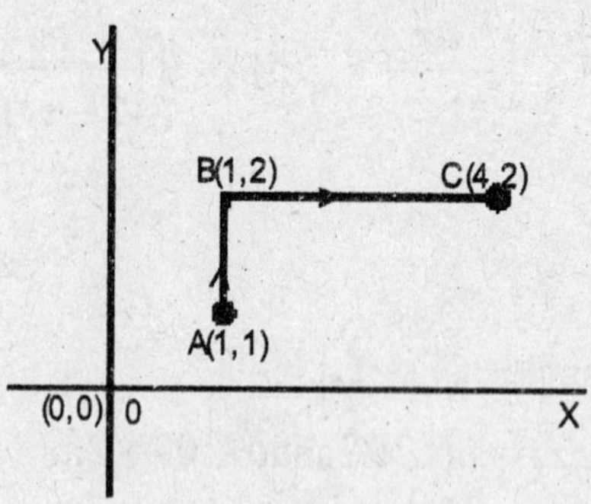

$$\int_{ABC} (x+y)dx+(y-x)dy = \int_{AB} (x+y)dx+(y-x)dy + \int_{BC} (x+y)dx+(y-x)dy \text{ ... (1)}$$

Along AB: joining A(1,1) B(1,2),

The equation of AB is $\dfrac{y-1}{2-1} = \dfrac{x-1}{1-1}$

Equation of AB is x = 1, dx = 0 $(y:1 \to 2)$

Equation of BC joining B(1, 2), C(4,2) is $\dfrac{y-2}{2-2} = \dfrac{x-1}{4-1}$

$\therefore$ Equation of BC is y = 2, dy = 0 $(x:1 \to 4)$

$$\therefore \quad \int_{ABC} (x+y)dx+(y-x)dy$$

$$\int_{y=1}^{2} (1+y)0+(y-1)dy + \int_{x=1}^{4} (x+2)dx+(2-x)0$$

$$= \left[\frac{y^2}{2} - y\right]_1^2 + \left[\frac{x^2}{2} + 2x\right]_1^4$$

$$= 14$$

(c) Along the curve $x = 2t^2 + t + 1, y = t^2 + 1$, $dx = (4t+1)dt$, $dy = 2tdt$

when $x = 1$, $y = 1$ the value of $t = 0$ and when $x = 4$, $y = 2$ the value of $t = 1$

$$\int_{(1,1)}^{(4,2)} (x+y)dx+(y-x)dy$$

$$= \int_{t=0}^{1} \left[(2t^2+t+1+t^2+1)(4t+1)+(t^2+1-2t^2-t-1)2t\right]dt$$

$$= \int_0^1 (10t^3+5t^2+9t+2)dt = \frac{32}{3}.$$

***Example* 22.6**: Evaluate $\int_C z^2 dz$ where C is the curve joining the points A(1, 1) and B(2, 4) along

(i) $y = x^2$

(ii) $y = 3x - 2$

Solution: (i) Along $y = x^2$, dy = 2x dx and x varies from 1 to 2.

$$\int_C z^2 dz = \int_{AB} (x+iy)^2 (dx+idy) = \int_{AB} (x^2-y^2+2xyi)(dx+idy)$$

Substituting $y = x^2$, $dy = 2x\,dx$

$$= \int_{x=1}^{2} \left(x^2 - x^4 + 2x^3 i\right)\left(dx + 2xidx\right)$$

$$= \int_{x=1}^{2} \left(x^2 - 5x^4\right)dx + i\int_{x=1}^{2} (4x^3 - 2x^5)dx$$

$$= \left[\frac{x^3}{3} - x^5\right]_1^2 + i\left[x^4 - \frac{x^6}{3}\right]_1^2 = \frac{-86-18i}{3}$$

(ii) Along $y=3x-2$, $dy=3dx$

$$\int_C z^2 dz = \int_{AB} \left(x^2 - y^2 + 2xyi\right)(dx + idy)$$

$$= \int_1^2 \{x^2 - (3x-2)^2 + 2x(3x-2)i\}(dx + 3idx)$$

$$= (1+3i)\int_1^2 \left[\left(-8x^2 + 12x - 4\right) + i(6x^2 - 4x)\right]dx$$

$$= \frac{-86-18i}{3}$$

Exercises - 22.1

I. Evaluate the following Integrals:

1. $\int_0^{1+i} \left(x^2 - iy\right)dz$ along $y = x^2$ from $(0,0)$ to $(1,1)$

2. $\int_0^{3+i} z^2 dz$ along the line $3y = x$ from $(0,0)$ to $(3,1)$

3. $\int \sin z\, dz$ along the line $z = 0, z = i$

4. $\int_C \bar{z}^2 dz$ where C is straight line path from $z = 0$ to $z = 3 + 2i$

5. $\int_C (x^2 - iy^2)dz$ where C is the path joining (1.2) to (2,8) on the parabola $y = 2x^2$

6. $\int_C e^z dz$ where C is straight line path from $z = \pi i$ to $z = 1$

7. $\int_C |z|^2\, dz$ where C is the boundary of the square with vertices at (0,0), (2,0), (2,2) and (0,2)

8. $\int_C \bar{z}dz$ where C is the upper half of the circle $|z| = 1$ from z = 1 to z = −1

9. $\int_C z^2 dz$ where C is the line x = 2y from the point (0,0) to (4,2)

10. Evaluate $\int_C (z^2 + 1)dz$ where C is the square with vertices at z = 0, z = 1, z = 1 + *i*, z = *i*

11. If O, A, B are points O, 2 + i, 2 respectively and OA, OB, BA are sides of a triangle OAB, show that $\int_{OABO} z^2 dz = 0$.

12. Evaluate $\int_0^{1+i} (x - y + ix^2)dz$ along the real axis from z = 0 to z = 1 and then along a line parallel to the imaginary axis from z = 1 to z = 1 + *i*

13. Find $\int_{1-i}^{2+3i} \left(z^2 + z\right) dz$ along line joining the points A(1,-1) B(2,3)

14. Evaluate $\int_2^{-2+i} (2+z)^2 dz$

Answers

1. $\left(\frac{1-i}{2}\right)$ 2. $\frac{18+26i}{3}$ 3. $1-\cos i$ 4. $29-(98/3)i$

5. $(1+4i)((7/3)-(124/5)i)$ 6. $1+e$ 7. 0 8. πi

9. $\frac{16+88i}{3}$ 10. 0 12. $\frac{-3+5i}{6}$ 13. $\frac{64i-103}{6}$ 14. $-i/3$

22.2 Curves and Regions

22.2.1 Definitions

1. A curve is called a simple closed curve if it does not cross itself.
2. A region D is said to be a connected region if any two points of the region 'D' can be connected by a curve which lies entirely within the region .
3. A connected region 'D' is said to be simply connected if all the points of the area bounded by any single closed curve 'C' drawn in the region D are the points of the region D

4. If all the points of the area bounded by two or more closed curves drawn in the region D are the points of the region D, then the region D is said to be multiply connected region.
5. The lines drawn in multiply connected region without intersecting with any one of the curves, which make a multiply connected region a simply connected one are called cross-cuts

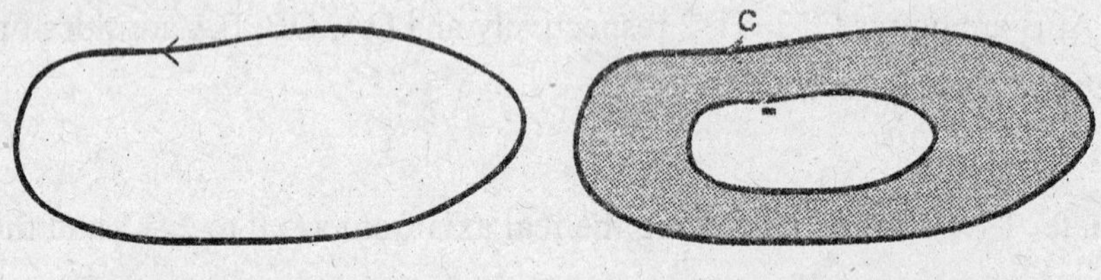

Simply Connected Region   Doubly Connected Region

Multiply connected region bounded by C and C_1, C_2,C_3 changed to simply connected with the help of the lines AB, PQ, RS which are called cross-cuts (i.e. now the region is bounded by single closed curve)

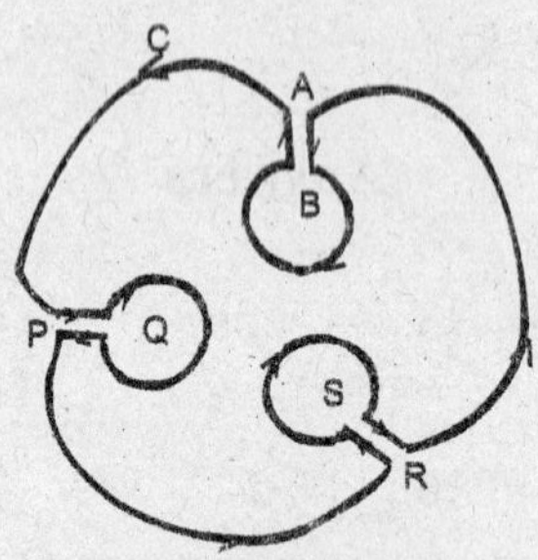

22.2.2 Cauchy's Integral Theorem (Cauchy's Theorem)

If f(z) is an analytic function of z and $f^1(z)$ is continuous at each point within and on a closed contour 'C', then $\int_C f(z)dz = 0$

Proof: Let D be the region bounded by the closed curve 'C', then

$$\int_C f(z)dz = \int_C (u+iv)(dx+idy)$$

where u and v are respectively the real and imaginary parts of f(z)

$$= \int_C (udx - vdy) + i\int_C vdx + udy \qquad .....(1)$$

Since $f^1(z)$ is continuous, the partial derivatives $\frac{\partial u}{\partial x}, \frac{\partial u}{\partial y}, \frac{\partial v}{\partial x}, \frac{\partial v}{\partial y}$ exist and are also continuous in D. Using Green's Theorem (1) gives

$$\int_C f(z)dz = \iint_D \left(\frac{-\partial v}{\partial x} - \frac{\partial u}{\partial y}\right)dxdy + i\iint_D \left(\frac{\partial u}{\partial x} - \frac{\partial v}{\partial y}\right)dxdy \qquad .....(2)$$

Since f(z) is analytic in D we have $\frac{\partial u}{\partial x} = \frac{\partial v}{\partial y}, \frac{\partial u}{\partial y} = \frac{-\partial v}{\partial x}$

Using Cauchy –Riemann equation in(2)

$$\therefore \quad \int_C f(z)dz = 0$$

An alternate form of the theorem is

If f(z) is analytic throughout a simply connected domain D, then for every closed contour C within D, $\int_C f(z)dz = 0$

22.2.3 Cauchy's Theorem for Doubly Connected Regions

If f(z) is analytic in the doubly connected region bounded by the curves C_1 and C_2 then

$$\int_{C_1} f(z)dz = \int_{C_2} f(z)dz$$

Proof : Let f(z) be analytic in the region D bounded by the curves C_1 and C_2. Draw a cross cut by joining a point A on C_1 to a point B on C_2. Now the doubly connected region becomes simply connected and by Cauchy's theorem

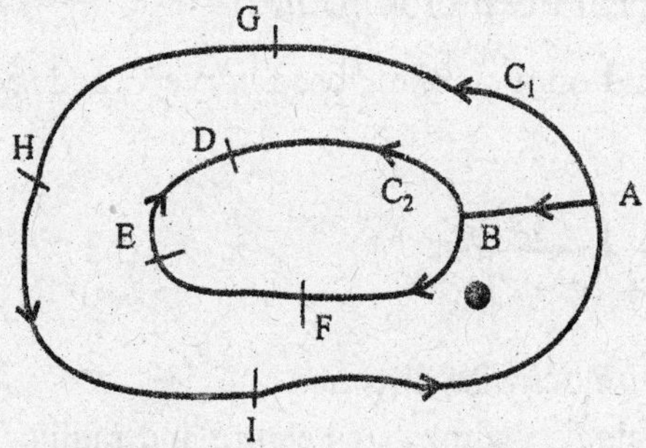

$$\int_{AGHIABFEDBA} f(z)dz = 0$$

$$\int_{AGHIA} f(z)dz + \int_{AB} f(z)dz + \int_{BFEDB} f(z)dz + \int_{BA} f(z)dz = 0 \qquad .....(1)$$

Since $\int_{AB} f(z)dz = -\int_{BA} f(z)dz$

(1) reduces to,

$$\int_{AGHIA} f(z)dz + \int_{BFEDB} f(z)dz = 0$$

$$\int_{C_1} f(z)dz - \int_{C_2} f(z)dz = 0$$

$$\therefore \quad \int_{C} f(z)dz = \int_{C_2} f(z)dz$$

***Note* 1**: In general this theorem can be extended to a multiply connected region bounded by an outer simple closed curve C and inner simple closed curves $C_1, C_2, C_3, ...C_n$ and it can be proved that

$$\int_{C} f(z)dz = \int_{C_1} f(z)dz + \int_{C_2} f(z)dz + ... + \int_{C_n} f(z)dz$$

***Note* 2**: Let C be a simple closed curve and let f(z) be analytic on and within C everywhere except at $z = z_0$. Suppose z_0 is enclosed within a small circle with centre at z_0 and radius r. Let C_0 be $|z - z_0| = r$. Then f(z) is analytic everywhere on and within C_0 and also out side C_0 (i.e., in D). From Note 1, it follows that

$$\int_{C} f(z)dz = \int_{C_0} f(z)dz$$

22.2.4 Cauchy's Integral Formulae (C.I.F.)

If f(z) is analytic within and on a simple closed curve C and z_0 is any point inside 'C' then

$$f(z_0) = \frac{1}{2\pi i}\int_{C} \frac{f(z)}{z - z_0} dz$$

where C is traversed in anti-clockwise direction

Proof : Draw a small circle C_1 with z_0 as centre and radius 'r' so small such that C_1 lies inside C. The only point where $\frac{f(z)}{z - z_0}$ is not analytic inside 'C' is z_0.

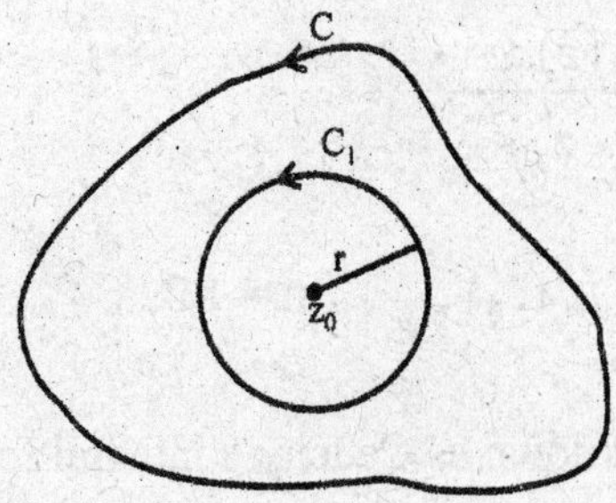

Hence $\dfrac{f(z)}{z-z_0}$ is analytic in the region between 'C' and 'C_1'. By using Cauchy's theorem for multiply connected regions,

$$\int_C \frac{f(z)}{z-z_0}dz = \int_{C_1} \frac{f(z)}{z-z_0}dz \qquad \text{.....(1)}$$

C_1 is the circle with centre at z_0 and radius 'r'

$\therefore$ $|z-z_0| = r$ is the equation of the circle .

Hence

$$z - z_0 = re^{i\theta} \quad \text{(i.e.) } z = z_0 + re^{i\theta},\ dz = re^{i\theta} i d\theta$$

Substituting these in R.H.S of (1)

$$\int_C \frac{f(z)}{z-z_0}dz = \int_0^{2\pi} \frac{f(z_0+re^{i\theta})}{re^{i\theta}} re^{i\theta} i d\theta$$

As $r \to 0$ the circle C_1 shrinks to the point z_0 and $f(z_0+re^{i\theta}) \to f(z_0)$

$$\therefore \quad \int_C \frac{f(z)}{z-z_0}dz = f(z_0)\int_0^{2\pi} i d\theta = f(z_0)2\pi i$$

$$\therefore \quad \int_C \frac{f(z)}{z-z_0}dz = 2\pi i f(z_0) \text{ i.e., } f(z_0) = \frac{1}{2\pi i}\int_C \frac{f(z)dz}{(z-z_0)} \qquad \text{.....(2)}$$

Differentiating (2) w.r.t to z_0 we get

$$f^1(z_0) = \frac{1}{2\pi i}\int_C \frac{f(z)dz}{(z-z_0)^2}$$

similarly, $$f^{11}(z_0) = \frac{2!}{2\pi i}\int_C \frac{f(z)dz}{(z-z_0)^3},\ f^{111}(z_0) = \frac{3!}{2\pi i}\int_C \frac{f(z)dz}{(z-z_0)^4}$$

....

$$f^n(z_0) = \frac{n!}{2\pi i}\int_C \frac{f(z)dz}{(z-z_0)^{n+1}} \quad .....(3)$$

i.e., $$\int_C \frac{f(z)dz}{(z-z_0)^{n+1}} = \frac{2\pi i}{n!} f^n(z_0), \qquad n = 1,2,....$$

The results (2) and (3) are known as Cauchy's integral formulae.

22.2.5 Cauchy's Integral Formula for Multiply Connected Region

In general if the region is bounded by exterior curve 'C' and interior curves C_1, C_2,...C_n then,

$$f^1(z_0) = \frac{1}{2\pi i}\int_C \frac{f(z)dz}{(z-z_0)^2} - \frac{1}{2\pi i}\int_{C_1} \frac{f(z)dz}{(z-z_0)^2} - \frac{1}{2\pi i}\int_{C_2} \frac{f(z)dz}{(z-z_0)^2} - \cdots \frac{1}{2\pi i}\int_{C_n} \frac{f(z)dz}{(z-z_0)^2}$$

This is called generalized Cauchy's integral formula.

Solved Examples

***Example* 22.7:** Evaluate $\int_C \frac{e^{2z}+1}{z} dz$ where C is $|z| = 1$

***Solution*:** The function $\frac{e^{2z}+1}{z}$ is not an analytic at z = 0 and the point z = 0 is inside the circle $|z| = 1$. The Cauchy's integral Formula is $\int_C \frac{f(z)}{z-z_0} dz = 2\pi i f(z_0)$

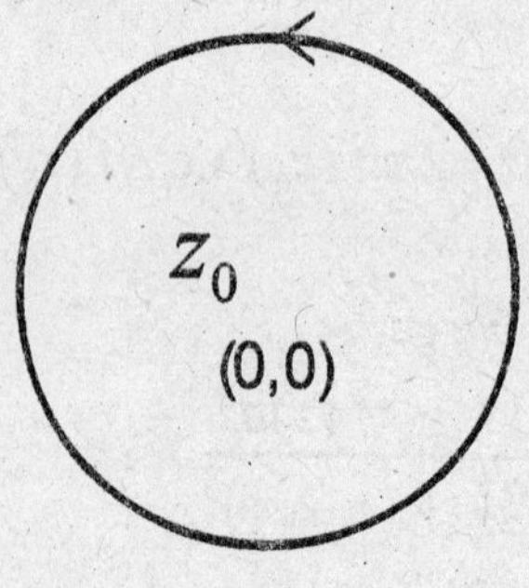

$$\int_C \frac{e^{2z}+1}{z-z_0} dz = 2\pi i f(z_0)$$

$$f(z) = e^{2z} + 1, z_0 = 0,\ f(z_0) = e^0 + 1 = 2$$

$$\int_C \frac{e^{2z} + 1}{z} dz = 2\pi i f(z_0) = 4\pi i$$

***Example* 22.8:** Evaluate $\int_C \frac{\cos z}{z(z^2+8)} dz$ where C is the boundary of the square whose sides lie along the lines x = ±2 and $y = \pm 2$

***Solution*:** The function $\int_C \frac{\cos z}{z(z^2+8)} dz$ is not analytic at z = 0 and $z = \pm 2\sqrt{2}$ i ; $z = 0$ lies inside *C*

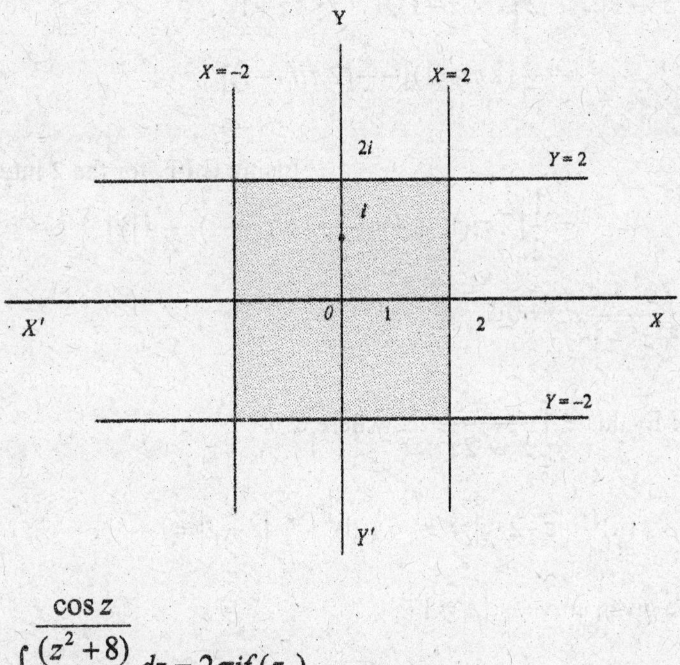

$$\int_C \frac{\frac{\cos z}{(z^2+8)}}{z-0} dz = 2\pi i f(z_0)$$

where $\quad f(z) = \frac{\cos z}{(z^2+8)},\ z_0 = 0$

$$f(0) = \frac{1}{8}$$

$$\int_C \frac{\cos z}{z(z^2+8)} dz = 2\pi i \frac{1}{8} = \frac{\pi i}{4}$$

***Example* 22.9:** Evaluate $\int_C \frac{z^2+4}{z^2-1}dz$ where C is $|z|=2$

Solution: $\frac{1}{z^2-1} = \frac{1}{2}\frac{1}{z-1} - \frac{1}{2}\frac{1}{z+1}$

$$\therefore \int_C \frac{z^2+4}{z^2-1}dz = \frac{1}{2}\int_C \frac{z^2+4}{z-1}dz - \frac{1}{2}\int_C \frac{z^2+4}{z+1}dz$$

the point z = 1 and z = –1 are inside C

Using Cauchy's integral formula

$$\int_C \frac{z^2+4}{z^2-1}dz = \frac{1}{2}\int_C \frac{(z^2+4)dz}{z-1} - \frac{1}{2}\int_C \frac{z^2+4}{z+1}dz$$

$$= \frac{1}{2}[2\pi i f(1)] - \frac{1}{2}[2\pi i f(-1)]$$

[using C.I.F. for the 2 integrals on R.H.S]

$$= \frac{1}{2}[2\pi i(1^2+4) - \frac{1}{2}\{2\pi i f[(-1)^2+4]\}]$$

$$\therefore \int_C \frac{z^2+4}{z^2-1}dz = 0$$

***Example* 22.10:** Evaluate $\int_C \frac{z+1}{z^3-2z^2}dz$ where C is

(i) $|z|=1$ (ii) $|z-2-i|=2$ (iii) $|z+2|=1$

Solution: (i) The given curve is $|z|=1$

Z = 0 is inside 'C' and z = 2 is out side 'C'

The given integral can be written as $\int_C \frac{z+1}{z^2(z-2)}dz = \int_C \frac{\frac{z+1}{(z-2)}}{(z-0)^2}dz$

which is in the form $\int_C \frac{f(z)dz}{(z-z_0)^{n+1}}$ .Here $f(z) = \frac{z+1}{(z-2)}, z_0 = 0, n = 1$

But by C.I.F $\int_C \frac{f(z)dz}{(z-z_0)^{n+1}} = \frac{2\pi i}{\underline{|n}} f^n(z_0)$

$$\therefore \int_C \frac{\frac{z+1}{(z-2)}}{(z-0)^2} dz = \frac{2\pi i}{1!} f^1(z_0) = 2\pi i \left. \frac{d\left(\frac{z+1}{z-2}\right)}{dz} \right|_{z=0}$$

$$= 2\pi i \left[\frac{-3}{(z-2)^2}\right]_{z=0} = \frac{-3\pi i}{2}.$$

(ii) The given curve C is $|z-2-i| = 2$ which is circle with centre (2, 1) and radius 2.

The point z = 0 is outside 'C' and the point z = 2 is inside 'C'.

$\int_C \frac{\left(\frac{z+1}{z^2}\right)}{z-2} dz$ is in the form $\int_C \frac{f(z)dz}{z-z_0}$

Where $f(z) = \left(\frac{z+1}{z^2}\right), z_0 = 2$ and $f(z)$ is analytic in C

$$\therefore \quad \int_C \frac{\left(\frac{z+1}{z^2}\right)}{z-2} dz = 2\pi i f(2) = 2\pi i \left[\frac{z+1}{z^2}\right]_{z=2} = 2\pi i \left(\frac{3}{4}\right) = \frac{3\pi i}{2} \quad \text{(using C.I.F)}$$

(iii) The given curve C is $|z+2| = 1$ which is a circle with centre (– 2, 0) and radius 1

Both the points z = 0, z = 2 are outside 'C'.

$$\therefore \quad \int_C \left(\frac{z+1}{z^2(z-2)}\right) dz = 0$$

***Example* 22.11:** Evaluate $\int_C \frac{e^{2z}}{(z-1)(z-2)} dz$ where C is $|z| = 4$

***Solution*:** Let $f(z) = e^{2z}$

$$\frac{1}{(z-1)(z-2)}=\frac{1}{z-2}-\frac{1}{z-1} \quad \text{(using partial fractions)}$$

z =1 and z = 2 are inside 'C'

$$\therefore \int_C \frac{e^{2z}}{(z-1)(z-2)}dz=\int_C \frac{e^{2z}}{z-2}dz-\int_C \frac{e^{2z}}{z-1}dz$$

In the first integral $f(z)=e^{2z}, z_0=2$

$$f(z_0)=f(2)=e^4$$

and in the second integral $f(z)=e^{2z}, z_0=1,\ f(z_0)=f(1)=e^2$

Using Cauchy's integral formula

$$\int_C \frac{e^{2z}}{z-2}dz-\int_C \frac{e^{2z}}{z-1}dz=2\pi i f(2)-2\pi i f(1)=2\pi i(e^4-e^2)$$

***Example* 22.12:** Evaluate $\int_C \frac{e^{2z}}{(z+1)^4}dz$ where C is $|z|=2$.

***Solution*:** Let f(z) = $e^{2z}, z_0=-1$, lies inside C; $n=3$; *f*(z) is analytic in C.

$$f^{111}(z)=8e^{2z}$$

Using Cauchy's integral formula, $\int_C \frac{f(z)dz}{(z-z_0)^{n+1}}=2\pi i\frac{f^n(z_0)}{n!}$

$$\int_C \frac{e^{2z}dz}{[z-(-1)]^{3+1}}=2\pi i\frac{f^{111}(-1)}{3!}=\frac{2\pi i}{6}8e^{-2}$$

$$\therefore \int_C \frac{e^{2z}}{(z+1)^4}dz=\frac{8\pi i}{3e^2}$$

***Example* 22.13:** Evaluate $\int_C \frac{z^3-z}{(z-2)^3}dz$ where C is $|z|=3$.

***Solution*:** Let *f*(z) = $z^3-z, z_0=2$, lies inside C $n=2$, and *f*(z) is analytic in C

$$f^1(z) = 3z^2 - 1, f^{11}(z) = 6z \quad \therefore f^{11}(2) = 12$$

Using Cauchy's integral formula

$$\int_C \frac{f(z)dz}{(z-z_0)^{n+1}} = \frac{2\pi i}{n!} f^n(z_0), \text{ we get}$$

$$\int_C \frac{(z^3 - z)dz}{(z-2)^{2+1}} = \frac{2\pi i}{2!} f^{11}(z_0) = \frac{2\pi i}{2!} f^{11}(2) = 12\pi i$$

Exercises – 22.2

I. Prove the following using Cauchy's integral formula.

1. $\int_C \frac{z^2 - 3z - i}{z - 2i} dz = 2\pi(7 - 4i)$ where C is $|z|=3$
2. $\int_C \frac{3z+5}{z^2+z} dz = 10\pi i$ where C is $|z| = 1/2$
3. $\int_C \frac{1}{z^2+4} dz = \frac{\pi}{2}$ where C is $|z - i| = 2$
4. $\int_C \frac{\sin \pi z^2 + \cos \pi z^2}{(z-1)(z-2)} dz = 4\pi i$ where C is $|z| = 3$
5. $\int_C \frac{z}{2z+1} dz = \frac{-\pi i}{2}$ where C is a square whose sides lie along the lines $x = \pm 2$ and $y = \pm 2$
6. $\int_C \frac{e^{-z}}{\left(z - \frac{\pi i}{6}\right)} dz = 2\pi$ where C is a square whose sides lie along the lines $x = \pm 2$ and $y = \pm 2$
7. $\int_C \frac{\sin^2 z}{\left(z - \frac{\pi}{6}\right)^3} dz = \pi i$ where C is $|z| = 1$
8. $\int_C \frac{e^z}{(z+1)^2} dz = 2\pi i / e$, where C is $|z - 1| = 3$
9. $\int_C \frac{e^{-2z}}{z^2} dz = -2\pi i$ where C is $|z| = 1$

10. $\int_C \frac{e^{-2z}}{(z+1)^6} dz = \frac{-8\pi i e^2}{15}$ where C is $|z| = 2$

11. $\int_C \frac{3z^2 + z}{z^2 - 1} dz = 2\pi i$ where C is $|z| = 2$

12. $\int_C \frac{\sin z}{\left(z - \frac{\pi}{2}\right)^5} dz = \frac{\pi i}{12}$ where C is $|z| = 2$

13. $\int_C \frac{e^{2z} dz}{(z-1)(z-2)} = 2\pi i(e^4 - e^2)$ where C is $|z| = 3$

Summary

1. *Line integral (or contour integral)*:
   Let C represent a simple curve and let *f*(*z*) be continuous at each point of C, then the integral $\int_c f(z)dz = \int_c f(x, y)[dx + i\,dy]$ is called a line integral.

2. *Cauchy's integral theorem (Cauchy's theorem)*:
   If *f*(*z*) is analytic and $f'(z)$ is continuous at each point within and on a closed contour C, then, $\int_c f(z)dz = 0$.

3. *Cauchy's integral formula* (C.I.F):
   If *f*(*z*) is analytic with in and on a closed curve C and z_0 is any point inside C, then,

$$\int_c \frac{f(z)dz}{z - z_0} = 2\pi i\, f(z_0)$$

4. Generalised C.I.F (or C.I.F for derivatives of *f*(*z*))

$$\int_c \frac{f(z)dz}{(z - z_0)^{n+1}} = \frac{2\pi i}{\underline{|n}} f^n(z_0), \text{ n} = 1, 2\ldots$$

Solved University Questions

1. $\int_c \frac{ze^z dz}{(z+a)^3}$ where *c* is any simple closed curve enclosing the point $z = -a$

***Solution*:** The point $z = -a$ lies inside 'C',

$$I=\int_c \frac{ze^z\,dz}{\left[z-(-a)\right]^{2+1}}$$

$f(z)=ze^z$, $z_0=-a$, $n=2$, $f^{11}(z)=(z+2)e^z$, $f^{11}(-a)=(2-a)e^{-a}$

and $f(z)$ is analytic in C.

$\therefore$ By Cauchy's integral formula $\dfrac{2\pi i f^n(z_0)}{n!}=\displaystyle\int_c \frac{f(z)\,dz}{(z-z_0)^{n+1}}$.

We have

$$\int_c \frac{ze^z\,dz}{\left[z-(-a)\right]^{2+1}}=\frac{2\pi i f^{11}(-a)}{2!}=\pi i(2-a)e^{-a}$$

2. $\displaystyle\int_c\left[\frac{e^z}{z^3}+\frac{z^4}{(z+i)^2}\right]dz$ where c is $|z|=2$ using Cauchy's integral formula.

Solution: Let $\displaystyle\int_c\left[\frac{e^z}{z^3}+\frac{z^4}{(z+i)^2}\right]dz=\text{I}$ (say)

$$I=\int_c\frac{e^z}{z^3}dz+\int_c\frac{z^4}{(z+i)^2}dz$$

when c is $|z|=2$.which is a circle with center at origin and radius 2.

Arranging both the integrals in the form of Cauchy's – Integral formula

$\dfrac{2\pi i f^n(z_0)}{n!}=\displaystyle\int_c\frac{f(z)\,dz}{(z-z_0)n+1}$,we have

$$I=\int_c\frac{e^z\,dz}{(z-0)^{2+1}}+\int_c\frac{z^4}{\left[z-(-i)\right]^{1+1}}$$

For the first integral $f(z)=e^z$, $z_0=0$ lies inside 'c', [$f(z)$ is analytic in C]

$$n=2, f^{11}(z)=e^z,\ f^{11}(0)=e^0=1$$

For the second integral $f(z)=z^4$, $z_0=-i$ lies inside 'c', $n=1$

$f^1(z) = 4z^3$, $f^1(-i) = 4(-i)^3 = 4i$ [$f(z)$ is analytic in C]

$$\therefore \quad I = \int_c \frac{e^2 dz}{(z-0)^{2+1}} + \int_c \frac{z^4 dz}{[z-(-i)]^{1+1}} = \frac{2\pi i f^{11}(0)}{2!} + \frac{2\pi i f^1(-i)}{1!}$$

$$I = \pi i(1) + 2\pi i(4i) \text{ i.e., } \quad I = \pi(i-8)$$

3. Evaluate $\int_c \frac{e^z}{(z^2+\pi^2)^2} dz$ where c is $|z| = 4$ using Cauchy's integral formula.

Solution: Given $\int_c \frac{e^z}{(z^2+\pi^2)^2} dz$ where c is $|z| = 4$ which is a circle with center at $z = 0$ and radius 4;

Let $f(z) = e^z$; $\therefore$ $f(z)$ is analytic in C.

The points $z = \pm\pi i$ lie inside c, consider $\frac{1}{(z^2+\pi^2)^2} = \frac{1}{(z+\pi i)^2(z-\pi i)^2}$

$$= \frac{A}{z+\pi i} + \frac{B}{(z+\pi i)^2} + \frac{C}{(z-\pi i)} + \frac{D}{(z-\pi i)^2}$$

Using partial fractions, solving for A, B, C, D we get,

$$A = \frac{7}{2\pi^3 i}, \quad B = \frac{-1}{4\pi^2}, \quad C = \frac{-7}{2\pi^3 i}, \quad D = \frac{-1}{4\pi^2}$$

$\therefore$

$$I = \int_c \frac{e^z dz}{(z^2+\pi^2)^2} = \frac{7}{2\pi^3 i}\int_c \frac{e^z dz}{z+\pi i} - \frac{1}{4\pi^2}\int_c \frac{e^z dz}{(z+\pi i)^2} - \frac{7}{2\pi^3 i}\int_c \frac{e^z dz}{z-\pi i} - \frac{1}{4\pi^2}\int_c \frac{e^z dz}{(z-\pi i)^2}$$

Using Cauchy's integral formula.

$$= \frac{7}{2\pi^3 i}[2\pi i f(-\pi i)] - \frac{1}{4\pi^2}[2\pi i f'(-\pi i)] - \frac{7}{2\pi^3 i}[2\pi i f(\pi i)] - \frac{i}{4\pi^2}[2\pi i f'(\pi i)]$$

$$= \frac{7}{\pi^2}e^{-\pi i} - \frac{i}{2\pi}e^{-i\pi} - \frac{7}{\pi^2}e^{i\pi} - \frac{i}{2\pi}e^{\pi i} = \frac{i}{\pi}$$

4 Evaluate $\int_c \frac{dz}{z^8(z+4)}$ when c is $|z| = 2$

***Solution*:** Given $I = \int_c \frac{dz}{z^8(z+4)}$, c is $|z| = 2$

The point z = 0 is inside '*c*' and z = – 4 out side '*c*' which is the circle $x^2 + y^2 = 4$

Let $f(z) = \frac{1}{z+4}$; which is analytic in C

$$I = \int_c \frac{\left(\frac{1}{z+4}\right)dz}{(z-0)^{7+1}} = \frac{2\pi i}{7!} f^7(z_0) \quad \text{using Cauchy's integral formula}$$

$$f(z) = \frac{1}{z+4},\ z_0 = 0,\ n = 7$$

$$f^7(z) = 7!(-1)^7(z+4)^{-8},\ f^7(0) = \frac{-7!}{(0+4)^8}$$

$$\therefore \quad I = \int_c \frac{\left(\frac{1}{z+4}\right)dz}{(z-0)^{7+1}} = \frac{2\pi i}{7!}\left[\frac{-7!}{4^8}\right] = \frac{-\pi i}{32768}$$

5. Evaluate $\int_c \frac{(z^2-z-1)}{z(z-i)^2} dz$ where c is | z- ½ | = 1 using Cauchy's integral formula.

***Solution*:** Given $I = \int_c \frac{(z^2-z-1)}{z(z-i)^2} dz$; c is | z- ½ | = 1

c is the circle $\left(x - \frac{1}{2}\right)^2 + y^2 = 1^2$ centre at (½ , 0), radius 1

The point z = 0 lies inside the circle and the point z = i lies outside '*c*'

∴ Using Cauchy's integral formula

$$I = \int_c \frac{\left[\frac{z^2-z-1}{(z-i)^2}\right]}{z-0} dz = 2\pi i f(0),\ f(0) = 1 \text{ where } f(z) = \frac{z^2-z-1}{(z-i)^2}$$

which is analytic in C

$$\therefore\ I = \int_c \frac{z^2 - z - 1}{z(z-i)^2}\,dz = 2\pi i(1) = 2\pi i \qquad \text{(by C.I.F.)}$$

6. Evaluate $\int_c \frac{z-2}{(z-1)^3(z+2)^3}\,dz$ where c is $|z| = 3$ using Cauchy's integral formula.

Solution: The integral $\frac{z-2}{(z-1)^3(z+2)^3}$ has singularities z = 1 and z = –2 inside the circle c:| z | = 3

$$\therefore \quad \int_c \frac{z-2}{(z-1)^3(z+2)^3}\,dz = \int_c \frac{\frac{z-2}{(z+2)^3}}{(z-1)^{2+1}}\,dz + \int \frac{\frac{z-2}{(z-1)^3}}{(z+2)^{2+1}}\,dz = I_1 + I_2 \text{ (say)}$$

Cauchy's integral formula is $\int_c \frac{f(z)}{(z-a)^{n+1}}\,dz = \frac{2\pi i f(a)}{n!}$

$$I_1 = \frac{2\pi i}{\underline{|2}}\frac{d^2}{dz^2}\left[\frac{z-2}{(z+2)^3}\right]_{z=+1} = \frac{2\pi i}{(2)}\frac{d^2}{dz^2}\left[(z-2)(z+2)^{-3}\right]_{z=1} = \left[\pi i D_1\right]_{z=1} \text{ (say)}$$

$$D_1 = \frac{d}{dz}\left[(z-2)(-3)(z+2)^{-4} + (z+2)^{-3}\right]$$

$$= \frac{d}{dz}\left[(z+2)^{-4}(-3z+6+z+2)\right] = \frac{d}{dz}\left[(z+2)^{-4}(8-2z)\right]$$

$$= 2\left[(z+2)^{-4}(-1) + (4-z)(z+2)^{-5}(-4)\right]$$

$$= 2\left[(z+2)^{-5}\right]\left[-16+4z-z-2\right] = \frac{6(z-6)}{(z+2)^5}$$

$$(D_1)_{z=1} = \frac{6\times -5}{3^5} = \frac{-10}{81}$$

$$I_2 = \frac{2\pi i}{2!}\frac{d^2}{dz^2}\left[\frac{z-2}{(z-1)^3}\right]_{z=-2} = \left[\pi i\, D_2\right]_{z=-2} \text{ (say)}$$

$$D_2 = \frac{d}{dz}\left[(z-1)^{-3}.1 + (z-2)(-3)(z-1)^{-4}\right]$$

$$= \frac{d}{dz}\left[(z-1)^{-4}(5-2z)\right] = \left[(z-1)^{-4}(-2)+(5-2z)(-4)(z-1)^{-5}\right]$$

$$= \frac{-2z+2-20+8z}{(z-1)^5} = \frac{6z-18}{(z-1)^5}$$

$$(D_2)_{z=-2} = \frac{-30}{(-3)^5} = \frac{10}{81}$$

$$\therefore \quad \int_c \frac{z-2}{(z-1)^3(z+2)^3}dz = I_1 + I_2 = \left[(D_1)_{z=1} + (D_2)_{z=-2}\right]\pi i$$

$$= \pi i\left[\frac{-10}{81} + \frac{10}{81}\right] = 0$$

7. Evaluate $\int_c \left(y^2+z^2\right)dx + \left(z^2+x^2\right)dy + \left(x^2+y^2\right)dz$ from (0,0 ,0) to (1,1,1) where c is the curve $x = t$, $y = t^2$, $z = t^3$ in the parametric form.

Solution: $x = t$; $dx = dt$; $y = t^2$; $dy = 2tdt$; $z = t^3 \Rightarrow dz = 3t^2dt$ and t varies from 0 to 1.

$$\therefore \text{ given integral} = \int_{t=0}^{1}\left[(t^4+t^6)+(t^6+t^2)2t+(t^2+t^4)3t^2\right]dt$$

$$\int_{t=0}^{1}\left[(t^4+t^6)+(t^6+t^2)2t+(t^2+t^4)3t^2\right]dt = \int_0^1\left[2t^7+4t^6+4t^4+2t^3\right]dt$$

$$= \frac{2}{8}+\frac{4}{7}+\frac{4}{5}+\frac{2}{4} = \frac{297}{140}$$

8. Evaluate $\int (x^2+ixy)\, dz$ from $A(1,1)$ to B(2,8) along $x = t,\ y = t^3$

Solution: $I = \int_{(1,1)}^{(2,8)} \left(x^2+ixy\right)\left(dx+idy\right)$

Equation of AB is $(y-1) = \frac{8-1}{2-1}(x-1)$

$y = 7x-6,\ dy = 7dx,\ x$ varies from 1 to 2

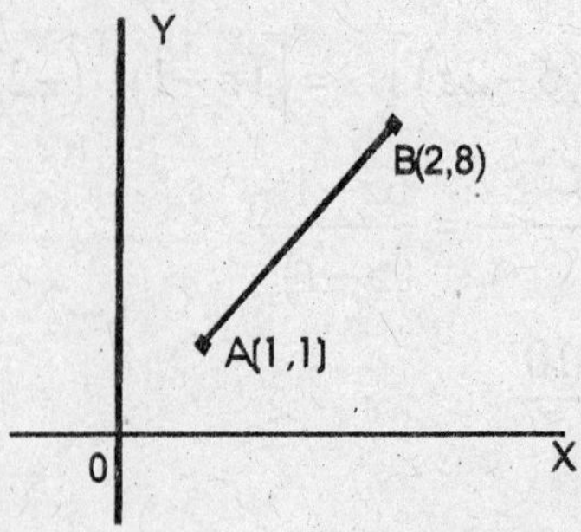

$$I = \int_{x=1}^{2} \left(x^2 + ix(7x-6)(dx + i7dx)\right) = (7i+1)\int_1^2 \left[(7i+1)x^2 - 6ix\right] dx$$

$$= (7i+1)\left\{(7i+1)\frac{x^3}{3} - 6i\frac{x^2}{2}\right\}_1^2 = (7i+1)\left\{(7i+1)\frac{8}{3} - 12i - \frac{7i+1}{3} + 3i\right\}$$

$$= \frac{(7i+1)}{3}[22i+7]$$

9. Using Cauchy's integral formula evaluate $\int_c \frac{z^4 dz}{(z+1)(z-i)^2}$ where c is $9x^2 + 4y^2 = 36$

Solution: Given $\frac{x^2}{2^2} + \frac{y^2}{3^2} = 1$ is the ellipse with centre (0, 0)

$\therefore$ The points $z = -1, i$ are inside 'c'

$$\frac{1}{(z+1)(z-i)^2} = \frac{A}{z+1} + \frac{B}{z-i} + \frac{C}{(z-i)^2}$$

using partial fractions , we get the values.

$$A = \frac{1}{(1+i)^2};\ B = \frac{-1}{(1+i)^2};\ C = \frac{1}{1+i}$$

$$\frac{1}{(z+1)(z-i)^2} = \frac{1}{(1+i)^2}\cdot\frac{1}{z+1} - \frac{1}{(1+i)^2}\cdot\frac{1}{(z-i)} + \frac{1}{1+i}\cdot\frac{1}{(z-i)^2}$$

Hence

$$I = \int_c \frac{z^4 dz}{(z+1)(z-i)^2} = \frac{1}{(1+i)^2}\int_c \frac{z^4}{z+1}dz - \frac{1}{(1+i)^2}\int_c \frac{z^4}{z-i}dz + \frac{1}{1+i}\int_c \frac{z^4}{(z-i)^2}dz$$

Using Cauchy's integral formula

$$=\frac{1}{(1+i)^2}[2\pi i f(-1)]-\frac{1}{(1+i)^2}[2\pi i f(i)]+\frac{1}{1+i}[2\pi i f'(i)]$$

(In all these integrals $f(z)=z^4$)

$$I=\frac{2\pi i}{(1+i)^2}.1-\frac{2\pi i}{(1+i)^2}+\frac{2\pi i}{1+i}4i^3=\frac{8\pi}{1+i}=4\pi(1-i)$$

10. Integrate z^2 along the straight lines along the path OLM when O is origin, L is the point z = 3 and M is z = 3+ i

Solution: $\int_{OLM} z^2dz=\int_{OL} z^2dz+\int_{LM} z^2dz$ ....(1)

Equation of OL is $y=0,\ dy=0,\ x$ varies from 0 to 3

Equation of LM is $x=3,\ dx=0,\ y$ varies from 0 to 1

Substituting these values in (1)

$$\int_{OLM} z^2dz=\int_0^3 (x+i0)^2(dx+i0)+\int_o^1 (3+iy)^2(0+idy)$$

$$=\left[\frac{x^3}{3}\right]_0^3+i\left[9y-\frac{y^3}{3}+6i\frac{y^2}{2}\right]_0^1$$

$$=9+i\left[9-\frac{1}{3}+3i\right]=\frac{18+26i}{3}$$

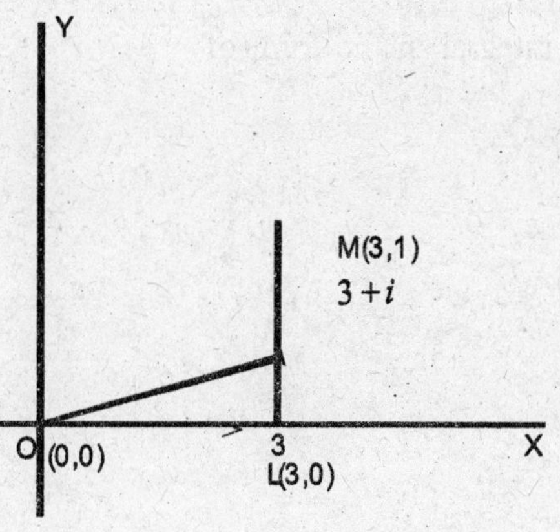

11. Evaluate $\int_c \frac{(z^3+z^2+2z-1)}{(z-1)^3}dz$ where c is $|z| = 3$ using Cauchy's integral formula.

***Solution*:** Given $\int_c \frac{(z^3+z^2+2z-1)}{(z-1)^3}dz$, c is $|z| = 3$

Arranging the given integral in the form of Cauchy's integral formula.

$$\int_c \frac{(z^3+z^2+2z-1)}{(z-1)^{2+1}}dz = 2\pi i\frac{f^{11}(z_0)}{2!} \qquad \text{......(1)}$$

where $f(z) = z^3+z^2+2z-1,\ z_0 = 1$ which lies inside 'c'

$$f'(z) = 3z^2+2z+2\ ;\ f''(z) = 6z+2;\ f''(1) = 8$$

$$\therefore \quad \int_c \frac{(z^3+z^2+2z-1)}{(z-1)^3}dz = 2\pi i.\frac{8}{2} = 8\pi i$$

12. Evaluate $\int_c \frac{(z^2-2z-2)}{z(z^2+1)^2}dz$ where c is $|z - i| = ½$ using Cauchy's integral formula.

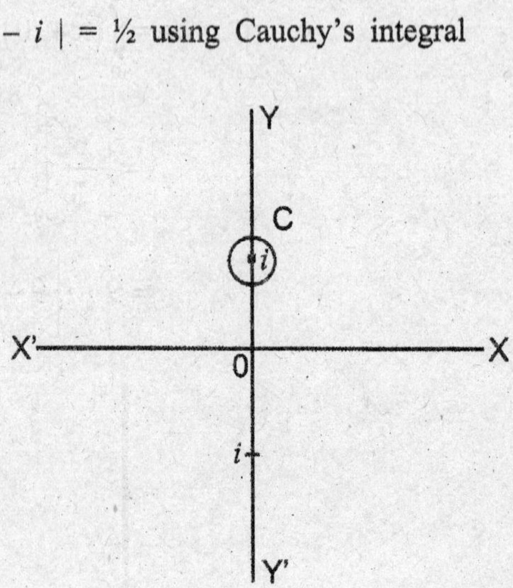

***Solution*:** Given $\int_c \frac{(z^2-2z-2)}{z(z^2+1)^2}dz$; c is $|z-i| = ½$

c is the circle centre at (0, 1) and radius ½ . $z(z^2+1)^2 = 0 \Rightarrow z = 0, \pm\ i$. Only $z = i$ lies inside 'c'

$\therefore$ Arranging the given integral in the form of Cauchy's integral formula

Given

$$\int_c \frac{\frac{(z^2-2z-2)}{z(z+i)^2}}{(z-i)^{1+1}}dz = 2\pi i f'(z_0) \qquad \text{....(1)}$$

where $z_0 = i$, $f(z) = \dfrac{z^2-2z-2}{z(z+i)^2}$

$$f'(z) = \frac{z(z+i)^2(2z-2)-(z^2-2z-2)\left[z.2(z+i)+(z+i)^2.1\right]}{z^2(z+i)^4}$$

$$f'(z) = \frac{2z(z-1)(z+i)-(z^2-2z-2)(3z+i)}{z^2(z+i)^3}$$

$$f'(z_0) = f'(i) = \frac{2i(i-1)i+i-(i^2-2i-2)(3i+i)}{i^2(i+i)^3}$$

$f'(i) = \dfrac{8i-4}{8i}$. Substituting in (1)

$$\int_c \frac{\dfrac{z^2-2z-2}{z(z+i)^2}}{(z-i)^2}dz = 2\pi i\left(\frac{8i-4}{8i}\right) = \pi(2i-1)$$

13. Evaluate $\displaystyle\int_{(0,0)}^{(1,1)} (3x^2+4xy+ix^2)\,dz$ along $y = x^2$

***Solution*:** $z = x+iy \Rightarrow dz = dx+idy$ Along the curve $y = x^2$, $dy = 2xdx$, x varies from 0 to 1

$$I = \int_{(0,0)}^{(1,1)} (3x^2+4xy+ix^2)\,dz$$

Substituting $y = x^2$, $dy = 2xdx$, $z = x+ix^2$, $dz = dx+i2xdx$

$$= \int_0^1 (3x^2+4x.x^2+ix^2)(dx+i2xdx)$$

$$I = \left[(3+i)\frac{x^3}{3}+2(3i+1)\frac{x^4}{4}+8i\frac{x^5}{5}\right]_0^1 = \frac{3+i}{3}+\frac{3i+1}{2}+\frac{8i}{5} = \frac{3}{2}+\frac{103i}{30}$$

14. Evaluate $\int_c \frac{e^{2z}}{\left(z^2+\pi^2\right)^3} dz$ where c is $|z| = 4$ using Cauchy's integral formula.

Solution: The integrand has singularities at $z = \pm i\pi$ inside $c : |z| = 4$.

$\therefore\ I = I_1 + I_2$ where

$$I_1 = \int_c \frac{e^{2z}/(z+i\pi)^3}{(z-i\pi)^3} dz = \frac{2\pi i}{2!}\frac{d^2}{dz^2}\left[e^{2z}(z+i\pi)^{-3}\right] \quad \text{at } z = i\pi$$

$$= \pi i\left[D_1\right]_{z=i\pi} \quad \text{(say)}$$

$$I_2 = \int_c \frac{e^{2z}/(z-i\pi)^3}{(z+i\pi)^3} = \frac{2\pi i}{2!}\frac{d^2}{dz^2}\left[e^{2z}(z-i\pi)^{-3}\right] \quad \text{at } z = -i\pi$$

$$= \pi i\left[D_2\right]_{z=-i\pi} \quad \text{(say)}$$

$$D_1 = \frac{d}{dz}\left[e^{2z}(-3)(z+i\pi)^{-4} + 2e^{2z}(z+i\pi)^{-3}\right]$$

$$= \frac{d}{dz}\left[e^{2z}(z+i\pi)^{-4}(-3+2z+2i\pi)\right]$$

$$= e^{2z}\left[2(z+i\pi)^{-4}(-3+2z+2i\pi)+(-4)(z+i\pi)^{-5}(-3+2z+2i\pi)+(z+i\pi)^{-4}.2\right]$$

$$= \frac{e^{2z}}{(z+i\pi)^5}\left[(-6+4z+4i\pi)+(12-8z-8i\pi)+2z+2i\pi\right]$$

$$= \frac{e^{2z}}{(z+i\pi)^5}\left[-6-2z-2i\pi\right] = \frac{e^{2z}}{(z+i\pi)^5}(3-z-i\pi)$$

$$\left[D_1\right]_{z=i\pi} = \frac{e^{2i\pi}}{32i\pi^5}\left[3-2i\pi\right] = \frac{e^{2i\pi}}{32\pi^5}(2\pi-3i) = \frac{1}{32\pi^5}(2\pi-3i)$$

Similarly, $\left[D_2\right]_{z=-i\pi} = \frac{1}{32\pi^5}(2\pi+3i)$

$$I = \pi i\left[\frac{1}{32\pi^5}(2\pi-3i+2\pi+3i)\right] = \frac{i}{32\pi^4}(4\pi) = \frac{i}{8\pi^3}$$

15. Evaluate $\int_c (y^2+2xy)dx+(x^2-2xy)dy$

where c is the boundary of the region $y=x^2$ and $x=y^2$

***Solution*:** The given two curves $y=x^2$ and $x=y^2$ intersect at the points (0, 0) and (1, 1)

$$I=\int_c (y^2+2xy)dx+(x^2-2xy)dy = \int_{opAqo} (y^2+2xy)dx+(x^2-2xy)dy$$

$$= \int_{opA} (y^2+2xy)dx+(x^2-2xy)dy + \int_{Aq0} (y^2+2xy)dx+(x^2-2xy)dy \quad \text{.....(1)}$$

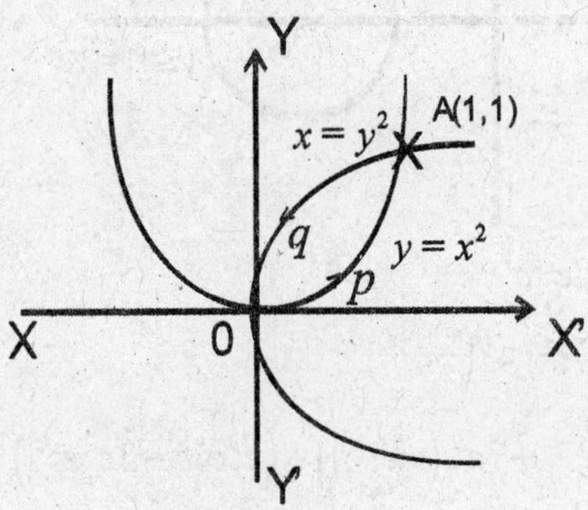

Along opA, $y=x^2$, $dy=2xdx$, x varies from 0 to 1 .

And along Aqo, $x=y^2$, dx =2ydy, y varies from 1 to 0

Substituting these in (1)

$$I=\int_{x=0}^{1} (x^4+2x.x^2)dx+(x^2-2x.x^2)2xdx+\int_{y=1}^{0} (y^2+2y^2-y)2y+(y^4-2y^2-y)dy$$

$$= \int_0^1 (4x^3-3x^4)dx+\int_{y=1}^{0} 5y^4dy$$

$$= \left[\frac{4x^4}{4}-\frac{3x^5}{5}\right]_{x=0}^{1}+\left[5\frac{y^5}{5}\right]_{y=1}^{0} = \left(1-\frac{3}{5}\right)+(0-1)=\frac{2}{5}-1=-\frac{3}{5}$$

16. Evaluate using Cauchy's Integral formula $\int_c \frac{z^3 e^{-z} dz}{(z-1)^3}$ where c is $|z-1| = ½$

Solution: Arranging the given integral in the form of Cauchy's integral formula

$$\int_c \frac{z^3 e^{-z}}{(z-1)^{2+1}} dz = 2\pi i \frac{1}{2!} f''(z_0) \text{ where } f(z) = z^3 e^{-z} \qquad \text{.......(1)}$$

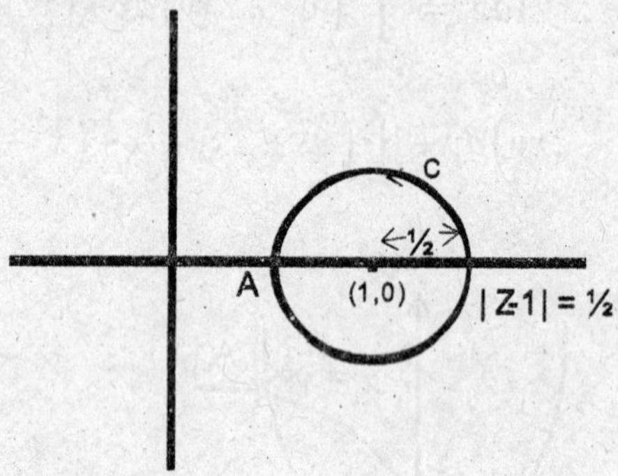

The point z = 1 lies inside '*c*'

$$f'(z) = 3z^2 e^{-z} + z^3(-e^{-z})$$

$$f''(z) = (6ze^{-z} + 3z^2(-e^{-z})) + (z^3 e^{-z} - 3z^2 e^{-z})$$

$$f''(1) = e^{-1} \qquad \therefore \text{ substituting in (1), we get}$$

$$\int_c \frac{z^3 e^{-z} dz}{(z-1)^{2+1}} = 2\pi i \times \frac{1}{2} e^{-1} = \frac{\pi i}{e}$$

17. Evaluate $\int_{(0,0)}^{(1,1)} [3x^2 + 5y + i(x^2 - y^2)] dz$ along $y^2 = x$

Solution: Given $I = \int_{0,0}^{(1,1)} [3x^2 + 5y + i(x^2 - y^2)] dz$ along $y^2 = x$

Substituting $x = y^2$, $dx = 2ydy$, (y varies from 0 to 1)

$$I = \int_0^1 [3y^4 + 5y + i(y^4 - y^2)](2ydy + idy)$$

$$= \int_0^1 (2y+i)\left[3y^4+5y+i\left(y^4-y^2\right)\right]dy$$

$$= \int_0^1 \left(6y^5+10y^2+i2y^5-i2y^3+3iy^4+5yi-y^4+y^2\right)dy$$

$$= \int_0^1 \left(6y^5+2iy^5+3iy^4-y^4-2iy^3+11y^2+5yi\right)dy$$

$$= 6\left(\frac{1}{6}\right)+2i\left(\frac{1}{6}\right)+3i\left(\frac{1}{5}\right)-\left(\frac{1}{5}\right)-2i\left(\frac{1}{4}\right)+11\left(\frac{1}{3}\right)+5i\left(\frac{1}{2}\right)$$

$$= 1+\frac{i}{3}+\frac{3i}{3}-\frac{1}{5}-\frac{i}{2}+\frac{11}{3}+\frac{5i}{2}=\frac{67}{15}+\frac{44}{15}i$$

18. Evaluate using Cauchy's integral theorem $\int_c \frac{zdz}{\left(z^2-6z+25\right)^2}$ when c is $|z-(3+4i)|$

Solution: Let I $= \int_c \frac{zdz}{\left(z^2-6z+25\right)^2}$; c is $|z-(3+4i)| = 4$ a circle centre at (3, 4) and radius 4.

$$z^2-6z+25=0 \;\Rightarrow z=\frac{6\pm\sqrt{36-100}}{2}=3\pm 4i$$

The point $z = 3 + 4i$ lies inside 'c' and $z = 3 - 4i$ lies outside 'c'

$$\therefore \quad \int_c \frac{\left\{\dfrac{z}{\left[z-(3-4i)\right]^2}\right\}dz}{\left[z-(3+4i)\right]^{1+1}} = 2\pi i\frac{f'(z_0)}{1!} \qquad \text{......(1)}$$

$$f(z)=\frac{z}{\left[z-(3-4i)\right]^2},\; z_0=3+4i$$

$$f'(z) = \frac{[z-(3-4i)]^2 .1 - z.2[z-(3-4i)]}{[z-(3-4i)]^4} = \frac{[z-(3-4i)]-2z}{[z-(3-4i)]^3}$$

$$f'(z_0) = f'(3+4i) = \frac{-z-(3-4i)}{[z-(3-4i)]^3}\Big|_{z=3+4i} = \frac{-3-4i-34i}{[(3+4i)-(3-4i)]^3}$$

$$f'(z_0) = \frac{-6}{(4i)^3} = \frac{3}{32i}.$$ Substituting in (1)

$$\int_c \frac{z}{(z^2-6z+25)^2} dz = 2\pi i\left[\frac{3}{32i}\right] = \frac{3\pi}{16}$$

19. Evaluate $\int_0^{1+i} z^2 dz$ along $y = x^2$

Solution: Given $\int_0^{1+i} z^2 dz$ along $y = x^2$

$\mathrm{I} = \int_{0A} z^2 dz$; Along $y = x^2$, $dy = 2xdx$

$dz = dx + i2xdx$ and x varies from 0 to 1

$$\mathrm{I} = \int_0^1 (x+ix^2)^2 (dx + i2xdx) = \int_0^1 (x^2 - x^4 + 2x^3 i)(1+2xi)\,dx$$

$$= \int_0^1 x^2 - 5x^4 + 2x^5 i + 4x^3 i\,dx$$

$$= \left[\frac{x^3}{3} - \frac{5x^5}{5} - \frac{2x^6}{6} i + \frac{4x^4}{4} i\right]_0^1 = \left(\frac{1}{3}-1\right) - \frac{i}{3} + i$$

$$= -\frac{2}{3} + \frac{2i}{3} = \frac{2}{3}(i-1)$$

20. Prove that $\int_c \frac{dz}{z-a} = 2\pi i$ where c is $|z-a| = r$

Solution: Let $\mathrm{I} = \int_c \frac{dz}{z-a}$ where c is $|z-a| = r$

$z - a = re^{i\theta}$; $dz = re^{i\theta} i d\theta$, θ varies from $0 \to 2\pi$

$$\int_c \frac{dz}{z-a} = \int_0^{2\pi} \frac{re^{i\theta} i d\theta}{re^{i\theta}} = i[\theta]_0^{2\pi} = 2\pi i$$

21. Evaluate $\int_c \frac{dz}{e^z (z-1)^3}$ where c is $|z| = 2$ using Cauchy's integral theorem.

Solution: Let $I = \int_c \frac{e^{-z} dz}{(z-1)^3}$, c is $|z| = 2$

The point $z = 1$ is inside 'c'

Let $$f(z) = e^{-z} \text{ ; and } z_0 = 1$$

$$f'(z) = -e^{-z} \quad f''(z) = e^{-z}$$

$$f''(1) = e^{-1}$$

Using Cauchy's integral formula.

$$I = \int_c \frac{e^{-z} dz}{(z-1)^{2+1}} = 2\pi i \frac{f''(z_0)}{2!} = 2\pi i \frac{e^{-1}}{2} = \frac{\pi i}{e}$$

22. Evaluate $\int_{-2+i}^{5+i} z^3 dz$ along, $y = x$

Solution: Let $I = \int_{-2+i}^{5+i} z^3 dz$

$$z = x + iy \quad , \quad y = x \quad , dy = dx$$

$$z = x + ix \Rightarrow \quad dz = (1+i)\,dx, \; x \text{ varies from -2 to 5}$$

$$I = \int_2^5 \left[(1+i)x\right]^3 (1+i)\,dx = (1+i)^4 \int_{-2}^5 x^3 dx = (1+i)^4 \left[\frac{x^4}{4}\right]_{-2}^5$$

$$= (1+i)^4 \left(\frac{1}{4}\right)\left[5^4 - 2^4\right] \text{ (Simplify yourself)}$$

23. $\int (x+y)\,dx + x^2 y dy$ along $y = x^2$ from (0,0) to (3,9)

Solution: Along 0 A, $y = x^2$; $dy = 2xdx$; x varies from 0 to 3

$$\int_{0A} (x+y)\,dx + x^2 y dy = \int_{x=0}^3 \left(x + x^2\right)dx + x^2.x^2 2xdx \;= \int_0^3 \left(2x^5 + x^2 + x\right)dx$$

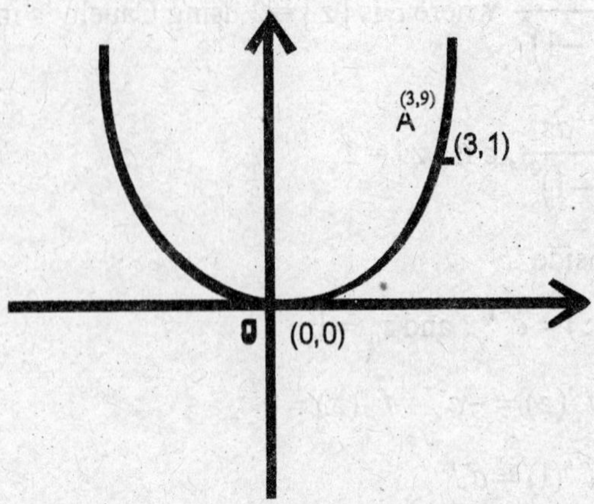

$$\left[2\frac{x^6}{6}+\frac{x^3}{3}+\frac{x^2}{2}\right]_0^3=\frac{1}{3}\left(3^6\right)+\frac{(3)^3}{3}+\frac{3^2}{2}=\frac{513}{2}$$

24. Evaluate $\int_c \frac{\cos z-\sin z}{(z+i)^3}dz$ with c is $|z|=2$ using Cauchy's integral formula.

Solution: c is $|z|=2$ which is a circle with centre at origin and radius 2

$\therefore$ The point $z=-i$ is inside c. Using Cauchy's integral formula.

$$\int_c \frac{(\cos z-\sin z)}{[z-(-i)]^{2+1}}dz = 2\pi i\frac{f''(z_0)}{2!} \quad .....(1)$$

where $z_0=-i; f(z)=\cos z-\sin z \quad , \quad f'(z)=-\sin z-\cos z$

$f''(z)=-\cos z+\sin z \quad , \quad f''(z_0)=f''(-i)=-\cos i-\sin i$

$f''(-i)=-[\cos i+\sin i]$. Substituting in (1)

$$\int_c \frac{(\cos z-\sin z)dz}{[z-(-i)]^{2+1}}=\frac{2\pi i[-(\cos i+\sin i)]}{2}=-\pi i(\cos i+\sin i)$$

25. Evaluate $\int_{1-i}^{2+i}(2x+1+iy)\,dz$ along $1-i$ to $2+i$.

Solution: The equation of the line joining the two points (1,-1) and (2,1) is

$$y+1=\frac{1-(-1)}{2-1}(x-1)$$

i.e., $y=2x-3$ ......(1)

$$z=x+iy=x+i(2x-3)=(1+2i)x-3i$$

$dz=(1+2i)\,dx$ and x varies from 1 to 2

$$\therefore \quad \int_{1-i}^{2+i}(2x+1+iy)\,dz=\int_{x=1}^{2}\left[(2x+1)+i(2x-3)\right][1+2i]\,dx$$

$$=(1+2i)\int_1^2\left[2(1+i)x+(1-3i)\right]dx=(1+2i)\left[(1+i)x^2+(1-3i)x\right]_1^2$$

$$=(1+2i)\left[(1+i)4+(1-3i)2-(1+i)-(1-3i)\right]=(1+2i)(4)=4+8i$$

26. Evaluate $\int_c \frac{z^3-\sin 3z}{\left(z-\frac{\pi}{2}\right)^3}dz$ with c is $|z|=2$ using Cauchy's integral formula.

***Solution*:** Using Cauchy's integral formula.

$$\int_c \frac{z^3-\sin 3z}{\left(z-\frac{\pi}{2}\right)^{2+1}}dz=2\pi i\frac{f''(z)}{2!}$$

where $f(z)=z^3-\sin 3z$, $z_0=\pi/2$ inside c

$$f'(z)=3z^2-3\cos 3z\,;\; f''(z)=6z+9\sin 3z$$

$$f''(z_0)=f''\left(\frac{\pi}{2}\right)=6\frac{\pi}{2}+9\sin 3\frac{\pi}{2};\; f''\left(\frac{\pi}{2}\right)=3\pi-9$$

$$\therefore \quad \int_c \frac{\left(z^3-\sin 3z\right)}{\left(z-\frac{\pi}{2}\right)^{2+1}}dz=2\pi i\frac{3(\pi-3)}{2}=3\pi i(\pi-3)$$

27. Evaluate $\int_c\left(z^2+3z+2\right)dz$ where c is the area of the cycloid

$x=a(\theta+\sin\theta),\, y=a(1-\cos\theta)$ between the points (0, 0), $(\pi a, 2a)$

Solution: Given integral, $I = \int_c \left(z^2+3z+2\right)dz = \int_{0AB} \left(z^2+3z+2\right)dz$

$$I = \int_{0A} \left(z^2+3z+2\right)dz + \int_{AB} \left(z^2+3z+2\right)dz \qquad ......(1)$$

Integration along 0A

Equation of 0A is $y=0,\ dy=0,\ x$ varies from 0 to $a\pi$

$\therefore \qquad z = x+iy \Rightarrow z = x,\ dz = dx$

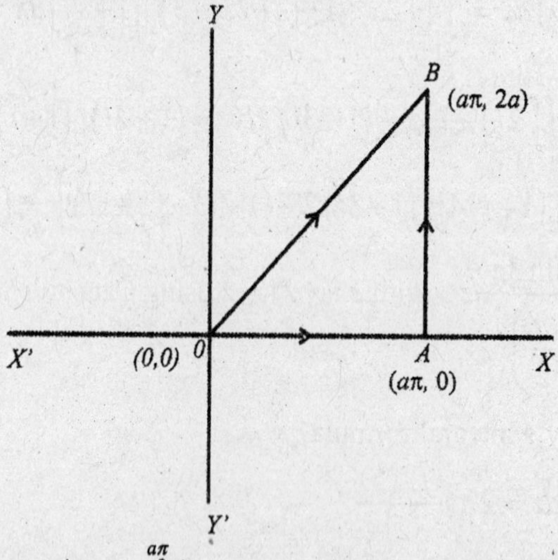

$$\int_{0A} \left(z^2+3z+2\right)dz = \int_0^{a\pi} \left(x^2+3x+2\right)dx$$

$$= \left[\frac{x^3}{3}+\frac{3x^2}{2}+2x\right]_0^{a\pi} = \frac{a^3\pi^3}{3}+\frac{3}{2}a^2\pi^2+2a\pi \qquad ......(2)$$

Integration along AB

Equation of AB is $x = a\pi$, $dx = 0$, y varies from 0 to $2a$

$\therefore \qquad z = x+iy = a\pi+iy \Rightarrow dz = idy$

$$\int_{BA} \left(z^2+3z+2\right)dz = \int_0^{2a} \left[\left(a\pi+iy\right)^2+3\left(a\pi+iy\right)+2\right]idy$$

$$= i\left[\frac{(a\pi + iy)^3}{3i} + 3\frac{(a\pi + iy)^2}{2i} + 2y\right]_0^{2a}$$

$$= \frac{1}{3}(a\pi + i2a)^3 + \frac{3}{2}(a\pi + 2ai)^2 + 4ia - \frac{(a\pi)^3}{3} - \frac{3}{2}(\pi a)^2 \quad \text{....(3)}$$

Substituting (2), (3) in (1) we get ,

$$\mathrm{I} = 2a\pi + \frac{1}{3}(a\pi + 2ai)^3 + \frac{3}{2}(a\pi + 2ai)^2 + 4ai$$

28. Evaluate $\int_c \frac{e^z \sin 2z - 1}{z^2(z+2)^2} dz$ when c is $|z| = ½$ using Cauchy's integral formula.

Solution: Given $\int_c \frac{e^z \sin 2z - 1}{z^2(z+2)^2} dz$, c is $|z| = ½$

$z = 0$ is in the only one point lies inside 'c'. Arranging the given integral in the form of Cauchy's integral formula,

$$\int_c \frac{\left[\frac{e^z \sin 2z - 1}{(z+2)^2}\right]}{(z-0)^{1+1}} dz = 2\pi i \frac{f'(z_0)}{1!} \quad \text{......(1)}$$

where $\quad f(z) = \frac{e^z \sin 2z - 1}{(z+2)^2}$ and $z_0 = 0$, $\quad n = 1$

$$f'(z) = \frac{(z+2)^2\left\{e^z \cos 2z.2 + \sin 2z.e^z\right\} - \left(e^z \sin 2z - 1\right)2(z+2)}{(z+2)^4}$$

$f'(0) = \frac{6}{8} = \frac{3}{4}$ Substituting in (1)

$$\int_c \frac{\left[\frac{e^z \sin 2z - 1}{(z+2)^2}\right]}{(z-0)^2} dz = 2\pi i \times \frac{3}{4} = \frac{3\pi i}{2}$$

29. Evaluate $\int_0^{1-i} \left(x - y^2 + ix^3\right) dz$ along the real axis from $z = 0$ to $z = 1$.

Solution: Let $z = 0$, $z = 1$ represent the points 0 and A

Along 0A, $y = 0$, $dy = 0$, $z = x + i.0 = x$, $dz = dx$, x varies from '0' to 1

$$\int_0^{1-i}\left(x-y^2+ix^3\right)dz = \int_{0A}\left(x-y^2+ix^3\right)dz$$

$$\int_0^1\left(x+ix^3\right)dx = \left[\frac{x^2}{2}+i\frac{x^4}{4}\right]_0^1 = \frac{1}{2}+\frac{1}{4}i$$

30. Evaluate $\int_c \frac{dz}{z^3(z+4)}$ where c is $|z| = 2$ using Cauchy's integral formula.

Solution: c is $|z| = 2$ which is a circle with center at origin and radius 2,

The point $z = 0$ lies inside 'c' and $z = -4$ lies outside 'c'

Arranging the integral in the form of Cauchy's integral formula, we have

$$\int_c \frac{\left(\frac{1}{z+4}\right)dz}{(z-0)^{2+1}} = 2\pi i\frac{f''(z_0)}{2!} \qquad \text{......(1)}$$

where $f(z) = \frac{1}{z+4}$, $z_0 = 0$, $n = 2$

$$f'(z) = \frac{-1}{(z+4)^2},\ f''(z) = \frac{2}{(z+4)^3},\ f''(0) = \frac{2}{4^3},\ f''(0) = \frac{1}{32}$$

$$\therefore \qquad \int_c \frac{\left(\frac{1}{z+4}\right)dz}{(z-0)^{2+1}} = 2\pi i\frac{1}{2!}.\frac{1}{32} = \frac{\pi i}{32}$$

31. Evaluate $\int_0^{1+i}\left(x-y+ix^2\right)dz$ along

(i) $z = 0$ to $1 + i$,

(ii) The real axis from $z = 0$ to $z = 1$ and then along a line parallel to the imaginary axis from $z = 1$ to $z = 1 + i$

Solution: Given $\int_0^{1+i}\left(x-y+ix^2\right)dz$

(i) Along the line $z = 0$ to $z = 1 + i$

Equation of 0A joining the points (0, 0) to (1,1) is $(y-0) = \frac{1-0}{1-0}(x-0)$

$y = x,\ dy = dx,\ x$ varies from 0 to 1

$$\int_0^{1+i}\left(x-y+ix^2\right)(dx+idy)=\int_0^1\left(x-x+ix^2\right)(dx+idx)=(1+i)i\left[\frac{x^3}{3}\right]_0^1=\frac{(i-1)}{3}$$

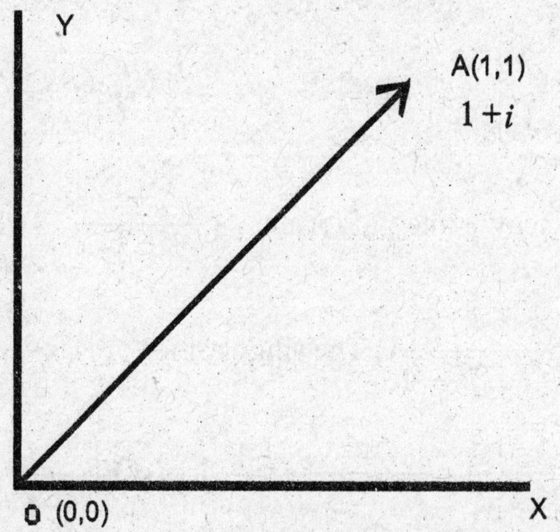

(ii) The real axis from z = 0 to z = 1 and then along a parallel to the imaginary axis from z = 1 to z = 1 + i

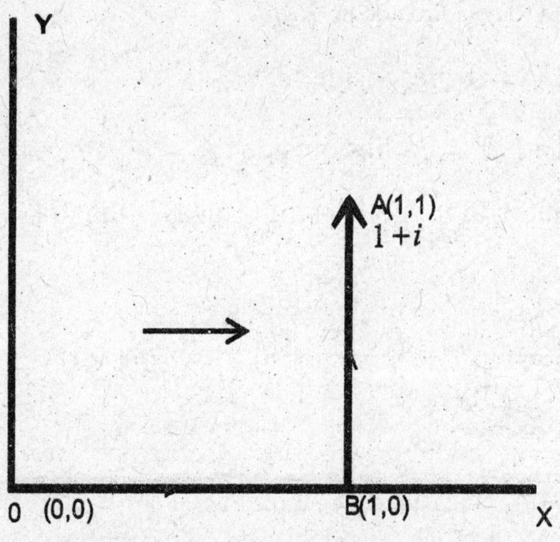

$$I=\int_0^{1+i}\left(x-y+ix^2\right)dz=\int_{0BA}\left[(x-y)+ix^2\right]dz$$

$$=\int_{0B}\left(x-y+ix^2\right)dz+\int_{BA}\left(x-y+ix^2\right)dz \qquad \text{......(1)}$$

Along 0B, $dy = 0$ and 'x' varies from 0 to 1

Along BA, $x = 1,\ dx = 0$, 'y' varies from 0 to 1

Substituting these values in (1)

$$I = \int_{x=0}^{1} \left(x - 0 + ix^2\right)\left(dx + i.0\right) + \int_{y=0}^{1} (1 - y + i)(0 + idy)$$

$$= \left[\frac{x^2}{2} + i\frac{x^3}{3}\right]_0^1 + i\left[y - \frac{y^2}{2} + iy\right]_0^1 = \frac{1}{2} + \frac{i}{3} + i - \frac{i}{2} - 1 = -\frac{1}{2} + \frac{5i}{6}$$

32. Evaluate using Cauchy's integral formula $\int_c \frac{(z+1)\,dz}{z^2 + 2z + 4}$ where c is $|z + 1 + i| = 2$

Solution: Let $f(z) = \frac{z+1}{z^2 + 2z + 4}$. The singularities of $f(z)$ are given by

$$z = \frac{-2 \pm \sqrt{4 - 16}}{2} = \frac{-2 \pm 2\sqrt{3}i}{2} = -1 \pm i\sqrt{3};$$

c is $|z + 1 + i| = 2 \Rightarrow (x+1)^2 + (y+1)^2 = 2^2$ which is a circle

Centre to the circle is $(-1, -1)$, radius = 2.

The point $z = -1 + i\sqrt{3} = \alpha$ lies inside 'c'

and the point $z = -1 - i\sqrt{3} = \beta$ lies outside 'c'

Arranging the given integral in the form of Cauchy's integral formula, we have

$$I = \int_c \frac{z+1}{z^2 + 2z + 4}dz = \int_c \frac{\left(\frac{z+1}{z-\beta}\right)}{z - \alpha}dz = 2\pi i f(\alpha) \qquad ....(1)$$

$$f(\alpha) = \frac{\alpha + 1}{\alpha - \beta} = \frac{-1 + i\sqrt{3} + 1}{\left(-1 + i\sqrt{3}\right) - \left(-1 - i\sqrt{3}\right)} = \frac{1}{2}$$

Substituting in (1)

$$I = 2\pi i\left(\frac{1}{2}\right) = \pi i$$

33. Evaluate $\int_c \bar{z}dz$ from $z = 0$ to $z = 4 + 2i$ along the curve C given by

(i) $z = t^2 + it$

(ii) along the line z = 0 to z = 2i and from there z = 2i to z = 4+ 2i

Solution: (i) Given curve is $z = t^2 + it$, $dz = (2t + i)\,dt$

$z = 0 \Rightarrow t = 0$, $z = 4 + 2i \Rightarrow t = 2$, and $\bar{z} = t^2 - it$

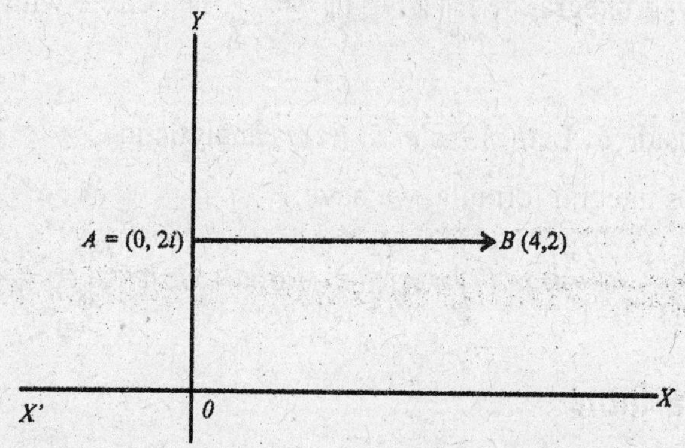

$$\int_c \bar{z}dz = \int_0^2 (t^2 - it)(2t + i)\,dt = \int_c (2t^3 - it^2 + t)\,dt = \left[2\frac{t^4}{4} - \frac{it^3}{3} + \frac{t^2}{2}\right]_0^2$$

$$= 8 - \frac{8i}{3} + 2 = 10 - \frac{8i}{3}$$

(ii) Along 0A, $x = 0$ $dx = 0$; and y varies from 0 to 2

$\therefore \quad z = iy \Rightarrow dz = idy$, $\bar{z} = -iy$

$$\int_c \bar{z}dz = \int_0^2 (-iy)(idy) = \int_0^2 ydy = \left[\frac{y^2}{2}\right]_0^2 = 2 \qquad \text{........(1)}$$

Along AB equation of AB is $y = 2$, $dy = 0$, x varies from 0 to 4

$z = x+iy$, $z = x + i2$, $\bar{z} = x - i2$, $dz = dx$

$$\int_c \bar{z}dz = \int_{AB} \bar{z}dz = \int_{x=0}^4 (x - i2)\,dx = \left[\frac{x^2}{2} - 2ix\right]_0^4 = 8(1 - i) \qquad \text{........(2)}$$

From (1) (2)

$$\therefore \quad \int_{0AB} \bar{z}dz = \int_{0A} \bar{z}dz + \int_{AB} \bar{z}dz = 2 + 8(1 - i) = 10 - 8i$$

34. Evaluate $\int_c \frac{z^3 e^{-z} dz}{(z-1)^3}$ where c is $|z-1| = \frac{1}{2}$ using Cauchy's integral formula.

Solution: Let the given integral be I; $|(z - 1)| = \frac{1}{2}$ is a circle with centre at (1, 0) and radius = 2;

$\therefore$ $z = 1$ lies inside C; Let $f(z) = z^3 e^{-z}$; $f(z)$ is analytic in C

$\therefore$ By cauchy's integral formula, we have,

$$I = \int_c \frac{f(z)dz}{(z-1)^3} = \frac{2\pi i}{\underline{|2}} f''(1) = \pi i\left[(z^3 - 6z^2 + 6z)e^{-z}\right]_{z=1} = \frac{\pi i}{e}$$

22.3 Complex Variable

22.3.1 Absolute Convergence

A series $\sum_{n=1}^{\infty} v_n(z)$ is said to be absolutely convergent if $\sum_{n=1}^{\infty} |v_n(z)|$ is convergent. If $\sum_{n=1}^{\infty} v_n$ is convergent and $\sum_{n=1}^{\infty} |v_n(z)|$ is divergent then $\sum v_n$ is said to be conditionally convergent.

22.3.2 Uniform Convergence

A series $\sum v_n$ is said to be uniformly convergent to $v(z)$ in a given region R of z-plane, if ,for any given $\in > 0$, there exists a positive number $N(\in)$ [i.e. N depends only on $\in$ but not on z] such that , $|S_n(z) - S(z)| < \in$, for all $n>N$, for every point $z \in R$.

22.3.3 Power Series

Let $z = a$ be any fixed point in the z-plane. Then the infinite series,

$$a_0 + a_1(z-a) + a_2(z-a)^2 + ... + a_n(z-a)^n + ... \qquad .....(1)$$

[$a_0, a_1, a_2, ...$ being constants] which is written as $\sum_0^{\infty} a_n(z-a)^n$ is called a power series in *(z-a)*. '*a*' is sometimes known as the centre of the series (1). $a_0 + a_1 z + a_2 z^2 + ... + a_n z^n + ...$ is a power series with centre at the origin $z = 0$.

Let the power series (1) be convergent at all points 'z' such that $|z-a|=R$ i.e. Geometrically , if $|z-a|=R$ denotes a circle 'C' then the power series (1) converges inside C and diverges outside C [on the circle it may or may not converge]

This circle 'C' is called the circle of convergence of the power series (1) and 'R' is the radius of convergence of (1)

22.3.4 Taylor's Theorem

If a function f(z) is analytic inside the circle 'C' centre at z_0 and radius R and z is any point inside 'C' then

$$f(z) = \sum_{n=0}^{\infty} a_n (z-z_0)^n \qquad \text{where} \qquad a_n = \frac{1}{n!} f^n(z_0)$$

Proof: Let $f(z)$ be analytic inside circle $C : |z-z_0|=R$. Let z be a point such that $|z-z_0|<R$. We can choose r > 0 such that $|z-z_0|<r<R$. $f(z)$ is analytic inside and on $C_1 : |z-z_0|=r$ .Using Cauchy's integral formula

$$f(z) = \frac{1}{2\pi\, \mathrm{i}} \int \frac{f(\zeta)}{\zeta - z} d\zeta$$

Now consider

$$\frac{1}{\zeta - z} = \frac{1}{(\zeta - z_0)-(z-z_0)} = \frac{1}{(\zeta - z_0)\left(1-\dfrac{z-z_0}{\zeta - z_0}\right)}$$

$$= \frac{1}{(\zeta - z_0)}\left[1+\frac{z-z_0}{\zeta - z_0}+\left(\frac{z-z_0}{\zeta - z_0}\right)^2+....\left(\frac{z-z_0}{\zeta - z_0}\right)^{n-1}+\left(\frac{z-z_0}{\zeta - z_0}\right)^n+...\right]$$

(Using the Binomial theorem)

(i.e.,) $$\frac{1}{(\zeta - z)} = \sum_0^{\infty} \frac{(z-z_0)^n}{(\zeta - z_0)^{n+1}}$$

Substituting the above in (1)

$$f(z) = \frac{1}{2\pi i} \int_{c_1} f(\zeta) \sum_0^{\infty} \frac{(z-z_0)^n}{(\zeta - z_0)^{n+1}} d\zeta$$

$$= \sum_0^{\infty} (z-z_0)^n \frac{1}{2\pi i} \int_{c_1} \frac{f(\zeta)}{(\zeta - z_0)^{n+1}} d\zeta$$

$$f(z)=\sum_{0}^{\infty}(z-z_0)^n\frac{f^n(z_0)}{n!}=\sum_{0}^{\infty}a_n(z-z_0)^n \quad \text{.....(2)}$$

where $a_n=\dfrac{f^n(z_0)}{n!}$ and z is any point inside the circle $|z-z_0|=R$.

Note : If the centre of the circle is $z_0=0$ (i.e., if the circle is $|z|=R$), then the series (2) reduces to

$$f(z)=\sum_{n=0}^{\infty}a_n z^n \text{ where } a_n=\frac{f^n(z)}{n!}$$

$$f(z)=f(0)+\frac{z}{1!}f^1(0)+\frac{z^2}{2!}f^2(0)+\ldots \text{ is called Maclaurin's series}$$

22.3.5 Laurent's Theorem

If a function $f(z)$ is analytic in the annulus (ring shaped region) between two concentric circles C and C_1 with centre z_0 and radii R and R_1 ($R_1>R$), then at any point z in the annulus

$$f(z)=\sum_{0}^{\infty}a_n(z-z_0)^n+\sum_{1}^{\infty}b_n(z-z_0)^{-n},$$

where $a_n=\dfrac{1}{2\pi i}\displaystyle\int_C\frac{f(\zeta)\,d\zeta}{(\zeta-z_0)^{n+1}}$; $b_n=\dfrac{1}{2\pi i}\displaystyle\int_{C_1}\frac{f(\zeta)\,d\zeta}{(\zeta-z_0)^{-n+1}}$

Proof : Let $f(z)$ be analytic in the ring shaped region bounded by two concentric circles C and C_1 with centre at z_0 and radii R and R_1 respectively. Let z be a point in the annulus. Using Cauchy's integral formula for doubly connected region

$$f(z)=\frac{1}{2\pi i}\int_C\frac{f(\zeta)}{(\zeta-z)}-\frac{1}{2\pi i}\int_{C_1}\frac{f(\zeta)}{(\zeta-z)}$$

From Taylor's theorem,

$$\frac{1}{2\pi i}\int_C\frac{f(\zeta)}{(\zeta-z)}d\zeta=\sum_{0}^{\infty}a_n(z-z_0)^n$$

$$\therefore \quad f(z)=\sum_{0}^{\infty}a_n(z-z_0)^n-\frac{1}{2\pi i}\int_{C_1}\frac{f(\zeta)}{(\zeta-z)}d\zeta \quad \text{.....(1)}$$

Now consider,

$$-\frac{1}{\zeta - z} = \frac{1}{z-\zeta} = \frac{1}{(z-z_0)-(\zeta-z_0)} = \frac{1}{(z-z_0)}\left[\frac{1}{1-\frac{\zeta-z_0}{z-z_0}}\right]$$

$$= \frac{1}{z-z_0}\left[1+\left(\frac{\zeta-z_0}{z-z_0}\right)+\left(\frac{\zeta-z_0}{z-z_0}\right)^2+\ldots.+\left(\frac{\zeta-z_0}{z-z_0}\right)^n+\ldots\right]$$

(Using Binomial theorem)

(i.e.,) $$-\frac{1}{\zeta - z} = \sum_1^{\infty}\frac{(\zeta-z_0)^{n-1}}{(z-z_0)^n}$$

Substituting in (1),

$$f(z) = \sum_0^{\infty} a_n (z-z_0)^n + \frac{1}{2\pi i}\int_{c_1} f(\zeta)\sum_1^{\infty}\frac{(\zeta-z_0)^{n-1}}{(z-z_0)^n}\,d\zeta$$

$$= \sum_0^{\infty} a_n (z-z_0)^n + \sum_1^{\infty}(z-z_0)^{-n}\,\frac{1}{2\pi i}\int_{c_1}\frac{f(\zeta)}{(\zeta-z_0)^{-n+1}}\,d\zeta$$

$$f(z) = \sum_0^{\infty} a_n (z-z_0)^n + \sum_1^{\infty} b_n (z-z_0)^{-n}$$

where $b_n = \dfrac{1}{2\pi i}\displaystyle\int_{C_1}\frac{f(\zeta)}{(\zeta-z_0)^{-n+1}}\,d\zeta$ is called Laurent's expansion of the function *f(z)* in ascending and descending powers of $(z-z_0)$

***Example* 22.14:** Expand $f(z) = \dfrac{1}{(z+1)(z+3)}$ in Laurents series, valid for

(a) $1 < |z| < 3$ (b) $|z| > 3$

(c) $0 < |z+1| < 2$ (d) $|z| < 1$

Solution: (a) Resolving the given function of z into partial fractions

$$f(z)=\frac{1}{(z+1)(z+3)}=\frac{1}{2}\left(\frac{1}{z+1}-\frac{1}{z+3}\right)$$

$$=\frac{1}{2}\left[\frac{1}{z}\left(1+\frac{1}{z}\right)^{-1}-\frac{1}{3}\left(1+\frac{2}{3}\right)^{-1}\right] \text{ since } |z|>1 \text{ and } |z|<3$$

$$=\frac{1}{2z}\left[1-\frac{1}{z}+\frac{1}{z^2}-\frac{1}{z^3}+\cdots\right]-\frac{1}{6}\left[1-\frac{z}{3}+\frac{z^2}{3^2}-\frac{z^3}{3^3}+\cdots\right]$$

(b) For $|z|>3$

$$f(z)=\frac{1}{2}\left[\frac{1}{z+1}-\frac{1}{z+3}\right]=\frac{1}{2^z}\left(1+\frac{1}{z}\right)^{-1}-\frac{1}{2z}\left(1+\frac{3}{z}\right)^{-1}$$

$$=\frac{1}{2z}\left(1-\frac{1}{z}+\frac{1}{z^2}-\frac{1}{z^3}+\cdots\right)-\frac{1}{2z}\left(1-\frac{3}{z}+\frac{9}{z^2}-\frac{27}{z^3}+\cdots\right)$$

$$=\frac{1}{2}\left(\frac{1}{z}-\frac{1}{z^2}+\frac{1}{z^3}-\frac{1}{z^4}+\cdots\right)-\frac{1}{2}\left(\frac{1}{z}-\frac{3}{z^2}+\frac{9}{z^3}-\frac{27}{z^4}+\cdots\right)$$

$$=\frac{1}{z^2}-\frac{4}{z^3}+\frac{13}{z^4}-\cdots$$

(c) For $|z+1|<2$

$$f(z)=\frac{1}{2}\left(\frac{1}{z+1}-\frac{1}{z+1+2}\right)=\frac{1}{z(z+1)}-\frac{1}{4}\left(1+\frac{z+1}{2}\right)^{-1}$$

$$=\frac{1}{2(z+1)}-\frac{1}{4}\left[1-\frac{z+1}{2}+\frac{(z+1)^2}{4}-\frac{(z+1)^3}{8}+\cdots\right]$$

$$=\frac{1}{2(z+1)}-\frac{1}{4}+\frac{z+1}{8}-\frac{(z+1)^2}{16}+\frac{(z+1)^3}{32}-\cdots$$

(d) For $|z|<1$

$$f(z)=\frac{1}{2}\left(\frac{1}{z+1}-\frac{1}{z+3}\right)=\frac{1}{2}(1+z)^{-1}-\frac{1}{6}\left(1+\frac{2}{3}\right)^{-1}$$

$$=\frac{1}{2}\left(1-z+z^2-z^3+\cdots\right)-\frac{1}{6}\left(1-\frac{z}{3}+\frac{z^2}{9}-\frac{z^3}{27}+\cdots\right)$$

$$=\frac{1}{3}-\frac{4}{9}z+\frac{13}{27}z^2+\ldots+(-1)^n\frac{1}{2}\left(1-3^{-n-1}\right)z^n+\ldots$$

Since $|z|<1$ the series becomes a Taylor's series.

***Example* 22.15:** Find the Laurent's expansion of $f(z)=\dfrac{1}{z(z-1)^2}$ at the point $z=1$.

Solution: To obtain the expansion of $f(z)$ about the point $z=1$, we put $z_1=z-1$ so that $z=1+z_1$, then we have

$$\begin{aligned} f(z) &= \frac{1}{(1+z_1)z_1^{\,2}} = \frac{1}{z_1^{\,2}}(1+z_1)^{-1} = \frac{1}{z_1^{\,2}}\left(1-z_1+z_1^{\,2}-z_1^{\,3}+\ldots\right) \\ &= \frac{1}{z_1^{\,2}} - \frac{1}{z_1} + 1 - z_1 + z_1^{\,2} - z_1^{\,3} + \ldots \\ &= \frac{1}{(z-1)^2} - \frac{1}{z-1} + 1 - (z-1) + (z-1)^2 - (z-1)^3 + \ldots \\ &= \sum_{n=0}^{\infty} (-1)^n (z-1)^{n-2} \text{ which is valid for } |z-1|<1. \end{aligned}$$

22.4 Zeros of an Analytic Function

Let $f(z)$ be an analytic function. If $f(z_0)=0$ then z_0 is called the zero of $f(z)$.

If $f(z_0)=0$ and $f'(z_0)\neq 0$ then z_0 is called a simple zero or a zero of first order. If $f(z_0)=f'(z_0)=0$ and $f''(z_0)\neq 0$ then z_0 is called a zero of second order. Extending it, we say that z_0 is a zero of order n if

$$f(z_0)=f'(z_0)=f''(z_0)=\ldots=f^{(n)}(z_0)=0 \text{ but } f^{(n+1)}(z_0)\neq 0.$$

For example, $z^2 \sin z$ has a zero of third order at $z=0$ and simple zeros at $z=\pm\pi, \pm 2\pi, \ldots$ and $(1-\cos z)$ has zeros of order two at $z=0, \pm 2\pi, \pm 4\pi, \ldots$.

Next suppose that the analytic function $f(z)$ has a zero of order n at $z=z_0,$ then its Taylor's series is of the form

$$\begin{aligned} f(z) &= b_n(z-z_0)^n + b_{n+1}(z-z_0)^{n+1} + b_{n+2}(z-z_0)^{n+2} + \ldots \\ &= (z-z_0)^n\left[b_n + (z-z_0)b_{n+1} + (z-z_0)^2 b_{n+2} + \ldots\right] \\ &= (z-z_0)^n\left[b_n + S\right], \text{ say.} \end{aligned}$$

By taking $|z-z_0|$ sufficiently small we can make

$$|b_n| > |z-z_0||S|.$$

Thus, neighbourhood of z_0 can be found where

$b_n + (z - z_0)S$ is not zero. It implies that $f(z)$ cannot have another zero in this neighbourhood. Thus we have the fact that zeros of an analytic function are isolated.

For example, $\frac{1}{1-z}$ has a simple zero at infinity, $\cos z$ has simple zeros at $z = \pm\pi/2, \pm 3\pi/2, \ldots$ and $\left(\frac{z+1}{z^2+1}\right)^3$ has a zero at $z = -1$ or order three.

22.4.1 Singularity of a Function

A point z_0 at which a function $f(z)$ is not analytic is known as a singular point or singularity of $f(z)$. There are different types of singularities as discussed below.

(a) ***Isolated Singularity*:** If the function $f(z)$ is analytic at every point in the neighbourhood of a point z_0 except at z_0 itself then z_0 is called an isolated singular point or isolated singularity. For example, $f(z) = 1/z$ then we have $f'(z) = -\frac{1}{z^2}$ it follows that $f(z)$ is analytic at every point except at $z = 0$, hence $z = 0$ is an isolated singularity. Also, $f(z) = \frac{1}{z^3(z^2+1)}$ has three isolated singularities at $z = 0$, $z = \pm i$.

Next, let $f(z)$ be analytic in a domain D except at the point $z = z_0$, then z_0 is an isolated singularity of $f(z)$ and we can draw two concentric circles with centre z_0 such that $f(z)$ can be expanded in a Laurent's series and we shall write

$$f(z) = \sum_{n=0}^{\infty} a_n (z - z_0)^n + \sum_{n=1}^{\infty} \frac{b_n}{(z - z_0)^n} \quad \text{.....(1)}$$

In (1), $\sum_{n=0}^{\infty} a_n (z - z_0)^n$ is called the regular part of $f(z)$ and $\sum_{n=1}^{\infty} \frac{b_n}{(z - z_0)^n}$ is called its principal part in the neighbourhood of z_0.

(b) Removable Singularity: In the equation (1) above if $b_n = 0$ for all n, then the singularity at z_0 is called a removable singularity of $f(z)$.

For example, $f(z)=\frac{\sin h(z-a)}{z-a}$ has a removable singularity at $z=a$, since we can write

$$f(z)=\frac{1}{z-a}\left[z-a+\frac{(z-a)^3}{3!}+\frac{(z-a)^5}{5!}+...\right]$$

$$=1+\frac{(z-a)^2}{3!}+\frac{(z-a)^4}{5!}+...$$

Here $f(z)$ does not contain any negative power of $(z-a)$. It is easy to see that the singularity in this case can be removed by redefining the function as follows:

$$f(z)=\frac{\sin h(z-a)}{z-a} \qquad \text{where } z\neq a$$

$$=1 \qquad \text{when } z=a.$$

(c) **Essential Singularity:** In the equation (1) above, if the principal part contains infinite number of terms then $z=z_0$ is called an isolated essential singularity of $f(z)$.

For example $f(z)=\sin\frac{1}{z}$ has an essential singularity at $z=0$, since

$$\sin\frac{1}{z}=\frac{1}{z}-\frac{1}{3!z^3}+\frac{1}{5!z^5}-...$$

(d) **Poles:** If in equation (1) above, the principal part has all the coefficients $b_{n+1}, b_{n+2},...$ as zero after a particular term b_n then the Laurents series of $f(z)$ reduces to

$$f(z)=\sum_{n=0}^{\infty}a_n(z-z_0)^2+\frac{b_1}{z-z_0}+\frac{b_2}{(z-z_0)^2}+...+\frac{b_n}{(z-z_0)^n}$$

then the singularity is called a pole of order n.

A pole of order one is called a simple pole

For example, $\frac{z^2-2z+5}{z-2}$ has a simple pole at $z=2$,

since $$\frac{z^2-2z+5}{z-2}=z+\frac{5}{z-2}=2+(z-2)+\frac{5}{z-2}$$

Similarly, $f(z)=\dfrac{1-e^{2z}}{z^3}$ has a pole of order 2 at $z=0$ since

$$f(z)=\frac{1}{z^3}\left(1-e^{2z}\right)=\frac{1}{z^3}\left[1-1-\frac{2z}{1!}+\frac{4z^2}{2!}+\frac{8z^3}{3!}+\frac{16z^4}{4!}+...\right]$$

$$=\frac{-2}{z^2}+\frac{2}{z}+\frac{4}{3}+\frac{2z}{3}+...$$

which shows that $f(z)$ has pole of order 2.

22.5 Contour Integration

22.5.1 Introduction – Residue of f(z)

In this section we define 'residue' at a point z = z_0 of a function f(z) and some methods for calculating residues.

Let c be a circle with centre at z_0 and f(z) be single valued and analytic inside and on c. Then f(z) can be expanded by Laurent series about z = z_0, as,

$$f(z)=\sum_{n=0}^{\infty}a_n(z-z_0)^n+\sum_{n=1}^{\infty}b_n(z-z_0)^{-n} \quad .....(1)$$

where $$a_n=\frac{1}{2\pi i}\oint_C\frac{f(z)dz}{(z-z_0)^{n+1}}, \quad n=0,1,2,3,4,......$$

$$b_n=\frac{1}{2\pi i}\oint_C\frac{f(z)dz}{(z-z_0)^{-n+1}}, \quad n=0,1,2,3,4,......$$

In particular, $b_1=\dfrac{1}{2\pi i}\oint_C f(z)dz$

If z = z_0 is an isolated singularity, the coefficient of $\dfrac{1}{z-z_0}$, that is, 'b_1' is called the residue of f(z) at z = z_0.

22.5.2 Residue at a Pole of Order m

Let z = z_0 be a pole of order 'm' of f(z), then (2) of 22.5.1 gives

$$f(z) = a_0 + a_1(z-z_0) + a_2(z-z_0)^2 + ... +$$

$$+\frac{b_1}{(z-z_0)} + \frac{b_2}{(z-z_0)^2} + ... + \frac{b_{m-1}}{(z-z_0)^{m-1}} + \frac{b_m}{(z-z_0)^m}$$

$$(z-z_0)^m f(z) = a_0(z-z_0)^m + a_1(z-z_0)^{m+1} +$$

$$...... + b_1(z-z_0)^{m-1} + b_2(z-z_0)^{m-2} + ... + b_{m-1}(z-z_0) + b_m \quad .....(1)$$

Differentiation of (1) (m–1) times w.r. to z gives

$$\frac{d^{m-1}\left[(z-z_0)^m f(z)\right]}{dz^{m-1}}$$

$$= \left[\text{Terms containing } (z-z_0) \text{ and its higher powers}\right] + [b_1(m-1)! + 0 + 0 + ...]$$

Applying the limit $z \to z_0$ and dividing with (m–1)!

We get $\displaystyle \frac{1}{(m-1)!}\lim_{z\to z_0}\frac{d^{m-1}\left[(z-z_0)^m f(z)\right]}{dz^{m-1}} = [0+0+......] + [b_1+0+0]$

∴ The Residue of f(z) at z = z_0 is given by

$$b_1 = \frac{1}{(m-1)!}\lim_{z\to z_0}\frac{d^{m-1}\left[(z-z_0)^m f(z)\right]}{dz^{m-1}} \quad .....(2)$$

22.5.3 Residue at a Simple Pole (m = 1)

b_1 = residue of $f(z)$ at $z = z_0$ from (2)

$$= \lim_{z\to z_0}\left[(z-z_0)\right].f(z)$$

Residue at infinity: If f(z) has an isolated singular point at $z=\infty$, or is analytic there, then the Residue of f(z) at z = ∞ is defined as lt $\int_{C_R} f(z)dz \to 0$ as $R \to \infty$.

For example consider f(z) = $\dfrac{b}{z-z_0}$ (the function may be regular at z = ∞)

Residue of f(z) at z = ∞ (using definition)

$$= -\frac{1}{2\pi i}\int_c \frac{b}{z-z_0}dz = -\frac{1}{2\pi i}[b.2\pi i] = -b \qquad .....(1)$$

Also z = z_0 is a simple pole of f(z) and its residue there is

$$= \frac{1}{2\pi i}\int_c f(z)dz = \frac{1}{2\pi i}\int_c \frac{b}{z-z_0}dz = \frac{1}{2\pi i}[b.2\pi i] = b$$

i.e., residue of f(z) at z = z_0 is $\lim_{z\to z_0}(z-z_0)f(z) = b$ .....(2)

From (1) and (2)

Residue of f(z) at z = z_0 = – residue of f(z) at z = ∞

∴ The residue of f(z) at z = ∞ is the negative of the coefficient of $\frac{1}{z}$ in the expansion of f(z) for values of z in the neighbourhood of z = ∞

The Residue of f(z) at z = ∞ is $-\underset{z\to\infty}{Lt}[-zf(z)]$ provided f(z) is analytic at z = ∞

22.5.4 Evaluation of Integrals using Residues

If $z = z_0$ is a simple pole of f(z), then f(z) is in the form $\frac{\phi(z)}{z-z_0}$

Residue of f(z) at simple pole at $z = z_0 = \lim_{z\to z_0}\left[(z-z_0).f(z)\right] = \phi(z_0)$ .....(3)

But, $\phi(z_0) = \frac{1}{2\pi i}\oint_C \frac{\phi(z)dz}{z-z_0}$ (using Cauchy's integral formula)

$$= \frac{1}{2\pi i}\oint_C f(z)dz \qquad .....(4)$$

From (3) and (4) we get

$$\oint_C f(z)dz = 2\pi i \text{ [Residue of f(z) at a simple pole } z = z_0\text{]} \qquad ....(5)$$

Similarly

$$\oint_C f(z)dz = 2\pi i \text{ [Residue of f(z) at } z = z_0 \text{ at a pole of order n]} \quad .....(6)$$

22.5.5 Cauchy's Residue Theorem

If f(z) is single valued and analytic within and on a closed contour c except possibly at a finite number of singularities $z_1, z_2, ..., z_n$ and $R_1, R_2, R_3, ..., R_n$ are respectively the residues of f(z) at these points , then

$$\oint_C f(z)dz = 2\pi i\left[R_1 + R_2 + R_3 + ... + R_n\right] = 2\pi i$$

(sum of the residues at the singularities within c)

Proof : Let $c_1, c_2, ..., c_n$ be the circles with centers $z_1, z_2, ..., z_n$ respectively with radii so small that they lie entirely within the closed curve 'c' and do not overlap, then f(z) is analytic within the multiply connected region bounded by c and $c_1, c_2, ..., c_n$. $R_1, R_2, ..., R_n$ are the residues at $z_1, z_2, ..., z_n$ respectively.

Using Cauchy's Integral Theorem for multiply connected region,

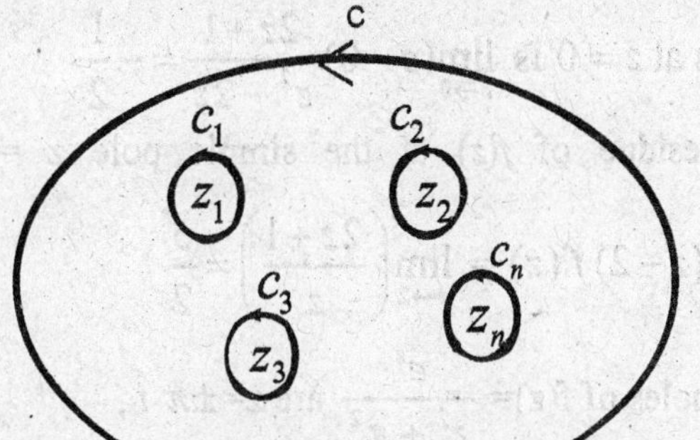

$$\oint_C f(z)dz = \oint_{C_1} f(z)dz + \oint_{C_2} f(z)dz + ... + \oint_{C_n} f(z)dz$$

Now c_1 is the circle with centre at z_1 and z_1 is the singular point of f(z) inside c_1.

Using (3) of 22.5.4.

$$\oint_{C_1} f(z)dz = 2\pi i[\text{Residue of } f(z) \text{ at the simple pole } z = z_1] \text{ is } 2\pi i R_1$$

Similarly for $\oint_{C_2} f(z)dz, \oint_{C_3} f(z)dz, \ldots\ldots$

Thus $$\oint_{C} f(z)dz = 2\pi i[R_1 + R_2 + \ldots\ldots\ldots\ldots + R_n]$$

Solved Examples

***Example* 22.16:** Find the singularities and also the residues at the poles of the following functions;

1. $\dfrac{2z+1}{z^2-2z}$ 2. $\dfrac{e^z}{z^2+\pi^2}$

***Solution*:**

1. f(z) = $\dfrac{2z+1}{z^2-2z}$. The roots of $z^2-2z = 0$ are the singular points of f(z). Thus the singular points are z = 0, 2. The Residue of f(z) at z = 0, a simple pole, is $\lim\limits_{z\to 0}(z-0)f(z)$

$$R_1 = \text{Res at } z = 0 \text{ is } \lim_{z\to 0}(z-0)\frac{2z+1}{z^2-2z} = -\frac{1}{2}$$

The residue of $f(z)$ at the simple pole z = 2 is $\lim\limits_{z\to 2}(z-2)f(z)$

$$= \lim_{z\to 2}(z-2)f(z) = \lim_{z\to 2}\left(\frac{2z+1}{z}\right) = \frac{5}{2}$$

(2) The simple poles of f(z)= $\dfrac{e^z}{z^2+\pi^2}$ are z=$\pm\pi\, i$,

Resi $$R_1 = \lim_{z\to\pi i}(z-\pi i)\frac{e^z}{z^2+\pi^2} = \lim_{z\to\pi i}\frac{e^z}{z+\pi i} = \frac{e^{\pi i}}{2\pi i} = -\frac{1}{2\pi i} = \frac{i^2}{2\pi i} = \frac{i}{2\pi}$$

Resi $$R_2 = \lim_{z\to-\pi i}(z+\pi i)\frac{e^z}{z^2+\pi^2} = \lim_{z\to-\pi i}\frac{e^z}{z-\pi i} = \frac{e^{-\pi i}}{-2\pi i} = \frac{-1}{-2\pi i} = \frac{i^2}{-2\pi i} = -\frac{i}{2\pi}$$

***Example* 22.17:** Evaluate $\int_c \frac{e^{2z}}{(z+1)^2} dz$ around the circle $|z-1|=3$

***Solution*:** $f(z) = \frac{e^{2z}}{(z+1)^2}$, c is the circle $|z-1|=3$

$z = -1$ is the singular point which is a pole of order 2, and this lies inside the circle $|z-1|=3$.

Resi. of f(z) at $z=-1$, a pole of order 2,= $\lim_{z\to -1}\left\{\frac{1}{(2-1)!}\frac{d^{2-1}(z+1)^2 f(z)}{dz^{2-1}}\right\}$

$$\text{Resi} = \lim_{z\to -1}\frac{1}{1!}\frac{d\left[(z+1)^2\frac{e^{2z}}{(z+1)^2}\right]}{dz}$$

$$\text{Resi} = \lim_{z\to -1} 2e^{2z} = 2e^{-2} = \frac{2}{e^2}$$

Using Cauchy's Residue theorem,

$$\int_c \frac{e^{2z}}{(z+1)^2} dz = 2\pi i(\text{Resi}) = 2\pi i\left(\frac{2}{e^2}\right) = \frac{4\pi i}{e^2}$$

***Example* 22.18:** Find the poles and residues of the following

(i) $\frac{3z+1}{(z+1)(2z-1)}$ (ii) $\frac{2z^2-1}{z^2(3z-5)}$ (iii) $z^2 e^{\frac{1}{z^2}}$ (iv) cot z

***Solution*:**

(i) $f(z) = \frac{3z+1}{(z+1)(2z-1)}$, $z=-1$, $z=\frac{1}{2}$ are simple poles.

Residue at $z=-1$ is $\lim_{z\to(-1)}(z+1)f(z) = \lim_{z\to(-1)}(z+1)\frac{3z+1}{(z+1)(2z-1)} = \frac{2}{3}$

Residue at $z=\frac{1}{2}$ is $\lim_{z\to\frac{1}{2}}\left[z-\left(\frac{1}{2}\right)\right]f(z) = \lim_{z\to\frac{1}{2}}\left[z-\left(\frac{1}{2}\right)\right]\frac{3z+1}{(z+1)(2z-1)} = \frac{5}{6}$

(ii) $f(z) = \dfrac{2z^2-1}{z^2(3z-5)}$, z = 0 is a pole of order 2, $z = \dfrac{5}{3}$ is a simple pole.

Residue of f(z) at z = 0 (pole of order 2) is $\lim\limits_{z\to 0}\dfrac{1}{(2-1)!}\dfrac{d^{2-1}\{z^2.f(z)\}}{dz^{2-1}}$

$$\lim_{z\to 0}\frac{d\{z^2.\frac{2z^2-1}{z^2(3z-5)}\}}{dz} = \lim_{z\to 0}\frac{d\{\frac{2z^2-1}{3z-5}\}}{dz}$$

$$= \lim_{z\to 0}\frac{(3z-5)(4z)-(2z^2-1)(3)}{(3z-5)^2} = \frac{3}{25}$$

Residue of f(z) at $z = \dfrac{5}{3}$ is $\lim\limits_{z\to\frac{5}{3}}\left[z-\dfrac{5}{3}\right]f(z) = \lim\limits_{z\to\frac{5}{3}}\left[\dfrac{3z-5}{3}\right]\dfrac{2z^2-1}{z^2(3z-5)} = \dfrac{41}{75}$

(iii) $f(z) = z^2 e^{\frac{1}{z^2}}$

$$f(z) = z^2\left[1+\frac{\left(\frac{1}{z^2}\right)}{1!}+\frac{\left(\frac{1}{z^4}\right)}{2!}+\frac{\left(\frac{1}{z^6}\right)}{3!}+...\right] = z^2+1+\frac{1}{2!z^2}+\frac{1}{3!z^4}+...$$

The principal part is non terminating. Hence z = 0 is an isolated essential singularity.

In this expansion, since the coefficient of $\dfrac{1}{z}$ is zero, the residue at z = 0 is 0.

(iv) $f(z) = \cot z = \dfrac{\cos z}{\sin z}$

The function has singularities at z = 0, $n\pi (n = \pm 1, \ \pm 2, ...)$

Residue at z = 0 is $\lim\limits_{z\to 0} zf(z) = \lim\limits_{z\to 0} z\dfrac{\cos z}{\sin z} = \lim\limits_{z\to 0}\dfrac{\cos z}{\left(\dfrac{\sin z}{z}\right)} = 1$

Similarly we can show that Residue at z = $n\pi$ is also 1.

***Example* 22.19:** Find the residue of (i) $f(z)=\dfrac{z^3}{z^2-1}$ at $z = \infty$

(ii) $f(z)=\dfrac{z^2}{(z-a)(z-b)(z-c)}$ at $z=\infty$

***Solution*:**

(i) $f(z)=\dfrac{z^3}{z^2-1}$ .....(1)

Substituting $z=\dfrac{1}{\xi}$ then $f\left(\dfrac{1}{\xi}\right)=\dfrac{1}{\xi(1-\xi^2)}$ .....(2)

(2) shows that $f(\xi)$ has a simple pole at $\xi=0$ hence $f(z)$ has a simple pole at $z=\infty$

Residue of $f(z)$ at $z=\infty$ is $\lim_{z\to\infty}\{-zf(z)\} = \lim_{z\to\infty}\{-z.\dfrac{z^3}{z^2-1}\}$ which does not exit

Hence we proceed as follows

$$\therefore f(z)=\frac{z^3}{z^2}\left(1-\frac{1}{z^2}\right)^{-1} = z(1+\frac{1}{z^2}+\frac{1}{z^4}+...) = z+\frac{1}{z}+\frac{1}{z^3}+...$$

the coefficient of $\dfrac{1}{z}$ is 1, hence residue of $f(z)$ at $z=\infty$ is equal to 1.

(ii) The function $f(z) = \dfrac{z^2}{(z-a)(z-b)(z-c)}$ is analytic at $z=\infty$. Since $f\left(\dfrac{1}{\xi}\right)$ is analytic at $\xi=0$

Residue of $f(z)$ at $z=\infty$ is $\lim_{z\to\infty}\{-zf(z)\}= \lim_{z\to\infty}\{-z\dfrac{z^2}{(z-a)(z-b)(z-c)}\}=-1.$

***Example* 22.20:** Evaluate $\displaystyle\int_c \frac{z^2+2z-2}{z(z-4)(z-1)}dz$ where c is $|z|=1.5$

Solution $f(z)=\dfrac{z^2+2z-2}{z(z-4)(z-1)}$; c is the circle $|z|=\left(\dfrac{3}{2}\right)$

the poles of $f(z)$ are $z=0, z=1, z=4$

$z=0$ is a simple pole which lie inside 'c'

$z = 1$ is a simple pole which lie inside 'c'

$z = 4$ is a simple pole which lie outside 'c'

Residue of $f(z)$ at the simple pole $z = 0$ is $\lim_{z\to 0}(z-0)f(z)$

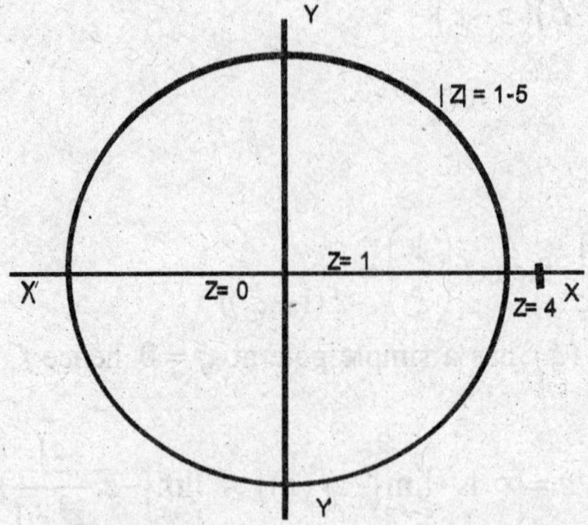

$$R_1 = \lim_{z\to 0}(z-0)\frac{z^2+2z-2}{z(z-4)(z-1)} = \frac{-2}{4} = \frac{-1}{2}$$

Residue of $f(z)$ at the simple pole $z=1$

$$R_2 = \lim_{z\to 1}(z-1)f(z) = \frac{-1}{3}$$

Using Cauchy's Residue Theorem

$$\int_c \frac{z^2+2z-2}{z(z-4)(z-1)}dz = 2\pi i\left[\left(\frac{-1}{2}\right)+\left(\frac{-1}{3}\right)\right] = \frac{-5\pi i}{3}.$$

Exercises – 22.3

I. Prove the following Integrals using Residues

1. $\int_c \frac{2z-1}{z(z+1)(z-3)}dz = \frac{-5\pi i}{6}$ where c is $|z| = 2$;
2. $\int_c \frac{\sin \pi z^2 + \cos \pi z^2}{(z-1)^2(z-2)}dz = 4\pi i(1+\pi)$ where c is $|z| = 3$;

3. $\int_c \frac{3z^2+2}{(z-1)(z^2+9)}dz = \pi i$ where c is $|z-2|=2$;

4. $\int_c \frac{3z+4}{(2z-1)(3z-1)(z^2+4)}dz = \frac{98\pi i}{629}$, where c is $|z|=1$;

5. $\int_c \frac{1-\cos z}{z^2}dz = 0$ where c is $|z|=1$;

6. $\int_c \frac{e^z}{\sin z}dz = 2\pi i$ where c is $|z|=1$

7. $\int_c \frac{z^2-2z}{(z+1)^2(z^2+4)}dz = \pi\left[\frac{4}{7}-\frac{28i}{25}\right]$ where c is $|z|=3$;

8. $\int_c \frac{e^z(z^2+4)}{(z-1)^3}dz = 11\pi ie$ where c is $|z|=2$;

9. $\int_c \tanh z dz = 4\pi i$ where c is $|z|=2$;

22.6 Application of Residue Theorem to Evaluate Some Real Definite Integrals

22.6.1 Type I: Integration around a Unit Circle

Integrals of the form $\int_0^{2\pi} f(\cos\theta,\sin\theta)d\theta$ where f is a rational function of $\cos\theta,\sin\theta$

can be converted into complex line integrals by the substitution $z=e^{i\theta}$ (i.e. $|z|=1$)

$$dz = ie^{i\theta}d\theta \quad \therefore \; d\theta = \frac{dz}{iz}$$

$$\cos\theta = \frac{e^{i\theta}+e^{-i\theta}}{2} = \frac{z+\frac{1}{z}}{2}, \quad \sin\theta = \frac{e^{i\theta}-e^{-i\theta}}{2i} = \frac{z-\frac{1}{z}}{2i}$$

As θ varies from 0 to 2π, the complex variable z traverses once around a unit circle in the anti clock wise direction.

The integral $\int_0^{2\pi} f(\cos\theta,\sin\theta)d\theta$ reduces to

$$\int_c f\left[\frac{z+\frac{1}{z}}{2}, \frac{z-\frac{1}{z}}{2i}\right]\frac{dz}{iz}, \text{where } c \text{ is } |z|=1$$

And this can be evaluated using residue theorem.

Solved Examples

***Example* 22.21**: Evaluate $\int_0^{2\pi} \frac{d\theta}{a+b\sin\theta}$, where c is $|z|=1$

Solution:

$$I = \int_c \frac{\frac{dz}{iz}}{a+b\left(\frac{z-\frac{1}{z}}{2i}\right)}$$

$$I = 2\oint_C \frac{dz}{bz^3+2iaz-b} \qquad .....(1)$$

Have $f(z) = \frac{1}{bz^3+2iaz-b}$ The poles of $f(z)$ are $z = i\left(\frac{-a\pm\sqrt{a^2-b^2}}{b}\right)$ of which only $i\left(\frac{-a+\sqrt{a^2-b^2}}{b}\right)$ lies inside c, $|z|=1$

Let $$i\frac{\left(-a+\sqrt{a^2-b^2}\right)}{b} = \alpha$$

Residue of f(z) at $z=\alpha$ is $\lim_{z\to\alpha}(z-\alpha)\frac{1}{bz^2+2iaz-b}$ (of the form 0/0)

Applying L' Hospital's Rule,

$$R = \lim_{z\to\alpha}\frac{1}{2(bz+ai)} = \frac{1}{2i\sqrt{a^2-b^2}}$$

Using Residue Theorem,

$$I = 2\oint_c \frac{dz}{bz^3 + 2iaz - b} = 2.2\pi i\left[\frac{1}{2i\sqrt{a^2-b^2}}\right] = \frac{2\pi}{\sqrt{a^2-b^2}}$$

***Example* 22.22:** Prove that $\int_0^{2\pi} \frac{\cos 2\theta d\theta}{5+4\cos\theta} = \frac{\pi}{6}$

***Solution*:**

Substituting $z = e^{i\theta}$ $d\theta = \frac{dz}{iz}, \sin\theta = \frac{z-z^{-1}}{2i}, \cos\theta = \frac{z+z^{-1}}{2}$, $\cos 2\theta = \frac{z^2+z^{-2}}{2}$ in the given integral I, we get

$$I = \int_0^{2\pi} \frac{\cos 2\theta d\theta}{5+4\cos\theta} = \oint_C \frac{\frac{1}{2}\left(z^2+\frac{1}{z^2}\right)}{5+2\left(z+\frac{1}{z}\right)}\frac{dz}{iz} \qquad \text{where } c \text{ is } |z| = 1$$

$$= \frac{1}{2i}\oint_C \frac{z^4+1}{z^2(2z^2+5z+2)}dz$$

Now $\qquad f(z) = \frac{z^4+1}{z^2(2z^2+5z+2)} = \frac{z^4+1}{z^2(2z+1)(z+1)}$

The poles of $f(z)$ are $z = 0$, a pole of order 2, inside 'c'

$z = \frac{-1}{2}$ simple pole inside 'c', $z = -2$ simple pole outside'c'

Resi of $f(z)$ at $z = 0$ the double pole,

$$R_1 = \lim_{z\to 0}\frac{1}{(2-1)!}\frac{d^{2-1}\{(z-0)^2 f(z)\}}{dz^{2-1}}$$

$$= \lim_{z\to 0}\frac{d\left\{z^2.\frac{(z^4+1)}{z^2(2z+1)(z+1)}\right\}}{dz} = \lim_{z\to 0}\frac{d\left\{\frac{(z^4+1)}{2z^2+5z+2}\right\}}{dz}.$$

$$R_1 = \lim_{z\to 0}\frac{(2z^2+5z+2)(4z^3)-(1+z^4)(4z+5)}{(2z^2+5z+2)^2} = \frac{-5}{4}$$

Resi of $f(z)$ at $z = \frac{-1}{2}$ a simple pole

$$= \lim_{z \to -\frac{1}{2}} \left[z - \left(\frac{-1}{2} \right) \right] . f(z) = \lim_{z \to -\frac{1}{2}} \left(\frac{2z+1}{2} \right) \frac{(1+z^4)}{z^2(2z+1)(z+2)} = \frac{17}{12}$$

$$R_1 + R_2 = \frac{-5}{4} + \frac{17}{12} = \frac{1}{6} \quad .....(1)$$

Using Cauchy's Residue Theorem

$$I = \frac{1}{2i} \left[2\pi i . \frac{1}{6} \right] = \frac{\pi}{6} \text{ from (1)}$$

***Example* 22.23:** Evaluate $\int_0^{\pi} \frac{d\theta}{1 + a\cos\theta}$, $0 < a < 1$ using residue theorem.

***Solution*:**

Consider the integral $I = \int_0^{2\pi} \frac{d\theta}{1 + a\cos\theta}$. Substituting $z = e^{i\theta}$, $d\theta = \frac{dz}{iz}$, $\cos\theta = \frac{\left(z + \frac{1}{z} \right)}{2}$

$$I = \frac{2}{i} \int_c \frac{\frac{dz}{iz}}{1 + a \frac{\left(z + \frac{1}{z} \right)}{2}} \quad \text{where c is unit circle } |z| = 1$$

$$I = \frac{2}{i} \int_c \frac{dz}{az^2 + 2z + a} = \frac{2}{i} [2\pi i \times \text{Residue}] \quad .....(1)$$

The singularities of $f(z) = \frac{1}{az^2 + 2z + a}$ are given by

$$az^2 + 2z + a \quad \text{i.e.,} \quad z = \frac{-1 \pm \sqrt{1 - a^2}}{a}$$

Since $0 < a < 1$, $\left| \frac{1 + \sqrt{1 - a^2}}{a} \right| > 1$ and $\beta = \frac{-1 - \sqrt{1 - a^2}}{a}$ lies outside $|z| = 1$

Since the product of the two roots is 1, the simple pole $\alpha = \dfrac{-1+\sqrt{1-a^2}}{a}$ lies inside $|z| = 1$

Residue at simple pole $\alpha = \lim\limits_{z \to \alpha}(z-\alpha).\dfrac{1}{az^2+2z+a} = \lim\limits_{z \to \alpha}(z-\alpha).\dfrac{1}{a(z-\alpha)(z-\beta)}$

$$\text{Residue} = \frac{1}{a(\alpha-\beta)} = \frac{1}{2\sqrt{1-a^2}}$$

Substituting in (1)

$$I = \frac{2}{i}\left[2\pi i\left(\frac{1}{2\sqrt{1-a^2}}\right)\right], I = \frac{2\pi}{\sqrt{1-a^2}}$$

$$\therefore \quad \int_0^{2\pi} \frac{d\theta}{1+a\cos\theta} = \frac{2\pi}{\sqrt{1-a^2}},$$

$$\therefore \quad \int_0^{\pi} \frac{d\theta}{1+a\cos\theta} = \frac{\pi}{\sqrt{1-a^2}} \ (\text{since} \cos(2\pi-\theta) = \cos\theta)$$

***Example* 22.24:** Show that $\displaystyle\int_0^{\pi} \frac{d\theta}{(a+b\cos\theta)^2} = \frac{\pi a}{(a^2-b^2)^{\frac{3}{2}}}, \ 0 < b < a$

***Solution*:** We will first evaluate $I = \displaystyle\int_0^{2\pi} \frac{d\theta}{(a+b\cos\theta)}$

Substituting $z = e^{i\theta}, d\theta = \dfrac{dz}{iz}, \cos\theta = \left(z+\dfrac{1}{z}\right)/2$

$$I = \int_c \frac{\frac{dz}{iz}}{a+b\frac{\left(z+\frac{1}{z}\right)}{2}} \qquad \text{where } c \text{ is the unit circle } |z| = 1$$

$$= \frac{2}{i}\int_c \frac{dz}{bz^2+2az+b}$$

The poles of $\dfrac{1}{bz^2+2az+b}$ are given by $bz^2+2az+b = 0$

$$z=\frac{-a\pm\sqrt{a^2-b^2}}{b},\ (i.e.)\ \alpha=\frac{-a+\sqrt{a^2-b^2}}{b},\ \beta=\frac{-a-\sqrt{a^2-b^2}}{b}$$

are the poles $|\beta|>1,$ since $0<b<a$

Hence β is outside c. Since $\alpha\beta=1$, $|\alpha|<1$ and hence α lies inside c.

Residue at $\alpha=\lim\limits_{z\to\alpha}(z-\alpha).\dfrac{1}{bz^2+2az+b}=\lim\limits_{z\to\alpha}(z-\alpha).\dfrac{1}{b(z-\alpha)(z-\beta)}=\dfrac{1}{b(\alpha-\beta)}$

$$=\frac{1}{2\sqrt{a^2-b^2}}$$

Using Residue Theorem

$$I=\frac{2}{i}\left[2\pi i\times\frac{1}{2\sqrt{a^2-b^2}}\right]=\frac{2\pi}{\sqrt{a^2-b^2}}.$$

$$\therefore\quad \int_0^{2\pi}\frac{d\theta}{a+b\cos\theta}=\frac{2\pi}{\sqrt{a^2-b^2}}$$

Differentiating both sides w.r.t 'a'

$$\int_0^{2\pi}\frac{-d\theta}{(a+b\cos\theta)^2}=2\pi\left[\frac{-1}{2}\left(a^2-b^2\right)^{\frac{-3}{2}}.2a\right]$$

i.e.,
$$\int_0^{2\pi}\frac{d\theta}{(a+b\cos\theta)^2}=\frac{2a\pi}{\left(a^2-b^2\right)^{\frac{3}{2}}}$$

$$\therefore\quad \int_0^{\pi}\frac{d\theta}{(a+b\cos\theta)^2}=\frac{a\pi}{\left(a^2-b^2\right)^{\frac{3}{2}}}$$

***Example* 22.25:** Evaluate : $\displaystyle\int_0^{2\pi}\frac{d\theta}{\left(1-2a\sin\theta+a^2\right)},0<a<1$

***Solution*:** Substituting $z=e^{i\theta}$, $d\theta=\dfrac{dz}{iz}$, $\sin\theta=\dfrac{\left(z-\dfrac{1}{z}\right)}{2i}$

$$I=\int_0^{2\pi}\frac{d\theta}{1-2a\sin\theta+a^2}=\int_c\frac{\frac{dz}{iz}}{1-2a\frac{\left(z-\frac{1}{z}\right)}{2i}+a^2}$$ where c is the unit circle $|z|=1$

$$=-\int_c\frac{dz}{az^2-i(a^2+1)z-a}$$

The poles of $\frac{1}{az^2-i(a^2+1)z-a}$ are given by $az^2-i(a^2+1)z-a=0$

$$z=\frac{(a^2+1)i\pm\sqrt{-(a^2+1)^2+4a^2}}{2a}=\frac{(a^2+1)i\pm\sqrt{-(a^4-2a^2+1)}}{2a}$$

$$z=\frac{(a^2+1)i\pm i(a^2-1)}{2a}=ai,\ \frac{i}{a}.$$

$z=$ ai is a pole of order one, inside c, (since $a<1$),

$z=i/a$ is a pole of order 1 out side c($a<1$).

Residue at ai $=\lim_{z\to ai}(z-ai).\frac{1}{az^2-i(a^2+1)z-a}=\lim_{z\to ai}(z-ai).\frac{1}{a(z-ai)\left(z-\frac{i}{a}\right)}$

Residue at $z=ai$ $=\frac{1}{i(a^2-1)}$.

Substituting in (1) and using Residue theorem

$$I=-2\pi i\left[\frac{1}{i(a^2-1)}\right]$$

$$\int_0^{2\pi}\frac{d\theta}{1-2a\sin\theta+a^2}=\frac{2\pi}{1-a^2}.$$

***Example* 22.26:** Evaluate $\int_0^{\pi} \frac{a\,d\phi}{a^2+\sin^2\phi}, a>0$

***Solution*:**

Let

$$I = \int_0^{\pi} \frac{a\,d\phi}{a^2+\sin^2\phi}, a>0$$

$$I = \int_0^{\pi} \frac{a\,d\phi}{a^2+\frac{1-\cos 2\phi}{2}} \qquad \left(\text{since } \sin^2\phi = \frac{1-\cos 2\phi}{2}\right)$$

$$= \int_0^{\pi} \frac{2a\,d\phi}{2a^2+1-\cos 2\phi}$$

Replacing 2ϕ by θ , $2d\phi = d\theta$. when $\phi = 0,\ \theta = 0$, when $\phi = \pi,\ \theta = 2\pi$

Substituting $z = e^{i\theta}, d\theta = \frac{dz}{iz}, \cos\theta = \frac{z+(1/z)}{2}$ in $I = \int_0^{2\pi} \frac{a\,d\theta}{2a^2+1-\cos\theta}$

$$\mathrm{I} = \int_c \frac{a\,.\frac{dz}{iz}}{2a^2+1-\frac{1}{2}\left(z+\frac{1}{z}\right)} = \frac{-2a}{i}\int_c \frac{dz}{z^2-(4a^2+2)z+1}$$

where c is unit circle $|z| = 1$

$$= 2ai\int_c \frac{dz}{z^2-(4a^2+2)z+1}$$

The poles of $\frac{1}{z^2-(4a^2+2)z+1}$ are solutions of $z^2-(4a^2+2)z+1=0$

$$z = \frac{(4a^2+2)\pm\sqrt{(4a^2+2)^2-4}}{2} = 2a^2+1\pm 2a\sqrt{(a^2+1)}$$

Simple pole $z = 2a^2+1+2a\sqrt{(a^2+1)} = \alpha$, say , > 1 since a > 0

$\therefore\ \alpha$ is out side c

Simple pole $2a^2+1-2a\sqrt{(a^2+1)} = \beta$, say, < 1 lies inside c (since $\alpha\beta = 1$)

Now, residue at $z = \beta = \lim_{z\to\beta}(z-\beta).\dfrac{1}{z^2-(4a^2+2)z+1}$

$$= \lim_{z\to\beta}(z-\beta).\frac{1}{(z-\alpha)(z-\beta)} = \frac{1}{(\beta-\alpha)}$$

$$= \frac{1}{\left[2a^2+1-2a\sqrt{a^2+1}\right]-\left[2a^2+1+2a\sqrt{a^2+1}\right]}$$

Residue $= \dfrac{1}{-4a\sqrt{a^2+1}}$

Using Cauchy's residue theorem $I = 2ai\left[2\pi i \times \dfrac{1}{-4a\sqrt{a^2+1}}\right]$

$$\therefore \qquad \int_0^{\pi}\left[\frac{a\,d\phi}{a^2+\sin^2\phi}\right] = \frac{\pi}{\sqrt{a^2+1}}$$

Exercise – 22.4

I. Prove the following using Residue theorem

1. $\displaystyle\int_0^{2\pi}\frac{d\theta}{a+b\cos\theta},\ a>b>0 = \frac{2\pi}{\sqrt{a^2-b^2}}$

2. $\displaystyle\int_0^{\pi}\frac{d\theta}{a+\cos\theta},\ a>1 = \frac{\pi}{\sqrt{a^2-1}}$

CHAPTER 23

Fourier Transforms

23.1 Integral Transforms

23.1.1 Definition of Integral Transform

The integral transform of a function $f(x)$ is defined by $\int_a^b f(x).k(s,x)dx$ where $k(s, x)$ is a known function of s, x and is called the kernel of the transform. It is denoted by I [f(x)] or F(s)

(i.e.,) $$I\,[f(x)] = \int_a^b f(x)\,.\,k\,(s, x)\,dx = F(s) \qquad \text{.....(1)}$$

where s is called the parameter of the transform and $f(x)$ is called the inverse transform of [f(x)].

If k (s, x) = e^{isx} then $\int_{-\infty}^{\infty} f(x)e^{isx}dx$ is called the Fourier transform and is denoted by F[$f(x)$].

(i.e.,) $$F[f(x)] = \int_{-\infty}^{\infty} f(x)e^{isx}dx = F(s)$$

If k $(s, x) = e^{-sx}$ in (1) then $\int_0^{\infty} f(x)e^{-sx}dx$ is called the Laplace transform of $f(x)$ and is denoted by $L[f(x)]$

(i.e.,) $$L[f(x)] = \int_0^{\infty} f(x)e^{-sx}dx = F(S)$$

If $k(s, x) = xJ_n(sx)$ in (1) ($J_n(sx)$ is a Bessel Function).

Then $\int_0^{\infty} f(x)xJ_n(sx)dx$ is called Hankel Transform denoted by H(S)

(i.e.,) $$H(S) = \int_0^{\infty} f(x)xJ_n(sx)dx$$

If $k(s, x) = x^{s-1}$ in (1)

Then $\int_0^{\infty} f(x)x^{S-1}dx$ is called Mellin Transform denoted by M(S)

$$M(S) = \int_0^{\infty} f(x)x^{s-1}dx$$

23.1.2 Fourier Integral Theorem

If $f(x)$ is a given function defined in $(-l, l)$ and satisfies Dirichlet's conditions, then

$$f(x) = \frac{1}{\pi}\int_0^{\infty}\int_{-\infty}^{\infty} f(t)\cos\lambda(t-x)dxd\lambda \text{ (is called Fourier Integal Formula)}$$

Proof : The Fourier series of $f(x)$ in $(-l, l)$ is

$$f(x) = \frac{a_0}{2} + \sum_{n=1}^{\infty}\left(a_n\cos\frac{n\pi x}{l} + b_n\sin\frac{n\pi x}{l}\right) \quad .....(1)$$

where $$a_0 = \frac{1}{l}\int_{-l}^{l} f(t)dt$$

$$\left.\begin{aligned} a_n &= \frac{1}{l}\int_{-l}^{l} f(t)\cos\frac{n\pi t}{l}\,dt \\ b_n &= \frac{1}{l}\int_{-l}^{l} f(t)\sin\frac{n\pi t}{l}\,dt \end{aligned}\right\} \quad \text{.....(2)}$$

substituting (2) in (1)

$$f(x) = \frac{1}{2l}\int_{-l}^{l} f(t)dt + \sum_{n=1}^{\infty}\frac{1}{l}\int_{-l}^{l} f(t)\ \cos\frac{n\pi t}{l}\ .\ \cos\frac{n\pi x}{l}\,dt$$

$$+ \sum_{n=1}^{\infty}\frac{1}{l}\int_{-l}^{l} f(t)\sin\frac{n\pi t}{l}.\sin\frac{n\pi x}{l}\,dt$$

(i.e.,) $$f(x) = \frac{1}{2l}\int_{-l}^{l} f(t)dt + \frac{1}{l}\int_{-l}^{l} f(t)\left\{\sum_{n=1}^{\infty}\cos\frac{n\pi t}{l}\cos\frac{n\pi x}{l} + \sin\frac{n\pi t}{l}\sin\frac{n\pi x}{l}\right\}dt$$

(i.e.,) $$f(x) = \frac{1}{2l}\int_{-l}^{l} f(t)\left\{1 + 2\sum_{n=1}^{\infty}\cos\frac{n\pi}{l}(t-x)\right\}dt \quad \text{.....(3)}$$

But cosine functions are even functions

$$\therefore \quad \sum_{n=-\infty}^{\infty}\cos\frac{n\pi}{l}(t-x) = \sum_{n=-\infty}^{1}\cos\frac{n\pi}{l}(t-x) + 1 + \sum_{n=1}^{\infty}\cos\frac{n\pi}{l}(t-x)$$

$$\sum_{n=-\infty}^{\infty}\cos\frac{n\pi}{l}(t-x) = 1 + 2\sum_{n=1}^{\infty}\cos\frac{n\pi}{l}(t-x) \quad \text{.....(4)}$$

Substituting in (3)

$$f(x) = \frac{1}{2l}\int_{-l}^{l} f(t)\left\{\sum_{n=-\infty}^{\infty}\cos\frac{n\pi}{l}(t-x)\right\}dt$$

$$f(x) = \frac{1}{2\pi}\int_{-l}^{l} f(t)\left\{\frac{\pi}{l}\sum_{n=-\infty}^{\infty}\cos\frac{n\pi}{l}(t-x)\right\}dt \quad \text{.....(5)}$$

Now assume l increases infinitely so that

$$\frac{n\pi}{l} = \lambda \text{ and } \frac{\pi}{l} = d\lambda$$

$$\therefore \quad \underset{l\to\infty}{Lt}\ \frac{\pi}{l}\sum_{n=-\infty}^{\infty}\cos\frac{n\pi}{l}(t-x) = \int_{-\infty}^{\infty}\cos\lambda(t-x)d\lambda$$

$$= 2\int_{0}^{\infty}\cos\lambda(t-x)d\lambda \quad .....(6)$$

Substituting (6) in (5)

$$f(x) = \frac{1}{2\pi}\int_{-\infty}^{\infty}f(t)\left\{2\int_{0}^{\infty}\cos\lambda(t-x)d\lambda\right\}dt$$

$$\therefore \quad f(x) = \frac{1}{\pi}\int_{0}^{\infty}\int_{-\infty}^{\infty}f(t)\cos\lambda(t-x)dtd\lambda$$ is called Fourier integral Formula.

At a point of discontinuity the value of the integral on the L. H. S of above formula is $\frac{1}{2}\{f(x+0)-f(x-0)\}$

$$\therefore \quad \frac{1}{2}\{f(x+0)-f(x-0)\} = \frac{1}{\pi}\int_{0}^{\infty}\int_{-\infty}^{\infty}f(t)\cos\lambda(t-x)dtd\lambda \quad .....(7)$$

23.1.3 Fourier Sine and Cosine Integrals

The Fourier integral of $f(x)$ is

$$f(x) = \frac{1}{\pi}\int_{0}^{\infty}\int_{-\infty}^{\infty}f(t)\cos\lambda(t-x)dtd\lambda$$

$$f(x) = \frac{1}{\pi}\int_{0}^{\infty}\int_{-\infty}^{\infty}f(t)[\cos\lambda t\cos\lambda x+\sin\lambda t\sin\lambda x]dtd\lambda$$

$$f(x) = \frac{1}{\pi}\int_{0}^{\infty}\cos\lambda x\int_{-\infty}^{\infty}f(t)\cos\lambda t\,dtd\lambda + \frac{1}{\pi}\int_{0}^{\infty}\sin\lambda x\int_{-\infty}^{\infty}f(t)\sin\lambda t\,dt\,d\lambda \quad .....(1)$$

Case (i) :

If $f(t)$ is an odd function, then $f(t)\cos\lambda t$ also is an odd function and hence the first integral in (1) becomes zero.

$$\therefore \qquad f(x) = \frac{1}{\pi}\int_0^{\infty}\int_{-\infty}^{\infty} f(t)\sin\lambda t \sin\lambda x \, dt\, d\lambda \qquad .....(2)$$

Since $f(t)$ is odd and sin λt is also odd $f(t)$sin λt is an even function. Hence equation (2) can be written as

$$f(x) = \frac{2}{\pi}\int_0^{\infty}\sin\lambda x \int_0^{\infty} f(t)\sin\lambda t \, dt\, d\lambda$$

This is known as the Fourier sine integral .

***Case (ii)* :**

If $f(t)$ is an even function, then $f(t)$ sin λt is an odd function. Hence the second integral in (1) becomes zero

$$\therefore \qquad f(x) = \frac{1}{\pi}\int_0^{\infty}\int_{-\infty}^{\infty} f(t)\cos\lambda t \cos\lambda x \, dt\, d\lambda \qquad .....(3)$$

$f(t)$ is even, cosλt is also even

$\therefore$ $f(t)$ cos λt is an even function. Hence equation (3) can be written as

$$f(x) = \frac{2}{\pi}\int_0^{\infty}\cos\lambda x \int_0^{\infty} f(t)\cos\lambda t \, dt\, d\lambda$$

This is known as Fourier cosine integral.

23.1.4 Complex Form of Fourier Integrals

The Fourier integral of $f(x)$ is

$$f(x) = \frac{1}{\pi}\int_0^{\infty}\int_{-\infty}^{\infty} f(t)\cos\lambda(t-x)\, dt\, d\lambda$$

$$\text{(i.e.)} \qquad f(x) = \frac{1}{2\pi}\int_{-\infty}^{\infty}\int_{-\infty}^{\infty} f(t)\cos\lambda(t-x)\, dt\, d\lambda \qquad .....(1)$$

Since cos λ $(t - x)$ is an even function and sin λ $(t - x)$ is an odd function of λ, therefore

$$\int_{-\infty}^{\infty}\sin\lambda(t-x)\, d\lambda = 0$$

$$\therefore \qquad \frac{1}{2\pi}\int_{-\infty}^{\infty}\int_{-\infty}^{\infty} f(t)\sin\lambda(t-x)\, dt\, d\lambda = 0 \qquad .....(2)$$

Multiplying (2) by i

$$\frac{i}{2\pi}\int_{-\infty}^{\infty}\int_{-\infty}^{\infty} f(t)\sin\lambda(t-x)dt\,d\lambda = 0 \qquad .....(3)$$

Adding (1) and (3)

$$f(x) = \frac{1}{2\pi}\int_{-\infty}^{\infty}\int_{-\infty}^{\infty} f(t)[\cos\lambda(t-x)+i\sin\lambda(t-x)]dt\,d\lambda$$

$$= \frac{1}{2\pi}\int_{-\infty}^{\infty}\int_{-\infty}^{\infty} f(t)e^{i\lambda(t-x)}dt\,d\lambda$$

$$f(x) = \frac{1}{2\pi}\int_{-\infty}^{\infty} e^{-i\lambda x}\int_{-\infty}^{\infty} f(t)e^{i\lambda t}dt\,d\lambda$$

This is known as complex form of Fourier integral.

Solved Examples

***Example* 23.1**: Find the Fourier integral of the function

$$f(x) = \begin{cases} 0, & x < 0 \\ \frac{1}{2}, & x = 0 \\ e^{-x}, & x > 0 \end{cases}$$

Solution: Fourier integral of $f(x)$ is

$$f(x) = \frac{1}{\pi}\int_{0}^{\infty}\int_{-\infty}^{\infty} f(t)\cos\lambda(t-x)dt\,d\lambda$$

$$= \frac{1}{\pi}\int_{0}^{\infty}\left\{\int_{-\infty}^{0} f(t)\cos\lambda(t-x)dt + \int_{0}^{\infty} e^{-t}\cos\lambda(t-x)dt\right\}d\lambda$$

$$= \frac{1}{\pi}\int_{0}^{\infty}\left\{\int_{-\infty}^{\infty} 0.\cos\lambda(t-x)dt + \int_{0}^{\infty} e^{-t}\cos\lambda(t-x)dt\right\}d\lambda$$

$$= \frac{1}{\pi}\int_0^\infty \left\{0 + \frac{e^{-t}}{1+\lambda^2}\left[-\cos\lambda(t-x) + \lambda\sin\lambda(t-x)\right]_0^\infty\right\} d\lambda$$

$$f(x) = \frac{1}{\pi}\int_0^\infty \frac{\cos\lambda x + \lambda\sin\lambda x}{1+\lambda^2} d\lambda$$

is the required Fourier integral of the given function $f(x)$.

Verification : Substituting $x = 0$ in the above integral

$$f(0) = \frac{1}{\pi}\int_0^\infty \frac{\cos 0 + \lambda\sin 0}{1+\lambda^2} d\lambda$$

$$f(0) = \frac{1}{\pi}\int_0^\infty \frac{d\lambda}{1+\lambda^2} = \frac{1}{\pi}\left[\tan^{-1}(\lambda)\right]_0^\infty$$

$$f(0) = \frac{1}{\pi}\left[\tan^{-1}(\infty) - \tan^{-1}(0)\right]$$

$$f(0) = \frac{1}{\pi}\left[\frac{\pi}{2}\right] = \frac{1}{2}$$

$\therefore$ The value of $f(x) = \frac{1}{2}$ at $x = 0$ is given.

Hence verified.

***Example* 23.2:** Obtain the Fourier integral representation of the function defined by

$$f(t) = 0, \quad t > -t_0$$
$$f(t) = 1, \quad -t_0 < t < t_0$$
$$f(t) = 0, \quad t > t_0$$

***Solution*:** The given function is

$$f(t) = 0, \quad t < -t_0$$
$$= 1, \quad -t_0 < t < t_0$$
$$= 0, \quad t > t_0$$

The Fourier integral of the given function $f(x)$ is

$$f(x) = \frac{1}{\pi}\int_0^{\infty}\int_{-\infty}^{\infty} f(t)\cos\lambda(t-x)\,dt\,d\lambda$$

$$= \frac{1}{\pi}\int_0^{\infty}\int_{-t_0}^{t_0} 1.\cos\lambda(t-x)\,dt\,d\lambda$$

$$= \frac{1}{\pi}\int_0^{\infty}\left[\frac{\sin\lambda(t-x)}{\lambda}\right]_{-t_0}^{t_0} d\lambda$$

$$= \frac{1}{\pi}\int_0^{\infty}\frac{1}{\lambda}\{\sin\lambda(t_0-x)-\sin\lambda(-t_0-x)\}d\lambda$$

$$f(x) = \frac{2}{\pi}\int_0^{\infty}\frac{1}{\lambda}\sin\lambda t_0\cos\lambda x\,d\lambda$$

***Example* 23.3:** Express the function

$$f(x) = 1 \quad \text{for } |x| \le 1$$
$$= 0 \quad \text{for } |x| > 1$$

as a Fourier integral. Hence evaluate $\int_0^{\infty}\frac{\sin\lambda\cos\lambda x}{\lambda}d\lambda$ and also find the value of $\int_0^{\infty}\frac{\sin\lambda}{\lambda}d\lambda$.

***Solution*:** The Fourier integral of $f(x)$ is

$$f(x) = \frac{1}{\pi}\int_0^{\infty}\int_{-\infty}^{\infty} f(t)\cos\lambda(t-x)\,dt\,d\lambda$$

$$= \frac{1}{\pi}\int_0^{\infty}\int_{-1}^{1} 1\cos\lambda(t-x)\,dt\,d\lambda$$

$$= \frac{1}{\pi}\int_0^{\infty}\left[\frac{\sin\lambda(t-x)}{\lambda}\right]_{-1}^{1} d\lambda$$

$$f(x) = \frac{1}{\pi}\int_0^{\infty}\frac{\sin\lambda(1-x)+\sin\lambda(1+x)}{\lambda}d\lambda$$

$$f(x) = \frac{2}{\pi}\int_0^{\infty}\frac{\sin\lambda\cos\lambda x}{\lambda}d\lambda$$

$$\therefore \qquad \int_0^{\infty}\frac{\sin\lambda\cos\lambda x}{\lambda}d\lambda = \frac{\pi}{2}f(x) \qquad .....(1)$$

But it is given that

$$f(x) = 1 \quad \text{for } |x| < 1$$
$$= 0 \quad \text{for } |x| > 1$$

$$\therefore \qquad \left.\begin{aligned}\int_0^{\infty}\frac{\sin\lambda\cos\lambda x}{\lambda}d\lambda &= \frac{\pi}{2}.1 \quad \text{for } |x| < 1\\ &= 0 \quad \text{for } |x| > 1\end{aligned}\right\} \qquad .....(2)$$

For $|x| = 1$ which is a point of discontinuity of $f(x)$, the integral has the value

$$\frac{\pi}{2}\frac{(1+0)}{2} = \frac{\pi}{4} \qquad \text{(using (7) of 23.1.2)}$$

From the result (2)

$$\int_0^{\infty}\frac{\sin\lambda\cos\lambda x}{\lambda}d\lambda = \frac{\pi}{2},\ |x| < 1$$

Substituting $x = 0$

$$\int_0^{\infty}\frac{\sin\lambda}{\lambda}d\lambda = \frac{\pi}{2}$$

***Example* 23.4:** Using Fourier integral, show that

$$\int_0^{\infty}\frac{\sin\pi\lambda\sin x\lambda}{1-\lambda^2}d\lambda = \begin{cases}\frac{1}{2}\pi\sin x & \text{when } 0 \le x \le \pi\\ 0 & \text{when } x > \pi\end{cases}$$

***Solution*:** Let $f(x) = \frac{\pi}{2}\sin x, \qquad 0 \le x \le p$

$= 0 \qquad x > \pi$

The Fourier sine integral of $f(x)$ is

$$f(x) = \frac{2}{\pi}\int_0^\infty \sin \lambda x \int_0^\infty f(t)\sin \lambda t\, dt\, d\lambda$$

$$f(x) = \frac{2}{\pi}\int_0^\infty \sin \lambda x \int_0^\pi \frac{\pi}{2} \sin t \sin \lambda t\, dt\, d\lambda$$

$$= \int_0^\infty \sin \lambda x \int_0^\pi \frac{1}{2}\left[\cos(\lambda-1)t - \cos(\lambda+1)t\right] dt\, d\lambda$$

$$= \frac{1}{2}\int_0^\infty \sin \lambda x \left[\frac{\sin(\lambda-1)t}{\lambda-1} - \frac{\sin(\lambda+1)t}{\lambda+1}\right]_0^\pi d\lambda$$

$$= \frac{1}{2}\int_0^\infty \sin \lambda x \frac{[2\sin \lambda\pi \cos\pi]}{\lambda^2-1} d\lambda$$

$$= \int_0^\infty \frac{\sin \lambda x \sin \lambda\pi}{1-\lambda^2} d\lambda$$

Thus $$\int_0^\infty \frac{\sin \lambda x \sin \lambda\pi}{1-\lambda^2} d\lambda = f(x) = \begin{cases} \frac{1}{2}\pi \sin x & \text{when } 0 \le x \le \pi \\ 0 & \text{when } x > \pi \end{cases}$$

***Example* 23.5:** Using Fourier integrals show that

$$\int_0^\infty \frac{1-\cos\pi\lambda}{\lambda} \sin \lambda x\, d\lambda = \begin{cases} \frac{\pi}{2}, & 0 < x < \pi \\ 0, & x > \pi \end{cases}$$

***Solution*:** Let $f(x) = \frac{\pi}{2}, \quad 0 < x < p$

$= 0, \quad x > \pi$

The Fourier sine integral of $f(x)$ is

$$f(x) = \frac{2}{\pi}\int_0^\infty \sin \lambda x \int_0^\infty f(t)\sin \lambda t\, dt\, d\lambda$$

$$= \frac{2}{\pi}\int_0^\infty \sin \lambda x \int_0^\pi \frac{\pi}{2} \sin \lambda t\, dt\, d\lambda \qquad (\because f(x) = 0 \text{ when } x > \pi)$$

$$= \int_0^\infty \sin \lambda x \left[\frac{-\cos \lambda t}{\lambda}\right]_0^\pi d\lambda$$

$$= \int_0^\infty \left[\frac{1-\cos \pi\lambda}{\lambda}\right] \sin \lambda x \, d\lambda$$

Hence the result.

***Example* 23.6:** Using Fourier integrals, show that

$$\int_0^\infty \frac{\lambda \sin \lambda x}{k^2+\lambda^2} d\lambda = \frac{\pi}{2} e^{-kx}, \qquad (x > 0, k > 0)$$

***Solution*:** Let $\quad f(x) = \frac{\pi}{2} e^{-kx}, \qquad (x > 0, k > 0)$

The Fourier sine integral of $f(x)$ is

$$f(x) = \frac{2}{\pi} \int_0^\infty \sin \lambda x \int_0^\infty f(t) \sin \lambda t \, dt \, d\lambda$$

$$= \frac{2}{\pi} \int_0^\infty \sin \lambda x \int_0^\infty \frac{\pi}{2} e^{-kt} \sin \lambda t \, dt \, d\lambda$$

$$= \int_0^\infty \sin \lambda x \left\{\frac{e^{-kt}}{k^2+\lambda^2}(-k \sin \lambda t - \lambda \cos \lambda t)\right\}_0^\infty d\lambda$$

$$= \int_0^\infty \frac{\lambda \sin \lambda x}{k^2+\lambda^2} d\lambda$$

Hence proved.

Exercise - 23.1

Using Fourier Integrals, show that.

1. $\displaystyle\int_0^\infty \frac{\cos \lambda x}{1+\lambda^2} d\lambda = \frac{\pi}{2} e^{-x}, \quad (x \geq 0)$

2. $\int_0^\infty \frac{\cos \lambda x}{\lambda^2 + k^2} d\lambda = \frac{\pi}{2} e^{-x},$ $(x \geq 0)$

3. $\int_{-\infty}^\infty \frac{\lambda^2 + 2}{\lambda^2 + 4} \cos \lambda x \, d\lambda = \frac{\pi}{2} e^{-x} \cos x,$ $(x \geq 0)$

4. $\int_0^\infty \frac{\sin \lambda \cos \lambda x}{\lambda} d\lambda = \frac{\pi}{2},$ $0 \leq x < 1$

23.2 Fourier Transforms

23.2.1 Definitions

The complex form of the Fourier integral of $f(x)$ is

$$f(x) = \frac{1}{2\pi} \int_{-\infty}^\infty e^{-i\lambda x} \int_{-\infty}^\infty f(t) e^{i\lambda t} \, dt \, d\lambda$$

Replacing λ by s $f(x) = \frac{1}{2\pi} \int_{-\infty}^\infty e^{-isx} \int_{-\infty}^\infty f(t) e^{ist} \, dt ds$

$$f(x) = \frac{1}{\sqrt{2\pi}} \int_{-\infty}^\infty e^{-isx} \frac{1}{\sqrt{2\pi}} \int_{-\infty}^\infty f(t) e^{ist} \, dt ds$$

Denoting $\frac{1}{\sqrt{2\pi}} \int_{-\infty}^\infty f(t) e^{ist} \, dt = F[f(x)] = F(s)$ .....(1)

Then $f(x) = \frac{1}{\sqrt{2\pi}} \int_{-\infty}^\infty e^{-isx} F[f(x)] ds$ .....(2)

From (1) and (2) ; $F[f(t)] = \frac{1}{\sqrt{2\pi}} \int_{-\infty}^\infty f(t) e^{ist} dt$ **is called the complex Fourier**

Transform of $f(t)$ and $f(x) = \frac{1}{\sqrt{2\pi}} \int_{-\infty}^\infty F[f(x)] e^{-isx}$ *ds* **is called the inverse formula**

for the complex Fourier Transform of $F[f(x)]$.

23.2.2 Fourier Cosine Transforms

The Fourier cosine integral of $f(x)$ is

$$f(x) = \frac{2}{\pi}\int_0^\infty \cos \lambda x \int_0^\infty f(t)\cos \lambda t \, dt \, d\lambda$$

writing s for λ

$$f(x) = \frac{2}{\pi}\int_0^\infty \cos sx \int_0^\infty f(t)\cos st \, dt \, ds$$

$$f(x) = \sqrt{\frac{2}{\pi}}\int_0^\infty \cos sx \sqrt{\frac{2}{\pi}}\int_0^\infty f(t)\cos st \, dt \, ds$$

Denoting $\sqrt{\frac{2}{\pi}}\int_0^\infty f(t)\cos st \, dt = F_c[f(x)]$

$$f(x) = \sqrt{\frac{2}{\pi}}\int_0^\infty \cos sx . F_c[f(x)] ds$$

$F_c[f(x)] = \sqrt{\frac{2}{\pi}}\int_0^\infty f(t)\cos st \, dt$ is called the Fourier cosine transform of $f(x)$, and

$f(x) = \sqrt{\frac{2}{\pi}}\int_0^\infty F_c[f(x)]\cos sx \, ds$ is called the inverse for the Fourier cosine transform.

23.2.3 Fourier Sine Transforms

The Fourier sine integral of $f(x)$ is

$$f(x) = \frac{2}{\pi}\int_0^\infty \sin \lambda x \int_0^\infty f(t)\sin \lambda t \, dt \, d\lambda$$

Replacing λ by s $\quad f(x) = \frac{2}{\pi}\int_0^\infty \sin sx \int_0^\infty f(t)\sin st \, dt \, ds$

$$f(x) = \sqrt{\frac{2}{\pi}}\int_0^\infty \sin sx \sqrt{\frac{2}{\pi}}\int_0^\infty f(t)\sin st \, dt \, ds$$

Denoting $\sqrt{\frac{2}{\pi}}\int_0^\infty f(t)\sin st\,dt = F_s[f(x)]$

$$f(x) = \sqrt{\frac{2}{\pi}}\int_0^\infty \sin sx\, F_s[f(x)]\,ds$$

The Function $F_s[f(x)] = \sqrt{\frac{2}{\pi}}\int_0^\infty f(t)\sin st\,dt$ is called **the Fourier sine transform** of the function $f(x)$ and $f(x) = \sqrt{\frac{2}{\pi}}\int_0^\infty F_s[f(x)]\sin sx\,ds$ is called the **inverse formula for the Fourier sine transform.**

Note: Some authors follow alternate definitions for Fourier infinite transforms, Fourier sine and cosine transforms in place of those given in 23.2.1, 23.2.2 and 23.2.3. (But in our book we have mostly followed the definitions given in 23.2.1 to 23.2.3). The alternate definitions are listed below

1. **Fourier Transform F(s) of f(x) and inverse Fourier transform f(x) of F(s):**

Definition (1) :- $F[f(x)] = F(s) = \int_{-\infty}^{\infty} f(x)e^{isx}dx$ and $F(x) = \frac{1}{2\pi}\int_{-\infty}^{\infty} F(s)e^{-isx}ds$

Defirnition (2):- $F[f(x)] = F(s) = \int_{-\infty}^{\infty} f(x)e^{-isx}dx$ and $f(x) = \frac{1}{2\pi}\int_{-\infty}^{\infty} F(s)e^{-isx}ds$

2. **Fourier cosine transform and its inversion formula:-**

$$F_c[f(x)] = F_c(s) = \int_0^\infty f(x)\cos sx\,dx$$

and $$f(x)=\frac{2}{\pi}\int_0^{\infty}F_c(s)\cos sx\,ds$$

3. Fourier sine transform and its inversion formula:-

$$F_s\left[f(x)\right]=F_s(s)=\int_0^{\infty}f(x)\sin sx\,dx$$

and $$f(x)=\frac{2}{\pi}\int F_s(s)\sin sx\,dx$$

(It is better to remember these formulae too as they are to be used in some problems)

23.2.4 Properties of Fourier Transforms

(a) *Linear property :*

$$F[af_1(x)+bf_2(x)]=aF_1(s)+bF_2(s)$$

Proof :

$$F[af_1(x)+bf_2(x)]=\frac{1}{\sqrt{2\pi}}\int_{-\infty}^{\infty}\left[af_1(t)+bf_2(t)\right]e^{ist}dt$$

$$=\frac{1}{\sqrt{2\pi}}\int_{-\infty}^{\infty}af_1(t)e^{ist}dt+\frac{1}{\sqrt{2\pi}}\int_{-\infty}^{\infty}bf_2(t)e^{ist}dt$$

$$=a\frac{1}{\sqrt{2\pi}}\int_{-\infty}^{\infty}f_1(t)e^{ist}dt+b\frac{1}{\sqrt{2\pi}}\int_{-\infty}^{\infty}f_2(t)e^{ist}dt$$

$$=a\,F_1(s)+bF_2(s)$$

(b) Shifting Theorem :

Show that $F[f(x-a)]=e^{ias}F(s)$

proof:

$$F[f(x-a)] = \frac{1}{\sqrt{2\pi}} \int_{-\infty}^{\infty} f(t-a)e^{ist}\,dt$$

Substituting $t - a = z \qquad t = -\infty \Rightarrow z = -\infty$

$dt = dz \qquad t = \infty \Rightarrow z = \infty$

$$= \frac{1}{\sqrt{2\pi}} \int_{-\infty}^{\infty} f(z)e^{is(z+a)}\,dz$$

$$= \frac{e^{ias}}{\sqrt{2\pi}} \int_{-\infty}^{\infty} f(z)e^{isz}\,dz$$

$$= e^{ias} \frac{1}{\sqrt{2\pi}} \int_{-\infty}^{\infty} f(t)e^{ist}\,dt$$

$$\therefore \qquad F[f(x-a)] = e^{ias}F(s)$$

(c) Show that $F[e^{iax}f(x)] = F(s + a)$

Proof:

$$F[e^{iax}f(x)] = \frac{1}{\sqrt{2\pi}} \int_{-\infty}^{\infty} e^{iat} f(t).e^{ist}\,dt$$

$$= \frac{1}{\sqrt{2\pi}} \int_{-\infty}^{\infty} f(t)e^{i(s+a)t}\,dt$$

$$\therefore \qquad F[e^{iax} f(x)] = F(s + a)$$

(d) ***Change of scale property :***

Show that $F[f(ax)] = \frac{1}{a}F\left(\frac{s}{a}\right)$ $(a > 0)$

Proof: $F[f(ax)] = \frac{1}{\sqrt{2\pi}}\int_{-\infty}^{\infty} f(at)e^{ist}\,dt$

Substituting $at = z$

$$dt = \frac{dz}{a},$$

As t changes from $-\infty$ to ∞, z changes from $-\infty$ to ∞

$$= \frac{1}{\sqrt{2\pi}}\int_{-\infty}^{\infty} f(z)e^{\frac{isz}{a}}.\frac{dz}{a}$$

$$= \frac{1}{a}\frac{1}{\sqrt{2\pi}}\int_{-\infty}^{\infty} f(t)e^{i\left(\frac{s}{a}\right)t}\,dt$$

$$= \frac{1}{a}F\left(\frac{s}{a}\right)$$

$$\therefore \quad F[f(ax)] = \frac{1}{a}F\left(\frac{s}{a}\right)$$

(e) Show that $F[x^n f(x)] = (-i)^n \frac{d^n F}{ds^n}$

Proof: $F(s) = \frac{1}{\sqrt{2\pi}}\int_{-\infty}^{\infty} f(t)e^{ist}\,dt$

$$\frac{dF(s)}{ds} = \frac{d\left[\frac{1}{\sqrt{2\pi}}\int_{-\infty}^{\infty} f(t)e^{ist}\,dt\right]}{ds}$$

$$= \frac{1}{\sqrt{2\pi}}\int_{-\infty}^{\infty} f(t)\frac{de^{ist}}{ds}\,dt$$

$$\frac{dF(s)}{ds} = \frac{(i)}{\sqrt{2\pi}} \int_{-\infty}^{\infty} (t) f(t).e^{ist} dt \qquad .....(1)$$

$$\frac{d^2F(s)}{ds^2} = \frac{1}{\sqrt{2\pi}} \frac{d\left[\int_{-\infty}^{\infty} f(t)(it)e^{ist} dt\right]}{ds}$$

$$= \frac{1}{\sqrt{2\pi}} \int_{-\infty}^{\infty} f(t)(it)^2 e^{ist} dt$$

$$= (i)^2 \frac{1}{\sqrt{2\pi}} \int_{-\infty}^{\infty} (t^2) f(t).e^{ist} dt$$

Similarly, we can show that

$$\frac{d^nF(s)}{ds^n} = (i)^n \frac{1}{\sqrt{2\pi}} \int_{-\infty}^{\infty} t^n f(t) e^{ist} dt$$

$$= (i)^n F\left[t^n f(t)\right]$$

$$\therefore \qquad F[t^n \, . \, f(t)] = \frac{1}{(i)^n} \frac{d^nF(s)}{ds^n}$$

$$F[x^n \, . \, f(x)] = (-i)^n \frac{d^nF(s)}{ds^n}$$

(f) Show that $F\left[\frac{d^n f(x)}{dx^n}\right] = (-is)^n F(s)$

proof: $F\left[\frac{df(x)}{dx}\right] = \frac{1}{\sqrt{2\pi}} \int_{-\infty}^{\infty} \frac{df(t)}{dt} e^{ist} dt$

$$= \frac{1}{\sqrt{2\pi}} \left\{ e^{ist} f(t) \Big|_{-\infty}^{\infty} - \int_{-\infty}^{\infty} f(t).e^{ist} (is) dt \right\}$$

$$= \frac{1}{\sqrt{2\pi}} \left\{ e^{ist} f(t) \Big|_{-\infty}^{\infty} - \int_{-\infty}^{\infty} f(t).is.e^{ist} dt \right\}$$

$$= \frac{-is}{\sqrt{2\pi}} \int_{-\infty}^{\infty} f(t) e^{ist}\, dt$$

Assuming $f(t) \to 0$ as $t \to \pm\infty$

$$F\left[\frac{df(x)}{dx}\right] = -isF(s)$$

$$F\left[\frac{d^2 f(x)}{dx^2}\right] = \frac{1}{\sqrt{2\pi}} \int_{-\infty}^{\infty} \frac{d^2 f(t)}{dt^2} e^{ist}\, dt$$

$$= \frac{1}{\sqrt{2\pi}} \left\{ e^{ist} f'(t) \Big|_{-\infty}^{\infty} - \int_{-\infty}^{\infty} \frac{df(t)}{dt} . e^{ist} (is) dt \right\}$$

$$= \frac{-is}{\sqrt{2\pi}} \int_{-\infty}^{\infty} e^{ist} \frac{df(t)}{dt}$$

(Assuming $\frac{df(t)}{dt} \to 0$ as $t \to \pm\infty$)

$$= \frac{-is}{\sqrt{2\pi}} \left\{ e^{ist} f(t) \Big|_{-\infty}^{\infty} - \int_{-\infty}^{\infty} f(t) e^{ist} (is) dt \right.$$

$$F\left[\frac{d^2 f(x)}{dx^2}\right] = \frac{(-is)^2}{\sqrt{2\pi}} \int_{-\infty}^{\infty} f(t) e^{ist}\, dt$$

(Assuming $f(t) \to 0$ as $t \to \pm\infty$)

Similarly
$$F\left[\frac{d^n f(x)}{dx^n}\right] = \frac{(-is)^n}{\sqrt{2\pi}} \int_{-\infty}^{\infty} f(t) e^{ist}\, dt$$

$$F\left[\frac{d^n f(x)}{dx^n}\right] = (-is)^n F(s)$$

(g) Show that $F[f(-x)] = F(-s)$

Proof:
$$F[f(-x)] = \frac{1}{\sqrt{2\pi}} \int_{-\infty}^{\infty} f(-t) e^{ist}\, dt$$

Substituting $-t = z$ when $t \to -\infty, z \to \infty$

$dt = -dz$ when $t \to -\infty, z \to -\infty$

$$= \frac{1}{\sqrt{2\pi}} \int_{\infty}^{-\infty} e^{-isz} f(z)(-dz)$$

$$= -\frac{1}{\sqrt{2\pi}} \int_{\infty}^{-\infty} f(z) e^{-isz} dz$$

$$= \frac{1}{\sqrt{2\pi}} \int_{-\infty}^{\infty} f(t) e^{-ist} dt$$

$$= \frac{1}{\sqrt{2\pi}} \int_{-\infty}^{\infty} f(t) e^{i(-s)t} dt$$

$$F[f(-x)] = F(-s)$$

(h) Modulation Theorem :

Show that $F[f(x) \cos ax] = \frac{1}{2}[f(s+a) + f(s-a)]$

when $f(s) = F[f(x)]$

Proof: $F[f(x) \cos ax] = \frac{1}{\sqrt{2\pi}} \int_{-\infty}^{\infty} f(t) \cos at\, e^{ist} dt$

$$= \frac{1}{\sqrt{2\pi}} \int_{-\infty}^{\infty} f(t) e^{ist} \left[\frac{e^{iat} + e^{-iat}}{2}\right] dt$$

$$= \frac{1}{2}\frac{1}{\sqrt{2\pi}} \int_{-\infty}^{\infty} f(t) e^{i(s+a)t} dt + \frac{1}{2}\frac{1}{\sqrt{2\pi}} \int_{-\infty}^{\infty} f(t) e^{i(s-a)t} dt$$

$$= \frac{1}{2}[f(s+a) + f(s-a)]$$

(i) Show that $\overline{F[f(x)]} = \overline{F(-s)}$

Proof: $F(s) = \frac{1}{\sqrt{2\pi}} \int_{-\infty}^{\infty} f(t) e^{ist} dt$

$$F(-s) = \frac{1}{\sqrt{2\pi}} \int_{-\infty}^{\infty} f(t) e^{-ist} dt$$

$$\overline{F(-s)} = \frac{1}{\sqrt{2\pi}} \int_{-\infty}^{\infty} \overline{f(t)} e^{-ist} \, dt$$

$$\therefore \qquad \overline{F(-s)} = \overline{F[f(x)]}$$

(j) Show that $F\left[\overline{f(-x)}\right] = \overline{F(s)}$

$$F(s) = \frac{1}{\sqrt{2\pi}} \int_{-\infty}^{\infty} f(t) e^{ist} \, dt$$

$$\overline{F(s)} = \frac{1}{\sqrt{2\pi}} \int_{-\infty}^{\infty} \overline{f(t)} \, \bar{e}^{\,ist} \, dt$$

Substituting $t = -z$ when $t \to -\infty,\ z \to \infty$

$dt = -dz$ $\quad t \to \infty,\ z \to -\infty$

$$\overline{F(s)} = \frac{1}{\sqrt{2\pi}} \int_{\infty}^{-\infty} \overline{f(-z)} \, e^{isz} (-dz)$$

$$= \frac{1}{\sqrt{2\pi}} \int_{-\infty}^{\infty} \overline{f(-t)} \, e^{ist} \, dt$$

$$\overline{F(s)} = F \, \overline{f(-x)}$$

If $F_s[f(x)]$ and $F_c[f(x)]$ are the Fourier sine and cosine transforms of $f(x)$ respectively, we can prove that

(i) $F_s[f(x)\cos ax] = \frac{1}{2}[F_s(s+a) + F_s(s-a)]$

(ii) $F_c[f(x)\cos ax] = \frac{1}{2}[F_c(s+a) + F_c(s-a)]$

(iii) $F_s[f(x)\sin ax] = \frac{1}{2}[F_c(s-a) - F_c(s+a)]$

(iv) $F_c[f(ax)] = \frac{1}{a} F_c\left(s/_a\right)$

(v) $F_s[f(ax)] = \frac{1}{a} F_s\left(\frac{s}{a}\right)$

(vi) $F_c[f(ax)] = \frac{1}{a} F_c\left(s/_a\right)$

(vii) $F_s[f'(x)] = -sF_c(s)$

(viii) $F_c[f'(x)] = -\sqrt{\frac{2}{\pi}} f(0) + sF_s(s)$

(ix) $\int_0^\infty F_c[f(x)]G_c[g(x)]ds = \int_0^\infty f(x)g(x)dx$

(x) $\int_0^\infty F_s[f(x)].G_s[g(x)]ds = \int_0^\infty f(x)g(x)dx$

(xi) $F_c[x.f(x)] = \frac{dF_s[f(x)]}{ds}$

(xii) $F_s[x.f(x)] = -\frac{dF_c[f(x)]}{ds}$

Solved Examples

***Example* 23.7:** Find the Fourier Transform of

$$f(x) = 1 - x^2, \quad |x| \le 1$$
$$= 0 \quad , \quad |x| > 1$$

***Solution*:** The Fourier Transform of $f(x)$ is

$$F[f(x)] = F(s) = \frac{1}{\sqrt{2\pi}} \int_{-\infty}^{\infty} f(t)e^{ist} dt$$

$$F(s) = \frac{1}{\sqrt{2\pi}} \int_{-1}^{1} (1 - t^2)e^{ist} dt$$

$$= \frac{1}{\sqrt{2\pi}}\left[\left(1-t^2\right)\frac{e^{ist}}{is} - (-2t)\frac{e^{ist}}{(is)^2} + (-2)\frac{e^{ist}}{(is)^3}\right]_{-1}^{1}$$

$$= \frac{1}{\sqrt{2\pi}}\left[0 - \frac{2}{s^2}\left(e^{is} + e^{-is}\right) + \frac{2}{is^3}\left(e^{is} - e^{-is}\right)\right]$$

$$= \frac{1}{\sqrt{2\pi}}\left[\frac{4}{s^3}\sin s - \frac{4}{s^2}\cos s\right]$$

where $\sin s = \dfrac{e^{is} - e^{-is}}{2i}$ $\cos s = \dfrac{e^{is} + e^{-is}}{2}$

$$= \frac{2\sqrt{2}}{s^3\sqrt{\pi}}\left[\sin s - s\cos s\right]$$

***Example* 23.8:** Find the Fourier transform of

$$f(x) = x^2, \quad |x| < a$$
$$= 0, \quad |x| > a$$

***Solution*:** The Fourier transform of $f(x)$ is

$$F[f(x)] = \frac{1}{\sqrt{2\pi}}\int_{-\infty}^{\infty} f(t)e^{ist}\,dt$$

$$F(s) = \frac{1}{\sqrt{2\pi}}\int_{-a}^{a} t^2 e^{ist}\,dt$$

$$= \frac{1}{\sqrt{2\pi}}\left[\left(t^2\right)\frac{e^{ist}}{is} - (2t)\frac{e^{ist}}{(is)^2} + (2)\frac{\left(e^{ist}\right)}{(is)^3}\right]_{-a}^{a}$$

$$= \frac{1}{\sqrt{2\pi}}\left[\frac{a^2e^{isa} - a^2e^{-isa}}{is} + \frac{2ae^{isa} + 2ae^{-isa}}{s^2} - \frac{2\left(e^{isa} - e^{-isa}\right)}{is^3}\right]$$

$$= \frac{1}{\sqrt{2\pi}}\left[\frac{a^2}{is}\{(\cos sa + i\sin sa) - (\cos sa - i\sin sa)\}\right.$$

$$+ \frac{2a}{s^2}\{\cos sa + i\sin sa + \cos sa - i\sin sa\}$$

$$\left.\frac{-2}{is^3}\{\cos sa + i\sin sa - \cos sa + i\sin sa\}\right]$$

$$= \frac{1}{\sqrt{2\pi}}\left[\frac{a^2}{is}2i\sin sa + \frac{2a}{s^2}2\cos sa - \frac{2}{is^3}2i\sin sa\right]$$

$$= \sqrt{\frac{2}{\pi}}\left[\frac{a^2s^2\sin as + 2as\cos as - 2\sin as}{s^3}\right]$$

$$= \frac{1}{s^3}\sqrt{\frac{2}{\pi}}\left[\left(a^2s^2 - 2\right)\sin(as) + 2as\cos as\right]$$

***Example* 23.9:** Find the Fourier sine and cosine transforms of

$$f(x) = x, \quad 0 < x < 1$$
$$= 2 - x, \quad 1 < x < 2$$
$$= 0, \quad x > 2$$

***Solution*:** The Fourier sine transform of $f(x)$ is

$$F_s(s) = \sqrt{\frac{2}{\pi}}\int_0^\infty f(t)\sin st\,dt$$

$$= \sqrt{\frac{2}{\pi}}\int_0^1 t.\sin st\,dt + \int_1^2 (2-t)\sin st\,dt + \int_2^\infty 0.\sin st\,dt$$

$$= \sqrt{\frac{2}{\pi}}\left\{\left[(t)\frac{(-\cos t)}{s} - (1)\frac{(-\sin st)}{s^2}\right]_0^1\right.$$

$$\left. + \left[(2-t)\left(\frac{-\cos st}{s}\right) - (-1)\left(\frac{-\sin st}{s^2}\right)\right]_1^2 + 0\right\}$$

$$= \sqrt{\frac{2}{\pi}}\left\{\frac{-\cos s}{s} + \frac{\sin s}{s^2} + \frac{\cos s}{s} - \left(\frac{\sin 2s - \sin s}{s^2}\right)\right\}$$

$$= \sqrt{\frac{2}{\pi}}.\frac{1}{s^2}\{2\sin s - \sin 2s\}$$

$$= \frac{2}{s^2}\sqrt{\frac{2}{\pi}}[\sin s(1 - \cos s)]$$

Now the Fourier cosine Transform of $f(x)$ is

$$F_c(s) = \sqrt{\frac{2}{\pi}}\int_0^\infty f(t)\cos st\,dt$$

$$= \sqrt{\frac{2}{\pi}}\left\{\int_0^1 t\cos st\,dt + \int_1^2 (2-t)\cos st\,dt + \int_2^\infty 0.\cos st\,dt\right.$$

$$= \sqrt{\frac{2}{\pi}}\left\{\left[(t)\left(\frac{\sin st}{s}\right) - (1)\left(\frac{-\cos st}{s^2}\right)\right]_0^1\right.$$

$$\left. + \left[(2-t)\left(\frac{\sin st}{s}\right) - (-1)\left(\frac{-\cos st}{s^2}\right)\right]_1^2\right\}$$

$$= \sqrt{\frac{2}{\pi}}\left\{\frac{\sin s}{s} + \frac{\cos s}{s^2} - \frac{1}{s^2} - \frac{\sin s}{s} - \left(\frac{\cos 2s - \cos s}{s^2}\right)\right\}$$

$$= \sqrt{\frac{2}{\pi}}\left\{\frac{2\cos s - \cos 2s - 1}{s^2}\right\}$$

$$= \sqrt{\frac{2}{\pi}}\left\{\frac{2\cos s - 2\cos^2 s + 1 - 1}{s^2}\right\}$$

$$= \frac{2}{s^2}\sqrt{\frac{2}{\pi}}\cos s(1-\cos s)$$

***Example* 23.10:** Find Fourier cosine and sine transform of $f(x) = e^{-ax}$.

***Solution*:** The Fourier cosine transform of $f(x)$ is

$$F_c[f(x)] = \sqrt{\frac{2}{\pi}}\int_0^\infty f(t)\cos st\,dt$$

$$= \sqrt{\frac{2}{\pi}}\int_0^\infty e^{-at}\cos st\,dt$$

$$= \sqrt{\frac{2}{\pi}}\left[\frac{e^{-at}}{a^2+s^2}(-a\cos st + s\sin st)\right]_0^\infty$$

$$= \sqrt{\frac{2}{\pi}}\left[0 - \frac{1}{a^2+s^2}(-a)\right]$$

$$= \sqrt{\frac{2}{\pi}}\frac{a}{a^2+s^2}$$

$$\therefore \qquad F_c[e^{-ax}] = \sqrt{\frac{2}{\pi}}\frac{a}{a^2+s^2}$$

The Fourier sine transform of $f(x)$ is

$$F_s[f(x)] = \sqrt{\frac{2}{\pi}}\int_0^\infty f(t)\sin st\, dt$$

$$= \sqrt{\frac{2}{\pi}}\int_0^\infty e^{-at} \sin st\, dt$$

$$= \sqrt{\frac{2}{\pi}}\left[\frac{e^{-at}}{a^2+s^2}(-a\sin st - s\cos st)\right]_0^\infty$$

$$= \sqrt{\frac{2}{\pi}}\left[0 - \frac{1}{a^2+s^2}(0-s)\right]$$

$$= \sqrt{\frac{2}{\pi}}\frac{s}{a^2+s^2}$$

$$\therefore \qquad F_s[e^{-ax}] = \sqrt{\frac{2}{\pi}}\frac{s}{a^2+s^2}$$

***Example* 23.11:** Find Fourier sine transform of the function $f(x) = \dfrac{e^{-ax}}{x}$

***Solution*:** The Fourier sine transform of $f(x)$ is

$$F_s[f(x)] = \sqrt{\frac{2}{\pi}}\int_0^\infty f(t)\sin st\, dt$$

$$= \sqrt{\frac{2}{\pi}}\int_0^\infty \frac{e^{-at}}{t}\sin st\, dt$$

Let $\qquad F_s[f(x)] = F(s)$

$$\therefore \qquad F(s) = \sqrt{\frac{2}{\pi}}\int_0^{\infty}\frac{e^{-at}}{t}\sin st\,dt$$

Differentiating both sides w.r.t to 's'

$$\frac{dF(s)}{ds} = \frac{d}{ds}\sqrt{\frac{2}{\pi}}\int_0^{\infty}\frac{e^{-at}}{t}\sin st\,dt$$

$$= \sqrt{\frac{2}{\pi}}\int_0^{\infty}\frac{e^{-at}}{t}\frac{d\sin st}{ds}dt$$

$$= \sqrt{\frac{2}{\pi}}\int_0^{\infty}\frac{e^{-at}}{t}t\cos st\,dt$$

$$\frac{dF(s)}{ds} = \sqrt{\frac{2}{\pi}}\int_0^{\infty}e^{-at}\cos st\,dt$$

$$= \sqrt{\frac{2}{\pi}}\left[\frac{e^{-at}}{a^2+s^2}(-a\cos st + s\sin st)\right]_0^{\infty}$$

$$\frac{dF(s)}{ds} = \sqrt{\frac{2}{\pi}}\frac{a}{a^2+s^2}$$

Integrating both sides w.r.to 's'

$$F(s) = \sqrt{\frac{2}{\pi}}\int\frac{a}{a^2+s^2}ds$$

$$= \sqrt{\frac{2}{\pi}}\; a \,.\, \frac{1}{a}\tan^{-1}\left(\frac{s}{a}\right)$$

$$F(s) = \sqrt{\frac{2}{\pi}}\tan^{-1}\left(\frac{s}{a}\right)$$

$$\therefore \qquad F_s[f(x)] = F(s) = \sqrt{\frac{2}{\pi}}\tan^{-1}\left(\frac{s}{a}\right)$$

$$\therefore \qquad F_s\left[\frac{e^{-ax}}{x}\right] = \sqrt{\frac{2}{\pi}}\tan^{-1}\left(s/a\right)$$

***Example* 23.12:** Find Fourier sine and cosine transforms of xe^{-ax}

Solution: We know that $F_s[xf(x)] = -\frac{d}{ds}F_c[f(x)]$

$$f(x) = e^{-ax}$$

$$F_s\left[xe^{-ax}\right] = \frac{-d}{ds}F_c\left[e^{-ax}\right] \quad .....(1)$$

But $$F_c\left[e^{-ax}\right] = \sqrt{\frac{2}{\pi}}\int_0^\infty e^{-at}\cos st\,dt$$

$$= \sqrt{\frac{2}{\pi}}\left[\frac{e^{-at}}{a^2+s^2}(-a\cos st + s\sin st)\right]_0^\infty$$

$$F_c[e^{-ax}] = \sqrt{\frac{2}{\pi}}\frac{a}{a^2+s^2}$$

Substituting in (1)

$$F_s[xe^{-ax}] = \frac{-d\left[F_c\left(e^{-ax}\right)\right]}{ds}$$

$$= \frac{-d\sqrt{\frac{2}{\pi}}\frac{a}{a^2+s^2}}{ds}$$

$$= -\sqrt{\frac{2}{\pi}}a\left[\frac{-2s}{\left(a^2+s^2\right)^2}\right]$$

$$F_s\left[xe^{-ax}\right] = \sqrt{\frac{2}{\pi}}\frac{2as}{\left(a^2+s^2\right)^2}$$

Similarly $$F_c[xf(x)] = \frac{dF_s[f(x)]}{ds}$$

$$\therefore \quad F_c\left[xe^{-ax}\right] = \frac{dF_s\left[e^{-ax}\right]}{ds} \quad .....(1)$$

But $$F_s[e^{-ax}] = \sqrt{\frac{2}{\pi}}\int_0^\infty e^{-at}\sin st\,dt$$

$$= \sqrt{\frac{2}{\pi}}\left\{\frac{e^{-at}}{a^2+s^2}[-a\sin st - s\cos st]\right\}_0^\infty$$

$$= \sqrt{\frac{2}{\pi}}\frac{s}{a^2+s^2} \qquad .....(2)$$

$$F_c\left[xe^{-ax}\right] = \frac{dF_s\left[e^{-ax}\right]}{ds}$$

$$= \frac{d\sqrt{\frac{2}{\pi}}\frac{s}{a^2+s^2}}{ds} \qquad \text{From (2)}$$

$$= \sqrt{\frac{2}{\pi}}\left[\frac{\left(a^2+s^2\right)1 - s.2s}{\left(a^2+s^2\right)^2}\right]$$

$$F_c\left[xe^{-ax}\right] = \sqrt{\frac{2}{\pi}}\left[\frac{a^2-s^2}{\left(a^2+s^2\right)^2}\right]$$

23.3 Finite Fourier Transforms

If $f(x)$ is a function defined in the interval $(0, l)$ then the finite Fourier sine transform of $f(x)$ in $0 < x < l$ is defined as $F_s[f(x)] = \int_0^l f(x)\sin\left(\frac{n\pi x}{l}\right)dx$ where n is an integer. The inverse finite Fourier sine transform of $F_s[f(x)]$ is $f(x)$ and is given by

$$f(x) = \frac{2}{l}\sum_{n=1}^{\infty} F_s[f(x)]\sin\left(\frac{n\pi x}{l}\right).$$

The finite Fourier cosine transform of $f(x)$ in $0 < x < l$ is defined as

$$F_c[f(x)] = \int_0^l f(x)\cos\frac{n\pi x}{l}dx,$$

where 'n' is an integer. The inverse finite Fourier cosine transform of $F_c[f(x)]$ is $f(x)$ and is given by

$$f(x) = \frac{1}{l}F_c(0) + \frac{2}{l}\sum_{n=1}^{\infty} F_c[f(x)]\cos\frac{n\pi x}{l}$$

Solved Examples

***Example* 23.13:** Find the finite Fourier sine and cosine transforms of $f(x) = x^2$ in $0 < x < l$.

***Solution*:** The Fourier sine transform is

$$F_s[f(x)] = \int_0^l f(x)\sin\frac{n\pi x}{l}\,dx$$

$$F_s[x^2] = \int_0^l x^2 \sin\frac{n\pi x}{l}\,dx$$

$$= \left[(x^2)\left(-\cos\frac{n\pi x}{l}\right)\frac{l}{n\pi} - (2x)\left(-\sin\frac{n\pi x}{l}\right)\frac{l^2}{n^2\pi^2}\right.$$

$$\left. + (2)\left(\cos\frac{n\pi x}{l}\right)\frac{l^3}{n^3\pi^3}\right]_0^l$$

$$= \frac{-l^3}{n\pi}\cos n\pi + \frac{2l^3}{n^3\pi^3}\cos n\pi - \frac{2l^3}{n^3\pi^3}$$

$$\therefore \quad F_s(x^2) = \frac{l^3}{n\pi}(-1)^{n+1} + \frac{2l^3}{n^3\pi^3}\left[(-1)^n - 1\right]$$

The Fourier cosine transform is

$$F_c[f(x)] = \int_0^l f(x)\cos\frac{n\pi x}{l}\,dx$$

$$F_c[x^2] = \int_0^l x^2 \cos\frac{n\pi x}{l}\,dx$$

$$= \left[(x^2)\left(\sin\frac{n\pi x}{l}\right)\left(\frac{l}{n\pi}\right) - (2x)\left(-\cos\frac{n\pi x}{l}\right) \times \frac{l^2}{n^2\pi^2}\right.$$

$$\left. + (2)\left(-\sin\frac{n\pi x}{l}\right) \times \frac{l^3}{n^3\pi^3}\right]_0^l$$

$$= \frac{2l^3}{n^2\pi^2}\cos n\pi$$

$$F_c[x^2] = \frac{2l^3(-1)^n}{n^2\pi^2}$$

***Example* 23.14:** Find the finite Fourier sine transform of $f(x) = e^{ax}$ in $(0, l)$

***Solution*:** The finite Fourier sine transform of $f(x)$ is

$$F_s[f(x)] = \int_0^l f(x)\sin\frac{n\pi x}{l}\,dx$$

$$= \int_0^l e^{ax}\sin\frac{n\pi x}{l}\,dx$$

$$= \left\{\frac{e^{ax}}{a^2+\frac{n^2\pi^2}{l^2}}\left[a\sin\frac{n\pi x}{l}-\frac{n\pi}{l}\cos\frac{n\pi x}{l}\right]\right\}_0^l$$

$$= \frac{e^{al}}{a^2+\frac{n^2\pi^2}{l^2}}\left(-\frac{n\pi}{l}\cos n\pi\right)+\frac{\left(\frac{n\pi}{l}\right)}{a^2+\frac{n^2\pi^2}{l^2}}$$

$$= \frac{n\pi l}{a^2l^2+n^2\pi^2}\left[(-1)^{n+1}e^{al}+1\right]$$

***Example* 23.15:** Find the finite Fourier cosine transforms of $f(x) = \sin ax$ in $(0, p)$.

***Solution*:** The Fourier cosine transform of $f(x)$ is

$$F_c[f(x)] = \int_0^\pi f(x)\cos nx\,dx$$

$$F_c[\sin ax] = \int_0^\pi \sin ax\cos nx\,dx$$

$$= \frac{1}{2}\int_0^\pi[\sin(a+n)x+\sin(a-n)x]dx$$

$$= \frac{1}{2}\left[\frac{-\cos(a+n)x}{a+n}-\frac{\cos(a-n)x}{a-n}\right]_0^\pi$$

$$= -\frac{1}{2}\left[\frac{\cos(a+n)\pi}{(a+n)} + \frac{\cos(a-n)\pi}{(a-n)} - \frac{1}{a+n} - \frac{1}{a-n}\right]$$

$$F_c[\sin ax] = -\frac{1}{2}\left[\frac{(-1)^{a+n}}{a+n} + \frac{(-1)^{a-n}}{a-n} - \frac{1}{a+n} - \frac{1}{a-n}\right]$$

***Example* 23.16:** If the finite sine transform of $f(x)$ is given by $\frac{2\pi(-1)^{s-1}}{s^3}$ where n is positive integer and $0 < x < p$ find $f(x)$.

***Solution*:** The inverse Fourier Sine Transform $F_s[f(x)]$ is

$$f(x) = \frac{2}{\pi}\sum_{s=1}^{\infty} F_s[f(x)]\sin sx \qquad .....(1)$$

Given $\qquad F_s(s) = F_s[f(x)] = \frac{2\pi(-1)^{s-1}}{s^3} \qquad .....(2)$

Substituting (2) in (1)

$$f(x) = \frac{2}{\pi}\sum_{s=1}^{\infty} \frac{2\pi(-1)^{s-1}}{s^3}\sin sx$$

$$\therefore \qquad f(x) = 4\sum_{s=1}^{\infty} \frac{(-1)^{s-1}}{s^3}\sin sx$$

***Example* 23.17:** Find $f(x)$, if its sine transform is

$$F_s(s) = F_s[f(x)] = \frac{1-\cos s\pi}{s^2\pi^2}, \qquad 0 < x < \pi, \quad s = 1, 2, 3, ...$$

***Solution*:** The inverse finite Fourier sine transform of $F_s[f(x)]$ in $0 < x < p$ is

$$f(x) = \frac{2}{\pi}\sum_{s=1}^{\infty} F_s[f(x)]\sin sx,$$

$$= \frac{2}{\pi}\sum_{s=1}^{\infty} \frac{1-\cos s\pi}{s^2\pi^2}\sin sx$$

$$\therefore \qquad f(x) = \frac{2}{\pi^3}\sum_{s=1}^{\infty} \frac{1-\cos s\pi}{s^2}\sin sx$$

***Example* 23.18:** Find $f(x)$, if its cosine transform is

$$F_c(s) = F_c[f(x)] = \frac{\cos\left(\frac{2s\pi}{3}\right)}{(2s+1)^2}, \; 0 < x < 1$$

***Solution*:** The inverse finite Fourier cosine transform of $F_c[f(x)]$ is

$$f(x) = \frac{1}{l}F_c(0) + \frac{2}{l}\sum_{s=1}^{\infty} F_c[f(x)]\frac{\cos s\pi x}{l}$$

given $$F_c(s) = F_c[f(x)] = \frac{\cos\left(\frac{2s\pi}{3}\right)}{(2s+1)^2} \text{ in } (0, 1)$$

and $$F_c(0) = \frac{\cos(0)}{(2.0+1)^2} = 1$$

$\therefore$ $$f(x) = \frac{1}{l} + \frac{2}{l}\sum_{s=1}^{\infty} \frac{\cos\left(\frac{2s\pi}{3}\right)}{2s+1}\cos(s\pi x); \; (l = 1)$$

Exercise - 23.2

1. Find the Fourier transform of $f(x) = 1, \; |x| < a$
$0, \; |x| > a$

Ans : $\left(\frac{2\sin(sa)}{s\sqrt{2\pi}}\right)$

2. Find the Fourier transform of $f(x) = x, \; |x| < a$
$= 0, \; |x| > a$

Ans : $\frac{2}{s}\left\{\frac{a}{i}\cos as + \frac{i}{s}\sin as\right\}$

3. Find the Fourier transform of $f(x)$ defined by $f(x) = 0, \quad x < a$

$$= 1, \quad a < x < b$$
$$= 0, \quad x > b$$

Ans : $\left\{\frac{1}{\sqrt{2\pi}}\left[\frac{e^{ibs} - e^{ias}}{is}\right]\right\}$

4. Find Fourier sine and cosine transform of $f(x) = 1, \quad 0 < x < a$

$$= 0, \quad x > a$$

Ans : $\left\{\sqrt{\frac{2}{\pi}}\left(\frac{1-\cos as}{s}\right) \text{ and } \sqrt{\frac{2}{\pi}}\frac{\sin as}{s}\right\}$

5. Find Fourier cosine transform of $\frac{e^{-ax}}{x}$

Ans : $\left\{\frac{-1}{\sqrt{2\pi}}\log\left(s^2 + a^2\right)\right\}$

6. Find the sine transform of the function $f(x) = \sin x, \quad 0 \le x < a$

$$= 0, \quad x > a$$

Ans : $\left\{\frac{1}{\sqrt{2\pi}}\left[\frac{\sin(1-s)a}{1-s} - \frac{\sin(1+s)a}{1+s}\right]\right\}$

7. Find the Fourier cosine transform of $\frac{e^{-ax} - e^{-bx}}{x}$

Ans : $\left\{\frac{1}{\sqrt{2\pi}}\log\left(\frac{s^2+b^2}{s^2+a^2}\right)\right\}$

8. Find the finite Fourier sine and cosine transforms of $f(x) = x$ in $(0, \pi)$

Ans : $(-1)^{n+1}\frac{\pi}{s}$ and $\frac{1}{s^2}(-1)^s - 1$

9. Find the finite cosine transform of $f(x) = \left(1 - \frac{x}{\pi}\right)^2$

Ans : $\left\{\frac{2}{\pi s^2}, (s > 0); \frac{\pi}{3}, (s = 0)\right\}$

10. Find $f(x)$, if its finite cosine transform is

$$F_s(s) = F_c\,[f(x)] = \frac{\left(6\sin\frac{s\pi}{2} - \cos s\pi\right)}{(2s+1)\pi}, \; s = 1, 2, 3...$$

$$= \frac{2}{\pi}, \qquad s = 0 \text{ in } 0 < x < \pi$$

$$\textbf{Ans : } f(x) = \frac{1}{2\pi} + \frac{1}{2\pi}\sum_{s=1}^{\infty}\left[\frac{6\sin s\frac{\pi}{2} - \cos s\pi}{(2s+1)\pi}\right]\frac{\cos s\pi x}{4}$$

11. Find $f(x)$ if its finite sine transform is

$$F_s(s) = F_s[f(x)] = \frac{1+\cos s\pi}{s\pi}, \; s = 1, 2, 3.... \text{ and } 0 < x < \pi$$

$$\textbf{Ans : } f(x) = \frac{2}{\pi^2}\sum_{s=1}^{\infty}\left[\frac{1+\cos s\pi}{s\pi}\right]\sin sx$$

23.4 Convolution Theorem

23.4.1 Convolution of Two Functions

If $f(x)$ and $g(x)$ are any two functions defined in $(-\infty, \infty)$ then the convolution of two functions is denoted by $f * g$ and is defined by $\frac{1}{\sqrt{2\pi}}\int_{-\infty}^{\infty} f(t)g(x-t)dt$.

$$\therefore \qquad f * g = \frac{1}{\sqrt{2\pi}}\int_{-\infty}^{\infty} f(t)g(x-t)dt$$

Convolution theorem for Fourier Transforms :

If $F(s)$ and $G(s)$ are the Fourier transforms of $f(x)$ and $g(x)$ respectively then the Fourier Transform of the convolution of $f(x)$ and $g(x)$ is the product of their Fourier transforms.

i.e., $\qquad F[(f * g)x] = F(s) \,.\, G(s)$

i.e.,

$$\frac{1}{\sqrt{2\pi}}\int_{-\infty}^{\infty}(f+g)(x)e^{isx}dx = \left\{\frac{1}{\sqrt{2\pi}}\int_{-\infty}^{\infty}f(x)e^{isx}dx\right\}$$

$$\times\left\{\frac{1}{\sqrt{2\pi}}\int_{-\infty}^{\infty}g(x)e^{isx}dx\right\}$$

Proof :

$$F[(f*g)\,x] = \frac{1}{\sqrt{2\pi}}\int_{-\infty}^{\infty}(f*g)(x)e^{isx}dx$$

$$= \frac{1}{\sqrt{2\pi}}\int_{-\infty}^{\infty}\left[\frac{1}{\sqrt{2\pi}}\int_{-\infty}^{\infty}f(t)g(x-t)dt\right]e^{isx}dx$$

$$= \frac{1}{\sqrt{2\pi}}\int_{-\infty}^{\infty}f(t)\left[\frac{1}{\sqrt{2\pi}}\int_{-\infty}^{\infty}g(x-t)e^{isx}dx\right]^{dt}$$

$$= \frac{1}{\sqrt{2\pi}}\int_{-\infty}^{\infty}f(t)\left[\frac{1}{\sqrt{2\pi}}\int_{-\infty}^{\infty}g(x-t)e^{isx}.e^{ist}.e^{-ist}dx\right]dt$$

$$= \frac{1}{\sqrt{2\pi}}\int_{-\infty}^{\infty}f(t)\left[\frac{1}{\sqrt{2\pi}}\int_{-\infty}^{\infty}g(x-t)e^{is(x-t)}d(x-t)\right]e^{ist}dt$$

$$= \frac{1}{\sqrt{2\pi}}\int_{-\infty}^{\infty}f(t).G(s)e^{ist}dt$$

$$= G(s)\frac{1}{\sqrt{2\pi}}\int_{-\infty}^{\infty}f(t)e^{ist}dt$$

$$= G(s)\,.\,F(s)$$

$$\therefore \qquad F[(f*g)x] = F(s)\,.\,G(s)$$

23.4.2 Parseval's Identity

If $f(x)$ is a given function defined in $(-\infty, \infty)$ and $F(s)$ is the Fourier Transform of $f(x)$, then

$$\int_{-\infty}^{\infty}|f(x)|^2\,dx = \int_{-\infty}^{\infty}|F(s)|^2\,ds$$

Proof :

We have $F[f * g] = F(s).\ G(s)$

$\therefore$ $F^{-1}[F(s)\ .\ G(s)] = f * g$

$$\frac{1}{\sqrt{2\pi}}\int_{-\infty}^{\infty} F(s)G(s)e^{-isx}ds = \frac{1}{\sqrt{2\pi}}\int_{-\infty}^{\infty} f(t)g(x-t)dt$$

Substituting $x = 0$

$$\int_{-\infty}^{\infty} F(s).G(s)ds = \int_{-\infty}^{\infty} f(t)g(-t)dt \qquad .....(1)$$

Considering $g(-t) = \overline{f(t)}$

$\Rightarrow$ $g(t) = \overline{f(-t)}$

$\therefore$ $G(s) = F[g(x)] = F[g(t)] = F\overline{f(-t)}$

$G(s) = F\overline{f(-x)} = \overline{F(s)}$, (using $F\left[\overline{f(-x)}\right] = \overline{F(s)}$)

Substituting $g(-t) = \overline{f(t)}$ and $G(s) = \overline{F(s)}$ in (1)

$$\int_{-\infty}^{\infty} F(s)\overline{F(s)}ds = \int_{-\infty}^{\infty} f(t)\overline{f(t)}dt$$

$$\int_{-\infty}^{\infty} |F(s)|^2 ds = \int_{-\infty}^{\infty} |f(t)|^2 dt\ .\ \left(F(s)\overline{F(s)} = |F(s)|^2\right)$$

$$\therefore \quad \int_{-\infty}^{\infty} |f(x)|^2 dx = \int_{-\infty}^{\infty} |F(s)|^2 ds$$

***Note*:** Some authors define parseval's identify as $\frac{1}{2\pi}\int_{-\infty}^{\infty} |f(s)|^2 ds = \int_{-\infty}^{\infty} |f(x)|^2 dx$

Solved Examples

***Example* 23.19:** Show that the Fourier transform of

$$f(x) = \begin{cases} a^2 - x^2, & |x| < a \\ 0, & |x| > a > 0 \end{cases} \text{ is } 2\sqrt{\frac{2}{\pi}}\left(\frac{\sin as - as\cos as}{s^3}\right)$$

Hence deduce that $\int_0^{\infty} \frac{\sin t - t\cos t}{t^3} dt = \frac{\pi}{4}$, using Parseval's identity show that $\int_0^\infty \left(\frac{\sin t - t\cos t}{t^3}\right)^2 dt = \frac{\pi}{15}$

Solution: $$F[f(x)] = \frac{1}{\sqrt{2\pi}}\int_0^\infty f(t)e^{ist}\,dt$$

$$= \frac{1}{\sqrt{2\pi}}\left\{\int_{-\infty}^{-a} f(t)e^{ist}\,dt + \int_{-a}^{a} f(t)e^{ist}\,dt + \int_a^\infty f(t)e^{ist}\,dt\right\}$$

$$= \frac{1}{\sqrt{2\pi}}\left\{0 + \int_{-a}^{a}\left(a^2 - t^2\right)e^{ist}\,dt + 0\right\}$$

$$= \frac{1}{\sqrt{2\pi}}\left\{\left(a^2 - t^2\right)\frac{\left(e^{ist}\right)}{is} - (-2t)\left(\frac{e^{ist}}{i^2 s^2}\right) + (-2)\left(\frac{e^{ist}}{i^3 s^3}\right)\right\}_{-a}^{a}$$

$$= \frac{1}{\sqrt{2\pi}}\left\{0 - \frac{2a}{s^2}e^{isa} - \frac{2a}{s^2}e^{-isa} - \frac{2ie^{isa}}{s^3} + \frac{2ie^{-isa}}{s^3}\right\}$$

$$= \frac{1}{\sqrt{2\pi}}\left\{-\frac{2a}{s^2}\left(e^{isa} + e^{-isa}\right) + \frac{2i}{s^3}\left(e^{-isa} - e^{isa}\right)\right\}$$

$$= \frac{1}{\sqrt{2\pi}}\left\{-\frac{2a}{s^2}(2\cos as) + \frac{4}{s^3}\sin as\right\}$$

$$= \frac{4}{\sqrt{2\pi}}\left[\frac{\sin as - as\cos as}{s^3}\right]$$

$$\therefore \quad F(s) = 2\sqrt{\frac{2}{\pi}}\left[\frac{\sin as - as\cos as}{s^3}\right] \qquad .....(1)$$

The inverse Fourier Transform is

$$f(x) = \frac{1}{\sqrt{2\pi}}2\sqrt{\frac{2}{\pi}}\int_{-\infty}^{\infty}\frac{1}{s^3}[\sin as - as\cos as]e^{-isx}\,ds$$

$$= \frac{2}{\pi}\int_{-\infty}^{\infty}\left(\frac{\sin as - as\cos as}{s^3}\right)(\cos sx - i\sin sx)\,ds$$

$$= \frac{2}{\pi}\int_{-\infty}^{\infty}\left(\frac{\sin as - as\cos as}{s^3}\right)\cos sx\,ds + \frac{2}{\pi}$$

$$i\int_{-\infty}^{\infty}\left(\frac{\sin as - as\cos as}{s^3}\right)\sin sx\,as$$

$$= \frac{2}{\pi}\int_{-\infty}^{\infty}\left(\frac{\sin as - as\cos as}{s^3}\right)\cos sx\,ds$$

(The second integral is zero, since the function is an odd function)

$$\therefore \quad f(x) = \frac{4}{\pi}\int_0^{\infty}\frac{\sin as - as\cos as}{s^3}\cos sx\,ds$$

Substituting $a = 1$

$$f(x) = \frac{4}{\pi}\int_0^{\infty}\frac{\sin t - t\cos t}{t^3}\cos t\,x\,dt$$

Substituting $x = 0$

$$f(0) = \frac{4}{\pi}\int_0^{\infty}\frac{\sin t - t\cos t}{t^3}\cos 0.dt\text{, but } f(0) = 1$$

$$1 = \frac{4}{\pi}\int_0^{\infty}\frac{\sin t - t\cos t}{t^3}dt$$

$$\therefore \quad \int_0^{\infty}\frac{\sin t - t\cos t}{t^3} = \frac{\pi}{4}$$

Now Parseval's identity is $\int_0^{\infty}|F(s)|^2 ds = \int_{-\infty}^{\infty}|f(x)|^2\,dx$

From (1) $\quad F(s) = 2\sqrt{\frac{2}{\pi}}\left[\frac{\sin as - as\cos as}{s^3}\right]$

$$\int_{-\infty}^{\infty}\left[2\sqrt{\frac{2}{\pi}}\left(\frac{\sin as - as\cos as}{s^3}\right)\right]^2 ds = \int_{-a}^{a}\left[a^2 - x^2\right]^2 dx$$

Substituting $a = 1$

$$\frac{8}{\pi}\int_{-\infty}^{\infty}\left(\frac{\sin s - s\cos s}{s^3}\right)^2 ds = \int_{-1}^{1}\left(1-x^2\right)^2 dx \quad \text{writing} \quad x = \sin\theta$$

$dx = \cos\theta\, d\theta$

θ varies from 0 to $\frac{\pi}{2}$

$$\frac{16}{\pi}\int_0^{\infty}\left(\frac{\sin s - s\cos s}{s^3}\right)^2 ds = 2\int_0^1 \left(1-x^2\right)^2 dx$$

$$\frac{16}{\pi}\int_0^{\infty}\left(\frac{\sin s - s\cos s}{s^3}\right)^2 ds = 2\int_0^{\pi/2} \left(1-\sin^2\theta\right)^2 \cos\theta\, d\theta$$

$$= 2\int_0^{\pi/2}\cos^5\theta\, d\theta = 2\frac{(5-1)}{5}\frac{(5-3)}{3}$$

$$\therefore \quad \frac{16}{\pi}\int_0^{\infty}\left(\frac{\sin s - \cos s}{s^3}\right)^2 ds = \frac{2.8}{15}$$

$$\therefore \quad \int_0^{\infty}\left(\frac{\sin s - s\cos s}{s^3}\right)^2 ds = \frac{\pi}{15}$$

***Example* 23.20:** Evaluate $\int_0^{\infty}\frac{dx}{\left(x^2+a^2\right)\left(x^2+b^2\right)}$ using Fourier Transform.

***Solution*:** The Fourier cosine transform of $f(x)$ is $f(x) = e^{-ax}$ is $\sqrt{\frac{2}{\pi}}\left(\frac{a}{s^2+a^2}\right)$ and the Fourier cosine transform of $f(x) = e^{-bx}$ is $\sqrt{\frac{2}{\pi}}\left(\frac{b}{s^2+b^2}\right)$

we have $$\int_0^{\infty} F_c[f(x)]F_c[g(x)]ds = \int_0^{\infty} f(x)g(x)dx$$

$$\int_0^{\infty}\sqrt{\frac{2}{\pi}}\left(\frac{a}{s^2+a^2}\right)\sqrt{\frac{2}{\pi}}\left(\frac{b}{s^2+b^2}\right)ds = \int_0^{\infty} e^{-ax}.e^{-bx}dx$$

$$\frac{2}{\pi}\int_0^{\infty}\frac{ab\,ds}{\left(s^2+a^2\right)\left(s^2+b^2\right)} = \int_0^{\infty} e^{-(a+b)x}dx$$

$$= \left[\frac{e^{-(a+b)x}}{-(a+b)}\right]_0^{\infty}$$

$$= -\frac{1}{a+b}\left[-e^0\right]$$

$$= \frac{1}{a+b}$$

$$\therefore \quad \int_0^\infty \frac{ds}{\left(s^2+a^2\right)\left(s^2+b^2\right)} = \frac{\pi}{2ab(a+b)}$$

***Example* 23.21:** Using Parseval's identity calculate

$$\int_0^\infty \frac{x^2}{\left(a^2+x^2\right)^2} dx (a > 0)$$

***Solution*:** The Fourier sine transform of $f(x) = e^{-ax}$ is

$$F_s[f(x)] = \sqrt{\frac{2}{\pi}}\left(\frac{s}{s^2+a^2}\right) = F(s)$$

Parseval's identity is

$$\int_{-\infty}^{\infty} |f(x)|^2 dx = \int_{-\infty}^{\infty} |F(s)|^2 ds$$

$$f(x) = e^{-ax} \text{ and } F(s) = \sqrt{\frac{2}{\pi}}\left(\frac{s}{s^2+a^2}\right)$$

$$\int_{-\infty}^{\infty} \left|e^{-ax}\right|^2 dx = \int_{-\infty}^{\infty} \left\{\sqrt{\frac{2}{\pi}} \frac{s}{s^2+a^2}\right\}^2 ds$$

$$2\int_0^\infty e^{-2ax} dx = \frac{2}{\pi}\int_{-\infty}^{\infty} \frac{s^2}{\left(s^2+a^2\right)} ds$$

$$\int_0^\infty e^{-2ax} dx = \frac{1}{\pi}.2\int_0^\infty \frac{s^2 ds}{\left(s^2+a^2\right)^2}$$

$$\therefore \quad \int_0^\infty \frac{s^2 ds}{\left(s^2+a^2\right)} = \frac{\pi}{2}\int_0^\infty e^{-2ax} dx$$

$$= \frac{\pi}{2}\left[\frac{e^{-2ax}}{-2a}\right]_0^\infty = \frac{\pi}{2}\left[0+\frac{1}{2a}\right] = \frac{\pi}{4a}$$

$$\int_0^\infty \frac{x^2 dx}{\left(x^2+a^2\right)^2} = \frac{\pi}{4a}$$

***Example* 23.22:** Evaluate $\int_0^\infty \frac{dx}{(a^2+x^2)^2}$ using Parseval's identity.

***Solution*:** The Fourier cosine transform of $f(x) = e^{-ax}$ is

$$F_c[f(x)] = \sqrt{\frac{2}{\pi}}\left[\frac{a}{s^2+a^2}\right] = F(s)$$

The Parseval's identity is

$$\int_{-\infty}^{\infty} |f(x)|^2\, dx = \int_{-\infty}^{\infty} |F(s)|^2\, ds$$

$$\int_{-\infty}^{\infty} |e^{-ax}|^2\, dx = \int_{-\infty}^{\infty}\left[\sqrt{\frac{2}{\pi}}\left(\frac{a}{s^2+a^2}\right)\right]^2 ds$$

$$\frac{2}{\pi}\int_{-\infty}^{\infty}\frac{a^2}{(s^2+a^2)^2}ds = \int_{-\infty}^{\infty}|e^{-ax}|^2\, dx$$

$$\frac{2}{\pi}.2\int_0^\infty \frac{a^2 ds}{(s^2+a^2)^2} = 2\int_0^\infty e^{-2ax}dx$$

$$\int_0^\infty \frac{ds}{(s^2+a^2)} = \frac{\pi}{2a^2}\cdot\left[\frac{e^{-2ax}}{-2a}\right]_0^\infty = = \frac{\pi}{4a^3}$$

$$= \frac{\pi}{2a^2}\left[0+\frac{1}{2a}\right]$$

***Example* 23.23:** Solve the integral equation $\int_0^\infty f(x)\cos\lambda x\, dx = e^{-\lambda}$ using Fourier Transform.

***Solution*:** $\int_0^\infty f(x)\cos\lambda x\, dx = e^{-\lambda}$

$$\therefore \quad \int_0^\infty f(x)\cos s x\, dx = e^{-s} \qquad .....(1)$$

From the definition of Fourier cosine transform of $f(x)$

$$F_c[f(x)] = \sqrt{\frac{2}{\pi}}\int_0^\infty f(x)\cos sx\, dx$$

$$\therefore \quad \int_0^\infty f(x)\cos sx\, dx = \sqrt{\frac{\pi}{2}}F_c[f(x)] \qquad .....(2)$$

From (1) & (2)

$$\therefore \quad \sqrt{\frac{\pi}{2}}F_c[f(x)] = e^{-s}$$

$$F_c[f(x)] = \sqrt{\frac{2}{\pi}}e^{-s}$$

$\therefore$ using inverse Fourier transform

$$f(x) = \sqrt{\frac{2}{\pi}}\int_0^\infty F_c[f(x)]\cos sx\, ds$$

$$= \sqrt{\frac{2}{\pi}}\int_0^\infty \sqrt{\frac{2}{\pi}}e^{-s}\cos sx\, ds$$

$$= \frac{2}{\pi}\int_0^\infty e^{-s}\cos sx\, ds$$

$$= \frac{2}{\pi}\left[\frac{e^{-s}}{1^2+x^2}(-1\cos sx + x\sin sx)\right]_0^\infty = \frac{2}{\pi(x^2+1)}$$

***Example* 23.24:** Solve the integral equation $\int_0^\infty f(x)\cos \lambda x\, dx = 1 - l,\ 0 \le l \le 1$
$= 0,\ \lambda > 1$

Hence deduce that $\int_0^\infty \frac{\sin^2 t}{t^2}dt = \frac{\pi}{2}$

***Solution*:** The Fourier cosine transform of $f(x)$ is

$$F_c[f(x)] = \sqrt{\frac{2}{\pi}}\int_0^\infty f(x)\cos sx\, dx \qquad .....(1)$$

But it is given that

$$\int_0^\infty f(x)\cos sx\,dx = 1 - s, \quad 0 \le s \le 1$$
$$= 0, \quad s > 1 \qquad .....(2)$$

from (1) and (2)

$$F_c[f(x)] = \sqrt{\frac{2}{\pi}}\begin{cases} 1-s, & 0 \le s \le 1 \\ 0, & s > 1 \end{cases}$$

$\therefore$ Inverse Fourier cosine transform is

$$f(x) = \sqrt{\frac{2}{\pi}}\left[\int_0^\infty F_c[f(x)]\cos sx\,ds\right]$$

$$f(x) = \sqrt{\frac{2}{\pi}}\left\{\int_0^1 F_c f(x)\cos sx\,ds + \int_1^\infty F_c[f(x)]\cos sx\,ds\right.$$

$$= \sqrt{\frac{2}{\pi}}\left\{\sqrt{\frac{2}{\pi}}\int_0^1 (1-s)\cos sx\,ds + \sqrt{\frac{2}{\pi}}\int_0^\infty 0.\cos sx\,ds\right.$$

$$= \frac{2}{\pi}\int_0^1 (1-s)\cos sx\,ds$$

$$= \frac{2}{\pi}\left[(1-s)\left(\frac{\sin sx}{x}\right) - (-1)\left(\frac{-\cos sx}{x^2}\right)\right]_0^1$$

$$= \frac{2}{\pi}\left[\frac{-\cos x}{x^2} + \frac{1}{x^2}\right]$$

$$\therefore \quad f(x) = \frac{2}{\pi x^2}(1-\cos x) \qquad .....(3)$$

Now given that $\displaystyle\int_0^\infty f(x)\cos \lambda x\,dx = 1 - \lambda, \quad 0 \le \lambda \le 1$

$$= 0, \quad \lambda > 1$$

substituting $f(x) = \frac{2}{\pi x^2}(1-\cos x)$ and $\lambda = 0$

$$\int_0^\infty \frac{2}{\pi x^2}(1-\cos x)\cos 0.dx = 1$$

$$\int_0^\infty \frac{2.2\sin^2 x/2}{\pi x^2}dx = 1$$

$$\int_0^\infty \frac{\sin^2(x/2)dx}{x^2} = \frac{\pi}{4}$$

put $\frac{x}{2} = t \qquad dx = 2dt$

$x = 0 \quad \Rightarrow t = 0$

$x \to \infty \Rightarrow t \to \infty$

$$\int_0^\infty \frac{\sin^2 t2dt}{4t^2} = \frac{\pi}{4}$$

$$\int_0^\infty \frac{\sin^2 tdt}{t^2} = \frac{\pi}{2}$$

23.5 Applications of Fourier Transforms in Initial and Boundary Value Problems

The problems in which one of the variables ranges from $-\infty$ to ∞ are solved by taking the infinite Fourier transform of both the sides and then taking the corresponding inverse Fourier transform of the solution of the differential equation choosing the sine and cosine transform depends on the given boundary conditions at the lower limit of the variable selected for exclusion.

(i) To remove $\frac{\partial^2 u}{\partial x^2}$ from a partial differential equation we require the value of u when $x = 0$ in sine transform and the value of $\frac{\partial u}{\partial x}$ when $x = 0$ in cosine transform.

(ii) Exclusion of odd order derivatives from the given differential equation is not possible with the help of the Fourier sine and cosine transform.

(iii) If one variable in a differential equation varies in the range $-\infty$ to ∞ then that variable can be excluded by the complex Fourier transform.

Solved Examples

***Example* 23.25:** Solve $\frac{\partial u}{\partial t} = \frac{\partial^2 u}{\partial x^2}$ if $\frac{\partial u}{\partial x} = 0$ at $x = 0$

$$u(x, 0) = \begin{cases} x, & 0 \le x \le 1 \\ 0, & x > 1 \end{cases}$$

and $u(x, t)$ is bounded where $x > 0$, $t > 0$.

***Solution*:** The given equation is $\frac{\partial u}{\partial t} = \frac{\partial^2 u}{\partial x^2}$.

Taking Fourier cosine transform of both the sides.

$$\sqrt{\frac{2}{\pi}}\int_0^\infty \frac{\partial u}{\partial t}.\cos sx\, dx = \sqrt{\frac{2}{\pi}}\int_0^\infty \frac{\partial^2 u}{\partial x^2}.\cos sx\, dx$$

$$\frac{\partial}{\partial t}\left\{\sqrt{\frac{2}{\pi}}\int_0^\infty u\cos sx\, dx\right\} = \sqrt{\frac{2}{\pi}}\left\{\left(\frac{\partial u}{\partial x}.\cos sx\right)_0^\infty + s\int_0^\infty \frac{\partial u}{\partial x}\sin sx\, dx\right\}$$

Since $\frac{\partial u}{\partial x} \to 0$ as $x \to \infty$ we have

$$\frac{\partial u_c}{\partial t} = \left\{0 - \sqrt{\frac{2}{\pi}}\left(\frac{\partial u}{\partial x}\right)_{x=0}\right\} + s\sqrt{\frac{2}{\pi}}\left\{(u\sin sx)_0^\infty + s\int_0^\infty u\cos sx\, dx\right\}$$

where $u_c = \sqrt{\frac{2}{\pi}}\int_0^\infty u\cos s\,x\, dx$

Since $u \to 0$ as $x \to \infty$ we have

$$\frac{\partial u_c}{\partial t} = -\sqrt{\frac{2}{\pi}}\left(\frac{\partial u}{\partial x}\right)_{x=0} + 0 + 0 - s^2 u_c$$

$$\frac{\partial u_c}{\partial t}+\sqrt{\frac{2}{\pi}}\left(\frac{\partial u}{\partial x}\right)_{x=0}+s^2u_c=0$$

But $$\left(\frac{\partial u}{\partial x}\right)_{x=0}=0$$

$$\therefore \quad \frac{\partial u_c}{\partial t}+s^2u_c=0$$

$$(D'+s^2)u_c=0 \text{ where } D'=\frac{\partial}{\partial t}$$

$\therefore$ The solution is $u_c = c_1 e^{-s^2 t}$ .....(1)

$$u(x, 0) = x, \quad 0 \le x \le 1$$
$$0, \quad x > 1$$

$$\therefore \quad u_c=\sqrt{\frac{2}{\pi}}\int_0^{\infty}u(x,0)\cos sx\ dx$$

$$=\sqrt{\frac{2}{\pi}}\int_0^1 x.\cos sx\ dx$$

$$=\sqrt{\frac{2}{\pi}}\left\{(x)\frac{\sin sx}{s}-(1)\frac{(-\cos sx)}{s^2}+0\right\}_0^1$$

$$u_c=\sqrt{\frac{2}{\pi}}\left\{\frac{\sin s}{s}+\frac{\cos s-1}{s^2}\right\} \quad .....(2)$$

from (1) when $t = 0,\ u_c = c_1$ .....(3)

from (2) & (3)

$$c_1 = \sqrt{\frac{2}{\pi}}\left[\frac{\sin s}{s} + \frac{\cos -1}{s^2}\right]$$

substituting in (2)

$$u_c = \sqrt{\frac{2}{\pi}}\left[\frac{\sin s}{s} + \frac{\cos s - 1}{s^2}\right]e^{-s^2 t}$$

Now applying inverse Fourier cosine transform

$$u(x,\ t) = \sqrt{\frac{2}{\pi}}\int_0^\infty u_c \cos sx\ dx$$

$$= \frac{2}{\pi}\int_0^\infty \left[\frac{\sin s}{s} + \frac{\cos s - 1}{s^2}\right]e^{-s^2 t}.\cos sx\ ds$$

***Example* 23.26:** Solve $\frac{\partial u}{\partial t} = 2\frac{\partial^2 u}{\partial x^2}$, if $u(0, t) = 0$, $u(x, 0) = e^{-x}$, $x > 0$ and $u(x, t)$ is bounded where $x > 0$, $t > 0$.

***Solution*:** The given equation is $\frac{\partial u}{\partial t} = 2\frac{\partial^2 u}{\partial x^2}$. Taking the Fourier sine transform of both sides.

$$\sqrt{\frac{2}{\pi}}\int_0^\infty \frac{\partial u}{\partial t}\sin sx\ dx = \sqrt{\frac{2}{\pi}}\int_0^\infty 2\frac{\partial^2 u}{\partial x^2}\sin sx\ dx$$

$$\frac{\partial}{\partial t}\left\{\sqrt{\frac{2}{\pi}}\int_0^\infty u \sin sx\ dx\right\} = 2\sqrt{\frac{2}{\pi}}\left\{\left(\frac{\partial u}{\partial x}\sin sx\right)_0^\infty - s\int_0^\infty \frac{\partial u}{\partial x}\cos sx\ dx\right\}$$

if $\frac{\partial u}{\partial x} \to 0$ as $x \to \infty$

$$\frac{\partial\{u_s\}}{\partial t} = -2s\sqrt{\frac{2}{\pi}}\left\{(u\cos sx)_0^\infty + s\int_0^\infty u\sin sx\, dx\right\}$$

if $u \to 0$ as $x \to \infty$

$$= 2s\sqrt{\frac{2}{\pi}}u(0, t)\cos 0 - 2s^2 u_s$$

$$\therefore \quad \frac{\partial u_s}{\partial t} + 2s^2 u_s = 2s\sqrt{\frac{2}{\pi}}u(0, t)$$

But from given condition $u(0, t) = 0$, we get

$$\frac{\partial u_s}{\partial t} + 2s^2 u_s = 0$$

(i.e.) $\quad (D' + 2s^2)\, u_s = 0$ .....(1)

The solution of (1) is

$$u_s = C_1\, e^{-2s^2 t} \quad .....(2)$$

But $\quad u(x, 0) = e^{-x}$

$\therefore$ when $t = 0$, the Fourier sine transform is

$$u_s = \sqrt{\frac{2}{\pi}}\int_0^\infty e^{-x}\sin sx\, dx = \sqrt{\frac{2}{\pi}}\left\{\frac{e^{-x}}{(-1)^2 + s^2}(-\sin sx - s\cos sx)\right\}_0^\infty$$

$$u_s = \sqrt{\frac{2}{\pi}}\left(\frac{s}{1+s^2}\right) \quad .....(3)$$

From (2) when $t = 0$, $u_s = c_1$ .....(4)

From (3) and (4)

$$c_1 = \sqrt{\frac{2}{\pi}}\left(\frac{s}{1+s^2}\right)$$

Substituting in the solution (2)

$$u_s = \sqrt{\frac{2}{\pi}}\left(\frac{s}{1+s^2}\right)e^{-2s^2 t}$$

Now applying inverse Fourier sine transform

$$u(x, t) = \sqrt{\frac{2}{\pi}} \int_0^\infty u_s . \sin sx \, ds$$

$$\therefore \qquad u(x, t) = \frac{2}{\pi} \int_0^\infty \left(\frac{s}{1+s^2} \right) e^{-2s^2 t} \sin sx \; ds$$

***Example* 23.27**: Solve $\frac{\partial v}{\partial t} = \frac{\partial^2 v}{\partial x^2}$ by using complex form of the Fourier transform under the boundary conditions $-\infty < x < \infty$, $t > 0$; $v = f(x)$, when $t = 0$.

Solution: The given equation is $\frac{\partial v}{\partial t} = \frac{\partial^2 v}{\partial x^2}$

Applying the complex Fourier transform of both the sides of the given equation,

$$\frac{1}{\sqrt{2\pi}} \int_{-\infty}^{\infty} \frac{\partial v}{\partial t} e^{isx} dx = \frac{1}{\sqrt{2\pi}} \int_{-\infty}^{\infty} \frac{\partial^2 v}{\partial x^2} e^{isx} dx$$

using Fourier transform of the derivatives of a function

that $\qquad F\left\{ \frac{\partial^n f}{\partial x^n} \right\} = (-is)^n F\{f(x)\}$, we have

$$\frac{d\left\{ \frac{1}{\sqrt{2\pi}} \int_{-\infty}^{\infty} \frac{\partial v}{\partial t} e^{isx} dx \right\}}{dt} = (-is)^2 \, v(s)$$

(i.e.) $\qquad \frac{dv(s)}{dt} = -s^2 v(s)$

$\therefore \qquad \frac{dv(s)}{dt} + s^2 \, v(s) = 0$

(or) $\qquad (D + s^2) \, v(s) = 0$ where $D = \frac{d}{dt}$

$\therefore$ The solution is $v(s) = Ae^{-s^2t}$ .....(1)

But when $t = 0$, $v(s) = Ae^0 \Rightarrow A = v(s)$ .....(2)

And at $t = 0$, $v(s) = \dfrac{1}{\sqrt{2\pi}} \int_{-\infty}^{\infty} f(u) e^{isu} du$ .....(3)

(given $v = f(x)$ when $t = 0$)

Using (2) and (3) in (1)

$$\therefore \quad V(s) = \left\{ \frac{1}{\sqrt{2\pi}} \int_{-\infty}^{\infty} f(u) e^{isu} du \right\} e^{-s^2t}$$

Applying the inverse Fourier transform

$$V(x, t) = \frac{1}{\sqrt{2\pi}} \int_{-\infty}^{\infty} \left\{ \frac{e^{-s^2t}}{\sqrt{2\pi}} \int_{-\infty}^{\infty} f(u).e^{isu} du \right\} e^{-isx} ds$$

$$V(x, t) = \frac{1}{2\pi} \int_{-\infty}^{\infty} f(u) \left\{ \int_{-\infty}^{\infty} e^{-s^2t - is(x-u)} ds \right\} du$$

This can be written as

$$V(x, t) = \frac{1}{2\pi} \int_{-\infty}^{\infty} f(u) \left\{ \int_{-\infty}^{\infty} e^{-t\left[s + \frac{i(x-u)}{2t}\right]^2 - \frac{(x-u)^2}{4t}} ds \right\} du$$

$$= \frac{1}{2\pi} \int_{-\infty}^{\infty} f(u) \left\{ \int_{-\infty}^{\infty} e^{-t\left[s + i\frac{(x-u)}{2t}\right]^2} ds \right\} e^{-\frac{(x-u)^2}{4t}} du$$

substituting $\sqrt{t}\left\{ s + i\dfrac{(x-u)}{2t} \right\} = y$

$$= \frac{1}{2\pi} \int_{-\infty}^{\infty} f(u) \left\{ \frac{1}{\sqrt{t}} \int_{-\infty}^{\infty} e^{-y^2} dy \right\} e^{-\frac{(x-u)^2}{4t}} du \qquad \left(\because ds = \frac{dy}{\sqrt{t}} \right)$$

using $\displaystyle\int_{-\infty}^{\infty} e^{-x^2} dx = \sqrt{\pi}$

$$= \frac{1}{2\pi} \int_{-\infty}^{\infty} f(u) \frac{1}{\sqrt{t}} . \sqrt{\pi}\, e^{-\frac{(x-u)^2}{4t}}\, du$$

$$\therefore \qquad V(x, t) = \frac{1}{2\sqrt{\pi t}} \int_{-\infty}^{\infty} f(u) e^{\frac{-(x-u)^2}{4t}}\, du$$

***Example* 23.28:** Use finite Fourier transforms to solve

$$\frac{\partial u}{\partial t} = \frac{\partial^2 u}{\partial x^2},\ u(0, t) = 0,\ u(4, t) = 0,\ u(x, 0) = 2x$$

and $0 < x < 4$ when $t > 0$.

***Solution*:** Applying the finite Fourier sine transform both sides to the partial differential

equation $\dfrac{\partial u}{\partial t} = \dfrac{\partial^2 u}{\partial x^2}$

$$\int_0^4 \frac{\partial u}{\partial t} \sin\frac{s\pi x}{4}\, dx = \int_0^4 \frac{\partial^2 u}{\partial x^2} \sin\frac{s\pi x}{4}\, dx \qquad \text{(here } l = 4\text{)}$$

or
$$\frac{du_s}{dt} = \frac{-s^2\pi^2}{4^2} u_s + \frac{s\pi}{4} \qquad [u(0, t) - u(4, t) \cos s\pi]$$

where u_s is finite Fourier sine transform of u.

Using the given condition $u(0, t) = 0$, $u(4, t) = 0$ we get

$$\frac{du_s}{dt} = \frac{-s^2\pi^2}{4^2} u_s$$

$$\therefore \text{ Solution is } u_s = Ae^{-\frac{s^2\pi^2 t}{16}} \qquad .....(1)$$

Taking finite Fourier sine transforms

$$\text{At } t = 0 \qquad u_s = \int_0^4 2x \sin\frac{s\pi x}{4}\, dx \qquad (\text{since } u(x, 0) = 2x,\ 0 < x < 4)$$

$$u_s = \left[(2x)\left(-\cos\frac{s\pi x}{4}\right) . \frac{4}{s\pi} - (2)\left(-\sin\frac{s\pi x}{4}\right)\left(\frac{4^2}{s^2\pi^2}\right) \right]_0^4$$

$$u_s = \frac{32(-\cos s\pi)}{s\pi} = \frac{32(-1)^{s+1}}{s\pi}$$

Substituting $t = 0$ in (1) we get

$$A = \frac{32(-1)^{s+1}}{s\pi}$$

Substituting in (1) for A

$$u_s = \frac{32(-1)^{s+1}}{s\pi} e^{-\frac{s^2\pi^2 t}{16}}$$

Taking inverse finite Fourier sine transform

$$u(x, t) = \frac{2}{4}\sum_{s=1}^{\infty} . \frac{32(-1)^{s+1}}{3\pi} . e^{-\frac{s^2\pi^2 t}{16}} . \sin\frac{s\pi x}{4}$$

(i.e.) $$u(x, t) = \frac{16}{\pi}\sum_{s=1}^{\infty} \frac{(-1)^{s+1}}{s\pi} e^{-\frac{s^2\pi^2 t}{16}} . \sin\left(\frac{s\pi x}{4}\right)$$

***Example* 23.29:** Solve $\frac{\partial u}{\partial t} = \frac{\partial^2 u}{\partial x^2}$ $\quad 0 < x < 6, t > 0$

subject to the conditions $u(0, t) = 0$, $u(6, t) = 0$

$$u(x, 0) = 1, \quad 0 < x < 3$$
$$= 0, \quad 3 < x < 6$$

***Solution*:** The finite Fourier sine transform of the given partial differential equation is

$$\int_0^6 \frac{\partial u}{\partial t} \sin\left(\frac{s\pi x}{6}\right) dx = \int_0^6 \frac{\partial^2 u}{\partial x^2} \sin\frac{s\pi x}{6} dx$$

$$\frac{du_s}{dt} = -\frac{s^2\pi^2}{6^2} u_s + \frac{s\pi}{6}\left[u(0,t) - u(6,t)\cos s\pi\right]$$

Using the conditions $u(0, t) = 0$, $u(6, t) = 0$

$$\frac{du_s}{dt} = \frac{s^2\pi^2}{6^2} u_s$$

The solution of the above equation is

$$u_s = Ae^{-\frac{s^2\pi^2 t}{36}} \qquad .....(1)$$

To find the constant 'A' we use the conditions that at $t = 0$

$$u = 1, \quad 0 < x < 3$$
$$0, \quad 3 < x < 6$$

Now $$u_s = \int_0^6 u \sin\frac{s\pi x}{6} dx$$

$$= \int_0^3 1.\sin\frac{s\pi x}{6}\partial x + \int_3^6 0.\sin\frac{s\pi x}{6}\partial x$$

$$= \left[-\cos\left(\frac{s\pi x}{6}\right).\frac{6}{s\pi}\right]_0^3$$

$$u_s = \frac{6}{s\pi}\left(1-\cos\frac{s\pi}{2}\right) \qquad .....(2)$$

from (1) and (2) $A = \frac{6}{s\pi}\left(1-\cos\frac{s\pi}{2}\right)$

substituting for A in (1)

$$u_s = \frac{6}{s\pi}\left(1-\cos\frac{s\pi}{2}\right)e^{-\frac{s^2\pi^2 t}{36}}$$

Taking the inverse Fourier sine transform

$$u(x, t) = \frac{2}{6}\sum_{s=1}^{\infty}\frac{6}{s\pi}\left(1-\cos\frac{s\pi}{2}\right)e^{-\frac{s^2\pi^2 t}{36}}.\sin\left(\frac{s\pi x}{6}\right)$$

$$u(x, t) = \frac{2}{\pi}\sum_{s=1}^{\infty}\frac{6}{s}\left(1-\cos\frac{s\pi}{2}\right)e^{-\frac{s^2\pi^2 t}{36}}.\sin\left(\frac{s\pi x}{6}\right)$$

***Example* 23.30:** Using cosine transform show that the steady temperature in the semi-infinite solid $y > 0$ when the temperature on the surface $y = 0$ is kept at unity over the strip $|x| < a$ and at zero outside the strip is

$$\frac{1}{\pi}\left[\tan^{-1}\left(\frac{a+x}{y}\right) + \tan^{-1}\left(\frac{a-x}{y}\right)\right]$$

by assuming the result

$$\int_3^{\infty} e^{-sx}x^{-1}\sin(rx)dx = \tan^{-1}\left(\frac{r}{s}\right) \qquad r > 0, s > 0$$

***Solution*:** The two dimensional heat equation (Laplace equation) is

$$\frac{\partial^2 u}{\partial x^2}+\frac{\partial^2 u}{\partial y^2}=0,\ 0<y<\infty, \quad -\infty<x<\infty \qquad .....(1)$$

The given conditions are

$u = 1, y = 0 \qquad -a < x < a$

$u = 0, y = 0 \qquad \text{when } x > a$

Applying Fourier cosine transform of both sides of (1)

$$\sqrt{\frac{2}{\pi}}\int_0^{\infty}\frac{\partial^2 u}{\partial y^2}\cos(s\,x)\,dx+\sqrt{\frac{2}{\pi}}\int_0^{\infty}\frac{\partial^2 u}{\partial y^2}\cos(s\,x)\,dx=0$$

$$\sqrt{\frac{2}{\pi}}\left[\frac{\partial u}{\partial x}\cos(s\,x)\right]_0^{\infty}+s\sqrt{\frac{2}{\pi}}\int_0^{\infty}\frac{\partial u}{\partial x}\sin(s\,x)dx+\frac{\partial^2 u_c}{\partial y^2}=0$$

As $\quad \frac{\partial u}{\partial x}\to 0$ as $x\to\infty$ and $\frac{\partial u}{\partial x}\to 0$ as $x\to 0$

we get $\quad s\sqrt{\frac{2}{\pi x}}[u\sin sx]_0^{\infty}-s^2\sqrt{\frac{2}{\pi x}}\int_0^{\infty}u\cos(s\,x)dx+\frac{\partial^2 u_c}{\partial y^2}=0$

$\therefore \quad \frac{d^2 u_c}{dy^2}-s^2 u_c=0 \qquad$ (If $u\to 0$ as $x\to\infty$)

Solution of the above differential equation is

$$u_c = Ae^{sy}+Be^{-sy} \qquad .....(2)$$

Using the condition that as $y\to\infty$, $u_c\to 0$ in (2) gives $A = 0$

$\therefore \quad$ Solution (2) reduces to $\quad u_c = Be^{-sy} \qquad .....(3)$

Also when $y = 0$, $u_c = 1$ for $|x| < a$ and $u_c = 0$ for $x > a$

$$\therefore \quad u_c=\sqrt{\frac{2}{\pi}}\int_0^{a}1.\cos sx\,dx+\sqrt{\frac{2}{\pi}}\int_a^{\infty}0.\cos sx\,dx$$

$$=\sqrt{\frac{2}{\pi}}\left[\frac{\sin sx}{s}\right]_0^{a}=\sqrt{\frac{2}{\pi}}\left(\frac{\sin sa}{s}\right) \qquad .....(4)$$

From (3) $\quad u_c = B \qquad (\because y = 0)$

$$\therefore \qquad B = \sqrt{\frac{2}{\pi}}\frac{\sin sa}{s}$$

$\therefore$ (3) reduces to

$$u_c = \sqrt{\frac{2}{\pi}}\frac{\sin sa}{s}e^{-sy}$$

Taking the inverse Fourier cosine Transform

$$u(x, y) = \sqrt{\frac{2}{\pi}}\int_0^\infty \sqrt{\frac{2}{\pi}}\frac{\sin sa}{s}e^{-sy}\cos sx\, ds$$

$$= \frac{1}{\pi}\int_0^\infty \frac{e^{-sy}}{s}\left[\sin(a+x)s + \sin(a-x)s\right]ds$$

$$u(x, y) = \frac{1}{\pi}\int_0^\infty e^{-sy}s^{-1}\sin(a+x)s\, ds + \frac{1}{\pi}\int_0^\infty e^{-sy}s^{-1}\sin(a-x)s\, ds$$

Now assuming the result $\int_0^\infty e^{-sx}x^{-1}\sin rx\, dx = \tan^{-1}\left(\frac{r}{s}\right)$ for $r > 0, s > 0$

$$u(x, y) = \frac{1}{\pi}\left[\tan^{-1}\left(\frac{a+x}{y}\right) + \tan^{-1}\left(\frac{a-x}{y}\right)\right]$$

***Example* 23.31:** Show that the solution of Laplace's equation for u inside the semi-infinite strip $x > 0$, $0 < y < b$ such that

$u = f(x)$ when $y = 0, 0 < x < \infty$

$u = 0$ when $y = b, 0 < x < \infty$

$u = 0$ when $x = 0, 0 < y < b$

is given by $$u = \frac{2}{\pi}\int_0^\infty f(u)du\int_0^\infty \frac{\sinh(b-y)s}{\sinh(sb)}.\sin(sx)\sin(us)ds$$

***Solution*:** Two dimensional Laplace's equation is

$$\frac{\partial^2 u}{\partial x^2} + \frac{\partial^2 u}{\partial y^2} = 0, \quad 0 < x < \infty, \; 0 < y < b \qquad \text{.....(1)}$$

Taking Fourier sine transform of (1)

$$\sqrt{\frac{2}{\pi}}\int_0^\infty \frac{\partial^2 u}{\partial x^2}\sin sx\,dx+\sqrt{\frac{2}{\pi}}\int_0^\infty \frac{\partial^2 u}{\partial y^2}\sin sx\,dx=0$$

$$\sqrt{\frac{2}{\pi}}\left[\frac{\partial u}{\partial x}\sin(sx)\right]_0^\infty - s\sqrt{\frac{2}{\pi}}\int_0^\infty \frac{\partial u}{\partial x}\cos\,sx\,dx+\frac{\partial^2 u_s}{\partial y^2}=0$$

$$-s\sqrt{\frac{2}{\pi}}\left[u\cos sx\right]_0^\infty - s^2\sqrt{\frac{2}{\pi}}\int_0^\infty u\sin sx\,ds+\frac{\partial^2 u_s}{\partial y^2}=0$$

As $\quad \frac{\partial u}{\partial x}\to 0 as\ x\ \infty\ and\ u\to 0\ as\ x\to\infty$

$\therefore$ we get $\quad \frac{d^2u_s}{dy^2}-s^2u_s=0$

The solution of the above equation is

$$u_s = A\cosh sy + B\sinh sy \qquad .....(2)$$

Now taking Fourier sine transforms of given conditions that

when $\quad y=0$ we have $u_s=\sqrt{\frac{2}{\pi}}\int_0^\infty f(u)\sin su\,du$

and when $\quad y=b,\ u_s=0$

$\therefore$ from (2) we have $\sqrt{\frac{2}{\pi}}\int_0^\infty f(u)\sin su\,du=A$

and $\quad 0=A\cosh sb+B\sinh sb$

$$B=-\frac{\cosh sb}{\sinh sb}\sqrt{\frac{2}{\pi}}\int_0^\infty f(u)\sin su\,du$$

Hence $\quad u_s=\sqrt{\frac{2}{\pi}}\cosh sy\int_0^\infty f(u)\sin su\,du$

$$-\sqrt{\frac{2}{\pi}}\frac{\cosh sb}{\sinh sb}\sinh sy\int_0^\infty f(u)\sin(su)\,du$$

$$u_s = \sqrt{\frac{2}{\pi}} \int_0^\infty f(u) \sin su \left[\cosh sy - \frac{\cosh sb \sinh sy}{\sinh sb} \right] du$$

$$u_s = \sqrt{\frac{2}{\pi}} \int_0^\infty f(u) \sin su \left[\frac{\sinh sb \cosh sy - \cosh sb \sinh sy}{\sinh sb} \right] du$$

$$u_s = \sqrt{\frac{2}{\pi}} \int_0^\infty f(u) \sin su \frac{\sinh (b-y)s}{\sinh sb} du$$

Taking the inverse Fourier transform

$$u = \sqrt{\frac{2}{\pi}} \int_0^\infty u_s \sin sx \, ds$$

(i.e.)
$$u = \frac{2}{\pi} \int_0^\infty f(u) du \int_0^\infty \frac{\sinh (b-y)s}{\sinh sb} \sin(su) \sin(sx) ds$$

***Example* 23.32:** Use the method of Fourier Transform to determine the displacement $y(x, t)$ of an infinite string, given that the string is initially at rest and that the initial displacement is $f(x)$, $-\infty < x < \infty$. Show that the solution can also be put in the form

$$y(x, t) = \frac{1}{2}\{f(x+ct) + f(x-ct)\}.$$

***Solution*:** The one dimensional wave equation is

$$\frac{\partial^2 y}{\partial t^2} = c^2 \frac{\partial^2 y}{\partial x^2} \qquad .....(1)$$

where $y(x, t)$ is the displacement at any time 't', $-\infty < x < \infty$, $t > 0$

Taking the Fourier transform of both the sides of equation (1)

$$\frac{1}{\sqrt{2\pi}} \int_{-\infty}^{\infty} \frac{\partial^2 y}{\partial t^2} e^{isx} dx = c^2 . \frac{1}{\sqrt{2\pi}} \int_{-\infty}^{\infty} \frac{\partial^2 y}{\partial x^2} e^{isx} dx$$

$$\frac{\partial^2}{\partial t^2} \frac{1}{\sqrt{2\pi}} \int_{-\infty}^{\infty} y e^{isx} dx = c^2 (-i\, s)^2\, y(s, t) = 0$$

$$\frac{\partial^2 y(s,t)}{\partial t^2} = -c^2 s^2 y(s, t)$$

$$\therefore \qquad \frac{\partial^2 y(s,t)}{\partial t^2} + c^2 s^2 y(s,t) = 0$$

The solution of the above equation is

$$y(s, t) = A\cos cst + B \sin cst \qquad .....(2)$$

Initially the string is at rest

$$\Rightarrow \qquad \frac{\partial y}{\partial t} = 0, \text{ at } t = 0$$

$$\therefore \qquad \frac{1}{\sqrt{2\pi}} \int_{-\infty}^{\infty} \frac{\partial y}{dt} e^{isx} dx = \frac{\partial}{\partial t} \frac{1}{\sqrt{2\pi}} \int_{-\infty}^{\infty} y e^{isx} dx$$

$$= \frac{dy(s,t)}{dt} = 0$$

Using $\frac{dy(s,t)}{dt} = 0$ in (2) we have $B\,c\,s = 0 \Rightarrow B = 0$

Also at $t = 0, y = f(x)$

$\therefore$ at $t = 0, y(s, 0) = \frac{1}{\sqrt{2\pi}} \int_{-\infty}^{\infty} f(u) e^{isu} du = f(s)$

from (2) $f(s) = A$

Hence $y(s, t) = f(s) \cos(c\,s\,t)$

Taking the inverse Fourier Transform

$$y(x, t) = \frac{1}{\sqrt{2\pi}} \int_{-\infty}^{\infty} f(u) e^{isu} du \frac{1}{\sqrt{2\pi}} \int_{-\infty}^{\infty} \cos(c\,s\,t) e^{-isx} ds$$

$$= \frac{1}{2\pi} \int_{-\infty}^{\infty} \left[\int_{-\infty}^{\infty} f(u) e^{isu} du \right] \cos c\,s\,t\, e^{-isx} ds$$

$$= \frac{1}{4\pi} \int_{-\infty}^{\infty} \left[\int_{-\infty}^{\infty} f(u) e^{isu} du \right] \left[e^{icst} + e^{-icst} \right] e^{-isx} ds$$

$$y(x,\,t)=\frac{1}{2}\left\{\frac{1}{2\pi}\int_{-\infty}^{\infty}\left[\int_{-\infty}^{\infty}f(u)e^{-i\alpha u}\,du\right]\left(e^{-ic\alpha t}+e^{-ic\alpha t}\right)e^{i\alpha x}\,d\alpha\right\}$$

where $\alpha = -s$

$ds = -\,d\alpha$

$$y(x,\,t)=\frac{1}{2}\left\{\frac{1}{2\pi}\int_{-\infty}^{\infty}f(u)e^{-i\alpha u}\left[\int_{-\infty}^{-\infty}e^{i\alpha(x+ct)}\,d\alpha\right]du\right.$$

$$\left.+\frac{1}{2}\int_{-\infty}^{\infty}f(u)e^{-i\alpha u}\left[\int_{-\infty}^{\infty}e^{i\alpha(x-ct)}\,d\alpha\right]du\right\}$$

Hence $\quad y(x,\,t)=\frac{1}{2}\{f(x+ct)+f(x-ct)\}$

Exercise 23.3

1. Solve $\dfrac{\partial u}{\partial t}=\dfrac{\partial^2 u}{\partial x^2}$, $x > 0$, $t > 0$ subject to the conditions

   (i) $u = 0$ when $x = 0,\ t > 0$

   (ii) $u=\begin{cases}1 & 0<x<1\\ 0 & x\ge 1\end{cases}$ when $t = 0$

   and (iii) $u(x, t)$ is bounded.

   [Ans : $u(x,\,t)=\dfrac{2}{\pi}\displaystyle\int_0^{\infty}\frac{se^{-2s^2t}\sin sx}{1+s^2}\,ds$ **]**

2. Solve the partial differential equation $\dfrac{\partial v}{\partial t}=k\,\dfrac{\partial^2 v}{\partial x^2}$ for $x > 0$, $t > 0$, using the Fourier sine transform, under the boundary conditions $V = V_0$ when $x = 0$, $t > 0$, and the initial condition $V = 0$ when $t = 0$, $x > 0$.

   [Ans : $V(x,\,t)=V_0\left(1-\dfrac{2}{\pi}\displaystyle\int_0^{\infty}\frac{e^{-ks^2t}}{s}\sin sx\,ds\right)$ **]**

3. If the flow of heat is linear so that the variation of θ (temperature) with z and y may be neglected and if it is assumed that no heat is generated in the medium, then solve the

differential equation $\frac{\partial \theta}{\partial t} = k\frac{\partial^2 \theta}{\partial x^2}, -\infty < x < \infty$ and $\theta = f(x)$ when $t = 0$, $f(x)$ being a given function of x.

[**Ans :** $\theta(x, t) = \frac{1}{\sqrt{2\pi}} \int_{-\infty}^{\infty} f(s) e^{-ks^2 t - isx} ds$]

4. Use the finite cosine transform to solve $\frac{\partial v}{\partial t} = k\frac{\partial^2 v}{\partial x^2}$ with the boundary conditions $\frac{\partial v}{\partial x} = 0$ when $x = 0$ and $x = \pi$, $t > 0$ and the initial condition $v = f(x)$ when $t = 0$, $0 < x < \pi$.

[**Ans :** $V(x, t) = \frac{1}{2}\int_0^{\pi} f(y)dy + \frac{2}{\pi}\sum_{s=1}^{\infty}\left[e^{-ks^2 t}\cos sx . \int_0^{\pi} f(y)\cos sydy\right]$]

5. Solve the boundary value problem $\frac{\partial u}{\partial t} = \frac{\partial^2 u}{\partial x^2}$, $u(0, t) = 1$, $u(\pi, t) = 3$, $u(x, 0) = 1$, $0 < x < \pi$, $t > 0$

Ans : $u(x, t) = \frac{4}{\pi}\sum_{s=1}^{\infty}\frac{\cos s\pi}{s} e^{-s^2 t}\sin sx + 1 + \frac{2x}{\pi}$

6. Determine displacements $u(x, t)$ in a horizontal string stretched from origin to the point $(\pi, 0)$ when the motion is due to the weight of the string alone. The string may be taken to be initially at rest in the position $u = 0$.

Ans : $u(x, t) = \frac{2g}{\pi a^2}\sum_{s=1}^{\infty}\frac{x(\pi - x)}{2}(1 - \cos sat)\sin sx$

7. Determine a function $v(x, y)$ which is harmonic $\left(\frac{\partial^2 v}{dx^2} + \frac{\partial^2 v}{\partial y^2} = 0\right)$ in the open square $0 < x < \pi$, $0 < y < \pi$, takes a constant value V_0 on the edge $y = \pi$ and vanishes on the other edges of the square.

Ans : $v(x, y) = \frac{2}{\pi}\sum_{s=1}^{\infty}\frac{v_0}{\sinh s\pi}\left(\frac{1 - \cos s\pi}{s}\right)\sinh sy \sin sx$

8. Solve the boundary value problem $\frac{\partial^2 u}{\partial t^2} = 9\frac{\partial^2 u}{\partial x^2}$ subject to the conditions $u(0< t) = 0$,

$u(2, t) = 0$, $u(x, 0) = (0\,.\,05)\,x\,(2 - x)$, $\frac{\partial u}{\partial t} = 0$ at $t = 0$ where $0 < x < 2$, $t > 0$.

Ans : $u(x, t) = \frac{1.6}{\pi^3}\sum_{n=1}^{\infty}\frac{1}{(2n-1)^3}\cos\frac{3(2n-1)\pi t}{2}.\sin\frac{(2n-1)\pi x}{2}$

CHAPTER 24

Laplace Transforms

24.1 Introduction

The theory of Laplace transform is an essential part of the mathematical background required by engineers, physicists and mathematicians. It gives an easy and effective means for solving certain types of differential and integral equations. It is the foundation of the modern form of operational calculus, which was originated in an attempt to justify certain operational methods used by an electrical engineer Oliver Heavinide, in the later part of the 19th century for solving equations in electromagnetic theory.

The Laplace transform reduces the problem of solving a differential equation to an algebraic problem. It is particularly useful for solving problems where the mechanical or electrical driving force has discontinuities, acts for a short time only or is periodic but not merely a sine or cosine function.

24.1.1 Definition

A class of linear transformations of particular importance is that of the integral transforms. The Laplace transformation (L.T) is a special case of an integral transformation and is defined by

$$F(s) = L\,[f(t)] = \int_0^\infty e^{-st} f(t)\, dt$$

Def. Piecewise continuity of a function

A function f(t) is sectionally continuous (Piecewise continuous) in an interval $0 \leq t \leq b$ if f(t) is continuous over every finite interval $0 \leq t \leq b$ except possibly at a finite number of points, where there are jump discontinuities and at which the function approaches different limits from left and right.

Sectionally continuous function

Def. Functions of exponential Order

A function f(t) is said to be of exponential order σ, as $t \to \infty$ if $\exists\ M > 0$

$$\ni |f(t)| < M e^{\sigma t},\ \forall\ t \geq 0 \qquad \text{.... (1)}$$

or equivalently, $|e^{-\sigma t} f(t)| \leq M,\ \forall\ t \geq 0$.

Remark: $e^{\sigma t}$ and sinbt are of exponential order, while e^{t^2} is not of exponential order since

$$\left| e^{-\sigma t} e^{t^2} \right| = e^{t^2 - \sigma t}$$

which can be made larger than any given constant by increasing 't' indefinitely.

24.1.2 Conditions for L.T to Exist

By using above remarks we can prove the conditions that are sufficient for the existence of Laplace transforms.

Theorem: If f(t) is piecewise continuous in every finite integral [0, N] for N > 0, and of exponential order σ, that is $|f(t)| \leq M e^{\sigma t},\ \forall\ t \geq 0$ and M > 0. (from 1)

then the Laplace transform of F(t) exists for all s > σ

Proof: Let $I = |F(s)| = \left| \int_0^\infty e^{-st} f(t) dt \right| \leq \int_0^\infty | f(t) | e^{-st} dt \leq \int_0^\infty M e^{\sigma t} e^{-st} dt = \frac{M}{s - \sigma}$

Here, s > σ since I is non-negative everywhere. This shows that above conditions are sufficient for the existence of L [f(t)].

24.1.3 Linearity Property

Let $f_1(t)$ and $f_2(t)$ be two functions on [0, ∞] such that the Laplace transforms $L[f_1(t)]$ and $L[f_2(t)]$ exist. If k_1 and k_2 are two constants, then

$$L[k_1 f_1(t) + k_2 f_2(t)] = k_1 L[f_1(t)] + k_2 L[f_2(t)] \qquad \text{....(1)}$$

This property is valid since

$$\int_0^\infty [k_1 f_1(t) + k_2 f_2(t)] e^{-st} dt = \int_0^\infty k_1 f_1(t) e^{-st} dt + \int_0^\infty k_2 f_2(t) e^{-st} dt = k_1 L[f_1(t)] + k_2 L[f_2(t)]$$

This result can be extended to the linear combinations of more than two functions.

24.1.4 First Shifting, Second Shifting and Change of Scale Properties

If f(t) is piecewise continuous on [0, N] for each N > 0, and f(t) is of exponential order σ, that is by 24.1.2(1), $|f(t)| \le Me^{\sigma t}$, $\forall\ t \ge 0$ and M > 0 and if L [f(t)] = F(s) then we have

(1) $L[e^{at}f(t)] = F(s-a)\ \forall\ s > \sigma + a\ \forall\ a > 0$ (First shifting property) .....(1)

(2) $L[f_1(t)] = e^{-as}F(s)$ (Second shifting property)

where $f_1(t) = \begin{Bmatrix} f(t-a) & t > a \\ 0 & t < a \end{Bmatrix}$ (a > 0) (Change of scale property)

and (3) $L(f(at)) = \frac{1}{a}F\left(\frac{s}{a}\right)$ (for any a > 0)

Proof **(1)**

From definition, we have

$$L[f(t)] = \int_0^\infty e^{-st} f(t)\ dt = F(s)$$

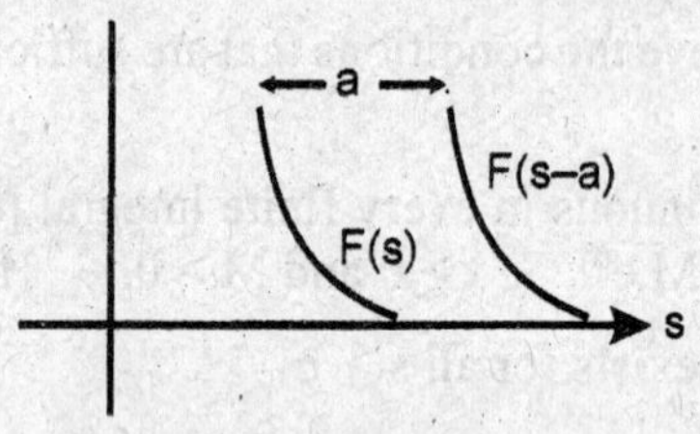

so that $L[e^{at}f(t)] = \int_0^\infty e^{-st}[e^{at} f(t)dt = \int_0^\infty e^{-(s-a)t} f(t)dt = F(s-a)$

$\therefore$ $L[e^{at}f(t)] = F(s-a)$

This property shows shifting on the s-axis and is called the *first shifting property* or first translation.

This means that replacing s by s – a in the Laplace transform corresponds to multiplying the original function by e^{at}

Similarly $L[e^{-at}f(t)] = F(s+a)$, we can have the following results with the help of the above property.

$L[e^{\pm at}\, t^n] = \frac{n!}{(s \mp a)^{n+1}}$; $\quad (\because L[t^n] = \frac{n!}{s^{n+1}})$

$L[e^{\pm at} \sin bt] = \frac{b}{(s \mp a)^2 + b^2}$ $\quad (\because L[\sin bt] = \frac{b}{s^2 + b^2})$

$$L[e^{\pm at}\cos bt] = \frac{(s\mp a)}{(s\mp a)^2+b^2}$$

$$L[e^{\pm at}\sinh bt] = \frac{b}{(s\mp a)^2-b^2}$$

$$L[e^{\pm at}\cosh bt] = \frac{(s\mp a)}{(s\mp a)^2-b^2}$$

(2) $$L(f_1(t)) = \int_0^{\infty} e^{-st} f_1(t)dt = \int_0^{a} e^{-st} f_1(t)dt + \int_a^{\infty} e^{-st} f_1(t)dt$$

$$= \int_0^{a} e^{-st}.0dt + \int_a^{\infty} e^{-st} f(t-a)\ dt \quad \text{from definition of } f_1(t) = \begin{cases} f(t-a) & t>a \\ 0, & t<a \end{cases}$$

$$= \int_a^{\infty} e^{-st} f(t-a)dt = \int_0^{\infty} e^{-s(p+a)} f(p)\ dp \quad \text{(where } t = p+a)$$

$$= e^{-as}\int_0^{\infty} e^{-ps} f(p)dp = e^{-as}F(s)$$

This property indicates shifting on the t-axis and is called the second shifting property or second translation.

(3) $$L[f(at)] = \int_0^{\infty} e^{-st} f(at)dt = \int_0^{\infty} e^{-s(p/a)} f(p)\frac{dp}{a} \quad \text{[taking } t = p/a]$$

$$= \frac{1}{a}\int_0^{\infty} e^{-sp/a} f(p)\ dp = \frac{1}{a}F\left(\frac{s}{a}\right)$$

This property is known as change of scale property.

24.1.5 Laplace Transforms of Standard Functions

(a) $$L(e^{-at}) = \frac{1}{s+a} \qquad (s > a) \qquad \text{......(1)}$$

Proof: we have $L(e^{-at}) = \int_0^{\infty} e^{-st}.e^{-at}dt = \int_0^{\infty} e^{-(s+a)t}dt = \frac{-1}{s+a}[e^{-(s+a)t}]_0^{\infty} = \frac{1}{s+a}$

if a = 0 the $L(1) = \frac{1}{s}$ (s > 0) .....(2)

Also changing the sign of 'a' we get

$$L(e^{at}) = \frac{1}{s-a} \qquad \text{.....(3)}$$

(b) $L(\sin at) = \dfrac{a}{s^2+a^2}$ and $L(\cos at) = \dfrac{s}{s^2+a^2}$ (s > 0)

Proof: $L(\cos at + i\sin at) = L(e^{iat}) = \dfrac{1}{s-ia} = \dfrac{s+ia}{s^2+a^2}$ from (a)

Equating real and imaginary parts we get

$$L(\sin at) = \frac{a}{s^2+a^2}, L(\cos at) = \frac{s}{s^2+a^2} \quad .....(4)$$

(c) $L(\sinh at) = \dfrac{a}{s^2-a^2}$ and $L(\cosh at) = \dfrac{s}{s^2-a^2}$ (s > |a|)

Proof: $L(\sinh at) = L\left(\dfrac{e^{at}-e^{-at}}{2}\right) = \dfrac{1}{2}L(e^{at}) - \dfrac{1}{2}L(e^{-at})$

$$= \frac{1}{2}\left[\frac{1}{s-a} - \frac{1}{s+a}\right] \quad \text{from (a)}$$

Thus $(\sinh at) = \dfrac{a}{s^2-a^2}$

similarly we can show that $L(\cosh at) = \dfrac{s}{s^2-a^2}$ ....(5)

(d) $L(t^n) = \dfrac{(n+1)}{s^{n+1}}$ [(n + 1) > 0 and s > 0]

Proof: $L(t^n) = \displaystyle\int_0^\infty e^{-st} t^n dt = \int_0^\infty e^{-p}.\frac{p^n}{s^n}\frac{dp}{s}$ where p = st

$$= \int_0^\infty \frac{e^{-p}.p^n}{s^{n+1}} dp = \frac{\Gamma(n+1)}{s^{n+1}}$$

Since $\Gamma(n+1) = \displaystyle\int_0^\infty e^{-x} x^n dx$ and $\Gamma(n+1) = n!$ if n is a +ve integer

$$\therefore L(t^n) = \frac{n!}{s^{n+1}} \quad .....(6)$$

From (6) we have

$$L(1) = \frac{1}{s}, \quad L(t) = \frac{1}{s^2}, \quad L(t^2) = \frac{2}{s^3}, \quad L(t^3) = \frac{6}{s^4}$$

Solved Examples

***Example* 24.1:** Find the Laplace transform of

1. $2t^3 + \cos 4t + e^{-2t}$
2. $e^{2t} + 4t^3 - \sin 2t \cos 3t$
3. $\cos(wt + \beta)$
4. $3t^2 + e^{-t} + \sin^3 2t$
5. $\cos^3 2t$
6. $e^{-bt} \cos at$
7. $t^3 e^{-4t}$
8. $e^{2t} \sin 3t$
9. $e^{-3t} \sin 3t$
10. $e^{-2t}[2 \cos 3t - 3 \sin 3t]$
11. If $L[f(t)] = \dfrac{3s}{s^2+9}$. find $L[f(3t)]$

***Solution*:**

1. $L(2t^3 + \cos 4t + e^{-2t}) = 2 \times \dfrac{6}{s^4} + \dfrac{s}{s^2+16} + \dfrac{1}{s+2} = \dfrac{12}{s^4} + \dfrac{s}{s^2+16} + \dfrac{1}{s+2}$

2. $L(e^{2t}) + 4L(t^3) - L(\sin 2t \cos 3t)$

$$\frac{1}{s-2} + 4\frac{6}{s^4} - \left(\frac{5}{s^2+25} - \frac{1}{n^2+1}\right)$$

$$\left[\because \sin 2t \cos 3t = \frac{1}{2}(\sin 5t - \sin t)\right]$$

3. $L \cos(wt + \beta) = L[\cos wt.\cos\beta - \sin wt \sin\beta]$

$$= \cos\beta\, L(\cos wt) - \sin\beta\, L(\sin wt) = \cos\beta \frac{s}{s^2+w^2} - \sin\beta \frac{w}{s^2+w^2}$$

4. $L[3t^2 + e^{-t} + \sin^3 2t] = 3L(t^2) + L(e^{-t}) + L(\sin^3 2t)$

$$= 3.\frac{2}{s^3} + \frac{1}{s+1} + \frac{3}{2(s^2+4)} - \frac{3}{2}\frac{1}{s^2+36}$$

$$= \frac{2}{s^3} + \frac{1}{s+1} + \frac{3}{2}\frac{1}{s^2+4} - \frac{3}{2}\frac{1}{s^2+36}$$

$$\left(\because L(\sin^3 2t) = \frac{1}{4} L[3\sin 2t - \sin 6t]\right]$$

5. $L(\cos^3 2t) = L\left[\dfrac{\cos 6t + 3\cos 2t}{4}\right]$

$$= \frac{1}{4}\left(\frac{s}{s^2+36} + \frac{3s}{s^2+4}\right)$$

6. $L(\cos at) = \dfrac{s}{s^2 + a^2}$

$\therefore$ $L(e^{-bt}\cos at) = \dfrac{s+b}{(s+b)^2 + a^2}$ (from 24.1.5(1) first shifting theorem)

7. $L(t^3) = \dfrac{6}{s^4}$; By shifting theorem 24.1.5(1) $L(e^{-4t}.t^3) = \dfrac{6}{(s+4)^4}$

8. $L(\sin 3t) = \dfrac{3}{s^2+9}$

by 24.1.5(1), $L(e^{2t}\sin 3t) = \dfrac{3}{(s-2)^2+9}$

9. $e^{-3t}.\sin^3 t$

$L(\sin^3 t) = L\left[\dfrac{1}{4}(3\sin t - \sin 3t)\right] = \dfrac{3}{4}L(\sin t) - \dfrac{1}{4}L(\sin 3t) = \dfrac{3}{4}\dfrac{1}{s^2+1} - \dfrac{1}{4}\dfrac{3}{s^2+9}$

By 24.1.5(1) $L(e^{-3t}\sin^3 t) = \dfrac{3}{4}.\dfrac{1}{(s+3)^2+1} - \dfrac{3}{4}.\dfrac{1}{(s+3)^2+9}$

10. $e^{-2t}(2\cos 3t - 3\sin 3t)$

$L(2\cos 3t) = 2.\dfrac{s}{s^2+9}$; $L(3\sin 3t) = \dfrac{9}{s^2+9}$

By shifting theorem 24.1.5(1) $2L(e^{-2t}\cos 3t) - 3L(e^{-2t}\sin 3t)$

$= \dfrac{2(s+2)}{(s+2)^2+9} - \dfrac{9}{(s+2)^2+9}$

11. $L[f(t)] = \dfrac{3s}{s^2+9}$; By 24.1.5(3), $L[f(3t)] = \dfrac{1}{3}\dfrac{3.\frac{s}{3}}{\left(\frac{s}{3}\right)^2+9} = \dfrac{3s}{s^2+27}$

Exercise 24.1

Find the Laplace transform of the following:

1. $4t^2 + \sin 3t + e^{-2t}$ **Ans.** $\dfrac{8}{s^3} + \dfrac{3}{s^2+9} + \dfrac{1}{s+2}$

2. $(\sin 2t - \cos 2t)^2$ **Ans.** $\dfrac{1}{s} - \dfrac{4}{s^2+16}$

3. $(\cosh^2 4t)$ **Ans.** $\frac{1}{2}\left(\frac{1}{s}+\frac{s}{s^2-64}\right)$

4. $t^2 - e^{-2t} + \cosh^2 4t$ **Ans.** $\frac{2}{s^3}-\frac{1}{s+2}+\frac{1}{2}\left(\frac{1}{s}+\frac{s}{s^2-64}\right)$

5. $t^4 + 2at^2 + a^2$ **Ans.** $\frac{a^2s^4+4as^2+24}{s^5}$

6. $\cos^2 (2bt)$ **Ans.** $\frac{1}{2}\left(\frac{1}{s}+\frac{s}{s^2+16b^2}\right)$

7. $\sinh^3 3t$ **Ans.** $\frac{162}{(s^2-81)(s^2-9)}$

8. $\sinh at - \sin bt$ **Ans.** $\frac{a}{s^2-a^2}-\frac{b}{s^2+b^2}$

9. $\cosh^3 2t$ **Ans.** $\frac{s(s^2-28)}{(s^2-36)(s^2-4)}$

10. $\cosh at - \cos bt$ **Ans.** $\frac{s}{s^2-a^2}-\frac{s}{s^2+b^2}$

11. $3t^4 + 4e^{-3t}\cos t,$ **Ans.** $\frac{72}{s^5}+\frac{4s+2}{s^2+6s+10}$

12. $t + t^2 + 2t^3$ **Ans.** $\frac{1}{s^2}+\frac{2}{s^3}+\frac{12}{s^4}$

13. $e^{-2t}(3\cos 4t - 2\sin 5t)$ **Ans.** $\frac{s(s+2)}{(s+2)^2+16}-\frac{10}{(s+2)^2+25}$

14. Find Laplace Transform f(t) where

$f(t) = \sin t, \quad 0 < t < \pi$

$= 0, \quad t > \pi$ **Ans.** $\frac{1+e^{-\pi s}}{s^2+1}$

15. Find Laplace Transform of f(t) where

$f(t) = \frac{t}{a}, \quad 0 < t < a$

$= 1, \quad t > a$ **Ans.** $\frac{1-e^{as}}{a s^2}$

16. $4c^{3t}\sin(2t+\alpha)$ **Ans.** $4\cos\alpha\frac{2}{(s-3)^2+16}+\sin\frac{s-3}{(s-3)^2+16}$

17. $e^{-4t}\sin 3t$ **Ans.** $\frac{3}{4}\frac{1}{(s+4)^2+1}-\frac{1}{4}\frac{3}{(s+4)^2+9}$

18. cosh2t.cos2t **Ans.** $\frac{1}{2}\left[\frac{s-2}{(s-2)^2+4}+\frac{s+2}{(s+2)^2+4}\right]$

19. coshat sinat **Ans.** $\frac{a(s^2+2a^2)}{s^4+2a^4}$

20. cos3t.cosh4t **Ans.** $\frac{s(s^2-7)}{s^4+625-14s^2}$

21. $e^{-2t}(2\cos 3t-3\sin 3t)$ **Ans.** $\frac{2s-5}{s^2+4s+13}$

22. coshat cosat **Ans.** $\frac{s^3}{s^4+4a^4}$

23. $e^t\sin 4t.\cos 2t$ **Ans.** $\frac{3}{(s-1)^2+36}$

24. $t^2-3t+5+e^{2t}.t^2$ **Ans.** $\frac{5s^2-3s+2}{s^3}$

25. $t^n e^{-at}+\sin^3 t$ **Ans.** $\frac{n+1}{(s+a)^{n+1}}+\frac{6}{(s^2+1)(s^2+a)}$

24.1.6 Multiplication by 't'

If $L[f(t)] = F(s)$ then $L[tf(t)] = -F^1(s)$

Proof: We know that $F(s) = \int_0^\infty e^{-st} f(t)\, dt.$

differentiating with respect to s

$$F'(s) = \int_0^\infty \frac{\partial}{\partial s}[e^{-st} f(t)]dt = \int_0^\infty e^{-st}(-t)f(t)dt$$

$$\therefore\ -F'(s) = \int_0^\infty e^{-st}[tf(t)]dt = L[tf(t)]$$

$\therefore$ L[t f(t)] = $-\dfrac{d\,F(s)}{ds}$

In general, multiplication by t^n

$$L[t^n f(t)] = (-1)^n \frac{d^n F(s)}{ds^n}, \qquad u = 1, 2, 3...$$

Examples and Solutions

1. $L(e^{-2t}) = \dfrac{1}{s+2}, L(te^{-2t}) = -\dfrac{d}{ds}\dfrac{1}{s+2} = \dfrac{1}{(s+2)^2}$

2. $L(t\cos at) = (-1)\dfrac{d}{ds}[L(\cos at)] = -\dfrac{d}{ds}\left(\dfrac{s}{s^2+a^2}\right) = \dfrac{s^2-a^2}{(s^2+a^2)^2}$

3. $L(t\sin t) = -\dfrac{d}{ds}\left(\dfrac{1}{s^2+1}\right) = \dfrac{2s}{(s^2+1)^2}$

4. L(atsinat – cosat) = aL(tsinat) – L(cosat) = $a\left[-\dfrac{d}{ds}\left(\dfrac{a}{s^2+a^2}\right)\right] - \dfrac{s}{s^2+a^2}$

$$= a\left[\frac{2as}{(s^2+a^2)^2}\right] - \frac{s}{(s^2+a^2)} = \frac{2a^2s}{(s^2+a^2)^2} - \frac{s}{(s^2+a^2)}$$

5. $L(1 + t\,e^t)^3 = L[1 + 3t\,e^t + 3t^2 e^{2t} + t^3 e^{3t}]$

$= L(1) + 3L(te^t) + 3L(t^2e^{2t}) + L(t^3e^{3t})$

$$= \frac{1}{s} - 3\frac{d}{ds}\frac{1}{(s-1)} + 3\frac{d^2}{ds^2}\frac{1}{(s-2)} - \frac{d^3}{ds^3}\left(\frac{1}{s-3}\right) = \frac{1}{s} + \frac{3}{(s-1)^2} + \frac{6}{(s-2)^3} + \frac{6}{(s-3)^4}$$

6. $L(t^2\cos t) = \dfrac{d^2}{ds^2}[L(\cos t)] = \dfrac{d^2}{ds^2}\left(\dfrac{s}{s^2+1}\right) = \dfrac{d}{ds}\left[\dfrac{1-s^2}{(s^2+1)^2}\right]$

$$= \left[\frac{-(s^2+1)^2.2s - 2(s^2+1).2s(1-s^2)}{(s^2+1)^4}\right] = \frac{-s(6+4s^2-2s^4)}{(s^2+1)^4}$$

7. $L(te^{4t}\sin 2t) = \dfrac{-d}{ds}L[e^{4t}\sin 2t] = -\dfrac{d}{ds}\left[\dfrac{2}{(s-4)^2+4}\right]$ by shifting property

$$= -\frac{d}{ds}\left[\frac{2}{s^2-8s+20}\right] = \frac{4s-16}{(s^2-8s+20)^2}$$

8. $L(t^2 e^t \cos 4t) = \frac{d^2}{ds^2} L(e^t \cos 4t) = \frac{d^2}{ds^2}\left[\frac{s-1}{(s-1)^2+16}\right]$ by shifting theorem

$$= -\frac{d}{ds}\left[\frac{s^2-2s-15}{(s^2-2s+17)^2}\right] = -\frac{(2s^3-6s^2-90s+94)}{(s^2-2s+17)^3}$$

9. Find $L[t \sin^2 t]$

We have $L[\sin^2 t] = L\,\frac{[1-\cos 2t]}{2} = \frac{1}{2}[L(1) - L(\cos 2t)] = \frac{1}{2}\left[\frac{1}{s} - \frac{s}{s^2+4}\right]$

$$\therefore L\left[t \sin^2 t\right] = (-1)^1 \frac{d\frac{1}{2}\left(\frac{1}{s} - \frac{s}{s^2+4}\right)}{ds} = -\frac{1}{2}\left[-\frac{1}{s^2} - \frac{(s+4)-2s^2}{(s^2+4)^2}\right]$$

$$L\left[t \sin^2 t\right] = \frac{6s^2+8}{s^2(s^2+4)^2}$$

10. Find $L\left[t^2 \sin 2t\right]$

We have $L[\sin 2t] = \frac{2}{s^2+4}$

$$\therefore L\left[t^2 \sin 2t\right] = (-1)^2 \frac{d^2\left(\frac{2}{s^2+4}\right)}{as^2} = \frac{d\left[\frac{-2.2s}{(s^2+4)^2}\right]}{ds}$$

$$= \frac{-4\left[(s^2+4)^2 - s2(s^2+4)2s\right]}{(s^2+4)^4} = \frac{-4(s^2+4-4s^2)}{(s^2+4)^3}$$

$$\therefore L[t \sin 2t] = \frac{4(3s^2-4)}{(s^2+4)^3}$$

Exercise 24.2

Find the Laplace transform of the following:

1. $te^{-3t}\sin 2t$ **Ans.** $\frac{1}{2}\left[\frac{1}{(s-3)^2} - \frac{(s^2-s+5)}{(s^2-6s+13)^2}\right]$

2. $te^{-3t}\sin 3t$ **Ans.** $\frac{6s-18}{(s^2-6s+18)^2}$

3. $t\, e^{-2t}\sin t$ **Ans.** $\frac{2s+4}{(s^2+4s+5)^2}$

4. 2tsin3t – 3tcos3t **Ans.** $\dfrac{9+12s-3s^2}{(s^2+9)^2}$

5. te^t sin2t cost **Ans.** $\dfrac{3s-3}{(s^2-2s+10)^2}$

24.1.7 Division By 't'

If $L[f(t)] = F(s)$, then $L\left[\frac{1}{t}f(t)\right] = \int_s^{\infty} F(s)\,ds$

Proof: We have $F(s) = \int_0^{\infty} e^{-st} f(t)\,dt$

Integrating bothsides w.r.t 's' between the limits s, ∞ and changing the order of integration on right hand side (R.H.S),

$$\int_s^{\infty} F(s)\,ds = \int_0^{\infty}\left[\int_s^{\infty} e^{-st} f(t)ds\right]dt = \int_0^{\infty} -\left[\frac{e^{-st}}{t} f(t)\right]_s^{\infty} dt = \int_0^{\infty} e^{-st}\frac{f(t)}{t}dt$$

$$= \int_0^{\infty} e^{-st}\frac{f(t)}{t}dt = L\left[\frac{1}{t}f(t)\right]$$

$$\therefore L[t f(t)] = \int_s^{\infty} F(s)ds$$

Solved Examples

***Example* 24.2:** Find the Laplace Transforms of the following

1. $\frac{1}{t}(1-\cos at)$ 2. $\frac{1}{t}(e^{-t}\sin at)$ 3. $\frac{1}{t}(1-e^{2t})$

4. $\frac{1}{t}(e^{-3t}\sin 2t)$ 5. $\frac{1}{t}(e^{at}-e^{bt})$ 6. $\dfrac{(1-\cos t)}{t^2}$

***Solution*:**

1. $L(1 - \cos at) = L(1) - L(\cos at) = \dfrac{1}{s} - \dfrac{s}{s^2+a^2}$

$$L\left(\frac{1}{t}(1-\cos at)\right) = \int_s^{\infty}\left[\frac{1}{s} - \frac{s}{s^2+a^2}\right]ds = \left[\log s - \frac{1}{2}\log(s^2+a^2)\right]_0^{\infty} = \frac{1}{2}\log\left(\frac{s^2+a^2}{s^2}\right)$$

2. $L(e^{-t}\sin at) = \dfrac{a}{(s+1)^2 + a^2}$

$$L\left[\frac{1}{t}(e^{-t}\sin at)\right] = \int_s^\infty \frac{a}{(s+1)^2 + a^2}\,ds = \left[\tan^{-1}\frac{(s+1)}{a}\right]_s^\infty = \frac{\pi}{2} - \tan^{-1}\frac{s+1}{a} = \cot^{-1}\left(\frac{s+1}{a}\right)$$

3. $L(1 - e^{2t}) = L(1) - L(e^{2t}) = \dfrac{1}{s} - \dfrac{1}{s-2}$

$$L\left[\frac{1}{t}(1 - e^{2t})\right] = \int_s^\infty \left(\frac{1}{s} - \frac{1}{s-2}\right) ds = [\log s - \log(s-2)]_s^\infty = \log\frac{(s-2)}{s}$$

4. $L(e^{-3t}\sin 2t) = \dfrac{2}{(s+3)^2 + 4}$

$$L\left(\frac{1}{t}e^{-3t}\sin 2t\right) = \int_s^\infty \frac{2}{(s+3)^2 + 4}\,ds = \tan^{-1}\left(\frac{s+3}{2}\right)_s^\infty$$

$$= \frac{\pi}{2} - \tan^{-1}\left(\frac{s+3}{2}\right) = \cot^{-1}\left(\frac{s+3}{2}\right)$$

5. $L(e^{at}) = \dfrac{1}{s-a}$, $\qquad L(e^{bt}) = \dfrac{1}{s-b}$

$$L\left[\frac{1}{t}(e^{at} - e^{bt})\right] = \int_s^\infty \left(\frac{1}{s-a} - \frac{1}{s-b}\right) ds = [\log(s-a) - \log(s-b)]_s^\infty$$

$$= \left[\log\left(\frac{s-a}{s-b}\right)\right]_s^\infty = -\log\left(\frac{s-a}{s-b}\right) = \log\left(\frac{s-b}{s-a}\right)$$

6. We have $L(1 - \cos t] = \dfrac{1}{s} - \dfrac{s}{s^2+1}$

$$\therefore\ L\left[\frac{1 - \cos t}{t}\right] = \int_s^\infty \left(\frac{1}{s} - \frac{s}{s^2+1}\right) ds = \log s - \frac{1}{2}\log\left(s^2+1\right)\Big]_s^\infty$$

$$= \left[\log\frac{s}{\sqrt{s^2+1}}\right]_s^\infty = \log\left[\frac{\sqrt{s^2+1}}{s}\right]$$

$$\text{Now } L\left[\frac{1-\cos t}{t^2}\right] = \int_s^\infty \log \frac{\sqrt{s^2+1}}{s} ds$$

$$= \frac{1}{2}\int_s^\infty \log\left[1+\frac{1}{s^2}\right] ds$$

$$= \frac{1}{2}\left\{\left[s \log\left(1+s^2\right)\right]_s^\infty - \int_s^\infty s.\frac{1}{1+\frac{1}{s^2}}.\left(\frac{-2}{s^3}\right) ds\right\}$$

$$= \frac{1}{2}\left\{\left[0 - s \log\left(1+s^2\right)\right] + 2\int_s^\infty \frac{ds}{s^2+1}\right\}$$

$$= \frac{1}{2}\left\{\left[-s \log\left(1+s^2\right)\right] + \left[2 \tan^{-1}(s)\right]_s^\infty\right\}$$

$$= \frac{-s}{2}\log\left(1+s^2\right) + \left(\frac{\pi}{2} - \tan^{-1} s\right)$$

$$\therefore L\left[\frac{1-\cos t}{t^2}\right] = \frac{-s^2}{2}\log\left(1+s^2\right) + \cot^{-1}(s)$$

Exercise 24.3

Find the Laplace transform of the following:

1. $\frac{1}{t}(1-\cos t)$ **Ans.** $\frac{1}{2}\log\left(\frac{s^2+1}{s}\right)$

2. $\frac{1}{t}e^{-t}\sin t$ **Ans.** $\cot^{-1}(s+1)$

3. $\frac{1}{t}(e^{-at} - e^{-bt})$ **Ans.** $\log\left(\frac{s+b}{s+a}\right)$

4. $\frac{1}{t}(e^{-3t}\sin 2t)$ **Ans.** $\cot^{-1}\left(\frac{s-3}{2}\right)$

5. $\frac{1}{t}(\sin 2t)$ **Ans.** $\frac{1}{4}\log\left(\frac{s^2+4}{s^2}\right)$

6. $\frac{1}{t}$ (co at – cosbt) **Ans.** $\frac{1}{2}\log\left(\frac{s^2+b^2}{s^2+a^2}\right)$

7. $\frac{1}{t}$ (1 – cos3t) **Ans.** $\frac{1}{2}\log\left(\frac{s^2+9}{s^2}\right)$

8. $\frac{1}{t}$ (sin3t) **Ans.** $\frac{1}{4}\left[3\cot^{-1}s-\cot^{-1}\frac{s}{3}\right]$

9. $t\,e^{-t}\cos t+\frac{\sin 2t}{t}$ **Ans.** $\frac{s^2+2s}{\left(s^2+2s+2\right)^2}+\cot^{-1}\left(\frac{s}{2}\right)$

10. $\frac{1-e^{2t}}{t}$ **Ans.** $\log\left(\frac{s-2}{s}\right)$

11. $\frac{\sinh t}{t}$ **Ans.** $-\frac{1}{2}\log\left(\frac{s-1}{s+1}\right)$

24.1.8 Laplace Transform of Derivatives

Theorem: If f(t) is continous ∀ $t \geq 0$ and of exponential order, say σ and has a derivative $f'(t)$ which is piecewise continuous on every finite interval [0, N] for each $N > 0$, then the Lapalce transform of the derivative $f'(t)$ exists for $s > \sigma$ and

$$L\{f'(t)\} = sLf(t) - f(0)$$

Proof:

$$L[f(t)] = \int_0^\infty e^{-st} f(t)dt$$

$$L[f'(t)] = \int_0^\infty e^{-st} f'(t)dt$$

Integration by parts gives $L[f'(t)] = \left[e^{-st} f(t)\right]_0^\infty + s\int_0^\infty e^{-st} f(t)dt$

Since f(t) is of an exponential order, the integrand in first integral on the R.H.S. is zero at the upper limit when $s > \sigma$ and if f(0) at the lower limit. Thus, we have

$$L[f'(t)] = 0 - f(0) + s\int_0^\infty e^{-st} f(t)dt = sLf(t) - f(0)$$

Similarly $L\left[f''(t)\right] = s^2 L\left[f(t)\right] - sf(0) - f'(0)$

$$L\left[f'''(t)\right] = s^3 L\left[f(t)\right] - s^2 f(0) - sf'(0) - f''(0)$$

$\underset{t\to 0}{Lt}\, f(t) = f(0+)$ exists, but is not equal to f(0), which may or may not exist then

$$L[f'(t)] = sLf(t) - f(0+).$$

Note: If f(t) and its derivatives $f'(t)$, $f''(t)$ .... $f^{n-1}(t)$ are continous $\forall\ t \geq 0$ and are functions of exponential order for some M > 0 and σ, and the derivative $f^n(t)$ is Piecewise continous on every finite interval in the range $t \geq 0$, then the Laplace transform of $f^n(t)$ exists when s > σ and is given by

$$L[f^n(t)] = s^n L[f(t)] - s^{n-1}f(0) - s^{n-2} f'(0) - s^{n-3} f''(0) - \ldots\ldots - f^{n-1}(0)]$$

Ex : Prove that (i) $L(\cos at) = \dfrac{s}{s^2+a^2}$ (ii) $e^{at} = \dfrac{1}{s-a}$ (iii) $t \sin at = \dfrac{2as}{s^2+a^2}$

Sol. (i) **We can show this with the help of above theorem**

Let $f(t) = \cos at \Rightarrow f(0) = 1$

then $f'(t) = -a \sin at \Rightarrow f'(0) = 0$

$f''(t) = -a^2 \cos at$

Hence, we have

$$L[f''(t)] = s^2Lf(t) - sf(0) - f'(0)$$

$$\therefore L[-a^2\cos at] = s^2L(\cos at) - s - 0 = s^2L(\cos at) - s$$

$$s^2L(\cos at) + a^2L(\cos at) = s$$

$\therefore L(\cos at) = \dfrac{s}{s^2+a^2}$ also we have same value from definition.

(ii) $f(t) = e^{at},\ f'(t) = a\,e^{at},\ f(0) = 1$

We have

$$L[f'(t)] = s\,L[f(t)] - f(0)$$

$$L[a\,e^{at}] = s\,L[e^{at}] - 1$$

$$a\,L[e^{at}] - s\,L[e^{at}] = -1$$

$$L[e^{at}]\{a - s\} = -1$$

$$\therefore L[e^{at}] = \frac{1}{s-a}$$

We have same value from definition

(iii) $f(t) = t \sin at, \; f'(t) = \sin at \cos at$

$$f''(t) = a \cos at + [\cos at - at \sin at] = 2a \cos at - a^2 t \sin at$$

$$f(0) = 0, \; f'(0) = 0$$

$$L[f''(t)] = s^2 L[f(t)] - s f(0) - f'(0)$$

$$L[2a \cos at - a^2 t \sin at] = s^2 L[t \sin at] - 0 - 0$$

$$2a\, L[\cos at] - a^2 L[t \sin at] - s^2 L[t \sin at] = 0$$

$$-(a^2 + s^2) L[t \sin at] = -2a\, L[\cos at]$$

$$L[t \sin at] = 2a.\frac{s}{s^2 + a^2}$$

$$\therefore \; L[t \sin at] = \frac{2as}{s^2 + a^2}$$

24.1.9 Laplace Transform of Integrals

Theorem: If f is Piecewise continous on [0, N] for each N > 0 and is of exponential order σ, then $L \int_0^t f(u)du = \frac{1}{s} L[f(t)] = \frac{1}{s} F(s)$ where L[f(t)] = F(s)

Proof: Let $\phi(t) = \int_0^t f(u)du$. Then we have $\phi'(t) = f(t)$ and f(0) = 0;

$$L[\varphi'(t)] = s.L[\varphi(t)] - \varphi(0) = s[L(\varphi(t))] = sL\left[\int_0^t f(u)du\right]$$

$$\therefore L\left[\int_0^t f(u)du = \frac{1}{s} L[\varphi'(t)] = \frac{1}{s}[L(f(t))]\right] = \frac{1}{s} F(s)$$

Similarly if L[f(t)] = F(s) then $L\left[\int_0^t \int_0^t f(u)\, du\, du\right] = \frac{1}{s^2} F(s)$

In general, $L\left[\int_0^t \int_0^t \ldots\ldots \int_0^t f(u)\, du\, du \ldots\ldots du \; n \; times\right] = \frac{1}{s^n} F(s)$

Solved Examples

***Example* 24.3:** Find the Laplace transforms of

1. $\int_0^t \sin 2p\,dp$ 2. $\int_0^t p\cosh p\,dp$ 3. $\int_0^t \frac{\sin p}{p}\,dp$

4. $\int_0^t e^p.\frac{\sin p}{p}\,dp$ 5. $\int_0^t e^{-p}\cos p\,dp$ 6. $\int_0^t \frac{e^p \sin p}{p}\,dp$

7. $\int_0^t \frac{e^p \sin p}{p}\,dp$

***Solution*:**

1. $L(\sin 2t) = \frac{2}{s^2+4} = f(s)$, say

$$\int_0^t \sin 2p\,dp = \frac{1}{s} f(s) = \frac{2}{s(s^2+4)}$$

2. $L(\cosh t) = \frac{s}{s^2-a^2}$ $L(t\cosh t) = \frac{(s^2+a^2)}{(s^2-a^2)^2} = F(s)$

then $L\int_0^t p\cosh p\,dp = \frac{1}{s} f(s) = -\frac{(s^2+a^2)}{(s^2-a^2)^2 s}$

3. $L(\sin t) = \frac{1}{s^2+1}$; $L\left(\frac{1}{t}\sin t\right) = \int_s^\infty \frac{1}{s^2+1}\,ds$

$$= \left[\tan^{-1}\right]_s^\infty = \frac{\pi}{2} - \tan^{-1} s = \cot^{-1} s = F(s) \text{ say}$$

$$L\int_0^t \frac{\sin p}{p}\,dp = \frac{1}{s} f(s) = \frac{1}{s}\cot^{-1} s$$

4. $L(\sin t) = \frac{1}{s^2+1}$ $\therefore L(e^t \sin t) = \frac{1}{(s-1)^2+1}$

$$L\left[\frac{1}{t}(e^t \sin t)\right] = \int_s^\infty \frac{ds}{(s-1)^2+1}$$

$$= \left[\tan^{-1}(s-1)\right]_s^\infty = \frac{\pi}{2} - \tan^{-1}(s-1) = \cot^{-1}(s-1) = F(s) \text{ say}$$

$$\therefore L\int_0^t e^p \frac{\sin p}{p} dp = \frac{1}{s} f(s) = \frac{1}{s}\cot^{-1}(s-1)$$

5. Consider $f(t) = e^{-t} \cos t$

We have $L\left[e^{-t}\cos t\right] = \dfrac{s+1}{(s+1)^2+1} = \dfrac{s+1}{s^2+2s+2} = F(s)$

Now $L\left[\int_0^t e^{-p} \cos p\, dp\right] = \dfrac{1}{s} F(s)$

$$\therefore \quad L\left[\int_0^t e^{-p} \cos p\, dp\right] = \frac{1}{s}\frac{s+1}{s^2+2s+2}$$

$$= \frac{s+1}{s\left(s^2+2s+2\right)}$$

6. Let $f(t) = \cos h\ at$

$$L[\cos h\ at] = \frac{s}{s^2-a^2} = F(s)$$

$$L\left[\int_0^t \cos h\ ap\, dp\right] = \frac{1}{s}\frac{s}{s^2-a^2} = \frac{1}{s^2-a^2}$$

Again applying Laplace transform of integral

$$L\left[\int_0^t\int_0^t \cos h\ ap\, dp\, dp\right] = \frac{1}{s}\frac{s}{s^2-a^2}$$

$$\therefore \quad L\left[\int_0^t\int_0^t \cos h\ ap\, dp\, dp\right] = \frac{1}{s\left(s^2-a^2\right)}$$

7. $L(\sin t) = \dfrac{1}{1+s^2}$ $\qquad \therefore L\left[\dfrac{\sin t}{t}\right] = \displaystyle\int_s^\infty \frac{ds}{s^2+1}$

$$L\left[\frac{\sin t}{t}\right] = \left[\tan^{-1}(s)\right]_s^\infty = \tan^{-1}\infty - \tan^{-1}s$$

$$L\left[\frac{\sin t}{t}\right] = \frac{\pi}{2} - \tan^{-1}s = \cot^{-1}s$$

Using first shifting theorem

$$L\left[\frac{e^t \sin t}{t}\right] = \cos^{-1}(s-1)$$

Now using Laplace transform of integral

$$L\left[\int_0^t \frac{e^p \sin p}{p} dp\right] = \frac{1}{s} \cdot \cot^{-1}(s-1)$$

Exercise 24.4

Find the Laplace Transform of

1. $\int_0^t \cos 2p\,dp$ 2. $\int_0^t pe^{-3}\sin 4p\,dp$ 3. $\int_0^t \frac{1}{p}\sin 3p\,dp$

4. $\int_0^t \frac{(1-e^{-p})}{p}dp$ 5. $\int_0^t\int_0^t \cos ap\,dp\,dp$ 6. $\int_0^t p\,e^{-p}\sin 4p\,dp$

Ans. 1. $\frac{s}{s(s^2+4)}$ 2. $\frac{8(s+3)}{s(s^2+6s+25)^2}$ 3. $\frac{1}{s}\cot^{-1}s$

4. $\frac{1}{s}\log\left(1+\frac{1}{s}\right)$ 5. $\frac{1}{s(s^2+a^2)}$ 6. $\frac{8(s+1)}{s(s^2+2s+17)^2}$

24.1.10 Laplace Transforms of Periodic Functions

A function f(t) is said to be periodic with period p> 0 if f(t + p) = f(t) for all t > 0

Since periodic functions appear in many practical problems and in most cases are more complicated than single sine or cosine functions, we find it useful to establish the following result:

$$L([f(t)] = \frac{1}{1-e^{-ps}}\int_0^p e^{-st} f(t)dt$$

Proof: Let f(t) be a periodic function of period 'p' so that f(t) = f(t + p) = f(t + 2p) = ...

$$L[f(t)] = \int_0^\infty e^{-st} f(t)dt = \int_0^p e^{-st} f(t)dt + \int_p^{2p} e^{-st} f(t)dt + \ldots\ldots\ldots\ldots$$

writing t = u + p

$$\text{now } \int_p^{2p} e^{-st} f(t)dt = \int_0^p e^{-s(u+p)} f(u+p)du$$

$$= e^{-ps}\int_0^p e^{-st} f(t+p)dt \qquad \text{(change u to t)}$$

$$= e^{-ps}\int_0^p e^{-st} f(t)dt \qquad [\because f(t+p) = f(t)]$$

similarly, $\int_{2p}^{3p} e^{-st} f(t)dt = e^{-2ps}\int_0^p e^{-st} f(t)dt$ and so on

$$L[f(t)] = (1 + e^{-ps} + e^{-2ps} + \ldots\infty) \int_0^p e^{-st} f(t)\,dt$$

The expression within parantheses is a G.P. with c.r. = e^{-ps}

$$\therefore L[f(t)] = \frac{1}{1-e^{-ps}}\int_0^p e^{-st} f(t)dt$$

Solved Examples

***Example* 24.4:** Find the Laplace Transform of

(a) $f(t) = k\,\dfrac{t}{p}$ for $0 < t < p$ and $f(t) = f(t+p)$

(b) $f(t) = 1$ for $0 < t < a$ and $f(t) = -1$, $a < t < 2a$ and f(t) is periodic with period 2a

***Solution*:**

(a) Since f(t) is periodic with period p.

$$L[f(t)] = \frac{1}{1-e^{-sp}}\int_0^p e^{-st} f(t)dt = \frac{1}{1-e^{-sp}}\int_0^p e^{-st}.\frac{kt}{p}dt$$

$$= \frac{1}{1-e^{-sp}}\frac{k}{p}\int_0^p e^{-st}t.dt = \frac{1}{1-e^{-sp}}\frac{k}{p}\left[-t\frac{e^{-st}}{s} - \frac{e^{-st}}{s^2}\right]_0^p$$

$$= \frac{1}{1-e^{-sp}}\frac{k}{p}\left[-p\frac{e^{-sp}}{s} - \frac{e^{-sp}}{s^2} + \frac{1}{s^2}\right] = \frac{1}{1-e^{-sp}}\frac{k}{p}\left[\frac{1}{s^2}(1-e^{-sp}) - \frac{pe^{-sp}}{s}\right]$$

(b) f(t) is peirodic with peirod 2a

$$L[f(t)] = \frac{1}{1-e^{-2as}}\int_0^{2a} e^{-st} f(t)dt = \frac{1}{1-e^{-2as}}\left[\int_0^a e^{-st}.1.dt + \int_0^{2a} e^{-st}(-1)dt\right]$$

$$= \frac{1}{1-e^{-2as}}\left[\left\{\frac{-e^{-st}}{s}\right\}_0^a - \left\{\frac{-e^{-st}}{s}\right\}_0^{2a}\right] = \frac{1}{s}\frac{1}{1-e^{-2as}}.(1-e^{-as})^2$$

$$= \frac{1}{s}.\left[\frac{1-e^{-2as}}{1+e^{-2as}}\right] = \frac{1}{s}.\left[\frac{e^{\frac{as}{2}} - e^{-\frac{as}{2}}}{e^{\frac{as}{2}} + e^{-\frac{as}{2}}}\right] = \frac{1}{s}\tanh\left[\frac{as}{2}\right]$$

***Example* 24.5:** Find the Laplace transform of the sqaure wave function period 2a defined as

$$f(t) = k, \qquad 0 < t < a$$
$$= -k, \qquad a < t < 2a$$

Solution: Given f(t) is a periodic function of period 2a

$$L[f(t)] = \frac{1}{1-e^{-2as}}\int_0^{2a} e^{-st} f(t)\, dt$$

$$= \frac{1}{1-e^{-2as}}\left[\int_0^a k\, e^{-st}\, dt + \int_a^{2a} (-k) e^{-st} dt\right]$$

$$= \frac{1}{1-e^{-2as}}\left[\left(k\frac{e^{-st}}{-s}\right)_0^a + \left(k\frac{e^{-st}}{s}\right)_a^{2a}\right]$$

$$= \frac{1}{1-e^{-2as}}\left[\frac{k}{s}\left(-e^{-as}+1\right) + \frac{k}{s}\left(e^{-2as} - e^{-as}\right)\right]$$

$$= \frac{1}{1-\left(e^{-as}\right)^2}\left(\frac{k}{s}\right)\left[1 - 2e^{-as} + e^{-2as}\right]$$

$$= \frac{1}{\left(1-e^{-as}\right)\left(1+e^{-as}\right)}\left(\frac{k}{s}\right)\left(1-e^{-as}\right)^2$$

$$= \frac{k}{s}\left(\frac{1-e^{-as}}{1+e^{-as}}\right)$$

***Example* 24.6:** Find the Laplace transmform of the rectified semi-wave function defined by

$$f(t) = \begin{cases} \sin \omega t & \text{for} \quad 0 < t < \pi/\omega \\ 0, & \textit{for} \quad \dfrac{\pi}{\omega} < t < 2\dfrac{\pi}{\omega} \end{cases}$$

Solution: Given f(t) is a periodic function of period $\frac{2\pi}{\omega}$

$$\therefore\ L[f(t)] = \frac{1}{1-e^{\frac{-2\pi s}{\omega}}}\int_0^{\frac{2\pi}{\omega}} e^{-st} f(t)\,dt$$

$$= \frac{1}{1-e^{\frac{-2\pi s}{\omega}}}\left[\int_0^{\frac{\pi}{\omega}} e^{-st} f(t)\,dt + \int_{\omega}^{\frac{2\pi}{\omega}} e^{-st} f(t)\,dt\right] = \frac{1}{1-e^{\frac{-2\pi s}{\omega}}}\left[\int_0^{\frac{\pi}{\omega}} e^{-st}\sin\omega t\,dt\right]$$

$$= \frac{1}{1-e^{\frac{-2\pi s}{\omega}}}\left[\frac{e^{\frac{-\pi s}{\omega}}}{s^2+\omega^2}(0+\omega) - \frac{1}{s^2+\omega^2}(0-\omega)\right] = \frac{\omega}{1-e^{\frac{-2\pi s}{\omega}}}\left[\frac{1+e^{\frac{\pi s}{\omega}}}{s^2+\omega^2}\right]$$

$$= \frac{\omega}{\left(1-e^{\frac{-\pi s}{\omega}}\right)\left(1+e^{-\frac{\pi s}{\omega}}\right)}\left[\frac{1+e^{-\frac{\pi s}{\omega}}}{s^2+\omega^2}\right] = \frac{\omega}{\left(1-e^{\frac{-\pi s}{\omega}}\right)\left(s^2+\omega^2\right)}$$

***Example* 24.7:** Find L[|sin t|]

Solution: f(t) = |sin t| is a periodic functions of period p.

$$L|\sin t| = \frac{1}{1-e^{-\pi s}}\int_0^{\pi} e^{-st}\,|\sin t|\,dt$$

$$= \frac{1}{1-e^{-\pi s}}\int_0^{\pi} e^{-st}\sin t\,dt \qquad \because\ |\sin t| = \sin t,\ 0 < t < \pi$$

$$= \frac{1}{1-e^{-\pi s}}\left[\frac{e^{-st}}{1+s^2}(-\sin t - \cos t)_0^{\pi}\right]$$

$$= \frac{1}{1-e^{-\pi s}}\left[\frac{e^{-\pi s}}{1+s^2}(0+1) - \frac{1}{1+s^2}(0-1)\right]$$

$$= \frac{1}{1-e^{-\pi s}}\left[\frac{1+e^{-\pi s}}{1+s^2}\right]$$

Example **24.8:** Find the Laplace transform of the rectangular wave shown in the figure:

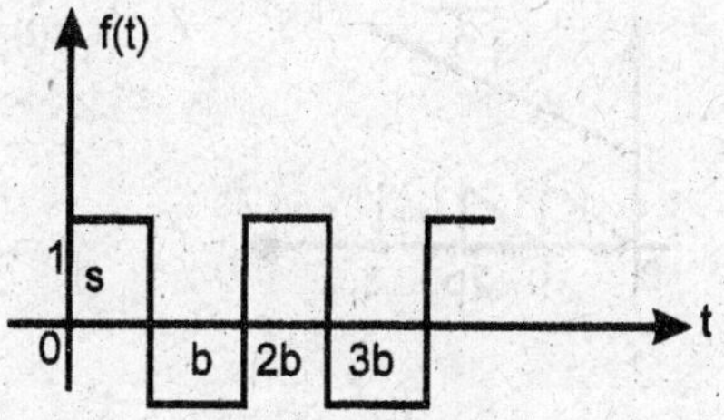

Rectangular wave

***Solution*:** The period of the given function is '2b'.

$$\text{Hence } L[f(t)] = \frac{1}{1-e^{-2bs}}\int_0^{2b} e^{-st} f(t)dt = \frac{1}{1-e^{-2bs}}\left[\int_0^b 1.e^{-st} f(t)dt + \int_b^{2b}(-1)e^{-st}dt\right]$$

$$= \frac{1}{1-e^{-2bs}}\left[\left\{\frac{-e^{-st}}{-s}\right\}_0^b - \left\{\frac{-e^{-st}}{-s}\right\}_0^{2b}\right] = \frac{1}{1-e^{-2bs}}\left(\frac{1-2e^{-bs}+e^{-2bs}}{s}\right) = \frac{(1-e^{-bs})^2}{s(1-e^{-bs})(1+e^{-bs})}$$

$$= \frac{1}{s}\left(\frac{1-e^{-bs}}{1+e^{-bs}}\right) = \frac{e^{\frac{bs}{2}} - e^{\frac{-bs}{2}}}{s\left(e^{\frac{bs}{2}} + e^{\frac{-bs}{2}}\right)} = \frac{1}{s}\tanh\frac{bs}{2}$$

Example **24.9:** What is the Laplace transform of the staircase function (Fig(a)) given by

f(t) = n + 1, np < t <(n + 1)p for n = 0, 1, 2 ....

***Solution*:** The easient way of getting the Laplace transform of the staircase function is to consider it as the difference between the two functions shown in Fig(b) the transform of the linear function $\frac{(t+p)}{p}$ can easily be found by using first shifting, second shifting and change of scale properties.

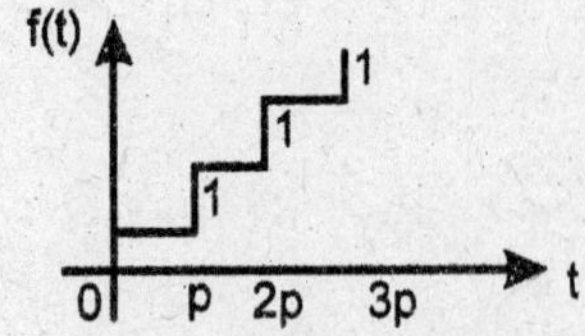

(a)

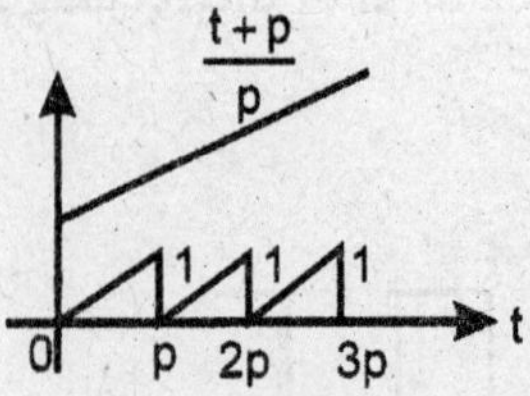

(b)

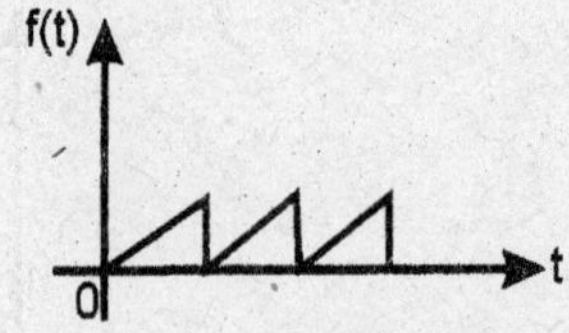

(c)

(a) Staircse function (b) solution of the stair case function (c) Saw-tooth function.

Thus, we have $$L\left(\frac{t+p}{p}\right)=\frac{1}{p}\left(\frac{1}{s^2}+\frac{p}{s}\right) \qquad .....(1)$$

The function $f_1(t) = \dfrac{t}{p}$ is called a saw-tooth function (Fig.(c)) and its transform is obtained as shown below.

$$L[f_1(t)]=\frac{1}{1-e^{-sp}}\int_0^p e^{-st}\,\frac{t}{p}\,dt=\frac{1}{1-e^{-sp}}\cdot\frac{1}{p}\left[\frac{e^{-st}}{s^2}(-st-1)\right]_0^p$$

$$=\frac{1-(1+ps)e^{-ps}}{ps^2(1-e^{-sp})}$$

This can be simplified to

$$L[f_1(t)]=\frac{1}{p}\left[\frac{1+ps}{s^2}-\frac{p}{s(1-e^{-ps})}\right] \qquad .....(2)$$

Thus from (1) and (2), we get

$$L[f(t)]=\frac{1}{p}\left(\frac{1}{s^2}+\frac{p}{s}\right)-\frac{1}{p}\left(\frac{1+ps}{s^2}-\frac{p}{s(1-e^{-ps})}\right)$$

$$=\frac{1}{s(1-e^{-ps})}$$

Exercise 24.5

1. **If Lf(t)** = $\phi(s)$, Prove that

   $L[f'''(t)] = s^3\phi(s) - s^2\,\phi(s) - s\,\phi'(s) - \phi''(s)$

2. If $L[f''(t)] = \text{Tan}^{-1}\left(\frac{1}{s}\right)$, f(0) = 2 and $f'(0) = -1$ find L[f(t)]

(**Ans.** $\frac{2s-1+\tan^{-1}\left(\frac{1}{s}\right)}{s^3}$)

3. A periodic function f(t) of period 2a is defind by

$f(t) = t, \quad \text{for } 0 \le t \le a,$

$= 2a - t, \quad \text{for } a < t < 2a$

show that $L[f(t)] = \frac{1}{s^2}\tanh\left(\frac{as}{2}\right)$

4. If $f(t) = t^2$, $0 < t < 2$ and $f(t+2) = f(t)$

find L[f(t)].

Ans. $\frac{2-2^{-2s}-4se^{-2s}-4s^2e^{-2s}}{s^3(1-e^{-2s})}$

5. If $F(s) = L(f(t)]$, Prove that

$L\left[\int_0^t dt\int_0^t f^{|}(u)du\right] = \frac{F(s)}{s^2}$ and hence generalise the result.

6. If $f(t) = \begin{cases} 3t, & 0 < t < 2 \\ 6, & 2 < t < 4 \end{cases}$ find Lf(t)

where f(t) has period 4

Ans: $\frac{3\left[1-e^{-2s}\left(1+e^{-2s}\right)\right]}{s^2\left(1-e^{-4s}\right)}$

7. Find L[f(t)] where f(t) is square wave function a period 'a'

$f(t) = 1, \quad 0 < t < \frac{a}{2}$

$= -1, \quad \frac{a}{2} < t < a.$

24.1.11 Laplace transform of the unit step function (Heavisides unit function)

The unit step function or the Heaviside function is defined as

$$H(t-a) = 0 \quad \text{if} \quad t < a,$$

$$H(t-a) = 1 \quad \text{if} \quad t > a$$

where a is a non-negative constant. In particular, when a = 0, we have

$$H(t) = \begin{cases} 0, & \text{if } t<0 \\ 1, & \text{if } t>0 \end{cases}$$

The unit step function is the basic building block of certain functions whose knowledge greately increases the usefulness of the Laplace transformation.

It is easy to show that the Laplace transform of H(t – a) is given by

$$L(H(t-a)) = \frac{e^{-as}}{s}$$

$$L[H(t-a)] = \int_0^\infty e^{-st} H(t-a)dt = \int_0^a e^{-st}\, 0.dt + \int_a^\infty e^{-st}\, 1.dt$$

$$= \left[\frac{-e^{-st}}{s}\right]_0^\infty = \frac{e^{-as}}{s}$$

In Particular, when a = 0, we get $L(H(t)) = \dfrac{1}{s}$

24.1.12 Heaviside Shift Theorem

If Lf(t) = F(s), then L[f(t–a)H(t–a)] = e^{-as} F(s).

Proof: By defnition, we have

$$L[f(t-a)H(t-a)] = \int_0^\infty e^{-st} f(t-a)H(t-a)dt = \int_0^a e^{-st} f(t-a)0.dt + \int_0^\infty e^{-st} f(t-a)1.dt$$

Using definition of unit step function

$$= \int_0^\infty e^{-s(p+a)} f(p)dp = e^{-sa} \int_0^\infty e^{-sp} f(p)(p)dp \qquad \text{(consider } p = t-a\text{)}$$

$$= e^{-as}\, F(s)$$

24.1.13 Dirac Delta Function (Unit Impulse Function)

The limiting form of unit impulse function is

$$f_\in(t-a)=\begin{cases}0, & \text{for } t<a\\ \frac{1}{\in}, & \text{for } a\le t\le a+\in\\ 0, & \text{for } t>a\end{cases}$$

or

$$f_\in(t-a)=\begin{cases}\frac{1}{\in}, & \text{for } a\le t\le a+\in\\ 0, & \text{otherwise}\end{cases}$$

as $\in>0$

From figure if $\in\to 0$, the height of the strip ABCD increases indefinitely and width decreases in such a way that its area is unity

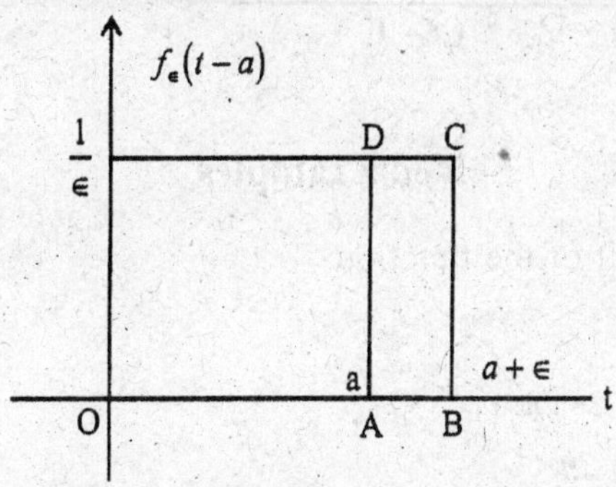

$$\therefore\ L\left[f_\in(t-a)\right]=\int_0^\infty e^{-st}f_\in(t-a)\,dt$$

$$=\int_a^{a+\in} e^{-st}f_\in(t-a)\,dt+\int_0^\infty e^{-st}f_\in(t-a)\,dt$$

$$=\int_a^{a+\in} e^{-st}.\frac{1}{\in}dt+\int_{a+\in}^\infty e^{-st}.0\,dt$$

$$=\frac{1}{\in}\left[\frac{e^{-st}}{-s}\right]_a^{a+\in}=\frac{e^{-as}\left(1-e^{-\in s}\right)}{\in s}$$

$\underset{\in\to 0}{Lt}\ f_\in(t-a)$ is denoted by $\delta(t-a)$

$\delta(t-a)$ is called the Dirac Delta function or the unit impulse function.

Laplace Transform of Dirac Delta Function

$$L[\delta(t-a)] = \underset{\epsilon\to 0}{Lt}\, L[f_\epsilon(t-a)]$$

$$= \underset{\epsilon\to 0}{Lt}\, \frac{e^{-as}\left(1-e^{-\epsilon s}\right)}{\epsilon s}(\%)$$

$$= e^{-as}.\underset{\epsilon\to 0}{Lt}\, \frac{0+e^{-\epsilon s}}{s} \qquad \text{(using L. Hospital's Rule)}$$

$$= e^{-as}.1$$

$\therefore$ $L[\delta(t-a)] = e^{-as}$

Thus the Dirac Delta function $\delta(t-a)$ is defined as

$$\delta(t-a) = \begin{cases} \infty, & \text{for } t = a \\ 0, & \text{for } t \neq a \end{cases}$$

such that $\int_0^\infty \delta(t-a)\, dt = 1, \qquad (a \geq 0)$

Solved Examples

***Example* 24.10:** Find the L.T. of the function

$$f(t) \quad \left.\begin{matrix} =1 \\ =0 \\ =\sin t \end{matrix}\right\} \begin{matrix} \text{if } 0< t<\pi \\ \text{if } \pi< t<2\pi \\ \text{if } t> 2\pi \end{matrix}$$

***Solution*:** Let us represent f(t) interms of unit step functions as

f(t) = H(t) – H(t – p) + H(t – 2p) sint

$$L[f(t)] = L[H(t)] - L[H(t-\pi)] + L[H(t-2\pi)\sin t] = \frac{1}{s} - \frac{e^{-\pi s}}{s} + \frac{e^{-2\pi s}}{s}$$

***Example* 24.11:** Find L.T. of $t^2H(t-3)$

***Solution*:** Let us express $f(t) = t^2$ as a function of (t – 3) by using Taylor's series, which states that

$$f(x) = f(a) + (x-a)\ f'(a) + \frac{(x-a)^2}{2!}\ f''(a) + \dots + \dots$$

$$\phi(t) = 9 + (t-3).6 + \frac{(t-3)^2}{2!}.2$$

Now, by the above result

$$L[f(t).H(t-a)] = L[9 + 6(t-3) + (t-3)^2H(t-3)]$$

$$= L[f(t-3).H(t-3)] = e^{-3s}Lf(t) \text{where } f(t) = 9 + 6t + t^2$$

$$L[t^2H(t-a)] = e^{-3s}\left[\frac{9}{s}+\frac{6}{s^2}+\frac{2}{s^3}+.........\right]$$

***Example* 24.12:** Express $F(t) = (t-2)^2$ when $t > 2$ and $f(t) = 0$ when $0 < t < 2$ in terms of Heaviside unit step function and find its Laplace transform

***Solution*:** It is easy to see that

$F(t) = (t-2)^2 = (t-2)^2.H(t-2)$ (since for $t < 2$, $H(t-2) = 0$ and for $t \geq 2$, $H(t-2) = 1$)

$L[f(t).H(t-2)] = L[(t-2)^2H(t-2)] = e^{-2s}.Lf(t)$ where $f(t) = t^2$

$$= e^{-2s}.\frac{2!}{s^3}$$

Exercise 24.6

1. Find the Laplace tranform of f(t) using the unit step function, where

$$f(t)=\begin{cases} if\ 1 & 0<t<1 \\ if\ 2 & 1<t<3 \\ if\ 4 & 3<t<4 \\ if\ -2 & 4<t \end{cases}$$

Ans. $\left[\frac{1}{s}+\frac{e^{-s}}{s}+\frac{2e^{-3s}}{s}-\frac{6e^{-4s}}{s}\right]$

2. Determine the Laplace transform of the following function:

(i) $f(t) = 6 - 2H(t-\pi) - 4H(t-2\pi)$ **Ans.** $\frac{6-2e^{-\pi s}-4e^{-2\pi s}}{s}$

(ii) $f(t) = e^{2t}H(t - \log 5)$ **Ans.** $\frac{5^{2-s}}{5-s}$

(iii) $f(t) = (t-a)H(t-a)$ **Ans.** $\frac{e^{-as}}{s^2}$

(iv) $f(t) = t^2H(t-2)$ **Ans.** $\frac{e^{-2s}(2+4s+4s^2)}{s^3}$

(v) $f(t) = H(t-e^t)$ **Ans.** $\frac{1}{s}$

(vi) $f(t) = tH(t-2)$ **Ans.** $e^{-2s}\left[\frac{2}{s}+\frac{1}{s^2}\right]$

(vii) $f(t) = (1+2t-3t^2+4t^3)H(t-2)$ **Ans.** $e^{-2s}\left[\frac{25}{s}+\frac{38}{s^2}+\frac{42}{s^3}+\frac{24}{s^4}\right]$

(viii) $f(t) = \sin t.H(t-\pi)$ **Ans.** $e^{-\pi s}\frac{1}{s^2+1}$

24.2 Inverse Laplace Transforms

If L[f(t)] = F(s) then f(t) is called the inverse Laplace transform of F(s) and is denoted by

$$L^{-1}[F(s)] = f(t)$$

If f_1 and f_2 are continuous functios on $[0, \infty]$ such that

$$L[f_1(t)] = L[f_2(t)] \text{ or } F_1(s) = F_2(s)\ (\forall\ s > 0) \text{ then } f_1(t) = f_2(t)\ (\forall\ t \geq 0)$$

Some of the important properties of the inverse laplace transforms are analogous to the corresponding properties of Laplace transforms.

1. ***Linearity Property:*** If k_1, k_2 are any constants and $F_1(s)$ and $F_2(s)$ are the Laplace transforms of $f_1(t)$ and $f_2(t)$ respectively, then

$$L^{-1}[k_1F_1(s) + k_2F_2(s)] = k_1f_1(t) + k_2f_2(t).$$

2. ***First shifting property:*** If $L^{-1}[F(s)] = f(t)$, then

$$L^{-1}[e^{-as}F(s)] = \begin{cases} f(t-a) & \text{for } t > a \\ 0 & \text{for } t < a \end{cases}$$

3. ***Change of scale property:*** if $L^{-1}[F(s)] = f(t)$, then

$$L^{-1}[F(as)] = \frac{1}{a}f\left(\frac{t}{a}\right)$$

4. ***Inverse Laplace transforms of derivatives:*** If $L^{-1}[F(s)] = f(t)$ then

$$L^{-1}[F^n(s)] = L^{-1}\left[\frac{d^n}{ds^n}F(s)\right] = (-1)^n t^n f(t)$$

5. ***Inverse Laplace transform of integrals:*** If $L^{-1}[F(s)] = f(t)$ then

$$L^{-1}\left\{\int_0^{\alpha} F(u)du\right\} = \frac{f(t)}{t}$$

6. ***Multiplication by 's':*** If $L^{-1}[F(s)] = f(t)$ and $f(0) = 0$, then

$L^{-}[sF(s)] = f^{1}(t)$

7. ***Division by 's':*** If $L^{-1}[F(s)] = f(t)$ then

$$L^{-1}\left(\frac{F(s)}{s}\right) = \int_0^t f(u)\,du.$$

This result can be extended to $L^{-1}\left(\frac{F(s)}{s^n}\right)$

$$= \int_0^t\int_0^t \dots\dots\dots\dots\dots \int_0^t f(u)\,du^n.$$

(the proof is left to the student)

24.2.1 Table of Inverse Transforms

Sr.No.	L[f(t)]	L^{-1}[F(s)]
1.	$L(1) = \frac{1}{s}$	$L^{-1}\left(\frac{1}{s}\right) = 1$
2.	$L(e^{-at}) = \frac{1}{s+a}$	$L^{-1}\left(\frac{1}{s+a}\right) = e^{-at}$
3.	$L(t^n) = \frac{n!}{s^{n+1}}$ if 'a' is a positive integer	$L^{-1}\left(\frac{1}{s^{n+1}}\right) = \frac{t^n}{(n+1)!}$
4.	$L(\sin at) = \frac{a}{s^2+a^2}$	$L^{-1}\left(\frac{1}{s^2+a^2}\right) = \frac{1}{a}\sin at$
5.	$L(\cos at) = \frac{s}{s^2+a^2}$	$L^{-1}\left(\frac{s}{s^2+a^2}\right) = \cos at$

Contd....

6.	$L(\sinh at) = \dfrac{a}{s^2 - a^2}$	$L^{-1}\left(\dfrac{1}{s^2 - a^2}\right) = \dfrac{1}{a}\sinh at$
7.	$L(\cosh at) = \dfrac{s}{s^2 - a^2}$	$L^{-1}\left(\dfrac{s}{s^2 - a^2}\right) = \cosh at$
8.	$L[e^{at}\, t^n] = \dfrac{n!}{(s-a)^{n+1}}$	$L^{-1}\left[\dfrac{1}{(s-a)^{n+1}}\right] = \dfrac{e^{at} t^n}{n!}$
9.	$L\left[e^{at} \sin bt\right] = \dfrac{b}{(s-a)^2 + b^2}$	$L^{-1}\left[\dfrac{1}{(s-a)^2 + b^2}\right] = \dfrac{e^{at} \sin bt}{b}$
10.	$L\left[e^{at} \cos bt\right] = \dfrac{(s-a)}{(s-a)^2 + b^2}$	$L^{-1}\left[\dfrac{s-a}{(s-a)^2 + b^2}\right] = e^{at} \cos bt$
11.	$L\left[e^{at} \sinh bt\right] = \dfrac{b}{(s-a)^2 - b^2}$	$L^{-1}\left[\dfrac{1}{(s-a)^2 - b^2}\right] = \dfrac{e^{at} \sinh bt}{b}$
12.	$L\left[e^{at} \cosh bt\right] = \dfrac{s-a}{(s-a)^2 - b^2}$	$L^{-1}\left[\dfrac{s-a}{(s-a)^2 - b^2}\right] = e^{at} \cosh bt$

Solved Examples

***Example* 24.13:** Find the inverse Laplace transform of:

1. $\dfrac{3+2s+s^2}{s^3}$ 2. $\dfrac{2s+3}{s^2+9}$ 3. $\dfrac{3s+5}{9s^2-25}$ 4. $\dfrac{1}{4s+5}$

5. $\dfrac{1}{4s-5}$ 6. $\dfrac{3}{s+4}$ 7. $\dfrac{2s+1}{(s^2+1)}$ 8. $\dfrac{s+3}{s^2+4}$

***Solution*:** From the above table, we get

1. $$L^{-1}\frac{3}{s^3} + L^{-1}\left(\frac{2}{s^2}\right) + L^{-1}\left(\frac{1}{s}\right) = 3L^{-1}\frac{1}{s^3} + 2L^{-1}\frac{1}{s^2} + L^{-1}\left(\frac{1}{s}\right) = 3\frac{t^2}{2} + 2t + 1$$

2. $$L^{-1}\left(\frac{2s+3}{s^2+9}\right) = L^{-1}\left(\frac{2s}{s^2+9}\right) + L^{-1}\left(\frac{3}{s^2+9}\right) = 2L^{-1}\left(\frac{s}{s^2+9}\right) + L^{-1}\left(\frac{3}{s^2+9}\right)$$

$$= 2\cos 3t + \sin 3t$$

3. $L^{-1}\left(\frac{3s+5}{9s^2-25}\right)=L^{-1}\left(\frac{3s}{9s^2-25}\right)+L^{-1}\left(\frac{5}{9s^2-25}\right)=\frac{1}{3}\cosh\left(\frac{5t}{3}\right)+\frac{1}{3}\sinh\left(\frac{5t}{3}\right)$

4. $L^{-1}\left(\frac{1}{4s+5}\right)=L^{-1}\frac{1}{4\left(s+\frac{5}{4}\right)}=\frac{1}{4}\ e^{-\frac{5}{4}t}$

5. $L^{-1}\left(\frac{1}{4s-5}\right)=L^{-1}\frac{1}{4\left(s-\frac{5}{4}\right)}=\frac{1}{4}\ e^{\frac{5}{4}t}$

6. $L^{-1}\left(\frac{3}{s+4}\right)=3L^{-1}\frac{1}{s+4}=3e^{-4t}$

7. $L^{-1}\frac{2s+1}{(s^2+1)}=2L^{-1}\left(\frac{s}{s^2+1}\right)+L^{-1}\left(\frac{1}{s^2+1}\right)=2\cos t+\sin t$

8. $L^{-1}\left(\frac{s+3}{s^2+4}\right)=L^{-1}\left(\frac{s}{s^2+4}\right)+L^{-1}\left(\frac{3}{s^2+4}\right)=\cos 2t+\frac{3}{2}\sin 2t$

Exercise 24.7

Find the invere Laplace transform of:

1. $\frac{2}{s}+\frac{1}{s^3}+\frac{1}{s+4}$ 2. $\frac{3s+4}{s^2+16}$ 3. $\frac{2s+3}{s^2-9}$ 4. $\frac{(s^2-1)^2}{s^5}$ 5. $\frac{4s+15}{16s^2-25}$

Ans. 1. $2+\frac{t^2}{2}+e^{-4t}$ 2. 3cos4t + sin4t 3. 2cosh3t+sinh3t

4. $1+t^2+\frac{t^4}{24}$ 5. $\frac{1}{4}\cosh\left(\frac{5t}{4}\right)=\frac{3}{4}\sinh\frac{5}{4}t$

Solved Examples

***Example* 24.14:** Find the inverse Laplace transform of:

1. $\frac{6s-4}{s^2-4s+20}$ 2. $\frac{1}{(s-3)^3}$ 3. $\frac{1}{(s-2)^4}$ 4. $\frac{1}{(s-2)^2}+\frac{s-2}{s^2-4s+5}$

Solution: 1. $L^{-1}\left(\dfrac{6s-4}{s^2-4s+20}\right)=L^{-1}\dfrac{6s-12+8}{(s-2)^2+16}=L^{-1}\dfrac{6s-12}{(s-2)^2+4^2}+L^{-1}\dfrac{8}{(s-2)^2+4^2}$

$$= 6L^{-1}\frac{(s-2)}{(s-2)^2+4^2}+2L^{-1}\frac{4}{(s-2)^2+4^2} = 6e^{2t}\cos 4t + 2e^{2t}\sin 4t$$

2. $L^{-1}\left(\dfrac{1}{(s-3)^3}\right)=\dfrac{1}{2}e^{3t}t^2$

3. $L^{-1}\left(\dfrac{1}{(s-2)^4}\right)=\dfrac{1}{6}e^{2t}t^3$

4. $L^{-1}\left(\dfrac{1}{(s-2)^2}\right)+L^{-1}\dfrac{(s-2)}{(s^2-4s+5)}=L^{-1}\left(\dfrac{1}{(s-2)^2}\right)+L^{-1}\left(\dfrac{(s-2)}{s^2-4s+4+1}\right)$

$$=L^{-1}\left(\frac{1}{(s-2)^2}\right)+L^{-1}\left(\frac{s-2}{(s-2)^2+1^2}\right) = e^t.t + e^{2t}.\cos t$$

Exercise 24.8

Find the inverse Laplace transform of:

(i) $\dfrac{s+4}{s^2-4s+13}$ (ii) $\dfrac{s+6}{s^2-8s+25}$ (iii) $\dfrac{4s+12}{s^2+8s+16}$

Ans. (i) $e^{2t}\cos 3t + 2e^{2t}\sin 3t$ (ii) $e^{4t}\cos 3t + e^{4t}\sin 3t$ (iii) $4e^{-4t}.1.-4e^{-4t}$ or $4e^{-4t}(1-t)$

24.2.2 Method of Partial Fractions

Whenever possible we express the given function F(s) into the sum of linear or quadratic partial fractions as

$F(s)=\dfrac{A}{(s+a)^r}+\dfrac{Bs+C}{(s^2+a^2)^r}$ and then use the standard results to find the inverse L.T.

Solved Examples

***Example* 24.15:** (1) Find $L^{-1}\left[\dfrac{s+4}{s^3+s}\right]$ for all s>0,

Solution: By partial fraction expansion, we have

$$\frac{s+1}{s(s^2+1)}=\frac{A}{s}+\frac{Bs+C}{s^2+1}$$

$$s + 1 = A(s^2 + 1) + (Bs + c)s = As^2 + A + Bs^2 + cs$$

comparing coefficients of different powers of s from both sides

$$0 = A + B, \qquad 1 = c, \qquad \therefore\ 1 = A,\ B = -1$$

$$\therefore \frac{s+1}{s(s^2+1)} = \frac{1}{s} + \frac{-s+1}{s^2+1}$$

By the linearity property of L^{-1}, we have

$$L^{-1}\left(\frac{s+1}{s^3+s}\right) = L^{-1}\left(\frac{1}{s}\right) - L^{-1}\left(\frac{s}{s^2+1}\right) + L^{-1}\left(\frac{1}{s^2+1}\right) = 1 - \cos t + \sin t$$

***Example* 24.16:** If $L[f(t)] = \dfrac{2s^2 - 6s + 5}{s^3 + 6s^2 + 11s - 6}$, find [f(t)]

***Solution*:** By partial fractions we have

$$f(s) = \frac{2s^2 - 6s + 5}{s^3 - 6s^2 + 11s - 6} = \frac{A}{(s-1)} + \frac{B}{(s-2)} + \frac{C}{(s-3)}$$

$$\therefore 2s^2 - 6s + 5 = A(s-2)(s-3) + B(s-1)(s-3) + C(s-1)(s-2)$$

put s = 1 we get, $\quad 1 = 2A, \quad A = 1/2$

put s = 2, $\quad B = -1$

Similarly s = 3 gives, $\quad C = 5/2$

$$\therefore L^{-1}[f(s)] = \frac{1}{2}L^{-1}\frac{1}{(s-1)} - L^{-1}\left(\frac{1}{s-2}\right) + \frac{5}{2}L^{-1}\left(\frac{1}{s-3}\right) = \frac{1}{2}e^{t} - e^{2t} + \frac{5}{2}e^{3t}$$

***Example* 24.17:** If $Lf(t) = \dfrac{3s+1}{(s-1)(s^2+1)}$, find f(t)

***Solution*:** By partial fractions we have

$$f(s) = \frac{A}{s-1} + \frac{Bs+C}{s^2+1}$$

$\therefore$ Solving for A, B C

we have A = 2, B = –2, C = 1

$$L^{-1}[f(s)] = 2L^{-1}\frac{1}{s-1} - 2L^{-1}\frac{s}{s^2+1} + L^{-1}\frac{1}{s^2+1}$$

$$= 2e^t - 2\cos t + \sin t$$

***Example* 24.18:** If L[f(t)] = $\frac{s+2}{(s+3)(s+1)^3}$, find f(t)

***Solution*:** $f(s) = \frac{s+2}{(s+3)(s+1)^3} = \frac{1}{8}\frac{1}{s+3} - \frac{1}{8}\frac{1}{s+1} + \frac{1}{4}\frac{1}{(s+1)^2} + \frac{1}{2}\frac{1}{(s+1)^3}$

(after expressing the function in partial fractions)

$$\therefore L^{-1}[f(s)] = \frac{1}{8}L^{-1}\frac{1}{s+3} - \frac{1}{8}L^{-1}\frac{1}{s+1} + \frac{1}{4}\frac{1}{(s+1)^2} + \frac{1}{2}\frac{1}{(s+1)^3}$$

$$= \frac{1}{8}e^{-3t} - \frac{1}{8}e^{-t} + \frac{1}{4}te^{-t} + t^2e^{-t}$$

***Example* 24.19:** If L[f(t)] = $\frac{(s^2+2s-4)}{(s^2+2s+5)(s^2+2s+2)}$, find f(t)

***Solution*:** Let f(s) = $\frac{as+b}{s^2+2s+5} + \frac{cs+d}{s^2+2s+2}$

$$\therefore s^2 + 2s - 4 = (as+b)(s^2+2s+2) + (cs+d)(s^2+2s+5)$$

Equating the coefficients of like powers of s we get a = 0, b = 3, c = 0, d = –2.

$$\therefore f(s) = \frac{3}{s^2+2s+5} - \frac{2}{s^2+2s+2} = \frac{3}{(s+1)^2+2^2} - \frac{2}{(s+1)^2+1^2}$$

$$\therefore L^{-1}[f(s)] = 3L^{-1}\left(\frac{1}{(s+1)^2+2^2}\right) - 2L^{-1}\left[\frac{1}{(s+1)^2+1^2}\right]$$

$$= \frac{3}{2}e^{-t}\sin 2t - 2e^{-t}\sin t.$$

***Example* 24.20:** If $L[f(t)] = \frac{2s}{4s^2+16}$, find f(t)

***Solution*:** By change of scale property we have

$$L^{-1}\left[\frac{2s}{4s^2+16}\right] = \frac{1}{2}\cos 2t \qquad \text{since } L^{-1}\left[\frac{s}{s^2+4}\right] = \cos 2t$$

***Example* 24.21:** If $L[f(t)] = \frac{e^{-5s}}{(s-2)^4}$, find f(t)

***Solution*:** We know $L^{-1}\left[\frac{1}{(s-2)^4}\right] = e^{2t}L^{-1}\left[\frac{1}{s^4}\right] = e^{2t}.\frac{t^3}{3!} = \frac{1}{6}t^3e^{2t}$

Thus by second shifting property

$$L^{-1}\left[\frac{e^{-5s}}{(s-2)^4}\right] = \begin{cases} \frac{1}{6}(t-5)^3 e^{2(t-5)} & \text{if } t > 5 \\ 0 & \text{if } t < 5 \end{cases}$$

Exercise 24.9

Determine the inverse Laplace transform of each of the following functions.

(i) $\frac{3}{s+2}$ (ii) $\frac{s}{s^2+2s+6}$ (iii) $\frac{2s+1}{s(s+1)}$ (iv) $\log\left(\frac{s+3}{s+2}\right)$

(v) $\frac{s-5}{s^2+6s+13}$ (vi) $\frac{e^{-2s}}{s(s+1)}$ (vii) $\frac{s^2}{(s^2+9)(s^2+16)}$

(viii) $\frac{2s^2+s-10}{(s-4)(s^2+2s+2)}$ (ix) $\log\left(1+\frac{1}{s^2}\right)$ (x) $\frac{1}{(s^4+64)}$

(xi) $\frac{1}{s(s+1)(s+2)(s+3)}$ (xii) $\frac{s}{(s^4+s^2+1)}$

(xiii) $\frac{s^2+10s+13}{s-10(s^2-5s+6)}$ (xiv) $\frac{s}{(s^2+1)(s^2+4)}$

(xv) $\dfrac{s+29}{(s+4)(s^2+9)}$ (xvi) $\dfrac{5s+3}{(s-1)(s^2+2s+5)}$

(xvii) $\dfrac{4s+4}{(s-1)^2(s+2)}$ (xviii) $\dfrac{s^2+1}{s^3+3s^2+2s}$

Ans. (i) $3e^{-2t}$ (ii) $e^{-1}\cos\sqrt{5t} - \dfrac{1}{\sqrt{5}}\, e^{-1}\sin\sqrt{5t}$ (iii) $1+e^{-t}$ (iv) $\dfrac{1}{t}(e^{-2t} - e^{-3t})$

(v) $e^{-3t}\cos 2t - 4e^{-3t}\sin 2t$ (vi) $4(1 - e^{-2t})\,(t - 2)$ (vii) $\left(\dfrac{4}{7}\sin 4t - \dfrac{3}{7}\sin 3t\right)$

(viii) $e^{4t} - e^{-t}(\cos 2t + 2\sin t)$ (ix) $\dfrac{2}{t}(1-\cos t)$ (x) $\dfrac{1}{8}$ sin2t sinh2t

(xi) $\dfrac{1}{6}.1 - \dfrac{1}{2}e^{-t} + \dfrac{1}{2}e^{-2t} - \dfrac{1}{6}e^{-3t}$ (xii) $\dfrac{2}{\sqrt{3}}\left(\sin\dfrac{\sqrt{5}}{2}t.\sinh\dfrac{t}{2}\right)$

(xiii) $12e^{t} - 37e^{2t} + 25e^{3t}$ (xiv) $\dfrac{1}{3}$ (cost – cos2t) (xv) $e^{-4t} - \cos 3t + \dfrac{5}{3}\sin 3t$

(xvi) $e^{t} - \dfrac{1}{2}\, e^{-t}\cos 2t + \dfrac{3}{2}\, e^{-t}\sin 2t$ (xvii) $\dfrac{4}{9}\,(e^{t} - e^{-2t} + 6te^{t})$

(xviii) $\dfrac{1}{2} - 2^{-et} + \dfrac{5}{2}\, e^{-2t}$

24.2.3 First Shifting Theorem

If $L^{-1}[F(s)] = f(t)$ then $L^{-1}[F(s-a)] = e^{at} f(t)$

Proof:

We have $L[e^{at} f(t)] = F(s-a)$

$\therefore$ $L^{-1}[s-a] = e^{at} f(t)$

similarly $L^{-1}[s+a] = e^{-at} f(t)$

Solved Examples

***Example* 24.22:** Find $L^{-1}\left[\frac{s}{s^4+4a^4}\right]$

Solution: Given $F(s) = \frac{s}{s^4+4a^4} = \frac{s}{(s^2+2a^2)^2-(2as)^2}$

$$= \frac{s}{(s^2+2a^2-2as)^2-(s^2+2as+2a^2)}$$

using partial fractions

$$= \frac{1}{4a}\left[\frac{1}{(s^2-2as+2a^2)} - \frac{1}{(s^2+2as+2a^2)}\right]$$

$$\frac{s}{s^4-4a^4} = \frac{1}{4a}\left[\frac{1}{(s-a)^2+a^2} - \frac{1}{(s+a)^2+a^2}\right]$$

$$L^{-1}\left(\frac{s}{s^4+4a^4}\right) = \frac{1}{4a}L^{-1}\left[\frac{1}{(s-a)^2+a^2}\right]\frac{-1}{4a}L^{-1}\left[\frac{1}{(s+a)^2+a^2}\right]$$

$$= \frac{1}{4a}\frac{e^{at}\,sin\,at}{a} - \frac{1}{4a}\frac{e^{-at}\,sin\,at}{a}$$

$$= \frac{1}{4a^2}sin\,at\left[e^{at} - e^{-at}\right]$$

$$= \frac{1}{2a^2}\sin at \sin h\, at$$

***Example* 24.23:** Find $L^{-1}\left[\frac{s+2}{(s-2)^3}\right]$

Solution : Given $F(s) = \frac{s+2}{(s-2)^3}$

$$= \frac{(s-2)+4}{(s-2)^3} = \frac{1}{(s-2)^2} + \frac{4}{(s-2)^3}$$

Now $L^{-1}\left[\frac{s+2}{(s-2)^3}\right]=L^{-1}\left[\frac{1}{(s-2)^2}\right]+4L^{-1}\left[\frac{1}{(s-2)^3}\right]$

$$= e^{2t}.t + 4\frac{e^{2t}}{2!}t^2 \qquad \because L(t)=\frac{1}{s^2}; \quad L(t^2)=\frac{2!}{s^3}$$

***Example* 24.24:** Find $L^{-1}\left[\frac{s+3}{s^2-10s+29}\right]$

Solution: Given $F(s)=\frac{s+3}{s^2-10s+29}$

$$=\frac{(s+3)}{(s-5)^2+2^2}=\frac{(s-5)+8}{(s-5)^2+2^2}$$

$$=\frac{(s-5)}{(s-5)^2+2^2}+\frac{8}{(s-5)^2+2^2}$$

$$\therefore \quad L^{-1}\left[\frac{s+3}{s^2-10s+29}\right]=L^{-1}\left[\frac{(s-5)}{(s-5)^2+2^2}\right]+8L^{-1}\left[\frac{1}{(s-5)^2+2^2}\right]$$

$$= e^{5t}\cos 2t + \frac{8}{2}e^{5t}\sin 2t$$

$$= e^{5t}[\cos 2t + 4\sin 2t]$$

***Example* 24.25:** Find $L^{-1}\left[\frac{s}{(s+3)^2}\right]$

Solution: $L^{-1}\left[\frac{1}{(s+3)^2}\right]=L^{-1}\left[\frac{(s+3)-3}{(s+3)^2}\right]$

$$=L^{-1}\left[\frac{1}{s+3}\right]-3L^{-1}\left[\frac{1}{(s+3)^2}\right]$$

$$= e^{-3t} - 3e^{-3t}t = e^{-3t}(1-3t)$$

Exercise 24.10

Find the inverse Laplace Transform of the following

1. $\dfrac{5s+3}{s^2+2s+5}$ [Ans: e^t (5 cos 2t – sin 2t)]

2. $\dfrac{1}{(s+1)(s^2+2s+2)}$ [Ans: e^{-t} (1–cos t)]

3. $\dfrac{16}{(s^2+2s+5)^2}$ [Ans: e^{-t} (sin 2t –2t cos 2t)]

4. $\left[\dfrac{s}{(s+7)^4}\right]$ [Ans: $e^{-7t}\left[\dfrac{t^2}{2}-\dfrac{7}{6}t^3\right]$]

5. $\dfrac{s^2+2s-4}{(s^2+2s+5)(s^2+2s+2)}$ [Ans: $\dfrac{e^{-t}}{2}[3\sin 2t-4\sin t]$)]

6. $\dfrac{2s^3+4s^2-s+1}{s^2(s^2-s+2)}$ [Ans: $\dfrac{e^{\frac{t}{2}}}{4}\left[9\cos\dfrac{\sqrt{7}}{2}t+5\sqrt{7}\,sin\dfrac{\sqrt{7}}{2}t+\dfrac{t}{2}-\dfrac{1}{4}\right]$]

7. $\dfrac{a^2}{s(s+a)^3}$ [Ans: $\dfrac{1}{a}-\dfrac{1}{a}e^{-at}-te^{-at}-a\dfrac{t^2}{2}e^{-at}$]

24.2.4 Second Shifting Theorem

If $L^{-1}[F(s)]=f(t)$; $L^{-1}[e^{-as}F(s)]=G(t)$

where G(t) = {f(t – a) ; t > a

{0 ; t < a

Proof:

$$L[G(t)] = \int_0^\infty e^{-st}G(t)dt = \int_0^a e^{-st}G(t)dt\ t\int_a^\infty e^{-st}G(t)dt$$

$$= \int_0^a e^{-st}.0\,dt + \int_a^\infty e^{-st}f(t-a)dt$$

$$= \int_a^\infty e^{-st}f(t-a)dt$$

Substituting t – a = x, dt = dx, t = a $\Rightarrow x = 0$, t = $\infty \Rightarrow$ $x = \infty$

$$\therefore \quad L[G(t)] = \int_0^{\infty} e^{-s(x+a)} f(x)\,dx = e^{-as}\int_0^{\infty} e^{-sx} f(x)\,dx$$

$$= e^{-as}\int_0^{\infty} e^{-st} f(t)\,dt = e^{-as} L[f(t)]$$

$$L[G(t)] = e^{-as} F(s)$$

$$\therefore \quad G(t) = L^{-1}[e^{-as} F(s)]$$

Note:

$$L^{-1}\left[e^{-as}F(s)\right] = \begin{cases} f(t-a), & t > a \\ 0, & t < a \end{cases}$$

or

$$L^{-1}[e^{-as} F(s)] = f(t-a)\, H(t-a)$$

where H(t – a) is Heaviside unit step function

Solved Examples

***Example* 24.26:** Find $L^{-1}\left[\frac{e^{-2s}(3+5s)}{s^2}\right]$

Solution: Consider $F(s) = \frac{3+5s}{s^2} = \frac{3}{s^2} + \frac{5}{s}$

$$f(t) = L^{-1}[F(s)]$$

$$= L^{-1}\left[\frac{3}{s^2} + \frac{5}{s}\right] = 3L^{-1}\left[\frac{1}{s^2}\right] + 5L^{-1}\left[\frac{1}{s}\right]$$

$$f(t) = 3t + 5$$

Using second shifting theorem

$$L^{-1}\left[e^{-2s}\frac{(3+5s)}{s^2}\right] = f(t-2)\,U(t-2)$$

$$= [3(t-2) + 5]\,[U(t-2)]$$

where U(t – 2) is the unit step function

$$L^{-1}\left[\frac{3+5s}{s^2 e^{2s}}\right] = \begin{cases} 0, & 0 < t < 2 \\ 3t-1, & t > 2 \end{cases}$$

***Example* 24.27:** Find $L^{-1}\left[\dfrac{e^{-\pi s}}{s^2+1}\right]$

Solution: We have $L^{-1}\left[\dfrac{1}{s^2+1}\right]=\sin t=f(t)$

using second shifting theorem

$$L^{-1}\left[\frac{e^{-\pi s}}{s^2+1}\right]=\begin{cases}\sin(t-\pi); & t>\pi\\ 0; & t<\pi\end{cases}$$

$$L^{-1}\left[\frac{e^{-\pi s}}{s^2+1}\right]=\sin(t-\pi)\,H(t-\pi)$$

where $H(t-\pi)$ is the Heaviside unit step function

***Example* 24.28:** Find $L^{-1}\left[\dfrac{e^{-3s}}{(s-4)^2}\right]$

Solution: We have $L^{-1}\left[\dfrac{1}{(s-4)^2}\right]=e^{4t}\,L^{-1}\left[\dfrac{1}{s^2}\right]$

$$= e^{4t}\,.\,t$$

$\therefore$ $f(t) = e^{4t}\,.\,t$

Using second shifting theorem

$$L^{-1}\left[\frac{e^{-3s}}{(s-4)^2}\right]=\begin{cases}e^{4(t-3)}(t-3); & t>3\\ 0\;; & t<3\end{cases}$$

or $$L^{-1}\left[\frac{e^{-3s}}{(s-4)^2}\right]=(t-3)e^{4(t-3)}H(t-3)$$

***Example* 24.29:** Find $L^{-1}\left[\dfrac{(s+1)e^{-\pi s}}{s^2+s+1}\right]$

Solution: Consider $f(t) = L^{-1}[F(s)]$

$$=L^{-1}\left[\frac{s+1}{s^2+s+1}\right]=L^{-1}\left[\frac{s+1}{\left(s+\frac{1}{2}\right)^2+\frac{3}{4}}\right]$$

$$= L^{-1}\left[\frac{\left(s+\frac{1}{2}\right)+\frac{1}{2}}{\left(s+\frac{1}{2}\right)^2+\left(\frac{\sqrt{3}}{2}\right)^2}\right]$$

$$f(t) = e^{\frac{-t}{2}} \cos\frac{\sqrt{3}}{2}t + \frac{1}{\sqrt{3}} e^{\frac{-t}{2}} \sin\frac{\sqrt{3}}{2}t$$

using second shifting theorem

$$L^{-1}\left[\frac{(s+1)e^{-\pi s}}{s^2+s+1}\right] = L^{-1}\left[e^{-\pi s}F(s)\right] = f(t-\pi)\, u\,(t-\pi)$$

$$L^{-1}\left[\frac{(s+1)e^{-\pi s}}{s^2+s+1}\right] = e^{\frac{-1}{2}(t-\pi)}\left[\cos\frac{\sqrt{3}}{2}(t-\pi) + \frac{1}{\sqrt{3}}\sin\frac{\sqrt{3}}{2}(t-\pi)\right]u(t-\pi)$$

where $u(t-\pi)$ is the unit step function

***Example* 24.30:** Find $L^{-1}\left[\frac{e^{-3s}}{(s-2)^4}\right]$

Solution: We have $L^{-1}\left[\frac{1}{(s-2)^4}\right] = e^{2t}\frac{t^3}{3!}$

$$L^{-1}\left[\frac{e^{-3s}}{(s-2)^4}\right] = \begin{cases} \frac{1}{6}\, e^{2(t-3)}(t-3)^3 ; t > 3 \\ 0 \qquad ; t < 3 \end{cases}$$

$$L^{-1}\left[\frac{e^{-3s}}{(s-2)^4}\right] = \frac{1}{6}(t-3)^3\, e^{2(t-3)}.\, H(t-3)$$

$H(t-3)$ is Heaviside unit step function

***Example* 24.31:** Find $L^{-1}\left[\frac{e^{6-5s}}{(s+4)^{\frac{5}{2}}}\right]$

Solution: We know that $L^{-1}\left[\frac{1}{(s+4)^{\frac{5}{2}}}\right] = e^{-4t}\frac{t^{\frac{5}{2}-1}}{\left\lfloor\frac{5}{2}\right.}$

$$= e^{-4t}\frac{t^{\frac{3}{2}}}{\frac{3}{2}\cdot\frac{1}{2}\sqrt{\pi}} = \frac{4}{3}\frac{e^{-4t}}{\sqrt{\pi}}t^{\frac{3}{2}}$$

$$L^{-1}\left[\frac{e^{6-5s}}{(s+4)^{\frac{5}{2}}}\right] = e^{6}\quad L^{-1}\left[\frac{e^{-5s}}{(s+4)^{\frac{5}{2}}}\right]$$

$$= e^{6}\cdot\frac{4}{3\sqrt{\pi}}e^{-4(t-5)}.(t-5)^{3/2};\ t>5$$

$$= 0 \qquad ;t<5$$

$$L^{-1}\left[\frac{e^{6-5s}}{(s+4)^{\frac{5}{2}}}\right] = \frac{4}{3\sqrt{\pi}}e^{-4(t-5)}.(t-5)^{\frac{3}{2}}H(t-5)$$

H(t – 5) is Heaviside unit step function

24.2.5 Change of Scale Property

***Proof*:** If L[*f*(t) = F(s), then L^{-1} [*f*(as)] = $\frac{1}{a}\ f\left(\frac{t}{a}\right)$, a > 0

Solution:

Definition of $L[f(t)] = \int_0^\infty e^{-st}f(t)dt = F(s)$

$$F(as) = \int_0^\infty e^{-ast}f(t)\,dt$$

substituting at = u, $= \int_0^\infty e^{-su}f\left(\frac{u}{a}\right)\frac{1}{a}\ du = \frac{1}{a}\int_0^\infty e^{-st}f\left(\frac{t}{a}\right)dt$

$$=\frac{1}{a}L\left[f\left(\frac{t}{a}\right)\right] \qquad\qquad =L\left[\frac{1}{a}f\left(\frac{t}{a}\right)\right]$$

$$\therefore \qquad L^{-1}[F(as)]=\frac{1}{a}f\left(\frac{t}{a}\right)$$

Solved Examples

***Example* 24.32:** If $L^{-1}\left[\frac{s^2-1}{(s^2+1)^2}\right]=t\cos t$ then find $L^{-1}\left[\frac{9s^2-1}{(9s^2+1)^2}\right]$

Solution: Given $L^{-1}\left[\frac{s^2-1}{(s^2+1)^2}\right]=t\cos t$

Replacing s by as

$$L^{-1}\left[\frac{a^2s^2-1}{(a^2s^2+1)^2}\right]=\frac{1}{a}\left[\frac{t}{a}\cos\left(\frac{t}{a}\right)\right]$$

Using the above result

$$L^{-1}\left[\frac{9s^2-1}{(9s^2+1)^2}\right]=\frac{1}{3}\left[\frac{t}{3}\cos\left(\frac{t}{3}\right)\right]$$

***Example* 24.33:** If $L^{-1}\left[\frac{e^{-\frac{1}{s}}}{s^{\frac{1}{2}}}\right]=\frac{\cos 2\sqrt{t}}{\sqrt{\pi t}}$, find $L^{-1}\left[\frac{e^{\frac{-a}{s}}}{s^{\frac{1}{2}}}\right]$

Solution: Given $L^{-1}\left[\frac{e^{-\frac{1}{s}}}{s^{\frac{1}{2}}}\right]=\frac{\cos 2\sqrt{t}}{\sqrt{\pi t}}$

Replacing s by bs

$$L^{-1}\left[\frac{e^{\frac{-1}{bs}}}{(bs)^{\frac{1}{2}}}\right]=\frac{1}{b}\frac{\cos 2\sqrt{\frac{t}{b}}}{\sqrt{\pi\frac{t}{b}}}$$

$$\frac{1}{b^{\frac{1}{2}}} L^{-1}\left[\frac{e^{-\frac{1}{bs}}}{s^{\frac{1}{2}}}\right] = \frac{1}{(b)^{\frac{1}{2}}} \frac{\cos 2\sqrt{\frac{t}{b}}}{\sqrt{\pi t}}$$

$$L^{-1}\left[\frac{e^{-\frac{1}{bs}}}{s^{\frac{1}{2}}}\right] = \frac{\cos 2\sqrt{\frac{t}{b}}}{\sqrt{\pi t}}$$

Substituting $b = \frac{1}{a}$, $L^{-1}\left[\frac{e^{-\frac{a}{s}}}{s^{\frac{1}{2}}}\right] = \frac{\cos\left(2\sqrt{at}\right)}{\sqrt{\pi t}}$

***Example* 24.34:** If $L^{-1}\left[\frac{s}{\left(s^2+1\right)^2}\right] = \frac{t}{2}\sin t$; Find $L^{-1}\left[\frac{8s}{\left(4s^2+1\right)^2}\right]$

Solution: Given $L^{-1}\left[\frac{s}{\left(s^2+1\right)^2}\right] = \frac{t}{2}\sin t$

Replacing s by as

$$L^{-1}\left[\frac{as}{\left(a^2s^2+1\right)^2}\right] = \frac{1}{2}\frac{1}{a}\left(\frac{t}{a}\right)\sin\left(\frac{t}{a}\right)$$

Substituting a = 2

$$L^{-1}\left[\frac{2s}{\left(4s^2+1\right)^2}\right] = \frac{1}{2}.\frac{1}{2}\left(\frac{t}{2}\right)\sin\left(\frac{t}{2}\right)$$

$$L^{-1}\left[\frac{8s}{\left(4s^2+1\right)^2}\right] = \frac{t}{2}\sin\left(\frac{t}{2}\right)$$

Exercise 24.11

Find the inverse Laplace transform of the following

1. $\frac{e^{-4s}}{(s-3)^4}$ **[Ans:** $\frac{1}{6}(t-4)^3 e^{3(t-4)} . H(t-4)$**]**

2. $\dfrac{s\,e^{\frac{-2\pi s}{3}}}{s^2+9}$ **[Ans:** $\cos 3\left(t-\dfrac{2\pi}{3}\right) H\left(t-\dfrac{2\pi}{3}\right)$ **]**

3. $\dfrac{3\left(1+e^{-\pi s}\right)}{s^2+9}$ **[Ans:** sin 3t – sin 3t H (t – π)**]**

4. $\dfrac{se^{-as}}{s^2-w^2}$ **[Ans:** cos h w (t – a) H(t – a), a > 0**]**

5. $\dfrac{e^{-3s}}{s^3}$ **[Ans:** $\dfrac{1}{2}(t-3)^2 H(t-3)$**]**

6. $\dfrac{8}{s}-\dfrac{5e^{-s}}{s^2}+\dfrac{4e^{-3s}}{s^2}$ **[Ans:** 8 – 5 (t – 1) H (t – 1) + 4 (t – 3) H (t – 3)**]**

7. $\dfrac{e^{4-3s}}{(s+4)^{\frac{5}{2}}}$ **[Ans:** $\dfrac{4}{3\sqrt{\pi}}(t-3)^{\frac{3}{2}} e^{-4(t-4)} H(t-3)$**]**

8. If $L^{-1}\left[\dfrac{s}{\left(s^2+1\right)^2}\right]=\dfrac{t}{2}\text{sint}$; show that $L^{-1}\left[\dfrac{32s}{\left(16s^2+1\right)^2}\right]=\dfrac{t}{4}\sin\left(\dfrac{t}{4}\right)$

9. If $L^{-1}\left[\dfrac{s^2-1}{\left(s^2+1\right)^2}\right]=t\cos t$; show that $L^{-1}\left[\dfrac{9s^2-1}{\left(9s^2+1\right)^2}\right]=\dfrac{t}{9}\cos\left(\dfrac{t}{3}\right)$

24.2.6 Inverse Laplace Transform of Derivatives

Theorem:

If $L^{-1}[F(s)] = f(t)$, then $L^{-1}[F^n(s)] = L^{-1}\left[\dfrac{d^n F(s)}{ds^n}\right]$

$$= (-1)^n t^n f(t),\ n = 1, 2, 3, \ldots$$

Proof:

We have $L\{t^n f(t)\} = (-1)^n \dfrac{d^n F(s)}{ds^n} = (-1)^n F^n(s)$

$$\therefore \quad L^{-1}[F^n(s)] = L^{-1}\left[\frac{d^n F(s)}{ds^n}\right] = (-1)^n t^n f(t)$$

24.2.7 Inverse Laplace Transform of Integrals

Theorem

If $L^{-1}[F(s)] = f(t)$, then $L^{-1}\left[\int_s^{\infty} F(x)dx\right] = \frac{f(t)}{t}$

Proof:

We have $L\left\{\frac{1}{t}f(t)\right\} = \int_s^{\infty} F(x)dx$ (Provided $\underset{t\to 0}{Lt}\left[\frac{f(t)}{t}\right]$ exists)

$$\therefore \quad L^{-1}\left[\int_s^{\infty} F(x)\,dx\right] = \frac{f(t)}{t}$$

24.2.8 Multiplication by powers of s

Theorem

If $L^{-1}[F(s)] = f(t)$, and $f(0) = 0$ then $L^{-1}[sF(s)] = f'(t)$

Proof:

In general, $L^{-1}[s^n F(s)] = f^n(t)$, if $f(0) = 0$, n = 2, 3, ...

We have $L[f'(t)] = s\,L\{f(t)\} - f(0) = s\,F(s)$ $\quad \because f(0) = 0$

$$\therefore \quad L^{-1}[s\,F(s)] = f'(t)$$

Note: If $f(0) \neq 0$ then $L^{-1}[s\,F(s) - f(0) = f'(t)\}$

$\therefore \quad L^{-1}[s\,F(s)] = f'(t) + f(0)\,\delta(t)$ where $\delta(t)$ is Dirac delta function or unit impulse function.

In general, $L^{-1}[s^n F(s)] = f^n(t)$, if $f(0) = 0$, n = 2, 3, ...

24.2.9 Division by powers of 's'

Theorem

If (i) $f(t)$ is sectionally continuous and of exponential order 'a'

(ii) $\underset{t\to 0}{Lim}\frac{f(t)}{t}$ exists then for s > a, $L^{-1}\left[\frac{F(s)}{s}\right] = \int_0^t f(x)\,dx$

Proof:

Let $g(t) = \int_0^t f(x)\,dx$, then $g'(t) = f(t)$ and $g(0) = 0$ ..... (1)

$\therefore$ $L\{g'(t)\} = s\,L\{g(t)\} - g(0)$ (Laplace transform of the derivative of a function)

$$L\{g'(t)\} = s\,L\{g(t)\} \qquad \because \quad g(0) = 0$$

But $L\{g'(t)\} = L\{f(t)\} = F(s)$

i.e., $F(s) = sL\{g(t)\}$

$$\therefore \quad L\{g(t)\} = \frac{F(s)}{s}$$

$$\therefore \quad L^{-1}\left[\frac{F(s)}{s}\right] = g(t)$$

$$L^{-1}\left[\frac{F(s)}{s}\right] = \int_0^t f(x)\,dx$$

24.2.10 Theorem

$$L^{-1}\left[\frac{F(s)}{s^2}\right] = \int_0^t\int_0^y f(x)\,dx\,dy$$

Proof: Let $g(t) = \int_0^t\int_0^y f(x)\,dx\,dy$

then $g'(t) = \int_0^t f(x)\,dx, \; g''(t) = f(t)$

Now $L\{g''(t)\} = s^2 L\{g(t)\} - s\,g(0) - g'(0)$

$$L\{g''(t)\} = s^2 L\{g(t)\}, \text{ since } g'(0) = g(0) = 0$$

$$F(s) = s^2 L\{g(t)\}$$

$$\therefore \quad L\{g(t)\} = \frac{F(s)}{s^2}$$

$$\therefore \quad L^{-1}\left[\frac{F(s)}{s^2}\right] = g(t) = \int_0^t\int_0^t f(x)\,dx\,dy$$

or
$$L^{-1}\left[\frac{F(s)}{s^2}\right] = \int_0^t\int_0^t f(t)\,(dt)^2$$

In general,

$$L^{-1}\left[\frac{F(s)}{s^n}\right] = \int_0^t\int_0^t\ldots\int_0^t f(t)\,(dt)^n.$$

24.2.11 Convolution

Definition: Let $f(t)$ and $g(t)$ be two functions of class A than the convolution of $f(t)$ and $g(t)$ is denoted by $f(t)*g(t)$ and is defined by

$$f(t) * g(t) = \int_0^t f(x)\,g(t-x)\,dx$$

Also $f(t) * g(t)$ is called the resultant or falting of $f(t)$ and $g(t)$.

Theorem : $f(t) * g(t)$ is (i) commutative (ii) distributive with respect to addition.

Proof : From definition $f(t) * g(t) = \int_0^t f(x)\,g(t-x)\,dx$

Substiuting $t - x = y$, $-dx = dy$

$$f(t) * g(t) = -\int_t^0 f(t-y)\,g(y)\,(-dy)$$

$$= \int_0^t g(y)\,f(t-y)\,dy = g(t) * f(t)$$

$\therefore \quad f(t)*g(t) = g(t)*f(t)$

$\therefore \quad f(t)*g(t)$ is commutative.

$f*g$ is distribution w.r. to addition means.

$f*(g+h) = f*g + f*h$

Consider $\quad f*(g+h) = \int_0^t f(t)\,\{g(t-x)+h(t-x)\}\,dx$

$$= \int_0^t f(t)\,g(t-x)\,dx + \int_0^t f(t)\,h(t-x)\,dx$$

$\therefore \quad f*(g+h) = f*g + f*h.$

24.2.12 Convolution Theorem (Convolution Property)

If $f(t)$ and $g(t)$ are two functions of class A ($t \geq 0$) and

$L^{-1}[F(s)] = f(t)$, $L^{-1}[G(s)] = g(t)$, then

$$L^{-1}[F(s).\,G(s)] = \int_0^t f(x)\,g(t-x)\;dx = f*g.$$

Proof : From definition of Laplace transform

$$L\left\{\int_0^t f(x)\,g(t-x)dx\right\} = \int_0^\infty e^{-st}\left\{\int_0^t f(x)\,g(t-x)dx\right\}dt$$

$$= \int_0^\infty\int_0^t e^{-st}\;f(x)\,g(t-x)\,dx\,dt$$

$$= \lim_{T\to\infty}\int_{t=0}^{T}\int_{x=0}^{t} e^{-st}f(x)\,g(t-x)dx\,dt \qquad \text{... (1)}$$

Dennte

$$I = \int_{t=0}^{T}\int_{x=0}^{t} e^{-st}f(x)\,g(t-x)dx\,dt$$

Substituting $t-x=y$

and by changing the order of integration, the shaded area A of the *tx*-Plane is transformed into A_1 of the *xy*-plane

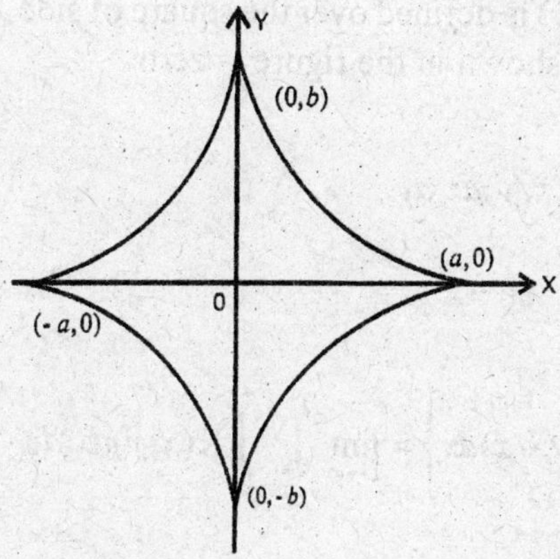

$$\therefore\ I = \int_A e^{-st} f(x)\, g(t-x)\, dx\, dt$$

$$= \int_{A_1} e^{-s(x+y)} f(x) g(y) \left|\frac{\partial(x,t)}{(x,y)}\right| dx\, dy$$

Where $\dfrac{\partial(x,t)}{(x,y)} = \begin{vmatrix} \dfrac{\partial x}{\partial x} & \dfrac{\partial x}{\partial y} \\ \dfrac{\partial t}{\partial x} & \dfrac{\partial t}{\partial y} \end{vmatrix} = \begin{vmatrix} 1 & 0 \\ 1 & 1 \end{vmatrix} = 1 = J$ which is the Jacobian of transformation

$$\therefore \qquad I = \int_{y=0}^{T} \int_{x=0}^{T-y} e^{-s(x+y)} f(x) g(y) dx\, dy \qquad \text{... (2)}$$

Now define a function

$$K(x, y) = \begin{cases} e^{-s(x+y)} f(x) g(y), & x + y \le T \\ 0 \qquad\qquad , & x + y > T \end{cases}$$

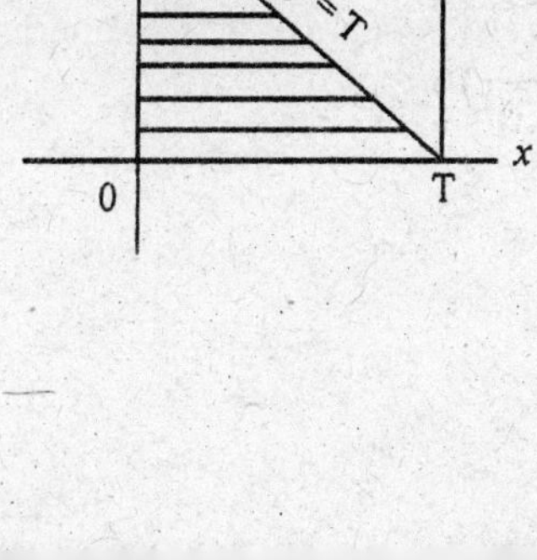

The function K(x, y) is defined over the square of side T, such that the imtegral I over the unshaded area shown in the figure is zero

$$\therefore \quad I = \int_{y=0}^{T} \int_{x=0}^{T} K(x, y)dx\, dy \qquad \dots (3)$$

From (1), (2), (3)

$$L\left\{\int_0^t f(x)g(t-x)dx\right\} = \lim_{T\to\infty} \int_{y=0}^{T} \int_{x=0}^{T} K(x, y)dx\, dy$$

$$= \int_{y=0}^{\infty} \int_{x=0}^{\infty} e^{-s(x+y)} f(x)g(y)dx\, dy$$

$$= \left[\int_{x=0}^{\infty} e^{-sx} f(x)dx\right]\left[\int_{y=0}^{\infty} e^{-sy} g(y)dy\right]$$

$$= L\{f(t)\}.\, L\{g(t)\} = F(s).\, G(s)$$

$$\therefore \quad L\left\{\int_0^t f(x)g(t-x)dx\right\} = F(s).\, G(s)$$

Hence $L^{-1}[F(s).\, G(s)] = \int_0^t f(x)g(t-x)dx = f * g$.

Solved Examples

***Example* 24.35:** Find $L^{-1}\left[\dfrac{s+3}{(s^2+6s+13)^2}\right]$.

***Solution*:** Taking $\dfrac{s+3}{(s^2+6s+13)^2} = \dfrac{dF(s)}{ds}$

$$F(s) = \int \frac{s+3}{(s^2+6s+13)^2} ds$$

$$s^2 + 6s + 13 = t$$

$$(2s + 6)\, ds = dt; \quad (s + 3)\, ds = \frac{dt}{2}$$

$$F(s) = \frac{1}{2}\int \frac{dt}{t^2} = -\frac{1}{2t} = -\frac{1}{2(s^2+6s+13)}$$

using $L^{-1}\{F'(s)\} = -t\, L^{-1}\{F(s)\}$

$$L^{-1}\frac{s+3}{(s^2+6s+13)^2} = -t\, L^{-1}\left[-\frac{1}{2(s^2+6s+13)}\right]$$

$$= \frac{t}{2}L^{-1}\left[\frac{1}{(s+3)^2+2^2}\right]$$

$$= \frac{t}{2}L^{-1}\left[\frac{1}{2}.\frac{1}{(s+3)^2+2^2}\right]$$

$$L^{-1}\left[\frac{s+3}{(s^2+6s+13)^2}\right] = \frac{t}{4}e^{-3t}\sin 2t.$$

***Example* 24.36:** Find $L^{-1}\left[\frac{s}{(s^2-9)^2}\right]$

***Solution*:** Let $\frac{d\,F(s)}{ds} = \frac{s}{(s^2-9)^2}$

$$F(s) = \int \frac{s}{(s^2-9)^2}ds = \frac{-1}{2(s^2-9)}$$

using $L^{-1}[F'(s)] = -t\, L^{-1}[F(s)]$

$$L^{-1}\left[\frac{s}{(s^2-9)^2}\right] = -t\, L^{-1}\left[\frac{-1}{2(s^2-9)}\right] = \frac{1}{2}t\, L^{-1}\left[\frac{1}{s^2-9}\right]$$

$$= \frac{t}{2}\, L^{-1}\left[\frac{3}{3(s^2-9)}\right] = \frac{t}{6}\, L^{-1}\left[\frac{3}{s^2-3^2}\right] = \frac{t}{6}\sin h3t.$$

Example 24.37: Find $L^{-1}\left[\log\left(\frac{s+6}{s+9}\right)\right]$

***Solution*:** Let $F(s) = \log\left(\frac{s+6}{s+8}\right) = \log(s+6) - \log(s+8)$

$$F'(s) = \frac{1}{s+6} - \frac{1}{s+8}$$

$$L^{-1}[F'(s)] = L^{-1}\left(\frac{1}{s+6}\right) - L^{-1}\left[\frac{1}{s+8}\right] = e^{-6t} - e^{-8t}$$

$$\Rightarrow \quad L^{-1} F(s) = -\frac{1}{t} L^{-1}[F'(s)]$$

$$\therefore \quad L^{-1}\left[\log\left(\frac{s+6}{s+8}\right)\right] = -\frac{1}{t}[e^{-6t} - e^{-8t}]$$

$$\therefore \quad = L^{-1}\left[\frac{e^{-8t} - e^{-6t}}{t}\right].$$

***Example* 24.38:** Find $L^{-1}\left[\log\sqrt{\frac{s^2+a^2}{s+b^2}}\right]$.

***Solution*:** Let $F(s) = \frac{1}{2}\log\frac{s^2+a^2}{s^2+b^2} = \frac{1}{2}\log(s^2+a^2) - \frac{1}{2}\log(s^2+b^2)$

$$F'(s) = \frac{1}{2}.\frac{2s}{s^2+a^2} - \frac{1}{2}.\frac{2s}{s^2+b^2}$$

$$L^{-1}[F'(s)] = L^{-1}\left[\frac{s}{s^2+a^2}\right] - L^{-1}\left[\frac{s}{s^2+b^2}\right]$$

$$L^{-1}[F'(s)] = \cos at - \cos bt$$

$$\Rightarrow \quad L^{-1}[F(s)] = -\frac{1}{t} L^{-1} F'(s)$$

$$L^{-1}\left[\frac{1}{2}\log\left(\frac{s^2+a^2}{s^2+b^2}\right)\right] = -\frac{1}{t}[\cos at - \cos bt]$$

***Example* 24.39:** Find $L^{-1}\left[\cot^{-1}\left(\frac{2}{s^2}\right)\right]$

***Solution*:** $F(s) = \cot^{-1}\left(\frac{2}{s^2}\right)$; $F'(s) = \frac{-1}{1+\frac{4}{s^4}} \times \frac{-4}{s^3} = \frac{4s}{s^4+4}$

$$F'(s) = \frac{4s}{(s^4+4s^2+4)-4s^2}$$

$$= \frac{4s}{(s^2+2)^2-(2s)^2} = \frac{4s}{(s^2+2s+2^2)(s^2-2s+2)}$$

$$= \left[\frac{1}{s^2-2s+2} - \frac{1}{s^2+2s+2}\right]$$

$$\therefore \quad L^{-1}\left[F'(s)\right] = L^{-1}\left[\frac{1}{(s-1)^2+1^2} - \frac{1}{(s+1)^2+1}\right]$$

$$\therefore \quad L^{-1}\left[F'(s)\right] = [e^t \sin t - e^{-t} \sin t]$$

$$\Rightarrow \quad L^{-1}\left[F(s) = -\frac{1}{t} L^{-1} F'(s)\right]$$

$$\therefore \quad L^{-1}\left[\cot\left(\frac{2}{s^2}\right)\right] = \frac{-1}{t}[e^t \sin t - e^{-t} \sin t]$$

$$= -\left(\frac{e^t - e^{-t}}{t}\right)\sin t = -\frac{2}{t}\sin t \sin ht.$$

***Example* 24.40:** Find $L^{-1}[\tan^{-1}(s+1)]$

***Solution*:** Let $F(s) = [\tan^{-1}(s+1)]$;

$$F'(s) = \frac{1}{1+(s+1)^2}$$

$$L^{-1}\left[F'(s)\right] = L^{-1}\left[\frac{1}{1+(s+1)^2}\right] = e^{-t}\sin t$$

$$\Rightarrow \quad L^{-1} F(s) - \frac{1}{t} L^{-1} F'(s)$$

$$L^{-1}[\tan^{-1}(s+1)] = -\frac{1}{t} e^{-t} \sin t.$$

***Example* 24.41:** Find $L^{-1}\left[\frac{s^2}{(s^2+a^2)^2}\right]$

Solution: $L^{-1}\left[\frac{s^2}{(s^2+a^2)^2}\right] = L^{-1}\left[s.\frac{s}{(s^2+a^2)^2}\right]$

$$= \frac{d}{dt}L'\left[\frac{s^2}{(s^2+a^2)^2}\right] \quad \text{using } L^{-1}\left[\frac{s^2}{(s^2+a^2)^2}\right] = \frac{t}{2a}\sin t$$

$$= \frac{d}{dt}\left[\frac{t}{2a}\sin at\right]$$

$$L^{-1}\left[\frac{s^2}{(s^2+a^2)^2}\right] = \frac{t}{2a}\ [at\cos at + \sin at]$$

***Example* 24.42:** Find $L^{-1}\left[\frac{s^2}{(s^2-2)^2}\right]$

Solution: $L^{-1}\left[\frac{s^2}{(s^2-2)^2}\right] = L^{-1}\left[s.\frac{s}{(s-2)^2}\right] = \frac{d}{dt}L^{-1}\left[\frac{s}{(s^2-2)^2}\right]$

$$= \frac{d}{dt}.\frac{d}{dt}L^{-1}\left[\frac{1}{(s-2)^2}\right] = \frac{d^2}{dt^2}e^{2t}L^{-1}\left[\frac{1}{(s^2)}\right]$$

$$= \frac{d^2}{dt^2}e^{2t}.t = \frac{d[e^{2t}.1+2te^{2t}]}{dt}$$

$$= 2e^{2t} + 2t.\ e^{2t}.\ 2 + 2\ e^{2t}$$

$$\therefore \quad L^{-1}\left[\frac{s^2}{(s-2)^2}\right] = 4e^{2t}(1+t).$$

***Example* 24.43:** Find $L^{-1}\left[\frac{s^2}{(s-a)^4}\right]$

Solution: $L^{-1}\left[s.\frac{s}{(s-a)^4}\right]$

$$= \frac{d\, L^{-1}\left[\frac{s}{(s-a)^4}\right]}{dt} = \frac{d\, L^{-1}\left[\frac{(s-a+a)}{(s-a)^4}\right]}{dt}$$

$$= \frac{d}{dt}\left\{L^{-1}\left[\frac{1}{(s-a)^3}\right] + L^{-1}\left[\frac{a}{(s-a)^4}\right]\right\}$$

$$= \frac{d}{dt}\left\{e^{at}\, L^{-1}\left[\frac{1}{s^3}\right] + ae^{at}\, L^{-1}\left[\frac{1}{s^4}\right]\right\} = \frac{d}{dt}\left\{e^{at}\, \frac{t^2}{2!} + ae^{at}.\frac{t^3}{3!}\right\}$$

$$= \frac{1}{2}\, [e^{at}.\, 2t + t^2\, ae^{at}] + \frac{a}{6}[e^{at}\, 3t^2 + t^3\, ae^{at}]$$

$$= \frac{e^{at}}{6}[at^3 + 6at^2 + 6at]$$

***Example* 24.44:** Find $L^{-1}\left[\frac{s}{(s-2)^2+9}\right]$

Solution: $L^{-1}\left[\frac{s}{(s-2)^2+9}\right] = L^{-1}\left[s.\frac{1}{(s-2)^2+9}\right]$

$$= \frac{d}{dt}L^{-1}\left[\frac{1}{(s-2)^2+9}\right] = \frac{d\, e^{2t} L^{-1}\left[\frac{1}{s^2+3^2}\right]}{dt}$$

$$= \frac{d\, e^{2t}\, \frac{1}{3}\sin 3t}{dt}$$

$$= \frac{1}{3}[e^{2t}.\, 3\cos 3t + 2\, e^{2t}\sin 3t]$$

$$\therefore \quad L^{-1}\left[\frac{s}{(s-2)^2+9}\right] = \frac{e^{2t}}{3}[3\cos 3t + 2\sin 3t]$$

Example **24.45:**

Find (i) $L^{-1}\left[\dfrac{1}{s^3(s^2+a^2)}\right]$ (ii) $L^{-1}\left[\dfrac{1}{s(s+2)^3}\right]$

(iii) $L^{-1}\left[\dfrac{1}{s(s^2+9)}\right]$ (iv) $L^{-1}\left[\dfrac{1}{(s^2+a^2)^2}\right]$

(v) $L^{-1}\left[\dfrac{1}{s(s^2-4s+5)}\right]$ (vi) $L^{-1}\left[\dfrac{1}{s}\log\left(\dfrac{s+3}{s+2}\right)\right]$

(vii) $L^{-1}\left[\dfrac{1}{s}\log\left(1+\dfrac{1}{s^2}\right)\right]$

Solution:

(i) $$L^{-1}\left[\frac{1}{s^3(s^2+a^2)}\right]$$

using $$L^{-1}\left[\frac{F(s)}{s}\right]=\int_0^t f(x)\,dx$$

i.e. $$L\left\{\int_0^t f(x)dx\right\}=\frac{1}{s}\,L\{f(t)\}$$

we have $$L^{-1}\left\{\frac{1}{s^2+a^2}\right\}=\frac{1}{a}\sin at$$

$$\therefore\quad L^{-1}\left[\frac{1}{s}\cdot\left(\frac{1}{s^2+a^2}\right)\right]=\int_0^t\frac{1}{a}\sin ax\,dx=\frac{1}{a}\left[-\frac{\cos ax}{a}\right]_0^t$$

$$=\frac{1}{a^2}[1-\cos at]$$

$$\therefore\quad L^{-1}\left[\frac{1}{s}\cdot\frac{1}{s(s^2+a^2)}\right]=\int_0^t\frac{1}{a^2}\,(1-\cos ax)\,dx \qquad \text{(using the same method)}$$

$$=\frac{1}{a^2}\left[x-\frac{\sin ax}{a}\right]_0^t=\frac{1}{a^2}\left[(t-0)-\frac{\sin at}{a}+0\right]$$

$$L^{-1}\left[\frac{1}{s^2(s^2+a^2)}\right]=\frac{1}{a^3}\ [at-\sin at]$$

Similarly

$$L^{-1}\left[\frac{1}{s^3(s^3+a^2)}\right]=L^{-1}\left[\frac{1}{s}.\frac{1}{s^2(s^2+a^2)}\right]$$

$$=\int_0^t\frac{1}{a^3}\ (ax-\sin ax)\,dx=\frac{1}{a^3}\left[a\frac{x^2}{2}+\frac{\cos\ ax}{a}\right]_0^t$$

$$=\frac{1}{a^3}\left[\frac{a}{2}(t^2-0)+\frac{1}{a}(\cos\ at-1)\right]_0^t=\left(\frac{1}{a^4}\cos\ at+\frac{t^2}{2a^2}-\frac{1}{a^4}\right)$$

$$\therefore\qquad L^{-1}\left[\frac{1}{s^3(s^2+a^2)}\right]=\frac{1}{2a^4}\ (2\cos at+a^2t^2-2)$$

(ii) $$L^{-1}\left[\frac{1}{s(s+2^3)}\right]=L^{-1}\left[\frac{1}{s}.\frac{1}{(s+2)^3}\right]$$

we have $L^{-1}\ \dfrac{1}{(s+2)^3}=e^{-2t}L^{-1}\ \dfrac{1}{s^3}=e^{-2t}\dfrac{t^2}{2}$

$$\therefore\qquad L^{-1}\left[\frac{1}{s.}\frac{1}{(s+2)^3}\right]=\int_0^t\frac{x^2}{2}e^{-2x}dx$$

$$=\frac{1}{2}\left[x^2\frac{e^{-2x}}{-2}-(2x)\left(\frac{e^{-2x}}{4}\right)+(2)\left(\frac{e^{-2x}}{-8}\right)\right]_0^t$$

$$=\frac{1}{2}\left[t^2\frac{e^{-2t}}{-2}-0-\frac{t}{2}e^{-2t}+0-\frac{1}{4}e^{-2t}-\frac{1}{4}\right]$$

$$=\frac{-e^{-2t}}{8}[1+2t+2t^2]+\frac{1}{8}$$

(iii) $$L^{-1}\left[\frac{1}{s(s^2+9)}\right]=L^{-1}\left[\frac{1}{s}.\frac{1}{s^2+9}\right]$$

we have $L^{-1}\left[\dfrac{1}{s^2+3^2)}\right]=\dfrac{1}{3}\ \sin 3t$

$$\therefore \quad L^{-1}\left[\frac{1}{s}.\frac{1}{s^2+9}\right]=\int_0^t \frac{1}{3}\sin 3x\,dx=\frac{1}{3}\left[-\frac{\cos 3x}{3}\right]_0^t=\frac{1}{9}(1-\cos 3t).$$

(iv) $L^{-1}\left[\frac{1}{(s^2+a^2)^2}\right]=L^{-1}\left[\frac{1}{s}.\frac{s}{(s^2+a^2)^2}\right]$

we have $\quad L^{-1}\left[\frac{s}{(s^2+a^2)^2}\right]=\frac{1}{2a}t\sin at$

$$\therefore \quad L^{-1}\left[\frac{1}{s}.\frac{s}{(s^2+a^2)^2}\right]=\int_0^t \frac{1}{2a}x\sin ax\,dx$$

$$=\frac{1}{2a}\left[x\left(\frac{-\cos ax}{a}\right)-(1)\left(\frac{-\sin ax}{a^2}\right)\right]_0^t$$

$$=\frac{1}{2a}\left[\frac{-t}{a}\cos at-0+\frac{\sin at}{a^2}\right]=\frac{1}{2a^3}[\sin at-at\cos at]$$

(v) $L^{-1}\left[\frac{1}{s(s^2-4s+5)}\right]=L^{-1}\left[\frac{1}{s}.\frac{1}{(s-2)^2+1}\right]$

we have $L^{-1}\left[\frac{1}{(s-2)^2+1}s\right]=e^{+2t}\,L^{-1}\frac{1}{s^2+1^2}=e^{2t}\sin t$

$$\therefore \quad L^{-1}\left[\frac{1}{s}.\frac{1}{(s-2)^2+1}\right]=\int_0^t e^{2x}\sin x\,dx=\left[\frac{e^{2x}}{2^2+1^2}(2\sin x+1\cos x)\right]_0^t$$

$$=\frac{e^{2t}}{5}[2\sin t+\cos t]-\frac{e^{2.0}}{5}(2.0+1)=\frac{e^{2t}}{5}(2\sin t+\cos t)-\frac{1}{5}.$$

(vi) From given $F(s)=\log\left(\frac{s+3}{s+2}\right)=\log(s+3)-\log(s+2)$

$$\therefore \quad F'(s)=\frac{1}{s+3}-\frac{1}{s+2}$$

$$L^{-1}[F'(s)]=e^{-3t}-e^{-2t}$$

$$-t\,L^{-1}[F(s)]=e^{-3t}-e^{-2t}$$

$$L^{-1}[F(s)] = \frac{e^{-3t} - e^{-2t}}{-t} = \frac{e^{-2t} - e^{-3t}}{t}$$

$$\therefore \quad L^{-1}\left[\frac{1}{s}.F(s)\right] = L^{-1}\left[\frac{1}{s}\log\left(\frac{s+3}{s+3}\right)\right] = \int_0^t \frac{e^{-2x} - e^{-3x}}{x}\, dx.$$

(vii) $L^{-1}\left[\frac{1}{s}\log\left(\frac{1}{s^2}\right)\right]$

Take $F(s) = \log\left(1+\frac{1}{s^2}\right) = \log(s^2+1) - \log s^2$. Then

$$F'(s) = \frac{2s}{s^2+1} - \frac{2}{s}$$

$$\therefore \quad L^{-1}\, F'(s) = 2\, L^{-1}\left(\frac{s}{s^2+1^2}\right) - 2\, L^{-1}\left(\frac{1}{s}\right)$$

$$L^{-1}\, F'(s) = 2\cos t - 2.$$

$$-t\, L^{-1}[F(s)] = 2\cos t - 2$$

$$L^{-1}[F(s)] = \frac{2(\cos t - 1)}{-t} \quad \frac{2(1-\cos t)}{t}.$$

$$L^{-1}\left[\frac{1}{s}F(s)\right] = \int_0^t \frac{2(1-\cos x)}{x}\, dx$$

$$\therefore \quad L^{-1}\left[\frac{1}{s}\log\left(1+\frac{1}{s^2}\right)\right] = \int_0^t 2\frac{(1-\cos x)}{x}\, dx.$$

***Example* 24.46:** Find $L^{-1}\left[\log\left(\frac{1+s}{s^2}\right)\right]$.

Solution: $L^{-1}\left[\log\left(\frac{1+s}{s^2}\right)\right] = f(t) \quad L\{f(t)\} = \log\left(\frac{1+s}{s^2}\right)$

$$L\{t\, f(t)\} = -\frac{d\log\left(\frac{1+s}{\frac{2}{s^2}}\right)}{ds} = -\frac{d}{ds}\log(1+s) + \frac{d(\log s^2)}{ds}$$

$$= -\frac{1}{1+s} + \frac{1}{s^2}.2s$$

$$L[t f(t)] = \frac{2}{s} - \frac{1}{1+s} \ ; \ t f(t) = L^{-1}\left\{\frac{2}{s} - \frac{1}{1+s}\right\}$$

$$= L^{-1}\left(\frac{2}{s}\right) - L^{-1}\left(\frac{1}{1+s}\right) = 2.1 - e^{-t}$$

$$\Rightarrow f(t) = \frac{2 - e^{-t}}{t}$$

$$L^{-1}\left[\log\left(\frac{1+s}{s^2}\right)\right] = \frac{2 - e^{-t}}{t}.$$

***Example* 24.47:** Find $L^{-1}\left[\log\left(\frac{s^2-4}{s^2+9}\right)\right]$

Solution: $L^{-1}\left[\log\left(\frac{s^2-4}{s^2+9}\right)\right] = f(t)\ L\{f(t)\} = \log\left(\frac{s^2-4}{s^2+9}\right)$

$$L\{t f(t)\} = -\frac{d\log\left(\frac{s^2-4}{s^2+9}\right)}{ds}$$

$$L\{t f(t)\} = -\frac{d\log(s^2-4)}{ds} + \frac{d\log(s^2+9)}{ds}$$

$$L\{t f(t)\} = -\frac{2s}{s^2-4} + \frac{2s}{s^2+9}$$

$$t f(t) = 2\,L^{-1}\left(\frac{s}{s^2+9}\right) - 2L^{-1}\left(\frac{s}{s^2-4}\right) = 2\cos 3t - 2\cos h\, 2t.$$

$$L^{-1}\left\{\log\left(\frac{s^2-4}{s^2+9}\right)\right\} = \frac{2(\cos 3t - \cos h2t)}{t}.$$

***Example* 24.48:** Find $L^{-1}\left[\log\frac{s(s+4)}{s^2+9}\right]$

Solution: $L^{-1}\left[\log\frac{s(s+4)}{s^2+9}\right] = f(t) \Rightarrow L\{f(t)\} = \log\frac{s(s+4)}{s^2+9}$

$$L\{t f(t)\} = -\frac{d\log\left[\frac{s(s+4)}{s^2+9}\right]}{ds} = -\frac{d\log s}{ds} - \frac{d\log(s+4)}{ds} + \frac{d(s^2+9)}{ds}$$

$$L\{t f(t)\} = -\frac{1}{s} - \frac{1}{s+4} + \frac{2s}{s^2+9}$$

$$t f(t) = -L^{-1}\left(\frac{1}{s}\right) - L^{-1}\left(\frac{1}{s+4}\right) + 2L^{-1}\left(\frac{2s}{s^2+3^2}\right)$$

$$= 1 - e^{-4t} + 2\cos 3t$$

$$\therefore \quad L^{-1}\left[\log\frac{s(s+4)}{s^2+9}\right] = \frac{2\cos 3t - e^{-4t} - 1}{t}$$

***Example* 24.49:** Find $L^{-1}\left[\log\left(1+\frac{w^2}{s^2}\right)\right]$.

***Solution*:** $L^{-1}\left[\log\left(1+\frac{w^2}{s^2}\right)\right] = f(t) \Rightarrow L\{f(t)\} = \log\frac{s^2+w^2}{s^2}$

$$L\{t f(t)\} = -\frac{d\log\left(\frac{s^2+w^2}{s^2}\right)}{ds}$$

$$= \frac{2s}{s^2+w^2} + \frac{2s}{s^2}$$

$$t f(t) = -2L^{-1}\left(\frac{s}{s^2+w^2}\right) + 2L^{-1}\left(\frac{1}{s}\right)$$

$$t f(t) = -2\cos wt + 2.$$

$$\therefore \quad L^{-1}\left[\log\left(1+\frac{w^2}{s^2}\right)\right] = 2\frac{(1-\cos wt)}{t}.$$

***Example* 24.50:** Find the inverse Laplace transforms using convolution theorem

(i) $\dfrac{1}{s(s^2+4)}$ (ii) $\dfrac{s}{(s^2+a^2)^2}$ (iii) $\dfrac{1}{(s^2+9)^2}$

(iv) $\dfrac{s^2}{(s^2+a^2)\,(s^2+b^2)}$ (v) $\dfrac{1}{(s^2+4)\,(s+2)}$ (vi) $\dfrac{1}{s(s^2+9)^2}$

(vii) $\dfrac{1}{(s+3)\,(s-4)}$ (viii) $\dfrac{1}{s\sqrt{s+4}}$.

Solution:

(i) $$L^{-1}\left[\frac{1}{s(s^2+4)}\right]=L^{-1}\left[\frac{1}{s}\right].\,L^{-1}\left[\frac{1}{s^2+4}\right]$$

$$=1*\frac{1}{2}\sin 2t \qquad \text{(By convolution theorem)}$$

$$=\frac{1}{2}\int_0^t 1.\sin 2(t-u)du$$

$$=\left[\frac{1}{2}\,\frac{-\cos 2(t-u)}{-2}\right]_0^t=\frac{1}{4}[1-\cos 2t].$$

(ii) $$L^{-1}\left[\frac{s}{(s^2+a^2)^2}\right]=L^{-1}\left[\frac{s}{s^2+a^2}\right].\,L^{-1}\left[\frac{1}{s^2+a^2}\right]$$

$$=\cos at * \sin at=\frac{1}{a}\int_0^t \cos au\ \sin a(t-u)\ du$$

$$=\frac{1}{a}\int_0^t\left[\frac{\sin(at-au+au)+\sin(at-au-au)}{2}\right]du$$

$$=\frac{1}{2a}\int_0^t[\sin at+\sin a(t-2u)]\ du$$

$$= \frac{1}{2a}\left\{\sin at.[u]_0^t + \left[-\frac{\cos a(t-2u)}{-2a}\right]_0^t\right\}$$

$$= \frac{1}{2a}\left\{\sin at.t + \frac{1}{2a}(\cos at - \cos at)\right\} = \frac{1}{2a}\sin at.$$

(iii) $L^{-1}\left[\frac{1}{(s^2+9)^2}\right] = L^{-1}\left[\frac{1}{s^2+9}\right].L^{-1}\left[\frac{1}{s^2+9}\right]$

$= \frac{1}{3}\sin 3t * \frac{1}{2}\sin 3t$ (By convolution theorem)

$$= \frac{1}{9}\int_0^t \sin 3u \sin 3(t-u)\, du$$

$$= \frac{1}{9}\int_0^t \left[\frac{[\cos(3u-3t+3u) - \cos(3u+3t-3u)]}{2}\right] du$$

$$= \frac{1}{18}\int_0^t [\cos 3\,(2u-t) - \cos 3t]\, du = \frac{1}{18}\left[\frac{\sin(6u-2t)}{6} - \cos 3t.u\right]_0^t$$

$$= \frac{1}{108}\sin 4t - \frac{t}{18}\cos 3t + 0 = \frac{1}{108}\sin 4t - \frac{t}{18}\cos 3t.$$

(iv) $L^{-1}\left[\frac{s^2}{(s^2+a^2)(s^2+b^2)}\right] = L^{-1}\left[\frac{s}{s^2+a^2}\right].L^{-1}\left[\frac{s}{s^2+b^2}\right]$

$\cos at * \cos bt$ (by convolution theorem)

$$= \int_0^t \cos au \cos b(t-u)\, du$$

$$= \int_0^t \frac{[\cos(au+bt-bu) + \cos(au-bt+bu)]\, du}{2}$$

$$= \frac{1}{2}\int_0^t \{\cos[(a-b)u+bt]+\cos[(a-b)u-bt]\}du$$

$$= \frac{1}{2}\left\{\frac{\sin[bt+(a-b)u]}{a-b}+\frac{\sin[\sin(a+b)u-bt]}{a+b}\right\}_0^t$$

$$= \frac{1}{2}\left\{\frac{\sin bt+at-bt}{a-b}+\frac{\sin at+bt-bt}{a+b}-\frac{\sin bt}{a-b}+\frac{\sin bt}{a+b}\right\}$$

$$= \frac{1}{2}\left[\frac{2a\sin at-2b\sin bt}{a^2-b^2}\right] = \frac{a\sin at-b\sin bt}{a^2-b^2}.$$

(v) $$L^{-1}\left[\frac{1}{(s^2+4)\,(s+2)}\right] = L^{-1}\left[\frac{1}{s^2+2^2}\right].\, L^{-1}\left[\frac{1}{s+2}\right]$$

$$= \frac{1}{2}\sin 2t * e^{-2t}.$$ (By convolution theorem)

$$= \frac{1}{2}\int_0^t \sin 2u e^{-2(t-u)}\, du$$

$$= \frac{1}{2}\left[\frac{e^{-2(t-u)}}{2^2+2^2}\{2\sin 2u-2\cos 2u\}\right]_0^t$$

$$= \frac{1}{8}\,[\sin 2t-\cos 2t+e^{-2t}].$$

(vi) $$L^{-1}\left[\frac{1}{s(s^2+9)^2}\right] = L^{-1}\left[\frac{1}{s^2}.\frac{s}{(s^2+9)^2}\right]$$

$$= L^{-1}\left[\frac{1}{s^2}\right].\, L^{-1}\left[\frac{s}{(s^2+9)^2}\right]$$ using $L^{-1}\left[\frac{s}{(s^2+a^2)^2}\right] = \frac{t}{2a}\sin at$

$$= t * \frac{t}{6}\sin 3t.$$ (By convolution theorem)

$$= \frac{1}{6}\int_0^t u \sin 3u.\ (t-u)\ du$$

$$= \frac{1}{6}\int_0^t u \sin 3u.\ du - \frac{1}{6}\int_0^t u^2 \sin\ 3u\ du$$

$$= \frac{1}{6}\left\{(u)\left(\frac{-\cos 3u}{3}\right) - (1)\left(\frac{-\sin\ 3u}{9}\right)\right\}_0^t$$

$$- \frac{1}{6}\left\{(u^2)\left(\frac{-\cos 3u}{3}\right) - (2u)\left(\frac{-\sin\ 3u}{9}\right) + (2)\left(\frac{\cos 3u}{27}\right)\right\}_0^t$$

$$= \frac{1}{81}(1-\cos 3t) - \frac{t}{54}\sin 3t.$$

(vii) $L^{-1}\left[\frac{1}{(s+3)\,(s-4)}\right] = L^{-1}\left[\frac{1}{s+3}\right].\ L^{-1}\left[\frac{1}{s-4}\right]$

$= e^{-3t} * e^{4t}$ (By convolution theorem)

$$= \int_0^t e^{-3u}.\ e^{4(t-u)} du$$

$$= e^{4t}\int_0^t e^{-7u} du = e^{4t}\left[\frac{e^{-7u}}{-7}\right]_0^t = -\frac{1}{7}(e^{-3t} - e^{4t}) = \frac{e^{4t} - e^{-3t}}{7}.$$

(viii) $L^{-1}\left[\frac{1}{s\sqrt{s+4}}\right] = L^{-1}\left[\frac{1}{s}\right].\ L^{-1}\left[\frac{1}{(s+4)^{1/2}}\right]$

$$= L^{-1}\left[\frac{1}{s}\right].\ e^{-4t}L^{-1}\left[\frac{1}{s^{1/2}}\right] = 1 * \frac{e^{-4t}}{}.\frac{t^{-1/2}}{\Gamma(1/2)}$$

$= 1 * \frac{e^{-4t}}{\sqrt{\pi t}}$ (By the convolution theorem)

$$= \int_0^t \frac{e^{-4u}}{\sqrt{\pi u}}.1.du$$

substituting $4u = x^2$, $du = \frac{1}{2}x\,dx, \frac{du}{\sqrt{u}} = dx$

$$= \frac{1}{\sqrt{\pi}}\int_0^{2\sqrt{t}} e^{-x^2}dx = \frac{1}{2}.\frac{2}{\sqrt{\pi}}\int_0^{2\sqrt{t}} e^{-x^2}dx$$

$$= \frac{1}{2} erf(2\sqrt{t}).$$

[using error function definition $erf(t) = \frac{2}{\sqrt{\pi}}\int_0^t e^{-x^2}dx$]

***Example* 24.51:** Using convolution theorem show that

$$b(m, n) = \int_0^t x^{m-1}(1-x)^{n-1}dx = \frac{\Gamma m\,\Gamma n}{\Gamma(m+n)}, \begin{matrix} m>0 \\ n>0 \end{matrix}$$

Solution : Let $f(x) = \int_0^t x^{m-1}(t-x)^{n-1}dx$ ... (1)

The functions of the integrand on R.H.S of (1) are in the form of the functions in the integrand of the convolution theorem.

i.e., $f(x) = \int_0^t f_1(x)\, f_2(t-x)\, dx$

where $f_1(x) = x^{m-1}, f_2(x) = x^{n-1}$

$$= f_1(x) * f_2(x)$$

$\therefore$ $L\{f(x)\} = L\{f_1 * f_2\} = L\{f_1(t)\}\ L\ \{f_2(t)\}$

$$= L\{t^{m-1}\}.\ L\{t^{n-1}\} = \frac{\Gamma m}{s^m}.\frac{\Gamma n}{s^n}$$

$$L\{f(t)\} = \frac{\Gamma m\,\Gamma n}{s^{m+n}}$$

$$\therefore \qquad f(x) = \Gamma m\, \Gamma n\, L^{-1}\left[\frac{1}{s^{m+n}}\right] \qquad \text{... (2)}$$

From (1) and (2)

$$f(x) = \int_0^t x^{m-1}(t-x)^{n-1}\,dx = \Gamma m\, \Gamma n\, L^{-1}\left[\frac{1}{s^{m+n}}\right]$$

$$f(x) = \Gamma m\, \Gamma n\left[\frac{t^{m+n-1}}{\Gamma(m+n)}\right]$$

i.e.,

$$f(x) = \int_0^t x^{m-1}(t-x)^{n-1}\,dx = \frac{\Gamma m\, \Gamma n}{\Gamma(m+n)}(t^{m+n-1})$$

taking $t = 1$

$$f(x) = \int_0^t x^{m-1}(1-x)^{n-1}\,dx = \frac{\Gamma m\, \Gamma n}{\Gamma(m+n)}$$

$$\therefore \qquad f(x) = \beta(m, n) = \frac{\Gamma m\, \Gamma n}{\Gamma(m+n)}$$

***Example* 24.52**

(1) Evaluate $= L^{-1}\left[\frac{1}{s^2(s+1)^2}\right]$, using the convolution theorem

(2) Evaluate $L^{-1}\left[\frac{s}{(s^2+2^2)^2}\right]$ using the convolution theorem.

(3) Evaluate $L^{-1}\left[\frac{1}{s\sqrt{s+4}}\right]$ using convolution theorem.

Solution:

(1) we have $L^{-1}\left(\frac{1}{s^2}\right) = t$ and $L^{-1}\left(\frac{1}{(s+1)^2}\right) = te^{-t}$

By the convolution theorem

$$= L^{-1}\left[\frac{1}{s^2(s+1)^2}\right] = \int_0^t (ue^{-u} - (t-u)\ du = \int_0^t (ut - u^2)e^{-u})du$$

$= [(ut - u^2)e^{-u} - (t - 2u)e^{-u} + (-2)(-e^{-u}) = t - e^{-t} + 2e^{-1} + t - 2$

(2) Let $F(s) = \frac{s}{s^2+4}$, $G(s) = \frac{1}{s^2+4}$ and Let $\phi(s) = F(s)\ G(s)$

$f(t) = L^{-1}[F(s)] = L^{-1}(\frac{s}{s^2+4}) = \cos 2t$

$g(t) = L-1[G(s)] = L^{-1}(\frac{1}{s^2+4}) = \frac{1}{2}\sin 2t$

$f(u) = \cos 2u$ and $g(t-u) = \frac{1}{2}\sin 2(t-u)$

$$L^{-1}[\varphi(s)] = \int_0^t \cos 2u \frac{1}{2}\sin 2(t-u)\,du$$

$$= \frac{1}{4}\int_0^t \sin 2t + \sin 2(t-2u)du$$

$$= \frac{1}{4}\left[\sin 2t.u + \frac{1}{2}\cos 2(t-2u\right]_0^t \frac{1}{4}t\sin 2u$$

(3) Let $F(s) = \frac{1}{\sqrt{s+4}}$; $G(s) = \frac{1}{s}$ and choose

$\phi(s) = F(s).G(s)$

$\therefore\ f(t) = L^{-1}[F(s)] = L^{-1}\frac{1}{(s+4)^{1/2}} = e^{-4t}\ L^{-1}\frac{1}{s^{1/2}} = e^{-4t}\frac{t^{1/2}}{\Gamma\frac{1}{2}} = \frac{e^{-4t}}{\sqrt{\pi t}}$

$g(t) = L^{-1}[G(s)] = L^{-1}(\frac{1}{s}) = 1$

$f(u) = \frac{e^{-4u}}{\sqrt{\pi u}}$ and $g(t-u) = 1$

By convolution theorem

$$L^{-1}[\phi(s)] = \int_0^t \frac{e^{-4u}}{\sqrt{\pi u}}\,du = \frac{1}{\sqrt{\pi}}\int_0^{2\sqrt{t}} e^{-x^2}\,dx$$

where $4u = x^2$ $\qquad 2du = xdx \qquad \sqrt{u} = \frac{x}{2}$

$$L^{-1}[\phi(s)] = \frac{1}{2} erf(2\sqrt{t}) \qquad \text{(where } erf(t) = \frac{2}{\sqrt{\pi}}\int_0^t e^{-x^2}\,dx$$

erf(t) stands for error function)

Exercise 24.12

Find the inverse Laplace transform of the following functions using convolution theorem.

1. $\dfrac{1}{(s+1)(s-2)}$, **Ans:** $\dfrac{e^{2t}-e^{-t}}{2}$

2. $\dfrac{1}{s(s^2+4)^2}$ **Ans:** $\dfrac{1}{16}(1 - t\sin 2t - \cos 2t)$

3. $\dfrac{s}{(s^2+4)^3}$ **Ans:** $\dfrac{t}{64}(\sin 2t - 2t\cos 2t)$

4. $\dfrac{1}{\sqrt{s}(s+1)}$ **Ans:** $erf(t)$

5. $\dfrac{1}{(s+2)^2(s-2)}$ **Ans:** $\dfrac{1}{16}\left[e^{2t} - 4te^{-2t} - e^{-2t}\right]$

6. $\dfrac{s^2+s}{(s^2+1)(s^2+2s+2)}$ **Ans:** $\dfrac{1}{10}\left[e^{-t}(2\sin t - 6\cos t) + (2\sin t + 6\cos t)\right]$

7. $\dfrac{s^2}{(s^2-a^2)^2}$ **Ans:** $(e^{at}-1)$

8. $\dfrac{s^2}{(s^2-a^2)^2}$ **Ans:** $\dfrac{1}{2a}(\sin h\, at + at\cos h\, at)$

9. $\dfrac{s}{(s^2+9)^2}$ **Ans:** $\dfrac{t}{2}\sin 3t$

10. $\dfrac{s^2}{(s^2+1)(s^2+4)}$ **Ans:** $\frac{1}{3}(2\sin 2t - \sin t)$

11. $\dfrac{s^2}{(s^2+1)^2}$ **Ans:** $\frac{1}{3}(t \sin t)$

12. $\dfrac{1}{s}\log\left(\dfrac{s+2}{s+1}\right)$ **Ans:** $\int_0^t \frac{e^{-u}-e^{-2u}}{u}\,du$

13. $\dfrac{1}{(s+1)(s^2+1)}$ **Ans:** $\frac{1}{2}\left[e^{-t}-\cos t-\sin t\right]$

24.2.13 Application of Laplace Transform to Solutions of Ordinary Differential Equations

Laplace transformation is useful in solving ordinary as well as partial differential equations. The advantage of this method over the conventional one, as already pointedout, is that it reduces the problem of solving a differential equation to an algebraic problem. Moreover this method yields the particular integral directly without finding complementary functin and particular integral and then evaluating the arbitrary constants.

Let us consider a linear differential equation with constant coefficients

$$\frac{d^n y}{dt^n}+a_1\frac{d^{n-1}y}{dt^{n-1}}+a_2\frac{d^{n-2}y}{dt^{n-2}}+\ldots+a_{n-1}\frac{dy}{dt}+a_n t=f(t) \qquad \ldots\ldots(1)$$

and $y(0)=\alpha_1, y'(0)=\alpha_2, \ldots y^{n-1}(0)=\alpha_{n-1}$ be the given inital or boundary conditions where $\alpha_1, \alpha_2, \ldots \alpha_{n-1}$ are constants.

Taking Laplace transform of both sides of the differential equation with the help of the formula

$$L\{y^n(t)\}=s^n L\{y\}-s^{n-1}y(0)-s^{n-2}y'(0)-\ldots-y^{n-1}(0)$$

and using the given inital conditions, the equation 1 reduces to an algebraic equation of the form Y = F(s), which is called subsidary equation, from we can find the required solution by taking inverse Laplace trnasform of F(s).

i.e., $L^{-1}[y]=L^{-1}[F(s)]$ or $y=f(t)$ is the required solution.

Solved Examples

***Example* 24.53:** Solve by Laplace transform method

$$y''-3y'+2y=4 \text{ where } y(0)=2,\ y'(0)=3.$$

***Solution*:** Taking Laplace transform of both sides.

$$L\{y''\}-3\,L\{y'\}+2\,L\{y\}=L\{4\}$$

$$\left[s^2L\{y\}-s\,y(0)-y'(0)\right]-3\left[s\,L\{y\}-y(0)\right]2+L\{y\}=4.\frac{1}{s}$$

Using $y(0)=2,\ y'(0)=3$

$$\left[s^2L\{y\}-s.2-3\right]-3\left[s\,L\{y\}-2\right]+2\,L\{y\}=\frac{4}{s}$$

$$L\{y\}\left[s^2-3s+2\right]=\frac{4}{s}\times 2s-3$$

$$L\{y\}=\frac{2s^2-3s+4}{s\left(s^2-3s+2\right)}$$

$$L\{y\}=\frac{2}{s}-\frac{3}{s-1}+\frac{3}{s-2} \qquad \text{(resolving into partial fractions)}$$

$$y=2\,L^{-1}\,\frac{1}{s}-3\,L^{-1}\left(\frac{1}{s-1}\right)+3\,L^{-1}\left[\frac{1}{s-2}\right]$$

$$y=2-3\,e^{t}+3\,e^{2t}$$

***Example* 24.54:** Solve $(D^2-D-2)\,y=20\sin 2t$ where $y(0)=1,\ y'(0)=2$

***Solution*:** Given differential equation is $(D^2-D-2)\,y=20\sin 2t$ applying Laplace transform on both sides

$$L\{y''(t)\}-L\{y'(t)\}-2\,L\{y(t)\}=20\,L\{\sin 2t\}$$

$$\left[s^2L\{y\}-s\,y(0)-y'(0)\right]-\left[s\,L\{y\}-y(0)\right]-2L\{y\}=20\frac{2}{s^2+2^2}$$

Using $y(0)=1,\ y'(0)=2$

$$L\{y\}\left[s^2-s-2\right]-s.1-2+1=\frac{40}{s^2+4}$$

$$L\{y\} = \left[\frac{40}{s^2+4} + s.1\right] . \frac{1}{s^2-s-2} = \frac{s^3+s^2+4s+44}{(s^2+4)(s^2-s-2)}$$

using paritial fractions

$$L\{y\} = -\frac{8}{3(s+1)} + \frac{8}{3(s-2)} + \frac{s-6}{s^2+4}$$

$$y = \frac{-8}{3} L^{-1}\left[\frac{1}{s+1}\right] + \frac{8}{3} L^{-1}\left[\frac{1}{s-2}\right] + L^{-1}\left[\frac{s}{s^2+2^2}\right] - 3\, L^{-1}\left[\frac{2}{s^2+2^2}\right]$$

$$y = \frac{-8}{3} e^{-t} + \frac{8}{3} e^{2t} + \cos 2t - 3\sin 2t$$

***Example* 24.55:** Solve $y''-3y'+2y=e^{2t}$, $y(0)=-3$ $y'(0)=5$

***Solution*:** Given $\frac{d^2y}{dt^2} - 3\frac{dy}{dy} + 2y = e^{2t}$

Applying Laplace transform on both sides

$$L\{y''\} - 3\, L\{y'\} + 2L\{y\} = L\{e^{2t}\}$$

$$\left[s^2 L\{y\} - s(0) - y'(0)\right] - 3\left[s\, L\{y\} - y(0)\right] + 2\, L\{y\} = \frac{1}{s-2}$$

Using $y(0)=-3$, $y'(0)=5$

$$L\{y\}\left[s^2+3s+2\right] - s(-3) - 5 + 3(-3) = \frac{1}{s-2}$$

$$L\{y\}\left[s^2+3s+2\right] = \frac{1}{s-2} - 3s + 14 = \frac{-3s^2+20s+27}{s-2}$$

$$L\{y\} = \frac{-3s^2+20s+27}{(s-2)(s^2-3s+2)}$$ using partial fractions

$$L\{y\} = \frac{-10}{s-1} + \frac{7}{s-2} + \frac{1}{(s-2)^2}$$

$$y = -10\, L^{-1}\left(\frac{1}{s-1}\right) + 7\, L^{-1}\left[\frac{1}{s-2}\right] + L^{-1}\left[\frac{1}{(s-2)^2}\right]$$

$$y = t\, e^{2t} + 7\, e^{2t} - 10\, e^{t}$$

***Example* 24.56**: Solve $\frac{d^2x}{dt^2} - 3\frac{dx}{dt} + 2x = 1 - e^{2t}$, $x(0) = 1$, $x'(0) = 0$

Solution: Given $\quad x'' - 3x' + 2x = 1 - e^{2t}$

Applying Laplace transform on both sides

$$L\{x''\} - 3\,L\{x'\} + 2\,L\{x\} = L\{1\} - \{e^{2t}\}$$

$$\left[s^2 L\{x\} - s\,x(0) - x'(0)\right] - 3\left[s\,L\{x\} - x(0)\right] + 2L\{x\} = \frac{1}{s} - \frac{1}{s-2}$$

using $x(0) = 1,\ x'(0) = 0$

$$L\{x\}\left[s^2 - 3s + 2\right] - s.1 - 0 + 3.1 = \frac{1}{s} - \frac{1}{s-2}$$

$$L\{x\}\left[s^2 - 3s + 2\right] = \frac{1}{s} - \frac{1}{s-2} + s - 3 = \frac{s^3 - 5s^2 + 6s - 2}{s(s-2)}$$

$$L\{x\} = \frac{s^3 - 5s^2 + 6s - 2}{s(s-2)(s^2 - 3s + 2)} = \frac{s^3 - 5s^2 + 6s - 2}{s(s-1)(s-2)^2} = \frac{(s-1)(s^2 - 4s + 2)}{s(s-1)(s-2)^2}$$

$$L\{x\} = \frac{s^3 - 4s + 2}{s(s-2)^2}$$

$$L\{x\} = \frac{1}{2}.\frac{1}{s} + \frac{1}{2}.\frac{1}{(s-2)} - \frac{1}{(s-2)^2} \qquad \text{using partial fractions}$$

$$x = \frac{1}{2}L^{-1}\left[\frac{1}{s}\right] + \frac{1}{2}L^{-1}\left[\frac{1}{s-2}\right] - L^{-1}\left[\frac{1}{(s-2)^2}\right]$$

$$x = \frac{1}{2}.1 + \frac{1}{2}e^{2t} - e^{2t}L^{-1}\left[\frac{1}{s^2}\right] = \frac{1}{2} + \frac{1}{2}e^{2t} - t\,e^{2t}$$

***Example* 24.57:** Solve $\frac{d^2y}{dt^2} + 2\frac{dy}{dt} + 5y = e^{-t}$ sin t, $y(0) = 0,\ y'(0) = 1$.

Solution: Given $y'' + 2y' + 5y = e^{-t}\sin t$

Applying Laplace transformation on both sides

$$L\{y''\} + 2L\{y'\} + 5L\{y\} = L\{e^{-t}\sin t\}$$

$$\left[s^2L\{y\} - s\,y(0) - y'(0)\right] + 2\left[s\,L\{y\} - y(0)\right] + 5L\{y\} = \frac{1}{(s+1)^2 + 1}$$

Using $y(0) = 0,\ y'(0) = 1$

$$L\{y\}\left[s^2 + 2s + 5\right] - s.0 - 1 - 2.0 = \frac{1}{s^2 + 2s + 2}$$

$$L\{y\} = \frac{1}{s^2 + 2s + 5}\left[\frac{s^2 + 2s + 3}{(s^2 + 2s + 2)}\right]$$

Resolving into partial fractions

$$L\{y\} = \frac{1}{3}.\frac{1}{s^2 + 2s + 2} + \frac{2}{3}.\frac{1}{s^2 + 2s + 5}$$

$$y = \frac{1}{3}L^{-1}\left[\frac{1}{s^2 + 2s + 2}\right] + \frac{2}{3}L^{-1}\left[\frac{1}{s^2 + 2s + 5}\right]$$

$$y = \frac{1}{3}L^{-1}\left[\frac{1}{(s+1)^2 + 1}\right] + \frac{1}{3}L^{-1}\left[\frac{2}{(s+1)^2 + 2^2}\right]$$

$$y = \frac{1}{3}e^{-t}\sin t + \frac{1}{3}e^{-t}\sin 2t$$

***Example* 24.58:** Find the solution of $\frac{d^2y}{dt^2}+\frac{dy}{dt}-2y$ = sint which satisfies the initial conditions y = 0. y' = 0 when t = 0

Solution: Applying the Laplace transforms to both sides of the given equation and by linearity, we have

$$L(y'')+L(y')-2L(y)=L(\sin t)$$

As we know that $L(\text{sint})=\frac{1}{s^2+1}$; $s>0$

$$[s^2L(y)-sy(0)-y'(0)]+[sL(y)-y(0)]-2L(y)=\frac{1}{s^2+1}$$

Given conditions y(0) = 0, y' (0) = 0

$$\Rightarrow s^2L(y)+sL(y)-2L(y)=\frac{1}{s^2+1}$$

$$L(y)=\frac{1}{(s^2+s-2)(s^2+1)}=\frac{1}{(s-1)(s+2)(s^2+1)}$$

we have the partial fraction expansion

$$\frac{1}{(s-1)(s+2)(s^2+1)}=\frac{A}{s-1}+\frac{B}{s+2}+\frac{Cs+D}{s^2+1}$$

where $\quad A=\frac{1}{6}, B=\frac{-1}{15}. C=\frac{-1}{10}$ and $D=\frac{-3}{10}$

Therefore we get

$$y(t)=L^{-1}\left|\frac{1}{(s-1)(s+2)(s^2+1)}\right|$$

$$=\frac{1}{6}L^{-1}\left[\frac{1}{s-1}\right]-\frac{1}{15}L^{-1}\left[\frac{1}{s+2}\right]-\frac{1}{10}L^{-1}\left[\frac{s}{s^2+1}\right]-\frac{3}{10}L^{-1}\left[\frac{1}{s^2+1}\right]$$

$$=\frac{1}{6}e^{t}-\frac{1}{15}e^{-2t}-\frac{1}{10}\cos t-\frac{3}{10}\sin t$$

as the solution to the given initial value problem.

***Example* 24.59:** Solve the initial value problem

$$y''-3y'+2y=e^{3t},\ y=1,\ y'=0 \text{ where } t=0$$

***Solution*:** Applying Laplace transform to both sides of the differential equation, we get

$$[s^2Ly(t)-sy(0)-y'(0)]-3[sLy(t)-y(0)]+2L(t)=L(e^{3t})=\frac{1}{s-3}$$

Applying the given conditions, we get

$$(s^2-3s+2)\,Ly(t)=s-3+\frac{1}{s-3}$$

$$\therefore \quad L[y(t)]=\frac{s-3}{(s^2-3s+2)}+\frac{1}{(s-3)(s^2-3s+2)}$$

$$\therefore \quad L(y(t)]=\frac{1}{(s-3)(s^2-3s+2)}+\frac{s-3}{(s^2-3s+2)}$$

$$=\frac{1}{(s-1)(s-2)(s-3)}+\frac{s-3}{(s-1)(s-2)} \qquad ....(1)$$

To find y(t), we expand each term on the right hand side of (1) in partial functions, Thus

$$\frac{1}{(s-1)(s-2)(s-3)}=\frac{A}{s-1}-\frac{B}{s-2}+\frac{C}{s-3}$$

$$1=A(s-2)(s-3)+B(s-1)(s-3)+C(s-1)(s-2)$$

substituting s = 1 gives $A=\frac{1}{2}$;

similarly substituting s = 2 gets B = –1 and substituting

s = 3 gives $C=\frac{1}{2}$

$$\therefore \quad \frac{1}{(s-1)(s-2)(s-3)} = \frac{1}{2(s-1)} - \frac{1}{s-2} + \frac{1}{2(s-3)}$$

Similarly we can also write

$$\frac{s-3}{(s-1)(s-2)} = \frac{D}{s-1} + \frac{E}{s-2}$$

$$s - 3 = D\,(s - 2)\; E(s - 1)$$

Taking s = 2, we get E = –1

similarly s = 1 we get D = 2

Hence $\quad L[y(t)] = \dfrac{1}{2(s-1)} - \dfrac{1}{s-2} + \dfrac{1}{2(s-3)} + \dfrac{1}{s-1} - \dfrac{1}{s-2}$

$$= \frac{2}{2(s-1)} - \frac{2}{s-2} + \frac{1}{2(s-3)}$$

(or) $\quad L[y(t)] = L[\frac{5}{2}e^{t} - 2e^{2t} + \frac{1}{2}e^{3t}]$

$$y(t) = \frac{5}{2}e^{t} - 2e^{2t} + \frac{1}{2}e^{3t}$$

***Example* 24.60:** Solve $y'' + 3y' + 2y = 2t^3 + 2t + 2$ with $y(0) = 2,\ y'(0)$

Solution: Applying Laplace transform to both sides of the given equation we get

$$L(y'') + 3L(y') + 2L(y) = 2L(t^2) + 2L(t) + 2$$

$$[s^2Ly(t) - sy(0) - y'(0)] + 3[sL[y(t)] - y(0)] + 2Ly(t) = \frac{4}{s^3} + \frac{2}{s^2} + \frac{2}{s} \qquad ....(1)$$

using the given conditions (1) gives

$$(s^2 + 3s + 2)Ly(t) = \frac{2(2+s+s^2)}{s^3} + 2s + 6 = \frac{2(s^4+3s^3+s^2+s+2)}{s^3}$$

$$\therefore \quad L[y(t)] = \frac{2(s^4+3s^3+s^2+s+2)}{s^3(s^2+s+2)}$$

By partial fractions,

$$y(t) = 3L^{-1}\left(\frac{1}{s}\right) - 2L^{-1}\left(\frac{1}{s^2}\right) + 2L^{-1}\left(\frac{1}{s^3}\right) - L^{-1}\left(\frac{1}{s+2}\right)$$

$$y(t) = 3 - 2t + t^2 - e^{-2t}$$

Exercise 24.13

Solve by Laplace Transform Method

1. $\frac{d^2y}{dt^2} + \frac{dy}{dt} - 2y = 3\cos 3t - 11\sin 3t$; $y(0) = 0,\ y'(0) = 6$

[Ans: ($y = e^t - e^{-2t} + \sin 3t$)]

2. $\frac{d^2x}{dt^2} + x = t\cos 2t,\ x(0) = x'(0) = 0$

[Ans: ($x = \frac{1}{9}(4\sin 2t - 5\sin t - 3t\cos 2t)$)]

3. $y'' + y' - 2y = t,\ y(0) = 1,\ y'(0) = 0$

[Ans: ($y = \frac{1}{4}(4e^t + e^{-2t} - 2t - 1)$)]

4. $y'' + 3y' + 2y = 2(t^2 + t + 1)$

[Ans: ($y = 3 - 2t + t^2 - e^{-2t}$)]

5. $y''' - 2y'' + y' = 0, y(0) = 0, y'(0) = 0, y''(0) = 1$

[Ans: ($y = \frac{1}{10}e^t \sin 2t - \frac{1}{5}e^t \cos 2t + \frac{1}{5}$)]

6. $(D^3 - 3D^2 + 3D - 1)y = t^2e^t,\ \ y(0) = 1, y'(0) = 0, y''(0) = -2$

[Ans: ($y = e^t - t\,e^t - \frac{t^2e^t}{2} + \frac{t^5e^t}{60}$)]

7. $(D^3 + D)y = 2,\ y(0) = 3,\ y'(0) = 1,\ y''(0) = -2$

Summary

Definition: $L[f(t)] = \int_0^\infty e^{-st} f(t)\,dt = F(s)$

Table 1 Laplace Transforms of Simple Functions

S.No.	Function $f(t)$	Laplace transform $L[f(t)$ $= F(s) = \int_0^\infty e^{-st} f(t)\,dt\,]$	Function f(t)	Laplace transform F(s)
1.	1	$\frac{1}{s}$	sin *at*	a/s^2+a^2
2.	t	$\frac{1}{s^2}$	cos *at*	s/s^2+a^2
3.	t^n	$\frac{n!}{s^{n+1}} = \frac{\sqrt{n+1}}{s^{n+1}}$	sinh *at*	a/s^2-a^2
4.	e^{at}	$\frac{1}{s-a}$	cosh *at*	s/s^2-a^2

CHAPTER 25

Z-Transforms and Applications

25.1 Introduction

The Z-transform is useful for the manipulation of discrete sequences and has acquired a new significance in the formulation and analysis of discrete-time systems. It is used extensively today in the areas of applied mathematics, digital signal processing, control theory, population science and economics. These discrete models are solved with difference equations in a manner that is analogous to solving continuous models with differential equations. The role played by Z-transform in the solution of difference equations corresponds to that played by Laplace transforms in the solution of differential equations.

25.2 Z-transform

Definition: Given the sequence {f(k)} the z-transform is defined as follows:

$$\begin{aligned} Z[\{f(k)\}] &= F(z) \\ &= \sum_{n=-\infty}^{\infty} f(k).z^{-k} \end{aligned}$$

Here Z is an operator of Z-transform and Z is a complex number; F(z) is the Z-transform of {f(x)}.

***Example* 25.1:** If f(k) = {15, 10, 7, 4, 1, –1, 0, 3, 6}, then $Z[\{f(k)\}] = f(z) = 15z^3 + 10z^2 + 7z + 4 + \frac{1}{z} - \frac{1}{z^2} + 0 + \frac{3}{z^4} + \frac{6}{z^5}$

25.3 Properties of Z-Transforms

Linearity

Theorem 1: If f(n) and g(n) are such that they can be added and a and b are constants then

$$Z\{a\, f(n) + b\, g(n)\} = a\, Z[f(n)] + bZ[g(n)]$$

Proof: $Z[\{a\, f(n) + bg(n)\}] = \sum_{k=-\infty}^{\infty} [af(n) + bg(n)]z^{-k}$ [By definition]

$$= \sum_{k=-\infty}^{\infty} [af(n)z^{-k} + bg(n)z^{-k}] = a\sum_{k=-\infty}^{\infty} f(k)z^{-k} + b\sum_{k=-\infty}^{\infty} g(k)z^{-4}$$

$= aZ[f(n) + bZ[g(n)]$ Proved

***Example* 25.2:** Find the Z transform of {f(k)} where

$$f(k) = \begin{cases} 5^k, & k < 0 \\ 3^k, & k \geq 0 \end{cases}$$

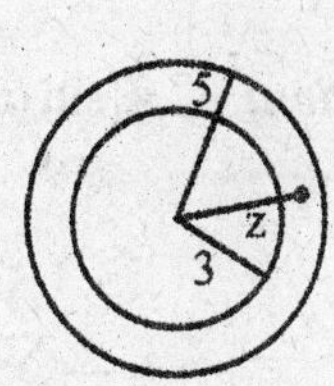

Solution: $Z\{f(k)\} = \sum_{k=-\infty}^{-1} 5^k z^{-k} + \sum_{k=0}^{\infty} 3^k z^{-k}$

$$= \left[\ldots + 5^{-3}z^3 + 5^{-2}z^2 + 5^{-1}z^1\right] + \left[1 + \frac{3}{z^{-1}} + \frac{9}{z^{-2}} + \frac{27}{z^{-3}} + \ldots\right] \quad [\text{G.P}]$$

$$= \frac{5^{-1}z}{1 - 5^{-1}z} + \frac{1}{1 - \frac{3}{z^{-1}}} = \frac{z}{5 - z} + \frac{z}{z - 3} \qquad [\text{G.P}] \qquad \left[S = \frac{a}{1 - r}\right]$$

$$= \frac{z^2 - 3z + 5z - z^2}{(5 - z)(z - 3)} = \frac{-2z}{z^2 - 2x + 15} \qquad \left|\frac{z}{5}\right| < 1,\ \left|\frac{3}{z}\right| < 1$$

Two series are converging in annulus. Here $3 < |z|$ and $|z| < 5$

***Example* 25.3:** Find the Z-transform of unit impulse

$$\delta(k) = \begin{cases} 1 & k = 0 \\ 0 & k \neq 0 \end{cases}$$

Solution: $Z\{f(k)\} = \sum_{k=-\infty}^{\infty} \delta(k)z^{-k}$

$$= [\ldots.\ 0 + 0 + 1 + 0\ \ldots] = 1$$

***Example* 25.4:** Find the Z-transform of discrete unit step

$$U(k) = \begin{cases} 0 & k < 0 \\ 1 & k \geq 0 \end{cases}$$

Solution: $$Z = \{[U(k)]\} = \sum_{k=0}^{\infty} U(k) z^{-k} = \left[1 + z^{-1} + z^{-2} + z^{-3} + \ldots\right]$$

This is G.P, its sum is $\dfrac{a}{1-r}$

$$= \frac{1}{1-z^{-1}} = \frac{1}{1-\frac{1}{z}} = \frac{z}{z-1}$$

***Example* 25.5:** Find the Z-transform of sin αk, k ≥ 0.

Solution: $$Z[\sin \alpha k] = \sum_{k=0}^{\infty} \sin \alpha\, k z^{-k} = \sum_{k=0}^{\infty} \frac{e^{i\alpha k} - e^{-i\alpha k}}{2i} z^{-k}$$

$$= \frac{1}{2i}\sum_{k=0}^{\infty} e^{i\alpha k} z^{-k} - \frac{1}{2i}\sum_{k=0}^{\infty} e^{-i\alpha k} z^{-k} = \frac{1}{2i}\sum_{k=0}^{\infty}\left(e^{i\alpha} z^{-1}\right)^k - \frac{1}{2i}\sum_{k=0}^{\infty}\left(e^{i\alpha} z^{-1}\right)^k$$

$$= \frac{1}{2i}\left[1 + \left(e^{i\alpha} z^{-1}\right) + \left(e^{i\alpha} z^{-1}\right)^2 + \ldots\right] - \frac{1}{2i}\left[1 + \left(e^{-i\alpha} z^{-1}\right) + \left(e^{-i\alpha} z^{-1}\right)^2 + \ldots\right]$$

These are G.P.

$$= \frac{1}{2i}\frac{1}{1-e^{i\alpha} z^{-1}} - \frac{1}{2i}\frac{1}{1-e^{-i\alpha} z^{-1}} = \frac{1}{2i}\frac{z}{z-e^{i\alpha}} - \frac{1}{2i}\frac{z}{z-e^{-i\alpha}}$$

$$= \frac{1}{2i}\left[\frac{z}{z-e^{i\alpha}} - \frac{z}{z-e^{-i\alpha}}\right] = \frac{1}{2i}\frac{z\left(z-e^{-i\alpha}\right) - z\left(z-e^{i\alpha}\right)}{\left(z-e^{-i\alpha}\right)\left(z-e^{-i\alpha}\right)}$$

$$= \frac{1}{2i}\frac{z\left(e^{i\alpha} - e^{-i\alpha}\right)}{z^2 - z\left(e^{i\alpha} + e^{-i\alpha}\right) + 1} = \frac{z \sin \alpha}{z^2 - 2z\cos\alpha + 1}$$

***Example* 25.6:** Find the Z-transform of ck cosh αk (k ≥ 0)

Solution: $$z[\{c^k \cosh \alpha k\}] = \sum_{k=0}^{\infty}\left(c^k \cosh \alpha k\right) z^{-k} = \sum_{k=0}^{\infty} \frac{c^k}{2}\left(e^{\alpha k} + e^{-\alpha k}\right) z^{-k}$$

$$=\frac{1}{2}\sum_{k=0}^{\infty}\left(ce^{\alpha}z^{-1}\right)^{k}+\frac{1}{2}\sum_{k=0}^{\infty}\left(ce^{-\alpha}z^{-1}\right)^{k}$$

$$=\frac{1}{2}\left[1+\left(c\,e^{\alpha}z^{-1}\right)+\left(ce^{\alpha}z^{-1}\right)^{2}+\ldots\right]+\frac{1}{2}\left[1+\left(ce^{-\alpha}z^{-1}\right)+\left(ce^{-\alpha}z^{-1}\right)^{2}+\ldots\right]\left[|z|>\left|ce^{\alpha}\right|,|z|>\left|ce^{-\alpha}\right|\right]$$

$$=\frac{1}{2}\frac{1}{1-c\,e^{\alpha}z^{-1}}+\frac{1}{2}\frac{1}{1-c\,e^{-\alpha}z^{-1}}=\frac{1}{2}\frac{1-ce^{-\alpha}z^{-1}+1-ce^{\alpha}z^{-1}}{\left(1-ce^{\alpha}z^{-1}\right)\left(1-ce^{-\alpha}z^{-1}\right)}$$

$$=\frac{1-c\left(\dfrac{e^{\alpha}+e^{-\alpha}}{2}\right)z^{-1}}{1+c^{2}z^{-2}-ce^{\alpha}z^{-1}-ce^{-\alpha}z^{-1}}=\frac{1-(c\cosh\alpha)z^{-1}}{1+c^{2}\,z^{-2}-2cz^{-1}\cosh\alpha}\qquad\text{(Multiply by } z^{2})$$

$$=\frac{z(z-c\cosh\alpha)}{z^{2}+c^{2}-2c\,z\cosh\alpha}$$

Corollary If $c = 1$

$$Z[\cosh\alpha k]=\frac{z(z-\cosh\alpha)}{z^{2}-2z\cosh\alpha+1}$$

***Example* 25.7:** Find the Z-transform of $\cosh\left(\dfrac{k\pi}{2}+\alpha\right)$

***Solution*:** $F(z)=\sum_{k=0}^{\infty}\cosh\left(\dfrac{k\pi}{2}+\alpha\right)z^{-k}=\sum_{k=0}^{\infty}\dfrac{e^{\frac{k\pi}{2}+\alpha}+e^{-\left(\frac{k\pi}{2}+\alpha\right)}}{2}z^{-k}$

$$=\frac{1}{2}\sum_{k=0}^{\infty}e^{\frac{k\pi}{2}+\alpha}z^{-k}+\frac{1}{2}\sum_{k=0}^{\infty}e^{-\frac{k\pi}{2}-\alpha}z^{-k}=\frac{1}{2}e^{\alpha}\sum_{k=0}^{\infty}\left(e^{\frac{\pi}{2}}z^{-1}\right)^{k}+\frac{1}{2}e^{-\alpha}\sum_{k=0}^{\infty}\left(e^{\frac{-\pi}{2}}z^{-1}\right)^{k}$$

$$=\frac{1}{2}e^{\alpha}\left[1+\left(e^{\frac{\pi}{2}}z^{-1}\right)+\left(e^{\frac{\pi}{2}}z^{-1}\right)^{2}+\ldots\right]+\frac{1}{2}e^{-\alpha}\left[1+\left(e^{-\frac{\pi}{2}}z^{-1}\right)+\left(e^{-\frac{\pi}{2}}z^{-1}\right)^{2}+\ldots\right]$$

(Geometrical series sum $=\dfrac{a}{1-r}$)

$$=\frac{1}{2}e^{\alpha}\frac{1}{1-e^{\frac{\pi}{2}}z^{-1}}+\frac{1}{2}e^{-\alpha}\frac{1}{1-e^{-\frac{\pi}{2}}z^{-1}}=\frac{1}{2}\frac{e^{\alpha}\left(1-e^{-\frac{\pi}{2}}z^{-1}\right)+e^{-\alpha}\left(1-e^{\frac{\pi}{2}}z^{-1}\right)}{\left(a-e^{\frac{\pi}{2}}z^{-1}\right)\left(1-e^{\frac{-\pi}{2}}z^{-1}\right)}$$

$$=\frac{\dfrac{e^{a}+e^{-\alpha}}{2}-\dfrac{e^{\alpha-\frac{\pi}{2}}+e^{-\alpha+\frac{\pi}{2}}}{2}.z^{-1}}{1-e^{\frac{\pi}{2}}z^{-1}-e^{-\frac{\pi}{2}}z^{-1}+z^{-2}}=\frac{\cosh\alpha-\cosh\left(\alpha-\dfrac{\pi}{2}\right).z^{-1}}{1-\left(2\cosh\dfrac{\pi}{2}\right)z^{-1}+z^{-2}}$$

$$=\frac{z^{2}\cosh\alpha-z\cosh\left(\dfrac{\pi}{2}-\alpha\right)}{z^{2}-2z\cosh\dfrac{\pi}{2}+1}$$

***Example* 25.8:** Find Z-transform of sin(3k + 5)

Solution: $F(z)=\sum_{k=0}^{\infty}\sin(3k+5)z^{-k}=\sum_{k=0}^{\infty}\frac{e^{i(3k+5)}-e^{-i(3k+5)}}{2i}z^{-k}$

$$=\frac{1}{2i}\sum_{k=0}^{\infty}e^{i(3k+5)}z^{-k}-\frac{1}{2i}\sum_{k=0}^{\infty}e^{-i(3k+5)}z^{-k}$$

$$=\frac{1}{2i}e^{i5}\sum_{k=0}^{\infty}\left(e^{3i}z^{-1}\right)^{k}-\frac{1}{2i}e^{-5i}\sum\left(e^{-3i}z^{-1}\right)^{k}$$

$$=\frac{1}{2i}e^{i5}\left[1+\left(e^{3i}z^{-1}\right)+\left(e^{3i}z^{-1}\right)^{2}+\ldots\right]-\frac{1}{2i}e^{-5i}\left[1+\left(e^{-3i}z^{-1}\right)+\left(e^{-3i}z^{-1}\right)^{2}+\ldots\right]$$

$$=\frac{e^{i5}}{2i}\frac{1}{1-e^{3i}z^{-1}}-\frac{1}{2i}e^{-i5}\frac{1}{1-e^{3i}z^{-1}}\qquad\left[S=\frac{a}{1-r}\right]$$

$$=\frac{1}{2}\frac{e^{i5}\left(1-e^{3i}z^{-1}\right)-e^{-5i}\left(1-e^{3i}z^{-1}\right)}{\left(1-e^{3i}z^{-1}\right)\left(1-e^{-3i}z^{-1}\right)}=\frac{1}{2i}\frac{\left(e^{i5}-e^{-5i}\right)-e^{2i}z^{-1}+e^{-2i}z^{-1}}{1-e^{3i}z^{-1}-e^{-3i}z^{-1}+z^{-2}}$$

$$=\frac{\dfrac{e^{i5}-e^{-5i}}{2i}-z^{-1}\dfrac{e^{2i}-e^{-2i}}{2i}}{1-\left(e^{3i}+e^{-3i}\right)z^{-1}+z^{-2}}=\frac{\sin5-z^{-1}\sin2}{1-(2\cos3)z^{-1}+z^{-2}}$$

$$=\frac{z^{2}\sin5-z\sin2}{z^{2}-2z\cos3+1}\qquad|z|>|$$

***Example* 25.9:** Find the Z-transform of $c^{k}\cos\alpha k,\ k\geq0$

Solution: $F(z)=\sum_{k=0}^{\infty}[c^{k}\cos\alpha k]z^{-k}$

$$= \sum_{k=0}^{\infty} c^k \left[\frac{e^{i\alpha k}+e^{-i\alpha k}}{2}\right] z^{-k} = \sum_{k=0}^{\infty}\left[\frac{1}{2}c^k e^{i\alpha k}\right]z^{-k} + \sum_{k=0}^{\infty}\left[\frac{1}{2}c^k e^{-i\alpha k}\right]z^{-k}$$

$$= \frac{1}{2}\sum_{k=0}^{\infty}\left(ce^{i\alpha}z^{-1}\right)^k + \frac{1}{2}\sum_{k=0}^{\infty}(ce^{-i\alpha}z^{-1})^k$$

$$= \frac{1}{2}\left[1+\left(ce^{i\alpha}z^{-1}\right)+\left(ce^{i\alpha}z^{-1}\right)^2+\ldots\right]+\frac{1}{2}\left[1+\left(ce^{-i\alpha}z^{-1}\right)+\left(ce^{-i\alpha}z^{-1}\right)^2+\ldots\right]$$

This is a Geometric series whose sum is $\frac{a}{1-r}$

$$= \frac{1}{2}\left[\frac{1}{1-ce^{i\alpha}z^{-1}}\right]+\frac{1}{2}\left[\frac{1}{1-ce^{-i\alpha}z^{-1}}\right], \quad |z|>|c|$$

$$= \frac{1}{2}\frac{1-ce^{-i\alpha}z^{-1}+1-ce^{i\alpha}z^{-1}}{\left(1-ce^{i\alpha}z^{-1}\right)\left(1-ce^{-i\alpha}z^{-1}\right)} = \frac{1}{2}\frac{2-c\left(e^{i\alpha}+e^{-i\alpha}\right)z^{-1}}{\left[1-ce^{-i\alpha}z^{-1}-ce^{i\alpha}z^{-1}+c^2z^{-2}\right]}$$

$$= \frac{1-c\cos\alpha . z^{-1}}{1-2cz^{-1}\cos\alpha+c^2z^{-2}}, \qquad |z|>|c|$$

$$= \frac{z^2-c\,z\cos\alpha}{z^2-2cz\cos\alpha+c^2}$$

***Example* 25.10:** Find the Z-transform of $\{f(k)\}$ where $f(k) = \begin{cases} 6^k, & k<0 \\ 4^k, & k\geq 0 \end{cases}$

Solution: By definition of Z-transform, we have

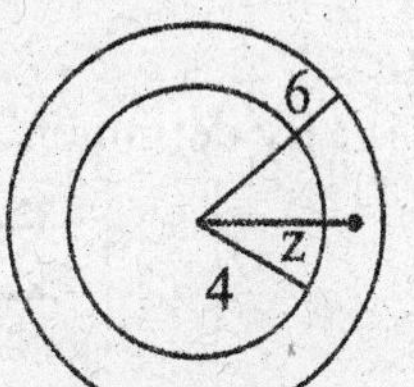

$$Z\left[f(k)\right] = \sum_{k=-\infty}^{-1} 6^k z^{-k} + \sum_{k=0}^{\infty} 4^k z^{-k}$$

$$= \left[\ldots+6^{-3}z^3+6^{-2}z^2+6^{-1}z^1\right]+\left[1+\frac{4}{z}+\frac{4^2}{z^2}+\frac{4^3}{z^3}+\ldots\right]$$

$$= \left[\ldots+\left(\frac{z}{6}\right)^3+\left(\frac{z}{6}\right)^2+\left(\frac{z}{6}\right)\right]+\left[1+\frac{4}{z}+\left(\frac{4}{z}\right)^2+\left(\frac{4}{z}\right)^3+\ldots\right] \qquad \text{[G.P]}$$

$$\left[s=\frac{a}{1-r}\right]$$

$$= \frac{z/6}{1-(z/6)} + \frac{1}{1-(4/z)} \text{ so that } \left|\frac{z}{6}\right| < 1 \text{ and } \left|\frac{4}{z}\right| < 1$$

$$= \frac{z}{6-z} + \frac{z}{z-4}$$

$$= \frac{-2z}{z^2 - 10z + 24}$$

Note that two series are convergent in annulus. Here $4 < |z|$ and $|z| < 6$.

***Example* 25.11:** Find the Z-transform of $\{a^{|k|}\}$

Solution: By definition,

$$Z\left[\alpha^{|k|}\right] = \sum_{k=-\infty}^{\infty} \alpha^{|k|} z^{-k}$$

$$= \sum_{k=-\infty}^{-1} \alpha^{-k} z^{-k} + \sum_{k=0}^{\infty} a^k z^{-k}$$

$$= \left[\ldots + a^3 z^3 + a^2 z^2 + az\right] + \left[1 + \alpha z^{-1} + a^2 z^{-2} + a^3 z^{-3} + \ldots\right]$$

These are two G.P and sum of G.P. $= \dfrac{a}{1-r}$

$$\therefore \quad z\left[a^{|k|}\right] = \frac{az}{1-az} + \frac{1}{1-az^{-1}} \qquad \text{so that } |az| < 1 \text{ and } |az^{-1}| < 1$$

$$= \frac{az}{1-az} + \frac{z}{z-a} = \frac{z - a^2 z}{(1-az)(z-a)}$$

***Example* 25.12:** Find the Z-transform of $\left\{\dfrac{a^k}{k!}\right\}$, $(k \geq 0)$

Solution: By definition

$$z\left[\frac{a^k}{k!}\right] = \sum_{k=0}^{\infty} \frac{a^k}{k!} z^{-k}$$

$$= \sum_{k=0}^{\infty} \frac{\left(az^{-1}\right)^k}{k!} = 1 + \frac{az^{-1}}{1!} + \frac{\left(az^{-1}\right)^2}{2!} + \frac{\left(az^{-1}\right)^3}{3!} + \ldots$$

This is an exponential series,

$$\therefore \quad z\left[\frac{a^k}{k!}\right] = e^{az^{-1}} = e^{a/z}$$

Exercise 25.1

Find the Z-transform of the following for ($k \geq 0$):

1. 2^k **Ans.** $\frac{z}{z-2}, |z| > 2$

2. Sin 2k **Ans.** $\frac{z \sin 2}{z^2 - 2z\cos 2 + 1}$

25.4 Theorem

If {f (k) } = F(z), {g(k)} = G (z), and a and b are constant, then

Z^{-1} [a F (z) + b G(z)] = aZ^{-1} [F (z)] + b Z^{-1} [G(z)]

Proof: we know that

$$Z[a\, f(k) + b\, g(k)] = a\, Z[f(k)] + b\, Z[g(k)]$$
$$= a\, F(z) + b\, G(z)$$
$$\therefore \quad Z^{-1}[a\, F(z) + b\, G(z)] = [a\, f(k) + b\, g(k)] = a\,[f(k)] + b\,[g(k)]$$
$$= a\, Z^{-1}[F(z)] + b\, Z^{-1}[G(z)]$$

25.5 Change of Scale

Theorem If Z[f(k)] = F(z) then $Z\left[\left|a^k f(k)\right|\right] = F\left(\frac{z}{a}\right)$

Proof: $F(z) = Z[f(k)] = \sum_{k=-\infty}^{\infty} f(k) z^{-k}$

Substituting $\frac{z}{a}$ for z we get

$$F\left(\frac{z}{a}\right) = \sum_{k=-\infty}^{\infty} f(k)\left(\frac{z}{a}\right)^{-k} \quad .....(1)$$

But $$Z[a^k f(k)] = \sum_{k=-\infty}^{\infty} a^k f(k) z^{-k} = \sum_{k=-\infty}^{\infty} f(k)\left(\frac{z}{a}\right)^{-k} \quad ....(2)$$

From (1) and (2)

$$Z[a^k f(k)] = F\left(\frac{z}{a}\right)$$ **Proved**

***Example* 25.13:** Find the Z-transform of a^k, $k \geq 0$

Solution: we know that

$$Z[(1)] = \frac{z}{z-1} \qquad \text{(see example 12)}$$

For the given sequence, by the scale change formula the Z-transform

$$Z\left[\left|a^k.1\right|\right] = \frac{\frac{z}{a}}{\frac{z}{a}-1} = \frac{z}{z-a}$$

***Example* 25.14:** Find the Z-transform of $c^k \sin \alpha k$, $k \geq 0$

Solution: We know that

$$Z[\sin \alpha k] = \frac{z \sin \alpha}{z^2 - 2z\cos\alpha + 1}$$

By applying the formula of change of scale we get

$$Z\left[c^k \sin \alpha k\right] = \frac{\frac{z}{c}\sin\alpha}{\left(\frac{z}{c}\right)^2 - 2\left(\frac{z}{c}\right)\cos\alpha + 1}$$

$$= \frac{cz\sin\alpha}{z^2 - 2c\, z\cos\alpha + c^2}$$

25.6 Shifting Property

Theorem: If $Z[f(k)] = F(z)$

then $Z[f(k \pm n)] = z^{\pm n} F(z)$

***Proof*:** $Z[f(k \pm n)] = \sum_{k=-\infty}^{\infty} f(k \pm n) z^{-k} = z^{\pm n} \sum_{k=-\infty}^{\infty} f(k \pm n) z^{-(k \pm n)} \qquad (r = k \pm n)$

$$= z^{\pm n} \sum_{k=-\infty}^{\infty} f(r) z^{-r}$$

$$= z^{\pm n} F(z)$$

Case I $Z[f(k+n)] = f(k+n)\, z^{-k}, \qquad k \geq 0$

$$= z^n \sum_{k=0}^{\infty} f(k+n)\, z^{-(k+n)}$$

Put $(r = k+n)$ $= z^n \sum_{r=n}^{\infty} f(r)\, z^{-r} = z^n \sum_{r=0}^{\infty} f(r) z^{-r} - z^n \sum_{r=0}^{n-1} f(r) z^{-r}$

$$= z^n F(z) - \sum_{r=0}^{n-1} f(r)\, z^{n-r}$$

Case II $Z[f(k-n)] = \sum_{k=0}^{\infty} f(k-n)\, z^{-k} \quad k \geq 0 = z^{-n} \sum_{k=0}^{\infty} f(k-n) z^{-(k-n)}$

Put $r = k - n$

$$= z^{-n} \sum_{r=-n}^{\infty} f(r) z^{-r}$$

$$= z^{-n} \sum_{r=0}^{\infty} f(r) z^{-r} + z^{-n} \sum_{r=-n}^{-1} f(r) z^{-r} = z^{-n} F(z) + \sum_{r=-n}^{-1} f(r) z^{-n-r} \quad \text{Put } r = -m$$

$$= z^{-n} F(z) + \sum_{m=1}^{n} f(-m) z^{-n+m}$$

Corollary 1. If $\{f(k)\}$ is casual sequence, then

$Z[f(k-n)] = z^{-n} F(z)$

Since $f(-1) = f(-2) = f(-3) = \ldots = f(-n) = 0$

Corollary 2. For casual sequence

$Z[f(k-1)] = z^{-1} F(z)$ as $f(-1) = 0$

$Z[f(k+1)] = z F(z) - z f(0)$

$Z[f(k+2)] = z^2 F(z) - z^2 f(0) - z f(1)$

25.7 Inverse Z-Transform

Finding the sequence $\{f(k)\}$ from $F(z)$ is defined as inverse Z-transform. It is denoted as

$Z^{-1} F(z) = \{f(k)\}$ $\quad$ Z^{-1} is the inverse Z-transform

***Example* 25.15:** Find the inverse Z-transform of $\frac{1}{z-2}$

Solution: $F(z) = \frac{1}{z-2}$

Case I If $\left|\frac{2}{z}\right| < 1$, $F(z) = \frac{1}{z}\frac{1}{1-2z^{-1}}$

$$= z^{-1}\left(1-2z^{-1}\right)^{-1} = z^{-1}\left[1+2z^{-1}+2^2z^{-2}+...\right]$$

$$= z^{-1} + 2z^{-2} + 2^2z^{-3} + ...$$

$$\{f(k)\} = \{2^{k-1}\}, \qquad k \leq 1$$

Case II If $\left|\frac{z}{2}\right| < 1$

$$F(z) = \frac{1}{z-2} = -\frac{1}{2}\frac{1}{1-\frac{z}{2}} = \frac{-1}{2}\left(1-\frac{z}{2}\right)^{-1} = -\frac{1}{2}\left[1+\frac{z}{2}+\frac{z^2}{z^2}+.....\right]$$

$$\{f(k)\} = (-2k-1), \quad k \leq 0$$

Note: The inverse Z-transform can only be settled when region of convergence (ROC) is given.

25.8 Solution of Difference Equations

***Example* 25.16:** Solve the difference equation

$$y_{k+1} - 2y_{k-1} = 0, k \geq 1, \quad y_{(0)} = 1$$

Solution: $y_{k+1} - 2y_{k-1} = 0$ ....(1)

Taking the Z-transform of both sides of (1), we get

$$Z[y_{k+1} - 2y_{k-1}] = 0$$

$$Z[y_{k+1}] - 2Z\,[y_{k-1}] = 0$$

$$z\,Y(z) - y_0\,z - 2Y(z) = 0 \qquad (y_0 = 1)$$

$$(z-2)\,Y(z) - z = 0$$

$$Y(z) = \frac{z}{z-2}$$

$$\left|y_{(k)}\right| = Z^{-1}\left[\frac{z}{z-2}\right] = Z^{-1}\left[\frac{1}{1-2z^{-1}}\right]$$

$$= Z^{-1}\,[1-2z^{-1}]^{-1} = 1 + 2z^{-1} + (2z^{-1})^2 + ...$$

$$= |\,2^k\,|, \qquad k \geq 0$$

25.9 Multiplication by K

Theorem: If $Z[f(k)] = F(z)$, then $Z[kf(k)] = -z\frac{d}{dz}F(z)$

Proof: If $Z[kf(k)] = \sum_{k=-\infty}^{\infty} kf(k)z^{-k} = -z\sum_{k=-\infty}^{\infty} -kf(k)z^{-k-1} = -z\sum_{k=-\infty}^{\infty} f(k)\left(-kz^{-k-1}\right)$

$$= -z\sum_{k=-\infty}^{\infty} f(k)\frac{d}{dz}\left(z^{-k}\right) = -z\frac{d}{dz}\sum_{k=-\infty}^{\infty} f(k)z^{-k}$$

$$= -z\frac{d}{dz}F(z)$$

In general. $Z[K^n f(k)] = \left(-z\frac{d}{dz}\right)^n F(z)$

Corollary 1. If $f(k) = 1$ then $Z[\{1\}] = (1-z^{-1})^{-1}$, $\quad k \geq 0$

Putting these values in the above theorem we get

$$Z[\{K\}] = -z\frac{d}{dz}\left(1-z^{-1}\right)^{-1} = z\left(1-z^{-1}\right)^{-2}\left(\frac{1}{z^2}\right)$$

$$= z^{-1}(1-z^{-1})^{-2}$$

Corollary 2. If $f(k) = 1$, then $Z[(1)] = (1-z^{-1})^{-1} = \frac{z}{z-1}$ and $n = 2$

$$= Z[|k^2|] = \left(-z\frac{d}{dz}\right)^2 \frac{z}{z-1} = \left(-z\frac{d}{dz}\right)\left(-z\frac{d}{dz}\right)\frac{z}{z-1}$$

$$= -Z\frac{d}{dz}\left[-z\frac{(z-1).1-z.1}{(z-1)^2}\right] = -z\frac{d}{dz}\left[\frac{z}{(z-1)^2}\right]$$

$$= -z\frac{(z-1)^2.1-z.2(z-1)}{(z-1)^4} = -z\frac{z-1-2z}{(z-1)^3}$$

$$= -z\frac{-z-1}{(z-1)^3} = \frac{z^2+z}{(z-1)^3} = \frac{z(z+1)}{(z-1)^3}$$

25.10 Division by K

Theorem: If Z [f(k)] = F(z) then $Z\left[\frac{f(k)}{k}\right]=-\int^{z} z^{-1}F(z)\,dz$

Proof: $Z\left[\frac{f(k)}{k}\right]=\sum_{k=-\infty}^{\infty}\frac{f(k)}{k}z^{-k}$

$$=\sum_{k=-\infty}^{\infty} f(k)\left(\frac{1}{k}z^{-k}\right)=-\sum_{k=-\infty}^{\infty} f(k)\int^{z} z^{-k-1}dz$$

$$=-\int^{z}\sum_{k=-\infty}^{\infty} f(k)z^{-k-1}dz=-\int^{z} z^{-1}\sum_{k=-\infty}^{\infty} f(k)z^{-1}dz=-\int^{z} z^{-1}F(z)\,dz$$

$$Z\left[\left\{\frac{f(k)}{k}\right\}\right]=-\int^{z} z^{-1}F(z)\,dz$$

25.11 Initial Value

Theorem: If Z[f (k)] = F(z), $k \geq 0$

then $f(0)=\lim_{z\to\infty} F(z)$

***Proof*:** $Z[f(k)]=\sum_{k=0}^{\infty} f(k)z^{-4}=F(z)$

$f(0) + f(1)\, z^{-1} + f(2)\, z^{-2} + \ldots = F(z)$

Taking the limit, $z \to \infty$, we get

$$f(0)=\lim_{z\to\infty} F(z)$$

25.12 Final Value

Theorem: $\lim_{k\to\infty} f(k)=\lim_{z\to 1}(z-1)F(z)$

***Proof*:** $Z[f(k+1)-f(k)]=\sum_{k=0}^{\infty}[f(k+1)-f(k)]z^{-k}$

$$zF(z)-f(0)-F(z)=\lim_{n\to\infty}\sum_{k=0}^{n}[f(k+1)-f(k)]z^{-k}$$

$$\lim_{z\to 1}(z-1)F(z)=f(0)+\lim_{z\to 1}\ \lim_{n\to\infty}\ f(k+1)-f(k)z^{-k}$$

By changing the order of limits, we get

$$\lim_{z\to 1}(z-1)F(z)=f(0)+\lim_{n\to\infty}\sum_{k=0}^{n}\lim_{z\to 1}[f(k+1)-f(k)]z^{-k}$$

$$=\lim_{n\to\infty}\left[f(0)+\sum_{k=0}^{n}\left(f(k+1)-f(k)\right)\right]$$

$$= [f(0) - f(0) + f(1) - f(1) + f(2) - f(2) + \ldots + f(n+1) - f(m)]$$

$$=\lim_{n\to\infty} f(n+1)=\lim_{n\to\infty} f(n)=\lim_{k\to\infty} f(k)$$

25.13 Partial Sum

Theorem: If $Z[\{ f(k) \}] = F(z)$

then
$$Z\left[\sum_{n\to-\infty}^{k} f(n)\right]=\frac{F(z)}{1-z^{-1}}$$

***Proof*:** Let [g (k)] be a sequence such that

$$g(k)=\sum_{n=-\infty}^{k} f(n)$$

We are required to find Z [g(k)]

We know that
$$g(k)-g(k-1)=\sum_{n=-\infty}^{k} f(n)-\sum_{n=-\infty}^{k-1} f(n)=f(k)$$

$$Z[\{g(k)\} - \{g(k-1)\} = Z[f(k)]$$

$$Z[\{g(k)\} - Z\{g(k-1)\}] = F(z)$$

$$G(z) - z^{-1} G(z) = F(Z)$$

$$\sum_{n=-\infty}^{k} f(k)=G(z)=\frac{F(z)}{1-z^{-1}}$$

25.14 Convolution

Let two sequence be [f(k)] and [g (k)] and the convolution of {f(k)} and {g(k)} be {h(k)} and denoted as

$$\{h(k)\} = \{f(k)\}\ \{g(k)\}$$

where $\{h(k)\} = \sum_{n=-\infty}^{k} f(n)g(k-n)$ .....(1)

$$= \sum_{n=-\infty}^{k} g(n)\, f(k-n) = [g(k)]\ [f(k)]$$

Proof: Z-transform of (1) is

$$Z\{h(k)\} = \sum_{k=-\infty}^{\infty}\left(\sum_{n=-\infty}^{\infty} f(n)g(k-n)\right)z^{-k} = \sum_{n=-\infty}^{\infty}\sum_{k=-\infty}^{\infty} f(n)g(k-n)z^{-k}$$

$$= \sum_{n=-\infty}^{\infty} f(n)z^{-n}\sum_{n=-\infty}^{\infty} g(k-n)z^{-(k-n)} = \left[\sum_{k=-\infty}^{\infty} f(n)z^{-k}\right]G(z) = F(z)G(z)$$

Note: Region of convergence of H(z) is the common region of convergence of F(z) and G(z)

25.15 Convolution Property of Casual Sequence

$$F(z) = \{f(0) + f(1)\, z^{-1} + f(2)\, z^{-2} + f(3)\, z^{-3} + \ldots.\}$$

$$G(z) = \{g(0) + g(1)\, z^{-1} + g(2)\, z^{-2} + g(3)\, z^{-3} \ldots\}$$

Now

$$F(z)\, G(z) = \{f(0) + f(1)\, z^{-1} + f(2)\, z^{-2} + f(3)\, z^{-3} + \ldots$$

$$\{g(0) + g(1)\, z^{-1} + g(2)\, z^{-2} + g(3)\, z^{-3} + \ldots\}$$

$$= f(0)\, g(0) + \{f(1)\, g(0) + f(0)\, g(1)\, z^{-1} +$$

$$\{f(0)\, g(2) + f(1)\, g(1) + f(2)\, g(0)\}\, z^{-2} + \ldots$$

$$= h(0) + h(1)\, z{-}1 + h(2)\, z - 2 + \ldots = Z\, \{h(k)\}$$

$$Z\{h(k)\} = Z\, \{f(k)\}\, \{g(k)\}$$

***Example* 25.17:** Evaluate the Z-transform of the sequence

$$\{f(k)\} = \sum_{k=0}^{\infty} 2^k \sum_{k=0}^{\infty} 3^k$$

Solution: $Z\left[2^k\right]=1+2z^{-1}+2^2z^{-2}+2^3z^{-3}+...=\dfrac{1}{1-2z^{-1}}$

Similarly $Z\left[3^k\right]=\dfrac{1}{1-3z^{-1}}$

$$Z\left[f(k)\right]=Z\left|2^k\right|\left|3^k\right|=\frac{1}{1-2z^{-1}}\cdot\frac{1}{1-3z^{-1}}$$

$$=\frac{1}{\left(1-2z^{-1}\right)\left(1-3z^{-1}\right)}$$

Exercise 25.2

Find the Z-transform of the following for (k > 0):

1. sin 5k **Ans.** $\dfrac{z\sin 5}{z^2-2z\cos 5+1}$

2. cos 3k **Ans.** $\dfrac{z(z-\cos 3)}{z^2-2z\cos 3+1}$

3. sinh 7k **Ans.** $\dfrac{z\sinh 7}{z^2-2z\cosh 7+1}$

4. cosh 9k **Ans.** $\dfrac{z(z-\cosh 9)}{z^2-2z\cosh 9+1}$

5. sin (5k + 3) **Ans.** $\dfrac{z^2\sin 3+z\sin 2}{z^2-2z\cos 5+1}$

6. $\cos\left(\dfrac{k\pi}{3}+5\right)$ **Ans.** $\dfrac{z^2\cos 5-z\cos\left(\dfrac{\pi}{3}-5\right)}{z^2-2z\cos\dfrac{\pi}{3}+1}$

7. 3^k cosh 5k **Ans.** $\dfrac{z(z-3\cosh 5)}{z^2-2z\cosh 5+9}$

25.16 Transform of Important Sequences

Sequence $k \geq 0$	Z-transform	
$\{f(k)\}$	$F(z)$	
$\delta(k)$	1	
U(k) or 1	$(1-z^{-1})^{-1}$	
k	$-z\frac{d}{dz}\left(1-z^{-1}\right)^{-1}$	$\|z\|>\|$

Solved Examples

***Example* 25.18:** Show that $Z\left[\frac{1}{n!}\right]=e^{1/z}$

Hence evaluate $Z\left[\frac{1}{(n+1)!}\right]$ and $Z\left[\frac{1}{(n+2)!}\right]$.

***Solution*:** By definition,

$$Z\left[\frac{1}{n!}\right]=\sum_{n=0}^{\infty}\frac{1}{n!}z^{-n}=1+\frac{z^{-1}}{1!}+\frac{z^{-2}}{2!}+\frac{z^{-3}}{3!}+\ldots=e^{1/z}.$$

Shifting $\frac{1}{n!}$ are unit to the left gives

$$Z\left[\frac{1}{(n+1)!}\right]=z\left[Z\left(\frac{1}{n!}\right)-1\right]=z\left(e^{1/z}-1\right).$$

Similarly shifting $\frac{1}{n!}$ two units to the left gives

$$Z\left[\frac{1}{(n+2)!}\right]=z^2\left(e^{1/z}-1-z^{-1}\right).$$

***Example* 25.19:** Find the Z-transform of

(i) $n\ \sin n\theta$

(ii) $n^2e^{n\theta}$.

***Solution*:** (i) Let $u_n = \sin n\theta$. Then $Z[u_n] = Z[\sin\ n\theta] = \dfrac{z\sin\theta}{z^2 - 2z\cos\theta + 1} = U(z)$, say.

We know that $Z[nu]_n = -z\dfrac{d}{dz}U(z)$.

$$\therefore \quad Z[n\ \sin n\theta] = -z\frac{d}{dz}Z[\sin\ n\theta] = -z\frac{d}{dz}\left(\frac{z\ \sin\theta}{z^2 - 2z\ \cos\theta + 1}\right)$$

$$= -z\frac{\sin\ \theta - z^2\sin\theta}{\left(z^2 - 2z\ \cos\theta + 1\right)^2} = \frac{z\left(z^2 - 1\right)\sin\theta}{\left(z^2 - 2z\cos\theta + 1\right)^2}.$$

(ii) Let $u_n = e^{n\theta}$. Then applying damping rule, we get

$$Z[u_n] = Z\left[e^{n\theta}\right]$$

$$= Z[(e^{-\theta})^{-n}.1] = \frac{ze^{-\theta}}{ze^{-\theta} - 1} \qquad \left[\because\ Z[1] = \frac{1}{z-1}\right]$$

$$= \frac{z}{z - e^{\theta}} = U(z), \text{ say.}$$

We know that $Z\left[n^2 u_n\right] = (-z)^2\dfrac{d^2U(z)}{dz^2}$.

$$\therefore \quad Z\ [n^2 e^{n\theta}] = z^2\frac{d^2}{dz^2}Z\left(e^{n\theta}\right) = z^2\frac{d^2}{dz^2}\left(\frac{z}{z - e^{\theta}}\right)$$

$$= z^2\frac{d}{dz}\left\{\frac{-e^{\theta}}{(z - e^{\theta})^2}\right\} = z^2\cdot\frac{2e^{\theta}}{(z - e^{\theta})^3} = \frac{2e^{\theta}\ z^2}{(z - e^{\theta})^3}.$$

***Example* 25.20:** If $Z\ [u_n] = \dfrac{z}{z-1} + \dfrac{z}{z^2+1}$, find the Z-transform of u_{n+2}.

***Solution*:** We have

$$Z\ [u_n] = \frac{z}{z-1} + \frac{z}{z^2+1} = \left(1 - \frac{1}{z}\right)^{-1} + \frac{1}{z}\left(1 + \frac{1}{z^2}\right)^{-1}$$

$$= (1 + z^{-1} + z^{-2} + \ldots\infty) + z^{-1}(1 - z^{-2} + z^{-4} - \ldots\infty)$$

$$= 1 + 2z^{-1} + z^{-2} + z^{-4} + 2z^{-5} + \ldots\infty. \qquad \ldots\ldots(i)$$

But we know that

$$U(z) = Z[u_n] = u_0 + u_1 z^{-1} + u_2 z^{-2} + u_3 z^{-3} + \ldots \infty. \qquad \ldots\text{(ii)}$$

Thus comparing (i) and (ii), we get

$$u_0 = 1,\ u_1 = 2,\ u_2 = 1 \text{ etc.}$$

Now shifting u_n two units to the left, we get

$$Z[u_{n+2}] = z^2 [U(z) - u_0 - u_1 z^{-1}]$$

$$= z^2\left[\frac{z}{z-1} + \frac{z}{z^2+1} - 1 - 2z^{-1}\right] = \frac{z^3}{z-1} + \frac{z^3}{z^2+1} - z^2 - 2z$$

$$= \frac{z^3(z^2+1) + z^3(z-1) - z^2(z^3 - z^2 + z - 1) - 2z\,(z^3 - z^2 + z - 1)}{(z-1)\,(z^2+1)}$$

$$= \frac{z^3 - z^2 + 2z}{(z-1)\,(z^2+1)}.$$

25.17 Final Value Theorem

If $\quad Z[U_n] = U(z)$, then

$$\operatorname*{Lt}_{n \to \infty} u_n = \operatorname*{Lt}_{z \to 1} (z-1)U(z)$$

Proof: By definition and linearity property of Z-transform, we have

$$Z[u_{n+1}] - Z[u_n] = Z[u_{n+1} - u_n] = \sum_{n=0}^{\infty} (u_{n+1} - u_n)\, z^{-n}$$

or $$z[U(z) - u_0] - U(z) = \sum_{n=0}^{\infty} (u_{n+1} - u_n)\, z^{-n}$$

or $$(z-1)\, U(z) - u_0 z = \sum_{n=0}^{\infty} (u_{n+1} - u_n)\, z^{-n}.$$

Taking limits of both sides as $z \to 1$, we get

$$\operatorname*{Lt}_{z \to 1} [(z-1)\, U(z)] - u_0 = \sum_{n=0}^{\infty} (u_{n+1} - u_n)$$

$$= \operatorname*{Lt}_{n \to \infty} [(u_1 - u_0) + (u_2 - u_1) + \ldots + (u_{n+1} - u_n)]$$

$$= \underset{n \to \infty}{\text{Lt}} [u_{n+1}] - u_0 = u_\infty - u_0.$$

Hence $\quad u_\infty = \underset{z \to 1}{\text{Lt}} [(z-1)\, U(z)].$

***Example* 25.21:** If $U(z) = \dfrac{2z^2 + 5z + 14}{(z'-1)^4}$, evaluate u_2

***Solution*:** Writing U (z) as follows:

$$U(z) = \frac{1}{z^2} \cdot \frac{z + 5z^{-1} + 14z^{-2}}{(1 - z^{-1})^4}.$$

By initial value theorem, we have

$$u_0 = \underset{z \to \infty}{\text{Lt}} U(z) = 0 \text{ as } \underset{z \to \infty}{\text{Lt}} \frac{1}{z} = 0 \text{ and } \underset{z \to \infty}{\text{Lt}} \frac{1}{z^2} = 0.$$

Similarly, $\quad u_1 = \underset{z \to \infty}{\text{Lt}} \{z\,[U(z) - u_0]\} = 0.$

Now $\quad u_2 = \underset{z \to \infty}{\text{Lt}} \{z^2\, [U(z) - u_0 - u_1\, z^{-1}]\}$

$$= \underset{z \to \infty}{\text{Lt}} \left\{ \frac{2 + 5z^{-1} + 14z^{-2}}{(1 - z^{-1})^4} \right\} = 2 - 0 - 0 = 2$$

k^n	$\left(-z\dfrac{d}{dz}\right)^n \left(1 - z^{-1}\right)^{-1}$	$\lvert z\rvert > 1$
${}^kC_n$	$z^{-n}\left(1 - z^{-1}\right)^{-(n+1)}$	$\lvert z\rvert > 1$
${}^{k+n}C_n.a^k$	$\left(1 - az^{-1}\right)^{-(n+1)}$	$\lvert z\rvert > \lvert a\rvert$
${}^nC_k$	$\left(1 - z^{-1}\right)^n$	$0 \le k \le n, \lvert z\rvert > 0$
a^k	$\left(1 - az^{-1}\right)^{-1}$	$\lvert z\rvert > \lvert a\rvert$
$k^n\, a^k$	$\left(-z\dfrac{d}{dz}\right)^n \left(1 - az^{-1}\right)^{-1}$	$\lvert z\rvert > \lvert a\rvert$
a^k	$-\left(1 - az^{-1}\right)^{-1}$	$\lvert z\rvert < \lvert a\rvert$

$k^n a^k$	$-\left(-z\frac{d}{dz}\right)^n\left(1-az^{-1}\right)^{-1}$	$\|z\| < \|a\|$
$a^{\|k\|}$	$\left(1-a^2\right)\left(1-az\right)\left(1-az^{-1}\right)^{-1}$	$\|a\| < z\| < \frac{1}{\|a\|}$
$\sin \alpha k$	$\frac{z\sin\alpha}{z^2-2z\cos\alpha+1}$	$k \geq 0$
$c^k \sin \alpha k$	$\frac{cz\sin\alpha}{z^2-2cz\cos\alpha+c^2}$	$k \geq 0$
$\cosh \alpha k$	$\frac{z^2-z\cos\alpha}{z^2-2z\cos\alpha+1}$	$k \geq 0$
$c^k \cosh \alpha k$	$\frac{z^2-cz\cos\alpha}{z^2-2cz\cos\alpha+c^2}$	$k \geq 0$
$\cosh \alpha k$	$\frac{z^2-z\cosh.\alpha}{z^2-2cz\cosh.\alpha+1}$	$k \geq 0$
$c^k \cosh \alpha k$	$\frac{z^2-cz\cos.\alpha}{z^2-2cz\cosh.\alpha+c^2}$	$k \geq 0$
$\sinh \alpha k$	$\frac{z\sinh\alpha}{z^2-2z\cosh\alpha+1}$	$k \geq 0$
$c^k \sinh \alpha k$	$\frac{cz\sinh\alpha}{z^2-2cz\cosh\alpha+c^2}$	$k \geq 0$
Change of scale	$a^k f(k)$	$f\left(\frac{z}{a}\right)$
Shifting	$f(k \pm n)$	$z^{\pm n}F(z)$
	$f(k+n)$	$z^nF(z)-\sum_{r=0}^{n-1}f(r)z^{n-r}$
	$f(k-n)$	$z^nF(z)\times\sum_{r=1}^{n}f(-r)z^{-n+r}, k<0$
Casual sequence	$f(k+n)$	$z^nF(z)-\sum_{r=0}^{n-1}f(r)z^{-n-r}$
	$f(k-n)$	$z^{-n}F(z)$

Multiplication by k	$kf(k)$	$-z\frac{d}{dz}F(z)$
	$k^n f(k)$	$\left(-z\frac{d}{dz}\right)^n F(z)$
Division by k	$\frac{1}{k}f(k)$	$-\int^z z^{-1}F(z)dz$
Initial value theorem	$f(0)$	$\lim_{z\to\infty} F(z)$
Final value theorem	$\lim_{k\to\infty} F(k)$	$\lim_{z\to 1}(z-1)F(z)$
Partial sum	$\sum_{n\to-\infty}^{k} f(n)$	$\frac{F(z)}{z-1}$
	$\sum_{n\to-\infty}^{\infty} f(n)$	$F(1)$
Convolution	$f(k)^* g(k)$	$F(z).G(z)$

Inverse Z-transforms

	Inverse Z-transform	
	$\|z\|>a\|,k>0$	$\|z\|<a\|,k<0$
$\frac{z}{z-a}$	$a^k U(k)$	$-a^k$
$\frac{z^2}{(z-a)^2}$	$(k+1)a^k$	$-(k+1)a^k$
$\frac{z^3}{(z-a)^3}$	$\frac{1}{\underline{\|2}}(k+1)(k+2)a^k U(k)$	$-\frac{1}{\underline{\|2}}(k+1)(k+2)a^k U(-k+2)$
$\frac{z^n}{(z-a)^n}$	$\frac{1}{\underline{\|n-1}}(k+1)....(k+n-1)$ $a^k U(k)$	$-\frac{1}{\underline{\|n-1}}(k+1)....(k+n-1)a^k$
$\frac{1}{z-a}$	$a^{k-1}U(k-1)$	$-a^{k-1}U(-k)$

$\frac{1}{(z-a)^2}$	$(k-1)a^{k-2}U(k-2)$	$-(k-1)a^{k-2}U(-k+1)$
$\frac{1}{(z-a)^3}$	$\frac{1}{2}(k-2)(k-1)a^{k-3}U(k-3)$	$-\frac{1}{2}(k-2)(k-1)a^{k-3}U(-k+2)$

25.18 By Binomial Expansion and Partial Fraction

***Example* 25.22:** Find the inverse Z-transform of $\frac{4z}{z-a}$

(i) $|z|>|a|$ (ii) $|z|<|a|$

Solution: **Case I.** $|z|>|a|$

$$\frac{4z}{z-a}=\frac{4z}{z}\frac{1}{1-\frac{a}{z}}=4\left(1-\frac{a}{z}\right)^{-1}-4\left[1+\frac{a}{z}+\frac{a^2}{z^2}+\frac{a^3}{z^3}+...\right]$$

$$=4+4az^{-1}+4a^2z^{-2}+4a^3z^{-3}+...+4a^kz^{-k}+...=\left\{4a^k\right\}z^{-k}$$

$$z^{-1}\left(\frac{4z}{z-a}\right)=\left\{4a^k\right\}$$

Case II $|z|<|a|$

$$\frac{4z}{z-a}=-\frac{1}{a}\frac{z}{\left(1-\frac{z}{a}\right)}=\frac{-z}{a}\left(1-\frac{z}{a}\right)^{-1}$$

$$=-\frac{z}{a}\left[1+\frac{z}{a}+\frac{z^2}{a^2}+\frac{z^3}{a^3}+...\right]=-\frac{z}{a}-\frac{z^2}{a^2}-\frac{z^3}{a^3}-\frac{z^4}{a^4}-...$$

$$\{f(k)\}=\left\{...-\frac{1}{a^4},-\frac{1}{a^3},-\frac{1}{a^2},\frac{1}{a}\right\}$$ **Ans.**

Exercise 25.3

Find the inverse of the Z-transform of the following:

1. $\dfrac{z^2}{\left(z-\frac{1}{4}\right)\left(z-\frac{1}{5}\right)}, \quad |z|<\frac{1}{5}$ **Ans.** $4\left(\frac{1}{5}\right)^k-5\left(\frac{1}{4}\right)^k$

2. $\dfrac{z}{\left(z-\frac{1}{4}\right)\left(z-\frac{1}{5}\right)}, \quad |z|>\frac{1}{4}$ **Ans.** $20\left[\left(\frac{1}{4}\right)^k-\left(\frac{1}{5}\right)^k\right]U(k)$

3. $\dfrac{z^3}{\left(z-\frac{1}{2}\right)^2(z-1)}, \quad |z|<\frac{1}{2}$ **Ans.** $-4(1)^k+(k+3)\left(\frac{1}{2}\right)^k, (k<0)$

4. $\dfrac{z+1}{z^2-2z+1}, \quad |z|>1$ **Ans.** $(2k+1)U(k)$

Solved Examples

***Example* 25.23:** Use convolution to evaluate $Z^{-1}\left\{\dfrac{z^2}{(z-a)(z-b)}\right\}$.

Solution: We know that

$$Z^{-1}\left\{\frac{z}{z-a}\right\}=a^n \quad \text{and} \quad Z^{-1}\left\{\frac{z}{z-b}\right\}=b^n$$

$$\therefore \quad Z^{-1}\left\{\frac{z^2}{(z-a)(z-b)}\right\}=Z^{-1}\left\{\frac{z}{z-a}\cdot\frac{z}{z-b}\right\}=a^n \times b^n$$

$$=\sum_{m=0}^{n} a^m \cdot b^{n-m}=b^n \sum_{m=0}^{n}\left(\frac{a}{b}\right)^m \quad \text{which is a G.P.}$$

$$=b^n \cdot \frac{(a/b)^{n+1}-1}{(a/b)-1}=\frac{a^{n+1}-b^{n+1}}{a-b}$$

***Example* 25.24:** Using convolution theorem, evaluate the inverse Z-transform of $\left(\frac{z}{z-a}\right)^2$.

Solution: By convolution theorem, we have

$$Z^{-1}\left[\left(\frac{z}{z-a}\right)^2\right]=Z^{-1}\left[\frac{z}{z-a}\cdot\frac{z}{z-a}\right]=u_n \times v_n$$

where $\quad u_n = Z^{-1}\left(\frac{z}{z-a}\right)=a^n$ and $v_n = Z^{-1}\left(\frac{z}{z-a}\right)=a^n$

$$=\sum_{m=0}^{n} u_n v_{n-m} = \sum_{m=0}^{n} a^n\ a^{n-m} = a^n \sum_{m=0}^{n} 1 = (n+a)a^n.$$

***Example* 25.25:** Show that $\frac{1}{n\,!} * \frac{1}{n\,!} = \frac{2^n}{n\,!}$

Solution: Let $u_n = v_n = \frac{1}{n\,!}$. Then by convolution theorem, we have

$$\frac{1}{n!}*\frac{1}{n!}=u_n * v_n = \sum_{m=0}^{n} u_m . v_{n-m}$$

$$=\sum_{m=0}^{n}\frac{1}{m!}.\frac{1}{(n-m)!}=\frac{1}{0!n!}+\frac{1}{1!(n-1)!}+\frac{1}{2!(n-2)!}+\ldots+\frac{1}{n!0!}$$

$$=\frac{1}{n!}\left[1+\frac{n}{1!}+\frac{n(n-1)}{2!}+\ldots+1\right]$$

$$=\frac{1}{n!}\left[{}^nC_0+{}^nC_1+{}^nC_2+\ldots+{}^nC_n\right]$$

$$=\frac{1}{n!}(1+1)^n=\frac{2^n}{n!}$$

***Example* 25.26:** Using convolution theorem, find inverse Z-transform of $\frac{18z^2}{(2z-1)(4z+1)}$

Solution: By convolution theorem, we have

$$Z^{-1}\left[\frac{18z^2}{(2z-1)(4z+1)}\right]=Z^{-1}\left[\frac{9}{4}.\frac{z}{z-\frac{1}{2}}.\frac{z}{z+\frac{1}{4}}\right]=\frac{9}{4}Z^{-1}\left[\frac{z}{z-\frac{1}{2}}.\frac{z}{z+\frac{1}{4}}\right]$$

$$= \frac{9}{4} u_n * v_n \quad \text{where } u_n = Z^{-1}\left(\frac{z}{z-\frac{1}{2}}\right) = \left(\frac{1}{2}\right)^n \text{ and } v_n = Z^{-1}\left(\frac{z}{z+\frac{1}{4}}\right) = \left(-\frac{1}{4}\right)^n$$

$$= \frac{9}{4}\sum_{m=0}^{n} u_n v_{n-m} = \frac{9}{4}\sum_{m=0}^{n}\left(\frac{1}{2}\right)^n\left(-\frac{1}{4}\right)^{n-m}$$

$$= \frac{9}{4}\left(\frac{1}{2}\right)^n \sum_{m=0}^{n}\left(-\frac{1}{4}\right)^{n-m} = \frac{9}{4}\left(\frac{1}{2}\right)^n\left[1+\left(-\frac{1}{4}\right)+\left(-\frac{1}{4}\right)^2+\ldots+\left(-\frac{1}{4}\right)^n\right]$$

$$= \frac{9}{4}\left(\frac{1}{2}\right)^n \cdot \frac{1-\left(-\frac{1}{4}\right)^{n+1}}{1-\left(-\frac{1}{4}\right)} = \frac{9}{5}\left(\frac{1}{2}\right)^n\left[1-\left(-\frac{1}{4}\right)^{n+1}\right]$$

Exercise 25.4

Using convolution theorem, find the inverse Z-transform of the following:

1. $\dfrac{z^2}{(z-1)(z-3)}$ **Ans.** $\frac{1}{2}(3^{k+1}-1)$

2. $\left(\dfrac{z}{z-1}\right)^3$ **Ans.** $\frac{1}{2}k(k+1)$

3. $\dfrac{2z^2-10z+13}{(z-3)^2(z-2)}$ **Ans.** $2^{k-1}, k \geq 1$ and $-(k+2)\, 3^{k-2}, k \leq 0$

25.19 Partial Fractions

Let $f(z) = \dfrac{R(z)}{D(z)}$ [if the degree of Numerator < the degree of Denominator]

and $F(z) = Q(z) + \dfrac{R(z)}{D(z)}$ [if the degree of Numerator > the degree of Denominator]

$\frac{R(z)}{D(z)}$ is then expressed into partial fractions.

We convert $\frac{F(z)}{z}$ into partial fractions and not that of F(z)

Let $\frac{F(z)}{z} = \frac{A}{z-a} + \frac{B}{z-b} + \frac{C}{z-c} + \frac{D}{(z-c)^2} + \frac{E}{(z-c)^3} + \frac{Mz+N}{z^2+pz+q}$

then

$$F(z) = A\frac{z}{z-a} + B\frac{z}{z-b} + C\frac{z}{z-c} + D\frac{z}{(z-c)^2} + E\frac{z}{(z-c)^3} + \frac{z(Mz+N)}{z^2+pz+q}$$

$$Z^{-1}F(z) = AZ^{-1}\frac{z}{z-a} + BZ^{-1}\frac{z}{z-b} + CZ^{-1}\frac{z}{z-c} + DZ^{-1}\frac{z}{(z-c)^2} + EZ^{-1}\frac{z}{(z-c)^3} + Z^{-1}\frac{z(Mz+N)}{z^2+pz+q}$$

(i) Linear non-repeated factor

Let the linear repeated factor be $\frac{z}{z-a}$

$$Z^{-1}\left(\frac{z}{z-a}\right) = Z^{-1}\frac{1}{1-az^{-1}} = \{a^k\} \qquad \text{if} \quad |z| > |a|$$

$$Z^{-1}\left(\frac{z}{z-a}\right) = -Z^{-1}\frac{\frac{z}{a}}{1-\frac{z}{a}} = \{-a^k\}, k < 0 \quad \text{if} \quad |z| < |a|$$

(ii) Linear repeated factor

Let the linear repeated factor be $\frac{z}{(z-b)^r}$, $r \geq 2, |z| > |b|$

$$Z^{-1}\frac{z}{(z-b)^r} = Z^{-1}\left[z^{-(r-1)}\frac{z^r}{(z-b)^r}\right]$$

$$= \frac{(k+2-r)(k+3-r).....kU(k-r+1)b^{(k-r+1)}}{(r-1)!} \quad \text{If} \quad |z| < |b|$$

$$Z^{-1}\frac{z}{(z-b)^r} = -\frac{(k+2-r)(k+3-r).....kb^{k-r+1}}{(r-1)!}U(-k-r-1)$$

(iii) Quadratic non-repeated factor

Let the quadratic non-repeated factor be

$$\frac{Mz^2+Nz}{z^2+pz+q}.... \qquad \text{If } |z|>0 \qquad .....(1)$$

Compare (1) with $Z[\{c^k \cos \alpha k\}] = \dfrac{z^2 - cz\cos\alpha}{z^2 - 2cz\cos\alpha + c^2}$

$p = -2c \cos \alpha,\ q = c^2$

or with $Z\left[c^k \cosh \alpha k\right] = \dfrac{Z^2 - cz\cosh\alpha}{Z^2 - 2cz\cosh\alpha + c^2}$

or $p = 2c\cosh\alpha,\ q = c^2$

$$\frac{p}{-2c} = \cos\alpha \Rightarrow \left|\frac{p}{2c}\right| < 1 \text{ or } > 1, \text{ is given by } (-2c\cos\alpha)$$

and α by $(-2c\cosh\alpha)$

Let $\left|\dfrac{p}{2c}\right| < 1$

$$\frac{Mz^2+Nz}{z^2+pz+q} = \frac{Mz(z-c\cos\alpha)+\dfrac{Mc\cos\alpha+N}{c\sin\alpha}(cz\sin\alpha)}{z^2-2cz\cos\alpha+c^2}$$

$$= \frac{M\left(z^2-cz\cos\alpha\right)}{z^2-2cz\cos\alpha+c^2} + \frac{Mc\cos\alpha+N}{c\sin\alpha}\cdot\frac{cz\sin\alpha}{z^2-2cz\cos\alpha+c^2}$$

$$Z^{-1}\frac{Mz^2+Nz}{z^2+pz+q} = M\left\{c^k\cos\alpha k\right\} + \left(\frac{Mc\cos\alpha+N}{c\sin\alpha}\right)\left\{c^k \sin\alpha k\right\}$$

If $\left|\dfrac{p}{2c}\right| > 1$

$$\frac{Mz^2+Nz}{z^2+pz+q} = \frac{Mz(z-c\cosh\alpha)+\dfrac{Mc\cosh\alpha+N}{c\sinh\alpha}(cz\sin\alpha)}{z^2-2cz\cosh\alpha+c^2}$$

$$= \frac{Mz(z - c\cosh\alpha)}{z^2 - 2cz\cosh\alpha + c^2} + \frac{Mc\cosh\alpha + N}{c\sinh\alpha} \cdot \frac{cz\sinh\alpha}{z^2 - 2cz\cosh\alpha + c^2}$$

$$= \frac{M(z^2 - cz\cosh\alpha)}{z^2 - 2cz\cosh\alpha + c^2} + \frac{Mc\cosh\alpha + N}{c\sinh\alpha}$$

$$Z^{-1}\left[\frac{Mz^2 + Nz}{z^2 + pz + q}\right] = M\{c^k \cosh\alpha k\} + \frac{Mc\cosh\alpha + N}{c\sinh\alpha}(c^k \sinh\alpha)$$

***Example* 25.27:** Find the inverse Z-transform of $\frac{1}{(z-3)(z-2)}$

(i) $|z| < 2$ (ii) $2 < |z| < 3$ (iii) $|z| > 3$

Solution: $F(z) = \frac{1}{(z-3)(z-2)} = \frac{1}{z-3} - \frac{1}{z-2}$

Case (i) $|z| < 2$

$$F(z) = -\frac{1}{3}\frac{1}{1-\frac{z}{3}} + \frac{1}{2}\frac{1}{1-\frac{z}{2}} = -\frac{1}{3}\left(1-\frac{z}{3}\right)^{-1} + \frac{1}{2}\left(1-\frac{z}{2}\right)^{-1}$$

$$= -\frac{1}{3}\left(1+\frac{z}{3}+\frac{z^2}{3^2}+\frac{z^3}{3^3}+\ldots\right) + \frac{1}{2}\left[1+\frac{z}{2}+\frac{z^2}{2}+\frac{z^3}{2^3}+\ldots\right]$$

$$= -\frac{1}{3} - \frac{z}{3^2} - \frac{z^2}{3^3} - \frac{z^3}{3^4} \ldots + \frac{1}{2} + \frac{z}{2^2} + \frac{z^2}{2^3} + \frac{z^3}{2^4} + \ldots$$

Case (ii) $2 < |z| < 3$.

$$F(z) = -\frac{1}{3}\frac{1}{1-\frac{z}{3}} - \frac{1}{z}\frac{1}{1-\frac{2}{z}} = -\frac{1}{3}\left[1-\frac{z}{3}\right]^{-1} - \frac{1}{z}\left[1-\frac{2}{z}\right]^{-1}$$

$$= -\frac{1}{3}\left[1+\frac{z}{3}+\frac{z^2}{3^2}+\frac{z^3}{3^3}+\ldots\ldots\right] - \frac{1}{z}\left[1+\frac{2}{z}+\frac{2^2}{z^2}+\frac{2^3}{z^3}\ldots\right]$$

$$= -\frac{1}{3} - \frac{z}{3^2} - \frac{z^2}{3^3} - \frac{z^3}{3^4} \ldots - \frac{1}{z} - \frac{2}{z^2} - \frac{2^2}{z^2} + \frac{2^3}{z^3} + \ldots$$

$$= \ldots\ldots - \frac{z^3}{3^4} - \frac{z^4}{3^3} - \frac{z}{3^2} - \frac{1}{3} - z^{-1} - 2z^{-2} - 2z^{-3} - 2^3 z^{-4} - \ldots$$

$$= \ldots - \frac{z^3}{3^4} - \frac{z^2}{3^3} - \frac{z}{3^2} - \frac{1}{3} - z^{-1} - 2z^{-2} - 2^2 z^{-3} - 2^3 z^{-4} - \ldots$$

$$\Rightarrow f(k) = \left\{-2^{k-1}\right\}, k > 0$$

$$\Rightarrow f(k) = \left\{-3^{k-1}\right\}, k \le 0$$

Case (iii) $|z| > 3$

$$F(z) = \frac{1}{z}\frac{1}{1-\frac{3}{2}} - \frac{1}{z}\frac{1}{1-\frac{2}{z}} = \frac{1}{z}\left(1-\frac{3}{z}\right)^{-1} - \frac{1}{z}\left(1-\frac{2}{z}\right)^{-1}$$

$$= \frac{1}{z}\left[1+\frac{3}{z}+\frac{3^2}{z^2}+\frac{3^3}{z^3}+\ldots\right] - \frac{1}{z}\left[1+\frac{2}{z}+\frac{2^2}{z^2}+\frac{2^3}{z^3}+\ldots\ldots\right]$$

$$= \frac{1}{z}+\frac{3}{z^2}+\frac{3^2}{z^3}+\frac{3^3}{z^4}+\ldots - \frac{1}{z} - \frac{2}{z^2} - \frac{4}{z^3} - \frac{8}{2^4} + \ldots\ldots$$

$$= \left\{3^{k-1} - 2^{k-1}\right\} z^{-k}, \quad k \ge 1$$

$$= 0, \quad k \le 0.$$

***Example* 25.28:** Find the inverse of Z-transform of $\frac{1}{(z-5)^3}$, $|z| > 5$.

***Solution*:** $$f(z) = \frac{1}{(z-5)^3} = \frac{1}{z^3}\frac{1}{\left(1-\frac{5}{z}\right)^3} = \frac{1}{z^3}\left(1-5\,z^{-1}\right)^{-3}$$

$$= z^{-3}\left[1+15z^{-1}+6\left(5z^{-1}\right)^2+10\left(5z^{-1}\right)^3+\ldots+\frac{(n+1)(n+2)}{2}\left(5z^{-1}\right)^n+\ldots\right]$$

$$= z^{-3}\left\{\frac{(k+1)(k+2)}{2}5^k\right\}z^{-k} = \left\{\frac{(k+1)(k+2)}{2}5^k\right\}z^{-k-3}$$

Replacing k by k – 3 we get

$$= \left\{\frac{(k-3+1)(k-3+2)}{2}5^{k-3}\right\}z^{-k} = \left\{\frac{(k-2)(k-1)}{2}5^{k-3}\right\}z^{-k}$$

$$z^{-1}F(z) = f(k) = \frac{(k-2)(k-1)}{2}5^{k-3}, \qquad k \geq 3$$

$$= 0, \qquad k < 3$$

***Example* 25.29**: Obtain $z^{-1}\dfrac{1}{\left(z-\frac{1}{2}\right)\left(z-\frac{1}{3}\right)}$

When (i) $\frac{1}{3} < |z| < \frac{1}{2}$ (ii) $\frac{1}{2} < |z|$.

Solution: $f(z) = \dfrac{1}{\left(z-\frac{1}{2}\right)\left(z-\frac{1}{3}\right)} = 6\left[\dfrac{1}{\left(z-\frac{1}{2}\right)} - \dfrac{1}{(z-3)}\right] = \dfrac{6}{z-\frac{1}{2}} - \dfrac{6}{z-\frac{1}{3}}$

(i) $F(z) = \dfrac{6}{-\frac{1}{2}(1-2^z)} - \dfrac{1}{z}\dfrac{6}{\left(1-\frac{1}{3}z^{-1}\right)}$

$$= -12(1-2z)^{-1} - \frac{6}{z}\left(1-\frac{1}{3}z^{-1}\right)^{-1}$$

$$= -12\left[(1+2z)+(2z)^2+(2z)^3+...\right] - \frac{6}{z}\left[1+\frac{1}{3z}+\frac{1}{(3z)^2}+\frac{1}{(3z^3)}+...\right]$$

$$= -12(2z)^k - \frac{6}{z}\left(\frac{1}{3z}\right)^k = -12(2z)^k - 6\left(\frac{1}{3^k z^{k+1}}\right)$$

$$f(k) = -\frac{6}{3^{k-1}} \quad \text{if } k > 0$$

$$f(k) = -12^{-k} \quad \text{if } k < 0$$

(ii) $\frac{1}{2}<|z|$.

$$f(z)=\frac{6}{z-\frac{1}{2}}-\frac{6}{z-\frac{1}{3}}=\frac{1}{z}\frac{6}{\left(1-\frac{1}{2z}\right)}-\frac{1}{z}\frac{6}{\left(1-\frac{1}{3z}\right)}$$

$$=6z^{-1}\left(1-\frac{1}{2}z^{-1}\right)^{-1}-6z^{-1}\left(1-\frac{1}{3}z^{-1}\right)^{-1}$$

$$=6z^{-1}\left[1+\frac{1}{2}z^{-1}+\left(\frac{1}{2}\right)^{2}z^{-2}+\left(\frac{1}{2}\right)^{3}z^{-3}+...\right]$$

$$-6z^{-1}\left[1+\frac{1}{3}z^{-1}+\left(\frac{1}{3}\right)z^{-2}+\left(\frac{1}{3}\right)^{3}z^{-3}+...\right]$$

$$=6\left[z^{-1}+\frac{1}{2}z^{-2}\left(\frac{1}{2}\right)^{2}z^{-3}+\left(\frac{1}{2}\right)^{3}z^{-4}+...\right]$$

$$-6\left[z^{-1}+\frac{1}{3}z^{-2}+\left(\frac{1}{3}\right)^{2}z^{-3}+\left(\frac{1}{3}\right)^{3}z^{-4}+...\right]$$

$$=6\left\{\frac{1}{2^{k+1}}\right\}z^{-k}-6\left\{\frac{1}{3^{k+1}}\right\}z^{-k}\Rightarrow Z^{-1}F(z)=f(k)=6\left[\frac{1}{2^{k-1}}-\frac{1}{3^{k-1}}\right],k\geq 1$$

***Example* 25.30:** Obtain $Z^{-1}\frac{2z^2-10z+13}{(z-3)^2(z-2)}$, when $2<|z|<3$

***Solution*:** Let $\frac{2z^2-10z+13}{(z-3)^2(z-2)}=\frac{A}{(z-3)^2}+\frac{B}{z-3}+\frac{C}{z-2}$

Converting into partial fractions we get.

$$=\frac{1}{(z-3)^2}+\frac{1}{z-3}+\frac{1}{z-2}$$

$$=\frac{1}{9}\frac{1}{\left(1-\frac{z}{3}\right)^2}-\frac{1}{3}\frac{1}{1-\frac{z}{3}}+\frac{1}{z}\frac{1}{1-\frac{2}{z}}=\frac{1}{9}\left(1-\frac{z}{3}\right)^{-2}-\frac{1}{3}\left(1-\frac{z}{3}\right)^{-1}+\frac{1}{z}\left(1-\frac{2}{z}\right)^{-1}$$

$$= \frac{1}{9}\left[1+\frac{2z}{3}+\frac{3z^2}{9}+\frac{4z^3}{27}+......\right]-\frac{1}{3}\left[1+\frac{z}{3}+\frac{z^2}{9}+\frac{z^3}{27}+......\right]+\frac{1}{z}\left[1+\frac{2}{z}+\frac{4}{z^2}+\frac{8}{z^3}+......\right]$$

$$= \frac{1}{3^2}+\frac{2z}{3^3}+\frac{3z^2}{3^4}+\frac{4z^3}{3^5}...-\frac{1}{3}-\frac{z}{3^2}-\frac{z^2}{3^3}-\frac{z^3}{3^4}......+\frac{1}{z}+\frac{2}{z^2}+\frac{4}{z^3}+\frac{8}{z^4}+...$$

$$= \left[2^{k-1}\right]z^{-k}, \; k \geq 1 \qquad .....(1)$$

$$= \left[\frac{k+1}{3^{k+2}}-\frac{1}{3^{k+1}}\right]z^k, \; k<0 \text{ or } = \frac{-k-2}{3^{-k+2}}, \; k \leq 0 \qquad .....(2)$$

Hence, $Z^{-1}\left[F(z)\right] = 2^{k-1}$ if $k \geq 1$, and $Z^{-1}F(z) = -(k+2)3^{k-2}$, $k \leq 0$

***Example* 25.31:** Find $Z^{-1}\dfrac{3z^2+4z}{z^2-z+1}$. $\quad |z| > 1$

Solution: Let $c^2 = 1$, $\quad \therefore c = \pm 1$

If $\;c = 1$, $\left|\dfrac{p}{2c}\right| = \left|\dfrac{-1}{2\times 1}\right| = \dfrac{1}{2} < 1$

So $-1 = -2c\cos\alpha$ or $-1 = -2\times 1\times\cos\alpha$, $\cos\alpha = \dfrac{1}{2}$

$\alpha = \dfrac{\pi}{3}$, $\sin\alpha = \sin\dfrac{\pi}{3} = \dfrac{\sqrt{3}}{2}$.

$$\frac{3z^2+4z}{z^2-z+1} = \frac{3z(z-c\cos\alpha)+\dfrac{(3c\cos\alpha+4)}{c\sin\alpha}cz\sin\alpha}{z^2-2cz\cos\alpha+c^2}$$

$$\frac{3z^2+4z}{z^2-z+1} = \frac{3z(z-c\cos\alpha)}{z^2-2cz\cos\alpha+c^2}+3z(z-c\cos\alpha)+\frac{\dfrac{3c\cos\alpha+4}{c\sin\alpha}cz\sin\alpha}{z^2-2cz\cos\alpha+c^2}$$

Putting the values of c and α, for the coefficient of $cz\sin\alpha$, we get

$$= \frac{3z(z-c\cos\alpha)}{z^2-2cz\cos\alpha+c^2}+\frac{\dfrac{3\times 1\times\frac{1}{2}+4}{1\times\frac{\sqrt{3}}{2}}cz\sin\alpha}{z^2-2cz\cos\alpha+c^2}$$

$$Z^{-1}\left[\frac{3z^2+4z}{z^2-z+1}\right]=\left[\left\{3c^k\cos\alpha k\right\}+\left\{\frac{11}{\sqrt{3}}c^k\sin ak\right\}\right]U_k$$

Putting the values of c and α, we get

$$Z^{-1}\left[\frac{3z^2+4z}{z^2-z+1}\right]=\left[3(1)^k\left\{\cos\frac{\pi}{3}k\right\}+\frac{11}{3}(1)^k\left\{\sin\frac{\pi}{3}k\right\}\right]U_k$$

$$=\left[3\left\{\cos\frac{\pi k}{3}\right\}+\frac{11}{\sqrt{3}}\left\{\sin\frac{\pi}{3}k\right\}\right]U_k$$

Example 25.32: Obtain $Z^{-1}\left[\dfrac{2z^2+3z}{z^2+z+\frac{1}{9}}\right]\quad\left(|z|>\dfrac{1}{3}\right)$.

Solution: $c^2=\frac{1}{9}\Rightarrow c=\pm\frac{1}{3}, p=1$

If $\quad c=-\frac{1}{3}, \left|\frac{p}{2c}\right|=\left|\frac{1}{2\times-\frac{1}{3}}\right|=\frac{3}{2}>1$

Hence $\quad 1=-2c\cosh\alpha$ or $1=-2\left(-\frac{1}{3}\right)\cosh\alpha$

or $\quad \cosh\alpha=\frac{3}{2},\ \ \sinh\alpha=\frac{\sqrt{5}}{2} \qquad (\cosh^2\alpha-\sinh^2\alpha=1)$

$$\frac{2z^2+3z}{z^2+z+\frac{1}{9}}=\frac{2z(z-c\cosh\alpha)+\frac{2c\cosh\alpha+3}{c\sin\ h\alpha}cz\ \sin\ h\alpha}{z^2-2cz\ \cosh\ \alpha+c^2}$$

Putting the values of c and α in coefficient of cz sin α we get

$$=2\frac{z(z-c\cosh\alpha)}{z^2-2cz\cosh a+c^2}+\frac{\dfrac{\left[2\left(-\frac{1}{3}\right)\left(\frac{3}{2}\right)+3\right]}{\frac{1}{3}+\frac{\sqrt{5}}{2}}cz\sinh\alpha}{z^2-2cz\ \cosh\ \alpha+c^2}$$

$$= 2\frac{z(z - c\cosh\alpha)}{z^2 - 2cz\cosh a + c^2} + \left(\frac{\frac{-1+3}{-\sqrt{5}}}{6}\right)\frac{cz\sinh\alpha}{z^2 - 2cz\cosh a + c^2}$$

$$= 2\frac{z(z - c\cosh\alpha)}{z^2 - 2cz\cosh\alpha + c^2} - \frac{12}{\sqrt{5}}\frac{cz\sinh\alpha}{z^2 - 2cz\cosh\alpha + c^2}$$

$$Z^{-1}\frac{2z^2 + 3z}{z^2 + z + \frac{1}{9}} = 2\left(-\frac{1}{3}\right)^k \cosh\alpha\, k - \frac{12}{\sqrt{5}}\left(-\frac{1}{3}\right)^k \sinh\alpha, k \geq 0$$

where $\cosh\alpha = 3/2$

Exercise 25.5

Evaluate inverse Z-transform of the following:

1. $\frac{z}{3-z}, \quad |z| < 3$ **Ans.** $\{3^k\}, k < 0$

2. $\frac{2z^2 - 5z}{(z-2)(z-3)}, \quad |z| > 3$ **Ans.** $\{2^k + 3^k\}, k \geq 0$

3. $\frac{z}{z - e^a}$ **Ans.** $\{e^{k\alpha}\}$

4. $\frac{z^3}{z^3 - 27}, \quad |z| > 3$ **Ans.** $\{f(k)\} = \{3^k\} = 0$ for k =0, 3, 6, 9,

5. $\frac{2z}{(z-2)^2}, \quad |z| > 2$ **Ans.** $\{k2^k\}, k \geq 0$

6. $\frac{z^2 + z\cos\alpha}{z^2 - 2z\cos\alpha + 1}, \quad |z| > 1$ **Ans.** $\{\cos\alpha k\}, k \geq 0$

7. $-\log(1 - 2z^{-1}), \quad |z| > 2$ **Ans.** $\left\{\frac{2^k}{k}\right\}, k \geq 1$

8. $\frac{4z}{4z+1}+\frac{5z}{5z+1},\ \frac{1}{5}<\ |z|\ <\frac{1}{4}$ **Ans.** $f(k)=\begin{cases}-\left(\frac{1}{4}\right)^k & k<0\\ \left(-\frac{1}{5}\right)^k & k\geq 0\end{cases}$

9. $\frac{z(z^2+4z+1)}{(z-1)^4},\ |z|>|$ **Ans.** $\{k^3\}$

10. $\frac{ze^{-a}}{(z-e^{-a})^2},\ |z|>|e^{-a}|$ **Ans.** $\{k\,e^{-ak}\}$

11. $\frac{ze^{-a}\sin b}{z^2-2e^{-a}z\cos b+e^{-2a}},\ |z|>|e^{-a}|$ **Ans.** $\{e^{-ak}\sin bk\}$

12. $\frac{z(z-e^{-a}\cos b)}{z^2-2e^{-a}z\cos b+e^{-2a}},$ **Ans.** $\{e^{-ak}\cos bk\}$

13. $\frac{z^2\sin\beta+\frac{z}{\sqrt{2}}(\cos\beta-\sin\beta)}{z^2-\sqrt{2}z+1},\ |z|>1$ **Ans.** $\left\{\sin\left(\frac{k\pi}{4}+\beta\right)\right\},\ k\geq 0$

14. $\frac{z(z+a)}{(z-a)^3},|z|>a$ **Ans.** $\{k^2a^{k-1}U(k-1),\}\ k\geq 0$

15. $\frac{15z}{(4-z)(4z-1)}\ \frac{1}{4}<|z|<4$ **Ans.** $\left\{\left(\frac{1}{4}\right)^{|k|}\right\}$

25.20 Inversion by Residue Method

Take the contour c such that all the poles of the function z lie within the contour.

Then by Residue Method

$f(k) = \Sigma$ Residue of $z^{k-1}F(z)$ at its poles.

where Residue for simple pole $z = z_i$ is

$$=\left[(z-z_i)z^{k-1}F(z)\right]_{z=z_i}$$

Residue of order r at the pole $z = z_i$

$$\left[\frac{1}{(r-1)!}\frac{d^{r-1}}{dz^{r-1}}(z-z_i)^r z^{k-1}F(z)\right]_{z=z_i}$$

***Example* 25.33:** Obtain $Z^{-1}\dfrac{z}{(z-2)(z-3)}$

***Solution*:** The poles are determined by

$$(z-2)(z-3)=0 \Rightarrow z=2,3$$

There are two poles. Let us consider the contour $|z|>3$.

Residue at $\quad (z=2)=\left[(z-2)z^{k-1}\dfrac{z}{(z-2)(z-3)}\right]_{z=2}=\left[\dfrac{z^k}{z-3}\right]=\dfrac{2^k}{-1}$

Residue at $\quad (z=3)=\left[(z-3)z^{k-1}\dfrac{z}{(z-2)(z-3)}\right]_{z=3}=\left[\dfrac{z^k}{z-2}\right]_{z=3}=\dfrac{3^k}{1}$

Hence f(k) = Sum of the residues = 3^k-2^k

***Example* 25.34:** Obtain $Z^{-1}\dfrac{3z^2-18z+26}{(z-2)(z-3)(z-4)}$

Solution: The poles are determined by

$$(z-2)(z-3)(z-4)=0 \Rightarrow z=2,\ 3,\ 4.$$

There are three poles. Let us consider the contour $|z|>4$.

Residue at $(z=2)=\left[\dfrac{(z-2)z^{k-1}[3z^2-18z+26]}{(z-2)(z-3)(z-4)}\right]_{z=2}$

$$=\left[\frac{3z^{k+1}-18z^k+26z^{k-1}}{(z-3)(z-4)}\right]_{z=2}=\frac{3.2^{k+1}-18.2^k+26.2^{k-1}}{(-1)(-2)}$$

$$=3.2^k-9.2^k+13.2^{k-1}$$

$$=-6.2^{k+1}\times 13.2^{k-1}=-12.2^{k-1}+13.2^{k-1}=2^{k-1}$$

Residue at $(z=3)=\left[\dfrac{(z-3)z^{k-1}(3z^2-18z+26)}{(z-2)(z-3)(z-4)}\right]_{z=3}=\left[\dfrac{3z^{k+1}-18z^k+26z^{k-1}}{(z-2)(z-4)}\right]_{z=3}$

$$=\frac{3.3^{k+1}-18.3^k+26.3^{k-1}}{1(-1)}=-3.3^{k+1}+18.3^k-26.3^{k-1}$$

$$=-273^{k-1}+54.3^{k-1}-26.3^{k-1}=3^{k-1}$$

$$\text{Residue at } (z=4) = \left[(z-4)\frac{z^{k-1}(3z^2-18z+26)}{(z-2)(z-3)(z-4)}\right]_{z=4} = \left[\frac{3z^{k+1}-18z^k+26z^{k-1}}{(z-2)(z-3)}\right]_{z=4}$$

$$= \frac{3.4^{k+1}-18.4^k+26.4^{k-1}}{(2)\ (1)} = \frac{3}{2}.4^{k+1}-9.4^k+13\ 4^{k-1}$$

$$= 24.4^{k-1}-36.4^{k-1}+13.4^{k-1} = 4^{k-1}$$

Hence f(k) = sum of the residues

$$= 2^{k-1}+3^{k-1}+4^{k-1},\ k>0 \qquad \textbf{Ans.}$$

***Example* 25.35:** Obtain $Z^{-1}\dfrac{z(3z^2-6z+4)}{(z-1)^2(z-2)}$

Solution: The poles are determined by $(z-1)^2(z-2)=0, \Rightarrow z=1,1,2$

There are two poles, simple pole at z = 2 and pole of order 2 at z = 1. Let us consider the contour $|z|>2$.

Residue at (z = 2) is

$$\left[\frac{(z-2).z^{k-1}.z(3z^2-6z+4)}{(z-1)^2(z-2)}\right]_{z=2} = \left[\frac{3z^{k+2}-6z^{k+1}+4z^k}{(z-1)^2}\right]_{z=2}$$

$$= 3.2^{k+2}-6.2^{k+1}+4.2^k$$

$$= 12.2^k-12.2^k+4.2^k = 4.2^k = 2^{k+2}$$

$$\text{Residue at } (z=1) = \frac{d}{dz}(z-1)^2\left.\frac{z^{k-1}.z(3z^2-6z+4)}{(z-1)^2(z-2)}\right]_{z=1} = \frac{d}{dz}\left[\frac{3z^{k+2}-6z^{k+1}+4z^k}{z-2}\right]_{z=1}$$

$$= \left[\frac{(z-2)[3(k+2)z^{k+1}-6(k+1)z^k+4kz^{k-1}]-(3z^{k+2}-6z^{k+1}+4z^k).1}{(z-2)^2}\right]_{z=1}$$

$$= -\{(3k+6)-6k-6+4k\}-\{3-6+4\}$$

$$= -3k-6+6k+6-4k-3+6-4 = -k-1$$

$$f(k) = \text{sum of the residues} = \left[2^{k+2}-k-1\right]U(k) \qquad \textbf{Ans.}$$

***Example* 25.36:** Find $Z^{-1}\dfrac{9z^3}{(3z-1)^2(z-2)}$

***Solution*:** The poles are determined by $(3z-1)^2(z-2)=0, z=\frac{1}{3},\frac{1}{3},2$

There are two poles i.e., one simple pole at z = 2 and second pole of order 2 at $z=\frac{1}{3}$. Let us consider the contour $|z|>2$.

$$\text{Residue at } (z=2)=\left[(z-2).z^{k-1}\frac{9z^3}{(3z-1)^2(z-2)}\right]_{z=2}=\left[\frac{9z^{k+2}}{(3z-1)^2}\right]_{z=2}=\frac{9.2^{k+2}}{25}$$

$$\text{Residue at } \left(z=\frac{1}{3}\right)=\frac{d}{dz}\left[\frac{\left(z-\frac{1}{3}\right)^2.z^{k-1}.9z^3}{(3z-1)^2.(z-2)}\right]_{z=\frac{1}{3}}=\frac{d}{dz}\left[\frac{9z^{k+2}}{9(z-2)}\right]_{z=\frac{1}{3}}=\frac{d}{dz}\left(\frac{z^{k+2}}{z-2}\right)_{z=\frac{1}{3}}$$

$$\frac{d}{dz}\left(\frac{z^{k+2}}{z-2}\right)=\left[\frac{(z-2)(k+2)z^{k+1}-z^{k+2}}{(z-2)^2}\right]_{z=\frac{1}{3}}=\left[\frac{(k+2)z^{k+2}-2(k+2)z^{k+1}-z^{k+2}}{(z-2)^2}\right]_{z=\frac{1}{3}}$$

$$=\left[\frac{(k+1)z^{k+2}-2(k+2)z^{k+1}}{(z-2)^2}\right]_{z=\frac{1}{3}}=\left[\frac{z^{k+1}\{(k+1)z-2k-4}{(z-2)^2}\right]_{z=\frac{1}{3}}$$

$$=\frac{\left(\frac{1}{3}\right)^{k+1}\left\{(k+1)\frac{1}{3}-2k-4\right\}}{\frac{25}{9}}=\frac{\left(\frac{1}{3}\right)^{k+1}.9.\frac{1}{3}[k+1-6k-12]}{25}$$

$$=\frac{\left(\frac{1}{3}\right)^{k}(-5k-11)}{25}$$

Hence, f(k) = sum of residues $=\frac{9}{25}.2^{k+2}-\left(\frac{1}{3}\right)^k\frac{(5k+11)}{25}$ **Ans.**

***Example* 25.37:** Find $Z^{-1}\dfrac{z^2}{z^2+4}$. $|z|>2$

***Solution*:** The poles are determined by $z^2+4=0 \Rightarrow z=\pm 2i$

There are two poles at z = 2i and z = –2i

Let us consider a contour | z | > 2

$$\text{Residue at } (z = 2i) = \left[\frac{(z-2i)z^{k-1}.z^2}{z^2+4}\right]_{z=2i} = \left[\frac{z^{k+1}}{z+2i}\right]_{z=2i} = \frac{(2i)^{k+1}}{4i} = 2^{k-1}(i)^k$$

$$= 2^{k-1}\left[\cos\frac{\pi}{2} + i\sin\frac{\pi}{2}\right]^k = 2^{k-1}\left[\cos\frac{k\pi}{2} + i\sin\frac{k\pi}{2}\right]$$

$$\text{Residue at } (z = -2i) = \left[(z+2i)z^{k-1}\frac{z^2}{z^2+4}\right]_{z=-2} = \left[\frac{z^{k+1}}{z-2i}\right]_{z=-2i} = \frac{(-2i)^{k+1}}{-4i}$$

$$= 2^{k+1}(-i)^k = 2^{k-1}\left[\cos\frac{\pi}{2} - i\sin\frac{\pi}{2}\right]^k = 2^{k-1}\left[\cos\frac{k\pi}{2} - i\sin\frac{k\pi}{2}\right]$$

$$f(k) = \text{sum of the residues} = 2^{k-1}\left[\cos\frac{k\pi}{2} + i\sin\frac{k\pi}{2}\right] + 2^{k-1}\left[\cos\frac{k\pi}{2} - i\sin\frac{k\pi}{2}\right]$$

$$= 2^{k-1}\left[2\cos\frac{k\pi}{2}\right] = 2^k\cos\frac{k\pi}{2}$$

Exercise 25.6

Find the inverse Z transform of the following functions by residue method:

1. $\dfrac{z}{(z-1)(z-2)}$ **Ans.** $2^k - 1,\ k \geq 0$

2. $\dfrac{3z^2+2z}{z^2-3z+2}, < |z| < 2$ **Ans.** $\{-5\}\ (k \geq 0), -8\,(2)^k, k < 0$

3. $\dfrac{2z^2+3z}{z^2+z+1},\ |z| > 1$ **Ans.** $\left[2\left\{\cos\dfrac{2\pi k}{3}\right\} + \dfrac{4}{\sqrt{3}}\left\{\sin\dfrac{2\pi k}{3}\right\}\right]U(k)$

4. $\dfrac{z(z+1)}{(z-1)(z^2+z+1)}, |z| > 1$ **Ans.** $\dfrac{2}{3}\left[1-\cos\dfrac{2\pi k}{3}\right]U(k)$

5. $\dfrac{16z^3}{(4z-1)^2(z-1)}$ **Ans.** $\left[\dfrac{16}{9} - \dfrac{1}{9}(3k+7)\left(\dfrac{1}{4}\right)^k\right]U(k)$

6. $\dfrac{z^2}{z^2+1}$ **Ans.** $\cos\dfrac{k\pi}{2}$

25.21 Solution of Difference Equation

***Example* 25.38:** Solve the difference equation

$$6y_{k+2} - y_{k-1} - y_k = 0, \quad y(0) = 0,\ y(1) = 1 \text{ by Z-transform}$$

Solution: $6y_{k+2} - y_{k-1} - y_k = 0,$ .....(1)

Taking the Z-transform of both sides of (1), we get

$$Z[6y_{k+2} - y_{k+1} - y_k] = 0$$

$$Z(6y_{k+2}) - Z(y_{k+1}) - Z(y_k) = 0$$

$$6[z^2Y(z) - z^2y(0) - 2y(1)] - [zY(z) - zy(0)] - Y(z) = 0$$

On putting the values of y (0) and y(1) we get

$$6z^2Y(z) - 6z - zY(z) - Y(z) = 0$$

$$(6z^2 - z - 1)Y(z) = 6z$$

$$Y(z) = \frac{6z}{6z^2 - z - 1} = \frac{6z}{(3z+1)(2z-1)} = \frac{z^{-1}}{\left(1 + \frac{z^{-1}}{3}\right)\left(1 - \frac{z^{-1}}{2}\right)} = \frac{\frac{6}{5}}{1 - \frac{z^{-1}}{2}} - \frac{\frac{6}{5}}{1 + \frac{z^{-1}}{3}}$$

$$y_k = Z^{-1}\left[\frac{\frac{6}{5}}{1 - \frac{z^{-1}}{2}}\right] - Z^{-1}\left[\frac{\frac{6}{5}}{1 + \frac{z^{-1}}{3}}\right] = \frac{6}{5}\left(\frac{1}{2}\right)^k - \frac{6}{5}\left(-\frac{1}{3}\right)^k$$

$$= \frac{6}{5}\left[\left(\frac{1}{2}\right)^k - \left(-\frac{1}{3}\right)^k\right]$$

***Example* 25.39:** Solve the difference equation

$$y_{k+3} - 3y_{k+2} + 3y_{k+1} - y_k = U(k)$$

$$y(0) = y(1) = y(2) = 0 \text{ by Z-transforms.}$$

Solution: $y_{k+3} - 3y_{k+2} + 3y_{k+1} - y_k = U(k)$ .....(1)

Taking the Z-transform of both sides of (1), we get

$$Z[y_{k+3} - 3y_{k+2} + 3y_{k+1} - y_k] = Z\,U(k),$$

$$\Rightarrow Z[y_{k+3}] - 3Z[y_{k+2}] + 3Z[y_{k+1}] - Z[y_k] = ZU(k)$$

$$\Rightarrow [z^3Y(z) - z^3y(0) - z^2y(1) - zy(2)] - 3[z^2Y(z) - z^2y(0) - zy(1)]$$

$$+3[zY(z) - zy(0)] - Y(z) = ZU(k)$$

Putting the values of $y(0) = y(1) = y(2) = 0$ in the above equation, we get

$$z^3Y(z) - 3z^2Y(z) + 3zY(z) - Y(z) = \frac{1}{1 - z^{-1}}$$

$$[z^3 - 3z^2 + 3z - 1]Y(z) = \frac{1}{1 - z^{-1}}$$

$$(z-1)^3 Y(z) = \frac{1}{1 - z^{-1}}$$

$$Y(z) = \frac{1}{(z-1)^3(1 - z^{-1})} = \frac{1}{z^3(1 - z^{-1})^3(1 - z^{-1})} = Z^{-3}(1 - z^{-1})^{-4}$$

$$y_k = \text{coeff.of } z^{-k} \text{ in } z^{-3}(1 - z^{-1})^{-4} = \text{coeff.of } z^{-k-3} \text{ in } (1 - z^{-1})^{-4}$$

$$= \frac{(k-2)(k-1)k}{6}, \quad k \geq 3$$

***Example* 25.40:** Solve by Z-transform

$$y_{k+1} + \frac{1}{4}y_k = \left(\frac{1}{4}\right)^k \qquad (k \geq 0),\ y(0) = 0$$

***Solution*:** $y_{k+1} + \frac{1}{4}y_k = \left(\frac{1}{4}\right)^k$ .....(1)

Taking Z transform of both sides of (1), we get

$$Z\left[y_{k+1} + \frac{1}{4}y_k\right] = Z\left[\left(\frac{1}{4}\right)^k\right]$$

$$Z[y_{k+1}] + Z\left[\frac{1}{4}y_k\right] = Z\left[\left(\frac{1}{4}\right)^k\right]$$

$$ZY(z) - zy(0) + \frac{1}{4}Y(z) = \frac{1}{1 - \frac{1}{4}z^{-1}} \qquad |z| > \frac{1}{4}$$

$$zY(z)-0+\frac{1}{4}Y(z)=\frac{1}{1-\frac{1}{4}z^{-1}}$$

$$\left(z+\frac{1}{4}\right)Y(z)=\frac{1}{1-\frac{1}{4}z^{-1}}$$

$$Y(z)=\frac{1}{z+\frac{1}{4}}\times\frac{1}{1-\frac{1}{4}z^{-1}}=\frac{z^{-1}}{1+\frac{1}{4}z^{-1}}\times\frac{1}{1-\frac{1}{4}z^{-1}}=\frac{-2}{1+\frac{1}{4}z^{-1}}+\frac{2}{1-\frac{1}{4}z^{-1}}$$

$$y(k)=z^{-1}\left[\frac{-2}{1+\frac{1}{4}z^{-1}}\right]+Z^{-1}\left[\frac{2}{1-\frac{1}{4}z^{-1}}\right]=Z^{-1}\left[-2\left(1+\frac{1}{4}z^{-1}\right)^{-1}\right]+Z^{-1}\left[2\left(1-\frac{1}{4}z^{-1}\right)^{-1}\right]$$

$$=-2\left(-\frac{1}{4}\right)^{k}+2\left(\frac{1}{4}\right)^{k}$$

***Example* 25.41:** Solve $y_k+\frac{1}{4}y_{k-1}=U_{(k)}+\frac{1}{3}U_{(k-1)}$

***Solution*:** $y_k+\frac{1}{4}y_{k-1}=U[k]+\frac{1}{3}U_{(k-1)}$ .....(1)

Taking the Z-transform of both sides of (1) we get

$$Z[\{y_k\}]+\frac{1}{4}Z[\{y_{k-1}\}]=Z[\{U_k\}]+\frac{1}{3}Z[\{U_{k-1}\}]$$

$$Y(z)+\frac{1}{4}z^{-1}Y(z)=\left[1+\frac{1}{3}z^{-1}\right]U(k)$$

$$\left[1+\frac{1}{4}z^{-1}\right]Y(z)=\left(1+\frac{1}{3}z^{-1}\right)U(k)$$

$$Y(z)=\frac{1+\frac{1}{3}z^{-1}}{1+\frac{1}{4}z^{-1}}=\frac{z+\frac{1}{3}}{z+\frac{1}{4}}$$

There is only one simple pole at $z=-\frac{1}{4}$.

Let us consider the contour $|z| > \frac{1}{4}$.

Residue at $\left(z = -\frac{1}{4}\right) = \left[\left(z+\frac{1}{4}\right)z^{k-1}\frac{z+\frac{1}{3}}{z+\frac{1}{4}}\right]_{z=-\frac{1}{4}} = \left[z^k + \frac{z^{k-1}}{3}\right]_{z=-\frac{1}{4}}$

$$\left(-\frac{1}{4}\right)^k + \frac{1}{3}\left(-\frac{1}{4}\right)^{k-1} = -\frac{1}{4}\left(-\frac{1}{4}\right)^{k-1} + \frac{1}{3}\left(-\frac{1}{4}\right)^{k-1} = \frac{1}{12}\left(-\frac{1}{4}\right)^{k-1}$$

Hence $y_k = \text{Residue} = \frac{1}{12}\left(-\frac{1}{4}\right)^{k-1}$

Solved Examples

***Example* 25.42:** Using Z-transform solve $U_{k+1} + U_k = 1$ if $u_0 = 0$.

Solution: The given difference equation is

$$u_{k+1} + u_k = 1 \qquad \text{.....(1)}$$

If $Z(u_k) = U(z)$, then $Z(u_{k+1}) = z\,[U(z) - u_0]$ Also $Z(1) = \frac{z}{z-1}$

∴ Taking the Z-transform of both sides of (1), we get

$$Z[u_{k+1} + u_k] = Z(1)$$

or $$Z(u_{k+1}) + Z(u_k) = Z(1)$$

or $$z(U(z) - u_0] + U(z) = \frac{z}{z-1} \qquad \text{.....(2)}$$

On putting the value of u_0 in (2), we get

$$U(z) = \frac{z}{(z-1)(z+1)}$$

∴ $$\frac{U(z)}{z} = \frac{1}{(z-1)(z+1)} = \frac{1}{2}.\frac{1}{z-1} - \frac{1}{2}.\frac{1}{z+1}$$ are breaking into partial fractions

or $$U(z) = \frac{1}{2}.\frac{z}{z-1} - \frac{1}{2}.\frac{z}{z+1}$$

On inversion, we obtain

$$u_k = \frac{1}{2}.Z^{-1}\left(\frac{z}{z-1}\right) - \frac{1}{2}.Z^{-1}\left(\frac{z}{z+1}\right)$$

$$u_k = \frac{1}{2} - \frac{1}{2}.(-1)^k$$

Hence the required solution of (1) is

$$u_k = \frac{1}{2} - \frac{1}{2}.(-1)^k$$

***Example* 25.43:** Using Z-transform, solve $y_{k+2} - 4y_k = 0$ given that $y_0 = 0$, $y_1 = 2$

***Solution*:** The given difference equation is

$$y_{k+2} - 4y_k = 0 \qquad .....(1)$$

Taking the Z-transform of both sides, we get

$$Z[y_{k+2} - 4y_k] = 0$$

or $\quad Z(y_{k+2} - 4Z = 0$

or $\quad z^2[Y(z) - y_0 - y_1 z^{-1}] - 4\,[Y(z) - y_0] = 0 \qquad ....(2)$

on putting the values of y_0 and y_1 in (2), we get

Taking the Z-transform of both sides, we get

$$z^2\,[Y(z) - 2z^{-1} - 4\,Y(z) = 0$$

or $\quad (z^2 - 4)\,Y(z) = 2z$

or $\quad Y(z) = \dfrac{2z}{z^2 - 4}$

or $\quad Y(z) = \dfrac{2z}{(z-2)(z+2)}$

Its poles are determined by

$$(z-2)\,(z+2) = 0 \Rightarrow z = 2, -2$$

Therefore there are two simple poles at $z = 2$ and $z = -2$

Let us consider a contour $|\,z\,| > 2$

Now Res $\quad [Y(z)\,z^{k-1}]_{z=2} = \underset{z\to 2}{\text{Lt}}\left\{(z-2).\dfrac{z^{k-1}.2z}{(z-2)(z+2)}\right\}$

$$= \underset{z \to 2}{\text{Lt}} \left\{ \frac{2z^k}{z+2} \right\} = \frac{2.2^k}{4} = 2^{k-1}$$

$$\text{Res} \quad \left[Y(z) z^{k-1} \right]_{z=-2} = \underset{z \to -2}{\text{Lt}} \left\{ (z+2) . \frac{z^{k-1} . 2z}{(z-2)(z+2)} \right\}$$

$$= \underset{z \to -2}{\text{Lt}} \left\{ \frac{2z^k}{z-2} \right\} = \frac{2.(-2)^k}{-4} = (-2)^{k-1}$$

Hence $\quad y_k$ = sum of residues

or $\quad y_k = 2^{k-1} + (-2)^{k-1}$

***Example* 25.44:** Using Z-transform, solve

$$u_{k+2} + 4u_{k+1} + 3u_k = 3^k$$

given that $\quad u_0 = 0, u_1 = 1$

***Solution*:** The given difference equation is

$$u_{k+2} + 4u_{k+1} + 3u_k = 3^k \qquad \text{.....(1)}$$

If $Z[u_k] = U(z)$, then $Z[u_{k+1}] = z[U(z) - u_0]$, $Z[u_{k+2}] = z^2 [U(z) - u_0 - u_1 z^{-1}]$

Also $\quad Z\left[3^k\right] = \dfrac{z}{z-3}$

Taking the Z-transform of both sides of (1), we get

$$Z[u_{k+2}] + 4Z[u_{k+1}] + 3Z[u_k] = Z[3^k]$$

or $$Z^2[U(z) - u_0 - u_1 z^{-1}] + 4z[U(z) - u_0] + 3U(z) = \frac{z}{z-3} \qquad \text{.....(2)}$$

Putting the values of u_0 and u_1 in (2), we get

$$z^2 [U(z) - 0 - z^{-1}] + 4z[U(z) - 0] + 3U(z) = \frac{z}{z-3}$$

or $$(z^2 + 4z + 3) U(z) = z + \frac{z}{z-3}$$

$$\therefore \quad \frac{U(z)}{z} = \frac{1}{(z+1)(z+3)} + \frac{1}{(z-3)(z+1)(z+3)}$$

$$= \frac{3}{8}\frac{1}{z+1} + \frac{1}{24}\frac{1}{z-3} - \frac{5}{12}\frac{1}{z+3}, \text{ on breaking into partial fractions}$$

$$U(z) = \frac{3}{8}\frac{z}{z+1} + \frac{1}{24}\frac{z}{z-3} - \frac{5}{12}\frac{z}{z+3}$$

On inversion, we obtain

$$u_k = \frac{3}{8}Z^{-1}\left(\frac{z}{z+1}\right) + \frac{1}{24}Z^{-1}\left(\frac{z}{z-3}\right) - \frac{5}{12}Z^{-1}\left(\frac{z}{z+3}\right)$$

$$= \frac{3}{8}(-1)^k + \frac{1}{24}3^k - \frac{5}{12}(-3)^k$$

***Example* 25.45:** Using Z-transform, solve $y_{k+2} + 6y_{k+1} + 9y_k = 2^k$ given that $y_0 = y_1 = 0$.

Solution: The given difference equation is

$$y_{k+2} + 6y_{k+1} + 9y_k = 2^k \qquad .....(1)$$

Taking the Z-transforms of both sides, we get

$$Z[y_{k+2}] + 6Z\,[y_{k+1}] + 9Z\,[y_k] = Z\,[2^k]$$

or $$z^2[Y(z) - y_0 - y_1\, z^{-1}] + 6z\,[Y(z) - y_0] + 9Y(z) = \frac{z}{z-2} \qquad .....(2)$$

Putting the values of y_0 and y_1 in (2), we get

$$(z^2 + 6z + 9)\, Y(z) = \frac{z}{z-2}$$

or $$Y(z) = \frac{z}{(z-2)(z^2+6z+9)}$$

or $$\frac{Y(z)}{z} = \frac{1}{(z-2)(z+3)^2} = \frac{1}{25}\left[\frac{1}{z-2} - \frac{1}{z+3} - \frac{5}{(z+3)^2}\right]$$

on splitting into partial fractions

or $$Y(z) = \frac{1}{25}\left[\frac{z}{z-2} - \frac{z}{z+3} - 5\frac{z}{(z+3)^2}\right]$$

on taking inverse Z-transforms of both sides, we obtain

$$y_k = \frac{1}{25}\left[Z^{-1}\left(\frac{z}{z-2}\right) - Z^{-1}\left(\frac{z}{z+3}\right) + \frac{5}{3}Z^{-1}\left(\frac{3z}{(z+3)^2}\right)\right]$$

$$= \frac{1}{2}\left[2^k - (-3)^k + \frac{5}{3}k(-3)^k\right] \qquad \left[\because Z^{-1}\left\{\frac{az}{(z-a)^2}\right\} = ka^k\right]$$

***Example* 25.46:** Using Z-transform, solve $6y_{k+2} + 5y_{k+1} - y_k = 6U(k)$ given that $y(0) = y(1) = 0$

Solution: The given difference equation is

$$6y_{k+2} + 5y_{k+1} - y_k = 6U(k) \qquad .....(1)$$

where U(k) is unit step sequence

Taking the Z-transform of both sides of (1), we get

$$Z[6y_{k+2} + 5y_{k+1} - y_k] = Z\,[6U(k)]$$

or $$6Z[y_{k+2}] + 5Z\,[y_{k+1}] - Z\,[y_k] = 6Z\,[U(k)]$$

or $$6z^2\,[Y(z) - y(0) - y(1)\,z^{-1}] + 5z\,[Y(z) - y(0)] - Y(z) = 6,\ \frac{z}{z-1}, \qquad .....(2)$$

$$k \geq 0 \quad \text{and} \quad 0, k < 0$$

On putting the values of y(0) and y(1), we get

$$6z^2Y(z) - 5z\,Y(z) - Y(z) = \frac{6z}{z-1}$$

or $$(6z^2 - 5z - 1)\,Y(z) = \frac{6z}{z-1}$$

or $$Y(z) = \frac{6z}{(6z+1)(z-1)^2},\quad k \geq 0 \ \text{ and } \ 0, k < 0$$

its poles are determined by

$$(6z + 1)\,(z - 1)^2 = 0 \Rightarrow z = -\frac{1}{6}.1,1$$

There are two poles; i.e., one simple pole at $z = -\frac{1}{6}$ and second pole of order 2 at $z = 1$

Let us consider the contour $|\,z\,| > 1$

Now Res $$\left[Y(z)z^{k-1}\right]_{z=-\frac{1}{6}} = \underset{z \to -\frac{1}{6}}{\text{Lt}} \left\{\left(z+\frac{1}{6}\right), \frac{z^{k-1}.6z}{(6z+1)(z-1)^2}\right\}$$

$$= \underset{z \to -\frac{1}{6}}{\text{Lt}} \left\{\frac{z^k}{(z-1)^2}\right\} = \frac{36}{49}\left(-\frac{1}{6}\right)^k,\quad k \geq 0 \text{ and } 0, k < 0$$

$$= \frac{36}{49}\left(-\frac{1}{6}\right)^k \ U(k)$$

$$\text{Res }[Y(z)\, zk-1]_{z=1} = \frac{d}{dz}\left[\frac{(z-1)^2.6z^k}{(6z+1)(z-1)^2}\right] = \left[\frac{d}{dz}\frac{6z^k}{6z+1}\right]$$

$$= \left[\frac{(6z+1).6kz^{k-1} - 6z^k.6}{(6z+1)^2}\right]_{z=1} = \left[\frac{6kz^{k-1}}{6z+1} - \frac{36z^k}{(6z+1)^2}\right]_{z=1}$$

$$= \left[\frac{6}{7}k - \frac{36}{49}\right] = \frac{6}{7}(k+1) - \frac{78}{49}, \quad k \geq 0 \text{ and } 0,\ k < 0$$

$$= \left[\frac{6}{7}(k+1) - \frac{78}{49}\right] U(k)$$

y_k = sum of residues

$$= \left[\frac{6}{7}(k+1) - \frac{78}{49} + \frac{36}{49}\left(-\frac{1}{6}\right)^k\right] U(k)$$

***Example* 25.47:** Using Z-transform, solve $y_{k+3} - 3y_{k+2} + 3y_{k+1} - y_k = U(k)$ given that $y(0) = y(1) = y(2) = 0$

Solution: The given difference equation is

$$y_{k+3} - 3y_{k+2} + 3y_{k+1} - y_k = U(k) \qquad \text{.....(1)}$$

Taking the Z-transform of both sides of (1), we get

$$Z[y_{k+3} - 3y_{k+2} + 3y_{k+1} - y_k] = Z[U(k)]$$

$$\Rightarrow \quad Z[y_{k+3}] - 3Z[y_{k+2}] + 3Z[y_{k+1}] - Z[y_k] = Z[U(k)]$$

$$\Rightarrow \quad z^3\left[Y(z) - y(0) - y(1)z^{-1} - y(2)z^{-2}\right] - 3z^2[Y(z) - y(0) - y(1)z^{-1}]$$

$$+3z\left[Y(z) - y(0)\right] - Y(z) = Z[U(k)]$$

Putting the values of $y(0) = y(1) = y(2) = 0$ in the above equation, we get

$$z^3Y(z) - 3z^2Y(z) + 3zY(z) - Y(z) = \frac{z}{z-1}$$

or $\quad (z^3 - 3z^2 + 3z - 1)Y(z) = \dfrac{z}{z-1}$

or $\quad (z-1)^3\, Y(z) = \dfrac{z}{z-1}$

or $\quad Y(z) = \dfrac{z}{(z-1)^4} = z^{-3}\left(1-z^{-1}\right)^{-4}$

$$= \sum y_k z^{-k}, \text{ say}$$

y_k = coefficient of z^{-k} in $z^{-3}\,(1 - z^{-1})^{-4}$

$$= \text{coefficient of } z^{-k-3} \text{ in } \left(1-z^{-1}\right)^{-4}$$

$$= \frac{(k-2)(k-1)k}{6}, k \geq 3$$

***Example* 25.48:** Solve $y_{k+1} - 5y_k = \begin{cases} \sin k & \text{for } k \geq 0 \\ 0 & \text{for } k < 0 \end{cases}$ by residue method.

Solution: Taking Z transform, we get

$$Z[y_{k+1} - 5y_k] = Z\,[\sin k] \qquad k \geq 0$$

$$Y(0) - 5\,Y(z) = \frac{z\sin 1}{z^2 - 2z\cos 1 + 1} \qquad \text{......(1)}$$

K = –1 in the given equation we get

$$y(0) - 5y_{(-1)} = 0, \qquad \text{But } y_{(-1)} = 0;\ y_{(0)} = 0$$

substituting the value of y(0) in (1), we get

$$(z-5)\,Y(z) = \frac{z\sin 1}{z^2 - 2z\cos 1 + 1}$$

$$Y(z) = \frac{z\sin 1}{(z-5)\left(z^2 - 2z\cos 1 + 1\right)}$$

$$= \frac{A}{z-5} - \frac{Az - \dfrac{A - \sin 1}{5}}{z^2 - 2z\cos 1 + 1} \quad \text{where } A = \frac{\sin 1}{26 - 10\cos 1}$$

$$Y(z) = A\frac{z}{z-5} - \frac{Az(z-\cos 1) + \dfrac{A\cos 1 - \dfrac{A-\sin 1}{5}z\sin 1}{\sin 1}}{z^2 - 2z\cos 1 + 1}$$

$$y_k = A(5)^k - A\cos k - A\frac{5-\cos 1}{\sin 1}\sin k$$

$$= A[(5)^k - \cos k - \frac{5-\cos 1}{\sin 1}\sin k], \text{ where } A = \frac{\sin 1}{26 - 10\cos 1}$$

Exercise 25.7

Solve the difference equations by Z-transform

1. $y_k - \frac{5}{6}y_{k-1} + \frac{1}{6}y_{k-2} = U_{(k)}$ **Ans.** $y_k = \left[3 - 3\left(\frac{1}{2}\right)^k + \left(\frac{1}{3}\right)^k\right]U_{(k)}$

2. $6y_{k+2} + 5y_{k+1} - y_k = 6U(k)$ **Ans.** $y_k = \left[\frac{6}{7}(k+1) - \frac{78}{49} + \frac{36}{49}\left(\frac{1}{6}\right)^k\right]U(k)$

3. $y_{k+1} - y_{k-1} = U(k)$, $y(0) = 0$ **Ans.** $\left[\frac{k}{2} + \frac{k}{4}\right]U(k)$

4. $yk+1 - 2y_k + y_{k-1} = a^k$. $a \neq 1$

Ans. $y_k = \frac{1}{a}(k+1)U(k) - \frac{a}{(a-1)^2}U(k) + \frac{a}{(a-1)^2}a^kU(k) + \frac{1}{1-a}U(k-1)$